Ordered Pairs

A pair of numbers enclosed in parentheses and separated by a comma, such as (2, 1) is called an ordered pair of numbers. The first number in the pair is called the *x-coordinate* of the ordered pair, while the second number is called the *y-coordinate*. For the ordered pair (2, 1), the *x*-coordinate is 2 and the *y*-coordinate is 1.

The Rectangular Coordinate System

We graph ordered pairs on a rectangular coordinate system. A rectangular coordinate system is made by drawing two real number lines at right angles to each other. The two number lines, called *axes*, cross each other at 0. This point is called the *origin*. The horizontal number line is the *x-axis* and the vertical number line is the *y-axis*.

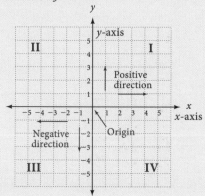

FIGURE 4 The rectangular coordinate system.

Graphing Ordered Pairs

To graph the ordered pair (*a*, *b*) on a rectangular coordinate system, we start at the origin and move *a* units right or left (right if *a* is positive and left if *a* is negative). We then move *b* units up or down (up if *b* is positive and down if *b* is negative). The point where we end up is the graph of the ordered pair (*a*, *b*). Figure 5 shows the graphs of the ordered pairs (2, 5), (−2, 5), (−2, −5), and (2, −5).

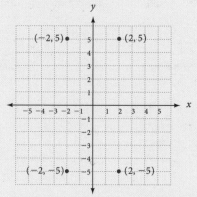

FIGURE 5 The graphs of four ordered pairs.

Graphing Functions

There are many ways to graph functions. Figure 6 is a graph constructed by substituting 0, 10, 20, 30, and 40 for *x* in the formula $f(x) = 7.5x$, and then using the formula to find the corresponding values of *y*. Each pair of numbers corresponds to a point in Figure 6. The points are connected with straight lines to form the graph.

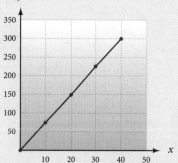

FIGURE 6 The graph of $f(x) = 7.5x$, $0 \leq x \leq 40$.

Function Notation

The notation $f(x)$ is used to denote elements in the range of a function. For example, if the rule for a function is given by $y = 7.5x$, then the rule can also be written as $f(x) = 7.5x$. If we ask for $f(20)$, we are asking for the value of *y* that comes from $x = 20$. That is

$$\text{if} \qquad f(x) = 7.5x$$
$$\text{then} \qquad f(20) = 7.5(20) = 150$$

Figure 7 is a diagram called a function map that gives a visual representation of this particular function:

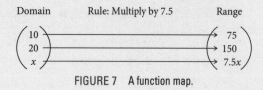

FIGURE 7 A function map.

A function can also be thought of as a machine. We put values of *x* into the machine which transforms them into values of $f(x)$, which are then output by the machine. The diagram is shown in Figure 8.

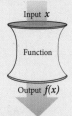

FIGURE 8 A function machine.

We can organize our work with paired data, graphing, and functions into a more standard form by using ordered pairs and the rectangular coordinate system.

(Continued on inside back cover)

Fourth Edition

ELEMENTARY AND INTERMEDIATE ALGEBRA

A Combined Course

Charles P. McKeague

CUESTA COLLEGE

BROOKS/COLE
CENGAGE Learning™

Australia • Brazil • Japan • Korea • Mexico • Singapore • Spain • United Kingdom • United States

BROOKS/COLE
CENGAGE Learning™

Elementary and Intermediate Algebra,
Fourth Edition
Charles P. McKeague

Acquisitions Editor: Marc Bove

Developmental Editor: Carolyn Crockett

Assistant Editor: Shaun Williams

Editorial Assistant: Zachary Crockett

Media Editors: Heleny Wong, Guanglei Zhang

Marketing Manager: Gordon Lee

Marketing Assistant: Shannon Myers

Marketing Communications Manager:
    Darlene Macanan

Content Project Manager: Jennifer Risden

Design Director: Rob Hugel

Art Director: Vernon Boes

Print Buyer: Karen Hunt

Rights Acquisitions Specialist: Roberta Broyer

Production Service: XYZ Textbooks

Text Designer: Diane Beasley

Photo Researcher: Bill Smith Group

Copy Editor: Katherine Shields, XYZ Textbooks

Illustrator: Kristina Chung, XYZ Textbooks

Cover Designer: Irene Morris

Cover Image: Pete McArthur

Compositor: Donna Looper, XYZ Textbooks

For product information and technology assistance, contact us at
Cengage Learning Customer & Sales Support, 1-800-354-9706

For permission to use material from this text or product,
submit all requests online at **www.cengage.com/permissions**
Further permissions questions can be emailed to
permissionrequest@cengage.com

Library of Congress Control Number: 2010933321

ISBN-13: 978-0-8400-6419-6

ISBN-10: 0-8400-6419-5

Brooks/Cole
20 Davis Drive
Belmont, CA 94002-3098
USA

Cengage Learning is a leading provider of customized learning solutions with office locations around the globe, including Singapore, the United Kingdom, Australia, Mexico, Brazil, and Japan. Locate your local office at:
www.cengage.com/global

Cengage Learning products are represented in Canada by Nelson Education, Ltd.

To learn more about Brooks/Cole, visit **www.cengage.com/brookscole**
Purchase any of our products at your local college store or at our preferred online store **www.cengagebrain.com**

Printed in the United States of America
1 2 3 4 5 6 7 14 13 12 11 10

Brief Contents

Contents

1 The Basics 1

Image © 2010 DigitalGlobe
Image © 2010 IGN France

2 Linear Equations and Inequalities 81

Image © 2010 DigitalGlobe
Image © 2010 TerraMetrics

5 Factoring   303

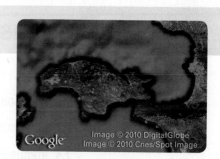

6 Rational Expressions   359

9 Rational Exponents and Roots 563

10 Quadratic Functions 627

13   Sequences and Series   785

The Elementary and Intermediate Algebra Course as a Bridge to Further Success

Elementary and Intermediate algebra is a bridge course. The course and its syllabus bring the student to the level of ability required of college students, while getting them ready to make a successful start in college algebra or precalculus.

New to This Edition

OBJECTIVES

In order to help organize topics within each section, we have greatly enhanced our use of objectives in both the sections and in the accompanying problem sets. Objectives for each chapter are shown immediately preceding the first section in each chapter. Section objectives are shown at the start of each section.

TICKET TO SUCCESS

Previously included as Getting Ready for Class, these questions require written responses from the student. They reinforce the idea of reading the section before coming to class, as their answers to these questions can truly be their ticket into class.

MOVING TOWARD SUCCESS

Each problem set now starts with this new feature that includes a motivational quote followed by a few questions to focus the students on their own success. Because good study habits are essential to the success of math students, this feature is prominently displayed at the start of every problem set.

CHAPTER TESTS

Each chapter now includes a comprehensive one-page chapter test.

CUMULATIVE REVIEW

Beginning in Chapter 2, each chapter now has a one-page cumulative review to reinforce material covered previously.

Our Proven Commitment to Student Success

After three successful editions, we have developed several interlocking, proven features that will improve students' chances of success in the course. Here are some of the important success features of the book.

Integration of Functions   Functions are introduced in Chapter 8 and then integrated in the rest of the intermediate algebra portion of the text. This feature forms a bridge to college algebra by requiring students to work with functions and function notation throughout the course.

Getting Ready for Chapter X   This is a set of problems from previous chapters that students need to be able to solve in order to be successful in the current chapter. These are review problems intended to reinforce the idea that all topics in the course are built on previous topics.

Blueprint for Problem Solving   Found in the main text, this feature is a detailed outline of steps required to successfully attempt application problems. Intended as a guide to problem solving in general, the blueprint takes the student through the solution process of various kinds of applications.

Organization of Problem Sets

The problem sets begin with drill problems that are linked to the section objectives. This is where you will find most of the problems that are new to this edition. Following the drill problems we have the categories of problems discussed below.

Applying the Concepts   Students are always curious about how the mathematics they are learning can be applied, so we have included applied problems in most of the problem sets in the book and have labeled them to show students the array of uses of mathematics. These applied problems are written in an inviting way, many times accompanied by new interesting illustrations to help students overcome some of the apprehension associated with application problems.

Getting Ready for the Next Section   Many students think of mathematics as a collection of discrete, unrelated topics. Their instructors know that this is not the case. The new Getting Ready for the Next Section problems reinforce the cumulative, connected nature of this course by showing how the concepts and techniques flow one from another throughout the course. These problems review all of the material that students will need in order to be successful in the next section, gently preparing students to move forward.

Maintaining Your Skills   One of the major themes of our book is continuous review. We strive to continuously hone techniques learned earlier by keeping the important concepts in the forefront of the course. The Maintaining Your Skills problems review material from the previous chapter, or they review problems that form the foundation of the course—the problems that you expect students to be able to solve when they get to the next course.

End-of-Chapter Summary, Review, and Assessment

We have learned that students are more comfortable with a chapter that sums up what they have learned thoroughly and accessibly, and reinforces concepts and techniques well. To help students grasp concepts and get more practice, each chapter ends with the following features that together give a comprehensive reexamination of the chapter.

Chapter Summary   The chapter summary recaps all main points from the chapter in a visually appealing grid. In the margin, next to each topic where appropriate, is an example that illustrates the type of problem associated with the topic being reviewed. When students prepare for a test, they can use the chapter summary as a guide to the main concepts of the chapter.

Chapter Review   Following the chapter summary in each chapter is the chapter review. It contains an extensive set of problems that review all the main topics in the chapter. This feature can be used flexibly, as assigned review, as a recommended self-test for students as they prepare for examinations, or as an in-class quiz or test.

Cumulative Review   Starting in Chapter 2, following the chapter review is a set of problems that reviews material from preceding chapters. This keeps students current with past topics and helps them retain the information they study.

Chapter Test   This set of problems is representative of all the main points of the chapter. These don't contain as many problems as the chapter review, and should be completed in 50 minutes.

Chapter Projects   Each chapter closes with a pair of projects. One is a group project, suitable for students to work on in class. The second project is a research project for students to do outside of class and tends to be open ended.

Additional Features of the Book

Facts from Geometry   Many of the important facts from geometry are listed under this heading. In most cases, an example or two accompanies each of the facts to give students a chance to see how topics from geometry are related to the algebra they are learning.

Supplements

If you are interested in any of the supplements below, please contact your sales representative.

For the Instructor

Annotated Instructor's Edition   ISBN-10: 1-111-42784-9 | ISBN-13: 978-1-111-42784-4
This special instructor's version of the text contains answers next to exercises and instructor notes at the appropriate location.

Complete Solutions Manual   ISBN-10: 1-111-57492-8 | ISBN-13: 978-1-111-57492-5
This manual contains complete solutions for all problems in the text.

Cengage Instructor's Resource Binder for Algebra Activities
ISBN-10: 0-538-73675-5 | ISBN-13: 978-0-538-73675-6
NEW! Each section of the main text is discussed in uniquely designed Teaching Guides containing instruction tips, examples, activities, worksheets, overheads, assessments, and solutions to all worksheets and activities.

Enhanced WebAssign   ISBN-10: 0-538-73810-3 | ISBN-13: 978-0-538-73810-1
Used by over 1 million students at more than 1,100 institutions, Enhanced WebAssign enables you to assign, collect, grade, and record homework assignments via the web. This proven and reliable homework system includes thousands of algorithmically generated homework problems, links to relevant textbook sections, video examples, problem-specific tutorials, and more. Diagnostic quizzing for each chapter identifies concepts that students still need to master and directs them to the appropriate review material. Students will appreciate the interactive eBook, which offers searching, highlighting, and note-taking functionality, as well as links to multimedia resources— all available to students when you choose Enhanced WebAssign.

PowerLecture with ExamView®   ISBN-10: 1-111-57486-3 | ISBN-13: 978-1-111-57486-4
This CD-ROM provides the instructor with dynamic media tools for teaching. Create, deliver, and customize tests (both print and online) in minutes with ExamView® computerized Testing Featuring Algorithmic Equations. Easily build solution sets for

homework or exams using Solution Builder's online solutions manual. Microsoft® PowerPoint® lecture slides, figures from the book, and Test Bank, in electronic format, are also included on this CD-Rom.

Solution Builder   This online instructor database offers complete worked solutions to all exercises in the text, allowing you to create customized, secure solutions printouts (in PDF format) matched exactly to the problems you assign in class. Visit http://www.cengage.com/solutionbuilder.

Text-Specific DVDs   ISBN-10: 1-111-57163-5 | ISBN-13: 978-1-111-57163-4
This set of text-specific DVDs features segments taught by the author and worked-out solutions to many examples in the book. Available to instructors only.

For the Student

Enhanced WebAssign   ISBN-10: 0-538-73810-3 | ISBN-13: 978-0-538-73810-1
Used by over 1 million students at more than 1,100 institutions, Enhanced WebAssign enables you to assign, collect, grade, and record homework assignments via the web. This proven and reliable homework system includes thousands of algorithmically generated homework problems, links to relevant textbook sections, video examples, problem-specific tutorials, and more.

Student Solutions Manual   ISBN-10: 1-111-57508-8 | ISBN-13: 978-1-111-57508-3
The Student Solutions Manual provides worked-out solutions to the odd-numbered problems in the textbook. The pretest and chapter test sections now include all solutions.

Student Workbook   ISBN-10: 1-111-57509-6 | ISBN-13: 978-1-111-57509-0
The Student Workbook contains all of the Assessments, Activities, and Worksheets from the Instructor's Resource Binder for classroom discussions, in-class activities, and group work.

Acknowledgments

I would like to thank my editor at Cengage Learning, Marc Bove, for his help and encouragement with this project, and Jennifer Risden, our production manager at Cengage Learning, for keeping our team on track. Donna Looper, the head of production at XYZ Textbooks, along with Staci Truelson did a tremendous job in organizing and planning the details of the production of the fourth edition of *Elementary and Intermediate Algebra*. Special thanks to our other team members Mary Gentilucci, Matthew Hoy, Kristina Chung and Katherine Shields; all of whom played important roles in the production of this book.

Pat McKeague
September 2010

Preface to the Student

I often find my students asking themselves the question "Why can't I understand this stuff the first time?" The answer is "You're not expected to." Learning a topic in mathematics isn't always accomplished the first time around. There are many instances when you will find yourself reading over new material a number of times before you can begin to work problems. That's just the way things are in mathematics. If you don't understand a topic the first time you see it, that doesn't mean there is something wrong with you. Understanding mathematics takes time. The process of understanding requires reading the book, studying the examples, working problems, and getting your questions answered.

How to Be Successful in Mathematics

1. If you are in a lecture class, be sure to attend all class sessions on time.   You cannot know exactly what goes on in class unless you are there. Missing class and then expecting to find out what went on from someone else is not the same as being there yourself.

2. Read the book.   It is best to read the section that will be covered in class beforehand. Reading in advance, even if you do not understand everything you read, is still better than going to class with no idea of what will be discussed.

3. Work problems every day and check your answers.   The key to success in mathematics is working problems. The more problems you work, the better you will become at working them. The answers to the odd-numbered problems are given in the back of the book. When you have finished an assignment, be sure to compare your answers with those in the book. If you have made a mistake, find out what it is, and correct it.

4. Do it on your own.   Don't be misled into thinking someone else's work is your own. Having someone else show you how to work a problem is not the same as working the same problem yourself. It is okay to get help when you are stuck. As a matter of fact, it is a good idea. Just be sure you do the work yourself.

5. Review every day.   After you have finished the problems your instructor has assigned, take another 15 minutes and review a section you have already completed. The more you review, the longer you will retain the material you have learned.

6. Don't expect to understand every new topic the first time you see it.   Sometimes you will understand everything you are doing, and sometimes you won't. Expecting to understand each new topic the first time you see it can lead to disappointment and frustration. The process of understanding takes time. You will need to read the book, work problems, and get your questions answered.

7. Spend as much time as it takes for you to master the material.   No set formula exists for the exact amount of time you need to spend on mathematics to master it. You will find out as you go along what is or isn't enough time for you. If you end up spending 2 or more hours on each section in order to master the material there, then that's how much time it takes; trying to get by with less will not work.

8. Relax.   It's probably not as difficult as you think.

1

The Basics

Image © 2010 DigitalGlobe
Image © 2010 IGN France

Introduction

In 1170 AD, the famous mathematician Fibonacci was born in Pisa, Italy. During this time, Europeans were still using Roman numerals to compute mathematical problems. A young Fibonacci often traveled with his father, a wealthy Italian merchant, across the Mediterranean Sea to Northern Africa. Working alongside his father in a busy port, Fibonacci learned of the simpler Hindu-Arabic numeral system that used ten digits, 0–9, instead of the more complicated Roman numerals. These ten digits fascinated Fibonacci and fueled his desire to study mathematics. By age 32, he had published a book of his mathematical knowledge called *Liber Abaci* (Book of Calculation) that was widely acclaimed by European scholars. It is in this book that Fibonacci first introduced Europe to the following number sequence, later named for him.

Fibonacci sequence: 1, 1, 2, 3, 5, 8, 13, 21, . . .

In this chapter, we will learn how to manipulate real numbers through addition, subtraction, multiplication, and division. Then using the characteristics and properties of real numbers, we will analyze number sequences, such as the Fibonacci sequence.

Getting Ready for Chapter 1

To get started in this book, we assume that you can do simple addition and multiplication problems with whole numbers and decimals. To check to see that you are ready for this chapter, work each of the problems below.

1. $5 \cdot 5 \cdot 5$

2. $12 \div 4$

3. $1.7 - 1.2$

4. $10 - 9.5$

5. $7(0.2)$

6. $0.3(6)$

7. $14 - 9$

8. $39 \div 13$

9. $5 \cdot 9$

10. $2 \cdot 2 \cdot 2 \cdot 2 \cdot 2$

11. $15 + 14$

12. $28 \div 7$

13. $26 \div 2$

14. $125 - 81$

15. $630 \div 63$

16. $210 \div 10$

17. $2 \cdot 3 \cdot 5 \cdot 7$

18. $3 \cdot 7 \cdot 11$

19. $2 \cdot 3 \cdot 3 \cdot 5 \cdot 7$

20. $24 \div 8$

Chapter Outline

1.1 Variables, Notation, and Symbols

A Translate between phrases written in English and expressions written in symbols.

B Simplify expressions containing exponents.

C Simplify expressions using the rule for order of operations.

D Recognize the pattern in a sequence of numbers.

1.2 Real Numbers

A Locate and label points on the number line.

B Change a fraction to an equivalent fraction with a new denominator.

C Identify the absolute value, opposite, and reciprocal of a number.

D Find the value of an algebraic expression.

E Find the perimeter and area of squares, rectangles, and triangles.

1.3 Addition and Subtraction of Real Numbers

A Add any combination of positive and negative numbers.

B Extend an arithmetic sequence.

C Subtract any combination of positive and negative numbers.

D Simplify expressions using the rule for order of operations.

E Find the complement and the supplement of an angle.

1.4 Multiplication of Real Numbers

A Multiply any combination of positive and negative numbers.

B Simplify expressions using the rule for order of operations.

C Multiply positive and negative fractions.

D Multiply using the multiplication property of zero.

E Extend a geometric sequence.

1.5 Division of Real Numbers

A Divide any combination of positive and negative numbers.

B Divide positive and negative fractions.

C Simplify expressions using the rule for order of operations.

1.6 Properties of Real Numbers

A Rewrite expressions using the commutative and associative properties.

B Multiply using the distributive property.

C Identify properties used to rewrite an expression.

1.7 Subsets of Real Numbers

A Associate numbers with subsets of the real numbers.

B Factor whole numbers into the product of prime factors.

C Reduce fractions to lowest terms using prime factorization.

1.8 Addition and Subtraction of Fractions with Variables

A Add or subtract two or more fractions with the same denominator.

B Find the least common denominator for a set of fractions.

C Add or subtract fractions with different denominators.

D Extend a sequence of numbers containing fractions.

1.1 Variables, Notation, and Symbols

OBJECTIVES

A Translate between phrases written in English and expressions written in symbols.

B Simplify expressions containing exponents.

C Simplify expressions using the rule for order of operations.

D Recognize the pattern in a sequence of numbers.

TICKET TO SUCCESS

Each section of the book will begin with some problems and questions like the ones below. Think about them while you read through the following section. Before you go to class, answer each problem or question with a written response using complete sentences. Writing about mathematics is a valuable exercise. As with all problems in this course, approach these writing exercises with a positive point of view. You will get better at giving written responses to questions as the course progresses. Even if you never feel comfortable writing about mathematics, just attempting the process will increase your understanding and ability in this course.

Keep these questions in mind as you read through the section. Then respond in your own words and in complete sentences.

1. What is a variable?
2. Why is it important to translate expressions written in symbols into the English language?
3. Write the four steps in the rule for order of operations.
4. What is inductive reasoning?

Image copyright © Perkus. Used under license from Shutterstock.com

A Variables: An Intuitive Look

When you filled out the application for the school you are attending, there was a space to fill in your first name. "First name" is a variable quantity because the value it takes depends on who is filling out the application. For example, if your first name is Manuel, then the value of "First Name" is Manuel. However, if your first name is Christa, then the value of "First Name" is Christa.

If we denote "First Name" as *FN*, "Last Name" as *LN*, and "Whole Name" as *WN*, then we take the concept of a variable further and write the relationship between the names this way:

$$FN + LN = WN$$

We use the + symbol loosely here to represent writing the names together with a space between them. This relationship we have written holds for all people who have only a first name and a last name. For those people who have a middle name, the relationship between the names is

$$FN + MN + LN = WN$$

where MN is "Middle Name."

A similar situation exists in algebra when we let a letter stand for a number or a group of numbers. For instance, if we say "let a and b represent numbers," then a and b are called *variables* because the values they take on vary. Furthermore, much of what we do in algebra involves comparison of quantities. We will begin by listing some symbols used to compare mathematical quantities. The comparison symbols fall into two major groups: equality symbols and inequality symbols.

> **Definition**
> A **variable** is a letter that stands for (represents) a mathematical quantity.

We use the variables a and b in the following lists so that the relationships shown there are true for all numbers that we will encounter in this book. By using variables, the following statements are general statements about all numbers, rather than specific statements about only a few numbers.

Comparison Symbols		
Equality:	$a = b$	a is equal to b (a and b represent the same number)
	$a \neq b$	a is not equal to b
Inequality:	$a < b$	a is less than b
	$a \not< b$	a is not less than b
	$a > b$	a is greater than b
	$a \not> b$	a is not greater than b
	$a \geq b$	a is greater than or equal to b
	$a \leq b$	a is less than or equal to b

The symbols for inequality, $<$ and $>$, always point to the smaller of the two quantities being compared. For example, $3 < x$ means 3 is smaller than x. In this case we can say "3 is less than x" or "x is greater than 3"; both statements are correct. Similarly, the expression $5 > y$ can be read as "5 is greater than y" or as "y is less than 5" because the inequality symbol is pointing to y, meaning y is the smaller of the two quantities.

Next, we consider the symbols used to represent the four basic operations: addition, subtraction, multiplication, and division.

> **NOTE**
> In the past you may have used the notation 3×5 to denote multiplication. In algebra it is best to avoid this notation if possible, because the multiplication symbol $\times$ can be confused with the variable x when written by hand.

Operation Symbols		
Addition:	$a + b$	The *sum* of a and b
Subtraction:	$a - b$	The *difference* of a and b
Multiplication:	$a \cdot b, (a)(b), a(b), (a)b, ab$	The *product* of a and b
Division:	$a \div b, a/b, \dfrac{a}{b}, b\overline{)a}$	The *quotient* of a and b

When we encounter the word *sum*, the implied operation is addition. To find the sum of two numbers, we simply add them. *Difference* implies subtraction, *product* implies multiplication, and *quotient* implies division. Notice also that there is more than one way to write the product or quotient of two numbers.

Grouping Symbols

Parentheses ( ) and brackets [ ] are the symbols used for grouping numbers together. Occasionally, braces { } are also used for grouping, although they are usually reserved for set notation, as we shall see later in this chapter.

The following examples illustrate the relationship between the symbols for comparing, operating, and grouping and the English language.

EXAMPLES

Mathematical Expression	*English Equivalent*
1. $4 + 1 = 5$	The sum of 4 and 1 is 5.
2. $8 - 1 < 10$	The difference of 8 and 1 is less than 10.
3. $2(3 + 4) = 14$	Twice the sum of 3 and 4 is 14.
4. $3x \geq 15$	The product of 3 and x is greater than or equal to 15.
5. $\dfrac{y}{2} = y - 2$	The quotient of y and 2 is equal to the difference of y and 2.

B Exponents

The last type of notation we need to discuss is the notation that allows us to write repeated multiplications in a more compact form—*exponents*. In the expression 2^3, the 2 is called the *base* and the 3 is called the *exponent*. The exponent 3 tells us the number of times the base appears in the product; that is,

$$2^3 = 2 \cdot 2 \cdot 2 = 8$$

The expression 2^3 is said to be in exponential form, whereas $2 \cdot 2 \cdot 2$ is said to be in expanded form.

Notation and Vocabulary Here is how we read expressions containing exponents.

Mathematical Expression	Written Equivalent
5^2	five to the second power
5^3	five to the third power
5^4	five to the fourth power
5^5	five to the fifth power
5^6	five to the sixth power

We have a shorthand vocabulary for second and third powers because the area of a square with a side of 5 is 5^2, and the volume of a cube with a side of 5 is 5^3.

5^2 can be read "five squared." 5^3 can be read "five cubed."

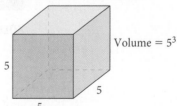

Here are some additional examples of expressions involving exponents.

EXAMPLES Expand and multiply.

6. $5^2 = 5 \cdot 5 = 25$ Base 5, exponent 2
7. $2^5 = 2 \cdot 2 \cdot 2 \cdot 2 \cdot 2 = 32$ Base 2, exponent 5
8. $10^3 = 10 \cdot 10 \cdot 10 = 1,000$ Base 10, exponent 3

C Order of Operations

The symbols for comparing, operating, and grouping are to mathematics what punctuation marks are to English. These symbols are the punctuation marks for mathematics.

Consider the following sentence:

Paul said John is tall.

It can have two different meanings, depending on how it is punctuated.

1. "Paul," said John, "is tall."
2. Paul said, "John is tall."

Let's take a look at a similar situation in mathematics. Consider the following mathematical statement:

$5 + 2 \cdot 7$

If we add the 5 and 2 first and then multiply by 7, we get an answer of 49. However, if we multiply the 2 and the 7 first and then add 5, we are left with 19. We have a problem that seems to have two different answers, depending on whether we add first or multiply first. We would like to avoid this type of situation. Every problem like $5 + 2 \cdot 7$ should have only one answer. Therefore, we have developed the following rule for the order of operations:

Rule: Order of Operations

When evaluating a mathematical expression, we will perform the operations in the following order:

1. Begin with the expression in the innermost parentheses or brackets first and work our way out.

2. Simplify all numbers with exponents, working from left to right if more than one of these expressions is present.

3. Work all multiplications and divisions left to right.

4. Perform all additions and subtractions left to right.

These next examples involve using the rule for order of operations.

EXAMPLE 9 Simplify $5 + 8 \cdot 2$.

SOLUTION $5 + 8 \cdot 2 = 5 + 16$ Multiply $8 \cdot 2$ first
                               $= 21$

EXAMPLE 10   Simplify $12 \div 4 \cdot 2$.

SOLUTION
$$12 \div 4 \cdot 2 = 3 \cdot 2$$
$$= 6$$

Work left to right

EXAMPLE 11   Simplify $2[5 + 2(6 + 3 \cdot 4)]$.

SOLUTION $2[5 + 2(6 + 3 \cdot 4)] = 2[5 + 2(6 + 12)]$ } Simplify within the innermost
$$= 2[5 + 2(18)]$$ grouping symbols first
$$= 2[5 + 36]$$ }
$$= 2[41]$$ } Next, simplify inside the brackets
$$= 82$$ Multiply

EXAMPLE 12   Simplify $10 + 12 \div 4 + 2 \cdot 3$.

SOLUTION $10 + 12 \div 4 + 2 \cdot 3 = 10 + 3 + 6$ Multiply and divide left to right
$$= 19$$ Add left to right

EXAMPLE 13   Simplify $2^4 + 3^3 \div 9 - 4^2$.

SOLUTION $2^4 + 3^3 \div 9 - 4^2 = 16 + 27 \div 9 - 16$ Simplify numbers with exponents
$$= 16 + 3 - 16$$ Then, divide
$$= 19 - 16$$ Finally, add and subtract
$$= 3$$ left to right

D Number Sequences and Inductive Reasoning

Suppose someone asks you to give the next number in the sequence of numbers below. (The dots mean that the sequence continues in the same pattern forever.)

$$2, 5, 8, 11, \ldots$$

If you notice that each number is 3 more than the number before it, you would say the next number in the sequence is 14 because $11 + 3 = 14$. When we reason in this way, we are using what is called *inductive reasoning*.

> **Definition**
>
> In mathematics, we use **inductive reasoning** when we notice a pattern to a sequence of numbers and then use the pattern to extend the sequence.

EXAMPLE 14   Use inductive reasoning to find the next number in each sequence.
  a. $3, 8, 13, 18, \ldots$
  b. $2, 10, 50, 250, \ldots$
  c. $2, 4, 7, 11, \ldots$

SOLUTION   To find the next number in each sequence, we need to look for a pattern or relationship.
  a. For the first sequence, each number is 5 more than the number before it; therefore, the next number will be $18 + 5 = 23$.

b. For the second sequence, each number is 5 times the number before it; therefore, the next number in the sequence will be $5 \cdot 250 = 1{,}250$.

c. For the third sequence, there is no number to add each time or multiply by each time. However, the pattern becomes apparent when we look at the differences between consecutive numbers:

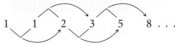

Proceeding in the same manner, we would add 5 to get the next number, giving us $11 + 5 = 16$. ■

In the introduction to this chapter we mentioned the mathematician known as Fibonacci. There is a special sequence in mathematics named for Fibonacci. Here it is.

Fibonacci sequence = 1, 1, 2, 3, 5, 8, . . .

Can you see the relationship among the numbers in this sequence? Start with two 1's, then add two consecutive members of the sequence to get the next number. Here is a diagram.

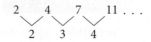

Sometimes we refer to the numbers in a sequence as *terms* of the sequence.

EXAMPLE 15   Write the first 10 terms of the Fibonacci sequence.

SOLUTION   The first six terms are given above. We extend the sequence by adding 5 and 8 to obtain the seventh term, 13. Then we add 8 and 13 to obtain 21. Continuing in this manner, the first 10 terms in the Fibonacci sequence are

1, 1, 2, 3, 5, 8, 13, 21, 34, 55   ■

Problem Set 1.1

Moving Toward Success

"Failing to prepare is preparing to fail."

—John Wooden, 1910–2010,
former UCLA basketball coach

1. What grade are you going to earn in this course? What will you have to do to achieve it?

2. How many hours a day outside of class are you setting aside for this course? Is that enough?

A For each sentence below, write an equivalent expression in symbols. [Examples 1–5]

1. The sum of x and 5 is 14.

2. The difference of x and 4 is 8.

3. The product of 5 and y is less than 30.

4. The product of 8 and y is greater than 16.

5. The product of 3 and y is less than or equal to the sum of y and 6.

6. The product of 5 and y is greater than or equal to the difference of y and 16.

7. The quotient of x and 3 is equal to the sum of x and 2.

8. The quotient of x and 2 is equal to the difference of x and 4.

B Expand and multiply. [Examples 6–8]

9. 3^2

10. 4^2

11. 7^2

12. 9^2

13. 2^3

14. 3^3

15. 4^3

16. 5^3

17. 2^4

18. 3^4

19. 10^2

20. 10^4

21. 11^2

22. 111^2

Simplify each expression.

23. $20 \div 2 \cdot 10$

24. $40 \div 4 \cdot 5$

25. $24 \div 8 \cdot 3$

26. $24 \div 4 \cdot 6$

27. $36 \div 6 \cdot 3$

28. $36 \div 9 \cdot 2$

29. $48 \div 12 \cdot 2$

30. $48 \div 8 \cdot 3$

31. $16 - 8 + 4$

32. $16 - 4 + 8$

33. $16 - 4 + 6$

34. $24 - 16 + 6$

35. $36 - 6 + 12$

36. $36 - 9 + 20$

37. $48 - 12 + 17$

38. $48 - 13 + 15$

C Use the rule for order of operations to simplify each expression as much as possible. [Examples 9–13]

39. $2 \cdot 3 + 5$

40. $8 \cdot 7 + 1$

41. $2(3 + 5)$

42. $8(7 + 1)$

43. $5 + 2 \cdot 6$

44. $8 + 9 \cdot 4$

45. $(5 + 2) \cdot 6$

46. $(8 + 9) \cdot 4$

47. $5 \cdot 4 + 5 \cdot 2$

48. $6 \cdot 8 + 6 \cdot 3$

49. $5(4 + 2)$

50. $6(8 + 3)$

51. $8 + 2(5 + 3)$

52. $7 + 3(8 - 2)$

53. $(8 + 2)(5 + 3)$

54. $(7 + 3)(8 - 2)$

55. $20 + 2(8 - 5) + 1$

56. $10 + 3(7 + 1) + 2$

57. $5 + 2(3 \cdot 4 - 1) + 8$

58. $11 - 2(5 \cdot 3 - 10) + 2$

59. $8 + 10 \div 2$

60. $16 - 8 \div 4$

61. $4 + 8 \div 4 - 2$

62. $6 + 9 \div 3 + 2$

63. $3 + 12 \div 3 + 6 \cdot 5$

64. $18 + 6 \div 2 + 3 \cdot 4$

65. $3 \cdot 8 + 10 \div 2 + 4 \cdot 2$

66. $5 \cdot 9 + 10 \div 2 + 3 \cdot 3$

67. $(5 + 3)(5 - 3)$

68. $(7 + 2)(7 - 2)$

69. $5^2 - 3^2$

70. $7^2 - 2^2$

71. $(4 + 5)^2$

72. $(6 + 3)^2$

73. $4^2 + 5^2$

74. $6^2 + 3^2$

75. $3 \cdot 10^2 + 4 \cdot 10 + 5$

76. $6 \cdot 10^2 + 5 \cdot 10 + 4$

77. $2 \cdot 10^3 + 3 \cdot 10^2 + 4 \cdot 10 + 5$

78. $5 \cdot 10^3 + 6 \cdot 10^2 + 7 \cdot 10 + 8$

79. $10 - 2(4 \cdot 5 - 16)$

80. $15 - 5(3 \cdot 2 - 4)$

81. $4[7 + 3(2 \cdot 9 - 8)]$

82. $5[10 + 2(3 \cdot 6 - 10)]$

83. $5(7 - 3) + 8(6 - 4)$

84. $3(10 - 4) + 6(12 - 10)$

85. $3(4 \cdot 5 - 12) + 6(7 \cdot 6 - 40)$

86. $6(8 \cdot 3 - 4) + 5(7 \cdot 3 - 1)$

87. $3^4 + 4^2 \div 2^3 - 5^2$

88. $2^5 + 6^2 \div 2^2 - 3^2$

89. $5^2 + 3^4 \div 9^2 + 6^2$

90. $6^2 + 2^5 \div 4^2 + 7^2$

D Find the next number in each sequence. [Examples 14–15]

91. 1, 2, 3, 4, . . . (The sequence of counting numbers)

92. 0, 1, 2, 3, . . . (The sequence of whole numbers)

93. 2, 4, 6, 8, . . . (The sequence of even numbers)

94. 1, 3, 5, 7, . . . (The sequence of odd numbers)

95. 1, 4, 9, 16, . . . (The sequence of squares)

96. 1, 8, 27, 64, . . . (The sequence of cubes)

97. 2, 2, 4, 6, . . . (A Fibonacci-like sequence)

98. 5, 5, 10, 15, . . . (A Fibonacci-like sequence)

Applying the Concepts

Food Labels In 1993 the government standardized the way in which nutrition information was presented on the labels of most packaged food products. Figure 1 shows a standardized food label from a package of cookies that I ate at lunch the day I was writing the problems for this problem set. Use the information in Figure 1 to answer the following questions.

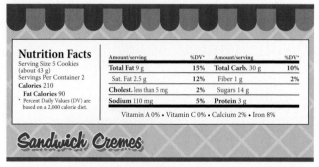

FIGURE 1

99. How many cookies are in the package?

100. If I paid $0.50 for the package of cookies, how much did each cookie cost?

101. If the "calories" category stands for calories per serving, how many calories did I consume by eating the whole package of cookies?

102. Suppose that, while swimming, I burn 11 calories each minute. If I swim for 20 minutes, will I burn enough calories to cancel out the calories I added by eating 5 cookies?

Education The chart shows the average income for people with different levels of education. Use the information in the chart to answer the following questions.

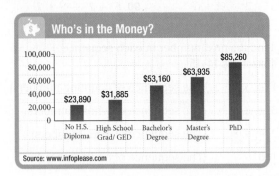

103. At a high-school reunion seven friends get together. One of them has a Ph.D., two have master's degrees, three have bachelor's degrees, and one did not go to college. Use order of operations to find their combined annual income.

104. At a dinner party for six friends, one has a master's degree, two have bachelor's degrees, two have high school diplomas, and one did not finish high school. Use order of operations to find their combined annual income.

105. **Reading Tables and Charts** The following bar chart gives the number of calories burned by a 150-pound person during 1 hour of various exercises. The accompanying table displays the same information. Use the bar chart to complete the table.

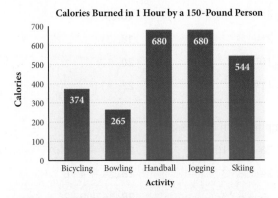

CALORIES BURNED BY 150-POUND PERSON	
Activity	**Calories Burned in 1 Hour**
Bicycling	374
Bowling	
Handball	
Jogging	
Skiing	

Real Numbers

OBJECTIVES

A Locate and label points on the number line.

B Change a fraction to an equivalent fraction with a new denominator.

C Identify the absolute value, opposite, and reciprocal of a number.

D Find the value of an algebraic expression.

E Find the perimeter and area of squares, rectangles, and triangles.

TICKET TO SUCCESS

Keep these questions in mind as you read through the section. Then respond in your own words and in complete sentences.

1. What is a real number?
2. Find the absolute value, the opposite, and the reciprocal of the number 4.
3. How do you find the value of the algebraic expression $4x + 6$ if $x = 3$?
4. Explain how you find the perimeter and the area of a rectangle.

Table 1 and Figure 1 give the record low temperature, in degrees Fahrenheit, for each month of the year in the city of Jackson, Wyoming. Notice that some of these temperatures are represented by negative numbers.

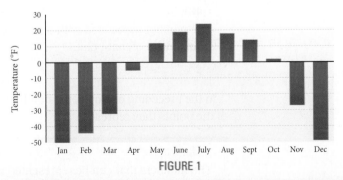

FIGURE 1

TABLE 1			
RECORD LOW TEMPERATURES FOR JACKSON, WYOMING			
Month	**Temperature (Degrees Fahrenheit)**	**Month**	**Temperature (Degrees Fahrenheit)**
January	−50	July	24
February	−44	August	18
March	−32	September	14
April	−5	October	2
May	12	November	−27
June	19	December	−49

A The Number Line

In this section we start our work with negative numbers. To represent negative numbers in algebra, we use what is called the *real number line*. Here is how we construct a real number line: We first draw a straight line and label a convenient point on the line with 0. Then we mark off equally spaced distances in both directions from 0. Label the points to the right of 0 with the numbers 1, 2, 3, . . . (the dots mean "and so on"). The points to the left of 0 we label in order, −1, −2, −3, ... Here is what it looks like.

> **NOTE**
> If there is no sign (+ or −) in front of a number, the number is assumed to be positive (+).

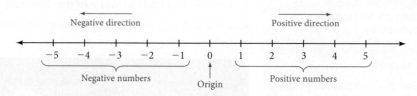

> **NOTE**
> There are other numbers on the number line that you may not be as familiar with. They are irrational numbers such as π, $\sqrt{2}$, $\sqrt{3}$. We will introduce these numbers later in the chapter.

The numbers increase in value going from left to right. If we "move" to the right, we are moving in the positive direction. If we move to the left, we are moving in the negative direction. When we compare two numbers on the number line, the number on the left is always smaller than the number on the right. For instance, −3 is smaller than −1 because it is to the left of −1 on the number line.

EXAMPLE 1   Locate and label the points on the real number line associated with the numbers -3.5, $-1\frac{1}{4}$, $\frac{1}{2}$, $\frac{3}{4}$, 2.5.

SOLUTION   We draw a real number line from −4 to 4 and label the points in question.

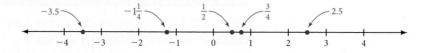

> **Definition**
> The number associated with a point on the real number line is called the **coordinate** of that point.

In the preceding example, the numbers $\frac{1}{2}$, $\frac{3}{4}$, 2.5, -3.5, and $-1\frac{1}{4}$ are the coordinates of the points they represent.

> **Definition**
> The numbers that can be represented with points on the real number line are called **real numbers**.

Real numbers include whole numbers, fractions, decimals, and other numbers that are not as familiar to us as these.

B Equivalent Fractions on the Number Line

As we proceed through Chapter 1, from time to time we will review some of the major concepts associated with fractions. To begin, here is the formal definition of a fraction:

Definition

If a and b are real numbers, then the expression

$$\frac{a}{b} \qquad b \neq 0$$

is called a **fraction.** The top number a is called the **numerator,** and the bottom number b is called the **denominator.** The restriction $b \neq 0$ keeps us from writing an expression that is undefined. (As you will see, division by zero is not allowed.)

The number line can be used to visualize fractions. Recall that for the fraction $\frac{a}{b}$, a is called the numerator and b is called the denominator. The denominator indicates the number of equal parts in the interval from 0 to 1 on the number line. The numerator indicates how many of those parts we have. If we take that part of the number line from 0 to 1 and divide it into *three equal parts,* we say that we have divided it into *thirds* (Figure 2). Each of the three segments is $\frac{1}{3}$ (one third) of the whole segment from 0 to 1.

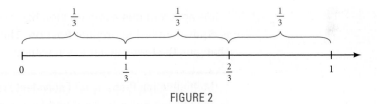

FIGURE 2

Two of these smaller segments together are $\frac{2}{3}$ (two thirds) of the whole segment, three of them would be $\frac{3}{3}$ (three thirds), or the whole segment.

Let's do the same thing again with six equal divisions of the segment from 0 to 1 (Figure 3). In this case we say each of the smaller segments has a length of $\frac{1}{6}$ (one sixth).

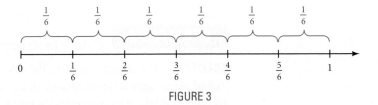

FIGURE 3

The same point we labeled with $\frac{1}{3}$ in Figure 2 is labeled with $\frac{2}{6}$ in Figure 3. Likewise, the point we labeled earlier with $\frac{2}{3}$ is now labeled $\frac{4}{6}$. It must be true then that

$$\frac{2}{6} = \frac{1}{3} \qquad \text{and} \qquad \frac{4}{6} = \frac{2}{3}$$

Actually, there are many fractions that name the same point as $\frac{1}{3}$. If we were to divide the segment between 0 and 1 into 12 equal parts, 4 of these 12 equal parts $\left(\frac{4}{12}\right)$ would be the same as $\frac{2}{6}$ or $\frac{1}{3}$; that is,

$$\frac{4}{12} = \frac{2}{6} = \frac{1}{3}$$

Even though these three fractions look different, each names the same point on the number line, as shown in Figure 4. All three fractions have the same value because they all represent the same number.

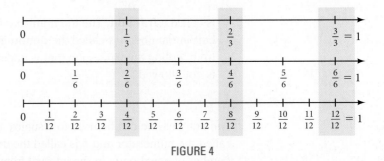

FIGURE 4

Definition
Fractions that represent the same number are said to be **equivalent**.
Equivalent fractions may look different, but they must have the same value.

It is apparent that every fraction has many different representations, each of which is equivalent to the original fraction. The following two properties give us a way of changing the terms of a fraction without changing its value.

Multiplication Property for Equivalent Fractions
Multiplying the numerator and denominator of a fraction by the same nonzero number never changes the value of the fraction.

Division Property for Equivalent Fractions
Dividing the numerator and denominator of a fraction by the same nonzero number never changes the value of the fraction.

EXAMPLE 2 Write $\frac{3}{4}$ as an equivalent fraction with denominator 20.

SOLUTION The denominator of the original fraction is 4. The fraction we are trying to find must have a denominator of 20. We know that if we multiply 4 by 5, we get 20. The multiplication property for equivalent fractions indicates that we are free to multiply the denominator by 5 as long as we do the same to the numerator.

$$\frac{3}{4} = \frac{3 \cdot 5}{4 \cdot 5} = \frac{15}{20}$$

The fraction $\frac{15}{20}$ is equivalent to the fraction $\frac{3}{4}$.

C Absolute Values, Opposites, and Reciprocals

Representing numbers on the number line lets us give each number two important properties: a direction from zero and a distance from zero. The direction from zero is represented by the sign in front of the number. (A number without a sign is understood to be positive.) The distance from zero is called the *absolute value* of the number, as the following definition indicates:

Definition

The **absolute value** of a real number is its distance from zero on the number line. If x represents a real number, then the absolute value of x is written $|x|$.

EXAMPLE 3 Write the expression $|5|$ without absolute value bars.

SOLUTION $|5| = 5$ The number 5 is 5 units from zero ■

EXAMPLE 4 Write the expression $|-5|$ without absolute value bars.

SOLUTION $|-5| = 5$ The number -5 is 5 units from zero ■

EXAMPLE 5 Write the expression $\left|-\dfrac{1}{2}\right|$ without absolute value bars.

SOLUTION $\left|-\dfrac{1}{2}\right| = \dfrac{1}{2}$ The number $-\frac{1}{2}$ is $\frac{1}{2}$ unit from zero ■

The absolute value of a number is *never* negative. It is the distance the number is from zero without regard to which direction it is from zero. When working with the absolute value of sums and differences, we must simplify the expression inside the absolute value symbols first and then find the absolute value of the simplified expression.

EXAMPLES Simplify each expression.

6. $|8 - 3| = |5| = 5$

7. $|3 \cdot 2^3 + 2 \cdot 3^2| = |3 \cdot 8 + 2 \cdot 9| = |24 + 18| = |42| = 42$

8. $|9 - 2| - |6 - 8| = |7| - |-2| = 7 - 2 = 5$ ■

Another important concept associated with numbers on the number line is that of *opposites*. Here is the definition:

Definition

Numbers the same distance from zero but in opposite directions from zero are called **opposites.**

EXAMPLE 9 Give the opposite of 5.

	Number	Opposite	
SOLUTION	5	-5	5 and -5 are opposites ■

EXAMPLE 10 Give the opposite of -3.

	Number	Opposite	
SOLUTION	-3	3	-3 and 3 are opposites ■

EXAMPLE 11   Give the opposite of $\frac{1}{4}$.

	Number	Opposite
SOLUTION	$\frac{1}{4}$	$-\frac{1}{4}$

$\frac{1}{4}$ and $-\frac{1}{4}$ are opposites

EXAMPLE 12   Give the opposite of −2.3.

	Number	Opposite
SOLUTION	−2.3	2.3

−2.3 and 2.3 are opposites

Each negative number is the opposite of some positive number, and each positive number is the opposite of some negative number. The opposite of a negative number is a positive number. In symbols, if a represents a positive number, then

$$-(-a) = a$$

Opposites always have the same absolute value and when you add any two opposites, the result is always zero:

$$a + (-a) = 0$$

Another concept we want to cover in this section is the concept of *reciprocals*. In order to understand reciprocals, we must use our knowledge of multiplication with fractions. To multiply two fractions, remember we simply multiply numerators and multiply denominators.

EXAMPLE 13   Multiply $\frac{3}{4} \cdot \frac{5}{7}$.

SOLUTION   The product of the numerators is 15, and the product of the denominators is 28.

$$\frac{3}{4} \cdot \frac{5}{7} = \frac{3 \cdot 5}{4 \cdot 7} = \frac{15}{28}$$

> **NOTE**
> In past math classes, you may have written fractions like $\frac{7}{3}$ (improper fractions) as mixed numbers, such as $2\frac{1}{3}$. In algebra, it is usually better to write them as improper fractions rather than mixed numbers.

EXAMPLE 14   Multiply $7\left(\frac{1}{3}\right)$.

SOLUTION   The number 7 can be thought of as the fraction $\frac{7}{1}$.

$$7\left(\frac{1}{3}\right) = \frac{7}{1}\left(\frac{1}{3}\right) = \frac{7 \cdot 1}{1 \cdot 3} = \frac{7}{3}$$

EXAMPLE 15   Expand and multiply $\left(\frac{2}{3}\right)^3$.

SOLUTION   Using the definition of exponents from the previous section, we have

$$\left(\frac{2}{3}\right)^3 = \frac{2}{3} \cdot \frac{2}{3} \cdot \frac{2}{3} = \frac{8}{27}$$

We are now ready for the definition of reciprocals.

> **Definition**
> Two numbers whose product is 1 are called **reciprocals.**

EXAMPLES Give the reciprocal of each number.

	Number	Reciprocal	
16.	5	$\dfrac{1}{5}$	Because $5\left(\dfrac{1}{5}\right) = \dfrac{5}{1}\left(\dfrac{1}{5}\right) = \dfrac{5}{5} = 1$
17.	2	$\dfrac{1}{2}$	Because $2\left(\dfrac{1}{2}\right) = \dfrac{2}{1}\left(\dfrac{1}{2}\right) = \dfrac{2}{2} = 1$
18.	$\dfrac{1}{3}$	3	Because $\dfrac{1}{3}(3) = \dfrac{1}{3}\left(\dfrac{3}{1}\right) = \dfrac{3}{3} = 1$
19.	$\dfrac{3}{4}$	$\dfrac{4}{3}$	Because $\dfrac{3}{4}\left(\dfrac{4}{3}\right) = \dfrac{12}{12} = 1$

Although we will not develop multiplication with negative numbers until later in the chapter, you should know that the reciprocal of a negative number is also a negative number. For example, the reciprocal of -4 is $-\dfrac{1}{4}$.

D The Value of an Algebraic Expression

Previously we mentioned that a variable is a letter used to represent a number or a group of numbers. An expression that contains variables with other numbers and symbols is called an *algebraic expression* and is further defined below.

> **Definition**
> An **algebraic expression** is an expression that contains any combination of numbers, variables, operation symbols, and grouping symbols. This definition includes the use of exponents and fractions.

Each of the following is an algebraic expression.

$$3x + 5 \qquad 4t^2 - 9 \qquad x^2 - 6xy + y^2 \qquad -15x^2y^4z^5 \qquad \frac{a^2 - 9}{a - 3} \qquad \frac{(x - 3)(x + 2)}{4x}$$

In the last two expressions, the fraction bar separates the numerator from the denominator and is treated the same as a pair of grouping symbols; it groups the numerator and denominator separately.

An expression such as $3x + 5$ will take on different values depending on what x is. If we were to let x equal 2, the expression $3x + 5$ would become 11. On the other hand, if x is 10, the same expression has a value of 35.

When →	$x = 2$	When →	$x = 10$
the expression →	$3x + 5$	the expression →	$3x + 5$
becomes →	$3(2) + 5$	becomes →	$3(10) + 5$
	$= 6 + 5$		$= 30 + 5$
	$= 11$		$= 35$

Table 2 lists some other algebraic expressions. It also gives some specific values for the variables and the corresponding value of the expression after the variable has been replaced with the given number.

TABLE 2

Original Expression	Value of the Variable	Value of the Expression
$5x + 2$	$x = 4$	$5(4) + 2 = 20 + 2$ $= 22$
$3x - 9$	$x = 2$	$3(2) - 9 = 6 - 9$ $= -3$
$4t^2 - 9$	$t = 5$	$4(5^2) - 9 = 4(25) - 9$ $= 100 - 9$ $= 91$
$\dfrac{a^2 - 9}{a - 3}$	$a = 8$	$\dfrac{8^2 - 9}{8 - 3} = \dfrac{64 - 9}{8 - 3}$ $= \dfrac{55}{5}$ $= 11$

EXAMPLE 20   Find the value of $3t^2 - 2$ when $t = 2$.

SOLUTION

$$\text{When} \rightarrow \quad t = 2$$
$$\text{the expression} \rightarrow \quad 3t^2 - 2$$
$$\text{becomes} \rightarrow \quad 3(2^2) - 2$$
$$= 3(4) - 2$$
$$= 12 - 2$$
$$= 10$$

E   Perimeter and Area of Geometric Figures

FACTS FROM GEOMETRY: Formulas for Area and Perimeter

A square, rectangle, and triangle are shown in the following figures. Note that we have labeled the dimensions of each with variables. The formulas for the perimeter and area of each object are given in terms of its dimensions.

NOTE
The vertical line labeled h in the triangle is its height, or altitude. It extends from the top of the triangle down to the base, meeting the base at an angle of 90°. The altitude of a triangle is always perpendicular to the base. The small square shown where the altitude meets the base is used to indicate that the angle formed is 90°.

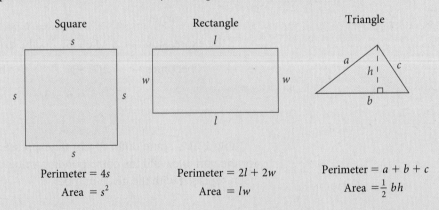

Square
Perimeter $= 4s$
Area $= s^2$

Rectangle
Perimeter $= 2l + 2w$
Area $= lw$

Triangle
Perimeter $= a + b + c$
Area $= \frac{1}{2}bh$

The formula for perimeter gives us the distance around the outside of the object along its sides, whereas the formula for area gives us a measure of the amount of surface the object occupies.

EXAMPLE 21 Find the perimeter and area of each figure.

a.
5 ft

b.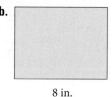
6 in.
8 in.

c.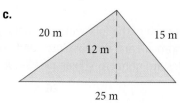
20 m   12 m   15 m
25 m

SOLUTION We use the preceding formulas to find the perimeter and the area. In each case, the units for perimeter are linear units, whereas the units for area are square units.

a. Perimeter = $4s = 4 \cdot 5$ feet = 20 feet
Area = $s^2 = (5 \text{ feet})^2 = 25$ square feet

b. Perimeter = $2l + 2w = 2(8 \text{ inches}) + 2(6 \text{ inches}) = 28$ inches
Area = $lw = (8 \text{ inches})(6 \text{ inches}) = 48$ square inches

c. Perimeter = $a + b + c$
  $= (20 \text{ meters}) + (25 \text{ meters}) + (15 \text{ meters})$
  $= 60$ meters
Area = $\frac{1}{2}bh = \frac{1}{2}(25 \text{ meters})(12 \text{ meters}) = 150$ square meters

Problem Set 1.2

Moving Toward Success

"Do not let what you cannot do interfere with what you can do."

—John Wooden, 1910–2010,
former UCLA basketball coach

1. Which of the following is most important for success in this course?

   a. Intelligence and natural ability

   b. Reading the book and working problems every day

   c. A good teacher

2. Which of the three items above do you have the most control over? Why is that important to understand?

A Draw a number line that extends from −5 to +5. Label the points with the following coordinates. [Example 1]

1. 5

2. −2

3. −4

4. −3

5. 1.5

6. −1.5

7. $\frac{9}{4}$

8. $\frac{8}{3}$

B Write each fraction of the following as an equivalent fraction with denominator 24. [Example 2]

9. $\frac{3}{4}$

10. $\frac{5}{6}$

11. $\frac{1}{2}$

12. $\frac{1}{8}$

13. $\frac{5}{8}$

14. $\frac{7}{12}$

B Write each fraction as an equivalent fraction with denominator 60. [Example 2]

15. $\frac{3}{5}$

16. $\frac{5}{12}$

17. $\frac{11}{30}$

18. $\frac{9}{10}$

Fill in the missing numerator so the fractions are equal. [Example 2]

19. $\frac{1}{-} = \frac{2}{4}$

20. $\frac{1}{5} = \frac{}{20}$

21. $\frac{5}{9} = \frac{}{45}$

22. $\frac{2}{5} = \frac{}{45}$

23. $\dfrac{3}{4} = \dfrac{}{8}$

24. $\dfrac{1}{2} = \dfrac{}{8}$

C For each of the following numbers, give the opposite, the reciprocal, and the absolute value. (Assume all variables are nonzero.) [Examples 3–19]

25. 10

26. 8

27. $\dfrac{3}{4}$

28. $\dfrac{5}{7}$

29. $\dfrac{11}{2}$

30. $\dfrac{16}{3}$

31. -3

32. -5

33. $-\dfrac{2}{5}$

34. $-\dfrac{3}{8}$

35. x

36. a

Place one of the symbols < or > between each of the following to make the resulting statement true.

37. $-5 \quad -3$

38. $-8 \quad -1$

39. $-3 \quad -7$

40. $-6 \quad 5$

41. $|-4| \quad -|-4|$

42. $3 \quad -|-3|$

43. $7 \quad -|-7|$

44. $-7 \quad |-7|$

45. $-\dfrac{3}{4} \quad -\dfrac{1}{4}$

46. $-\dfrac{2}{3} \quad -\dfrac{1}{3}$

47. $-\dfrac{3}{2} \quad -\dfrac{3}{4}$

48. $-\dfrac{8}{3} \quad -\dfrac{17}{3}$

C Simplify each expression. [Examples 3–8]

49. $|8 - 2|$

50. $|6 - 1|$

51. $|5 \cdot 2^3 - 2 \cdot 3^2|$

52. $|2 \cdot 10^2 + 3 \cdot 10|$

53. $|7 - 2| - |4 - 2|$

54. $|10 - 3| - |4 - 1|$

55. $10 - |7 - 2(5 - 3)|$

56. $12 - |9 - 3(7 - 5)|$

57. $15 - |8 - 2(3 \cdot 4 - 9)| - 10$

58. $25 - |9 - 3(4 \cdot 5 - 18)| - 20$

Multiply the following. [Examples 13–15]

59. $\dfrac{2}{3} \cdot \dfrac{4}{5}$

60. $\dfrac{1}{4} \cdot \dfrac{3}{5}$

61. $\dfrac{1}{2}(3)$

62. $\dfrac{1}{3}(2)$

63. $\dfrac{1}{4}(5)$

64. $\dfrac{1}{5}(4)$

65. $\dfrac{4}{3} \cdot \dfrac{3}{4}$

66. $\dfrac{5}{7} \cdot \dfrac{7}{5}$

67. $6\left(\dfrac{1}{6}\right)$

68. $8\left(\dfrac{1}{8}\right)$

69. $3 \cdot \dfrac{1}{3}$

70. $4 \cdot \dfrac{1}{4}$

Find the next number in each sequence.

71. $1, \dfrac{1}{3}, \dfrac{1}{5}, \dfrac{1}{7}, \ldots$ (Reciprocals of odd numbers)

72. $\dfrac{1}{2}, \dfrac{1}{4}, \dfrac{1}{6}, \dfrac{1}{8}, \ldots$ (Reciprocals of even numbers)

73. $1, \dfrac{1}{4}, \dfrac{1}{9}, \dfrac{1}{16}, \ldots$ (Reciprocals of squares)

74. $1, \dfrac{1}{8}, \dfrac{1}{27}, \dfrac{1}{64}, \ldots$ (Reciprocals of cubes)

D Problems here involve finding the value of an algrebraic expression. [Example 20]

75. Find the value of $2x - 6$ when
   a. $x = 5$
   b. $x = 10$
   c. $x = 15$
   d. $x = 20$

76. Find the value of $2(x - 3)$ when
   a. $x = 5$
   b. $x = 10$
   c. $x = 15$
   d. $x = 20$

77. Find the value of each expression when x is 10.
   a. $x + 2$
   b. $2x$
   c. x^2
   d. 2^x

78. Find the value of each expression when x is 3.
   a. $x + 3$
   b. $3x$
   c. x^2
   d. 3^x

79. Find the value of each expression when x is 4.
   a. $x^2 + 1$
   b. $(x + 1)^2$
   c. $x^2 + 2x + 1$

80. Find the value of $b^2 - 4ac$ when

    a. $a = 2, b = 6, c = 3$

    b. $a = 1, b = 5, c = 6$

    c. $a = 1, b = 2, c = 1$

E Find the perimeter and area of each figure. [Example 21]

81.

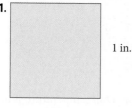

1 in.

1 in.

82.

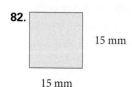

15 mm

15 mm

83.

0.75 in.

1.5 in.

84.

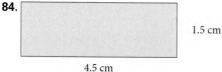

1.5 cm

4.5 cm

85.

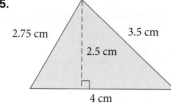

2.75 cm   3.5 cm

2.5 cm

4 cm

86.

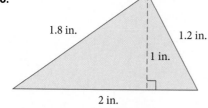

1.8 in.   1.2 in.

1 in.

2 in.

Applying the Concepts

87. Football Yardage A football team gains 6 yards on one play and then loses 8 yards on the next play. To what number on the number line does a loss of 8 yards correspond? The total yards gained or lost on the two plays corresponds to what negative number?

88. Checking Account Balance A woman has a balance of $20 in her checking account. If she writes a check for $30, what negative number can be used to represent the new balance in her checking account?

Temperature In the United States, temperature is measured on the Fahrenheit temperature scale. On this scale, water boils at 212 degrees and freezes at 32 degrees. To denote a temperature of 32 degrees on the Fahrenheit scale, we write

    32°F, which is read "32 degrees Fahrenheit"

Use this information for Problems 89 and 90.

89. Temperature and Altitude Marilyn is flying from Seattle to San Francisco on a Boeing 737 jet. When the plane reaches an altitude of 35,000 feet, the temperature outside the plane is 64 degrees below zero Fahrenheit. Represent the temperature with a negative number. If the temperature outside the plane gets warmer by 10 degrees, what will the new temperature be?

90. Temperature Change At 10:00 in the morning in White Bear Lake, Minnesota, John notices the temperature outside is 10 degrees below zero Fahrenheit. Write the temperature as a negative number. An hour later it has warmed up by 6 degrees. What is the temperature at 11:00 that morning?

91. Google Earth The Google image shows the Stratosphere Tower in Las Vegas, Nevada. The Stratosphere Tower is $\frac{23}{12}$ the height of the Space Needle in Seattle, Washington. If the Space Needle is about 600 feet tall, how tall is the Stratosphere Tower?

92. Skyscrapers The chart shows the heights of the three tallest buildings in the world. Bloomberg Tower in New York City is half the height of the Shanghai World Finance Center. How tall is the Bloomberg Tower?

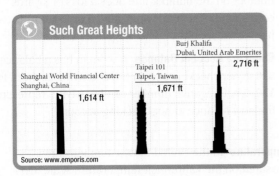

Such Great Heights

Burj Khalifa
Dubai, United Arab Emerites
2,716 ft

Taipei 101
Taipei, Taiwan
1,671 ft

Shanghai World Financial Center
Shanghai, China
1,614 ft

Source: www.emporis.com

Wind Chill Table 3 is a table of wind chill temperatures. The top row gives the air temperature, and the first column is wind speed in miles per hour. The numbers within the table indicate how cold the weather will feel. For example, if the thermometer reads 30°F and the wind is blowing at 15 miles per hour, the wind chill temperature is 9°F. Use Table 3 to answer Questions 93 through 96.

TABLE 3

WIND CHILL TEMPERATURES

Wind Speed (mph)	Air Temperature (°F)							
	30°	25°	20°	15°	10°	5°	0°	−5°
10	16°	10°	3°	−3°	−9°	−15°	−22°	−27°
15	9°	2°	−5°	−11°	−18°	−25°	−31°	−38°
20	4°	−3°	−10°	−17°	−24°	−31°	−39°	−46°
25	1°	−7°	−15°	−22°	−29°	−36°	−44°	−51°
30	−2°	−10°	−18°	−25°	−33°	−41°	−49°	−56°

93. Reading Tables Find the wind chill temperature if the thermometer reads 20°F and the wind is blowing at 25 miles per hour.

94. Reading Tables Which will feel colder: a day with an air temperature of 10°F with a 25-mile-per-hour wind, or a day with an air temperature of −5° F and a 10-mile-per-hour wind?

95. Reading Tables Find the wind chill temperature if the thermometer reads 5°F and the wind is blowing at 15 miles per hour.

96. Reading Tables Which will feel colder: a day with an air temperature of 5°F with a 20-mile-per-hour wind, or a day with an air temperature of −5° F and a 10-mile-per-hour wind?

1.3 Addition and Subtraction of Real Numbers

OBJECTIVES

A Add any combination of positive and negative numbers.

B Extend an arithmetic sequence.

C Subtract any combination of positive and negative numbers.

D Simplify expressions using the rule for order of operations.

E Find the complement and the supplement of an angle.

TICKET TO SUCCESS

Keep these questions in mind as you read through the section. Then respond in your own words and in complete sentences.

1. Explain how you would add 3 and −5 on the number line.

2. What is an arithmetic sequence?

3. Define subtraction in terms of addition.

4. What are complementary and supplementary angles?

Suppose that you are playing a friendly game of poker with some friends, and you lose $3 on the first hand and $4 on the second hand. If you represent winning with positive numbers and losing with negative numbers, how can you translate this situation into symbols? Because you lost $3 and $4 for a total of $7, one way to represent this situation is with addition of negative numbers:

$$(-\$3) + (-\$4) = -\$7$$

From this equation, we see that the sum of two negative numbers is a negative number. To generalize addition with positive and negative numbers, we use the number line.

Because real numbers have both a distance from zero (absolute value) and a direction from zero (sign), we can think of addition of two numbers in terms of distance and direction from zero.

A Adding Positive and Negative Numbers

Let's look at a problem for which we know the answer. Suppose we want to add the numbers 3 and 4. The problem is written 3 + 4. To put it on the number line, we read the problem as follows:

1. The 3 tells us to "start at the origin and move 3 units in the positive direction."
2. The + sign is read "and then move."
3. The 4 means "4 units in the positive direction."

To summarize, 3 + 4 means to start at the origin, move 3 units in the *positive* direction, and then move 4 units in the *positive* direction.

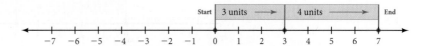

We end up at 7, which is the answer to our problem: 3 + 4 = 7.

Let's try adding other combinations of positive and negative 3 and 4 on the number line.

EXAMPLE 1   Add 3 + (−4).

SOLUTION Starting at the origin, move 3 units in the *positive* direction and then 4 units in the *negative* direction.

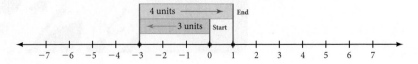

We end up at -1; therefore, $3 + (-4) = -1$.

EXAMPLE 2 Add $-3 + 4$.

SOLUTION Starting at the origin, move 3 units in the *negative* direction and then 4 units in the *positive* direction.

We end up at $+1$; therefore, $-3 + 4 = 1$.

EXAMPLE 3 Add $-3 + (-4)$.

SOLUTION Starting at the origin, move 3 units in the *negative* direction and then 4 units in the *negative* direction.

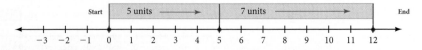

We end up at -7; therefore, $-3 + (-4) = -7$.

Here is a summary of what we have just completed:

$$3 + 4 = 7$$
$$3 + (-4) = -1$$
$$-3 + 4 = 1$$
$$-3 + (-4) = -7$$

Let's do four more problems on the number line and then summarize our results into a rule we can use to add any two real numbers.

EXAMPLE 4 Show that $5 + 7 = 12$.

SOLUTION

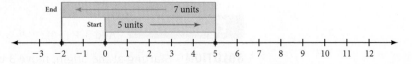

EXAMPLE 5 Show that $5 + (-7) = -2$.

SOLUTION

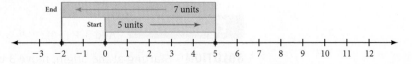

EXAMPLE 6   Show that $-5 + 7 = 2$.

SOLUTION

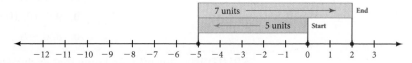

EXAMPLE 7   Show that $-5 + (-7) = -12$.

SOLUTION

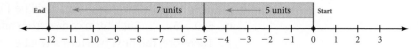

If we look closely at the results of the preceding addition problems, we can see that they support (or justify) the following rule:

> **NOTE**
> This rule is what we have been working toward. The rule is very important. Be sure that you understand it and can use it. The problems we have done up to this point have been done simply to justify this rule. Now that we have the rule, we no longer need to do our addition problems on the number line.

Rule: Adding Real Numbers

To add two real numbers with

1. The *same* sign: Simply add their absolute values and use the common sign. (Both numbers are positive, the answer is positive. Both numbers are negative, the answer is negative.)
2. *Different* signs: Subtract the smaller absolute value from the larger. The answer will have the sign of the number with the larger absolute value.

This rule covers all possible combinations of addition with real numbers. You must memorize it. After you have worked a number of problems, it will seem almost automatic.

EXAMPLE 8   Add all combinations of positive and negative 10 and 13.

SOLUTION   Rather than work these problems on the number line, we use the rule for adding positive and negative numbers to obtain our answers:

$$10 + 13 = 23$$
$$10 + (-13) = -3$$
$$-10 + 13 = 3$$
$$-10 + (-13) = -23$$

B Arithmetic Sequences

The pattern in a sequence of numbers is easy to identify when each number in the sequence comes from the preceding number by adding the same amount.

Definition

An **arithmetic sequence** is a sequence of numbers in which each number (after the first number) comes from adding the same amount to the number before it.

EXAMPLE 9   Each sequence below is an arithmetic sequence. Find the next two numbers in each sequence.

  a. 7, 10, 13, . . .
  b. 9.5, 10, 10.5, . . .
  c. 5, 0, −5, . . .

SOLUTION   Because we know that each sequence is arithmetic, we know to look for the number that is added to each term to produce the next consecutive term.

  a. 7, 10, 13, . . . : Each term is found by adding 3 to the term before it. Therefore, the next two terms will be 16 and 19.
  b. 9.5, 10, 10.5, . . . : Each term comes from adding 0.5 to the term before it. Therefore, the next two terms will be 11 and 11.5.
  c. 5, 0, −5, . . . : Each term comes from adding −5 to the term before it. Therefore, the next two terms will be $-5 + (-5) = -10$ and $-10 + (-5) = -15$.

C Subtracting Positive and Negative Numbers

Suppose that the temperature at noon is 20° Fahrenheit and 12 hours later, at midnight, it has dropped to −15° Fahrenheit. What is the difference between the temperature at noon and the temperature at midnight? Intuitively, we know the difference in the two temperatures is 35°. We also know that the word difference indicates subtraction. The difference between 20 and −15 is written

$$20 - (-15)$$

It must be true that $20 - (-15) = 35$. In the following examples, we will see how our definition for subtraction confirms that this last statement is in fact correct.

At the beginning of this section, we spent some time developing the rule for addition of real numbers. Because we want to make as few rules as possible, we can define subtraction in terms of addition. By doing so, we can then use the rule for addition to solve our subtraction problems.

Rule: Subtracting Real Numbers

To subtract one real number from another, simply add its opposite.

Algebraically, the rule is written like this: If a and b represent two real numbers, then it is always true that

$$a - b = a + (-b)$$

To subtract b add the opposite of b

This is how subtraction is defined in algebra. This definition of subtraction will not conflict with what you already know about subtraction, but it will allow you to do subtraction using negative numbers.

EXAMPLE 10   Subtract all possible combinations of positive and negative 7 and 2.

SOLUTION

$$\left. \begin{array}{l} 7 - 2 = 7 + (-2) = 5 \\ -7 - 2 = -7 + (-2) = -9 \end{array} \right\}$$ Subtracting 2 is the same as adding −2

$$7 - (-2) = 7 + 2 = 9$$
$$-7 - (-2) = -7 + 2 = -5$$

Subtracting -2 is the same as adding 2

Notice that each subtraction problem is first changed to an addition problem. The rule for addition is then used to arrive at the answer.

We have defined subtraction in terms of addition, and we still obtain answers consistent with the answers we are used to getting with subtraction. Moreover, we now can do subtraction problems involving both positive and negative numbers.

As you work more problems like these, you will begin to notice shortcuts you can use in working them. You will not always have to change subtraction to addition of the opposite to be able to get answers quickly. Use all the shortcuts you wish as long as you consistently get the correct answers.

EXAMPLE 11   Subtract all combinations of positive and negative 8 and 13.

SOLUTION

$$8 - 13 = 8 + (-13) = -5$$
$$-8 - 13 = -8 + (-13) = -21$$

Subtracting $+13$ is the same as adding -13

$$8 - (-13) = 8 + 13 = 21$$
$$-8 - (-13) = -8 + 13 = 5$$

Subtracting -13 is the same as adding 13

D Order of Operations

EXAMPLE 12   Add $-3 + 2 + (-4)$.

SOLUTION   Applying the rule for order of operations, we add left to right.

$$-3 + 2 + (-4) = -1 + (-4)$$
$$= -5$$

EXAMPLE 13   Add $-8 + [2 + (-5)] + (-1)$.

SOLUTION   Adding inside the brackets first and then left to right, we have

$$-8 + [2 + (-5)] + (-1) = -8 + (-3) + (-1)$$
$$= -11 + (-1)$$
$$= -12$$

EXAMPLE 14   Simplify $-10 + 2(-8 + 11) + (-4)$.

SOLUTION   First, we simplify inside the parentheses. Then, we multiply. Finally, we add left to right.

$$-10 + 2(-8 + 11) + (-4) = -10 + 2(3) + (-4)$$
$$= -10 + 6 + (-4)$$
$$= -4 + (-4)$$
$$= -8$$

EXAMPLE 15 Simplify the expression $7 + (-3) - 5$ as much as possible.

SOLUTION
$$7 + (-3) - 5 = 7 + (-3) + (-5)$$

Begin by changing subtraction to addition

$$= 4 + (-5)$$
$$= -1$$

Then add left to right

EXAMPLE 16 Simplify the expression $8 - (-2) - 6$ as much as possible.

SOLUTION
$$8 - (-2) - 6 = 8 + 2 + (-6)$$

Begin by changing all subtractions to additions

$$= 10 + (-6)$$
$$= 4$$

Then add left to right

EXAMPLE 17 Simplify the expression $-2 - (-3 + 1) - 5$ as much as possible.

SOLUTION
$$-2 - (-3 + 1) - 5 = -2 - (-2) - 5$$

Do what is in the parentheses first

$$= -2 + 2 + (-5)$$
$$= -5$$

The next two examples involve multiplication and exponents as well as subtraction. Remember, according to the rule for order of operations, we evaluate the numbers containing exponents and multiply before we subtract.

EXAMPLE 18 Simplify $2 \cdot 5 - 3 \cdot 8 - 4 \cdot 9$.

SOLUTION First, we multiply left to right, and then we subtract.

$$2 \cdot 5 - 3 \cdot 8 - 4 \cdot 9 = 10 - 24 - 36$$
$$= -14 - 36$$
$$= -50$$

EXAMPLE 19 Simplify $3 \cdot 2^3 - 2 \cdot 4^2$.

SOLUTION We begin by evaluating each number that contains an exponent. Then we multiply before we subtract.

$$3 \cdot 2^3 - 2 \cdot 4^2 = 3 \cdot 8 - 2 \cdot 16$$
$$= 24 - 32$$
$$= -8$$

E Complementary and Supplementary Angles

We can apply our knowledge of algebra to help solve some simple geometry problems. Before we do, however, we need to review some of the vocabulary associated with angles.

Definition
In geometry, two angles that add to 90° are called **complementary angles.**
In a similar manner, two angles that add to 180° are called **supplementary angles.** The diagrams that follow illustrate the relationships between angles that are complementary and between angles that are supplementary.

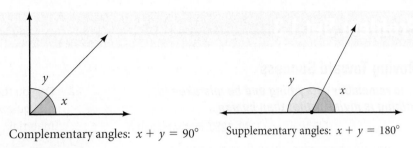

Complementary angles: $x + y = 90°$ Supplementary angles: $x + y = 180°$

EXAMPLE 20   Find x in each of the following diagrams.

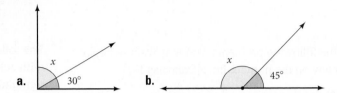

a.   30°   **b.**   45°

SOLUTION   We use subtraction to find each angle.

a. Because the two angles are complementary, we can find x by subtracting 30° from 90°:
$$x = 90° - 30° = 60°$$

We say 30° and 60° are complementary angles. The complement of 30° is 60°.

b. The two angles in the diagram are supplementary. To find x, we subtract 45° from 180°:
$$x = 180° - 45° = 135°$$

We say 45° and 135° are supplementary angles. The supplement of 45° is 135°.

Subtraction and Taking Away

For some people taking algebra for the first time, subtraction of positive and negative numbers can be a problem. These people may believe that $-5 - 9$ should be -4 or 4, not -14. If this is happening to you, you probably are thinking of subtraction in terms of taking one number away from another. Thinking of subtraction in this way works well with positive numbers if you always subtract the smaller number from the larger. In algebra, however, we encounter many situations other than this. The definition of subtraction, that $a - b = a + (-b)$, clearly indicates the correct way to use subtraction; that is, when working subtraction problems, you should think "addition of the opposite," not "take one number away from another." To be successful in algebra, you need to apply properties and definitions exactly as they are presented here.

Problem Set 1.3

Moving Toward Success

"...to remember everything and be mistaken in nothing is divine rather than human..."

—Fibonacci, 1170–1250, mathematician

1. Name two resources, other than the book and your instructor, that will help you get the grade you want in this course.

2. Do you think you will make mistakes on tests and quizzes, even when you understand the material?

A Work the following problems. You may want to begin by doing a few on the number line. [Examples 1–7]

1. $6 + (-3)$
2. $7 + (-8)$
3. $18 + (-32)$
4. $6 + (-9)$
5. $-6 + 3$
6. $-8 + 7$
7. $-30 + 5$
8. $-18 + 6$
9. $-6 + (-6)$
10. $-5 + (-5)$
11. $-10 + (-15)$
12. $-18 + (-30)$
13. $3.9 + 7.1$
14. $4.7 + 4.3$
15. $8.1 + 2.7$
16. $2.4 + 7.3$

A [Examples 8]

17. Add all combinations of positive and negative 3 and 5.

18. Add all combinations of positive and negative 6 and 4.

19. Add all combinations of positive and negative 15 and 20.

20. Add all combinations of positive and negative 18 and 12.

B Each sequence below is an arithmetic sequence. In each case, find the next two numbers in the sequence. [Example 9]

21. 3, 8, 13, 18, . . .
22. 1, 5, 9, 13, . . .
23. 10, 15, 20, 25, . . .
24. 10, 16, 22, 28, . . .
25. 6, 0, −6, . . .
26. 1, 0, −1, . . .

27. Is the sequence of odd numbers an arithmetic sequence?

28. Is the sequence of squares an arithmetic sequence?

The following problems are intended to give you practice with subtraction of positive and negative numbers. Remember in algebra subtraction is not taking one number away from another. Instead, subtracting a number is equivalent to adding its opposite.

C Subtract. [Examples 10–11}

29. $5 - 8$
30. $6 - 7$
31. $5 - 5$
32. $8 - 8$
33. $-8 - 2$
34. $-6 - 3$
35. $-4 - 12$
36. $-3 - 15$
37. $15 - (-20)$
38. $20 - (-5)$
39. $-4 - (-4)$
40. $-5 - (-5)$
41. $-3.4 - 7.9$
42. $-3.5 - 2.3$
43. $3.3 - 6.9$
44. $2.2 - 7.5$

D Work the following problems using the rule for addition of real numbers. You may want to refer back to the rule for order of operations. [Examples 12–17]

45. $5 + (-6) + (-7)$
46. $6 + (-8) + (-10)$
47. $5 + [6 + (-2)] + (-3)$
48. $10 + [8 + (-5)] + (-20)$
49. $[6 + (-2)] + [3 + (-1)]$
50. $[18 + (-5)] + [9 + (-10)]$
51. $-3 + (-2) + [5 + (-4)]$

D Simplify each expression by applying the rule for order of operations. [Examples 12–17]

52. $3 - 2 - 5$
53. $4 - 8 - 6$
54. $-6 - 8 - 10$
55. $-5 - 7 - 9$
56. $-22 + 4 - 10$
57. $-13 + 6 - 5$
58. $10 - (-20) - 5$
59. $15 - (-3) - 20$
60. $8 - (2 - 3) - 5$
61. $10 - (4 - 6) - 8$
62. $7 - (3 - 9) - 6$
63. $4 - (3 - 7) - 8$

64. $-(5 - 7) - (2 - 8)$ **65.** $-(4 - 8) - (2 - 5)$

66. $-(3 - 10) - (6 - 3)$ **67.** $-(3 - 7) - (1 - 2)$

68. $5 - [(2 - 3) - 4]$ **69.** $6 - [(4 - 1) - 9]$

70. $21 - [-(3 - 4) - 2] - 5$

71. $30 - [-(10 - 5) - 15] - 25$

D   The following problems involve multiplication and exponents. Use the rule for order of operations to simplify each expression as much as possible. [Examples 18–19]

72. $3 \cdot 5 - 2 \cdot 7$ **73.** $6 \cdot 10 - 5 \cdot 20$

74. $3 \cdot 8 - 2 \cdot 4 - 6 \cdot 7$ **75.** $5 \cdot 9 - 3 \cdot 8 - 4 \cdot 5$

76. $2 \cdot 3^2 - 5 \cdot 2^2$ **77.** $3 \cdot 7^2 - 2 \cdot 8^2$

78. $4 \cdot 3^3 - 5 \cdot 2^3$

79. $3 \cdot 6^2 - 2 \cdot 3^2 - 8 \cdot 6^2$

E   Find x in each of the following diagrams. [Example 20]

80.   **81.**

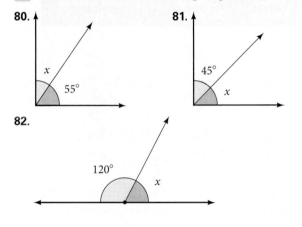

82.

83.

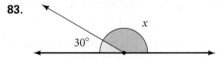

Applying the Concepts

84. Music The chart shows the country music artists that earned the most in 2008-2009. How much did Toby Keith and Tim McGraw make combined?

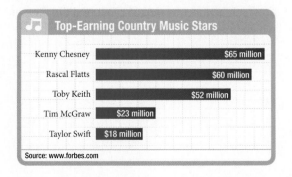

85. Internet The chart shows the three countries with the most internet users. What is the total number of internet users that can be found in the three countries?

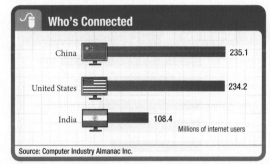

86. Temperature Change The temperature at noon is 12 degrees below 0 Fahrenheit. By 1:00 it has risen 4 degrees. Write an expression using the numbers -12 and 4 to describe this situation.

87. Temperature Change The temperature inside a space shuttle is 73°F before reentry. During reentry the temperature inside the craft increases 10°. On landing it drops 8°F. Write an expression using the numbers 73, 10, and 8 to describe this situation. What is the temperature inside the shuttle on landing?

88. Checkbook Balance Suppose that you balance your checkbook and find that you are overdrawn by $30; that is, your balance is $-\$30$. Then you go to the bank and deposit $40. Translate this situation into an addition problem, the answer to which gives the new balance in your checkbook.

89. Checkbook Balance The balance in your checkbook is $-\$25$. If you make a deposit of $75, and then write a check for $18, what is the new balance?

90. Checkbook Balance Bob has $98 in his checking account when he writes a check for $65 and then another check for $53. Write a subtraction problem that gives the new balance in Bob's checkbook. What is his new balance?

91. Pitchers The chart shows the number of career strikeouts for starting pitchers through the 2009 season. How many more strikeouts does Pedro Martinez have than Andy Pettitte?

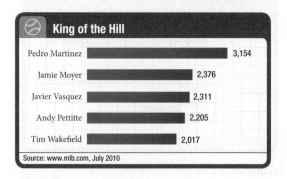

King of the Hill

Pedro Martinez	3,154
Jamie Moyer	2,376
Javier Vasquez	2,311
Andy Pettitte	2,205
Tim Wakefield	2,017

Source: www.mlb.com, July 2010

92. Gambling On three consecutive hands of draw poker, a gambler wins $10, loses $6, and then loses another $8. Write an expression using positive and negative numbers and addition to describe this situation and then simplify.

93. Gambling A man who has lost $35 playing roulette in Las Vegas wins $15 playing blackjack. He then loses $20 playing the wheel of fortune. Write an expression using the numbers −35, 15, and 20 to describe this situation and then simplify it.

94. Stock Value On Monday a certain stock gains 2 points. On Tuesday it loses 3 points. Write an expression using positive and negative numbers with addition to describe this situation and then simplify.

95. Number Problem You know from your past experience with numbers that subtracting 5 from 8 results in 3 (8 − 5 = 3). What addition problem that starts with the number 8 gives the same result?

96. Altitude Change An airplane flying at 10,000 feet lowers its altitude by 1,500 feet to avoid other air traffic. Then it increases its altitude by 3,000 feet to clear a mountain range. Write an expression that describes this situation and then simplify it.

97. Depreciation Stacey buys a used car for $4,500. With each year that passes, the car drops $550 in value. Write a sequence of numbers that gives the value of the car at the beginning of each of the first 5 years she owns it. Can this sequence be considered an arithmetic sequence?

98. Depreciation Wade buys a computer system for $6,575. Each year after that he finds that the system is worth $1,250 less than it was the year before.

Write a sequence of numbers that gives the value of the computer system at the beginning of each of the first four years he owns it. Can this sequence be considered an arithmetic sequence?

Drag Racing In the sport of drag racing, two cars at the starting line race to the finish line $\frac{1}{4}$ mile away. The car that crosses the finish line first wins the race. Jim Rizzoli owns and races an alcohol dragster. On board the dragster is a computer that records data during each of Jim's races. Table 1 gives some of the data from a race Jim was in. In addition to showing the time and speed of Jim Rizzoli's dragster during a race, it also shows the distance past the starting line that his dragster has traveled. Use the information in the table shown here to answer Problems 99–104.

TABLE 1		
SPEED AND DISTANCE FOR A RACE CAR		
Time in Seconds	**Speed in Miles/Hour**	**Distance Traveled in Feet**
0	0	0
1	72.7	69
2	129.9	231
3	162.8	439
4	192.2	728
5	212.4	1,000
6	228.1	1,373

99. Find the difference in the distance traveled by the dragster after 5 seconds and after 2 seconds.

100. How much faster is he traveling after 4 seconds than he is after 2 seconds?

101. How far from the starting line is he after 3 seconds?

102. How far from the starting line is he when his speed is 192.2 miles per hour?

103. How many seconds have gone by between the time his speed is 162.8 miles per hour and the time at which he has traveled 1,000 feet?

104. How many seconds have gone by between the time at which he has traveled 231 feet and the time at which his speed is 228.1 miles per hour?

1.4 Multiplication of Real Numbers

OBJECTIVES

A Multiply any combination of positive and negative numbers.

B Simplify expressions using the rule for order of operations.

C Multiply positive and negative fractions.

D Multiply using the multiplication property of zero.

E Extend a geometric sequence.

TICKET TO SUCCESS

Keep these questions in mind as you read through the section. Then respond in your own words and in complete sentences.

1. Give an example of how a multiplication problem is repeated addition.
2. How do you multiply two numbers with different signs?
3. How do you multiply two negative numbers?
4. What is a geometric sequence?

Suppose that you own 5 shares of a stock and the price per share drops $3. How much money have you lost? Intuitively, we know the loss is $15. Because it is a loss, we can express it as −$15. To describe this situation with numbers, we would write

5 shares each lose $3 for a total of $15

$$5(-3) = -15$$

Reasoning in this manner, we conclude that the product of a positive number with a negative number is a negative number. Let's look at multiplication in more detail.

A Multiplication of Real Numbers

From our experience with counting numbers, we know that multiplication is simply repeated addition; that is, $3(5) = 5 + 5 + 5$. We will use this fact, along with our knowledge of negative numbers, to develop the rule for multiplication of any two real numbers. The following examples illustrate multiplication with all of the possible combinations of positive and negative numbers.

EXAMPLES   Multiply.

1. Two positives:
$$3(5) = 5 + 5 + 5$$
$$= 15 \qquad \text{Positive answer}$$

2. One positive:
$$3(-5) = -5 + (-5) + (-5)$$
$$= -15 \qquad \text{Negative answer}$$

3. One negative:
$$-3(5) = -15 \qquad \text{Negative answer}$$

To understand why multiplying a negative by a positive (or a positive by a negative) results in a negative answer, we consider the number line. Recall that, on the number line, the negative sign means to move to the left of the zero. If we think of -3 in the problem $-3(5)$ as moving 5 units to the left of the zero 3 times, then we have.

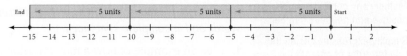

$$-5 + (-5) + (-5) = -15$$

4. Two negatives: $-3(-5) = ?$

With two negatives, $-3(-5)$, it is not possible to work the problem in terms of repeated addition. (It doesn't "make sense" to write -5 down a -3 number of times.) The answer is probably $+15$ (that's just a guess), but we need some justification for saying so. We will solve a different problem and in so doing get the answer to the problem $(-3)(-5)$.

Here is a problem to which we know the answer. We will work it two different ways.

$$-3[5 + (-5)] = -3(0) = 0$$

The answer is zero. We also can work the problem by distributing the -3 over both terms inside the parenthesis. The distributive property is one of the properties we will be covering more in depth later in the chapter.

$$-3[5 + (-5)] = -3(5) + (-3)(-5) \qquad \text{Distributive property}$$

$$= -15 + ?$$

Because the answer to the problem is 0, our ? must be $+15$. (What else could we add to -15 to get 0? Only $+15$.)

Here is a summary of the results we have obtained from the first four examples.

Original Numbers Have		The Answer is
the same sign	$3(5) = 15$	positive
different signs	$3(-5) = -15$	negative
different signs	$-3(5) = -15$	negative
the same sign	$-3(-5) = 15$	positive

By examining Examples 1 through 4 and the preceding table, we can use the information there to write the following rule. This rule tells us how to multiply any two real numbers.

> **Rule: Multiplying Real Numbers**
>
> To multiply any two real numbers, simply multiply their absolute values. The sign of the answer is
>
> **1.** *Positive* if both numbers have the same sign (both + or both −).
>
> **2.** *Negative* if the numbers have opposite signs (one +, the other −).

The following examples illustrate how we use the preceding rule to multiply real numbers.

NOTE
You may have to read the explanation for Example 4 several times before you understand it completely. The purpose of the explanation in Example 4 is simply to justify the fact that the product of two negative numbers is a positive number.

NOTE
Some students have trouble with the expression $-8(-3)$ because they want to subtract rather than multiply. Because we are very precise with the notation we use in algebra, the expression $-8(-3)$ has only one meaning—multiplication. A subtraction problem that uses the same numbers is $-8 - 3$. Compare the two following lists.

All Multiplication	No Multiplication
$5(4)$	$5 + 4$
$-5(4)$	$-5 + 4$
$5(-4)$	$5 - 4$
$-5(-4)$	$-5 - 4$

EXAMPLES   Multiply.

5. $-8(-3) = 24$
6. $-10(-5) = 50$ If the two numbers in the product have the same sign, the answer is positive
7. $-4(-7) = 28$

8. $5(-7) = -35$
9. $-4(8) = -32$ If the two numbers in the product have different signs, the answer is negative
10. $-6(10) = -60$

B Using Order of Operations

In the following examples, we combine the rule for order of operations with the rule for multiplication to simplify expressions. Remember, the rule for order of operations specifies that we are to work inside the parentheses first and then simplify numbers containing exponents. After this, we multiply and divide, left to right. The last step is to add and subtract, left to right.

EXAMPLE 11   Simplify $-5(-3)(-4)$ as much as possible.

SOLUTION $-5(-3)(-4) = 15(-4)$ Multiply
$\qquad\qquad\qquad = -60$

EXAMPLE 12   Simplify $4(-3) + 6(-5) - 10$ as much as possible.

SOLUTION $4(-3) + 6(-5) - 10 = -12 + (-30) - 10$ Multiply
$\qquad\qquad\qquad\qquad\quad = -42 - 10$ Add
$\qquad\qquad\qquad\qquad\quad = -52$ Subtract

EXAMPLE 13   Simplify $(-2)^3$ as much as possible.

SOLUTION $(-2)^3 = (-2)(-2)(-2)$ Definition of exponents
$\qquad\qquad = -8$ Multiply, left to right

EXAMPLE 14   Simplify $-3(-2)^3 - 5(-4)^2$ as much as possible.

SOLUTION $-3(-2)^3 - 5(-4)^2 = -3(-8) - 5(16)$ Exponents first
$\qquad\qquad\qquad\qquad = 24 - 80$ Multiply
$\qquad\qquad\qquad\qquad = -56$ Subtract

EXAMPLE 15   Simplify $6 - 4(7 - 2)$ as much as possible.

SOLUTION $6 - 4(7 - 2) = 6 - 4(5)$ Inside parentheses first
$\qquad\qquad\qquad = 6 - 20$ Multiply
$\qquad\qquad\qquad = -14$ Subtract

EXAMPLE 16   Figure 1 gives the calories that are burned in 1 hour for a variety of forms of exercise by a person weighing 150 pounds. Figure 2 gives the calories that are consumed by eating some popular fast foods. Find the net change in calories for a 150-pound person playing handball for 2 hours and then eating a Whopper.

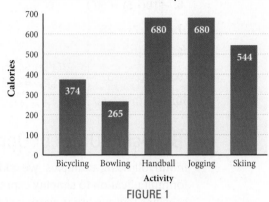

Calories Burned in 1 Hour by a 150-Pound Person

FIGURE 1

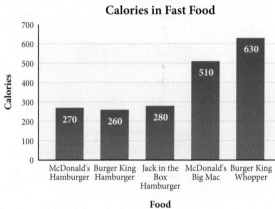

Calories in Fast Food

FIGURE 2

SOLUTION   The net change in calories will be the difference of the calories gained from eating and the calories lost from exercise.

$$\text{Net change in calories} = 630 - 2(680) = -730 \text{ calories}$$

C   Multiplying Fractions

Previously, we mentioned that to multiply two fractions we multiply numerators and multiply denominators. We can apply the rule for multiplication of positive and negative numbers to fractions in the same way we apply it to other numbers. We multiply absolute values: The product is positive if both fractions have the same sign and negative if they have different signs. Here are some examples.

EXAMPLE 17   Multiply $-\dfrac{3}{4}\left(\dfrac{5}{7}\right)$.

SOLUTION
$$-\frac{3}{4}\left(\frac{5}{7}\right) = -\frac{3 \cdot 5}{4 \cdot 7}$$ Different signs give a negative answer
$$= -\frac{15}{28}$$

EXAMPLE 18 Multiply $-6\left(\dfrac{1}{2}\right)$.

SOLUTION

$$-6\left(\frac{1}{2}\right) = -\frac{6}{1}\left(\frac{1}{2}\right) \qquad \text{Different signs give a negative answer}$$

$$= -\frac{6}{2}$$

$$= -3$$

EXAMPLE 19 Multiply $-\dfrac{2}{3}\left(-\dfrac{3}{2}\right)$.

SOLUTION

$$-\frac{2}{3}\left(-\frac{3}{2}\right) = \frac{2 \cdot 3}{3 \cdot 2} \qquad \text{Same signs give a positive answer}$$

$$= \frac{6}{6}$$

$$= 1$$

D The Multiplication Property of Zero

As previously mentioned, we will cover properties of real numbers later in the chapter. For now we will consider the multiplication property of zero. It is of great importance, as you will see, throughout our study of algebra and deserves special attention.

> **Rule: Multiplication Property of Zero**
> The product of any real number a and zero is 0.
>
> $$a(0) = 0 \qquad a \text{ is a real number}$$
>
> $$-a(0) = 0 \qquad a \text{ is a real number}$$
>
> $$0(0) = 0$$

EXAMPLE 20 Multiply.

   a. $5(0)$ **b.** $0(6)$ **c.** $5{,}398{,}423(0)$ **d.** $(-1)(-6)(0)$

SOLUTION

   a. $5(0) = 0$ **b.** $0(6) = 0$

   c. $5{,}398{,}423(0) = 0$ **d.** $(-1)(-6)(0) = 0$

Whenever you see a 0 as one of the numbers to be multiplied, you can simply write the answer down without any calculations. It doesn't matter what the other numbers multiply to be or if the numbers are positive, negative, fractions, very large, or very small.

EXAMPLE 21 Multiply.

   a. $\dfrac{1}{12}(0)$ **b.** $25^3(0)$ **c.** $(-3)\left(\dfrac{9}{13}\right)(0)$ **d.** $(465{,}832)(0.00026)(0)$

SOLUTION

   a. $\dfrac{1}{12}(0) = 0$ **b.** $25^3(0) = 0$

   c. $(-3)\left(\dfrac{9}{13}\right)(0) = 0$ **d.** $(465{,}832)(0.00026)(0) = 0$

E   Geometric Sequences

> **Definition**
> A **geometric sequence** is a sequence of numbers in which each number (after the first number) comes from the number before it by multiplying by the same amount each time.

For example, the sequence

$$2, 6, 18, 54, \ldots$$

is a geometric sequence because each number is obtained by multiplying the number before it by 3.

EXAMPLE 22   Each sequence below is a geometric sequence. Find the next number in each sequence.

a. $5, 10, 20, \ldots$ **b.** $3, -15, 75, \ldots$ **c.** $\frac{1}{8}, \frac{1}{4}, \frac{1}{2}, \ldots$

SOLUTION   Because each sequence is a geometric sequence, we know that each term is obtained from the previous term by multiplying by the same number each time.

a. $5, 10, 20, \ldots$: The sequence starts with 5. After that, each number is obtained from the previous number by multiplying by 2 each time. The next number will be $20 \cdot 2 = 40$.

b. $3, -15, 75, \ldots$: The sequence starts with 3. After that, each number is obtained by multiplying by -5 each time. The next number will be $75(-5) = -375$.

c. $\frac{1}{8}, \frac{1}{4}, \frac{1}{2}, \ldots$: This sequence starts with $\frac{1}{8}$. Multiplying each number in the sequence by 2 produces the next number in the sequence. To extend the sequence, we multiply $\frac{1}{2}$ by 2: $\frac{1}{2} \cdot 2 = 1$. The next number in the sequence is 1.

Problem Set 1.4

Moving Toward Success

"The quality of a person's life is in direct proportion to their commitment to excellence, regardless of their chosen field of endeavor."
—Vince Lombardi, 1913–1970, American football coach

1. What do successful students do when they are stuck on a problem?
2. If you make a mistake on a test or a quiz, how do you plan on learning from it?

A Use the rule for multiplying two real numbers to find each of the following products. [Examples 1–10]

1. $7(-6)$ **2.** $8(-4)$

3. $-8(2)$ **4.** $-16(3)$

5. $-3(-1)$ **6.** $-7(-1)$

7. $-11(-11)$ **8.** $-12(-12)$

B Use the rule for order of operations to simplify each expression as much as possible. [Examples 11–16]

9. $-3(2)(-1)$ **10.** $-2(3)(-4)$

11. $-3(-4)(-5)$ **12.** $-5(-6)(-7)$

13. $-2(-4)(-3)(-1)$ **14.** $-1(-3)(-2)(-1)$

15. $(-7)^2$ **16.** $(-8)^2$

17. $(-3)^3$ **18.** $(-2)^4$

19. $-2(2 - 5)$ **20.** $-3(3 - 7)$

21. $-5(8 - 10)$ **22.** $-4(6 - 12)$

23. $(4 - 7)(6 - 9)$ **24.** $(3 - 10)(2 - 6)$

25. $(-3 - 2)(-5 - 4)$ **26.** $(-3 - 6)(-2 - 8)$

27. $-3(-6) + 4(-1)$ **28.** $-4(-5) + 8(-2)$

29. $2(3) - 3(-4) + 4(-5)$ **30.** $5(4) - 2(-1) + 5(6)$

31. $4(-3)^2 + 5(-6)^2$ **32.** $2(-5)^2 + 4(-3)^2$

33. $7(-2)^3 - 2(-3)^3$ **34.** $10(-2)^3 - 5(-2)^4$

35. $6 - 4(8 - 2)$ **36.** $7 - 2(6 - 3)$

37. $9 - 4(3 - 8)$ **38.** $8 - 5(2 - 7)$

39. $-4(3 - 8) - 6(2 - 5)$ **40.** $-8(2 - 7) - 9(3 - 5)$

41. $7 - 2[-6 - 4(-3)]$ **42.** $6 - 3[-5 - 3(-1)]$

43. $7 - 3[2(-4 - 4) - 3(-1 - 1)]$

44. $5 - 3[7(-2 - 2) - 3(-3 + 1)]$

45. Simplify each expression.
  a. $5(-4)(-3)$
  b. $5(-4) - 3$
  c. $5 - 4(-3)$
  d. $5 - 4 - 3$

46. Simplify each expression.
  a. $-2(-3)(-5)$
  b. $-2(-3) - 5$
  c. $-2 - 3(-5)$

C Multiply the following fractions. [Examples 17–19]

47. $-\dfrac{2}{3} \cdot \dfrac{5}{7}$ **48.** $-\dfrac{6}{5} \cdot \dfrac{2}{7}$

49. $-8\left(\dfrac{1}{2}\right)$ **50.** $-12\left(\dfrac{1}{3}\right)$

51. $\left(-\dfrac{3}{4}\right)^2$ **52.** $\left(-\dfrac{2}{5}\right)^2$

53. Simplify each expression.
  a. $\dfrac{5}{8}(24) + \dfrac{3}{7}(28)$
  b. $\dfrac{5}{8}(24) - \dfrac{3}{7}(28)$
  c. $\dfrac{5}{8}(-24) + \dfrac{3}{7}(-28)$
  d. $-\dfrac{5}{8}(24) - \dfrac{3}{7}(28)$

54. Simplify each expression.
  a. $\dfrac{5}{6}(18) + \dfrac{3}{5}(15)$
  b. $\dfrac{5}{6}(18) - \dfrac{3}{5}(15)$
  c. $\dfrac{5}{6}(-18) + \dfrac{3}{5}(-15)$
  d. $-\dfrac{5}{6}(18) - \dfrac{3}{5}(15)$

Simplify.

55. $\left(\dfrac{1}{2} \cdot 6\right)^2$ **56.** $\left(\dfrac{1}{2} \cdot 10\right)^2$

57. $\left(\dfrac{1}{2} \cdot 5\right)^2$ **58.** $\left[\dfrac{1}{2}(0.8)\right]^2$

59. $\left[\dfrac{1}{2}(-4)\right]^2$ **60.** $\left[\dfrac{1}{2}(-12)\right]^2$

61. $\left[\dfrac{1}{2}(-3)\right]^2$ **62.** $\left[\dfrac{1}{2}(-0.8)\right]^2$

D Find the following products. [Examples 20–21]

63. $-2(0)$ **64.** $2x(0)$

65. $-7x(0)$ **66.** $\dfrac{1}{5}(0)$

67. $(3x)(-5)(0)$ **68.** $(16^5)\left(\dfrac{4}{19}\right)(0)(x)$

E Each of the following is a geometric sequence. In each case, find the next number in the sequence. [Example 22]

69. $1, 2, 4, \ldots$ **70.** $1, 5, 25, \ldots$

71. $10, -20, 40, \ldots$ **72.** $10, -30, 90, \ldots$

73. $1, \dfrac{1}{2}, \dfrac{1}{4}, \ldots$ **74.** $1, \dfrac{1}{3}, \dfrac{1}{9}, \ldots$

75. $3, -6, 12, \ldots$ **76.** $-3, 6, -12, \ldots$

Here are some problems you will see later in the book. Simplify.

77. $2(5 - 5) + 4$ **78.** $5x(2) + 3x(0) + 4$

79. $6(3 + 1) - 5(0)$ **80.** $2(3) + 4(5) - 5(2)$

81. $\left(\dfrac{1}{2} \cdot 18\right)^2$ **82.** $\left[\dfrac{1}{2}(-10)\right]^2$

83. $\left(\dfrac{1}{2} \cdot 3\right)^2$ **84.** $\left(\dfrac{1}{2} \cdot 5\right)^2$

Applying the Concepts

85. Picture Messaging Jan got a new phone in March that allowed picture messaging. The graph shows the number of picture messages she sent each of the first nine months of the year. If she gets charged $0.25 per message, how much did she get charged in April?

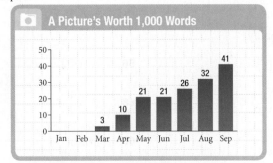

A Picture's Worth 1,000 Words

86. Google Earth The Google Earth map shows Yellowstone National Park. The park covers about 3,472 square miles. If there is an average of 2.3 moose per square mile, about how many moose live in Yellowstone? Round to the nearest moose.

87. Stock Value Suppose you own 20 shares of a stock. If the price per share drops $3, how much money have you lost?

88. Stock Value Imagine that you purchase 50 shares of a stock at a price of $18 per share. If the stock is selling for $11 a share a week after you purchased it, how much money have you lost?

89. Temperature Change The temperature is 25°F at 5:00 in the afternoon. If the temperature drops 6°F every hour after that, what is the temperature at 9:00 in the evening?

90. Temperature Change The temperature is −5°F at 6:00 in the evening. If the temperature drops 3°F every hour after that, what is the temperature at midnight?

91. Nursing Problem A patient's prescription requires him to take two 25 mg tablets twice a day. What is the total dosage for the day?

92. Nursing Problem A patient takes three 50 mg capsules a day. How many milligrams is he taking daily?

1.5 Division of Real Numbers

OBJECTIVES

A Divide any combination of positive and negative numbers.

B Divide positive and negative fractions.

C Simplify experssions using the rule for order of operations.

TICKET TO SUCCESS

Keep these questions in mind as you read through the section. Then respond in your own words and in complete sentences.

1. Why would you define division in terms of multiplication?
2. How would you divide fractions?
3. Why is it important to use the order of operations when dividing real numbers?
4. Why is division by zero not allowed with real numbers?

Image copyright © bokan. Used under license from Shutterstock.com

Suppose that you and four friends bought equal shares of an investment for a total of $15,000 and then sold it later for only $13,000. How much did each person lose? Because the total amount of money that was lost can be represented by −$2,000, and there are 5 people with equal shares, we can represent each person's loss with division:

$$\frac{-\$2,000}{5} = -\$400$$

From this discussion, it seems reasonable to say that a negative number divided by a positive number is a negative number. Here is a more detailed discussion of division with positive and negative numbers.

We will use the same approach to define division as we used for subtraction; that is, we will define division in terms of rules we already know.

Recall that we developed the rule for subtraction of real numbers by defining subtraction in terms of addition. We changed our subtraction problems to addition problems and then added to get our answers. Because we already have a rule for multiplication of real numbers, and division is the inverse operation of multiplication, we will simply define division in terms of multiplication.

We know that division by the number 2 is the same as multiplication by $\frac{1}{2}$; that is, 6 divided by 2 is 3, which is the same as 6 times $\frac{1}{2}$. Similarly, dividing a number by 5 gives the same result as multiplying by $\frac{1}{5}$. We can extend this idea to all real numbers with the following rule:

> **Rule: Dividing Real Numbers**
> If a and b represent any two real numbers ($b \neq 0$), then it is always true that
> $$a \div b = \frac{a}{b} = a\left(\frac{1}{b}\right)$$

NOTE

We are defining division this way simply so that we can use what we already know about multiplication to do division problems. We actually want as few rules as possible. Defining division in terms of multiplication allows us to avoid writing a separate rule for division.

A Dividing Positive and Negative Numbers

Division by a number is the same as multiplication by its reciprocal. Because every division problem can be written as a multiplication problem and because we already know the rule for multiplication of two real numbers, we do not have to write a new rule for division of real numbers. We will simply replace our division problem with multiplication and use the rule we already have.

EXAMPLES Write each division problem as an equivalent multiplication problem, and then multiply.

1. $\dfrac{6}{2} = 6\left(\dfrac{1}{2}\right) = 3$ The product of two positives is positive

2. $\dfrac{6}{-2} = 6\left(-\dfrac{1}{2}\right) = -3$ ⎫

3. $\dfrac{-6}{2} = -6\left(\dfrac{1}{2}\right) = -3$ ⎬ The product of a positive and a negative is a negative

4. $\dfrac{-6}{-2} = -6\left(-\dfrac{1}{2}\right) = 3$ The product of two negatives is positive ▪

NOTE
What we are saying here is that the work shown in Examples 1 through 4 is shown simply to justify the answers we obtain. In the future we won't show the middle step in these kinds of problems. Even so, we need to know that division is defined to be multiplication by the reciprocal.

The second step in the previous examples is used only to show that we *can* write division in terms of multiplication. [In actual practice we wouldn't write $\dfrac{6}{2}$ as $6\left(\dfrac{1}{2}\right)$.] The answers, therefore, follow from the rule for multiplication; that is, like signs produce a positive answer, and unlike signs produce a negative answer.

Here are some examples. This time we will not show division as multiplication by the reciprocal. We will simply divide. If the original numbers have the same signs, the answer will be positive. If the original numbers have different signs, the answer will be negative.

EXAMPLE 5 Divide $\dfrac{12}{6}$.

SOLUTION $\dfrac{12}{6} = 2$ Like signs give a positive answer ▪

EXAMPLE 6 Divide $\dfrac{12}{-6}$.

SOLUTION $\dfrac{12}{-6} = -2$ Unlike signs give a negative answer ▪

EXAMPLE 7 Divide $\dfrac{-12}{6}$.

SOLUTION $\dfrac{-12}{6} = -2$ Unlike signs give a negative answer ▪

EXAMPLE 8 Divide $\dfrac{-12}{-6}$.

SOLUTION $\dfrac{-12}{-6} = 2$ Like signs give a positive answer ▪

EXAMPLE 9   Divide $\dfrac{15}{-3}$.

SOLUTION $\dfrac{15}{-3} = -5$ Unlike signs give a negative answer

EXAMPLE 10   Divide $\dfrac{-40}{-5}$.

SOLUTION $\dfrac{-40}{-5} = 8$ Like signs give a positive answer

EXAMPLE 11   Divide $\dfrac{-14}{2}$.

SOLUTION $\dfrac{-14}{2} = -7$ Unlike signs give a negative answer

B Division with Fractions

We can apply the definition of division to fractions. Because dividing by a fraction is equivalent to multiplying by its reciprocal, we can divide a number by the fraction $\dfrac{3}{4}$ by multiplying it by the reciprocal of $\dfrac{3}{4}$, which is $\dfrac{4}{3}$. For example,

$$\frac{2}{5} \div \frac{3}{4} = \frac{2}{5} \cdot \frac{4}{3} = \frac{8}{15}$$

You may have learned this rule in previous math classes. In some math classes, multiplication by the reciprocal is referred to as "inverting the divisor and multiplying" or "flip that guy and multiply." No matter how you say it, division by any number (except zero) is always equivalent to multiplication by its reciprocal. Here are additional examples that involve division by fractions.

EXAMPLE 12   Divide $\dfrac{2}{3} \div \dfrac{5}{7}$.

SOLUTION $\dfrac{2}{3} \div \dfrac{5}{7} = \dfrac{2}{3} \cdot \dfrac{7}{5}$ Rewrite as multiplication by the reciprocal

$= \dfrac{14}{15}$ Multiply

EXAMPLE 13   Divide $-\dfrac{3}{4} \div \dfrac{7}{9}$.

SOLUTION $-\dfrac{3}{4} \div \dfrac{7}{9} = -\dfrac{3}{4} \cdot \dfrac{9}{7}$ Rewrite as multiplication by the reciprocal

$= -\dfrac{27}{28}$ Multiply

EXAMPLE 14   Divide $8 \div \left(-\dfrac{4}{5}\right)$.

SOLUTION $8 \div \left(-\dfrac{4}{5}\right) = \dfrac{8}{1}\left(-\dfrac{5}{4}\right)$ Rewrite as multiplication by the reciprocal

$= -\dfrac{40}{4}$ Multiply

$= -10$ Divide 40 by 4 to simplify

C   Using Order of Operations

As in Example 14, the last step in each of the following examples involves reducing a fraction to lowest terms. To reduce a fraction to lowest terms, we divide the numerator and denominator by the largest number that divides each of them exactly. For example, to reduce $\frac{15}{20}$ to lowest terms, we divide 15 and 20 by 5 to get $\frac{3}{4}$.

EXAMPLE 15   Simplify as much as possible: $\frac{-4(5)}{6}$

SOLUTION $\frac{-4(5)}{6} = \frac{-20}{6}$ Simplify numerator

$= -\frac{10}{3}$ Reduce to lowest terms by dividing numerator and denominator by 2   ■

EXAMPLE 16   Simplify as much as possible: $\frac{30}{-4 - 5}$

SOLUTION $\frac{30}{-4 - 5} = \frac{30}{-9}$ Simplify denominator

$= -\frac{10}{3}$ Reduce to lowest terms by dividing numerator and denominator by 3   ■

In the examples that follow, the numerators and denominators contain expressions that are somewhat more complicated than those we have seen thus far. To apply the rule for order of operations to these examples, we treat fraction bars the same way we treat grouping symbols; that is, fraction bars separate numerators and denominators so that each will be simplified separately.

EXAMPLE 17   Simplify $\frac{-8 - 8}{-5 - 3}$.

SOLUTION $\frac{-8 - 8}{-5 - 3} = \frac{-16}{-8}$ Simplify numerator and denominator separately

$= 2$ Division   ■

EXAMPLE 18   Simplify $\frac{2(-3) + 4}{12}$.

SOLUTION $\frac{2(-3) + 4}{12} = \frac{-6 + 4}{12}$ In the numerator, we multiply before we add

$= \frac{-2}{12}$ Addition

$= -\frac{1}{6}$ Reduce to lowest terms by dividing numerator and denominator by 2   ■

EXAMPLE 19   Simplify $\frac{5(-4) + 6(-1)}{2(3) - 4(1)}$.

SOLUTION $\frac{5(-4) + 6(-1)}{2(3) - 4(1)} = \frac{-20 + (-6)}{6 - 4}$ Multiplication before addition

$= \frac{-26}{2}$ Simplify numerator and denominator

$= -13$ Divide -26 by 2   ■

We must be careful when we are working with expressions such as $(-5)^2$ and -5^2 that we include the negative sign with the base only when parentheses indicate we are to do so.

Unless there are parentheses to indicate otherwise, we consider the base to be only the number directly below and to the left of the exponent. *If we want to include a negative sign with the base, we must use parentheses.*

To simplify a more complicated expression, we follow the same rule. For example,

$$7^2 - 3^2 = 49 - 9 \qquad \text{The bases are 7 and 3; the sign between the two terms is a subtraction sign}$$

$$= 40$$

For another example,

$$5^3 - 3^4 = 125 - 81 \qquad \text{We simplify exponents first, then subtract}$$

$$= 44$$

EXAMPLE 20   Simplify $\dfrac{5^2 - 3^2}{-5 + 3}$.

SOLUTION $\dfrac{5^2 - 3^2}{-5 + 3} = \dfrac{25 - 9}{-2}$ Simplify numerator and denominator separately

$$= \frac{16}{-2}$$

$$= -8$$

EXAMPLE 21   Simplify $\dfrac{(3 + 2)^2}{-3^2 - 2^2}$.

SOLUTION $\dfrac{(3 + 2)^2}{-3^2 - 2^2} = \dfrac{5^2}{-9 - 4}$ Simplify numerator and denominator separately

$$= \frac{25}{-13}$$

$$= -\frac{25}{13}$$

Division with the Number 0

For every division problem there is an associated multiplication problem involving the same numbers. For example, the following two problems say the same thing about the numbers 2, 3, and 6:

Division	Multiplication
$\dfrac{6}{3} = 2$	$6 = 2(3)$

We can use this relationship between division and multiplication to clarify division involving the number 0.

First, dividing 0 by a number other than 0 is allowed and always results in 0. To see this, consider dividing 0 by 5. We know the answer is 0 because of the relationship between multiplication and division. This is how we write it:

$$\frac{0}{5} = 0 \qquad \text{because} \qquad 0 = 0(5)$$

However, dividing a nonzero number by 0 is not allowed in the real numbers. Suppose we were attempting to divide 5 by 0. We don't know if there is an answer to this problem, but if there is, let's say the answer is a number that we can represent with the letter n. If 5 divided by 0 is a number n, then

$$\frac{5}{0} = n \qquad \text{and} \qquad 5 = n(0)$$

This is impossible, however, because no matter what number n is, when we multiply it by 0 the answer must be 0. It can never be 5. In algebra, we say expressions like $\frac{5}{0}$ are undefined because there is no answer to them; that is, division by 0 is not allowed in the real numbers.

The only other possibility for division involving the number 0 is 0 divided by 0. We will treat problems like $\frac{0}{0}$ as if they were undefined also. Here is a rule that summarizes this information:

Rule: Dividing with Zero

1. Division by 0 is not defined.

$$a \div 0 \text{ or } \frac{a}{0} \text{ is not defined, for all real numbers, } a.$$

2. Zero divided by any number is 0.

$$\frac{0}{a} = 0 \text{ for all real numbers, } a \text{ so long as } a \neq 0.$$

Problem Set 1.5

Moving Toward Success

"The way to get started is to quit talking and begin doing."

—Walt Disney, 1901–1966, American film producer/director/animator

1. Which of the following is the best time to seek help with the problem set?

a. After you have tried the corresponding examples in the book

b. Before you look at the examples in the book

2. Which of the times above is the least effective time to seek help? Why?

A Find the following quotients (divide). [Examples 1–11]

1. $\dfrac{8}{-4}$

2. $\dfrac{10}{-5}$

3. $\dfrac{-48}{16}$

4. $\dfrac{-32}{4}$

5. $\dfrac{-7}{21}$

6. $\dfrac{-25}{100}$

7. $\dfrac{-39}{-13}$

8. $\dfrac{-18}{-6}$

9. $\dfrac{-6}{-42}$

10. $\dfrac{-4}{-28}$

11. $\dfrac{0}{-32}$

12. $\dfrac{0}{17}$

The following problems review all four operations with positive and negative numbers. Perform the indicated operations.

13. $-3 + 12$

14. $5 + (-10)$

15. $-3 - 12$

16. $5 - (-10)$

17. $-3(12)$

18. $5(-10)$

19. $-3 \div 12$

20. $5 \div (-10)$

B Divide and reduce all answers to lowest terms. [Examples 12–14]

21. $\dfrac{4}{5} \div \dfrac{3}{4}$

22. $\dfrac{6}{8} \div \dfrac{3}{4}$

23. $-\dfrac{5}{6} \div \left(-\dfrac{5}{8}\right)$

24. $-\dfrac{7}{9} \div \left(-\dfrac{1}{6}\right)$

25. $\dfrac{10}{13} \div \left(-\dfrac{5}{4}\right)$

26. $\dfrac{5}{12} \div \left(-\dfrac{10}{3}\right)$

27. $-\dfrac{5}{6} \div \dfrac{5}{6}$

28. $-\dfrac{8}{9} \div \dfrac{8}{9}$

29. $-\dfrac{3}{4} \div \left(-\dfrac{3}{4}\right)$

30. $-\dfrac{6}{7} \div \left(-\dfrac{6}{7}\right)$

C The following problems involve more than one operation. Simplify as much as possible. [Examples 15–21]

31. $\dfrac{3(-2)}{-10}$

32. $\dfrac{4(-3)}{24}$

33. $\dfrac{-5(-5)}{-15}$

34. $\dfrac{-7(-3)}{-35}$

35. $\dfrac{-8(-7)}{-28}$

36. $\dfrac{-3(-9)}{-6}$

37. $\dfrac{27}{4-13}$

38. $\dfrac{27}{13-4}$

39. $\dfrac{20-6}{5-5}$

40. $\dfrac{10-12}{3-3}$

41. $\dfrac{-3+9}{2\cdot5-10}$

42. $\dfrac{-4+8}{2\cdot4-8}$

43. $\dfrac{15(-5)-25}{2(-10)}$

44. $\dfrac{10(-3)-20}{5(-2)}$

45. $\dfrac{27-2(-4)}{-3(5)}$

46. $\dfrac{20-5(-3)}{10(-3)}$

47. $\dfrac{12-6(-2)}{12(-2)}$

48. $\dfrac{3(-4)+5(-6)}{10-6}$

49. $\dfrac{5^2-2^2}{-5+2}$

50. $\dfrac{7^2-4^2}{-7+4}$

51. $\dfrac{8^2-2^2}{8^2+2^2}$

52. $\dfrac{4^2-6^2}{4^2+6^2}$

53. $\dfrac{(5+3)^2}{-5^2-3^2}$

54. $\dfrac{(7+2)^2}{-7^2-2^2}$

55. $\dfrac{(8-4)^2}{8^2-4^2}$

56. $\dfrac{(6-2)^2}{6^2-2^2}$

57. $\dfrac{-4\cdot3^2-5\cdot2^2}{-8(7)}$

58. $\dfrac{-2\cdot5^2+3\cdot2^3}{-3(13)}$

59. $\dfrac{3\cdot10^2+4\cdot10+5}{345}$

60. $\dfrac{5\cdot10^2+6\cdot10+7}{567}$

61. $\dfrac{7-[(2-3)-4]}{-1-2-3}$

62. $\dfrac{2-[(3-5)-8]}{-3-4-5}$

63. $\dfrac{6(-4)-2(5-8)}{-6-3-5}$

64. $\dfrac{3(-4)-5(9-11)}{-9-2-3}$

65. $\dfrac{3(-5-3)+4(7-9)}{5(-2)+3(-4)}$

66. $\dfrac{-2(6-10)-3(8-5)}{6(-3)-6(-2)}$

67. $\dfrac{|3-9|}{3-9}$

68. $\dfrac{|4-7|}{4-7}$

69. $\dfrac{2+0.15(10)}{10}$

70. $\dfrac{5(5)+250}{640(5)}$

71. $\dfrac{1-3}{3-1}$

72. $\dfrac{25-16}{16-25}$

73. Simplify.

  a. $\dfrac{5-2}{3-1}$

  b. $\dfrac{2-5}{1-3}$

74. Simplify.

  a. $\dfrac{6-2}{3-5}$

  b. $\dfrac{2-6}{5-3}$

75. Simplify.

  a. $\dfrac{-4-1}{5-(-2)}$

  b. $\dfrac{1-(-4)}{(-2)-5}$

76. Simplify.

  a. $\dfrac{-6-1}{4-(-5)}$

  b. $\dfrac{1-(-6)}{-5-4}$

77. Simplify.

  a. $\dfrac{3+2.236}{2}$

  b. $\dfrac{3-2.236}{2}$

  c. $\dfrac{3+2.236}{2}+\dfrac{3-2.236}{2}$

78. Simplify.

  a. $\dfrac{1+1.732}{2}$

  b. $\dfrac{1-1.732}{2}$

  c. $\dfrac{1+1.732}{2}+\dfrac{1-1.732}{2}$

79. Simplify each expression.

  a. $20\div4\cdot5$

  b. $-20\div4\cdot5$

  c. $20\div(-4)\cdot5$

  d. $20\div4(-5)$

  e. $-20\div4(-5)$

80. Simplify each expression.

  a. $32\div8\cdot4$

  b. $-32\div8\cdot4$

  c. $32\div(-8)\cdot4$

  d. $32\div8(-4)$

  e. $-32\div8(-4)$

81. Simplify each expression.

  a. $8\div\dfrac{4}{5}$

  b. $8\div\dfrac{4}{5}-10$

  c. $8\div\dfrac{4}{5}(-10)$

  d. $8\div\left(-\dfrac{4}{5}\right)-10$

82. Simplify each expression.

  a. $10\div\dfrac{5}{6}$

  b. $10\div\dfrac{5}{6}-12$

  c. $10\div\dfrac{5}{6}(-12)$

d. $10 \div \left(-\dfrac{5}{6}\right) - 12$

Answer the following questions.

83. What is the quotient of -12 and -4?

84. The quotient of -4 and -12 is what number?

85. What number do we divide by -5 to get 2?

86. What number do we divide by -3 to get 4?

87. Twenty-seven divided by what number is -9?

88. Fifteen divided by what number is -3?

89. If the quotient of -20 and 4 is decreased by 3, what number results?

90. If -4 is added to the quotient of 24 and -8, what number results?

Applying the Concepts

91. Golf The chart shows the lengths of the longest golf courses used for the U.S. Open. If the length of the Torrey Pines South course is 7,643 yards, what is the average length of each hole on the 18-hole course? Round to the nearest yard.

Longest Courses in U.S. Open History

Torrey Pines, South Course (2008)	7,643 yds
Bethpage State Park, Black Course (2009)	7,426 yds
Winged Foot Golf Course, West Course (2006)	7,264 yds
Oakmont Country Club (2007)	7,230 yds
Pinehurst, No2 Course (2005)	7,214 yds

Source: http://www.usopen.com

92. Broadway The chart shows the number of plays performed by the longest running Broadway musicals. If *Les Miserables* ran for 16 years, what was the mean number of shows per year? Round to the nearest show.

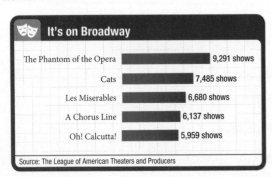

It's on Broadway

The Phantom of the Opera	9,291 shows
Cats	7,485 shows
Les Miserables	6,680 shows
A Chorus Line	6,137 shows
Oh! Calcutta!	5,959 shows

Source: The League of American Theaters and Producers

93. Investment Suppose that you and 3 friends bought equal shares of an investment for a total of $15,000

and then sold it later for only $13,600. How much did each person lose?

94. Investment If 8 people invest $500 each in a stamp collection and after a year the collection is worth $3,800, how much did each person lose?

95. Temperature Change Suppose that the temperature outside is dropping at a constant rate. If the temperature is 75°F at noon and drops to 61°F by 4:00 in the afternoon, by how much did the temperature change each hour?

96. Temperature Change In a chemistry class, a thermometer is placed in a beaker of hot water. The initial temperature of the water is 165°F. After 10 minutes the water has cooled to 72°F. If the water temperature drops at a constant rate, by how much does the water temperature change each minute?

97. Nursing Problem A patient is required to take 75 mg of a drug. If the capsule strength is 25 mg, how many capsules should she take?

98. Nursing Problem A patient is given prescribed a dosage of 25 mg, and the strength of each tablet is 50 mg. How many tablets should he take?

99. Nursing Problem A patient is required to take 1.2 mg of drug and is told to take 4 tablets. What is the strength of each tablet?

100. Nursing Problem A patient is required to take two tablets for a 3.6 mg dosage. What is the dosage strength of each tablet?

101. Internet Mailing Lists A company sells products on the Internet through an email list. They predict that they sell one $50 product for every 25 people on their mailing list.

  a. What is their projected revenue if their list contains 10,000 email addresses?

  b. What is their projected revenue if their list contains 25,000 email addresses?

  c. They can purchase a list of 5,000 email addresses for $5,000. Is this a wise purchase?

102. Internet Mailing Lists A new band has a following on the Internet. They sell their CDs through an email list. They predict that they sell one $15 CD for every 10 people on their mailing list.

  a. What is their projected revenue if their list contains 5,000 email addresses?

  b. What is their projected revenue if their list contains 20,000 email addresses?

  c. If they need to make $45,000, how many people do they need on their email list?

1.6 Properties of Real Numbers

OBJECTIVES

A Rewrite expressions using the commutative and associative properties.

B Multiply using the distributive property.

C Identify properties used to rewrite an expression.

TICKET TO SUCCESS

Keep these questions in mind as you read through the section. Then respond in your own words and in complete sentences.

1. What is the commutative property of addition?
2. Write the commutative property of multiplication in symbols and words.
3. Use the associative property of multiplication to simplify $4(3x)$.
4. How do you use the distributive property to find the area of a rectangle?

Image copyright © John Wollwerth. Used under license from Shutterstock.com

In this section, we will list facts (properties) that you know from past experience are true about numbers in general. We will give each property a name so we can refer to it later in this book. Mathematics is very much like a game. The game involves numbers. The rules of the game are the properties and rules we are developing in this chapter. The goal of the game is to extend the basic rules to as many situations as possible.

You know from past experience with numbers that it makes no difference in which order you add two numbers; that is, $3 + 5$ is the same as $5 + 3$. This fact about numbers is called the *commutative property of addition*. We say addition is a commutative operation. Changing the order of the numbers does not change the answer.

Another basic operation is commutative. Because $3(5)$ is the same as $5(3)$, we say multiplication is a commutative operation. Changing the order of the two numbers you are multiplying does not change the answer.

A The Commutative and Associative Properties

For all properties listed in this section, a, b, and c represent real numbers.

Commutative Property of Addition

In symbols: $a + b = b + a$

In words: Changing the *order* of the numbers in a sum will not change the result.

Commutative Property of Multiplication

In symbols: $a \cdot b = b \cdot a$

In words: Changing the *order* of the numbers in a product will not change the result.

EXAMPLES

1. The statement $5 + 8 = 8 + 5$ is an example of the commutative property of addition.

2. The statement $2 \cdot y = y \cdot 2$ is an example of the commutative property of multiplication.

3. The expression $5 + x + 3$ can be simplified using the commutative property of addition.

$$5 + x + 3 = x + 5 + 3 \qquad \text{Commutative property of addition}$$
$$= x + 8$$

The other two basic operations, subtraction and division, are not commutative. The order in which we subtract or divide two numbers makes a difference in the answer.

Another property of numbers that you have used many times has to do with grouping. You know that when we add three numbers it makes no difference which two we add first. When adding $3 + 5 + 7$, we can add the 3 and 5 first and then the 7, or we can add the 5 and 7 first and then the 3. Mathematically, it looks like this: $(3 + 5) + 7 = 3 + (5 + 7)$. This property is true of multiplication as well. Operations that behave in this manner are called *associative* operations. The answer will not change when we change the association (or grouping) of the numbers.

Associative Property of Addition

In symbols: $a + (b + c) = (a + b) + c$

In words: Changing the *grouping* of the numbers in a sum will not change the result.

Associative Property of Multiplication

In symbols: $a(bc) = (ab)c$

In words: Changing the *grouping* of the numbers in a product will not change the result.

The following examples illustrate how the associative properties can be used to simplify expressions that involve both numbers and variables.

EXAMPLE 4
Simplify $4 + (5 + x)$.

SOLUTION
$$4 + (5 + x) = (4 + 5) + x \qquad \text{Associative property of addition}$$
$$= 9 + x$$

EXAMPLE 5
Simplify $5(2x)$.

SOLUTION
$$5(2x) = (5 \cdot 2)x \qquad \text{Associative property of multiplication}$$
$$= 10x$$

EXAMPLE 6 Simplify $\frac{1}{5}(5x)$.

SOLUTION

$$\frac{1}{5}(5x) = \left(\frac{1}{5} \cdot 5\right)x \qquad \text{Associative property of multiplication}$$
$$= 1x$$
$$= x$$

EXAMPLE 7 Simplify $3\left(\frac{1}{3}x\right)$.

SOLUTION

$$3\left(\frac{1}{3}x\right) = \left(3 \cdot \frac{1}{3}\right)x \qquad \text{Associative property of multiplication}$$
$$= 1x$$
$$= x$$

EXAMPLE 8 Simplify $12\left(\frac{2}{3}x\right)$.

SOLUTION

$$12\left(\frac{2}{3}x\right) = \left(12 \cdot \frac{2}{3}\right)x \qquad \text{Associative property of multiplication}$$
$$= 8x$$

B The Distributive Property

The associative and commutative properties apply to problems that are either all multiplication or all addition. There is a third basic property that involves both addition and multiplication. It is called the *distributive property* and looks like this:

Distributive Property

In symbols: $a(b + c) = ab + ac$

In words: Multiplication *distributes* over addition.

> **NOTE**
> Because subtraction is defined in terms of addition, it is also true that the distributive property applies to subtraction as well as addition; that is, $a(b - c) = ab - ac$ for any three real numbers a, b, and c.

You will see as we progress through the book that the distributive property is used very frequently in algebra. We can give a visual justification to the distributive property by finding the areas of rectangles. Figure 1 shows a large rectangle that is made up of two smaller rectangles. We can find the area of the large rectangle two different ways.

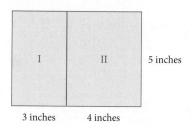

FIGURE 1

METHOD 1 We can calculate the area of the large rectangle directly by finding its length and width. The width is 5 inches, and the length is $(3 + 4)$ inches.

$$\text{Area of large rectangle} = 5(3 + 4)$$
$$= 5(7)$$
$$= 35 \text{ square inches}$$

METHOD 2 Because the area of the large rectangle is the sum of the areas of the two smaller rectangles, we find the area of each small rectangle and then add to find the area of the large rectangle.

$$\text{Area of large rectangle} = \text{Area of rectangle I} + \text{Area of rectangle II}$$
$$= \qquad 5(3) \qquad + \qquad 5(4)$$
$$= \qquad 15 \qquad + \qquad 20$$
$$= \qquad 35 \text{ square inches}$$

In both cases the result is 35 square inches. Because the results are the same, the two original expressions must be equal. Stated mathematically, $5(3 + 4) = 5(3) + 5(4)$. We can either add the 3 and 4 first and then multiply that sum by 5, or we can multiply the 3 and the 4 separately by 5 and then add the products. In either case we get the same answer.

Here are some examples that illustrate how we use the distributive property using algebraic expressions.

EXAMPLE 9 Apply the distributive property to the expression $2(x + 3)$, and then simplify the result.

SOLUTION
$$2(x + 3) = 2(x) + 2(3) \qquad \text{Distributive property}$$
$$= 2x + 6$$

EXAMPLE 10 Apply the distributive property to the expression $5(2x - 8)$, and then simplify the result.

SOLUTION
$$5(2x - 8) = 5(2x) - 5(8) \qquad \text{Distributive property}$$
$$= 10x - 40$$

Notice in this example that multiplication distributes over subtraction as well as addition.

EXAMPLE 11 Apply the distributive property to the expression $4(x + y)$, and then simplify the result.

SOLUTION
$$4(x + y) = 4x + 4y \qquad \text{Distributive property}$$

EXAMPLE 12 Apply the distributive property to the expression $5(2x + 4y)$, and then simplify the result.

SOLUTION
$$5(2x + 4y) = 5(2x) + 5(4y) \qquad \text{Distributive property}$$
$$= 10x + 20y$$

EXAMPLE 13 Apply the distributive property to the expression $\frac{1}{2}(3x + 6)$, and then simplify the result.

SOLUTION
$$\frac{1}{2}(3x + 6) = \frac{1}{2}(3x) + \frac{1}{2}(6) \qquad \text{Distributive property}$$
$$= \frac{3}{2}x + 3$$

EXAMPLE 14 Apply the distributive property to the expression $4(2a + 3) + 8$, and then simplify the result.

SOLUTION
$$4(2a + 3) + 8 = 4(2a) + 4(3) + 8 \qquad \text{Distributive property}$$
$$= 8a + 12 + 8$$
$$= 8a + 20$$

EXAMPLE 15 Apply the distributive property to the expression $a\left(1 + \frac{1}{a}\right)$, and then simplify the result.

SOLUTION
$$a\left(1 + \frac{1}{a}\right) = a \cdot 1 + a \cdot \frac{1}{a} \qquad \text{Distributive property}$$
$$= a + 1$$

EXAMPLE 16 Apply the distributive property to the expression $3\left(\frac{1}{3}x + 5\right)$, and then simplify the result.

SOLUTION $\qquad 3\left(\frac{1}{3}x + 5\right) = 3 \cdot \frac{1}{3}x + 3 \cdot 5 \qquad$ Distributive property

$$= x + 15$$

EXAMPLE 17 Apply the distributive property to the expression $12\left(\frac{2}{3}x + \frac{1}{2}y\right)$, and then simplify the result.

SOLUTION $\qquad 12\left(\frac{2}{3}x + \frac{1}{2}y\right) = 12 \cdot \frac{2}{3}x + 12 \cdot \frac{1}{2}y \quad$ Distributive property

$$= 8x + 6y$$

Special Numbers

In addition to the three properties mentioned so far, we want to include in our list two special numbers that have unique properties. They are the numbers 0 and 1.

Additive Identity Property

There exists a unique number 0 such that

In symbols: $a + 0 = a$ and $0 + a = a$

In words: Zero preserves identities under addition. (The identity of the number is unchanged after addition with 0.)

Multiplicative Identity Property

There exists a unique number 1 such that

In symbols: $a(1) = a$ and $1(a) = a$

In words: The number 1 preserves identities under multiplication. (The identity of the number is unchanged after multiplication by 1.)

Additive Inverse Property

For each real number a, there exists a unique number $-a$ such that

In symbols: $a + (-a) = 0$

In words: Opposites add to 0.

Multiplicative Inverse Property

For every real number a, except 0, there exists a unique real number $\frac{1}{a}$ such that

In symbols: $a\left(\frac{1}{a}\right) = 1$

In words: Reciprocals multiply to 1.

C Identifying Properties

Of all the basic properties listed, the commutative, associative, and distributive properties are the ones we will use most often. They are important because they will be used as justifications or reasons for many of the things we will do.

The following examples illustrate how we use the preceding properties. Each one contains an algebraic expression that has been changed in some way. The property that justifies the change is written to the right.

EXAMPLE 18   State the property that justifies $x + 5 = 5 + x$.

SOLUTION $x + 5 = 5 + x$ Commutative property of addition

EXAMPLE 19   State the property that justifies $(2 + x) + y = 2 + (x + y)$.

SOLUTION $(2 + x) + y = 2 + (x + y)$ Associative property of addition

EXAMPLE 20   State the property that justifies $6(x + 3) = 6x + 18$.

SOLUTION $6(x + 3) = 6x + 18$ Distributive property

EXAMPLE 21   State the property that justifies $2 + (-2) = 0$.

SOLUTION $2 + (-2) = 0$ Additive inverse property

EXAMPLE 22   State the property that justifies $3\left(\dfrac{1}{3}\right) = 1$.

SOLUTION $3\left(\dfrac{1}{3}\right) = 1$ Multiplicative inverse property

EXAMPLE 23   State the property that justifies $(2 + 0) + 3 = 2 + 3$.

SOLUTION $(2 + 0) + 3 = 2 + 3$ Additive identity property

EXAMPLE 24   State the property that justifies $(2 + 3) + 4 = 3 + (2 + 4)$.

SOLUTION $(2 + 3) + 4 = 3 + (2 + 4)$ Commutative and associative properties of addition

EXAMPLE 25   State the property that justifies $(x + 2) + y = (x + y) + 2$.

SOLUTION $(x + 2) + y = (x + y) + 2$ Commutative and associative properties of addition

As a final note on the properties of real numbers, we should mention that although some of the properties are stated for only two or three real numbers, they hold for as many numbers as needed. For example, the distributive property holds for expressions like $3(x + y + z + 5 + 2)$; that is,

$$3(x + y + z + 5 + 2) = 3x + 3y + 3z + 15 + 6$$

It is not important how many numbers are contained in the sum, only that it is a sum. Multiplication, you see, distributes over addition, whether there are two numbers in the sum or two hundred.

Problem Set 1.6

Moving Toward Success

"When everything seems to be going against you, remember that the airplane takes off against the wind, not with it."

—Henry Ford, 1863–1947, American industrialist and founder of Ford Motor Company

1. When things are not going well for you, who or what is usually the reason?

2. What can you do to stay focused in this course if things are not going well for you?

A Use the associative property to rewrite each of the following expressions, and then simplify the result. [Examples 4–8]

1. $4 + (2 + x)$

2. $5 + (6 + x)$

3. $(x + 2) + 7$

4. $(x + 8) + 2$

5. $3(5x)$ 6. $5(3x)$

7. $9(6y)$ 8. $6(9y)$

9. $\frac{1}{2}(3a)$ 10. $\frac{1}{3}(2a)$

11. $\frac{1}{3}(3x)$ 12. $\frac{1}{4}(4x)$

13. $\frac{1}{2}(2y)$ 14. $\frac{1}{7}(7y)$

15. $\frac{3}{4}\left(\frac{4}{3}x\right)$ 16. $\frac{3}{2}\left(\frac{2}{3}x\right)$

17. $\frac{6}{5}\left(\frac{5}{6}a\right)$ 18. $\frac{2}{5}\left(\frac{5}{2}a\right)$

B Apply the distributive property to each of the following expressions. Simplify when possible. [Examples 9–14]

19. $8(x + 2)$ 20. $5(x + 3)$

21. $8(x - 2)$ 22. $5(x - 3)$

23. $4(y + 1)$ 24. $4(y - 1)$

25. $3(6x + 5)$ 26. $3(5x + 6)$

27. $2(3a + 7)$ 28. $5(3a + 2)$

29. $9(6y - 8)$ 30. $2(7y - 4)$

31. $\frac{1}{2}(3x - 6)$ 32. $\frac{1}{3}(2x - 6)$

33. $\frac{1}{3}(3x + 6)$ 34. $\frac{1}{2}(2x + 4)$

35. $3(x + y)$ 36. $2(x - y)$

37. $8(a - b)$ 38. $7(a + b)$

39. $6(2x + 3y)$ 40. $8(3x + 2y)$

41. $4(3a - 2b)$ 42. $5(4a - 8b)$

43. $\frac{1}{2}(6x + 4y)$ 44. $\frac{1}{3}(6x + 9y)$

45. $4(a + 4) + 9$ 46. $6(a + 2) + 8$

47. $2(3x + 5) + 2$ 48. $7(2x + 1) + 3$

49. $7(2x + 4) + 10$ 50. $3(5x + 6) + 20$

B Here are some problems you will see later in the book. Apply the distributive property and simplify, if possible.

51. $0.09(x + 2{,}000)$

52. $0.04(x + 7{,}000)$

53. $0.05(3x + 1{,}500)$

54. $0.08(4x + 3{,}000)$

55. $6\left(\frac{1}{2}x - \frac{1}{3}y\right)$ 56. $12\left(\frac{1}{4}x + \frac{2}{3}y\right)$

57. $12\left(\frac{1}{3}x + \frac{1}{4}y\right)$ 58. $15\left(\frac{2}{5}x - \frac{1}{3}y\right)$

59. $\frac{1}{2}(4x + 2)$ 60. $\frac{1}{3}(6x + 3)$

61. $\frac{3}{4}(8x - 4)$ 62. $\frac{2}{5}(5x + 10)$

63. $\frac{5}{6}(6x + 12)$ 64. $\frac{2}{3}(9x - 3)$

65. $10\left(\frac{3}{5}x + \frac{1}{2}\right)$ 66. $8\left(\frac{1}{4}x - \frac{5}{8}\right)$

67. $15\left(\frac{1}{3}x + \frac{2}{5}\right)$ 68. $12\left(\frac{1}{12}m + \frac{1}{6}\right)$

69. $12\left(\frac{1}{2}m - \frac{5}{12}\right)$ 70. $8\left(\frac{1}{8} + \frac{1}{2}m\right)$

71. $21\left(\frac{1}{3} + \frac{1}{7}x\right)$ 72. $6\left(\frac{3}{2}y + \frac{1}{3}\right)$

73. $12\left(\dfrac{1}{2}x - \dfrac{1}{3}y\right)$ **74.** $24\left(\dfrac{1}{4}x + \dfrac{2}{3}y\right)$

75. $0.15(x + 600)$ **76.** $0.17(x + 400)$

77. $0.12(x + 500)$ **78.** $0.06(x + 800)$

79. $a\left(1 + \dfrac{1}{a}\right)$ **80.** $a\left(1 - \dfrac{1}{a}\right)$

81. $a\left(\dfrac{1}{a} - 1\right)$ **82.** $a\left(\dfrac{1}{a} + 1\right)$

C State the property or properties that justify the following. [Examples 18–25]

83. $3 + 2 = 2 + 3$

84. $5 + 0 = 5$

85. $4\left(\dfrac{1}{4}\right) = 1$

86. $10(0.1) = 1$

87. $4 + x = x + 4$

88. $3(x - 10) = 3x - 30$

89. $2(y + 8) = 2y + 16$

90. $3 + (4 + 5) = (3 + 4) + 5$

91. $(3 + 1) + 2 = 1 + (3 + 2)$

92. $(5 + 2) + 9 = (2 + 5) + 9$

93. $(8 + 9) + 10 = (8 + 10) + 9$

94. $(7 + 6) + 5 = (5 + 6) + 7$

95. $3(x + 2) = 3(2 + x)$

96. $2(7y) = (7 \cdot 2)y$

97. $x(3y) = 3(xy)$

98. $a(5b) = 5(ab)$

99. $4(xy) = 4(yx)$

100. $3[2 + (-2)] = 3(0)$

101. $8[7 + (-7)] = 8(0)$

102. $7(1) = 7$

Each of the following problems has a mistake in it. Correct the right-hand side.

103. $3(x + 2) = 3x + 2$

104. $5(4 + x) = 4 + 5x$

105. $9(a + b) = 9a + b$

106. $2(y + 1) = 2y + 1$

107. $3(0) = 3$ **108.** $5\left(\dfrac{1}{5}\right) = 5$

109. $3 + (-3) = 1$ **110.** $8(0) = 8$

111. $10(1) = 0$ **112.** $3 \cdot \dfrac{1}{3} = 0$

Applying the Concepts

113. Getting Dressed While getting dressed for work, a man puts on his socks and puts on his shoes. Are the two statements "put on your socks" and "put on your shoes" commutative? That is, will changing the order of the events always produce the same result?

114. Getting Dressed Are the statements "put on your left shoe" and "put on your right shoe" commutative?

115. Skydiving A skydiver flying over the jump area is about to do two things: jump out of the plane and pull the rip cord. Are the two events "jump out of the plane" and "pull the rip cord" commutative?

116. Commutative Property Give an example of two events in your daily life that are commutative.

117. Solar and Wind Energy The chart shows the cost to install either solar panels or a wind turbine. A homeowner buys 3 solar modules, and each module is $100 off the original price. Use the distributive property to calculate the total cost.

🔆 **Solar Versus Wind Energy Costs**			
Solar Energy Equipment Cost:		Wind Energy Equipment Cost:	
Modules	$6200	Turbine	$3300
Fixed Rack	$1570	Tower	$3000
Charge Controller	$971	Cable	$715
Cable	$440		
TOTAL	**$9181**	**TOTAL**	**$7015**

Source: a Limited 2006

118. Pitchers The chart shows the number of saves for relief pitchers. Which property of real numbers can be used to find the sum of the saves of Troy Percival and Billy Wagner in either order?

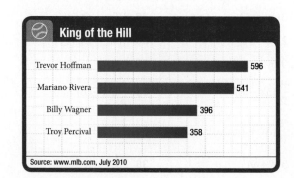

King of the Hill

Trevor Hoffman 596
Mariano Rivera 541
Billy Wagner 396
Troy Percival 358

Source: www.mlb.com, July 2010

119. Division Give an example that shows that division is not a commutative operation; that is, find two numbers for which changing the order of division gives two different answers.

120. Subtraction Simplify the expression $10 - (5 - 2)$ and the expression $(10 - 5) - 2$ to show that subtraction is not an associative operation.

121. Take-Home Pay Jose works at a winery. His monthly salary is $2,400. To cover his taxes and retirement, the winery withholds $480 from each check. Calculate his yearly "take-home" pay using the numbers 2,400, 480, and 12. Do the calculation two different ways so that the results give further justification for the distributive property.

122. Hours Worked Carlo works as a waiter. He works double shifts 4 days a week. The lunch shift is 2 hours and the dinner shift is 3 hours. Find the total number of hours he works per week using the numbers 2, 3, and 4. Do the calculation two different ways so that the results give further justification for the distributive property.

1.7 Subsets of Real Numbers

OBJECTIVES

A Associate numbers with subsets of the real numbers.

B Factor whole numbers into the product of prime factors.

C Reduce fractions to lowest terms using prime factorization.

TICKET TO SUCCESS

Keep these questions in mind as you read through the section. Then respond in your own words and in complete sentences.

1. What is a whole number?
2. Is every integer also a rational number? Explain.
3. What is a prime number?
4. Factor the number 480 into a product of primes.

In Section 1.2, we introduced the real numbers and defined them as the numbers associated with points on the real number line. At that time, we said the real numbers include whole numbers, fractions, and decimals, as well as other numbers that are not as familiar to us as these numbers. In this section we take a more detailed look at the kinds of numbers that make up the set of real numbers.

The numbers that make up the set of real numbers can be classified as *counting numbers, whole numbers, integers, rational numbers,* and *irrational numbers;* each is said to be a *subset* of the real numbers.

Definition

Set A is called a **subset** of set B if set A is contained in set B, that is, if each and every element in set A is also a member of set B.

A   Subsets of Real Numbers

Here is a detailed description of the major subsets of the real numbers.

The *counting numbers* are the numbers with which we count. For example, the numbers we use to count down a shuttle launch (10, 9, 8, 7...) are counting numbers. They are the numbers 1, 2, 3, and so on. The notation we use to specify a group of numbers like this is *set notation*. We use the symbols { and } to enclose the members of the set.

$$\text{Counting numbers} = \{1, 2, 3, \dots\}$$

EXAMPLE 1   Which of the numbers in the following set are not counting numbers?

$$\left\{-3, 0, \frac{1}{2}, 1, 1.5, 3\right\}$$

SOLUTION   The numbers $-3, 0, \frac{1}{2}$, and 1.5 are not counting numbers.   ∎

Here are definitions of a few more important subsets:

Definition

The **whole numbers** include the counting numbers and the number 0.

$$\text{Whole numbers} = \{0, 1, 2, \dots\}$$

Definition

The set of **integers** includes the whole numbers and the opposites of all the counting numbers.

$$\text{Integers} = \{\dots, -3, -2, -1, 0, 1, 2, 3, \dots\}$$

When we refer to *positive integers*, we are referring to the numbers $1, 2, 3, \dots$. Likewise, the *negative integers* are $-1, -2, -3, \dots$. The number 0 is neither positive nor negative.

EXAMPLE 2   Which of the numbers in the following set are not integers?

$$\left\{-5, -1.75, 0, \frac{2}{3}, 1, \pi, 3\right\}$$

SOLUTION   The only numbers in the set that are not integers are $-1.75, \frac{2}{3}$, and π.   ∎

Definition

The set of **rational numbers** is the set of numbers commonly called "fractions" together with the integers.

The set of rational numbers is difficult to list in the same way we have listed the other sets, so we will use a different kind of notation:

$$\text{Rational numbers} = \left\{ \frac{a}{b} \,\middle|\, a \text{ and } b \text{ are integers } (b \neq 0) \right\}$$

This notation is read "The set of elements $\frac{a}{b}$ such that a and b are integers (and b is not 0)." If a number can be put in the form $\frac{a}{b}$, where a and b are both from the set of integers, then it is called a rational number.

Rational numbers include any number that can be written as the *ratio* of two integers; that is, rational numbers are numbers that can be put in the form

$$\frac{\text{integer}}{\text{integer}}$$

EXAMPLE 3   Show why each of the numbers in the following set is a rational number.

$$\left\{ -3, -\frac{2}{3}, 0, 0.333\ldots, 0.75 \right\}$$

SOLUTION   The number -3 is a rational number because it can be written as the ratio of -3 to 1; that is,

$$-3 = \frac{-3}{1}$$

Similarly, the number $-\frac{2}{3}$ can be thought of as the ratio of -2 to 3, whereas the number 0 can be thought of as the ratio of 0 to 1.

Any repeating decimal, such as $0.333\ldots$ (the dots indicate that the 3's repeat forever), can be written as the ratio of two integers. In this case $0.333\ldots$ is the same as the fraction $\frac{1}{3}$.

Finally, any decimal that terminates after a certain number of digits can be written as the ratio of two integers. The number 0.75 is equal to the fraction $\frac{3}{4}$ and is therefore a rational number.   ■

Still other numbers exist, each of which is associated with a point on the real number line, that cannot be written as the ratio of two integers. In decimal form, they never terminate and never repeat a sequence of digits indefinitely. They are called *irrational numbers* (because they are not rational):

$$\text{Irrational numbers} = \{\text{nonrational numbers}\}$$

$$= \{\text{nonrepeating, nonterminating decimals}\}$$

Here is a formal definition:

Definition
Irrational numbers are numbers on the number line with nonrepeating, nonterminating decimals and cannot be written as the ratio of two intergers.

We cannot write irrational numbers in a form that is familiar to us because they are all nonrepeating, nonterminating decimals. They have to be represented in other ways. One irrational number you have probably seen before is π. It is not 3.14. Rather, 3.14 is an approximation of π, or *3.141592653589...* It cannot be written as a terminating decimal number. Other representations for irrational numbers are $\sqrt{2}$, $\sqrt{3}$, $\sqrt{5}$, $\sqrt{6}$, and, in general, the square root of any number that is not itself a perfect square. (If you are not familiar with square roots, you will be later in the book.) Right now it is enough to know that some numbers on the number line—irrational numbers—cannot be written as the ratio of two integers or in decimal form.

> **Definition**
>
> The set of **real numbers** is the set of numbers that are either rational or irrational; that is, a real number is either rational or irrational.
>
> Real numbers = {all rational numbers and all irrational numbers}

B Prime Numbers and Factoring

The following diagram shows the relationship between multiplication and factoring:

$$\text{Multiplication}$$
$$\text{Factors} \longrightarrow 3 \cdot 4 = 12 \longleftarrow \text{Product}$$
$$\text{Factoring}$$

When we read the problem from left to right, we say the product of 3 and 4 is 12. Or we multiply 3 and 4 to get 12. When we read the problem in the other direction, from right to left, we say we have *factored* 12 into 3 times 4, or 3 and 4 are *factors* of 12.

The number 12 can be factored still further:

$$12 = 4 \cdot 3$$
$$= 2 \cdot 2 \cdot 3$$
$$= 2^2 \cdot 3$$

The numbers 2 and 3 are called *prime factors* of 12 because neither of them can be factored any further.

> **Definition**
>
> If a and b represent integers, then a is said to be a **factor** (or divisor) of b if a divides b evenly, that is, if a divides b with no remainder.

> **Definition**
>
> A **prime number** is any positive integer larger than 1 whose only positive factors (divisors) are itself and 1.

Here is a list of the first few prime numbers.

Prime numbers = {2, 3, 5, 7, 11, 13, 17, 19, 23, 29, 31, 37, 41, . . . }

When a number is not prime, we can factor it into the product of prime numbers. The number 15 is not a prime number because it has factors of 3 and 5; that is, $15 = 3 \cdot 5$. When a whole number larger than 1 is not prime, it is said to be *composite*. To factor a number into the product of primes, we simply factor it until it cannot be factored further.

EXAMPLE 4 Factor the number 60 into the product of prime numbers.

SOLUTION We begin by writing 60 as the product of any two positive integers whose product is 60, like 6 and 10.

$$60 = 6 \cdot 10$$

Then we factor these numbers:

NOTE

It is customary to write prime factors in order from smallest to largest.

$$60 = 6 \cdot 10$$
$$= (2 \cdot 3) \cdot (2 \cdot 5)$$
$$= 2 \cdot 2 \cdot 3 \cdot 5$$
$$= 2^2 \cdot 3 \cdot 5$$

NOTE

There are some "tricks" to finding the divisors of a number. For instance, if a number ends in 0 or 5, then it is divisible by 5. If a number ends in an even number (0, 2, 4, 6, or 8), then it is divisible by 2. A number is divisible by 3 if the sum of its digits is divisible by 3. For example, 921 is divisible by 3 because the sum of its digits is $9 + 2 + 1 = 12$, which is divisible by 3.

EXAMPLE 5 Factor the number 630 into the product of its primes.

SOLUTION Let's begin by writing 630 as the product of 63 and 10.

$$630 = 63 \cdot 10$$
$$= (7 \cdot 9) \cdot (2 \cdot 5)$$
$$= 7 \cdot 3 \cdot 3 \cdot 2 \cdot 5$$
$$= 2 \cdot 3^2 \cdot 5 \cdot 7$$

It makes no difference which two numbers we start with, as long as their product is 630. We always will get the same result because a number has only one set of prime factors.

$$630 = 18 \cdot 35$$
$$= 3 \cdot 6 \cdot 5 \cdot 7$$
$$= 3 \cdot 2 \cdot 3 \cdot 5 \cdot 7$$
$$= 2 \cdot 3^2 \cdot 5 \cdot 7$$

If we factor 210 into its prime factors, we have $210 = 2 \cdot 3 \cdot 5 \cdot 7$, which means that 2, 3, 5, and 7 divide 210, as well as any combination of products of 2, 3, 5, and 7; that is, because 3 and 7 divide 210, then so does their product 21. Because 3, 5, and 7 each divide 210, then so does their product 105.

$$\begin{array}{c} 21 \text{ divides } 210 \\ 210 = 2 \cdot 3 \cdot 5 \cdot 7 \\ 105 \text{ divides } 210 \end{array}$$

C Reducing to Lowest Terms

Recall that we reduce fractions to lowest terms by dividing the numerator and denominator by the same number. We can use the prime factorization of numbers to help us reduce fractions with large numerators and denominators.

EXAMPLE 6 Reduce $\dfrac{210}{231}$ to lowest terms.

SOLUTION First, we factor 210 and 231 into the product of prime factors. Then we reduce to lowest terms by dividing the numerator and denominator by any factors they have in common.

NOTE

The small lines we have drawn through the factors that are common to the numerator and denominator are used to indicate that we have divided the numerator and denominator by those factors.

$$\frac{210}{231} = \frac{2 \cdot 3 \cdot 5 \cdot 7}{3 \cdot 7 \cdot 11} \qquad \text{Factor the numerator and denominator completely}$$

$$= \frac{2 \cdot 3 \cdot 5 \cdot 7}{3 \cdot 7 \cdot 11} \qquad \text{Divide the numerator and denominator by } 3 \cdot 7$$

$$= \frac{2 \cdot 5}{11} = \frac{10}{11}$$

Problem Set 1.7

Moving Toward Success

"Don't find fault, find a remedy."

—Henry Ford, 1863–1947, American industrialist and founder of Ford Motor Company

1. Is it important to look at the answers at the back of the book? Why?

2. When will you use the answers in the back of the book?

a. Before starting the problem

b. To check if an answer you have reached is correct

c. When you are stuck on a problem

d. Never

A Given the numbers in the set $\{-3, -2.5, 0, 1, \frac{3}{2}, \sqrt{15}\}$: [Examples 1–3]

1. List all the whole numbers.

2. List all the integers.

3. List all the rational numbers.

4. List all the irrational numbers.

5. List all the real numbers.

A Given the numbers in the set $\{-10, -8, -0.333\ldots, -2, 9, \frac{25}{3}, \pi\}$:

6. List all the whole numbers.

7. List all the integers.

8. List all the rational numbers.

9. List all the irrational numbers.

10. List all the real numbers.

Identify the following statements as either true or false.

11. Every whole number is also an integer.

12. The set of whole numbers is a subset of the set of integers.

13. A number can be both rational and irrational.

14. The set of rational numbers and the set of irrational numbers have some elements in common.

15. Some whole numbers are also negative integers.

16. Every rational number is also a real number.

17. All integers are also rational numbers.

18. The set of integers is a subset of the set of rational numbers.

B Label each of the following numbers as *prime* or *composite*. If a number is composite, then factor it completely. [Examples 4–5]

19. 48

20. 72

21. 37

22. 23

23. 1,023

24. 543

B Factor the following into the product of primes. When the number has been factored completely, write its prime factors from smallest to largest.

25. 144

26. 288

27. 38

28. 63

29. 105

30. 210

31. 180

32. 900

33. 385

34. 1,925

35. 121

36. 546

37. 420

38. 598

39. 620

40. 2,310

C Reduce each fraction to lowest terms by first factoring the numerator and denominator into the product of prime factors and then dividing out any factors they have in common. [Example 6]

41. $\dfrac{105}{165}$

42. $\dfrac{165}{385}$

43. $\dfrac{525}{735}$

44. $\dfrac{550}{735}$

45. $\dfrac{385}{455}$

46. $\dfrac{385}{735}$

47. $\dfrac{322}{345}$

48. $\dfrac{266}{285}$

49. $\dfrac{205}{369}$

50. $\dfrac{111}{185}$

51. $\dfrac{215}{344}$ **52.** $\dfrac{279}{310}$

The next two problems are intended to give you practice reading, and paying attention to, the instructions that accompany the problems you are working. You will see a number of problems like this throughout the book. Working these problems is an excellent way to get ready for a test or a quiz.

53. Work each problem according to the instructions given. (Note that each of these instructions could be replaced with the instruction *Simplify*.)

   a. Add: $50 + (-80)$

   b. Subtract: $50 - (-80)$

   c. Multiply: $50(-80)$

   d. Divide: $\dfrac{50}{-80}$

54. Work each problem according to the instructions given. (Note that each of these instructions could be replaced with the instruction *Simplify*.)

   a. Add: $-2.5 + 7.5$

   b. Subtract: $-2.5 - 7.5$

   c. Multiply: $-2.5(7.5)$

   d. Divide: $\dfrac{-2.5}{7.5}$

Simplify each expression without using a calculator.

55. $\dfrac{6.28}{9(3.14)}$ **56.** $\dfrac{12.56}{4(3.14)}$

57. $\dfrac{9.42}{2(3.14)}$ **58.** $\dfrac{12.56}{2(3.14)}$

59. $\dfrac{32}{0.5}$ **60.** $\dfrac{16}{0.5}$

61. $\dfrac{5,599}{11}$ **62.** $\dfrac{840}{80}$

63. Find the value of $\dfrac{2 + 0.15x}{x}$ for each of the values of x given below. Write your answers as decimals, to the nearest hundreth.

   a. $x = 10$

   b. $x = 15$

   c. $x = 20$

64. Find the value of $\dfrac{5x + 250}{640x}$ for each of the values of x given below. Write your answers as decimals, to the nearest thousandth.

   a. $x = 10$

   b. $x = 15$

   c. $x = 20$

65. Factor 6^3 into the product of prime factors by first factoring 6 and then raising each of its factors to the third power.

66. Factor 12^2 into the product of prime factors by first factoring 12 and then raising each of its factors to the second power.

67. Factor $9^4 \cdot 16^2$ into the product of prime factors by first factoring 9 and 16 completely.

68. Factor $10^2 \cdot 12^3$ into the product of prime factors by first factoring 10 and 12 completely.

69. Simplify the expression $3 \cdot 8 + 3 \cdot 7 + 3 \cdot 5$, and then factor the result into the product of primes. (Notice one of the factors of the answer is 3.)

70. Simplify the expression $5 \cdot 4 + 5 \cdot 9 + 5 \cdot 3$, and then factor the result into the product of primes.

Applying the Concepts

71. Cars The chart shows the fastest cars in the world. Factor the speed of the SSC Ultimate Aero into the product of primes.

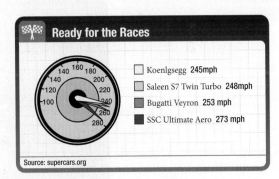

72. Energy Use The chart shows how much energy is used by gaming consoles. Write the energy used by a Wii over the energy used by the PS3 and then reduce to lowest terms.

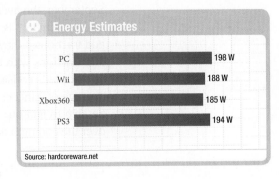

Recall the Fibonacci sequence we introduced earlier in this chapter.

Fibonacci sequence = 1, 1, 2, 3, 5, 8, . . .

Any number in the Fibonacci sequence is a *Fibonacci number.*

73. The Fibonacci numbers are not a subset of which of the following sets: real numbers, rational numbers, irrational numbers, whole numbers?

74. Name three Fibonacci numbers that are prime numbers.

75. Name three Fibonacci numbers that are composite numbers.

76. Is the sequence of odd numbers a subset of the Fibonacci numbers?

1.8 Addition and Subtraction of Fractions with Variables

OBJECTIVES

A Add or subtract two or more fractions with the same denominator.

B Find the least common denominator for a set of fractions.

C Add or subtract fractions with different denominators.

D Extend a sequence of numbers containing fractions.

TICKET TO SUCCESS

Keep these questions in mind as you read through the section. Then respond in your own words and in complete sentences.

1. Use symbols to show how you would add two fractions with common denominators.

2. What is a least common denominator?

3. What is the first step in adding two fractions that have different denominators?

4. What is the last thing you do when subtracting two fractions?

In May, the average monthly rainfall for San Luis Obispo, CA is $\frac{1}{2}$ inches. In September, the average is $\frac{1}{3}$ inches. What if we needed to find the total rainfall for May and September? Based on our knowledge from previous math classes, we can only add fractions if they have common denominators. In this section, we will learn how to add fractions with different denominators including those that contain variables. But to start, let's consider some equivalent quantities for our rainfall total. If it rained an average of $\frac{3}{6}$ inches in May and an average of $\frac{2}{6}$ inches in September, then we would be able to find our total by simply adding the two numerators and putting the result over the common denominator.

$$\frac{3}{6} + \frac{2}{6} = \frac{(3+2)}{6} = \frac{5}{6} \text{ inches}$$

NOTE
Most people who have done any work with adding fractions know that you add fractions that have the same denominator by adding their numerators but not their denominators. However, most people don't know why this works. The reason why we add numerators but not denominators is because of the distributive property. That is what the discussion at the right is all about. If you really want to understand addition of fractions, pay close attention to this discussion.

The reason we add numerators but do not add denominators is that we must follow the distributive property. To see this, you first have to recall that $\frac{3}{6}$ can be written as $3 \cdot \frac{1}{6}$, and $\frac{2}{6}$ can be written as $2 \cdot \frac{1}{6}$ (dividing by 6 is equivalent to multiplying by $\frac{1}{6}$). Here is the addition problem again, this time showing the use of the distributive property:

$$\frac{3}{6} + \frac{2}{6} = 3 \cdot \frac{1}{6} + 2 \cdot \frac{1}{6}$$

$$= (3 + 2) \cdot \frac{1}{6} \qquad \text{Distributive property}$$

$$= 5 \cdot \frac{1}{6}$$

$$= \frac{5}{6}$$

A Adding and Subtracting with the Same Denominator

What we have here is the sum of the numerators placed over the *common denominator*. In symbols, we have the following:

Addition and Subtraction of Fractions
If a, b, and c are integers and c is not equal to zero, then

$$\frac{a}{c} + \frac{b}{c} = \frac{a + b}{c}$$

This rule holds for subtraction as well; that is,

$$\frac{a}{c} - \frac{b}{c} = \frac{a - b}{c}$$

In Examples 1–4, find the sum or difference. (Add or subtract as indicated.) Reduce all answers to lowest terms. (Assume all variables represent nonzero numbers.)

EXAMPLE 1 Add $\frac{3}{8} + \frac{1}{8}$.

SOLUTION $\frac{3}{8} + \frac{1}{8} = \frac{3 + 1}{8}$ Add numerators; keep the same denominator

$$= \frac{4}{8} \qquad \text{The sum of 3 and 1 is 4}$$

$$= \frac{1}{2} \qquad \text{Reduce to lowest terms}$$

EXAMPLE 2 Subtract: $\frac{a + 5}{8} - \frac{3}{8}$.

SOLUTION $\frac{a + 5}{8} - \frac{3}{8} = \frac{a + 5 - 3}{8}$ Combine numerators; keep the same denominator

$$= \frac{a + 2}{8}$$

EXAMPLE 3 Subtract: $\frac{9}{x} - \frac{3}{x}$.

SOLUTION $\frac{9}{x} - \frac{3}{x} = \frac{9 - 3}{x}$ Subtract numerators; keep the same denominator

$$= \frac{6}{x} \qquad \text{The difference of 9 and 3 is 6}$$

EXAMPLE 4 Simplify: $\dfrac{3}{7} + \dfrac{2}{7} - \dfrac{9}{7}$.

SOLUTION $\dfrac{3}{7} + \dfrac{2}{7} - \dfrac{9}{7} = \dfrac{3 + 2 - 9}{7}$

$$= \dfrac{-4}{7}$$

$$= -\dfrac{4}{7} \qquad \text{Unlike signs give a negative answer}$$

> **NOTE**
> The ability to find least common denominators is very important in mathematics. The discussion here is a detailed explanation of how to do it.

B Finding the Least Common Denominator

We will now turn our attention to the process of adding fractions that have different denominators. To get started, we need the following definition:

> **Definition**
> The **least common denominator (LCD)** for a set of denominators is the smallest number that is exactly divisible by each denominator, also called the *least common multiple*.
>
> In other words, all the denominators of the fractions in a problem must divide into the least common denominator without a remainder.

EXAMPLE 5 Find the LCD for the fractions $\dfrac{5}{12}$ and $\dfrac{7}{18}$.

SOLUTION The least common denominator for the denominators 12 and 18 must be the smallest number divisible by both 12 and 18. We can factor 12 and 18 completely and then build the LCD from these factors. Factoring 12 and 18 completely gives us

$$12 = 2 \cdot 2 \cdot 3 \qquad\qquad 18 = 2 \cdot 3 \cdot 3$$

Now, if 12 is going to divide the LCD exactly, then the LCD must have factors of $2 \cdot 2 \cdot 3$. If 18 is to divide it exactly, it must have factors of $2 \cdot 3 \cdot 3$. We don't need to repeat the factors that 12 and 18 have in common:

$$\left.\begin{array}{l} 12 = 2 \cdot 2 \cdot 3 \\ 18 = 2 \cdot 3 \cdot 3 \end{array}\right\} \qquad \overset{\text{12 divides the LCD}}{\text{LCD} = 2 \cdot 2 \cdot 3 \cdot 3 = 36} \atop {\text{18 divides the LCD}}$$

In other words, first we write down the factors of 12, then we attach the factors of 18 that do not already appear as factors of 12.

The LCD for 12 and 18 is 36. It is the smallest number that is divisible by both 12 and 18; 12 divides it exactly three times, and 18 divides it exactly two times.

EXAMPLE 6 Find the LCD for $\dfrac{3}{4}$ and $\dfrac{1}{6}$.

SOLUTION We factor 4 and 6 into products of prime factors and build the LCD from these factors:

$$\left.\begin{array}{l} 4 = 2 \cdot 2 \\ 6 = 2 \cdot 3 \end{array}\right\} \qquad \text{LCD} = 2 \cdot 2 \cdot 3 = 12$$

The LCD is 12. Both denominators divide it exactly; 4 divides 12 exactly three times, and 6 divides 12 exactly two times.

C Adding and Subtracting with Different Denominators

We can use the results of Example 5 to find the sum of the fractions $\frac{5}{12}$ and $\frac{7}{18}$.

EXAMPLE 7   Add $\frac{5}{12} + \frac{7}{18}$.

SOLUTION   We can add fractions only when they have the same denominators. In Example 5 we found the LCD for $\frac{5}{12}$ and $\frac{7}{18}$ to be 36. We change $\frac{5}{12}$ and $\frac{7}{18}$ to equivalent fractions that each have 36 for a denominator by applying the multiplication proprety for fractions:

$$\frac{5}{12} = \frac{5 \cdot 3}{12 \cdot 3} = \frac{15}{36} \qquad \frac{5}{12} \text{ is equivalent to } \frac{15}{36}$$

$$\frac{7}{18} = \frac{7 \cdot 2}{18 \cdot 2} = \frac{14}{36} \qquad \frac{7}{18} \text{ is equivalent to } \frac{14}{36}$$

The fraction $\frac{15}{36}$ is equivalent to $\frac{5}{12}$, because it was obtained by multiplying both the numerator and denominator by 3. Likewise, $\frac{14}{36}$ is equivalent to $\frac{7}{18}$ because it was obtained by multiplying the numerator and denominator by 2. All we have left to do is to add numerators:

$$\frac{15}{36} + \frac{14}{36} = \frac{29}{36}$$

The sum of $\frac{5}{12}$ and $\frac{7}{18}$ is the fraction $\frac{29}{36}$. Let's write the complete solution again step-by-step.

$$\frac{5}{12} + \frac{7}{18} = \frac{5 \cdot 3}{12 \cdot 3} + \frac{7 \cdot 2}{18 \cdot 2} \qquad \text{Rewrite each fraction as an equivalent fraction with denominator 36}$$

$$= \frac{15}{36} + \frac{14}{36}$$

$$= \frac{29}{36} \qquad \text{Add numerators; keep the common denominator}$$

EXAMPLE 8   Add $\frac{3}{4} + \frac{1}{6}$.

SOLUTION   In Example 6 we found that the LCD for these two fractions is 12. We begin by changing $\frac{3}{4}$ and $\frac{1}{6}$ to equivalent fractions with denominator 12:

$$\frac{3}{4} = \frac{3 \cdot 3}{4 \cdot 3} = \frac{9}{12} \qquad \frac{3}{4} \text{ is equivalent to } \frac{9}{12}$$

$$\frac{1}{6} = \frac{1 \cdot 2}{6 \cdot 2} = \frac{2}{12} \qquad \frac{1}{6} \text{ is equivalent to } \frac{2}{12}$$

To complete the problem, we add numerators:

$$\frac{9}{12} + \frac{2}{12} = \frac{11}{12}$$

The sum of $\frac{3}{4}$ and $\frac{1}{6}$ is $\frac{11}{12}$. Here is how the complete problem looks:

$$\frac{3}{4} + \frac{1}{6} = \frac{3 \cdot 3}{4 \cdot 3} + \frac{1 \cdot 2}{6 \cdot 2} \qquad \text{Rewrite each fraction as an equivalent fraction with denominator 12}$$

$$= \frac{9}{12} + \frac{2}{12} = \frac{11}{12} \qquad \text{Add numerators; keep the same denominator}$$

EXAMPLE 9   Subtract $\dfrac{7}{15} - \dfrac{3}{10}$.

SOLUTION   Let's factor 15 and 10 completely and use these factors to build the LCD.

$$
\begin{array}{c}
\text{15 divides the LCD} \\
\\
\left.\begin{array}{l} 15 = 3 \cdot 5 \\ 10 = 2 \cdot 5 \end{array}\right\} \quad \text{LCD} = 2 \cdot 3 \cdot 5 = 30 \\
\\
\text{10 divides the LCD}
\end{array}
$$

Changing to equivalent fractions and subtracting, we have

$$\dfrac{7}{15} - \dfrac{3}{10} = \dfrac{7 \cdot \mathbf{2}}{15 \cdot \mathbf{2}} - \dfrac{3 \cdot \mathbf{3}}{10 \cdot \mathbf{3}}$$ Rewrite as equivalent fractions with the LCD for denominator

$$= \dfrac{14}{30} - \dfrac{9}{30}$$

$$= \dfrac{5}{30}$$ Subtract numerators; keep the LCD

$$= \dfrac{1}{6}$$ Reduce to lowest terms

As a summary of what we have done so far and as a guide to working other problems, we will now list the steps involved in adding and subtracting fractions with different denominators.

Strategy   To Add or Subtract Any Two Fractions

Step 1: Factor each denominator completely and use the factors to build the LCD. (Remember, the LCD is the smallest number divisible by each of the denominators in the problem.)

Step 2: Rewrite each fraction as an equivalent fraction that has the LCD for its denominator. This is done by multiplying both the numerator and denominator of the fraction in question by the appropriate whole number.

Step 3: Add or subtract the numerators of the fractions produced in step 2. This is the numerator of the sum or difference. The denominator of the sum or difference is the LCD.

Step 4: Reduce the fraction produced in step 3 to lowest terms if it is not already in lowest terms.

The idea behind adding or subtracting fractions is really very simple. We can add or subtract only fractions that have the same denominators. If the fractions we are trying to add or subtract do not have the same denominators, we rewrite each of them as an equivalent fraction with the LCD for a denominator.

Here are some further examples of sums and differences of fractions.

EXAMPLE 10   Add $\dfrac{1}{6} + \dfrac{1}{8} + \dfrac{1}{4}$.

SOLUTION   We begin by factoring the denominators completely and building the LCD from the factors that result. We then change to equivalent fractions and add as usual:

$$6 = 2 \cdot 3$$
$$8 = 2 \cdot 2 \cdot 2$$
$$4 = 2 \cdot 2$$

8 divides the LCD

$$LCD = 2 \cdot 2 \cdot 2 \cdot 3 = 24$$

4 divides the LCD   6 divides the LCD

$$\frac{1}{6} + \frac{1}{8} + \frac{1}{4} = \frac{1 \cdot \mathbf{4}}{6 \cdot \mathbf{4}} + \frac{1 \cdot \mathbf{3}}{8 \cdot \mathbf{3}} + \frac{1 \cdot \mathbf{6}}{4 \cdot \mathbf{6}}$$

$$= \frac{4}{24} + \frac{3}{24} + \frac{6}{24}$$

$$= \frac{13}{24}$$

EXAMPLE 11 Subtract $3 - \frac{5}{6}$.

SOLUTION The denominators are 1 $\left(\text{because } 3 = \frac{3}{1}\right)$ and 6. The smallest number divisible by both 1 and 6 is 6.

$$3 - \frac{5}{6} = \frac{3}{1} - \frac{5}{6}$$

$$= \frac{3 \cdot \mathbf{6}}{1 \cdot \mathbf{6}} - \frac{5}{6}$$

$$= \frac{18}{6} - \frac{5}{6}$$

$$= \frac{13}{6}$$

EXAMPLE 12 Subtract: $\frac{x}{5} - \frac{1}{6}$

SOLUTION The LCD for 5 and 6 is their product, 30. We begin by rewriting each fraction with this common denominator:

$$\frac{x}{5} - \frac{1}{6} = \frac{x \cdot \mathbf{6}}{5 \cdot \mathbf{6}} - \frac{1 \cdot \mathbf{5}}{6 \cdot \mathbf{5}}$$

$$= \frac{6x}{30} - \frac{5}{30}$$

$$= \frac{6x - 5}{30}$$

EXAMPLE 13 Add: $\frac{4}{x} + \frac{2}{3}$

NOTE
In Example 13, it is understood that x cannot be 0. Do you know why? Revisit "Division with the Number 0" in Section 5, if needed.

SOLUTION The LCD for x and 3 is $3x$. We multiply the numerator and the denominator of the first fraction by 3 and the numerator and the denominator of the second fraction by x to get two fractions with the same denominator. We then add the numerators:

$$\frac{4}{x} + \frac{2}{3} = \frac{4 \cdot \mathbf{3}}{x \cdot \mathbf{3}} + \frac{2 \cdot \mathbf{x}}{3 \cdot \mathbf{x}} \qquad \text{Change to equivalent fractions}$$

$$= \frac{12}{3x} + \frac{2x}{3x}$$

$$= \frac{12 + 2x}{3x} \qquad \text{Add the numerators}$$

When we are working with fractions, we can change the form of a fraction without changing its value. There will be times when one form is easier to work with than another form. Look over the material below and be sure you see that the pairs of expressions are equal.

The expressions $\dfrac{x}{2}$ and $\dfrac{1}{2}x$ are equal.

The expressions $\dfrac{3a}{4}$ and $\dfrac{3}{4}a$ are equal.

The expressions $\dfrac{7y}{3}$ and $\dfrac{7}{3}y$ are equal.

EXAMPLE 14 Add: $\dfrac{x}{3} + \dfrac{5x}{6}$.

SOLUTION We can do the problem two ways. One way probably seems easier, but both ways are valid methods of finding this sum. You should understand both of them.

METHOD 1:
$$\frac{x}{3} + \frac{5x}{6} = \frac{\mathbf{2} \cdot x}{\mathbf{2} \cdot 3} + \frac{5x}{6} \qquad \text{LCD}$$
$$= \frac{2x + 5x}{6} \qquad \text{Add numerators}$$
$$= \frac{7x}{6}$$

METHOD 2:
$$\frac{x}{3} + \frac{5x}{6} = \frac{1}{3}x + \frac{5}{6}x$$
$$= \left(\frac{1}{3} + \frac{5}{6}\right)x \qquad \text{Distributive property}$$
$$= \left(\frac{\mathbf{2} \cdot 1}{\mathbf{2} \cdot 3} + \frac{5}{6}\right)x$$
$$= \left(\frac{2}{6} + \frac{5}{6}\right)x$$
$$= \frac{7}{6}x$$

D Sequences Containing Fractions

EXAMPLE 15 Find the next number in each sequence.

a. $\dfrac{1}{2}, 0, -\dfrac{1}{2}, \ldots$ **b.** $\dfrac{1}{2}, 1, \dfrac{3}{2}, \ldots$ **c.** $\dfrac{1}{2}, \dfrac{1}{4}, \dfrac{1}{8}, \ldots$

SOLUTION **a.** $\dfrac{1}{2}, 0, -\dfrac{1}{2}, \ldots$: Adding $-\dfrac{1}{2}$ to each term produces the next term. The fourth term will be $-\dfrac{1}{2} + \left(-\dfrac{1}{2}\right) = -1$. This is an arithmetic sequence.

b. $\dfrac{1}{2}, 1, \dfrac{3}{2}, \ldots$: Each term comes from the term before it by adding $\dfrac{1}{2}$. The fourth term will be $\dfrac{3}{2} + \dfrac{1}{2} = 2$. This sequence is also an arithmetic sequence.

c. $\dfrac{1}{2}, \dfrac{1}{4}, \dfrac{1}{8}, \ldots$: This is a geometric sequence in which each term comes from the term before it by multiplying by $\dfrac{1}{2}$ each time. The next term will be $\dfrac{1}{8} \cdot \dfrac{1}{2} = \dfrac{1}{16}$.

Problem Set 1.8

Moving Toward Success

"Mistakes are the portals of discovery."
—James Joyce, 1882–1941, Irish writer and poet

1. Why is making mistakes important to the process of learning mathematics?

2. If you complete an assignment, what should you do?
   a. Read the next section in the book
   b. Work more problems
   c. Both of the above

A Find the following sums and differences, and reduce to lowest terms. Add and subtract as indicated. Assume all variables represent nonzero numbers. [Examples 1–4]

1. $\dfrac{3}{6} + \dfrac{1}{6}$

2. $\dfrac{2}{5} + \dfrac{3}{5}$

3. $\dfrac{3}{8} - \dfrac{5}{8}$

4. $\dfrac{1}{7} - \dfrac{6}{7}$

5. $-\dfrac{1}{4} + \dfrac{3}{4}$

6. $-\dfrac{4}{9} + \dfrac{7}{9}$

7. $\dfrac{x}{3} - \dfrac{1}{3}$

8. $\dfrac{x}{8} - \dfrac{1}{8}$

9. $\dfrac{1}{4} + \dfrac{2}{4} + \dfrac{3}{4}$

10. $\dfrac{2}{5} + \dfrac{3}{5} + \dfrac{4}{5}$

11. $\dfrac{x+7}{2} - \dfrac{1}{2}$

12. $\dfrac{x+5}{4} - \dfrac{3}{4}$

13. $\dfrac{1}{10} - \dfrac{3}{10} - \dfrac{4}{10}$

14. $\dfrac{3}{20} - \dfrac{1}{20} - \dfrac{4}{20}$

15. $\dfrac{1}{a} + \dfrac{4}{a} + \dfrac{5}{a}$

16. $\dfrac{5}{a} + \dfrac{4}{a} + \dfrac{3}{a}$

B **C** Find the LCD for each of the following. Then use the methods developed in this section to add and subtract as indicated. [Examples 5–11]

17. $\dfrac{1}{8} + \dfrac{3}{4}$

18. $\dfrac{1}{6} + \dfrac{2}{3}$

19. $\dfrac{3}{10} - \dfrac{1}{5}$

20. $\dfrac{5}{6} - \dfrac{1}{12}$

21. $\dfrac{4}{9} + \dfrac{1}{3}$

22. $\dfrac{1}{2} + \dfrac{1}{4}$

23. $2 + \dfrac{1}{3}$

24. $3 + \dfrac{1}{2}$

25. $-\dfrac{3}{4} + 1$

26. $-\dfrac{3}{4} + 2$

27. $\dfrac{1}{2} + \dfrac{2}{3}$

28. $\dfrac{2}{3} + \dfrac{1}{4}$

29. $\dfrac{5}{12} - \left(-\dfrac{3}{8}\right)$

30. $\dfrac{9}{16} - \left(-\dfrac{7}{12}\right)$

31. $-\dfrac{1}{20} + \dfrac{8}{30}$

32. $-\dfrac{1}{30} + \dfrac{9}{40}$

33. $\dfrac{17}{30} + \dfrac{11}{42}$

34. $\dfrac{19}{42} + \dfrac{13}{70}$

35. $\dfrac{25}{84} + \dfrac{41}{90}$

36. $\dfrac{23}{70} + \dfrac{29}{84}$

37. $\dfrac{13}{126} - \dfrac{13}{180}$

38. $\dfrac{17}{84} - \dfrac{17}{90}$

39. $\dfrac{3}{4} + \dfrac{1}{8} + \dfrac{5}{6}$

40. $\dfrac{3}{8} + \dfrac{2}{5} + \dfrac{1}{4}$

41. $\dfrac{1}{2} + \dfrac{1}{3} + \dfrac{1}{4} + \dfrac{1}{6}$

42. $\dfrac{1}{8} + \dfrac{1}{4} + \dfrac{1}{5} + \dfrac{1}{10}$

43. $1 - \dfrac{5}{2}$

44. $1 - \dfrac{5}{3}$

45. $1 + \dfrac{1}{2}$

46. $1 + \dfrac{2}{3}$

D Find the fourth term in each sequence. [Example 15]

47. $\dfrac{1}{3}, 0, -\dfrac{1}{3}, \ldots$

48. $\dfrac{2}{3}, 0, -\dfrac{2}{3}, \ldots$

49. $\dfrac{1}{3}, 1, \dfrac{5}{3}, \ldots$

50. $1, \dfrac{3}{2}, 2, \ldots$

51. $1, \dfrac{1}{5}, \dfrac{1}{25}, \ldots$

52. $1, -\dfrac{1}{2}, \dfrac{1}{4}, \ldots$

Use the rule for order of operations to simplify each expression.

53. $9 - 3\left(\dfrac{5}{3}\right)$

54. $6 - 4\left(\dfrac{7}{2}\right)$

55. $-\dfrac{1}{2} + 2\left(-\dfrac{3}{4}\right)$

56. $\dfrac{5}{4} - 3\left(\dfrac{7}{12}\right)$

57. $\dfrac{3}{5}(-10) + \dfrac{4}{7}(-21)$

58. $-\dfrac{3}{5}(10) - \dfrac{4}{7}(21)$

59. $16\left(-\dfrac{1}{2}\right)^2 - 125\left(-\dfrac{2}{5}\right)^2$

60. $16\left(-\dfrac{1}{2}\right)^3 - 125\left(-\dfrac{2}{5}\right)^3$

61. $-\dfrac{4}{3} \div 2 \cdot 3$ **62.** $-\dfrac{8}{7} \div 4 \cdot 2$

63. $-\dfrac{4}{3} \div 2(-3)$ **64.** $-\dfrac{8}{7} \div 4(-2)$

65. $-6 \div \dfrac{1}{2} \cdot 12$ **66.** $-6 \div \left(-\dfrac{1}{2}\right) \cdot 12$

67. $-15 \div \dfrac{5}{3} \cdot 18$ **68.** $-15 \div \left(-\dfrac{5}{3}\right) \cdot 18$

Add or subtract the following fractions. (Assume all variables represent nonzero numbers.) [Examples 12–14]

69. $\dfrac{x}{4} + \dfrac{1}{5}$ **70.** $\dfrac{x}{3} + \dfrac{1}{5}$

71. $\dfrac{1}{3} + \dfrac{a}{12}$ **72.** $\dfrac{1}{8} + \dfrac{a}{32}$

73. $\dfrac{x}{2} + \dfrac{1}{3} + \dfrac{x}{4}$ **74.** $\dfrac{x}{3} + \dfrac{1}{4} + \dfrac{x}{5}$

75. $\dfrac{2}{x} + \dfrac{3}{5}$ **76.** $\dfrac{3}{x} - \dfrac{2}{5}$

77. $\dfrac{3}{7} + \dfrac{4}{y}$ **78.** $\dfrac{2}{9} + \dfrac{5}{y}$

79. $\dfrac{3}{a} + \dfrac{3}{4} + \dfrac{1}{5}$ **80.** $\dfrac{4}{a} + \dfrac{2}{3} + \dfrac{1}{2}$

81. $\dfrac{1}{2}x + \dfrac{1}{6}x$ **82.** $\dfrac{2}{3}x + \dfrac{5}{6}x$

83. $\dfrac{1}{2}x - \dfrac{3}{4}x$ **84.** $\dfrac{2}{3}x - \dfrac{5}{6}x$

85. $\dfrac{1}{3}x + \dfrac{3}{5}x$ **86.** $\dfrac{2}{3}x - \dfrac{3}{5}x$

87. $\dfrac{3x}{4} + \dfrac{x}{6}$ **88.** $\dfrac{3x}{4} - \dfrac{2x}{3}$

89. $\dfrac{2x}{5} + \dfrac{5x}{8}$ **90.** $\dfrac{3x}{5} - \dfrac{3x}{8}$

Simplify.

91. $1 - \dfrac{1}{x}$ **92.** $1 + \dfrac{1}{x}$

The next two problems are intended to give you practice reading, and paying attention to, the instructions that accompany the problems you are working. As we mentioned previously, working these problems is an excellent way to get ready for a test or a quiz.

93. Work each problem according to the instructions given.

     a. Add: $\dfrac{3}{4} + \left(-\dfrac{1}{2}\right)$

     b. Subtract: $\dfrac{3}{4} - \left(-\dfrac{1}{2}\right)$

     c. Multiply: $\dfrac{3}{4}\left(-\dfrac{1}{2}\right)$

     d. Divide: $\dfrac{3}{4} \div \left(-\dfrac{1}{2}\right)$

94. Work each problem according to the instructions given.

     a. Add: $-\dfrac{5}{8} + \left(-\dfrac{1}{2}\right)$

     b. Subtract: $-\dfrac{5}{8} - \left(-\dfrac{1}{2}\right)$

     c. Multiply: $-\dfrac{5}{8}\left(-\dfrac{1}{2}\right)$

     d. Divide: $-\dfrac{5}{8} \div \left(-\dfrac{1}{2}\right)$

Simplify.

95. $\left(1 - \dfrac{1}{2}\right)\left(1 - \dfrac{1}{3}\right)$ **96.** $\left(1 + \dfrac{1}{2}\right)\left(1 + \dfrac{1}{3}\right)$

97. $\left(1 + \dfrac{1}{2}\right)\left(1 - \dfrac{1}{2}\right)$ **98.** $\left(1 + \dfrac{1}{3}\right)\left(1 - \dfrac{1}{3}\right)$

99. Find the value of $1 + \dfrac{1}{x}$ when x is

     a. 2

     b. 3

     c. 4

100. Find the value of $1 - \dfrac{1}{x}$ when x is

     a. 2

     b. 3

     c. 4

101. Find the value of $2x + \dfrac{6}{x}$ when x is

     a. 1

     b. 2

     c. 3

102. Find the value of $x + \dfrac{4}{x}$ when x is

     a. 1

     b. 2

     c. 3

Chapter 1 Summary

The number(s) in brackets next to each heading indicates the section(s) in which that topic is discussed.

NOTE
We will use the margins in the chapter summaries to give examples that correspond to the topic being reviewed whenever it is appropriate.

Symbols [1.1]

$a = b$	a is equal to b
$a \neq b$	a is not equal to b
$a < b$	a is less than b
$a \not< b$	a is not less than b
$a > b$	a is greater than b
$a \not> b$	a is not greater than b
$a \geq b$	a is greater than or equal to b
$a \leq b$	a is less than or equal to b

EXAMPLES

1. $2^5 = 2 \cdot 2 \cdot 2 \cdot 2 \cdot 2 = 32$
$5^2 = 5 \cdot 5 = 25$
$10^3 = 10 \cdot 10 \cdot 10 = 1{,}000$
$1^4 = 1 \cdot 1 \cdot 1 \cdot 1 = 1$

Exponents [1.1]

Exponents are notation used to indicate repeated multiplication. In the expression 3^4, 3 is the *base* and 4 is the *exponent*.

$$3^4 = 3 \cdot 3 \cdot 3 \cdot 3 = 81$$

Order of Operations [1.1]

2. $10 + (2 \cdot 3^2 - 4 \cdot 2)$
$= 10 + (2 \cdot 9 - 4 \cdot 2)$
$= 10 + (18 - 8)$
$= 10 + 10$
$= 20$

When evaluating a mathematical expression, we will perform the operations in the following order, beginning with the expression in the innermost parentheses or brackets and working our way out.

1. Simplify all numbers with exponents, working from left to right if more than one of these numbers is present.
2. Then do all multiplications and divisions left to right.
3. Finally, perform all additions and subtractions left to right.

Absolute Value [1.2]

3. $|5| = 5$
$|-5| = 5$

The absolute value of a real number is its distance from zero on the real number line. Absolute value is never negative.

Opposites [1.2]

4. The numbers 3 and -3 are opposites; their sum is 0.
$3 + (-3) = 0$

Any two real numbers the same distance from zero on the number line but in opposite directions from zero are called opposites. Opposites always add to zero.

Reciprocals [1.2]

5. The numbers 2 and $\frac{1}{2}$ are reciprocals; their product is 1.

$$2\left(\frac{1}{2}\right) = 1$$

Any two real numbers whose product is 1 are called reciprocals. Every real number has a reciprocal except 0.

Areas and Perimeters [1.2]

Rectangle

l

w w

l

Perimeter $= 2l + 2w$

Area $= lw$

The formula for perimeter gives the distance around the outside of an object, along its sides.

The formula for area gives a measurement of the amount of surface an object has.

Addition of Real Numbers [1.3]

6. Add all combinations of positive and negative 10 and 13.

$$10 + 13 = 23$$
$$10 + (-13) = -3$$
$$-10 + 13 = 3$$
$$-10 + (-13) = -23$$

To add two real numbers with

1. The same sign: Simply add their absolute values and use the common sign.
2. Different signs: Subtract the smaller absolute value from the larger absolute value. The answer has the same sign as the number with the larger absolute value.

Subtraction of Real Numbers [1.3]

7. Subtracting 2 is the same as adding -2.

$$7 - 2 = 7 + (-2) = 5$$

To subtract one number from another, simply add the opposite of the number you are subtracting; that is, if a and b represent real numbers, then

$$a - b = a + (-b)$$

Multiplication of Real Numbers [1.4]

8.
$$3(5) = 15$$
$$3(-5) = -15$$
$$-3(5) = -15$$
$$-3(-5) = 15$$

To multiply two real numbers, simply multiply their absolute values. Like signs give a positive answer. Unlike signs give a negative answer.

Division of Real Numbers [1.5]

9. $-\dfrac{6}{2} = -6\left(\dfrac{1}{2}\right) = -3$

$\dfrac{-6}{-2} = -6\left(-\dfrac{1}{2}\right) = 3$

Division by a number is the same as multiplication by its reciprocal. Like signs give a positive answer. Unlike signs give a negative answer.

Properties of Real Numbers [1.6]

	For Addition	*For Multiplication*
Commutative:	$a + b = b + a$	$a \cdot b = b \cdot a$
Associative:	$a + (b + c) = (a + b) + c$	$a \cdot (b \cdot c) = (a \cdot b) \cdot c$
Identity:	$a + 0 = a$	$a \cdot 1 = a$
Inverse:	$a + (-a) = 0$	$a\left(\dfrac{1}{a}\right) = 1$
Distributive:	$a(b + c) = ab + ac$	

10. a. 7 and 100 are counting numbers, but 0 and -2 are not.

b. 0 and 241 are whole numbers, but -4 and $\frac{1}{2}$ are not.

c. $-15, 0$, and 20 are integers.

d. $-4, -\frac{1}{2}, 0.75$, and $0.666\ldots$ are rational numbers.

e. $-\pi, \sqrt{3}$, and π are irrational numbers.

f. All the numbers listed above are real numbers.

Subsets of the Real Numbers [1.7]

Counting numbers:	$\{1, 2, 3, \ldots\}$
Whole numbers:	$\{0, 1, 2, 3, \ldots\}$
Integers:	$\{\ldots, -3, -2, -1, 0, 1, 2, 3, \ldots\}$
Rational numbers:	{all numbers that can be expressed as the ratio of two integers}
Irrational numbers:	{all numbers on the number line that cannot be expressed as the ratio of two integers}
Real numbers:	{all numbers that are either rational or irrational}

Factoring [1.7]

11. The number 150 can be factored into the product of prime numbers.
$$150 = 15 \cdot 10$$
$$= (3 \cdot 5)(2 \cdot 5)$$
$$= 2 \cdot 3 \cdot 5^2$$

Factoring is the reverse of multiplication.

Multiplication

$$\text{Factors} \rightarrow 3 \cdot 5 = 15 \leftarrow \text{Product}$$

Factoring

Least Common Denominator (LCD) [1.8]

12. The LCD for $\frac{5}{12}$ and $\frac{7}{18}$ is 36.

The *least common denominator* (LCD) for a set of denominators is the smallest number that is exactly divisible by each denominator.

Addition and Subtraction of Fractions [1.8]

13. $\frac{5}{12} + \frac{7}{18} = \frac{5}{12} \cdot \frac{3}{3} + \frac{7}{18} \cdot \frac{2}{2}$
$$= \frac{15}{36} + \frac{14}{36}$$
$$= \frac{29}{36}$$

To add (or subtract) two fractions with a common denominator, add (or subtract) numerators and use the common denominator.

$$\frac{a}{c} + \frac{b}{c} = \frac{a+b}{c} \quad \text{and} \quad \frac{a}{c} - \frac{b}{c} = \frac{a-b}{c}$$

⊘ COMMON MISTAKES

1. Interpreting absolute value as changing the sign of the number inside the absolute value symbols. $|-5| = +5, |+5| = -5$. (The first expression is correct; the second one is not.) To avoid this mistake, remember: Absolute value is a distance and distance is always measured in positive units.

2. Using the phrase "two negatives make a positive." This works only with multiplication and division. With addition, two negative numbers produce a negative answer. It is best not to use the phrase "two negatives make a positive" at all.

Write the numerical expression that is equivalent to each phrase, and then simplify. [1.3, 1.4, 1.5]

1. The sum of -7 and -10

2. Five added to the sum of -7 and 4

3. The sum of -3 and 12 increased by 5

4. The difference of 4 and 9

5. The difference of 9 and -3

6. The difference of -7 and -9

7. The product of -3 and -7 decreased by 6

8. Ten added to the product of 5 and -6

9. Twice the product of -8 and $3x$

10. The quotient of -25 and -5

Simplify. [1.2]

11. $|-1.8|$

12. $-|-10|$

For each number, give the opposite and the reciprocal. [1.2]

13. 6

14. $-\dfrac{12}{5}$

Multiply. [1.2, 1.4]

15. $\dfrac{1}{2}(-10)$

16. $\left(-\dfrac{4}{5}\right)\left(\dfrac{25}{16}\right)$

Add. [1.3]

17. $-9 + 12$

18. $-18 + (-20)$

19. $-2 + (-8) + [-9 + (-6)]$

20. $(-21) + 40 + (-23) + 5$

Subtract. [1.3]

21. $6 - 9$

22. $14 - (-8)$

23. $-12 - (-8)$

24. $4 - 9 - 15$

Find the products. [1.4]

25. $(-5)(6)$

26. $4(-3)$

27. $-2(3)(4)$

28. $(-1)(-3)(-1)(-4)$

Find the following quotients. [1.5]

29. $\dfrac{12}{-3}$

30. $-\dfrac{8}{9} \div \dfrac{4}{3}$

Simplify. [1.1, 1.4, 1.5]

31. $4 \cdot 5 + 3$

32. $9 \cdot 3 + 4 \cdot 5$

33. $2^3 - 4 \cdot 3^2 + 5^2$

34. $12 - 3(2 \cdot 5 + 7) + 4$

35. $20 + 8 \div 4 + 2 \cdot 5$

36. $2(3 - 5) - (2 - 8)$

37. $30 \div 3 \cdot 2$

38. $(-2)(3) - (4)(-3) - 9$

39. $3(4 - 7)^2 - 5(3 - 8)^2$

40. $(-5 - 2)(-3 - 7)$

41. $\dfrac{4(-3)}{-6}$

42. $\dfrac{3^2 + 5^2}{(3 - 5)^2}$

43. $\dfrac{15 - 10}{6 - 6}$

44. $\dfrac{2(-7) + (-11)(-4)}{7 - (-3)}$

State the property or properties that justify the following. [1.6]

45. $9(3y) = (9 \cdot 3)y$

46. $8(1) = 8$

47. $(4 + y) + 2 = (y + 4) + 2$

48. $5 + (-5) = 0$

Use the associative property to rewrite each expression, and then simplify the result. [1.6]

49. $7 + (5 + x)$

50. $4(7a)$

51. $\dfrac{1}{9}(9x)$

52. $\dfrac{4}{5}\left(\dfrac{5}{4}y\right)$

Apply the distributive property to each of the following expressions. Simplify when possible. [1.6]

53. $7(2x + 3)$

54. $3(2a - 4)$

55. $\dfrac{1}{2}(5x - 6)$

56. $-\dfrac{1}{2}(3x - 6)$

For the set $\{\sqrt{7}, -\dfrac{1}{3}, 0, 5, -4.5, \dfrac{2}{5}, \pi, -3\}$ list all the [1.7]

57. rational numbers

58. whole numbers

59. irrational numbers

60. integers

Factor into the product of primes. [1.7]

61. 90

62. 840

Combine. [1.8]

63. $\dfrac{18}{35} + \dfrac{13}{42}$

64. $\dfrac{x}{6} + \dfrac{7}{12}$

Translate into symbols. [1.1]

1. The sum of x and 3 is 8.

2. The product of 5 and y is 15.

Simplify according to the rule for order of operations. [1.1]

3. $5^2 + 3(9 - 7) + 3^2$ **4.** $10 - 6 \div 3 + 2^3$

For each number, name the opposite, reciprocal, and absolute value. [1.2]

5. -4 **6.** $\dfrac{3}{4}$

Add. [1.3]

7. $3 + (-7)$

8. $|-9 + (-6)| + |-3 + 5|$

Subtract. [1.3]

9. $-4 - 8$ **10.** $9 - (7 - 2) - 4$

Match each expression below with the letter of the property that justifies it. [1.6]

11. $(x + y) + z = x + (y + z)$ **12.** $3(x + 5) = 3x + 15$

13. $5(3x) = (5 \cdot 3)x$ **14.** $(x + 5) + 7 = 7 + (x + 5)$

   a. Commutative property of addition

   b. Commutative property of multiplication

   c. Associative property of addition

   d. Associative property of multiplication

   e. Distributive property

Multiply. [1.4]

15. $-3(7)$ **16.** $-4(8)(-2)$

17. $8\left(-\dfrac{1}{4}\right)$ **18.** $\left(-\dfrac{2}{3}\right)^3$

Simplify using the rule for order of operations. [1.1, 1.4, 1.5]

19. $-3(-4) - 8$ **20.** $5(-6)^2 - 3(-2)^3$

21. $7 - 3(2 - 8)$

22. $4 - 2[-3(-1 + 5) + 4(-3)]$

23. $\dfrac{4(-5) - 2(7)}{-10 - 7}$ **24.** $\dfrac{2(-3 - 1) + 4(-5 + 2)}{-3(2) - 4}$

Apply the associative property, and then simplify. [1.6]

25. $3 + (5 + 2x)$ **26.** $-2(-5x)$

Multiply by applying the distributive property. [1.6]

27. $2(3x + 5)$ **28.** $-\dfrac{1}{2}(4x - 2)$

From the set of numbers $\{-8, \frac{3}{4}, 1, \sqrt{2}, 1.5\}$ list all the elements that are in the following sets. [1.7]

29. Integers

30. Rational numbers

31. Irrational numbers

32. Real numbers

Factor into the product of primes. [1.7]

33. 592   **34.** 1,340

Combine. [1.8]

35. $\dfrac{5}{15} + \dfrac{11}{42}$ **36.** $\dfrac{5}{x} + \dfrac{3}{x}$

Write an expression in symbols that is equivalent to each English phrase, and then simplify it.

37. The sum of 8 and -3 [1.1, 1.3]

38. The difference of -24 and 2 [1.1, 1.3]

39. The product of -5 and -4 [1.1, 1.4]

40. The quotient of -24 and -2 [1.1, 1.5]

Find the next number in each sequence. [1.1, 1.2, 1.3, 1.4, 1.8]

41. $-8, -3, 2, 7, \ldots$ **42.** $8, -4, 2, -1, \ldots$

Use the illustration below to answer the following questions. [1.1, 1.3]

43. How many people speak Spanish?

44. What is the total number of people who speak Mandarin Chinese or Russian?

45. How many more people speak English than Hindustani?

Chapter 1 Projects

THE BASICS

GROUP PROJECT

Binary Numbers

Students and Instructors: The end of each chapter in this book will contain a section like this one containing two projects. The group project is intended to be done in class. The research projects are to be completed outside of class. They can be done in groups or individually. In my classes, I use the research projects for extra credit. I require all research projects to be done on a word processor and to be free of spelling errors.

Number of People	2 or 3
Time Needed	10 minutes
Equipment	Paper and pencil
Background	Our decimal number system is a base-10 number system. We have 10 digits—0, 1, 2, 3, 4, 5, 6, 7, 8, and 9—which we use to write all the numbers in our number system. The number 10 is the first number that is written with a combination of digits. Although our number system is very useful, there are other number systems that are more appropriate for some disciplines. For example, computers and programmers use both the binary number system, which is base 2, and the hexadecimal number system, which is base 16. The binary number system has only digits 0 and 1, which are used to write all the other numbers. Every number in our base 10 number system can be written in the base 2 number system as well.
Procedure	To become familiar with the binary number system, we first learn to count in base 2. Imagine that the odometer on your car had only

0's and 1's. Here is what the odometer would look like for the first 6 miles the car was driven.

Odometer Reading	mileage
0 0 0 0 0 0	0
0 0 0 0 0 1	1
0 0 0 0 1 0	2
0 0 0 0 1 1	3
0 0 0 1 0 0	4
0 0 0 1 0 1	5
0 0 0 1 1 0	6

Continue the table above to show the odometer reading for the first 32 miles the car is driven. At 32 miles, the odometer should read

1 0 0 0 0 0

Chapter 1 Projects

RESEARCH PROJECT

Sophie Germain

The photograph at the right shows the street sign in Paris named for the French mathematician Sophie Germain (1776–1831). Among her contributions to mathematics is her work with prime numbers. In this chapter we had an introductory look at some of the classifications for numbers, including the prime numbers. Within the prime numbers themselves, there are still further classifications. In fact, a Sophie Germain prime is a prime number P, for which both P and $2P + 1$ are

Cheryl Slaughter

primes. For example, the prime number 2 is the first Sophie Germain prime because both 2 and $2 \cdot 2 + 1 = 5$ are prime numbers. The next Germain prime is 3 because both 3 and $2 \cdot 3 + 1 = 7$ are primes.

Sophie Germain was born on April 1, 1776, in Paris, France. She taught herself mathematics by reading the books in her father's library at home. Today she is recognized most for her work in number theory, which includes her work with prime numbers. Research the life of Sophie Germain. Write a short essay that includes information on her work with prime numbers and how her results contributed to solving Fermat's Last Theorem almost 200 years later.

2 Linear Equations and Inequalities

Google   Image © 2010 DigitalGlobe
Image © 2010 TerraMetrics

Introduction

In the 1950s, scientists studied the effects of high altitude on the human body to guarantee the safety of eventual space travel. Joseph Kittinger spent much of his life piloting military jets for the U.S. Air Force. With his vast knowledge and own body's tolerance of high altitude's harsh conditions, government researchers recruited him to float thousands of feet up into the stratosphere in open-gondola

helium balloons. During these balloon flights, Kittinger wore a pressure suit that protected his body from the extremely cold temperatures and low oxygen levels of high altitude. In order to ensure his safety, engineers who designed the suit needed to know what temperatures Kittinger would encounter as the balloon rose. They may have begun by using the linear formula

$$T = -0.0035A + 70$$

to determine the temperature of the troposphere at any specific altitude, from sea level to 55,000 feet, when the temperature on the ground is 70°F.

After Kittinger completed his tests at high altitude, he jumped out of the balloon and free-fell back to Earth. He set a record for one fall starting at an altitude over 102,000 feet! If you are having difficulty picturing how high that is, then imagine commercial jets flying around only 30,000 feet. Quite a feat!

In this chapter, we will learn how to solve linear equations and formulas, like the one above.

Getting Ready for Chapter 2

Simplify each of the following.

1. $-3 + 7$

2. $-10 - 4$

3. $9 - (-24)$

4. $-6(5) - 5$

5. $-3(2y + 1)$

6. $8.1 + 2.7$

7. $-\frac{3}{4} + \left(-\frac{1}{2}\right)$

8. $-\frac{7}{10} + \left(-\frac{1}{2}\right)$

9. $\frac{1}{5}(5x)$

10. $4\left(\frac{1}{4}x\right)$

11. $\frac{3}{2}\left(\frac{2}{3}x\right)$

12. $\frac{5}{2}\left(\frac{2}{5}x\right)$

13. $-1(3x + 4)$

14. $-2.4 + (-7.3)$

15. $0.04(x + 7,000)$

16. $0.09(x + 2,000)$

Apply the distributive property and then simplify if possible.

17. $3(x - 5) + 4$

18. $5(x - 3) + 2$

19. $5(2x - 8) - 3$

20. $4(3x - 5) - 2$

Chapter Outline

2.1 Simplifying Expressions

A Simplify expressions by combining similar terms.

B Simplify expressions by applying the distributive property and then combining similar terms.

C Calculate the value of an expression for a given value of the variable.

2.2 Addition Property of Equality

A Check the solution to an equation by substitution.

B Use the addition property of equality to solve an equation.

2.3 Multiplication Property of Equality

A Use the multiplication property of equality to solve an equation.

B Use the addition and multiplication properties of equality together to solve an equation.

2.4 Solving Linear Equations

A Solve a linear equation in one variable.

2.5 Formulas

A Find the value of a variable in a formula given replacements for the other variables.

B Solve a formula for one of its variables.

C Find the complement and supplement of an angle.

D Solve basic percent problems.

2.6 Applications

A Apply the Blueprint for Problem Solving to a variety of application problems.

2.7 More Applications

A Apply the Blueprint for Problem Solving to a variety of application problems.

2.8 Linear Inequalities

A Use the addition property for inequalities to solve an inequality and graph the solution set.

B Use the multiplication property for inequalities to solve an inequality.

C Use both the addition and multiplication properties to solve an inequality.

D Translate and solve application problems involving inequalities.

2.1 Simplifying Expressions

OBJECTIVES

A Simplify expressions by combining similar terms.

B Simplify expressions by applying the distributive property and then combining similar terms.

C Calculate the value of an expression for a given value of the variable.

TICKET TO SUCCESS

Keep these questions in mind as you read through the section. Then respond in your own words and in complete sentences.

1. What is the first step in solving an equation?
2. What are similar terms?
3. Explain how you would use the distributive property to combine similar terms.
4. Explain how you would find the value of $5x + 3$ when x is 6.

If a cellular phone company charges \$35 per month plus \$0.25 for each minute, or fraction of a minute, that you use one of their cellular phones, then the amount of your monthly bill is given by the expression $35 + 0.25t$. To find the amount you will pay for using that phone 30 minutes in one month, you substitute 30 for t and simplify the resulting expression. This process is one of the topics we will study in this section.

A Combining Similar Terms

As you will see in the next few sections, the first step in solving an equation is to simplify both sides as much as possible. In the first part of this section, we will practice simplifying expressions by combining what are called *similar* (or *like*) terms.

For our immediate purposes, a term is a number, or a number and one or more variables multiplied together. For example, the number 5 is a term, as are the expressions $3x$, $-7y$, and $15xy$.

> **Definition**
>
> Two or more terms with the same variable part are called **similar** (or **like**) terms.

The terms $3x$ and $4x$ are similar because their variable parts are identical. Likewise, the terms $18y$, $-10y$, and $6y$ are similar terms.

To simplify an algebraic expression, we simply reduce the number of terms in the expression. We accomplish this by applying the distributive property along with our knowledge of addition and subtraction of positive and negative real numbers. The following examples illustrate the procedure.

EXAMPLE 1 Simplify $3x + 4x$ by combining similar terms.

SOLUTION

$$3x + 4x = (3 + 4)x \qquad \text{Distributive property}$$
$$= 7x \qquad\qquad \text{Addition of 3 and 4}$$

EXAMPLE 2 Simplify $7a - 10a$ by combining similar terms.

SOLUTION

$$7a - 10a = (7 - 10)a \qquad \text{Distributive property}$$
$$= -3a \qquad\qquad \text{Addition of 7 and } -10$$

EXAMPLE 3 Simplify $18y - 10y + 6y$ by combining similar terms.

SOLUTION

$$18y - 10y + 6y = (18 - 10 + 6)y \qquad \text{Distributive property}$$
$$= 14y \qquad\qquad\qquad\quad \text{Addition of 18, } -10, \text{ and 6}$$

When the expression we intend to simplify is more complicated, we use the commutative and associative properties first.

EXAMPLE 4 Simplify the expression $3x + 5 + 2x - 3$.

SOLUTION

$$3x + 5 + 2x - 3 = 3x + 2x + 5 - 3 \qquad \text{Commutative property}$$
$$= (3x + 2x) + (5 - 3) \qquad \text{Associative property}$$
$$= (3 + 2)x + (5 - 3) \qquad \text{Distributive property}$$
$$= 5x + 2 \qquad\qquad\qquad \text{Addition}$$

EXAMPLE 5 Simplify the expression $4a - 7 - 2a + 3$.

SOLUTION

$$4a - 7 - 2a + 3 = (4a - 2a) + (-7 + 3) \qquad \text{Commutative and associative properties}$$
$$= (4 - 2)a + (-7 + 3) \qquad \text{Distributive property}$$
$$= 2a - 4 \qquad\qquad\qquad\quad \text{Addition}$$

EXAMPLE 6 Simplify the expression $5x + 8 - x - 6$.

SOLUTION

$$5x + 8 - x - 6 = (5x - x) + (8 - 6) \qquad \text{Commutative and associative properties}$$
$$= (5 - 1)x + (8 - 6) \qquad \text{Distributive property}$$
$$= 4x + 2 \qquad\qquad\qquad\quad \text{Addition}$$

B Simplifying Expressions Containing Parentheses

If an expression contains parentheses, it is often necessary to apply the distributive property to remove the parentheses before combining similar terms.

EXAMPLE 7 Simplify the expression $5(2x - 8) - 3$.

SOLUTION We begin by distributing the 5 across $2x - 8$. We then combine similar terms:

$$5(2x - 8) - 3 = 10x - 40 - 3 \qquad \text{Distributive property}$$
$$= 10x - 43$$

EXAMPLE 8 Simplify $7 - 3(2y + 1)$.

SOLUTION By the rule for order of operations, we must multiply before we add or subtract. For that reason, it would be incorrect to subtract 3 from 7 first. Instead, we multiply -3 and $2y + 1$ to remove the parentheses and then combine similar terms.

$$7 - 3(2y + 1) = 7 - 6y - 3 \qquad \text{Distributive property}$$
$$= -6y + 4$$

EXAMPLE 9   Simplify $5(x - 2) - (3x + 4)$.

SOLUTION   We begin by applying the distributive property to remove the parentheses. The expression $-(3x + 4)$ can be thought of as $-1(3x + 4)$. Thinking of it in this way allows us to apply the distributive property.

$$-1(3x + 4) = -1(3x) + (-1)(4)$$
$$= -3x - 4$$

The complete solution looks like this:

$$5(x - 2) - (3x + 4) = 5x - 10 - 3x - 4 \qquad \text{Distributive property}$$
$$= 2x - 14 \qquad \text{Combine similar terms}$$

As you can see from the explanation in Example 9, we use the distributive property to simplify expressions in which parentheses are preceded by a negative sign. In general we can write

$$-(a + b) = -1(a + b)$$
$$= -a + (-b)$$
$$= -a - b$$

The negative sign outside the parentheses ends up changing the sign of each term within the parentheses. In words, we say "the opposite of a sum is the sum of the opposites."

C More About the Value of an Expression

Recall from Chapter 1, an expression like $3x + 2$ has a certain value depending on what number we assign to x. For instance, when x is 4, $3x + 2$ becomes $3(4) + 2$, or 14. When x is -8, $3x + 2$ becomes $3(-8) + 2$, or -22. The value of an expression is found by replacing the variable with a given number.

EXAMPLES   Find the value of the following expressions by replacing the variable with the given number.

Expression	Value of the Variable	Value of the Expression
10. $3x - 1$	$x = 2$	$3(2) - 1 = 6 - 1 = 5$
11. $7a + 4$	$a = -3$	$7(-3) + 4 = -21 + 4 = -17$
12. $2x - 3 + 4x$	$x = -1$	$2(-1) - 3 + 4(-1)$ $= -2 - 3 + (-4) = -9$
13. $2x - 5 - 8x$	$x = 5$	$2(5) - 5 - 8(5)$ $= 10 - 5 - 40 = -35$
14. $y^2 - 6y + 9$	$y = 4$	$4^2 - 6(4) + 9 = 16 - 24 + 9 = 1$

Simplifying an expression should not change its value; that is, if an expression has a certain value when x is 5, then it will always have that value no matter how much it has been simplified, as long as x is 5. If we were to simplify the expression in Example 13 first, it would look like

$$2x - 5 - 8x = -6x - 5$$

When x is 5, the simplified expression $-6x - 5$ is

$$-6(5) - 5 = -30 - 5 = -35$$

It has the same value as the original expression when x is 5.

We also can find the value of an expression that contains two variables if we know the values for both variables.

EXAMPLE 15 Find the value of the expression $2x - 3y + 4$ when x is -5 and y is 6.

SOLUTION Substituting -5 for x and 6 for y, the expression becomes

$$2(-5) - 3(6) + 4 = -10 - 18 + 4$$
$$= -28 + 4$$
$$= -24$$

EXAMPLE 16 Find the value of the expression $x^2 - 2xy + y^2$ when x is 3 and y is -4.

SOLUTION Replacing each x in the expression with the number 3 and each y in the expression with the number -4 gives us

$$3^2 - 2(3)(-4) + (-4)^2 = 9 - 2(3)(-4) + 16$$
$$= 9 - (-24) + 16$$
$$= 33 + 16$$
$$= 49$$

More about Sequences

As the next example indicates, when we substitute the counting numbers, in order, into algebraic expressions, we form some of the sequences of numbers that we studied in Chapter 1. To review, recall that the sequence of counting numbers (also called the sequence of positive integers) is

$$\text{Counting numbers} = 1, 2, 3, \ldots$$

EXAMPLE 17 Substitute 1, 2, 3, and 4 for n in the expression $2n - 1$.

SOLUTION Substituting as indicated, we have

When $n = 1$, $2n - 1 = 2 \cdot 1 - 1 = 1$
When $n = 2$, $2n - 1 = 2 \cdot 2 - 1 = 3$
When $n = 3$, $2n - 1 = 2 \cdot 3 - 1 = 5$
When $n = 4$, $2n - 1 = 2 \cdot 4 - 1 = 7$

As you can see, substituting the first four counting numbers into the expression $2n - 1$ produces the first four numbers in the sequence of odd numbers.

Problem Set 2.1

Moving Toward Success

"Our attitude toward life determines life's attitude towards us."

—Earl Nightingale, 1921–1989, American motivational author and radio broadcaster

1. What was the most important study skill you used while working through Chapter 1?

2. Why should you continue to place an importance on study skills as you work through Chapter 2?

A Simplify the following expressions. [Examples 1–6]

1. $3x - 6x$

2. $7x - 5x$

3. $-2a + a$

4. $3a - a$

5. $7x + 3x + 2x$

6. $8x - 2x - x$

7. $3a - 2a + 5a$

8. $7a - a + 2a$

9. $4x - 3 + 2x$

10. $5x + 6 - 3x$

11. $3a + 4a + 5$

12. $6a + 7a + 8$

13. $2x - 3 + 3x - 2$

14. $6x + 5 - 2x + 3$

15. $3a - 1 + a + 3$

16. $-a + 2 + 8a - 7$

17. $-4x + 8 - 5x - 10$

18. $-9x - 1 + x - 4$

19. $7a + 3 + 2a + 3a$

20. $8a - 2 + a + 5a$

B Apply distributive property, then simplify. [Examples 7–9]

21. $5(2x - 1) + 4$

22. $2(4x - 3) + 2$

23. $7(3y + 2) - 8$

24. $6(4y + 2) - 7$

25. $-3(2x - 1) + 5$

26. $-4(3x - 2) - 6$

27. $5 - 2(a + 1)$

28. $7 - 8(2a + 3)$

B Simplify the following expressions. [Examples 7–9]

29. $6 - 4(x - 5)$

30. $12 - 3(4x - 2)$

31. $-9 - 4(2 - y) + 1$

32. $-10 - 3(2 - y) + 3$

33. $-6 + 2(2 - 3x) + 1$

34. $-7 - 4(3 - x) + 1$

35. $(4x - 7) - (2x + 5)$

36. $(7x - 3) - (4x + 2)$

37. $8(2a + 4) - (6a - 1)$

38. $9(3a + 5) - (8a - 7)$

39. $3(x - 2) + (x - 3)$

40. $2(2x + 1) - (x + 4)$

41. $4(2y - 8) - (y + 7)$

42. $5(y - 3) - (y - 4)$

43. $-9(2x + 1) - (x + 5)$

44. $-3(3x - 2) - (2x + 3)$

C Evaluate the following expressions when x is 2. [Examples 10–14]

45. $3x - 1$

46. $4x + 3$

47. $-2x - 5$

48. $-3x + 6$

49. $x^2 - 8x + 16$

50. $x^2 - 10x + 25$

51. $(x - 4)^2$

52. $(x - 5)^2$

C Find the value of each expression when x is -5. Then simplify the expression, and check to see that it has the same value for $x = -5$. [Examples 15–16]

53. $7x - 4 - x - 3$

54. $3x + 4 + 7x - 6$

55. $5(2x + 1) + 4$

56. $2(3x - 10) + 5$

C Find the value of each expression when x is -3 and y is 5. [Examples 15–16]

57. $x^2 - 2xy + y^2$

58. $x^2 + 2xy + y^2$

59. $(x - y)^2$

60. $(x + y)^2$

61. $x^2 + 6xy + 9y^2$

62. $x^2 + 10xy + 25y^2$

63. $(x + 3y)^2$

64. $(x + 5y)^2$

C Find the value of $12x - 3$ for each of the following values of x. [Example 17]

65. $\dfrac{1}{2}$

66. $\dfrac{1}{3}$

67. $\dfrac{1}{4}$

68. $\dfrac{1}{6}$

69. $\dfrac{3}{2}$

70. $\dfrac{2}{3}$

71. $\dfrac{3}{4}$

72. $\dfrac{5}{6}$

73. Fill in the tables below to find the sequences formed by substituting the first four counting numbers into the expressions $3n$ and n^3.

a.

n	1	2	3	4
$3n$				

b.
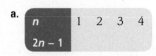

n	1	2	3	4
n^3				

74. Fill in the tables below to find the sequences formed by substituting the first four counting numbers into the expressions $2n - 1$ and $2n + 1$.

a.

n	1	2	3	4
$2n - 1$				

b.
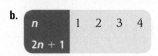

n	1	2	3	4
$2n + 1$				

C Find the sequences formed by substituting the counting numbers, in order, into the following expressions.
[Example 17]

75. $3n - 2$

76. $2n - 3$

77. $n^2 - 2n + 1$

78. $(n - 1)^2$

Here are some problems you will see later in the book. Simplify.

79. $7 - 3(2y + 1)$

80. $4(3x - 2) - (6x - 5)$

81. $0.08x + 0.09x$

82. $0.04x + 0.05x$

83. $(x + y) + (x - y)$

84. $(-12x - 20y) + (25x + 20y)$

85. $3x + 2(x - 2)$

86. $2(x - 2) + 3(5x)$

87. $4(x + 1) + 3(x - 3)$

88. $5(x + 2) + 3(x - 1)$

89. $x + (x + 3)(-3)$

90. $x - 2(x + 2)$

91. $3(4x - 2) - (5x - 8)$

92. $2(5x - 3) - (2x - 4)$

93. $-(3x + 1) - (4x - 7)$

94. $-(6x + 2) - (8x - 3)$

95. $(x + 3y) + 3(2x - y)$

96. $(2x - y) - 2(x + 3y)$

97. $3(2x + 3y) - 2(3x + 5y)$

98. $5(2x + 3y) - 3(3x + 5y)$

99. $-6\left(\dfrac{1}{2}x - \dfrac{1}{3}y\right) + 12\left(\dfrac{1}{4}x + \dfrac{2}{3}y\right)$

100. $6\left(\dfrac{1}{3}x + \dfrac{1}{2}y\right) - 4\left(x + \dfrac{3}{4}y\right)$

101. $0.08x + 0.09(x + 2{,}000)$

102. $0.06x + 0.04(x + 7{,}000)$

103. $0.10x + 0.12(x + 500)$

104. $0.08x + 0.06(x + 800)$

105. Find a so the expression $(5x + 4y) + a(2x - y)$ simplifies to an expression that does not contain y. Using that value of a, simplify the expression.

106. Find a so the expression $(5x + 4y) - a(x - 2y)$ simplifies to an expression that does not contain x. Using that value of a, simplify the expression.

Find the value of $b^2 - 4ac$ for the given values of a, b, and c. (You will see these problems later in the book.)

107. $a = 1, b = -5, c = -6$

108. $a = 1, b = -6, c = 7$

109. $a = 2, b = 4, c = -3$

110. $a = 3, b = 4, c = -2$

Applying the Concepts

111. Temperature and Altitude If the temperature on the ground is 70°F, then the temperature at A feet above the ground can be found from the expression $-0.0035A + 70$. Find the temperature at the following altitudes.
  a. 8,000 feet   **b.** 12,000 feet   **c.** 24,000 feet

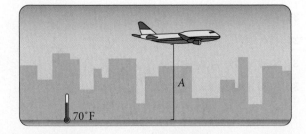

112. Perimeter of a Rectangle The expression $2l + 2w$ gives the perimeter of a rectangle with length l and width w. Find the perimeter of the rectangles with the following lengths and widths.

   a. Length = 8 meters, Width = 5 meters

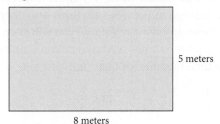

5 meters

8 meters

   b. Length = 10 feet, Width = 3 feet

3 feet

10 feet

113. Cellular Phone Rates A cellular phone company charges $35 per month plus $0.25 for each minute, or fraction of a minute, that you use one of their cellular phones. The expression $35 + 0.25t$ gives the amount of money you will pay for using one of their phones for t minutes a month. Find the monthly bill for using one of their phones.

   a. 10 minutes in a month
   b. 20 minutes in a month
   c. 30 minutes in a month

114. Cost of Bottled Water A water bottling company charges $7.00 per month for their water dispenser and $1.10 for each gallon of water delivered. If you have g gallons of water delivered in a month, then the expression $7 + 1.1g$ gives the amount of your bill for that month. Find the monthly bill for each of the following deliveries.

   a. 10 gallons
   b. 20 gallons
   c. 30 gallons

WBC WATER BOTTLE CO.		
MONTHLY BILL		
234 5th Street Glendora, CA 91740		DUE 07/23/10
Water dispenser	1	$7.00
Gallons of water	8	$1.10
		$15.80

115. Taxes We all have to pay taxes. Suppose that 21% of your monthly pay is withheld for federal income taxes and another 8% is withheld for Social Security, state income tax, and other miscellaneous items. If G is your monthly pay before any money is deducted

(your gross pay), then the amount of money that you take home each month is given by the expression $G - 0.21G - 0.08G$. Simplify this expression and then find your take-home pay if your gross pay is $1,250 per month.

116. Taxes If you work H hours a day and you pay 21% of your income for federal income tax and another 8% for miscellaneous deductions, then the expression $0.21H + 0.08H$ tells you how many hours a day you are working to pay for those taxes and miscellaneous deductions. Simplify this expression. If you work 8 hours a day under these conditions, how many hours do you work to pay for your taxes and miscellaneous deductions?

117. Cars The chart shows the fastest cars in the world. How fast is a car traveling if its speed is 56 miles per hour less than half the speed of the Saleen S7 Twin Turbo?

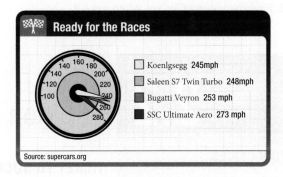

Ready for the Races

Koenlgsegg **245mph**
Saleen S7 Twin Turbo **248mph**
Bugatti Veyron **253 mph**
SSC Ultimate Aero **273 mph**

Source: supercars.org

118. Sound The chart shows the decibel level of various sounds. The loudness of a loud rock concert is 5 decibels less than twice the loudness of normal conversation. Find the sound level of a rock concert.

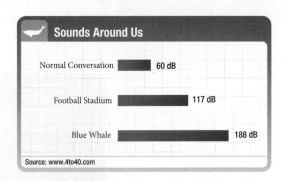

Sounds Around Us

Normal Conversation **60 dB**
Football Stadium **117 dB**
Blue Whale **188 dB**

Source: www.4to40.com

Getting Ready for the Next Section

Problems under this heading, *Getting Ready for the Next Section,* are problems that you must be able to work to understand the material in the next section. The problems below are exactly the types of problems you will see in the explanations and examples in the next section.

Simplify.

119. $17 - 5$

120. $12 + (-2)$

121. $2 - 5$

122. $25 - 20$

123. $-2.4 + (-7.3)$

124. $8.1 + 2.7$

125. $-\dfrac{1}{2} + \left(-\dfrac{3}{4}\right)$

126. $-\dfrac{1}{6} + \left(-\dfrac{2}{3}\right)$

127. $4(2 \cdot 9 - 3) - 7 \cdot 9$

128. $5(3 \cdot 45 - 4) - 14 \cdot 45$

129. $4(2a - 3) - 7a$

130. $5(3a - 4) - 14a$

131. $-3 - \dfrac{1}{2}$

132. $-5 - \dfrac{1}{3}$

133. $\dfrac{4}{5} + \dfrac{1}{10} + \dfrac{3}{8}$

134. $\dfrac{3}{10} + \dfrac{7}{25} + \dfrac{3}{4}$

135. Find the value of $2x - 3$ when x is 5.

136. Find the value of $3x + 4$ when x is -2.

Maintaining Your Skills

From this point on, each problem set will contain a number of problems under the heading *Maintaining Your Skills*. These problems cover the most important skills you have learned in previous sections and chapters. By working these problems regularly, you will keep yourself current on all the topics we have covered and possibly need less time to study for tests and quizzes.

137. $\dfrac{1}{8} - \dfrac{1}{6}$

138. $\dfrac{x}{8} - \dfrac{x}{6}$

139. $\dfrac{5}{9} - \dfrac{4}{3}$

140. $\dfrac{x}{9} - \dfrac{x}{3}$

141. $-\dfrac{7}{30} + \dfrac{5}{28}$

142. $-\dfrac{11}{105} + \dfrac{11}{30}$

2.2 Addition Property of Equality

OBJECTIVES

A Check the solution to an equation by substitution.

B Use the addition property of equality to solve an equation.

TICKET TO SUCCESS

Keep these questions in mind as you read through the section. Then respond in your own words and in complete sentences.

1. What is a solution set for an equation?

2. What are equivalent equations?

3. Explain in words the addition property of equality.

4. How do you check a solution to an equation?

When light comes into contact with any object, it is reflected, absorbed, and transmitted, as shown here.

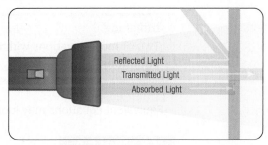

Reflected Light
Transmitted Light
Absorbed Light

For a certain type of glass, 88% of the light hitting the glass is transmitted through to the other side, whereas 6% of the light is absorbed into the glass. To find the percent of light that is reflected by the glass, we can solve the equation

$$88 + R + 6 = 100$$

Solving equations of this type is what we will study in this section. To solve an equation we must find all replacements for the variable that make the equation a true statement.

A Solutions to Equations

Definition
The **solution set** for an equation is the set of all numbers that when used in place of the variable make the equation a true statement.

For example, the equation $x + 2 = 5$ has solution set {3} because when x is 3 the equation becomes the true statement $3 + 2 = 5$, or $5 = 5$.

EXAMPLE 1   Is 5 a solution to $2x - 3 = 7$?

SOLUTION   We substitute 5 for x in the equation, and then simplify to see if a true statement results. A true statement means we have a solution; a false statement indicates the number we are using is not a solution.

$$\begin{aligned}
\text{When} \rightarrow \qquad\qquad & x = 5 \\
\text{the equation} \rightarrow \qquad\qquad & 2x - 3 = 7 \\
\text{becomes} \rightarrow \qquad\qquad & 2(5) - 3 \overset{?}{=} 7 \\
& 10 - 3 \overset{?}{=} 7 \\
& 7 = 7 \qquad\qquad \text{A true statement}
\end{aligned}$$

Because $x = 5$ turns the equation into the true statement $7 = 7$, we know 5 is a solution to the equation. ■

> **NOTE**
> We can use a question mark over the equal signs to show that we don't know yet whether the two sides of the equation are equal.

EXAMPLE 2   Is -2 a solution to $8 = 3x + 4$?

SOLUTION   Substituting -2 for x in the equation, we have

$$\begin{aligned}
8 &\overset{?}{=} 3(-2) + 4 \\
8 &\overset{?}{=} -6 + 4 \\
8 &= -2 \qquad\qquad \text{A false statement}
\end{aligned}$$

Substituting -2 for x in the equation produces a false statement. Therefore, $x = -2$ is not a solution to the equation. ■

The important thing about an equation is its solution set. Therefore, we make the following definition to classify together all equations with the same solution set.

> **Definition**
> Two or more equations with the same solution set are said to be **equivalent equations.**

Equivalent equations may look different but must have the same solution set.

EXAMPLE 3

a. $x + 2 = 5$ and $x = 3$ are equivalent equations because both have solution set {3}.

b. $a - 4 = 3$, $a - 2 = 5$, and $a = 7$ are equivalent equations because they all have solution set {7}.

c. $y + 3 = 4$, $y - 8 = -7$, and $y = 1$ are equivalent equations because they all have solution set {1}.

B Addition Property of Equality

If two numbers are equal and we increase (or decrease) both of them by the same amount, the resulting quantities are also equal. We can apply this concept to equations. Adding the same amount to both sides of an equation always produces an equivalent equation—one with the same solution set. This fact about equations is called the *addition property of equality* and can be stated more formally as follows:

> **Addition Property of Equality**
> For any three algebraic expressions A, B, and C,
> *In symbols:*   if $A = B$
> then $A + C = B + C$
>
> *In words:* Adding the same quantity to both sides of an equation will not change the solution set.

NOTE
We will use this property many times in the future. Be sure you understand it completely by the time you finish this section.

This property is just as simple as it seems. We can add any amount to both sides of an equation and always be sure we have not changed the solution set.

Consider the equation $x + 6 = 5$. We want to solve this equation for the value of x that makes it a true statement. We want to end up with x on one side of the equal sign and a number on the other side. Because we want x by itself, we will add -6 to both sides.

$$x + 6 + (\mathbf{-6}) = 5 + (\mathbf{-6}) \qquad \text{Addition property of equality}$$
$$x + 0 = -1 \qquad \text{Addition}$$
$$x = -1$$

All three equations say the same thing about x. They all say that x is -1. All three equations are equivalent. The last one is just easier to read.

Here are some further examples of how the addition property of equality can be used to solve equations.

EXAMPLE 4 Solve the equation $x - 5 = 12$ for x.

SOLUTION Because we want x alone on the left side, we choose to add $+5$ to both sides:

$$x - 5 + \mathbf{5} = 12 + \mathbf{5} \qquad \text{Addition property of equality}$$
$$x + 0 = 17$$
$$x = 17$$

To check our solution, we substitute 17 for x in the original equation:

$$\text{When} \rightarrow \qquad x = 17$$
$$\text{the equation} \rightarrow \quad x - 5 = 12$$
$$\text{becomes} \rightarrow \qquad 17 - 5 \overset{?}{=} 12$$
$$12 = 12 \qquad \text{A true statement}$$

As you can see, our solution checks. The purpose for checking a solution to an equation is to catch any mistakes we may have made in the process of solving the equation.

EXAMPLE 5 Solve for a: $a + \dfrac{3}{4} = -\dfrac{1}{2}$.

SOLUTION Because we want a by itself on the left side of the equal sign, we add the opposite of $\dfrac{3}{4}$ to each side of the equation.

$$a + \frac{3}{4} + \left(-\frac{\mathbf{3}}{\mathbf{4}}\right) = -\frac{1}{2} + \left(-\frac{\mathbf{3}}{\mathbf{4}}\right) \qquad \text{Addition property of equality}$$
$$a + 0 = -\frac{1}{2} \cdot \frac{\mathbf{2}}{\mathbf{2}} + \left(-\frac{3}{4}\right) \qquad \text{LCD on the right side is 4}$$
$$a = -\frac{2}{4} + \left(-\frac{3}{4}\right) \qquad \tfrac{2}{4} \text{ is equivalent to } \tfrac{1}{2}$$
$$a = -\frac{5}{4} \qquad \text{Add fractions}$$

The solution is $a = -\dfrac{5}{4}$. To check our result, we replace a with $-\dfrac{5}{4}$ in the original equation. The left side then becomes $-\dfrac{5}{4} + \dfrac{3}{4}$, which simplifies to $-\dfrac{1}{2}$, so our solution checks.

EXAMPLE 6 Solve for x: $7.3 + x = -2.4$.

SOLUTION Again, we want to isolate x, so we add the opposite of 7.3 to both sides.

$$7.3 + (-\mathbf{7.3}) + x = -2.4 + (-\mathbf{7.3}) \qquad \text{Addition property of equality}$$
$$0 + x = -9.7$$
$$x = -9.7$$

The addition property of equality also allows us to add variable expressions to each side of an equation.

EXAMPLE 7 Solve for x: $3x - 5 = 4x$.

SOLUTION Adding $-3x$ to each side of the equation gives us our solution.

$$3x - 5 = 4x$$
$$3x + (-\mathbf{3x}) - 5 = 4x + (-\mathbf{3x}) \qquad \text{Distributive property}$$
$$-5 = x$$

Sometimes it is necessary to simplify each side of an equation before using the addition property of equality. The reason we simplify both sides first is that we want as few terms as possible on each side of the equation before we use the addition property of equality. The following examples illustrate this procedure.

EXAMPLE 8 Solve $4(2a - 3) - 7a = 2 - 5$.

SOLUTION We must begin by applying the distributive property to separate terms on the left side of the equation. Following that, we combine similar terms and then apply the addition property of equality.

$4(2a - 3) - 7a = 2 - 5$	Original equation
$8a - 12 - 7a = 2 - 5$	Distributive property
$a - 12 = -3$	Simplify each side
$a - 12 + \mathbf{12} = -3 + \mathbf{12}$	Add 12 to each side
$a = 9$	

To check our solution, we replace the variable a with 9 in the original equation.

$$4(2 \cdot 9 - 3) - 7 \cdot 9 \overset{?}{=} 2 - 5$$
$$4(15) - 63 \overset{?}{=} -3$$
$$60 - 63 \overset{?}{=} -3$$
$$-3 = -3 \qquad \text{A true statement}$$

NOTE
Again, we place a question mark over the equal sign because we don't know yet whether the expressions on the left and right side of the equal sign will be equal.

EXAMPLE 9 Solve $3x - 5 = 2x + 7$.

SOLUTION We can solve this equation in two steps. First, we add $-2x$ to both sides of the equation. When this has been done, x appears on the left side only. Second, we add 5 to both sides.

$3x + (\mathbf{-2x}) - 5 = 2x + (\mathbf{-2x}) + 7$	Add $-2x$ to both sides
$x - 5 = 7$	Simplify each side
$x - 5 + \mathbf{5} = 7 + \mathbf{5}$	Add 5 to both sides
$x = 12$	Simplify each side

NOTE
In my experience teaching algebra, I find that students make fewer mistakes if they think in terms of addition rather than subtraction. So, you are probably better off if you continue to use the addition property just the way we have used it in the examples in this section. But, if you are curious as to whether you can subtract the same number from both sides of an equation, the answer is yes.

A Note on Subtraction
Although the addition property of equality is stated for addition only, we can subtract the same number from both sides of an equation as well. Because subtraction is defined as addition of the opposite, subtracting the same quantity from both sides of an equation does not change the solution.

$x + 2 = 12$	Original equation
$x + 2 - \mathbf{2} = 12 - \mathbf{2}$	Subtract 2 from each side
$x = 10$	

Problem Set 2.2

Moving Toward Success

"The big secret in life is that there is no big secret. Whatever your goal, you can get there if you're willing to work."

—Oprah Winfrey, 1954–present, American television host and philanthropist

1. Why do you think your work should look like the work you see in this book?

2. Do you think your work should imitate the work of people (e.g., your instructor) who have been successful? Explain.

B Find the solution for the following equations. Be sure to show when you have used the addition property of equality. [Examples 4–7]

1. $x - 3 = 8$

2. $x - 2 = 7$

3. $x + 2 = 6$

4. $x + 5 = 4$

5. $a + \dfrac{1}{2} = -\dfrac{1}{4}$

6. $a + \dfrac{1}{3} = -\dfrac{5}{6}$

7. $x + 2.3 = -3.5$

8. $x + 7.9 = -3.4$

9. $y + 11 = -6$

10. $y - 3 = -1$

11. $x - \dfrac{5}{8} = -\dfrac{3}{4}$

12. $x - \dfrac{2}{5} = -\dfrac{1}{10}$

13. $m - 6 = 2m$

14. $3m - 10 = 4m$

15. $6.9 + x = 3.3$

16. $7.5 + x = 2.2$

17. $5a = 4a - 7$

18. $12a = -3 + 11a$

19. $-\dfrac{5}{9} = x - \dfrac{2}{5}$

20. $-\dfrac{7}{8} = x - \dfrac{4}{5}$

B Simplify both sides of the following equations as much as possible, and then solve. [Example 7]

21. $4x + 2 - 3x = 4 + 1$

22. $5x + 2 - 4x = 7 - 3$

23. $8a - \dfrac{1}{2} - 7a = \dfrac{3}{4} + \dfrac{1}{8}$

24. $9a - \dfrac{4}{5} - 8a = \dfrac{3}{10} - \dfrac{1}{5}$

25. $-3 - 4x + 5x = 18$

26. $10 - 3x + 4x = 20$

27. $-11x + 2 + 10x + 2x = 9$

28. $-10x + 5 - 4x + 15x = 0$

29. $-2.5 + 4.8 = 8x - 1.2 - 7x$

30. $-4.8 + 6.3 = 7x - 2.7 - 6x$

31. $2y - 10 + 3y - 4y = 18 - 6$

32. $15 - 21 = 8x + 3x - 10x$

B The following equations contain parentheses. Apply the distributive property to remove the parentheses, then simplify each side before using the addition property of equality. [Example 8]

33. $2(x + 3) - x = 4$

34. $5(x + 1) - 4x = 2$

35. $-3(x - 4) + 4x = 3 - 7$

36. $-2(x - 5) + 3x = 4 - 9$

37. $5(2a + 1) - 9a = 8 - 6$

38. $4(2a - 1) - 7a = 9 - 5$

39. $-(x + 3) + 2x - 1 = 6$

40. $-(x - 7) + 2x - 8 = 4$

41. $4y - 3(y - 6) + 2 = 8$

42. $7y - 6(y - 1) + 3 = 9$

43. $-3(2m - 9) + 7(m - 4) = 12 - 9$

44. $-5(m - 3) + 2(3m + 1) = 15 - 8$

A **B** Solve the following equations by the method used in Example 9 in this section. Check each solution in the original equation. [Example 9]

45. $4x = 3x + 2$

46. $6x = 5x - 4$

47. $8a = 7a - 5$

48. $9a = 8a - 3$

49. $2x = 3x + 1$

50. $4x = 3x + 5$

51. $2y + 1 = 3y + 4$

52. $4y + 2 = 5y + 6$

53. $2m - 3 = m + 5$

54. $8m - 1 = 7m - 3$

55. $4x - 7 = 5x + 1$

56. $3x - 7 = 4x - 6$

57. $4x + \dfrac{4}{3} = 5x - \dfrac{2}{3}$

58. $2x + \dfrac{1}{4} = 3x - \dfrac{5}{4}$

59. $8a - 7.1 = 7a + 3.9$

60. $10a - 4.3 = 9a + 4.7$

61. $12x - 5.8 = 11x + 42$

62. $4y - 8.4 = 3y + 3.6$

Applying the Concepts

63. Light When light comes into contact with any object, it is reflected, absorbed, and transmitted, as shown in the following figure. If *T* represents the percent of light transmitted, *R* the percent of light reflected, and *A* the percent of light absorbed by a surface, then the equation $T + R + A = 100$ shows one way these quantities are related.

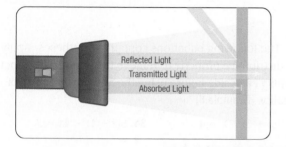

a. For glass, $T = 88$ and $A = 6$, meaning that 88% of the light hitting the glass is transmitted and 6% is absorbed. Substitute $T = 88$ and $A = 6$ into the equation $T + R + A = 100$ and solve for *R* to find the percent of light that is reflected.

b. For flat black paint, $A = 95$ and no light is transmitted, meaning that $T = 0$. What percent of light is reflected by flat black paint?

c. A pure white surface can reflect 98% of light, so $R = 98$. If no light is transmitted, what percent of light is absorbed by the pure white surface?

d. Typically, shiny gray metals reflect 70–80% of light. Suppose a thick sheet of aluminum absorbs 25% of light. What percent of light is reflected by this shiny gray metal? (Assume no light is transmitted.)

64. Movie Tickets A movie theater has a total of 300 seats. For a special Saturday night preview, they reserve 20 seats to give away to their VIP guests at no charge. If *x* represents the number of tickets they can sell for the preview, then *x* can be found by solving the equation $x + 20 = 300$.

a. Solve the equation for *x*.

b. If tickets for the preview are $7.50 each, what is the maximum amount of money they can make from ticket sales?

65. Geometry The three angles shown in the triangle at the front of the tent in the following figure add up to 180°. Use this fact to write an equation containing

x, and then solve the equation to find the number of degrees in the angle at the top of the triangle.

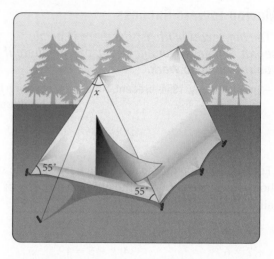

66. Geometry The figure shows part of a room. From a point on the floor, the angle of elevation to the top of the door is 47°, whereas the angle of elevation to the ceiling above the door is 59°. Use this diagram to write an equation involving *x*, and then solve the equation to find the number of degrees in the angle that extends from the top of the door to the ceiling.

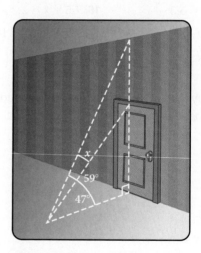

67. Movie Rentals The chart shows the market share as a percent for the years 2008 and 2009 for the three largest movie rental companies.

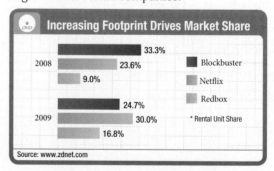

a. What was the combined market share for these companies in 2008?.

b. What was the combined market share for these companies in 2009?

c. What is the total market share of all movie rental companies not shown in the chart for 2008?

d. What is the total market share of all movie rental companies not shown in the chart for 2009?

68. Skyscrapers The chart shows the heights of three of the tallest buildings in the world. The height of the Sears Tower is 358 feet taller than the Tokyo Tower in Japan. How tall is the Tokyo Tower?

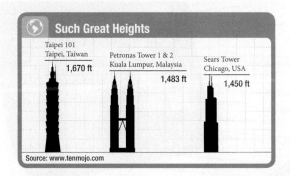

Getting Ready for the Next Section

To understand all of the explanations and examples in the next section you must be able to work the problems below.

Simplify.

69. $\dfrac{3}{2}\left(\dfrac{2}{3}y\right)$

70. $\dfrac{5}{2}\left(\dfrac{2}{5}y\right)$

71. $\dfrac{1}{7}(7x)$

72. $-\dfrac{1}{4}(-4a)$

73. $\dfrac{1}{5}(30)$

74. $-\dfrac{1}{4}(24)$

75. $\dfrac{3}{2}(4)$

76. $\dfrac{1}{26}(13)$

77. $12\left(-\dfrac{3}{4}\right)$

78. $12\left(\dfrac{1}{2}\right)$

79. $\dfrac{3}{2}\left(-\dfrac{5}{4}\right)$

80. $\dfrac{5}{3}\left(-\dfrac{6}{5}\right)$

81. $-13 + (-5)$

82. $-14 + (-3)$

83. $-\dfrac{3}{4} + \left(-\dfrac{1}{2}\right)$

84. $-\dfrac{7}{10} + \left(-\dfrac{1}{2}\right)$

85. $7x + (-4x)$

86. $5x + (-2x)$

Maintaining Your Skills

The problems that follow review some of the more important skills you have learned in previous sections and chapters. You can consider the time you spend working these problems as time spent studying for exams.

87. $3(6x)$

88. $5(4x)$

89. $\dfrac{1}{5}(5x)$

90. $\dfrac{1}{3}(3x)$

91. $8\left(\dfrac{1}{8}y\right)$

92. $6\left(\dfrac{1}{6}y\right)$

93. $-2\left(-\dfrac{1}{2}x\right)$

94. $-4\left(-\dfrac{1}{4}x\right)$

95. $-\dfrac{4}{3}\left(-\dfrac{3}{4}a\right)$

96. $-\dfrac{5}{2}\left(-\dfrac{2}{5}a\right)$

OBJECTIVES

A Use the multiplication property of equality to solve an equation.

B Use the addition and multiplication properties of equality together to solve an equation.

TICKET TO SUCCESS

Keep these questions in mind as you read through the section. Then respond in your own words and in complete sentences.

1. Explain in words the multiplication property of equality.

2. If an equation contains fractions, how do you use the multiplication property of equality to clear the equation of fractions?

3. Why is it okay to divide both sides of an equation by the same nonzero number?

4. Give an example of using an LCD with the multiplication property of equality to simplify an equation.

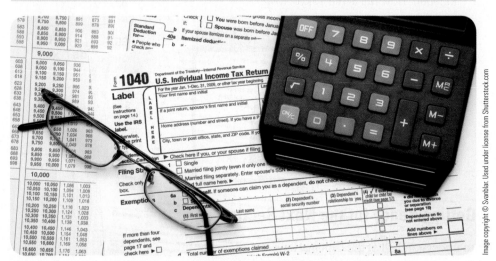

As we have mentioned before, we all have to pay taxes. According to Figure 1, people have been paying taxes for quite a long time.

FIGURE 1 **Collection of taxes, ca. 3000 BC. Clerks and scribes appear at the right, with pen and papyrus, and officials and taxpayers appear at the left.**

If 21% of your monthly pay is withheld for federal income taxes and another 8% is withheld for Social Security, state income tax, and other miscellaneous items, leaving you with $987.50 a month in take-home pay, then the amount you earned before the deductions were removed from your check is given by the equation

$$G - 0.21G - 0.08G = 987.5$$

In this section we will learn how to solve equations of this type.

A Multiplication Property of Equality

In the previous section, we found that adding the same number to both sides of an equation never changed the solution set. The same idea holds for multiplication by numbers other than zero. We can multiply both sides of an equation by the same nonzero number and always be sure we have not changed the solution set. (The reason we cannot multiply both sides by zero will become apparent later.) This fact about equations is called the *multiplication property of equality,* which can be stated formally as follows:

Multiplication Property of Equality

For any three algebraic expressions A, B, and C, where $C \neq 0$,

In symbols: if $A = B$
then $AC = BC$

In words: Multiplying both sides of an equation by the same nonzero number will not change the solution set.

Suppose we want to solve the equation $5x = 30$. We have $5x$ on the left side but would like to have just x. We choose to multiply both sides by $\frac{1}{5}$ because $\left(\frac{1}{5}\right)(5) = 1$. Here is the solution:

$$5x = 30$$

$$\frac{1}{5}(5x) = \frac{1}{5}(30) \qquad \text{Multiplication property of equality}$$

$$\left(\frac{1}{5} \cdot 5\right)x = \frac{1}{5}(30) \qquad \text{Associative property of multiplication}$$

$$1x = 6$$

$$x = 6 \qquad \text{Note: } 1x \text{ is the same as } x$$

We chose to multiply by $\frac{1}{5}$ because it is the reciprocal of 5. We can see that multiplication by any number except zero will not change the solution set. If, however, we were to multiply both sides by zero, the result would always be $0 = 0$ because multiplication by zero always results in zero. Although the statement $0 = 0$ is true, we have lost our variable and cannot solve the equation. This is the only restriction of the multiplication property of equality. We are free to multiply both sides of an equation by any number except zero.

Here are some more examples that use the multiplication property of equality.

EXAMPLE 1   Solve for a: $-4a = 24$.

SOLUTION   Because we want the variable a alone on the left side, we choose to multiply both sides by $-\frac{1}{4}$:

$$-\frac{1}{4}(-4a) = -\frac{1}{4}(24) \qquad \text{Multiplication property of equality}$$

$$\left[-\frac{1}{4}(-4)\right]a = -\frac{1}{4}(24) \qquad \text{Associative property}$$

$$a = -6$$

EXAMPLE 2   Solve for t: $-\frac{t}{3} = 5$.

SOLUTION   Because division by 3 is the same as multiplication by $\frac{1}{3}$, we can write $-\frac{t}{3}$ as $-\frac{1}{3}t$. To solve the equation, we multiply each side by the reciprocal of $-\frac{1}{3}$, which is -3.

$$-\frac{t}{3} = 5 \qquad \text{Original equation}$$

$$-\frac{1}{3}t = 5 \qquad \text{Dividing by 3 is equivalent to multiplying by } \frac{1}{3}$$

$$-3\left(-\frac{1}{3}t\right) = -3(5)$$ Multiply each side by -3

$$t = -15$$

NOTE
Notice in Examples 1 through 4 that if the variable is being multiplied by a number like -4 or $\frac{2}{3}$, we always multiply by the number's reciprocal, $-\frac{1}{4}$ or $\frac{3}{2}$, to end up with just the variable on one side of the equation.

EXAMPLE 3 Solve $\frac{2}{3}y = 4$.

SOLUTION We can multiply both sides by $\frac{3}{2}$ and have $1y$ on the left side.

$$\frac{3}{2}\left(\frac{2}{3}y\right) = \frac{3}{2}(4)$$ Multiplication property of equality

$$\left(\frac{3}{2} \cdot \frac{2}{3}\right)y = \frac{3}{2}(4)$$ Associative property

$$y = 6$$ Simplify $\frac{3}{2}(4) = \frac{3}{2}\left(\frac{4}{1}\right) = \frac{12}{2} = 6$

EXAMPLE 4 Solve $5 + 8 = 10x + 20x - 4x$.

SOLUTION Our first step will be to simplify each side of the equation.

$$13 = 26x$$ Simplify both sides first

$$\frac{1}{26}(13) = \frac{1}{26}(26x)$$ Multiplication property of equality

$$\frac{13}{26} = x$$

$$\frac{1}{2} = x$$ Reduce to lowest terms

B Solving Equations

In the next three examples, we will use both the addition property of equality and the multiplication property of equality.

EXAMPLE 5 Solve for x: $6x + 5 = -13$.

SOLUTION We begin by adding -5 to both sides of the equation.

$$6x + 5 + (-5) = -13 + (-5)$$ Add -5 to both sides

$$6x = -18$$ Simplify

$$\frac{1}{6}(6x) = \frac{1}{6}(-18)$$ Multiply both sides by $\frac{1}{6}$

$$x = -3$$

NOTE
Notice that in Example 6, we used the addition property of equality first to combine all the terms containing x on the left side of the equation. Once this had been done, we used the multiplication property to isolate x on the left side.

EXAMPLE 6 Solve for x: $5x = 2x + 12$.

SOLUTION We begin by adding $-2x$ to both sides of the equation.

$$5x + (-2x) = 2x + (-2x) + 12$$ Add $-2x$ to both sides

$$3x = 12$$ Simplify

$$\frac{1}{3}(3x) = \frac{1}{3}(12)$$ Multiply both sides by $\frac{1}{3}$

$$x = 4$$ Simplify

EXAMPLE 7 Solve for x: $3x - 4 = -2x + 6$.

SOLUTION We begin by adding $2x$ to both sides.

$$3x + 2x - 4 = -2x + 2x + 6$$ Add $2x$ to both sides

$$5x - 4 = 6$$ Simplify

Now we add 4 to both sides.

$$5x - 4 + \mathbf{4} = 6 + \mathbf{4}$$ Add 4 to both sides
$$5x = 10$$ Simplify
$$\frac{\mathbf{1}}{\mathbf{5}}(5x) = \frac{\mathbf{1}}{\mathbf{5}}(10)$$ Multiply by $\frac{1}{5}$
$$x = 2$$ Simplify ◼

The next example involves fractions. You will see that the properties we use to solve equations containing fractions are the same as the properties we used to solve the previous equations. Also, the LCD that we used previously to add fractions can be used with the multiplication property of equality to simplify equations containing fractions.

EXAMPLE 8   Solve $\frac{2}{3}x + \frac{1}{2} = -\frac{3}{4}$.

SOLUTION   We can solve this equation by applying our properties and working with the fractions, or we can begin by eliminating the fractions.

METHOD 1 Working with the fractions.

$$\frac{2}{3}x + \frac{1}{2} + \left(-\frac{\mathbf{1}}{\mathbf{2}}\right) = -\frac{3}{4} + \left(-\frac{\mathbf{1}}{\mathbf{2}}\right)$$ Add $-\frac{1}{2}$ to each side

$$\frac{2}{3}x = -\frac{5}{4}$$ Note that $-\frac{3}{4} + \left(-\frac{1}{2}\right) = -\frac{3}{4} + \left(-\frac{2}{4}\right)$

$$\frac{\mathbf{3}}{\mathbf{2}}\left(\frac{2}{3}x\right) = \frac{\mathbf{3}}{\mathbf{2}}\left(-\frac{5}{4}\right)$$ Multiply each side by $\frac{3}{2}$

$$x = -\frac{15}{8}$$

> **NOTE**
> Our original equation has denominators of 3, 2, and 4. The LCD for these three denominators is 12, and it has the property that all three denominators will divide it evenly. Therefore, if we multiply both sides of our equation by 12, each denominator will divide into 12 and we will be left with an equation that does not contain any denominators other than 1.

METHOD 2 Eliminating the fractions in the beginning.

$$\mathbf{12}\left(\frac{2}{3}x + \frac{1}{2}\right) = \mathbf{12}\left(-\frac{3}{4}\right)$$ Multiply each side by the LCD 12

$$\mathbf{12}\left(\frac{2}{3}x\right) + \mathbf{12}\left(\frac{1}{2}\right) = \mathbf{12}\left(-\frac{3}{4}\right)$$ Distributive property on the left side

$$8x + 6 = -9$$ Multiply

$$8x = -15$$ Add -6 to each side

$$x = -\frac{15}{8}$$ Multiply each side by $\frac{1}{8}$

As the third line in Method 2 indicates, multiplying each side of the equation by the LCD eliminates all the fractions from the equation.

As you can see, both methods yield the same solution. ◼

More About the Multiplication Property of Equality

Because division is defined as multiplication by the reciprocal, multiplying both sides of an equation by the same number is equivalent to dividing both sides of the equation by the reciprocal of that number; that is, multiplying each side of an equation by $\frac{1}{3}$ and dividing each side of the equation by 3 are equivalent operations. If we were to solve the equation $3x = 18$ using division instead of multiplication, the steps would look like this:

$$3x = 18 \qquad \text{Original equation}$$

$$\frac{3x}{\mathbf{3}} = \frac{18}{\mathbf{3}} \qquad \text{Divide each side by 3}$$

$$x = 6$$

Using division instead of multiplication on a problem like this may save you some writing. However, with multiplication, it is easier to explain "why" we end up with just one x on the left side of the equation. (The "why" has to do with the associative property of multiplication.) Continue to use multiplication to solve equations like this one until you understand the process completely. Then, if you find it more convenient, you can use division instead of multiplication.

Problem Set 2.3

Moving Toward Success

"It doesn't matter if you try and try and try again, and fail. It does matter if you try and fail, and fail to try again."

—Charles F. Kettering, 1876–1958, American engineer and inventor

1. What is it going to take for you to be successful in mathematics?

2. Do you think it is important to take notes as you read this book? Explain.

A Solve the following equations. Be sure to show your work. [Examples 1–3]

1. $5x = 10$

2. $6x = 12$

3. $7a = 28$

4. $4a = 36$

5. $-8x = 4$

6. $-6x = 2$

7. $8m = -16$

8. $5m = -25$

9. $-3x = -9$

10. $-9x = -36$

11. $-7y = -28$

12. $-15y = -30$

13. $2x = 0$

14. $7x = 0$

15. $-5x = 0$

16. $-3x = 0$

17. $\frac{x}{3} = 2$

18. $\frac{x}{4} = 3$

19. $-\frac{m}{5} = 10$

20. $-\frac{m}{7} = 1$

21. $-\frac{x}{2} = -\frac{3}{4}$

22. $-\frac{x}{3} = \frac{5}{6}$

23. $\frac{2}{3}a = 8$

24. $\frac{3}{4}a = 6$

25. $-\frac{3}{5}x = \frac{9}{5}$

26. $-\frac{2}{5}x = \frac{6}{15}$

27. $-\frac{5}{8}y = -20$

28. $-\frac{7}{2}y = -14$

A Simplify both sides as much as possible, and then solve. [Example 4]

29. $-4x - 2x + 3x = 24$

30. $7x - 5x + 8x = 20$

31. $4x + 8x - 2x = 15 - 10$

32. $5x + 4x + 3x = 4 + 8$

33. $-3 - 5 = 3x + 5x - 10x$

34. $10 - 16 = 12x - 6x - 3x$

35. $18 - 13 = \frac{1}{2}a + \frac{3}{4}a - \frac{5}{8}a$

36. $20 - 14 = \frac{1}{3}a + \frac{5}{6}a - \frac{2}{3}a$

Solve the following equations by multiplying both sides by −1.

37. $-x = 4$

38. $-x = -3$

39. $-x = -4$

40. $-x = 3$

41. $15 = -a$

42. $-15 = -a$

43. $-y = \dfrac{1}{2}$

44. $-y = -\dfrac{3}{4}$

B Solve each of the following equations.
[Examples 5–8]

45. $3x - 2 = 7$

46. $2x - 3 = 9$

47. $2a + 1 = 3$

48. $5a - 3 = 7$

49. $\dfrac{1}{8} + \dfrac{1}{2}x = \dfrac{1}{4}$

50. $\dfrac{1}{3} + \dfrac{1}{7}x = -\dfrac{8}{21}$

51. $6x = 2x - 12$

52. $8x = 3x - 10$

53. $2y = -4y + 18$

54. $3y = -2y - 15$

55. $-7x = -3x - 8$

56. $-5x = -2x - 12$

57. $2x - 5 = 8x + 4$

58. $3x - 6 = 5x + 6$

59. $x + \dfrac{1}{2} = \dfrac{1}{4}x - \dfrac{5}{8}$

60. $\dfrac{1}{3}x + \dfrac{2}{5} = \dfrac{1}{5}x - \dfrac{2}{5}$

61. $m + 2 = 6m - 3$

62. $m + 5 = 6m - 5$

63. $\dfrac{1}{2}m - \dfrac{1}{4} = \dfrac{1}{12}m + \dfrac{1}{6}$

64. $\dfrac{1}{2}m - \dfrac{5}{12} = \dfrac{1}{12}m + \dfrac{5}{12}$

65. $6y - 4 = 9y + 2$

66. $2y - 2 = 6y + 14$

67. $\dfrac{3}{2}y + \dfrac{1}{3} = y - \dfrac{2}{3}$

68. $\dfrac{3}{2}y + \dfrac{7}{2} = \dfrac{1}{2}y - \dfrac{1}{2}$

69. $5x + 6 = 2$

70. $2x + 15 = 3$

71. $\dfrac{x}{2} = \dfrac{6}{12}$

72. $\dfrac{x}{4} = \dfrac{6}{8}$

73. $\dfrac{3}{x} = \dfrac{6}{7}$

74. $\dfrac{2}{9} = \dfrac{8}{x}$

75. $\dfrac{a}{3} = \dfrac{5}{12}$

76. $\dfrac{a}{2} = \dfrac{7}{20}$

77. $\dfrac{10}{20} = \dfrac{20}{x}$

78. $\dfrac{15}{60} = \dfrac{60}{x}$

79. $\dfrac{2}{x} = \dfrac{6}{7}$

80. $\dfrac{4}{x} = \dfrac{6}{7}$

Applying the Concepts

81. Break-Even Point Movie theaters pay a certain price for the movies that you and I see. Suppose a theater pays $1,500 for each showing of a popular movie. If they charge $7.50 for each ticket they sell, then the equation $7.5x = 1{,}500$ gives the number of tickets they must sell to equal the $1,500 cost of showing the movie. This number is called the break-even point. Solve the equation for x to find the break-even point.

82. Basketball Laura plays basketball for her community college. In one game she scored 13 points total, with a combination of free throws, field goals, and three-pointers. Each free throw is worth 1 point, each field goal is 2 points, and each three-pointer is worth 3 points. If she made 1 free throw and 3 field goals, then solving the equation

$$1 + 3(2) + 3x = 13$$

will give us the number of three-pointers she made. Solve the equation to find the number of three-point shots Laura made.

83. Taxes Suppose 21% of your monthly pay is withheld for federal income taxes and another 8% is withheld for Social Security, state income tax, and other miscellaneous items and you are left with $987.61 a month in take-home pay. The amount you earned before the deductions were removed from your check is given by the equation

$$G - 0.21G - 0.08G = 987.61$$

Solve this equation to find your gross income.

84. Rhind Papyrus The *Rhind Papyrus* is an ancient document that contains mathematical riddles. One problem asks the reader to find a quantity such that when it is added to one-fourth of itself the sum is 15. The equation that describes this situation is

$$x + \dfrac{1}{4}x = 15$$

Solve this equation.

85. Skyscrapers The chart shows the heights of three of the tallest buildings in the world. Two times the height of the Hoover Dam is the height of the Sears Tower. How tall is the Hoover Dam?

Such Great Heights

Taipei 101
Taipei, Taiwan
1,670 ft

Petronas Tower 1 & 2
Kuala Lumpur, Malaysia
1,483 ft

Sears Tower
Chicago, USA
1,450 ft

Source: www.tenmojo.com

86. Cars The chart shows the fastest cars in the world. The speed of the SSC Ultimate Aero in miles per hour is $\frac{15}{22}$ of its speed in feet per second. What is its speed in feet per second? Round to the nearest foot per second.

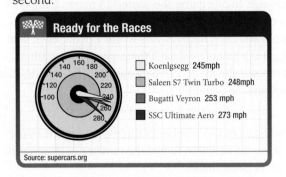

Ready for the Races

☐ Koenlgsegg **245mph**
☐ Saleen S7 Twin Turbo **248mph**
☐ Bugatti Veyron **253 mph**
☐ SSC Ultimate Aero **273 mph**

Source: supercars.org

Getting Ready for the Next Section

To understand all of the explanations and examples in the next section you must be able to work the problems below.

Solve each equation.

87. $2x = 4$

88. $3x = 24$

89. $30 = 5x$

90. $0 = 5x$

91. $0.17x = 510$

92. $0.1x = 400$

Apply the distributive property and then simplify if possible.

93. $3(x - 5) + 4$

94. $5(x - 3) + 2$

95. $0.09(x + 2,000)$

96. $0.04(x + 7,000)$

97. $5 - 2(3y + 1)$

98. $4 - 2(3y + 1)$

99. $3(2x - 5) - (2x - 4)$

100. $4(3x - 2) - (6x - 5)$

Simplify.

101. $10x + (-5x)$

102. $12x + (-7x)$

103. $0.08x + 0.09x$

104. $0.06x + 0.04x$

Maintaining Your Skills

The problems that follow review some of the more important skills you have learned in previous sections and chapters. You can consider the time you spend working these problems as time spent studying for exams.

Apply the distributive property, and then simplify each expression as much as possible.

105. $2(3x - 5)$

106. $4(2x - 6)$

107. $\frac{1}{2}(3x + 6)$

108. $\frac{1}{4}(2x + 8)$

109. $\frac{1}{3}(-3x + 6)$

110. $\frac{1}{2}(-2x + 6)$

111. $5(2x - 8) - 3$

112. $4(3x - 1) + 7$

113. $-2(3x + 5) + 3(x - 1)$

114. $6(x + 3) - 2(2x + 4)$

115. $7 - 3(2y + 1)$

116. $8 - 5(3y - 4)$

2.4 Solving Linear Equations

OBJECTIVES

A Solve a linear equation in one variable.

TICKET TO SUCCESS

Keep these questions in mind as you read through the section. Then respond in your own words and in complete sentences.

1. What is the first step in solving a linear equation containing parentheses?

2. When solving a linear equation, why should you get all variable terms on one side and all constant terms on the other before using the multiplication property of equality?

3. What is the last step in solving a linear equation?

4. If an equation contains decimals, what can you do to eliminate the decimals?

A Solving Linear Equations

Study the graph in Figure 1. The world's capacity for generating power from wind energy has visibly increased since the year 2002. How would this graph help us find the year that produced 34,000 megawatts of wind power? If we used the linear equation $7x - 13,980 = 34$, where x represents the year that produced 34,000 megawatts, then we would need to know how to solve the equation for x to get our answer.

We will now use the material we have developed in the first three sections of this chapter to build a method for solving any linear equation.

> **Definition**
>
> A **linear equation in one variable** is any equation that can be put in the form $ax + b = 0$, where a and b are real numbers and a is not zero.

Each of the equations we will solve in this section is a linear equation in one variable. The steps we use to solve a linear equation in one variable are listed here:

> **Strategy**   Solving Linear Equations in One Variable
>
> **Step 1a:** Use the distributive property to separate terms, if necessary.
>
> > **1b:** If fractions are present, consider multiplying both sides by the LCD to eliminate the fractions. If decimals are present, consider multiplying both sides by a power of 10 to clear the equation of decimals.
> >
> > **1c:** Combine similar terms on each side of the equation.
>
> **Step 2:** Use the addition property of equality to get all variable terms on one side of the equation and all constant terms on the other side. A variable term is a term that contains the variable (for example, 5x). A constant term is a term that does not contain the variable (the number 3, for example).
>
> **Step 3:** Use the multiplication property of equality to get x (that is, $1x$) by itself on one side of the equation.
>
> **Step 4:** Check your solution in the original equation to be sure that you have not made a mistake in the solution process.

As you work through the examples in this section, it is not always necessary to use all four steps when solving equations. The number of steps used depends on the

---

World Wind Electricity-Generating Capacity

Megawatts

250,000
200,000
150,000
100,000
50,000

31,000   48,000   74,000   121,000   204,000

2002   2004   2006   2008   2010

Source: World Wind Energy Association

FIGURE 1

NOTE

You may have some previous experience solving equations. Even so, you should solve the equations in this section using the method developed here. Your work should look like the examples in the text. If you have learned shortcuts or a different method of solving equations somewhere else, you always can go back to them later. What is important now is that you are able to solve equations by the methods shown here.

equation. In Example 1, there are no fractions or decimals in the original equation, so step 1b will not be used. Likewise, after applying the distributive property to the left side of the equation in Example 1, there are no similar terms to combine on either side of the equation, making step 1c also unnecessary.

EXAMPLE 1   Solve $2(x + 3) = 10$.

SOLUTION   To begin, we apply the distributive property to the left side of the equation to separate terms.

Step 1a:	$2x + 6 = 10$	Distributive property
Step 2:	$\begin{cases} 2x + 6 + (\mathbf{-6}) = 10 + (\mathbf{-6}) \\ 2x = 4 \end{cases}$	Addition property of equality
Step 3:	$\begin{cases} \dfrac{\mathbf{1}}{\mathbf{2}}(2x) = \dfrac{\mathbf{1}}{\mathbf{2}}(4) \\ x = 2 \end{cases}$	Multiply each side by $\frac{1}{2}$ The solution is 2

The solution to our equation is 2. We check our work (to be sure we have not made either a mistake in applying the properties or an arithmetic mistake) by substituting 2 into our original equation and simplifying each side of the result separately.

CHECK:

Step 4:
$$\begin{cases} \text{When} \rightarrow & x = 2 \\ \text{the equation} \rightarrow & 2(x + 3) = 10 \\ \text{becomes} \rightarrow & 2(2 + 3) \stackrel{?}{=} 10 \\ & 2(5) \stackrel{?}{=} 10 \\ & 10 = 10 \quad \text{A true statement} \end{cases}$$

Our solution checks.

The general method of solving linear equations is actually very simple. It is based on the properties we developed in Chapter 1 and on two very simple new properties. We can add any number to both sides of the equation and multiply both sides by any nonzero number. The equation may change in form, but the solution set will not. If we look back to Example 1, each equation looks a little different from each preceding equation. What is interesting and useful is that each equation says the same thing about x. They all say x is 2. The last equation, of course, is the easiest to read, and that is why our goal is to end up with x by itself.

EXAMPLE 2   Solve for x: $3(x - 5) + 4 = 13$.

SOLUTION   Our first step will be to apply the distributive property to the left side of the equation.

Step 1a:	$3x - 15 + 4 = 13$	Distributive property
Step 1c:	$3x - 11 = 13$	Simplify the left side
Step 2:	$\begin{cases} 3x - 11 + \mathbf{11} = 13 + \mathbf{11} \\ 3x = 24 \end{cases}$	Add 11 to both sides
Step 3:	$\begin{cases} \dfrac{\mathbf{1}}{\mathbf{3}}(3x) = \dfrac{\mathbf{1}}{\mathbf{3}}(24) \\ x = 8 \end{cases}$	Multiply both sides by $\frac{1}{3}$ The solution is 8

CHECK:

Step 4:
$$\begin{cases} \text{When} \rightarrow & x = 8 \\ \text{the equation} \rightarrow & 3(x - 5) + 4 = 13 \\ \text{becomes} \rightarrow & 3(8 - 5) + 4 \stackrel{?}{=} 13 \\ & 3(3) + 4 \stackrel{?}{=} 13 \\ & 9 + 4 \stackrel{?}{=} 13 \\ & 13 = 13 \quad \text{A true statement} \end{cases}$$

EXAMPLE 3   Solve $5(x - 3) + 2 = 5(2x - 8) - 3$.

SOLUTION   In this case, we first apply the distributive property to each side of the equation.

Step 1a:	$5x - 15 + 2 = 10x - 40 - 3$	Distributive property
Step 1c:	$5x - 13 = 10x - 43$	Simplify each side

Step 2:
$$\begin{cases} 5x + (\mathbf{-5x}) - 13 = 10x + (\mathbf{-5x}) - 43 & \text{Add } -5x \text{ to both sides} \\ -13 = 5x - 43 \\ -13 + \mathbf{43} = 5x - 43 + \mathbf{43} & \text{Add } 43 \text{ to both sides} \\ 30 = 5x \end{cases}$$

Step 3:
$$\begin{cases} \dfrac{1}{5}(30) = \dfrac{1}{5}(5x) & \text{Multiply both sides by } \frac{1}{5} \\ 6 = x & \text{The solution is } 6 \end{cases}$$

CHECK: Replacing x with 6 in the original equation, we have

Step 4:
$$\begin{cases} 5(6 - 3) + 2 \overset{?}{=} 5(2 \cdot 6 - 8) - 3 \\ 5(3) + 2 \overset{?}{=} 5(12 - 8) - 3 \\ 5(3) + 2 \overset{?}{=} 5(4) - 3 \\ 15 + 2 \overset{?}{=} 20 - 3 \\ 17 = 17 \qquad\qquad \text{A true statement} \end{cases}$$

> **NOTE**
> It makes no difference on which side of the equal sign x ends up. Most people prefer to have x on the left side because we read from left to right, and it seems to sound better to say x is 6 rather than 6 is x. Both expressions, however, have exactly the same meaning.

EXAMPLE 4   Solve the equation $0.08x + 0.09(x + 2{,}000) = 690$.

SOLUTION   We can solve the equation in its original form by working with the decimals, or we can eliminate the decimals first by using the multiplication property of equality and solving the resulting equation. Both methods follow.

METHOD 1   Working with the decimals

	$0.08x + 0.09(x + 2{,}000) = 690$	Original equation
Step 1a:	$0.08x + 0.09x + 0.09(2{,}000) = 690$	Distributive property
Step 1c:	$0.17x + 180 = 690$	Simplify the left side

Step 2:
$$\begin{cases} 0.17x + 180 + (\mathbf{-180}) = 690 + (\mathbf{-180}) & \text{Add } -180 \text{ to each side} \\ 0.17x = 510 \end{cases}$$

Step 3:
$$\begin{cases} \dfrac{0.17x}{\mathbf{0.17}} = \dfrac{510}{\mathbf{0.17}} & \text{Divide each side by } 0.17 \\ x = 3{,}000 & \text{The solution is } 3{,}000 \end{cases}$$

Note that we divided each side of the equation by 0.17 to obtain the solution. This is still an application of the multiplication property of equality because dividing by 0.17 is equivalent to multiplying by $\frac{1}{0.17}$.

METHOD 2   Eliminating the decimals in the beginning

	$0.08x + 0.09(x + 2{,}000) = 690$	Original equation
Step 1a:	$0.08x + 0.09x + 180 = 690$	Distributive property

Step 1b:
$$\begin{cases} \mathbf{100}(0.08x + 0.09x + 180) = \mathbf{100}(690) & \text{Multiply both sides by } 100 \\ 8x + 9x + 18{,}000 = 69{,}000 \end{cases}$$

Step 1c:	$17x + 18{,}000 = 69{,}000$	Simplify the left side
Step 2:	$17x = 51{,}000$	Add $-18{,}000$ to each side

Step 3:
$$\begin{cases} \dfrac{17x}{\mathbf{17}} = \dfrac{51{,}000}{\mathbf{17}} & \text{Divide each side by } 17 \end{cases}$$

$$x = 3{,}000 \qquad\qquad \text{The solution is } 3{,}000$$

CHECK: Substituting 3,000 for x in the original equation, we have

Step 4:
$$
\begin{cases}
0.08(3,000) + 0.09(3,000 + 2,000) \stackrel{?}{=} 690 \\
0.08(3,000) + 0.09(5,000) \stackrel{?}{=} 690 \\
240 + 450 \stackrel{?}{=} 690 \\
690 = 690 \qquad \text{A true statement}
\end{cases}
$$

EXAMPLE 5   Solve $7 - 3(2y + 1) = 16$.

SOLUTION   We begin by multiplying -3 times the sum of $2y$ and 1.

Step 1a:	$7 - 6y - 3 = 16$	Distributive property
Step 1c:	$-6y + 4 = 16$	Simplify the left side

Step 2: $\begin{cases} -6y + 4 + (\mathbf{-4}) = 16 + (\mathbf{-4}) & \text{Add } -4 \text{ to both sides} \\ -6y = 12 \end{cases}$

Step 3: $\begin{cases} -\dfrac{1}{6}(-6y) = -\dfrac{1}{6}(12) & \text{Multiply both sides by } -\dfrac{1}{6} \\ y = -2 & \text{The solution is } -2 \end{cases}$

Step 4: Replacing y with -2 in the original equation yields a true statement.

There are two things to notice about the example that follows: first, the distributive property is used to remove parentheses that are preceded by a negative sign; second, the addition property and the multiplication property are not shown in as much detail as in the previous examples.

EXAMPLE 6   Solve $3(2x - 5) - (2x - 4) = 6 - (4x + 5)$.

SOLUTION   When we apply the distributive property to remove the grouping symbols and separate terms, we have to be careful with the signs. Remember, we can think of $-(2x - 4)$ as $-1(2x - 4)$, so that

$$-(2x - 4) = -1(2x - 4) = -2x + 4$$

It is not uncommon for students to make a mistake with this type of simplification and write the result as $-2x - 4$, which is incorrect. Here is the complete solution to our equation:

$3(2x - 5) - (2x - 4) = 6 - (4x + 5)$	Original equation
$6x - 15 - 2x + 4 = 6 - 4x - 5$	Distributive property
$4x - 11 = -4x + 1$	Simplify each side
$8x - 11 = 1$	Add $4x$ to each side
$8x = 12$	Add 11 to each side
$x = \dfrac{12}{8}$	Multiply each side by $\dfrac{1}{8}$
$x = \dfrac{3}{2}$	Reduce to lowest terms

The solution, $\dfrac{3}{2}$, checks when replacing x in the original equation.

Problem Set 2.4

Moving Toward Success

"Our greatest weakness lies in giving up. The most certain way to succeed is always to try just one more time."

—Thomas Edison, 1847–1931, American inventor and entrepreneur

1. What should you do when you come across a difficult problem?
   a. Make a guess for the answer and move on to the next problem

   b. Add it to an ongoing list of difficult problems to address promptly
   c. Ask someone on your resources list for guidance
   d. Answers (b) and (c)

2. Why is making a list of difficult problems important?

A Solve each of the following equations using the four steps shown in this section. [Examples 1–2]

1. $2(x + 3) = 12$ **2.** $3(x - 2) = 6$

3. $6(x - 1) = -18$ **4.** $4(x + 5) = 16$

5. $2(4a + 1) = -6$ **6.** $3(2a - 4) = 12$

7. $14 = 2(5x - 3)$ **8.** $-25 = 5(3x + 4)$

9. $-2(3y + 5) = 14$ **10.** $-3(2y - 4) = -6$

11. $-5(2a + 4) = 0$ **12.** $-3(3a - 6) = 0$

13. $1 = \frac{1}{2}(4x + 2)$ **14.** $1 = \frac{1}{3}(6x + 3)$

15. $3(t - 4) + 5 = -4$ **16.** $5(t - 1) + 6 = -9$

A Solve each equation. [Examples 2–6]

17. $4(2y + 1) - 7 = 1$ **18.** $6(3y + 2) - 8 = -2$

19. $\frac{1}{2}(x - 3) = \frac{1}{4}(x + 1)$ **20.** $\frac{1}{3}(x - 4) = \frac{1}{2}(x - 6)$

21. $-0.7(2x - 7) = 0.3(11 - 4x)$

22. $-0.3(2x - 5) = 0.7(3 - x)$

23. $-2(3y + 1) = 3(1 - 6y) - 9$

24. $-5(4y - 3) = 2(1 - 8y) + 11$

25. $\frac{3}{4}(8x - 4) + 3 = \frac{2}{5}(5x + 10) - 1$

26. $\frac{5}{6}(6x + 12) + 1 = \frac{2}{3}(9x - 3) + 5$

27. $0.06x + 0.08(100 - x) = 6.5$

28. $0.05x + 0.07(100 - x) = 6.2$

29. $6 - 5(2a - 3) = 1$ **30.** $-8 - 2(3 - a) = 0$

31. $0.2x - 0.5 = 0.5 - 0.2(2x - 13)$

32. $0.4x - 0.1 = 0.7 - 0.3(6 - 2x)$

33. $2(t - 3) + 3(t - 2) = 28$

34. $-3(t - 5) - 2(2t + 1) = -8$

35. $5(x - 2) - (3x + 4) = 3(6x - 8) + 10$

36. $3(x - 1) - (4x - 5) = 2(5x - 1) - 7$

37. $2(5x - 3) - (2x - 4) = 5 - (6x + 1)$

38. $3(4x - 2) - (5x - 8) = 8 - (2x + 3)$

39. $-(3x + 1) - (4x - 7) = 4 - (3x + 2)$

40. $-(6x + 2) - (8x - 3) = 8 - (5x + 1)$

41. $x + (2x - 1) = 2$ **42.** $x + (5x + 2) = 20$

43. $x - (3x + 5) = -3$ **44.** $x - (4x - 1) = 7$

45. $15 = 3(x - 1)$ **46.** $12 = 4(x - 5)$

47. $4x - (-4x + 1) = 5$ **48.** $-2x - (4x - 8) = -1$

49. $5x - 8(2x - 5) = 7$ **50.** $3x + 4(8x - 15) = 10$

51. $7(2y - 1) - 6y = -1$ **52.** $4(4y - 3) + 2y = 3$

53. $0.2x + 0.5(12 - x) = 3.6$ **54.** $0.3x + 0.6(25 - x) = 12$

55. $0.5x + 0.2(18 - x) = 5.4$

56. $0.1x + 0.5(40 - x) = 32$

57. $x + (x + 3)(-3) = x - 3$ **58.** $x - 2(x + 2) = x - 2$

59. $5(x + 2) + 3(x - 1) = -9$

60. $4(x + 1) + 3(x - 3) = 2$

61. $3(x - 3) + 2(2x) = 5$ **62.** $2(x - 2) + 3(5x) = 30$

63. $5(y + 2) = 4(y + 1)$ **64.** $3(y - 3) = 2(y - 2)$

65. $3x + 2(x - 2) = 6$ **66.** $5x - (x - 5) = 25$

67. $50(x - 5) = 30(x + 5)$ **68.** $34(x - 2) = 26(x + 2)$

69. $0.08x + 0.09(x + 2,000) = 860$

70. $0.11x + 0.12(x + 4,000) = 940$

71. $0.10x + 0.12(x + 500) = 214$

72. $0.08x + 0.06(x + 800) = 104$

73. $5x + 10(x + 8) = 245$ **74.** $5x + 10(x + 7) = 175$

75. $5x + 10(x + 3) + 25(x + 5) = 435$

76. $5(x + 3) + 10x + 25(x + 7) = 390$

The next two problems are intended to give you practice reading, and paying attention to, the instructions that accompany the problems you are working. Working these problems is a excellent way to get ready for a test or a quiz.

77. Work each problem according to the instructions.

   a. Solve: $4x - 5 = 0$

   b. Solve: $4x - 5 = 25$

   c. Add: $(4x - 5) + (2x + 25)$

   d. Solve: $4x - 5 = 2x + 25$

   e. Multiply: $4(x - 5)$

   f. Solve: $4(x - 5) = 2x + 25$

78. Work each problem according to the instructions.

   a. Solve: $3x + 6 = 0$

   b. Solve: $3x + 6 = 4$

   c. Add: $(3x + 6) + (7x + 4)$

   d. Solve: $3x + 6 = 7x + 4$

   e. Multiply: $3(x + 6)$

   f. Solve: $3(x + 6) = 7x + 4$

Applying the Concepts

79. Wind Energy The graph shows the world's capacity for generating power from wind energy. The year that produced 48,000 megawatts is given by the equation $48 = 7x - 13,980$. What year is this?

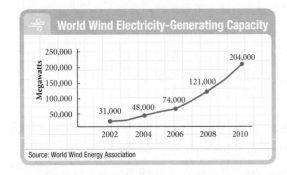

World Wind Electricity-Generating Capacity

Source: World Wind Energy Association

80. Camera Phones The chart shows the estimated number of camera phones and non-camera phones sold from 2004 to 2010. The year in which 730 million phones were sold can be given by the equation $730 = 10(5x - 9,952)$. What year is this?

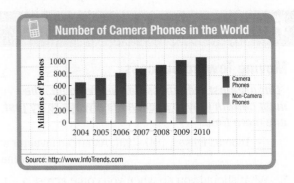

Number of Camera Phones in the World

Source: http://www.InfoTrends.com

Getting Ready for the Next Section

To understand all of the explanations and examples in the next section you must be able to work the problems below.

Solve each equation.

81. $40 = 2x + 12$ **82.** $80 = 2x + 12$

83. $12 + 2y = 6$ **84.** $3x + 18 = 6$

85. $24x = 6$ **86.** $45 = 0.75x$

87. $70 = x \cdot 210$ **88.** $15 = x \cdot 80$

Apply the distributive property.

89. $\dfrac{1}{2}(-3x + 6)$ **90.** $-\dfrac{1}{4}(-5x + 20)$

Maintaining Your Skills

The problems that follow review some of the more important skills you have learned in previous sections and chapters. You can consider the time you spend working these problems as time spent studying for exams.

Multiply.

91. $\dfrac{1}{2}(3)$ **92.** $\dfrac{1}{3}(2)$

93. $\dfrac{2}{3}(6)$ **94.** $\dfrac{3}{2}(4)$

95. $\dfrac{5}{9} \cdot \dfrac{9}{5}$ **96.** $\dfrac{3}{7} \cdot \dfrac{7}{3}$

Fill in the tables by finding the value of each expression for the given values of the variables.

97.

x	$3(x + 2)$	$3x + 2$	$3x + 6$
0			
1			
2			
3			

98.

x	$7(x-5)$	$7x-5$	$7x-35$
-3			
-2			
-1			
0			

99.

a	$(2a+1)^2$	$4a^2+4a+1$
1		
2		
3		

100.

a	$(a+1)^3$	a^3+3a^2+3a+1
1		
2		
3		

2.5 Formulas

OBJECTIVES

A Find the value of a variable in a formula given replacements for the other variables.

B Solve a formula for one of its variables.

C Find the complement and supplement of an angle.

D Solve basic percent problems.

TICKET TO SUCCESS

Keep these questions in mind as you read through the section. Then respond in your own words and in complete sentences.

1. What is a formula?
2. How do you solve a formula for one of its variables?
3. What are complementary angles?
4. Translate the following question into an equation: what number is 40% of 20?

A *formula* is a mathematical sentence that describes how one variable relates to one or more other variables. For example, let's say you want to buy a donut and a coffee. One donut costs $1.25 and a coffee costs $2.50. Therefore, your cost is $1.25 + $2.50 = $3.75. But if you want to buy donuts and coffee for the 5 people in your study group (including yourself), you need $6.25 for 5 donuts and $12.50 for 5 coffees. The formula in words for any number of treats is therefore, the cost of treats is equal to $1.25 multiplied by the number of people plus $2.50 multiplied by the same number of people. In math we would simply substitute variables for each item in the formula.

If T = total cost of treats

$$D = \text{number of donuts}$$

$$C = \text{number of coffees}$$

then the formula now becomes

$$T = 1.25D + 2.50C$$

If you know the values of any two of these three variables, you can figure out the value of the third.

Here is the formal definition of a formula:

> **Definition**
>
> In mathematics, a **formula** is an equation that contains more than one variable.

The equation $P = 2l + 2w$, which tells us how to find the perimeter of a rectangle, is an example of a formula.

A   Solving Formulas

To begin our work with formulas, we will consider some examples in which we are given numerical replacements for all but one of the variables.

EXAMPLE 1   The perimeter P of a rectangular livestock pen is 40 feet. If the width w is 6 feet, find the length.

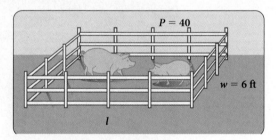

SOLUTION   First, we substitute 40 for P and 6 for w in the formula $P = 2l + 2w$. Then we solve for l.

When → $P = 40$ and $w = 6$

the formula → $P = 2l + 2w$

becomes → $40 = 2l + 2(6)$

$40 = 2l + 12$ Multiply 2 and 6

$28 = 2l$ Add -12 to each side

$14 = l$ Multiply each side by $\frac{1}{2}$

To summarize our results, if a rectangular pen has a perimeter of 40 feet and a width of 6 feet, then the length must be 14 feet.

EXAMPLE 2   Find y when $x = 4$ in the formula $3x + 2y = 6$.

SOLUTION   We substitute 4 for x in the formula and then solve for y.

When → $x = 4$

the formula → $3x + 2y = 6$

becomes → $3(4) + 2y = 6$

$12 + 2y = 6$ Multiply 3 and 4

$2y = -6$ Add -12 to each side

$y = -3$ Multiply each side by $\frac{1}{2}$ ∎

B Solving for a Variable

In the next examples we will solve a formula for one of its variables without being given numerical replacements for the other variables.

Consider the formula for the area of a triangle:

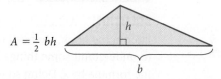

$A = \frac{1}{2} bh$

where A = area, b = length of the base, and h = height of the triangle.

Suppose we want to solve this formula for h. What we must do is isolate the variable h on one side of the equal sign. We begin by multiplying both sides by 2.

$$2 \cdot A = 2 \cdot \frac{1}{2}bh$$

$$2A = bh$$

Then we divide both sides by b.

$$\frac{2A}{b} = \frac{bh}{b}$$

$$h = \frac{2A}{b}$$

The original formula $A = \frac{1}{2}bh$ and the final formula $h = \frac{2A}{b}$ both give the same relationship among A, b, and h. The first one has been solved for A and the second one has been solved for h.

Rule: Solving for a Variable

To solve a formula for one of its variables, we must isolate that variable on either side of the equal sign. All other variables and constants will appear on the other side.

EXAMPLE 3   Solve $3x + 2y = 6$ for y.

SOLUTION   To solve for y, we must isolate y on one side of the equation. To begin, we use the addition property of equality to add $-3x$ to each side.

$3x + 2y = 6$ Original formula

$3x + (\mathbf{-3x}) + 2y = (\mathbf{-3x}) + 6$ Add $-3x$ to each side

$2y = -3x + 6$ Simplify the left side

$\frac{\mathbf{1}}{\mathbf{2}}(2y) = \frac{\mathbf{1}}{\mathbf{2}}(-3x + 6)$ Multiply each side by $\frac{1}{2}$

$y = -\frac{3}{2}x + 3$ ∎

EXAMPLE 4   Solve $h = vt - 16t^2$ for v.

SOLUTION   Let's begin by interchanging the left and right sides of the equation. That way, the variable we are solving for, v, will be on the left side.

$$vt - 16t^2 = h \qquad \text{Exchange sides}$$

$$vt - 16t^2 + \mathbf{16t^2} = h + \mathbf{16t^2} \qquad \text{Add } 16t^2 \text{ to each side}$$

$$vt = h + 16t^2$$

$$\frac{vt}{\boldsymbol{t}} = \frac{h + 16t^2}{\boldsymbol{t}} \qquad \text{Divide each side by } t$$

$$v = \frac{h + 16t^2}{t}$$

We know we are finished because we have isolated the variable we are solving for on the left side of the equation and it does not appear on the other side. ▪

EXAMPLE 5   Solve for y: $\dfrac{y - 1}{x} = \dfrac{3}{2}$.

SOLUTION   Although we will do more extensive work with formulas of this form later in the book, we need to know how to solve this particular formula for y in order to understand some things in the next chapter. We begin by multiplying each side of the formula by x. Doing so will simplify the left side of the equation and make the rest of the solution process simple.

$$\frac{y - 1}{x} = \frac{3}{2} \qquad \text{Original formula}$$

$$x \cdot \frac{y - 1}{x} = \frac{3}{2} \cdot x \qquad \text{Multiply each side by } x$$

$$y - 1 = \frac{3}{2}x \qquad \text{Simplify each side}$$

$$y = \frac{3}{2}x + 1 \qquad \text{Add 1 to each side}$$

This is our solution. If we look back to the first step, we can justify our result on the left side of the equation this way: Dividing by x is equivalent to multiplying by its reciprocal $\frac{1}{x}$. Here is what it looks like when written out completely:

$$x \cdot \frac{y - 1}{x} = x \cdot \frac{1}{x} \cdot (y - 1) = 1(y - 1) = y - 1 \qquad ▪$$

C Complementary and Supplementary Angles

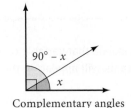
Complementary angles

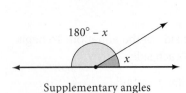
Supplementary angles

FACTS FROM GEOMETRY: More on Complementary and Supplementary Angles
In Chapter 1 we defined complementary angles as angles that add to 90°; that is, if x and y are complementary angles, then

$$x + y = 90°$$

If we solve this formula for y, we obtain a formula equivalent to our original formula:

$$y = 90° - x$$

Because y is the complement of x, we can generalize by saying that the complement of angle x is the angle $90° - x$. By a similar reasoning process, we can say that the supplement of angle x is the angle $180° - x$. To summarize, if x is an angle, then

The complement of x is $90° - x$, and

The supplement of x is $180° - x$

If you go on to take a trigonometry class, you will see this formula again.

EXAMPLE 6 Find the complement and the supplement of 25°.

SOLUTION We can use the formulas with $x = 25°$.
The complement of 25° is $90° - 25° = 65°$.
The supplement of 25° is $180° - 25° = 155°$.

D Basic Percent Problems

The last examples in this section show how basic percent problems can be translated directly into equations. To understand these examples, you must recall that *percent* means "per hundred." That is, 75% is the same as $\frac{75}{100}$, 0.75, and, in reduced fraction form, $\frac{3}{4}$. Likewise, the decimal 0.25 is equivalent to 25%. To change a decimal to a percent, we move the decimal point two places to the right and write the % symbol. To change from a percent to a decimal, we drop the % symbol and move the decimal point two places to the left. The table that follows gives some of the most commonly used fractions and decimals and their equivalent percents.

Fraction	Decimal	Percent
$\frac{1}{2}$	0.5	50%
$\frac{1}{4}$	0.25	25%
$\frac{3}{4}$	0.75	75%
$\frac{1}{3}$	$0.\overline{3}$	$33\frac{1}{3}\%$
$\frac{2}{3}$	$0.\overline{6}$	$66\frac{2}{3}\%$
$\frac{1}{5}$	0.2	20%
$\frac{2}{5}$	0.4	40%
$\frac{3}{5}$	0.6	60%
$\frac{4}{5}$	0.8	80%

EXAMPLE 7 What number is 25% of 60?

SOLUTION To solve a problem like this, we let x = the number in question (that is, the number we are looking for). Then, we translate the sentence directly into an equation by using an equal sign for the word "is" and multiplication for the word "of." Here is how it is done:

$$\text{What number is 25\% of 60?}$$
$$x \quad = 0.25 \cdot 60$$
$$x = 15$$

Notice that we must write 25% as a decimal to do the arithmetic in the problem. The number 15 is 25% of 60.

EXAMPLE 8 What percent of 24 is 6?

SOLUTION Translating this sentence into an equation, as we did in Example 7, we have

$$\text{What percent of 24 is 6?}$$
$$x \quad \cdot 24 = 6$$
$$\text{or} \quad 24x = 6$$

Next, we multiply each side by $\frac{1}{24}$. (This is the same as dividing each side by 24.)

$$\begin{aligned}\mathbf{\frac{1}{24}}(24x) &= \mathbf{\frac{1}{24}}(6)\\ x &= \frac{6}{24}\\ &= \frac{1}{4}\\ &= 0.25, \text{ or } 25\%\end{aligned}$$

The number 6 is 25% of 24.

EXAMPLE 9 45 is 75% of what number?

SOLUTION Again, we translate the sentence directly:

$$\begin{array}{ccccc} 45 & \text{is} & 75\% & \text{of} & \underbrace{\text{what number?}} \\ \downarrow & \downarrow & \downarrow & \downarrow & \downarrow \\ 45 & = & 0.75 & \cdot & x \end{array}$$

Next, we multiply each side by $\frac{1}{0.75}$ (which is the same as dividing each side by 0.75).

$$\begin{aligned}\mathbf{\frac{1}{0.75}}(45) &= \mathbf{\frac{1}{0.75}}(0.75x)\\ \frac{45}{0.75} &= x\\ 60 &= x\end{aligned}$$

The number 45 is 75% of 60.

We can solve application problems involving percent by translating each problem into one of the three basic percent problems shown in Examples 7, 8, and 9.

EXAMPLE 10 The American Dietetic Association (ADA) recommends eating foods in which the calories from fat are less than 30% of the total calories. The nutrition labels from two kinds of granola bars are shown in Figure 1. For each bar, what percent of the total calories come from fat?

SOLUTION The information needed to solve this problem is located toward the top of each label. Each serving of Bar I contains 210 calories, of which 70 calories come from fat. To find the percent of total calories that come from fat, we must answer this question:

70 is what percent of 210?

Bar I

Nutrition Facts	
Serving Size 2 bars (47g)	
Servings Per Container 6	
Amount Per Serving	
Calories	210
Calories from Fat	**70**
	% Daily Value*
Total Fat 8g	**12%**
Saturated Fat 1g	**5%**
Cholesterol 0mg	**0%**
Sodium 150mg	**6%**
Total Carbohydrate 32g	**11%**
Dietary Fiber 2g	**10%**
Sugars 12g	
Protein 4g	

* Percent Daily Values are based on a 2,000 calorie diet. Your daily values may be higher or lower depending on your calorie needs.

Bar II

Nutrition Facts	
Serving Size 1 bar (21g)	
Servings Per Container 8	
Amount Per Serving	
Calories	80
Calories from Fat	**15**
	% Daily Value*
Total Fat 1.5g	**2%**
Saturated Fat 0g	**0%**
Cholesterol 0mg	**0%**
Sodium 60mg	**3%**
Total Carbohydrate 16g	**5%**
Dietary Fiber 1g	**4%**
Sugars 5g	
Protein 2g	

* Percent Daily Values are based on a 2,000 calorie diet. Your daily values may be higher or lower depending on your calorie needs.

FIGURE 1

For Bar II, one serving contains 80 calories, of which 15 calories come from fat. To find the percent of total calories that come from fat, we must answer this question:

15 is what percent of 80?

Translating each equation into symbols, we have

70 is what percent of 210	15 is what percent of 80
$70 = x \cdot 210$	$15 = x \cdot 80$
$x = \dfrac{70}{210}$	$x = \dfrac{15}{80}$
$x = 0.33$ to the nearest hundredth	$x = 0.19$ to the nearest hundredth
$x = 33\%$	$x = 19\%$

Comparing the two bars, 33% of the calories in Bar I are fat calories, whereas 19% of the calories in Bar II are fat calories. According to the ADA, Bar II is the healthier choice.

Problem Set 2.5

Moving Toward Success

"You cannot create experience. You must undergo it."

—Albert Camus, 1913–1960, French author and philosopher

1. What will you do if you notice you are repeatedly making mistakes on certain types of problems?

2. Why is confidence important in mathematics?

A For the next two problems, use the formula $P = 2l + 2w$ to find the length l of a rectangular lot if [Examples 1–2]

1. The width w is 50 feet and the perimeter P is 300 feet

2. The width w is 75 feet and the perimeter P is 300 feet

3. For the equation $2x + 3y = 6$
   a. Find y when x is 0.
   b. Find x when y is 1.
   c. Find y when x is 3.

4. For the equation $2x - 5y = 20$
   a. Find y when x is 0.
   b. Find x when y is 0.
   c. Find x when y is 2.

5. For the equation $y = -\dfrac{1}{3}x + 2$
   a. Find y when x is 0.
   b. Find x when y is 3.
   c. Find y when x is 3.

6. For the equation $y = -\dfrac{2}{3}x + 1$
   a. Find y when x is 0.
   b. Find x when y is -1.
   c. Find y when x is -3.

A Use the formula $3x + 3y = 6$ to find y if

7. x is 3
8. x is -2
9. x is 0
10. x is -3

A Use the formula $2x - 2y = 20$ to find x if

11. y is 2
12. y is -4
13. y is 0
14. y is -6

A Use the equation $y = (x + 1)^2 - 3$ to find the value of y when

15. $x = -2$
16. $x = -1$
17. $x = 1$
18. $x = 2$

19. Use the formula $y = \dfrac{20}{x}$ to find y when
    a. $x = 10$
    b. $x = 5$

20. Use the formula $y = 2x^2$ to find y when
   a. $x = 5$
   b. $x = -6$

21. Use the formula $y = Kx$ to find K when
   a. $y = 15$ and $x = 3$
   b. $y = 72$ and $x = 4$

22. Use the formula $y = Kx^2$ to find K when
   a. $y = 32$ and $x = 4$
   b. $y = 45$ and $x = 3$

23. If $y = \dfrac{K}{x}$, find K if
   a. x is 5 and y is 4.
   b. x is 5 and y is 15.

24. If $I = \dfrac{K}{d^2}$, find K if
   a. $I = 200$ and $d = 10$.
   b. $I = 200$ and $d = 5$.

B Solve each of the following for the indicated variable.
[Examples 3–5]

25. $A = lw$ for l **26.** $d = rt$ for r

27. $V = lwh$ for h **28.** $PV = nRT$ for P

29. $P = a + b + c$ for a

30. $P = a + b + c$ for b

31. $x - 3y = -1$ for x

32. $x + 3y = 2$ for x

33. $-3x + y = 6$ for y

34. $2x + y = -17$ for y

35. $2x + 3y = 6$ for y

36. $3x + 5y = 20$ for y

37. $y - 3 = -2(x + 4)$ for y

38. $y + 5 = 2(x + 2)$ for y

39. $y - 3 = -\dfrac{2}{3}(x + 3)$ for y

40. $y - 1 = -\dfrac{1}{2}(x + 4)$ for y

41. $P = 2l + 2w$ for w

42. $P = 2l + 2w$ for l

43. $h = vt + 16t^2$ for v

44. $h = vt - 16t^2$ for v

45. $A = \pi r^2 + 2\pi rh$ for h

46. $A = 2\pi r^2 + 2\pi rh$ for h

47. Solve for y.
   a. $y - 3 = -2(x + 4)$

   b. $y - 5 = 4(x - 3)$

48. Solve for y.
   a. $y + 1 = -\dfrac{2}{3}(x - 3)$
   b. $y - 4 = -\dfrac{2}{4}(x + 4)$

49. Solve for y.
   a. $y - 1 = \dfrac{3}{4}(x - 1)$
   b. $y + 2 = \dfrac{3}{4}(x - 4)$

50. Solve for y.
   a. $y + 3 = \dfrac{3}{2}(x - 2)$
   b. $y + 4 = \dfrac{4}{3}(x - 3)$

51. Solve for y.
   a. $\dfrac{y - 1}{x} = \dfrac{3}{5}$ **b.** $\dfrac{y - 2}{x} = \dfrac{1}{2}$

52. Solve for y.
   a. $\dfrac{y + 1}{x} = -\dfrac{3}{5}$ **b.** $\dfrac{y + 2}{x} = -\dfrac{1}{2}$

B Solve each formula for y.

53. $\dfrac{x}{7} - \dfrac{y}{3} = 1$ **54.** $\dfrac{x}{4} - \dfrac{y}{9} = 1$

55. $-\dfrac{1}{4}x + \dfrac{1}{8}y = 1$

56. $-\dfrac{1}{9}x + \dfrac{1}{3}y = 1$

The next two problems are intended to give you practice reading, and paying attention to, the instructions that accompany the problems you are working. As we have mentioned previously, working these problems is an excellent way to get ready for a test or a quiz.

57. Work each problem according to the instructions.
   a. Solve: $4x + 5 = 20$
   b. Find the value of $4x + 5$ when x is 3.
   c. Solve for y: $4x + 5y = 20$
   d. Solve for x: $4x + 5y = 20$

58. Work each problem according to the instructions.
   a. Solve: $-2x + 1 = 4$
   b. Find the value of $-2x + 1$ when x is 8.
   c. Solve for y: $-2x + y = 20$
   d. Solve for x: $-2x + y = 20$

C Find the complement and the supplement of each angle. [Example 6]

59. 30°   **60.** 60°

61. 45°   **62.** 15°

D Translate each of the following into an equation, and then solve that equation. [Examples 7–9]

63. What number is 25% of 40?

64. What number is 75% of 40?

65. What number is 12% of 2,000?

66. What number is 9% of 3,000?

67. What percent of 28 is 7?

68. What percent of 28 is 21?

69. What percent of 40 is 14?

70. What percent of 20 is 14?

71. 32 is 50% of what number?

72. 16 is 50% of what number?

73. 240 is 12% of what number?

74. 360 is 12% of what number?

Applying the Concepts

More About Temperatures As we mentioned in Chapter 1, in the U.S. system, temperature is measured on the Fahrenheit scale. In the metric system, temperature is measured on the Celsius scale. On the Celsius scale, water boils at 100 degrees and freezes at 0 degrees. To denote a temperature of 100 degrees on the Celsius scale, we write

100°C, which is read "100 degrees Celsius"

Table 1 is intended to give you an intuitive idea of the relationship between the two temperature scales. Table 2 gives the formulas, in both symbols and words, that are used to convert between the two scales.

TABLE 1		
	Temperature	
Situation	**Fahrenheit**	**Celsius**
Water freezes	32°F	0°C
Room temperature	68°F	20°C
Normal body temperature	98.6°F	37°C
Water boils	212°F	100°C
Bake cookies	365°F	185°C

TABLE 2		
To Convert from	**Formula in Symbols**	**Formula in Words**
Fahrenheit to Celsius	$C = \frac{5}{9}(F - 32)$	Subtract 32, multiply by 5, then divide by 9.
Celsius to Fahrenheit	$F = \frac{9}{5}C + 32$	Multiply by $\frac{9}{5}$, then add 32.

75. Let $F = 212$ in the formula $C = \frac{5}{9}(F - 32)$, and solve for C. Does the value of C agree with the information in Table 1?

76. Let $C = 100$ in the formula $F = \frac{9}{5}C + 32$, and solve for F. Does the value of F agree with the information in Table 1?

77. Let $F = 68$ in the formula $C = \frac{5}{9}(F - 32)$, and solve for C. Does the value of C agree with the information in Table 1?

78. Let $C = 37$ in the formula $F = \frac{9}{5}C + 32$, and solve for F. Does the value of F agree with the information in Table 1?

79. Solve the formula $F = \frac{9}{5}C + 32$ for C.

80. Solve the formula $C = \frac{5}{9}(F - 32)$ for F.

81. Population The map shows the most populated cities in the United States. If the population of California in 2007 was about 36.6 million, what percentage of the population is found in Los Angeles and San Diego? Round to the nearest tenth of a percent.

Where Is Everyone?

Los Angeles, CA — 3.8
San Diego, CA — 1.3
Phoeniz, AZ — 1.6
San Antonio, TX — 1.3
Houston, TX — 2.2
Chicago, IL — 2.8
Philadelphia, PA — 1.4
New York City, NY — 8.3

* Population in Millions

Source: U.S. Census Bureau

82. Seasonal Employees The bar chart shows which traits American employers look for when hiring seasonal workers. Use the chart to find how many more employers look for positive attitude than experience.

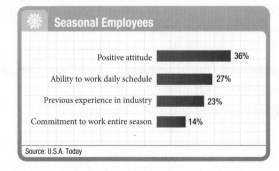

Seasonal Employees

Positive attitude — 36%
Ability to work daily schedule — 27%
Previous experience in industry — 23%
Commitment to work entire season — 14%

Source: U.S.A. Today

Nutrition Labels The nutrition label in Figure 2 is from a quart of vanilla ice cream. The label in Figure 3 is from a pint of vanilla frozen yogurt. Use the information on these labels for problems 83–86. Round your answers to the nearest tenth of a percent. [Example 10]

Nutrition Facts
Serving Size 1/2 cup (65g)
Servings Per Container 8

Amount Per Serving

Calories 150	Calories from Fat 90

	% Daily Value*
Total Fat 10g	**16%**
Saturated Fat 6g	**32%**
Cholesterol 35mg	**12%**
Sodium 30mg	**1%**
Total Carbohydrate 14g	**5%**
Dietary Fiber 0g	**0%**
Sugars 11g	
Protein 2g	

Vitamin A 6%	•	Vitamin C 0%
Calcium 6%	•	Iron 0%

* Percent Daily Values are based on a 2,000 calorie diet.

FIGURE 2   Vanilla ice cream

Nutrition Facts
Serving Size 1/2 cup (98g)
Servings Per Container 4

Amount Per Serving

Calories 160	Calories from Fat 25

	% Daily Value*
Total Fat 2.5g	**4%**
Saturated Fat 1.5g	**7%**
Cholesterol 45mg	**15%**
Sodium 55mg	**2%**
Total Carbohydrate 26g	**9%**
Dietary Fiber 0g	**0%**
Sugars 19g	
Protein 8g	

Vitamin A 0%	•	Vitamin C 0%
Calcium 25%	•	Iron 0%

* Percent Daily Values are based on a 2,000 calorie diet.

FIGURE 3   Vanilla frozen yogurt

83. What percent of the calories in one serving of the vanilla ice cream are fat calories?

84. What percent of the calories in one serving of the frozen yogurt are fat calories?

85. One serving of frozen yogurt is 98 grams, of which 26 grams are carbohydrates. What percent of one serving are carbohydrates?

86. One serving of vanilla ice cream is 65 grams. What percent of one serving is sugar?

Circumference The circumference of a circle is given by the formula $C = 2\pi r$. Find r if

87. The circumference C is 44 meters and π is $\frac{22}{7}$

88. The circumference C is 176 meters and π is $\frac{22}{7}$

89. The circumference is 9.42 inches and π is 3.14

90. The circumference is 12.56 inches and π is 3.14

Volume The volume of a cylinder is given by the formula $V = \pi r^2 h$. Find the height h to the nearest hundreth if

91. The volume V is 42 cubic feet, the radius is $\frac{7}{22}$ feet, and π is $\frac{22}{7}$

92. The volume V is 84 cubic inches, the radius is $\frac{7}{11}$ inches, and π is $\frac{22}{7}$

93. The volume is 6.28 cubic centimeters, the radius is 3 centimeters, and π is 3.14

94. The volume is 12.56 cubic centimeters, the radius is 2 centimeters, and π is 3.14

Getting Ready for the Next Section

To understand all of the explanations and examples in the next section you must be able to work the problems below.

Write an equivalent expression in English. Include the words *sum* and *difference* when possible.

95. $4 + 1 = 5$

96. $7 + 3 = 10$

97. $6 - 2 = 4$

98. $8 - 1 = 7$

99. $x - 5 = -12$

100. $2x + 3 = 7$

101. $x + 3 = 4(x - 3)$

102. $2(2x - 5) = 2x - 34$

For each of the following expressions, write an equivalent equation.

103. Twice the sum of 6 and 3 is 18.

104. Four added to the product of 3 and −1 is 1.

105. The sum of twice 5 and 3 is 13.

106. Twice the difference of 8 and 2 is 12.

107. The sum of a number and five is thirteen.

108. The difference of ten and a number is negative eight.

109. Five times the sum of a number and seven is thirty.

110. Five times the difference of twice a number and six is negative twenty.

Maintaining Your Skills

The problems that follow review some of the more important skills you have learned in previous sections and chapters. You can consider the time you spend working these problems as time spent studying for exams.

111.
  a. $27 - (-68)$
  b. $27 + (-68)$
  c. $-27 - 68$
  d. $-27 + 68$

112.
  a. $55 - (-29)$
  b. $55 + (-29)$
  c. $-55 - 29$
  d. $-55 + 29$

113.
  a. $-32 - (-41)$
  b. $-32 + (-41)$
  c. $-32 + 41$
  d. $-32 - 41$

114.
  a. $-56 - (-35)$
  b. $-56 + (-35)$
  c. $-56 + 35$
  d. $-56 - 35$

2.6 Applications

OBJECTIVES

A Apply the Blueprint for Problem Solving to a variety of application problems.

TICKET TO SUCCESS

Keep these questions in mind as you read through the section. Then respond in your own words and in complete sentences.

1. Why is the first step in the Blueprint for Problem Solving done mentally?

2. What is the last thing you do when solving an application problem?

3. Why is it important to still use the blueprint and show your work even if you can solve a problem without using algebra?

4. Write an application problem whose solution depends on solving the equation $4x + 2 = 10$.

The title of the section is "Applications," but many of the problems here don't seem to have much to do with real life. Example 3 is what we refer to as an "age problem." Imagine a conversation in which you ask a mother how old her children are and she replies, "Bill is 6 years older than Tom. Three years ago the sum of their ages was 21. You figure it out." Although many of the application problems in this section are contrived, they are also good for practicing the strategy we will use to solve all application problems.

A Blueprint for Problem Solving

To begin this section, we list the steps used in solving application problems. We call this strategy the *Blueprint for Problem Solving.* It is an outline that will overlay the solution process we use on all application problems.

> **Strategy** Blueprint for Problem Solving
>
> **Step 1:** *Read* the problem, and then mentally *list* the items that are known and the items that are unknown.
>
> **Step 2:** *Assign a variable* to one of the unknown items. (In most cases this will amount to letting x = the item that is asked for in the problem.) Then *translate* the other *information* in the problem to expressions involving the variable.
>
> **Step 3:** *Reread* the problem, and then *write an equation,* using the items and variables listed in steps 1 and 2, that describes the situation.
>
> **Step 4:** *Solve the equation* found in step 3.
>
> **Step 5:** *Write* your *answer* using a complete sentence.
>
> **Step 6:** *Reread* the problem, and *check* your solution with the original words in the problem.

There are a number of substeps within each of the steps in our blueprint. For instance, with steps 1 and 2 it is always a good idea to draw a diagram or picture if it helps visualize the relationship between the items in the problem. In other cases, a table helps organize the information. As you gain more experience using the blueprint to solve application problems, you will find additional techniques that expand the blueprint.

To help with problems of the type shown next in Example 1, here are some common English words and phrases and their mathematical translations.

English	Algebra
The sum of a and b	$a + b$
The difference of a and b	$a - b$
The product of a and b	$a \cdot b$
The quotient of a and b	$\dfrac{a}{b}$
of	$\cdot$ (multiply)
is	= (equals)
A number	x
4 more than x	$x + 4$
4 times x	$4x$
4 less than x	$x - 4$

Number Problems

EXAMPLE 1   The sum of twice a number and three is seven. Find the number.

SOLUTION   Using the Blueprint for Problem Solving as an outline, we solve the problem as follows.

Step 1:   **Read** the problem, and then mentally **list** the items that are known and the items that are unknown.

Known items: The numbers 3 and 7

Unknown items: The number in question

Step 2:   **Assign a variable** to one of the unknown items. Then **translate** the other **information** in the problem to expressions involving the variable.

Let x = the number asked for in the problem.
Then "The sum of twice a number and three" translates to $2x + 3$.

Step 3:   **Reread** the problem, and then **write an equation,** using the items and variables listed in steps 1 and 2, that describes the situation.
With all word problems, the word *is* translates to =.

$$\underbrace{\text{The sum of twice } x \text{ and 3}}\ \text{is}\ 7$$
$$2x + 3 \qquad\qquad = 7$$

Step 4:   **Solve the equation** found in step 3.

$$2x + 3 = 7$$
$$2x + 3 + (\mathbf{-3}) = 7 + (\mathbf{-3})$$
$$2x = 4$$
$$\frac{1}{2}(2x) = \frac{1}{2}(4)$$
$$x = 2$$

Step 5:   **Write** your **answer** using a complete sentence.

The number is 2.

Step 6:   **Reread** the problem, and **check** your solution with the original words in the problem.

The sum of twice 2 and 3 is 7. A true statement.   ∎

You may find some examples and problems in this section that you can solve without using algebra or our blueprint. It is very important that you still solve these problems using the methods we are showing here. The purpose behind these problems is to give you experience using the blueprint as a guide to solving problems written in words. Your answers are much less important than the work that you show to obtain your answer. You will be able to condense the steps in the blueprint later in the course. For now, though, you need to show your work in the same detail that we are showing in the examples in this section.

EXAMPLE 2   One number is three more than twice another; their sum is eighteen. Find the numbers.

SOLUTION

Step 1:   Read and list.

Known items: Two numbers that add to 18. One is 3 more than twice the other.

Unknown items: The numbers in question.

Step 2:   Assign a variable, and translate information.
Let x = the first number. The other is $2x + 3$.

Step 3:   Reread and write an equation.

$$\overbrace{\text{Their sum is } 18}$$

$$x + (2x + 3) = 18$$

Step 4:   Solve the equation.

$$x + (2x + 3) = 18$$
$$3x + 3 = 18$$
$$3x + 3 + (-3) = 18 + (-3)$$
$$3x = 15$$
$$x = 5$$

Step 5:   Write the answer.
The first number is 5. The other is $2 \cdot 5 + 3 = 13$.

Step 6:   Reread and check.
The sum of 5 and 13 is 18, and 13 is 3 more than twice 5.

Age Problems

Remember, as you read through the solutions to the examples in this section, step 1 is done mentally. Read the problem, and then mentally list the items that you know and the items that you don't know. The purpose of step 1 is to give you direction as you begin to work application problems. Finding the solution to an application problem is a process; it doesn't happen all at once. The first step is to read the problem with a purpose in mind. That purpose is to mentally note the items that are known and the items that are unknown.

EXAMPLE 3   Bill is 6 years older than Tom. Three years ago Bill's age was four times Tom's age. Find the age of each boy now.

SOLUTION   We apply the Blueprint for Problem Solving.

Step 1:   Read and list.

Known items: Bill is 6 years older than Tom. Three years ago Bill's age was four times Tom's age.

Unknown items: Bill's age and Tom's age

Step 2:   Assign a variable, and translate information.
Let x = Tom's age now. That makes Bill $x + 6$ years old now. A table like the

one shown here can help organize the information in an age problem. Notice how we placed the x in the box that corresponds to Tom's age now.

	Three Years Ago	Now
Bill		$x + 6$
Tom		x

If Tom is x years old now, 3 years ago he was $x - 3$ years old. If Bill is $x + 6$ years old now, 3 years ago he was $x + 6 - 3 = x + 3$ years old. We use this information to fill in the remaining entries in the table.

	Three Years Ago	Now
Bill	$x + 3$	$x + 6$
Tom	$x - 3$	x

Step 3: *Reread and write an equation.*

Reading the problem again, we see that 3 years ago Bill's age was four times Tom's age. Writing this as an equation, we have Bill's age 3 years ago = 4 · (Tom's age 3 years ago):

$$x + 3 = 4(x - 3)$$

Step 4: *Solve the equation.*

$$x + 3 = 4(x - 3)$$
$$x + 3 = 4x - 12$$
$$x + (\mathbf{-x}) + 3 = 4x + (\mathbf{-x}) - 12$$
$$3 = 3x - 12$$
$$3 + \mathbf{12} = 3x - 12 + \mathbf{12}$$
$$15 = 3x$$
$$x = 5$$

Step 5: *Write the answer.*

Tom is 5 years old. Bill is 11 years old.

Step 6: *Reread and check.*

If Tom is 5 and Bill is 11, then Bill is 6 years older than Tom. Three years ago Tom was 2 and Bill was 8. At that time, Bill's age was four times Tom's age. As you can see, the answers check with the original problem.

Geometry Problems

To understand Example 4 completely, you need to recall from Chapter 1 that the perimeter of a rectangle is the sum of the lengths of the sides. The formula for the perimeter is $P = 2l + 2w$.

EXAMPLE 4   The length of a rectangle is 5 inches more than twice the width. The perimeter is 34 inches. Find the length and width.

SOLUTION   When working problems that involve geometric figures, a sketch of the figure helps organize and visualize the problem.

Step 1: *Read and list.*

Known items: The figure is a rectangle. The length is 5 inches more than twice the width. The perimeter is 34 inches.

Unknown items: The length and the width

Step 2: *Assign a variable, and translate information.*

Because the length is given in terms of the width (the length is 5 more than twice the width), we let x = the width of the rectangle. The length is 5 more than twice the width, so it must be $2x + 5$. The diagram below is a visual description of the relationships we have listed so far.

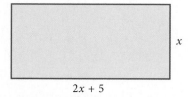

$2x + 5$

Step 3: *Reread and write an equation.*

The equation that describes the situation is

Twice the length + twice the width is the perimeter

$$2(2x + 5) \quad + \quad 2x \quad = \quad 34$$

Step 4: *Solve the equation.*

$2(2x + 5) + 2x = 34$	Original equation
$4x + 10 + 2x = 34$	Distributive property
$6x + 10 = 34$	Add $4x$ and $2x$
$6x = 24$	Add -10 to each side
$x = 4$	Divide each side by 6

Step 5: *Write the answer.*

The width x is 4 inches. The length is $2x + 5 = 2(4) + 5 = 13$ inches.

Step 6: *Reread and check.*

If the length is 13 and the width is 4, then the perimeter must be $2(13) + 2(4) = 26 + 8 = 34$, which checks with the original problem.

Coin Problems

EXAMPLE 5 Jennifer has $2.45 in dimes and nickels. If she has 8 more dimes than nickels, how many of each coin does she have?

SOLUTION

Step 1: *Read and list.*

Known items: The type of coins, the total value of the coins, and that there are 8 more dimes than nickels

Unknown items: The number of nickels and the number of dimes

Step 2: *Assign a variable, and translate information.*

If we let x = the number of nickels, then $x + 8$ = the number of dimes. Because the value of each nickel is 5 cents, the amount of money in nickels is 5x. Similarly, because each dime is worth 10 cents, the amount of money in dimes is $10(x + 8)$. Here is a table that summarizes the information we have so far:

	Nickels	Dimes
Number	x	x + 8
Value (in cents)	5x	10(x + 8)

Step 3: ***Reread, and write an equation.***
Because the total value of all the coins is 245 cents, the equation that describes this situation is

$$\underset{\text{in nickels}}{\text{Amount of money}} + \underset{\text{in dimes}}{\text{Amount of money}} = \underset{\text{of money}}{\text{Total amount}}$$
$$5x + 10(x+8) = 245$$

Step 4: ***Solve the equation.***
To solve the equation, we apply the distributive property first.

$5x + 10x + 80 = 245$	Distributive property
$15x + 80 = 245$	Add $5x$ and $10x$
$15x = 165$	Add -80 to each side
$x = 11$	Divide each side by 15

Step 5: ***Write the answer.***
The number of nickels is $x = 11$.
The number of dimes is $x + 8 = 11 + 8 = 19$.

Step 6: ***Reread and check.***
To check our results

11 nickels are worth $5(11) = $ 55 cents
19 dimes are worth $10(19) = $ 190 cents
The total value is 245 cents = $2.45

When you begin working the problems in the problem set that follows, there are a couple of things to remember. The first is that you may have to read the problems a number of times before you begin to see how to solve them. The second thing to remember is that word problems are not always solved correctly the first time you try them. Sometimes it takes a couple of attempts and some wrong answers before you can set up and solve these problems correctly. Don't give up.

Problem Set 2.6

Moving Toward Success

"What is defeat? Nothing but education; nothing but the first step to something better."

—Wendell Phillips, 1811–1884, American abolitionist and orator

1. Why should you keep track of the additional techniques you find that help you expand the Blueprint for Problem Solving?

2. Should you always expect to solve a word problem on the first try? Explain.

A Solve the following word problems. Follow the steps given in the Blueprint for Problem Solving.

Number Problems [Examples 1–2]

1. The sum of a number and five is thirteen. Find the number.

2. The difference of ten and a number is negative eight. Find the number.

3. The sum of twice a number and four is fourteen. Find the number.

4. The difference of four times a number and eight is sixteen. Find the number.

5. Five times the sum of a number and seven is thirty. Find the number.

6. Five times the difference of twice a number and six is negative twenty. Find the number.

7. One number is two more than another. Their sum is eight. Find both numbers.

8. One number is three less than another. Their sum is fifteen. Find the numbers.

9. One number is four less than three times another. If their sum is increased by five, the result is twenty-five. Find the numbers.

10. One number is five more than twice another. If their sum is decreased by ten, the result is twenty-two. Find the numbers.

Age Problems [Example 3]

11. Shelly is 3 years older than Michele. Four years ago the sum of their ages was 67. Find the age of each person now.

	Four Years Ago	Now
Shelly	$x - 1$	$x + 3$
Michele	$x - 4$	x

12. Cary is 9 years older than Dan. In 7 years the sum of their ages will be 93. Find the age of each man now.

	Now	In Seven Years
Cary	$x + 9$	
Dan	x	$x + 7$

13. Cody is twice as old as Evan. Three years ago the sum of their ages was 27. Find the age of each boy now.

	Three Years Ago	Now
Cody		
Evan	$x - 3$	x

14. Justin is 2 years older than Ethan. In 9 years the sum of their ages will be 30. Find the age of each boy now.

	Now	In Nine Years
Justin		
Ethan	x	

15. Fred is 4 years older than Barney. Five years ago the sum of their ages was 48. How old are they now?

	Five Years Ago	Now
Fred		
Barney		x

16. Tim is 5 years older than JoAnn. Six years from now the sum of their ages will be 79. How old are they now?

	Now	Six Years From Now
Tim		
JoAnn	x	

17. Jack is twice as old as Lacy. In 3 years the sum of their ages will be 54. How old are they now?

18. John is 4 times as old as Martha. Five years ago the sum of their ages was 50. How old are they now?

19. Pat is 20 years older than his son Patrick. In 2 years Pat will be twice as old as Patrick. How old are they now?

20. Diane is 23 years older than her daughter Amy. In 6 years Diane will be twice as old as Amy. How old are they now?

Geometry Problems [Example 4]

21. The perimeter of a square is 36 inches. Find the length of one side.

22. The perimeter of a square is 44 centimeters. Find the length of one side.

23. The perimeter of a square is 60 feet. Find the length of one side.

24. The perimeter of a square is 84 meters. Find the length of one side.

25. One side of a triangle is three times the shortest side. The third side is 7 feet more than the shortest side. The perimeter is 62 feet. Find all three sides.

26. One side of a triangle is half the longest side. The third side is 10 meters less than the longest side. The perimeter is 45 meters. Find all three sides.

27. One side of a triangle is half the longest side. The third side is 12 feet less than the longest side. The perimeter is 53 feet. Find all three sides.

28. One side of a triangle is 6 meters more than twice the shortest side. The third side is 9 meters more than the shortest side. The perimeter is 75 meters. Find all three sides.

29. The length of a rectangle is 5 inches more than the width. The perimeter is 34 inches. Find the length and width.

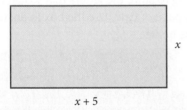

$x + 5$

30. The width of a rectangle is 3 feet less than the length. The perimeter is 10 feet. Find the length and width.

31. The length of a rectangle is 7 inches more than twice the width. The perimeter is 68 inches. Find the length and width.

32. The length of a rectangle is 4 inches more than three times the width. The perimeter is 72 inches. Find the length and width.

33. The length of a rectangle is 6 feet more than three times the width. The perimeter is 36 feet. Find the length and width.

34. The length of a rectangle is 3 feet less than twice the width. The perimeter is 54 feet. Find the length and width.

Coin Problems [Example 5]

35. Marissa has $4.40 in quarters and dimes. If she has 5 more quarters than dimes, how many of each coin does she have?

	Dimes	Quarters
Number	x	$x + 5$
Value (cents)	$10(x)$	$25(x + 5)$

36. Kendra has $2.75 in dimes and nickels. If she has twice as many dimes as nickels, how many of each coin does she have?

	Nickels	Dimes
Number	x	$2x$
Value (cents)	$5(x)$	

37. Tanner has $4.35 in nickels and quarters. If he has 15 more nickels than quarters, how many of each coin does he have?

	Nickels	Quarters
Number	$x + 15$	x
Value (cents)		

38. Connor has $9.00 in dimes and quarters. If he has twice as many quarters as dimes, how many of each coin does he have?

	Dimes	Quarters
Number	x	$2x$
Value (cents)		

39. Sue has $2.10 in dimes and nickels. If she has 9 more dimes than nickels, how many of each coin does she have?

40. Mike has $1.55 in dimes and nickels. If he has 7 more nickels than dimes, how many of each coin does he have?

41. Katie has a collection of nickels, dimes, and quarters with a total value of $4.35. There are 3 more dimes than nickels and 5 more quarters than nickels. How many of each coin is in her collection?

	Nickels	Dimes	Quarters
Number	x		
Value			

42. Mary Jo has $3.90 worth of nickels, dimes, and quarters. The number of nickels is 3 more than the number of dimes. The number of quarters is 7 more than the number of dimes. How many of each coin is in her collection?

	Nickels	Dimes	Quarters
Number			
Value			

43. Cory has a collection of nickels, dimes, and quarters with a total value of $2.55. There are 6 more dimes than nickels and twice as many quarters as nickels. How many of each coin is in her collection?

	Nickels	Dimes	Quarters
Number	x		
Value			

44. Kelly has a collection of nickels, dimes, and quarters with a total value of $7.40. There are four more nickels than dimes and twice as many quarters as nickels. How many of each coin is in her collection?

	Nickels	Dimes	Quarters
Number			
Value			

Getting Ready for the Next Section

To understand all of the explanations and examples in the next section you must be able to work the problems below.

Simplify the following expressions.

45. $x + 2x + 2x$

46. $x + 2x + 3x$

47. $x + 0.075x$

48. $x + 0.065x$

49. $0.09(x + 2,000)$

50. $0.06(x + 1,500)$

Solve each of the following equations.

51. $0.05x + 0.06(x - 1,500) = 570$

52. $0.08x + 0.09(x + 2,000) = 690$

53. $x + 2x + 3x = 180$

54. $2x + 3x + 5x = 180$

Maintaining Your Skills

Write an equivalent statement in English.

55. $4 < 10$

56. $4 \le 10$

57. $9 \ge -5$

58. $x - 2 > 4$

Place the symbol < or the symbol > between the quantities in each expression.

59. 12   20

60. -12 20

61. -8 -6

62. -10 -20

Simplify.

63. $|8 - 3| - |5 - 2|$

64. $|9 - 2| - |10 - 8|$

65. $15 - |9 - 3(7 - 5)|$

66. $10 - |7 - 2(5 - 3)|$

Extending the Concepts

67. The Alphabet and Area Have you ever wondered which of the letters in the alphabet is the most popular? No? The box shown here is called a *typecase,* or *printer's tray.* It was used in early typesetting to store the letters that would be used to lay out a page of type. Use the box to answer the questions below.

a. What letter is used most often in printed material?

b. Which letter is printed more often, the letter *i* or the letter *f*?

c. Regardless of what is shown in the figure, what letter is used most often in problems?

d. What is the relationship between area and how often a letter in the alphabet is printed?

2.7 More Applications

OBJECTIVES

A Apply the Blueprint for Problem Solving to a variety of application problems.

TICKET TO SUCCESS

Keep these questions in mind as you read through the section. Then respond in your own words and in complete sentences.

1. Write an equation for "The sum of two consecutive integers is 13."
2. Write an application problem whose solution depends on solving the equation $x + 0.075x = 500$.
3. How do you label a triangle?
4. What rule is always true about the three angles in a triangle?

Image copyright © Rafael Ramirez Lee. Used under license from Shutterstock.com

In this section, we will continue to use our knowledge of the Blueprint for Problem Solving to solve more application problems.

A More Applications

Our next example involves consecutive integers; the definition follows.

Consecutive Integers

> **Definition**
> **Consecutive integers** are integers that are next to each other on the number line, like 5 and 6, or 13 and 14, or −4 and −3.

In the dictionary, consecutive is defined as following one another in uninterrupted order. If we ask for consecutive *odd* integers, then we mean odd integers that follow one another on the number line. For example, 3 and 5, 11 and 13, and −9 and −7 are consecutive odd integers. As you can see, to get from one odd integer to the next consecutive odd integer we add 2.

If we are asked to find two consecutive integers and we let x equal the first integer, the next one must be $x + 1$, because consecutive integers always differ by 1. Likewise, if we are asked to find two consecutive odd or even integers, and we let x equal the first integer, then the next one will be $x + 2$, because consecutive even or odd integers always differ by 2. Here is a table that summarizes this information.

In Words	Using Algebra	Example
Two consecutive integers	$x, x + 1$	The sum of two consecutive integers is 15. $x + (x + 1) = 15$ or $7 + 8 = 15$
Three consecutive integers	$x, x + 1, x + 2$	The sum of three consecutive integers is 24. $x + (x + 1) + (x + 2) = 24$ or $7 + 8 + 9 = 24$
Two consecutive odd integers	$x, x + 2$	The sum of two consecutive odd integers is 16. $x + (x + 2) = 16$ or $7 + 9 = 16$
Two consecutive even integers	$x, x + 2$	The sum of two consecutive even integers is 18. $x + (x + 2) = 18$ or $8 + 10 = 18$

EXAMPLE 1 The sum of two consecutive odd integers is 28. Find the two integers.

SOLUTION

Step 1: **Read and list.**

Known items: Two consecutive odd integers. Their sum is equal to 28.

Unknown items: The numbers in question

Step 2: **Assign a variable, and translate information.**
If we let $x =$ the first of the two consecutive odd integers, then $x + 2$ is the next consecutive one.

Step 3: **Reread and write an equation.**
Their sum is 28.

$$x + (x + 2) = 28$$

Step 4: **Solve the equation.**

$2x + 2 = 28$	Simplify the left side
$2x = 26$	Add -2 to each side
$x = 13$	Multiply each side by $\frac{1}{2}$

Step 5: **Write the answer.**
The first of the two integers is 13. The second of the two integers will be two more than the first, which is 15.

Step 6: **Reread and check.**
Suppose the first integer is 13. The next consecutive odd integer is 15. The sum of 15 and 13 is 28. ∎

> **NOTE**
> Now that you have worked through a number of application problems using our blueprint, you probably have noticed that step 3, in which we write an equation that describes the situation, is the key step. Anyone with experience solving application problems will tell you that there will be times when your first attempt at step 3 results in the wrong equation. Remember, mistakes are part of the process of learning to do things correctly.

Interest Problems

EXAMPLE 2 Suppose you invest a certain amount of money in an account that earns 8% in annual interest. At the same time, you invest $2,000 more than that in an account that pays 9% in annual interest. If the total interest from both accounts at the end of the year is $690, how much is invested in each account?

SOLUTION

Step 1: **Read and list.**

Known items: The interest rates, the total interest earned, and how much more is invested at 9%

Unknown items: The amounts invested in each account

Step 2: **Assign a variable, and translate information.**

Let x = the amount of money invested at 8%. From this, $x + 2,000$ = the amount of money invested at 9%. The interest earned on x dollars invested at 8% is $0.08x$. The interest earned on $x + 2,000$ dollars invested at 9% is $0.09(x + 2,000)$.

Here is a table that summarizes this information:

	Dollars Invested at 8%	Dollars Invested at 9%
Number of	x	$x + 2,000$
Interest on	$0.08x$	$0.09(x + 2,000)$

Step 3: **Reread and write an equation.**

Because the total amount of interest earned from both accounts is $690, the equation that describes the situation is

Interest earned at 8%	+	Interest earned at 9%	=	Total interest earned
$0.08x$	+	$0.09(x + 2,000)$	=	690

Step 4: **Solve the equation.**

$$0.08x + 0.09(x + 2,000) = 690$$
$$0.08x + 0.09x + 180 = 690 \quad \text{Distributive property}$$
$$0.17x + 180 = 690 \quad \text{Add } 0.08x \text{ and } 0.09x$$
$$0.17x = 510 \quad \text{Add } -180 \text{ to each side}$$
$$x = 3,000 \quad \text{Divide each side by } 0.17$$

Step 5: **Write the answer.**

The amount of money invested at 8% is $3,000, whereas the amount of money invested at 9% is $x + 2,000 = 3,000 + 2,000 = \$5,000$.

Step 6: **Reread and check.**

$$\text{The interest at 8\% is 8\% of } 3,000 = 0.08(3,000) = \$240$$
$$\text{The interest at 9\% is 9\% of } 5,000 = 0.09(5,000) = \$450$$
$$\text{The total interest is } \$690$$

Triangle Problems

FACTS FROM GEOMETRY: Labeling Triangles and the Sum of the Angles in a Triangle
One way to label the important parts of a triangle is to label the vertices with
capital letters and the sides with small letters, as shown in Figure 1.

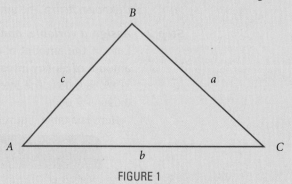

FIGURE 1

In Figure 1, notice that side a is opposite vertex A, side b is opposite vertex B,
and side c is opposite vertex C. Also, because each vertex is the vertex of one of
the angles of the triangle, we refer to the three interior angles as A, B, and C.

In any triangle, the sum of the interior angles is 180°. For the triangle shown
in Figure 1, the relationship is written

$$A + B + C = 180°$$

EXAMPLE 3   The angles in a triangle are such that one angle is twice the
smallest angle, whereas the third angle is three times as large as the smallest angle.
Find the measure of all three angles.

SOLUTION

Step 1:   Read and list.

Known items: The sum of all three angles is 180°; one angle is twice the
smallest angle; the largest angle is three times the smallest angle.

Unknown items: The measure of each angle

Step 2:   Assign a variable, and translate information.
Let x be the smallest angle, then $2x$ will be the measure of another angle
and $3x$ will be the measure of the largest angle.

Step 3:   Reread and write an equation.
When working with geometric objects, drawing a generic diagram
sometimes will help us visualize what it is that we are asked to find. In
Figure 2, we draw a triangle with angles A, B, and C.

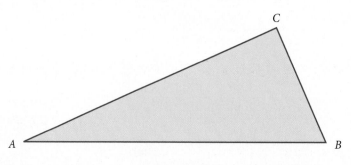

FIGURE 2

We can let the value of $A = x$, the value of $B = 2x$, and the value of $C = 3x$. We know that the sum of angles A, B, and C will be 180°, so our equation becomes

$$x + 2x + 3x = 180°$$

Step 4:   Solve the equation.

$$x + 2x + 3x = 180°$$
$$6x = 180°$$
$$x = 30°$$

Step 5:   Write the answer.

The smallest angle A measures 30°
Angle B measures $2x$, or $2(30°) = 60°$
Angle C measures $3x$, or $3(30°) = 90°$

Step 6:   Reread and check.

The angles must add to 180°:

$$A + B + C = 180°$$
$$30° + 60° + 90° \stackrel{?}{=} 180°$$
$$180° = 180° \quad \text{Our answers check}$$

Problem Set 2.7

Moving Toward Success

"In every difficult situation is potential value. Believe this, then begin looking for it."

—Norman Vincent Peale, 1898–1993, American preacher and author of "The Power of Positive Thinking"

1. Why are word problems important in mathematics?

2. Why should you be actively involved in a word problem?

A   Consecutive Integer Problems

[Example 1]

1. The sum of two consecutive integers is 11. Find the numbers.

2. The sum of two consecutive integers is 15. Find the numbers.

3. The sum of two consecutive integers is −9. Find the numbers.

4. The sum of two consecutive integers is −21. Find the numbers.

5. The sum of two consecutive odd integers is 28. Find the numbers.

6. The sum of two consecutive odd integers is 44. Find the numbers.

7. The sum of two consecutive even integers is 106. Find the numbers.

8. The sum of two consecutive even integers is 66. Find the numbers.

9. The sum of two consecutive even integers is −30. Find the numbers.

10. The sum of two consecutive odd integers is −76. Find the numbers.

11. The sum of three consecutive odd integers is 57. Find the numbers.

12. The sum of three consecutive odd integers is −51. Find the numbers.

13. The sum of three consecutive even integers is 132. Find the numbers.

14. The sum of three consecutive even integers is −108. Find the numbers.

A   Interest Problems [Example 2]

15. Suppose you invest money in two accounts. One of the accounts pays 8% annual interest, whereas the other pays 9% annual interest. If you have $2,000 more invested at 9% than you have invested at 8%, how much do you have invested in each account

if the total amount of interest you earn in a year is $860? (Begin by completing the following table.)

	Dollars Invested at 8%	Dollars Invested at 9%
Number of	x	
Interest on		

16. Suppose you invest a certain amount of money in an account that pays 11% interest annually, and $4,000 more than that in an account that pays 12% annually. How much money do you have in each account if the total interest for a year is $940?

	Dollars Invested at 11%	Dollars Invested at 12%
Number of	x	
Interest on		

17. Tyler has two savings accounts that his grandparents opened for him. The two accounts pay 10% and 12% in annual interest; there is $500 more in the account that pays 12% than there is in the other account. If the total interest for a year is $214, how much money does he have in each account?

18. Travis has a savings account that his parents opened for him. It pays 6% annual interest. His uncle also opened an account for him, but it pays 8% annual interest. If there is $800 more in the account that pays 6%, and the total interest from both accounts is $104, how much money is in each of the accounts?

19. A stockbroker has money in three accounts. The interest rates on the three accounts are 8%, 9%, and 10%. If she has twice as much money invested at 9% as she has invested at 8%, three times as much at 10% as she has at 8%, and the total interest for the year is $280, how much is invested at each rate? (*Hint:* Let x = the amount invested at 8%.)

20. An accountant has money in three accounts that pay 9%, 10%, and 11% in annual interest. He has twice as much invested at 9% as he does at 10% and three times as much invested at 11% as he does at 10%. If the total interest from the three accounts is $610 for the year, how much is invested at each rate? (*Hint:* Let x = the amount invested at 10%.)

A **Triangle Problems** [Example 3]

21. Two angles in a triangle are equal and their sum is equal to the third angle in the triangle. What are the measures of each of the three interior angles?

22. One angle in a triangle measures twice the smallest angle, whereas the largest angle is six times the smallest angle. Find the measures of all three angles.

23. The smallest angle in a triangle is $\frac{1}{5}$ as large as the largest angle. The third angle is twice the smallest angle. Find the three angles.

24. One angle in a triangle is half the largest angle but three times the smallest. Find all three angles.

25. A right triangle has one 37° angle. Find the other two angles.

26. In a right triangle, one of the acute angles is twice as large as the other acute angle. Find the measure of the two acute angles.

27. One angle of a triangle measures 20° more than the smallest, while a third angle is twice the smallest. Find the measure of each angle.

28. One angle of a triangle measures 50° more than the smallest, while a third angle is three times the smallest. Find the measure of each angle.

Miscellaneous Problems

29. **Ticket Prices** Miguel is selling tickets to a barbecue. Adult tickets cost $6.00 and children's tickets cost $4.00. He sells six more children's tickets than adult tickets. The total amount of money he collects is $184. How many adult tickets and how many children's tickets did he sell?

	Adult	Child
Number	x	
Income		

30. **Working Two Jobs** Maggie has a job working in an office for $10 an hour and another job driving a tractor for $12 an hour. One week she works in the office twice as long as she drives the tractor. Her total income for that week is $416. How many hours did she spend at each job?

Job	Office	Tractor
Hours Worked		x
Wages Earned		

31. Phone Bill The cost of a long-distance phone call is $0.41 for the first minute and $0.32 for each additional minute. If the total charge for a long-distance call is $5.21, how many minutes was the call?

32. Phone Bill Danny, who is 1 year old, is playing with the telephone when he accidentally presses one of the buttons his mother has programmed to dial her friend Sue's number. Sue answers the phone and realizes Danny is on the other end. She talks to Danny, trying to get him to hang up. The cost for a call is $0.23 for the first minute and $0.14 for every minute after that. If the total charge for the call is $3.73, how long did it take Sue to convince Danny to hang up the phone?

33. Hourly Wages JoAnn works in the publicity office at the state university. She is paid $12 an hour for the first 35 hours she works each week and $18 an hour for every hour after that. If she makes $492 one week, how many hours did she work?

34. Hourly Wages Diane has a part-time job that pays her $6.50 an hour. During one week she works 26 hours and is paid $178.10. She realizes when she sees her check that she has been given a raise. How much per hour is that raise?

35. Office Numbers Professors Wong and Gil have offices in the mathematics building at Miami Dade College. Their office numbers are consecutive odd integers with a sum of 14,660. What are the office numbers of these two professors?

36. Cell Phone Numbers Diana and Tom buy two cell phones. The phone numbers assigned to each are consecutive integers with a sum of 11,109,295. If the smaller number is Diana's, what are their phone numbers?

37. Age Marissa and Kendra are 2 years apart in age. Their ages are two consecutive even integers. Kendra is the younger of the two. If Marissa's age is added to twice Kendra's age, the result is 26. How old is each girl?

38. Age Justin's and Ethan's ages form two consecutive odd integers. What is the difference of their ages?

39. Arrival Time Jeff and Carla Cole are driving separately from San Luis Obispo, California, to the north shore of Lake Tahoe, a distance of 425 miles. Jeff leaves San Luis Obispo at 11:00 AM and averages 55 miles per hour on the drive, Carla leaves later, at 1:00 PM but averages 65 miles per hour. Which person arrives in Lake Tahoe first?

40. Chores Tyler's parents pay him $0.50 to do the laundry and $1.25 to mow the lawn. In one month, he does the laundry 6 more times than he mows the lawn. If his parents pay him $13.50 that month, how many times did he mow the lawn?

At one time, the Texas Junior College Teachers Association annual conference was held in Austin. At that time a taxi ride in Austin was $1.25 for the first $\frac{1}{5}$ of a mile and $0.25 for each additional $\frac{1}{5}$ of a mile. Use this information for Problems 41 and 42.

41. Cost of a Taxi Ride If the distance from one of the convention hotels to the airport is 7.5 miles, how much will it cost to take a taxi from that hotel to the airport?

42. Cost of a Taxi Ride Suppose the distance from one of the hotels to one of the western dance clubs in Austin is 12.4 miles. If the fare meter in the taxi gives the charge for that trip as $16.50, is the meter working correctly?

43. Geometry The length and width of a rectangle are consecutive even integers. The perimeter is 44 meters. Find the length and width.

44. Geometry The length and width of a rectangle are consecutive odd integers. The perimeter is 128 meters. Find the length and width.

45. Geometry The angles of a triangle are three consecutive integers. Find the measure of each angle.

46. Geometry The angles of a triangle are three consecutive even integers. Find the measure of each angle.

Ike and Nancy give western dance lessons at the Elks Lodge on Sunday nights. The lessons cost $3.00 for members of the lodge and $5.00 for nonmembers. Half of the money collected for the lesson is paid to Ike and Nancy. The Elks Lodge keeps the other half. One Sunday night Ike counts 36 people in the dance lesson. Use this information to work Problems 47 through 50.

47. Dance Lessons What is the least amount of money Ike and Nancy will make?

48. Dance Lessons What is the largest amount of money Ike and Nancy will make?

49. Dance Lessons At the end of the evening, the Elks Lodge gives Ike and Nancy a check for $80 to cover half of the receipts. Can this amount be correct?

50. Dance Lessons Besides the number of people in the dance lesson, what additional information does Ike need to know to always be sure he is being paid the correct amount?

Getting Ready for the Next Section

To understand all of the explanations and examples in the next section you must be able to work the problems below.

Solve the following equations.

51. a. $x - 3 = 6$
   b. $x + 3 = 6$
   c. $-x - 3 = 6$
   d. $-x + 3 = 6$

52. a. $x - 7 = 16$
   b. $x + 7 = 16$
   c. $-x - 7 = 16$
   d. $-x + 7 = 16$

53. a. $\dfrac{x}{4} = -2$
   b. $-\dfrac{x}{4} = -2$
   c. $\dfrac{x}{4} = 2$
   d. $-\dfrac{x}{4} = 2$

54. a. $3a = 15$
   b. $3a = -15$
   c. $-3a = 15$
   d. $-3a = -15$

55. $2.5x - 3.48 = 4.9x + 2.07$

56. $2(1 - 3x) + 4 = 4x - 14$

57. $3(x - 4) = -2$

58. Solve $2x - 3y = 6$ for y.

Maintaining Your Skills

The problems that follow review some of the more important skills you have learned in previous sections and chapters. You can consider the time you spend working these problems as time spent studying for exams.

Simplify the expression $36x - 12$ for each of the following values of x.

59. $\dfrac{1}{4}$ **60.** $\dfrac{1}{6}$

61. $\dfrac{1}{9}$ **62.** $\dfrac{3}{2}$

63. $\dfrac{1}{3}$ **64.** $\dfrac{5}{12}$

65. $\dfrac{5}{9}$ **66.** $\dfrac{2}{3}$

Find the value of each expression when $x = -4$.

67. $3(x - 4)$ **68.** $-3(x - 4)$

69. $-5x + 8$ **70.** $5x + 8$

71. $\dfrac{x - 14}{36}$ **72.** $\dfrac{x - 12}{36}$

73. $\dfrac{16}{x} + 3x$ **74.** $\dfrac{16}{x} - 3x$

75. $7x - \dfrac{12}{x}$ **76.** $7x + \dfrac{12}{x}$

77. $8\left(\dfrac{x}{2} + 5\right)$ **78.** $-8\left(\dfrac{x}{2} + 5\right)$

2.8 Linear Inequalities

OBJECTIVES

A Use the addition property for inequalities to solve an inequality and graph the solution set.

B Use the multiplication property for inequalities to solve an inequality.

C Use both the addition and multiplication properties to solve an inequality.

D Translate and solve application problems involving inequalities.

TICKET TO SUCCESS

Keep these questions in mind as you read through the section. Then respond in your own words and in complete sentences.

1. State the addition property for inequalities.
2. When do you darken the circle on the graph of a solution set?
3. How is the multiplication property for inequalities different from the multiplication property of equality?
4. When would you reverse the direction of an inequality symbol?

Based on the map in Figure 1, we can write the following inequality for the number of organic stores in New Mexico.

$$x \leq 10$$

FIGURE 1

If we know for certain of three stores in New Mexico, we can add to the original inequality and say $x + 3 \leq 10$. Up to now we have solved equalities. For example, when we say $x + 3 = 10$, we know x must be 7 since the statement is only true when $x = 7$. But if we say that $x + 3 \leq 10$, then there are many numbers that make this statement true. If x is 3, $3 + 3$ is less than 10. If x is 2.5, $2.5 + 3$ is less than 10. If x is any real number less than or equal to 7, then the statement $x + 3 \leq 10$ is true.

We also know that any number greater than 7 makes the statement false. If x is 8, then $x + 3$ is not less than 10.

When we solve an inequality, we look for all possible solutions—and there will usually be many of them. Any number that makes an inequality true is a *solution* to that inequality. All the numbers that make the inequality true are called the *solution set*.

We can express a solution set symbolically. If the solution set to an inequality is all the real numbers less than 7, we say the solution is $x < 7$. If the solution set to this inequality is all numbers less than or equal to 7, we say the solution is $x \leq 7$.

We can also express the solution set graphically using the number line. When a solution set is all numbers less than a certain number, we mark that number with an open circle:

If we want to include the number (7, in this case) in the solution set, we mark that number with a closed circle:

Linear inequalities are solved by a method similar to the one used in solving linear equations. The only real differences between the methods are in the multiplication property for inequalities and in graphing the solution set.

A   Addition Property for Inequalities

The addition property for inequalities is almost identical to the addition property for equality.

Addition Property for Inequalities

For any three algebraic expressions A, B, and C,

$$\text{if} \qquad A < B$$
$$\text{then} \qquad A + C < B + C$$

In words: Adding the same quantity to both sides of an inequality will not change the solution set.

It makes no difference which inequality symbol we use to state the property. Adding the same amount to both sides always produces an inequality equivalent to the original inequality. Also, because subtraction can be thought of as addition of the opposite, this property holds for subtraction as well as addition.

EXAMPLE 1   Solve the inequality $x + 5 < 7$.

SOLUTION   To isolate x, we add -5 to both sides of the inequality.

$$x + 5 < 7$$
$$x + 5 + (\mathbf{-5}) < 7 + (\mathbf{-5}) \qquad \text{Addition property for inequalities}$$
$$x < 2$$

We can go one step further here and graph the solution set. The solution set is all real numbers less than 2. To graph this set, we simply draw a straight line and label the center 0 (zero) for reference. Then we label the 2 on the right side of zero and extend an arrow beginning at 2 and pointing to the left. We use an open circle at 2 because it is not included in the solution set. Here is the graph.

EXAMPLE 2   Solve $x - 6 \leq -3$.

SOLUTION   Adding 6 to each side will isolate x on the left side.

$$x - 6 \leq -3$$
$$x - 6 + \mathbf{6} \leq -3 + \mathbf{6} \qquad \text{Add 6 to both sides}$$
$$x \leq 3$$

The graph of the solution set is

Notice that the dot at the 3 is darkened because 3 is included in the solution set.

B   Multiplication Property for Inequalities

To see the idea behind the multiplication property for inequalities, we will consider three true inequality statements and explore what happens when we multiply both

NOTE
This discussion is intended to show why the multiplication property for inequalities is written the way it is. You may want to look ahead to the property itself and then come back to this discussion if you are having trouble making sense out of it.

sides by a positive number and then what happens when we multiply by a negative number.

Consider the following three true statements

$$3 < 5 \qquad -3 < 5 \qquad -5 < -3$$

Now multiply both sides by the positive number 4.

$$4(3) < 4(5) \qquad 4(-3) < 4(5) \qquad 4(-5) < 4(-3)$$
$$12 < 20 \qquad\quad -12 < 20 \qquad\quad -20 < -12$$

In each case, the inequality symbol in the result points in the same direction it did in the original inequality. We say the "sense" of the inequality doesn't change when we multiply both sides by a positive quantity.

Notice what happens when we go through the same process but multiply both sides by −4 instead of 4.

$$3 < 5 \qquad\qquad -3 < 5 \qquad\qquad -5 < -3$$
$$-4(3) > -4(5) \quad -4(-3) > -4(5) \quad -4(-5) > -4(-3)$$
$$-12 > -20 \qquad\quad 12 > -20 \qquad\quad 20 > 12$$

In each case, we have to change the direction in which the inequality symbol points to keep each statement true. Multiplying both sides of an inequality by a negative quantity *always* reverses the sense of the inequality. Our results are summarized in the multiplication property for inequalities.

NOTE
Because division is defined in terms of multiplication, this property is also true for division. We can divide both sides of an inequality by any nonzero number we choose. If that number happens to be negative, we must also reverse the direction of the inequality symbol.

Multiplication Property for Inequalities

For any three algebraic expressions A, B, and C,

$$\text{if} \qquad A < B$$
$$\text{then} \qquad AC < BC \qquad \text{when } C \text{ is positive}$$
$$\text{and} \qquad AC > BC \qquad \text{when } C \text{ is negative}$$

In words: Multiplying both sides of an inequality by a positive number does not change the solution set. When multiplying both sides of an inequality by a negative number, it is necessary to reverse the direction of inequality symbol to produce an equivalent inequality.

We can multiply both sides of an inequality by any nonzero number we choose. If that number happens to be negative, we must also reverse the sense of the inequality.

EXAMPLE 3   Solve $3a < 15$ and graph the solution.

SOLUTION   We begin by multiplying each side by $\frac{1}{3}$. Because $\frac{1}{3}$ is a positive number, we do not reverse the direction of the inequality symbol.

$$3a < 15$$
$$\frac{1}{3}(3a) < \frac{1}{3}(15) \qquad\qquad \text{Multiply each side by } \tfrac{1}{3}$$
$$a < 5$$

EXAMPLE 4   Solve $-3a \le 18$, and graph the solution.

SOLUTION   We begin by multiplying both sides by $-\frac{1}{3}$. Because $-\frac{1}{3}$ is a negative number, we must reverse the direction of the inequality symbol at the same time that we multiply by $-\frac{1}{3}$.

$$-3a \le 18$$

$$-\frac{1}{3}(-3a) \ge -\frac{1}{3}(18)$$ Multiply both sides by $-\frac{1}{3}$ and reverse the direction of the inequality symbol

$$a \ge -6$$

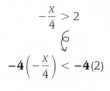

EXAMPLE 5   Solve $-\frac{x}{4} > 2$ and graph the solution.

SOLUTION   To isolate x, we multiply each side by -4. Because -4 is a negative number, we also must reverse the direction of the inequality symbol.

$$-\frac{x}{4} > 2$$

$$-4\left(-\frac{x}{4}\right) < -4(2)$$ Multiply each side by -4, and reverse the direction of the inequality symbol

$$x < -8$$

C Solving Linear Inequalities in One Variable

To solve more complicated inequalities, we use the following steps.

Strategy   Solving Linear Inequalities in One Variable

Step 1a: Use the distributive property to separate terms, if necessary.

   1b: If fractions are present, consider multiplying both sides by the LCD to eliminate the fractions. If decimals are present, consider multiplying both sides by a power of 10 to clear the inequality of decimals.

   1c: Combine similar terms on each side of the inequality.

Step 2: Use the addition property for inequalities to get all variable terms on one side of the inequality and all constant terms on the other side.

Step 3: Use the multiplication property for inequalities to get x by itself on one side of the inequality.

Step 4: Graph the solution set.

EXAMPLE 6   Solve $2.5x - 3.48 < -4.9x + 2.07$.

SOLUTION   We have two methods we can use to solve this inequality. We can simply apply our properties to the inequality the way it is currently written and work with the decimal numbers, or we can eliminate the decimals to begin with and solve the resulting inequality.

METHOD 1 Working with the decimals.

$$2.5x - 3.48 < -4.9x + 2.07$$ Original inequality
$$2.5x + \mathbf{4.9x} - 3.48 < -4.9x + \mathbf{4.9x} + 2.07$$ Add $4.9x$ to each side

$$7.4x - 3.48 < 2.07$$
$$7.4x - 3.48 + \mathbf{3.48} < 2.07 + \mathbf{3.48} \qquad \text{Add 3.48 to each side}$$
$$7.4x < 5.55$$
$$\frac{7.4x}{\mathbf{7.4}} < \frac{5.55}{\mathbf{7.4}} \qquad \text{Divide each side by 7.4}$$
$$x < 0.75$$

METHOD 2 Eliminating the decimals in the beginning.

Because the greatest number of places to the right of the decimal point in any of the numbers is 2, we can multiply each side of the inequality by 100 and we will be left with an equivalent inequality that contains only whole numbers.

$2.5x - 3.48 < -4.9x + 2.07$	Original inequality
$\mathbf{100}(2.5x - 3.48) < \mathbf{100}(-4.9x + 2.07)$	Multiply each side by 100
$\mathbf{100}(2.5x) - \mathbf{100}(3.48) < \mathbf{100}(-4.9x) + \mathbf{100}(2.07)$	Distributive property
$250x - 348 < -490x + 207$	Multiplication
$740x - 348 < 207$	Add 490x to each side
$740x < 555$	Add 348 to each side
$\dfrac{740x}{\mathbf{740}} < \dfrac{555}{\mathbf{740}}$	Divide each side by 740
$x < 0.75$	

The solution by either method is $x < 0.75$. Here is the graph:

EXAMPLE 7   Solve $3(x - 4) \geq -2$.

SOLUTION

$3x - 12 \geq -2$	Distributive property
$3x - 12 + \mathbf{12} \geq -2 + \mathbf{12}$	Add 12 to both sides
$3x \geq 10$	
$\dfrac{\mathbf{1}}{\mathbf{3}}(3x) \geq \dfrac{\mathbf{1}}{\mathbf{3}}(10)$	Multiply both sides by $\frac{1}{3}$
$x \geq \dfrac{10}{3}$	

EXAMPLE 8   Solve and graph $2(1 - 3x) + 4 < 4x - 14$.

SOLUTION

$2 - 6x + 4 < 4x - 14$	Distributive property
$-6x + 6 < 4x - 14$	Simplify
$-6x + 6 + (\mathbf{-6}) < 4x - 14 + (\mathbf{-6})$	Add -6 to both sides
$-6x < 4x - 20$	
$-6x + (\mathbf{-4x}) < 4x + (\mathbf{-4x}) - 20$	Add $-4x$ to both sides
$-10x < -20$	
$\left(-\dfrac{\mathbf{1}}{\mathbf{10}}\right)(-10x) > \left(-\dfrac{\mathbf{1}}{\mathbf{10}}\right)(-20)$	Multiply by $-\frac{1}{10}$, reverse the sense of the inequality

$x > 2$

Inequalities with Two Variables

EXAMPLE 9   Solve $2x - 3y < 6$ for y.

SOLUTION   We can solve this formula for y by first adding $-2x$ to each side and then multiplying each side by $-\frac{1}{3}$. When we multiply by $-\frac{1}{3}$ we must reverse the direction of the inequality symbol.

$$2x - 3y < 6 \qquad \text{Original inequality}$$
$$2x + (\mathbf{-2x}) - 3y < (\mathbf{-2x}) + 6 \qquad \text{Add } -2x \text{ to each side}$$
$$-3y < -2x + 6$$
$$-\frac{\mathbf{1}}{\mathbf{3}}(-3y) > -\frac{\mathbf{1}}{\mathbf{3}}(-2x + 6) \qquad \text{Multiply each side by } -\frac{1}{3}$$
$$y > \frac{2}{3}x - 2$$

Because this is an inequality in two variables, we will not graph the solution until later in the book.

D   Applications

When working application problems that involve inequalities, the phrases "at least" and "at most" translate as follows:

In Words	In Symbols
x is at least 30	$x \geq 30$
x is at most 20	$x \leq 20$

Our next example is similar to an example done earlier in this chapter. This time it involves an inequality instead of an equation.

We can modify our Blueprint for Problem Solving to solve application problems whose solutions depend on writing and then solving inequalities.

EXAMPLE 10   The sum of two consecutive odd integers is at most 28. What are the possibilities for the first of the two integers?

SOLUTION   When we use the phrase "their sum is at most 28," we mean that their sum is less than or equal to 28.

Step 1:   Read and list.

Known items: Two consecutive odd integers. Their sum is less than or equal to 28.

Unknown items: The numbers in question

Step 2:   Assign a variable, and translate information.
If we let $x =$ the first of the two consecutive odd integers, then $x + 2$ is the next consecutive one.

Step 3: **Reread and write an inequality.**

Their sum is at most 28.

$$x + (x + 2) \leq 28$$

Step 4: **Solve the inequality.**

$2x + 2 \leq 28$	Simplify the left side
$2x \leq 26$	Add -2 to each side
$x \leq 13$	Multiply each side by $\frac{1}{2}$

Step 5: **Write the answer.**

The first of the two integers must be an odd integer that is less than or equal to 13. The second of the two integers will be two more than whatever the first one is.

Step 6: **Reread and check.**

Suppose the first integer is 13. The next consecutive odd integer is 15. The sum of 15 and 13 is 28. If the first odd integer is less than 13, the sum of it and the next consecutive odd integer will be less than 28.

EXAMPLE 11   Monica wants to invest more money into an account that pays $6\frac{1}{4}$% annual interest. She wants to be able to count on at least $500 in interest at the end of the year. How much money should she invest to get $500 or more at the end of the year?

SOLUTION   The phrase "or more" tells us this is an inequality problem. Once we find out the amount needed to get $500 interest, we can solve the problem and answer the question.

Step 1: **Read and list.**

Known items: The interest rate is $6\frac{1}{4}$%. The minimum acceptable interest amount is $500.

Unknown items: The investment amount.

Step 2: **Assign a variable, and translate information.**

We will let $x =$ the amount to invest.

In words: $6\frac{1}{4}$% *of* x must be at least $500.

Step 3: **Reread and write an inequality.**

Our sentence from step 2 can be translated directly into the inequality

$$0.0625x \geq 500.$$

Step 4: **Solve the inequality.**

$0.0625x \geq 500$	Original inequality
$x \geq 8,000$	Divide both sides by 0.0625

Step 5: **Write the answer.**

We now know that if Monica invests $8,000, her interest will be $500. Any more than $8,000 will give her more than $500 so the answer is that Monica must invest at least $8,000.

Step 6: **Reread and check.**

Put in any numbers greater than or equal to 8,000 in the inequality and check to make sure the result is at least 500.

$$x = 9,000$$
$$0.0625(9,000) = 562.5, \text{ which is } > \$500.$$

EXAMPLE 12 Jack is on a diet and wants to make sure not to eat more than 2,000 calories a day. If Jack's breakfast had 550 calories, lunch had 625 calories, and a afternoon snack had 180 calories, what is the minimum and maximum amount of calories Jack can eat for dinner?

SOLUTION We need to form an inequality to solve the problem.

Step 1: **Read and list.**

Known items: Calories consumed all day, except for dinner, and maximum calories allowed.

Unknown items: The least amount of calories Jack can have for dinner and the most amount to stay at or under 2,000.

Step 2: **Assign a variable, and translate information.**
We will let x = number of possible calories eaten at dinner. We want to find out how $x + 550 + 625 + 180$ can be less than or equal to 2,000.

Step 3: **Reread and write an inequality.**

$$x + 550 + 625 + 180 \leq 2,000$$

Step 4: **Solve the inequality.**

$x + 550 + 625 + 180 \leq 2,000$	Original inequality
$x + 1355 \leq 2,000$	Combine like terms
$x \leq 645$	Add -1355 to both sides

Step 5: **Write the answer.**
Our solution to Step 4 is really not the answer. This says $x \leq 645$ which means x can be any real number less than 645. But in reality, Jack cannot eat a negative quantity of calories. The actual answer is that Jack can eat no dinner (calories = 0) up to and including a dinner of 645 calories to stay within the required 2,000 calories limit.

Step 6: **Reread and check.**
By putting any number between 0 and 645 into the original inequality and doing the addition, we can check our answer. ■

Problem Set 2.8

Moving Toward Success

"What the mind of man can conceive and believe, it can achieve."

—Napoleon Hill, 1883–1970, American speaker and motivational author

1. Why is making the decision to be successful important to solving word problems?

2. Why should you mentally list the items that are known and unknown in a word problem?

A Solve the following inequalities using the addition property of inequalities. Graph each solution set.
[Examples 1–2]

1. $x - 5 < 7$

2. $x + 3 < -5$

3. $a - 4 \leq 8$

4. $a + 3 \leq 10$

5. $x - 4.3 > 8.7$

6. $x - 2.6 > 10.4$

7. $y + 6 \geq 10$

8. $y + 3 \geq 12$

9. $2 < x - 7$

10. $3 < x + 8$

B Solve the following inequalities using the multiplication property of inequalities. If you multiply both sides by a negative number, be sure to reverse the direction of the inequality symbol. Graph the solution set.
[Examples 3–5]

11. $3x < 6$

12. $2x < 14$

13. $5a \le 25$

14. $4a \le 16$

15. $\dfrac{x}{3} > 5$

16. $\dfrac{x}{7} > 1$

17. $-2x > 6$

18. $-3x \ge 9$

19. $-3x \ge -18$

20. $-8x \ge -24$

21. $-\dfrac{x}{5} \le 10$

22. $-\dfrac{x}{9} \ge -1$

23. $-\dfrac{2}{3}y > 4$

24. $-\dfrac{3}{4}y > 6$

C Solve the following inequalities. Graph the solution set in each case. [Examples 6–9]

25. $2x - 3 < 9$

26. $3x - 4 < 17$

27. $-\dfrac{1}{5}y - \dfrac{1}{3} \le \dfrac{2}{3}$

28. $-\dfrac{1}{6}y - \dfrac{1}{2} \le \dfrac{2}{3}$

29. $-7.2x + 1.8 > -19.8$

30. $-7.8x - 1.3 > 22.1$

31. $\dfrac{2}{3}x - 5 \le 7$

32. $\dfrac{3}{4}x - 8 \le 1$

33. $-\dfrac{2}{5}a - 3 > 5$

34. $-\dfrac{4}{5}a - 2 > 10$

35. $5 - \dfrac{3}{5}y > -10$

36. $4 - \dfrac{5}{6}y > -11$

37. $0.3(a + 1) \le 1.2$

38. $0.4(a - 2) \le 0.4$

39. $2(5 - 2x) \le -20$

40. $7(8 - 2x) > 28$

41. $3x - 5 > 8x$

42. $8x - 4 > 6x$

43. $\dfrac{1}{3}y - \dfrac{1}{2} \le \dfrac{5}{6}y + \dfrac{1}{2}$

44. $\dfrac{7}{6}y + \dfrac{4}{3} \le \dfrac{11}{6}y - \dfrac{7}{6}$

45. $-2.8x + 8.4 < -14x - 2.8$

46. $-7.2x - 2.4 < -2.4x + 12$

47. $3(m - 2) - 4 \ge 7m + 14$

48. $2(3m - 1) + 5 \ge 8m - 7$

49. $3 - 4(x - 2) \le -5x + 6$

50. $8 - 6(x - 3) \le -4x + 12$

Solve each of the following inequalities for y.

51. $3x + 2y < 6$

52. $-3x + 2y < 6$

53. $2x - 5y > 10$

54. $-2x - 5y > 5$

55. $-3x + 7y \le 21$

56. $-7x + 3y \le 21$

57. $2x - 4y \ge -4$

58. $4x - 2y \ge -8$

The next two problems are intended to give you practice reading, and paying attention to, the instructions that accompany the problems you are working.

59. Work each problem according to the instructions given.

   a. Evaluate when $x = 0$: $-5x + 3$

   b. Solve: $-5x + 3 = -7$

   c. Is 0 a solution to $-5x + 3 < -7$

   d. Solve: $-5x + 3 < -7$

60. Work each problem according to the instructions given.

   a. Evaluate when $x = 0$: $-2x - 5$

   b. Solve: $-2x - 5 = 1$

   c. Is 0 a solution to $-2x - 5 > 1$

   d. Solve: $-2x - 5 > 1$

For each graph below, write an inequality whose solution is the graph.

61.

62.

63.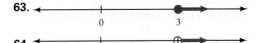

64.

D Applying the Concepts [Examples 10–12]

65. Organic Groceries The map shows a range of how many organic stores are found in each state. Write an inequality that describes the maximum number of organic stores found in Florida.

Organic Stores in the U.S.

- 0-10
- 11-20
- 21-30
- 31-40 Number of organic
- 41-50 grocery stores
- 51-60 per state
- 61-70
- 71-80
- 81-90

Source: organic.org

66. Population The map shows the most populated cities in the United States. Write an inequality that describes the population of Chicago relative to San Diego.

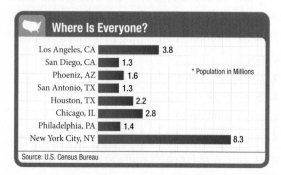

Where Is Everyone?

Los Angeles, CA	3.8
San Diego, CA	1.3
Phoeniz, AZ	1.6
San Antonio, TX	1.3
Houston, TX	2.2
Chicago, IL	2.8
Philadelphia, PA	1.4
New York City, NY	8.3

* Population in Millions

Source: U.S. Census Bureau

67. Consecutive Integers The sum of two consecutive integers is at least 583. What are the possibilities for the first of the two integers?

68. Consecutive Integers The sum of two consecutive integers is at most 583. What are the possibilities for the first of the two integers?

69. Number Problems The sum of twice a number and six is less than ten. Find all solutions.

70. Number Problems Twice the difference of a number and three is greater than or equal to the number increased by five. Find all solutions.

71. Number Problems The product of a number and four is greater than the number minus eight. Find the solution set.

72. Number Problems The quotient of a number and five is less than the sum of seven and two. Find the solution set.

73. Geometry Problems The length of a rectangle is 3 times the width. If the perimeter is to be at least 48 meters, what are the possible values for the width? (If the perimeter is at least 48 meters, then it is greater than or equal to 48 meters.)

74. Geometry Problems The length of a rectangle is 3 more than twice the width. If the perimeter is to be at least 51 meters, what are the possible values for the width? (If the perimeter is at least 51 meters, then it is greater than or equal to 51 meters.)

75. Geometry Problems The numerical values of the three sides of a triangle are given by three consecutive even integers. If the perimeter is greater than 24 inches, what are the possibilities for the shortest side?

76. Geometry Problems The numerical values of the three sides of a triangle are given by three consecutive odd integers. If the perimeter is greater than 27 inches, what are the possibilities for the shortest side?

77. Car Heaters If you have ever gotten in a cold car early in the morning you know that the heater does not work until the engine warms up. This is because the heater relies on the heat coming off the engine. Write an equation using an inequality sign to express when the heater will work if the heater works only after the engine is 100°F.

78. Exercise When Kate exercises, she either swims or runs. She wants to spend a minimum of 8 hours a week exercising, and she wants to swim 3 times the amount she runs. What is the minimum amount of time she must spend doing each exercise?

79. Profit and Loss Movie theaters pay a certain price for the movies that you and I see. Suppose a theater pays $1,500 for each showing of a popular movie. If they charge $7.50 for each ticket they sell, then they will lose money if ticket sales are less than $1,500. However, they will make a profit if ticket sales are greater than $1,500. What is the range of tickets they can sell and still lose money? What is the range of tickets they can sell and make a profit?

80. Stock Sales Suppose you purchase x shares of a stock at $12 per share. After 6 months you decide to sell all your shares at $20 per share. Your broker charges you $15 for the trade. If your profit is at least $3,985, how many shares did you purchase in the first place?

Maintaining Your Skills

The problems that follow review some of the more important skills you have learned in previous sections and chapters. You can consider the time you spend working these problems as time spent studying for exams.

Apply the distributive property, then simplify.

81. $\dfrac{1}{6}(12x + 6)$

82. $\dfrac{3}{5}(15x - 10)$

83. $\dfrac{2}{3}(-3x - 6)$

84. $\dfrac{3}{4}(-4x - 12)$

85. $3\left(\dfrac{5}{6}a + \dfrac{4}{9}\right)$

86. $2\left(\dfrac{3}{4}a - \dfrac{5}{6}\right)$

87. $-3\left(\dfrac{2}{3}a + \dfrac{5}{6}\right)$

88. $-4\left(\dfrac{5}{6}a + \dfrac{4}{9}\right)$

Find the LCD and simplify.

89. $\dfrac{1}{2}x + \dfrac{1}{6}x$

90. $\dfrac{1}{2}x - \dfrac{3}{4}x$

91. $\dfrac{2}{3}x - \dfrac{5}{6}x$

92. $\dfrac{1}{3}x + \dfrac{3}{5}x$

93. $\dfrac{3}{4}x + \dfrac{1}{6}x$

94. $\dfrac{3}{2}x - \dfrac{2}{3}x$

95. $\dfrac{2}{5}x + \dfrac{5}{8}x$

96. $\dfrac{3}{5}x - \dfrac{3}{8}x$

EXAMPLES

Similar Terms [2.1]

1. The terms $2x$, $5x$, and $-7x$ are all similar because their variable parts are the same.

A *term* is a number or a number and one or more variables multiplied together. *Similar terms* are terms with the same variable part.

Simplifying Expressions [2.1]

2. Simplify $3x + 4x$.
$$3x + 4x = (3 + 4)x$$
$$= 7x$$

In this chapter we simplified expressions that contained variables by using the distributive property to combine similar terms.

Solution Set [2.2]

3. The solution set for the equation $x + 2 = 5$ is $\{3\}$ because when x is 3 the equation is $3 + 2 = 5$, or $5 = 5$.

The *solution set* for an equation (or inequality) is all the numbers that, when used in place of the variable, make the equation (or inequality) a true statement.

Equivalent Equations [2.2]

4. The equation $a - 4 = 3$ and $a - 2 = 5$ are equivalent because both have solution set $\{7\}$.

Two equations are called *equivalent* if they have the same solution set.

Addition Property of Equality [2.2]

5. Solve $x - 5 = 12$.
$$x - 5 + 5 = 12 + 5$$
$$x + 0 = 17$$
$$x = 17$$

When the same quantity is added to both sides of an equation, the solution set for the equation is unchanged. Adding the same amount to both sides of an equation produces an equivalent equation.

Multiplication Property of Equality [2.3]

6. Solve $3x = 18$.
$$\frac{1}{3}(3x) = \frac{1}{3}(18)$$
$$x = 6$$

If both sides of an equation are multiplied by the same nonzero number, the solution set is unchanged. Multiplying both sides of an equation by a nonzero quantity produces an equivalent equation.

Linear Equation [2.4]

A linear equation in one variable is any equation that can be put in the form $ax + b = 0$, where a and b are real numbers and a is not zero.

Strategy for Solving Linear Equations in One Variable [2.4]

7. Solve $2(x + 3) = 10$.
$$2x + 6 = 10$$
$$2x + 6 + (-6) = 10 + (-6)$$
$$2x = 4$$
$$\frac{1}{2}(2x) = \frac{1}{2}(4)$$
$$x = 2$$

Step 1a: Use the distributive property to separate terms, if necessary.

1b: If fractions are present, consider multiplying both sides by the LCD to eliminate the fractions. If decimals are present, consider multiplying both sides by a power of 10 to clear the equation of decimals.

1c: Combine similar terms on each side of the equation.

Step 2: Use the addition property of equality to get all variable terms on one side of the equation and all constant terms on the other side. A variable term is a term that contains the variable (for example, $5x$). A constant term is a term that does not contain the variable (the number 3, for example.)

Step 3: Use the multiplication property of equality to get x (that is, $1x$) by itself on one side of the equation.

Step 4: Check your solution in the original equation to be sure that you have not made a mistake in the solution process.

Formulas [2.5]

8. Solving $P = 2l + 2w$ for l, we have
$$P - 2w = 2l$$
$$\frac{P - 2w}{2} = l$$

A formula is an equation with more than one variable. To solve a formula for one of its variables, we use the addition and multiplication properties of equality to move everything except the variable in question to one side of the equal sign so the variable in question is alone on the other side.

Blueprint for Problem Solving [2.6, 2.7]

Step 1: **Read** the problem, and then mentally **list** the items that are known and the items that are unknown.

Step 2: **Assign a variable** to one of the unknown items. (In most cases this will amount to letting $x =$ the item that is asked for in the problem.) Then **translate** the other **information** in the problem to expressions involving the variable.

Step 3: **Reread** the problem, and then **write an equation,** using the items and variables listed in steps 1 and 2, that describes the situation.

Step 4: **Solve the equation** found in step 3.

Step 5: **Write** your **answer** using a complete sentence.

Step 6: **Reread** the problem, and **check** your solution with the original words in the problem.

Addition Property for Inequalities [2.8]

9. Solve $x + 5 < 7$.
$$x + 5 + (-5) < 7 + (-5)$$
$$x < 2$$

Adding the same quantity to both sides of an inequality produces an equivalent inequality, one with the same solution set.

■ Multiplication Property for Inequalities [2.8]

10. Solve $-3a \le 18$.

$$-\frac{1}{3}(-3a) \ge -\frac{1}{3}(18)$$

$$a \ge -6$$

Multiplying both sides of an inequality by a positive number never changes the solution set. If both sides are multiplied by a negative number, the sense of the inequality must be reversed to produce an equivalent inequality.

■ Strategy for Solving Linear Inequalities in One Variable [2.8]

11. Solve $3(x - 4) \ge -2$.

$$3x - 12 \ge -2$$

$$3x - 12 + \mathbf{12} \ge -2 + \mathbf{12}$$

$$3x \ge 10$$

$$\frac{1}{3}(3x) \ge \frac{1}{3}(10)$$

$$x \ge \frac{10}{3}$$

Step 1a: Use the distributive property to separate terms, if necessary.

1b: If fractions are present, consider multiplying both sides by the LCD to eliminate the fractions. If decimals are present, consider multiplying both sides by a power of 10 to clear the inequality of decimals.

1c: Combine similar terms on each side of the inequality.

Step 2: Use the addition property for inequalities to get all variable terms on one side of the inequality and all constant terms on the other side.

Step 3: Use the multiplication property for inequalities to get x by itself on one side of the inequality.

Step 4: Graph the solution set.

⃠ COMMON MISTAKES

1. Trying to subtract away coefficients (the number in front of variables) when solving equations. For example:

$$4x = 12$$

$$4x - \mathbf{4} = 12 - \mathbf{4}$$

$$x = 8 \leftarrow \text{Mistake}$$

It is not incorrect to add (-4) to both sides; it's just that $4x - 4$ is not equal to x. Both sides should be multiplied by $\frac{1}{4}$ to solve for x.

2. Forgetting to reverse the direction of the inequality symbol when multiplying both sides of an inequality by a negative number. For instance:

$$-3x < 12$$

$$-\frac{1}{3}(-3x) < -\frac{1}{3}(12) \leftarrow \text{Mistake}$$

$$x < -4$$

It is not incorrect to multiply both sides by $-\frac{1}{3}$. But if we do, we must also reverse the sense of the inequality.

Simplify each expression as much as possible. [2.1]

1. $5x - 8x$

2. $6x - 3 - 8x$

3. $-a + 2 + 5a - 9$

4. $5(2a - 1) - 4(3a - 2)$

5. $6 - 2(3y + 1) - 4$

6. $4 - 2(3x - 1) - 5$

Find the value of each expression when x is 3. [2.1]

7. $7x - 2$

8. $-4x - 5 + 2x$

9. $-x - 2x - 3x$

Find the value of each expression when x is -2. [2.1]

10. $5x - 3$

11. $-3x + 2$

12. $7 - x - 3$

Solve each equation. [2.2, 2.3]

13. $x + 2 = -6$

14. $x - \dfrac{1}{2} = \dfrac{4}{7}$

15. $10 - 3y + 4y = 12$

16. $-3 - 4 = -y - 2 + 2y$

17. $2x = -10$

18. $3x = 0$

19. $\dfrac{x}{3} = 4$

20. $-\dfrac{x}{4} = 2$

21. $3a - 2 = 5a$

22. $\dfrac{7}{10}a = \dfrac{1}{5}a + \dfrac{1}{2}$

23. $3x + 2 = 5x - 8$

24. $6x - 3 = x + 7$

25. $0.7x - 0.1 = 0.5x - 0.1$

26. $0.2x - 0.3 = 0.8x - 0.3$

Solve each equation. Be sure to simplify each side first. [2.4]

27. $2(x - 5) = 10$

28. $12 = 2(5x - 4)$

29. $\dfrac{1}{2}(3t - 2) + \dfrac{1}{2} = \dfrac{5}{2}$

30. $\dfrac{3}{5}(5x - 10) = \dfrac{2}{3}(9x + 3)$

31. $2(3x + 7) = 4(5x - 1) + 18$

32. $7 - 3(y + 4) = 10$

Use the formula $4x - 5y = 20$ to find y if [2.5]

33. x is 5

34. x is 0

35. x is -5

36. x is 10

Solve each of the following formulas for the indicated variable. [2.5]

37. $2x - 5y = 10$ for y

38. $5x - 2y = 10$ for y

39. $V = \pi r^2 h$ for h

40. $P = 2l + 2w$ for w

41. What number is 86% of 240? [2.5]

42. What percent of 2,000 is 180? [2.5]

Solve each of the following word problems. In each case, be sure to show the equation that describes the situation. [2.6, 2.7]

43. Number Problem The sum of twice a number and 6 is 28. Find the number.

44. Geometry The length of a rectangle is 5 times as long as the width. If the perimeter is 60 meters, find the length and the width.

45. Investing A man invests a certain amount of money in an account that pays 9% annual interest. He invests $300 more than that in an account that pays 10% annual interest. If his total interest after a year is $125, how much does he have invested in each account?

46. Coin Problem A collection of 15 coins is worth $1.00. If the coins are dimes and nickels, and there are twice as many nickels as dimes, how many of each coin are there?

Solve each inequality. [2.8]

47. $-2x < 4$

48. $-5x > -10$

49. $-\dfrac{a}{2} \le -3$

50. $-\dfrac{a}{3} > 5$

Solve each inequality, and graph the solution. [2.8]

51. $-4x + 5 > 37$

52. $2x + 10 < 5x - 11$

53. $2(3t + 1) + 6 \ge 5(2t + 4)$

Simplify.

1. $7 - 9 - 12$

2. $-11 + 17 + (-13)$

3. $8 - 4 \cdot 5$

4. $30 \div 3 \cdot 10$

5. $6 + 3(6 + 2)$

6. $-6(5 - 11) - 4(13 - 6)$

7. $\left(-\dfrac{2}{3}\right)^3$

8. $\left[\dfrac{1}{2}(-8)\right]^2$

9. $\dfrac{-30}{120}$

10. $\dfrac{234}{312}$

11. $-\dfrac{3}{4} \div \dfrac{15}{16}$

12. $\dfrac{2}{5} \cdot \dfrac{4}{7}$

13. $\dfrac{-4(-6)}{-9}$

14. $\dfrac{2(-9) + 3(6)}{-7 - 8}$

15. $\dfrac{5}{9} + \dfrac{1}{3}$

16. $\dfrac{4}{21} - \dfrac{9}{35}$

17. $\dfrac{1}{5}(10x)$

18. $-8(9x)$

19. $\dfrac{1}{4}(8x - 4)$

20. $3(2x - 1) + 5(x + 2)$

Solve each equation.

21. $7x = 6x + 4$

22. $x + 12 = -4$

23. $-\dfrac{3}{5}x = 30$

24. $\dfrac{x}{5} = -\dfrac{3}{10}$

25. $3x - 4 = 11$

26. $5x - 7 = x - 1$

27. $4(3x - 8) + 5(2x + 7) = 25$

28. $15 - 3(2t + 4) = 1$

29. $3(2a - 7) - 4(a - 3) = 15$

30. $\dfrac{1}{3}(x - 6) = \dfrac{1}{4}(x + 8)$

31. Solve $P = a + b + c$ for c.

32. Solve $3x + 4y = 12$ for y.

Solve each inequality, and graph the solution.

33. $3(x - 4) \le 6$

34. $-5x + 9 < -6$

35. $2x + 1 > 7$

36. $x + 1 < 5$

Translate into symbols.

37. The difference of x and 5 is 12.

38. The sum of x and 7 is 4.

For the set $\{-5, -3, -1.7, 2.3, \frac{12}{7}, \pi\}$ list all the

39. Integers

40. Rational numbers

41. Evaluate $x^2 - 8x - 9$ when $x = -2$.

42. Evaluate $a^2 - 2ab + b^2$ when $a = 3$, $b = -2$.

43. What is 30% of 50?

44. What percent of 36 is 27?

45. Number Problem   The sum of a number and 9 is 23. Find the number.

46. Number Problem   Twice a number increased by 7 is 31. Find the number.

47. Geometry   A right triangle has one 42° angle. Find the other two angles.

48. Geometry   Two angles are complementary. If one is 25°, find the other one.

49. Hourly Pay   Carol tutors in the math lab. She gets paid $8 per hour for the first 15 hours and $10 per hour for each hour after that. She made $150 one week. How many hours did she work?

50. Cost of a Letter   The cost of mailing a letter was 44¢ for the first ounce and 17¢ for each additional ounce. If the cost of mailing a letter was 78¢, how many ounces did the letter weigh?

The illustration shows the annual circulation (in millions) of the top five magazines. Use the illustration to answer the following questions.

Top 5 Magazine Sales

AARP — 24.1M
Readers Digest — 8.3M
Better Homes & Gardens — 7.7M
National Geographic — 5.1M
Good Housekeeping — 4.7M

Copies Sold

Source: Magazine Publishers of America

51. Find the total annual circulation *Reader's Digest* and *AARP*.

52. How many more copies of *Better Homes & Gardens* are sold compared to *National Geographic*?

Simplify each of the following expressions. [2.1]

1. $3x + 2 - 7x + 3$

2. $4a - 5 - a + 1$

3. $7 - 3(y + 5) - 4$

4. $8(2x + 1) - 5(x - 4)$

5. Find the value of $2x - 3 - 7x$ when $x = -5$. [2.1]

6. Find the value of $x^2 + 2xy + y^2$ when $x = 2$ and $y = 3$. [2.1]

7. Fill in the tables below to find the sequences formed by substituting the first four counting numbers into the expressions $(n + 1)^2$ and $n^2 + 1$. [2.1]

a.

n	$(n + 1)^2$
1	
2	
3	
4	

b.

n	$n^2 + 1$
1	
2	
3	
4	

Solve the following equations. [2.2, 2.3, 2.4]

8. $2x - 5 = 7$

9. $2y + 4 = 5y$

10. $\dfrac{1}{2}x - \dfrac{1}{10} = \dfrac{1}{5}x + \dfrac{1}{2}$

11. $\dfrac{2}{5}(5x - 10) = -5$

12. $-5(2x + 1) - 6 = 19$

13. $0.04x + 0.06(100 - x) = 4.6$

14. $2(t - 4) + 3(t + 5) = 2t - 2$

15. $2x - 4(5x + 1) = 3x + 17$

16. What number is 15% of 38? [2.5]

17. 240 is 12% of what number? [2.5]

18. If $2x - 3y = 12$, find x when $y = -2$. [2.5]

19. If $2x - 3y = 12$, find y when $x = -6$. [2.5]

20. Solve $2x + 5y = 20$ for y. [2.5]

21. Solve $h = x + vt + 16t^2$ for v. [2.5]

Solve each word problem. [2.6, 2.7]

22. Age Problem Dave is twice as old as Rick. Ten years ago the sum of their ages was 40. How old are they now?

23. Geometry A rectangle is twice as long as it is wide. The perimeter is 60 inches. What are the length and width?

24. Coin Problem A man has a collection of dimes and quarters with a total value of $3.50. If he has 7 more dimes than quarters, how many of each coin does he have?

25. Investing A woman has money in two accounts. One account pays 7% annual interest, whereas the other pays 9% annual interest. If she has $600 more invested at 9% than she does at 7% and her total interest for a year is $182, how much does she have in each account?

Solve each inequality, and graph the solution. [2.8]

26. $2x + 3 < 5$

27. $-5a > 20$

28. $0.4 - 0.2x \geq 1$

29. $4 - 5(m + 1) \leq 9$

Use the illustration below to find the percent of the hours in one week (168 hours) that a person from the given country spends on social media sites. Round your answers to the nearest tenth of a percent. [2.5]

Checking out Social Media Sites

Country	Hours per Week
Australia	6.9
United States	6.2
United Kingdom	6.1
Italy	6
Spain	5.4

Source: The Nielsen Company

30. Australia

31. Spain

LINEAR EQUATIONS AND INEQUALITIES

GROUP PROJECT

Tables and Numbers

Number of People	2-3
Time Needed	5–10 minutes
Equipment	Pencil and graph paper
Background	Building tables is a method of visualizing information. We can build a table from a situation (as below) or from an equation. In this project, we will first build a table and then write an equation from the information in the table.
Procedure	A parking meter, which accepts only dimes and quarters, is emptied at the end of each day. The amount of money in the meter at the end of one particular day is $3.15.

1. Complete the following table so that all possible combinations of dimes and quarters, along with the total number of coins, is shown. Remember, although the number of coins will vary, the value of the dimes and quarters must total $3.15.

Number of Dimes	Number of Quarters	Total Coins	Value
29	1	30	$3.15
24			$3.15
			$3.15
			$3.15
			$3.15
			$3.15

2. From the information in the table, answer the following questions.

a. What is the maximum possible number of coins taken from the meter?

b. What is the minimum possible number of coins taken from the meter?

c. When is the number of dimes equal to the number of quarters?

3. Let x = the number of dimes and y = the number of quarters. Write an equation in two variables such that the value of the dimes added to the value of the quarters is $3.15.

RESEARCH PROJECT

Stand and Deliver

The 1988 film *Stand and Deliver* starring Edward James Olmos and Lou Diamond Phillips is based on a true story. Olmos, in his portrayal of high-school math teacher Jaime Escalante, earned an Academy Award nomination for best actor.

Watch the movie *Stand and Deliver*. After briefly describing the movie, explain how Escalante's students became successful in math. Make a list of specific things you observe that the students had to do to become successful. Indicate which items on this list you think will also help you become successful.

WARNER BROS/ THE KOBAL COLLECTION

3

Equations and Inequalities in Two Variables

Introduction

The remains of the French philosopher René Descartes (1596-1650) rest in the Abbey of Saint-Germain-des-Prés (shown above) in Paris, France. In the 17th century, mathematicians and scientists treated algebra and geometry as separate subjects. In 1637, Descartes discovered a distinct link between algebra and geometry by associating geometric shapes with algebraic equations. He plotted this connection on a graph using an x- and a y-axis drawn perpendicular to each other. This graphing sysem was later named the Cartesian coordinate system after him.

Getty Images

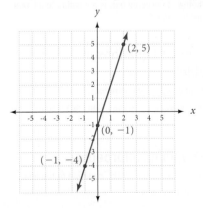

Our ability to graph an equation, such as $y = 3x - 1$ whose graph is shown in the diagram, is due to Descartes' invention of the Cartesian coordinate system. This coordinate system, also known as the rectangular coordinate system, is the foundation upon which we build our study of graphing in this chapter.

Simplify

1. $2(2) + (-1)$ **2.** $2(5) - 1$ **3.** $-\frac{1}{3}(3) + 2$ **4.** $\frac{3}{2}(2) - 3$ **5.** $-\frac{1}{2}(-3x + 6)$ **6.** $-\frac{2}{3}(x - 3)$

Solve.

7. $6 + 3y = 6$ **8.** $2x - 10 = 20$ **9.** $7 = 2x - 1$ **10.** $3 = 2x - 1$

11. If $3x - 5y = 10$, find y when x is 0.   **12.** If $y = \frac{2}{3}x + 5$, find y when x is 6

Find the value of each expression when $x = -2$ and $y = 3$.

13. $\frac{y - 1}{x + 4}$ **14.** $\frac{y + 7}{x - 3}$ **15.** $\frac{4 - y}{-1 - x}$ **16.** $\frac{3 - y}{x + 8}$

Solve each of the following equations for y.

17. $7x - 2y = 14$ **18.** $-2x - 5y = 4$ **19.** $y - 3 = -2(x + 4)$ **20.** $y + 1 = -\frac{2}{3}x + 2$

Chapter Outline

3.1 Paired Data and Graphing Ordered Pairs
A   Create a scatter diagram or line graph from a table of data.
B   Graph ordered pairs on a rectangular coordinate system.

3.2 Solutions to Linear Equations in Two Variables
A   Find solutions to linear equations in two variables.
B   Determine whether an ordered pair is a solution to a linear equation in two variables.

3.3 Graphing Linear Equations in Two Variables
A   Graph a linear equation in two variables.
B   Graph horizontal lines, vertical lines, and lines through the origin.

3.4 More on Graphing: Intercepts
A   Find the intercepts of a line from the equation of the line.
B   Use intercepts to graph a line.

3.5 The Slope of a Line
A   Find the slope of a line from two points on the line.
B   Graph a line given the slope and y-intercept.

3.6 Graphing Linear Inequalities in Two Variables
A   Graph linear inequalities in two variables.

3.7 Solving Inequalities in Two Variables
A   Solve a system of linear equation in two variables by graphing.

3.8 The Elimination Method
A   Use the elimination method to solve a system of linear equations in two variables.

3.9 The Substitution Method
A   Use the substitution method to solve a system of linear equations in two variables.

3.1 Paired Data and Graphing Ordered Pairs

OBJECTIVES

A Create a scatter diagram or line graph from a table of data.

B Graph ordered pairs on a rectangular coordinate system.

TICKET TO SUCCESS

Keep these questions in mind as you read through the section. Then respond in your own words and in complete sentences.

1. What is an ordered pair of numbers?
2. Explain in words how you would graph the ordered pair (3,4).
3. How does a scatter diagram differ from a line graph?
4. Where is the origin on a rectangular coordinate system and why is it important?

Visually

FIGURE 1

TABLE 1	
SPEED OF A RACECAR	
Time in Seconds	**Speed in Miles per Hour**
0	0
1	72.7
2	129.9
3	162.8
4	192.2
5	212.4
6	228.1

In a previous math class, you may have learned about the relationship between a table of values and its corresponding bar chart. We will review this concept now using Table 1 and Figure 1. In Figure 1, the horizontal line that shows the elapsed time in seconds of a race car is called the *horizontal axis,* and the vertical line that shows the speed in miles per hour is called the *vertical axis.*

The data in Table 1 are called *paired data* because the information is organized so that each number in the first column is paired with a specific number in the second column. Each pair of numbers is associated with one of the solid bars in Figure 1. For example, the third bar in the bar chart is associated with the pair of numbers 3 seconds and 162.8 miles per hour. The first number, 3 seconds, is associated with the horizontal axis, and the second number, 162.8 miles per hour, is associated with the vertical axis.

A Scatter Diagrams and Line Graphs

The information in Table 1 can be visualized with a *scatter diagram* and *line graph* as well. Figure 2 is a scatter diagram of the information in Table 1. We use dots instead of the bars shown in Figure 1 to show the speed of the racecar at each second during the race. Figure 3 is called a *line graph.* It is constructed by taking the dots in Figure 2 and connecting each one to the next with a straight line. Notice that we have labeled the axes in these two figures a little differently than we did with the bar chart by making the axes intersect at the number 0.

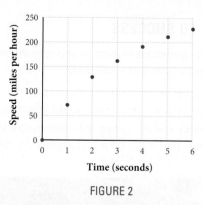

FIGURE 2

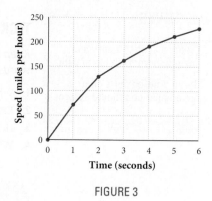

FIGURE 3

The number sequences we have worked with in the past can also be written as paired data by associating each number in the sequence with its position in the sequence. For instance, in the sequence of odd numbers

$$1, 3, 5, 7, 9, \ldots$$

the number 7 is the fourth number in the sequence. Its position is 4 and its value is 7. Here is the sequence of odd numbers written so that the position of each term is noted:

| Position | $1, 2, 3, 4, 5, \ldots$ |
| Value | $1, 3, 5, 7, 9, \ldots$ |

EXAMPLE 1 Tables 2 and 3 give the first five terms of the sequence of odd numbers and the sequence of squares as paired data. In each case construct a scatter diagram.

TABLE 2	
ODD NUMBERS	
Position	**Value**
1	1
2	3
3	5
4	7
5	9

TABLE 3	
SQUARES	
Position	**Value**
1	1
2	4
3	9
4	16
5	25

SOLUTION The two scatter diagrams are based on the data from Tables 2 and 3 shown here. Notice how the dots in Figure 4 seem to line up in a straight line, whereas the dots in Figure 5 give the impression of a curve. We say the points in Figure 4 suggest a *linear* relationship between the two sets of data, whereas the points in Figure 5 suggest a *nonlinear* relationship.

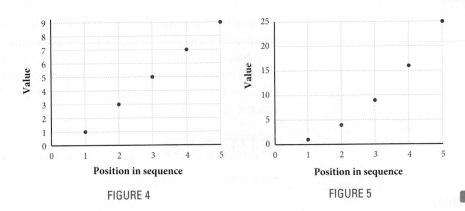

FIGURE 4   FIGURE 5

B Graphing Ordered Pairs

As you know, each dot in Figures 4 and 5 corresponds to a pair of numbers, one of which is associated with the horizontal axis and the other with the vertical axis. Paired data play a very important role in the equations we will solve in the next section. To prepare ourselves for those equations, we need to expand the concept of paired data to include negative numbers. At the same time, we want to standardize the position of the axes in the diagrams that we use to visualize paired data.

> **Definition**
> A pair of numbers enclosed in parentheses and separated by a comma, such as $(-2, 1)$, is called an **ordered pair** of numbers. The first number in the pair is called the **x-coordinate** of the ordered pair; the second number is called the **y-coordinate.** For the ordered pair $(-2, 1)$, the x-coordinate is -2 and the y-coordinate is 1.

Ordered pairs of numbers are important in the study of mathematics because they give us a way to visualize solutions to equations. To see the visual component of ordered pairs, we need the diagram shown in Figure 6. It is called the *rectangular coordinate system.*

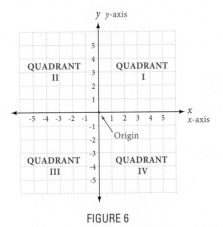

FIGURE 6

The rectangular coordinate system is built from two number lines oriented perpendicular to each other. The horizontal number line is exactly the same as our real number line and is called the *x-axis.* The vertical number line is also the same as our real number line with the positive direction up and the negative direction down. It is called the *y-axis.* The point where the two axes intersect is called the *origin.* As you can see from Figure 6, the axes divide the plane into four quadrants, which are numbered I through IV in a counterclockwise direction.

To graph the ordered pair (a, b), we start at the origin and move a units right or left (right if a is positive and left if a is negative). Then we move b units up or down (up if b is positive, down if b is negative). The point where we end up is the graph of the ordered pair (a, b).

EXAMPLE 2 Graph the ordered pairs $(3, 4)$, $(3, -4)$, $(-3, 4)$, and $(-3, -4)$.

SOLUTION

> **NOTE**
> It is very important that you graph ordered pairs quickly and accurately. Remember, the first coordinate goes with the horizontal axis and the second coordinate goes with the vertical axis.

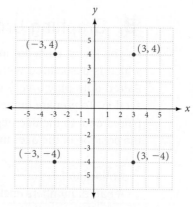

FIGURE 7

We can see in Figure 7 that when we graph ordered pairs, the x-coordinate corresponds to movement parallel to the x-axis (horizontal) and the y-coordinate corresponds to movement parallel to the y-axis (vertical). ■

EXAMPLE 3 Graph the ordered pairs $(-1, 3)$, $(2, 5)$, $(0, 0)$, $(0, -3)$, and $(4, 0)$.

SOLUTION

> **NOTE**
> If we do not label the axes of a coordinate system, we assume that each square is one unit long and one unit wide.

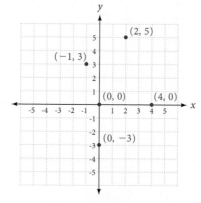

FIGURE 8

■

Problem Set 3.1

Moving Toward Success

"Achievement is not always success while reputed failure often is. It is honest endeavor, persistent effort to do the best possible under any and all circumstances."

—Orison Swett Marden, 1850–1924, writer

1. Should you take notes in class? Why or why not?

2. If your instructor writes something on the board in class, should you write it in your notes? Why or why not?

B Graph the following ordered pairs. [Examples 2–3]

1. $(3, 2)$
2. $(3, -2)$
3. $(-3, 2)$
4. $(-3, -2)$
5. $(5, 1)$
6. $(5, -1)$
7. $(1, 5)$
8. $(1, -5)$
9. $(-1, 5)$
10. $(-1, -5)$
11. $\left(2, \frac{1}{2}\right)$
12. $\left(3, \frac{3}{2}\right)$
13. $\left(-4, -\frac{5}{2}\right)$
14. $\left(-5, -\frac{3}{2}\right)$
15. $(3, 0)$
16. $(-2, 0)$
17. $(0, 5)$
18. $(0, 0)$

19–28. Give the coordinates of each numbered point in the figure.

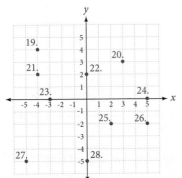

Graph the points $(4, 3)$ and $(-4, -1)$, and draw a straight line that passes through both of them. Then answer the following questions.

29. Does the graph of $(2, 2)$ lie on the line?

30. Does the graph of $(-2, 0)$ lie on the line?

31. Does the graph of $(0, -2)$ lie on the line?

32. Does the graph of $(-6, 2)$ lie on the line?

Graph the points $(-2, 4)$ and $(2, -4)$, and draw a straight line that passes through both of them. Then answer the following questions.

33. Does the graph of $(0, 0)$ lie on the line?

34. Does the graph of $(-1, 2)$ lie on the line?

35. Does the graph of $(2, -1)$ lie on the line?

36. Does the graph of $(1, -2)$ lie on the line?

Draw a straight line that passes through the points $(3, 4)$ and $(3, -4)$. Then answer the following questions.

37. Is the graph of $(3, 0)$ on this line?

38. Is the graph of $(0, 3)$ on this line?

39. Is there any point on this line with an x-coordinate other than 3?

40. If you extended the line, would it pass through a point with a y-coordinate of 10?

Draw a straight line that passes through the points $(3, 4)$ and $(-3, 4)$. Then answer the following questions.

41. Is the graph of $(4, 0)$ on this line?

42. Is the graph of $(0, 4)$ on this line?

43. Is there any point on this line with a y-coordinate other than 4?

44. If you extended the line, would it pass through a point with an x-coordinate of 10?

A Applying the Concepts [Example 1]

45. **Light Bulbs** The chart shows a comparison of power usage between incandescent and LED light bulbs.

Using the graph, estimate the number of watts used by the following bulbs.

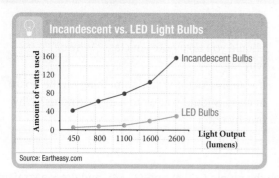

Source: Eartheasy.com

a. 800 lumen incandescent bulb

b. 1,600 lumen LED bulb

c. 1,600 lumen incandescent bulb

46. Health Care The graph shows the rising cost of health care. Write the five ordered pairs that describe the information on the chart.

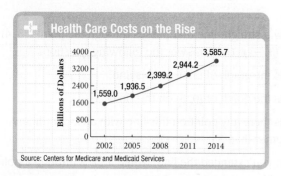

Source: Centers for Medicare and Medicaid Services

47. Hourly Wages Jane takes a job at the local Marcy's department store. Her job pays $8.00 per hour. The graph shows how much Jane earns for working from 0 to 40 hours in a week.

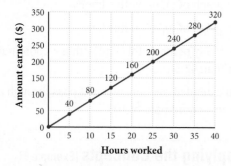

a. List three ordered pairs that lie on the line graph.

b. How much will she earn for working 40 hours?

c. If her check for one week is $240, how many hours did she work?

d. She works 35 hours one week, but her paycheck before deductions are subtracted is for $260. Is this correct? Explain.

48. Hourly Wages Judy takes a job at Gigi's boutique. Her job pays $6.00 per hour plus $50 per week in commission. The graph shows how much Judy earns for working from 0 to 40 hours in a week.

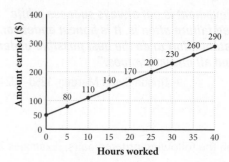

a. List three ordered pairs that lie on the line graph.

b. How much will she earn for working 40 hours?

c. If her check for one week is $230, how many hours did she work?

d. She works 35 hours one week, but her paycheck before deductions are subtracted is for $260. Is this correct? Explain.

49. Right triangle ABC has legs of length 5. Point C is the ordered pair (6, 2). Find the coordinates of A and B.

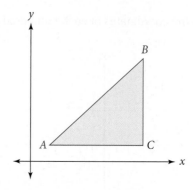

50. Right triangle ABC has legs of length 7. Point C is the ordered pair (−8, −3). Find the coordinates of A and B.

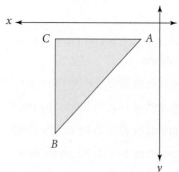

51. Rectangle *ABCD* has a length of 5 and a width of 3. Point *D* is the ordered pair (7, 2). Find points *A*, *B*, and *C*.

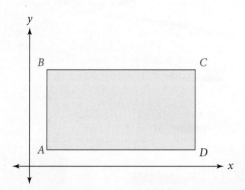

52. Rectangle *ABCD* has a length of 5 and a width of 3. Point *D* is the ordered pair (−1, 1). Find points *A*, *B*, and *C*.

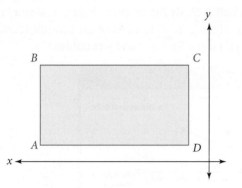

Getting Ready for the Next Section

53. Let $2x + 3y = 6$.

    a. Find x if $y = 4$.   **b.** Find x if $y = -2$.

c. Find y if $x = 3$.   **d.** Find y if $x = 9$.

54. Let $2x - 5y = 20$.

    a. Find x if $y = 0$.   **b.** Find x if $y = -6$.

    c. Find y if $x = 0$.   **d.** Find y if $x = 5$.

55. Let $y = 2x - 1$.

    a. Find x if $y = 7$.   **b.** Find x if $y = 3$.

    c. Find y if $x = 0$.   **d.** Find y if $x = 5$.

56. Let $y = 3x - 2$.

    a. Find x if $y = 4$.   **b.** Find x if $y = 7$.

    c. Find y if $x = 0$.   **d.** Find y if $x = -3$.

Maintaining Your Skills

Add or subtract as indicated.

57. $\dfrac{x}{5} + \dfrac{3}{5}$ **58.** $\dfrac{x}{5} + \dfrac{3}{4}$

59. $\dfrac{2}{7} - \dfrac{a}{7}$ **60.** $\dfrac{2}{7} - \dfrac{a}{5}$

61. $\dfrac{1}{14} - \dfrac{y}{7}$ **62.** $\dfrac{3}{4} + \dfrac{x}{5}$

63. $\dfrac{1}{2} + \dfrac{3}{x}$ **64.** $\dfrac{2}{3} - \dfrac{6}{y}$

65. $\dfrac{5 + x}{6} - \dfrac{5}{6}$ **66.** $\dfrac{3 - x}{3} + \dfrac{2}{3}$

67. $\dfrac{4}{x} + \dfrac{1}{2}$ **68.** $\dfrac{3}{y} + \dfrac{2}{3}$

3.2 Solutions to Linear Equations in Two Variables

OBJECTIVES

A Find solutions to linear equations in two variables.

B Determine whether an ordered pair is a solution to a linear equation in two variables.

TICKET TO SUCCESS

Keep these questions in mind as you read through the section. Then respond in your own words and in complete sentences.

1. How can you tell if an ordered pair is a solution to an equation?

2. How would you find a solution to $y = 3x - 5$?

3. Why is (3, 2) not a solution to $y = 3x - 5$?

4. How many solutions are there to an equation that contains two variables?

Suppose you want to switch all of the light bulbs in your employer's office from standard incandescent bulbs to energy-efficient bulbs. In order to convince your employer that the change would be cost effective, you need to provide a graph that compares power usage of the two bulbs. Study the graph in Figure 1. If you know the equation of the line segment from 1600 to 2600 lumens of an incandescent bulb is given by $y = \frac{1}{20}x + 20$, how would you solve for the x and y variables?

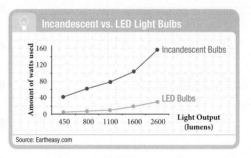

FIGURE 1

In this section, we will begin to investigate equations in two variables. As you will see, equations in two variables have pairs of numbers for solutions. Because we know how to use paired data to construct tables, line graphs, and other charts, we can take our work with paired data further by using equations in two variables to construct tables of paired data.

A Solving Linear Equations

Let's begin this section by reviewing the relationship between equations in one variable and their solutions.

If we solve the equation $3x - 2 = 10$, the solution is $x = 4$. If we graph this solution, we simply draw the real number line and place a dot at the point whose coordinate is 4. The relationship between linear equations in one variable, their solutions, and the graphs of those solutions look like this:

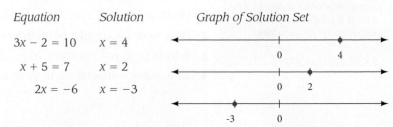

Equation	Solution	Graph of Solution Set
$3x - 2 = 10$	$x = 4$	
$x + 5 = 7$	$x = 2$	
$2x = -6$	$x = -3$	

NOTE

If this discussion seems a little long and confusing, you may want to look over some of the examples first and then come back and read this. Remember, it isn't always easy to read material in mathematics. What is important is that you understand what you are doing when you work problems. The reading is intended to assist you in understanding what you are doing. It is important to read everything in the book, but you don't always have to read it in the order it is written.

When the equation has one variable, the solution is a single number whose graph is a point on a line.

Now, consider the equation $2x + y = 3$. The first thing we notice is that there are two variables instead of one. Therefore, a solution to the equation $2x + y = 3$ will be not a single number but a pair of numbers, one for x and one for y, that makes the equation a true statement. One pair of numbers that works is $x = 2$, $y = -1$ because when we substitute them for x and y in the equation, we get a true statement.

$$2(2) + (-1) \stackrel{?}{=} 3$$
$$4 - 1 = 3$$
$$3 = 3 \qquad \text{A true statement}$$

The pair of numbers $x = 2$, $y = -1$ is written as $(2, -1)$. As you know from Section 3.1, $(2, -1)$ is called an *ordered pair* because it is a pair of numbers written in a specific order. The first number is always associated with the variable x, and the second number is always associated with the variable y. We call the first number in the ordered pair the *x-coordinate* and the second number the *y-coordinate* of the ordered pair.

Let's look back to the equation $2x + y = 3$. The ordered pair $(2, -1)$ is not the only solution. Another solution is $(0, 3)$ because when we substitute 0 for x and 3 for y we get

$$2(0) + 3 \stackrel{?}{=} 3$$
$$0 + 3 = 3$$
$$3 = 3 \qquad \text{A true statement}$$

Still another solution is the ordered pair $(5, -7)$ because

$$2(5) + (-7) \stackrel{?}{=} 3$$
$$10 - 7 = 3$$
$$3 = 3 \qquad \text{A true statement}$$

As a matter of fact, for any number we want to use for x, there is another number we can use for y that will make the equation a true statement. There is an infinite number of ordered pairs that satisfy (are solutions to) the equation $2x + y = 3$; we have listed just a few of them.

EXAMPLE 1 Given the equation $2x + 3y = 6$, complete the following ordered pairs so they will be solutions to the equation: $(0, y)$, $(x, 1)$, $(3, y)$.

SOLUTION To complete the ordered pair $(0, y)$, we substitute 0 for x in the equation and then solve for y:

$$2(0) + 3y = 6$$
$$3y = 6$$
$$y = 2$$

The ordered pair is $(0, 2)$.

To complete the ordered pair $(x, 1)$, we substitute 1 for y in the equation and solve for x:

$$2x + 3(1) = 6$$
$$2x + 3 = 6$$
$$2x = 3$$
$$x = \frac{3}{2}$$

The ordered pair is $\left(\frac{3}{2}, 1\right)$.

To complete the ordered pair $(3, y)$, we substitute 3 for x in the equation and solve for y.

$$2(3) + 3y = 6$$
$$6 + 3y = 6$$
$$3y = 0$$
$$y = 0$$

The ordered pair is $(3, 0)$.

Notice in each case in Example 1 that once we have used a number in place of one of the variables, the equation becomes a linear equation in one variable. We then use the method explained in Chapter 2 to solve for that variable.

EXAMPLE 2 Complete the following table for the equation $2x - 5y = 20$.

x	y
0	
	2
	0
−5	

SOLUTION Filling in the table is equivalent to completing the following ordered pairs: $(0, y)$, $(x, 2)$, $(x, 0)$, $(−5, y)$. So we proceed as in Example 1.

When $x = 0$, we have
$$2(0) - 5y = 20$$
$$0 - 5y = 20$$
$$-5y = 20$$
$$y = -4$$

When $y = 2$, we have
$$2x - 5(2) = 20$$
$$2x - 10 = 20$$
$$2x = 30$$
$$x = 15$$

When $y = 0$, we have
$$2x - 5(0) = 20$$
$$2x - 0 = 20$$
$$2x = 20$$
$$x = 10$$

When $x = -5$, we have
$$2(-5) - 5y = 20$$
$$-10 - 5y = 20$$
$$-5y = 30$$
$$y = -6$$

The completed table looks like this:

x	y
0	−4
15	2
10	0
−5	−6

The above table is equivalent to the ordered pairs $(0, −4)$, $(15, 2)$, $(10, 0)$, and $(−5, −6)$.

EXAMPLE 3 Complete the following table for the equation $y = 2x - 1$.

x	y
0	
5	
	7
	3

SOLUTION   When $x = 0$, we have   When $x = 5$, we have

$$y = 2(0) - 1 \qquad\qquad y = 2(5) - 1$$
$$y = 0 - 1 \qquad\qquad\quad y = 10 - 1$$
$$y = -1 \qquad\qquad\qquad y = 9$$

When $y = 7$, we have   When $y = 3$, we have

$$7 = 2x - 1 \qquad\qquad 3 = 2x - 1$$
$$8 = 2x \qquad\qquad\quad 4 = 2x$$
$$4 = x \qquad\qquad\qquad 2 = x$$

The completed table is

x	y
0	-1
5	9
4	7
2	3

Therefore, the ordered pairs $(0, -1)$, $(5, 9)$, $(4, 7)$, and $(2, 3)$ are among the solutions to the equation $y = 2x - 1$.   ■

B   Testing Solutions

EXAMPLE 4   Which of the ordered pairs $(2, 3)$, $(1, 5)$, and $(-2, -4)$ are solutions to the equation $y = 3x + 2$?

SOLUTION   If an ordered pair is a solution to the equation, then it must satisfy the equation; that is, when the coordinates are used in place of the variables in the equation, the equation becomes a true statement.

Try $(2, 3)$ in $y = 3x + 2$:

$$3 \stackrel{?}{=} 3(2) + 2$$
$$3 = 6 + 2$$
$$3 = 8 \qquad\qquad \text{A false statement}$$

Try $(1, 5)$ in $y = 3x + 2$:

$$5 \stackrel{?}{=} 3(1) + 2$$
$$5 = 3 + 2$$
$$5 = 5 \qquad\qquad \text{A true statement}$$

Try $(-2, -4)$ in $y = 3x + 2$:

$$-4 \stackrel{?}{=} 3(-2) + 2$$
$$-4 = -6 + 2$$
$$-4 = -4 \qquad\qquad \text{A true statement}$$

The ordered pairs $(1, 5)$ and $(-2, -4)$ are solutions to the equation $y = 3x + 2$, and $(2, 3)$ is not.   ■

Problem Set 3.2

Moving Toward Success

"It isn't the mountain ahead that wears you out; it's the grain of sand in your shoe."

—Robert W. Service, 1874–1958, poet and writer

the book, even if you have already taken notes in class?

1. Why is it important to take notes while you read

2. How will you mark things in your notes that are difficult for you?

A For each equation, complete the given ordered pairs. [Examples 1–3]

1. $2x + y = 6$ $(0, \ )$, $(\ , 0)$, $(\ , -6)$

2. $3x - y = 5$ $(0, \ )$, $(1, \ )$, $(\ , 5)$

3. $3x + 4y = 12$ $(0, \ )$, $(\ , 0)$, $(-4, \ )$

4. $5x - 5y = 20$ $(0, \ )$, $(\ , -2)$, $(1, \ )$

5. $y = 4x - 3$ $(1, \ )$, $(\ , 0)$, $(5, \ )$

6. $y = 3x - 5$ $(\ , 13)$, $(0, \ )$, $(-2, \ )$

7. $y = 7x - 1$ $(2, \ )$, $(\ , 6)$, $(0, \ )$

8. $y = 8x + 2$ $(3, \ )$, $(\ , 0)$, $(\ , -6)$

9. $x = -5$ $(\ , 4)$, $(\ , -3)$, $(\ , 0)$

10. $y = 2$ $(5, \ )$, $(-8, \ )$, $\left(\dfrac{1}{2}, \ \right)$

A For each of the following equations, complete the given table. [Examples 2–3]

11. $y = 3x$

x	y
1	3
-3	
	12
	18

12. $y = -2x$

x	y
-4	
0	
	10
	12

13. $y = 4x$

x	y
0	
	-2
-3	
	12

14. $y = -5x$

x	y
3	
	0
-2	
	-20

15. $x + y = 5$

x	y
2	
3	
	0
	-4

16. $x - y = 8$

x	y
0	
4	
	-3
	-2

17. $2x - y = 4$

x	y
	0
	2
1	
-3	

18. $3x - y = 9$

x	y
	0
	-9
5	
-4	

19. $y = 6x - 1$

x	y
0	
	-7
-3	
	8

20. $y = 5x + 7$

x	y
0	
-2	
-4	
	-8

B For the following equations, tell which of the given ordered pairs are solutions. [Example 4]

21. $2x - 5y = 10$ $(2, 3)$, $(0, -2)$, $\left(\dfrac{5}{2}, 1\right)$

22. $3x + 7y = 21$ $(0, 3)$, $(7, 0)$, $(1, 2)$

23. $y = 7x - 2$ $(1, 5)$, $(0, -2)$, $(-2, -16)$

24. $y = 8x - 3$ $(0, 3)$, $(5, 16)$, $(1, 5)$

25. $y = 6x$ $(1, 6)$, $(-2, -12)$, $(0, 0)$

26. $y = -4x$ $(0, 0)$, $(2, 4)$, $(-3, 12)$

27. $x + y = 0$ $(1, 1)$, $(2, -2)$, $(3, 3)$

28. $x - y = 1$ $(0, 1)$, $(0, -1)$, $(1, 2)$

29. $x = 3$ $(3, 0)$, $(3, -3)$, $(5, 3)$

30. $y = -4$ $(3, -4), (-4, 4), (0, -4)$

Applying the Concepts

31. Wind Energy The graph shows the world's capacity for generating power from wind energy. The line segment from 2008 to 2010 has the equation $y = 40,000x - 80,200,000$. Use it to complete the ordered pairs.

(, 120,000), (2009,   ), (, 200,000)

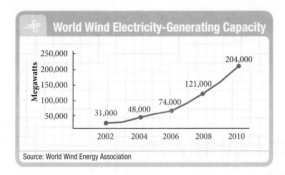

World Wind Electricity-Generating Capacity

Source: World Wind Energy Association

32. Light Bulbs The chart shows a comparison of power usage between incandescent and LED light bulbs. The equation of the line segment from 1600 to 2600 lumens of an incandescent bulb is given by $y = \left(\frac{1}{20}\right)x + 20$. Use it to complete the ordered pairs.

(1800,   ), (, 120), (2400,   )

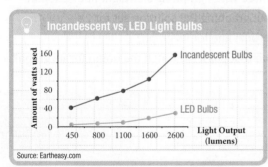

Incandescent vs. LED Light Bulbs

Source: Eartheasy.com

33. Perimeter If the perimeter of a rectangle is 30 inches, then the relationship between the length l and the width w is given by the equation

$$2l + 2w = 30$$

What is the length when the width is 3 inches?

34. Perimeter The relationship between the perimeter P of a square and the length of its side s is given by the formula $P = 4s$. If each side of a square is 5 inches, what is the perimeter? If the perimeter of a square is 28 inches, how long is a side?

35. Janai earns $12 per hour working as a math tutor. We can express the amount she earns each week, y, for working x hours with the equation $y = 12x$. Indicate with a *yes* or *no* which of the following could be one of Janai's paychecks. If you answer no, explain your answer.

  a. $60 for working 5 hours

  b. $100 for working nine hours

  c. $80 for working 7 hours

  d. $168 for working 14 hours

36. Erin earns $15 per hour working as a graphic designer. We can express the amount she earns each week, y, for working x hours with the equation $y = 15x$. Indicate with a *yes* or *no* which of the following could be one of Erin's paychecks. If you answer no, explain your answer.

  a. $75 for working 5 hours

  b. $125 for working 9 hours.

  c. $90 for working 6 hours

  d. $500 for working 35 hours.

37. The equation $V = -45,000t + 600,000$ can be used to find the value, V, of a small crane at the end of t years.

  a. What is the value of the crane at the end of 5 years?

  b. When is the crane worth $330,000?

  c. Is it true that the crane will be worth $150,000 after 9 years?

  d. How much did the crane cost?

38. The equation $V = -400t + 2,500$, can be used to find the value, V, of a notebook computer at the end of t years.

  a. What is the price of the notebook computer at the end of 4 years?

  b. When is the notebook computer worth $1,700?

  c. Is it true that the notebook computer will be worth $100 after 5 years?

  d. How much did the notebook computer cost?

Getting Ready for the Next Section

39. Find y when x is 4 in the formula $3x + 2y = 6$.

40. Find y when x is 0 in the formula $3x + 2y = 6$.

41. Find y when x is 0 in $y = -\frac{1}{3}x + 2$.

42. Find y when x is 3 in $y = -\frac{1}{3}x + 2$.

43. Find y when x is 2 in $y = \frac{3}{2}x - 3$.

44. Find y when x is 4 in $y = \frac{3}{2}x - 3$.

45. Solve $5x + y = 4$ for y.

46. Solve $-3x + y = 5$ for y.

47. Solve $3x - 2y = 6$ for y.

48. Solve $2x - 3y = 6$ for y.

Maintaining Your Skills

49. $\dfrac{11(-5) - 17}{2(-6)}$

50. $\dfrac{12(-4) + 15}{3(-11)}$

51. $\dfrac{13(-6) + 18}{4(-5)}$

52. $\dfrac{9^2 - 6^2}{-9 - 6}$

53. $\dfrac{7^2 - 5^2}{(7 - 5)^2}$

54. $\dfrac{7^2 - 2^2}{-7 - 2}$

55. $\dfrac{-3 \cdot 4^2 - 3 \cdot 2^4}{-3(8)}$

56. $\dfrac{-4(8 - 13) - 2(6 - 11)}{-5(3) + 5}$

3.3 Graphing Linear Equations in Two Variables

OBJECTIVES

A Graph a linear equation in two variables.

B Graph horizontal lines, vertical lines, and lines through the origin.

TICKET TO SUCCESS

Keep these questions in mind as you read through the section. Then respond in your own words and in complete sentences.

1. Should you use two points or three when graphing a straight line? Explain.

2. Explain how you would go about graphing the solution set for $2x - y = 4$.

3. What kind of equation has a vertical line for its graph?

4. What kind of equation has a line that passes through the origin for its graph?

A Graphing

One sunny day, you decide to test your golfing skills on a 9-hole golf course. After you sink your ball in the ninth hole, you realize that the number of strokes you used per hole was the hole number plus 2. The equation of the line that represents your afternoon of golf is given by $y = x + 2$. Aside from questioning your future career as a professional golfer, how do we know what this line looks like when we draw it on a rectangular coordinate system?

In this section, we will use the rectangular coordinate system introduced in Section 3.1 to obtain a visual picture of *all* solutions to a linear equation in two variables. The process we use to obtain a visual picture of all solutions to an equation is called *graphing*. The picture itself is called the *graph* of the equation.

EXAMPLE 1   Graph the solution set for $x + y = 5$.

SOLUTION   We know from the previous section that an infinite number of ordered pairs are solutions to the equation $x + y = 5$. We can't possibly list them all. What we can do is list a few of them and see if there is any pattern to their graphs.

Some ordered pairs that are solutions to $x + y = 5$ are (0, 5), (2, 3), (3, 2), (5, 0). The graph of each is shown in Figure 1.

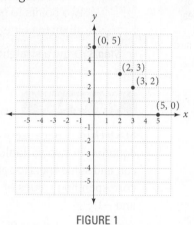

FIGURE 1

Now, by passing a straight line through these points, we can graph the solution set for the equation $x + y = 5$. Linear equations in two variables always have graphs that are straight lines. The graph of the solution set for $x + y = 5$ is shown in Figure 2.

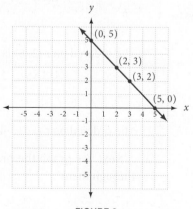

FIGURE 2

Every ordered pair that satisfies $x + y = 5$ has its graph on the line, and any point on the line has coordinates that satisfy the equation. So, there is a one-to-one correspondence between points on the line and solutions to the equation.

Here is the precise definition for a linear equation in two variables.

> **Definition**
>
> Any equation that can be put in the form $ax + by = c$, where a, b, and c are real numbers and a and b are not both 0, is called a **linear equation in two variables.** The graph of any equation of this form is a straight line (that is why these equations are called "linear"). The form $ax + by = c$ is called **standard form.**

To graph a linear equation in two variables, as we did in Example 1, we simply graph its solution set; that is, we draw a line through all the points whose coordinates satisfy the equation. Here are the steps to follow:

> **Strategy** Graphing a Linear Equation in Two Variables
>
> **Step 1:** Find any three ordered pairs that satisfy the equation. This can be done by using a convenient number for one variable and solving for the other variable.
>
> **Step 2:** Graph the three ordered pairs found in step 1. Actually, we need only two points to graph a straight line. The third point serves as a check. If all three points do not line up, there is a mistake in our work.
>
> **Step 3:** Draw a straight line through the three points graphed in step 2.

NOTE
The meaning of the *convenient numbers* referred to in step 1 of the strategy for graphing a linear equation in two variables will become clear as you read the next two examples.

EXAMPLE 2 Graph the equation $y = 3x - 1$.

SOLUTION Because $y = 3x - 1$ can be put in the form $ax + by = c$, it is a linear equation in two variables. Hence, the graph of its solution set is a straight line. We can find some specific solutions by substituting numbers for x and then solving for the corresponding values of y. We are free to choose any numbers for x, so let's use 0, 2, and -1.

NOTE
It may seem that we have simply picked the numbers 0, 2, and -1 out of the air and used them for x. In fact, we have done just that. Could we have used numbers other than these? The answer is yes. We can substitute any number for x. There will always be a value of y to go with it.

Let $x = 0$:
$$y = 3(0) - 1$$
$$y = 0 - 1$$
$$y = -1$$

The ordered pair $(0, -1)$ is one solution.

Let $x = 2$:
$$y = 3(2) - 1$$
$$y = 6 - 1$$
$$y = 5$$

The ordered pair $(2, 5)$ is a second solution.

Let $x = -1$:
$$y = 3(-1) - 1$$
$$y = -3 - 1$$
$$y = -4$$

The ordered pair $(-1, -4)$ is a third solution.

In table form

x	y
0	-1
2	5
-1	-4

Next, we graph the ordered pairs $(0, -1)$, $(2, 5)$, $(-1, -4)$ and draw a straight line through them.

The line we have drawn in Figure 3 is the graph of $y = 3x - 1$.

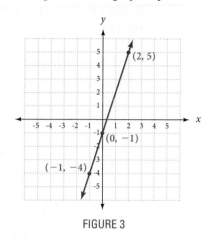

FIGURE 3

Example 2 illustrates the connection between algebra and geometry that we mentioned in this chapter introduction. Descartes's rectangular coordinate system allows us to associate the equation $y = 3x - 1$ (an algebraic concept) with a specific straight line (a geometric concept). The study of the relationship between equations in algebra and their associated geometric figures is called *analytic geometry*. The rectangular coordinate system is often referred to as the *Cartesian coordinate system* in honor of Descartes.

EXAMPLE 3   Graph the equation $y = -\frac{1}{3}x + 2$.

SOLUTION   We need to find three ordered pairs that satisfy the equation. To do so, we can let x equal any numbers we choose and find corresponding values of y. But every value of x we substitute into the equation is going to be multiplied by $-\frac{1}{3}$. For our convenience, let's use numbers for x that are divisible by 3, like -3, 0, and 3. That way, when we multiply them by $-\frac{1}{3}$, the result will be an integer.

> **NOTE**
> In Example 3 the values of x we used, -3, 0, and 3, are referred to as convenient values of x because they are easier to work with than some other numbers. For instance, if we let $x = 2$ in our original equation, we would have to add $-\frac{2}{3}$ and 2 to find the corresponding value of y. Not only would the arithmetic be more difficult, but also the ordered pair we obtained would have a fraction for its y-coordinate, making it more difficult to graph accurately.

Let $x = -3$: $y = -\frac{1}{3}(-3) + 2$

$y = 1 + 2$

$y = 3$

The ordered pair $(-3, 3)$ is one solution.

In table form

x	y
−3	3
0	2
3	1

Let $x = 0$: $y = -\frac{1}{3}(0) + 2$

$y = 0 + 2$

$y = 2$

The ordered pair $(0, 2)$ is a second solution.

Let $x = 3$: $y = -\frac{1}{3}(3) + 2$

$y = -1 + 2$

$y = 1$

The ordered pair $(3, 1)$ is a third solution.

Graphing the ordered pairs $(-3, 3)$, $(0, 2)$, and $(3, 1)$ and drawing a straight line through their graphs, we have the graph of the equation $y = -\frac{1}{3}x + 2$, as shown in Figure 4.

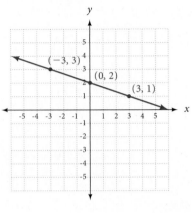

FIGURE 4

EXAMPLE 4   Graph the solution set for $3x - 2y = 6$.

SOLUTION   It will be easier to find convenient values of x to use in the equation if we first solve the equation for y. To do so, we add $-3x$ to each side, and then we multiply each side by $-\frac{1}{2}$.

$3x - 2y = 6$	Original equation
$-2y = -3x + 6$	Add $-3x$ to each side
$-\dfrac{1}{2}(-2y) = -\dfrac{1}{2}(-3x + 6)$	Multiply each side by $-\dfrac{1}{2}$
$y = \dfrac{3}{2}x - 3$	Simplify each side

Now, because each value of x will be multiplied by $\frac{3}{2}$, it will be to our advantage to choose values of x that are divisible by 2. That way, we will obtain values of y that do not contain fractions. This time, let's use 0, 2, and 4 for x.

When $x = 0$: $y = \dfrac{3}{2}(0) - 3$

$y = 0 - 3$

$y = -3$ $(0, -3)$ is one solution

When $x = 2$: $y = \dfrac{3}{2}(2) - 3$

$y = 3 - 3$

$y = 0$ $(2, 0)$ is a second solution

When $x = 4$: $y = \dfrac{3}{2}(4) - 3$

$y = 6 - 3$

$y = 3$ $(4, 3)$ is a third solution

Graphing the ordered pairs $(0, -3)$, $(2, 0)$, and $(4, 3)$ and drawing a line through them, we have the graph shown in Figure 5.

NOTE
After reading through Example 4, many students ask why we didn't use -2 for x when we were finding ordered pairs that were solutions to the original equation. The answer is, we could have. If we were to let $x = -2$, the corresponding value of y would have been -6. As you can see by looking at the graph in Figure 5, the ordered pair $(-2, -6)$ is on the graph.

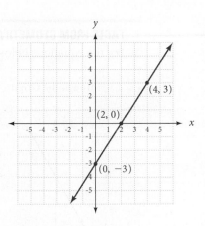

FIGURE 5

B Special Equations

EXAMPLE 5 Graph each of the following lines.

a. $y = \dfrac{1}{2}x$ **b.** $x = 3$ **c.** $y = -2$

SOLUTION

a. The line $y = \frac{1}{2}x$ passes through the origin because (0, 0) satisfies the equation. To sketch the graph, we need at least one more point on the line. When x is 2, we obtain the point (2, 1), and when x is −4, we obtain the point (−4, −2). The graph of $y = \frac{1}{2}x$ is shown in Figure 6A.

b. The line $x = 3$ is the set of all points whose x-coordinate is 3. The variable y does not appear in the equation, so the y-coordinate can be any number. Note that we can write our equation as a linear equation in two variables by writing it as $x + 0y = 3$. Because the product of 0 and y will always be 0, y can be any number. The graph of $x = 3$ is the vertical line shown in Figure 6B.

c. The line $y = -2$ is the set of all points whose y-coordinate is −2. The variable x does not appear in the equation, so the x-coordinate can be any number. Again, we can write our equation as a linear equation in two variables by writing it as $0x + y = -2$. Because the product of 0 and x will always be 0, x can be any number. The graph of $y = -2$ is the horizontal line shown in Figure 6C.

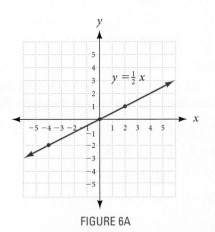

FIGURE 6A

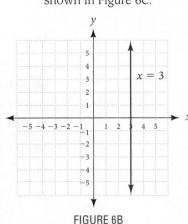

FIGURE 6B

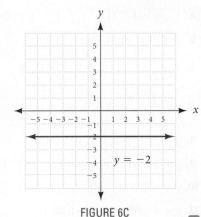

FIGURE 6C

FACTS FROM GEOMETRY: Special Equations and Their Graphs

For the equations below, m, a, and b are real numbers.

Through the Origin

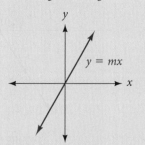

Vertical Line

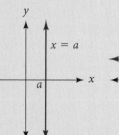

Horizontal Line

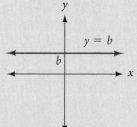

FIGURE 7A   **Any equation of the form $y = mx$ is a line that passes through the origin.**

FIGURE 7B   **Any equation of the form $x = a$ has a vertical line for its graph.**

FIGURE 7C   **Any equation of the form $y = b$ has a horizontal line for its graph.**

Problem Set 3.3

Moving Toward Success

"Without goals, and plans to reach them, you are like a ship that has set sail with no destination."

—Fitzhugh Dodson, 1923–1993, American psychologist and author

1. Why should you pay attention to concepts in your notes that are difficult for you?

2. When should you review your notes?
   a. Immediately after class
   b. Each day
   c. When studying for a quiz or exam
   d. All of the above

A **B** For the following equations, complete the given ordered pairs, and use the results to graph the solution set for the equation. [Examples 1–3, 5]

1. $x + y = 4$ (0,), (2,), (, 0)

2. $x - y = 3$ (0,), (2,), (, 0)

3. $x + y = 3$ (0,), (2,), (, −1)

4. $x - y = 4$ (1,), (−1,), (, 0)

5. $y = 2x$ (0,), (−2,), (2,)

6. $y = \dfrac{1}{2}x$ (0,), (−2,), (2,)

7. $y = \dfrac{1}{3}x$ (−3,), (0,), (3,)

8. $y = 3x$ (−2,), (0,), (2,)

9. $y = 2x + 1$ (0,), (−1,), (1,)

10. $y = -2x + 1$ (0,), (−1,), (1,)

11. $y = 4$ (0,), (−1,), (2,)

12. $x = 3$ (, −2), (, 0), (, 5)

13. $y = \dfrac{1}{2}x + 3$ (−2,), (0,), (2,)

14. $y = \dfrac{1}{2}x - 3$ (−2,), (0,), (2,)

15. $y = -\dfrac{2}{3}x + 1$ (−3,), (0,), (3,)

16. $y = -\dfrac{2}{3}x - 1$ (−3,), (0,), (3,)

A Solve each equation for y. Then, complete the given ordered pairs, and use them to draw the graph. [Example 4]

17. $2x + y = 3$ (−1,), (0,), (1,)

18. $3x + y = 2$ (−1,), (0,), (1,)

19. $3x + 2y = 6$ (0,), (2,), (4,)

20. $2x + 3y = 6$ (0,), (3,), (6,)

21. $-x + 2y = 6$ $(-2,\ ), (0,\ ), (2,\ )$

22. $-x + 3y = 6$ $(-3,\ ), (0,\ ), (3,\ )$

A **B** Find three solutions to each of the following equations, and then graph the solution set. [Examples 2–5]

23. $y = -\dfrac{1}{2}x$

24. $y = -2x$

25. $y = 3x - 1$

26. $y = -3x - 1$

27. $-2x + y = 1$

28. $-3x + y = 1$

29. $3x + 4y = 8$

30. $3x - 4y = 8$

31. $x = -2$

32. $y = 3$

33. $y = 2$

34. $x = -3$

A Graph each equation. [Examples 2–3]

35. $y = \dfrac{3}{4}x + 1$

36. $y = \dfrac{2}{3}x + 1$

37. $y = \dfrac{1}{3}x + \dfrac{2}{3}$

38. $y = \dfrac{1}{2}x + \dfrac{1}{2}$

39. $y = \dfrac{2}{3}x + \dfrac{2}{3}$

40. $y = -\dfrac{3}{4}x + \dfrac{3}{2}$

B For each equation in each table below, indicate whether the graph is horizontal (H), or vertical (V), or whether it passes through the origin (O). [Example 5]

41.

Equation	H, V, and/or O
$x = 3$	
$y = 3$	
$y = 3x$	
$y = 0$	

42.

Equation	H, V, and/or O
$x = \dfrac{1}{2}$	
$y = \dfrac{1}{2}$	
$y = \dfrac{1}{2}x$	
$x = 0$	

43.

Equation	H, V, and/or O
$x = -\dfrac{3}{5}$	
$y = -\dfrac{3}{5}$	
$y = -\dfrac{3}{5}x$	
$x = 0$	

44.

Equation	H, V, and/or O
$x = -4$	
$y = -4$	
$y = -4x$	
$y = 0$	

The next two problems are intended to give you practice reading, and paying attention to, the instructions that accompany the problems you are working. Working these problems is an excellent way to get ready for a test or a quiz.

45. Work each problem according to the instructions given.

    a. Solve: $2x + 5 = 10$

    b. Find x when y is 0: $2x + 5y = 10$

    c. Find y when x is 0: $2x + 5y = 10$

    d. Graph: $2x + 5y = 10$

    e. Solve for y: $2x + 5y = 10$

46. Work each problem according to the instructions given.

    a. Solve: $x - 2 = 6$

    b. Find x when y is 0: $x - 2y = 6$

    c. Find y when x is 0: $x - 2y = 6$

    d. Graph: $x - 2y = 6$

    e. Solve for y: $x - 2y = 6$

Applying the Concepts

47. Solar Energy The graph shows the rise in solar thermal collectors from 1997 to 2008. Use the chart to answer the following questions.

    a. Does the graph contain the point (2000, 7,500)?

    b. Does the graph contain the point (2004, 15,000)?

    c. Does the graph contain the point (2007, 15,000)?

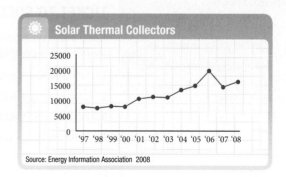

Solar Thermal Collectors

Source: Energy Information Association 2008

48. Health Care Costs The graph shows the projected rise in the cost of health care from 2002 to 2014. Using years as x and billions of dollars as y, write five ordered pairs that describe the information in the graph.

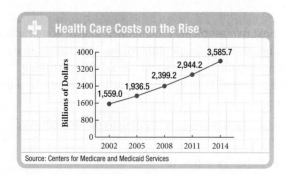

Health Care Costs on the Rise

Source: Centers for Medicare and Medicaid Services

Getting Ready for the Next Section

49. Let $3x - 2y = 6$.
  a. Find x when $y = 0$.
  b. Find y when $x = 0$.

50. Let $2x - 5y = 10$.
  a. Find x when $y = 0$.
  b. Find y when $x = 0$.

51. Let $-x + 2y = 4$.
  a. Find x when $y = 0$.
  b. Find y when $x = 0$.

52. Let $3x - y = 6$.
  a. Find x when $y = 0$.
  b. Find y when $x = 0$.

53. Let $y = -\dfrac{1}{3}x + 2$.
  a. Find x when $y = 0$.
  b. Find y when $x = 0$.

54. Let $y = \dfrac{3}{2}x - 3$.
  a. Find x when $y = 0$.
  b. Find y when $x = 0$.

Maintaining Your Skills

Apply the distributive property.

55. $\dfrac{1}{2}(4x + 10)$ **56.** $\dfrac{1}{2}(6x - 12)$

57. $\dfrac{2}{3}(3x - 9)$ **58.** $\dfrac{1}{3}(2x + 12)$

59. $\dfrac{3}{4}(4x + 10)$ **60.** $\dfrac{3}{4}(8x - 6)$

61. $\dfrac{3}{5}(10x + 15)$ **62.** $\dfrac{2}{5}(5x - 10)$

63. $5\left(\dfrac{2}{5}x + 10\right)$ **64.** $3\left(\dfrac{2}{3}x + 5\right)$

65. $4\left(\dfrac{3}{2}x - 7\right)$ **66.** $4\left(\dfrac{3}{4}x + 5\right)$

67. $\dfrac{3}{4}(2x + 12y)$

68. $\dfrac{3}{4}(8x - 16y)$

69. $\dfrac{1}{2}(5x - 10y) + 6$

70. $\dfrac{1}{3}(5x - 15y) - 5$

3.4 More on Graphing: Intercepts

OBJECTIVES

A Find the intercepts of a line from the equation of the line.

B Use intercepts to graph a line.

TICKET TO SUCCESS

Keep these questions in mind as you read through the section. Then respond in your own words and in complete sentences.

1. What is the x-intercept for a graph?

2. What is the y-intercept for a graph?

3. How would you find the y-intercept for a line from the equation $3x - y = 6$?

4. How would you graph a line for $y = -\dfrac{1}{2}x + 3$ using its intercepts?

A Intercepts

Suppose you were training a horse to race in an upcoming amateur event. You have paid a local college student $15 per hour to ride the horse during the first few days of training. The equation that gives the number of hours the student rides the horse x and the total amount of money you paid the student y is written as $y = 15x$. How would we graph this equation using the x- and y- intercepts?

In this section, we continue our work with graphing lines by finding the points where a line crosses the axes of our coordinate system. To do so, we use the fact that any point on the x-axis has a y-coordinate of 0 and any point on the y-axis has an x-coordinate of 0. We begin with the following definition:

> **Definition**
> The **x-intercept** of a straight line is the x-coordinate of the point where the graph crosses the x-axis. The **y-intercept** is defined similarly. It is the y-coordinate of the point where the graph crosses the y-axis.

B Using Intercepts to Graph Lines

If the x-intercept is a, then the point $(a, 0)$ lies on the graph. (This is true because any point on the x-axis has a y-coordinate of 0.)

If the y-intercept is b, then the point $(0, b)$ lies on the graph. (This is true because any point on the y-axis has an x-coordinate of 0.)

Graphically, the relationship is shown in Figure 1.

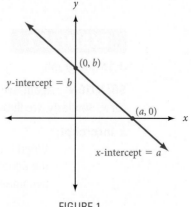

FIGURE 1

EXAMPLE 1 Find the x- and y-intercepts for $3x - 2y = 6$, and then use them to draw the graph.

SOLUTION To find where the graph crosses the x-axis, we let $y = 0$. (The y-coordinate of any point on the x-axis is 0.)

x-intercept:

$$\text{When} \rightarrow \qquad y = 0$$
$$\text{the equation} \rightarrow \quad 3x - 2y = 6$$
$$\text{becomes} \rightarrow \quad 3x - 2(0) = 6$$
$$3x - 0 = 6$$
$$3x = 6$$
$$x = 2 \qquad \text{Multiply each side by } \tfrac{1}{3}$$

The graph crosses the x-axis at (2, 0), which means the x-intercept is 2.

y-intercept:

$$\text{When} \rightarrow \qquad x = 0$$
$$\text{the equation} \rightarrow \quad 3x - 2y = 6$$
$$\text{becomes} \rightarrow \quad 3(0) - 2y = 6$$
$$0 - 2y = 6$$
$$-2y = 6$$
$$y = -3 \qquad \text{Multiply each side by } -\tfrac{1}{2}$$

The graph crosses the y-axis at (0, −3), which means the y-intercept is −3.

Plotting the x- and y-intercepts and then drawing a line through them, we have the graph of $3x - 2y = 6$, as shown in Figure 2.

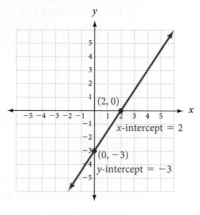

FIGURE 2

EXAMPLE 2 Graph $-x + 2y = 4$ by finding the intercepts and using them to draw the graph.

SOLUTION Again, we find the x-intercept by letting $y = 0$ in the equation and solving for x. Similarly, we find the y-intercept by letting $x = 0$ and solving for y.

x-intercept:

$$\text{When} \rightarrow \qquad y = 0$$
$$\text{the equation} \rightarrow \quad -x + 2y = 4$$
$$\text{becomes} \rightarrow \quad -x + 2(0) = 4$$

$$-x + 0 = 4$$
$$-x = 4$$
$$x = -4 \qquad \text{Multiply each side by } -1$$

The x-intercept is -4, indicating that the point $(-4, 0)$, is on the graph of $-x + 2y = 4$.

y-intercept:

$$\text{When} \rightarrow \qquad x = 0$$
$$\text{the equation} \rightarrow -x + 2y = 4$$
$$\text{becomes} \rightarrow \quad -0 + 2y = 4$$
$$2y = 4$$
$$y = 2 \qquad \text{Multiply each side by } \frac{1}{2}$$

The y-intercept is 2, indicating that the point $(0, 2)$ is on the graph of $-x + 2y = 4$.

Plotting the intercepts and drawing a line through them, we have the graph of $-x + 2y = 4$, as shown in Figure 3.

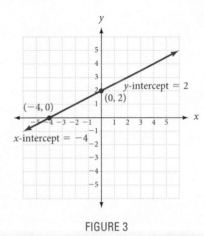

FIGURE 3

Graphing a line by finding the intercepts, as we have done in Examples 1 and 2, is an easy method of graphing if the equation has the form $ax + by = c$ and both the numbers a and b divide the number c evenly.

In our next example, we use the intercepts to graph a line in which y is given in terms of x.

EXAMPLE 3   Use the intercepts for $y = -\dfrac{1}{3}x + 2$ to draw its graph.

SOLUTION   We graphed this line previously in Example 3 of Section 3.3 by substituting three different values of x into the equation and solving for y. This time we will graph the line by finding the intercepts.

x-intercept:

$$\text{When} \rightarrow \qquad y = 0$$

$$\text{the equation} \rightarrow \quad y = -\frac{1}{3}x + 2$$

$$\text{becomes} \rightarrow \qquad 0 = -\frac{1}{3}x + 2$$

$$-2 = -\frac{1}{3}x \qquad \text{Add } -2 \text{ to each side}$$

$$6 = x \qquad \text{Multiply each side by } -3$$

The x-intercept is 6, which means the graph passes through the point $(6, 0)$.

y-intercept:

When → $x = 0$

the equation → $y = -\dfrac{1}{3}x + 2$

becomes → $y = -\dfrac{1}{3}(0) + 2$

$y = 2$

The y-intercept is 2, which means the graph passes through the point (0, 2).

The graph of $y = -\dfrac{1}{3}x + 2$ is shown in Figure 4. Compare this graph, and the method used to obtain it, with Example 3 in Section 3.3.

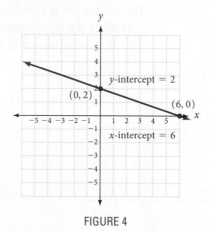

FIGURE 4

Problem Set 3.4

Moving Toward Success

"Success is the sum of small efforts, repeated day in and day out."

—Robert Collier, 1885–1950, author

1. Should you do all of the assigned homework problems or just a few? Explain.

2. When should you review material for the next exam?

   a. Only at the last minute before the exam

   b. After you see it on the exam

   c. Each day

   d. Only after you learn it

A **B** Find the x- and y-intercepts for the following equations. Then use the intercepts to graph each equation. [Examples 1–3]

1. $2x + y = 4$
2. $2x + y = 2$
3. $-x + y = 3$
4. $-x + y = 4$
5. $-x + 2y = 2$
6. $-x + 2y = 4$
7. $5x + 2y = 10$
8. $2x + 5y = 10$
9. $-4x + 5y = 20$
10. $-5x + 4y = 20$
11. $3x - 4y = -4$
12. $-2x + 3y = 3$
13. $x - 3y = 2$
14. $x - 2y = 1$
15. $2x - 3y = -2$
16. $3x + 4y = 6$
17. $y = 2x - 6$
18. $y = 2x + 6$
19. $y = 2x - 1$
20. $y = -2x - 1$
21. $y = \dfrac{1}{2}x + 3$
22. $y = \dfrac{1}{2}x - 3$
23. $y = -\dfrac{1}{3}x - 2$
24. $y = -\dfrac{1}{3}x + 2$

For each of the following lines, the x-intercept and the y-intercept are both 0, which means the graph of each will go through the origin, (0, 0). Graph each line by finding a point on each, other than the origin, and then drawing a line through that point and the origin.

25. $y = -2x$

26. $y = \frac{1}{2}x$

27. $y = -\frac{1}{3}x$

28. $y = -3x$

29. $y = \frac{2}{3}x$

30. $y = \frac{3}{2}x$

A Complete each table.

31.

Equation	x-intercept	y-intercept
$3x + 4y = 12$		
$3x + 4y = 4$		
$3x + 4y = 3$		
$3x + 4y = 2$		

32.

Equation	x-intercept	y-intercept
$-2x + 3y = 6$		
$-2x + 3y = 3$		
$-2x + 3y = 2$		
$-2x + 3y = 1$		

33.

Equation	x-intercept	y-intercept
$x - 3y = 2$		
$y = \frac{1}{3}x - \frac{2}{3}$		
$x - 3y = 0$		
$y = \frac{1}{3}x$		

34.

Equation	x-intercept	y-intercept
$x - 2y = 1$		
$y = \frac{1}{2}x - \frac{1}{2}$		
$x - 2y = 0$		
$y = \frac{1}{2}x$		

The next two problems are intended to give you practice reading, and paying attention to, the instructions that accompany the problems you are working. Working these problems is an excellent way to get ready for a test or a quiz.

35. Work each problem according to the instructions given.

    a. Solve: $2x - 3 = -3$

    b. Find the x-intercept: $2x - 3y = -3$

    c. Find y when x is 0: $2x - 3y = -3$

    d. Graph: $2x - 3y = -3$

    e. Solve for y: $2x - 3y = -3$

36. Work each problem according to the instructions given.

    a. Solve: $3x - 4 = -4$

    b. Find the y-intercept: $3x - 4y = -4$

    c. Find x when y is 0: $3x - 4y = -4$

    d. Graph: $3x - 4y = -4$

    e. Solve for y: $3x - 4y = -4$

37. Graph the line that passes through the points $(-2, 5)$ and $(5, -2)$. What are the x- and y-intercepts for this line?

38. Graph the line that passes through the points $(5, 3)$ and $(-3, -5)$. What are the x- and y-intercepts for this line?

From the graphs below, find the x- and y-intercepts for each line.

39.

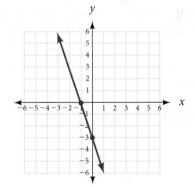

40.

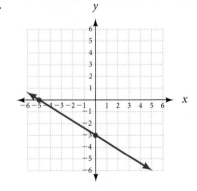

41. Use the graph to complete the following table.

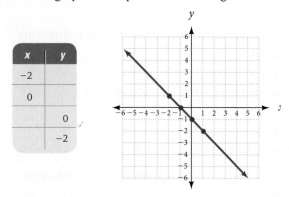

x	y
−2	
0	
	0
	−2

42. Use the graph to complete the following table.

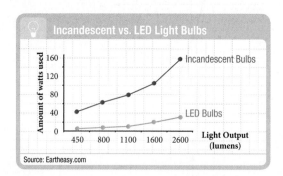

x	y
−2	
0	
	0
	6

Applying the Concepts

43. Light Bulbs The chart shows a comparison of power usage between incandescent and LED light bulbs. The equation of the line segment from 450 to 800 lumens of an LED bulb can be given by $y = \frac{1}{175}x + \frac{17}{7}$. Find the x- and y-intercepts of this equation.

Incandescent vs. LED Light Bulbs

Source: Eartheasy.com

44. Horse Racing The graph shows the total amount of money wagered on the Kentucky Derby. The equation of the line segment from 1985 to 1990 is given by the equation $y = 2.84x - 5{,}617.2$. Find the x- and y-intercepts of this equation to the nearest tenth.

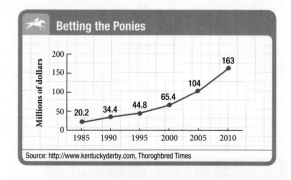

Betting the Ponies

Source: http://www.kentuckyderby.com, Thoroghbred Times

45. Working Two Jobs Maggie has a job working in an office for $10 an hour and another job driving a tractor for $12 an hour. Maggie works at both jobs and earns $480 in one week.

　a. Write an equation that gives the number of hours, x, she worked for $10 per hour and the number of hours, y, she worked for $12 per hour.

　b. Find the x- and y-intercepts for this equation.

　c. Graph this equation from the intercepts, using only the first quadrant.

　d. From the graph, find how many hours she worked at $12 if she worked 36 hours at $10 per hour.

　e. From the graph, find how many hours she worked at $10 if she worked 25 hours at $12 per hour.

46. Ticket prices Devin is selling tickets to a barbecue. Adult tickets cost $6.00 and children's tickets cost $4.00. The total amount of money he collects is $240.

　a. Write an equation that gives the number of adult tickets, x, he sold for $6 and the children's tickets, y, he sold for $4.

　b. Find the x- and y-intercepts for this equation.

　c. Graph this equation from the intercepts, using only the first quadrant.

　d. From the graph, find how many children's tickets were sold if 10 adult tickets were sold.

　e. From the graph, find how many adult tickets were sold if 30 children's tickets were sold.

47. Evaluate.

   a. $\dfrac{5-2}{3-1}$ **b.** $\dfrac{2-5}{1-3}$

48. Evaluate.

   a. $\dfrac{-4-1}{5-(-2)}$ **b.** $\dfrac{1+4}{-2-5}$

49. Evaluate the following expressions when $x = 3$ and $y = 5$.

   a. $\dfrac{y-2}{x-1}$ **b.** $\dfrac{2-y}{1-x}$

50. Evaluate the following expressions when $x = 3$ and $y = 2$.

   a. $\dfrac{-4-y}{5-x}$ **b.** $\dfrac{y+4}{x-5}$

Solve each equation.

51. $-12y - 4 = -148$ **52.** $-2x - 18 = 4$

53. $-5y - 4 = 51$ **54.** $-2y + 18 = -14$

55. $11x - 12 = -78$ **56.** $21 + 9y = -24$

57. $9x + 3 = 66$ **58.** $-11 - 15a = -71$

59. $-9c - 6 = 12$ **60.** $-7a + 28 = -84$

61. $4 + 13c = -9$ **62.** $-3x + 15 = -24$

63. $3y - 12 = 30$ **64.** $9x + 11 = -16$

65. $-11y + 9 = 75$ **66.** $9x - 18 = -72$

3.5 The Slope of a Line

OBJECTIVES

A   Find the slope of a line from two points on the line.

B   Graph a line given the slope and y-intercept.

TICKET TO SUCCESS

Keep these questions in mind as you read through the section. Then respond in your own words and in complete sentences.

1. Using x- and y-coordinates, how do you find the slope of a line?
2. Would you rather climb a hill with a slope of 1 or a slope of 3? Explain why.
3. Describe how you would graph a line from its slope and y-intercept.
4. Describe how to obtain the slope of a line if you know the coordinates of two points on the line.

In 1990, the United States produced 205 million tons of garbage. Ten years later, the amount of garbage produced had increased to 224 million tons. We can find the slope of a line that represents this data by creating a graph, with the years on the x-axis and the garbage production in millions of tons on the *y*-axis. In this section, we will learn how to find the slope of a line from a linear equation in two variables.

In defining the slope of a straight line, we are looking for a number to associate with that line that does two things. First of all, we want the slope of a line to measure the "steepness" of the line; that is, in comparing two lines, the slope of the steeper line should have the larger numerical absolute value. Second, we want a line that *rises* going from left to right to have a *positive* slope. We want a line that *falls* going from left to right to have a *negative* slope. (A line that neither rises nor falls going from left to right must, therefore, have 0 slope.) These are illustrated in Figure 1.

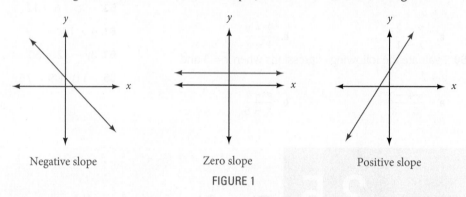

Negative slope   Zero slope   Positive slope

FIGURE 1

NOTE

The 2 in x_2 is called a subscript. It is notation that allows us to distinguish between the variables x_1 and x_2, while still showing that they are both *x*-coordinates.

Suppose we know the coordinates of two points on a line. Because we are trying to develop a general formula for the slope of a line, we will use general points. Let's call the two points $P_1(x_1, y_1)$ and $P_2(x_2, y_2)$. They represent the coordinates of any two different points on our line. We define the *slope* of our line to be the ratio of the vertical change of *y*-coordinates to the horizontal change of *x*-coordinates as we move from point (x_1, y_1) to point (x_2, y_2) on the line. (See Figure 2.)

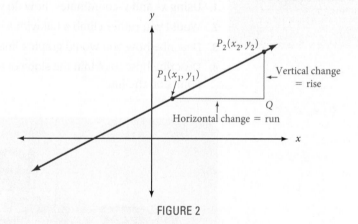

FIGURE 2

We call the vertical change the *rise* in the graph and the horizontal change the *run* in the graph. The slope, then, is

$$\text{Slope} = \frac{\text{vertical change}}{\text{horizontal change}} = \frac{\text{rise}}{\text{run}}$$

We would like to have a numerical value to associate with the rise in the graph and a numerical value to associate with the run in the graph. A quick study of Figure 2 shows that the coordinates of point Q must be (x_2, y_1), because Q is directly below point P_2 and right across from point P_1. We can draw our diagram again in the manner shown in Figure 3. It is apparent from this graph that the rise can be expressed as $(y_2 - y_1)$ and the run as $(x_2 - x_1)$. We usually denote the slope of a line by the letter *m*.

The complete definition of slope follows along with a diagram (Figure 3) that illustrates the definition.

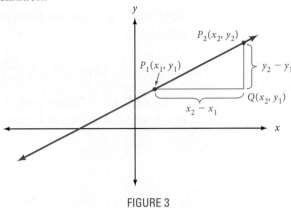

FIGURE 3

> **Definition**
>
> If points (x_1, y_1) and (x_2, y_2) are any two different points, then the **slope** of the line on which they lie is
>
> $$\text{Slope} = m = \frac{\text{rise}}{\text{run}} = \frac{y_2 - y_1}{x_2 - x_1}$$

This definition of the slope of a line does just what we want it to do. If the line rises going from left to right, the slope will be positive. If the line falls from left to right, the slope will be negative. Also, the steeper the line, the larger numerical absolute value the slope will have.

A Finding the Slope Given Two Points on a Line

EXAMPLE 1 Find the slope of the line between the points $(1, 2)$ and $(3, 5)$.

SOLUTION We can let

$$(x_1, y_1) = (1, 2)$$

and

$$(x_2, y_2) = (3, 5)$$

then

$$m = \frac{y_2 - y_1}{x_2 - x_1} = \frac{5 - 2}{3 - 1} = \frac{3}{2}$$

The slope is $\frac{3}{2}$. For every vertical change of 3 units, there will be a corresponding horizontal change of 2 units. (See Figure 4.)

NOTE
If we let $(x_1, y_1) = (3, 5)$ and $(x_2, y_2) = (1, 2)$, the slope is:
$$m = \frac{y_2 - y_1}{x_2 - x_1} = \frac{2 - 5}{1 - 3}$$
$$= \frac{-3}{-2}$$
$$= \frac{3}{2}$$
The slope is the same no matter which point we use first.

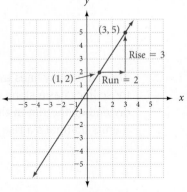

FIGURE 4

EXAMPLE 2 Find the slope of the line through $(-2, 1)$ and $(5, -4)$.

SOLUTION It makes no difference which ordered pair we call (x_1, y_1) and which we call (x_2, y_2).

$$\text{Slope} = m = \frac{y_2 - y_1}{x_2 - x_1} = \frac{-4 - 1}{5 - (-2)} = \frac{-5}{7}$$

The slope is $-\dfrac{5}{7}$. Every vertical change of -5 units (down 5 units) is accompanied by a horizontal change of 7 units (to the right 7 units). (See Figure 5.)

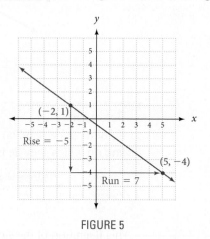

FIGURE 5

B Graphing a Line from Its Slope and y-intercept

EXAMPLE 3 Graph the line with slope $\dfrac{3}{2}$ and y-intercept 1.

SOLUTION Because the y-intercept is 1, we know that one point on the line is $(0, 1)$. So, we begin by plotting the point $(0, 1)$, as shown in Figure 6.

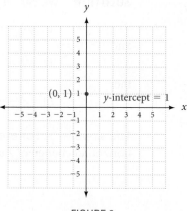

FIGURE 6

There are many lines that pass through the point shown in Figure 6, but only one of those lines has a slope of $\dfrac{3}{2}$. The slope, $\dfrac{3}{2}$, can be thought of as the rise in the graph divided by the run in the graph. Therefore, if we start at the point $(0, 1)$ and move 3 units up (that's a rise of 3) and then 2 units to the right (a run of 2), we will be at another point on the graph. Figure 7 shows that the point we reach by doing so is the point (2, 4).

$$\text{Slope} = m = \frac{\text{rise}}{\text{run}} = \frac{3}{2}$$

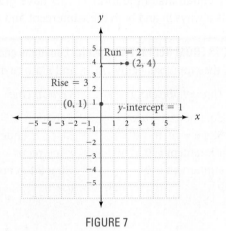

FIGURE 7

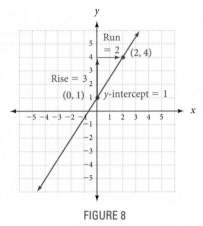

FIGURE 8

To graph the line with slope $\frac{3}{2}$ and y-intercept 1, we simply draw a line through the two points in Figure 7 to obtain the graph shown in Figure 8.

EXAMPLE 4   Find the slope of the line containing (3, −1) and (3, 4).

SOLUTION   Using the definition for slope, we have

$$m = \frac{-1 - 4}{3 - 3} = \frac{-5}{0}$$

The expression $\frac{-5}{0}$ is undefined; that is, there is no real number to associate with it. In this case, we say the *slope is undefined.*

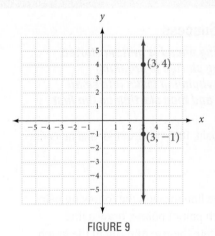

FIGURE 9

The graph of our line is shown in Figure 9. Our line is a vertical line. All vertical lines have an undefined slope. (And all horizontal lines, as we mentioned earlier, have 0 slope.)

As a final note, the summary below reminds us that all horizontal lines have equations of the form $y = b$ and slopes of 0. Because they cross the y-axis at b, the y-intercept is b; there is no x-intercept. Vertical lines have an undefined slope and equations of the form $x = a$. Each will have an x-intercept at a and no y-intercept. Finally, equations of the form $y = mx$ have graphs that pass through the origin. The slope is always m and both the x-intercept and the y-intercept are 0.

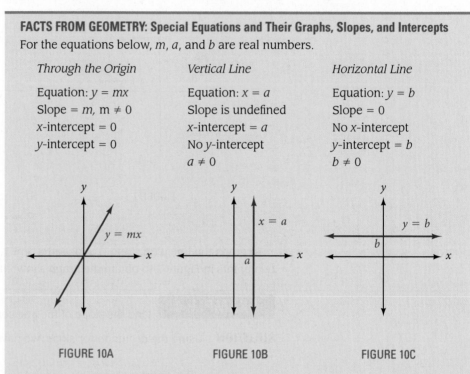

FACTS FROM GEOMETRY: Special Equations and Their Graphs, Slopes, and Intercepts
For the equations below, m, a, and b are real numbers.

Through the Origin

Equation: $y = mx$
Slope = m, m ≠ 0
x-intercept = 0
y-intercept = 0

Vertical Line

Equation: $x = a$
Slope is undefined
x-intercept = a
No y-intercept
a ≠ 0

Horizontal Line

Equation: $y = b$
Slope = 0
No x-intercept
y-intercept = b
b ≠ 0

FIGURE 10A   FIGURE 10B   FIGURE 10C

Problem Set 3.5

Moving Toward Success

"The secret of getting ahead is getting started. The secret of getting started is breaking your complex overwhelming tasks into small manageable tasks, and then starting on the first one."

—Mark Twain, 1835–1910, author and humorist

1. Why should you review material that will be covered on the exam each day?

2. Should you work problems when you review material? Why or why not?

A Find the slope of the line through the following pairs of points. Then plot each pair of points, draw a line through them, and indicate the rise and run in the graph. [Examples 1–2]

1. (2, 1), (4, 4)

2. (3, 1), (5, 4)

3. (1, 4), (5, 2)

4. (1, 3), (5, 2)

5. (1, −3), (4, 2)

6. (2, −3), (5, 2)

7. (−3, −2), (1, 3)

8. (−3, −1), (1, 4)

9. (−3, 2), (3, −2)

10. (−3, 3), (3, −1)

11. (2, −5), (3, −2)

12. (2, −4), (3, −1)

B In each of the following problems, graph the line with the given slope and y-intercept b. [Example 3]

13. $m = \dfrac{2}{3}, b = 1$

14. $m = \dfrac{3}{4}, b = -2$

15. $m = \dfrac{3}{2}, b = -3$

16. $m = \dfrac{4}{3}, b = 2$

17. $m = -\dfrac{4}{3}, b = 5$ **18.** $m = -\dfrac{3}{5}, b = 4$

19. $m = 2, b = 1$ **20.** $m = -2, b = 4$

21. $m = 3, b = -1$ **22.** $m = 3, b = -2$

Find the slope and y-intercept for each line.

23.

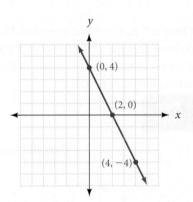

24.

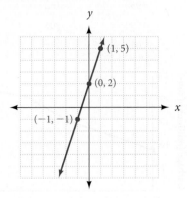

25.

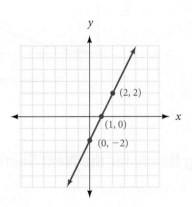

26.

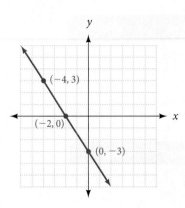

27. Graph the line that has an x-intercept of 3 and a y-intercept of -2. What is the slope of this line?

28. Graph the line that has an x-intercept of 2 and a y-intercept of -3. What is the slope of this line?

29. Graph the line with x-intercept 4 and y-intercept 2. What is the slope of this line?

30. Graph the line with x-intercept -4 and y-intercept -2. What is the slope of this line?

31. Graph the line $y = 2x - 3$, then name the slope and y-intercept by looking at the graph.

32. Graph the line $y = -2x + 3$, then name the slope and y-intercept by looking at the graph.

33. Graph the line $y = \dfrac{1}{2}x + 1$, then name the slope and y-intercept by looking at the graph.

34. Graph the line $y = -\dfrac{1}{2}x - 2$, then name the slope and y-intercept by looking at the graph.

35. Find y if the line through $(4, 2)$ and $(6, y)$ has a slope of 2.

36. Find y if the line through $(1, y)$ and $(7, 3)$ has a slope of 6.

For each equation in each table, give the slope of the graph.

37.

Equation	slope
$x = 3$	
$y = 3$	
$y = 3x$	

38.

Equation	slope
$y = \dfrac{3}{2}$	
$x = \dfrac{3}{2}$	
$y = \dfrac{3}{2}x$	

39.

Equation	slope
$y = -\frac{2}{3}$	
$x = -\frac{2}{3}$	
$y = -\frac{2}{3}x$	

40.

Equation	slope
$x = -2$	
$y = -2$	
$y = -2x$	

Applying the Concepts

41. Health Care The graph shows the rising cost of health care. What is the slope between 2005 and 2008? Round to the nearest hundredth.

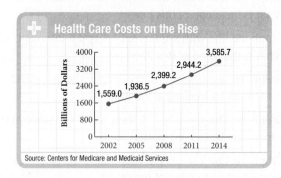

Health Care Costs on the Rise

1,559.0 1,936.5 2,399.2 2,944.2 3,585.7

2002 2005 2008 2011 2014

Source: Centers for Medicare and Medicaid Services

42. Wind Energy The graph shows the world's capacity for generating power from wind energy. Estimate the slope from 2008 to 2010.

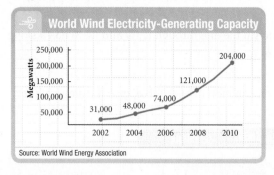

World Wind Electricity-Generating Capacity

31,000 48,000 74,000 121,000 204,000

2002 2004 2006 2008 2010

Source: World Wind Energy Association

43. Garbage Production The table and completed line graph shown here give the annual production of garbage in the United States for some specific years.

Find the slope of each of the four line segments, A, B, C, and D.

Year	Garbage (million of tons)
1960	88
1970	121
1980	152
1990	205
2000	224

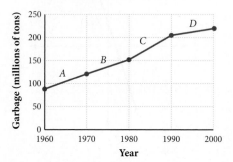

44. Plant Height The table and completed line graph shown here give the growth rate of a certain species of plant over time. Find the slopes of the line segments labeled A, B, and C.

Day	Plant Height
0	0
1	0.5
2	1
3	1.5
4	3
5	4
6	6
7	9
8	13
9	18
10	23

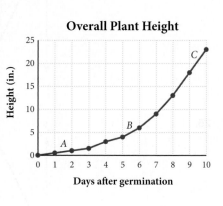

Overall Plant Height

Getting Ready for the Next Section

Solve each equation for y.

45. $-2x + y = 4$ **46.** $-4x + y = -2$

47. $3x + y = 3$ **48.** $3x + 2y = 6$

49. $4x - 5y = 20$

50. $-2x - 5y = 10$

51. $y - 3 = -2(x + 4)$

52. $y + 5 = 2(x + 2)$

53. $y - 3 = -\frac{2}{3}(x + 3)$

54. $y - 1 = -\frac{1}{2}(x + 4)$

55. $\frac{y - 1}{x} = \frac{3}{2}$

56. $\frac{y + 1}{x} = \frac{3}{2}$

Maintaining Your Skills

Solve each equation.

57. $\frac{1}{2}(4x + 10) = 11$ **58.** $\frac{1}{2}(6x - 12) = -18$

59. $\frac{2}{3}(3x - 9) = 24$ **60.** $\frac{1}{3}(2x + 15) = -29$

61. $\frac{3}{4}(4x + 8) = -12$ **62.** $\frac{3}{4}(8x - 16) = -36$

63. $\frac{3}{5}(10x + 15) = 45$ **64.** $\frac{2}{5}(5x - 10) = -10$

65. $5\left(\frac{2}{5}x + 10\right) = -28$ **66.** $3\left(\frac{2}{3}x + 5\right) = -13$

67. $4\left(\frac{3}{2}x - 7\right) = -4$ **68.** $4\left(\frac{3}{4}x + 5\right) = -40$

69. $\frac{3}{4}(2x + 12) = 24$ **70.** $-\frac{3}{4}(12x - 16) = -42$

71. $\frac{1}{2}(5x - 10) + 6 = -49$

72. $\frac{1}{3}(5x - 15) - 5 = 20$

TICKET TO SUCCESS

Keep these questions in mind as you read through the section. Then respond in your own words and in complete sentences.

1. When graphing a linear inequality in two variables, how do you find the equation of the boundary line?
2. What is the significance of a broken line in the graph of an inequality?
3. When graphing a linear inequality in two variables, how do you know which side of the boundary line to shade?
4. Why is the coordinate $(0, 0)$ a convenient test point?

When statistics are gathered, a random sampling error is calculated and applied to the result. This margin of error can be represented by a linear inequality. The graph in Figure 1 shows the rising cost of health care projected up to the year 2014. If we said the margin of error for this graph is ±300 billion dollars, how would we redraw the graph to include this margin? In this section, we will learn how to graph linear inequalities in two variables.

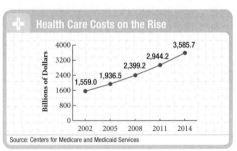

FIGURE 1

A Graphing Linear Inequalities in Two Variables

A linear inequality in two variables is any expression that can be put in the form

$$ax + by < c$$

where a, b, and c are real numbers (a and b not both 0). The inequality symbol can be any of the following four: $<$, $\leq$, $>$, $\geq$.

Some examples of linear inequalities are

$$2x + 3y < 6 \qquad y \geq 2x + 1 \qquad x - y \leq 0$$

Although not all of these inequalities have the form $ax + by < c$, each one can be put in that form.

The solution set for a linear inequality is a section of the coordinate plane. The boundary for the section is found by replacing the inequality symbol with an equal sign and graphing the resulting equation using the methods summarized in the previous section. The boundary is included in the solution set (and represented with a solid line) if the inequality symbol used originally is $\leq$ or $\geq$. The boundary is not included (and is represented with a broken line) if the original symbol is $<$ or $>$.

Let's look at some examples.

EXAMPLE 1　Graph the solution set for $x + y \leq 4$.

SOLUTION　The boundary for the graph is the graph of $x + y = 4$. The boundary (a solid line) is included in the solution set because the inequality symbol is $\leq$.

The graph of the boundary is shown in Figure 2.

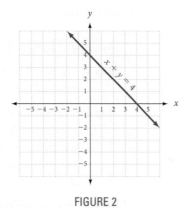

FIGURE 2

The boundary separates the coordinate plane into two sections, or regions: the region above the boundary and the region below the boundary. The solution set for $x + y \leq 4$ is one of these two regions along with the boundary. To find the correct region, we simply choose any convenient point that is *not* on the boundary. We then substitute the coordinates of the point into the original inequality $x + y \leq 4$. If the point we choose satisfies the inequality, then it is a member of the solution set, and we can assume that all points on the same side of the boundary as the chosen point are also in the solution set. If the coordinates of our point do not satisfy the original inequality, then the solution set lies on the other side of the boundary.

In this example, a convenient point not on the boundary is the origin. Substituting $(0, 0)$ into $x + y \leq 4$ gives us

$$0 + 0 \overset{?}{\leq} 4$$
$$0 \leq 4 \qquad\qquad \text{A true statement}$$

Because the origin is a solution to the inequality $x + y \leq 4$, and the origin is below the boundary, all other points below the boundary are also solutions.

The graph of $x + y \leq 4$ is shown in Figure 3.

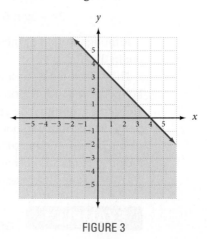

FIGURE 3

The region above the boundary is described by the inequality $x + y > 4$.

Here is a list of steps to follow when graphing the solution set for linear inequalities in two variables.

> **Strategy**   To Graph the Solution Set for Linear Inequalities in Two Variables
>
> **Step 1:** Replace the inequality symbol with an equal sign. The resulting equation represents the boundary for the solution set.
>
> **Step 2:** Graph the boundary found in step 1 using a *solid line* if the boundary is included in the solution set (that is, if the original inequality symbol was either ≤ or ≥). Use a *broken line* to graph the boundary if it is *not* included in the solution set, that is if the original inequality was either < or >.
>
> **Step 3:** Choose any convenient point not on the boundary and substitute the coordinates into the *original* inequality. If the resulting statement is *true*, the graph lies on the *same* side of the boundary as the chosen point. If the resulting statement is *false*, the solution set lies on the *opposite* side of the boundary.

EXAMPLE 2   Graph the solution set for $y < 2x - 3$.

SOLUTION   The boundary is the graph of $y = 2x - 3$. The boundary is not included because the original inequality symbol is <. We therefore use a broken line to represent the boundary, as shown in Figure 4.

A convenient test point is again the origin. Using $(0, 0)$ in $y < 2x - 3$, we have

$$0 \overset{?}{<} 2(0) - 3$$

$$0 < -3 \qquad\qquad \text{A false statement}$$

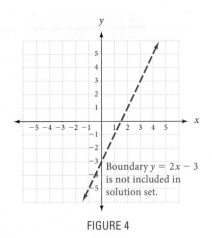

FIGURE 4

Because our test point gives us a false statement and it lies above the boundary, the solution set must lie on the other side of the boundary, as shown in Figure 5.

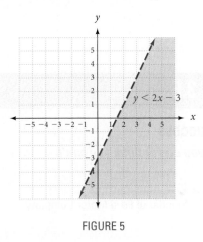

FIGURE 5

EXAMPLE 3   Graph the inequality $2x + 3y \leq 6$.

SOLUTION   We begin by graphing the boundary $2x + 3y = 6$. The boundary is included in the solution because the inequality symbol is $\leq$.

If we use $(0, 0)$ as our test point, we see that it yields a true statement when its coordinates are substituted into $2x + 3y \leq 6$. The graph, therefore, lies below the boundary, as shown in Figure 6.

The ordered pair $(0, 0)$ is a solution to $2x + 3y \leq 6$; all points on the same side of the boundary as $(0, 0)$ also must be solutions to the inequality $2x + 3y \leq 6$.

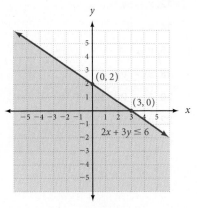

FIGURE 6

EXAMPLE 4   Graph the solution set for $x \leq 5$.

SOLUTION   The boundary is $x = 5$, which is a vertical line. All points to the left have x-coordinates less than 5, and all points to the right have x-coordinates greater than 5, as shown in Figure 7.

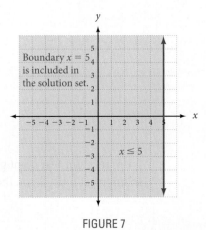

FIGURE 7

Problem Set 3.6

Moving Toward Success

"Whether you think you can or think you can't, you're right."

—Henry Ford, 1863–1947, American industrialist and founder of Ford Motor Company

1. Why should you pay attention to instructions?
2. How can you prevent yourself from getting confused when reading instructions for an exam problem?

A Graph the following linear inequalities.
[Examples 1–4]

1. $2x - 3y < 6$ **2.** $3x + 2y \geq 6$

3. $x - 2y \leq 4$ **4.** $2x + y > 4$

5. $x - y \leq 2$ **6.** $x - y \leq 1$

7. $3x - 4y \geq 12$ **8.** $4x + 3y < 12$

9. $5x - y \leq 5$ **10.** $4x + y > 4$

11. $2x + 6y \leq 12$ **12.** $x - 5y > 5$

13. $x \geq 1$ **14.** $x < 5$

15. $x \geq -3$ **16.** $y \leq -4$

17. $y < 2$ **18.** $3x - y > 1$

19. $2x + y > 3$ **20.** $5x + 2y < 2$

21. $y \leq 3x - 1$ **22.** $y \geq 3x + 2$

23. $y \leq -\dfrac{1}{2}x + 2$ **24.** $y < \dfrac{1}{3}x + 3$

The next two problems are intended to give you practice reading, and paying attention to, the instructions that accompany the problems you are working.

25. Work each problem according to the instructions given.

 a. Solve: $4 + 3y < 12$

 b. Solve: $4 - 3y < 12$

 c. Solve for y: $4x + 3y = 12$

 d. Graph: $y < -\dfrac{4}{3}x + 4$

26. Work each problem according to the instructions given.

 a. Solve: $3x + 2 \geq 6$

 b. Solve: $-3x + 2 \geq 6$

 c. Solve for y: $3x + 2y = 6$

 d. Graph: $y \geq -\dfrac{3}{2}x + 3$

27. The equation of the line shown in the graph is $y = \frac{2}{5}x + 2$. Use this information to find the inequalities for the graphs in parts a and b.

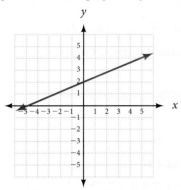

a.

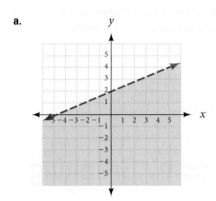

b.

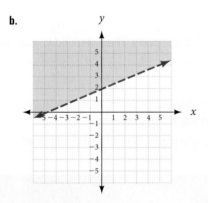

28. The equation of the line shown in the graph is $y = -\frac{3}{2}x + 3$. Use this information to find the inequalities for the graphs in parts a and b.

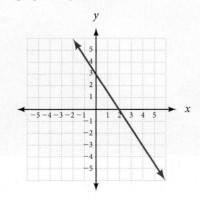

a.

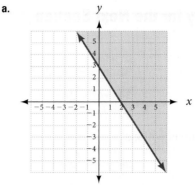

b.

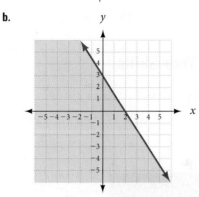

Applying the Concepts

29. Horse Racing The graph shows the total amount of money wagered on the Kentucky Derby. Suppose that the margin of error for this information was ±7 million. Redraw the graph to take into account the margin of error.

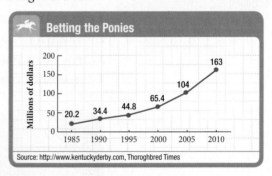

Betting the Ponies

Source: http://www.kentuckyderby.com, Thoroghbred Times

30. Health Care The graph shows the rising cost of health care. The information for 2008 to 2014 is projected costs. Suppose that there is a margin of error on this data of ±300 billion dollars. Redraw the graph from 2008 to 2014 so that it reflects the margin of error.

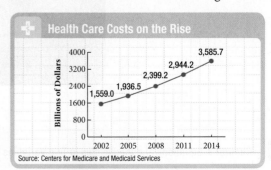

Health Care Costs on the Rise

Source: Centers for Medicare and Medicaid Services

Getting Ready for the Next Section

Solve for y.

31. $2x + y = 5$

32. $3x - 6y = 12$

33. $x + 3y = 6$

34. $-4x + 2y = 6$

Graph the following equations.

35. $4x - 2y = 8$

36. $y = -\frac{2}{3}x + 4$

37. $x - 3y = 6$

38. $y = 3x - 2$

Maintaining Your Skills

39. Simplify the expression $7 - 3(2x - 4) - 8$.

40. Find the value of $x^2 - 2xy + y^2$ when $x = 3$ and $y = -4$.

Solve each equation.

41. $-\frac{3}{2}x = 12$

42. $2x - 4 = 5x + 2$

43. $8 - 2(x + 7) = 2$

44. $3(2x - 5) - (2x - 4) = 6 - (4x + 5)$

45. Solve the formula $P = 2l + 2w$ for w.

Solve each inequality, and graph the solution.

46. $-4x < 20$

47. $3 - 2x > 5$

48. $3 - 4(x - 2) \geq -5x + 6$

49. Solve the formula $3x - 2y \leq 12$ for y.

50. What number is 12% of 2,000?

51. Geometry The length of a rectangle is 5 inches more than 3 times the width. If the perimeter is 26 inches, find the length and width.

3.7 Solving Linear Systems by Graphing

OBJECTIVES

A Solve a system of linear equations in two variables by graphing.

TICKET TO SUCCESS

Keep these questions in mind as you read through the section. Then respond in your own words and in complete sentences.

1. What is a system of linear equations?
2. What is a solution to a system of linear equations?
3. How would you solve a system of linear equations by graphing?
4. Under what conditions will a system of linear equations not have a solution?

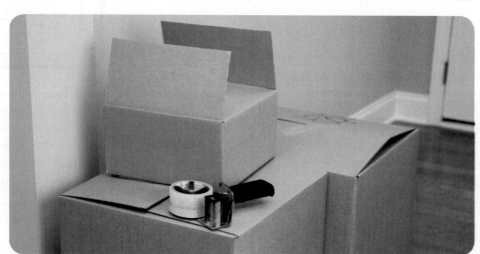

Image copyright © Rob Marmion. Used under license from Shutterstock.com

Your friend is moving to a new apartment and needs to rent a truck to move all her furniture. She asked for your help to decide between two rental companies. Rider Moving Trucks charges $50 per day and $0.50 per mile, given by $y = 0.50x + 50$ where x represents the number of miles and y represents the total rental cost. UMove Trucks charges $45 per day and $0.75 per mile, given by $y = 0.75x + 45$. We could graph these two linear equations to find out which rental company will be the least expensive for your friend. Doing so allows us to consider the two linear equations at the same time and learn how they relate to one another.

A Solving a System of Linear Equations by Graphing

Two linear equations considered at the same time make up what is called a *system of linear equations*. Both equations contain two variables and, of course, have graphs that are straight lines. The following are systems of linear equations:

$$\begin{array}{ccc} x + y = 3 & y = 2x + 1 & 2x - y = 1 \\ 3x + 4y = 2 & y = 3x + 2 & 3x - 2y = 6 \end{array}$$

The solution set for a system of linear equations is all ordered pairs that are solutions to both equations. Because each linear equation has a graph that is a straight line, we can expect the intersection of the graphs to be a point whose coordinates are solutions to the system; that is, if we graph both equations on the same coordinate system, we can read the coordinates of the point of intersection and have the solution to our system. Here is an example.

EXAMPLE 1   Solve the following system by graphing.

$$\begin{array}{l} x + y = 4 \\ x - y = -2 \end{array}$$

SOLUTION   On the same set of coordinate axes, we graph each equation separately. Figure 1 shows both graphs, without showing the work necessary to get them. We can see from the graphs that they intersect at the point (1, 3). Therefore, the point (1, 3) must be the solution to our system because it is the only ordered pair whose graph lies on both lines. Its coordinates satisfy both equations.

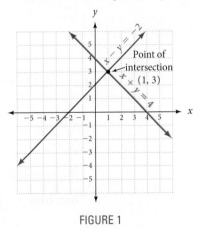

FIGURE 1

We can check our results by substituting the coordinates $x = 1$, $y = 3$ into both equations to see if they work.

| When → | $x = 1$ | When → | $x = 1$ |
| and → | $y = 3$ | and → | $y = 3$ |

$$\text{the equation} \rightarrow \quad x + y = 4 \qquad\qquad \text{the equation} \rightarrow \quad x - y = -2$$
$$\text{becomes} \rightarrow \quad 1 + 3 \overset{?}{=} 4 \qquad\qquad \text{becomes} \rightarrow \quad 1 - 3 \overset{?}{=} -2$$
$$4 = 4 \qquad\qquad\qquad\qquad -2 = -2$$

The point $(1, 3)$ satisfies both equations.

Here are some steps to follow when solving linear systems by graphing.

Strategy Solving a Linear System by Graphing

Step 1: Graph the first equation by the methods described in Section 3.3 or 3.4.

Step 2: Graph the second equation on the same set of axes used for the first equation.

Step 3: Read the coordinates of the point of intersection of the two graphs.

Step 4: Check the solution in both equations.

EXAMPLE 2 Solve the following system by graphing.
$$x + 2y = 8$$
$$2x - 3y = 2$$

SOLUTION Graphing each equation on the same coordinate system, we have the lines shown in Figure 2.

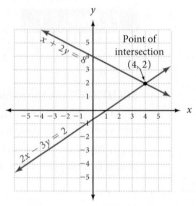

FIGURE 2

From Figure 2, we can see the solution for our system is $(4, 2)$. We check this solution as follows:

$$\text{When} \rightarrow \qquad x = 4 \qquad\qquad \text{When} \rightarrow \qquad x = 4$$
$$\text{and} \rightarrow \qquad y = 2 \qquad\qquad \text{and} \rightarrow \qquad y = 2$$
$$\text{the equation} \rightarrow x + 2y = 8 \qquad \text{the equation} \rightarrow \quad 2x - 3y = 2$$
$$\text{becomes} \rightarrow \quad 4 + 2(2) \overset{?}{=} 8 \qquad \text{becomes} \rightarrow \quad 2(4) - 3(2) \overset{?}{=} 2$$
$$4 + 4 = 8 \qquad\qquad\qquad\qquad 8 - 6 = 2$$
$$8 = 8 \qquad\qquad\qquad\qquad\qquad 2 = 2$$

The point $(4, 2)$ satisfies both equations and, therefore, must be the solution to our system.

EXAMPLE 3   Solve this system by graphing.

$$y = 2x - 3$$
$$x = 3$$

SOLUTION   Graphing both equations on the same set of axes, we have Figure 3.

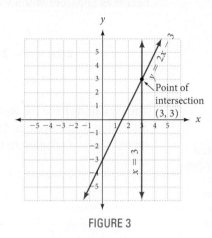

Point of intersection (3, 3)

FIGURE 3

The solution to the system is the point (3, 3).

EXAMPLE 4   Solve by graphing.

$$y = x - 2$$
$$y = x + 1$$

SOLUTION   Graphing both equations produces the lines shown in Figure 4. We can see in Figure 4 that the lines are parallel and therefore do not intersect. Our system has no ordered pair as a solution because there is no ordered pair that satisfies both equations. We say the solution set is the empty set and write $\varnothing$.

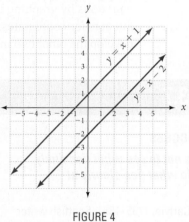

FIGURE 4

Example 4 is one example of two special cases associated with linear systems. The other special case happens when the two graphs coincide. Here is an example.

EXAMPLE 5 Graph the system.

$$2x + y = 4$$
$$4x + 2y = 8$$

SOLUTION Both graphs are shown in Figure 5. The two graphs coincide. The reason becomes apparent when we multiply both sides of the first equation by 2:

$$2x + y = 4$$

$$\mathbf{2}(2x + y) = \mathbf{2}(4) \qquad \text{Multiply both sides by 2}$$

$$4x + 2y = 8$$

NOTE

We sometimes use special vocabulary to describe the cases shown in Examples 4 and 5. When a system of equations has no solution because the lines are parallel (as in Example 4), we say the system is *inconsistent*. When the lines coincide (as in Example 5), we say the system is *dependent*.

The equations have the same solution set. Any ordered pair that is a solution to one is a solution to the system. The system has an infinite number of solutions. (Any point on the line is a solution to the system.)

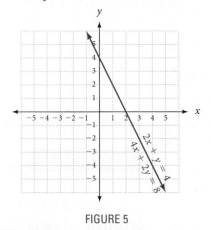

FIGURE 5

The special cases illustrated in the previous two examples do not happen often. Usually, a system has a single ordered pair as a solution. Solving a system of linear equations by graphing is useful only when the ordered pair in the solution set has integers for coordinates. Two other solution methods work well in all cases. We will develop the other two methods in the next two sections.

Problem Set 3.7

Moving Toward Success

"Our main business is not to see what lies dimly at a distance, but to do what lies clearly at hand."

—Thomas Carlyle, 1795–1881, Scottish writer and historian

1. What is mathematical intuition and how is it important?

2. Do you think mathematical intuition gets in the way of success? Why or why not?

A Solve the following systems of linear equations by graphing. [Examples 1–5]

1. $x + y = 3$
$x - y = 1$

2. $x + y = 2$
$x - y = 4$

3. $x + y = 1$
$-x + y = 3$

4. $x + y = 1$
$x - y = -5$

5. $x + y = 8$
$-x + y = 2$

6. $x + y = 6$
$-x + y = -2$

7. $3x - 2y = 6$
   $x - y = 1$

8. $5x - 2y = 10$
   $x - y = -1$

9. $6x - 2y = 12$
   $3x + y = -6$

10. $4x - 2y = 8$
   $2x + y = -4$

11. $4x + y = 4$
   $3x - y = 3$

12. $5x - y = 10$
   $2x + y = 4$

13. $x + 2y = 0$
   $2x - y = 0$

14. $3x + y = 0$
   $5x - y = 0$

15. $3x - 5y = 15$
   $-2x + y = 4$

16. $2x - 4y = 8$
   $2x - y = -1$

17. $y = 2x + 1$
   $y = -2x - 3$

18. $y = 3x - 4$
   $y = -2x + 1$

19. $x + 3y = 3$
   $y = x + 5$

20. $2x + y = -2$
   $y = x + 4$

21. $x + y = 2$
   $x = -3$

22. $x + y = 6$
   $y = 2$

23. $x = -4$
   $y = 6$

24. $x = 5$
   $y = -1$

25. $x + y = 4$
   $2x + 2y = -6$

26. $x - y = 3$
   $2x - 2y = 6$

27. $4x - 2y = 8$
   $2x - y = 4$

28. $3x - 6y = 6$
   $x - 2y = 4$

The next two problems are intended to give you practice reading, and paying attention to, the instructions that accompany the problems you are working.

29. Work each problem according to the instructions given.
   a. Simplify: $(3x - 4y) + (x - y)$
   b. Find y when x is 4 in $3x - 4y = 8$
   c. Find the y-intercept: $3x - 4y = 8$
   d. Graph: $3x - 4y = 8$
   e. Find the point where the graphs of $3x - 4y = 8$ and $x - y = 2$ cross.

30. Work each problem according to the instructions given.
   a. Simplify: $(x + 4y) + (-2x + 3y)$
   b. Find y when x is 3 in $-2x + 3y = 3$
   c. Find the y-intercept: $-2x + 3y = 3$
   d. Graph: $-2x + 3y = 3$
   e. Find the point where the graphs of $-2x + 3y = 3$ and $x + 4y = 4$ cross.

31. As you probably have guessed by now, it can be difficult to solve a system of equations by graphing if the solution to the system contains a fraction. The solution to the following system is $\left(\frac{1}{2}, 1\right)$. Solve the system by graphing.
   $$y = -2x + 2$$
   $$y = 4x - 1$$

32. The solution to the following system is $\left(\frac{1}{3}, -2\right)$. Solve the system by graphing.
   $$y = 3x - 3$$
   $$y = -3x - 1$$

33. A second difficulty can arise in solving a system of equations by graphing if one or both of the equations is difficult to graph. The solution to the following system is (2, 1). Solve the system by graphing.
   $$3x - 8y = -2$$
   $$x - y = 1$$

34. The solution to the following system is (−3, 2). Solve the system by graphing.
   $$2x + 5y = 4$$
   $$x - y = -5$$

Applying the Concepts

35. Light Bulbs The chart shows a comparison of power usage between incandescent and LED light bulbs. Would the line segments from 800 to 1100 lumens intersect in the first quadrant, if you were to extend the lines?

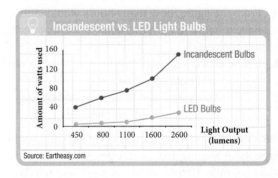

36. Television The graph shows how much television certain groups of people watch. In which years do the amounts of time that teens and children watch TV intersect?

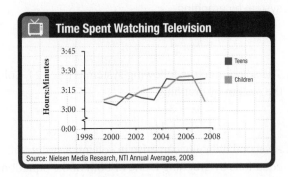

37. Job Comparison Jane is deciding between two sales positions. She can work for Marcy's and receive $8.00 per hour or for Gigi's, where she earns $6.00 per hour but also receives a $50 commission per week. The two lines in the following figure represent the money Jane will make for working at each of the jobs.

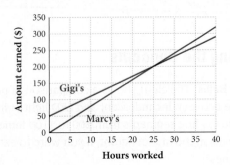

a. From the figure, how many hours would Jane have to work to earn the same amount at each of the positions?

b. If Jane expects to work less than 20 hours a week, which job should she choose?

c. If Jane expects to work more than 30 hours a week, which job should she choose?

38. Truck Rental You need to rent a moving truck for two days. Rider Moving Trucks charges $50 per day and $0.50 per mile. UMove Trucks charges $45 per day and $0.75 per mile. The following figure represents the cost of renting each of the trucks for two days.

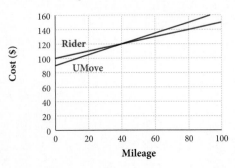

a. From the figure, after how many miles would the trucks cost the same?

b. Which company will give you a better deal if you drive less than 30 miles?

c. Which company will give you a better deal if you drive more than 60 miles?

Getting Ready for the Next Section

Simplify each of the following.

39. $(x + y) + (x - y)$ **40.** $(x + 2y) + (-x + y)$

41. $(6x - 3y) + (x + 3y)$

42. $(6x + 9y) + (-6x - 10y)$

43. $(-12x - 20y) + (25x + 20y)$

44. $(-3x + 2y) + (3x + 8y)$

45. $-4(3x + 5y)$ **46.** $6\left(\frac{1}{2}x - \frac{1}{3}y\right)$

47. $12\left(\frac{1}{4}x + \frac{2}{3}y\right)$ **48.** $5(5x + 4y)$

49. $-2(2x - y)$ **50.** $-2(4x - 3y)$

51. Let $x + y = 4$. If $x = 3$, find y.

52. Let $x + 2y = 4$. If $x = 3$, find y.

53. Let $x + 3y = 3$. If $x = 3$, find y.

54. Let $2x + 3y = -1$. If $y = -1$, find x.

55. Let $3x + 5y = -7$. If $x = 6$, find y.

56. Let $3x - 2y = 12$. If $y = 6$, find x.

Maintaining Your Skills

Simplify each expression.

57. $6x + 100(0.04x + 0.75)$

58. $5x + 100(0.03x + 0.65)$

59. $13x - 1,000(0.002x + 0.035)$

60. $9x - 1,000(0.023x + 0.015)$

61. $16x - 10(1.7x - 5.8)$

62. $43x - 10(3.1x - 2.7)$

63. $0.04x + 0.06(100 - x)$

64. $0.07x + 0.03(100 - x)$

65. $0.025x - 0.028(1,000 + x)$

66. $0.065x - 0.037(1,000 + x)$

67. $2.56x - 1.25(100 + x)$

68. $8.42x - 6.68(100 + x)$

The Elimination Method

OBJECTIVES

A Use the elimination method to solve a system of linear equations in two variables.

TICKET TO SUCCESS

Keep these questions in mind as you read through the section. Then respond in your own words and in complete sentences.

1. How would you use the addition property of equality in the elimination method of solving a system of linear equations?

2. What happens when you use the elimination method to solve a system of linear equations consisting of two parallel lines?

3. How would you use the multiplication property of equality to solve a system of linear equations?

4. What is the first step in solving a system of linear equations that contains fractions?

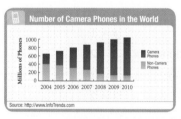

FIGURE 1

Consider the chart in Figure 1. How do we determine the last year that the sale of non-camera phones accounted for the sale of all phones? We would need to consider a system of equations that represented the information in the chart. Let's say the total sale of phones is given by the equation $60x - y = 119{,}560$, where x is the year and y is the total sale, and $40x + y = 80{,}560$ gives the sale of non-camera phones. In this section, we will learn how to set up these two equations so that we can eliminate one variable in order to solve for the other. This method is called the elimination method.

A The Elimination Method

The addition property states that if equal quantities are added to both sides of an equation, the solution set is unchanged. In the past, we have used this property to help solve equations in one variable. We will now use it to solve systems of linear equations. Here is another way to state the addition property of equality.

Let A, B, C, and D represent algebraic expressions.

$$
\begin{aligned}
\text{If} &\to & A &= B \\
\text{and} &\to & C &= D \\
\hline
\text{then} &\to & A + C &= B + D
\end{aligned}
$$

Because C and D are equal (that is, they represent the same number), what we have done is added the same amount to both sides of the equation $A = B$. Let's see how we can use this form of the addition property of equality to solve a system of linear equations.

EXAMPLE 1 Solve the following system.

$$x + y = 4$$
$$x - y = 2$$

SOLUTION The system is written in the form of the addition property of equality as shown above. It looks like this:

$$A = B$$
$$C = D$$

where A is $x + y$, B is 4, C is $x - y$, and D is 2.

We use the addition property of equality to add the left sides together and the right sides together.

$$
\begin{array}{r}
x + y = 4 \\
x - y = 2 \\
\hline
2x + 0 = 6
\end{array}
$$

We now solve the resulting equation for x.

$$2x + 0 = 6$$
$$2x = 6$$
$$x = 3$$

The value we get for x is the value of the x-coordinate of the point of intersection of the two lines $x + y = 4$ and $x - y = 2$. To find the y-coordinate, we simply substitute $x = 3$ into either of the two original equations. Using the first equation, we get

$$3 + y = 4$$
$$y = 1$$

The solution to our system is the ordered pair (3, 1). It satisfies both equations.

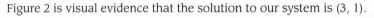

When →	$x = 3$	When →	$x = 3$
and →	$y = 1$	and →	$y = 1$
the equation →	$x + y = 4$	the equation →	$x - y = 2$
becomes →	$3 + 1 \overset{?}{=} 4$	becomes →	$3 - 1 \overset{?}{=} 2$
	$4 = 4$		$2 = 2$

> **NOTE**
> The graphs shown with our first three examples are not part of their solutions. The graphs are there simply to show you that the results we obtain by the elimination method are consistent with the results we would obtain by graphing.

Figure 2 is visual evidence that the solution to our system is (3, 1).

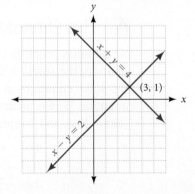

FIGURE 2

The most important part of this method of solving linear systems is eliminating one of the variables when we add the left and right sides together. In our first example, the equations were written so that the y variable was eliminated when we added the left and right sides together. If the equations are not set up this way to begin with, we have to work on one or both of them separately before we can add them together to eliminate one variable.

EXAMPLE 2   Solve the following system.

$$x + 2y = 4$$
$$x - y = -5$$

SOLUTION   Notice that if we were to add the equations together as they are, the resulting equation would have terms in both x and y. Let's eliminate the variable x by multiplying both sides of the second equation by -1 before we add the equations together. (As you will see, we can choose to eliminate either the x or the y variable.)

$$x + 2y = 4 \xrightarrow{\text{No change}} x + 2y = 4$$

$$x - y = -5 \xrightarrow[\text{Multiply by } -1]{} -x + y = 5$$

$$0 + 3y = 9 \quad \text{Add left and right sides to get}$$

$$3y = 9$$

$$y = 3 \quad \left\{ \begin{array}{l} \text{y-coordinate of the} \\ \text{point of intersection} \end{array} \right.$$

Substituting $y = 3$ into either of the two original equations, we get $x = -2$. The solution to the system is $(-2, 3)$. It satisfies both equations. Figure 3 shows the solution to the system as the point where the two lines cross.

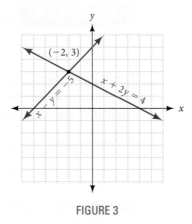

FIGURE 3

EXAMPLE 3   Solve the following system.

$$2x - y = 6$$
$$x + 3y = 3$$

SOLUTION   Let's eliminate the y variable from the two equations. We can do this by multiplying the first equation by 3 and leaving the second equation unchanged.

$$2x - y = 6 \xrightarrow[\text{}]{\text{Multiply by 3}} 6x - 3y = 18$$

$$x + 3y = 3 \xrightarrow[\text{No change}]{} x + 3y = 3$$

The important thing about our system now is that the *coefficients* (the numbers in front) of the y variables are opposites. When we add the terms on each side of the equal sign, then the terms in y will add to zero and be eliminated.

$$6x - 3y = 18$$
$$\underline{x + 3y = 3}$$
$$7x \quad\; = 21 \qquad \text{Add corresponding terms}$$

This gives us $x = 3$. Using this value of x in the second equation of our original system, we have

$$3 + 3y = 3$$
$$3y = 0$$
$$y = 0$$

We could substitute $x = 3$ into any of the equations with both x and y variables and also get $y = 0$. The solution to our system is the ordered pair $(3, 0)$. Figure 4 is a picture of the system of equations showing the solution $(3, 0)$.

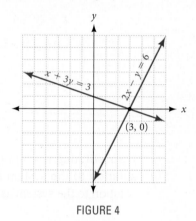

FIGURE 4

EXAMPLE 4   Solve the system.
$$2x + 3y = -1$$
$$3x + 5y = -2$$

SOLUTION   Let's eliminate x from the two equations. If we multiply the first equation by 3 and the second by -2, the coefficients of x will be 6 and -6, respectively. The x terms in the two equations will then add to zero.

$$2x + 3y = -1 \xrightarrow{\text{Multiply by 3}} 6x + 9y = -3$$
$$3x + 5y = -2 \xrightarrow[\text{Multiply by } -2]{} -6x - 10y = 4$$

We now add the left and right sides of our new system together.

$$6x + 9y = -3$$
$$\underline{-6x - 10y = \;\;4}$$
$$-y = \;\;1$$
$$y = -1$$

Substituting $y = -1$ into the first equation in our original system, we have

$$2x + 3(-1) = -1$$
$$2x - 3 = -1$$
$$2x = \;\;2$$
$$x = \;\;1$$

NOTE
If you are having trouble understanding this method of solution, it is probably because you can't see why we chose to multiply by 3 and -2 in the first step of Example 4. Look at the result of doing so: the $6x$ and $-6x$ will add to 0. We chose to multiply by 3 and -2 because they produce $6x$ and $-6x$, which will add to 0.

The solution to our system is $(1, -1)$. It is the only ordered pair that satisfies both equations.

EXAMPLE 5   Solve the system.

$$3x + 5y = -7$$
$$5x + 4y = 10$$

SOLUTION   Let's eliminate y by multiplying the first equation by -4 and the second equation by 5.

$$3x + 5y = -7 \xrightarrow{\text{Multiply by } -4} -12x - 20y = 28$$
$$5x + 4y = 10 \xrightarrow{\text{Multiply by } 5} \underline{25x + 20y = 50}$$
$$13x = 78$$
$$x = 6$$

Substitute $x = 6$ into either equation in our original system, and the result will be $y = -5$. Therefore, the solution is $(6, -5)$. ■

EXAMPLE 6   Solve the system.

$$\frac{1}{2}x - \frac{1}{3}y = 2$$
$$\frac{1}{4}x + \frac{2}{3}y = 6$$

SOLUTION   Although we could solve this system without clearing the equations of fractions, there is probably less chance for error if we have only integer coefficients to work with. So let's begin by multiplying both sides of the top equation by 6 and both sides of the bottom equation by 12, to clear each equation of fractions.

$$\frac{1}{2}x - \frac{1}{3}y = 2 \xrightarrow{\text{Multiply by } 6} 3x - 2y = 12$$
$$\frac{1}{4}x + \frac{2}{3}y = 6 \xrightarrow{\text{Multiply by } 12} 3x + 8y = 72$$

Now we can eliminate x by multiplying the top equation by -1 and leaving the bottom equation unchanged.

$$3x - 2y = 12 \xrightarrow{\text{Multiply by } -1} -3x + 2y = -12$$
$$3x + 8y = 72 \xrightarrow{\text{No change}} \underline{3x + 8y = 72}$$
$$10y = 60$$
$$y = 6$$

We can substitute $y = 6$ into any equation that contains both x and y. Let's use $3x - 2y = 12$.

$$3x - 2(6) = 12$$
$$3x - 12 = 12$$
$$3x = 24$$
$$x = 8$$

The solution to the system is $(8, 6)$. ■

   Our next two examples will show what happens when we apply the elimination method to a system of equations consisting of parallel lines and to a system in which the lines coincide.

EXAMPLE 7   Solve the system.

$$2x - y = 2$$
$$4x - 2y = 12$$

SOLUTION   Let's choose to eliminate y from the system. We can do this by multiplying the first equation by -2 and leaving the second equation unchanged.

$$2x - y = 2 \xrightarrow{\text{Multiply by } -2} -4x + 2y = -4$$

$$4x - 2y = 12 \xrightarrow[\text{No change}]{} 4x - 2y = 12$$

If we add both sides of the resulting system, we have

$$\begin{array}{r} -4x + 2y = -4 \\ \underline{4x - 2y = 12} \\ 0 + 0 = 8 \end{array}$$

or $0 = 8$ A false statement

Both variables have been eliminated and we end up with the false statement $0 = 8$. We have tried to solve a system that consists of two parallel lines. There is no solution, and that is the reason we end up with a false statement. Figure 5 is a visual representation of the situation and is conclusive evidence that there is no solution to our system.

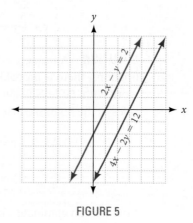

FIGURE 5

EXAMPLE 8   Solve the system.

$$4x - 3y = 2$$
$$8x - 6y = 4$$

SOLUTION   Multiplying the top equation by -2 and adding, we can eliminate the variable x.

$$4x - 3y = 2 \xrightarrow{\text{Multiply by } -2} -8x + 6y = -4$$

$$8x - 6y = 4 \xrightarrow[\text{No change}]{} \underline{8x - 6y = 4}$$

$$0 = 0$$

Both variables have been eliminated, and the resulting statement $0 = 0$ is true. In this case the lines coincide because the equations are equivalent. The solution set consists of all ordered pairs that satisfy either equation.

The preceding two examples illustrate the two special cases in which the graphs of the equations in the system either coincide or are parallel.

Here is a summary of our results from these two examples:

Both variables are eliminated and   ↔   The lines are parallel and there
the resulting statement is false.   is no solution to the system.

Both variables are eliminated and   ↔   The lines coincide and there
the resulting statement is true.   is an infinite number of
                                                 solutions to the system.

The main idea in solving a system of linear equations by the elimination method is to use the multiplication property of equality on one or both of the original equations, if necessary, to make the coefficients of either variable opposites. The following box shows some steps to follow when solving a system of linear equations by the elimination method.

Strategy   Solving a System of Linear Equations by the Elimination Method

Step 1: Decide which variable to eliminate. (In some cases one variable will be easier to eliminate than the other. With some practice you will notice which one it is.)

Step 2: Use the multiplication property of equality on each equation separately to make the coefficients of the variable that is to be eliminated opposites.

Step 3: Add the respective left and right sides of the system together.

Step 4: Solve for the variable remaining.

Step 5: Substitute the value of the variable from step 4 into an equation containing both variables and solve for the other variable.

Step 6: Check your solution in both equations, if necessary.

Problem Set 3.8

Moving Toward Success

"It is common sense to take a method and try it. If it fails, admit it frankly and try another. But above all, try something."

        —Franklin D. Roosevelt, 1882–1945, 32nd President of the United States

1. Why should you test properties about which you are unsure?

2. Does substituting numbers for variables help you test a property? Explain.

A Solve the following systems of linear equations by elimination. [Examples 1–3, 7–8]

1. $x + y = 3$
   $x - y = 1$

2. $x + y = -2$
   $x - y = 6$

3. $x + y = 10$
   $-x + y = 4$

4. $x - y = 1$
   $-x - y = -7$

5. $x - y = 7$
   $-x - y = 3$

6. $x - y = 4$
   $2x + y = 8$

7. $x + y = -1$
   $3x - y = -3$

8. $2x - y = -2$
   $-2x - y = 2$

9. $3x + 2y = 1$
   $-3x - 2y = -1$

10. $-2x - 4y = 1$
   $2x + 4y = -1$

A Solve each of the following systems by eliminating the y variable. [Examples 3, 5, 7]

11. $3x - y = 4$
   $2x + 2y = 24$

12. $2x + y = 3$
   $3x + 2y = 1$

13. $5x - 3y = -2$
   $10x - y = 1$

14. $4x - y = -1$
   $2x + 4y = 13$

15. $11x - 4y = 11$
   $5x + y = 5$

16. $3x - y = 7$
   $10x - 5y = 25$

A Solve each of the following systems by eliminating the x variable. [Examples 4, 6, 8]

17. $3x - 5y = 7$
$-x + y = -1$

18. $4x + 2y = 32$
$x + y = -2$

19. $-x - 8y = -1$
$-2x + 4y = 13$

20. $-x + 10y = 1$
$-5x + 15y = -9$

21. $-3x - y = 7$
$6x + 7y = 11$

22. $-5x + 2y = -6$
$10x + 7y = 34$

A Solve each of the following systems of linear equations by the elimination method. [Examples 3–8]

23. $6x - y = -8$
$2x + y = -16$

24. $5x - 3y = -3$
$3x + 3y = -21$

25. $x + 3y = 9$
$2x - y = 4$

26. $x + 2y = 0$
$2x - y = 0$

27. $x - 6y = 3$
$4x + 3y = 21$

28. $8x + y = -1$
$4x - 5y = 16$

29. $2x + 9y = 2$
$5x + 3y = -8$

30. $5x + 2y = 11$
$7x + 8y = 7$

31. $\frac{1}{3}x + \frac{1}{4}y = \frac{7}{6}$
$\frac{3}{2}x - \frac{1}{3}y = \frac{7}{3}$

32. $\frac{7}{12}x - \frac{1}{2}y = \frac{1}{6}$
$\frac{2}{5}x - \frac{1}{3}y = \frac{11}{15}$

33. $3x + 2y = -1$
$6x + 4y = 0$

34. $8x - 2y = 2$
$4x - y = 2$

35. $11x + 6y = 17$
$5x - 4y = 1$

36. $3x - 8y = 7$
$10x - 5y = 45$

37. $\frac{1}{2}x + \frac{1}{6}y = \frac{1}{3}$
$-x - \frac{1}{3}y = -\frac{1}{6}$

38. $-\frac{1}{3}x - \frac{1}{2}y = -\frac{2}{3}$
$-\frac{2}{3}x - y = -\frac{4}{3}$

39. $x + y = 22$
$5x + 10y = 170$

40. $x + y = 14$
$10x + 25y = 185$

41. $x + y = 14$
$5x + 25y = 230$

42. $x + y = 11$
$5x + 10y = 95$

43. $x + y = 15,000$
$6x + 7y = 98,000$

44. $x + y = 10,000$
$6x + 7y = 63,000$

45. $x + y = 11,000$
$4x + 7y = 68,000$

46. $x + y = 20,000$
$8x + 6y = 138,000$

47. $x + y = 23$
$5x + 10y = 175$

48. $x + y = 45$
$25x + 5y = 465$

49. Multiply both sides of the second equation in the following system by 100, and then solve as usual.

$$x + y = 22$$
$$0.05x + 0.10y = 1.70$$

50. Multiply both sides of the second equation in the following system by 100, and then solve as usual.

$$x + y = 15,000$$
$$0.06x + 0.07y = 980$$

Applying the Concepts

51. Camera Phones The chart shows the estimated number of camera phones and non-camera phones sold from 2004 to 2010. Suppose the equation that describes the total sales in phones is $60x - y = 119,560$, and the equation that describes the sale of non-camera phones is $40x + y = 80,560$. What was the last year that the sale of non-camera phones accounted for the sale of all phones?

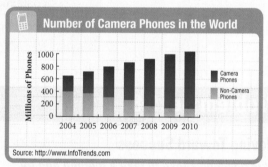

Source: http://www.InfoTrends.com

52. Light Bulbs The chart shows a comparison of power usage between incandescent and LED light bulbs. Suppose that any output less than 450 lumens is just an extension of the line segment from 450 to 800 lumens. If the equations are $35y - 2x = 500$ for the incandescent bulb and $175y - x = 425$ for the LED bulb, where do the lines intersect?

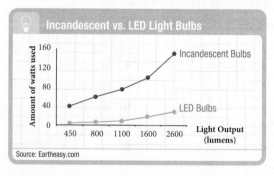

Source: Eartheasy.com

Getting Ready for the Next Section

Solve.

53. $x + (2x - 1) = 2$ **54.** $2x - 3(2x - 8) = 12$

55. $2(3y - 1) - 3y = 4$ **56.** $-2x + 4(3x + 6) = 14$

57. $-2x + 3(5x - 1) = 10$

58. $1.5x + 15 = 0.75x + 24.95$

Solve each equation for the indicated variable.

59. $x - 3y = -1$ for x

60. $-3x + y = 6$ for y

61. Let $y = 2x - 1$. If $x = 1$, find y.

62. Let $y = 2x - 8$. If $x = 5$, find y.

63. Let $x = 3y - 1$. If $y = 2$, find x.

64. Let $x = 4y - 5$. If $y = 2$, find x.

Let $y = 1.5x + 15$.

65. If $x = 13$, find y.   **66.** If $x = 14$, find y.

Let $y = 0.75x + 24.95$.

67. If $x = 12$, find y.   **68.** If $x = 16$, find y.

Maintaining Your Skills

For each of the equations, determine the slope and y-intercept.

69. $3x - y = 3$

70. $2x + y = -2$

71. $2x - 5y = 25$

72. $-3x + 4y = -12$

Find the slope of the line through the given points.

73. $(-2, 3)$ and $(6, -5)$

74. $(2, -4)$ and $(8, -2)$

75. $(5, 3)$ and $(2, -3)$

76. $(-1, -4)$ and $(-4, -1)$

77. $(-3, -5)$, $(3, 1)$

78. $(-1, -5)$, $(2, 1)$

79. Find y if the line through $(-2, 5)$ and $(-4, y)$ has a slope of -3.

80. Find y if the line through $(-2, 4)$ and $(6, y)$ has a slope of -2.

81. Find y if the line through $(3, -6)$ and $(6, y)$ has a slope of 5.

82. Find y if the line through $(3, 4)$ and $(-2, y)$ has a slope of -4.

3.9 The Substitution Method

OBJECTIVES

A Use the substitution method to solve a system of linear equations in two variables.

TICKET TO SUCCESS

Keep these questions in mind as you read through the section. Then respond in your own words and in complete sentences.

1. What is the first step in solving a system of linear equations by substitution?
2. When would substitution be more efficient than the elimination method in solving two linear equations?
3. What does it mean when you solve a system of linear equations by the substitution method and you end up with the statement 8 = 8?
4. How would you begin solving the following system using the substitution method?

$$x + y = 2$$
$$y = 2x - 1$$

If you are a business owner, you need to analyze the daily cost and daily revenue (the amount of money earned each day from sales) for running your business. What if you wanted to own a flower shop? The daily wholesale cost for x flowers could be $y = \frac{2}{3}x + 5$; whereas the daily revenue for selling flowers could be $y = 3x$. In this chapter, we will learn another method for solving a system of equations called the substitution method. For the flower shop example, you can use the substitution method to determine what your minimum revenue goal should be to break even; that is, when your x input (number of flowers you purchased) yields the same y output (the total revenue earned) for both equations. Afterward, any additional flowers purchased will be profit for your business.

A Substitution Method

The substitution method, like the elimination method, can be used on any system of linear equations. Some systems, however, lend themselves more to the substitution method than others.

EXAMPLE 1 Solve the following system.

$$x + y = 2$$
$$y = 2x - 1$$

SOLUTION If we were to solve this system by the methods used in the previous section, we would have to rearrange the terms of the second equation so that similar terms would be in the same column. There is no need to do this, however, because the second equation tells us that y is $2x - 1$. We can replace the y variable in the first equation with the expression $2x - 1$ from the second equation; that is, we *substitute* $2x - 1$ from the second equation for y in the first equation. Here is what it looks like:

$$x + (2x - 1) = 2$$

The equation we end up with contains only the variable x. The y variable has been eliminated by substitution.

Solving the resulting equation, we have

$$x + (2x - 1) = 2$$
$$3x - 1 = 2$$
$$3x = 3$$
$$x = 1$$

This is the x-coordinate of the solution to our system. To find the y-coordinate, we substitute $x = 1$ into the second equation of our system. (We could substitute $x = 1$ into the first equation also and have the same result.)

$$y = 2(1) - 1$$
$$y = 2 - 1$$
$$y = 1$$

The solution to our system is the ordered pair $(1, 1)$. It satisfies both of the original equations. Figure 1 provides visual evidence that the substitution method yields the correct solution.

NOTE

Sometimes this method of solving systems of equations is confusing the first time you see it. If you are confused, you may want to read through this first example more than once and try it on your own.

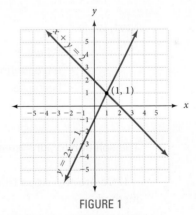

FIGURE 1

EXAMPLE 2 Solve the following system by the substitution method.

$$2x - 3y = 12$$
$$y = 2x - 8$$

SOLUTION Again, the second equation says y is $2x - 8$. Because we are looking for the ordered pair that satisfies both equations, the y in the first equation must also be $2x - 8$. Substituting $2x - 8$ from the second equation for y in the first equation, we have

$$2x - 3(2x - 8) = 12$$

This equation can still be read as $2x - 3y = 12$ because $2x - 8$ is the same as y. Solving the equation, we have

$$2x - 3(2x - 8) = 12$$
$$2x - 6x + 24 = 12$$
$$-4x + 24 = 12$$

$$-4x = -12$$
$$x = 3$$

To find the y-coordinate of our solution, we substitute $x = 3$ into the second equation in the original system.

When $\rightarrow$ $x = 3$
the equation $\rightarrow$ $y = 2x - 8$
becomes $\rightarrow$ $y = 2(3) - 8$
                 $y = 6 - 8 = -2$

The solution to our system is $(3, -2)$.

EXAMPLE 3   Solve the following system by solving the first equation for x and then using the substitution method:

$$x - 3y = -1$$
$$2x - 3y = 4$$

SOLUTION   We solve the first equation for x by adding $3y$ to both sides to get

$$x = 3y - 1$$

Using this value of x in the second equation, we have

$$2(3y - 1) - 3y = 4$$
$$6y - 2 - 3y = 4$$
$$3y - 2 = 4$$
$$3y = 6$$
$$y = 2$$

Next, we find x.

When $\rightarrow$ $y = 2$
the equation $\rightarrow x = 3y - 1$
becomes $\rightarrow$ $x = 3(2) - 1$
                 $x = 6 - 1$
                 $x = 5$

The solution to our system is $(5, 2)$.

Here are the steps to use in solving a system of equations by the substitution method.

Strategy   Solving a System of Equations by the Substitution Method

Step 1: Solve either one of the equations for x or y. (This step is not necessary if one of the equations is already in the correct form, as in Examples 1 and 2.)

Step 2: Substitute the expression for the variable obtained in step 1 into the other equation and solve it.

Step 3: Substitute the solution from step 2 into any equation in the system that contains both variables and solve it.

Step 4: Check your results, if necessary.

EXAMPLE 4   Solve by substitution.

$$-2x + 4y = 14$$
$$-3x + y = 6$$

SOLUTION We can solve either equation for either variable. If we look at the system closely, it becomes apparent that solving the second equation for y is the easiest way to go. If we add $3x$ to both sides of the second equation, we have

$$y = 3x + 6$$

Substituting the expression $3x + 6$ back into the first equation in place of y yields the following result.

$$-2x + 4(3x + 6) = 14$$
$$-2x + 12x + 24 = 14$$
$$10x + 24 = 14$$
$$10x = -10$$
$$x = -1$$

Substituting $x = -1$ into the equation $y = 3x + 6$ leaves us with

$$y = 3(-1) + 6$$
$$y = -3 + 6$$
$$y = 3$$

The solution to our system is $(-1, 3)$.

EXAMPLE 5   Solve by substitution.
$$4x + 2y = 8$$
$$y = -2x + 4$$

SOLUTION Substituting the expression $-2x + 4$ for y from the second equation into the first equation, we have

$$4x + 2(-2x + 4) = 8$$
$$4x - 4x + 8 = 8$$
$$8 = 8 \qquad \text{A true statement}$$

Both variables have been eliminated, and we are left with a true statement. Recall from the last section that a true statement in this situation tells us the lines coincide; that is, the equations $4x + 2y = 8$ and $y = -2x + 4$ have exactly the same graph. Any point on that graph has coordinates that satisfy both equations and is a solution to the system.

EXAMPLE 6   The following table shows two contract rates charged by a cellular phone company. How many text messages must a person send so that the cost will be the same, regardless of which plan is chosen?

	Flat Monthly Rate	plus	Per Text Charge
Plan 1	$40		$0.15
Plan 2	$70		$0.05

SOLUTION If we let $y = $ the monthly charge for x number of text messages, then the equations for each plan are

$$\text{Plan 1: } \quad y = 0.15x + 40$$
$$\text{Plan 2: } \quad y = 0.05x + 70$$

We can solve this system using the substitution method by replacing the variable y in Plan 2 with the expression $0.15x + 40$ from Plan 1. If we do so, we have

$$0.15x + 40 = 0.05x + 70$$
$$0.10x + 40 = 70$$
$$0.10x = 30$$
$$x = 300$$

The monthly bill is based on the number of text messages you send. We can use this value of x to gather more information by plugging it into either plan equation to find y.

Plan 1: $y = 0.15x + 40$ or   Plan 2: $y = 0.05x + 70$
$\quad\quad\quad y = 0.15(300) + 40$ $\quad\quad\quad\quad\quad\quad y = 0.05(300) + 70$
$\quad\quad\quad y = 45 + 40$ $\quad\quad\quad\quad\quad\quad\quad\quad y = 15 + 70$
$\quad\quad\quad y = 85$ $\quad\quad\quad\quad\quad\quad\quad\quad\quad y = 85$

Therefore, when you send 300 text messages in a month, the total cost for that month will be $85 regardless of which plan you used. ■

Problem Set 3.9

Moving Toward Success

"Even if you are on the right track, you will get run over if you just sit there."

—Will Rogers, 1879–1935, actor and humorist

1. Why do you think eating healthy balanced meals leads to success in math?

2. Why do you think scheduling physical activity during your day can help you be successful in math?

A Solve the following systems by substitution. Substitute the expression in the second equation into the first equation and solve. [Examples 1–2]

1. $x + y = 11$
   $y = 2x - 1$

2. $x - y = -3$
   $y = 3x + 5$

3. $x - y = -14$
   $y = 5x + 2$

4. $3x - y = -1$
   $x = 2y - 7$

5. $-2x + y = -1$
   $y = -4x + 8$

6. $4x - y = 5$
   $y = -4x + 1$

7. $3x - 2y = -2$
   $x = -y + 6$

8. $2x - 3y = 17$
   $x = 2y + 9$

9. $5x - 4y = -16$
   $y = 4$

10. $6x + 2y = 18$
    $x = 3$

11. $5x + 4y = 7$
    $y = 3$

12. $10x + 2y = -6$
    $y = -5x$

A Solve the following systems by solving one of the equations for x or y and then using the substitution method. [Examples 3–4]

13. $x + 3y = 4$
    $x - 2y = -1$

14. $x - y = 5$
    $x + 2y = -1$

15. $2x + y = 1$
    $x - 5y = 17$

16. $2x - 2y = 2$
    $x - 3y = -7$

17. $3x + 5y = -3$
    $x - 5y = -5$

18. $2x - 4y = -4$
    $x + 2y = 8$

19. $5x + 3y = 0$
    $x - 3y = -18$

20. $x - 3y = -5$
    $x - 2y = 0$

21. $-3x - 9y = 7$
    $x + 3y = 12$

22. $2x + 6y = -18$
    $x + 3y = -9$

A Solve the following systems using the substitution method. [Examples 1–2, 5]

23. $5x - 8y = 7$
    $y = 2x - 5$

24. $3x + 4y = 10$
    $y = 8x - 15$

25. $7x - 6y = -1$
    $x = 2y - 1$

26. $4x + 2y = 3$
    $x = 4y - 3$

27. $-3x + 2y = 6$
    $y = 3x$

28. $-2x - y = -3$
    $y = -3x$

29. $3x - 5y = -8$
    $y = x$

30. $2x - 4y = 0$
    $y = x$

31. $3x + 3y = 9$
    $y = 2x - 12$

32. $7x + 6y = -9$
    $y = -2x + 1$

33. $7x - 11y = 16$
    $y = 10$

34. $9x - 7y = -14$
    $x = 7$

35. $-4x + 4y = -8$
        $y = x - 2$

36. $-4x + 2y = -10$
        $y = 2x - 5$

Solve each system.

37. $2x + 5y = 36$
        $y = 12 - x$

38. $3x + 6y = 120$
        $y = 25 - x$

39. $5x + 2y = 54$
        $y = 18 - x$

40. $10x + 5y = 320$
        $y = 40 - x$

41. $2x + 2y = 96$
        $y = 2x$

42. $x + y = 22$
        $y = x + 9$

Solve each system by substitution. You can eliminate the decimals if you like, but you don't have to. The solution will be the same in either case.

43. $0.05x + 0.10y = 1.70$
        $y = 22 - x$

44. $0.20x + 0.50y = 3.60$
        $y = 12 - x$

The next two problems are intended to give you practice reading, and paying attention to, the instructions that accompany the problems you are working. Working these problems is an excellent way to get ready for a test or quiz.

45. Work each problem according to the instructions given.

   a. Solve: $4y - 5 = 20$

   b. Solve for y: $4x - 5y = 20$

   c. Solve for x: $x - y = 5$

   d. Solve the system: $4x - 5y = 20$
                        $x - y = 5$

46. Work each problem according to the instructions given.

   a. Solve: $2x - 1 = 4$

   b. Solve for y: $2x - y = 4$

   c. Solve for x: $x + 3y = 9$

   d. Solve the system: $2x - y = 4$
                        $x + 3y = 9$

Applying the Concepts [Example 6]

47. Camera Phones The chart shows the estimated number of camera phones and non-camera phones sold from 2004 to 2010. In the last section we solved this problem using the method of elimination. Now use substitution to solve for the last year that the sale

of non-camera phones accounted for the sale of all phones. The equation that describes the total sales in phones is $60x - y = 119,560$ and the equation that describes the sale of non-camera phones is $40x + y = 80,560$.

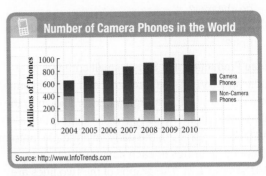

Number of Camera Phones in the World

Source: http://www.InfoTrends.com

48. Light Bulbs The chart shows a comparison of power usage between incandescent and LED light bulbs. The line segments from 1,600 to 2,600 lumens can be given by the equations $y = \left(\frac{1}{20}\right)x + 20$ and $y = \left(\frac{9}{1000}\right)x + \frac{8}{5}$. Find where these two equations intersect.

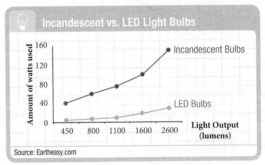

Incandescent vs. LED Light Bulbs

Source: Eartheasy.com

49. Gas Mileage Daniel is trying to decide whether to buy a car or a truck. The truck he is considering will cost him $150 a month in loan payments, and it gets 20 miles per gallon in gas mileage. The car will cost $180 a month in loan payments, but it gets 35 miles per gallon in gas mileage. Daniel estimates that he will pay $2.50 per gallon for gas. This means that the monthly cost to drive the truck x miles will be $y = \frac{2.50}{20}x + 150$. The total monthly cost to drive the car x miles will be $y = \frac{2.50}{35}x + 180$. The following figure shows the graph of each equation.

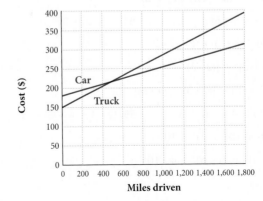

a. At how many miles do the car and the truck cost the same to operate?

b. If Daniel drives more than 800 miles, which will be cheaper?

c. If Daniel drives fewer than 400 miles, which will be cheaper?

d. Why do the graphs appear in the first quadrant only?

50. Video Production Pat runs a small company that duplicates DVD's. The daily cost and daily revenue for a company duplicating DVD's are shown in the following figure. The daily cost for duplicating x DVD's is $y = \frac{6}{5}x + 20$; the daily revenue (the amount of money he brings in each day) for duplicating x DVD's is $y = 1.7x$. The graphs of the two lines are shown in the following figure.

a. Pat will "break even" when his cost and his revenue are equal. How many DVD's does he need to duplicate to break even?

b. Pat will incur a loss when his revenue is less than his cost. If he duplicates 30 DVD's in one day, will he incur a loss?

c. Pat will make a profit when his revenue is larger than his costs. For what values of x will Pat make a profit?

d. Why does the graph appear in the first quadrant only?

62. $5 \cdot 10^3 + 2 \cdot 10^2 + 8 \cdot 10^1$

63. $1 \cdot 10^3 + 7 \cdot 10^2 + 6 \cdot 10^1 + 0$

64. $4(2 - 1) + 5(3 - 2)$

65. $4 \cdot 2 - 1 + 5 \cdot 3 - 2$ **66.** $2^3 + 3^2 \cdot 4 - 5$

67. $(2^3 + 3^2) \cdot 4 - 5$ **68.** $4^2 - 2^4 + (2 \cdot 2)^2$

69. $2(2^2 + 3^2) + 3(3^2)$ **70.** $2 \cdot 2^2 + 3^2 + 3 \cdot 3^2$

Maintaining Your Skills

51. $6(3 + 4) + 5$

52. $[(1 + 2)(2 + 3)] + (4 \div 2)$

53. $1^2 + 2^2 + 3^2$ **54.** $(1 + 2 + 3)^2$

55. $5(6 + 3 \cdot 2) + 4 + 3 \cdot 2$

56. $(1 + 2)^3 + [(2 \cdot 3) + (4 \cdot 5)]$

57. $(1^3 + 2^3) + [(2 \cdot 3) + (4 \cdot 5)]$

58. $[2(3 + 4 + 5)] \div 3$ **59.** $(2 \cdot 3 + 4 + 5) \div 3$

60. $10^4 + 10^3 + 10^2 + 10^1$

61. $6 \cdot 10^3 + 5 \cdot 10^2 + 4 \cdot 10^1$

■ Linear Equation in Two Variables [3.3]

EXAMPLES

1. The equation $3x + 2y = 6$ is an example of a linear equation in two variables.

A linear equation in two variables is any equation that can be put in the form $ax + by = c$. The graph of every linear equation is a straight line.

2. The graph of $y = -\frac{2}{3}x - 1$ is shown below.

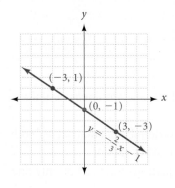

■ Strategy for Graphing Linear Equations in Two Variables [3.3]

Step 1 Find any three ordered pairs that satisfy the equation. This can be done by using a convenient number for one variable and solving for the other variable.

Step 2: Graph the three ordered pairs found in step 1. Actually, we need only two points to graph a straight line. The third point serves as a check. If all three points do not line up, there is a mistake in our work.

Step 3: Draw a straight line through the three points graphed in step 2.

■ Intercepts [3.4]

3. To find the x-intercept for $3x + 2y = 6$, we let $y = 0$ and get

$$3x = 6$$
$$x = 2$$

In this case the x-intercept is 2, and the graph crosses the x-axis at $(2, 0)$.

The x-intercept of an equation is the x-coordinate of the point where the graph crosses the x-axis. The y-intercept is the y-coordinate of the point where the graph crosses the y-axis. We find the y-intercept by substituting $x = 0$ into the equation and solving for y. The x-intercept is found by letting $y = 0$ and solving for x.

■ Slope of a Line [3.5]

4. The slope of the line through $(3, -5)$ and $(-2, 1)$ is

$$m = \frac{-5 - 1}{3 - (-2)} = \frac{-6}{5} = -\frac{6}{5}$$

The *slope* of the line containing the points (x_1, y_1) and (x_2, y_2) is given by

$$\text{Slope} = m = \frac{y_2 - y_1}{x_2 - x_1} = \frac{\text{rise}}{\text{run}}$$

To Graph a Linear Inequality in Two Variables [3.6]

5. Graph $x - y \geq 3$.

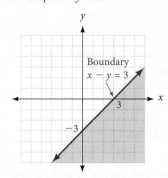

Step 1: Replace the inequality symbol with an equal sign. The resulting equation represents the boundary for the solution set.

Step 2: Graph the boundary found in step 1, using a *solid line* if the original inequality symbol was either $\leq$, or $\geq$. Use a *broken line* otherwise.

Step 3: Choose any convenient point not on the boundary and substitute the coordinates into the *original* inequality. If the resulting statement is *true,* the graph lies on the *same* side of the boundary as the chosen point. If the resulting statement is *false,* the solution set lies on the *opposite* side of the boundary.

Definitions [3.7]

6. The solution to the system
$$x + 2y = 4$$
$$x - y = 1$$
is the ordered pair (2, 1). It is the only ordered pair that satisfies both equations.

1. A *system of linear equations,* as the term is used in this book, is two linear equations that each contain the same two variables.

2. The *solution set* for a system of equations is the set of all ordered pairs that satisfy *both* equations. The solution set to a system of linear equations will contain:

Case I One ordered pair when the graphs of the two equations intersect at only one point (this is the most common situation)

Case II No ordered pairs when the graphs of the two equations are parallel lines

Case III An infinite number of ordered pairs when the graphs of the two equations coincide (are the same line)

Strategy for Solving a System by Graphing [3.7]

7. Solving the system in Example 1 by graphing looks like

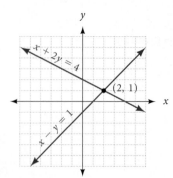

Step 1: Graph the first equation.

Step 2: Graph the second equation on the same set of axes.

Step 3: Read the coordinates of the point where the graphs cross each other (the coordinates of the point of intersection).

Step 4: Check the solution to see that it satisfies *both* equations.

8. We can eliminate the y variable from the system in Example 1 by multiplying both sides of the second equation by 2 and adding the result to the first equation

$$
\begin{array}{r}
x + 2y = 4 \qquad\quad x + 2y = 4 \\
x - y = 1 \xrightarrow[\text{Multiply by 2}]{\quad} 2x - 2y = 2 \\
\hline
3x \quad\;\; = 6 \\
x \quad\;\; = 2
\end{array}
$$

Substituting $x = 2$ into either of the original two equations gives $y = 1$. The solution is $(2, 1)$.

◼ Strategy for Solving a System by the Elimination Method [3.8]

Step 1: Look the system over to decide which variable will be easier to eliminate.

Step 2: Use the multiplication property of equality on each equation separately to ensure that the coefficients of the variable to be eliminated are opposites.

Step 3: Add the left and right sides of the system produced in step 2, and solve the resulting equation.

Step 4: Substitute the solution from step 3 back into any equation with both x and y variables, and solve.

Step 5: Check your solution in both equations, if necessary.

9. We can apply the substitution method to the system in Example 1 by first solving the second equation for x to get $x = y + 1$. Substituting this expression for x into the first equation, we have

$$
\begin{aligned}
(y + 1) + 2y &= 4 \\
3y + 1 &= 4 \\
3y &= 3 \\
y &= 1
\end{aligned}
$$

Using $y = 1$ in either of the original equations gives $x = 2$.

◼ Strategy for Solving a System by the Substitution Method [3.9]

Step 1: Solve either of the equations for one of the variables (this step is not necessary if one of the equations has the correct form already).

Step 2: Substitute the results of step 1 into the other equation, and solve.

Step 3: Substitute the results of step 2 into an equation with both x and y variables, and solve. (The equation produced in step 1 is usually a good one to use.)

Step 4: Check your solution, if necessary.

◼ Special Cases [3.7, 3.8, 3.9]

In some cases, using the elimination or substitution method eliminates both variables. The situation is interpreted as follows.

1. If the resulting statement is *false,* then the lines are parallel and there is no solution to the system.

2. If the resulting statement is *true,* then the equations represent the same line (the lines coincide). In this case any ordered pair that satisfies either equation is a solution to the system.

> 🚫 **COMMON MISTAKES**
>
> The most common mistake encountered in solving linear systems is the failure to complete the problem. Here is an example:
>
> $$
> \begin{aligned}
> x + y &= 8 \\
> x - y &= 4 \\
> \hline
> 2x &= 12 \\
> x &= 6
> \end{aligned}
> $$
>
> This is only half the solution. To find the other half, we must substitute the 6 back into one of the original equations and then solve for y.
>
> Remember, solutions to systems of linear equations always consist of ordered pairs. We need an x-coordinate and a y-coordinate; $x = 6$ can never be a solution to a system of linear equations.

For each equation, complete the given ordered pairs. [3.2]

1. $3x + y = 6$ (4,), (0,), (, 3), (, 0)

2. $2x - 5y = 20$ (5,), (0,), (, 2), (, 0)

3. $y = 2x - 6$ (4,), (, −2), (, 3)

4. $y = 5x + 3$ (2,), (, 0), (, −3)

5. $y = -3$ (2,), (−1,), (−3,)

6. $x = 6$ (, 5), (, 0), (, −1)

For the following equations, tell which of the given ordered pairs are solutions. [3.2]

7. $3x - 4y = 12$ $\left(-2, \dfrac{9}{2}\right)$, (0, 3), $\left(2, -\dfrac{3}{2}\right)$

8. $y = 3x + 7$ $\left(-\dfrac{8}{3}, -1\right)$, $\left(\dfrac{7}{3}, 0\right)$, (−3, −2)

Graph the following ordered pairs. [3.1]

9. (4, 2)

10. (−3, 1)

11. (0, 5)

12. (−2, −3)

13. (−3, 0)

14. $\left(5, -\dfrac{3}{2}\right)$

For the following equations, complete the given ordered pairs, and use the results to graph the solution set for the equations. [3.3]

15. $x + y = -2$ (, 0), (0,), (1,)

16. $y = 3x$ (−1,), (1,), (, 0)

17. $y = 2x - 1$ (1,), (0,), (, −3)

18. $x = -3$ (, 0), (, 5), (, −5)

Graph the following equations. [3.3]

19. $3x - y = 3$

20. $x - 2y = 2$

21. $y = -\dfrac{1}{3}x$

22. $y = \dfrac{3}{4}x$

23. $y = 2x + 1$

24. $y = -\dfrac{1}{2}x + 2$

25. $x = 5$

26. $y = -3$

27. $2x - 3y = 3$

28. $5x - 2y = 5$

Find the x- and y-intercepts for each equation. [3.4]

29. $3x - y = 6$

30. $2x - 6y = 24$

31. $y = x - 3$

32. $y = 3x - 6$

33. $y = -5$

34. $x = 4$

Find the slope of the line through the given pair of points. [3.5]

35. (2, 3), (3, 5)

36. (−2, 3), (6, −5)

37. (−1, −4), (−3, −8)

38. $\left(\dfrac{1}{2}, 4\right)$, $\left(-\dfrac{1}{2}, 2\right)$

39. Find x if the line through (3, 3) and (x, 9) has slope 2.

40. Find y if the line through (5, −5) and (−5, y) has slope 2.

Graph the following linear inequalities. [3.6]

41. $x - y < 3$

42. $x \geq -3$

43. $y \leq -4$

44. $y \leq -2x + 3$

Solve the following systems by graphing. [3.7]

45. $x + y = 2$
$x - y = 6$

46. $x + y = -1$
$-x + y = 5$

47. $2x - 3y = 12$
$-2x + y = -8$

48. $4x - 2y = 8$
$3x + y = 6$

49. $y = 2x - 3$
$y = -2x + 5$

50. $y = -x - 3$
$y = 3x + 1$

Solve the following systems by the elimination method. [3.8]

51. $x - y = 4$
$x + y = -2$

52. $-x - y = -3$
$2x + y = 1$

53. $5x - 3y = 2$
$-10x + 6y = -4$

54. $2x + 3y = -2$
$3x - 2y = 10$

55. $-3x + 4y = 1$
$-4x + y = -3$

56. $-4x - 2y = 3$
$2x + y = 1$

57. $-2x + 5y = -11$
$7x - 3y = -5$

58. $-2x + 5y = -1$
$3x - 4y = 19$

Solve the following systems by substitution. [3.9]

59. $x + y = 5$
  $y = -3x + 1$

60. $x - y = -2$
  $y = -2x - 10$

61. $4x - 3y = -16$
  $y = 3x + 7$

62. $5x + 2y = -2$
  $y = -8x + 10$

63. $x - 4y = 2$
  $-3x + 12y = -8$

64. $4x - 2y = 8$
  $3x + y = -19$

65. $10x - 5y = 20$
  $x + 6y = -11$

66. $3x - y = 2$
  $-6x + 2y = -4$

Simplify.

1. $3 \cdot 4 + 5$

2. $8 - 6(5 - 9)$

3. $4 \cdot 3^2 + 4(6 - 3)$

4. $7[8 + (-5)] + 3(-7 + 12)$

5. $\dfrac{12 - 3}{8 - 8}$

6. $\dfrac{5(4 - 12) - 8(14 - 3)}{3 - 5 - 6}$

7. $\dfrac{11}{60} - \dfrac{13}{84}$

8. $\dfrac{2}{3} + \dfrac{3}{4} - \dfrac{1}{6}$

9. $2(x - 5) + 8$

10. $7 - 5(2a - 3) + 7$

Solve each equation.

11. $-5 - 6 = -y - 3 + 2y$

12. $-2x = 0$

13. $3(x - 4) = 9$

14. $8 - 2(y + 4) = 12$

Solve each inequality, and graph the solution.

15. $0.3x + 0.7 \leq -2$

16. $5x + 10 \leq 7x - 14$

Graph on a rectangular coordinate system.

17. $y = -2x + 1$

18. $y = -\dfrac{2}{3}x$

Solve each system by graphing.

19. $2x + 3y = 3$
$4x + 6y = -4$

20. $3x - 3y = 3$
$2x + 3y = 2$

Solve each system.

21. $x + y = 7$
$2x + 2y = 14$

22. $2x + y = -1$
$2x - 3y = 11$

23. $2x + 3y = 13$
$x - y = -1$

24. $x + y = 13$
$0.05x + 0.10y = 1$

25. $2x + 5y = 33$
$x - 3y = 0$

26. $3x + 4y = 8$
$x - y = 5$

27. $3x - 7y = 12$
$2x + y = 8$

28. $5x + 6y = 9$
$x - 2y = 5$

29. $2x - 3y = 7$
$y = 5x + 2$

30. $3x - 6y = 9$
$x = 2y + 3$

Find the next number in each sequence.

31. $4, 1, -2, -5, \ldots$

32. $2, 4, 8, 16, \ldots$

33. What is the quotient of -30 and 6?

34. Subtract 8 from -9.

Factor into primes.

35. 180

36. 300

Find the value of each expression when x is 3.

37. $-3x + 7 + 5x$

38. $-x - 4x - 2x$

39. Given $4x - 5y = 12$, complete the ordered pair $(-2, \quad)$.

40. Given $y = \dfrac{3}{5}x - 2$, complete the ordered pairs $(2, \quad)$ and $(\quad, -1)$.

41. Find the x- and y-intercepts for the line $3x - 4y = 12$.

42. Find the slope of the line $x = 3y - 6$.

Find the slope of the line through the given pair of points.

43. $(-1, 1), (-5, -4)$

44. $\left(6, \dfrac{1}{2}\right), \left(-3, \dfrac{5}{4}\right)$

45. Graph the equation of the line with slope $\dfrac{2}{3}$ and y-intercept 3.

46. Graph the equation of the line with slope 2 if it passes through $(4, 1)$.

47. Find the slope of the line through $(4, 3)$ and $(6, 6)$.

48. Find the x- and y-intercepts for the line that passes through the points $(-2, 3)$ and $(6, -1)$.

49. Coin Problem Joy has one more nickel than she has dimes. If she has $\$1.10$ in change, how many of each coin does she have?

50. Investing I have invested money in two accounts. One account pays 5% annual interest, and the other pays 6%. I have $\$200$ more in the 6% account than I have in the 5% account. If the total amount of interest was $\$56$, how much do I have in each account?

Use the pie chart to answer the following questions.

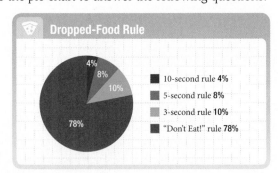

Dropped-Food Rule

4%
8%
10%
78%

- 10-second rule **4%**
- 5-second rule **8%**
- 3-second rule **10%**
- "Don't Eat!" rule **78%**

51. If 200 people were surveyed, how many said they don't eat food dropped on the floor?

52. If 20 people said they would eat food that stays on the floor for five seconds, how many people were surveyed?

Graph the ordered pairs. [3.1]

1. $(3, -4)$

2. $(-1, -2)$

3. $(4, 0)$

4. $(0, -1)$

5. Fill in the following ordered pairs for the equation
$2x - 5y = 10$. [3.2]
$$(0, \quad) \, (\quad, 0) \, (10, \quad) \, (\quad, -3)$$

6. Which of the following ordered pairs are solutions to
$y = 4x - 3$? [3.2]
$$(2, 5) \, (0, -3) \, (3, 0) \, (-2, 11)$$

Graph each line. [3.3]

7. $y = 3x - 2$

8. $x = -2$

Find the x- and y-intercepts. [3.4]

9. $3x - 5y = 15$

10. $y = \dfrac{3}{2}x + 1$

11.

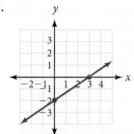

Find the slope of the line through each pair of points. [3.5]

12. $(2, -3), (4, -7)$

13. $(-3, 5), (2, -8)$

Find the slope of each line. [3.5]

14.

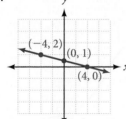

15.

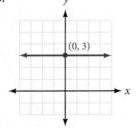

16.

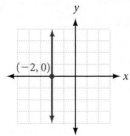

Graph each linear inequality in two variables. [3.6]

17. $y < -x + 4$

18. $3x - 4y \geq 12$

19. Write the solution to the system which is graphed
below. [4.1]

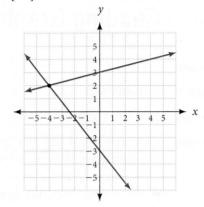

Solve each system by graphing. [3.7]

20. $x + 2y = 5$
$y = 2x$

21. $x - y = 5$
$x = 3$

22. $2x - y = 1$
$4x - 2y = -2$

Solve each system by the elimination method. [3.8]

23. $x - y = 1$
$2x + y = -10$

24. $2x + y = 7$
$3x + y = 12$

25. $7x + 8y = -2$
$3x - 2y = 10$

26. $6x - 10y = 6$
$9x - 15y = 9$

Solve each system by the substitution method. [3.9]

27. $3x + 2y = 20$
$y = 2x + 3$

28. $3x - 6y = -6$
$x = y + 1$

29. $7x - 2y = -4$
$-3x + y = 3$

30. $2x - 3y = -7$
$x + 3y = -8$

Chapter 3 Projects

EQUATIONS AND INEQUALITIES IN TWO VARIABLES

Reading Graphs

Number of People	2-3
Time Needed	5–10 minutes
Equipment	Pencil and paper
Background	Although most of the graphs we have encountered in this chapter have been straight lines, many of the graphs that describe the world around us are not straight lines. In this group project we gain experience working with graphs that are not straight lines.
Procedure	Read the introduction to each problem below. Then use the graphs to answer the questions.

1. A patient is taking a prescribed dose of a medication every 4 hours during the day to relieve the symptoms of a cold. Figure 1 shows how the concentration of that medication in the patient's system changes over time. The 0 on the horizontal axis corresponds to the time the patient takes the first dose of medication. (The units of concentration on the vertical axis are nanograms per milliliter.)

    a. Explain what the steep vertical line segments show with regard to the patient and his medication.

    b. What has happened to make the graph fall off on the right?

    c. What is the maximum concentration of the medication in the patient's system during the time period shown in Figure 1?

    d. Find the values of A, B, and C.

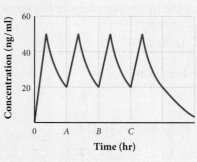

FIGURE 1

2. **Reading Graphs.** Figure 2 shows the number of people in line at a theater box office to buy tickets for a movie that starts at 7:30. The box office opens at 6:45.

    a. How many people are in line at 6:30?

    b. How many people are in line when the box office opens?

    c. How many people are in line when the show starts?

    d. At what times are there 60 people in line?

    e. How long after the show starts is there no one left in line?

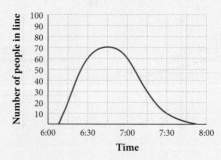

FIGURE 2

RESEARCH PROJECT

Least Squares Curve Fitting

In 1929, the astronomer Edwin Hubble (shown in Figure 1) announced his discovery that the other galaxies in the universe are moving away from us at velocities that increase with distance. The relationship between velocity and distance is described by the linear equation

$$v = Hr$$

where r is the distance of the galaxy from us, v is its velocity away from us, and H is "Hubble's constant." Figure 2 shows a plot of velocity versus distance, where each point represents a galaxy. The fact that the dots all lie approximately on a straight line is the basis of "Hubble's law."

As you can imagine, there are many lines that could be drawn through the dots in Figure 2. The line shown in Figure 2 is called the *line of best fit*

Bettmann/Corbis

FIGURE 1

for the points shown in the figure. The method used most often in mathematics to find the line of best fit is called the *least squares method.* Research the least squares method of finding the line of best fit and write an essay that describes the method. Your essay should answer the question: "Why is this method of curve fitting called the *least squares method*?"

Hubble's Law:

velocity = Hubble's constant × distance

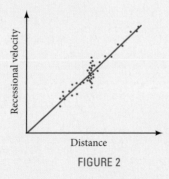

FIGURE 2

4

Exponents and Polynomials

Introduction

The French mathematician and philosopher, Blaise Pascal, was born in France in 1623. Both a scientist and a mathematician, he is credited for defining the scientific method. In the image below, Pascal carries a barometer to the top of the bell tower at the church of Saint-Jacques-de-la-Boucherie, overlooking Paris, to test a scientific theory.

Mary Evans Picture Library/Alamy

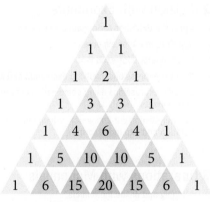

The triangular array of numbers to the right of the painting is called Pascal's triangle. Pascal's triangle is connected to the work we will do in this chapter when we find increasing powers of binomials.

Simplify.

1. $\dfrac{1}{2} \cdot \dfrac{5}{7}$

2. $\dfrac{9.6}{3}$

3. $9 - 20$

4. $1 - 8$

5. $-8 - 2(3)$

6. $2(3) - 4 - 3(-4)$

7. $2(3) + 4(5) - 5(2)$

8. $3(-2)^2 - 5(-2) + 4$

9. $-(4x^2 - 2x - 6)$

10. $3x \cdot 3x$

11. $2(2x)(-3)$

12. $(3x)(2x)$

13. $-3x - 10x$

14. $x - 2x$

15. $(2x^3 + 0x^2) - (2x^3 - 10x^2)$

16. $(4x - 14) - (4x - 10)$

17. $-4x(x + 5)$

18. $-2x(2x + 7)$

19. $2x^2(x - 5)$

20. $10x(x - 5)$

Chapter Outline

4.1 Multiplication with Exponents

A Use the definition of integer exponents to evaluate expressions containing exponents.

B Use Property 1 for exponents.

C Use Property 2 for exponents.

D Use Property 3 for exponents.

E Write numbers in scientific notation and expanded form.

4.2 Division with Exponents

A Apply the definition for negative exponents.

B Use Property 4 for exponents.

C Use Property 5 for exponents.

D Simplify expressions involving exponents of 0 and 1.

E Simplify expressions using combinations of the properties of exponents.

F Use negative exponents when writing numbers in scientific notation and expanded form.

4.3 Operations with Monomials

A Multiply and divide monomials.

B Multiply and divide numbers written in scientific notation.

C Add and subtract monomials.

4.4 Addition and Subtraction of Polynomials

A Add and subtract polynomials.

B Find the value of a polynomial for a given value of the variable.

4.5 Multiplication of Polynomials

A Multiply a monomial with a polynomial.

B Multiply two binomials using the FOIL method.

C Multiply two polynomials using the Column method.

4.6 Binomial Squares and Other Special Products

A Find the square of a binomial.

B Multiply expressions of the form $(a + b)(a - b)$.

4.7 Dividing a Polynomial by a Monomial

A Divide a polynomial by a monomial.

4.8 Dividing a Polynomial by a Polynomial

A Divide a polynomial by a polynomial.

4.1 Multiplication with Exponents

OBJECTIVES

A Use the definition of integer exponents to evaluate expressions containing exponents.

B Use Property 1 for exponents.

C Use Property 2 for exponents.

D Use Property 3 for exponents.

E Write numbers in scientific notation and expanded form.

TICKET TO SUCCESS

Keep these questions in mind as you read through the section. Then respond in your own words and in complete sentences.

1. Explain the difference between -5^2 and $(-5)^2$.

2. How do you multiply two expressions containing exponents when they each have the same base?

3. Explain in words how you would use Property 2 and Property 3 for exponents to simplify $(2x^3)^2$.

4. How would you write 5,190,000 in scientific notation?

During a race, a thirty-year-old marathon runner's heart beat 145 times per minute. If he ran the marathon in 4.5 hours, his heart beat roughly 39,150 times during the race. In this chapter, we will learn how to write this large number in scientific notation. For instance, 39,150 can be written as 3.915×10^4. But first, we must learn to evaluate exponents and work with the properties of exponents to further understand their meaning in this notation.

A Evaluating Exponents

Recall that an *exponent* is a number written just above and to the right of another number, which is called the *base*. In the expression 5^2, for example, the exponent is 2 and the base is 5. The expression 5^2 is read "5 to the second power" or "5 squared." The meaning of the expression is

$$5^2 = 5 \cdot 5 = 25$$

In the expression 5^3, the exponent is 3 and the base is 5. The expression 5^3 is read "5 to the third power" or "5 cubed." The meaning of the expression is

$$5^3 = 5 \cdot 5 \cdot 5 = 125$$

Here are some further examples.

EXAMPLE 1

$$4^3 = 4 \cdot 4 \cdot 4 = 16 \cdot 4 = 64 \qquad \text{Exponent 3, base 4}$$

EXAMPLE 2

$$-3^4 = -3 \cdot 3 \cdot 3 \cdot 3 = -81$$ Exponent 4, base 3 ■

EXAMPLE 3

$$(-2)^5 = (-2)(-2)(-2)(-2)(-2) = -32$$ Exponent 5, base -2 ■

EXAMPLE 4

$$\left(-\frac{3}{4}\right)^2 = \left(-\frac{3}{4}\right)\left(-\frac{3}{4}\right) = \frac{9}{16}$$ Exponent 2, base $-\frac{3}{4}$ ■

QUESTION: In what way are $(-5)^2$ and -5^2 different?

ANSWER: In the first case, the base is -5. In the second case, the base is 5. The answer to the first is 25. The answer to the second is -25. Can you tell why? Would there be a difference in the answers if the exponent in each case were changed to 3?

We can simplify our work with exponents by developing some properties of exponents. We want to list the things we know are true about exponents and then use these properties to simplify expressions that contain exponents.

B Property 1 for Exponents

The first property of exponents applies to products with the same base. We can use the definition of exponents, as indicating repeated multiplication, to simplify expressions like $7^4 \cdot 7^2$.

$$7^4 \cdot 7^2 = (7 \cdot 7 \cdot 7 \cdot 7)(7 \cdot 7)$$

$$= (7 \cdot 7 \cdot 7 \cdot 7 \cdot 7 \cdot 7)$$

$$= 7^6 \qquad \text{Note: } 4 + 2 = 6$$

As you can see, multiplication with the same base resulted in addition of exponents. We can summarize this result with the following property.

Property 1 for Exponents

If a is any real number and r and s are integers, then

$$a^r \cdot a^s = a^{r+s}$$

In words: To multiply two expressions with the same base, add exponents and use the common base.

Here are some examples using Property 1.

EXAMPLES   Use Property 1 to simplify the following expressions. Leave your answers in terms of exponents:

5. $5^3 \cdot 5^6 = 5^{3+6} = 5^9$

6. $x^7 \cdot x^8 = x^{7+8} = x^{15}$

7. $3^4 \cdot 3^8 \cdot 3^5 = 3^{4+8+5} = 3^{17}$ ■

NOTE
In Examples 5, 6, and 7, notice that each base in the original problem is the same base that appears in the answer and that it is written only once in the answer. A very common mistake that people make when they first begin to use Property 1 is to write a 2 in front of the base in the answer. For example, people making this mistake would get $2x^{15}$ or $(2x)^{15}$ as the result in Example 6. To avoid this mistake, you must be sure you understand the meaning of Property 1 exactly as it is written.

C   Property 2 for Exponents

Another common type of expression involving exponents is one in which an expression containing an exponent is raised to another power. The expression $(5^3)^2$ is an example.

$$(5^3)^2 = (5^3)(5^3)$$
$$= 5^{3+3}$$
$$= 5^6 \qquad \text{Note: } 3 \cdot 2 = 6$$

This result offers justification for the second property of exponents.

Property 2 for Exponents

If a is any real number and r and s are integers, then
$$(a^r)^s = a^{r \cdot s}$$

In words: A power raised to another power is the base raised to the product of the powers.

EXAMPLES   Simplify the following expressions:

8. $(4^5)^6 = 4^{5 \cdot 6} = 4^{30}$

9. $(x^3)^5 = x^{3 \cdot 5} = x^{15}$

D   Property 3 for Exponents

The third property of exponents applies to expressions in which the product of two or more numbers or variables is raised to a power. Let's look at how the expression $(2x)^3$ can be simplified.

$$(2x)^3 = (2x)(2x)(2x)$$
$$= (2 \cdot 2 \cdot 2)(x \cdot x \cdot x)$$
$$= 2^3 \cdot x^3 \qquad \text{Note: The exponent 3 distributes over the product } 2x$$
$$= 8x^3$$

We can generalize this result into a third property of exponents.

Property 3 for Exponents

If a and b are any two real numbers and r is an integer, then
$$(ab)^r = a^r b^r$$

In words: The power of a product is the product of the powers.

Here are some examples using Property 3 to simplify expressions.

EXAMPLE 10   Simplify the following expression $(5x)^2$.

SOLUTION $(5x)^2 = 5^2 \cdot x^2$ $\qquad\qquad$ Property 3
$$= 25x^2$$

EXAMPLE 11   Simplify the following expression $(2xy)^3$.

SOLUTION $(2xy)^3 = 2^3 \cdot x^3 \cdot y^3$ $\qquad\qquad$ Property 3
$$= 8x^3y^3$$

EXAMPLE 12   Simplify the following expression $(3x^2)^3$.

SOLUTION $(3x^2)^3 = 3^3(x^2)^3$ Property 3
$= 27x^6$ Property 2

EXAMPLE 13   Simplify the following expression $\left(-\dfrac{1}{4}x^2y^3\right)^2$.

SOLUTION $\left(-\dfrac{1}{4}x^2y^3\right)^2 = \left(-\dfrac{1}{4}\right)^2(x^2)^2(y^3)^2$ Property 3

$= \dfrac{1}{16}x^4y^6$ Property 2

EXAMPLE 14   Simplify the following expression $(x^4)^3(x^2)^5$.

SOLUTION $(x^4)^3(x^2)^5 = x^{12} \cdot x^{10}$ Property 2
$= x^{22}$ Property 1

EXAMPLE 15   Simplify the following expression $(2y)^3(3y^2)$.

SOLUTION $(2y)^3(3y^2) = 2^3y^3(3y^2)$ Property 3
$= 8 \cdot 3(y^3 \cdot y^2)$ Commutative and
                                                associative properties

$= 24y^5$ Property 1

EXAMPLE 16   Simplify the following expression $(2x^2y^5)^3(3x^4y)^2$.

SOLUTION $(2x^2y^5)^3(3x^4y)^2 = 2^3(x^2)^3(y^5)^3 \cdot 3^2(x^4)^2y^2$ Property 3
$= 8x^6y^{15} \cdot 9x^8y^2$ Property 2
$= (8 \cdot 9)(x^6x^8)(y^{15}y^2)$ Commutative and
                                                associative properties

$= 72x^{14}y^{17}$ Property 1

NOTE

If we include units with the dimensions of the diagrams, then the units for the area will be square units and the units for volume will be cubic units. More specifically:

If a square has a side 5 inches long, then its area will be $A = (5 \text{ inches})^2 = 25 \text{ inches}^2$ where the unit inches^2 stands for square inches.

If a cube has a single side 5 inches long, then its volume will be $V = (5 \text{ in.})^3 = 125 \text{ in.}^3$ where the unit inches^3 stands for cubic inches.

If a rectangular solid has a length of 5 inches, a width of 4 inches, and a height of 3 inches, then its volume is $V = (5 \text{ in.})(4 \text{ in.})(3 \text{ in.}) = 60 \text{ in.}^3$

FACTS FROM GEOMETRY: Volume of a Rectangular Solid

It is easy to see why the phrase "five squared" is associated with the expression 5^2. Simply find the area of the square shown in Figure 1 with a side of 5.

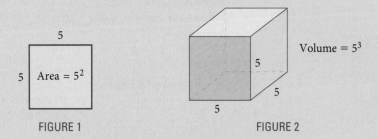

FIGURE 1   FIGURE 2

To see why the phrase "five cubed" is associated with the expression 5^3, we have to find the *volume* of a cube for which all three dimensions are 5 units long. The volume of a cube is a measure of the space occupied by the cube. To calculate the volume of the cube shown in Figure 2, we multiply the three dimensions together to get $5 \cdot 5 \cdot 5 = 5^3$.

The cube shown in Figure 2 is a special case of a general category of three-dimensional geometric figures called *rectangular solids*. Rectangular solids have rectangles for sides, and all connecting sides meet at right angles. The three dimensions are length, width, and height. To find the volume of a rectangular solid, we find the product of the three dimensions, as shown in Figure 3.

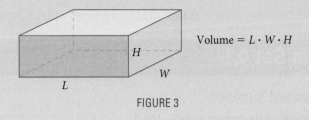

FIGURE 3

E   Scientific Notation

Many branches of science require working with very large numbers. In astronomy, for example, distances commonly are given in light-years. A light-year is the distance light travels in a year. It is approximately

5,880,000,000,000 miles

This number is difficult to use in calculations because of the number of zeros it contains. Scientific notation provides a way of writing very large numbers in a more manageable form.

Definition

A number is in **scientific notation** when it is written as the product of a number between 1 and 10 and an integer power of 10. A number written in scientific notation has the form

$$n \times 10^r$$

where $1 \le n < 10$ and $r =$ an integer.

EXAMPLE 17 Write 376,000 in scientific notation.

SOLUTION We must rewrite 376,000 as the product of a number between 1 and 10 and a power of 10. To do so, we move the decimal point 5 places to the left so that it appears between the 3 and the 7. Then we multiply this number by 10^5. The number that results has the same value as our original number and is written in scientific notation.

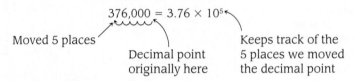

Moved 5 places

$$376,000 = 3.76 \times 10^5$$

Decimal point originally here

Keeps track of the 5 places we moved the decimal point

EXAMPLE 18 Write 4.52×10^3 in expanded form.

SOLUTION Since 10^3 is 1,000, we can think of this as simply a multiplication problem; that is,

$$4.52 \times 10^3 = 4.52 \times 1,000 = 4,520$$

On the other hand, we can think of the exponent 3 as indicating the number of places we need to move the decimal point to write our number in expanded form. Since our exponent is positive 3, we move the decimal point three places to the right:

$$4.52 \times 10^3 = 4,520$$

In the next section, we will see how a negative power of 10 affects our answer.

Problem Set 4.1

Moving Toward Success

"Nothing worthwhile comes easily. Work, continuous work and hard work, is the only way to accomplish results that last."

—Hamilton Holt, 1872–1951, American educator and author

1. Why is it important to stay focused when learning mathematics?

2. How can distraction impede your success in this class?

A Name the base and exponent in each of the following expressions. Then use the definition of exponents as repeated multiplication to simplify. [Examples 1–4]

1. 4^2

2. 6^2

3. $(0.3)^2$

4. $(0.03)^2$

5. 4^3

6. 10^3

7. $(-5)^2$

8. -5^2

9. -2^3

10. $(-2)^3$

11. 3^4

12. $(-3)^4$

13. $\left(\dfrac{2}{3}\right)^2$

14. $\left(\dfrac{2}{3}\right)^3$

15. $\left(\dfrac{1}{2}\right)^4$

16. $\left(\dfrac{4}{5}\right)^2$

17. a. Complete the following table.

Number x	Square x^2
1	
2	
3	
4	
5	
6	
7	

b. Using the results of part a, fill in the blank in the following statement: For numbers larger than 1, the square of the number is _____ than the number.

18. a. Complete the following table.

Number x	Square x^2
$\frac{1}{2}$	
$\frac{1}{3}$	
$\frac{1}{4}$	
$\frac{1}{5}$	
$\frac{1}{6}$	
$\frac{1}{7}$	
$\frac{1}{8}$	

b. Using the results of part a, fill in the blank in the following statement: For numbers between 0 and 1, the square of the number is _____ than the number.

B Use Property 1 for exponents to simplify each expression. Leave all answers in terms of exponents. [Examples 5–7]

19. $x^4 \cdot x^5$

20. $x^7 \cdot x^3$

21. $y^{10} \cdot y^{20}$

22. $y^{30} \cdot y^{30}$

23. $2^5 \cdot 2^4 \cdot 2^3$

24. $4^2 \cdot 4^3 \cdot 4^4$

25. $x^4 \cdot x^6 \cdot x^8 \cdot x^{10}$

26. $x^{20} \cdot x^{18} \cdot x^{16} \cdot x^{14}$

C Use Property 2 for exponents to write each of the following problems with a single exponent. (Assume all variables are positive numbers.) [Examples 8–9]

27. $(x^2)^5$

28. $(x^5)^2$

29. $(5^4)^3$

30. $(5^3)^4$

31. $(y^3)^3$

32. $(y^2)^2$

33. $(2^5)^{10}$

34. $(10^5)^2$

35. $(a^3)^x$

36. $(a^5)^x$

37. $(b^x)^y$

38. $(b^r)^s$

D Use Property 3 for exponents to simplify each of the following expressions. [Examples 10–16]

39. $(4x)^2$

40. $(2x)^4$

41. $(2y)^5$

42. $(5y)^2$

43. $(-3x)^4$

44. $(-3x)^3$

45. $(0.5ab)^2$

46. $(0.4ab)^2$

47. $(4xyz)^3$

48. $(5xyz)^3$

Simplify the following expressions by using the properties of exponents.

49. $(2x^4)^3$

50. $(3x^5)^2$

51. $(4a^3)^2$

52. $(5a^2)^2$

53. $(x^2)^3(x^4)^2$

54. $(x^5)^2(x^3)^5$

55. $(a^3)^1(a^2)^4$

56. $(a^4)^1(a^1)^3$

57. $(2x)^3(2x)^4$

58. $(3x)^2(3x)^3$

59. $(3x^2)^3(2x)^4$

60. $(3x)^3(2x^3)^2$

61. $(4x^2y^3)^2$

62. $(9x^3y^5)^2$

63. $\left(\frac{2}{3}a^4b^5\right)^3$

64. $\left(\frac{3}{4}ab^7\right)^3$

Write each expression as a perfect square.

65. $x^4 = (\ )^2$

66. $x^6 = (\ )^2$

67. $16x^2 = (\ )^2$

68. $256x^4 = (\ \ )^2$

Write each expression as a perfect cube.

69. $8 = (\ )^3$

70. $27 = (\ )^3$

71. $64x^3 = (\ )^3$

72. $27x^6 = (\ )^3$

73. Let $x = 2$ in each of the following expressions and simplify.

a. x^3x^2

b. $(x^3)^2$

c. x^5

d. x^6

74. Let $x = -1$ in each of the following expressions and simplify.

a. x^3x^4

b. $(x^3)^4$

c. x^7

d. x^{12}

75. Complete the following table, and then construct a line graph of the information in the table.

Number x	Square x^2
-3	
-2	
-1	
0	
1	
2	
3	

76. Complete the table, and then construct a line graph of the information in the table.

Number x	Cube x^3
−3	
−2	
−1	
0	
1	
2	
3	

77. Complete the table. When you are finished, notice how the points in this table could be used to refine the line graph you created in Problem 75.

Number x	Square x^2
−2.5	
−1.5	
−0.5	
0	
0.5	
1.5	
2.5	

78. Complete the following table. When you are finished, notice that this table contains exactly the same entries as the table from Problem 77. This table uses fractions, whereas the table from Problem 77 uses decimals.

Number x	Square x^2
$-\frac{5}{2}$	
$-\frac{3}{2}$	
$-\frac{1}{2}$	
0	
$\frac{1}{2}$	
$\frac{3}{2}$	
$\frac{5}{2}$	

E Write each number in scientific notation. [Example 17]

79. 43,200

80. 432,000

81. 570

82. 5,700

83. 238,000

84. 2,380,000

E Write each number in expanded form. [Example 18]

85. 2.49×10^3

86. 2.49×10^4

87. 3.52×10^2

88. 3.52×10^5

89. 2.8×10^4

90. 2.8×10^3

Applying the Concepts

91. Google Earth This Google Earth image is of the Luxor Hotel in Las Vegas, Nevada. The casino has a square base with sides of 525 feet. What is the area of the casino floor?

92. Google Earth This is a three dimensional model created by Google Earth of the Louvre Museum in Paris, France. The pyramid that dominates the Napoleon Courtyard has a square base with sides of 35.50 meters. What is the area of the base of the pyramid?

93. Volume of a Cube Find the volume of a cube if each side is 3 inches long.

94. Volume of a Cube Find the volume of a cube if each side is 0.3 feet long.

95. Volume of a Cube A bottle of perfume is packaged in a box that is in the shape of a cube. Find the volume of the box if each side is 2.5 inches long. Round to the nearest tenth.

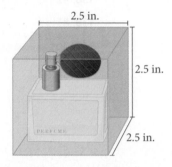

96. Volume of a Cube A television set is packaged in a box that is in the shape of a cube. Find the volume of the box if each side is 18 inches long.

97. Volume of a Box A rented videotape is in a plastic container that has the shape of a rectangular solid. Find the volume of the container if the length is 8 inches, the width is 4.5 inches, and the height is 1 inch.

98. Volume of a Box Your textbook is in the shape of a rectangular solid. Find the volume in cubic inches.

99. Age in seconds If you are 21 years old, you have been alive for more than 650,000,000 seconds. Write this last number in scientific notation.

100. Distance Around the Earth The distance around the Earth at the equator is more than 130,000,000 feet. Write this number in scientific notation.

101. Lifetime Earnings If you earn at least $12 an hour and work full-time for 30 years, you will make at least 7.4×10^5 dollars. Write this last number in expanded form.

102. Heart Beats per Year If your pulse is 72, then in one year your heart will beat at least 3.78×10^7 times. Write this last number in expanded form.

103. Investing If you put $1,000 into a savings account every year from the time you are 25 years old until you are 55 years old, you will have more than 1.8×10^5 dollars in the account when you reach 55 years of age (assuming 10% annual interest). Write 1.8×10^5 in expanded form.

104. Investing If you put $20 into a savings account every month from the time you are 20 years old until you are 30 years old, you will have more than 3.27×10^3 dollars in the account when

you reach 30 years of age (assuming 6% annual interest compounded monthly). Write 3.27×10^3 in expanded form.

Displacement The displacement, in cubic inches, of a car engine is given by the formula
$$d = \pi \cdot s \cdot c \cdot \left(\frac{1}{2} \cdot b\right)^2$$
where s is the stroke and b is the bore, as shown in the figure, and c is the number of cylinders. (Note: The bore is the diameter of a cylinder in a piston engine, and the stroke is the distance the cylinder travels.)

Calculate the engine displacement for each of the following cars. Use 3.14 to approximate π. Round your answers to the nearest cubic inch.

105. Ferrari Modena 8 cylinders, 3.35 inches of bore, 3.11 inches of stroke

106. Audi A8 8 cylinders, 3.32 inches of bore, 3.66 inches of stroke

107. Mitsubishi Eclipse 6 cylinders, 3.59 inches of bore, 2.99 inches of stroke

108. Porsche 911 GT3 6 cylinders, 3.94 inches of bore, 3.01 inches of stroke

Getting Ready for the Next Section

Subtract.

109. $4 - 7$

110. $-4 - 7$

111. $4 - (-7)$

112. $-4 - (-7)$

113. $15 - 20$

114. $15 - (-20)$

115. $-15 - (-20)$

116. $-15 - 20$

Simplify.

117. $2(3) - 4$

118. $5(3) - 10$

119. $4(3) - 3(2)$

120. $-8 - 2(3)$

121. $2(5 - 3)$

122. $2(3) - 4 - 3(-4)$

123. $5 + 4(-2) - 2(-3)$

124. $2(3) + 4(5) - 5(2)$

Maintaining Your Skills

Factor each of the following into its product of prime factors.

125. 128   **126.** 200

127. 250   **128.** 512

129. 720   **130.** 555

131. 820   **132.** 1,024

Factor the following by first factoring the base and then raising each of its factors to the third power.

133. 6^3 **134.** 10^3

135. 30^3 **136.** 42^3

137. 25^3 **138.** 8^3

139. 12^3 **140.** 36^3

4.2 Division with Exponents

OBJECTIVES

A Apply the definition for negative exponents.

B Use Property 4 for exponents.

C Use Property 5 for exponents.

D Simplify expressions involving exponents of 0 and 1.

E Simplify expressions using combinations of the properties of exponents.

F Use negative exponents when writing numbers in scientific notation and expanded form.

TICKET TO SUCCESS

Keep these questions in mind as you read through the section. Then respond in your own words and in complete sentences.

1. How do you divide two expressions containing exponents when they each have the same base?
2. Explain the difference between 3^2 and 3^{-2}.
3. Explain what happens when we use 0 as an exponent.
4. What does a negative exponent mean in scientific notation?

Image copyright © Graham Tomlin. Used under license from Shutterstock.com

Previously, we found that multiplication with the same base results in addition of exponents; that is, $a^r \cdot a^s = a^{r+s}$. Since division is the inverse operation of multiplication, we can expect division with the same base to result in subtraction of exponents.

Suppose you need to figure out how many small cube-shaped boxes with a side of x inches long will fit into a larger shipping box, also shaped like a cube but with a side of $5x$ inches long. First, you would need to find the volume of each box.

Volume for the smaller box: $x \cdot x \cdot x = x^3$

Volume for the larger box: $5x \cdot 5x \cdot 5x = (5x)^3$

To find how many smaller boxes fit in the larger box, we use division.

$$\frac{(5x)^3}{x^3}$$

In this chapter, we will learn a property of exponents under division that will help us simplify this problem.

A Negative Exponents

To develop the properties for exponents under division, we again apply the definition of exponents.

$$\frac{x^5}{x^3} = \frac{x \cdot x \cdot x \cdot x \cdot x}{x \cdot x \cdot x} \qquad\qquad \frac{2^4}{2^7} = \frac{2 \cdot 2 \cdot 2 \cdot 2}{2 \cdot 2 \cdot 2 \cdot 2 \cdot 2 \cdot 2 \cdot 2}$$

$$= \frac{x \cdot x \cdot x}{x \cdot x \cdot x}(x \cdot x) \qquad\qquad = \frac{2 \cdot 2 \cdot 2 \cdot 2}{2 \cdot 2 \cdot 2 \cdot 2} \cdot \frac{1}{2 \cdot 2 \cdot 2}$$

$$= 1(x \cdot x) \qquad\qquad\qquad = \frac{1}{2 \cdot 2 \cdot 2}$$

$$= x^2 \quad \text{Note: } 5 - 3 = 2 \qquad\qquad = \frac{1}{2^3} \quad \text{Note: } 7 - 4 = 3$$

In both cases, division with the same base resulted in subtraction of the smaller exponent from the larger. The problem is deciding whether the answer is a fraction. The problem is resolved easily by the following definition.

> **Definition**
> If r is a positive integer, then $a^{-r} = \dfrac{1}{a^r} = \left(\dfrac{1}{a}\right)^r \qquad (a \neq 0)$

The following examples illustrate how we use this definition to simplify expressions that contain negative exponents.

EXAMPLES   Write each expression with a positive exponent and then simplify.

1. $2^{-3} = \dfrac{1}{2^3} = \dfrac{1}{8}$ Note: Negative exponents do not indicate negative numbers. They indicate reciprocals

2. $5^{-2} = \dfrac{1}{5^2} = \dfrac{1}{25}$

3. $3x^{-6} = 3 \cdot \dfrac{1}{x^6} = \dfrac{3}{x^6}$

B Property 4 for Exponents

Now let us look back to one of our original problems and try to work it again with the help of a negative exponent. We know that $\frac{2^4}{2^7} = \frac{1}{2^3}$. Let's decide now that with division of the same base, we will always subtract the exponent in the denominator from the exponent in the numerator and see if this conflicts with what we know is true.

$$\frac{2^4}{2^7} = 2^{4-7} \qquad \text{Subtracting the bottom exponent from the top exponent}$$

$$= 2^{-3} \qquad \text{Subtraction}$$

$$= \frac{1}{2^3} \qquad \text{Definition of negative exponents}$$

Subtracting the exponent in the denominator from the exponent in the numerator and then using the definition of negative exponents gives us the same result we obtained previously. We can now continue the list of properties of exponents we started in Section 4.1.

Property 4 for Exponents

If a is any real number and r and s are integers, then

$$\frac{a^r}{a^s} = a^{r-s} \quad (a \neq 0)$$

In words: To divide with the same base, subtract the exponent in the denominator from the exponent in the numerator and raise the base to the exponent that results.

The following examples show how we use Property 4 and the definition for negative exponents to simplify expressions involving division.

EXAMPLES Simplify the following expressions:

4. $\dfrac{x^9}{x^6} = x^{9-6} = x^3$

5. $\dfrac{x^4}{x^{10}} = x^{4-10} = x^{-6} = \dfrac{1}{x^6}$

6. $\dfrac{2^{15}}{2^{20}} = 2^{15-20} = 2^{-5} = \dfrac{1}{2^5} = \dfrac{1}{32}$

C Property 5 for Exponents

Our final property of exponents is similar to Property 3 from Section 4.1, but it involves division instead of multiplication. After we have stated the property, we will give a proof of it. The proof shows why this property is true.

Property 5 for Exponents

If a and b are any two real numbers ($b \neq 0$) and r is an integer, then

$$\left(\frac{a}{b}\right)^r = \frac{a^r}{b^r}$$

In words: A quotient raised to a power is the quotient of the powers.

Proof

$$\left(\frac{a}{b}\right)^r = \left(a \cdot \frac{1}{b}\right)^r \qquad \text{Definition of division}$$

$$= a^r \cdot \left(\frac{1}{b}\right)^r \qquad \text{Property 3}$$

$$= a^r \cdot b^{-r} \qquad \text{Definition of negative exponents}$$

$$= a^r \cdot \frac{1}{b^r} \qquad \text{Definition of negative exponents}$$

$$= \frac{a^r}{b^r} \qquad \text{Definition of division}$$

EXAMPLES Simplify the following expressions:

7. $\left(\dfrac{x}{2}\right)^3 = \dfrac{x^3}{2^3} = \dfrac{x^3}{8}$

8. $\left(\dfrac{5}{y}\right)^2 = \dfrac{5^2}{y^2} = \dfrac{25}{y^2}$

9. $\left(\dfrac{2}{3}\right)^4 = \dfrac{2^4}{3^4} = \dfrac{16}{81}$

D Zero and One as Exponents

We have two special exponents left to deal with before our rules for exponents are complete: 0 and 1. To obtain an expression for x^1, we will solve a problem two different ways:

$$\left.\begin{array}{l} \dfrac{x^3}{x^2} = \dfrac{x \cdot x \cdot x}{x \cdot x} = x \\[2em] \dfrac{x^3}{x^2} = x^{3-2} = x^1 \end{array}\right\} \quad \text{Hence } x^1 = x$$

Stated generally, this rule says that $a^1 = a$. This seems reasonable and we will use it since it is consistent with our property of division using the same base.

We use the same procedure to obtain an expression for x^0.

$$\left.\begin{array}{l} \dfrac{5^2}{5^2} = \dfrac{25}{25} = 1 \\[2em] \dfrac{5^2}{5^2} = 5^{2-2} = 5^0 \end{array}\right\} \quad \text{Hence } 5^0 = 1$$

It seems, therefore, that the best definition of x^0 is 1 for all bases equal to x except $x = 0$. In the case of $x = 0$, we have 0^0, which we will not define. This definition will probably seem awkward at first. Most people would like to define x^0 as 0 when they first encounter it. Remember, the zero in this expression is an exponent, so x^0 does not mean to multiply by zero. Thus, we can make the general statement that $a^0 = 1$ for all real numbers except $a = 0$.

Here are some examples involving the exponents 0 and 1.

EXAMPLES   Simplify the following expressions:

10. $8^0 = 1$

11. $8^1 = 8$

12. $4^0 + 4^1 = 1 + 4 = 5$

13. $(2x^2y)^0 = 1$

E Combinations of Properties

Here is a summary of the definitions and properties of exponents we have developed so far. For each definition or property in the list, a and b are real numbers, and r and s are integers.

Definitions of Exponents	Properties of Exponents
$a^{-r} = \dfrac{1}{a^r} = \left(\dfrac{1}{a}\right)^r \quad a \neq 0$	**1.** $a^r \cdot a^s = a^{r+s}$
$a^1 = a$	**2.** $(a^r)^s = a^{rs}$
$a^0 = 1 \quad a \neq 0$	**3.** $(ab)^r = a^r b^r$
	4. $\dfrac{a^r}{a^s} = a^{r-s} \quad a \neq 0$
	5. $\left(\dfrac{a}{b}\right)^r = \dfrac{a^r}{b^r} \quad b \neq 0, r \geq 0$

Here are some additional examples. These examples use a combination of the preceding properties and definitions.

EXAMPLE 14 Simplify the expression $\dfrac{(5x^3)^2}{x^4}$. Write the answer with a positive exponent.

SOLUTION

$$\dfrac{(5x^3)^2}{x^4} = \dfrac{25x^6}{x^4} \qquad\qquad \text{Properties 2 and 3}$$

$$= 25x^2 \qquad\qquad \text{Property 4}$$

EXAMPLE 15 Simplify the expression $\dfrac{x^{-8}}{(x^2)^3}$. Write the answer with a positive exponent.

SOLUTION

$$\dfrac{x^{-8}}{(x^2)^3} = \dfrac{x^{-8}}{x^6} \qquad\qquad \text{Property 2}$$

$$= x^{-8-6} \qquad\qquad \text{Property 4}$$

$$= x^{-14} \qquad\qquad \text{Subtraction}$$

$$= \dfrac{1}{x^{14}} \qquad\qquad \text{Definition of negative exponents}$$

EXAMPLE 16 Simplify the expression $\left(\dfrac{y^5}{y^3}\right)^2$. Write the answer with a positive exponent.

SOLUTION

$$\left(\dfrac{y^5}{y^3}\right)^2 = \dfrac{(y^5)^2}{(y^3)^2} \qquad\qquad \text{Property 5}$$

$$= \dfrac{y^{10}}{y^6} \qquad\qquad \text{Property 2}$$

$$= y^4 \qquad\qquad \text{Property 4}$$

Notice that we could have simplified inside the parentheses first and then raised the result to the second power.

$$\left(\dfrac{y^5}{y^3}\right)^2 = (y^2)^2 = y^4$$

EXAMPLE 17 Simplify the expression $(3x^5)^{-2}$. Write the answer with a positive exponent.

SOLUTION

$$(3x^5)^{-2} = \dfrac{1}{(3x^5)^2} \qquad\qquad \text{Definition of negative exponents}$$

$$= \dfrac{1}{9x^{10}} \qquad\qquad \text{Properties 2 and 3}$$

EXAMPLE 18 Simplify the expression $x^{-8} \cdot x^5$. Write the answer with a positive exponent.

SOLUTION

$$x^{-8} \cdot x^5 = x^{-8+5} \qquad\qquad \text{Property 1}$$

$$= x^{-3} \qquad\qquad \text{Addition}$$

$$= \dfrac{1}{x^3} \qquad\qquad \text{Definition of negative exponents}$$

EXAMPLE 19 Simplify the expression $\dfrac{(a^3)^2 a^{-4}}{(a^{-4})^3}$. Write the answer with a positive exponent.

SOLUTION　$\dfrac{(a^3)^2 a^{-4}}{(a^{-4})^3} = \dfrac{a^6 a^{-4}}{a^{-12}}$　　　　　Property 2

$= \dfrac{a^2}{a^{-12}}$　　　　　Property 1

$= a^{14}$　　　　　Property 4 ■

In the next two examples we use division to compare the area and volume of geometric figures.

EXAMPLE 20　Suppose you have two squares, one of which is larger than the other. If the length of a side of the larger square is 3 times as long as the length of a side of the smaller square, how many of the smaller squares will it take to cover up the larger square?

SOLUTION　If we let x represent the length of a side of the smaller square, then the length of a side of the larger square is $3x$. The area of each square, along with a diagram of the situation, is given in Figure 1.

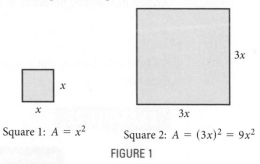

Square 1: $A = x^2$　　　　Square 2: $A = (3x)^2 = 9x^2$

FIGURE 1

To find out how many smaller squares it will take to cover up the larger square, we divide the area of the larger square by the area of the smaller square.

$$\frac{\text{Area of square 2}}{\text{Area of square 1}} = \frac{9x^2}{x^2} = 9$$

It will take 9 of the smaller squares to cover the larger square. ■

EXAMPLE 21　Suppose you have two boxes, each of which is a cube. If the length of a side in the second box is 3 times as long as the length of a side of the first box, how many of the smaller boxes will fit inside the larger box?

SOLUTION　If we let x represent the length of a side of the smaller box, then the length of a side of the larger box is $3x$. The volume of each box, along with a diagram of the situation, is given in Figure 2.

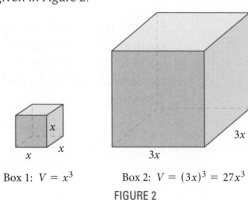

Box 1: $V = x^3$　　　　Box 2: $V = (3x)^3 = 27x^3$

FIGURE 2

To find out how many smaller boxes will fit inside the larger box, we divide the volume of the larger box by the volume of the smaller box.

$$\frac{\text{Volume of box 2}}{\text{Volume of box 1}} = \frac{27x^3}{x^3} = 27$$

We can fit 27 of the smaller boxes inside the larger box.

F More on Scientific Notation

Now that we have completed our list of definitions and properties of exponents, we can expand the work we did previously with scientific notation.

Recall that a number is in scientific notation when it is written in the form

$$n \times 10^r$$

where $1 \leq n < 10$ and r is an integer.

Since negative exponents give us reciprocals, we can use negative exponents to write very small numbers in scientific notation. For example, the number 0.00057, when written in scientific notation, is equivalent to 5.7×10^{-4}. Here's why:

$$5.7 \times 10^{-4} = 5.7 \times \frac{1}{10^4} = 5.7 \times \frac{1}{10,000} = \frac{5.7}{10,000} = 0.00057$$

The table below lists some other numbers in both scientific notation and expanded form.

EXAMPLE 22

Number Written the Expanded Form		Number Written Again in Scientific Notation
376,000	=	3.76×10^5
49,500	=	4.95×10^4
3,200	=	$3.2 \ \times 10^3$
591	=	5.91×10^2
46	=	$4.6 \ \times 10^1$
8	=	$8 \ \ \times 10^0$
0.47	=	$4.7 \ \times 10^{-1}$
0.093	=	$9.3 \ \times 10^{-2}$
0.00688	=	6.88×10^{-3}
0.0002	=	$2 \ \ \times 10^{-4}$
0.000098	=	$9.8 \ \times 10^{-5}$

Notice that in each case in Example 22, when the number is written in scientific notation, the decimal point in the first number is placed so that the number is between 1 and 10. The exponent on 10 in the second number keeps track of the number of places we moved the decimal point in the original number to get a number between 1 and 10.

$$376,000 = 3.76 \times 10^5$$

Moved 5 places

Decimal point originally here

Keeps track of the 5 places we moved the decimal point

$$0.00688 = 6.88 \times 10^{-3}$$

Moved 3 places

Keeps track of the 3 places we moved the decimal point

Problem Set 4.2

Moving Toward Success

"Thoughts lead on to purposes; purposes go forth in action; actions form habits; habits decide character; and character fixes our destiny."

—Tyron Edwards, 1809–1894, American theologian

1. How do questions like "why am I taking this class?" and "when am I ever going to use this stuff?" distract you?

2. How will you focus your energies toward success in this class?

A Write each of the following with positive exponents, and then simplify, when possible. [Examples 1–3]

1. 3^{-2}

2. 3^{-3}

3. 6^{-2}

4. 2^{-6}

5. 8^{-2}

6. 3^{-4}

7. 5^{-3}

8. 9^{-2}

9. $2x^{-3}$

10. $5x^{-1}$

11. $(2x)^{-3}$

12. $(5x)^{-1}$

13. $(5y)^{-2}$

14. $5y^{-2}$

15. 10^{-2}

16. 10^{-3}

17. Complete the following table.

Number x	Square x^2	Power of 2 2^x
−3		
−2		
−1		
0		
1		
2		
3		

18. Complete the following table.

Number x	Cube x^3	Power of 3 3^x
−3		
−2		
−1		
0		
1		
2		
3		

B Use Property 4 to simplify each of the following expressions. Write all answers that contain exponents with positive exponents only. [Examples 4–6]

19. $\dfrac{5^1}{5^3}$

20. $\dfrac{7^6}{7^8}$

21. $\dfrac{x^{10}}{x^4}$

22. $\dfrac{x^4}{x^{10}}$

23. $\dfrac{4^3}{4^0}$

24. $\dfrac{4^0}{4^3}$

25. $\dfrac{(2x)^7}{(2x)^4}$

26. $\dfrac{(2x)^4}{(2x)^7}$

27. $\dfrac{6^{11}}{6}$

28. $\dfrac{8^7}{8}$

29. $\dfrac{6}{6^{11}}$

30. $\dfrac{8}{8^7}$

31. $\dfrac{2^{-5}}{2^3}$

32. $\dfrac{2^{-5}}{2^{-3}}$

33. $\dfrac{2^5}{2^{-3}}$

34. $\dfrac{2^{-3}}{2^{-5}}$

35. $\dfrac{(3x)^{-5}}{(3x)^{-8}}$

36. $\dfrac{(2x)^{-10}}{(2x)^{-15}}$

E Simplify the following expressions. Any answers that contain exponents should contain positive exponents only. [Examples 7–19]

37. $(3xy)^4$

38. $(4xy)^3$

39. 10^0

40. 10^1

41. $(2a^2b)^1$

42. $(2a^2b)^0$

43. $(7y^3)^{-2}$

44. $(5y^4)^{-2}$

45. $x^{-3} \cdot x^{-5}$

46. $x^{-6} \cdot x^8$

47. $y^7 \cdot y^{-10}$

48. $y^{-4} \cdot y^{-6}$

49. $\dfrac{(x^2)^3}{x^4}$

50. $\dfrac{(x^5)^3}{x^{10}}$

51. $\dfrac{(a^4)^3}{(a^3)^2}$

52. $\dfrac{(a^5)^3}{(a^5)^2}$

53. $\dfrac{y^7}{(y^2)^8}$

54. $\dfrac{y^2}{(y^3)^4}$

55. $\left(\dfrac{y^7}{y^2}\right)^8$

56. $\left(\dfrac{y^2}{y^3}\right)^4$

57. $\dfrac{(x^{-2})^3}{x^{-5}}$

58. $\dfrac{(x^2)^{-3}}{x^{-5}}$

59. $\left(\dfrac{x^{-2}}{x^{-5}}\right)^3$

60. $\left(\dfrac{x^2}{x^{-5}}\right)^{-3}$

61. $\dfrac{(a^3)^2(a^4)^5}{(a^5)^2}$

62. $\dfrac{(a^4)^8(a^2)^5}{(a^3)^4}$

63. $\dfrac{(a^{-2})^3(a^4)^2}{(a^{-3})^{-2}}$

64. $\dfrac{(a^{-5})^{-3}(a^7)^{-1}}{(a^{-3})^5}$

65. Let $x = 2$ in each of the following expressions and simplify.

   a. $\dfrac{x^7}{x^2}$ **b.** x^5

   c. $\dfrac{x^2}{x^7}$ **d.** x^{-5}

66. Let $x = -1$ in each of the following expressions and simplify.

   a. $\dfrac{x^{12}}{x^9}$ **b.** x^3

   c. $\dfrac{x^{11}}{x^9}$ **d.** x^2

67. Write each expression as a perfect square.

   a. $\dfrac{1}{25} = \left(-\!\!\!-\right)^2$ **b.** $\dfrac{1}{64} = \left(-\!\!\!-\right)^2$

   c. $\dfrac{1}{x^2} = \left(-\!\!\!-\right)^2$ **d.** $\dfrac{1}{x^4} = \left(-\!\!\!-\right)^2$

68. Write each expression as a perfect cube.

   a. $\dfrac{1}{125} = \left(-\!\!\!-\right)^3$ **b.** $\dfrac{1}{27} = \left(-\!\!\!-\right)^3$

   c. $\dfrac{x^6}{125} = \left(-\!\!\!-\right)^3$ **d.** $\dfrac{x^3}{27} = \left(-\!\!\!-\right)^3$

69. Complete the following table, and then construct a line graph of the information in the table.

Number x	Power of 2 2^x
−3	
−2	
−1	
0	
1	
2	
3	

70. Complete the following table, and then construct a line graph of the information in the table.

Number x	Power of 3 3^x
−3	
−2	
−1	
0	
1	
2	
3	

F Write each of the following numbers in scientific notation. [Example 22]

71. 0.0048

72. 0.000048

73. 25

74. 35

75. 0.000009

76. 0.0009

77. Complete the following table.

Expanded Form	Scientific Notation $n \times 10^r$
0.000357	3.57×10^{-4}
0.00357	
0.0357	
0.357	
3.57	
35.7	
357	
3,570	
35,700	

78. Complete the following table.

Expanded Form	Scientific Notation $n \times 10^r$
0.000123	1.23×10^{-4}
	1.23×10^{-3}
	1.23×10^{-2}
	1.23×10^{-1}
	1.23×10^{0}
	1.23×10^{1}
	1.23×10^{2}
	1.23×10^{3}
	1.23×10^{4}

F Write each of the following numbers in expanded form.

79. 4.23×10^{-3}

80. 4.23×10^{3}

81. 8×10^{-5}

82. 8×10^{5}

83. 4.2×10^{0}

84. 4.2×10^{1}

Applying the Concepts [Examples 20–21]

85. Music The chart shows the country music singers that earned the most in 2009. Use the chart and write the earnings of Toby Keith in scientific notation. Remember the sales numbers are in millions.

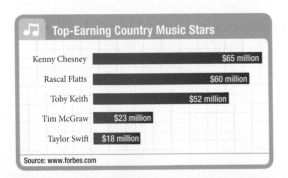

Top-Earning Country Music Stars

Kenny Chesney — $65 million
Rascal Flatts — $60 million
Toby Keith — $52 million
Tim McGraw — $23 million
Taylor Swift — $18 million

Source: www.forbes.com

86. Cars The map shows the highest producers of cars for 2009. Find, to the nearest thousand, how many more cars China produced than the United States. Then write the answer in scientific notation.

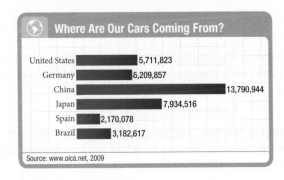

Where Are Our Cars Coming From?

United States — 5,711,823
Germany — 5,209,857
China — 13,790,944
Japan — 7,934,516
Spain — 2,170,078
Brazil — 3,182,617

Source: www.oica.net, 2009

87. Some home computers can do a calculation in 2×10^{-3} seconds. Write this number in expanded form.

88. Some of the cells in the human body have a radius of 3×10^{-5} inches. Write this number in expanded form.

89. The number 25×10^{3} is not in scientific notation because 25 is larger than 10. Write 25×10^{3} in scientific notation.

90. The number 0.25×10^{3} is not in scientific notation because 0.25 is less than 1. Write 0.25×10^{3} in scientific notation.

91. The number 23.5×10^{4} is not in scientific notation because 23.5 is not between 1 and 10. Rewrite 23.5×10^{4} in scientific notation.

92. The number 375×10^{3} is not in scientific notation because 375 is not between 1 and 10. Rewrite 375×10^{3} in scientific notation.

93. The number 0.82×10^{-3} is not in scientific notation because 0.82 is not between 1 and 10. Rewrite 0.82×10^{-3} in scientific notation.

94. The number 0.93×10^{-2} is not in scientific notation because 0.93 is not between 1 and 10. Rewrite 0.93×10^{-2} in scientific notation.

Comparing Areas Suppose you have two squares, one of which is larger than the other, and the side of the larger square is twice as long as the side of the smaller square.

95. If the length of the side of the smaller square is 10 inches, give the area of each square. Then find the number of smaller squares it will take to cover the larger square.

96. How many smaller squares will it take to cover the larger square if the length of the side of the smaller square is 1 foot?

97. If the length of the side of the smaller square is x, find the area of each square. Then find the number of smaller squares it will take to cover the larger square.

98. Suppose the length of the side of the larger square is 1 foot. How many smaller squares will it take to cover the larger square?

Comparing Volumes Suppose you have two boxes, each of which is a cube, and the length of a side of the second box is twice as long as the length of a side of the first box.

99. If the length of a side of the first box is 6 inches, give the volume of each box. Then find the number of smaller boxes that will fit inside the larger box.

100. How many smaller boxes can be placed inside the larger box if the length of a side of the second box is 1 foot?

101. If the length of a side of the first box is x, find the volume of each box. Then find the number of smaller boxes that will fit inside the larger box.

102. Suppose the length of a side of the larger box is 12 inches. How many smaller boxes will fit inside the larger box?

Getting Ready for the Next Section

Simplify.

103. $3(4.5)$

104. $\dfrac{1}{2} \cdot \dfrac{5}{7}$

105. $\dfrac{4}{5}(10)$

106. $\dfrac{9.6}{3}$

107. $6.8(3.9)$

108. $9 - 20$

109. $-3 + 15$

110. $2x \cdot x \cdot \dfrac{1}{2}x$

111. $x^5 \cdot x^3$

112. $y^2 \cdot y$

113. $\dfrac{x^3}{x^2}$

114. $\dfrac{x^2}{x}$

115. $\dfrac{y^3}{y^5}$

116. $\dfrac{x^2}{x^5}$

Write in expanded form.

117. 3.4×10^2

118. 6.0×10^{-4}

Maintaining Your Skills

Simplify the following expressions.

119. $4x + 3x$

120. $9x + 7x$

121. $5a - 3a$

122. $10a - 2a$

123. $4y + 5y + y$

124. $6y - y + 2y$

4.3 Operations with Monomials

OBJECTIVES

A Multiply and divide monomials.

B Multiply and divide numbers written in scientific notation.

C Add and subtract monomials.

TICKET TO SUCCESS

Keep these questions in mind as you read through the section. Then respond in your own words and in complete sentences.

1. What is a monomial?
2. Describe how you would multiply $3x^2$ and $5x^2$.
3. Describe how you would add $3x^2$ and $5x^2$.
4. Describe how you would multiply two numbers written in scientific notation.

We have developed all the tools necessary to perform the four basic operations on the simplest of polynomials: monomials.

Definition

A **monomial** is a one-term expression that is either a constant (number) or the product of a constant and one or more variables raised to whole number exponents.

The following are examples of monomials:

$$-3 \qquad 15x \qquad -23x^2y \qquad 49x^4y^2z^4 \qquad \frac{3}{4}a^2b^3$$

The numerical part of each monomial is called the *numerical coefficient,* or just *coefficient.* Monomials are also called *terms.*

Let's consider a carpet installer who may need to create a formula to know how much carpet two bedrooms in a house needs. The first room is in the shape of a square with walls the length of x. The second room is the shape of a rectangle, with one wall the same length as the first room but a second wall twice as long. Using our knowledge of area, we can write two monomials that each represent the area of the rooms.

Square room area: $x \cdot x = x^2$

Rectangular room area: $x \cdot 2x = 2x^2$

Later in this section, we can add these monomials to determine how much carpet the installer will need. But first, we must practice more problems that involve the multiplication and division of monomials.

A Multiplication and Division of Monomials

There are two basic steps involved in the multiplication of monomials. First, we rewrite the products using the commutative and associative properties. Then, we simplify by multiplying coefficients and adding exponents of like bases.

EXAMPLE 1 Multiply $(-3x^2)(4x^3)$.

SOLUTION
$$(-3x^2)(4x^3) = (-3 \cdot 4)(x^2 \cdot x^3) \qquad \text{Commutative and associative properties}$$
$$= -12x^5 \qquad \text{Multiply coefficients, add exponents}$$

EXAMPLE 2 Multiply $\left(\frac{4}{5}x^5 \cdot y^2\right)(10x^3 \cdot y)$.

SOLUTION $\left(\frac{4}{5}x^5 \cdot y^2\right)(10x^3 \cdot y) = \left(\frac{4}{5} \cdot 10\right)(x^5 \cdot x^3)(y^2 \cdot y)$ Commutative and associative exponents

$$= 8x^8y^3$$ Multiply coefficients, add exponents

You can see that in each example above the work was the same—multiply coefficients and add exponents of the same base. We can expect division of monomials to proceed in a similar way. Since our properties are consistent, division of monomials will result in division of coefficients and subtraction of exponents of like bases.

EXAMPLE 3 Divide $\frac{15x^3}{3x^2}$.

SOLUTION $\frac{15x^3}{3x^2} = \frac{15}{3} \cdot \frac{x^3}{x^2}$ Write as separate fractions

$$= 5x$$ Divide coefficients, subtract exponents

EXAMPLE 4 Divide $\frac{39x^2y^3}{3xy^5}$.

SOLUTION $\frac{39x^2y^3}{3xy^5} = \frac{39}{3} \cdot \frac{x^2}{x} \cdot \frac{y^3}{y^5}$ Write as separate fractions

$$= 13x \cdot \frac{1}{y^2}$$ Divide coefficients, subtract exponents

$$= \frac{13x}{y^2}$$ Write answer as a single fraction

In Example 4, the expression $\frac{y^3}{y^5}$ simplifies to $\frac{1}{y^2}$ because of Property 4 for exponents and the definition of negative exponents. If we were to show all the work in this simplification process, it would look like this:

$$\frac{y^3}{y^5} = y^{3-5}$$ Property 4 for exponents

$$= y^{-2}$$ Subtraction

$$= \frac{1}{y^2}$$ Definition of negative exponents

The point of the explanation is this: Even though we may not show all the steps when simplifying an expression involving exponents, the result we obtain still can be justified using the properties of exponents. We have not introduced any new properties in Example 4. We have just not shown the details of each simplification.

EXAMPLE 5 Divide $\frac{25a^5b^3}{50a^2b^7}$.

SOLUTION $\frac{25a^5b^3}{50a^2b^7} = \frac{25}{50} \cdot \frac{a^5}{a^2} \cdot \frac{b^3}{b^7}$ Write as separate fractions

$$= \frac{1}{2} \cdot a^3 \cdot \frac{1}{b^4}$$ Divide coefficients, subtract exponents

$$= \frac{a^3}{2b^4}$$ Write answer as a single fraction

Notice in Example 5 that dividing 25 by 50 results in $\frac{1}{2}$. This is the same result we would obtain if we reduced the fraction $\frac{25}{50}$ to lowest terms, and there is no harm in thinking of it that way. Also, notice that the expression $\frac{b^3}{b^7}$ simplifies to $\frac{1}{b^4}$ by Property 4 for exponents and the definition of negative exponents, even though we have not shown the steps involved in doing so.

B   Multiplication and Division of Numbers Written in Scientific Notation

We multiply and divide numbers written in scientific notation using the same steps we used to multiply and divide monomials.

EXAMPLE 6   Multiply $(4 \times 10^7)(2 \times 10^{-4})$.

SOLUTION   Because multiplication is commutative and associative, we can rearrange the order of these numbers and group them as follows:

$$(4 \times 10^7)(2 \times 10^{-4}) = (4 \times 2)(10^7 \times 10^{-4})$$
$$= 8 \times 10^3$$

Notice that we add exponents, $7 + (-4) = 3$, when we multiply with the same base. ■

EXAMPLE 7   Divide $\dfrac{9.6 \times 10^{12}}{3 \times 10^4}$.

SOLUTION   We group the numbers between 1 and 10 separately from the powers of 10 and proceed as we did in Example 6.

$$\frac{9.6 \times 10^{12}}{3 \times 10^4} = \frac{9.6}{3} \times \frac{10^{12}}{10^4}$$
$$= 3.2 \times 10^8$$

■

Notice that the procedure we used in both of the examples above is very similar to multiplication and division of monomials, for which we multiplied or divided coefficients and added or subtracted exponents.

C   Addition and Subtraction of Monomials

Addition and subtraction of monomials will be almost identical since subtraction is defined as addition of the opposite. With multiplication and division of monomials, the key was rearranging the numbers and variables using the commutative and associative properties. With addition, the key is application of the distributive property. We sometimes use the phrase *combine monomials* to describe addition and subtraction of monomials.

Definition

Two terms (monomials) with the same variable part (same variables raised to the same powers) are called **similar** (or *like*) **terms.**

You can add only similar terms. This is because the distributive property (which is the key to addition of monomials) cannot be applied to terms that are not similar.

EXAMPLE 8 Combine the monomial: $-3x^2 + 15x^2$.

SOLUTION $-3x^2 + 15x^2 = (-3 + 15)x^2$ Distributive property

$= 12x^2$ Add coefficients

EXAMPLE 9 Subtract $9x^2y - 20x^2y$.

SOLUTION $9x^2y - 20x^2y = (9 - 20)x^2y$ Distributive property

$= -11x^2y$ Add coefficients

EXAMPLE 10 Add: $5x^2 + 8y^2$.

SOLUTION $5x^2 + 8y^2$ In this case we cannot apply the distributive property, so we cannot add the monomials

The next examples show how we simplify expressions containing monomials when more than one operation is involved.

EXAMPLE 11 Simplify $\dfrac{(6x^4y)(3x^7y^5)}{9x^5y^2}$.

SOLUTION We begin by multiplying the two monomials in the numerator.

$\dfrac{(6x^4y)(3x^7y^5)}{9x^5y^2} = \dfrac{18x^{11}y^6}{9x^5y^2}$ Simplify numerator

$= 2x^6y^4$ Divide

EXAMPLE 12 Simplify $\dfrac{(6.8 \times 10^5)(3.9 \times 10^{-7})}{7.8 \times 10^{-4}}$.

SOLUTION We group the numbers between 1 and 10 separately from the powers of 10.

$\dfrac{(6.8)(3.9)}{7.8} \times \dfrac{(10^5)(10^{-7})}{10^{-4}} = 3.4 \times 10^{5+(-7)-(-4)}$

$= 3.4 \times 10^2$

EXAMPLE 13 Simplify $\dfrac{14x^5}{2x^2} + \dfrac{15x^8}{3x^5}$.

SOLUTION Simplifying each expression separately and then combining similar terms gives

$\dfrac{14x^5}{2x^2} + \dfrac{15x^8}{3x^5} = 7x^3 + 5x^3$ Divide

$= 12x^3$ Add

EXAMPLES Apply the distributive property, then simplify, if possible.

14. $x^2\left(1 - \dfrac{6}{x}\right) = x^2 \cdot 1 - x^2 \cdot \dfrac{6}{x} = x^2 - \dfrac{6x^2}{x} = x^2 - 6x$

15. $ab\left(\dfrac{1}{b} - \dfrac{1}{a}\right) = ab \cdot \dfrac{1}{b} - ab \cdot \dfrac{1}{a} = \dfrac{ab}{b} - \dfrac{ab}{a} = a - b$

EXAMPLE 16   A rectangular solid is twice as long as it is wide and one-half as high as it is wide. Write an expression for the volume in terms of the width, x.

SOLUTION   We begin by making a diagram of the object (Figure 1) with the dimensions labeled as given in the problem.

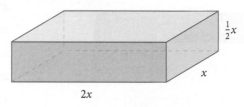

FIGURE 1

The volume is the product of the three dimensions:

$$V = 2x \cdot x \cdot \frac{1}{2}x = x^3$$

The box has the same volume as a cube with side x, as shown in Figure 2.

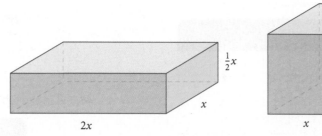

Equal Volumes

FIGURE 2

Problem Set 4.3

Moving Toward Success

"Don't dwell on what went wrong. Instead, focus on what to do next. Spend your energies on moving forward toward finding the answer."

—Denis Waitley, 1933–present, motivational speaker and writer

1. Why is resiliency a good quality to have when taking a math course?

2. Do poor scores on tests or difficulties with problems typically derail your focus in this class? Why or why not?

A Multiply. [Examples 1–2]

1. $(3x^4)(4x^3)$
2. $(6x^5)(-2x^2)$
3. $(-2y^4)(8y^7)$
4. $(5y^{10})(2y^5)$
5. $(8x)(4x)$
6. $(7x)(5x)$
7. $(10a^3)(10a)(2a^2)$
8. $(5a^4)(10a)(10a^4)$
9. $(6ab^2)(-4a^2b)$
10. $(-5a^3b)(4ab^4)$
11. $(4x^2y)(3x^3y^3)(2xy^4)$
12. $(5x^6)(-10xy^4)(-2x^2y^6)$

A Divide. Write all answers with positive exponents only. [Examples 3–5]

13. $\dfrac{15x^3}{5x^2}$
14. $\dfrac{25x^5}{5x^4}$
15. $\dfrac{18y^9}{3y^{12}}$
16. $\dfrac{24y^4}{-8y^7}$
17. $\dfrac{32a^3}{64a^4}$
18. $\dfrac{25a^5}{75a^6}$

19. $\dfrac{21a^2b^3}{-7ab^5}$

20. $\dfrac{32a^5b^6}{8ab^5}$

21. $\dfrac{3x^3y^2z}{27xy^2z^3}$

22. $\dfrac{5x^5y^4z}{30x^3yz^2}$

23. Fill in the table.

a	b	ab	$\dfrac{a}{b}$	$\dfrac{b}{a}$
10	$5x$			
$20x^3$	$6x^2$			
$25x^5$	$5x^4$			
$3x^{-2}$	$3x^2$			
$-2y^4$	$8y^7$			

24. Fill in the table.

a	b	ab	$\dfrac{a}{b}$	$\dfrac{b}{a}$
$10y$	$2y^2$			
$10y^2$	$2y$			
$5y^3$	15			
5	$15y^3$			
$4y^{-3}$	$4y^3$			

B Find each product. Write all answers in scientific notation. [Example 6]

25. $(3 \times 10^3)(2 \times 10^5)$

26. $(4 \times 10^8)(1 \times 10^6)$

27. $(3.5 \times 10^4)(5 \times 10^{-6})$

28. $(7.1 \times 10^5)(2 \times 10^{-8})$

29. $(5.5 \times 10^{-3})(2.2 \times 10^{-4})$

30. $(3.4 \times 10^{-2})(4.5 \times 10^{-6})$

B Find each quotient. Write all answers in scientific notation. [Example 7]

31. $\dfrac{8.4 \times 10^5}{2 \times 10^2}$

32. $\dfrac{9.6 \times 10^{20}}{3 \times 10^6}$

33. $\dfrac{6 \times 10^8}{2 \times 10^{-2}}$

34. $\dfrac{8 \times 10^{12}}{4 \times 10^{-3}}$

35. $\dfrac{2.5 \times 10^{-6}}{5 \times 10^{-4}}$

36. $\dfrac{4.5 \times 10^{-8}}{9 \times 10^{-4}}$

C Combine by adding or subtracting as indicated. [Examples 8–10]

37. $3x^2 + 5x^2$

38. $4x^3 + 8x^3$

39. $8x^5 - 19x^5$

40. $75x^6 - 50x^6$

41. $2a + a - 3a$

42. $5a + a - 6a$

43. $10x^3 - 8x^3 + 2x^3$

44. $7x^5 + 8x^5 - 12x^5$

45. $20ab^2 - 19ab^2 + 30ab^2$

46. $18a^3b^2 - 20a^3b^2 + 10a^3b^2$

47. Fill in the table.

a	b	ab	$a + b$
$5x$	$3x$		
$4x^2$	$2x^2$		
$3x^3$	$6x^3$		
$2x^4$	$-3x^4$		
x^5	$7x^5$		

48. Fill in the table.

a	b	ab	$a - b$
$2y$	$3y$		
$-2y$	$3y$		
$4y^2$	$5y^2$		
y^3	$-3y^3$		
$5y^4$	$7y^4$		

A Simplify. Write all answers with positive exponents only. [Example 11]

49. $\dfrac{(3x^2)(8x^5)}{6x^4}$

50. $\dfrac{(7x^3)(6x^8)}{14x^5}$

51. $\dfrac{(9a^2b)(2a^3b^4)}{18a^5b^7}$

52. $\dfrac{(21a^5b)(2a^8b^4)}{14ab}$

53. $\dfrac{(4x^3y^2)(9x^4y^{10})}{(3x^5y)(2x^6y)}$

54. $\dfrac{(5x^4y^4)(10x^3y^3)}{(25xy^5)(2xy^7)}$

B Simplify each expression, and write all answers in scientific notation. [Example 12]

55. $\dfrac{(6 \times 10^8)(3 \times 10^5)}{9 \times 10^7}$

4.5 Multiplication of Polynomials

OBJECTIVES

A Multiply a monomial with a polynomial.

B Multiply two binomials using the FOIL method.

C Multiply two polynomials using the Column method.

TICKET TO SUCCESS

Keep these questions in mind as you read through the section. Then respond in your own words and in complete sentences.

1. Describe how you would use the FOIL method to multiply two binomials.
2. Describe how the distributive property is used to multiply a monomial and a polynomial.
3. Describe how you would use the Column method to multiply two polynomials.
4. Show how the product of two binomials can be a trinomial.

Image copyright © muzsy. Used under license from Shutterstock.com

A Multiplying Monomials with Polynomials

We begin our discussion of multiplication of polynomials by finding the product of a monomial and a trinomial.

EXAMPLE 1 Multiply $3x^2(2x^2 + 4x + 5)$.

SOLUTION Applying the distributive property gives us

$$3x^2(2x^2 + 4x + 5) = 3x^2(2x^2) + 3x^2(4x) + 3x^2(5) \quad \text{Distributive property}$$
$$= 6x^4 + 12x^3 + 15x^2 \quad \text{Multiplication}$$

B Multiplying Binomials

Suppose a volleyball coach had coached a college team for x years. He plans to spend two more years coaching the team before retiring. If the amount of money he spends on uniforms each year can be given by the binomial $30x + 10$, how much will he have spent by the end of his coaching career?

We can set up this problem as the multiplication of two binomials. Therefore, if $x + 2$ equals the number of total coaching years, then we have

$$(x + 2)(30x + 10)$$

The distributive property is the key to multiplication of polynomials. We can use it to find the product of any two polynomials. There are some shortcuts we can use in certain situations, however. Let's look at an example that involves the product of two binomials.

EXAMPLE 2 Multiply $(3x - 5)(2x - 1)$.

SOLUTION

$$
\begin{aligned}
(3x - 5)(2x - 1) &= 3x(2x - 1) - 5(2x - 1) \\
&= 3x(2x) + 3x(-1) + (-5)(2x) + (-5)(-1) \\
&= 6x^2 - 3x - 10x + 5 \\
&= 6x^2 - 13x + 5
\end{aligned}
$$

If we look closely at the second and third lines of work in the previous example, we can see that the terms in the answer come from all possible products of terms in the first binomial with terms in the second binomial. This result is generalized as follows.

Rule: Multiplying Two Polynomials

To multiply any two polynomials, multiply each term in the first with each term in the second.

There are two ways we can put this rule to work.

FOIL Method

If we look at the original problem in Example 2 and then at the answer, we see that the first term in the answer came from multiplying the first terms in each binomial.

$$3x \cdot 2x = 6x^2 \qquad \text{First}$$

The middle term in the answer came from adding the product of the two outside terms in each binomial and the product of the two inside terms in each binomial.

$$
\begin{aligned}
3x(-1) &= -3x \qquad \text{Outside} \\
-5(2x) &= \underline{-10x} \qquad \text{Inside} \\
&\quad\ -13x
\end{aligned}
$$

The last term in the answer came from multiplying the two last terms:

$$-5(-1) = 5 \qquad \text{Last}$$

To summarize the FOIL method, we will multiply another two binomials.

EXAMPLE 3 Multiply $(2x + 3)(5x - 4)$.

SOLUTION $(2x + 3)(5x - 4) = 2x(5x) + 2x(-4) + 3(5x) + 3(-4)$

First

Outside

Inside

Last

$$
\begin{aligned}
&= 10x^2 - 8x + 15x - 12 \\
&= 10x^2 + 7x - 12
\end{aligned}
$$

With practice, adding the products of the outside and inside terms, $-8x + 15x = 7x$ can be done mentally.

C Column Method

The FOIL method can be applied only when multiplying two binomials. To find products of polynomials with more than two terms, we use what is called the Column method.

The Column method of multiplying two polynomials is very similar to long multiplication with whole numbers. It is just another way of finding all possible products of terms in one polynomial with terms in another polynomial.

EXAMPLE 4   Multiply $(2x + 3)(3x^2 - 2x + 1)$.

SOLUTION

$$
\begin{array}{r}
3x^2 - 2x + 1 \\
2x + 3 \\
\hline
6x^3 - 4x^2 + 2x \qquad \leftarrow 2x(3x^2 - 2x + 1) \\
9x^2 - 6x + 3 \quad \leftarrow \quad 3(3x^2 - 2x + 1) \\
\hline
6x^3 + 5x^2 - 4x + 3
\end{array}
$$

Add similar terms

It will be to your advantage to become very fast and accurate at multiplying polynomials. You should be comfortable using either method. The following examples illustrate different types of multiplication.

EXAMPLES   Multiply.

5. $4a^2(2a^2 - 3a + 5)$

$$
\begin{array}{r}
2a^2 - 3a + 5 \\
4a^2 \\
\hline
8a^4 - 12a^3 + 20a^2
\end{array}
$$

6. $(x - 2)(y + 3) = \underset{F}{x(y)} + \underset{O}{x(3)} + \underset{I}{(-2)(y)} + \underset{L}{(-2)(3)}$

$= xy + 3x - 2y - 6$

7. $(x + y)(a - b) = \underset{F}{x(a)} + \underset{O}{x(-b)} + \underset{I}{y(a)} + \underset{L}{y(-b)}$

$= xa - xb + ya - yb$

8. $(5x - 1)(2x + 6) = \underset{F}{5x(2x)} + \underset{O}{5x(6)} + \underset{I}{(-1)(2x)} + \underset{L}{(-1)(6)}$

$= 10x^2 + 30x + (-2x) + (-6)$

$= 10x^2 + 28x - 6$

EXAMPLE 9   The length of a rectangle is 3 more than twice the width. Write an expression for the area of the rectangle.

SOLUTION   We begin by drawing a rectangle and labeling the width with x. Since the length is 3 more than twice the width, we label the length with $2x + 3$.

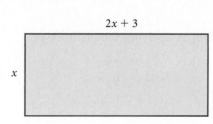

Since the area A of a rectangle is the product of the length and width, we write our formula for the area of this rectangle as

$$A = x(2x + 3)$$
$$A = 2x^2 + 3x \quad \text{Multiply}$$

Revenue

Suppose that a store sells x items at p dollars per item. The total amount of money obtained by selling the items is called the *revenue*. It can be found by multiplying the number of items sold, x, by the price per item, p. For example, if 100 items are sold for $6 each, the revenue is $100(6) = \$600$. Similarly, if 500 items are sold for $8 each, the total revenue is $500(8) = \$4,000$. If we denote the revenue with the letter R, then the formula that relates R, x, and p is

Revenue = (number of items sold)(price of each item)

In symbols: $R = xp$.

EXAMPLE 10 A store selling flash drives for home computers knows from past experience that it can sell x flash drives each day at a price of p dollars per flash drive, according to the equation $x = 800 - 100p$. Write a formula for the daily revenue that involves only the variables R and p.

SOLUTION From our previous discussion we know that the revenue R is given by the formula

$$R = xp$$

But, since $x = 800 - 100p$, we can substitute $800 - 100p$ for x in the revenue equation to obtain

$$R = (800 - 100p)p$$
$$R = 800p - 100p^2$$

This last formula gives the revenue, R, in terms of the price, p.

Problem Set 4.5

Moving Toward Success

"Concentrate all your thoughts upon the work at hand. The sun's rays do not burn until brought to a focus."

—Alexander Graham Bell, 1847–1922, scientist and inventor

1. Why is it important to intend to master the material in this class, not just go through the motions?

2. What are things you can do to make sure you study with intention?

A Multiply the following by applying the distributive property. [Example 1]

1. $2x(3x + 1)$

2. $4x(2x - 3)$

3. $2x^2(3x^2 - 2x + 1)$

4. $5x(4x^3 - 5x^2 + x)$

5. $2ab(a^2 - ab + 1)$

6. $3a^2b(a^3 + a^2b^2 + b^3)$

7. $y^2(3y^2 + 9y + 12)$

8. $5y(2y^2 - 3y + 5)$

9. $4x^2y(2x^3y + 3x^2y^2 + 8y^3)$

10. $6xy^3(2x^2 + 5xy + 12y^2)$

B **C** Multiply the following binomials. You should do about half the problems using the FOIL method and the other half using the Column method. Remember, you want to be comfortable using both methods. [Examples 2–8]

11. $(x + 3)(x + 4)$

12. $(x + 2)(x + 5)$

13. $(x + 6)(x + 1)$ **14.** $(x + 1)(x + 4)$

15. $\left(x + \dfrac{1}{2}\right)\left(x + \dfrac{3}{2}\right)$

16. $\left(x + \dfrac{3}{5}\right)\left(x + \dfrac{2}{5}\right)$

17. $(a + 5)(a - 3)$

18. $(a - 8)(a + 2)$

19. $(x - a)(y + b)$

20. $(x + a)(y - b)$

21. $(x + 6)(x - 6)$ **22.** $(x + 3)(x - 3)$

23. $\left(y + \dfrac{5}{6}\right)\left(y - \dfrac{5}{6}\right)$ **24.** $\left(y - \dfrac{4}{7}\right)\left(y + \dfrac{4}{7}\right)$

25. $(2x - 3)(x - 4)$

26. $(3x - 5)(x - 2)$

27. $(a + 2)(2a - 1)$

28. $(a - 6)(3a + 2)$

29. $(2x - 5)(3x - 2)$

30. $(3x + 6)(2x - 1)$

31. $(2x + 3)(a + 4)$

32. $(2x - 3)(a - 4)$

33. $(5x - 4)(5x + 4)$

34. $(6x + 5)(6x - 5)$

35. $\left(2x - \dfrac{1}{2}\right)\left(x + \dfrac{3}{2}\right)$

36. $\left(4x - \dfrac{3}{2}\right)\left(x + \dfrac{1}{2}\right)$

37. $(1 - 2a)(3 - 4a)$

38. $(1 - 3a)(3 + 2a)$

B For each of the following problems, fill in the area of each small rectangle and square, and then add the results together to find the indicated product. [Example 9]

39. $(x + 2)(x + 3)$

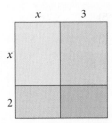

40. $(x + 4)(x + 5)$

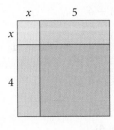

41. $(x + 1)(2x + 2)$

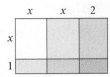

42. $(2x + 1)(2x + 2)$

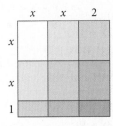

B **C** Multiply the following. [Examples 2–8]

43. $(a - 3)(a^2 - 3a + 2)$

44. $(a + 5)(a^2 + 2a + 3)$

45. $(x + 2)(x^2 - 2x + 4)$

46. $(x + 3)(x^2 - 3x + 9)$

47. $(2x + 1)(x^2 + 8x + 9)$

48. $(3x - 2)(x^2 - 7x + 8)$

49. $(5x^2 + 2x + 1)(x^2 - 3x + 5)$

50. $(2x^2 + x + 1)(x^2 - 4x + 3)$

Multiply.

51. $(x^2 + 3)(2x^2 - 5)$

52. $(4x^3 - 8)(5x^3 + 4)$

53. $(3a^4 + 2)(2a^2 + 5)$

54. $(7a^4 - 8)(4a^3 - 6)$

55. $(x + 3)(x + 4)(x + 5)$

56. $(x - 3)(x - 4)(x - 5)$

Simplify.

57. $(x - 3)(x - 2) + 2$

58. $(2x - 5)(3x + 2) - 4$

59. $(2x - 3)(4x + 3) + 4$

60. $(3x + 8)(5x - 7) + 52$

61. $(x + 4)(x - 5) + (-5)(2)$

62. $(x + 3)(x - 4) + (-4)(2)$

63. $2(x - 3) + x(x + 2)$

64. $5(x + 3) + 1(x + 4)$

65. $3x(x + 1) - 2x(x - 5)$

66. $4x(x - 2) - 3x(x - 4)$

67. $x(x + 2) - 3$

68. $2x(x - 4) + 6$

69. $a(a - 3) + 6$ **70.** $a(a - 4) + 8$

71. Find each product.
 a. $(x + 1)(x - 1)$
 b. $(x + 1)(x + 1)$
 c. $(x + 1)(x^2 + 2x + 1)$
 d. $(x + 1)(x^3 + 3x^2 + 3x + 1)$

72. Find each product.
 a. $(x + 1)(x^2 - x + 1)$
 b. $(x + 2)(x^2 - 2x + 4)$
 c. $(x + 3)(x^2 - 3x + 9)$
 d. $(x + 4)(x^2 - 4x + 16)$

73. Find each product.
 a. $(x + 1)(x - 1)$
 b. $(x + 1)(x - 2)$
 c. $(x + 1)(x - 3)$
 d. $(x + 1)(x - 4)$

74. Find each product.
 a. $(x + 2)(x - 2)$
 b. $(x - 2)(x^2 + 2x + 4)$
 c. $(x^2 + 4)(x^2 - 4)$
 d. $(x^3 + 8)(x^3 - 8)$

If the product of two expressions is 0, then one or both of the expressions must be zero. That is, the only way to multiply and get 0, is to multiply by 0. For each expression below, find all values of x that make the expression 0. (If the expression cannot be 0, say so.)

75. $5x$ **76.** $3x^2$

77. $x + 5$ **78.** $x^2 + 5$

79. $(x - 3)(x + 2)$ **80.** $x(x - 5)$

81. $x^2 + 16$ **82.** $x^2 - 100$

Applying the Concepts [Example 10]

Solar and Wind Energy The chart shows the cost to install either solar panels or a wind turbine. Use the chart to work Problems 83 and 84.

Solar Versus Wind Energy Costs

Solar Energy Equipment Cost:		Wind Energy Equipment Cost:	
Modules	$6200	Turbine	$3300
Fixed Rack	$1570	Tower	$3000
Charge Controller	$971	Cable	$715
Cable	$440		
TOTAL	**$9181**	**TOTAL**	**$7015**

Source: a Limited 2006

83. A homeowner is buying a certain number of solar panel modules. He is going to get a discount on each module that is equal to 25 times the number of modules he buys. Write an equation that describes this situation and simplify. Then find the cost if he buys 3 modules.

84. A farmer is replacing several turbines in his field. He is going to get a discount on each turbine that is equal to 50 times the number of turbines he buys. Write an expression that describes this situation and simplify. Then find the cost if he replaces 5 turbines.

85. Area The length of a rectangle is 5 units more than twice the width. Write an expression for the area of the rectangle.

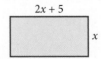

86. Area The length of a rectangle is 2 more than three times the width. Write an expression for the area of the rectangle.

87. Area The width and length of a rectangle are given by two consecutive integers. Write an expression for the area of the rectangle.

88. Area The width and length of a rectangle are given by two consecutive even integers. Write an expression for the area of the rectangle.

89. Revenue A store selling smart phones knows that the number of phones it can sell each week, x, is related to the price per phone, p, by the equation $x = 1,200 - 100p$. Write an expression for the weekly revenue that involves only the variables R and p. (*Remember:* The equation for revenue is $R = xp$.)

90. Revenue A store selling MP3 players knows from past experience that the number of MP3 players it can sell each week, x, is related to the price per MP3 players, p, by the equation $x = 1,300 - 100p$. Write an expression for the weekly revenue that involves only the variables R and p.

Getting Ready for the Next Section

Simplify.

91. $13 \cdot 13$ **92.** $3x \cdot 3x$

93. $2(x)(-5)$ **94.** $2(2x)(-3)$

95. $6x + (-6x)$ **96.** $3x + (-3x)$

97. $(2x)(-3) + (2x)(3)$ **98.** $(2x)(-5y) + (2x)(5y)$

Multiply.

99. $-4(3x - 4)$ **100.** $-2x(2x + 7)$

101. $(x - 1)(x + 2)$

102. $(x + 5)(x - 6)$

103. $(x + 3)(x + 3)$

104. $(3x - 2)(3x - 2)$

Maintaining Your Skills

105. Solve this system by graphing: $x + y = 4$
$x - y = 2$

Solve each system by the elimination method.

106. $3x + 2y = 12$
$x + y = 3$

107. $2x + 3y = -1$
$3x + 5y = -2$

Solve each system by the substitution method.

108. $x + y = 20$
$y = 5x + 2$

109. $2x - 6y = 2$
$y = 3x + 1$

110. Investing A total of $1,200 is invested in two accounts. One of the accounts pays 8% interest annually and the other pays 10% interest annually. If the total amount of interest earned from both accounts for the year is $104, how much is invested in each account?

111. Coin Problem Amy has $1.85 in dimes and quarters. If she has a total of 11 coins, how many of each coin does she have?

4.6

Binomial Squares and Other Special Products

OBJECTIVES

A Find the square of a binomial.

B Multiply expressions of the form $(a + b)(a - b)$.

TICKET TO SUCCESS

Keep these questions in mind as you read through the section. Then respond in your own words and in complete sentences.

1. Describe how you would square the binomial $a + b$.

2. Explain why $(x + 3)^2$ cannot be $x^2 + 9$.

3. Why is the middle term in the product of $(a + b)$ and $(a - b)$ equal to zero?

4. When multiplied out, how will $(x + 3)^2$ and $(x - 3)^2$ differ?

Image copyright © Mana Photo. Used under license from Shutterstock.com

A Squaring Binomials

Consider the following word problem:

> For a surfer, the number of successful rides in two hours may be equal to $(x + 2)^2$. How would we square this binomial to help find a solution?

In this section, we will combine the results of the last section with our definition of exponents to find some special products.

> **EXAMPLE 1** Find the square of $(3x - 2)$.

SOLUTION To square $(3x - 2)$, we multiply it by itself.

$$
\begin{aligned}
(3x - 2)^2 &= (3x - 2)(3x - 2) && \text{Definition of exponents} \\
&= 9x^2 - 6x - 6x + 4 && \text{FOIL method} \\
&= 9x^2 - 12x + 4 && \text{Combine similar terms}
\end{aligned}
$$

Notice in Example 1 that the first and last terms in the answer are the squares of the first and last terms in the original problem and that the middle term is twice the product of the two terms in the original binomial.

EXAMPLES

2. $(a + b)^2 = (a + b)(a + b)$
$$= a^2 + 2ab + b^2$$
3. $(a - b)^2 = (a - b)(a - b)$
$$= a^2 - 2ab + b^2$$

Binomial squares having the form of Examples 2 and 3 occur very frequently in algebra. It will be to your advantage to memorize the following rule for squaring a binomial.

Rule: Binomial Squares

The square of a binomial is the sum of the square of the first term, the square of the last term, and twice the product of the two original terms. In symbols this rule is written as follows:

$$(x + y)^2 \quad = \quad x^2 \quad + \quad 2xy \quad + \quad y^2$$

Square of first term ⸱ Twice the product of the two terms ⸱ Square of last term

---

EXAMPLES Multiply using the preceding rule.

		First term squared		Twice their product		Last term squared	Answer
4.	$(x - 5)^2 =$	x^2	$+$	$2(x)(-5)$	$+$	25	$= x^2 - 10x + 25$
5.	$(x + 2)^2 =$	x^2	$+$	$2(x)(2)$	$+$	4	$= x^2 + 4x + 4$
6.	$(2x - 3)^2 =$	$4x^2$	$+$	$2(2x)(-3)$	$+$	9	$= 4x^2 - 12x + 9$
7.	$(5x - 4)^2 =$	$25x^2$	$+$	$2(5x)(-4)$	$+$	16	$= 25x^2 - 40x + 16$

B More Special Products

Another special product that occurs frequently is $(a + b)(a - b)$. The only difference in the two binomials is the sign between the two terms. The interesting thing about this type of product is that the middle term is always zero. Here are some examples.

EXAMPLES Multiply using the FOIL method.

8. $(2x - 3)(2x + 3) = 4x^2 + 6x - 6x - 9$ FOIL method
$$= 4x^2 - 9$$

9. $(x - 5)(x + 5) = x^2 + 5x - 5x - 25$ FOIL method
$$= x^2 - 25$$

10. $(3x - 1)(3x + 1) = 9x^2 + 3x - 3x - 1$ FOIL method
$$= 9x^2 - 1$$

Notice that in each case in the examples above, the middle term is zero and therefore doesn't appear in the answer. The answers all turn out to be the difference of two squares. Here is a rule to help you memorize the result.

Rule: Difference of Two Squares

When multiplying two binomials that differ only in the sign between their terms, subtract the square of the last term from the square of the first term.
$$(a - b)(a + b) = a^2 - b^2$$

Here are some problems that result in the difference of two squares.

EXAMPLES Multiply using the preceding rule.

11. $(x + 3)(x - 3) = x^2 - 9$
12. $(a + 2)(a - 2) = a^2 - 4$
13. $(9a + 1)(9a - 1) = 81a^2 - 1$
14. $(2x - 5y)(2x + 5y) = 4x^2 - 25y^2$
15. $(3a - 7b)(3a + 7b) = 9a^2 - 49b^2$

Although all the problems in this section can be worked correctly using the methods in the previous section, they can be done much faster if the two rules are *memorized*. Here is a summary of the two rules:

$$
\left.
\begin{aligned}
(a + b)^2 &= (a + b)(a + b) = a^2 + 2ab + b^2 \\
(a - b)^2 &= (a - b)(a - b) = a^2 - 2ab + b^2
\end{aligned}
\right\} \quad \text{Binomial Squares}
$$

$$(a - b)(a + b) = a^2 - b^2 \qquad \text{Difference of Two Squares}$$

EXAMPLE 16 Write an expression in symbols for the sum of the squares of three consecutive even integers. Then, simplify that expression.

SOLUTION If we let $x =$ the first of the even integers, then $x + 2$ is the next consecutive even integer, and $x + 4$ is the one after that. An expression for the sum of their squares is

$$
\begin{aligned}
x^2 + (x + 2)^2 + (x + 4)^2 & \qquad \text{Sum of squares} \\
= x^2 + (x^2 + 4x + 4) + (x^2 + 8x + 16) & \qquad \text{Expand squares} \\
= 3x^2 + 12x + 20 & \qquad \text{Add similar terms}
\end{aligned}
$$

Problem Set 4.6

Moving Toward Success

"You can't depend on your eyes when your imagination is out of focus."

—Mark Twain, 1835–1910, American author and humorist

1. Why do you think it is important to create pictures in your head as you learn mathematics?

2. Why can it be helpful to sometimes read this book or your notes out loud as you study?

A Perform the indicated operations. [Examples 1–7]

1. $(x - 2)^2$
2. $(x + 2)^2$
3. $(a + 3)^2$
4. $(a - 3)^2$
5. $(x - 5)^2$
6. $(x - 4)^2$
7. $\left(a - \dfrac{1}{2}\right)^2$
8. $\left(a + \dfrac{1}{2}\right)^2$
9. $(x + 10)^2$
10. $(x - 10)^2$
11. $(a + 0.8)^2$
12. $(a - 0.4)^2$
13. $(2x - 1)^2$
14. $(3x + 2)^2$
15. $(4a + 5)^2$
16. $(4a - 5)^2$
17. $(3x - 2)^2$
18. $(2x - 3)^2$
19. $(3a + 5b)^2$
20. $(5a - 3b)^2$
21. $(4x - 5y)^2$
22. $(5x + 4y)^2$
23. $(7m + 2n)^2$
24. $(2m - 7n)^2$
25. $(6x - 10y)^2$

26. $(10x + 6y)^2$

27. $(x^2 + 5)^2$ **28.** $(x^2 + 3)^2$

29. $(a^2 + 1)^2$ **30.** $(a^2 - 2)^2$

31. $\left(y + \dfrac{3}{2}\right)^2$ **32.** $\left(y - \dfrac{3}{2}\right)^2$

33. $\left(a + \dfrac{1}{2}\right)^2$ **34.** $\left(a - \dfrac{5}{2}\right)^2$

35. $\left(x + \dfrac{3}{4}\right)^2$ **36.** $\left(x - \dfrac{3}{8}\right)^2$

37. $\left(t + \dfrac{1}{5}\right)^2$ **38.** $\left(t - \dfrac{3}{5}\right)^2$

Comparing Expressions Fill in each table.

39.

x	$(x + 3)^2$	$x^2 + 9$	$x^2 + 6x + 9$
1			
2			
3			
4			

40.

x	$(x - 5)^2$	$x^2 + 25$	$x^2 - 10x + 25$
1			
2			
3			
4			

41.

a	b	$(a + b)^2$	$a^2 + b^2$	$a^2 + ab + b^2$	$a^2 + 2ab + b^2$
1	1				
3	5				
3	4				
4	5				

42.

a	b	$(a - b)^2$	$a^2 - b^2$	$a^2 - 2ab + b^2$
2	1			
5	2			
2	5			
4	3			

B Multiply. [Examples 8–15]

43. $(a + 5)(a - 5)$ **44.** $(a - 6)(a + 6)$

45. $(y - 1)(y + 1)$ **46.** $(y - 2)(y + 2)$

47. $(9 + x)(9 - x)$ **48.** $(10 - x)(10 + x)$

49. $(2x + 5)(2x - 5)$ **50.** $(3x + 5)(3x - 5)$

51. $\left(4x + \dfrac{1}{3}\right)\left(4x - \dfrac{1}{3}\right)$

52. $\left(6x + \dfrac{1}{4}\right)\left(6x - \dfrac{1}{4}\right)$

53. $(2a + 7)(2a - 7)$

54. $(3a + 10)(3a - 10)$

55. $(6 - 7x)(6 + 7x)$

56. $(7 - 6x)(7 + 6x)$

57. $(x^2 + 3)(x^2 - 3)$ **58.** $(x^2 + 2)(x^2 - 2)$

59. $(a^2 + 4)(a^2 - 4)$ **60.** $(a^2 + 9)(a^2 - 9)$

61. $(5y^4 - 8)(5y^4 + 8)$

62. $(7y^5 + 6)(7y^5 - 6)$

Multiply and simplify.

63. $(x + 3)(x - 3) + (x - 5)(x + 5)$

64. $(x - 7)(x + 7) + (x - 4)(x + 4)$

65. $(2x + 3)^2 - (4x - 1)^2$

66. $(3x - 5)^2 - (2x + 3)^2$

67. $(a + 1)^2 - (a + 2)^2 + (a + 3)^2$

68. $(a - 1)^2 + (a - 2)^2 - (a - 3)^2$

69. $(2x + 3)^3$

70. $(3x - 2)^3$

71. Find the value of each expression when x is 6.
    a. $x^2 - 25$
    b. $(x - 5)^2$
    c. $(x + 5)(x - 5)$

72. Find the value of each expression when x is 5.
    a. $x^2 - 9$
    b. $(x - 3)^2$
    c. $(x + 3)(x - 3)$

73. Evaluate each expression when x is -2.
    a. $(x + 3)^2$
    b. $x^2 + 9$
    c. $x^2 + 6x + 9$

74. Evaluate each expression when x is -3.
    a. $(x + 2)^2$
    b. $x^2 + 4$
    c. $x^2 + 4x + 4$

Applying the Concepts [Example 16]

75. Comparing Expressions Evaluate the expression $(x + 3)^2$ and the expression $x^2 + 6x + 9$ for $x = 2$.

76. Comparing Expressions Evaluate the expression $x^2 - 25$ and the expression $(x - 5)(x + 5)$ for $x = 6$.

77. Number Problem Write an expression for the sum of the squares of two consecutive integers. Then, simplify that expression.

78. Number Problem Write an expression for the sum of the squares of two consecutive odd integers. Then, simplify that expression.

79. Number Problem Write an expression for the sum of the squares of three consecutive integers. Then, simplify that expression.

80. Number Problem Write an expression for the sum of the squares of three consecutive odd integers. Then, simplify that expression.

81. Area We can use the concept of area to further justify our rule for squaring a binomial. The length of each side of the square shown in the figure is $a + b$. (The longer line segment has length a and the shorter line segment has length b.) The area of the whole square is $(a + b)^2$. However, the whole area is the sum of the areas of the two smaller squares and the two smaller rectangles that make it up. Write the area of the two smaller squares and the two smaller rectangles and then add them together to verify the formula $(a + b)^2 = a^2 + 2ab + b^2$.

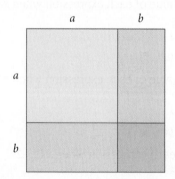

82. Area The length of each side of the large square shown in the figure is $x + 5$. Therefore, its area is $(x + 5)^2$. Find the area of the two smaller squares and the two smaller rectangles that make up the large square, then add them together to verify the formula $(x + 5)^2 = x^2 + 10x + 25$.

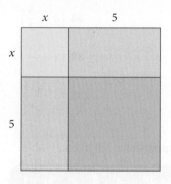

Getting Ready for the Next Section

Simplify each expression (divide).

83. $\dfrac{10x^3}{5x}$　　　　　　　**84.** $\dfrac{15x^2}{5x}$

85. $\dfrac{3x^2}{3}$　　　　　　　　**86.** $\dfrac{4x^2}{2}$

87. $\dfrac{9x^2}{3x}$　　　　　　　　**88.** $\dfrac{3x^4}{9x^2}$

89. $\dfrac{24x^3y^2}{8x^2y}$　　　　　**90.** $-\dfrac{4x^2y^3}{8x^2y}$

91. $\dfrac{15x^2y}{3xy}$　　　　　　**92.** $\dfrac{21xy^2}{3xy}$

93. $\dfrac{35a^6b^8}{70a^2b^{10}}$　　　　**94.** $\dfrac{75a^2b^6}{25a^4b^3}$

95. Biggest Hits The chart shows the number of hits for the five best charting artists in the United States. Write the number of hits Madonna had over the numbers of hits the Beatles had, and then reduce to lowest terms.

96. Text Messaging The graph shows the number of text messages sent each month in 2008. Write the number of text messages during August over the number of text messages sent during April, and then reduce to lowest terms.

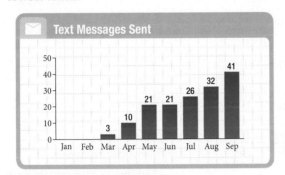

Text Messages Sent

Jan Feb Mar Apr May Jun Jul Aug Sep
3 10 21 21 26 32 41

Maintaining Your Skills

Solve each system by graphing.

97. $x + y = 2$
$x - y = 4$

98. $x + y = 1$
$x - y = -3$

99. $y = 2x + 3$
$y = -2x - 1$

100. $y = 2x - 1$
$y = -2x + 3$

4.7 Dividing a Polynomial by a Monomial

OBJECTIVES

A Divide a polynomial by a monomial.

TICKET TO SUCCESS

Keep these questions in mind as you read through the section. Then respond in your own words and in complete sentences.

1. What property of real numbers is the key to dividing a polynomial by a monomial?

2. Describe how you would divide the polynomial $x^2 + 10x$ by $5x$.

3. Describe how you would divide a polynomial by a negative monomial.

4. In your own words, explain the mistake in the problem $\frac{x+2}{2} = x + 1$.

A Divide a Polynomial by a Monomial

The number of tubes an artist used to paint his masterpieces in one month may be given by $3x^2 + 2x$. If $3x$ represents the number of paintings he completed in that month, how would we find the number of tubes used per painting? We would set this up as a division problem. To solve it, we need to understand how to divide a polynomial by a monomial.

To divide a polynomial by a monomial, we will use the definition of division and apply the distributive property. Follow the steps in this example closely.

EXAMPLE 1 Divide $10x^3 - 15x^2$ by $5x$.

SOLUTION

$$\frac{10x^3 - 15x^2}{5x} = (10x^3 - 15x^2)\frac{1}{5x}$$ 　　Division by $5x$ is the same as multiplication by $\frac{1}{5x}$

$$= 10x^3\left(\frac{1}{5x}\right) - 15x^2\left(\frac{1}{5x}\right)$$ 　　Distribute $\frac{1}{5x}$ to both terms

$$= \frac{10x^3}{5x} - \frac{15x^2}{5x}$$ 　　Multiplication by $\frac{1}{5x}$ is the same as division by $5x$

$$= 2x^2 - 3x$$ 　　Division of monomials as done in Section 4.3

If we were to leave out the first steps in Example 1, the problem would look like this:

$$\frac{10x^3 - 15x^2}{5x} = \frac{10x^3}{5x} - \frac{15x^2}{5x}$$

$$= 2x^2 - 3x$$

The problem is much shorter and clearer this way. You may leave out the first two steps from Example 1 when working problems in this section. They are part of Example 1 only to help show you why the following rule is true.

Rule: Dividing a Polynomial by a Monomial

To divide a polynomial by a monomial, simply divide each term in the polynomial by the monomial.

Here are some further examples using our rule for division of a polynomial by a monomial.

EXAMPLE 2 Divide $\frac{3x^2 - 6}{3}$.

SOLUTION We begin by writing the 3 in the denominator under each term in the numerator. Then we simplify the result.

$$\frac{3x^2 - 6}{3} = \frac{3x^2}{3} - \frac{6}{3}$$ 　　Divide each term in the numerator by 3

$$= x^2 - 2$$ 　　Simplify

EXAMPLE 3 Divide $\frac{4x^2 - 2}{2}$.

SOLUTION Dividing each term in the numerator by 2, we have

$$\frac{4x^2 - 2}{2} = \frac{4x^2}{2} - \frac{2}{2}$$ 　　Divide each term in the numerator by 2

$$= 2x^2 - 1$$ 　　Simplify

EXAMPLE 4 Find the quotient of $27x^3 - 9x^2$ and $3x$.

SOLUTION We again are asked to divide the first polynomial by the second one.

$$\frac{27x^3 - 9x^2}{3x} = \frac{27x^3}{3x} - \frac{9x^2}{3x} \qquad \text{Divide each term by } 3x$$

$$= 9x^2 - 3x \qquad \text{Simplify}$$

EXAMPLE 5 Divide $(15x^2y - 21xy^2) \div (-3xy)$.

SOLUTION This is the same type of problem we have shown in the first four examples; it is just worded a little differently. Note that when we divide each term in the first polynomial by $-3xy$, we must take the negative sign into account.

$$\frac{15x^2y - 21xy^2}{-3xy} = \frac{15x^2y}{-3xy} - \frac{21xy^2}{-3xy} \qquad \text{Divide each term by } -3xy$$

$$= -5x - (-7y) \qquad \text{Simplify}$$

$$= -5x + 7y \qquad \text{Simplify}$$

EXAMPLE 6 Divide $\dfrac{24x^3y^2 + 16x^2y^2 - 4x^2y^3}{8x^2y}$.

SOLUTION Writing $8x^2y$ under each term in the numerator and then simplifying, we have

$$\frac{24x^3y^2 + 16x^2y^2 - 4x^2y^3}{8x^2y} = \frac{24x^3y^2}{8x^2y} + \frac{16x^2y^2}{8x^2y} - \frac{4x^2y^3}{8x^2y}$$

$$= 3xy + 2y - \frac{y^2}{2}$$

From the examples in this section, it is clear that to divide a polynomial by a monomial, we must divide each term in the polynomial by the monomial. Often, students taking algebra for the first time will make the following mistake:

$$\frac{x + \cancel{2}}{\cancel{2}} = x + 1 \qquad \text{Mistake}$$

The mistake here is in not dividing both terms in the numerator by 2. The correct way to divide $x + 2$ by 2 looks like this:

$$\frac{x + 2}{2} = \frac{x}{2} + \frac{2}{2} = \frac{x}{2} + 1 \qquad \text{Correct}$$

Problem Set 4.7

Moving Toward Success

"I have become my own version of an optimist. If I can't make it through one door, I'll go through another door; or I'll make a door..."

—Rabindranath Tagore, 1861–1941, writer and musician

1. A mnemonic device is a mental tool that uses an acronym or a short verse to help a person remember. Why do you think a mnemonic device is a helpful learning tool?

2. Have you used a mnemonic device to help you study mathematics? If so, what was it? If not, create one for this section.

A Divide the following polynomials by $5x$. [Example 1]

1. $5x^2 - 10x$

2. $10x^3 - 15x$

3. $15x - 10x^3$

4. $50x^3 - 20x^2$

5. $25x^2y - 10xy$

6. $15xy^2 + 20x^2y$

7. $35x^5 - 30x^4 + 25x^3$

8. $40x^4 - 30x^3 + 20x^2$

9. $50x^5 - 25x^3 + 5x$

10. $75x^6 + 50x^3 - 25x$

A Divide the following by $-2a$. [Example 1]

11. $8a^2 - 4a$

12. $a^3 - 6a^2$

13. $16a^5 + 24a^4$

14. $30a^6 + 20a^3$

15. $8ab + 10a^2$

16. $6a^2b - 10ab^2$

17. $12a^3b - 6a^2b^2 + 14ab^3$

18. $4ab^3 - 16a^2b^2 - 22a^3b$

19. $a^2 + 2ab + b^2$

20. $a^2b - 2ab^2 + b^3$

A Perform the following divisions (find the following quotients). [Examples 1–6]

21. $\dfrac{6x + 8y}{2}$

22. $\dfrac{9x - 3y}{3}$

23. $\dfrac{7y - 21}{-7}$

24. $\dfrac{14y - 12}{2}$

25. $\dfrac{2x^2 + 16x - 18}{2}$

26. $\dfrac{3x^2 - 3x - 18}{3}$

27. $\dfrac{3y^2 - 9y + 3}{3}$

28. $\dfrac{2y^2 - 8y + 2}{2}$

29. $\dfrac{10xy - 8x}{2x}$

30. $\dfrac{12xy^2 - 18x}{-6x}$

31. $\dfrac{x^2y - x^3y^2}{x}$

32. $\dfrac{x^2y - x^3y^2}{x^2}$

33. $\dfrac{x^2y - x^3y^2}{-x^2y}$

34. $\dfrac{ab + a^2b^2}{ab}$

35. $\dfrac{a^2b^2 - ab^2}{-ab^2}$

36. $\dfrac{a^2b^2c + ab^2c^2}{abc}$

37. $\dfrac{x^3 - 3x^2y + xy^2}{x}$

38. $\dfrac{x^2 - 3xy^2 + xy^3}{x}$

39. $\dfrac{10a^2 - 15a^2b + 25a^2b^2}{5a^2}$

40. $\dfrac{11a^2b^2 - 33ab}{-11ab}$

41. $\dfrac{26x^2y^2 - 13xy}{-13xy}$

42. $\dfrac{6x^2y^2 - 3xy}{6xy}$

43. $\dfrac{4x^2y^2 - 2xy}{4xy}$

44. $\dfrac{6x^2a + 12x^2b - 6x^2c}{36x^2}$

45. $\dfrac{5a^2x - 10ax^2 + 15a^2x^2}{20a^2x^2}$

46. $\dfrac{12ax - 9bx + 18cx}{6x^2}$

47. $\dfrac{16x^5 + 8x^2 + 12x}{12x^3}$

48. $\dfrac{27x^2 - 9x^3 - 18x^4}{-18x^3}$

A Divide. Assume all variables represent positive numbers. [Example 6]

49. $\dfrac{9a^{5m} - 27a^{3m}}{3a^{2m}}$

50. $\dfrac{26a^{3m} - 39a^{5m}}{13a^{3m}}$

51. $\dfrac{10x^{5m} - 25x^{3m} + 35x^m}{5x^m}$

52. $\dfrac{18x^{2m} + 24x^{4m} - 30x^{6m}}{6x^{2m}}$

A Simplify each numerator, and then divide.

53. $\dfrac{2x^3(3x + 2) - 3x^2(2x - 4)}{2x^2}$

54. $\dfrac{5x^2(6x - 3) + 6x^3(3x - 1)}{3x}$

55. $\dfrac{(x + 2)^2 - (x - 2)^2}{2x}$

56. $\dfrac{(x - 3)^2 - (x + 3)^2}{3x}$

57. $\dfrac{(x + 5)^2 + (x + 5)(x - 5)}{2x}$

58. $\dfrac{(x - 4)^2 + (x + 4)(x - 4)}{2x}$

59. Find the value of each expression when x is 2.

a. $2x + 3$

b. $\dfrac{10x + 15}{5}$

c. $10x + 3$

60. Find the value of each expression when x is 5.

  a. $3x + 2$

  b. $\dfrac{6x^2 + 4x}{2x}$

  c. $6x^2 + 2$

61. Evaluate each expression for $x = 10$.

  a. $\dfrac{3x + 8}{2}$

  b. $3x + 4$

  c. $\dfrac{3}{2}x + 4$

62. Evaluate each expression for $x = 10$.

  a. $\dfrac{5x - 6}{2}$

  b. $5x - 3$

  c. $\dfrac{5}{2}x - 3$

63. Find the value of each expression when x is 2.

  a. $2x^2 - 3x$

  b. $\dfrac{10x^3 - 15x^2}{5x}$

64. Find the value of each expression when a is 3.

  a. $-4a + 2$

  b. $\dfrac{8a^2 - 4a}{-2a}$

65. Comparing Expressions Evaluate the expression $\dfrac{10x + 15}{5}$ and the expression $2x + 3$ when $x = 2$.

66. Comparing Expressions Evaluate the expression $\dfrac{6x^2 + 4x}{2x}$ and the expression $3x + 2$ when $x = 5$.

67. Comparing Expressions Show that the expression $\dfrac{3x + 8}{2}$ is not the same as the expression $3x + 4$ by replacing x with 10 in both expressions and simplifying the results.

68. Comparing Expressions Show that the expression $\dfrac{x + 10}{x}$ is not equal to 10 by replacing x with 5 and simplifying.

Getting Ready for the Next Section

Divide.

69. $27\overline{)3{,}962}$ **70.** $13\overline{)18{,}780}$

71. $\dfrac{2x^2 + 5x}{x}$ **72.** $\dfrac{7x^2 + 9x^3 + 3x^7}{x^3}$

Multiply

73. $(x - 3)x$ **74.** $(x - 3)(-2)$

75. $2x^2(x - 5)$ **76.** $10x(x - 5)$

Subtract.

77. $(x^2 - 5x) - (x^2 - 3x)$

78. $(2x^3 + 0x^2) - (2x^3 - 10x^2)$

79. $(-2x + 8) - (-2x + 6)$

80. $(4x - 14) - (4x - 10)$

Maintaining Your Skills

81. Movie Rentals The chart shows the market share as a percent for the years 2008 and 2009 for the three largest movie rental companies. These figures are based on units, or DVDs, rented rather than revenue.

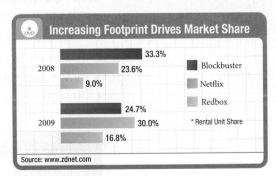

a. If Redbox rented 365 million DVDs in 2009, how many total DVDs were rented that year?

b. From your answer in part a, how many DVDs did Blockbuster rent in 2009?

82. Skyscrapers The chart shows the heights of the three tallest buildings in the world. Five times the height of Big Ben in London, England, and 70 feet is the height of Taipei 101 tower. How tall is Big Ben? Round to the nearest foot.

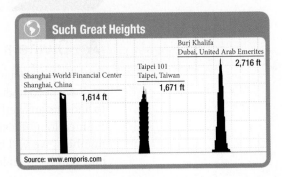

Solve each system of equations by the elimination method.

83. $x + y = 6$
$x - y = 8$

84. $2x + y = 5$
$-x + y = -4$

85. $2x - 3y = -5$
$x + y = 5$

86. $2x - 4y = 10$
$3x - 2y = -1$

Solve each system by the substitution method.

87. $x + y = 2$
$y = 2x - 1$

88. $2x - 3y = 4$
$x = 3y - 1$

89. $4x + 2y = 8$
$y = -2x + 4$

90. $4x + 2y = 8$
$y = -2x + 5$

4.8 Dividing a Polynomial by a Polynomial

OBJECTIVES

A Divide a polynomial by a polynomial.

TICKET TO SUCCESS

Keep these questions in mind as you read through the section. Then respond in your own words and in complete sentences.

1. What are the four steps used in long division with whole numbers?
2. How is division of two polynomials similar to long division with whole numbers?
3. How would you check your answer when dividing two polynomials?
4. How do we use 0 when dividing the polynomial $2x^3 - 3x + 2$ by $x - 5$?

A Divide a Polynomial by a Polynomial

Suppose the annual cost to live in Los Angeles is $117,726. To determine the monthly cost, we need to divide 117,726 by 12. How would you set up a long division problem to find this answer?

Since long division for polynomials is very similar to long division with whole numbers, we will begin by reviewing a division problem with whole numbers. You may realize when looking at Example 1 that you don't have a very good idea why you proceed as you do with long division. What you do know is that the process always works. We are going to approach the explanations in this section in much the same manner; that is, we won't always be sure why the steps we will use are important, only that they always produce the correct result.

EXAMPLE 1   Divide $27\overline{)3{,}962}$.

SOLUTION

$$
\begin{array}{r}
1 \\
27\overline{)3{,}962} \\
2\,7 \\
\hline
1\,2
\end{array}
$$
    Estimate 27 into 39

    Multiply $1 \times 27 = 27$
    Subtract $39 - 27 = 12$

$$
\begin{array}{r}
1 \\
27\overline{)3{,}962} \\
2\,7\downarrow \\
\hline
1\,26
\end{array}
$$

    Bring down the 6

These are the four basic steps in long division. Estimate, multiply, subtract, and bring down the next term. To finish the problem, we simply perform the same four steps again.

$$
\begin{array}{r}
14 \\
27\overline{)3{,}962} \\
2\,7\downarrow \\
\hline
1\,26 \\
1\,08 \\
\hline
182
\end{array}
$$

    4 is the estimate

    Multiply to get 108
    Subtract to get 18, then bring down the 2

One more time.

$$
\begin{array}{r}
146 \\
27\overline{)3{,}962} \\
2\,7\downarrow \\
\hline
1\,26 \\
1\,08 \\
\hline
182 \\
162 \\
\hline
20
\end{array}
$$

    6 is the estimate

    Multiply to get 162
    Subtract to get 20

Since there is nothing left to bring down, we have our answer.

$$
\frac{3{,}962}{27} = 146 + \frac{20}{27} \quad \text{or} \quad 146\frac{20}{27}
$$

Here is how it works with polynomials.

EXAMPLE 2   Divide $\dfrac{x^2 - 5x + 8}{x - 3}$.

SOLUTION

$$
\begin{array}{r}
x \\
x - 3\overline{)\; x^2 - 5x + 8\;} \\
\underset{-}{\cancel{+}}\; x^2 \underset{+}{\cancel{-}}\; 3x \\
\hline
-2x
\end{array}
$$

    Estimate $x^2 \div x = x$

    Multiply $x(x - 3) = x^2 - 3x$
    Subtract $(x^2 - 5x) - (x^2 - 3x) = -2x$

$$
\begin{array}{r}
x \phantom{{}- 5x + 8} \\
x - 3 \overline{)\ x^2 - 5x + 8} \\
\underline{\ \ \not{-}\ x^2\ \not{+}\ 3x \quad \downarrow} \\
-2x + 8
\end{array}
$$

Bring down the 8

Notice that to subtract one polynomial from another, we add its opposite. That is why we change the signs on $x^2 - 3x$ and add what we get to $x^2 - 5x$. (To subtract the second polynomial, simply change the signs and add.)

We perform the same four steps again.

$$
\begin{array}{r}
x - 2 \phantom{{}+ 8} \\
x - 3 \overline{)\ x^2 - 5x + 8} \\
\underline{\ \ \not{-}\ x^2\ \not{+}\ 3x \quad \downarrow} \\
-2x + 8 \\
\underline{\ \not{+}\ 2x\ \not{-}\ 6} \\
2
\end{array}
$$

-2 is the estimate $(-2x \div x = -2)$

Multiply $-2(x - 3) = -2x + 6$
Subtract $(-2x + 8) - (-2x + 6) = 2$

Since there is nothing left to bring down, we have our answer:

$$
\frac{x^2 - 5x + 8}{x - 3} = x - 2 + \frac{2}{x - 3}
$$

To check our answer, we multiply $(x - 3)(x - 2)$ to get $x^2 - 5x + 6$. Then, adding on the remainder, 2, we have $x^2 - 5x + 8$. ▪

EXAMPLE 3 Divide $\dfrac{6x^2 - 11x - 14}{2x - 5}$.

SOLUTION

$$
\begin{array}{r}
3x + 2 \phantom{{}- 14} \\
2x - 5 \overline{)\ 6x^2 - 11x - 14} \\
\underline{\ \not{-}\ 6x^2\ \not{+}\ 15x \quad \downarrow} \\
+4x - 14 \\
\underline{\ \not{-}\ 4x\ \not{+}\ 10} \\
-4
\end{array}
$$

$$
\frac{6x^2 - 11x - 14}{2x - 5} = 3x + 2 + \frac{-4}{2x - 5}
$$
▪

One last step is sometimes necessary. The two polynomials in a division problem must both be in descending powers of the variable and cannot skip any powers from the highest power down to the constant term.

EXAMPLE 4 Divide $\dfrac{2x^3 - 3x + 2}{x - 5}$.

SOLUTION The problem will be much less confusing if we write $2x^3 - 3x + 2$ as $2x^3 + 0x^2 - 3x + 2$. Adding $0x^2$ does not change our original problem.

$$
\begin{array}{r}
2x^2 \phantom{{}- 3x + 2} \\
x - 5 \overline{)\ 2x^3 + 0x^2 - 3x + 2} \\
\underline{\ \not{-}\ 2x^3\ \not{+}\ 10x^2 \quad \downarrow} \\
+10x^2 - 3x
\end{array}
$$

Estimate $2x^3 \div x = 2x^2$

Multiply $2x^2(x - 5) = 2x^3 - 10x^2$
Subtract:
$(2x^3 + 0x^2) - (2x^3 - 10x^2) = 10x^2$
Bring down the next term

Adding the term $0x^2$ gives us a column in which to write $10x^2$. (Remember, you can add and subtract only similar terms.)

Here is the completed problem:

$$
\begin{array}{r}
2x^2 + 10x + 47 \\
x - 5 \overline{)\ 2x^3 + 0x^2 - 3x + 2} \\
\mp 2x^3 \pm 10x^2 \\
\hline
+ 10x^2 - 3x \\
\mp 10x^2 \pm 50x \\
\hline
+ 47x + 2 \\
\mp 47x \pm 235 \\
\hline
237
\end{array}
$$

Our answer is $\dfrac{2x^3 - 3x + 2}{x - 5} = 2x^2 + 10x + 47 + \dfrac{237}{x - 5}$.

As you can see, long division with polynomials is a mechanical process. Once you have done it correctly a couple of times, it becomes very easy to produce the correct answer.

Problem Set 4.8

Moving Toward Success

"Tomorrow is the most important thing in life. Comes into us at midnight very clean. It's perfect when it arrives and puts itself in our hands. It hopes we've learned something from yesterday."

—John Wayne, 1907–1979, American actor and director

1. Where do you prefer to study?
2. Is your chosen location conducive to quality studying? Why or why not?

A Divide. [Examples 2–4]

1. $\dfrac{x^2 - 5x + 6}{x - 3}$

2. $\dfrac{x^2 - 5x + 6}{x - 2}$

3. $\dfrac{a^2 + 9a + 20}{a + 5}$

4. $\dfrac{a^2 + 9a + 20}{a + 4}$

5. $\dfrac{x^2 - 6x + 9}{x - 3}$

6. $\dfrac{x^2 + 10x + 25}{x + 5}$

7. $\dfrac{2x^2 + 5x - 3}{2x - 1}$

8. $\dfrac{4x^2 + 4x - 3}{2x - 1}$

9. $\dfrac{2a^2 - 9a - 5}{2a + 1}$

10. $\dfrac{4a^2 - 8a - 5}{2a + 1}$

11. $\dfrac{x^2 + 5x + 8}{x + 3}$

12. $\dfrac{x^2 + 5x + 4}{x + 3}$

13. $\dfrac{a^2 + 3a + 2}{a + 5}$

14. $\dfrac{a^2 + 4a + 3}{a + 5}$

15. $\dfrac{x^2 + 2x + 1}{x - 2}$

16. $\dfrac{x^2 + 6x + 9}{x - 3}$

17. $\dfrac{x^2 + 5x - 6}{x + 1}$

18. $\dfrac{x^2 - x - 6}{x + 1}$

19. $\dfrac{a^2 + 3a + 1}{a + 2}$

20. $\dfrac{a^2 - a + 3}{a + 1}$

21. $\dfrac{2x^2 - 2x + 5}{2x + 4}$

22. $\dfrac{15x^2 + 19x - 4}{3x + 8}$

23. $\dfrac{6a^2 + 5a + 1}{2a + 3}$

24. $\dfrac{4a^2 + 4a + 3}{2a + 1}$

25. $\dfrac{6a^3 - 13a^2 - 4a + 15}{3a - 5}$

26. $\dfrac{2a^3 - a^2 + 3a + 2}{2a + 1}$

Fill in the missing terms in the numerator, and then use long division to find the quotients. [Example 4]

27. $\dfrac{x^3 + 4x + 5}{x + 1}$

28. $\dfrac{x^3 + 4x^2 - 8}{x + 2}$

29. $\dfrac{x^3 - 1}{x - 1}$

30. $\dfrac{x^3 + 1}{x + 1}$

31. $\dfrac{x^3 - 8}{x - 2}$

32. $\dfrac{x^3 + 27}{x + 3}$

33. Find the value of each expression when x is 3.

  a. $x^2 + 2x + 4$

  b. $\dfrac{x^3 - 8}{x - 2}$

  c. $x^2 - 4$

34. Find the value of each expression when x is 2.

  a. $x^2 - 3x + 9$

  b. $\dfrac{x^3 + 27}{x + 3}$

  c. $x^2 + 9$

35. Find the value of each expression when x is 4.

  a. $x + 3$

  b. $\dfrac{x^2 + 9}{x + 3}$

  c. $x - 3 + \dfrac{18}{x + 3}$

36. Find the value of each expression when x is 2.

  a. $x + 1$

  b. $\dfrac{x^2 + 1}{x + 1}$

  c. $x - 1 + \dfrac{2}{x + 1}$

Applying the Concepts

37. Golf The chart shows the lengths of the longest golf courses used for the U.S. Open. If the length of the Winged Foot course is 7,264 yards, what is the average length of each hole on the 18-hole course? Round to the nearest yard.

38. Car Thefts The chart shows the U.S. cities with the highest car theft rates in a year. How many cars were stolen in one month in Modesto, CA? Round to the nearest car.

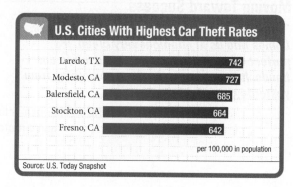

Maintaining Your Skills

Use systems of equations to solve the following word problems.

39. Number Problem The sum of two numbers is 25. One of the numbers is 4 times the other. Find the numbers.

40. Number Problem The sum of two numbers is 24. One of the numbers is 3 more than twice the other. Find the numbers.

41. Investing Supposed you have two savings accounts. One of the accounts pays 8% annual interest and the other pays 9% annual interest. Suppose you invest twice as much at 8% than at 9% If your total interest for the year is $100, how much money did you invest in each of the accounts?

42. **Investing** If you invest twice as much money in an account that pays 12% annual interest as you do in an account that pays 11% annual interest, how much do you have in each account if your total interest for a year is $210?

43. **Money Problem** If you have a total of $160 in $5 bills and $10 bills, how many of each type of bill do you have if you have 4 more $10 bills than $5 bills?

44. **Coin Problem** Suppose you hae two more nickels than quarters and you have a total of $2.80 in change. How many of each type of coin do you have?

Exponents: Definition and Properties [4.1, 4.2]

EXAMPLES

Integer exponents indicate repeated multiplications.

1. a. $2^3 = 2 \cdot 2 \cdot 2 = 8$

$a^r \cdot a^s = a^{r+s}$ To multiply with the same base, you add exponents

b. $x^5 \cdot x^3 = x^{5+3} = x^8$

$\dfrac{a^r}{a^s} = a^{r-s}$ To divide with the same base, you subtract exponents

c. $\dfrac{x^5}{x^3} = x^{5-3} = x^2$

d. $(3x)^2 = 3^2 \cdot x^2 = 9x^2$

$(ab)^r = a^r \cdot b^r$ Exponents distribute over multiplication

e. $\left(\dfrac{2}{3}\right)^3 = \dfrac{2^3}{3^3} = \dfrac{8}{27}$

$\left(\dfrac{a}{b}\right)^r = \dfrac{a^r}{b^r}$ Exponents distribute over division

f. $(x^5)^3 = x^{5 \cdot 3} = x^{15}$

$(a^r)^s = a^{r \cdot s}$ A power of a power is the product of the powers

g. $3^{-2} = \dfrac{1}{3^2} = \dfrac{1}{9}$

$a^{-r} = \dfrac{1}{a^r}$ Negative exponents imply reciprocals

Multiplication of Monomials [4.3]

2. $(5x^2)(3x^4) = 15x^6$

To multiply two monomials, multiply coefficients and add exponents.

Division of Monomials [4.3]

3. $\dfrac{12x^9}{4x^5} = 3x^4$

To divide two monomials, divide coefficients and subtract exponents.

Scientific Notation [4.1, 4.2, 4.3]

4. $768,000 = 7.68 \times 10^5$
$0.00039 = 3.9 \times 10^{-4}$

A number is in scientific notation when it is written as the product of a number between 1 and 10 and an integer power of 10.

Addition of Polynomials [4.4]

5. $(3x^2 - 2x + 1) + (2x^2 + 7x - 3)$
$\quad = 5x^2 + 5x - 2$

To add two polynomials, add coefficients of similar terms.

Subtraction of Polynomials [4.4]

6. $(3x + 5) - (4x - 3)$
$\quad = 3x + 5 - 4x + 3$
$\quad = -x + 8$

To subtract one polynomial from another, add the opposite of the second to the first.

7. a. $2a^2(5a^2 + 3a - 2)$
    $= 10a^4 + 6a^3 - 4a^2$

b. $(x + 2)(3x - 1)$
    $= 3x^2 - x + 6x - 2$
    $= 3x^2 + 5x - 2$

c.
$$
\begin{array}{r}
2x^2 - 3x + 4 \\
3x - 2 \\
\hline
6x^3 - 9x^2 + 12x \\
-4x^2 + 6x - 8 \\
\hline
6x^3 - 13x^2 + 18x - 8
\end{array}
$$

◼ Multiplication of Polynomials [4.5]

To multiply a polynomial by a monomial, we apply the distributive property. To multiply two binomials, we use the FOIL method. In other situations we use the Column method. Each method achieves the same result: To multiply any two polynomials, we multiply each term in the first polynomial by each term in the second polynomial.

◼ Special Products [4.6]

8. $(x + 3)^2 = x^2 + 6x + 9$
    $(x - 3)^2 = x^2 - 6x + 9$
$(x + 3)(x - 3) = x^2 - 9$

$\left.\begin{array}{l}(a + b)^2 = a^2 + 2ab + b^2 \\ (a - b)^2 = a^2 - 2ab + b^2\end{array}\right\}$ Binomial squares

$(a + b)(a - b) = a^2 - b^2$ Difference of two squares

◼ Dividing a Polynomial by a Monomial [4.7]

9. $\dfrac{12x^3 - 18x^2}{6x} = 2x^2 - 3x$

To divide a polynomial by a monomial, divide each term in the polynomial by the monomial.

◼ Long Division with Polynomials [4.8]

10.
$$
\begin{array}{r}
x - 2 \\
x - 3 \overline{)\, x^2 - 5x + 8} \\
\underline{\mp x^2 \pm 3x} \\
-2x + 8 \\
\underline{\pm 2x \mp 6} \\
2
\end{array}
$$

$x^2 - 5x + 8 \div x - 3$

$= x - 2 + \dfrac{2}{x - 3}$

Division with polynomials is similar to long division with whole numbers. The steps in the process are estimate, multiply, subtract, and bring down the next term. The divisors in all the long-division problems in this chapter were binomials.

1. If a term contains a variable that is raised to a power, then the exponent on the variable is associated only with that variable, unless there are parentheses; that is, the expression $3x^2$ means $3 \cdot x \cdot x$, not $3x \cdot 3x$. It is a mistake to write $3x^2$ as $9x^2$. The only way to end up with $9x^2$ is to start with $(3x)^2$.

2. It is a mistake to add nonsimilar terms. For example, $2x$ and $3x^2$ are non-similar terms and therefore cannot be combined; that is, $2x + 3x^2 \neq 5x^3$. If you were to substitute 10 for x in the preceding expression, you would see that the two sides are not equal.

3. It is a mistake to distribute exponents over sums and differences; that is, $(a + b)^2 \neq a^2 + b^2$. Convince yourself of this by letting $a = 2$ and $b = 3$ and then simplifying both sides.

4. Another common mistake can occur when dividing a polynomial by a monomial. Here is an example:

$$\frac{x + \cancel{2}}{\cancel{2}} = x + 1 \qquad \text{Mistake}$$

The mistake here is in not dividing both terms in the numerator by 2. The correct way to divide $x + 2$ by 2 looks like this:

$$\frac{x + 2}{2} = \frac{x}{2} + \frac{2}{2} \qquad \text{Correct}$$

$$= \frac{x}{2} + 1$$

Chapter 4   Review

Simplify. [4.1]

1. $(-1)^3$

2. -8^2

3. $\left(\dfrac{3}{7}\right)^2$

4. $y^3 \cdot y^9$

5. $x^{15} \cdot x^7 \cdot x^5 \cdot x^3$

6. $(x^7)^5$

7. $(2^6)^4$

8. $(3y)^3$

9. $(-2xyz)^3$

Simplify each expression. Any answers that contain exponents should contain positive exponents only. [4.2]

10. 7^{-2}

11. $4x^{-5}$

12. $(3y)^{-3}$

13. $\dfrac{a^9}{a^3}$

14. $\left(\dfrac{x^3}{x^5}\right)^2$

15. $\dfrac{x^9}{x^{-6}}$

16. $\dfrac{x^{-7}}{x^{-2}}$

17. $(-3xy)^0$

18. $3^0 - 5^1 + 5^0$

Simplify. Any answers that contain exponents should contain positive exponents only. [4.1, 4.2]

19. $(3x^3y^2)^2$

20. $(2a^3b^2)^4(2a^5b^6)^2$

21. $(-3xy^2)^{-3}$

22. $\dfrac{(b^3)^4(b^2)^5}{(b^7)^3}$

23. $\dfrac{(x^{-3})^3(x^6)^{-1}}{(x^{-5})^{-4}}$

Simplify. Write all answers with positive exponents only. [4.3]

24. $\dfrac{(2x^4)(15x^9)}{(6x^6)}$

25. $\dfrac{(10x^3y^5)(21x^2y^6)}{(7xy^3)(5x^9y)}$

26. $\dfrac{21a^{10}}{3a^4} - \dfrac{18a^{17}}{6a^{11}}$ **27.** $\dfrac{8x^8y^3}{2x^3y} - \dfrac{10x^6y^9}{5xy^7}$

Simplify, and write all answers in scientific notation. [4.3]

28. $(3.2 \times 10^3)(2 \times 10^4)$

29. $\dfrac{4.6 \times 10^5}{2 \times 10^{-3}}$

30. $\dfrac{(4 \times 10^6)(6 \times 10^5)}{3 \times 10^8}$

Perform the following additions and subtractions. [4.4]

31. $(3a^2 - 5a + 5) + (5a^2 - 7a - 8)$

32. $(-7x^2 + 3x - 6) - (8x^2 - 4x + 7) + (3x^2 - 2x - 1)$

33. Subtract $8x^2 + 3x - 2$ from $4x^2 - 3x - 2$.

34. Find the value of $2x^2 - 3x + 5$ when $x = 3$.

Multiply. [4.5]

35. $3x(4x - 7)$

36. $8x^3y(3x^2y - 5xy^2 + 4y^3)$

37. $(a + 1)(a^2 + 5a - 4)$

38. $(x + 5)(x^2 - 5x + 25)$

39. $(3x - 7)(2x - 5)$

40. $\left(5y + \dfrac{1}{5}\right)\left(5y - \dfrac{1}{5}\right)$

41. $(a^2 - 3)(a^2 + 3)$

Perform the indicated operations. [4.6]

42. $(a - 5)^2$

43. $(3x + 4)^2$

44. $(y^2 + 3)^2$

45. Divide $10ab + 20a^2$ by $-5a$. [4.7]

46. Divide $40x^5y^4 - 32x^3y^3 - 16x^2y$ by $-8xy$.

Divide using long division. [4.8]

47. $\dfrac{x^2 + 15x + 54}{x + 6}$

48. $\dfrac{6x^2 + 13x - 5}{3x - 1}$

49. $\dfrac{x^3 + 64}{x + 4}$

50. $\dfrac{3x^2 - 7x + 10}{3x + 2}$

51. $\dfrac{2x^3 - 7x^2 + 6x + 10}{2x + 1}$

Simplify.

1. $-\left(-\dfrac{3}{4}\right)$

2. $-|-9|$

3. $2 \cdot 7 + 10$

4. $6 \cdot 7 + 7 \cdot 9$

5. $10 + (-15)$

6. $-15 - (-3)$

7. $-7\left(-\dfrac{1}{7}\right)$

8. $(-9)(-7)$

9. $6\left(-\dfrac{1}{2}x + \dfrac{1}{6}y\right)$

10. $6(4a + 2) - 3(5a - 1)$

11. $y^{10} \cdot y^6$

12. $(2y)^4$

13. $(3a^2b^5)^3(2a^6b^7)^2$

14. $\dfrac{(12xy^5)^3(16x^2y^2)}{(8x^3y^3)(3x^5y)}$

Simplify and write your answer in scientific notation.

15. $(3.5 \times 10^3)(8 \times 10^{-6})$

16. $\dfrac{3.5 \times 10^{-7}}{7 \times 10^{-3}}$

Perform the indicated operations.

17. $10a^2b - 5a^2b + a^2b$

18. $(-4x^2 - 5x + 2) + (3x^2 - 6x + 1) - (-x^2 + 2x - 7)$

19. $(5x - 1)^2$

20. $(x - 1)(x^2 + x + 1)$

Solve each equation.

21. $8 - 2y + 3y = 12$

22. $6a - 5 = 4a$

23. $18 = 3(2x - 2)$

24. $2(3x + 5) + 8 = 2x + 10$

Solve the inequality, and graph the solution set.

25. $-\dfrac{a}{6} > 4$

26. $3(2t - 5) - 7 \le 5(3t + 1) + 5$

27. $2x + 3 < 5$

28. $-5 \le 4x + 3$

Divide.

29. $\dfrac{4x^2 + 8x - 10}{2x - 3}$

30. $\dfrac{15x^5 - 10x^2 + 20x}{5x^5}$

Graph on a rectangular coordinate system.

31. $x + y = 3$

32. $x = 2$

Solve each system by graphing.

33. $x + 2y = 1$
$x + 2y = 2$

34. $3x + 2y = 4$
$3x + 6y = 12$

Solve each system.

35. $x + 2y = 5$
$x - y = 2$

36. $x + 2y = 5$
$3x + 6y = 14$

37. $2x + y = -3$
$x - 3y = -5$

38. $\dfrac{1}{6}x + \dfrac{1}{4}y = 1$
$\dfrac{6}{5}x - y = \dfrac{8}{5}$

39. $x + y = 9$
$y = x + 1$

40. $4x + 5y = 25$
$2y = x - 3$

41. Find the value of $8x - 3$ when $x = 3$.

42. Find x when y is 8 in the equation $y = 3x - 1$.

43. Solve $A = \dfrac{1}{2}bh$ for h.

44. Solve $\dfrac{y - 1}{x} = -2$ for y.

For the set $\{-3, -\sqrt{2}, 0, \frac{3}{4}, 1.5, \pi\}$ list all of the

45. Whole numbers

46. Irrational numbers

Give the property that justifies the statement.

47. $5 + (-5) = 0$

48. $a + 2 = 2 + a$

49. Find the x- and y-intercepts for the equation $y = 2x + 4$.

50. Find the slope of the line $2x - 5y = 7$.

Use the illustration to answer the following questions.

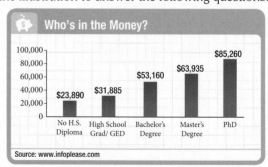

Who's in the Money?

No H.S. Diploma: $23,890
High School Grad/ GED: $31,885
Bachelor's Degree: $53,160
Master's Degree: $63,935
PhD: $85,260

Source: www.infoplease.com

51. How much more is the average income of a worker with a master's degree than that of a worker with a bachelor's degree?

52. On average, how much less does a high school graduate earn than a worker with a bachelor's degree?

Simplify each of the following expressions. [4.1]

1. $(-3)^4$

2. $\left(\dfrac{3}{4}\right)^2$

3. $(3x^3)^2(2x^4)^3$

Simplify each expression. Write all answers with positive exponents only. [4.2]

4. 3^{-2}

5. $(3a^4b^2)^0$

6. $\dfrac{a^{-3}}{a^{-5}}$

7. $\dfrac{(x^{-2})^3(x^{-3})^{-5}}{(x^{-4})^{-2}}$

8. Write 0.0278 in scientific notation. [4.2]

9. Write 2.43×10^5 in expanded form. [4.1]

Simplify. Write all answers with positive exponents only. [4.3]

10. $\dfrac{35x^2y^4z}{70x^6y^2z}$

11. $\dfrac{(6a^2b)(9a^3b^2)}{18a^4b^3}$

12. $\dfrac{24x^7}{3x^2} + \dfrac{14x^9}{7x^4}$

13. $\dfrac{(2.4 \times 10^5)(4.5 \times 10^{-2})}{1.2 \times 10^{-6}}$

Add or subtract as indicated. [4.4]

14. $8x^2 - 4x + 6x + 2$

15. $(5x^2 - 3x + 4) - (2x^2 - 7x - 2)$

16. Subtract $3x - 4$ from $6x - 8$. [5.4]

17. Find the value of $2y^2 - 3y - 4$ when y is -2.

Multiply. [4.5]

18. $2a^2(3a^2 - 5a + 4)$

19. $\left(x + \dfrac{1}{2}\right)\left(x + \dfrac{1}{3}\right)$

20. $(4x - 5)(2x + 3)$

21. $(x - 3)(x^2 + 3x + 9)$

Multiply. [4.6]

22. $(x + 5)^2$

23. $(3a - 2b)^2$

24. $(3x - 4y)(3x + 4y)$

25. $(a^2 - 3)(a^2 + 3)$

26. Divide $10x^3 + 15x^2 - 5x$ by $5x$. [5.7]

Divide. [4.8]

27. $\dfrac{8x^2 - 6x - 5}{2x - 3}$

28. $\dfrac{3x^3 - 2x + 1}{x - 3}$

29. **Volume** Find the volume of a cube if the length of a side is 2.5 centimeters. [4.1]

30. **Volume** Find the volume of a rectangular solid if the length is five times the width, and the height is one fifth the width. [4.3]

The illustration below shows the amount of money (in millions) wagered at the Belmont Stakes.

Betting for the Belmont Stakes

2004	$63.7M
2008	$56.2M
2002	$54.2M
2009	$51.1M
2003	$48.1M
2005	$48.0M

Millions of dollars

Source: USA Today Snapshot

Write in scientific notation the amount of money bet in the following years. [4.1]

31. 2004

32. 2009

The illustration below shows the top five margins of victory for Nascar.

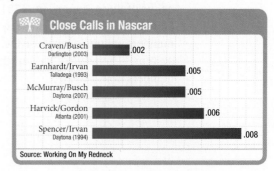

Close Calls in Nascar

Race	Margin
Craven/Busch Darlington (2003)	.002
Earnhardt/Irvan Talladega (1993)	.005
McMurray/Busch Daytona (2007)	.005
Harvick/Gordon Atlanta (2001)	.006
Spencer/Irvan Daytona (1994)	.008

Source: Working On My Redneck

Write in scientific notation the margin of victory for the given race. [4.2]

33. Earnhardt/Irvan

34. Craven/Busch

Chapter 4 Projects

EXPONENTS AND POLYNOMIALS

GROUP PROJECT

Discovering Pascal's Triangle

Number of People 3

Time Needed 20 minutes

Equipment Paper and pencils

Background The triangular array of numbers shown here is known as Pascal's triangle, after the French philosopher Blaise Pascal (1623–1662).

Look at Pascal's triangle and discover how the numbers in each row of the triangle are obtained from the numbers in the row above it.

Procedure **1.** Once you have discovered how to extend the triangle, write the next two rows.

2. Pascal's triangle can be linked to the Fibonacci sequence by rewriting Pascal's triangle so that the 1's on the left side of the triangle line up under one another, and the other columns are equally spaced to the right of the first column. Rewrite Pascal's triangle as indicated and then look along the diagonals of the new array until you discover how the Fibonacci sequence can be obtained from it.

3. The second diagram shows Pascal's triangle as written in Japan in 1781. Use your knowledge of Pascal's triangle to translate the numbers written in Japanese into our number system. Then write down the Japanese numbers from 1 to 20.

```
        1
      1   1
    1   2   1
   1   3   3   1
  1   4   6   4   1
 1  5  10  10   5   1
```

Pascal's triangle in Japanese (1781)

RESEARCH PROJECT
Working Mathematicians

It may seem at times as if all the mathematicians of note lived 100 or more years ago. However, that is not the case. There are mathematicians doing research today who are discovering new mathematical ideas and extending what is known about mathematics.

Use the Internet to find a mathematician working in the field today. Find out what drew them to mathematics in the first place, what it took for them to be successful, and what they like about their career in mathematics. Then summarize your results into an essay that gives anyone reading it a profile of a working mathematician.

5

Factoring

Introduction

In 570 BC, the Ionian Greek philosopher Pythagoras was born on the Island of Samos in the North Aegean Sea. In 540 BC, he formed a secret society to study mathematics. Society members, known as the Pythagoreans, kept no written record of their work; everything was handed down by spoken word. They influenced not only mathematics, but religion, science, medicine, and music as well. Among other things, they discovered the correlation between musical notes and the reciprocals of counting numbers, $\frac{1}{2}$, $\frac{1}{3}$, $\frac{1}{4}$, and so on.

In this chapter, we will introduce the Pythagorean Theorem. This theorem, named for the often credited discoverer Pythagoras, uses variables to represent the three sides of a triangle. Once we manipulate these variables, the theorem often translates into a quadratic equation we can factor and solve. Factoring and quadratic equations are the primary focus of this chapter.

Divide.

1. $\dfrac{y^3 - 16y^2 + 64y}{y}$ **2.** $\dfrac{5x^3 + 35x^2 + 60x}{5x}$ **3.** $\dfrac{-12x^4 + 48x^3 + 144x^2}{-12x^2}$ **4.** $\dfrac{16x^5 + 20x^4 + 60x^3}{4x^3}$

Multiply

5. $(x - 4)(x + 4)$ **6.** $(x - 6)(x + 6)$ **7.** $(x - 8)(x + 8)$ **8.** $(x^2 - 4)(x^2 + 4)$

9. $x(x^2 + 2x + 4)$ **10.** $2x(4x^2 - 10x + 25)$ **11.** $-2(x^2 + 2x + 4)$ **12.** $5(4x^2 - 10x + 25)$

13. $(x - 2)(x^2 + 2x + 4)$ **14.** $(2x + 5)(4x^2 - 10x + 25)$ **15.** $3x^2(x + 3)(x - 3)$ **16.** $2x^3(x + 5)^2$

17. $y^3(y^2 + 36)$ **18.** $(3a - 4)(2a - 1)$ **19.** $6x(x - 4)(x + 2)$ **20.** $2a^3b(a^2 + 3a + 1)$

Chapter Outline

5.1 The Greatest Common Factor and Factoring by Grouping

OBJECTIVES

A Factor the greatest common factor from a polynomial.

B Factor by grouping.

TICKET TO SUCCESS

Keep these questions in mind as you read through the section. Then respond in your own words and in complete sentences.

1. What is the greatest common factor for a polynomial?
2. How would you use the distributive property when factoring a polynomial?
3. When would you try to factor by grouping?
4. After factoring a polynomial, how can you check your result?

NOTE

As you will see as we progress through the book, factoring is a tool that is used in solving a number of problems. Before seeing how it is used, however, we first must learn how to do it. So, in this section and the two sections that follow, we will be developing our factoring skills.

In Chapter 1, we used the following diagram to illustrate the relationship between multiplication and factoring.

$$\text{Multiplication}$$
$$\text{Factors} \rightarrow 3 \cdot 5 = 15 \leftarrow \text{Product}$$
$$\text{Factoring}$$

A similar relationship holds for multiplication of polynomials. Reading the following diagram from left to right, we say the product of the binomials $x + 2$ and $x + 3$ is the trinomial $x^2 + 5x + 6$. However, if we read in the other direction, we can say that $x^2 + 5x + 6$ factors into the product of $x + 2$ and $x + 3$.

$$\text{Multiplication}$$
$$\text{Factors} \rightarrow (x + 2)(x + 3) = x^2 + 5x + 6 \leftarrow \text{Product}$$
$$\text{Factoring}$$

In this chapter, we develop a systematic method of factoring polynomials.

A Factoring Out the Greatest Common Factor

Picture an archer shooting an arrow into the air. The curved path that the arrow takes from the bow to the ground creates the shape of a parabola, which we will discuss further later in the book. For now, let's say the arrow's path can be represented by the polynomial $-16x^2 + 62x + 8$. To begin factoring this polynomial, we need to find the greatest common factor.

In this section we will apply the distributive property to polynomials to factor from them what is called the *greatest common factor*.

> **Definition**
> The **greatest common factor** for a polynomial is the largest monomial that evenly divides (is a factor of) each term of the polynomial.

We use the term *largest monomial* to mean the monomial with the greatest coefficient and highest power of the variable.

EXAMPLE 1 Find the greatest common factor for the polynomial:

$$3x^5 + 12x^2$$

SOLUTION The terms of the polynomial are $3x^5$ and $12x^2$. The largest number that divides the coefficients is 3, and the highest power of x that is a factor of x^5 and x^2 is x^2. Therefore, the greatest common factor for $3x^5 + 12x^2$ is $3x^2$; that is, $3x^2$ is the largest monomial that divides each term of $3x^5 + 12x^2$.

EXAMPLE 2 Find the greatest common factor for

$$8a^3b^2 + 16a^2b^3 + 20a^3b^3$$

SOLUTION The largest number that divides each of the coefficients is 4. The highest power of the variable that is a factor of a^3b^2, a^2b^3, and a^3b^3 is a^2b^2. The greatest common factor for $8a^3b^2 + 16a^2b^3 + 20a^3b^3$ is $4a^2b^2$. It is the largest monomial that is a factor of each term.

Once we have recognized the greatest common factor of a polynomial, we can apply the distributive property and factor it out of each term. We rewrite the polynomial as the product of its greatest common factor and the polynomial that remains after the greatest common factor has been factored from each term in the original polynomial.

EXAMPLE 3 Factor the greatest common factor from $3x - 15$.

SOLUTION The greatest common factor for the terms $3x$ and 15 is 3. We can rewrite both $3x$ and 15 so that the greatest common factor 3 is showing in each term. It is important to realize that $3x$ means $3 \cdot x$. The 3 and the x are not "stuck" together.

$$3x - 15 = 3 \cdot x - 3 \cdot 5$$

Now, applying the distributive property, we have

$$3 \cdot x - 3 \cdot 5 = 3(x - 5)$$

To check a factoring problem like this, we can multiply 3 and $x - 5$ to get $3x - 15$, which is what we started with. Factoring is simply a procedure by which we change sums and differences into products. In this case, we changed the difference $3x - 15$ into the product $3(x - 5)$. Note, however, that we have not changed the meaning or value of the expression. The expression we end up with is equivalent to the expression we started with.

EXAMPLE 4   Factor the greatest common factor from $5x^3 - 15x^2$.

SOLUTION   The greatest common factor is $5x^2$. We rewrite the polynomial as

$$5x^3 - 15x^2 = 5x^2 \cdot x - 5x^2 \cdot 3$$

Then we apply the distributive property to get

$$5x^2 \cdot x - 5x^2 \cdot 3 = 5x^2(x - 3)$$

To check our work, we simply multiply $5x^2$ and $(x - 3)$ to get $5x^3 - 15x^2$, which is our original polynomial.

EXAMPLE 5   Factor the greatest common factor from $16x^5 - 20x^4 + 8x^3$.

SOLUTION   The greatest common factor is $4x^3$. We rewrite the polynomial so we can see the greatest common factor $4x^3$ in each term. Then we apply the distributive property to factor it out.

$$16x^5 - 20x^4 + 8x^3 = 4x^3 \cdot 4x^2 - 4x^3 \cdot 5x + 4x^3 \cdot 2$$
$$= 4x^3(4x^2 - 5x + 2)$$

EXAMPLE 6   Factor the greatest common factor from $6x^3y - 18x^2y^2 + 12xy^3$.

SOLUTION   The greatest common factor is $6xy$. We rewrite the polynomial in terms of $6xy$ and then apply the distributive property as follows:

$$6x^3y - 18x^2y^2 + 12xy^3 = 6xy \cdot x^2 - 6xy \cdot 3xy + 6xy \cdot 2y^2$$
$$= 6xy(x^2 - 3xy + 2y^2)$$

EXAMPLE 7   Factor the greatest common factor from $3a^2b - 6a^3b^2 + 9a^3b^3$.

SOLUTION   The greatest common factor is $3a^2b$

$$3a^2b - 6a^3b^2 + 9a^3b^3 = 3a^2b(1) - 3a^2b(2ab) + 3a^2b(3ab^2)$$
$$= 3a^2b(1 - 2ab + 3ab^2)$$

B Factoring by Grouping

To develop our next method of factoring, called *factoring by grouping*, we start by examining the polynomial $xc + yc$. The greatest common factor for the two terms is c. Factoring c from each term we have

$$xc + yc = c(x + y)$$

But suppose that c itself was a more complicated expression, such as $a + b$, so that the expression we were trying to factor was $x(a + b) + y(a + b)$, instead of $xc + yc$. The greatest common factor for $x(a + b) + y(a + b)$ is $(a + b)$. Factoring this common factor from each term looks like this:

$$x(a + b) + y(a + b) = (a + b)(x + y)$$

To see how all of this applies to factoring polynomials, consider the polynomial

$$xy + 3x + 2y + 6$$

There is no greatest common factor other than the number 1. However, if we group the terms together two at a time, we can factor an x from the first two terms and a 2 from the last two terms

$$xy + 3x + 2y + 6 = x(y + 3) + 2(y + 3)$$

The expression on the right can be thought of as having two terms: $x(y + 3)$ and $2(y + 3)$. Each of these expressions contains the common factor $y + 3$, which can be factored out using the distributive property

$$x(y + 3) + 2(y + 3) = (y + 3)(x + 2)$$

This last expression is in factored form. Here are some additional examples using the process called factoring by grouping.

EXAMPLE 8   Factor $ax + bx + ay + by$.

SOLUTION   We begin by factoring the greatest common factor x from the first two terms and the greatest common factor y from the last two terms.

$$ax + bx + ay + by = x(a + b) + y(a + b)$$
$$= (a + b)(x + y)$$

To convince yourself that this is factored correctly, multiply the two factors $(a + b)$ and $(x + y)$.   ∎

EXAMPLE 9   Factor by grouping: $3ax - 2a + 15x - 10$

SOLUTION   First, we factor a from the first two terms and 5 from the last two terms. Then, we factor $3x - 2$ from the remaining two expressions.

$$3ax - 2a + 15x - 10 = a(3x - 2) + 5(3x - 2)$$
$$= (3x - 2)(a + 5)$$

Again, multiplying $(3x - 2)$ and $(a + 5)$ will convince you that these are the correct factors.   ∎

EXAMPLE 10   Factor $6x^2 - 3x - 4x + 2$ by grouping.

SOLUTION   The first two terms have $3x$ in common, and the last two terms have either a 2 or a −2 in common. Suppose we factor $3x$ from the first two terms and 2 from the last two terms. We get

$$6x^2 - 3x - 4x + 2 = 3x(2x - 1) + 2(-2x + 1)$$

We can't go any further because there is no common factor that will allow us to factor further. However, if we factor −2, instead of 2, from the last two terms, our problem is solved.

$$6x^2 - 3x - 4x + 2 = 3x(2x - 1) - 2(2x - 1)$$
$$= (2x - 1)(3x - 2)$$

In this case, factoring −2 from the last two terms gives us an expression that can be factored further.   ∎

EXAMPLE 11 Factor $2x^2 + 5ax - 2xy - 5ay$.

SOLUTION From the first two terms we factor x. From the second two terms we must factor $-y$ so that the binomial that remains after we do so matches the binomial produced by the first two terms.

$$2x^2 + 5ax - 2xy - 5ay = x(2x + 5a) - y(2x + 5a)$$

$$= (2x + 5a)(x - y)$$

Another way to accomplish the same result is to use the commutative property to interchange the middle two terms, and then factor by grouping.

$$2x^2 + 5ax - 2xy - 5ay = 2x^2 - 2xy + 5ax - 5ay \qquad \text{Commutative}$$
property

$$= 2x(x - y) + 5a(x - y)$$

$$= (x - y)(2x + 5a)$$

This is the same result we obtained previously.

Problem Set 5.1

Moving Toward Success

"One important key to success is self confidence. An important key to self confidence is preparation."

—Arthur Ashe, 1943–1993, professional tennis player

1. Will becoming overconfident help or hurt your pursuit of success in this class? Explain.
2. Should you still read the chapter even if you understand the concepts being taught in class? Why or why not?

A Factor the following by taking out the greatest common factor. [Examples 1–7]

1. $15x + 25$
2. $14x + 21$
3. $6a + 9$
4. $8a + 10$
5. $4x - 8y$
6. $9x - 12y$
7. $3x^2 - 6x - 9$
8. $2x^2 + 6x + 4$
9. $3a^2 - 3a - 60$
10. $2a^2 - 18a + 28$
11. $24y^2 - 52y + 24$
12. $18y^2 + 48y + 32$
13. $9x^2 - 8x^3$
14. $7x^3 - 4x^2$
15. $13a^2 - 26a^3$
16. $5a^2 - 10a^3$
17. $21x^2y - 28xy^2$
18. $30xy^2 - 25x^2y$

19. $22a^2b^2 - 11ab^2$
20. $15x^3 - 25x^2 + 30x$
21. $7x^3 + 21x^2 - 28x$
22. $16x^4 - 20x^2 - 16x$
23. $121y^4 - 11x^4$
24. $25a^4 - 5b^4$
25. $100x^4 - 50x^3 + 25x^2$
26. $36x^5 + 72x^3 - 81x^2$
27. $8a^2 + 16b^2 + 32c^2$
28. $9a^2 - 18b^2 - 27c^2$
29. $4a^2b - 16ab^2 + 32a^2b^2$
30. $5ab^2 + 10a^2b^2 + 15a^2b$

A Factor the following by taking out the greatest common factor.

31. $121a^3b^2 - 22a^2b^3 + 33a^3b^3$

32. $20a^4b^3 - 18a^3b^4 + 22a^4b^4$

33. $12x^2y^3 - 72x^5y^3 - 36x^4y^4$

34. $49xy - 21x^2y^2 + 35x^3y^3$

B Factor by grouping. [Examples 8–11]

35. $xy + 5x + 3y + 15$

36. $xy + 2x + 4y + 8$

37. $xy + 6x + 2y + 12$

38. $xy + 2y + 6x + 12$

39. $ab + 7a - 3b - 21$

40. $ab + 3b - 7a - 21$

41. $ax - bx + ay - by$

42. $ax - ay + bx - by$

43. $2ax + 6x - 5a - 15$

44. $3ax + 21x - a - 7$

45. $3xb - 4b - 6x + 8$

46. $3xb - 4b - 15x + 20$

47. $x^2 + ax + 2x + 2a$

48. $x^2 + ax + 3x + 3a$

49. $x^2 - ax - bx + ab$

50. $x^2 + ax - bx - ab$

B Factor by grouping. You can group the terms together two at a time or three at a time. Either way will produce the same result.

51. $ax + ay + bx + by + cx + cy$

52. $ax + bx + cx + ay + by + cy$

B Factor the following polynomials by grouping the terms together two at a time. [Examples 8–11]

53. $6x^2 + 9x + 4x + 6$

54. $6x^2 - 9x - 4x + 6$

55. $20x^2 - 2x + 50x - 5$

56. $20x^2 + 25x + 4x + 5$

57. $20x^2 + 4x + 25x + 5$

58. $20x^2 + 4x - 25x - 5$

59. $x^3 + 2x^2 + 3x + 6$

60. $x^3 - 5x^2 - 4x + 20$

61. $6x^3 - 4x^2 + 15x - 10$

62. $8x^3 - 12x^2 + 14x - 21$

63. The greatest common factor of the binomial $3x + 6$ is 3. The greatest common factor of the binomial $2x + 4$ is 2. What is the greatest common factor of their product $(3x + 6)(2x + 4)$ when it has been multiplied out?

64. The greatest common factors of the binomials $4x + 2$ and $5x + 10$ are 2 and 5, respectively. What is the greatest common factor of their product $(4x + 2)$ $(5x + 10)$ when it has been multiplied out?

65. The following factorization is incorrect. Find the mistake, and correct the right-hand side.

$$12x^2 + 6x + 3 = 3(4x^2 + 2x)$$

66. Find the mistake in the following factorization, and then rewrite the right-hand side correctly.
$$10x^2 + 2x + 6 = 2(5x^2 + 3)$$

Getting Ready for the Next Section

Multiply each of the following.

67. $(x - 7)(x + 2)$

68. $(x - 7)(x - 2)$

69. $(x - 3)(x + 2)$ **70.** $(x + 3)(x - 2)$

71. $(x + 3)(x^2 - 3x + 9)$

72. $(x - 2)(x^2 + 2x + 4)$

73. $(2x + 1)(x^2 + 4x - 3)$

74. $(3x + 2)(x^2 - 2x - 4)$

75. $3x^4(6x^3 - 4x^2 + 2x)$

76. $2x^4(5x^3 + 4x^2 - 3x)$

77. $\left(x + \dfrac{1}{3}\right)\left(x + \dfrac{2}{3}\right)$

78. $\left(x + \dfrac{1}{4}\right)\left(x + \dfrac{3}{4}\right)$

79. $(6x + 4y)(2x - 3y)$

80. $(8a - 3b)(4a - 5b)$

81. $(9a + 1)(9a - 1)$

82. $(7b + 1)(7b + 1)$

83. $(x - 9)(x - 9)$

84. $(x - 8)(x - 8)$

85. $(x + 2)(x^2 - 2x + 4)$

86. $(x - 3)(x^2 + 3x + 9)$

Divide.

87. $\dfrac{y^3 - 16y^2 + 64y}{y}$

88. $\dfrac{5x^3 + 35x^2 + 60x}{5x}$

89. $\dfrac{-12x^4 + 48x^3 + 144x^2}{-12x^2}$

90. $\dfrac{16x^5 + 20x^4 + 60x^3}{4x^3}$

91. $\dfrac{-18y^5 + 63y^4 - 108y^3}{-9y^3}$

92. $\dfrac{36y^6 - 66y^5 + 54y^4}{6y^4}$

Subtract.

93. $(5x^2 + 5x - 4) - (3x^2 - 2x + 7)$

94. $(7x^4 - 4x^2 - 5) - (2x^4 - 4x^2 + 5)$

95. Subtract $4x - 5$ from $7x + 3$.

96. Subtract $3x + 2$ from $-6x + 1$.

97. Subtract $2x^2 - 4x$ from $5x^2 - 5$.

98. Subtract $6x^2 + 3$ from $2x^2 - 4x$.

5.2 Factoring Trinomials

OBJECTIVES

A Factor a trinomial whose leading coefficient is the number 1.

B Factor a polynomial by first factoring out the greatest common factor and then factoring the polynomial that remains.

TICKET TO SUCCESS

Keep these questions in mind as you read through the section. Then respond in your own words and in complete sentences.

1. When the leading coefficient of a trinomial is 1, what is the relationship between the other two coefficients and the factors of the trinomial?

2. In words, explain how you would factor a trinomial if the leading coefficient is a number other than one.

3. How can you check to see that you have factored a trinomial correctly?

4. Describe how you would find the factors of $x^2 + 8x + 12$.

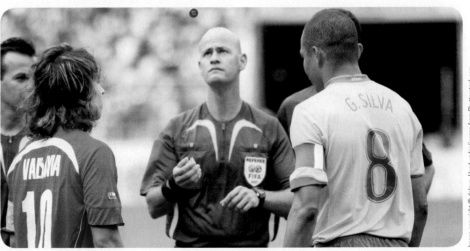

At the beginning of many athletic events, a referee flips a coin to determine which competitor or team goes first. The path that the coin takes can be given by the trinomial $-16x^2 + 29x + 6$. In this section, we will learn how to factor similar trinomials. You will also see this problem again in a later problem set.

Now we will factor trinomials in which the coefficient of the squared term is 1. The more familiar we are with multiplication of binomials, the easier factoring trinomials will be.

Recall multiplication of binomials from Chapter 4:

$$(x + 3)(x + 4) = x^2 + 7x + 12$$
$$(x - 5)(x + 2) = x^2 - 3x - 10$$

Using the FOIL method we learned in Chapter 4, the first term in the answer is the product of the first terms in each binomial. The last term in the answer is the product of the last terms in each binomial. The middle term in the answer comes from adding the product of the outside terms to the product of the inside terms.

Let's have a and b represent real numbers and look at the product of $(x + a)$ and $(x + b)$.

$$(x + a)(x + b) = x^2 + ax + bx + ab$$
$$= x^2 + (a + b)x + ab$$

The coefficient of the middle term is the sum of a and b. The last term is the product of a and b. Writing this as a factoring problem, we have

$$x^2 + \underset{\text{Sum}}{(a + b)}x + \underset{\text{Product}}{ab} = (x + a)(x + b)$$

A Factoring with a Leading Coefficient of 1

To factor a trinomial in which the coefficient of x^2 is 1, we need only find the numbers a and b whose sum is the coefficient of the middle term and whose product is the constant term (last term).

EXAMPLE 1   Factor $x^2 + 8x + 12$.

SOLUTION   The coefficient of x^2 is 1. We need two numbers whose sum is 8 and whose product is 12. The numbers are 6 and 2.

$$x^2 + 8x + 12 = (x + 6)(x + 2)$$

We can easily check our work by multiplying $(x + 6)$ and $(x + 2)$.

$$\text{Check:}\quad (x + 6)(x + 2) = x^2 + 6x + 2x + 12$$
$$= x^2 + 8x + 12$$

NOTE
Again, we can check our results by multiplying our factors to see if their product is the original polynomial.

EXAMPLE 2   Factor $x^2 - 2x - 15$.

SOLUTION   The coefficient of x^2 is again 1. We need to find a pair of numbers whose sum is -2 and whose product is -15. It can be helpful to create a table of products and sums to find our correct pair of numbers. Here are all the possibilities for products that are -15.

Products	Sums
$-1(15) = -15$	$-1 + 15 = 14$
$1(-15) = -15$	$1 + (-15) = -14$
$-5(3) = -15$	$-5 + 3 = -2$
$5(-3) = -15$	$5 + (-3) = 2$

The third line gives us what we want. The factors of $x^2 - 2x - 15$ are $(x - 5)$ and $(x + 3)$.

$$x^2 - 2x - 15 = (x - 5)(x + 3)$$

B Factoring Out the Greatest Common Factor

The first step in factoring any trinomial is to look for the greatest common factor. If the trinomial in question has a greatest common factor other than 1, we factor it out first and then try to factor the trinomial that remains.

EXAMPLE 3 Factor $2x^2 + 10x - 28$.

SOLUTION The coefficient of x^2 is 2. We begin by factoring out the greatest common factor, which is 2.

$$2x^2 + 10x - 28 = 2(x^2 + 5x - 14)$$

Now, we factor the remaining trinomial by finding a pair of numbers whose sum is 5 and whose product is -14. Here are the possibilities:

Products	Sums
$-1(14) = -14$	$-1 + 14 = 13$
$1(-14) = -14$	$1 + (-14) = -13$
$-7(2) = -14$	$-7 + 2 = -5$
$7(-2) = -14$	$7 + (-2) = 5$

From the last line, we see that the factors of $x^2 + 5x - 14$ are $(x + 7)$ and $(x - 2)$. Here is the complete solution:

$$2x^2 + 10x - 28 = 2(x^2 + 5x - 14)$$

$$= 2(x + 7)(x - 2)$$

EXAMPLE 4 Factor $3x^3 - 3x^2 - 18x$.

SOLUTION We begin by factoring out the greatest common factor, which is $3x$. Then we factor the remaining trinomial. Without showing the table of products and sums as we did in Examples 2 and 3, here is the complete solution:

$$3x^3 - 3x^2 - 18x = 3x(x^2 - x - 6)$$

$$= 3x(x - 3)(x + 2)$$

This approach may also be used to factor trinomials with two variables.

EXAMPLE 5 Factor $x^2 + 8xy + 12y^2$.

SOLUTION This time we need two expressions whose product is $12y^2$ and whose sum is $8y$. The two expressions are $6y$ and $2y$ (see Example 1 in this section).

$$x^2 + 8xy + 12y^2 = (x + 6y)(x + 2y)$$

You should convince yourself that these factors are correct by finding their product.

NOTE
Trinomials in which the coefficient of the second-degree term is 1 are the easiest to factor. Success in factoring any type of polynomial is directly related to the amount of time spent working the problems. The more we practice, the more accomplished we become at factoring.

Problem Set 5.2

Moving Toward Success

"Great things are not done by impulse, but by a series of small things brought together."

—Vincent Van Gogh, 1853–1890, Dutch painter

1. Why are the Ticket To Success questions helpful to read before reading the section?

2. Should you read the notes in the side columns of the section pages? Why or why not?

A Factor the following trinomials. [Examples 1–2]

1. $x^2 + 7x + 12$

2. $x^2 + 7x + 10$

3. $x^2 + 3x + 2$

4. $x^2 + 7x + 6$

5. $a^2 + 10a + 21$

6. $a^2 - 7a + 12$

7. $x^2 - 7x + 10$

8. $x^2 - 3x + 2$

9. $y^2 - 10y + 21$

10. $y^2 - 7y + 6$

11. $x^2 - x - 12$

12. $x^2 - 4x - 5$

13. $y^2 + y - 12$

14. $y^2 + 3y - 18$

15. $x^2 + 5x - 14$

16. $x^2 - 5x - 24$

17. $r^2 - 8r - 9$

18. $r^2 - r - 2$

19. $x^2 - x - 30$

20. $x^2 + 8x + 12$

21. $a^2 + 15a + 56$

22. $a^2 - 9a + 20$

23. $y^2 - y - 42$

24. $y^2 + y - 42$

25. $x^2 + 13x + 42$

26. $x^2 - 13x + 42$

B Factor the following problems completely. First, factor out the greatest common factor, and then factor the remaining trinomial. [Examples 3–4]

27. $2x^2 + 6x + 4$

28. $3x^2 - 6x - 9$

29. $3a^2 - 3a - 60$

30. $2a^2 - 18a + 28$

31. $100x^2 - 500x + 600$

32. $100x^2 - 900x + 2,000$

33. $100p^2 - 1,300p + 4,000$

34. $100p^2 - 1,200p + 3,200$

35. $x^4 - x^3 - 12x^2$

36. $x^4 - 11x^3 + 24x^2$

37. $2r^3 + 4r^2 - 30r$

38. $5r^3 + 45r^2 + 100r$

39. $2y^4 - 6y^3 - 8y^2$

40. $3r^3 - 3r^2 - 6r$

41. $x^5 + 4x^4 + 4x^3$

42. $x^5 + 13x^4 + 42x^3$

43. $3y^4 - 12y^3 - 15y^2$

44. $5y^4 - 10y^3 + 5y^2$

45. $4x^4 - 52x^3 + 144x^2$

46. $3x^3 - 3x^2 - 18x$

B Factor the following trinomials. [Example 5]

47. $x^2 + 5xy + 6y^2$

48. $x^2 - 5xy + 6y^2$

49. $x^2 - 9xy + 20y^2$

50. $x^2 + 9xy + 20y^2$

51. $a^2 + 2ab - 8b^2$

52. $a^2 - 2ab - 8b^2$

53. $a^2 - 10ab + 25b^2$

54. $a^2 + 6ab + 9b^2$

55. $a^2 + 10ab + 25b^2$

56. $a^2 - 6ab + 9b^2$

57. $x^2 + 2xa - 48a^2$

58. $x^2 - 3xa - 10a^2$

59. $x^2 - 5xb - 36b^2$

60. $x^2 - 13xb + 36b^2$

61. $x^4 - 5x^2 + 6$

62. $x^6 - 2x^3 - 15$

63. $x^2 - 80x - 2{,}000$

64. $x^2 - 190x - 2{,}000$

65. $x^2 - x + \dfrac{1}{4}$

66. $x^2 - \dfrac{2}{3}x + \dfrac{1}{9}$

67. $x^2 + 0.6x + 0.08$

68. $x^2 + 0.8x + 0.15$

69. If one of the factors of $x^2 + 24x + 128$ is $x + 8$, what is the other factor?

70. If one factor of $x^2 + 260x + 2{,}500$ is $x + 10$, what is the other factor?

71. What polynomial, when factored, gives $(4x + 3)(x - 1)$?

72. What polynomial factors to $(4x - 3)(x + 1)$?

Getting Ready for the Next Section

Multiply using the FOIL method.

73. $(6a + 1)(a + 2)$

74. $(6a - 1)(a - 2)$

75. $(3a + 2)(2a + 1)$

76. $(3a - 2)(2a - 1)$

77. $(6a + 2)(a + 1)$

78. $(3a + 1)(2a + 2)$

Maintaining Your Skills

Simplify each expression. Write using only positive exponents.

79. $\left(-\dfrac{2}{5}\right)^2$

80. $\left(-\dfrac{3}{8}\right)^2$

81. $(3a^3)^2(2a^2)^3$

82. $(-4x^4)^2(2x^5)^4$

83. $\dfrac{(4x)^{-7}}{(4x)^{-5}}$

84. $\dfrac{(2x)^{-3}}{(2x)^{-5}}$

85. $\dfrac{12a^5b^3}{72a^2b^5}$

86. $\dfrac{25x^5y^3}{50x^2y^7}$

87. $\dfrac{15x^{-5}y^3}{45x^2y^5}$

88. $\dfrac{25a^2b^7}{75a^5b^3}$

89. $(-7x^3y)(3xy^4)$

90. $(9a^6b^4)(6a^4b^3)$

91. $(-5a^3b^{-1})(4a^{-2}b^4)$

92. $(-3a^2b^{-4})(6a^5b^{-2})$

93. $(9a^2b^3)(-3a^3b^5)$

94. $(-7a^5b^8)(6a^7b^4)$

5.3 More Trinomials to Factor

OBJECTIVES

A Use trial and error to factor a trinomial whose leading coefficient is other than 1.

B Use the grouping method to factor a trinomial whose leading coefficient is other than 1.

TICKET TO SUCCESS

Keep these questions in mind as you read through the section. Then respond in your own words and in complete sentences.

1. Why would you use the trial and error method of factoring a trinomial?

2. Describe the criteria you would use to set up a table of possible factors of a trinomial.

3. What does it mean if you factor a trinomial and one of your factors has a greatest common factor of 3?

4. Describe how you would use the grouping method to find the factors of $6a^2 + 7a + 2$.

Image copyright © plantongkoh. Used under license from Shutterstock.com

Imagine yourself tossing a rock off a bridge into the river 32 feet below. If the equation that gives the height of a rock above the surface of the water at any time t is

$$h = 32 + 16t - 16t^2$$

then how would you factor the right side of the equation before finding the height at time t?

Notice that the coefficient of the squared term is -16. In the last section, we worked on factoring a trinomial with a leading coefficient of 1. In this section, we will focus more on considering trinomials whose greatest common factor is 1 and whose leading coefficient (the coefficient of the squared term) is a number other than 1. We present two methods for factoring trinomials of this type. The first method involves listing possible factors until the correct pair of factors is found. This requires a certain amount of trial and error. The second method is based on the factoring by grouping process that we covered previously. Either method can be used to factor trinomials whose leading coefficient is a number other than 1.

A Method 1: Factoring $ax^2 + bx + c$ by Trial and Error

Suppose we want to factor the trinomial $2x^2 - 5x - 3$. We know that the factors (if they exist) will be a pair of binomials. The product of their first terms is $2x^2$ and the product of their last term is -3. Let's list all the possible factors along with the trinomial that would result if we were to multiply them together. Remember, the middle term comes from the product of the inside terms plus the product of the outside terms.

Binomial Factors	First Term	Middle Term	Last Term
$(2x - 3)(x + 1)$	$2x^2$	$-x$	-3
$(2x + 3)(x - 1)$	$2x^2$	$+x$	-3
$(2x - 1)(x + 3)$	$2x^2$	$+5x$	-3
$(2x + 1)(x - 3)$	$2x^2$	$-5x$	-3

We can see from the last line that the factors of $2x^2 - 5x - 3$ are $(2x + 1)$ and $(x - 3)$. There is no straightforward way, as there was in the previous section, to find the factors, other than by trial and error or by simply listing all the possibilities. We look for possible factors that, when multiplied, will give the correct first and last terms, and then we see if we can adjust them to give the correct middle term.

EXAMPLE 1   Factor $6a^2 + 7a + 2$.

SOLUTION   We list all the possible pairs of factors that, when multiplied together, give a trinomial whose first term is $6a^2$ and whose last term is $+2$.

Binomial Factors	First Term	Middle Term	Last Term
$(6a + 1)(a + 2)$	$6a^2$	$+13a$	$+2$
$(6a - 1)(a - 2)$	$6a^2$	$-13a$	$+2$
$(3a + 2)(2a + 1)$	$6a^2$	$+7a$	$+2$
$(3a - 2)(2a - 1)$	$6a^2$	$-7a$	$+2$

The factors of $6a^2 + 7a + 2$ are $(3a + 2)$ and $(2a + 1)$ since these give the correct middle term.

$$\text{Check:}\quad (3a + 2)(2a + 1) = 6a^2 + 7a + 2$$

NOTE
Remember, we can always check our results by multiplying the factors we have and comparing that product with our original polynomial.

Notice that in the preceding list, we did not include the factors $(6a + 2)$ and $(a + 1)$. We do not need to try these since the first factor has a 2 common to each term and so could be factored again, giving $2(3a + 1)(a + 1)$. Since our original trinomial, $6a^2 + 7a + 2$, did *not* have a greatest common factor of 2, neither of its factors will.

EXAMPLE 2   Factor $4x^2 - x - 3$.

SOLUTION   We list all the possible factors that, when multiplied, give a trinomial whose first term is $4x^2$ and whose last term is -3.

Binomial Factors	First Term	Middle Term	Last Term
$(4x + 1)(x - 3)$	$4x^2$	$-11x$	-3
$(4x - 1)(x + 3)$	$4x^2$	$+11x$	-3
$(4x + 3)(x - 1)$	$4x^2$	$-x$	-3
$(4x - 3)(x + 1)$	$4x^2$	$+x$	-3
$(2x + 1)(2x - 3)$	$4x^2$	$-4x$	-3
$(2x - 1)(2x + 3)$	$4x^2$	$+4x$	-3

The third line shows that the factors are $(4x + 3)$ and $(x - 1)$.

$$\text{Check:}\quad (4x + 3)(x - 1) = 4x^2 - x - 3$$

You will find that the more practice you have at factoring this type of trinomial, the faster you will get the correct factors. You will pick up some shortcuts along the way, or you may come across a system of eliminating some factors as possibilities. Whatever works best for you is the method you should use. Factoring is a very important tool, and you must be good at it.

Factoring Out the Greatest Common Factor

EXAMPLE 3   Factor $12y^3 + 10y^2 - 12y$.

SOLUTION   We begin by factoring out the greatest common factor, $2y$:

$$12y^3 + 10y^2 - 12y = 2y(6y^2 + 5y - 6)$$

NOTE
Once again, the first step in any factoring problem is to factor out the greatest common factor if it is a number other than 1.

We now list all possible factors of a trinomial with the first term $6y^2$ and last term -6, along with the associated middle terms.

Possible Factors	Middle Term When Multiplied
$(3y + 2)(2y - 3)$	$-5y$
$(3y - 2)(2y + 3)$	$+5y$
$(6y + 1)(y - 6)$	$-35y$
$(6y - 1)(y + 6)$	$+35y$

The second line gives the correct factors. The complete problem is

$$12y^3 + 10y^2 - 12y = 2y(6y^2 + 5y - 6)$$
$$= 2y(3y - 2)(2y + 3)$$

EXAMPLE 4   Factor $30x^2y - 5xy^2 - 10y^3$.

SOLUTION   The greatest common factor is $5y$.

$$30x^2y - 5xy^2 - 10y^3 = 5y(6x^2 - xy - 2y^2)$$
$$= 5y(2x + y)(3x - 2y)$$

B   Method 2: Factoring $ax^2 + bx + c$ by Grouping

Recall that previously we used factoring by grouping to factor the polynomial $6x^2 - 3x - 4x + 2$. We began by factoring $3x$ from the first two terms and -2 from the last two terms. For review, here is the complete problem:

$$6x^2 - 3x - 4x + 2 = 3x(2x - 1) - 2(2x - 1)$$
$$= (2x - 1)(3x - 2)$$

Now, let's back up a little and notice that our original polynomial $6x^2 - 3x - 4x + 2$ can be simplified to $6x^2 - 7x + 2$ by adding $-3x$ and $-4x$. This means that $6x^2 - 7x + 2$ can be factored to $(2x - 1)(3x - 2)$ by the grouping method shown previously. The key to using this process is to rewrite the middle term $-7x$ as $-3x - 4x$.

To generalize this discussion, here are the steps we use to factor trinomials by grouping.

Strategy   Factoring $ax^2 + bx + c$ by Grouping

Step 1: Form the product ac.

Step 2: Find a pair of numbers whose product is ac and whose sum is b.

Step 3: Rewrite the polynomial to be factored so that the middle term bx is written as the sum of two terms whose coefficients are the two numbers found in step 2.

Step 4: Factor by grouping.

EXAMPLE 5   Factor $3x^2 - 10x - 8$ using these steps.

SOLUTION   The trinomial $3x^2 - 10x - 8$ has the form $ax^2 + bx + c$, where $a = 3$, $b = -10$, and $c = -8$.

Step 1:   The product ac is $3(-8) = -24$.

Step 2:   We need to find two numbers whose product is -24 and whose sum is -10. Let's systematically begin to list all the pairs of numbers whose product is -24 to find the pair whose sum is -10.

Product	Sum
$-24(1) = -24$	$-24 + 1 = -23$
$-12(2) = -24$	$-12 + 2 = -10$

We stop here because we have found the pair of numbers whose product is -24 and whose sum is -10. The numbers are -12 and 2.

Step 3:   We now rewrite our original trinomial so the middle term $-10x$ is written as the sum of $-12x$ and $2x$.

$$3x^2 - 10x - 8 = 3x^2 - 12x + 2x - 8$$

Step 4:   Factoring by grouping, we have

$$3x^2 - 12x + 2x - 8 = 3x(x - 4) + 2(x - 4)$$
$$= (x - 4)(3x + 2)$$

We can check our work by multiplying $x - 4$ and $3x + 2$ to get $3x^2 - 10x - 8$.   ■

EXAMPLE 6   Factor $4x^2 - x - 3$.

SOLUTION   In this case, $a = 4$, $b = -1$, and $c = -3$. The product ac is $4(-3) = -12$. We need a pair of numbers whose product is -12 and whose sum is -1. We begin listing pairs of numbers whose product is -12 and whose sum is -1.

Product	Sum
$-12(1) = -12$	$-12 + 1 = -11$
$-6(2) = -12$	$-6 + 2 = -4$
$-4(3) = -12$	$-4 + 3 = -1$

We stop here because we have found the pair of numbers for which we are looking. They are -4 and 3. Next, we rewrite the middle term $-x$ as the sum $-4x + 3x$ and proceed to factor by grouping.

$$4x^2 - x - 3 = 4x^2 - 4x + 3x - 3$$
$$= 4x(x - 1) + 3(x - 1)$$
$$= (x - 1)(4x + 3)$$

Compare this procedure and the result with those shown in Example 2 of this section.   ■

EXAMPLE 7   Factor $8x^2 - 2x - 15$.

SOLUTION   The product ac is $8(-15) = -120$. There are many pairs of numbers with the product of -120. We are looking for the pair with the sum that is also -2. The numbers are -12 and 10. Writing $-2x$ as $-12x + 10x$ and then factoring by grouping, we have

$$8x^2 - 2x - 15 = 8x^2 - 12x + 10x - 15$$
$$= 4x(2x - 3) + 5(2x - 3)$$

$$= (2x - 3)(4x + 5)$$

EXAMPLE 8 A ball is tossed into the air with an upward velocity of 16 feet per second from the top of a building 32 feet high. The equation that gives the height of the ball above the ground at any time t is

$$h = 32 + 16t - 16t^2$$

Factor the right side of this equation and then find h when t is 2.

SOLUTION We begin by factoring out the greatest common factor, 16. Then, we factor the trinomial that remains.

$$h = 32 + 16t - 16t^2$$
$$h = 16(2 + t - t^2)$$
$$h = 16(2 - t)(1 + t)$$

Letting $t = 2$ in the equation, we have

$$h = 16(0)(3) = 0$$

When t is 2, h is 0.

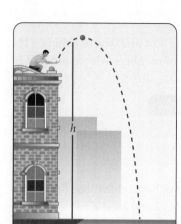

Problem Set 5.3

Moving Toward Success

"Most people never run far enough on their first wind to find out they've got a second. Give your dreams all you've got and you'll be amazed at the energy that comes out of you."

—William James, 1842–1910, American psychologist and philosopher

1. Should you put less time into your homework if you are doing well in this class?

2. How will keeping the same studying schedule you started this class with benefit you?

A Factor the following trinomials. [Examples 1–2, 5–7]

1. $2x^2 + 7x + 3$

2. $2x^2 + 5x + 3$

3. $2a^2 - a - 3$

4. $2a^2 + a - 3$

5. $3x^2 + 2x - 5$

6. $3x^2 - 2x - 5$

7. $3y^2 - 14y - 5$

8. $3y^2 + 14y - 5$

9. $6x^2 + 13x + 6$

10. $6x^2 - 13x + 6$

11. $4x^2 - 12xy + 9y^2$

12. $4x^2 + 12xy + 9y^2$

13. $4y^2 - 11y - 3$

14. $4y^2 + y - 3$

15. $20x^2 - 41x + 20$

16. $20x^2 + 9x - 20$

17. $20a^2 + 48ab - 5b^2$

18. $20a^2 + 29ab + 5b^2$

19. $20x^2 - 21x - 5$

20. $20x^2 - 48x - 5$

21. $12m^2 + 16m - 3$

22. $12m^2 + 20m + 3$

23. $20x^2 + 37x + 15$

24. $20x^2 + 13x - 15$

25. $12a^2 - 25ab + 12b^2$

26. $12a^2 + 7ab - 12b^2$

27. $3x^2 - xy - 14y^2$

28. $3x^2 + 19xy - 14y^2$

29. $14x^2 + 29x - 15$

30. $14x^2 + 11x - 15$

31. $6x^2 - 43x + 55$

32. $6x^2 - 7x - 55$

33. $15t^2 - 67t + 38$

34. $15t^2 - 79t - 34$

B Factor each of the following completely. Look first for the greatest common factor. [Examples 3–7]

35. $4x^2 + 2x - 6$

36. $6x^2 - 51x + 63$

37. $24a^2 - 50a + 24$

38. $18a^2 + 48a + 32$

39. $10x^3 - 23x^2 + 12x$

40. $10x^4 + 7x^3 - 12x^2$

41. $6x^4 - 11x^3 - 10x^2$

42. $6x^3 + 19x^2 + 10x$

43. $10a^3 - 6a^2 - 4a$

44. $6a^3 + 15a^2 + 9a$

45. $15x^3 - 102x^2 - 21x$

46. $2x^4 - 24x^3 + 64x^2$

47. $35y^3 - 60y^2 - 20y$

48. $14y^4 - 32y^3 + 8y^2$

49. $15a^4 - 2a^3 - a^2$

50. $10a^5 - 17a^4 + 3a^3$

51. $12x^2y - 34xy^2 + 14y^3$

52. $12x^2y - 46xy^2 + 14y^3$

53. Evaluate the expression $2x^2 + 7x + 3$ and the expression $(2x + 1)(x + 3)$ for $x = 2$.

54. Evaluate the expression $2a^2 - a - 3$ and the expression $(2a - 3)(a + 1)$ for $a = 5$.

55. What polynomial factors to $(2x + 3)(2x - 3)$?

56. What polynomial factors to $(5x + 4)(5x - 4)$?

57. What polynomial factors to $(x + 3)(x - 3)(x^2 + 9)$?

58. What polynomial factors to $(x + 2)(x - 2)(x^2 + 4)$?

59. One factor of $12x^2 - 71x + 105$ is $x - 3$. Find the other factor.

60. One factor of $18x^2 + 121x - 35$ is $x + 7$. Find the other factor.

61. One factor of $54x^2 + 111x + 56$ is $6x + 7$. Find the other factor.

62. One factor of $63x^2 + 110x + 48$ is $7x + 6$. Find the other factor.

63. One factor of $16t^2 - 64t + 48$ is $t - 1$. Find the other factors, then write the polynomial in factored form.

64. One factor of $16t^2 - 32t + 12$ is $2t - 1$. Find the other factors, then write the polynomial in factored form.

Applying the Concepts [Example 8]

Skyscrapers The chart shows the heights of some of the tallest buildings in the world. Use the chart to answer Problems 65 and 66.

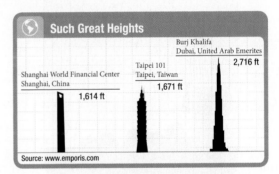

65. If you drop an object off the top of the Burj Khalifa tower, the height that the object is from the ground after t seconds is given by the equation $h = 2{,}716 - 16t^2$. Factor the right hand side of this equation. How far from the ground is the object after 6 seconds?

66. If you throw an object off the top of the Shanghai World Financial Center with an initial upward velocity of 6 feet per second, the approximate height that the object is from the ground after t seconds is given by the equation $h = 1{,}614 + 6t - 16t^2$. Factor this equation.

67. Archery Margaret shoots an arrow into the air. The equation for the height (in feet) of the tip of the arrow is

$$h = 8 + 62t - 16t^2$$

Time t (seconds)	Height h (feet)
0	
1	
2	
3	
4	

a. Factor the right side of this equation.

b. Fill in the table for various heights of the arrow, using the factored form of the equation.

68. Coin Toss At the beginning of every football game, the referee flips a coin to see who will kick off. The equation that gives the height (in feet) of the coin tossed in the air is:

$$h = 6 + 29t - 16t^2$$

a. Factor the right side of this equation.

b. Use the factored form of the equation to find the height of the quarter after 0 seconds, 1 second, and 2 seconds.

Getting Ready for the Next Section

Multiply each of the following.

69. $(x + 3)(x - 3)$
70. $(x - 4)(x + 4)$
71. $(x + 5)(x - 5)$
72. $(x - 6)(x + 6)$
73. $(x + 7)(x - 7)$
74. $(x - 8)(x + 8)$
75. $(x + 9)(x - 9)$
76. $(x - 10)(x + 10)$

77. $(2x - 3y)(2x + 3y)$
78. $(5x - 6y)(5x + 6y)$
79. $(x^2 + 4)(x + 2)(x - 2)$
80. $(x^2 + 9)(x + 3)(x - 3)$
81. $(x + 3)^2$
82. $(x - 4)^2$
83. $(x + 5)^2$
84. $(x - 6)^2$
85. $(x + 7)^2$
86. $(x - 8)^2$
87. $(x + 9)^2$
88. $(x - 10)^2$
89. $(2x + 3)^2$
90. $(3x - y)^2$
91. $(4x - 2y)^2$
92. $(5x - 6y)^2$

Maintaining Your Skills

Perform the following additions and subtractions.

93. $(6x^3 - 4x^2 + 2x) + (9x^2 - 6x + 3)$
94. $(6x^3 - 4x^2 + 2x) - (9x^2 - 6x + 3)$
95. $(-7x^4 + 4x^3 - 6x) + (8x^4 + 7x^3 - 9)$
96. $(-7x^4 + 4x^3 - 6x) - (8x^4 + 7x^3 - 9)$
97. $(2x^5 + 3x^3 + 4x) + (5x^3 - 6x - 7)$
98. $(2x^5 + 3x^3 + 4x) - (5x^3 - 6x - 7)$
99. $(-8x^5 - 5x^4 + 7) + (7x^4 + 2x^2 + 5)$
100. $(-8x^5 - 5x^4 + 7) - (7x^4 + 2x^2 + 5)$

101. $\dfrac{24x^3y^7}{6x^{-2}y^4} + \dfrac{27x^{-2}y^{10}}{9x^{-7}y^7}$

102. $\dfrac{15x^8y^4}{5x^2y^2} - \dfrac{4x^7y^5}{2xy^3}$

103. $\dfrac{18a^5b^9}{3a^3b^6} - \dfrac{48a^{-3}b^{-1}}{16a^{-5}b^{-4}}$

104. $\dfrac{54a^{-3}b^5}{6a^{-7}b^{-2}} - \dfrac{32a^6b^5}{8a^2b^{-2}}$

TICKET TO SUCCESS

Keep these questions in mind as you read through the section. Then respond in your own words and in complete sentences.

1. Describe how you factor the difference of two squares.
2. What is a perfect square trinomial?
3. How do you know when you've factored completely?
4. Describe how you would factor $25x^2 + 60x + 36$.

Suppose a bakery can bake x hundred cookies for a total cost of $C = 100x^2 + 600x + 900$. Upon factoring out the greatest common factor of 100, we find the remaining equation contains a special trinomial, which we will discuss in this section, known as a perfect square trinomial.

$$C = 100x^2 + 600x + 900$$

$$C = 100(x^2 + 6x + 9)$$

But first, we will examine another special product called the difference of two squares.

In Chapter 4, we listed the following three special products:

$$(a + b)^2 = (a + b)(a + b) = a^2 + 2ab + b^2$$

$$(a - b)^2 = (a - b)(a - b) = a^2 - 2ab + b^2$$

$$(a + b)(a - b) = a^2 - b^2$$

Since factoring is the reverse of multiplication, we can also consider the three special products as three special factorings:

$$a^2 + 2ab + b^2 = (a + b)^2$$

$$a^2 - 2ab + b^2 = (a - b)^2$$

$$a^2 - b^2 = (a + b)(a - b)$$

Any trinomial of the form $a^2 + 2ab + b^2$ or $a^2 - 2ab + b^2$ can be factored by the methods of Section 5.3. The last line is the factoring to obtain the difference of two squares. The difference of two squares always factors in this way. Again, these are patterns you must be able to recognize on sight.

A The Difference of Two Squares

EXAMPLE 1 Factor $16x^2 - 25$.

SOLUTION We can see that the first term is a perfect square, and the last term is also. This fact becomes even more obvious if we rewrite the problem as

$$16x^2 - 25 = (4x)^2 - (5)^2$$

The first term is the square of the quantity $4x$, and the last term is the square of 5. The completed problem looks like this:

$$16x^2 - 25 = (4x)^2 - (5)^2$$
$$= (4x + 5)(4x - 5)$$

To check our results, we multiply.

$$(4x + 5)(4x - 5) = 16x^2 + 20x - 20x - 25$$
$$= 16x^2 - 25$$ ∎

EXAMPLE 2 Factor $36a^2 - 1$.

SOLUTION We rewrite the two terms to show they are perfect squares and then factor. Remember, 1 is its own square, $1^2 = 1$.

$$36a^2 - 1 = (6a)^2 - (1)^2$$
$$= (6a + 1)(6a - 1)$$

To check our results, we multiply.

$$(6a + 1)(6a - 1) = 36a^2 + 6a - 6a - 1$$
$$= 36a^2 - 1$$ ∎

EXAMPLE 3 Factor $x^4 - y^4$.

SOLUTION x^4 is the perfect square $(x^2)^2$, and y^4 is $(y^2)^2$.

$$x^4 - y^4 = (x^2)^2 - (y^2)^2$$
$$= (x^2 - y^2)(x^2 + y^2)$$

NOTE
If you think the sum of two squares $x^2 + y^2$ factors, you should try it. Write down the factors you think it has, and then multiply them using the FOIL method. You won't get $x^2 + y^2$.

Notice the factor $(x^2 - y^2)$ is itself the difference of two squares and therefore can be factored again. The factor $(x^2 + y^2)$ is the *sum* of two squares and cannot be factored again. The complete solution is this:

$$x^4 - y^4 = (x^2)^2 - (y^2)^2$$
$$= (x^2 - y^2)(x^2 + y^2)$$
$$= (x + y)(x - y)(x^2 + y^2)$$ ∎

B Perfect Square Trinomials

EXAMPLE 4 Factor $25x^2 - 60x + 36$.

SOLUTION Although this trinomial can be factored by the grouping or trial and error method we used in Section 5.3, we notice that the first and last terms are the perfect squares $(5x)^2$ and $(6)^2$. Before going through the method for factoring trinomials by

NOTE

As we have indicated before, perfect square trinomials like the ones in Examples 4 and 5 can be factored by the methods developed in previous sections. Recognizing that they factor to binomial squares simply saves time in factoring.

listing all possible factors, we can check to see if $25x^2 - 60x + 36$ factors to $(5x - 6)^2$. We need only multiply to check.

$$(5x - 6)^2 = (5x - 6)(5x - 6)$$
$$= 25x^2 - 30x - 30x + 36$$
$$= 25x^2 - 60x + 36$$

The trinomial $25x^2 - 60x + 36$ factors to $(5x - 6)(5x - 6) = (5x - 6)^2$. ■

EXAMPLE 5   Factor $5x^2 + 30x + 45$.

SOLUTION   We begin by factoring out the greatest common factor, which is 5. Then we notice that the trinomial that remains is a perfect square trinomial:

$$5x^2 + 30x + 45 = 5(x^2 + 6x + 9)$$
$$= 5(x + 3)^2$$ ■

EXAMPLE 6   Factor $(x - 3)^2 - 25$.

SOLUTION   This example has the form $a^2 - b^2$, where a is $x - 3$ and b is 5. We factor it according to the formula for the difference of two squares.

$$(x - 3)^2 - 25 = (x - 3)^2 - 5^2 \qquad \text{Write 25 as } 5^2$$
$$= [(x - 3) - 5][(x - 3) + 5] \qquad \text{Factor}$$
$$= (x - 8)(x + 2) \qquad \text{Simplify}$$

Notice in this example we could have expanded $(x - 3)^2$, subtracted 25, and then factored to obtain the same result.

$$(x - 3)^2 - 25 = x^2 - 6x + 9 - 25 \qquad \text{Expand } (x - 3)^2$$
$$= x^2 - 6x - 16 \qquad \text{Simplify}$$
$$= (x - 8)(x + 2) \qquad \text{Factor}$$ ■

Problem Set 5.4

Moving Toward Success

"We are what we repeatedly do. Excellence, then, is not an act, but a habit."

—Aristotle, 384 BC–322 BC, Greek philosopher

1. Why will increasing the effectiveness of the time you spend learning help you?

2. Should you still add to your list of difficult work problems this far into the class? Why or why not?

A Factor the following. [Examples 1–3]

1. $x^2 - 9$

2. $x^2 - 25$

3. $a^2 - 36$

4. $a^2 - 64$

5. $x^2 - 49$

6. $x^2 - 121$

7. $4a^2 - 16$

8. $4a^2 + 16$

9. $9x^2 + 25$

10. $16x^2 - 36$

11. $25x^2 - 169$

12. $x^2 - y^2$

13. $9a^2 - 16b^2$

14. $49a^2 - 25b^2$

15. $9 - m^2$

16. $16 - m^2$

17. $25 - 4x^2$

18. $36 - 49y^2$

19. $2x^2 - 18$

20. $3x^2 - 27$

21. $32a^2 - 128$

22. $3a^3 - 48a$

23. $8x^2y - 18y$

24. $50a^2b - 72b$

25. $a^4 - b^4$

26. $a^4 - 16$

27. $16m^4 - 81$

28. $81 - m^4$

29. $3x^3y - 75xy^3$

30. $2xy^3 - 8x^3y$

B Factor the following. [Examples 4–6]

31. $x^2 - 2x + 1$ **32.** $x^2 - 6x + 9$

33. $x^2 + 2x + 1$ **34.** $x^2 + 6x + 9$

35. $a^2 - 10a + 25$ **36.** $a^2 + 10a + 25$

37. $y^2 + 4y + 4$ **38.** $y^2 - 8y + 16$

39. $x^2 - 4x + 4$ **40.** $x^2 + 8x + 16$

41. $m^2 - 12m + 36$ **42.** $m^2 + 12m + 36$

43. $4a^2 + 12a + 9$ **44.** $9a^2 - 12a + 4$

45. $49x^2 - 14x + 1$ **46.** $64x^2 - 16x + 1$

47. $9y^2 - 30y + 25$ **48.** $25y^2 + 30y + 9$

49. $x^2 + 10xy + 25y^2$

50. $25x^2 + 10xy + y^2$ **51.** $9a^2 + 6ab + b^2$

52. $9a^2 - 6ab + b^2$

53. $y^2 - 3y + \dfrac{9}{4}$ **54.** $y^2 + 3y + \dfrac{9}{4}$

55. $a^2 + a + \dfrac{1}{4}$ **56.** $a^2 - 5a + \dfrac{25}{4}$

57. $x^2 - 7x + \dfrac{49}{4}$ **58.** $x^2 + 9x + \dfrac{81}{4}$

59. $x^2 - \dfrac{3}{4}x + \dfrac{9}{64}$ **60.** $x^2 - \dfrac{3}{2}x + \dfrac{9}{16}$

Factor the following by first factoring out the greatest common factor.

61. $3a^2 + 18a + 27$ **62.** $4a^2 - 16a + 16$

63. $2x^2 + 20xy + 50y^2$

64. $3x^2 + 30xy + 75y^2$

65. $x^3 + 4x^2 + 4x$ **66.** $a^3 - 10a^2 + 25a$

67. $y^4 - 8y^3 + 16y^2$

68. $x^4 + 12x^3 + 36x^2$

69. $5x^3 + 30x^2y + 45xy^2$

70. $12x^2y - 36xy^2 + 27y^3$

71. $12y^4 - 60y^3 + 75y^2$

72. $18a^4 - 12a^3b + 2a^2b^2$

Factor by grouping the first three terms together.

73. $x^2 + 6x + 9 - y^2$

74. $x^2 + 10x + 25 - y^2$

75. $x^2 + 2xy + y^2 - 9$

76. $a^2 + 2ab + b^2 - 25$

77. Find a value for b so that the polynomial $x^2 + bx + 49$ factors to $(x + 7)^2$.

78. Find a value of b so that the polynomial $x^2 + bx + 81$ factors to $(x + 9)^2$.

79. Find the value of c for which the polynomial $x^2 + 10x + c$ factors to $(x + 5)^2$.

80. Find the value of a for which the polynomial $ax^2 + 12x + 9$ factors to $(2x + 3)^2$.

Getting Ready for the Next Section

Multiply each of the following.

81. a. 1^3 **b.** 2^3
     c. 3^3 **d.** 4^3
     e. 5^3

82. a. $(-1)^3$ **b.** $(-2)^3$
     c. $(-3)^3$ **d.** $(-4)^3$
     e. $(-5)^3$

83. a. $x(x^2 - x + 1)$ **b.** $1(x^2 - x + 1)$
     c. $(x + 1)(x^2 - x + 1)$

84. a. $x(x^2 + x + 1)$
     b. $-1(x^2 + x + 1)$
     c. $(x - 1)(x^2 + x + 1)$

85. a. $x(x^2 - 2x + 4)$
     b. $2(x^2 - 2x + 4)$
     c. $(x + 2)(x^2 - 2x + 4)$

86. a. $x(x^2 + 2x + 4)$
     b. $-2(x^2 + 2x + 4)$
     c. $(x - 2)(x^2 + 2x + 4)$

87. a. $x(x^2 - 3x + 9)$
     b. $3(x^2 - 3x + 9)$
     c. $(x + 3)(x^2 - 3x + 9)$

88. a. $x(x^2 + 3x + 9)$
     b. $-3(x^2 + 3x + 9)$
     c. $(x - 3)(x^2 + 3x + 9)$

89. a. $x(x^2 - 4x + 16)$

　　b. $4(x^2 - 4x + 16)$

　　c. $(x + 4)(x^2 - 4x + 16)$

90. a. $x(x^2 + 4x + 16)$

　　b. $-4(x^2 + 4x + 16)$

　　c. $(x - 4)(x^2 + 4x + 16)$

91. a. $x(x^2 - 5x + 25)$

　　b. $5(x^2 - 5x + 25)$

　　c. $(x + 5)(x^2 - 5x + 25)$

92. a. $x(x^2 + 5x + 25)$

　　b. $-5(x^2 + 5x + 25)$

　　c. $(x - 5)(x^2 + 5x + 25)$

Maintaining Your Skills

93. Picture Messaging The graph shows the number of text messages Mike sent each month in 2008. If Verizon Wireless charged $6.05 for the month of August, how much were they charging Mike for picture messages? Round to the nearest cent.

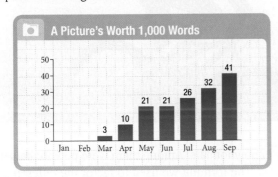

94. Marathon The chart shows the number of participants in the top five marathons in the United States. What was the average number of participants in the five races?

Divide.

95. $\dfrac{24y^3 - 36y^2 - 18y}{6y}$

96. $\dfrac{77y^3 + 35y^2 + 14y}{-7y}$

97. $\dfrac{48x^7 - 36x^5 + 12x^2}{4x^2}$

98. $\dfrac{-50x^5 + 15x^4 + 10x^2}{5x^2}$

99. $\dfrac{18x^7 + 12x^6 - 6x^5}{-3x^4}$

100. $\dfrac{-64x^5 - 18x^4 - 56x^3}{2x^3}$

101. $\dfrac{-42x^5 + 24x^4 - 66x^2}{6x^2}$

102. $\dfrac{63x^7 - 27x^6 - 99x^5}{-9x^4}$

Use long division to divide.

103. $\dfrac{x^2 - 5x + 8}{x - 3}$

104. $\dfrac{x^2 + 7x + 12}{x + 4}$

105. $\dfrac{6x^2 + 5x + 3}{2x + 3}$

106. $\dfrac{x^3 + 27}{x + 3}$

Extending the Concepts

Factor.

107. $t^2 - \dfrac{2}{5}t + \dfrac{1}{25}$

108. $t^2 + \dfrac{6}{5}t + \dfrac{9}{25}$

The Sum and Difference of Two Cubes

TICKET TO SUCCESS

Keep these questions in mind as you read through the section. Then respond in your own words and in complete sentences.

1. Compare and contrast the formula for the sum of two cubes with the formula for the difference of two cubes.
2. Why are the numbers 8, 27, 64, and 125 used so frequently in this section?
3. List the cubes of the numbers 1 through 10.
4. Why would you use the difference of two squares instead of the difference of two cubes to factor $x^6 - y^6$?

Image copyright © Rachael Goldberg. Used under license from Shutterstock.com

Previously, we factored a variety of polynomials. Among them were polynomials that were the difference of two squares. The formula we used to factor the difference of two squares looks like this:

$$a^2 - b^2 = (a + b)(a - b)$$

If we ran across a binomial that had the form of the difference of two squares, we factored it by applying this formula. For example, to factor $x^2 - 25$, we simply notice that it can be written in the form $x^2 - 5^2$, which looks like the difference of two squares. According to the formula above, this binomial factors into $(x + 5)(x - 5)$.

A Factoring the Sum and Difference of Two Cubes

A company that sells mountain biking equipment has just received an order for some new brake pads and pedals. The brake pads are packaged in a box with a volume of a^3 and the pedals are packaged in a box with a volume of b^3. In order to figure out the minimum volume needed for a shipping box to contain both parts, we need to set up the following expression:

$$a^3 + b^3$$

This expression can also be called the sum of two cubes.

In this section, we want to use two new formulas that will allow us to factor the sum and difference of two cubes. For example, we want to factor the binomial $x^3 - 8$, which is the difference of two cubes. To see that it is the difference of two cubes, notice that it can be written $x^3 - 2^3$. We also want to factor $y^3 + 27$, which is the sum of two cubes. (To see this, notice that $y^3 + 27$ can be written as $y^3 + 3^3$.)

The formulas that allow us to factor the sum of two cubes and the difference of two cubes are not as simple as the formula for factoring the difference of two squares. Here is what they look like:

$$a^3 + b^3 = (a + b)(a^2 - ab + b^2) \qquad \text{Sum of two cubes}$$
$$a^3 - b^3 = (a - b)(a^2 + ab + b^2) \qquad \text{Difference of two cubes}$$

Let's begin our work with these two formulas by showing that they are true. To do so, we multiply out the right side of each formula.

EXAMPLE 1   Verify the formula for a sum of two cubes and the formula for a difference of two cubes.

SOLUTION   We verify the formulas by multiplying the right sides and comparing the results with the left sides.

$$
\begin{array}{ll}
\begin{array}{r}
a^2 - ab + b^2 \\
a\ \ + b \\
\hline
a^3 - a^2b + ab^2 \\
a^2b - ab^2 + b^3 \\
\hline
a^3 \qquad\quad\ + b^3
\end{array}
&
\begin{array}{r}
a^2 + ab\ + b^2 \\
a\ \ - b \\
\hline
a^3 + a^2b + ab^2 \\
- a^2b - ab^2 - b^3 \\
\hline
a^3 \qquad\quad\ - b^3
\end{array}
\end{array}
$$

The first formula is correct.   The second formula is correct.   ∎

Here are some examples that use the formulas for factoring the sum and difference of two cubes.

EXAMPLE 2   Factor $x^3 - 8$.

SOLUTION   Since the two terms are perfect cubes, we write them as such and apply the formula.

$$
\begin{aligned}
x^3 - 8 &= x^3 - 2^3 \\
&= (x - 2)(x^2 + 2x + 2^2) \\
&= (x - 2)(x^2 + 2x + 4)
\end{aligned}
$$

EXAMPLE 3   Factor $y^3 + 27$.

SOLUTION   Proceeding as we did in Example 2, we first write 27 as 3^3. Then, we apply the formula for factoring the sum of two cubes, which is $a^3 + b^3 = (a + b)(a^2 - ab + b^2)$.

$$
\begin{aligned}
y^3 + 27 &= y^3 + 3^3 \\
&= (y + 3)(y^2 - 3y + 3^2) \\
&= (y + 3)(y^2 - 3y + 9)
\end{aligned}
$$

EXAMPLE 4   Factor $64 + t^3$.

SOLUTION   The first term is the cube of 4 and the second term is the cube of t.

$$
\begin{aligned}
64 + t^3 &= 4^3 + t^3 \\
&= (4 + t)(16 - 4t + t^2)
\end{aligned}
$$

EXAMPLE 5 Factor $27x^3 + 125y^3$.

SOLUTION Writing both terms as perfect cubes, we have

$$27x^3 + 125y^3 = (3x)^3 + (5y)^3$$
$$= (3x + 5y)(9x^2 - 15xy + 25y^2)$$

EXAMPLE 6 Factor $a^3 - \dfrac{1}{8}$.

SOLUTION The first term is the cube of a, whereas the second term is the cube of $\dfrac{1}{2}$.

$$a^3 - \frac{1}{8} = a^3 - \left(\frac{1}{2}\right)^3$$
$$= \left(a - \frac{1}{2}\right)\left(a^2 + \frac{1}{2}a + \frac{1}{4}\right)$$

EXAMPLE 7 Factor $x^6 - y^6$.

SOLUTION To begin, we have a choice of how we want to write the two terms. We can write the expression as the difference of two squares, $(x^3)^2 - (y^3)^2$, or as the difference of two cubes, $(x^2)^3 - (y^2)^3$. It is easier to use the difference of two squares if we have a choice.

$$x^6 - y^6 = (x^3)^2 - (y^3)^2$$
$$= (x^3 - y^3)(x^3 + y^3)$$
$$= (x - y)(x^2 + xy + y^2)(x + y)(x^2 - xy + y^2)$$

Problem Set 5.5

Moving Toward Success

"You've got to say, 'I think that if I keep working at this and want it badly enough I can have it.' It's called perseverance."

—Lee Iacocca, 1924–present, American businessman and automobile executive

1. How do you use your list of difficult problems?
2. Why is it important to go into an exam knowing you can work any problem on your list of difficult problems?

A Factor each of the following as the sum or difference of two cubes. [Examples 1–7]

1. $x^3 - y^3$

2. $x^3 + y^3$

3. $a^3 + 8$ 4. $a^3 - 8$

5. $27 + x^3$

6. $27 - x^3$

7. $y^3 - 1$ 8. $y^3 + 1$

9. $y^3 - 64$

10. $y^3 + 64$

11. $125h^3 - t^3$

12. $t^3 + 125h^3$

13. $x^3 - 216$

14. $216 + x^3$

15. $2y^3 - 54$

16. $81 + 3y^3$

17. $2a^3 - 128b^3$

18. $128a^3 + 2b^3$

19. $2x^3 + 432y^3$

20. $432x^3 - 2y^3$

21. $10a^3 - 640b^3$

22. $640a^3 + 10b^3$

23. $10r^3 - 1{,}250$

24. $10r^3 + 1{,}250$

25. $64 + 27a^3$

26. $27 - 64a^3$

27. $8x^3 - 27y^3$

28. $27x^3 - 8y^3$

29. $t^3 + \dfrac{1}{27}$

30. $t^3 - \dfrac{1}{27}$

31. $27x^3 - \dfrac{1}{27}$

32. $8x^3 + \dfrac{1}{8}$

33. $64a^3 + 125b^3$

34. $125a^3 - 27b^3$

35. $\dfrac{1}{8}x^3 - \dfrac{1}{27}y^3$

36. $\dfrac{1}{27}x^3 + \dfrac{1}{8}y^3$

37. $a^6 - b^6$

38. $x^6 - 64y^6$

39. $64x^6 - y^6$

40. $x^6 - (3y)^6$

41. $x^6 - (5y)^6$

42. $(4x)^6 - (7y)^6$

Getting Ready for the Next Section

Multiply each of the following.

43. $2x^3(x + 2)(x - 2)$

44. $3x^2(x + 3)(x - 3)$

45. $3x^2(x - 3)^2$

46. $2x^3(x + 5)^2$

47. $y(y^2 + 25)$ **48.** $y^3(y^2 + 36)$

49. $(5a - 2)(3a + 1)$

50. $(3a - 4)(2a - 1)$

51. $4x^2(x - 5)(x + 2)$

52. $6x(x - 4)(x + 2)$

53. $2ab^3(b^2 - 4b + 1)$

54. $2a^3b(a^2 + 3a + 1)$

Maintaining Your Skills

55. U.S. Energy The bar chart shows where Americans get their energy. If America uses 101.6 quadrillion Btu of energy, how many Btu did we use from Petroleum?

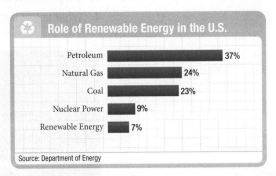

56. Gas Prices The chart shows the breakdown in gas prices. If gasoline costs $4.05, how much is paid for the actual oil? Round to the nearest cent.

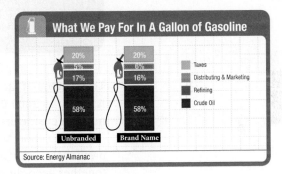

Solve each equation for x.

57. $2x - 6y = 8$

58. $-3x + 9y = 12$

59. $4x - 6y = 8$

60. $-20x + 15y = -10$

Solve each equation for y.

61. $3x - 6y = -18$

62. $-3x + 9y = -18$

63. $4x - 6y = 24$

64. $-20x + 5y = -10$

OBJECTIVES

A Factor a variety of polynomials.

TICKET TO SUCCESS

Keep these questions in mind as you read through the section. Then respond in your own words and in complete sentences.

1. What is the first step in factoring any polynomial?
2. If a polynomial has two terms, what method of factoring should you try?
3. If a polynomial has four terms, what method of factoring should you try?
4. What is the last step in factoring any polynomial?

The process of factoring has been used by mathematicians for thousands of years. As early as 2000 BC, the Babylonians were factoring polynomials by carving their numeric characters into stone tablets.

In this section, we will review the different methods of factoring that we presented previously. Prior to this section, the polynomials you worked with were grouped together according to the method used to factor them; for example, in Section 5.4 all the polynomials you factored were either the difference of two squares or perfect square trinomials. What usually happens in a situation like this is that you become proficient at factoring the kind of polynomial you are working with at the time but have trouble when given a variety of polynomials to factor.

We begin this section with a checklist that can be used in factoring polynomials of any type. When you have finished this section and the problem set that follows, you want to be proficient enough at factoring that the checklist is second nature.

> **Strategy**   Factoring a Polynomial
>
> **Step 1:** If the polynomial has a greatest common factor other than 1, then factor out the greatest common factor.
>
> **Step 2:** If the polynomial has two terms (a binomial), then see if it is the difference of two squares or the sum or difference of two cubes, and then factor accordingly. Remember, if it is the sum of two squares, it will not factor.
>
> **Step 3:** If the polynomial has three terms (a trinomial), then either it is a perfect square trinomial, which will factor into the square of a binomial, or it is not a perfect square trinomial, in which case you use the trial and error method or the factoring by grouping method.
>
> **Step 4:** If the polynomial has more than three terms, try to factor it by grouping.
>
> **Step 5:** As a final check, see if any of the factors you have written can be factored further. If you have overlooked a common factor, you can catch it here.

A Factoring a Variety of Polynomials

Here are some examples illustrating how we use the checklist.

EXAMPLE 1   Factor $2x^5 - 8x^3$.

SOLUTION   First, we check to see if the greatest common factor is other than 1. Since the greatest common factor is $2x^3$, we begin by factoring it out. Once we have done so, we notice that the binomial that remains is the difference of two squares.

$$
\begin{aligned}
2x^5 - 8x^3 &= 2x^3(x^2 - 4) && \text{Factor out the greatest common factor, } 2x^3 \\
&= 2x^3(x + 2)(x - 2) && \text{Factor the difference of two squares}
\end{aligned}
$$

Note that the greatest common factor $2x^3$ that we factored from each term in the first step of Example 1 remains as part of the answer to the problem, because it is one of the factors of the original binomial. Remember, the expression we end up with when factoring must be equal to the expression we start with. We can't just drop a factor and expect the resulting expression to equal the original expression.

EXAMPLE 2   Factor $3x^4 - 18x^3 + 27x^2$.

SOLUTION   Step 1 is to factor out the greatest common factor, $3x^2$. After we have done so, we notice that the trinomial that remains is a perfect square trinomial, which will factor as the square of a binomial.

$$
\begin{aligned}
3x^4 - 18x^3 + 27x^2 &= 3x^2(x^2 - 6x + 9) && \text{Factor out } 3x^2 \\
&= 3x^2(x - 3)^2 && x^2 - 6x + 9 \text{ is the square of } x - 3
\end{aligned}
$$

EXAMPLE 3 Factor $y^3 + 25y$.

SOLUTION We begin by factoring out the y that is common to both terms. The binomial that remains after we have done so is the sum of two squares, which does not factor, so after the first step we are finished.

$$y^3 + 25y = y(y^2 + 25)$$ Factor out the greatest common factor, y; then notice that $y^2 + 25$ cannot be factored further

EXAMPLE 4 Factor $6a^2 - 11a + 4$.

SOLUTION Here we have a trinomial that does not have a greatest common factor other than 1. Since it is not a perfect square trinomial, we factor it by trial and error; that is, we look for binomial factors the product of whose first terms is $6a^2$ and the product of whose last terms is 4. Then we look for the combination of these types of binomials whose product gives us a middle term of $-11a$. Without showing all the different possibilities, here is the answer:

$$6a^2 - 11a + 4 = (3a - 4)(2a - 1)$$

EXAMPLE 5 Factor $6x^3 - 12x^2 - 48x$.

SOLUTION This trinomial has a greatest common factor of $6x$. The trinomial that remains after the $6x$ has been factored from each term must be factored by trial and error.

$$6x^3 - 12x^2 - 48x = 6x(x^2 - 2x - 8)$$
$$= 6x(x - 4)(x + 2)$$

EXAMPLE 6 Factor $2ab^5 + 8ab^4 + 2ab^3$.

SOLUTION The greatest common factor is $2ab^3$. We begin by factoring it from each term. After that we find the trinomial that remains cannot be factored further.

$$2ab^5 + 8ab^4 + 2ab^3 = 2ab^3(b^2 + 4b + 1)$$

EXAMPLE 7 Factor $xy + 8x + 3y + 24$.

SOLUTION Since our polynomial has four terms, we try factoring by grouping.

$$xy + 8x + 3y + 24 = x(y + 8) + 3(y + 8)$$
$$= (y + 8)(x + 3)$$

Problem Set 5.6

Moving Toward Success

"If a man insisted always on being serious, and never allowed himself a bit of fun and relaxation, he would go mad or become unstable without knowing it."

—Herodotus, 484 BC– 425 BC, Greek historian

1. After you have adhered to your study schedule, what should you do?

a. Keep studying until you are exhausted and the words on the page are blurry

b. Take a break and do something fun

c. Eat well and get a good night's sleep

d. Answers (b) and (c) are correct

2. Think of some fun things you may rather be doing than studying. How can you work those things into your day and still meet your study schedule?

A Factor each of the following polynomials completely; that is, once you are finished factoring, none of the factors you obtain should be factorable. Also, note that the even-numbered problems are not necessarily similar to the odd-numbered problems that precede them in this problem set. [Examples 1–7]

1. $x^2 - 81$

2. $x^2 - 18x + 81$

3. $x^2 + 2x - 15$

4. $15x^2 + 11x - 6$

5. $x^2 + 6x + 9$

6. $12x^2 - 11x + 2$

7. $y^2 - 10y + 25$

8. $21y^2 - 25y - 4$

9. $2a^3b + 6a^2b + 2ab$

10. $6a^2 - ab - 15b^2$

11. $x^2 + x + 1$

12. $2x^2 - 4x + 2$

13. $12a^2 - 75$

14. $18a^2 - 50$

15. $9x^2 - 12xy + 4y^2$

16. $x^3 - x^2$

17. $4x^3 + 16xy^2$

18. $16x^2 + 49y^2$

19. $2y^3 + 20y^2 + 50y$

20. $3y^2 - 9y - 30$

21. $a^6 + 4a^4b^2$

22. $5a^2 - 45b^2$

23. $xy + 3x + 4y + 12$

24. $xy + 7x + 6y + 42$

25. $x^3 - 27$

26. $x^4 - 81$

27. $xy - 5x + 2y - 10$

28. $xy - 7x + 3y - 21$

29. $5a^2 + 10ab + 5b^2$

30. $3a^3b^2 + 15a^2b^2 + 3ab^2$

31. $x^2 + 49$

32. $16 - x^4$

33. $3x^2 + 15xy + 18y^2$

34. $3x^2 + 27xy + 54y^2$

35. $2x^2 + 15x - 38$

36. $2x^2 + 7x - 85$

37. $100x^2 - 300x + 200$

38. $100x^2 - 400x + 300$

39. $x^2 - 64$

40. $9x^2 - 4$

41. $x^2 + 3x + ax + 3a$

42. $x^2 + 4x + bx + 4b$

43. $49a^7 - 9a^5$

44. $8a^3 + 1$

45. $49x^2 + 9y^2$

46. $12x^4 - 62x^3 + 70x^2$

47. $25a^3 + 20a^2 + 3a$

48. $36a^4 - 100a^2$

49. $xa - xb + ay - by$

50. $xy - bx + ay - ab$

51. $48a^4b - 3a^2b$

52. $18a^4b^2 - 12a^3b^3 + 8a^2b^4$

53. $20x^4 - 45x^2$

54. $16x^3 + 16x^2 + 3x$

55. $3x^2 + 35xy - 82y^2$

56. $3x^2 + 37xy - 86y^2$

57. $16x^5 - 44x^4 + 30x^3$

58. $16x^2 + 16x - 1$

59. $2x^2 + 2ax + 3x + 3a$

60. $2x^2 + 2ax + 5x + 5a$

61. $y^4 - 1$

62. $a^7 + 8a^4b^3$

63. $12x^4y^2 + 36x^3y^3 + 27x^2y^4$

64. $16x^3y^2 - 4xy^2$

65. $16t^2 - 64t + 48$

66. $16t^2 - 32t + 12$

67. $54x^2 + 111x + 56$

68. $63x^2 + 110x + 48$

Getting Ready for the Next Section

Solve each equation.

69. $3x - 6 = 9$

70. $5x - 1 = 14$

71. $2x + 3 = 0$

72. $4x - 5 = 0$

73. $4x + 3 = 0$

74. $3x - 1 = 0$

Maintaining Your Skills

Solve each equation.

75. $-2(x + 4) = -10$

76. $\dfrac{3}{4}(-4x - 8) = 21$

77. $\dfrac{3}{5}x + 4 = 22$

78. $-10 = 4 - \dfrac{7}{4}x$

79. $6x - 4(9 - x) = -96$

80. $-2(x - 5) + 5x = 4 - 9$

81. $2x - 3(4x - 7) = -3x$

82. $\dfrac{3}{4}(8 + x) = \dfrac{1}{5}(5x - 15)$

83. $\dfrac{1}{2}x - \dfrac{5}{12} = \dfrac{1}{12}x + \dfrac{5}{12}$

84. $\dfrac{3}{10}x + \dfrac{5}{2} = \dfrac{3}{5}x - \dfrac{1}{2}$

5.7 Solving Quadratic Equations by Factoring

OBJECTIVES

A Solve an equation by writing it in standard form and then factoring.

TICKET TO SUCCESS

Keep these questions in mind as you read through the section. Then respond in your own words and in complete sentences.

1. When is an equation in standard form?

2. What is the first step in solving an equation by factoring?

3. Describe the zero-factor property in your own words.

4. Describe how you would solve $2x^2 - 5x = 12$ for x.

In this section, we will use the methods of factoring developed in previous sections, along with a special property of 0, to find solutions for quadratic equations.

NOTE
Notice that to solve a quadratic equation by this method, it must be possible to factor it. If we can't factor it, we can't solve it by this method. We will learn how to solve quadratic equations that do not factor later in the book.

A   Solving a Quadratic Equation by Factoring

Definition

Any equation that can be put in the form $ax^2 + bx + c = 0$, where a, b, and c are real numbers ($a \neq 0$), is called a **quadratic equation.** The equation $ax^2 + bx + c = 0$ is called **standard form** for a quadratic equation.

$$\underset{\text{an } x^2 \text{ term}}{a(\text{variable})^2} + \underset{\text{an } x \text{ term}}{b(\text{variable})} + \underset{\text{and a constant term}}{c(\text{absence of the variable})} = 0$$

The number 0 has a special property. If we multiply two numbers and the product is 0, then one or both of the original two numbers must be 0. In symbols, this property looks like this.

Zero-Factor Property

Let a and b represent real numbers. If $a \cdot b = 0$, then $a = 0$ or $b = 0$.

Suppose we want to solve the quadratic equation $x^2 + 5x + 6 = 0$. We can factor the left side into $(x + 2)(x + 3)$. Then we have

$$x^2 + 5x + 6 = 0$$
$$(x + 2)(x + 3) = 0$$

Now, $(x + 2)$ and $(x + 3)$ both represent real numbers. Their product is 0; therefore, either $(x + 3)$ is 0 or $(x + 2)$ is 0. Either way we have a solution to our equation. We use the property of 0 stated to finish the problem:

$$x^2 + 5x + 6 = 0$$
$$(x + 2)(x + 3) = 0$$
$$x + 2 = 0 \quad \text{or} \quad x + 3 = 0$$
$$x = -2 \quad \text{or} \quad x = -3$$

Our solution set is $\{-2, -3\}$. Our equation has two solutions. To check our solutions we have to check each one separately to see that they both produce a true statement when used in place of the variable:

$$\begin{aligned}
\text{When} \rightarrow \quad & x = -3 \\
\text{the equation} \rightarrow \quad & x^2 + 5x + 6 = 0 \\
\text{becomes} \rightarrow \quad & (-3)^2 + 5(-3) + 6 \stackrel{?}{=} 0 \\
& 9 + (-15) + 6 = 0 \\
& 0 = 0
\end{aligned}$$

$$\begin{aligned}
\text{When} \rightarrow \quad & x = -2 \\
\text{the equation} \rightarrow \quad & x^2 + 5x + 6 = 0 \\
\text{becomes} \rightarrow \quad & (-2)^2 + 5(-2) + 6 \stackrel{?}{=} 0 \\
& 4 + (-10) + 6 = 0 \\
& 0 = 0
\end{aligned}$$

We have solved a quadratic equation by replacing it with two linear equations in one variable. Based on this process, we have developed the following strategy to solve a quadratic equation by factoring.

Strategy **Solving a Quadratic Equation by Factoring**

Step 1: Put the equation in standard form; that is, 0 on one side and decreasing powers of the variable on the other.

Step 2: Factor completely.

Step 3: Use the zero-factor property to set each variable factor from step 2 to 0.

Step 4: Solve each equation produced in step 3.

Step 5: Check each solution, if necessary.

EXAMPLE 1   Solve the equation $2x^2 - 5x = 12$.

SOLUTION

Step 1:   Begin by adding -12 to both sides, so the equation is in standard form.

$$2x^2 - 5x = 12$$
$$2x^2 - 5x - 12 = 0$$

Step 2:   Factor the left side completely.

$$(2x + 3)(x - 4) = 0$$

Step 3:   Set each factor to 0.

$$2x + 3 = 0 \quad \text{or} \quad x - 4 = 0$$

Step 4:   Solve each of the equations from step 3.

$$2x + 3 = 0 \qquad x - 4 = 0$$
$$2x = -3 \qquad\qquad x = 4$$
$$x = -\frac{3}{2}$$

Step 5:   Substitute each solution into $2x^2 - 5x = 12$ to check.

$$\text{Check: } -\frac{3}{2} \qquad\qquad \text{Check: } 4$$

$$2\left(-\frac{3}{2}\right)^2 - 5\left(-\frac{3}{2}\right) \overset{?}{=} 12 \qquad 2(4)^2 - 5(4) \overset{?}{=} 12$$

$$2\left(\frac{9}{4}\right) + 5\left(\frac{3}{2}\right) = 12 \qquad 2(16) - 20 = 12$$

$$\frac{9}{2} + \frac{15}{2} = 12 \qquad 32 - 20 = 12$$

$$\frac{24}{2} = 12 \qquad\qquad 12 = 12$$

$$12 = 12$$

EXAMPLE 2   Solve for a: $16a^2 - 25 = 0$.

SOLUTION   The equation is already in standard form.

$$16a^2 - 25 = 0$$

$$(4a - 5)(4a + 5) = 0 \qquad \text{Factor left side}$$

$$4a - 5 = 0 \quad \text{or} \quad 4a + 5 = 0 \qquad \text{Set each factor to 0}$$

$$4a = 5 \qquad\qquad 4a = -5 \qquad \text{Solve the resulting equations}$$

$$a = \frac{5}{4} \qquad\qquad a = -\frac{5}{4}$$

EXAMPLE 3   Solve $4x^2 = 8x$.

SOLUTION   We begin by adding $-8x$ to each side of the equation to put it in standard form. Then we factor the left side of the equation by factoring out the greatest common factor.

$$4x^2 = 8x$$

$$4x^2 - 8x = 0 \qquad \text{Add } -8x \text{ to each side}$$

$$4x(x - 2) = 0 \qquad \text{Factor the left side}$$

$$4x = 0 \quad \text{or} \quad x - 2 = 0 \qquad \text{Set each factor to 0}$$

$$x = 0 \quad \text{or} \quad x = 2 \qquad \text{Solve the resulting equations}$$

The solutions are 0 and 2.

EXAMPLE 4   Solve $x(2x + 3) = 44$.

SOLUTION   We must multiply out the left side first and then put the equation in standard form.

$$x(2x + 3) = 44$$

$$2x^2 + 3x = 44 \qquad \text{Multiply out the left side}$$

$$2x^2 + 3x - 44 = 0 \qquad \text{Add } -44 \text{ to each side}$$

$$(2x + 11)(x - 4) = 0 \qquad \text{Factor the left side}$$

$$2x + 11 = 0 \quad \text{or} \quad x - 4 = 0 \qquad \text{Set each factor to 0}$$

$$2x = -11 \quad \text{or} \quad x = 4 \qquad \text{Solve the resulting equations}$$

$$x = -\frac{11}{2}$$

The two solutions are $-\dfrac{11}{2}$ and 4.

EXAMPLE 5   Solve for x: $5^2 = x^2 + (x + 1)^2$.

SOLUTION   Before we can put this equation in standard form we must square the binomial. Remember, to square a binomial, we use the formula $(a + b)^2 = a^2 + 2ab + b^2$.

$$5^2 = x^2 + (x + 1)^2$$

$$25 = x^2 + x^2 + 2x + 1 \qquad \text{Expand } 5^2 \text{ and } (x + 1)^2$$

$$25 = 2x^2 + 2x + 1 \qquad \text{Simplify the right side}$$

$$0 = 2x^2 + 2x - 24 \qquad \text{Add } -25 \text{ to each side}$$

$$0 = 2(x^2 + x - 12) \qquad \text{Factor out 2}$$

$$0 = 2(x + 4)(x - 3) \qquad \text{Factor completely}$$

$$x + 4 = 0 \quad \text{or} \quad x - 3 = 0 \qquad \text{Set each variable factor to 0}$$

$$x = -4 \quad \text{or} \quad x = 3$$

Note, in the second to last line, that we do not set 2 equal to 0. That is because 2 can never be 0. It is always 2. We only use the zero-factor property to set variable factors to 0 because they are the only factors that can possibly be 0.

Also notice that it makes no difference which side of the equation is 0 when we write the equation in standard form.

Although the equation in the next example is not a quadratic equation, it can be solved by the method shown in the first five examples.

EXAMPLE 6 Solve $24x^3 = -10x^2 + 6x$ for x.

SOLUTION First, we write the equation in standard form.

$$24x^3 + 10x^2 - 6x = 0 \qquad \text{Standard form}$$

$$2x(12x^2 + 5x - 3) = 0 \qquad \text{Factor out } 2x$$

$$2x(3x - 1)(4x + 3) = 0 \qquad \text{Factor remaining trinomial}$$

$$2x = 0 \quad \text{or} \quad 3x - 1 = 0 \quad \text{or} \quad 4x + 3 = 0 \qquad \text{Set factors to 0}$$

$$x = 0 \quad \text{or} \qquad x = \frac{1}{3} \quad \text{or} \qquad x = -\frac{3}{4} \quad \text{Solutions}$$

Problem Set 5.7

Moving Toward Success

You are the person who has to decide,
whether you'll do it or toss it aside;
You are the person who makes up your mind,
whether you'll lead or will linger behind,
Whether you'll try for the goal that's afar, or just
be contented to stay where you are.
— Edgar Guest, 1881–1959, poet

1. Should you ask questions in class? Why or why not?

2. What do you plan to do if you have questions about the section or a homework problem outside of class?

A The following equations are already in factored form. Use the zero-factor property to set the factors to 0 and solve.

1. $(x + 2)(x - 1) = 0$
2. $(x + 3)(x + 2) = 0$
3. $(a - 4)(a - 5) = 0$
4. $(a + 6)(a - 1) = 0$
5. $x(x + 1)(x - 3) = 0$
6. $x(2x + 1)(x - 5) = 0$
7. $(3x + 2)(2x + 3) = 0$
8. $(4x - 5)(x - 6) = 0$
9. $m(3m + 4)(3m - 4) = 0$
10. $m(2m - 5)(3m - 1) = 0$
11. $2y(3y + 1)(5y + 3) = 0$
12. $3y(2y - 3)(3y - 4) = 0$

Solve the following equations. [Examples 1–6]

13. $x^2 + 3x + 2 = 0$
14. $x^2 - x - 6 = 0$
15. $x^2 - 9x + 20 = 0$
16. $x^2 + 2x - 3 = 0$
17. $a^2 - 2a - 24 = 0$
18. $a^2 - 11a + 30 = 0$
19. $100x^2 - 500x + 600 = 0$
20. $100x^2 - 300x + 200 = 0$
21. $x^2 = -6x - 9$
22. $x^2 = 10x - 25$
23. $a^2 - 16 = 0$
24. $a^2 - 36 = 0$
25. $2x^2 + 5x - 12 = 0$
26. $3x^2 + 14x - 5 = 0$
27. $9x^2 + 12x + 4 = 0$
28. $12x^2 - 24x + 9 = 0$
29. $a^2 + 25 = 10a$
30. $a^2 + 16 = 8a$
31. $2x^2 = 3x + 20$
32. $6x^2 = x + 2$
33. $3m^2 = 20 - 7m$
34. $2m^2 = -18 + 15m$
35. $4x^2 - 49 = 0$
36. $16x^2 - 25 = 0$

37. $x^2 + 6x = 0$ **38.** $x^2 - 8x = 0$

39. $x^2 - 3x = 0$ **40.** $x^2 + 5x = 0$

41. $2x^2 = 8x$ **42.** $2x^2 = 10x$

43. $3x^2 = 15x$ **44.** $5x^2 = 15x$

45. $1{,}400 = 400 + 700x - 100x^2$

46. $2{,}700 = 700 + 900x - 100x^2$

47. $6x^2 = -5x + 4$ **48.** $9x^2 = 12x - 4$

49. $x(2x - 3) = 20$ **50.** $x(3x - 5) = 12$

51. $t(t + 2) = 80$ **52.** $t(t + 2) = 99$

53. $4{,}000 = (1{,}300 - 100p)p$

54. $3{,}200 = (1{,}200 - 100p)p$

55. $x(14 - x) = 48$ **56.** $x(12 - x) = 32$

57. $(x + 5)^2 = 2x + 9$ **58.** $(x + 7)^2 = 2x + 13$

59. $(y - 6)^2 = y - 4$ **60.** $(y + 4)^2 = y + 6$

61. $10^2 = (x + 2)^2 + x^2$

62. $15^2 = (x + 3)^2 + x^2$

63. $2x^3 + 11x^2 + 12x = 0$

64. $3x^3 + 17x^2 + 10x = 0$

65. $4y^3 - 2y^2 - 30y = 0$

66. $9y^3 + 6y^2 - 24y = 0$

67. $8x^3 + 16x^2 = 10x$

68. $24x^3 - 22x^2 = -4x$

69. $20a^3 = -18a^2 + 18a$

70. $12a^3 = -2a^2 + 10a$

71. $16t^2 - 32t + 12 = 0$

72. $16t^2 - 64t + 48 = 0$

Simplify each side as much as possible, then solve the equation.

73. $(a - 5)(a + 4) = -2a$

74. $(a + 2)(a - 3) = -2a$

75. $3x(x + 1) - 2x(x - 5) = -42$

76. $4x(x - 2) - 3x(x - 4) = -3$

77. $2x(x + 3) = x(x + 2) - 3$

78. $3x(x - 3) = 2x(x - 4) + 6$

79. $a(a - 3) + 6 = 2a$ **80.** $a(a - 4) + 8 = 2a$

81. $15(x + 20) + 15x = 2x(x + 20)$

82. $15(x + 8) + 15x = 2x(x + 8)$

83. $15 = a(a + 2)$ **84.** $6 = a(a - 5)$

Use factoring by grouping to solve the following equations.

85. $x^3 + 3x^2 - 4x - 12 = 0$

86. $x^3 + 5x^2 - 9x - 45 = 0$

87. $x^3 + x^2 - 16x - 16 = 0$

88. $4x^3 + 12x^2 - 9x - 27 = 0$

89. Find a quadratic equation that has two solutions; $x = 3$ and $x = 5$. Write your answer in standard form.

90. Find a quadratic equation that has two solutions; $x = 9$ and $x = 1$. Write your answer in standard form.

91. Find a quadratic equation that has the two given solutions.
   a. $x = 3$ and $x = 2$.
   b. $x = 1$ and $x = 6$.
   c. $x = 3$ and $x = -2$.

92. Find a quadratic equation that has the two given solutions.
   a. $x = 4$ and $x = 5$.
   b. $x = 2$ and $x = 10$.
   c. $x = -4$ and $x = 5$.

Getting Ready for the Next Section

Write each sentence as an algebraic equation.

93. The product of two consecutive integers is 72.

94. The product of two consecutive even integers is 80.

95. The product of two consecutive odd integers is 99.

96. The product of two consecutive odd integers is 63.

97. The product of two consecutive even integers is 10 less than 5 times their sum.

98. The product of two consecutive odd integers is 1 less than 4 times their sum.

The following word problems are taken from the book *Academic Algebra,* written by William J. Milne and published by the American Book Company in 1901. Solve each problem.

99. Cost of a Bicycle and a Suit A bicycle and a suit cost $90. How much did each cost, if the bicycle cost 5 times as much as the suit?

100. Cost of a Cow and a Calf A man bought a cow and a calf for $36, paying 8 times as much for the cow as for the calf. What was the cost of each?

101. Cost of a House and a Lot A house and a lot cost $3,000. If the house cost 4 times as much as the lot, what was the cost of each?

102. Daily Wages A plumber and two helpers together earned $7.50 per day. How much did each earn per day, if the plumber earned 4 times as much as each helper?

105. $\dfrac{x^5}{x^{-3}}$

106. $\dfrac{x^{-2}}{x^{-5}}$

107. $\dfrac{(x^2)^3}{(x^{-3})^4}$

108. $\dfrac{(x^2)^{-4}(x^{-2})^3}{(x^{-3})^{-5}}$

109. Write the number 0.0056 in scientific notation.

110. Write the number 2.34×10^{-4} in expanded form.

111. Write the number 5,670,000,000 in scientific notation.

112. Write the number 0.00000567 in scientific notation.

Maintaining Your Skills

Use the properties of exponents to simplify each expression.

103. 2^{-3}

104. 5^{-2}

5.8 Applications of Quadratic Equations

OBJECTIVES

A Solve number problems by factoring.

B Solve geometry problems by factoring.

C Solve business problems by factoring.

D Use the Pythagorean Theorem to solve problems.

TICKET TO SUCCESS

Keep these questions in mind as you read through the section. Then respond in your own words and in complete sentences.

1. What are consecutive integers?

2. How would you use factoring to solve a geometry problem?

3. Explain the Pythagorean theorem in words.

4. Write an application problem for which the solution depends on solving the equation $x(x + 1) = 12$.

Newscom

In this section, we will look at some application problems, the solutions to which require solving a quadratic equation. We will also introduce the Pythagorean theorem, one of the oldest theorems in the history of mathematics. As we mentioned in the introduction to this chapter, the person whose name we associate with the theorem, Pythagoras (of Samos), was a Greek philosopher and mathematician who lived from about 560 BC to 480 BC. According to the British philosopher Bertrand Russell, Pythagoras was "intellectually one of the most important men that ever lived."

Also in this section, the solutions to the examples show only the essential steps from our Blueprint for Problem Solving. Recall that step 1 is done mentally; we read the problem and mentally list the items that are known and the items that are unknown. This is an essential part of problem solving. However, now that you have had experience with application problems, you are doing step 1 automatically.

A　Number Problems

EXAMPLE 1　The product of two consecutive odd integers is 63. Find the integers.

SOLUTION　Let x = the first odd integer; then $x + 2$ = the second odd integer. An equation that describes the situation is

$$x(x + 2) = 63 \qquad \text{Their product is 63}$$

Then we solve the equation.

$$x(x + 2) = 63$$
$$x^2 + 2x = 63$$
$$x^2 + 2x - 63 = 0$$
$$(x - 7)(x + 9) = 0$$
$$x - 7 = 0 \quad \text{or} \quad x + 9 = 0$$
$$x = 7 \quad \text{or} \quad x = -9$$

If the first odd integer is 7, the next odd integer is $7 + 2 = 9$. If the first odd integer is -9, the next consecutive odd integer is $-9 + 2 = -7$. We have two pairs of consecutive odd integers that are solutions. They are 7, 9 and -9, -7.

We check to see that their products are 63.

$$7(9) = 63$$
$$-7(-9) = 63$$

Suppose we know that the sum of two numbers is 50. We want to find a way to represent each number using only one variable. If we let x represent one of the two numbers, how can we represent the other? Let's suppose for a moment that x turns out to be 30. Then the other number will be 20, because their sum is 50; that is, if two numbers add up to 50 and one of them is 30, then the other must be $50 - 30 = 20$. Generalizing this to any number x, we see that if two numbers have a sum of 50 and one of the numbers is x, then the other must be $50 - x$. The table that follows shows some additional examples.

If two numbers have a sum of	and one of them is	then the other must be
50	x	$50 - x$
100	x	$100 - x$
10	y	$10 - y$
12	n	$12 - n$

Now, let's look at an example that uses this idea.

EXAMPLE 2 The sum of two numbers is 13. Their product is 40. Find the numbers.

SOLUTION If we let x represent one of the numbers, then $13 - x$ must be the other number because their sum is 13. Since their product is 40, we can write:

$$x(13 - x) = 40 \qquad \text{The product of the two numbers is 40}$$

$$13x - x^2 = 40 \qquad \text{Multiply the left side}$$

$$x^2 - 13x = -40 \qquad \text{Multiply both sides by } -1 \text{ and reverse the order of the terms on the left side}$$

$$x^2 - 13x + 40 = 0 \qquad \text{Add 40 to each side}$$

$$(x - 8)(x - 5) = 0 \qquad \text{Factor the left side}$$

$$x - 8 = 0 \quad \text{or} \quad x - 5 = 0$$

$$x = 8 \qquad\qquad x = 5$$

The two solutions are 8 and 5. If x is 8, then the other number is $13 - x = 13 - 8 = 5$. Likewise, if x is 5, the other number is $13 - x = 13 - 5 = 8$. Therefore, the two numbers we are looking for are 8 and 5. Their sum is 13 and their product is 40.

B Geometry Problems

Many word problems dealing with area can best be described algebraically by quadratic equations.

EXAMPLE 3 The length of a rectangle is 3 more than twice the width. The area is 44 square inches. Find the dimensions (find the length and width).

SOLUTION As shown in Figure 1, let x = the width of the rectangle. Then $2x + 3$ = the length of the rectangle because the length is three more than twice the width.

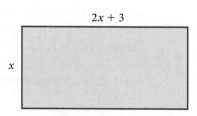

FIGURE 1

Since the area is 44 square inches, an equation that describes the situation is

$$x(2x + 3) = 44 \quad \text{Length} \cdot \text{width} = \text{area}$$

We now solve the equation.

$$x(2x + 3) = 44$$

$$2x^2 + 3x = 44$$

$$2x^2 + 3x - 44 = 0$$

$$(2x + 11)(x - 4) = 0$$

$$2x + 11 = 0 \quad \text{or} \quad x - 4 = 0$$

$$x = -\frac{11}{2} \quad \text{or} \quad x = 4$$

The solution $x = -\frac{11}{2}$ cannot be used since length and width are always given in positive units. The width is 4. The length is 3 more than twice the width or $2(4) + 3 = 11$.

$$\text{Width} = 4 \text{ inches}$$

$$\text{Length} = 11 \text{ inches}$$

The solutions check in the original problem since $4(11) = 44$. ∎

EXAMPLE 4   The numerical value of the area of a square is twice its perimeter. What is the length of its side?

SOLUTION   As shown in Figure 2, let x = the length of its side. Then x^2 = the area of the square and $4x$ = the perimeter of the square.

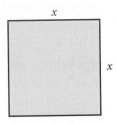

x

x

FIGURE 2

An equation that describes the situation is

$$x^2 = 2(4x) \quad \text{The area is 2 times the perimeter}$$

$$x^2 = 8x$$

$$x^2 - 8x = 0$$

$$x(x - 8) = 0$$

$$x = 0 \quad \text{or} \quad x = 8$$

Since $x = 0$ does not make sense in our original problem, we use $x = 8$. If the side has length 8, then the perimeter is $4(8) = 32$ and the area is $8^2 = 64$. Since 64 is twice 32, our solution is correct. ∎

C Business Problems

EXAMPLE 5 A company can manufacture x hundred items for a total profit P of $P = 300 + 500x - 100x^2$. How many items were manufactured if the total profit is $900?

SOLUTION We are looking for x when P is 900. We begin by substituting 900 for P in the profit equation. Then we solve for x.

$$\text{When} \rightarrow \quad P = 900$$
$$\text{the equation} \rightarrow \quad P = 300 + 500x - 100x^2$$
$$\text{becomes} \rightarrow \quad 900 = 300 + 500x - 100x^2$$

We can write this equation in standard form by adding -300, $-500x$, and $100x^2$ to each side. The result looks like this:

$$900 = 300 + 500x - 100x^2$$

$100x^2 - 500x + 600 = 0$	Standard form
$100(x^2 - 5x + 6) = 0$	Factor out 100
$100(x - 2)(x - 3) = 0$	Factor completely
$x - 2 = 0 \quad \text{or} \quad x - 3 = 0$	Set variable factors to 0
$x = 2 \quad \text{or} \quad\quad x = 3$	

Our solutions are two and three, which means that the company can manufacture 200 items or 300 items for a total profit of $900.

NOTE
If you are planning on taking finite mathematics, statistics, or business calculus in the future, Examples 5 and 6 will give you a head start on some of the problems you will see in those classes.

EXAMPLE 6 A manufacturer of small portable radios knows that the number of radios she can sell each week is related to the price of the radios by the equation $x = 1{,}300 - 100p$ (x is the number of radios and p is the price per radio). What price should she charge for the radios to have a weekly revenue of $4,000?

SOLUTION First, we must find the revenue equation. The equation for total revenue is $R = xp$, where x is the number of units sold and p is the price per unit. Since we want R in terms of p, we substitute $1{,}300 - 100p$ for x in the equation $R = xp$.

$$\text{If} \quad\quad R = xp$$
$$\text{and} \quad\quad x = 1{,}300 - 100p$$
$$\text{then} \quad\quad R = (1{,}300 - 100p)p$$

We want to find p when R is 4,000. Substituting 4,000 for R in the equation gives us.

$$4{,}000 = (1{,}300 - 100p)p$$

If we multiply out the right side, we have

$$4{,}000 = 1{,}300p - 100p^2$$

To write this equation in standard form, we add $100p^2$ and $-1{,}300p$ to each side.

$100p^2 - 1{,}300p + 4{,}000 = 0$	Add $100p^2$ and $-1{,}300p$ to each side
$100(p^2 - 13p + 40) = 0$	Factor out 100
$100(p - 5)(p - 8) = 0$	Factor completely
$p - 5 = 0 \quad \text{or} \quad p - 8 = 0$	Set variable factors to 0
$p = 5 \quad \text{or} \quad\quad p = 8$	

If she sells the radios for $5 each or for $8 each, she will have a weekly revenue of $4,000. ●

D   The Pythagorean Theorem

Next, we will work some problems involving the Pythagorean theorem, which uses variables to represent the three sides of a triangle. When calculating the lengths of these sides, the theorem often translates into a quadratic equation we can factor and solve.

Pythagorean Theorem
In any right triangle (Figure 3), the square of the longer side (called the hypotenuse) is equal to the sum of the squares of the other two sides (called legs).

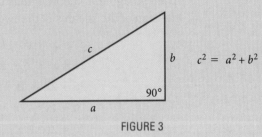

FIGURE 3

EXAMPLE 7   The three sides of a right triangle are three consecutive integers. Find the lengths of the three sides.

SOLUTION   Let x = the first integer (shortest side)
then $x + 1$ = the next consecutive integer
and $x + 2$ = the consecutive integer (longest side)

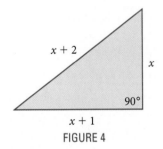

FIGURE 4

A diagram of the triangle is shown in Figure 4.

The Pythagorean theorem tells us that the square of the longest side $(x + 2)^2$ is equal to the sum of the squares of the two shorter sides, $(x + 1)^2 + x^2$. Here is the equation:

$$(x + 2)^2 = (x + 1)^2 + x^2$$

$$x^2 + 4x + 4 = x^2 + 2x + 1 + x^2 \qquad \text{Expand squares}$$

$$x^2 - 2x - 3 = 0 \qquad \text{Standard form}$$

$$(x - 3)(x + 1) = 0 \qquad \text{Factor}$$

$$x - 3 = 0 \quad \text{or} \quad x + 1 = 0 \qquad \text{Set factors to 0}$$

$$x = 3 \quad \text{or} \quad x = -1$$

Since a triangle cannot have a side with a negative number for its length, we must not use -1 for a solution to our original problem; therefore, the shortest side is 3. The other two sides are the next two consecutive integers, 4 and 5. ●

EXAMPLE 8   The hypotenuse of a right triangle is 5 inches, and the lengths of the two legs (the other two sides) are given by two consecutive integers. Find the lengths of the two legs.

SOLUTION   If we let x = the length of the shortest side, then the other side must be $x + 1$. A diagram of the triangle is shown in Figure 5.

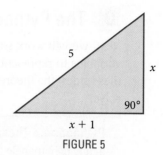

FIGURE 5

The Pythagorean theorem tells us that the square of the longest side, 5^2, is equal to the sum of the squares of the two shorter sides, $x^2 + (x + 1)^2$. Here is the equation:

$$5^2 = x^2 + (x + 1)^2 \qquad \text{Pythagorean theorem}$$
$$25 = x^2 + x^2 + 2x + 1 \qquad \text{Expand } 5^2 \text{ and } (x + 1)^2$$
$$25 = 2x^2 + 2x + 1 \qquad \text{Simplify the right side}$$
$$0 = 2x^2 + 2x - 24 \qquad \text{Add } -25 \text{ to each side}$$
$$0 = 2(x^2 + x - 12) \qquad \text{Factor out 2}$$
$$0 = 2(x + 4)(x - 3) \qquad \text{Factor completely}$$
$$x + 4 = 0 \quad \text{or} \quad x - 3 = 0 \qquad \text{Set variable factors to 0}$$
$$x = -4 \quad \text{or} \quad x = 3$$

Since a triangle cannot have a side with a negative number for its length, we cannot use -4; therefore, the shortest side must be 3 inches. The next side is $x + 1 = 3 + 1 = 4$ inches. Since the hypotenuse is 5, we can check our solutions with the Pythagorean theorem as shown in Figure 6.

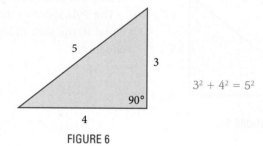

FIGURE 6

Problem Set 5.8

Moving Toward Success

"A successful life doesn't require that we've done the best, but that we've done our best."

—H. Jackson Brown, 1940–present, American author

1. What will you do if a classmate asks you a question about your assigned work to which you do not know the answer?

2. If you sought help from your instructor in the beginning of the course, should you continue to do so as the course continues? Why or why not?

A Number Problems [Examples 1–2]

Solve the following word problems. Be sure to show the equation used.

1. The product of two consecutive even integers is 80. Find the two integers.

2. The product of two consecutive integers is 72. Find the two integers.

3. The product of two consecutive odd integers is 99. Find the two integers.

4. The product of two consecutive integers is 132. Find the two integers.

5. The product of two consecutive even integers is 10 less than 5 times their sum. Find the two integers.

6. The product of two consecutive odd integers is 1 less than 4 times their sum. Find the two integers.

7. The sum of two numbers is 14. Their product is 48. Find the numbers.

8. The sum of two numbers is 12. Their product is 32. Find the numbers.

9. One number is 2 more than 5 times another. Their product is 24. Find the numbers.

10. One number is 1 more than twice another. Their product is 55. Find the numbers.

11. One number is 4 times another. Their product is 4 times their sum. Find the numbers.

12. One number is 2 more than twice another. Their product is 2 more than twice their sum. Find the numbers.

B Geometry Problems [Examples 3–4]

13. The length of a rectangle is 1 more than the width. The area is 12 square inches. Find the dimensions.

14. The length of a rectangle is 3 more than twice the width. The area is 44 square inches. Find the dimensions.

15. The height of a triangle is twice the base. The area is 9 square inches. Find the base.

16. The height of a triangle is 2 more than twice the base. The area is 20 square feet. Find the base.

17. The hypotenuse of a right triangle is 10 inches. The lengths of the two legs are given by two consecutive even integers. Find the lengths of the two legs.

18. The hypotenuse of a right triangle is 15 inches. One of the legs is 3 inches more than the other. Find the lengths of the two legs.

19. The shorter leg of a right triangle is 5 meters. The hypotenuse is 1 meter longer than the longer leg. Find the length of the longer leg.

20. The shorter leg of a right triangle is 12 yards. If the hypotenuse is 20 yards, how long is the other leg?

C Business Problems [Examples 5–6]

21. A company can manufacture x hundred items for a total profit of $p = 400 + 700x - 100x^2$. Find x if the total profit is $1,400.

22. If the total profit P of manufacturing x hundred items is given by the equation $P = 700 + 900x - 100x^2$, find x when P is $2,700.

23. The total profit P of manufacturing x hundred DVD's is given by the equation

$$P = 600 + 1{,}000x - 100x^2$$

Find x if the total profit is \$2,200.

24. The total profit P of manufacturing x hundred pen and pencil sets is given by the equation

$$P = 500 + 800x - 100x^2$$

Find x when P is \$1,700.

25. A company that manufactures hair ribbons knows that the number of ribbons it can sell each week, x, is related to the price p per ribbon by the equation $x = 1{,}200 - 100p$. At what price should the company sell the ribbons if it wants the weekly revenue to be \$3,200? (*Remember:* The equation for revenue is $R = xp$.)

26. A company manufactures small flash drives for home computers. It knows from experience that the number of flash drives it can sell each day, x, is related to the price p per flash drive by the equation $x = 800 - 100p$. At what price should the company sell the flash drives if it wants the daily revenue to be \$1,200?

27. The relationship between the number of calculators a company sells per week, x, and the price p of each calculator is given by the equation $x = 1{,}700 - 100p$. At what price should the calculators be sold if the weekly revenue is to be \$7,000?

28. The relationship between the number of pencil sharpeners a company can sell each week, x, and the price p of each sharpener is given by the equation $x = 1{,}800 - 100p$. At what price should the sharpeners be sold if the weekly revenue is to be \$7,200?

29. Pythagorean Theorem

A 13-foot ladder is placed so that it reaches to a point on the wall that is 2 feet higher than twice the distance from the base of the wall to the base of the ladder.

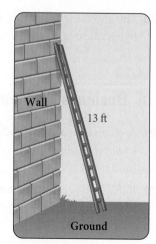

a. How far from the wall is the base of the ladder?

b. How high does the ladder reach?

30. Height of a Projectile If a rocket is fired vertically into the air with a speed of 240 feet per second, its height at time t seconds is given by $h(t) = -16t^2 + 240t$. Here is a graph of its height at various times, with the details left out:

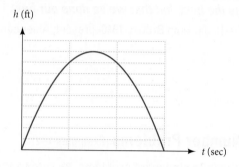

At what time(s) will the rocket be the following number of feet above the ground?

a. 704 feet

b. 896 feet

c. Why do parts a and b each have two answers?

d. How long will the rocket be in the air? (*Hint:* How high is it when it hits the ground?)

e. When the equation for part d is solved, one of the answers is $t = 0$ second. What does this represent?

31. Projectile Motion A gun fires a bullet almost straight up from the edge of a 100-foot cliff. If the bullet leaves the gun with a speed of 396 feet per second, its height at time t is given by $h(t) = -16t^2 + 396t + 100$, measured from the ground below the cliff.

a. When will the bullet land on the ground below the cliff? (*Hint:* What is its height when it lands? Remember that we are measuring from the ground below, not from the cliff.)

b. Make a table showing the bullet's height every five seconds, from the time it is fired ($t = 0$) to the time it lands. (*Note:* It is faster to substitute into the factored form.)

t (sec)	h (feet)
0	
5	
10	
15	
20	
25	

32. **Constructing a Box** I have a piece of cardboard that is twice as long as it is wide. If I cut a 2-inch by 2-inch square from each corner and fold up the resulting flaps, I get a box with a volume of 32 cubic inches. What are the dimensions of the cardboard?

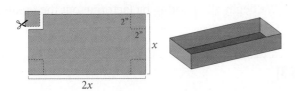

Skyscrapers The chart shows the heights of the three tallest buildings in the world. Use the chart to answer Problems 33 and 34.

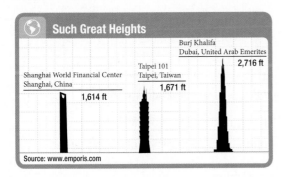

33. If you drop an object off the top of the Shanghai World Financial Center, the height that the object is from the ground after t seconds is given by the equation $h = 1,614 - 16t^2$.
    a. When is the object 1,550 feet above the ground?

    b. How far has the object fallen after 4 seconds?

34. If you throw an object off the top of the Taipei 101 tower with an initial upward velocity of 6 feet per second, the height that the object is from the ground after t seconds is given by the equation $h = 1,671 + 6t - 16t^2$.
    a. When is the object 1,661 feet above the ground?

    b. When is the object 929 feet above the ground?

Maintaining Your Skills

Simplify each expression. (Write all answers with positive exponents only.)

35. $(5x^3)^2(2x^6)^3$

36. 2^{-3}

37. $\dfrac{x^4}{x^{-3}}$

38. $\dfrac{(20x^2y^3)(5x^4y)}{(2xy^5)(10x^2y^3)}$

39. $(2 \times 10^{-4})(4 \times 10^5)$

40. $\dfrac{9 \times 10^{-3}}{3 \times 10^{-2}}$

41. $20ab^2 - 16ab^2 + 6ab^2$

42. Subtract $6x^2 - 5x - 7$ from $9x^2 + 3x - 2$.

Multiply.

43. $2x^2(3x^2 + 3x - 1)$

44. $(2x + 3)(5x - 2)$

45. $(3y - 5)^2$

46. $(a - 4)(a^2 + 4a + 16)$

47. $(2a^2 + 7)(2a^2 - 7)$

48. Divide $15x^{10} - 10x^8 + 25x^6$ by $5x^6$.

Greatest Common Factor [5.1]

EXAMPLES

1. $8x^4 - 10x^3 + 6x^2$
$= 2x^2 \cdot 4x^2 - 2x^2 \cdot 5x + 2x^2 \cdot 3$
$= 2x^2(4x^2 - 5x + 3)$

The largest monomial that divides each term of a polynomial is called the greatest common factor for that polynomial. We begin all factoring by factoring out the greatest common factor.

Factoring Trinomials [5.2, 5.3]

2. $x^2 + 5x + 6 = (x + 2)(x + 3)$
$x^2 - 5x + 6 = (x - 2)(x - 3)$
$6x^2 - x - 2 = (2x + 1)(3x - 2)$
$6x^2 + 7x + 2 = (2x + 1)(3x + 2)$

One method of factoring a trinomial is to list all pairs of binomials the product of whose first terms gives the first term of the trinomial, and the product of whose last terms gives the last term of the trinomial. Then we choose the pair that gives the correct middle term for the original trinomial.

Special Factorings [5.4]

3. $x^2 + 10x + 25 = (x + 5)^2$
$x^2 - 10x + 25 = (x - 5)^2$
$x^2 - 25 = (x + 5)(x - 5)$

$a^2 + 2ab + b^2 = (a + b)^2$ Binomial Square

$a^2 - 2ab + b^2 = (a - b)^2$ Binomial Square

$a^2 - b^2 = (a + b)(a - b)$ Difference of two squares

Sum and Difference of Two Cubes [5.5]

4. $x^3 - 27 = (x - 3)(x^2 + 3x + 9)$
$x^3 + 27 = (x + 3)(x^2 - 3x + 9)$

$a^3 - b^3 = (a - b)(a^2 + ab + b^2)$ Difference of two cubes

$a^3 + b^3 = (a + b)(a^2 - ab + b^2)$ Sum of two cubes

Strategy for Factoring a Polynomial [5.6]

5. **a.** $2x^5 - 8x^3 = 2x^3(x^2 - 4)$
           $= 2x^3(x + 2)(x - 2)$
b. $3x^4 - 18x^3 + 27x^2$
           $= 3x^2(x^2 - 6x + 9)$
           $= 3x^2(x - 3)^2$
c. $6x^3 - 12x^2 - 48x$
           $= 6x(x^2 - 2x - 8)$
           $= 6x(x - 4)(x + 2)$
d. $x^2 + ax + bx + ab$
           $= x(x + a) + b(x + a)$
           $= (x + a)(x + b)$

Step 1: If the polynomial has a greatest common factor other than 1, then factor out the greatest common factor.

Step 2: If the polynomial has two terms (it is a binomial), then see if it is the difference of two squares or the sum or difference of two cubes, and then factor accordingly. Remember, if it is the sum of two squares, it will not factor.

Step 3: If the polynomial has three terms (a trinomial), then it is either a perfect square trinomial that will factor into the square of a binomial, or it is not a perfect square trinomial, in which case you use the trial and error method or the factoring by grouping method.

Step 4: If the polynomial has more than three terms, then try to factor it by grouping.

Step 5: As a final check, see if any of the factors you have written can be factored further. If you have overlooked a common factor, you can catch it here.

■ Strategy for Solving a Quadratic Equation [5.7]

6. Solve $x^2 - 6x = -8$.
$$x^2 - 6x + 8 = 0$$
$$(x - 4)(x - 2) = 0$$
$$x - 4 = 0 \quad \text{or} \quad x - 2 = 0$$
$$x = 4 \quad \text{or} \quad x = 2$$
Both solutions check.

Step 1: Write the equation in standard form.
$$ax^2 + bx + c = 0$$

Step 2: Factor completely.

Step 3: Set each variable factor equal to 0.

Step 4: Solve the equations found in step 3.

Step 5: Check solutions, if necessary.

■ The Pythagorean Theorem [5.8]

7. The hypotenuse of a right triangle is 5 inches, and the lengths of the two legs (the other two sides) are given by two consecutive integers. Find the lengths of the two legs.

 If we let x = the length of the shortest side, then the other side must be $x + 1$. The Pythagorean theorem tells us that
$$5^2 = x^2 + (x + 1)^2$$
$$25 = x^2 + x^2 + 2x + 1$$
$$25 = 2x^2 + 2x + 1$$
$$0 = 2x^2 + 2x - 24$$
$$0 = 2(x^2 + x - 12)$$
$$0 = 2(x + 4)(x - 3)$$
$$x + 4 = 0 \quad \text{or} \quad x - 3 = 0$$
$$x = -4 \quad \text{or} \quad x = 3$$
Since a triangle cannot have a side with a negative number for its length, we cannot use -4. One leg is $x = 3$ and the other leg is $x + 1 = 3 + 1 = 4$.

In any right triangle, the square of the longest side (called the hypotenuse) is equal to the sum of the squares of the other two sides (called legs).

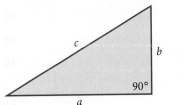

$$c^2 = a^2 + b^2$$

⊘ COMMON MISTAKES

It is a mistake to apply the zero-factor property to numbers other than zero. For example, consider the equation $(x - 3)(x + 4) = 18$. A fairly common mistake is to attempt to solve it with the following steps:
$$(x - 3)(x + 4) = 18$$
$$x - 3 = 18 \quad \text{or} \quad x + 4 = 18 \leftarrow \text{Mistake}$$
$$x = 21 \quad \text{or} \quad x = 14$$

These are obviously not solutions, as a quick check will verify.

Check: $x = 21$ | Check: $x = 14$
$$(21 - 3)(21 + 4) \stackrel{?}{=} 18 \qquad (14 - 3)(14 + 4) \stackrel{?}{=} 18$$
$$18 \cdot 25 = 18 \qquad\qquad 11 \cdot 18 = 18$$
$$450 = 18 \text{ false statements } 198 = 18$$

The mistake is in setting each factor equal to 18. It is not necessarily true that when the product of two numbers is 18, either one of them is itself 18. The correct solution looks like this:
$$(x - 3)(x + 4) = 18$$
$$x^2 + x - 12 = 18$$
$$x^2 + x - 30 = 0$$
$$(x + 6)(x - 5) = 0$$
$$x + 6 = 0 \quad \text{or} \quad x - 5 = 0$$
$$x = -6 \quad \text{or} \quad x = 5$$

To avoid this mistake, remember that before you factor a quadratic equation, you must write it in standard form. It is in standard form only when 0 is on one side and decreasing powers of the variable are on the other.

Factor the following by factoring out the greatest common factor. [5.1]

1. $10x - 20$

2. $4x^3 - 9x^2$

3. $5x - 5y$

4. $7x^3 + 2x$

5. $49a^3 - 14b^3$

6. $6ab^2 + 18a^3b^3 - 24a^2b$

Factor by grouping. [5.1]

7. $xy + bx + ay + ab$

8. $xy + 4x - 5y - 20$

9. $2xy + 10x - 3y - 15$

10. $5x^2 - 4ax - 10bx + 8ab$

Factor the following trinomials. [5.2]

11. $y^2 + 9y + 14$

12. $w^2 + 15w + 50$

13. $y^2 + 20y + 99$

14. $y^2 + 8y + 12$

Factor the following trinomials. [5.3]

15. $2x^2 + 13x + 15$

16. $4y^2 - 12y + 5$

17. $6r^2 + 5rt - 6t^2$

18. $10x^2 - 29x - 21$

Factor the following if possible. [5.4]

19. $n^2 - 81$

20. $4y^2 - 9$

21 $x^2 + 49$

22. $36y^2 - 121x^2$

Factor the following. [5.4]

23. $64t^2 + 16t + 1$

24. $16n^2 - 24n + 9$

25. $4r^2 - 12rt + 9t^2$

26. $9m^2 + 30mn + 25n^2$

Factor the following. [5.2]

27. $2x^2 + 20x + 48$

28. $a^3 - 10a^2 + 21a$

29. $3m^3 - 18m^2 - 21m$

30. $5y^4 + 10y^3 - 40y^2$

Factor the following trinomials. [5.3]

31. $8x^2 + 16x + 6$

32. $3a^3 - 14a^2 - 5a$

33. $20m^3 - 34m^2 + 6m$

34. $30x^2y - 55xy^2 + 15y^3$

Factor the following. [5.4]

35. $4x^2 + 40x + 100$

36. $4x^3 + 12x^2 + 9x$

37. $5x^2 - 45$

38. $12x^3 - 27xy^2$

Factor the following. [5.5]

39. $27x^3 + 8y^3$

40. $125x^3 - 64y^3$

Factor the following polynomials completely. [5.6]

41. $6a^3b + 33a^2b^2 + 15ab^3$

42. $x^5 - x^3$

43. $4y^6 + 9y^4$

44. $18a^3b^2 + 3a^2b^3 - 6ab^4$

Solve. [5.7]

45. $(x - 5)(x + 2) = 0$

46. $a^2 - 49 = 0$

47. $6y^2 = -13y - 6$

48. $9x^4 + 9x^3 = 10x^2$

Solve the following word problems. [5.8]

49. Number Problem The product of two consecutive even integers is 120. Find the two integers.

50. Number Problem The product of two consecutive integers is 110. Find the two integers.

51. Number Problem The sum of two numbers is 20. Their product is 75. Find the numbers.

52. Geometry The height of a triangle is 8 times the base. The area is 16 square inches. Find the base.

Simplify.

1. $9 + (-7) + (-8)$

2. $20 - (-9)$

3. $\dfrac{-63}{-7}$

4. $\dfrac{9(-2)}{-2}$

5. $(-4)^3$

6. 9^{-2}

7. $\dfrac{-3(4-7) - 5(7-2)}{-5 - 2 - 1}$

8. $\dfrac{4^2 - 8^2}{(4-8)^2}$

9. $-a + 3 + 6a - 8$

10. $6 - 2(4a + 2) - 5$

11. $(x^4)^{10}$

12. $(9xy)^0$

13. $(5x - 2)(3x + 4)$

14. $(a^2 + 7)(a^2 - 7)$

Solve each equation.

15. $3x = -18$

16. $\dfrac{x}{2} = 5$

17. $-\dfrac{x}{3} = 7$

18. $\dfrac{1}{2}(4t - 1) + \dfrac{1}{3} = -\dfrac{25}{6}$

19. $4m(m - 7)(2m - 7) = 0$

20. $16x^2 - 81 = 0$

Solve each inequality.

21. $-2x > -8$

22. $2x - 7 \geq 7$

Graph on a rectangular coordinate system.

23. $y = -3x$

24. $y = 2$

25. $2x + 3y \geq 6$

26. $x < -2$

27. Which of the ordered pairs $(0, 3)$, $(4, 0)$, and $\left(\dfrac{16}{3}, 1\right)$ are solutions to the equation $3x - 4y = 12$?

28. Find the x- and y-intercepts for the equation $10x - 3y = 30$.

29. Find the slope of the line that passes through the points $(7, 3)$ and $(-2, -4)$.

30. Find the equation for the line with slope $\dfrac{2}{3}$ that passes through $(-6, 4)$.

31. Find the equation for the line with slope $-\dfrac{2}{5}$ and y-intercept $-\dfrac{2}{3}$.

32. Find the slope and y-intercept for the equation $3x - 4y = -16$.

Solve each system by graphing.

33. $\begin{aligned} y &= x + 3 \\ x + y &= -1 \end{aligned}$

34. $\begin{aligned} 2x - 2y &= 3 \\ 2x - 2y &= 2 \end{aligned}$

Solve each system.

35. $\begin{aligned} -x + y &= 3 \\ x + y &= 7 \end{aligned}$

36. $\begin{aligned} 5x + 7y &= -18 \\ 8x + 3y &= 4 \end{aligned}$

37. $\begin{aligned} 2x + y &= 4 \\ x &= y - 1 \end{aligned}$

38. $\begin{aligned} x + y &= 5{,}000 \\ 0.04x + 0.06y &= 270 \end{aligned}$

Factor completely.

39. $n^2 - 5n - 36$

40. $14x^2 + 31xy - 10y^2$

41. $16 - a^2$

42. $49x^2 - 14x + 1$

43. $45x^2y - 30xy^2 + 5y^3$

44. $18x^3 - 3x^2y - 3xy^2$

45. $3xy + 15x - 2y - 10$

46. $a^3 + 64$

Give the opposite, reciprocal, and absolute value of the given number.

47. -2

48. $\dfrac{1}{5}$

Divide using long division.

49. $\dfrac{4x^2 - 7x - 13}{x - 3}$

50. $\dfrac{x^3 - 27}{x - 3}$

51. Carpentry A 72-inch board is to be cut into two pieces. One piece is to be 4 inches longer than the other. How long is each piece?

52. Hamburgers and Fries Sheila bought burgers and fries for her children and some friends. The burgers cost $2.05 each, and the fries are $0.85 each. She bought a total of 14 items, for a total cost of $19.10. How many of each did she buy?

Use the bar chart to answer problems 53 and 54.

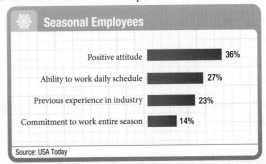

Seasonal Employees

Positive attitude — 36%
Ability to work daily schedule — 27%
Previous experience in industry — 23%
Commitment to work entire season — 14%

Source: USA Today

If 15,000 employers were surveyed, how many would list the following as their most valued attribute in seasonal employees? Round your answers to the nearest tenth.

53. Positive attitude

54. Commitment to work entire season

Factor out the greatest common factor. [5.1]

1. $5x - 10$

2. $18x^2y - 9xy - 36xy^2$

Factor by grouping. [5.1]

3. $x^2 + 2ax - 3bx - 6ab$

4. $xy + 4x - 7y - 28$

Factor the following completely. [5.2–5.6]

5. $x^2 - 5x + 6$ **6.** $x^2 - x - 6$

7. $a^2 - 16$ **8.** $x^2 + 25$

9. $x^4 - 81$

10. $27x^2 - 75y^2$

11. $x^3 + 5x^2 - 9x - 45$

12. $x^2 - bx + 5x - 5b$

13. $4a^2 + 22a + 10$

14. $3m^2 - 3m - 18$

15. $6y^2 + 7y - 5$

16. $12x^3 - 14x^2 - 10x$

17. $a^3 + 64b^3$

18. $54x^3 - 16$

Solve the following equations. [5.7]

19. $x^2 + 7x + 12 = 0$ **20.** $x^2 - 4x + 4 = 0$

21. $x^2 - 36 = 0$ **22.** $x^2 = x + 20$

23. $x^2 - 11x = -30$ **24.** $y^3 = 16y$

25. $2a^2 = a + 15$ **26.** $30x^3 - 20x^2 = 10x$

Solve the following word problems. Be sure to show the equation used. [5.8]

27. Number Problem Two numbers have a sum of 20. Their product is 64. Find the numbers.

28. Consecutive Integers The product of two consecutive odd integers is 7 more than their sum. Find the integers.

29. Geometry The length of a rectangle is 5 feet more than 3 times the width. The area is 42 square feet. Find the dimensions.

30. Geometry One leg of a right triangle is 2 meters more than twice the other. The hypotenuse is 13 meters. Find the lengths of the two legs.

31. Production Cost A company can manufacture x hundred items for a total profit P, given the equation $P = 200 + 500x - 100x^2$. How many items can be manufactured if the total profit is to be $800?

32. Price and Revenue A manufacturer knows that the number of items he can sell each week, x, is related to the price p of each item by the equation $x = 900 - 100p$. What price should he charge for each item to have a weekly revenue of $1,800? (*Remember:* $R = xp$.)

33. Pythagorean Theorem A 17 foot ladder is placed so that it leans against a wall at a point that is one foot less than twice the distance from the wall to the base of the ladder.

  a. How far from the wall is the ladder?

  b. How high does the ladder reach?

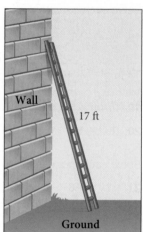

34. Projectile Motion A ball is thrown into the air with an upward velocity of 40 feet per second from a building that is 24 feet high. The equation that gives the height of the ball above the ground at time t is

$h = 24 + 40t - 16t^2$

  a. At what time(s) will the ball be 40 feet above the ground?

  b. When will the ball hit the ground?

Chapter 5 Projects

FACTORING

GROUP PROJECT

Visual Factoring

Number of People   2 or 3

Time Needed   10–15 minutes

Equipment   Pencil, graph paper, and scissors

Background   When a geometric figure is divided into smaller figures, the area of the original figure and the area of any rearrangement of the smaller figures must be the same. We can use this fact to help visualize some factoring problems.

Procedure   Use the diagram below to work the following problems.

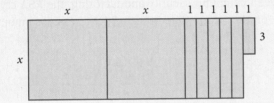

1. Write a polynomial involving x that gives the area of the diagram.

2. Factor the polynomial found in Part 1.

3. Copy the figure onto graph paper, then cut along the lines so that you end up with 2 squares and 6 rectangles.

4. Rearrange the pieces from Part 3 to show that the factorization you did in Part 2 is correct.

RESEARCH PROJECT

Factoring and Internet Security

The security of the information on computers is directly related to factoring of whole numbers. The key lies in the fact that multiplying whole numbers is a straightforward, simple task, whereas factoring can be very time-consuming. For example, multiplying the numbers 1,234 and 3,433 to obtain 4,236,322 takes very little time, even if done by hand. But given the number 4,236,322, finding its factors, even with a calculator or computer, is more than you want to try. The discipline that studies how to make and break codes is cryptography. The current Web browsers, such as Internet Explorer and Netscape, use a system called RSA public-key cryptosystem invented by Adi Shamir of Israel's Weizmann Institute of Science. In 1999, Shamir announced that he had found a method of factoring large numbers quickly that will put the current Internet security system at risk.

GABRIEL BOUYS/AFP/Getty Images

Research the connection between computer security and factoring, the RSA cryptosystem, and the current state of security on the Internet, and then write an essay summarizing your results.

6

Rational Expressions

Introduction

The Google Earth photograph shows the Bibliotheca Alexandria, a massive library in Alexandria, Egypt, which opened in 2003. Within the new library are rare artifacts from the city's ancient library, which existed from approximately 300 BC to 415 AD. One of the most famous books kept in the ancient library was *Euclid's Elements*, written in approximately 300 BC. The photograph to the right is from a later printing of *Euclid's Elements*.

Hypatia of Alexandria, the first woman mentioned in the history of mathematics, was the last librarian at the ancient library. She studied and revised *Elements* to make it more understandable for the other students at the library. Proportions, one of the topics we will study in this chapter, is also a topic in Euclid's *Elements*.

Simplify

1. $\dfrac{2}{9} \cdot \dfrac{15}{22}$

2. $\dfrac{3}{5} \div \dfrac{15}{7}$

3. $1 - \dfrac{5}{3}$

4. $1 + \dfrac{3}{5}$

5. $\dfrac{5}{0}$

6. $\dfrac{1}{10} + \dfrac{3}{14}$

7. $\dfrac{1}{21} + \dfrac{4}{15}$

Multiply.

8. $x(x + 2)$

9. $(x + 3)(x - 4)$

Factor completely.

10. $x^2 - 25$

11. $2x - 4$

12. $x^2 + 7x + 12$

13. $a^2 - 4a$

14. $xy^2 + y$

Solve.

15. $72 = 2x$

16. $15 = 3x - 3$

17. $a^2 - a - 20 = -2a$

18. $x^2 + 2x = 8$

19. Use the formula $y = \dfrac{20}{x}$ to find y when $x = 10$.

20. Use the formula $y = 2x^2$ to find x when $y = 72$.

Chapter Outline

6.1 Reducing Rational Expressions to Lowest Terms

A Find the restrictions on the variable in a rational expression.
B Reduce a rational expression to lowest terms.
C Work problems involving ratios.

6.2 Multiplication and Division of Rational Expressions

A Multiply and divide rational expressions by factoring and then dividing out common factors.
B Convert between units using unit analysis.

6.3 Addition and Subtraction of Rational Expressions

A Add and subtract rational expressions that have the same denominators.
B Add and subtract rational expressions that have different denominators.

6.4 Equations Involving Rational Expressions

A Solve equations that contain rational expressions.

6.5 Applications of Rational Expressions

A Solve applications whose solutions depend on solving an equation containing rational expressions.
B Graph an equation involving a rational expression.

6.6 Complex Fractions

A Simplify a complex fraction.

6.7 Proportions

A Solve a proportion.
B Solve application problems involving proportions.

OBJECTIVES

A Find the restrictions on the variable in a rational expression.

B Reduce a rational expression to lowest terms.

C Work problems involving ratios.

TICKET TO SUCCESS

Keep these questions in mind as you read through the section. Then respond in your own words and in complete sentences.

1. Explain what it means to have a restriction on a rational expression.
2. What properties would you use to manipulate a rational expression?
3. How do you reduce a rational expression to lowest terms?
4. What is a ratio?

In 1986, the Colossus Ferris Wheel was built at a staggering 180 feet tall in St. Louis, Missouri. The Ferris wheel has 40 gondolas that carry up to 8 passengers each. In this section, we will introduce the rate equation, which will allow us to calculate the speed of any given passenger as he travels around the wheel. To help, we will start by learning about rational expressions and how to reduce them to lowest terms.

We have defined the set of rational numbers to be the set of all numbers that could be put in the form $\frac{a}{b}$, where a and b are integers ($b \neq 0$).

$$\text{Rational numbers} = \left\{ \frac{a}{b} \,\middle|\, a \text{ and } b \text{ are integers}, b \neq 0 \right\}$$

> **Definition**
> A **rational expression** is any expression that can be put in the form $\frac{P}{Q}$, where P and Q are polynomials and $Q \neq 0$.
>
> $$\text{Rational expressions} = \left\{ \frac{P}{Q} \,\middle|\, P \text{ and } Q \text{ are polynomials}, Q \neq 0 \right\}$$

Each of the following is an example of a rational expression:

$$\frac{2x + 3}{x} \qquad \frac{x^2 - 6x + 9}{x^2 - 4} \qquad \frac{5}{x^2 + 6} \qquad \frac{2x^2 + 3x + 4}{2}$$

For the rational expression

$$\frac{x^2 - 6x + 9}{x^2 - 4} \quad \begin{array}{l} \leftarrow \text{numerator} \\ \leftarrow \text{denominator} \end{array}$$

the polynomial on top, $x^2 - 6x + 9$, is called the numerator, and the polynomial on the bottom, $x^2 - 4$, is called the denominator. The same is true of the other rational expressions.

A Restrictions on Variables

When working with rational expressions, we must be careful that we do not use a value of the variable that will give us a denominator of zero. Remember, division by zero is not defined.

EXAMPLES State the restrictions on the variable in the following rational expressions:

1. $\dfrac{x + 2}{x - 3}$

SOLUTION The variable x can be any real number except $x = 3$ since, when $x = 3$, the denominator is $3 - 3 = 0$. We state this restriction by writing $x \neq 3$.

2. $\dfrac{5}{x^2 - x - 6}$

SOLUTION If we factor the denominator, we have $x^2 - x - 6 = (x - 3)(x + 2)$. By applying our knowledge of the zero-factor property if either of the factors is zero, then the whole denominator is zero. Our restrictions are $x \neq 3$ and $x \neq -2$ since either one makes $x^2 - x - 6 = 0$.

We will not always list each restriction on a rational expression, but we should be aware of them and keep in mind that no rational expression can have a denominator of zero.

The two fundamental properties of rational expressions are listed next. We will use these two properties many times in this chapter.

Properties of Rational Expressions

Property 1
Multiplying the numerator and denominator of a rational expression by the same nonzero quantity will not change the value of the rational expression.

Property 2
Dividing the numerator and denominator of a rational expression by the same nonzero quantity will not change the value of the rational expression.

B Reducing Rational Expressions

We can use Property 2 to reduce rational expressions to lowest terms. Since this process is almost identical to the process of reducing fractions to lowest terms, let's recall how the fraction $\frac{6}{15}$ is reduced to lowest terms.

$$\frac{6}{15} = \frac{2 \cdot 3}{5 \cdot 3} \qquad \text{Factor numerator and denominator}$$

$$= \frac{2 \cdot \cancel{3}}{5 \cdot \cancel{3}} \qquad \text{Divide out the common factor, 3}$$

$$= \frac{2}{5} \qquad \text{Reduce to lowest terms}$$

The same procedure applies to reducing rational expressions to lowest terms. The process is summarized in the following rule:

> **Rule: Reducing Rational Expressions**
> To reduce a rational expression to lowest terms, first factor the numerator and denominator completely and then divide both the numerator and denominator by any factors they have in common.

EXAMPLE 3   Reduce $\dfrac{x^2 - 9}{x^2 + 5x + 6}$ to lowest terms.

SOLUTION   We begin by factoring.

$$\frac{x^2 - 9}{x^2 + 5x + 6} = \frac{(x - 3)(x + 3)}{(x + 2)(x + 3)}$$

Notice that both polynomials contain the factor $(x + 3)$. If we divide the numerator by $(x + 3)$, we are left with $(x - 3)$. If we divide the denominator by $(x + 3)$, we are left with $(x + 2)$. The complete solution looks like this:

$$\frac{x^2 - 9}{x^2 + 5x + 6} = \frac{(x - 3)\cancel{(x + 3)}}{(x + 2)\cancel{(x + 3)}} \qquad \begin{array}{l}\text{Factor the numerator and} \\ \text{denominator completely}\end{array}$$

$$= \frac{x - 3}{x + 2} \qquad \begin{array}{l}\text{Divide out the common} \\ \text{factor, } x + 3\end{array}$$

It is convenient to draw a line through the factors as we divide them out. It is especially helpful when the problems become longer.

EXAMPLE 4   Reduce to lowest terms $\dfrac{10a + 20}{5a^2 - 20}$.

SOLUTION   We begin by factoring out the greatest common factor from the numerator and denominator.

$$\frac{10a + 20}{5a^2 - 20} = \frac{10(a + 2)}{5(a^2 - 4)} \qquad \begin{array}{l}\text{Factor out the greatest common factor} \\ \text{from the numerator and denominator}\end{array}$$

$$= \frac{10\cancel{(a + 2)}}{5\cancel{(a + 2)}(a - 2)} \qquad \begin{array}{l}\text{Factor the denominator as the} \\ \text{difference of two squares}\end{array}$$

$$= \frac{2}{a - 2} \qquad \begin{array}{l}\text{Divide out the common factors 5 and} \\ a + 2\end{array}$$

EXAMPLE 5 Reduce $\dfrac{2x^3 + 2x^2 - 24x}{x^3 + 2x^2 - 8x}$ to lowest terms.

SOLUTION We begin by factoring the numerator and denominator completely. Then we divide out all factors common to the numerator and denominator. Here is what it looks like:

$$\frac{2x^3 + 2x^2 - 24x}{x^3 + 2x^2 - 8x} = \frac{2x(x^2 + x - 12)}{x(x^2 + 2x - 8)}$$ Factor out the greatest common factor first

$$= \frac{2\cancel{x}(x - 3)\cancel{(x + 4)}}{\cancel{x}(x - 2)\cancel{(x + 4)}}$$ Factor the remaining trinomials

$$= \frac{2(x - 3)}{x - 2}$$ Divide out the factors common to the numerator and denominator ∎

EXAMPLE 6 Reduce $\dfrac{x - 5}{x^2 - 25}$ to lowest terms.

SOLUTION We can factor the denominator, recognizing it as a difference of squares.

$$\frac{x - 5}{x^2 - 25} = \frac{\cancel{x - 5}}{\cancel{(x - 5)}(x + 5)}$$ Factor numerator and denominator completely

$$= \frac{1}{x + 5}$$ Divide out the common factor, $x - 5$ ∎

EXAMPLE 7 Reduce $\dfrac{x^3 + y^3}{x^2 - y^2}$ to lowest terms.

SOLUTION We begin by factoring the numerator and denominator completely. (Remember, we can only reduce to lowest terms when the numerator and denominator are in factored form. Trying to reduce before factoring will only lead to mistakes.)

$$\frac{x^3 + y^3}{x^2 - y^2} = \frac{\cancel{(x + y)}(x^2 - xy + y^2)}{\cancel{(x + y)}(x - y)}$$ Factor

$$= \frac{x^2 - xy + y^2}{x - y}$$ Divide out the common factor ∎

C Ratios

For the rest of this section we will concern ourselves with ratios, a topic closely related to reducing fractions and rational expressions to lowest terms. Let's start with a definition.

Definition

If a and b are any two numbers, $b \neq 0$, then the **ratio** of a and b is

$$\frac{a}{b}$$

As you can see, ratios are another name for fractions or rational numbers. They are a way of comparing quantities. Since we also can think of $\frac{a}{b}$ as the quotient of a and b, ratios are also quotients. The following table gives some ratios in words and as fractions.

Ratio	As a Fraction	In Lowest Terms
25 to 75	$\frac{25}{75}$	$\frac{1}{3}$
8 to 2	$\frac{8}{2}$	$\frac{4}{1}$
20 to 16	$\frac{20}{16}$	$\frac{5}{4}$

EXAMPLE 8   A solution of hydrochloric acid (HCl) and water contains 49 milliliters of water and 21 milliliters of HCl. Find the ratio of HCl to water and of HCl to the total volume of the solution.

SOLUTION   The ratio of HCl to water is 21 to 49, or

$$\frac{21}{49} = \frac{3}{7}$$

The amount of total solution volume is 49 + 21 = 70 milliliters. Therefore, the ratio of HCl to total solution is 21 to 70, or

$$\frac{21}{70} = \frac{3}{10}$$

Rate Equation

Many of the problems in this chapter will use what is called the *rate equation.* You use this equation on an intuitive level when you are estimating how long it will take you to drive long distances. For example, if you drive at 50 miles per hour for 2 hours, you will travel 100 miles. Here is the rate equation:

$$\text{Distance} = \text{rate} \cdot \text{time}$$

$$d = r \cdot t$$

The rate equation has two equivalent forms, the most common of which is obtained by solving for r. Here it is:

$$r = \frac{d}{t}$$

The rate r in the rate equation is the ratio of distance to time and also is referred to as *average speed.* The units for rate are miles per hour, feet per second, kilometers per hour, and so on.

EXAMPLE 9   The Forest Chair Lift at the Northstar Ski Resort in Lake Tahoe is 5,603 feet long. If a ride on this chair lift takes 11 minutes, what is the average speed of the lift in feet per minute?

SOLUTION   To find the speed of the lift, we find the ratio of distance covered to time. (Our answer is rounded to the nearest whole number.)

$$\text{Rate} = \frac{\text{distance}}{\text{time}} = \frac{5{,}603 \text{ feet}}{11 \text{ minutes}} = \frac{5{,}603}{11} \text{ feet/minute} = 509 \text{ feet/minute}$$

Note how we separate the numerical part of the problem from the units. In the next section, we will convert this rate to miles per hour.

l = 5,603 ft

The Forest Chair Lift
Northstar Ski Resort, Lake Tahoe

EXAMPLE 10 Recall the Ferris wheel named *Colossus* we discussed at the beginning of the section. The circumference of the wheel is 518 feet. It has 40 cars, each of which holds 8 passengers. A trip around the wheel takes 40 seconds. Find the average speed of a rider on *Colossus*.

Circumference

SOLUTION To find the average speed, we divide the distance traveled, which in this case is the circumference, by the time it takes to travel once around the wheel.

$$r = \frac{d}{t} = \frac{518 \text{ feet}}{40 \text{ seconds}} = 13.0 \text{ feet/second (rounded)}$$

The average speed of a rider on the *Colossus* is approximately 13.0 feet per second.

In the next section, you will convert the ratio into an equivalent ratio that gives the speed of the rider in miles per hour.

Problem Set 6.1

Moving Toward Success

"Nothing will work unless you do."

—John Wooden, 1910–2010, former UCLA basketball coach

1. What traits do you like in a study partner?
2. Have you possessed those traits when studying for this class? Why or why not?

1. Simplify each expression.

a. $\dfrac{5 + 1}{25 - 1}$

b. $\dfrac{x + 1}{x^2 - 1}$

c. $\dfrac{x^2 - x}{x^2 - 1}$

d. $\dfrac{x^3 - 1}{x^2 - 1}$

e. $\dfrac{x^3 - 1}{x^3 - x^2}$

2. Simplify each expression.

a. $\dfrac{25 - 30 + 9}{25 - 9}$

b. $\dfrac{x^2 - 6x + 9}{x^2 - 9}$

c. $\dfrac{x^2 - 10x + 9}{x^2 - 9x}$

d. $\dfrac{x^2 + 3x + ax + 3a}{x^2 - 9}$

e. $\dfrac{x^3 + 27}{x^3 - 9x}$

A **B** Reduce the following rational expressions to lowest terms, if possible. Also, specify any restrictions on the variable in Problems 3 through 12. [Examples 1–7]

3. $\dfrac{5}{5x - 10}$

4. $\dfrac{-4}{2x - 8}$

5. $\dfrac{a - 3}{a^2 - 9}$

6. $\dfrac{a + 4}{a^2 - 16}$

7. $\dfrac{x + 5}{x^2 - 25}$

8. $\dfrac{x - 2}{x^2 - 4}$

9. $\dfrac{2x^2 - 8}{4}$

10. $\dfrac{5x - 10}{x - 2}$

11. $\dfrac{2x - 10}{3x - 6}$

12. $\dfrac{4x - 8}{x - 2}$

13. $\dfrac{10a + 20}{5a + 10}$

14. $\dfrac{11a + 33}{6a + 18}$

15. $\dfrac{5x^2 - 5}{4x + 4}$

16. $\dfrac{7x^2 - 28}{2x + 4}$

17. $\dfrac{x - 3}{x^2 - 6x + 9}$

18. $\dfrac{x^2 - 10x + 25}{x - 5}$

19. $\dfrac{3x + 15}{3x^2 + 24x + 45}$

20. $\dfrac{5x + 15}{5x^2 + 40x + 75}$

21. $\dfrac{a^2 - 3a}{a^3 - 8a^2 + 15a}$

22. $\dfrac{a^2 + 3a}{a^3 - 2a^2 - 15a}$

23. $\dfrac{3x - 2}{9x^2 - 4}$

24. $\dfrac{2x - 3}{4x^2 - 9}$

25. $\dfrac{x^2 + 8x + 15}{x^2 + 5x + 6}$

26. $\dfrac{x^2 - 8x + 15}{x^2 - x - 6}$

27. $\dfrac{2m^3 - 2m^2 - 12m}{m^2 - 5m + 6}$

28. $\dfrac{2m^3 + 4m^2 - 6m}{m^2 - m - 12}$

29. $\dfrac{x^3 + 3x^2 - 4x}{x^3 - 16x}$

30. $\dfrac{3a^2 - 8a + 4}{9a^3 - 4a}$

31. $\dfrac{4x^3 - 10x^2 + 6x}{2x^3 + x^2 - 3x}$

32. $\dfrac{3a^3 - 8a^2 + 5a}{4a^3 - 5a^2 + 1a}$

33. $\dfrac{4x^2 - 12x + 9}{4x^2 - 9}$

34. $\dfrac{5x^2 + 18x - 8}{5x^2 + 13x - 6}$

35. $\dfrac{x + 3}{x^4 - 81}$

36. $\dfrac{x^2 + 9}{x^4 - 81}$

37. $\dfrac{3x^2 + x - 10}{x^4 - 16}$

38. $\dfrac{5x^2 - 16x + 12}{x^4 - 16}$

39. $\dfrac{42x^3 - 20x^2 - 48x}{6x^2 - 5x - 4}$

40. $\dfrac{36x^3 + 132x^2 - 135x}{6x^2 + 25x - 9}$

41. $\dfrac{x^3 - y^3}{x^2 - y^2}$

42. $\dfrac{x^3 + y^3}{x^2 - y^2}$

43. $\dfrac{x^3 + 8}{x^2 - 4}$

44. $\dfrac{x^3 - 125}{x^2 - 25}$

45. $\dfrac{x^3 + 8}{x^2 + x - 2}$

46. $\dfrac{x^2 - 2x - 3}{x^3 - 27}$

B To reduce each of the following rational expressions to lowest terms, you will have to use factoring by grouping. Be sure to factor each numerator and denominator completely before dividing out any common factors. (Remember, factoring by grouping takes two steps.)

47. $\dfrac{xy + 3x + 2y + 6}{xy + 3x + 5y + 15}$

48. $\dfrac{xy + 7x + 4y + 28}{xy + 3x + 4y + 12}$

49. $\dfrac{x^2 - 3x + ax - 3a}{x^2 - 3x + bx - 3b}$

50. $\dfrac{x^2 - 6x + ax - 6a}{x^2 - 7x + ax - 7a}$

51. $\dfrac{xy + bx + ay + ab}{xy + bx + 3y + 3b}$

52. $\dfrac{x^2 + 5x + ax + 5a}{x^2 + 5x + bx + 5b}$

The next two problems are intended to give you practice reading, and paying attention to, the instructions that accompany the problems you are working. Working these problems is an excellent way to get ready for a test or quiz.

53. Work each problem according to the instructions given.

   a. Add: $(x^2 - 4x) + (4x - 16)$

   b. Subtract: $(x^2 - 4x) - (4x - 16)$

   c. Multiply: $(x^2 - 4x)(4x - 16)$

   d. Reduce: $\dfrac{x^2 - 4x}{4x - 16}$

54. Work each problem according to the instructions given.

   a. Add: $(9x^2 - 3x) + (6x - 2)$

   b. Subtract: $(9x^2 - 3x) - (6x - 2)$

   c. Multiply: $(9x^2 - 3x)(6x - 2)$

   d. Reduce: $\dfrac{9x^2 - 3x}{6x - 2}$

C Write each ratio as a fraction in lowest terms. [Example 8]

55. 8 to 6

56. 6 to 8

57. 200 to 250

58. 250 to 200

59. 32 to 4

60. 4 to 32

Applying the Concepts [Examples 9–10]

61. Cars The chart shows the fastest cars in the world. Which car or cars could travel 375 miles in one-and-one-half hours?

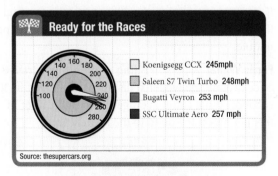

Ready for the Races

Koenigsegg CCX **245mph**

Saleen S7 Twin Turbo **248mph**

Bugatti Veyron **253 mph**

SSC Ultimate Aero **257 mph**

Source: thesupercars.org

62. Fifth Avenue Mile The chart shows the times of the five fastest runners for the 2009 Continental Airlines Fifth Avenue Mile. Insert < or > to make each statement regarding the runners' speeds true (based on the times given below).

   a. H. Lagat   Baddeley

   b. Lalang   Manzano

   c. B. Lagat   Lalang

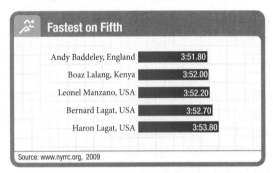

Fastest on Fifth

Andy Baddeley, England	3:51.80
Boaz Lalang, Kenya	3:52.00
Leonel Manzano, USA	3:52.20
Bernard Lagat, USA	3:52.70
Haron Lagat, USA	3:53.80

Source: www.nyrrc.org, 2009

63. Speed of a Car A car travels 122 miles in 3 hours. Find the average speed of the car in miles per hour. Round to the nearest tenth.

64. Speed of a Bullet A bullet fired from a gun travels a distance of 4,500 feet in 3 seconds. Find the average speed of the bullet in feet per second.

65. Ferris Wheel The first Ferris wheel was designed and built by George Ferris in 1893. It was a large wheel with a circumference of 785 feet. If one trip around the circumference of the wheel took 20 minutes, find the average speed of a rider in feet per minute.

66. Ferris Wheel In 1897 a large Ferris wheel was built in Vienna; it is still in operation today. Known as *The Great Wheel*, it has a circumference of 618 feet. If one trip around the wheel takes 15 minutes, find the average speed of a rider on this wheel in feet per minute.

67. Ferris Wheel A person riding a Ferris wheel travels once around the wheel, a distance of 188 feet, in 30 seconds. What is the average speed of the rider in feet per second? Round to the nearest tenth.

68. Average Speed Tina is training for a biathlon. As part of her training, she runs an 8-mile course, 2 miles of which is on level ground and 6 miles of which is downhill. It takes her 20 minutes to run the level part of the course and 40 minutes to run the downhill part of the course. Find her average speed in minutes per mile and in miles per minute for each part of the course. Round to the nearest hundredth, if rounding is necessary.

69. Fuel Consumption An economy car travels 168 miles on 3.5 gallons of gas. Give the average fuel consumption of the car in miles per gallon.

70. Fuel Consumption A luxury car travels 100 miles on 8 gallons of gas. Give the average fuel consumption of the car in miles per gallon.

71. Comparing Expressions Replace x with 5 and y with 4 in the expression
$$\frac{x^2 - y^2}{x - y}$$
and simplify the result. Is the result equal to $5 - 4$ or $5 + 4$?

72. Comparing Expressions Replace x with 2 in the expression
$$\frac{x^3 - 1}{x - 1}$$
and simplify the result. Your answer should be equal to what you would get if you replaced x with 2 in $x^2 + x + 1$.

73. Comparing Expressions Complete the following table; then show why the table turns out as it does.

x	$\dfrac{x - 3}{3 - x}$
−2	
−1	
0	
1	
2	

74. Comparing Expressions Complete the following table; then show why the table turns out as it does.

x	$\dfrac{25 - x^2}{x^2 - 25}$
−4	
−2	
0	
2	
4	

75. Comparing Expressions You know from reading through Example 6 in this section that $\frac{x-5}{x^2-25} = \frac{1}{x+5}$. Compare these expressions by completing the following table. (Be careful—not all the rows have equal entries.)

x	$\dfrac{x-5}{x^2-25}$	$\dfrac{1}{x+5}$
0		
2		
−2		
5		
−5		

76. Comparing Expressions You know from your work in this section that $\frac{x^2-6x+9}{x^2-9} = \frac{x-3}{x+3}$. Compare these expressions by completing the following table. (Be careful—not all the rows have equal entries.)

x	$\dfrac{x^2-6x+9}{x^2-9}$	$\dfrac{x-3}{x+3}$
−3		
−2		
−1		
0		
1		
2		
3		

Getting Ready for the Next Section

Perform the indicated operation.

77. $\dfrac{3}{4} \cdot \dfrac{10}{21}$

78. $\dfrac{2}{90} \cdot \dfrac{15}{22}$

79. $\dfrac{4}{5} \div \dfrac{8}{9}$

80. $\dfrac{3}{5} \div \dfrac{15}{7}$

Factor completely.

81. $x^2 - 9$

82. $x^2 - 25$

83. $3x - 9$

84. $2x - 4$

85. $x^2 - x - 20$

86. $x^2 + 7x + 12$

87. $a^2 + 5a$

88. $a^2 - 4a$

Reduce to lowest terms.

89. $\dfrac{a(a+5)(a-5)(a+4)}{a^2+5a}$

90. $\dfrac{a(a+2)(a-4)(a+5)}{a^2-4a}$

Multiply. Give the answers as decimals rounded to the nearest tenth.

91. $\dfrac{5,603}{11} \cdot \dfrac{1}{5,280} \cdot \dfrac{60}{1}$

92. $\dfrac{772}{2.2} \cdot \dfrac{1}{5,280} \cdot \dfrac{60}{1}$

Maintaining Your Skills

Simplify.

93. $\dfrac{27x^5}{9x^2} - \dfrac{45x^8}{15x^5}$

94. $\dfrac{36x^9}{4x} - \dfrac{45x^3}{5x^{-5}}$

95. $\dfrac{72a^3b^7}{9ab^5} + \dfrac{64a^5b^3}{8a^3b}$

96. $\dfrac{80a^5b^{11}}{10a^2b} + \dfrac{33a^6b^{12}}{11a^3b^2}$

Divide.

97. $\dfrac{38x^7 + 42x^5 - 84x^3}{2x^3}$

98. $\dfrac{49x^6 - 63x^4 - 35x^2}{7x^2}$

99. $\dfrac{28a^5b^5 + 36ab^4 - 44a^4b}{4ab}$

100. $\dfrac{30a^3b - 12a^2b^2 + 6ab^3}{6ab}$

101. Stock Market One method of comparing stocks on the stock market is the price to earnings ratio, or P/E.

$$P/E = \frac{\text{Current Stock Price}}{\text{Earnings per Share}}$$

Most stocks have a P/E between 25 and 40. A stock with a P/E of less than 25 may be undervalued, while a stock with a P/E greater than 40 may be overvalued. Fill in the P/E for each stock listed in the table below. Based on your results, are any of the stocks undervalued?

Stock	Price	Earnings per Share	P/E
Yahoo	37.80	1.07	
Google	381.24	4.51	
Disney	24.96	1.34	
Nike	85.46	4.88	
eBay	40.96	0.73	

TICKET TO SUCCESS

Keep these questions in mind as you read through the section. Then respond in your own words and in complete sentences.

1. How do we multiply rational expressions?
2. Explain the steps used to divide rational expressions.
3. Give an example of an application that uses unit analysis.
4. Why are all conversion factors the same as the number 1?

In this section, we will learn how to apply our knowledge of multiplying and dividing rational expressions to a new concept called *unit analysis*. Consider a skier riding a chair lift to the top of a snow-covered slope. The skier's riding speed is given by the rate equation $\left(\text{rate} = \frac{\text{distance}}{\text{time}}\right)$ in feet per minute. You will see later how to convert this feet-per-minute rate to that of one in miles per hour. Let's first work some problems to practice multiplying and dividing rational expressions.

A Multiplying and Dividing Rational Expressions

Recall that to multiply two fractions we simply multiply numerators and multiply denominators and then reduce to lowest terms, if possible.

$$\frac{3}{4} \cdot \frac{10}{21} = \frac{30}{84} \qquad \text{Multiply numerators} \\ \text{Multiply denominators}$$

$$= \frac{5}{14} \qquad \text{Reduce to lowest terms}$$

Recall also that the same result can be achieved by factoring numerators and denominators first and then dividing out the factors they have in common.

$$\frac{3}{4} \cdot \frac{10}{21} = \frac{3}{2 \cdot 2} \cdot \frac{2 \cdot 5}{3 \cdot 7} \qquad \text{Factor}$$

$$= \frac{3 \cdot 2 \cdot 5}{2 \cdot 2 \cdot 3 \cdot 7} \qquad \text{Multiply numerators} \\ \text{Multiply denominators}$$

$$= \frac{5}{14} \qquad \text{Divide out common factors}$$

We can apply the second process to the product of two rational expressions, as the following example illustrates.

EXAMPLE 1   Multiply $\dfrac{x - 2}{x + 3} \cdot \dfrac{x^2 - 9}{2x - 4}$.

SOLUTION   We begin by factoring numerators and denominators as much as possible. Then we multiply the numerators and denominators. The last step consists of dividing out all factors common to the numerator and denominator.

$$\frac{x - 2}{x + 3} \cdot \frac{x^2 - 9}{2x - 4} = \frac{x - 2}{x + 3} \cdot \frac{(x - 3)(x + 3)}{2(x - 2)} \qquad \text{Factor completely}$$

$$= \frac{(x - 2)(x - 3)(x + 3)}{(x + 3)(2)(x - 2)} \qquad \begin{array}{l}\text{Multiply numerators} \\ \text{and denominators}\end{array}$$

$$= \frac{x - 3}{2} \qquad \text{Divide out common factors}$$

In Chapter 1, we defined division as the equivalent of multiplication by the reciprocal. This is how it looks with fractions:

$$\frac{4}{5} \div \frac{8}{9} = \frac{4}{5} \cdot \frac{9}{8} \qquad \text{Division as multiplication by the reciprocal}$$

$$\left. \begin{array}{l} = \dfrac{2 \cdot 2 \cdot 3 \cdot 3}{5 \cdot 2 \cdot 2 \cdot 2} \\[2em] = \dfrac{9}{10} \end{array} \right\} \text{Factor and divide out common factors}$$

The same idea holds for division with rational expressions. The rational expression that follows the division symbol is called the *divisor;* to divide, we multiply by the reciprocal of the divisor.

EXAMPLE 2   Divide $\dfrac{3x - 9}{x^2 - x - 20} \div \dfrac{x^2 + 2x - 15}{x^2 - 25}$.

SOLUTION   We begin by taking the reciprocal of the divisor and writing the problem again in terms of multiplication. We then factor, multiply, and, finally, divide out all factors common to the numerator and denominator of the resulting expression. The complete solution looks like this:

$$\frac{3x - 9}{x^2 - x - 20} \div \frac{x^2 + 2x - 15}{x^2 - 25}$$

$$= \frac{3x - 9}{x^2 - x - 20} \cdot \frac{x^2 - 25}{x^2 + 2x - 15} \qquad \text{Multiply by the reciprocal of the divisor}$$

$$= \frac{3(x - 3)}{(x + 4)(x - 5)} \cdot \frac{(x - 5)(x + 5)}{(x + 5)(x - 3)} \qquad \text{Factor}$$

$$= \frac{3(x - 3)(x - 5)(x + 5)}{(x + 4)(x - 5)(x + 5)(x - 3)} \qquad \text{Multiply}$$

$$= \frac{3}{x + 4} \qquad \text{Divide out common factors}$$

As you can see, factoring is the single most important tool we use in working with rational expressions. It is easier to work with rational expressions if they are in factored form. Here are more examples of multiplication and division with rational expressions.

EXAMPLE 3 Multiply $\dfrac{3a + 6}{a^2} \cdot \dfrac{a}{2a + 4}$.

SOLUTION

$$\dfrac{3a + 6}{a^2} \cdot \dfrac{a}{2a + 4}$$

$$= \dfrac{3(a + 2)}{a^2} \cdot \dfrac{a}{2(a + 2)} \qquad \text{Factor completely}$$

$$= \dfrac{3\,\cancel{(a + 2)}\,\cancel{a}}{a^2(2)\,\cancel{(a + 2)}} \qquad \text{Multiply}$$

$$= \dfrac{3}{2a} \qquad \begin{array}{l}\text{Divide numerator and denominator}\\ \text{by common factors } a\,(a + 2)\end{array}$$

EXAMPLE 4 Divide $\dfrac{x^2 + 7x + 12}{x^2 - 16} \div \dfrac{x^2 + 6x + 9}{2x - 8}$.

SOLUTION

$$\dfrac{x^2 + 7x + 12}{x^2 - 16} \div \dfrac{x^2 + 6x + 9}{2x - 8}$$

$$= \dfrac{x^2 + 7x + 12}{x^2 - 16} \cdot \dfrac{2x - 8}{x^2 + 6x + 9} \qquad \begin{array}{l}\text{Division is multiplication}\\ \text{by the reciprocal}\end{array}$$

$$= \dfrac{\cancel{(x + 3)}\,\cancel{(x + 4)}\,(2)\,\cancel{(x - 4)}}{\cancel{(x - 4)}\,\cancel{(x + 4)}\,\cancel{(x + 3)}\,(x + 3)} \qquad \text{Factor and multiply}$$

$$= \dfrac{2}{x + 3} \qquad \text{Divide out common factors}$$

In this example, we factored and multiplied the two expressions in a single step. This saves writing the problem one extra time.

EXAMPLE 5 Multiply $(x^2 - 49)\left(\dfrac{x + 4}{x + 7}\right)$.

SOLUTION We can think of the polynomial $x^2 - 49$ as having a denominator of 1. Thinking of $x^2 - 49$ in this way allows us to proceed as we did in previous examples.

$$(x^2 - 49)\left(\dfrac{x + 4}{x + 7}\right) = \dfrac{x^2 - 49}{1} \cdot \dfrac{x + 4}{x + 7} \qquad \text{Write } x^2 - 49 \text{ with denominator 1}$$

$$= \dfrac{\cancel{(x + 7)}\,(x - 7)\,(x + 4)}{\cancel{x + 7}} \qquad \text{Factor and multiply}$$

$$= (x - 7)(x + 4) \qquad \text{Divide out common factors}$$

We can leave the answer in this form or multiply to get $x^2 - 3x - 28$. In this section, let's agree to leave our answers in factored form.

EXAMPLE 6 Multiply $a(a + 5)(a - 5)\left(\dfrac{a + 4}{a^2 + 5a}\right)$.

SOLUTION We can think of the expression $a(a + 5)(a - 5)$ as having a denominator of 1.

$$a(a + 5)(a - 5)\left(\dfrac{a + 4}{a^2 + 5a}\right)$$

$$= \dfrac{a(a + 5)(a - 5)}{1} \cdot \dfrac{a + 4}{a^2 + 5a}$$

$$= \frac{\cancel{d}\cancel{(a+5)}(a-5)(a+4)}{\cancel{d}\cancel{(a+5)}} \qquad \text{Factor and multiply}$$

$$= (a-5)(a+4) \qquad \text{Divide out common factors}$$

B Unit Analysis

Unit analysis is a method of converting between units of measure by multiplying by the number 1. Here is our first illustration: Suppose you are flying in a commercial airliner and the pilot tells you the plane has reached its cruising altitude of 35,000 feet. How many miles is the plane above the ground?

If you know that 1 mile is 5,280 feet, then it is simply a matter of deciding what to do with the two numbers, 5,280 and 35,000. By using unit analysis, this decision is unnecessary.

$$35,000 \text{ feet} = \frac{35,000 \text{ feet}}{1} \cdot \frac{1 \text{ mile}}{5,280 \text{ feet}}$$

We treat the units common to the numerator and denominator in the same way we treat factors common to the numerator and denominator; common units can be divided out, just as common factors are. In the previous expression, we have feet common to the numerator and denominator. Dividing them out leaves us with miles only. Here is the complete solution:

$$35,000 \text{ feet} = \frac{35,000 \cancel{\text{ feet}}}{1} \cdot \frac{1 \text{ mile}}{5,280 \cancel{\text{ feet}}}$$

$$= \frac{35,000}{5,280} \text{ miles}$$

$$= 6.6 \text{ miles to the nearest tenth of a mile}$$

The expression $\frac{1 \text{ mile}}{5,280 \text{ feet}}$ is called a *conversion factor*. It is simply the number 1 written in a convenient form. Because it is the number 1, we can multiply any other number by it and always be sure we have not changed that number. The key to unit analysis is choosing the right conversion factors.

EXAMPLE 7   The Mall of America in Minnesota covers 78 acres of land. If 1 square mile = 640 acres, how many square miles does the Mall of America cover? Round your answer to the nearest hundredth of a square mile.

SOLUTION   We are starting with acres and want to end up with square miles. We need to multiply by a conversion factor that will allow acres to divide out and leave us with square miles.

$$78 \text{ acres} = \frac{78 \cancel{\text{ acres}}}{1} \cdot \frac{1 \text{ square mile}}{640 \cancel{\text{ acres}}}$$

$$= \frac{78}{640} \text{ square miles}$$

$$= 0.12 \text{ square miles to the nearest hundredth}$$

EXAMPLE 8   The Forest chair lift at the Northstar ski resort in Lake Tahoe is 5,603 feet long. If a ride on this chair lift takes 11 minutes, what is the average speed of the lift in miles per hour?

The Forest Chair Lift
Northstar Ski Resort, Lake Tahoe

SOLUTION First, we find the speed of the lift in feet per second, as we did in Example 9 of Section 6.1, by taking the ratio of distance to time.

$$\text{Rate} = \frac{\text{distance}}{\text{time}} = \frac{5{,}603 \text{ feet}}{11 \text{ minutes}} = \frac{5{,}603}{11} \text{ feet per minute}$$
$$= 509 \text{ feet per minute}$$

Next, we convert feet per minute to miles per hour. To do this, we need to know that

1 mile = 5,280 feet
1 hour = 60 minutes

$$\text{Speed} = 509 \text{ feet per minute} = \frac{509 \text{ feet}}{1 \text{ minute}} \cdot \frac{1 \text{ mile}}{5{,}280 \text{ feet}} \cdot \frac{60 \text{ minutes}}{1 \text{ hour}}$$
$$= \frac{509 \cdot 60}{5{,}280} \text{ miles per hour}$$
$$= 5.8 \text{ miles per hour to the nearest tenth}$$

Problem Set 6.2

Moving Toward Success

"The superior man is modest in his speech, but exceeds in his actions."

—Confucius, 551–479 BC, Chinese philosopher

1. What would you do if your study partner is constantly complaining about the class or wanting to quit?

2. How do you cope with any frustration you may feel so that it does not interfere with your success in this class?

A Multiply or divide as indicated. Be sure to reduce all answers to lowest terms. (The numerator and denominator of the answer should not have any factors in common.) [Examples 1–4]

1. $\dfrac{x + y}{3} \cdot \dfrac{6}{x + y}$

2. $\dfrac{x - 1}{x + 1} \cdot \dfrac{5}{x - 1}$

3. $\dfrac{2x + 10}{x^2} \cdot \dfrac{x^3}{4x + 20}$

4. $\dfrac{3x^4}{3x - 6} \cdot \dfrac{x - 2}{x^2}$

5. $\dfrac{9}{2a - 8} \div \dfrac{3}{a - 4}$

6. $\dfrac{8}{a^2 - 25} \div \dfrac{16}{a + 5}$

7. $\dfrac{x + 1}{x^2 - 9} \div \dfrac{2x + 2}{x + 3}$

8. $\dfrac{11}{x - 2} \div \dfrac{22}{2x^2 - 8}$

9. $\dfrac{a^2 + 5a}{7a} \cdot \dfrac{4a^2}{a^2 + 4a}$

10. $\dfrac{4a^2 + 4a}{a^2 - 25} \cdot \dfrac{a^2 - 5a}{8a}$

11. $\dfrac{y^2 - 5y + 6}{2y + 4} \div \dfrac{2y - 6}{y + 2}$

12. $\dfrac{y^2 - 7y}{3y^2 - 48} \div \dfrac{y^2 - 9}{y^2 - 7y + 12}$

13. $\dfrac{2x - 8}{x^2 - 4} \cdot \dfrac{x^2 + 6x + 8}{x - 4}$

14. $\dfrac{x^2 + 5x + 1}{7x - 7} \cdot \dfrac{x - 1}{x^2 + 5x + 1}$

15. $\dfrac{x - 1}{x^2 - x - 6} \cdot \dfrac{x^2 + 5x + 6}{x^2 - 1}$

16. $\dfrac{x^2 - 3x - 10}{x^2 - 4x + 3} \cdot \dfrac{x^2 - 5x + 6}{x^2 - 3x - 10}$

17. $\dfrac{a^2 + 10a + 25}{a + 5} \div \dfrac{a^2 - 25}{a - 5}$

18. $\dfrac{a^2 + a - 2}{a^2 + 5a + 6} \div \dfrac{a - 1}{a}$

19. $\dfrac{y^3 - 5y^2}{y^4 + 3y^3 + 2y^2} \div \dfrac{y^2 - 5y + 6}{y^2 - 2y - 3}$

20. $\dfrac{y^2 - 5y}{y^2 + 7y + 12} \div \dfrac{y^3 - 7y^2 + 10y}{y^2 + 9y + 18}$

21. $\dfrac{2x^2 + 17x + 21}{x^2 + 2x - 35} \cdot \dfrac{x^3 - 125}{2x^2 - 7x - 15}$

22. $\dfrac{x^2 + x - 42}{4x^2 + 31x + 21} \cdot \dfrac{4x^2 - 5x - 6}{x^3 - 8}$

23. $\dfrac{2x^2 + 10x + 12}{4x^2 + 24x + 32} \cdot \dfrac{2x^2 + 18x + 40}{x^2 + 8x + 15}$

24. $\dfrac{3x^2 - 3}{6x^2 + 18x + 12} \cdot \dfrac{2x^2 - 8}{x^2 - 3x + 2}$

25. $\dfrac{2a^2 + 7a + 3}{a^2 - 16} \div \dfrac{4a^2 + 8a + 3}{2a^2 - 5a - 12}$

26. $\dfrac{3a^2 + 7a - 20}{a^2 + 3a - 4} \div \dfrac{3a^2 - 2a - 5}{a^2 - 2a + 1}$

27. $\dfrac{4y^2 - 12y + 9}{y^2 - 36} \div \dfrac{2y^2 - 5y + 3}{y^2 + 5y - 6}$

28. $\dfrac{5y^2 - 6y + 1}{y^2 - 1} \div \dfrac{16y^2 - 9}{4y^2 + 7y + 3}$

29. $\dfrac{x^2 - 1}{6x^2 + 42x + 60} \cdot \dfrac{7x^2 + 17x + 6}{x^3 + 1} \cdot \dfrac{6x + 30}{7x^2 - 11x - 6}$

30. $\dfrac{4x^2 - 1}{3x - 15} \cdot \dfrac{4x^2 - 17x - 15}{4x^2 - 9x - 9} \cdot \dfrac{3x - 9}{8x^3 - 1}$

31. $\dfrac{18x^3 + 21x^2 - 60x}{21x^2 - 25x - 4} \cdot \dfrac{28x^2 - 17x - 3}{16x^3 + 28x^2 - 30x}$

32. $\dfrac{56x^3 + 54x^2 - 20x}{8x^2 - 2x - 15} \cdot \dfrac{6x^2 + 5x - 21}{63x^3 + 129x^2 - 42x}$

The next two problems are intended to give you practice reading, and paying attention to, the instructions that accompany the problems you are working. Working these problems is an excellent way to get ready for a test or quiz.

33. Work each problem according to the instructions given.

   a. Simplify: $\dfrac{9 - 1}{27 - 1}$

   b. Reduce: $\dfrac{x^2 - 1}{x^3 - 1}$

   c. Multiply: $\dfrac{x^2 - 1}{x^3 - 1} \cdot \dfrac{x - 1}{x + 1}$

   d. Divide: $\dfrac{x^2 - 1}{x^3 - 1} \div \dfrac{x - 1}{x^2 + x + 1}$

34. Work each problem according to the instructions given.

   a. Simplify: $\dfrac{16 - 9}{16 + 24 + 9}$

   b. Reduce: $\dfrac{4x^2 - 9}{4x^2 + 12x + 9}$

   c. Multiply: $\dfrac{4x^2 - 9}{4x^2 + 12x + 9} \cdot \dfrac{2x + 3}{2x - 3}$

   d. Divide: $\dfrac{4x^2 - 9}{4x^2 + 12x + 9} \div \dfrac{2x + 3}{2x - 3}$

A Multiply the following expressions using the method shown in Examples 5 and 6 in this section.
[Examples 5–6]

35. $(x^2 - 9)\left(\dfrac{2}{x + 3}\right)$ **36.** $(x^2 - 9)\left(\dfrac{-3}{x - 3}\right)$

37. $a(a + 5)(a - 5)\left(\dfrac{2}{a^2 - 25}\right)$

38. $a(a^2 - 4)\left(\dfrac{a}{a + 2}\right)$

39. $(x^2 - x - 6)\left(\dfrac{x + 1}{x - 3}\right)$

40. $(x^2 - 2x - 8)\left(\dfrac{x + 3}{x - 4}\right)$

41. $(x^2 - 4x - 5)\left(\dfrac{-2x}{x + 1}\right)$

42. $(x^2 - 6x + 8)\left(\dfrac{4x}{x - 2}\right)$

A Each of the following problems involves some factoring by grouping. Remember, before you can divide out factors common to the numerators and denominators of a product, you must factor completely.

43. $\dfrac{x^2 - 9}{x^2 - 3x} \cdot \dfrac{2x + 10}{xy + 5x + 3y + 15}$

44. $\dfrac{x^2 - 16}{x^2 - 4x} \cdot \dfrac{3x + 18}{xy + 6x + 4y + 24}$

45. $\dfrac{2x^2 + 4x}{x^2 - y^2} \cdot \dfrac{x^2 + 3x + xy + 3y}{x^2 + 5x + 6}$

46. $\dfrac{x^2 - 25}{3x^2 + 3xy} \cdot \dfrac{x^2 + 4x + xy + 4y}{x^2 + 9x + 20}$

47. $\dfrac{x^3 - 3x^2 + 4x - 12}{x^4 - 16} \cdot \dfrac{3x^2 + 5x - 2}{3x^2 - 10x + 3}$

48. $\dfrac{x^3 - 5x^2 + 9x - 45}{x^4 - 81} \cdot \dfrac{5x^2 + 18x + 9}{5x^2 - 22x - 15}$

Simplify each expression. Work inside parentheses first, and then divide out common factors.

49. $\left(1 - \dfrac{1}{2}\right)\left(1 - \dfrac{1}{3}\right)\left(1 - \dfrac{1}{4}\right)\left(1 - \dfrac{1}{5}\right)$

50. $\left(1 + \dfrac{1}{2}\right)\left(1 + \dfrac{1}{3}\right)\left(1 + \dfrac{1}{4}\right)\left(1 + \dfrac{1}{5}\right)$

The dots in the following problems represent factors not written that are in the same pattern as the surrounding factors. Simplify.

51. $\left(1 - \dfrac{1}{2}\right)\left(1 - \dfrac{1}{3}\right)\left(1 - \dfrac{1}{4}\right) \cdots \left(1 - \dfrac{1}{99}\right)\left(1 - \dfrac{1}{100}\right)$

52. $\left(1 - \dfrac{1}{3}\right)\left(1 - \dfrac{1}{4}\right)\left(1 - \dfrac{1}{5}\right) \cdots \left(1 - \dfrac{1}{98}\right)\left(1 - \dfrac{1}{99}\right)$

B **Applying the Concepts** [Examples 7–8]

Cars The chart shows the fastest cars in the world. Use the chart to answer Problems 53 and 54.

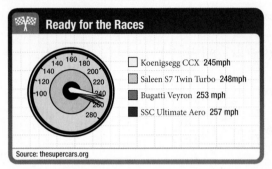

Ready for the Races

☐ Koenigsegg CCX **245mph**
☐ Saleen S7 Twin Turbo **248mph**
◼ Bugatti Veyron **253 mph**
◼ SSC Ultimate Aero **257 mph**

Source: thesupercars.org

53. Convert the speed of the Koenigsegg CCX to feet per second. Round to the nearest tenth.

54. Convert the speed of the Saleen S7 Twin Turbo to feet per second. Round to the nearest tenth.

55. Mount Whitney The top of Mount Whitney, the highest point in California, is 14,494 feet above sea level. Give this height in miles to the nearest tenth of a mile.

56. Motor Displacement The relationship between liters and cubic inches, both of which are measures of volume, is 0.0164 liters = 1 cubic inch. If a Ford Mustang has a engine with a displacement of 4.9 liters, what is the displacement in cubic inches? Round your answer to the nearest cubic inch.

57. Speed of Sound The speed of sound is 1,088 feet per second. Convert the speed of sound to miles per hour. Round your answer to the nearest whole number.

58. Average Speed A car travels 122 miles in 3 hours. Find the average speed of the car in feet per second. Round to the nearest whole number.

59. Ferris Wheel The first Ferris wheel was built in 1893. It was a large wheel with a circumference of 785 feet. If one trip around the circumference of the wheel took 20 minutes, find the average speed of a rider in miles per hour. Round to the nearest hundredth.

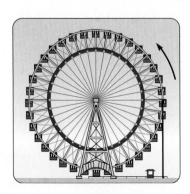

60. Unit Analysis The photograph shows the Cone Nebula as seen by the Hubble telescope in April 2002. The distance across the photograph is about 2.5 light-years. If we assume light travels 186,000 miles in one second, we can find the number of miles in one light-year by converting 186,000 miles/second to miles/year. Find the number of miles in one light-year. Write your answer in expanded form and in scientific notation.

NASA

61. Ferris Wheel A Ferris wheel called *Colossus* has a circumference of 518 feet. If a trip around the circumference of *Colossus* takes 40 seconds, find the average speed of a rider in miles per hour. Round to the nearest tenth.

62. Ferris Wheel A person riding a Ferris wheel travels once around the wheel, a distance of 188 feet, in 30 seconds. What is the average speed of the rider in miles per hour? Round to the nearest tenth.

Getting Ready for the Next Section

Perform the indicated operation.

63. $\dfrac{1}{5} + \dfrac{3}{5}$

64. $\dfrac{1}{7} + \dfrac{5}{7}$

65. $\dfrac{1}{10} + \dfrac{3}{14}$

66. $\dfrac{1}{21} + \dfrac{4}{15}$

67. $\dfrac{1}{10} - \dfrac{3}{14}$

68. $\dfrac{1}{21} - \dfrac{4}{15}$

Multiply.

69. $2(x - 3)$

70. $x(x + 2)$

71. $(x + 4)(x - 5)$

72. $(x + 3)(x - 4)$

Reduce to lowest terms.

73. $\dfrac{x + 3}{x^2 - 9}$

74. $\dfrac{x + 7}{x^2 - 49}$

75. $\dfrac{x^2 - x - 30}{2(x + 5)(x - 5)}$

76. $\dfrac{x^2 - x - 20}{2(x + 4)(x - 4)}$

Simplify.

77. $(x + 4)(x - 5) - 10$

78. $(x + 3)(x - 4) - 8$

Maintaining Your Skills

Add the following fractions.

79. $\dfrac{1}{2} + \dfrac{5}{2}$

80. $\dfrac{2}{3} + \dfrac{8}{3}$

81. $2 + \dfrac{3}{4}$

82. $1 + \dfrac{4}{7}$

Simplify each term, then add.

83. $\dfrac{10x^4}{2x^2} + \dfrac{12x^6}{3x^4}$

84. $\dfrac{32x^8}{8x^3} + \dfrac{27x^7}{3x^2}$

85. $\dfrac{12a^2b^5}{3ab^3} + \dfrac{14a^4b^7}{7a^3b^5}$

86. $\dfrac{16a^3b^2}{4ab} + \dfrac{25a^6b^5}{5a^4b^4}$

6.3 Addition and Subtraction of Rational Expressions

OBJECTIVES

A Add and subtract rational expressions that have the same denominators.

B Add and subtract rational expressions that have different denominators.

TICKET TO SUCCESS

Keep these questions in mind as you read through the section. Then respond in your own words and in complete sentences.

1. How do we add two rational expressions that have the same denominator?

2. What is the least common denominator for two fractions that include rational expressions?

3. What role does factoring play in finding a least common denominator when adding rational expressions?

4. Explain how you would reduce the solution to an addition problem of two rational expressions.

Imagine sitting in rush hour traffic through a big city. You drive two miles down one freeway, then take an interchange and drive 2 more miles. The complete trip takes 1 hour. If your speed after the interchange is 10 miles per hour faster than your speed before, how fast were you going during the first two miles?

To solve this problem, we will apply x to our unknown speed during the first two miles. Then, using the rate equation, you can set up a rational expression for each half of the trip. The following expression represents the addition problem we need to help us find our solution.

$$\frac{2}{x} + \frac{2}{(x + 10)}$$

After you learn how to add and subtract rational expressions in this section, return to this problem and see if you can work it out.

A Adding and Subtracting with the Same Denominator

In Chapter 1, we combined fractions having the same denominator by combining their numerators and putting the result over the common denominator. We use the same process to add two rational expressions with the same denominator.

EXAMPLE 1 Add $\frac{5}{x} + \frac{3}{x}$.

SOLUTION Adding numerators, we have

$$\frac{5}{x} + \frac{3}{x} = \frac{8}{x}$$

EXAMPLE 2 Add $\frac{x}{x^2 - 9} + \frac{3}{x^2 - 9}$.

SOLUTION Since both expressions have the same denominator, we add numerators and reduce to lowest terms.

$$\frac{x}{x^2 - 9} + \frac{3}{x^2 - 9} = \frac{x + 3}{x^2 - 9}$$

$$= \frac{\cancel{x + 3}}{\cancel{(x + 3)}(x - 3)} \quad \left\} \begin{array}{l} \text{Reduce to lowest terms by} \\ \text{factoring the denominator and} \\ \text{then dividing out the common} \\ \text{factor } x + 3 \end{array} \right.$$

$$= \frac{1}{x - 3}$$

B Adding and Subtracting with Different Denominators

Remember, that the distributive property allows us to add rational expressions by simply adding numerators. Because of this, we must begin all addition problems involving rational expressions by first making sure all the expressions have the least common denominator. Also remember, the least common denominator (LCD) for a set of denominators is the simplest quantity that is exactly divisible by all the denominators.

Use the strategy for adding or subtracting any two fractions we learned in Chapter 1 to work the following examples.

EXAMPLE 3 Add $\frac{1}{10} + \frac{3}{14}$.

SOLUTION

Step 1: Find the LCD for 10 and 14. To do so, we factor each denominator and build the LCD from the factors:

$$\left. \begin{array}{l} 10 = 2 \cdot 5 \\ 14 = 2 \cdot 7 \end{array} \right\} \quad \text{LCD} = 2 \cdot 5 \cdot 7 = 70$$

NOTE
If you have had a difficult time in the past with addition and subtraction of fractions with different denominators, this is the time to get it straightened out. Go over Example 3 as many times as is necessary for you to understand the process.

We know the LCD is divisible by 10 because it contains the factors 2 and 5. It is also divisible by 14 because it contains the factors 2 and 7.

Step 2: Change to equivalent fractions that each have denominator 70. To accomplish this task, we multiply the numerator and denominator of each fraction by the factor of the LCD that is not also a factor of its denominator.

Original Fractions		Denominators in Factored Form		Multiply by Factor Needed to Obtain LCD		These Have the Same Value as the Original Fractions
$\dfrac{1}{10}$	$=$	$\dfrac{1}{2 \cdot 5}$	$=$	$\dfrac{1}{2 \cdot 5} \cdot \dfrac{\mathbf{7}}{\mathbf{7}}$	$=$	$\dfrac{7}{70}$
$\dfrac{3}{14}$	$=$	$\dfrac{3}{2 \cdot 7}$	$=$	$\dfrac{3}{2 \cdot 7} \cdot \dfrac{\mathbf{5}}{\mathbf{5}}$	$=$	$\dfrac{15}{70}$

The fraction $\dfrac{7}{70}$ has the same value as the fraction $\dfrac{1}{10}$. Likewise, the fractions $\dfrac{15}{70}$ and $\dfrac{3}{14}$ are equivalent; they have the same value.

Step 3: Add numerators and put the result over the LCD.

$$\frac{7}{70} + \frac{15}{70} = \frac{7+15}{70} = \frac{22}{70}$$

Step 4: Reduce to lowest terms.

$$\frac{22}{70} = \frac{11}{35} \qquad \text{Divide numerator and denominator by 2}$$

The main idea in adding fractions is to write each fraction again with the LCD for a denominator. Once we have done that, we simply add numerators. The same process can be used to add rational expressions, as the next example illustrates.

EXAMPLE 4   Subtract $\dfrac{3}{x} - \dfrac{1}{2}$.

SOLUTION

Step 1: The LCD for x and 2 is $2x$. It is the smallest expression divisible by x and by 2.

Step 2: To change to equivalent expressions with the denominator $2x$, we multiply the first fraction by $\dfrac{2}{2}$ and the second by $\dfrac{x}{x}$.

$$\frac{3}{x} \cdot \frac{\mathbf{2}}{\mathbf{2}} = \frac{6}{2x}$$

$$\frac{1}{2} \cdot \frac{\mathbf{x}}{\mathbf{x}} = \frac{x}{2x}$$

Step 3: Subtracting numerators of the rational expressions in step 2, we have

$$\frac{6}{2x} - \frac{x}{2x} = \frac{6-x}{2x}$$

Step 4: Since $6 - x$ and $2x$ do not have any factors in common, we cannot reduce any further. Here is the complete solution:

$$\frac{3}{x} - \frac{1}{2} = \frac{3}{x} \cdot \frac{\mathbf{2}}{\mathbf{2}} - \frac{1}{2} \cdot \frac{\mathbf{x}}{\mathbf{x}}$$

$$= \frac{6}{2x} - \frac{x}{2x}$$

$$= \frac{6-x}{2x}$$

EXAMPLE 5 Add $\dfrac{5}{2x - 6} + \dfrac{x}{x - 3}$.

SOLUTION If we factor $2x - 6$, we have $2x - 6 = 2(x - 3)$. We only need to multiply the second rational expression in our problem by $\dfrac{2}{2}$ to have two expressions with the same denominator.

$$\dfrac{5}{2x - 6} + \dfrac{x}{x - 3} = \dfrac{5}{2(x - 3)} + \dfrac{x}{x - 3} \qquad \text{Factor the first denominator}$$

$$= \dfrac{5}{2(x - 3)} + \dfrac{\mathbf{2}}{\mathbf{2}}\left(\dfrac{x}{x - 3}\right) \qquad \text{Change to equivalent rational expressions}$$

$$= \dfrac{5}{2(x - 3)} + \dfrac{2x}{2(x - 3)} \qquad \text{Multiply}$$

$$= \dfrac{2x + 5}{2(x - 3)} \qquad \text{Add numerators}$$

EXAMPLE 6 Add $\dfrac{1}{x + 4} + \dfrac{8}{x^2 - 16}$.

SOLUTION After writing each denominator in factored form, we find that the least common denominator is $(x + 4)(x - 4)$. To change the first rational expression to an equivalent rational expression with the common denominator, we multiply its numerator and denominator by $x - 4$.

$$\dfrac{1}{x + 4} + \dfrac{8}{x^2 - 16}$$

$$= \dfrac{1}{x + 4} + \dfrac{8}{(x + 4)(x - 4)} \qquad \text{Factor each denominator}$$

$$= \dfrac{1}{x + 4} \cdot \dfrac{\mathbf{x - 4}}{\mathbf{x - 4}} + \dfrac{8}{(x + 4)(x - 4)} \qquad \text{Change to equivalent rational expressions}$$

$$= \dfrac{x - 4}{(x + 4)(x - 4)} + \dfrac{8}{(x + 4)(x - 4)} \qquad \text{Multiply}$$

$$= \dfrac{x + 4}{(x + 4)(x - 4)} \qquad \text{Add numerators}$$

$$= \dfrac{1}{x - 4} \qquad \text{Divide out common factor } x + 4$$

NOTE
In the last step we reduced the rational expression to lowest terms by dividing out the common factor of $x + 4$.

EXAMPLE 7 Add $\dfrac{2}{x^2 + 5x + 6} + \dfrac{x}{x^2 - 9}$.

SOLUTION

Step 1: We factor each denominator and build the LCD from the factors.

$$\left.\begin{array}{l} x^2 + 5x + 6 = (x + 2)(x + 3) \\ x^2 - 9 = (x + 3)(x - 3) \end{array}\right\} \qquad \text{LCD} = (x + 2)(x + 3)(x - 3)$$

Step 2: Change to equivalent rational expressions.

$$\dfrac{2}{x^2 + 5x + 6} = \dfrac{2}{(x + 2)(x + 3)} \cdot \dfrac{\mathbf{(x - 3)}}{\mathbf{(x - 3)}} = \dfrac{2x - 6}{(x + 2)(x + 3)(x - 3)}$$

$$\dfrac{x}{x^2 - 9} = \dfrac{x}{(x + 3)(x - 3)} \cdot \dfrac{\mathbf{(x + 2)}}{\mathbf{(x + 2)}} = \dfrac{x^2 + 2x}{(x + 2)(x + 3)(x - 3)}$$

Step 3:   Add numerators of the rational expressions produced in step 2.

$$\frac{2x - 6}{(x + 2)(x + 3)(x - 3)} + \frac{x^2 + 2x}{(x + 2)(x + 3)(x - 3)}$$

$$= \frac{x^2 + 4x - 6}{(x + 2)(x + 3)(x - 3)}$$

The numerator and denominator do not have any factors in common.

EXAMPLE 8   Subtract $\dfrac{x + 4}{2x + 10} - \dfrac{5}{x^2 - 25}$.

SOLUTION   We begin by factoring each denominator.

$$\frac{x + 4}{2x + 10} - \frac{5}{x^2 - 25} = \frac{x + 4}{2(x + 5)} - \frac{5}{(x + 5)(x - 5)}$$

NOTE
In the second step, we replaced subtraction by addition of the opposite. There seems to be less chance for error when this is done on longer problems.

The LCD is $2(x + 5)(x - 5)$. Completing the problem, we have

$$= \frac{x + 4}{2(x + 5)} \cdot \frac{(x - 5)}{(x - 5)} + \frac{-5}{(x + 5)(x - 5)} \cdot \frac{2}{2}$$

$$= \frac{x^2 - x - 20}{2(x + 5)(x - 5)} + \frac{-10}{2(x + 5)(x - 5)}$$

$$= \frac{x^2 - x - 30}{2(x + 5)(x - 5)}$$

To see if this expression will reduce, we factor the numerator into $(x - 6)(x + 5)$.

$$= \frac{(x - 6)\cancel{(x + 5)}}{2\cancel{(x + 5)}(x - 5)}$$

$$= \frac{x - 6}{2(x - 5)}$$

EXAMPLE 9   Write an expression for the sum of a number and its reciprocal, and then simplify that expression.

SOLUTION   If we let x = the number, then its reciprocal is $\dfrac{1}{x}$. To find the sum of the number and its reciprocal, we add them.

$$x + \frac{1}{x}$$

The first term x can be thought of as having a denominator of 1. Since the denominators are 1 and x, the least common denominator is x.

$$x + \frac{1}{x} = \frac{x}{1} + \frac{1}{x} \qquad \text{Write } x \text{ as } \frac{x}{1}$$

$$= \frac{x}{1} \cdot \frac{x}{x} + \frac{1}{x} \qquad \text{The LCD is } x$$

$$= \frac{x^2}{x} + \frac{1}{x}$$

$$= \frac{x^2 + 1}{x} \qquad \text{Add numerators}$$

Problem Set 6.3

Moving Toward Success

"If you can't feed a hundred people, then just feed one."

—Mother Teresa, 1910–1997, Albanian-born humanitarian and missionary

1. Is it helpful to share your study tools with your study partner? Why or why not?

2. What has been your most useful study tool for this class so far?

A Find the following sums and differences.
[Examples 1–2]

1. $\dfrac{3}{x} + \dfrac{4}{x}$

2. $\dfrac{5}{x} + \dfrac{3}{x}$

3. $\dfrac{9}{a} - \dfrac{5}{a}$

4. $\dfrac{8}{a} - \dfrac{7}{a}$

5. $\dfrac{1}{x+1} + \dfrac{x}{x+1}$

6. $\dfrac{x}{x-3} - \dfrac{3}{x-3}$

7. $\dfrac{y^2}{y-1} - \dfrac{1}{y-1}$

8. $\dfrac{y^2}{y+3} - \dfrac{9}{y+3}$

9. $\dfrac{x^2}{x+2} + \dfrac{4x+4}{x+2}$

10. $\dfrac{x^2-6x}{x-3} + \dfrac{9}{x-3}$

11. $\dfrac{x^2}{x-2} - \dfrac{4x-4}{x-2}$

12. $\dfrac{x^2}{x-5} - \dfrac{10x-25}{x-5}$

13. $\dfrac{x+2}{x+6} - \dfrac{x-4}{x+6}$

14. $\dfrac{x+5}{x+2} - \dfrac{x+3}{x+2}$

B [Examples 4–8]

15. $\dfrac{y}{2} - \dfrac{2}{y}$

16. $\dfrac{3}{y} + \dfrac{y}{3}$

17. $\dfrac{1}{2} + \dfrac{a}{3}$

18. $\dfrac{2}{3} + \dfrac{2a}{5}$

19. $\dfrac{x}{x+1} + \dfrac{3}{4}$

20. $\dfrac{x}{x-3} + \dfrac{1}{3}$

21. $\dfrac{x+1}{x-2} - \dfrac{4x+7}{5x-10}$

22. $\dfrac{3x+1}{2x-6} - \dfrac{x+2}{x-3}$

23. $\dfrac{4x-2}{3x+12} - \dfrac{x-2}{x+4}$

24. $\dfrac{6x+5}{5x-25} - \dfrac{x+2}{x-5}$

25. $\dfrac{6}{x(x-2)} + \dfrac{3}{x}$

26. $\dfrac{10}{x(x+5)} - \dfrac{2}{x}$

27. $\dfrac{4}{a} - \dfrac{12}{a^2+3a}$

28. $\dfrac{5}{a} + \dfrac{20}{a^2-4a}$

29. $\dfrac{2}{x+5} - \dfrac{10}{x^2-25}$

30. $\dfrac{6}{x^2-1} + \dfrac{3}{x+1}$

31. $\dfrac{x-4}{x-3} + \dfrac{6}{x^2-9}$

32. $\dfrac{x+1}{x-1} - \dfrac{4}{x^2-1}$

33. $\dfrac{a-4}{a-3} + \dfrac{5}{a^2-a-6}$

34. $\dfrac{a+2}{a+1} + \dfrac{7}{a^2-5a-6}$

35. $\dfrac{8}{x^2-16} - \dfrac{7}{x^2-x-12}$

36. $\dfrac{6}{x^2-9} - \dfrac{5}{x^2-x-6}$

37. $\dfrac{4y}{y^2+6y+5} - \dfrac{3y}{y^2+5y+4}$

38. $\dfrac{3y}{y^2+7y+10} - \dfrac{2y}{y^2+6y+8}$

39. $\dfrac{4x+1}{x^2+5x+4} - \dfrac{x+3}{x^2+4x+3}$

40. $\dfrac{2x-1}{x^2+x-6} - \dfrac{x+2}{x^2+5x+6}$

41. $\dfrac{1}{x} + \dfrac{x}{3x+9} - \dfrac{3}{x^2+3x}$

42. $\dfrac{1}{x} + \dfrac{x}{2x+4} - \dfrac{2}{x^2+2x}$

43. Work each problem according to the instructions given.

a. Multiply: $\dfrac{4}{9} \cdot \dfrac{1}{6}$

b. Divide: $\dfrac{4}{9} \div \dfrac{1}{6}$

c. Add: $\dfrac{4}{9} + \dfrac{1}{6}$

d. Multiply: $\dfrac{x+2}{x-2} \cdot \dfrac{3x+10}{x^2-4}$

e. Divide: $\dfrac{x+2}{x-2} \div \dfrac{3x+10}{x^2-4}$

f. Subtract: $\dfrac{x+2}{x-2} - \dfrac{3x+10}{x^2-4}$

44. Work each problem according to the instructions given.

a. Multiply: $\dfrac{9}{25} \cdot \dfrac{1}{15}$

b. Divide: $\dfrac{9}{25} \div \dfrac{1}{15}$

c. Subtract: $\dfrac{9}{25} - \dfrac{1}{15}$

d. Multiply: $\dfrac{3x - 2}{3x + 2} \cdot \dfrac{15x + 6}{9x^2 - 4}$

e. Divide: $\dfrac{3x - 2}{3x + 2} \div \dfrac{15x + 6}{9x^2 - 4}$

f. Subtract: $\dfrac{3x + 2}{3x - 2} - \dfrac{15x + 6}{9x^2 - 4}$

Complete the following tables.

45.

Number x	Reciprocal $\dfrac{1}{x}$	Sum $1 + \dfrac{1}{x}$	Sum $\dfrac{x + 1}{x}$
1			
2			
3			
4			

46.

Number x	Reciprocal $\dfrac{1}{x}$	Difference $1 - \dfrac{1}{x}$	Difference $\dfrac{x - 1}{x}$
1			
2			
3			
4			

47.

x	$x + \dfrac{4}{x}$	$\dfrac{x^2 + 4}{x}$	$x + 4$
1			
2			
3			
4			

48.

x	$2x + \dfrac{6}{x}$	$\dfrac{2x^2 + 6}{x}$	$2x + 6$
1			
2			
3			
4			

B Add or subtract as indicated. [Example 9]

49. $1 + \dfrac{1}{x + 2}$

50. $1 - \dfrac{1}{x + 2}$

51. $1 - \dfrac{1}{x + 3}$

52. $1 + \dfrac{1}{x + 3}$

Getting Ready for the Next Section

Simplify.

53. $6\left(\dfrac{1}{2}\right)$

54. $10\left(\dfrac{1}{5}\right)$

55. $\dfrac{0}{5}$

56. $\dfrac{0}{2}$

57. $\dfrac{5}{0}$

58. $\dfrac{2}{0}$

59. $1 - \dfrac{5}{2}$

60. $1 - \dfrac{5}{3}$

Use the distributive property to simplify.

61. $6\left(\dfrac{x}{3} + \dfrac{5}{2}\right)$

62. $10\left(\dfrac{x}{2} + \dfrac{3}{5}\right)$

63. $x^2\left(1 - \dfrac{5}{x}\right)$

64. $x^2\left(1 - \dfrac{3}{x}\right)$

Solve.

65. $2x + 15 = 3$

66. $15 = 3x - 3$

67. $-2x - 9 = x - 3$

68. $a^2 - a - 20 = -2a$

Maintaining Your Skills

Solve each equation.

69. $2x + 3(x - 3) = 6$

70. $4x - 2(x - 5) = 6$

71. $x - 3(x + 3) = x - 3$

72. $x - 4(x + 4) = x - 4$

73. $7 - 2(3x + 1) = 4x + 3$

74. $8 - 5(2x - 1) = 2x + 4$

Solve each quadratic equation.

75. $x^2 + 5x + 6 = 0$

76. $x^2 - 5x + 6 = 0$

77. $x^2 - x = 6$

78. $x^2 + x = 6$

79. $x^2 - 5x = 0$

80. $x^2 - 6x = 0$

81. Temperature The chart shows the temperatures for some of the world's hottest places. We can find the temperature in Celsius using the equation, $C = \left(\frac{5}{9}\right)(F - 32)$, where C is the temperature in Celsius and F is the temperature in Fahrenheit. Find the temperature of Al'Aziziyah, Libya, in Celsius.

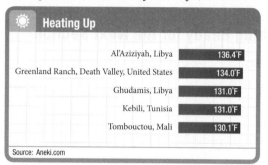

☀ **Heating Up**

Al'Aziziyah, Libya	136.4°F
Greenland Ranch, Death Valley, United States	134.0°F
Ghudamis, Libya	131.0°F
Kebili, Tunisia	131.0°F
Tombouctou, Mali	130.1°F

Source: Aneki.com

82. Population The map shows the average number of days spent commuting per year in the United States' largest cities. We can find the average number of hours per day spent in the car with the equation, $y = \left(\frac{24}{365}\right)x$, where y is hours per day and x is days per year. Find the hours per day spent commuting in Atlanta, GA. Round to the nearest hundredth.

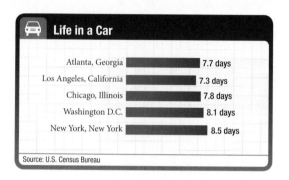

🚗 **Life in a Car**

Atlanta, Georgia	7.7 days
Los Angeles, California	7.3 days
Chicago, Illinois	7.8 days
Washington D.C.	8.1 days
New York, New York	8.5 days

Source: U.S. Census Bureau

6.4 Equations Involving Rational Expressions

OBJECTIVES

A Solve equations that contain rational expressions.

TICKET TO SUCCESS

Keep these questions in mind as you read through the section. Then respond in your own words and in complete sentences.

1. What is the first step in solving an equation that contains rational expressions?

2. Explain how you would find the LCD used to clear an equation of fractions.

3. When will an equation containing rational expressions have more than one possible solution?

4. How would you check for extraneous solutions to an equation containing rational expressions?

Picture a person kayaking on a river. The time it takes the person to go 1 mile down the river is given by $\frac{1}{(x+5)}$. When he paddles back upriver, the time it takes him is given by $\frac{1}{x}$. If x is his downriver speed and it takes him 1.5 hours to complete his trip, we can set up the following rational equation to solve for x.

$$\frac{1}{x} + \frac{1}{(x+5)} = 1.5$$

Let's practice working with some other rational equations, and then you can come back to this problem to solve it.

A Solving Rational Equations

The first step in solving an equation that contains one or more rational expressions is to find the LCD for all denominators in the equation. Once the LCD has been found, we multiply both sides of the equation by it. The resulting equation should be equivalent to the original one (unless we inadvertently multiplied by zero) and free from any denominators except the number 1.

EXAMPLE 1 Solve $\frac{x}{3} + \frac{5}{2} = \frac{1}{2}$ for x.

SOLUTION The LCD for 3 and 2 is 6. If we multiply both sides by 6, we have

$$6\left(\frac{x}{3} + \frac{5}{2}\right) = 6\left(\frac{1}{2}\right) \qquad \text{Multiply both sides by 6}$$

$$6\left(\frac{x}{3}\right) + 6\left(\frac{5}{2}\right) = 6\left(\frac{1}{2}\right) \qquad \text{Distributive property}$$

$$2x + 15 = 3$$

$$2x = -12$$

$$x = -6$$

We can check our solution by replacing x with -6 in the original equation.

$$-\frac{6}{3} + \frac{5}{2} \stackrel{?}{=} \frac{1}{2}$$

$$\frac{1}{2} = \frac{1}{2}$$

Multiplying both sides of an equation containing fractions by the LCD clears the equation of all denominators, because the LCD has the property that all denominators will divide it evenly.

EXAMPLE 2 Solve $\frac{3}{x-1} = \frac{3}{5}$ for x.

SOLUTION The LCD for $(x-1)$ and 5 is $5(x-1)$. Multiplying both sides by $5(x-1)$, we have

$$\frac{5(x-1) \cdot 3}{x-1} = \frac{5(x-1) \cdot 3}{5}$$

$$5 \cdot 3 = (x-1) \cdot 3$$

$$15 = 3x - 3$$

$$18 = 3x$$

$$6 = x$$

If we substitute $x = 6$ into the original equation, we have

$$\frac{3}{6-1} \stackrel{?}{=} \frac{3}{5}$$

$$\frac{3}{5} = \frac{3}{5}$$

The solution set is {6}.

EXAMPLE 3   Solve $1 - \dfrac{5}{x} = \dfrac{-6}{x^2}$.

SOLUTION   The LCD is x^2. Multiplying both sides by x^2, we have

$$x^2\left(1 - \frac{5}{x}\right) = x^2\left(\frac{-6}{x^2}\right) \qquad \text{Multiply both sides by } x^2$$

$$x^2(1) - x^2\left(\frac{5}{x}\right) = x^2\left(\frac{-6}{x^2}\right) \qquad \text{Apply distributive property to the left side}$$

$$x^2 - 5x = -6 \qquad \text{Simplify each side}$$

We have a quadratic equation, which we write in standard form, factor, and solve as we did in Chapter 5.

$$x^2 - 5x + 6 = 0 \qquad \text{Standard form}$$

$$(x - 2)(x - 3) = 0 \qquad \text{Factor}$$

$$x - 2 = 0 \quad \text{or} \quad x - 3 = 0 \qquad \text{Set factors equal to 0}$$

$$x = 2 \quad \text{or} \qquad x = 3$$

The two possible solutions are 2 and 3. Checking each in the original equation, we find they both give true statements. They are both solutions to the original equation.

Check $x = 2$ $\qquad\qquad$ Check $x = 3$

$$1 - \frac{5}{2} \stackrel{?}{=} \frac{-6}{4} \qquad\qquad 1 - \frac{5}{3} \stackrel{?}{=} \frac{-6}{9}$$

$$\frac{2}{2} - \frac{5}{2} = -\frac{3}{2} \qquad\qquad \frac{3}{3} - \frac{5}{3} = -\frac{2}{3}$$

$$-\frac{3}{2} = -\frac{3}{2} \qquad\qquad -\frac{2}{3} = -\frac{2}{3}$$

EXAMPLE 4   Solve $\dfrac{x}{x^2 - 9} - \dfrac{3}{x - 3} = \dfrac{1}{x + 3}$.

SOLUTION   The factors of $x^2 - 9$ are $(x + 3)(x - 3)$. The LCD, then, is $(x + 3)(x - 3)$.

$$(x + 3)(x - 3) \cdot \frac{x}{(x + 3)(x - 3)} + (x + 3)(x - 3) \cdot \frac{-3}{x - 3}$$

$$= (x + 3)(x - 3) \cdot \frac{1}{x + 3}$$

$$x + (x + 3)(-3) = (x - 3) \cdot 1$$

$$x + (-3x) + (-9) = x - 3$$

$$-2x - 9 = x - 3$$

$$-3x = 6$$

$$x = -2$$

The solution is $x = -2$. It checks when substituted for x in the original equation.

EXAMPLE 5 Solve $\dfrac{x}{x-3} + \dfrac{3}{2} = \dfrac{3}{x-3}$.

SOLUTION We begin by multiplying each term on both sides of the equation by the LCD, $2(x-3)$.

$$2(x-3) \cdot \frac{x}{x-3} + 2(x-3) \cdot \frac{3}{2} = 2(x-3) \cdot \frac{3}{x-3}$$

$$2x + (x-3) \cdot 3 = 2 \cdot 3$$

$$2x + 3x - 9 = 6$$

$$5x - 9 = 6$$

$$5x = 15$$

$$x = 3$$

Our only possible solution is $x = 3$. If we substitute $x = 3$ into our original equation, we get

$$\frac{3}{3-3} + \frac{3}{2} \overset{?}{=} \frac{3}{3-3}$$

$$\frac{3}{0} + \frac{3}{2} = \frac{3}{0}$$

Two of the terms are undefined, so the equation is meaningless. What has happened is that we have multiplied both sides of the original equation by zero. The equation produced by doing this is not equivalent to our original equation. We always must check our solution when we multiply both sides of an equation by an expression containing the variable to make sure we have not multiplied both sides by zero.

Our original equation has no solution; that is, there is no real number x such that

$$\frac{x}{x-3} + \frac{3}{2} = \frac{3}{x-3}$$

The solution set is $\varnothing$.

EXAMPLE 6 Solve $\dfrac{a+4}{a^2+5a} = \dfrac{-2}{a^2-25}$ for a.

SOLUTION Factoring each denominator, we have

$$a^2 + 5a = a(a+5)$$

$$a^2 - 25 = (a+5)(a-5)$$

The LCD is $a(a+5)(a-5)$. Multiplying both sides of the equation by the LCD gives us

$$\frac{a(a+5)(a-5) \cdot (a+4)}{a(a+5)} = \frac{-2}{(a+5)(a-5)} \cdot a(a+5)(a-5)$$

$$(a-5)(a+4) = -2a$$

$$a^2 - a - 20 = -2a$$

The result is a quadratic equation, which we write in standard form, factor, and solve.

$$a^2 + a - 20 = 0 \qquad \text{Add } 2a \text{ to both sides}$$

$$(a+5)(a-4) = 0 \qquad \text{Factor}$$

$$a + 5 = 0 \quad \text{ or } \quad a - 4 = 0 \qquad \text{Set each factor to 0}$$

$$a = -5 \quad \text{ or } \quad a = 4$$

The two possible solutions are -5 and 4. There is no problem with the 4. It checks when substituted for a in the original equation. However, -5 is not a solution. Substituting -5 into the original equation gives

$$\frac{-5 + 4}{(-5)^2 + 5(-5)} \stackrel{?}{=} \frac{-2}{(-5)^2 - 25}$$

$$\frac{-1}{0} = \frac{-2}{0}$$

This indicates -5 is not a solution. The solution is 4.

Problem Set 6.4

Moving Toward Success

"There are two primary choices in life; to accept conditions as they exist, or accept the responsibility for changing them."

—Denis Waitley, 1933–present, motivational speaker and writer

1. How might negative statements or thoughts about this class threaten your path to success?

2. How does coming to class prepared help to reduce stress?

A Solve the following equations. Be sure to check each answer in the original equation if you multiply both sides by an expression that contains the variable.
[Examples 1–6]

1. $\dfrac{x}{3} + \dfrac{1}{2} = -\dfrac{1}{2}$

2. $\dfrac{x}{2} + \dfrac{4}{3} = -\dfrac{2}{3}$

3. $\dfrac{4}{a} = \dfrac{1}{5}$

4. $\dfrac{2}{3} = \dfrac{6}{a}$

5. $\dfrac{3}{x} + 1 = \dfrac{2}{x}$

6. $\dfrac{4}{x} + 3 = \dfrac{1}{x}$

7. $\dfrac{3}{a} - \dfrac{2}{a} = \dfrac{1}{5}$

8. $\dfrac{7}{a} + \dfrac{1}{a} = 2$

9. $\dfrac{3}{x} + 2 = \dfrac{1}{2}$

10. $\dfrac{5}{x} + 3 = \dfrac{4}{3}$

11. $\dfrac{1}{y} - \dfrac{1}{2} = -\dfrac{1}{4}$

12. $\dfrac{3}{y} - \dfrac{4}{5} = -\dfrac{1}{5}$

13. $1 - \dfrac{8}{x} = \dfrac{-15}{x^2}$

14. $1 - \dfrac{3}{x} = \dfrac{-2}{x^2}$

15. $\dfrac{x}{2} - \dfrac{4}{x} = -\dfrac{7}{2}$

16. $\dfrac{x}{2} - \dfrac{5}{x} = -\dfrac{3}{2}$

17. $\dfrac{x - 3}{2} + \dfrac{2x}{3} = \dfrac{5}{6}$

18. $\dfrac{x - 2}{3} + \dfrac{5x}{2} = 5$

19. $\dfrac{x + 1}{3} + \dfrac{x - 3}{4} = \dfrac{1}{6}$

20. $\dfrac{x + 2}{3} + \dfrac{x - 1}{5} = -\dfrac{3}{5}$

21. $\dfrac{6}{x + 2} = \dfrac{3}{5}$

22. $\dfrac{4}{x + 3} = \dfrac{1}{2}$

23. $\dfrac{3}{y - 2} = \dfrac{2}{y - 3}$

24. $\dfrac{5}{y + 1} = \dfrac{4}{y + 2}$

25. $\dfrac{x}{x - 2} + \dfrac{2}{3} = \dfrac{2}{x - 2}$

26. $\dfrac{x}{x - 5} + \dfrac{1}{5} = \dfrac{5}{x - 5}$

27. $\dfrac{x}{x - 2} + \dfrac{3}{2} = \dfrac{9}{2(x - 2)}$

28. $\dfrac{x}{x + 1} + \dfrac{4}{5} = \dfrac{-14}{5(x + 1)}$

29. $\dfrac{5}{x + 2} + \dfrac{1}{x + 3} = \dfrac{-1}{x^2 + 5x + 6}$

30. $\dfrac{3}{x - 1} + \dfrac{2}{x + 3} = \dfrac{-3}{x^2 + 2x - 3}$

31. $\dfrac{8}{x^2 - 4} + \dfrac{3}{x + 2} = \dfrac{1}{x - 2}$

32. $\dfrac{10}{x^2 - 25} - \dfrac{1}{x - 5} = \dfrac{3}{x + 5}$

33. $\dfrac{a}{2} + \dfrac{3}{a - 3} = \dfrac{a}{a - 3}$

34. $\dfrac{a}{2} + \dfrac{4}{a - 4} = \dfrac{a}{a - 4}$

35. $\dfrac{6}{y^2 - 4} = \dfrac{4}{y^2 + 2y}$

36. $\dfrac{2}{y^2 - 9} = \dfrac{5}{y^2 - 3y}$

37. $\dfrac{2}{a^2 - 9} = \dfrac{3}{a^2 + a - 12}$

38. $\dfrac{2}{a^2 - 1} = \dfrac{6}{a^2 - 2a - 3}$

39. $\dfrac{3x}{x - 5} - \dfrac{2x}{x + 1} = \dfrac{-42}{x^2 - 4x - 5}$

40. $\dfrac{4x}{x - 4} - \dfrac{3x}{x - 2} = \dfrac{-3}{x^2 - 6x + 8}$

41. $\dfrac{2x}{x + 2} = \dfrac{x}{x + 3} - \dfrac{3}{x^2 + 5x + 6}$

42. $\dfrac{3x}{x-4} = \dfrac{2x}{x-3} + \dfrac{6}{x^2 - 7x + 12}$

43. Solve each equation.

  a. $5x - 1 = 0$

  b. $\dfrac{5}{x} - 1 = 0$

  c. $\dfrac{x}{5} - 1 = \dfrac{2}{3}$

  d. $\dfrac{5}{x} - 1 = \dfrac{2}{3}$

  e. $\dfrac{5}{x^2} + 5 = \dfrac{26}{x}$

44. Solve each equation.

  a. $2x - 3 = 0$

  b. $2 - \dfrac{3}{x} = 0$

  c. $\dfrac{x}{3} - 2 = \dfrac{1}{2}$

  d. $\dfrac{3}{x} - 2 = \dfrac{1}{2}$

  e. $\dfrac{1}{x} + \dfrac{3}{x^2} = 2$

45. Work each problem according to the instructions given.

  a. Divide: $\dfrac{7}{a^2 - 5a - 6} \div \dfrac{a+2}{a+1}$

  b. Add: $\dfrac{7}{a^2 - 5a - 6} + \dfrac{a+2}{a+1}$

  c. Solve: $\dfrac{7}{a^2 - 5a - 6} + \dfrac{a+2}{a+1} = 2$

46. Work each problem according to the instructions given.

  a. Divide: $\dfrac{6}{x^2 - 9} \div \dfrac{x-4}{x-3}$

  b. Add: $\dfrac{6}{x^2 - 9} + \dfrac{x-4}{x-3}$

  c. Solve: $\dfrac{6}{x^2 - 9} + \dfrac{x-4}{x-3} = \dfrac{3}{4}$

Getting Ready for the Next Section

Solve.

47. $\dfrac{1}{x} + \dfrac{1}{2x} = \dfrac{9}{2}$

48. $\dfrac{50}{x+5} = \dfrac{30}{x-5}$

49. $\dfrac{1}{10} - \dfrac{1}{15} = \dfrac{1}{x}$

50. $\dfrac{15}{x} + \dfrac{15}{x+20} = 2$

Find the value of $y = \dfrac{-6}{x}$ for the given value of x.

51. $x = -6$

52. $x = -3$

53. $x = 2$

54. $x = 1$

Maintaining Your Skills

55. Google Earth This Google Earth image is of the London Eye. The distance around the wheel is 420 meters. If it takes a person 27 minutes to complete a full revolution, what is the speed of the London Eye in centimeters per second? Round to the nearest whole number.

56. Google Earth The Google Earth image shows three cities in Colorado. If the distance between North Washington and Edgewater is 4.7 miles, and the distance from Edgewater to Denver is 4 miles, what is the distance from Denver to North Washington? Round to the nearest tenth.

57. Number Problem If twice the difference of a number and 3 were decreased by 5, the result would be 3. Find the number.

58. Number Problem If 3 times the sum of a number and 2 were increased by 6, the result would be 27. Find the number.

59. Geometry The length of a rectangle is 5 inches more than twice the width. The perimeter is 34 inches. Find the length and width.

60. Geometry The length of a rectangle is 2 feet more than 3 times the width. The perimeter is 44 feet. Find the length and width.

61. Number Problem The product of two consecutive even integers is 48. Find the two integers.

62. Number Problem The product of two consecutive odd integers is 35. Find the two integers.

63. Geometry The hypotenuse (the longest side) of a right triangle is 10 inches, and the lengths of the two legs (the other two sides) are given by two consecutive even integers. Find the lengths of the two legs.

64. Geometry One leg of a right triangle is 2 feet more than twice the other. If the hypotenuse is 13 feet, find the lengths of the two legs.

OBJECTIVES

A Solve applications whose solutions depend on solving an equation containing rational expressions.

B Graph an equation involving a rational expression.

TICKET TO SUCCESS

Keep these questions in mind as you read through the section. Then respond in your own words and in complete sentences.

1. Why is it useful to construct a table when working an application problem involving rational expressions?

2. Use rational expressions to show how the current of a river affects the speed of a motor boat.

3. Write an application problem for which the solution depends on solving the equation $\frac{1}{2} + \frac{1}{3} = \frac{1}{x}$.

4. Explain why the graph of the equation $y = -\frac{4}{x}$ would not cross the x- or the y-axis.

A Applications Involving Rational Expressions

In this section, we will solve some word problems whose equations involve rational expressions. Like the other word problems we have encountered, the more you work with them, the easier they become.

EXAMPLE 1 One number is twice another. The sum of their reciprocals is $\frac{9}{2}$. Find the two numbers.

SOLUTION Let x = the smaller number. The larger then must be $2x$. Their reciprocals are $\frac{1}{x}$ and $\frac{1}{2x}$, respectively. An equation that describes the situation is

$$\frac{1}{x} + \frac{1}{2x} = \frac{9}{2}$$

We can multiply both sides by the LCD, $2x$, and then solve the resulting equation.

$$2x\left(\frac{1}{x}\right) + 2x\left(\frac{1}{2x}\right) = 2x\left(\frac{9}{2}\right)$$

$$2 + 1 = 9x$$

$$3 = 9x$$

$$x = \frac{3}{9} = \frac{1}{3}$$

The smaller number is $\frac{1}{3}$. The other number is twice as large, or $\frac{2}{3}$. If we add their reciprocals, we have

$$\frac{3}{1} + \frac{3}{2} = \frac{6}{2} + \frac{3}{2} = \frac{9}{2}$$

The solutions check with the original problem. ■

EXAMPLE 2   A boat travels 30 miles up a river in the same amount of time it takes to travel 50 miles down the same river. If the current is 5 miles per hour, what is the speed of the boat in still water?

SOLUTION   The easiest way to work a problem like this is with a table. The top row of the table is labeled with d for distance, r for rate, and t for time. The left column of the table is labeled with the two trips: upstream and downstream. Here is what the table looks like:

	d	r	t
Upstream			
Downstream			

The next step is to read the problem over again and fill in as much of the table as we can with the information in the problem. The distance the boat travels upstream is 30 miles and the distance downstream is 50 miles. Since we are asked for the speed of the boat in still water, we will let that be x. If the speed of the boat in still water is x, then its speed upstream (against the current) must be $x - 5$, and its speed downstream (with the current) must be $x + 5$. Putting these four quantities into the appropriate positions in the table, we have

	d	r	t
Upstream	30	$x - 5$	
Downstream	50	$x + 5$	

The last positions in the table are filled in by using an equivalent form of the rate equation $t = \frac{d}{r}$.

	d	r	t
Upstream	30	$x - 5$	$\dfrac{30}{x - 5}$
Downstream	50	$x + 5$	$\dfrac{50}{x + 5}$

Reading the problem again, we find that the time for the trip upstream is equal to the time for the trip downstream. Setting these two quantities equal to each other, we have our equation:

Time (downstream) = Time (upstream)

$$\frac{50}{x + 5} = \frac{30}{x - 5}$$

NOTE
There are two things to note about this problem. The first is that to use the rate equation learned earlier in this chapter and solve the equation $d = r \cdot t$ for t, we divide each side by r, like this:

$$\frac{d}{r} = \frac{r \cdot t}{r}$$

$$\frac{d}{r} = t$$

The second thing is this: The speed of the boat in still water is the rate at which it would be traveling if there were no current; that is, it is the speed of the boat through the water. Since the water itself is moving at 5 miles per hour, the boat is going 5 miles per hour slower when it travels against the current and 5 miles per hour faster when it travels with the current.

The LCD is $(x + 5)(x - 5)$. We multiply both sides of the equation by the LCD to clear it of all denominators. Here is the solution:

$$(x + 5)(x - 5) \cdot \frac{50}{x + 5} = (x + 5)(x - 5) \cdot \frac{30}{x - 5}$$

$$50x - 250 = 30x + 150$$

$$20x = 400$$

$$x = 20$$

The speed of the boat in still water is 20 miles per hour.

EXAMPLE 3 Tina is training for a biathlon. To train for the bicycle portion, she rides her bike 15 miles uphill and then 15 miles back down. The complete trip takes her 2 hours. If her downhill speed is 20 miles per hour faster than her uphill speed, how fast does she ride uphill?

SOLUTION Again, we make a table. As in the previous example, we label the top row with distance, rate, and time. We label the left column with the two trips, uphill and downhill.

Total distance = 30 miles
Total time = 2 hours

	d	r	t
Uphill			
Downhill			

Next, we fill in the table with as much information as we can from the problem. We know the distance traveled is 15 miles uphill and 15 miles downhill, which allows us to fill in the distance column. To fill in the rate column, we first note that she rides 20 miles per hour faster downhill than uphill. Therefore, if we let x equal her rate uphill, then her rate downhill is $x + 20$. Filling in the table with this information gives us

	d	r	t
Uphill	15	x	
Downhill	15	$x + 20$	

Since time is distance divided by rate, $t = \dfrac{d}{r}$, we can fill in the last column in the table.

	d	r	t
Uphill	15	x	$\dfrac{15}{x}$
Downhill	15	$x + 20$	$\dfrac{15}{x + 20}$

Rereading the problem, we find that the total time (the time riding uphill plus the time riding downhill) is two hours. We write our equation as follows:

$$\text{Time (uphill)} + \text{Time (downhill)} = 2$$

$$\frac{15}{x} + \frac{15}{x + 20} = 2$$

We solve this equation for x by first finding the LCD and then multiplying each term in the equation by it to clear the equation of all denominators. Our LCD is $x(x + 20)$. Here is our solution:

$$x(x+20)\frac{15}{x} + x(x+20)\frac{15}{x+20} = 2 \cdot [x(x+20)]$$

$$15(x+20) + 15x = 2x(x+20)$$

$$15x + 300 + 15x = 2x^2 + 40x$$

$$0 = 2x^2 + 10x - 300$$

$$0 = x^2 + 5x - 150 \qquad \text{Divide both sides by 2}$$

$$0 = (x+15)(x-10)$$

$$x = -15 \quad \text{or} \quad x = 10$$

Since we cannot have a negative speed, our only solution is $x = 10$. Tina rides her bike at a rate of 10 miles per hour when going uphill. (Her downhill speed is $x + 20 = 30$ miles per hour.)

EXAMPLE 4   An inlet pipe can fill a water tank in 10 hours, while an outlet pipe can empty the same tank in 15 hours. By mistake, both pipes are left open. How long will it take to fill the water tank with both pipes open?

SOLUTION   Let x = amount of time to fill the tank with both pipes open.

One method of solving this type of problem is to think in terms of how much of the job is done by a pipe in 1 hour.

1. If the inlet pipe fills the tank in 10 hours, then in 1 hour the inlet pipe fills $\frac{1}{10}$ of the tank.

2. If the outlet pipe empties the tank in 15 hours, then in 1 hour the outlet pipe empties $\frac{1}{15}$ of the tank.

3. If it takes x hours to fill the tank with both pipes open, then in 1 hour the tank is $\frac{1}{x}$ full.

Here is how we set up the equation. In 1 hour,

$$\underset{\substack{\text{Amount of water let}\\\text{in by inlet pipe}}}{\frac{1}{10}} - \underset{\substack{\text{Amount of water let}\\\text{out by outlet pipe}}}{\frac{1}{15}} = \underset{\substack{\text{Total amount of}\\\text{water into tank}}}{\frac{1}{x}}$$

The LCD for our equation is $30x$. We multiply both sides by the LCD and solve.

$$30x\left(\frac{1}{10}\right) - 30x\left(\frac{1}{15}\right) = 30x\left(\frac{1}{x}\right)$$

$$3x - 2x = 30$$

$$x = 30$$

It takes 30 hours with both pipes open to fill the tank.

Inlet Pipe
10 hours
to fill

Outlet Pipe
15 hours
to empty

NOTE
In solving a problem of this type, we have to assume that the thing doing the work (whether it is a pipe, a person, or a machine) is working at a constant rate; that is, as much work gets done in the first hour as is done in the last hour and any other hour in between.

B Graphing Rational Expressions

EXAMPLE 5   Graph the equation $y = \frac{1}{x}$.

SOLUTION   Since this is the first time we have graphed an equation of this form, we will make a table of values for x and y that satisfy the equation. Before we do, let's make some generalizations about the graph (Figure 1).

First, notice that since y is equal to 1 divided by x, y will be positive when x is positive. (The quotient of two positive numbers is a positive number.) Likewise, when x is negative, y will be negative. In other words, x and y always will have the same sign. Thus, our graph will appear in quadrants I and III only because in those quadrants x and y have the same sign.

Next, notice that the expression $\frac{1}{x}$ will be undefined when x is 0, meaning that there is no value of y corresponding to $x = 0$. Because of this, the graph will not cross the y-axis. Further, the graph will not cross the x-axis either. If we try to find the x-intercept by letting $y = 0$, we have

$$0 = \frac{1}{x}$$

But there is no value of x to divide into 1 to obtain 0. Therefore, since there is no solution to this equation, our graph will not cross the x-axis.

To summarize, we can expect to find the graph in quadrants I and III only, and the graph will cross neither axis.

x	y
-3	$-\frac{1}{3}$
-2	$-\frac{1}{2}$
-1	-1
$-\frac{1}{2}$	-2
$-\frac{1}{3}$	-3
0	Undefined
$\frac{1}{3}$	3
$\frac{1}{2}$	2
1	1
2	$\frac{1}{2}$
3	$\frac{1}{3}$

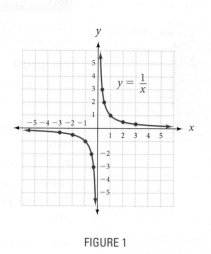

FIGURE 1

EXAMPLE 6 Graph the equation $y = \frac{-6}{x}$.

SOLUTION Since y is -6 divided by x, when x is positive, y will be negative (a negative divided by a positive is negative), and when x is negative, y will be positive (a negative divided by a negative). Thus, the graph (Figure 2) will appear in quadrants II and IV only. As was the case in Example 5, the graph will not cross either axis.

x	y
-6	1
-3	2
-2	3
-1	6
0	Undefined
1	-6
2	-3
3	-2
6	-1

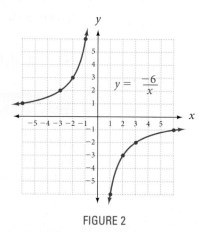

FIGURE 2

Problem Set 6.5

A Number Problems [Example 1]

1. One number is 3 times as large as another. The sum of their reciprocals is $\frac{16}{3}$. Find the two numbers.

2. If $\frac{3}{5}$ is added to twice the reciprocal of a number, the result is 1. Find the number.

3. The sum of a number and its reciprocal is $\frac{13}{6}$. Find the number.

4. The sum of a number and 10 times its reciprocal is 7. Find the number.

5. If a certain number is added to both the numerator and denominator of the fraction $\frac{7}{9}$, the result is $\frac{5}{7}$. Find the number.

6. The numerator of a certain fraction is 2 more than the denominator. If $\frac{1}{3}$ is added to the fraction, the result is 2. Find the fraction.

7. The sum of the reciprocals of two consecutive even integers is $\frac{5}{12}$. Find the integers.

8. The sum of the reciprocals of two consecutive integers is $\frac{7}{12}$. Find the two integers.

A Motion Problems [Examples 2–3]

9. A boat travels 26 miles up the river in the same amount of time it takes to travel 38 miles down the same river. If the current is 3 miles per hour, what is the speed of the boat in still water?

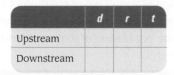

10. A boat can travel 9 miles up a river in the same amount of time it takes to travel 11 miles down the same river. If the current is 2 miles per hour, what is the speed of the boat in still water?

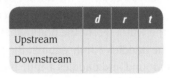

11. An airplane flying against the wind travels 140 miles in the same amount of time it would take the same plane to travel 160 miles with the wind. If the wind speed is a constant 20 miles per hour, how fast would the plane travel in still air?

12. An airplane flying against the wind travels 500 miles in the same amount of time that it would take to travel 600 miles with the wind. If the speed of the wind is 50 miles per hour, what is the speed of the plane in still air?

13. One plane can travel 20 miles per hour faster than another. One of them goes 285 miles in the same time it takes the other to go 255 miles. What are their speeds?

14. One car travels 300 miles in the same amount of time it takes a second car traveling 5 miles per hour slower than the first to go 275 miles. What are the speeds of the cars?

15. Tina, whom we mentioned in Example 3 of this section, is training for a biathlon. To train for the running portion of the race, she runs 8 miles each day, over the same course. The first 2 miles of the course are on level ground, while the last 6 miles are downhill. She runs 3 miles per hour slower on level ground than she runs downhill. If the complete course takes 1 hour, how fast does she run on the downhill part of the course?

16. Jerri is training for the same biathlon as Tina (Example 3 and Problem 15). To train for the bicycle portion of the race, she rides 24 miles out a straight road, then turns around and rides 24 miles back. The trip out is against the wind, whereas the trip

back is with the wind. If she rides 10 miles per hour faster with the wind then she does against the wind, and the complete trip out and back takes 2 hours, how fast does she ride when she rides against the wind?

17. To train for the running of a triathlon, Jerri jogs 1 hour each day over the same 9-mile course. Five miles of the course is downhill, whereas the other 4 miles is on level ground. Jerri figures that she runs 2 miles per hour faster downhill than she runs on level ground. Find the rate at which Jerri runs on level ground.

18. Travis paddles his kayak in the harbor at Morro Bay, California, where the incoming tide has caused a current in the water. From the point where he enters the water, he paddles 1 mile against the current, then turns around and paddles 1 mile back to where he started. His average speed when paddling with the current is 4 miles per hour faster than his speed against the current. If the complete trip (out and back) takes him 1.2 hours, find his average speed when he paddles against the current.

A **Work Problems** [Example 4]

19. An inlet pipe can fill a pool in 12 hours, while an outlet pipe can empty it in 15 hours. If both pipes are left open, how long will it take to fill the pool?

20. A water tank can be filled in 20 hours by an inlet pipe and emptied in 25 hours by an outlet pipe. How long will it take to fill the tank if both pipes are left open?

21. A bathtub can be filled by the cold water faucet in 10 minutes and by the hot water faucet in 12 minutes. How long does it take to fill the tub if both faucets are open?

22. A water faucet can fill a sink in 6 minutes, whereas the drain can empty it in 4 minutes. If the sink is full, how long will it take to empty if both the faucet and the drain are open?

23. A sink can be filled by the cold water faucet in 3 minutes. The drain can empty a full sink in 4 minutes. If the sink is empty and both the cold water faucet and the drain are open, how long will it take the sink to overflow?

24. A bathtub can be filled by the cold water faucet in 9 minutes and by the hot water faucet in 10 minutes. The drain can empty the tub in 5 minutes. Can the tub be filled if both faucets and the drain are open?

B Graph each of the following equations. [Examples 5–6]

25. $y = \dfrac{-4}{x}$

26. $y = \dfrac{4}{x}$

27. $y = \dfrac{8}{x}$

28. $y = \dfrac{-8}{x}$

29. Graph $y = \dfrac{3}{x}$ and $x + y = 4$ on the same coordinate system. At what points do the two graphs intersect?

30. Graph $y = \dfrac{4}{x}$ and $x - y = 3$ on the same coordinate system. At what points do the two graphs intersect?

Getting Ready for the Next Section

Simplify.

31. $\dfrac{1}{2} \div \dfrac{2}{3}$

32. $\dfrac{1}{3} \div \dfrac{3}{4}$

33. $1 + \dfrac{1}{2}$

34. $1 + \dfrac{2}{3}$

35. $y^5 \cdot \dfrac{2x^3}{y^2}$

36. $y^7 \cdot \dfrac{3x^5}{y^4}$

37. $\dfrac{2x^3}{y^2} \cdot \dfrac{y^5}{4x}$

38. $\dfrac{3x^5}{y^4} \cdot \dfrac{y^7}{6x^2}$

Factor.

39. $x^2y + x$

40. $xy^2 + y$

Reduce.

41. $\dfrac{2x^3y^2}{4x}$

42. $\dfrac{3x^5y^3}{6x^2}$

43. $\dfrac{x^2 - 4}{x^2 - x - 6}$

44. $\dfrac{x^2 - 9}{x^2 - 5x + 6}$

Maintaining Your Skills

45. **Population** The map shows the most populated cities in the United States. If the population of New York City is about 42% of the state's population, what is the approximate population of the state? Round to the nearest tenth.

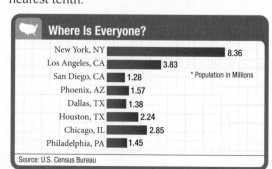

Where Is Everyone?

City	Population (Millions)
New York, NY	8.36
Los Angeles, CA	3.83
San Diego, CA	1.28
Phoenix, AZ	1.57
Dallas, TX	1.38
Houston, TX	2.24
Chicago, IL	2.85
Philadelphia, PA	1.45

* Population in Millions

Source: U.S. Census Bureau

46. Horse Racing The graph shows the total amount of money wagered on the Kentucky Derby. What was the percent increase in wagers from 1995 to 2005? Round to the nearest tenth of a percent.

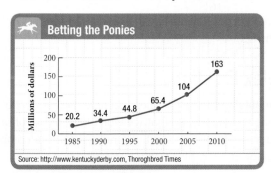

Betting the Ponies

Source: http://www.kentuckyderby.com, Thoroghbred Times

47. Factor out the greatest common factor for $15a^3b^3 - 20a^2b - 35ab^2$.

48. Factor by grouping $3ax - 2a + 15x - 10$.

Factor completely.

49. $x^2 - 4x - 12$

50. $4x^2 - 20xy + 25y^2$

51. $x^4 - 16$

52. $2x^2 + xy - 21y^2$

53. $5x^3 - 25x^2 - 30x$

6.6 | Complex Fractions

OBJECTIVES

A Simplify a complex fraction.

TICKET TO SUCCESS

Keep these questions in mind as you read through the section. Then respond in your own words and in complete sentences.

1. What is a complex fraction?

2. Explain the first method of simplifying complex fractions.

3. When would you use an LCD to simplify a complex fraction rather than multiply the numerator by the denominator's reciprocal?

4. Give an example of a complex fraction problem that can be rewritten as a division problem.

A *complex fraction* is a fraction or rational expression that contains other fractions in its numerator or denominator. Each of the following is a complex fraction:

$$\frac{\frac{1}{2}}{\frac{2}{3}} \qquad \frac{x + \frac{1}{y}}{y + \frac{1}{x}} \qquad \frac{\frac{a + 1}{a^2 - 9}}{\frac{2}{a + 3}}$$

In the United States, our speed limits are calculated in miles per hour. In many other countries, the limits appear in kilometers per hour. We can use the complex fraction

$$\dfrac{\dfrac{x}{100}}{161}$$

to convert miles per hour to kilometers per hour. You will use this complex fraction to do just that later on in the problem set.

A Simplifying Complex Fractions

We will begin this section by simplifying the first of these complex fractions. Before we do, though, let's agree on some vocabulary. So that we won't have to use phrases such as the "numerator of the denominator," let's call the numerator of a complex fraction the *top* and the denominator of a complex fraction the *bottom*.

EXAMPLE 1 Simplify $\dfrac{\dfrac{1}{2}}{\dfrac{2}{3}}$.

SOLUTION There are two methods we can use to solve this problem.

METHOD 1 We can multiply the top and bottom of this complex fraction by the LCD for both fractions. In this case, the LCD is 6.

$$\begin{matrix} \text{top} \\ \text{bottom} \end{matrix} \left\{ \quad \dfrac{\dfrac{1}{2}}{\dfrac{2}{3}} = \dfrac{\mathbf{6} \cdot \dfrac{1}{2}}{\mathbf{6} \cdot \dfrac{2}{3}} = \dfrac{3}{4} \right.$$

METHOD 2 We can treat this as a division problem. To divide by $\dfrac{2}{3}$, we multiply by its reciprocal $\dfrac{3}{2}$.

$$\dfrac{\dfrac{1}{2}}{\dfrac{2}{3}} = \dfrac{1}{2} \cdot \dfrac{3}{2} = \dfrac{3}{4}$$

Using either method, we obtain the same result.

EXAMPLE 2 Simplify $\dfrac{\dfrac{2x^3}{y^2}}{\dfrac{4x}{y^5}}$.

SOLUTION

METHOD 1 The LCD for each rational expression is y^5. Multiplying the top and bottom of the complex fraction by y^5, we have

$$\dfrac{\dfrac{2x^3}{y^2}}{\dfrac{4x}{y^5}} = \dfrac{\mathbf{y^5} \cdot \dfrac{2x^3}{y^2}}{\mathbf{y^5} \cdot \dfrac{4x}{y^5}} = \dfrac{2x^3y^3}{4x} = \dfrac{x^2y^3}{2}$$

METHOD 2 To divide by $\dfrac{4x}{y^5}$ we multiply by its reciprocal, $\dfrac{y^5}{4x}$

$$\dfrac{\dfrac{2x^3}{y^2}}{\dfrac{4x}{y^5}} = \dfrac{2x^3}{y^2} \cdot \dfrac{y^5}{4x} = \dfrac{x^2y^3}{2}$$

Again the result is the same, whether we use Method 1 or Method 2.

EXAMPLE 3 Simplify $\dfrac{x + \dfrac{1}{y}}{y + \dfrac{1}{x}}$.

SOLUTION To apply Method 2 as we did in the first two examples, we would have to simplify the top and bottom separately to obtain a single rational expression for both before we could multiply by the reciprocal. In this case, it is much easier to use Method 1 to multiply the top and bottom by the LCD xy.

$$\frac{x + \dfrac{1}{y}}{y + \dfrac{1}{x}} = \frac{\boldsymbol{xy}\left(x + \dfrac{1}{y}\right)}{\boldsymbol{xy}\left(y + \dfrac{1}{x}\right)} \qquad \text{Multiply top and bottom by } xy$$

$$= \frac{xy \cdot x + xy \cdot \dfrac{1}{y}}{xy \cdot y + xy \cdot \dfrac{1}{x}} \qquad \text{Distributive property}$$

$$= \frac{x^2 y + x}{xy^2 + y} \qquad \text{Simplify}$$

We can factor an x from $x^2 y + x$ and a y from $xy^2 + y$ and then reduce to lowest terms:

$$= \frac{x(\cancel{xy + 1})}{y(\cancel{xy + 1})}$$

$$= \frac{x}{y}$$

EXAMPLE 4 Simplify $\dfrac{1 - \dfrac{4}{x^2}}{1 - \dfrac{1}{x} - \dfrac{6}{x^2}}$.

SOLUTION Again, the easiest way to simplify this complex fraction is to multiply the top and bottom by the LCD, x^2.

$$\frac{1 - \dfrac{4}{x^2}}{1 - \dfrac{1}{x} - \dfrac{6}{x^2}} = \frac{\boldsymbol{x^2}\left(1 - \dfrac{4}{x^2}\right)}{\boldsymbol{x^2}\left(1 - \dfrac{1}{x} - \dfrac{6}{x^2}\right)} \qquad \text{Multiply top and bottom by } x^2$$

$$= \frac{x^2 \cdot 1 - x^2 \cdot \dfrac{4}{x^2}}{x^2 \cdot 1 - x^2 \cdot \dfrac{1}{x} - x^2 \cdot \dfrac{6}{x^2}} \qquad \text{Distributive property}$$

$$= \frac{x^2 - 4}{x^2 - x - 6} \qquad \text{Simplify}$$

$$= \frac{(x - 2)\cancel{(x + 2)}}{(x - 3)\cancel{(x + 2)}} \qquad \text{Factor}$$

$$= \frac{x - 2}{x - 3} \qquad \text{Reduce}$$

In our next example, we find the relationship between a sequence of complex fractions and the numbers in the Fibonacci sequence.

EXAMPLE 5 Simplify each term in the following sequence, and then explain how this sequence is related to the Fibonacci sequence.

$$1 + \cfrac{1}{1 + 1}, \; 1 + \cfrac{1}{1 + \cfrac{1}{1 + 1}}, \; 1 + \cfrac{1}{1 + \cfrac{1}{1 + \cfrac{1}{1 + 1}}} \;, \; \ldots$$

SOLUTION We can simplify our work somewhat if we notice that the first term $1 + \frac{1}{1+1}$ is the larger denominator in the second term, and that the second term is the largest denominator in the third term.

First term: $\quad 1 + \cfrac{1}{1 + 1} = 1 + \frac{1}{2} = \frac{2}{2} + \frac{1}{2} = \frac{3}{2}$

Second term: $\quad 1 + \cfrac{1}{1 + \cfrac{1}{1 + 1}} = 1 + \cfrac{1}{\frac{3}{2}} = 1 + \frac{2}{3} = \frac{3}{3} + \frac{2}{3} = \frac{5}{3}$

Third term: $\quad 1 + \cfrac{1}{1 + \cfrac{1}{1 + \cfrac{1}{1 + 1}}} = 1 + \cfrac{1}{\frac{5}{3}} = 1 + \frac{3}{5} = \frac{5}{5} + \frac{3}{5} = \frac{8}{5}$

Here are the simplified numbers for the first three terms in our sequence:

$$\frac{3}{2}, \frac{5}{3}, \frac{8}{5}, \ldots$$

Recall the Fibonacci sequence:

$$1, 1, 2, 3, 5, 8, 13, 21, \ldots$$

As you can see, each term in the sequence we have simplified is the ratio of two consecutive numbers in the Fibonacci sequence. If the pattern continues in this manner, the next number in our sequence will be $\frac{13}{8}$. ∎

Problem Set 6.6

Moving Toward Success

"Most of us serve our ideals by fits and starts. The person who makes a success of living is one who sees his goal steadily and aims for it unswervingly. That's dedication."

—Cecil B. DeMille, 1881–1959, American film director and producer

1. Do you find it difficult to put forth as much effort toward this class now compared to in the beginning? Why or why not?

2. How do you stay focused on the needs of this class when your mind wanders or distractions call you away from your work?

A Simplify each complex fraction. [Examples 1–4]

1. $\dfrac{\frac{3}{4}}{\frac{1}{8}}$

2. $\dfrac{\frac{1}{3}}{\frac{5}{6}}$

3. $\dfrac{\frac{2}{3}}{4}$

4. $\dfrac{5}{\frac{1}{2}}$

5. $\dfrac{\frac{x^2}{y}}{\frac{x}{y^3}}$

6. $\dfrac{\frac{x^5}{y^3}}{\frac{x^2}{y^8}}$

7. $\dfrac{\dfrac{4x^3}{y^6}}{\dfrac{8x^2}{y^7}}$

8. $\dfrac{\dfrac{6x^4}{y}}{\dfrac{2x}{y^5}}$

9. $\dfrac{y + \dfrac{1}{x}}{x + \dfrac{1}{y}}$

10. $\dfrac{y - \dfrac{1}{x}}{x - \dfrac{1}{y}}$

11. $\dfrac{1 + \dfrac{1}{a}}{1 - \dfrac{1}{a}}$

12. $\dfrac{\dfrac{1}{a} - 1}{\dfrac{1}{a} + 1}$

13. $\dfrac{\dfrac{x + 1}{x^2 - 9}}{\dfrac{2}{x + 3}}$

14. $\dfrac{\dfrac{3}{x - 5}}{\dfrac{x + 1}{x^2 - 25}}$

15. $\dfrac{\dfrac{1}{a + 2}}{\dfrac{1}{a^2 - a - 6}}$

16. $\dfrac{\dfrac{1}{a^2 + 5a + 6}}{\dfrac{1}{a + 3}}$

17. $\dfrac{1 - \dfrac{9}{y^2}}{1 - \dfrac{1}{y} - \dfrac{6}{y^2}}$

18. $\dfrac{1 - \dfrac{4}{y^2}}{1 - \dfrac{2}{y} - \dfrac{8}{y^2}}$

19. $\dfrac{\dfrac{1}{y} + \dfrac{1}{x}}{\dfrac{1}{xy}}$

20. $\dfrac{\dfrac{1}{xy}}{\dfrac{1}{y} - \dfrac{1}{x}}$

21. $\dfrac{1 - \dfrac{1}{a^2}}{1 - \dfrac{1}{a}}$

22. $\dfrac{1 + \dfrac{1}{a}}{1 - \dfrac{1}{a^2}}$

23. $\dfrac{\dfrac{1}{10x} - \dfrac{y}{10x^2}}{\dfrac{1}{10} - \dfrac{y}{10x}}$

24. $\dfrac{\dfrac{1}{2x} + \dfrac{y}{2x^2}}{\dfrac{1}{4} + \dfrac{y}{4x}}$

25. $\dfrac{\dfrac{1}{a + 1} + 2}{\dfrac{1}{a + 1} + 3}$

26. $\dfrac{\dfrac{2}{a + 1} + 3}{\dfrac{3}{a + 1} + 4}$

Although the following problems do not contain complex fractions, they do involve more than one operation. Simplify inside the parentheses first, then multiply. [Example 5]

27. $\left(1 - \dfrac{1}{x}\right)\left(1 - \dfrac{1}{x + 1}\right)\left(1 - \dfrac{1}{x + 2}\right)$

28. $\left(1 + \dfrac{1}{x}\right)\left(1 + \dfrac{1}{x + 1}\right)\left(1 + \dfrac{1}{x + 2}\right)$

29. $\left(1 + \dfrac{1}{x + 3}\right)\left(1 + \dfrac{1}{x + 2}\right)\left(1 + \dfrac{1}{x + 1}\right)$

30. $\left(1 - \dfrac{1}{x + 3}\right)\left(1 - \dfrac{1}{x + 2}\right)\left(1 - \dfrac{1}{x + 1}\right)$

31. Simplify each term in the following sequence.

$$2 + \dfrac{1}{2 + 1},\ 2 + \dfrac{1}{2 + \dfrac{1}{2 + 1}},\ 2 + \dfrac{1}{2 + \dfrac{1}{2 + \dfrac{1}{2 + 1}}},\ \ldots$$

32. Simplify each term in the following sequence.

$$2 + \dfrac{3}{2 + 3},\ 2 + \dfrac{3}{2 + \dfrac{3}{2 + 3}},\ 2 + \dfrac{3}{2 + \dfrac{3}{2 + \dfrac{3}{2 + 3}}},\ \ldots$$

Complete the following tables.

33.

Number	Reciprocal	Quotient	Square
x	$\dfrac{1}{x}$	$\dfrac{x}{\dfrac{1}{x}}$	x^2
1			
2			
3			
4			

34.

Number	Reciprocal	Quotient	Square
x	$\dfrac{1}{x}$	$\dfrac{\dfrac{1}{x}}{x}$	x^2
1			
2			
3			
4			

35.

Number	Reciprocal	Sum	Quotient
x	$\dfrac{1}{x}$	$1 + \dfrac{1}{x}$	$\dfrac{1 + \dfrac{1}{x}}{\dfrac{1}{x}}$
1			
2			
3			
4			

36.

Number	Reciprocal	difference	Quotient
x	$\dfrac{1}{x}$	$1 - \dfrac{1}{x}$	$\dfrac{1 - \dfrac{1}{x}}{\dfrac{1}{x}}$
1			
2			
3			
4			

Applying the Concepts

Cars The chart shows the fastest cars in the world. Use the chart to answer Problems 37 and 38.

37. To convert miles per hour to kilometers per hour, we can use the complex fraction $\dfrac{x}{\left(\frac{100}{161}\right)}$ where x is the speed in miles per hour. What is the top speed of the Bugatti Veyron in kilometers per hour? Round to the nearest whole number.

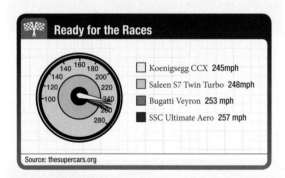

Ready for the Races

☐ Koenigsegg CCX **245mph**
☐ Saleen S7 Twin Turbo **248mph**
■ Bugatti Veyron **253 mph**
■ SSC Ultimate Aero **257 mph**

Source: thesupercars.org

38. To convert miles per hour to meters per second we use the complex fraction, $\dfrac{\left(\frac{x}{3,600}\right)}{\left(\frac{1}{1,609}\right)}$, where x is the speed in miles per hour. What is the top speed of the Saleen S7 Twin Turbo in meters per second? Round to the nearest whole number.

Getting Ready for the Next Section

Solve.

39. $21 = 6x$

40. $72 = 2x$

41. $x^2 + x = 6$

42. $x^2 + 2x = 8$

Maintaining Your Skills

Solve each inequality.

43. $2x + 3 < 5$

44. $3x - 2 > 7$

45. $-3x \le 21$

46. $-5x \ge -10$

47. $-2x + 8 > -4$

48. $-4x - 1 < 11$

49. $4 - 2(x + 1) \ge -2$

50. $6 - 2(x + 3) \le -8$

6.7 Proportions

OBJECTIVES

A Solve a proportion.

B Solve application problems involving proportions.

TICKET TO SUCCESS

Keep these questions in mind as you read through the section. Then respond in your own words and in complete sentences.

1. What is a proportion?

2. What are the means and extremes of a proportion?

3. What is the relationship between the means and the extremes in a proportion?

4. Write an application problem using the means-extremes property.

Picture a basketball player shooting free throws. During one game, the player makes 6 of his 8 free throws. During another game, he shoots a total of 12 free throws with the same accuracy that he shot in the first game. How many of his 12 free throws did he make? In this section, we solve similar proportions to help you answer that question.

Definition
A **proportion** is two equal ratios; that is, if $\frac{a}{b}$ and $\frac{c}{d}$ are ratios, then:

$$\frac{a}{b} = \frac{c}{d}$$

is a proportion.

Each of the four numbers in a proportion is called a *term* of the proportion. We number the terms as follows:

$$\text{First term} \rightarrow \frac{a}{b} = \frac{c}{d} \leftarrow \text{Third term}$$
$$\text{Second term} \rightarrow \phantom{\frac{a}{b} = \frac{c}{d}} \leftarrow \text{Fourth term}$$

The first and fourth terms are called the *extremes,* and the second and third terms are called the *means:*

$$\text{Means} \; \frac{a}{b} = \frac{c}{d} \; \text{Extremes}$$

For example, in the proportion:

$$\frac{3}{8} = \frac{12}{32}$$

the extremes are 3 and 32, and the means are 8 and 12.

Means-Extremes Property
If a, b, c, and d are real numbers with $b \neq 0$ and $d \neq 0$, then

$$\text{if} \quad \frac{a}{b} = \frac{c}{d}$$

$$\text{then} \quad ad = bc$$

In words: In any proportion, the product of the extremes is equal to the product of the means.

A Solving Proportions

This property of proportions comes from the multiplication property of equality. We can use it to solve for a missing term in a proportion.

EXAMPLE 1 Solve the proportion $\dfrac{3}{x} = \dfrac{6}{7}$ for x.

SOLUTION We could solve for x by using the method developed in Section 6.4; that is, multiplying both sides by the LCD $7x$. Instead, let's use our new means-extremes property.

$$\frac{3}{x} = \frac{6}{7}$$ Extremes are 3 and 7; means are x and 6

$$21 = 6x$$ Product of extremes = product of means

$$\frac{21}{6} = x$$ Divide both sides by 6

$$x = \frac{7}{2}$$ Reduce to lowest terms

EXAMPLE 2 Solve $\dfrac{x + 1}{2} = \dfrac{3}{x}$ for x.

SOLUTION Again, we want to point out that we could solve for x by using the method we used in Section 6.4. Using the means-extremes property is simply an alternative to the method developed in Section 6.4.

$$\frac{x + 1}{2} = \frac{3}{x}$$ Extremes are $x + 1$ and x; means are 2 and 3

$$x^2 + x = 6$$ Product of extremes = product of means

$$x^2 + x - 6 = 0$$ Standard form for a quadratic equation

$$(x + 3)(x - 2) = 0$$ Factor

$$x + 3 = 0 \quad \text{or} \quad x - 2 = 0$$ Set factors equal to 0

$$x = -3 \quad \text{or} \quad x = 2$$

This time we have two solutions: -3 and 2.

B Applications with Proportions

EXAMPLE 3 A manufacturer knows that during a production run, 8 out of every 100 parts produced by a certain machine will be defective. If the machine produces 1,450 parts, how many can be expected to be defective?

SOLUTION The ratio of defective parts to total parts produced is $\dfrac{8}{100}$. If we let x represent the number of defective parts out of the total of 1,450 parts, then we can write this ratio again as $\dfrac{x}{1,450}$. This gives us a proportion to solve.

Defective parts in numerator ⟶ $\dfrac{x}{1,450} = \dfrac{8}{100}$ Extremes are x and 100; means are 1,450 and 8

Total parts in denominator ⟶ $100x = 11,600$ Product of extremes = product of means

$$x = 116$$

The manufacturer can expect 116 defective parts out of the total of 1,450 parts if the machine usually produces 8 defective parts for every 100 parts it produces.

EXAMPLE 4   The scale on a map indicates that 1 inch on the map corresponds to an actual distance of 85 miles. Two cities are 3.5 inches apart on the map. What is the actual distance between the two cities?

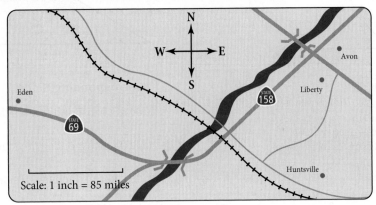

Scale: 1 inch = 85 miles

SOLUTION   We let x represent the actual distance between the two cities. The proportion is

$$\text{Miles} \rightarrow \frac{x}{3.5} = \frac{85}{1} \begin{array}{l} \leftarrow \text{Miles} \\ \leftarrow \text{Inches} \end{array}$$

$$x \cdot 1 = 3.5(85)$$
$$x = 297.5 \text{ miles}$$

EXAMPLE 5   A woman drives her car 270 miles in 6 hours. If she continues at the same rate, how far will she travel in 10 hours?

SOLUTION   We let x represent the distance traveled in 10 hours. Using x, we translate the problem into the following proportion:

$$\text{Miles} \rightarrow \frac{x}{10} = \frac{270}{6} \begin{array}{l} \leftarrow \text{Miles} \\ \leftarrow \text{Hours} \end{array}$$

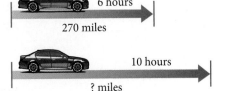

Notice that the two ratios in the proportion compare the same quantities. That is, both ratios compare miles to hours. In words this proportion says:

x miles is to 10 hours as 270 miles is to 6 hours

$$\frac{x}{10} = \frac{270}{6}$$

Next, we solve the proportion.

$$x \cdot 6 = 10 \cdot 270$$
$$x \cdot 6 = 2,700$$
$$\frac{x \cdot 6}{6} = \frac{2,700}{6}$$
$$x = 450 \text{ miles}$$

If the woman continues at the same rate, she will travel 450 miles in 10 hours.

NOTE
Recall from Chapter 2 that one way to label the important parts of a triangle is to label the vertices with capital letters and the sides with lower-case letters.

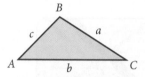

Notice that side a is opposite vertex A, side b is opposite vertex B, and side c is opposite vertex C. Also, because each vertex is the vertex of one of the angles of the triangle, we refer to the three interior angles as A, B, and C.

Similar Triangles

Two triangles that have the same shape are similar when their corresponding sides are proportional, or have the same ratio. The triangles below are similar.

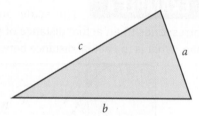
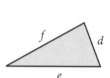

Corresponding Sides	**Ratio**
side a corresponds with side d	$\dfrac{a}{d}$
side b corresponds with side e	$\dfrac{b}{e}$
side c corresponds with side f	$\dfrac{c}{f}$

Because their corresponding sides are proportional, we write

$$\frac{a}{d} = \frac{b}{e} = \frac{c}{f}$$

EXAMPLE 6   The two triangles below are similar. Find side x.

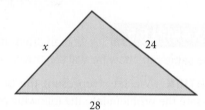
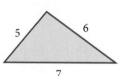

SOLUTION   To find the length x, we set up a proportion of equal ratios. The ratio of x to 5 is equal to the ratio of 24 to 6 and to the ratio of 28 to 7. Algebraically we have

$$\frac{x}{5} = \frac{24}{6} \quad \text{and} \quad \frac{x}{5} = \frac{28}{7}$$

We can solve either proportion to get our answer. The first gives us

$$\frac{x}{5} = 4 \qquad \frac{24}{6} = 4$$

$x = 4 \cdot 5$ 	Multiply both sides by 5

$x = 20$ 	Simplify

Problem Set 6.7

1. What does practice and repetition have to do
    with mathematics?

2. Why should you pay attention to the bold words
    or the information in colored boxes?

A Solve each of the following proportions.
[Examples 1–2]

1. $\dfrac{x}{2} = \dfrac{6}{12}$

2. $\dfrac{x}{4} = \dfrac{6}{8}$

3. $\dfrac{2}{5} = \dfrac{4}{x}$

4. $\dfrac{3}{8} = \dfrac{9}{x}$

5. $\dfrac{10}{20} = \dfrac{20}{x}$

6. $\dfrac{15}{60} = \dfrac{60}{x}$

7. $\dfrac{a}{3} = \dfrac{5}{12}$

8. $\dfrac{a}{2} = \dfrac{7}{20}$

9. $\dfrac{2}{x} = \dfrac{6}{7}$

10. $\dfrac{4}{x} = \dfrac{6}{7}$

11. $\dfrac{x+1}{3} = \dfrac{4}{x}$

12. $\dfrac{x+1}{6} = \dfrac{7}{x}$

13. $\dfrac{x}{2} = \dfrac{8}{x}$

14. $\dfrac{x}{9} = \dfrac{4}{x}$

15. $\dfrac{4}{a+2} = \dfrac{a}{2}$

16. $\dfrac{3}{a+2} = \dfrac{a}{5}$

17. $\dfrac{1}{x} = \dfrac{x-5}{6}$

18. $\dfrac{1}{x} = \dfrac{x-6}{7}$

B **Applying the Concepts** [Examples 3–5]

19. Google Earth The Google Earth image shows the
energy consumption for parts of Europe. In 2006, the
ratio of the oil consumption of Switzerland to the oil
consumption of the Czech Republic was 9 to 7. If the
oil consumption of the Czech Republic was 9.8
million metric tons, what was the oil consumption of
Switzerland?

20. Eiffel Tower The Eiffel Tower at the Paris Las Vegas
Hotel is a replica of the Eiffel Tower in France. The
heights of the Eiffel Tower in Las Vegas and the one
in France are 460 feet and 1,063 feet respectively. The
base of the Eiffel Tower in France is 410 feet wide.
What is the width of the base of the Eiffel Tower in
Las Vegas? Round to the nearest foot.

21. Baseball A baseball player gets 6 hits in the first 18
games of the season. If he continues hitting at the
same rate, how many hits will he get in the first 45
games?

22. Basketball A basketball player makes 8 of 12 free
throws in the first game of the season. If she shoots
with the same accuracy in the second game, how
many of the 15 free throws she attempts will she
make?

23. Mixture Problem A solution contains 12 milliliters of
alcohol and 16 milliliters of water. If another solution
is to have the same concentration of alcohol in water
but is to contain 28 milliliters of water, how much
alcohol must it contain?

24. Mixture Problem A solution contains 15 milliliters of
HCl and 42 milliliters of water. If another solution is
to have the same concentration of HCl in water but
is to contain 140 milliliters of water, how much HCl
must it contain?

25. Nutrition If 100 grams of ice cream contains 13
grams of fat, how much fat is in 350 grams of ice
cream?

26. Nutrition A 6-ounce serving of grapefruit juice contains 159 grams of water. How many grams of water are in 20 ounces of grapefruit juice?

27. Map Reading A map is drawn so that every 3.5 inches on the map corresponds to an actual distance of 100 miles. If the actual distance between the two cities is 420 miles, how far apart are they on the map?

28. Map Reading The scale on a map indicates that 1 inch on the map corresponds to an actual distance of 105 miles. Two cities are 4.5 inches apart on the map. What is the actual distance between the two cities?

On the map shown here, 0.5 inches on the map is equal to 5 miles. Use the information from the map to work Problems 29 through 32.

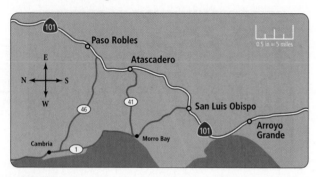

29. Map Reading Suppose San Luis Obispo is 1.25 inches from Arroyo Grande on the map. How far apart are the two cities?

30. Map Reading Suppose San Luis Obispo is 3.4 inches from Paso Robles on the map. How far apart are the two cities?

31. Driving Time If Ava drives from Paso Robles to San Luis Obispo in 46 minutes, how long will it take her to drive from San Luis Obispo to Arroyo Grande, if she drives at the same speed? Round to the nearest minute.

32. Driving Time If Brooke drives from Arroyo Grande to San Luis Obispo in 15 minutes, how long will it take her to drive from San Luis Obispo to Paso Robles, if she drives at the same speed? Round to the nearest minute.

33. Distance A man drives his car 245 miles in 5 hours. At this rate, how far will he travel in 7 hours?

34. Distance An airplane flies 1,380 miles in 3 hours. How far will it fly in 5 hours?

For each pair of similar triangles, set up a proportion in order to find the unknown. [Example 6]

35.

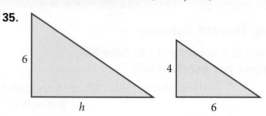

36.

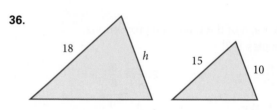

37.

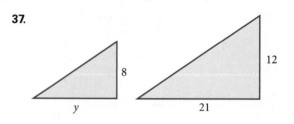

38.

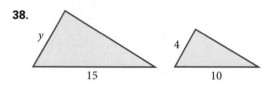

Maintaining Your Skills

Reduce to lowest terms.

39. $\dfrac{x^2 - x - 6}{x^2 - 9}$

40. $\dfrac{xy + 5x + 3y + 15}{x^2 + ax + 3x + 3a}$

Multiply or divide, as indicated.

41. $\dfrac{x^2 - 25}{x + 4} \cdot \dfrac{2x + 8}{x^2 - 9x + 20}$

42. $\dfrac{3x + 6}{x^2 + 4x + 3} \div \dfrac{x^2 + x - 2}{x^2 + 2x - 3}$

Add or subtract, as indicated.

43. $\dfrac{x}{x^2 - 16} + \dfrac{4}{x^2 - 16}$

44. $\dfrac{2}{x^2 - 1} - \dfrac{5}{x^2 + 3x - 4}$

45. Proportions in the News Searching Google News for articles that involve proportions produces a variety of items, some of which are shown here. Search the Internet for current news articles by searching on the word *proportion*. Then find an article that the material in this section helps you understand.

Confidence Falls in Higher Housing Prices

The latest consumer confidence survey from property website Rightmove suggested the **proportion** of people expecting prices to be higher next year has dropped from 50 percent to 41 percent . . .
Telegraph, UK, July 26, 2010

US Hot Stocks

. . . the Department of Education's gainful employment proposal will restrict a significant **proportion** of ITT's programs.
Wall Street Journal, NY, July 26, 2010

Forty Years of Less Society

Real household disposable income per head is $2\frac{1}{2}$ times more today than it was in 1970. We spend a small **proportion** on necessities like food and non-alcoholic drink, and more on recreation and housing.
The Guardian, UK, July 26, 2010

Rational Numbers [6.1]

EXAMPLES

1. We can reduce $\frac{6}{8}$ to lowest terms by dividing the numerator and denominator by their greatest common factor 2:

$$\frac{6}{8} = \frac{2 \cdot 3}{2 \cdot 4} = \frac{3}{4}$$

Any number that can be put in the form $\frac{a}{b}$, where a and b are integers ($b \neq 0$), is called a rational number.

Multiplying or dividing the numerator and denominator of a rational number by the same nonzero number never changes the value of the rational number.

Rational Expressions [6.1]

2. We reduce rational expressions to lowest terms by factoring the numerator and denominator and then dividing out any factors they have in common:

$$\frac{x - 3}{x^2 - 9} = \frac{x - 3}{(x - 3)(x + 3)} = \frac{1}{x + 3}$$

Any expression of the form $\frac{P}{Q}$, where P and Q are polynomials ($Q \neq 0$), is a rational expression.

Multiplying or dividing the numerator and denominator of a rational expression by the same nonzero quantity always produces a rational expression equivalent to the original one.

Multiplication of Rational Expressions [6.2]

3. $\dfrac{x - 1}{x^2 + 2x - 3} \cdot \dfrac{x^2 - 9}{x - 2}$

$= \dfrac{x - 1}{(x + 3)(x - 1)} \cdot \dfrac{(x - 3)(x + 3)}{x - 2}$

$= \dfrac{x - 3}{x - 2}$

To multiply two rational numbers or two rational expressions, multiply numerators, multiply denominators, and divide out any factors common to the numerator and denominator.

For rational numbers $\dfrac{a}{b}$ and $\dfrac{c}{d}$, $\qquad \dfrac{a}{b} \cdot \dfrac{c}{d} = \dfrac{ac}{bd}$

For rational expressions $\dfrac{P}{Q}$ and $\dfrac{R}{S}$, $\quad \dfrac{P}{Q} \cdot \dfrac{R}{S} = \dfrac{PR}{QS}$

Division of Rational Expressions [6.2]

4. $\dfrac{2x}{x^2 - 25} \div \dfrac{4}{x - 5}$

$= \dfrac{2x}{(x - 5)(x + 5)} \cdot \dfrac{(x - 5)}{4}$

$= \dfrac{x}{2(x + 5)}$

To divide by a rational number or rational expression, simply multiply by its reciprocal.

For rational numbers $\dfrac{a}{b}$ and $\dfrac{c}{d}$, $\qquad \dfrac{a}{b} \div \dfrac{c}{d} = \dfrac{a}{b} \cdot \dfrac{d}{c}$

For rational expressions $\dfrac{P}{Q}$ and $\dfrac{R}{S}$, $\quad \dfrac{P}{Q} \div \dfrac{R}{S} = \dfrac{P}{Q} \cdot \dfrac{S}{R}$

Addition of Rational Expressions [6.3]

5. $\dfrac{3}{x - 1} + \dfrac{x}{2}$

$= \dfrac{3}{x - 1} \cdot \dfrac{2}{2} + \dfrac{x}{2} \cdot \dfrac{x - 1}{x - 1}$

$= \dfrac{6}{2(x - 1)} + \dfrac{x^2 - x}{2(x - 1)}$

$= \dfrac{x^2 - x + 6}{2(x - 1)}$

To add two rational numbers or rational expressions, find a common denominator, change each expression to an equivalent expression having the common denominator, and then add numerators and reduce if possible.

For rational numbers $\dfrac{a}{c}$ and $\dfrac{b}{c}$, $\qquad \dfrac{a}{c} + \dfrac{b}{c} = \dfrac{a + b}{c}$

For rational expressions $\dfrac{P}{S}$ and $\dfrac{Q}{S}$, $\quad \dfrac{P}{S} + \dfrac{Q}{S} = \dfrac{P + Q}{S}$

Subtraction of Rational Expressions [6.3]

6. $\dfrac{x}{x^2 - 4} - \dfrac{2}{x^2 - 4}$

$= \dfrac{x - 2}{x^2 - 4}$

$= \dfrac{\cancel{x - 2}}{(\cancel{x - 2})(x + 2)}$

$= \dfrac{1}{x + 2}$

To subtract a rational number or rational expression, simply add its opposite

For rational numbers $\dfrac{a}{c}$ and $\dfrac{b}{c}$, $\dfrac{a}{c} - \dfrac{b}{c} = \dfrac{a}{c} + \left(\dfrac{-b}{c} \right)$

For rational expressions $\dfrac{P}{S}$ and $\dfrac{Q}{S}$, $\dfrac{P}{S} - \dfrac{Q}{S} = \dfrac{P}{S} + \left(\dfrac{-Q}{S} \right)$

Equations Involving Rational Expressions [6.4]

7. Solve $\dfrac{1}{2} + \dfrac{3}{x} = 5$.

$2x\left(\dfrac{1}{2}\right) + 2x\left(\dfrac{3}{x}\right) = 2x(5)$

$x + 6 = 10x$

$6 = 9x$

$x = \dfrac{2}{3}$

To solve equations involving rational expressions, first find the least common denominator (LCD) for all denominators. Then multiply both sides by the LCD and solve as usual. Check all solutions in the original equation to be sure there are no undefined terms.

8. $\dfrac{1 - \dfrac{4}{x}}{x - \dfrac{16}{x}} = \dfrac{x\left(1 - \dfrac{4}{x}\right)}{x\left(x - \dfrac{16}{x}\right)}$

$= \dfrac{x - 4}{x^2 - 16}$

$= \dfrac{\cancel{x - 4}}{(\cancel{x - 4})(x + 4)}$

$= \dfrac{1}{x + 4}$

Complex Fractions [6.6]

A rational expression that contains a fraction in its numerator or denominator is called a complex fraction. The most common method of simplifying a complex fraction is to multiply the top and bottom by the LCD for all denominators.

Ratio and Proportion [6.1, 6.7]

9. Solve $\dfrac{3}{x} = \dfrac{5}{20}$ for x.

$3 \cdot 20 = 5 \cdot x$

$60 = 5x$

$x = 12$

The ratio of a to b is:

$$\dfrac{a}{b}$$

Two equal ratios form a proportion. In the proportion

$$\dfrac{a}{b} = \dfrac{c}{d}$$

a and d are the *extremes*, and b and c are the *means*. In any proportion the product of the extremes is equal to the product of the means.

The numbers in brackets refer to the sections of the text in which similar problems can be found. Reduce to lowest terms. Also specify any restriction on the variable. [6.1]

1. $\dfrac{7}{14x - 28}$

2. $\dfrac{a + 6}{a^2 - 36}$

3. $\dfrac{8x - 4}{4x + 12}$

4. $\dfrac{x + 4}{x^2 + 8x + 16}$

5. $\dfrac{3x^3 + 16x^2 - 12x}{2x^3 + 9x^2 - 18x}$

6. $\dfrac{x + 2}{x^4 - 16}$

7. $\dfrac{x^2 + 5x - 14}{x + 7}$

8. $\dfrac{a^2 + 16a + 64}{a + 8}$

9. $\dfrac{xy + bx + ay + ab}{xy + 5x + ay + 5a}$

Multiply or divide as indicated. [6.2]

10. $\dfrac{3x + 9}{x^2} \cdot \dfrac{x^3}{6x + 18}$

11. $\dfrac{x^2 + 8x + 16}{x^2 + x - 12} \div \dfrac{x^2 - 16}{x^2 - x - 6}$

12. $(a^2 - 4a - 12)\left(\dfrac{a - 6}{a + 2}\right)$

13. $\dfrac{3x^2 - 2x - 1}{x^2 + 6x + 8} \div \dfrac{3x^2 + 13x + 4}{x^2 + 8x + 16}$

Find the following sums and differences. [6.3]

14. $\dfrac{2x}{2x + 3} + \dfrac{3}{2x + 3}$

15. $\dfrac{x^2}{x - 9} - \dfrac{18x - 81}{x - 9}$

16. $\dfrac{a + 4}{a + 8} - \dfrac{a - 9}{a + 8}$

17. $\dfrac{x}{x + 9} + \dfrac{5}{x}$

18. $\dfrac{5}{4x + 20} + \dfrac{x}{x + 5}$

19. $\dfrac{3}{x^2 - 36} - \dfrac{2}{x^2 - 4x - 12}$

20. $\dfrac{3a}{a^2 + 8a + 15} - \dfrac{2}{a + 5}$

Solve each equation. [6.4]

21. $\dfrac{3}{x} + \dfrac{1}{2} = \dfrac{5}{x}$

22. $\dfrac{a}{a - 3} = \dfrac{3}{2}$

23. $1 - \dfrac{7}{x} = \dfrac{-6}{x^2}$

24. $\dfrac{3}{x + 6} - \dfrac{1}{x - 2} = \dfrac{-8}{x^2 + 4x - 12}$

25. $\dfrac{2}{y^2 - 16} = \dfrac{10}{y^2 + 4y}$

26. Number Problem The sum of a number and 7 times its reciprocal is $\dfrac{16}{3}$. Find the number. [6.5]

27. Distance, Rate, and Time A boat travels 48 miles up a river in the same amount of time it takes to travel 72 miles down the same river. If the current is 3 miles per hour, what is the speed of the boat in still water? [6.5]

28. Filling a Pool An inlet pipe can fill a pool in 21 hours, whereas an outlet pipe can empty it in 28 hours. If both pipes are left open, how long will it take to fill the pool? [7.5]

Simplify each complex fraction. [6.6]

29. $\dfrac{\dfrac{x + 4}{x^2 - 16}}{\dfrac{2}{x - 4}}$

30. $\dfrac{1 - \dfrac{9}{y^2}}{1 + \dfrac{4}{y} - \dfrac{21}{y^2}}$

31. $\dfrac{\dfrac{1}{a - 2} + 4}{\dfrac{1}{a - 2} + 1}$

32. Write the ratio of 40 to 100 as a fraction in lowest terms. [6.7]

33. If there are 60 seconds in 1 minute, what is the ratio of 40 seconds to 3 minutes? [6.7]

Solve each proportion. [6.7]

34. $\dfrac{x}{9} = \dfrac{4}{3}$

35. $\dfrac{a}{3} = \dfrac{12}{a}$

36. $\dfrac{8}{x - 2} = \dfrac{x}{6}$

Simplify.

1. $8 - 11$

2. $-20 + 14$

3. $\dfrac{-48}{12}$

4. $\dfrac{1}{6}(-18)$

5. $5x - 4 - 9x$

6. $8 - x - 4$

7. $\dfrac{x^{-9}}{x^{-13}}$

8. $\left(\dfrac{x^5}{x^3}\right)^{-2}$

9. $4^1 + 9^0 + (-7)^0$

10. $\dfrac{(x^{-4})^{-3}(x^{-3})^4}{x^0}$

11. $4x - 7x$

12. $(4a^3 - 10a^2 + 6) - (6a^3 + 5a - 7)$

13. $\dfrac{x^2}{x - 7} - \dfrac{14x - 49}{x - 7}$

14. $\dfrac{6}{10x + 30} + \dfrac{x}{x + 3}$

Solve each equation.

15. $4x - 3 = 8x + 5$

16. $\dfrac{3}{4}(8x - 12) = \dfrac{1}{2}(4x + 4)$

17. $98r^2 - 18 = 0$

18. $6x^4 = 33x^3 - 42x^2$

19. $\dfrac{5}{x} - \dfrac{1}{3} = \dfrac{3}{x}$

20. $\dfrac{4}{x - 5} - \dfrac{3}{x + 2} = \dfrac{28}{x^2 - 3x - 10}$

Solve each system.

21. $\begin{aligned} x + 2y &= 1 \\ x - y &= 4 \end{aligned}$

22. $\begin{aligned} 9x + 14y &= -4 \\ 15x - 8y &= 9 \end{aligned}$

23. $\begin{aligned} x &= 4y - 3 \\ 2x - 5y &= 9 \end{aligned}$

24. $\begin{aligned} \dfrac{1}{2}x + \dfrac{1}{3}y &= -1 \\ \dfrac{1}{3}x &= \dfrac{1}{4}y + 5 \end{aligned}$

Graph each equation on a rectangular coordinate system.

25. $y = -3x + 2$

26. $y = \dfrac{1}{3}x$

Solve each inequality.

27. $-\dfrac{a}{3} \le -2$

28. $-3x < 9$

Factor completely.

29. $xy + 5x + ay + 5a$

30. $a^2 + 2a - 35$

31. $20y^2 - 27y + 9$

32. $4r^2 - 9t^2$

33. $3x^2 + 12y^2$

34. $16x^2 + 72xy + 81y^2$

Solve the following systems by graphing.

35. $\begin{aligned} 2x + y &= 4 \\ 3x - 2y &= 6 \end{aligned}$

36. $\begin{aligned} 4x + 7y &= -1 \\ 3x - 2y &= -8 \end{aligned}$

37. For the equation $2x + 5y = 10$, find the x- and y-intercepts.

38. For the equation $3x - y = 4$, find the slope and y-intercept.

39. Find the equation for the line with slope -5 and y-intercept -1.

40. Graph the equation for the line with slope $-\dfrac{2}{5}$ that passes through $(-2, -1)$.

41. Subtract -2 from 6.

42. Add -2 to the product of -3 and 4.

Give the property that justifies the statement.

43. $(2 + x) + 3 = 2 + (x + 3)$

44. $(2 + x) + 3 = (x + 2) + 3$

45. Divide $\dfrac{x^2 - 3x - 28}{x + 4}$.

46. Multiply $\dfrac{6x - 12}{6x + 12} \cdot \dfrac{3x + 3}{12x - 24}$.

Reduce to lowest terms.

47. $\dfrac{5a + 10}{10a + 20}$

48. $\dfrac{2xy + 10x + 3y + 15}{3xy + 15x + 2y + 10}$

49. $\dfrac{2x^2 + x - 6}{x^2 + 6x + 8}$

50. $\dfrac{2x^2 + 2x - 24}{4x^2 - 64}$

The illustration shows the sizes of several deserts, in square kilometers. (Note: 1 sq mi $\approx$ 2.59 square km.)

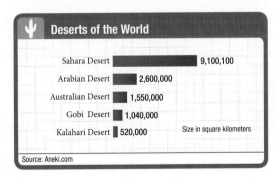

Deserts of the World

Sahara Desert — 9,100,100
Arabian Desert — 2,600,000
Australian Desert — 1,550,000
Gobi Desert — 1,040,000
Kalahari Desert — 520,000

Size in square kilometers

Source: Aneki.com

51. How many square miles is the Arabian Desert?

52. How many square miles is the Gobi Desert?

Reduce to lowest terms. [6.1]

1. $\dfrac{x^2 - 16}{x^2 - 8x + 16}$

2. $\dfrac{10a + 20}{5a^2 + 20a + 20}$

3. $\dfrac{xy + 7x + 5y + 35}{x^2 + ax + 5x + 5a}$

Multiply or divide as indicated. [6.2]

4. $\dfrac{3x - 12}{4} \cdot \dfrac{8}{2x - 8}$

5. $\dfrac{x^2 - 49}{x + 1} \div \dfrac{x + 7}{x^2 - 1}$

6. $\dfrac{x^2 - 3x - 10}{x^2 - 8x + 15} \div \dfrac{3x^2 + 2x - 8}{x^2 + x - 12}$

7. $(x^2 - 9)\left(\dfrac{x + 2}{x + 3}\right)$

Add or subtract as indicated. [6.3]

8. $\dfrac{3}{x - 2} - \dfrac{6}{x - 2}$

9. $\dfrac{x}{x^2 - 9} + \dfrac{4}{4x - 12}$

10. $\dfrac{2x}{x^2 - 1} + \dfrac{x}{x^2 - 3x + 2}$

Solve the following equations. [6.4]

11. $\dfrac{7}{5} = \dfrac{x + 2}{3}$

12. $\dfrac{10}{x + 4} = \dfrac{6}{x} - \dfrac{4}{x}$

13. $\dfrac{3}{x - 2} - \dfrac{4}{x + 1} = \dfrac{5}{x^2 - x - 2}$

Solve the following problems. [6.5]

14. Speed of a Boat A boat travels 26 miles up a river in the same amount of time it takes to travel 34 miles down the same river. If the current is 2 miles per hour, what is the speed of the boat in still water?

15. Emptying a Pool An inlet pipe can fill a pool in 15 hours, whereas an outlet pipe can empty it in 12 hours. If the pool is full and both pipes are open, how long will it take to empty?

Simplify each complex fraction. [6.6]

16. $\dfrac{1 + \dfrac{1}{x}}{1 - \dfrac{1}{x}}$

17. $\dfrac{1 - \dfrac{16}{x^2}}{1 - \dfrac{2}{x} - \dfrac{8}{x^2}}$

Solve the following problems involving ratio and proportion. [6.7]

18. Ratio A solution of alcohol and water contains 27 milliliters of alcohol and 54 milliliters of water. What is the ratio of alcohol to water and the ratio of alcohol to total volume?

19. Ratio A manufacturer knows that during a production run 8 out of every 100 parts produced by a certain machine will be defective. If the machine produces 1,650 parts, how many can be expected to be defective?

20. Ratio Use the illustration below to find the ratio of the sound emitted by a blue whale to the sound of normal conversation. Round to the nearest tenth. [6.1]

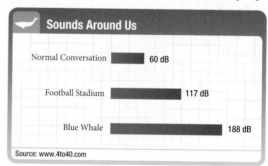

21. Unit Analysis The illustration below shows the top speeds of the four fastest cars in the world. Convert each speed to feet per second to the nearest tenth. [6.2]

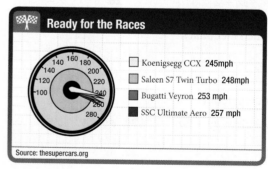

RATIONAL EXPRESSIONS

GROUP PROJECT

Kayak Race

Number of People 2-3

Time Needed 20 minutes

Equipment Paper and pencil

Background In a kayak race, the participants must paddle a kayak 450 meters down a river and then return 450 meters up the river to the starting point (see figure). Susan has deduced correctly that the total time t (in seconds) depends on the speed c (in meters per second) of the water according to the following expression:

$$t = \frac{450}{v + c} + \frac{450}{v - c}$$

where v is the speed of the kayak relative to the water (the speed of the kayak in still water).

Procedure **1.** Fill in the following table.

Time t (seconds)	Speed of Kayak Relative to the Water v (meters/second)	Current of the River c (meters/second)
240		1
300		2
	4	3
	3	1
540	3	
	3	3

2. If the kayak race were conducted in the still waters of a lake, do you think that the total time of a given participant would be greater than, equal to, or smaller than the time in the river? Justify your answer.

3. Suppose Peter can paddle his kayak at 4.1 meters per second and that the speed of the current is 4.1 meters per second. What will happen when Peter makes the turn and tries to come back up the river? How does this situation show up in the equation for total time?

RESEARCH PROJECT

Bertrand Russell

Here is a quote taken from the beginning of the first sentence in the book *Principles of Mathematics* by the British philosopher and mathematician, Bertrand Russell.

> Pure Mathematics is the class of all propositions of the form "*p* implies *q*," where *p* and *q* are propositions containing one or more variables . . .

He is using the phrase "*p* implies *q*" in the same way mathematicians use the phrase "If *A*, then *B*." Conditional statements are an introduction to the foundations on which all of mathematics is built.

Write an essay on the life of Bertrand Russell. In the essay, indicate what purpose he had for writing and publishing his book *Principles of Mathematics*. Write in complete sentences and organize your work just as you would if you were writing a paper for an English class.

Bertrand Russell

7

Transitions

Ulysses S. Grant High School

Google   Image © 2009 City of Glendale

Introduction

One of the most innovative and successful textbook authors of our time is Harold Jacobs. Mr. Jacobs taught mathematics, physics, and chemistry at Ulysses S. Grant High School in Los Angeles for 35 years. The page below is from the transparency masters that accompany his elementary algebra textbook. It is a page from a French mathematics book showing an example of how to solve a system of linear equations in two variables, which is one of the main topics of this chapter.

Based on W.H. Freeman and Company

SYSTÈME DE DEUX ÉQUATIONS À DEUX INCONNUES

1° **Elimination d'une inconnue par addition.**

Exemple $\left\{\begin{array}{ll} 2x + y = 1 & (1) \\ 3x - 4y = 29 & (2) \end{array}\right.$

Multiplions par 4 les deux membres de (1):
$$8x + 4y = 4 \qquad (3)$$

Ajoutons (2) et (3):
$$11x = 33$$
$$x = \frac{33}{11} = 3.$$

Portons x = 3 dans (1):
$$6 + y = 1$$
$$y = 1 - 6 = -5.$$

Règle pratique : par le jeu des multiplications, faire apparaître dans les équations deux coefficients opposés.

After you have finished this chapter, come back to this introduction and see if you can understand the example written in French. I think you will be surprised how easy it is to understand the French words, when you understand the mathematics they are describing.

Simplify.

1. $0.09(6{,}000)$ **2.** $1.5(500)$ **3.** $3(11) - 5(7)$ **4.** $6 - 12$

5. $3(3) - 5$ **6.** $4(5) - 3(-3)$ **7.** $1 + 2(2) - 3(3)$ **8.** $-4(-1) + 1(2) + 2(6)$

Apply the distributive property, then simplify if possible.

9. $-3(2x - 3y)$ **10.** $3(4x + 5y)$ **11.** $6\left(\dfrac{1}{2}x - \dfrac{1}{3}y\right)$

12. $12\left(\dfrac{1}{4}x + \dfrac{2}{3}y\right)$ **13.** $10(0.3x + 0.7y)$ **14.** $100(0.06x + 0.07y)$

Solve.

15. $2(1) + y = 4$ **16.** $2x - 3(3x - 5) = -6$ **17.** $2(2y + 6) + 3y = 5$

18. $5\left(\dfrac{19}{15}\right) + 5y = 9$ **19.** $4x - 2x = 8$ **20.** $20x + 9{,}300 > 18{,}000$

Chapter Outline

7.1 Review of Solving Equations

A   Solve a linear equation in one variable.
B   Solve equations by factoring.
C   Apply the Blueprint for Problem Solving to solve application problems whose solutions involve quadratic equations.

7.2 Equations with Absolute Value

A   Locate and label points on the number line.

7.3 Compound Inequalities and Interval Notation

A   Solve a linear inequality in one variable and graph the solution set.
B   Write solutions to inequalities using interval notation.
C   Solve a compound inequality and graph the solution set.
D   Solve application problems using inequalities.

7.4 Inequalities Involving Absolute Value

A   Solve inequalities with absolute value and graph the solution set.

7.5 Review of Systems of Equations in Two Variables

A   Solve systems of linear equations in two variables by graphing.
B   Solve systems of linear equations in two variables by the addition method.
C   Solve systems of linear equations in two variables by the substitution method.

7.6 Systems of Equations in Three Variables

A   Solve systems of linear equations in three variables.

7.7 Linear Inequalities in Two Variables

A   Graph linear inequalities in two variables.

7.8 Systems of Linear Inequalities

A   Graph the solution to a system of linear inequalities in two variables.

Review of Solving Equations

OBJECTIVES

A Solve a linear equation in one variable.

B Solve equations by factoring.

C Apply the Blueprint for Problem Solving to solve application problems whose solutions involve quadratic equations.

TICKET TO SUCCESS

Keep these questions in mind as you read through the section. Then respond in your own words and in complete sentences.

1. What are equivalent equations?
2. What is the addition property of equality?
3. What is the multiplication property of equality?
4. Briefly explain the strategy for solving linear equations in one variable.

Image copyright ©Adisa, 2009. Used under license from Shutterstock.com

The population density of a state is calculated by using the number of people in the state compared to the area of the state. If the population of Massachusetts is 6,497,967 and the population density is 828.82 people per square mile, how many square miles does Massachusetts cover? In this section, we will begin our work with linear equations to solve problems like this one.

This chapter marks the transition from introductory algebra to intermediate algebra. Some of the material here is a review of material we covered earlier, and some of the material here is new. If you cover all the sections in this chapter, you will review all the important points contained in the first six chapters of the book. So, it is a good idea to put some extra time and effort into this chapter to ensure that you get a good start with the rest of the course. Let's begin by reviewing the methods we use to solve equations.

A Solving Linear Equations

A *linear equation in one variable* is any equation that can be put in the form

$$ax + b = c$$

where a, b, and c are constants and $a \neq 0$. For example, each of the equations

$$5x + 3 = 2 \qquad 2x = 7 \qquad 2x + 5 = 0$$

is linear because they can be put in the form $ax + b = c$. In the first equation, $5x$, 3, and 2 are called *terms* of the equation: $5x$ is a variable term; 3 and 2 are constant terms.

Furthermore, we can find a solution for the equation by substituting a number for the variable.

> **Definition**
>
> The **solution set** for an equation is the set of all numbers that when used in place of the variable make the equation a true statement.

> **Definition**
>
> Two or more equations with the same solution set are called **equivalent equations.**

EXAMPLE 1 The equations $2x - 5 = 9$, $x - 1 = 6$, and $x = 7$ are all equivalent equations because the solution set for each is $\{7\}$.

Properties of Equality

The first property of equality states that adding the same quantity to both sides of an equation preserves equality. Or, more importantly, adding the same amount to both sides of an equation *never changes* the solution set. This property is called the *addition property of equality* and is stated in symbols as follows.

NOTE
Because subtraction is defined in terms of addition and division is defined in terms of multiplication, we do not need to introduce separate properties for subtraction and division. The solution set for an equation will never be changed by subtracting the same amount from both sides or by dividing both sides by the same nonzero quantity.

> **Addition Property of Equality**
>
> For any three algebraic expressions, A, B, and C,
>
> $$\text{if} \qquad A = B$$
> $$\text{then} \qquad A + C = B + C$$
>
> *In words:* Adding the same quantity to both sides of an equation will not change the solution set.

Our second new property is called the *multiplication property of equality* and is stated as follows.

> **Multiplication Property of Equality**
>
> For any three algebraic expressions A, B, and C, where $C \neq 0$,
>
> $$\text{if} \qquad A = B$$
> $$\text{then} \qquad AC = BC$$
>
> *In words:* Multiplying both sides of an equation by the same nonzero quantity will not change the solution set.

NOTE
From the previous chapter, we know that multiplication by a number and division by its reciprocal always produce the same result. Because of this fact, instead of multiplying each side of our equation by $\frac{1}{9}$, we could just as easily divide each side by 9. If we did so, the last two lines in our solution would look like this:

$$\frac{9a}{9} = \frac{6}{9}$$
$$a = \frac{2}{3}$$

EXAMPLE 2 Find the solution set for $3a - 5 = -6a + 1$.

SOLUTION To solve for a we must isolate it on one side of the equation. Let's decide to isolate a on the left side by adding $6a$ to both sides of the equation.

$$3a - 5 = -6a + 1$$
$$3a + \mathbf{6a} - 5 = -6a + \mathbf{6a} + 1 \qquad \text{Add } 6a \text{ to both sides}$$
$$9a - 5 = 1$$
$$9a - 5 + \mathbf{5} = 1 + \mathbf{5} \qquad \text{Add 5 to both sides}$$
$$9a = 6$$

$$\frac{1}{9}(9a) = \frac{1}{9}(6) \qquad \text{Multiply both sides by } \frac{1}{9}$$

$$a = \frac{2}{3} \qquad\qquad \frac{1}{9}(6) = \frac{6}{9} = \frac{2}{3}$$

The solution set is $\left\{\frac{2}{3}\right\}$.

We can check our solution in Example 2 by replacing a in the original equation with $\frac{2}{3}$.

$$\text{When} \rightarrow \qquad\qquad a = \frac{2}{3}$$

$$\text{the equation} \rightarrow \quad 3a - 5 = -6a + 1$$

$$\text{becomes} \rightarrow \quad 3\left(\frac{2}{3}\right) - 5 = -6\left(\frac{2}{3}\right) + 1$$

$$2 - 5 = -4 + 1$$

$$-3 = -3 \qquad \text{A true statement}$$

There will be times when we solve equations and end up with a negative sign in front of the variable. The next example shows how to handle this situation.

EXAMPLE 3   Solve each equation.

a. $-x = 4$ **b.** $-y = -8$

SOLUTION   Neither equation can be considered solved because of the negative sign in front of the variable. To eliminate the negative signs, we simply multiply both sides of each equation by -1.

a. $-x = 4$	**b.** $-y = -8$	
$-\mathbf{1}(-x) = -\mathbf{1}(4)$	$-\mathbf{1}(-y) = -\mathbf{1}(-8)$	Multiply each side by -1
$x = -4$	$y = 8$	

EXAMPLE 4   Solve $\frac{2}{3}x + \frac{1}{2} = -\frac{3}{8}$.

SOLUTION   We can solve this equation by applying our properties and working with fractions, or we can begin by eliminating the fractions. Let's use both methods.

METHOD 1 Working with the fractions.

$$\frac{2}{3}x + \frac{1}{2} + \left(-\frac{\mathbf{1}}{\mathbf{2}}\right) = -\frac{3}{8} + \left(-\frac{\mathbf{1}}{\mathbf{2}}\right) \qquad \text{Add } -\frac{1}{2} \text{ to each side}$$

$$\frac{2}{3}x = -\frac{7}{8} \qquad\qquad -\frac{3}{8} + \left(-\frac{1}{2}\right) = -\frac{3}{8} + \left(-\frac{4}{8}\right)$$

$$\frac{\mathbf{3}}{\mathbf{2}}\left(\frac{2}{3}x\right) = \frac{\mathbf{3}}{\mathbf{2}}\left(-\frac{7}{8}\right) \qquad \text{Multiply each side by } \frac{3}{2}$$

$$x = -\frac{21}{16}$$

METHOD 2 Eliminating the fractions in the beginning.

Our original equation has denominators of 3, 2, and 8. The least common denominator, abbreviated LCD, for these three denominators is 24, and it has the property that all three denominators will divide it evenly. If we multiply both sides of our equation by 24, each denominator will divide into 24, and we will be left with an equation that does not contain any denominators other than 1.

$$24\left(\frac{2}{3}x + \frac{1}{2}\right) = 24\left(-\frac{3}{8}\right)$$ Multiply each side by the LCD 24

$$24\left(\frac{2}{3}x\right) + 24\left(\frac{1}{2}\right) = 24\left(-\frac{3}{8}\right)$$ Distributive property on the left side

$$16x + 12 = -9$$ Multiply

$$16x = -21$$ Add -12 to each side

$$x = -\frac{21}{16}$$ Multiply each side by $\frac{1}{16}$

CHECK To check our solution, we substitute $x = -\frac{21}{16}$ back into our original equation to obtain

$$\frac{2}{3}\left(-\frac{21}{16}\right) + \frac{1}{2} \stackrel{?}{=} -\frac{3}{8}$$

$$-\frac{7}{8} + \frac{1}{2} \stackrel{?}{=} -\frac{3}{8}$$

$$-\frac{7}{8} + \frac{4}{8} \stackrel{?}{=} -\frac{3}{8}$$

$$-\frac{3}{8} = -\frac{3}{8}$$ A true statement

NOTE
We are placing a question mark over the equal sign because we don't know yet if the expression on the left will be equal to the expression on the right.

EXAMPLE 5 Solve the equation $0.06x + 0.05(10{,}000 - x) = 560$.

SOLUTION We can solve the equation in its original form by working with the decimals, or we can eliminate the decimals first by using the multiplication property of equality and solve the resulting equation. Here are both methods.

METHOD 1 Working with the decimals.

$$0.06x + 0.05(10{,}000 - x) = 560$$ Original equation

$$0.06x + 0.05(10{,}000) - 0.05x = 560$$ Distributive property

$$0.01x + 500 = 560$$ Simplify the left side

$$0.01x + 500 + (-\mathbf{500}) = 560 + (-\mathbf{500})$$ Add -500 to each side

$$0.01x = 60$$

$$\frac{0.01x}{\mathbf{0.01}} = \frac{60}{\mathbf{0.01}}$$ Divide each side by 0.01

$$x = 6{,}000$$

METHOD 2 Eliminating the decimals in the beginning: To move the decimal point two places to the right in $0.06x$ and 0.05, we multiply each side of the equation by 100.

$$0.06x + 0.05(10{,}000 - x) = 560$$ Original equation

$$0.06x + 500 - 0.05x = 560$$ Distributive property

$$\mathbf{100}(0.06x) + \mathbf{100}(500) - \mathbf{100}(0.05x) = \mathbf{100}(560)$$ Multiply each side by 100

$$6x + 50{,}000 - 5x = 56{,}000$$

$$x + 50{,}000 = 56{,}000$$ Simplify the left side

$$x = 6{,}000$$ Add $-50{,}000$ to each side

Using either method, the solution to our equation is 6,000.

CHECK We check our work (to be sure we have not made a mistake in applying the properties or in arithmetic) by substituting 6,000 into our original equation and simplifying each side of the result separately, as the following shows.

$$0.06(\mathbf{6,000}) + 0.05(10,000 - \mathbf{6,000}) \stackrel{?}{=} 560$$

$$0.06(6,000) + 0.05(4,000) \stackrel{?}{=} 560$$

$$360 + 200 \stackrel{?}{=} 560$$

$$560 = 560 \qquad \text{A true statement}$$

Here is a list of steps to use as a guideline for solving linear equations in one variable.

Strategy   Solving Linear Equations in One Variable

Step 1: **a.** Use the distributive property to separate terms, if necessary.

b. If fractions are present, consider multiplying both sides by the LCD to eliminate the fractions. If decimals are present, consider multiplying both sides by a power of 10 to clear the equation of decimals.

c. Combine similar terms on each side of the equation.

Step 2: Use the addition property of equality to get all variable terms on one side of the equation and all constant terms on the other side. A **variable term** is a term that contains the variable. A **constant term** is a term that does not contain the variable (the number 3, for example).

Step 3: Use the multiplication property of equality to get the variable by itself on one side of the equation.

Step 4: Check your solution in the original equation to be sure that you have not made a mistake in the solution process.

As you will see as you work through the problems in the problem set, it is not always necessary to use all four steps when solving equations. The number of steps used depends on the equation. In Example 6, there are no fractions or decimals in the original equation, so Step 1b will not be used.

EXAMPLE 6   Solve the equation $8 - 3(4x - 2) + 5x = 35$.

SOLUTION   We must begin by distributing the -3 across the quantity $4x - 2$.

NOTE
It would be a mistake to subtract 3 from 8 first because the rule for order of operations indicates we are to do multiplication before subtraction.

Step 1:	**a.** $8 - 3(4x - 2) + 5x = 35$	Original equation
	$8 - 12x + 6 + 5x = 35$	Distributive property
	c. $-7x + 14 = 35$	Simplify
Step 2:	$-7x = 21$	Add -14 to each side
Step 3:	$x = -3$	Multiply by $-\frac{1}{7}$

Step 4: When x is replaced by -3 in the original equation, a true statement results. Therefore, -3 is the solution to our equation.

Identities and Equations with No Solution

Two special cases are associated with solving linear equations in one variable, each of which is illustrated in the following examples.

EXAMPLE 7   Solve $2(3x - 4) = 3 + 6x$ for x.

SOLUTION   Applying the distributive property to the left side gives us

$$6x - 8 = 3 + 6x \qquad \text{Distributive property}$$

Now, if we add $-6x$ to each side, we are left with the following

$$-8 = 3$$

which is a false statement. This means that there is no solution to our equation. Any number we substitute for x in the original equation will lead to a similar false statement. ∎

EXAMPLE 8   Solve $-15 + 3x = 3(x - 5)$ for x.

SOLUTION   We start by applying the distributive property to the right side.

$$-15 + 3x = 3x - 15 \qquad \text{Distributive property}$$

If we add $-3x$ to each side, we are left with the true statement

$$-15 = -15$$

In this case, our result tells us that any number we use in place of x in the original equation will lead to a true statement. Therefore, all real numbers are solutions to our equation. We say the original equation is an *identity* because the left side is always identically equal to the right side. ∎

B   Solving Equations by Factoring

Next, we will use our knowledge of factoring to solve equations. To begin, here is the definition of a quadratic equation.

Definition

Any equation that can be written in the form

$$ax^2 + bx + c = 0$$

where a, b, and c are constants and a is not 0 ($a \neq 0$) is called a **quadratic equation**. The form $ax^2 + bx + c = 0$ is called **standard form** for quadratic equations.

NOTE
The third equation is clearly a quadratic equation since it is in standard form. (Notice that a is 4, b is -3, and c is 2.) The first two equations are also quadratic because they could be put in the form $ax^2 + bx + c = 0$ by using the addition property of equality.

Each of the following is a quadratic equation:

$$2x^2 = 5x + 3 \qquad 5x^2 = 75 \qquad 4x^2 - 3x + 2 = 0$$

NOTATION   For a quadratic equation written in standard form, the first term ax^2 is called the *quadratic term*; the second term bx is the *linear term*; and the third term c is called the *constant term*.

In the past we have noticed that the number 0 is a special number. There is another property of 0 that is the key to solving quadratic equations. It is called the *zero-factor property*.

NOTE
What the zero-factor property says in words is that we can't multiply and get 0 without multiplying by 0; that is, if we multiply two numbers and get 0, then one or both of the original two numbers we multiplied must have been 0.

Zero-Factor Property

For all real numbers r and s,

$$r \cdot s = 0 \quad \text{if and only if} \quad r = 0 \quad \text{or} \quad s = 0 \quad \text{(or both)}$$

EXAMPLE 9   Solve $x^2 - 2x - 24 = 0$.

SOLUTION   We begin by factoring the left side as $(x - 6)(x + 4)$ and get

$$(x - 6)(x + 4) = 0$$

Now both $(x - 6)$ and $(x + 4)$ represent real numbers. We notice that their product is 0. By the zero-factor property, one or both of them must be 0.

$$x - 6 = 0 \quad \text{or} \quad x + 4 = 0$$

We have used factoring and the zero-factor property to rewrite our original second-degree equation as two first-degree equations connected by the word *or*. Completing the solution, we solve the two first-degree equations.

$$x - 6 = 0 \quad \text{or} \quad x + 4 = 0$$
$$x = 6 \quad \text{or} \quad x = -4$$

NOTE
We are placing a question mark over the equal sign because we don't know yet if the expression on the left will be equal to the expression on the right.

We check our solutions in the original equation as follows:

Check $x = 6$ Check $x = -4$

$6^2 - 2(6) - 24 \stackrel{?}{=} 0$ $(-4)^2 - 2(-4) - 24 \stackrel{?}{=} 0$

$36 - 12 - 24 \stackrel{?}{=} 0$ $16 + 8 - 24 \stackrel{?}{=} 0$

$0 = 0$ $0 = 0$

In both cases the result is a true statement, which means that both 6 and -4 are solutions to the original equation.

Although the next equation is not quadratic, the method we use is similar.

EXAMPLE 10   Solve $\dfrac{1}{3}x^3 = \dfrac{5}{6}x^2 + \dfrac{1}{2}x$.

SOLUTION   We can simplify our work if we clear the equation of fractions. Multiplying both sides by the LCD, 6, we have

$$\mathbf{6} \cdot \frac{1}{3}x^3 = \mathbf{6} \cdot \frac{5}{6}x^2 + \mathbf{6} \cdot \frac{1}{2}x$$

$$2x^3 = 5x^2 + 3x$$

Next we add $-5x^2$ and $-3x$ to each side so that the right side will become 0.

$$2x^3 - 5x^2 - 3x = 0 \qquad \text{Standard form}$$

We factor the left side and then use the zero-factor property to set each factor to 0.

$$x(2x^2 - 5x - 3) = 0 \qquad \text{Factor out the greatest common factor}$$

$$x(2x + 1)(x - 3) = 0 \qquad \text{Continue factoring}$$

$$x = 0 \quad \text{or} \quad 2x + 1 = 0 \quad \text{or} \quad x - 3 = 0 \qquad \text{Zero-factor property}$$

Solving each of the resulting equations, we have

$$x = 0 \quad \text{or} \quad x = -\frac{1}{2} \quad \text{or} \quad x = 3$$

To generalize the preceding example, here are the steps used in solving a quadratic equation by factoring.

> ### Strategy   To Solve an Equation by Factoring
>
> **Step 1:** Write the equation in standard form.
>
> **Step 2:** Factor the left side.
>
> **Step 3:** Use the zero-factor property to set each factor equal to 0.
>
> **Step 4:** Solve the resulting linear equations.

EXAMPLE 11   Solve $100x^2 = 300x$.

SOLUTION   We begin by writing the equation in standard form and factoring.

$$100x^2 = 300x$$
$$100x^2 - 300x = 0 \quad \text{Standard form}$$
$$100x(x - 3) = 0 \quad \text{Factor}$$

Using the zero-factor property to set each factor to 0, we have

$$100x = 0 \quad \text{or} \quad x - 3 = 0$$
$$x = 0 \quad \text{or} \quad x = 3$$

The two solutions are 0 and 3.

EXAMPLE 12   Solve $(x - 2)(x + 1) = 4$.

SOLUTION   We begin by multiplying the two factors on the left side. (Notice that it would be incorrect to set each of the factors on the left side equal to 4. The fact that the product is 4 does not imply that either of the factors must be 4.)

$$(x - 2)(x + 1) = 4$$
$$x^2 - x - 2 = 4 \quad \text{Multiply the left side}$$
$$x^2 - x - 6 = 0 \quad \text{Standard form}$$
$$(x - 3)(x + 2) = 0 \quad \text{Factor}$$
$$x - 3 = 0 \quad \text{or} \quad x + 2 = 0 \quad \text{Zero-factor property}$$
$$x = 3 \quad \text{or} \quad x = -2$$

EXAMPLE 13   Solve $x^3 + 2x^2 - 9x - 18 = 0$ for x.

SOLUTION   We start with factoring by grouping.

$$x^3 + 2x^2 - 9x - 18 = 0$$
$$x^2(x + 2) - 9(x + 2) = 0$$
$$(x + 2)(x^2 - 9) = 0$$
$$(x + 2)(x - 3)(x + 3) = 0 \quad \text{The difference of two squares}$$
$$x + 2 = 0 \quad \text{or} \quad x - 3 = 0 \quad \text{or} \quad x + 3 = 0 \quad \text{Set factors to 0}$$
$$x = -2 \quad \text{or} \quad x = 3 \quad \text{or} \quad x = -3$$

We have three solutions: -2, 3, and -3.

C Application

EXAMPLE 14   The sum of the squares of two consecutive integers is 25. Find the two integers.

SOLUTION   We apply the Blueprint for Problem Solving to solve this application problem. Remember, step 1 in the blueprint is done mentally.

Step 1: *Read and list.*
Known items: Two consecutive integers. If we add their squares, the result is 25.
Unknown items: The two integers

Step 2: *Assign a variable and translate information.*
Let x = the first integer; then $x + 1$ = the next consecutive integer.

Step 3: *Reread and write an equation.*
Since the sum of the squares of the two integers is 25, the equation that describes the situation is

$$x^2 + (x + 1)^2 = 25$$

NOTE
The common factor can be divided out, since 0 divided by any number is 0.

Step 4: *Solve the equation.*

$$x^2 + (x + 1)^2 = 25$$
$$x^2 + (x^2 + 2x + 1) = 25$$
$$2x^2 + 2x - 24 = 0$$
$$x^2 + x - 12 = 0$$
$$(x + 4)(x - 3) = 0$$
$$x = -4 \quad \text{or} \quad x = 3$$

Step 5: *Write the answer.*
If $x = -4$, then $x + 1 = -3$. If $x = 3$, then $x + 1 = 4$. The two integers are -4 and -3, or the two integers are 3 and 4.

Step 6: *Reread and check.*
The two integers in each pair are consecutive integers, and the sum of the squares of either pair is 25.

Problem Set 7.1

Moving Toward Success

"One of the greatest discoveries a man makes, one of his great surprises, is to find he can do what he was afraid he couldn't do."

—Henry Ford, 1863–1947, American industrialist and founder of Ford Motor Company

1. What was the more important study skill you used while working through the first half of this book?

2. Why should you continue to place an importance on study skills as you work through the next half?

A Solve each of the following equations.
[Examples 1–6]

1. $x - 5 = 3$

2. $x + 2 = 7$

3. $2x - 4 = 6$

4. $3x - 5 = 4$

5. $7 = 4a - 1$

6. $10 = 3a - 5$

7. $3 - y = 10$

8. $5 - 2y = 11$

9. $-3 - 4x = 15$

10. $-8 - 5x = -6$

11. $-3 = 5 + 2x$

12. $-12 = 6 + 9x$

13. $-300y + 100 = 500$

14. $-20y + 80 = 30$

15. $160 = -50x - 40$

16. $110 = -60x - 50$

17. $-x = 2$

18. $-x = \dfrac{1}{2}$

19. $-a = -\dfrac{3}{4}$

20. $-a = -5$

21. $\dfrac{2}{3}x = 8$

22. $\dfrac{3}{2}x = 9$

23. $-\dfrac{3}{5}a + 2 = 8$

24. $-\dfrac{5}{3}a + 3 = 23$

25. $8 = 6 + \dfrac{2}{7}y$

26. $1 = 4 + \dfrac{3}{7}y$

27. $2x - 5 = 3x + 2$

28. $5x - 1 = 4x + 3$

29. $-3a + 2 = -2a - 1$

30. $-4a - 8 = -3a + 7$

31. $5 - 2x = 3x + 1$

32. $7 - 3x = 8x - 4$

33. $11x - 5 + 4x - 2 = 8x$

34. $2x + 7 - 3x + 4 = -2x$

35. $6 - 7(m - 3) = -1$

36. $3 - 5(2m - 5) = -2$

37. $7 + 3(x + 2) = 4(x - 1)$

38. $5 + 2(4x - 4) = 3(2x - 1)$

39. $5 = 7 - 2(3x - 1) + 4x$

40. $20 = 8 - 5(2x - 3) + 4x$

41. $\dfrac{1}{2}x + \dfrac{1}{4} = \dfrac{1}{3}x + \dfrac{5}{4}$

42. $\dfrac{2}{3}x - \dfrac{3}{4} = \dfrac{1}{6}x + \dfrac{21}{4}$

43. $-\dfrac{2}{5}x + \dfrac{2}{15} = \dfrac{2}{3}$

44. $-\dfrac{1}{6}x + \dfrac{2}{3} = \dfrac{1}{4}$

45. $\dfrac{3}{4}(8x - 4) = \dfrac{2}{3}(6x - 9)$

46. $\dfrac{3}{5}(5x + 10) = \dfrac{5}{6}(12x - 18)$

47. $\dfrac{1}{4}(12a + 1) - \dfrac{1}{4} = 5$

48. $\dfrac{2}{3}(6x - 1) + \dfrac{2}{3} = 4$

49. $0.35x - 0.2 = 0.15x + 0.1$

50. $0.25x - 0.05 = 0.2x + 0.15$

51. $0.42 - 0.18x = 0.48x - 0.24$

52. $0.3 - 0.12x = 0.18x + 0.06$

A Solve each equation, if possible. [Examples 7–8]

53. $3x - 6 = 3(x + 4)$

54. $7x - 14 = 7(x - 2)$

55. $4y + 2 - 3y + 5 = 3 + y + 4$

56. $7y + 5 - 2y - 3 = 6 + 5y - 4$

57. $2(4t - 1) + 3 = 5t + 4 + 3t$

58. $5(2t - 1) + 1 = 2t - 4 + 8t$

B Solve each equation. [Examples 9–13]

59. $x^2 - 5x - 6 = 0$

60. $x^2 + 5x - 6 = 0$

61. $x^3 - 5x^2 + 6x = 0$

62. $x^3 + 5x^2 + 6x = 0$

63. $3y^2 + 11y - 4 = 0$

64. $3y^2 - y - 4 = 0$

65. $60x^2 - 130x + 60 = 0$

66. $90x^2 + 60x - 80 = 0$

67. $\dfrac{1}{10}t^2 - \dfrac{5}{2} = 0$

68. $\dfrac{2}{7}t^2 - \dfrac{7}{2} = 0$

69. $100x^4 = 400x^3 + 2{,}100x^2$

70. $100x^4 = -400x^3 + 2{,}100x^2$

71. $\dfrac{1}{5}y^2 - 2 = -\dfrac{3}{10}y$

72. $\dfrac{1}{2}y^2 + \dfrac{5}{3} = \dfrac{17}{6}y$

73. $9x^2 - 12x = 0$

74. $4x^2 + 4x = 0$

75. $0.02r + 0.01 = 0.15r^2$

76. $0.02r - 0.01 = -0.08r^2$

77. $9a^3 = 16a$

78. $16a^3 = 25a$

79. $-100x = 10x^2$

80. $800x = 100x^2$

81. $(x + 6)(x - 2) = -7$

82. $(x - 7)(x + 5) = -20$

83. $(y - 4)(y + 1) = -6$

84. $(y - 6)(y + 1) = -12$

85. $(x + 1)^2 = 3x + 7$

86. $(x + 2)^2 = 9x$

87. $(2r + 3)(2r - 1) = -(3r + 1)$

88. $(3r + 2)(r - 1) = -(7r - 7)$

89. $x^3 + 3x^2 - 4x - 12 = 0$

90. $x^3 + 5x^2 - 4x - 20 = 0$

91. $x^3 + 2x^2 - 25x - 50 = 0$

92. $x^3 + 4x^2 - 9x - 36 = 0$

93. $2x^3 + 3x^2 - 8x - 12 = 0$

94. $3x^3 + 2x^2 - 27x - 18 = 0$

95. $4x^3 + 12x^2 - 9x - 27 = 0$

96. $9x^3 + 18x^2 - 4x - 8 = 0$

B Problems 97–106 are problems you will see later in the book. Solve each equation. [Examples 9–13]

97. $3x^2 + x = 10$ **98.** $y^2 + y - 20 = 2y$

99. $12(x + 3) + 12(x - 3) = 3(x^2 - 9)$

100. $8(x + 2) + 8(x - 2) = 3(x^2 - 4)$

101. $(y + 3)^2 + y^2 = 9$

102. $(2y + 4)^2 + y^2 = 4$

103. $(x + 3)^2 + 1^2 = 2$

104. $(x - 3)^2 + (-1)^2 = 10$

105. $(x + 2)(x) = 2^3$

106. $(x + 3)(x) = 2^2$

C Applying the Concepts [Example 14]

107. **Distance** Two cyclists leave from an intersection at the same time. One travels due north at a speed of 15 miles per hour, and the other travels due east at a speed of 20 miles per hour. How long until the distance between the two cyclists is 75 miles?

108. **Distance** Two airplanes leave from an airport at the same time. One travels due south at a speed of 480 miles per hour, and the other travels due west at a speed of 360 miles per hour. How long until the distance between the two airplanes is 2400 miles?

109. **Consecutive Integers** The square of the sum of two consecutive integers is 81. Find the two integers.

110. **Consecutive Integers** Find two consecutive even integers whose sum squared is 100.

111. **Right Triangle** A 25-foot ladder is leaning against a building. The base of the ladder is 7 feet from the side of the building. How high does the ladder reach along the side of the building?

25 ft

7 ft

112. **Right Triangle** Noreen wants to place her 13-foot ramp against the side of her house so that the top of the ramp rests on a ledge that is 5 feet above the ground. How far will the base of the ramp be from the house?

113. **Right Triangle** The lengths of the three sides of a right triangle are given by three consecutive even integers. Find the lengths of the three sides.

114. **Right Triangle** The longest side of a right triangle is 3 less than twice the shortest side. The third side measures 12 inches. Find the length of the shortest side.

115. **Geometry** The length of a rectangle is 2 feet more than 3 times the width. If the area is 16 square feet, find the width and the length.

116. **Geometry** The length of a rectangle is 4 yards more than twice the width. If the area is 70 square yards, find the width and the length.

117. **Geometry** The base of a triangle is 2 inches more than 4 times the height. If the area is 36 square inches, find the base and the height.

118. **Geometry** The height of a triangle is 4 feet less than twice the base. If the area is 48 square feet, find the base and the height.

119. **Projectile Motion** If an object is thrown straight up into the air with an initial velocity of 32 feet per second, then its height above the ground at any time t is given by the formula $h = 32t - 16t^2$. Find the times at which the object is on the ground by letting $h = 0$ in the equation and solving for t.

120. Projectile Motion An object is projected into the air with an initial velocity of 64 feet per second. Its height at any time t is given by the formula $h = 64t - 16t^2$. Find the times at which the object is on the ground.

C The formula $h = vt - 16t^2$ gives the height h, in feet, of an object projected into the air with an initial vertical velocity v, in feet per second, after t seconds. [Example 14]

121. Projectile Motion If an object is projected upward with an initial velocity of 48 feet per second, at what times will it reach a height of 32 feet above the ground?

122. Projectile Motion If an object is projected upward into the air with an initial velocity of 80 feet per second, at what times will it reach a height of 64 feet above the ground?

123. Projectile Motion An object is projected into the air with a vertical velocity of 24 feet per second. At what times will the object be on the ground? (It is on the ground when h is 0.)

124. Projectile Motion An object is projected into the air with a vertical velocity of 20 feet per second. At what times will the object be on the ground?

125. Height of a Bullet A bullet is fired into the air with an initial upward velocity of 80 feet per second from the top of a building 96 feet high. The equation that gives the height of the bullet at any time t is $h = 96 + 80t - 16t^2$. At what times will the bullet be 192 feet in the air?

126. Height of an Arrow An arrow is shot into the air with an upward velocity of 48 feet per second from a hill 32 feet high. The equation that gives the height of the arrow at any time t is $h = 32 + 48t - 16t^2$. Find the times at which the arrow will be 64 feet above the ground.

127. Price and Revenue A company that manufactures typewriter ribbons knows that the number of ribbons x it can sell each week is related to the price per ribbon p by the equation $x = 1,200 - 100p$. At what price should it sell the ribbons if it wants the weekly revenue to be $3,200? (*Remember:* The equation for revenue is $R = xp$.)

128. Price and Revenue A company manufactures CDs for home computers. It knows from past experience that the number of CDs x it can sell each day is related to the price per CD p by the equation $x = 800 - 100p$. At what price should it sell its CDs if it wants the daily revenue to be $1,200?

129. Price and Revenue The relationship between the number of calculators x a company sells per day and the price of each calculator p is given by the equation $x = 1,700 - 100p$. At what price should the calculators be sold if the daily revenue is to be $7,000?

130. Price and Revenue The relationship between the number of pencil sharpeners x a company can sell each week and the price of each sharpener p is given by the equation $x = 1,800 - 100p$. At what price should the sharpeners be sold if the weekly revenue is to be $7,200?

Maintaining Your Skills

Solve each system.

131. $2x - 5y = -8$
$\quad\ \ 3x + y = 5$

132. $\quad 4x - 7y = -2$
$\quad\quad\ -5x + 6y = -3$

133. $\dfrac{1}{3}x - \dfrac{1}{6}y = 3$
$\quad -\dfrac{1}{5}x + \dfrac{1}{4}y = 0$

134. $2x - 5y = 14$
$\quad\quad\ y = 3x + 8$

135. $2x - y + z = 9$
$\quad\ x + y - 3z = -2$
$\quad\ 3x + y - z = 6$

136. Number Problem A number is 1 less than twice another. Their sum is 14. Find the two numbers.

137. Investing John invests twice as much money at 6% as he does at 5%. If his investments earn a total of $680 in 1 year, how much does he have invested at each rate?

138. Speed of a Boat A boat can travel 20 miles downstream in 2 hours. The same boat can travel 18 miles upstream in 3 hours. What is the speed of the boat in still water, and what is the speed of the current?

Getting Ready for the Next Section

Solve the following equations.

139. $4x - 7 = 9$

140. $6 - 2x = 12$

141. $5 - 3x = -10$

142. $7y - 3 = 9$

143. $3a + 12 = 21$

144. $6m - 3 = 18$

7.2 Equations with Absolute Value

OBJECTIVES

A Locate and label points on the number line.

TICKET TO SUCCESS

Keep these questions in mind as you read through the section. Then respond in your own words and in complete sentences.

1. Why would some equations that involve absolute value have two solutions instead of one?

2. Translate $|x| = 6$ into words using the definition of absolute value.

3. Explain in words what the equation $|x - 3| = 4$ means with respect to distance on the number line.

4. When is the statement $|x| = x$ true?

Suppose you are going to a new friend's house for a barbecue. You begin by driving 4 blocks north and 10 blocks east, when you realize you are lost. You backtrack 5 blocks south and 8 blocks west before you stop for directions. How far are you away from your starting point? Since you are traveling in both positive and negative directions (backtracking) your answer may end up negative. Intuitively we know that distance cannot be negative, so what do you do with this information? After working through this section you will be able to use absolute value to determine the absolute value of a number and how it relates to the distance you have traveled.

A Solve Equations with Absolute Values

Previously, we defined the absolute value of x, $|x|$, to be the distance between x and 0 on the number line. The absolute value of a number measures its distance from 0.

EXAMPLE 1 Solve $|x| = 5$ for x.

SOLUTION Using the definition of absolute value, we can read the equation as, "The distance between x and 0 on the number line is 5." If x is 5 units from 0, then x can be 5 or −5.

$$\text{If } |x| = 5 \quad \text{then } x = 5 \quad \text{or} \quad x = -5$$

In general, then, we can see that any equation of the form $|a| = b$ is equivalent to the equations $a = b$ or $a = -b$, as long as $b > 0$.

EXAMPLE 2   Solve $|2a - 1| = 7$.

SOLUTION   We can read this question as "$2a - 1$ is 7 units from 0 on the number line." The quantity $2a - 1$ must be equal to 7 or -7:

$$|2a - 1| = 7$$
$$2a - 1 = 7 \quad \text{or} \quad 2a - 1 = -7$$

We have transformed our absolute value equation into two equations that do not involve absolute value. We can solve each equation separately.

$$2a - 1 = 7 \quad \text{or} \quad 2a - 1 = -7$$
$$2a = 8 \quad \text{or} \quad 2a = -6 \quad \text{Add 1 to both sides}$$
$$a = 4 \quad \text{or} \quad a = -3 \quad \text{Multiply by } \tfrac{1}{2}$$

Our solution set is $\{4, -3\}$.

To check our solutions, we put them into the original absolute value equation:

When $\rightarrow$ $\qquad a = 4$ When $\rightarrow$ $\qquad a = -3$

the equation $\rightarrow$ $|2a - 1| = 7$ the equation $\rightarrow$ $|2a - 1| = 7$

becomes $\rightarrow$ $|2(4) - 1| = 7$ becomes $\rightarrow$ $|2(-3) - 1| = 7$

$\qquad\qquad\qquad |7| = 7$ $\qquad\qquad\qquad\qquad |-7| = 7$

$\qquad\qquad\qquad 7 = 7$ $\qquad\qquad\qquad\qquad\quad 7 = 7$

EXAMPLE 3   Solve $\left|\frac{2}{3}x - 3\right| + 5 = 12$.

SOLUTION   To use the definition of absolute value to solve this equation, we must isolate the absolute value on the left side of the equal sign. To do so, we add -5 to both sides of the equation to obtain

$$\left|\frac{2}{3}x - 3\right| = 7$$

Now that the equation is in the correct form, we can write

$$\frac{2}{3}x - 3 = 7 \quad \text{or} \quad \frac{2}{3}x - 3 = -7$$
$$\frac{2}{3}x = 10 \quad \text{or} \quad \frac{2}{3}x = -4 \quad \text{Add 3 to both sides}$$
$$x = 15 \quad \text{or} \quad x = -6 \quad \text{Multiply by } \tfrac{3}{2}$$

The solution set is $\{15, -6\}$.

EXAMPLE 4   Solve $|3a - 6| = -4$.

SOLUTION   The solution set is $\varnothing$ because the left side cannot be negative and the right side is negative. No matter what we try to substitute for the variable a, the quantity $|3a - 6|$ will always be positive or zero. It can never be -4.

NOTE
Recall that $\varnothing$ is the symbol we use to denote the empty set. When we use it to indicate the solutions to an equation, then we are saying the equation has no solution.

Consider the statement $|a| = |b|$. What can we say about a and b? We know they are equal in absolute value. By the definition of absolute value, they are the same distance from 0 on the number line. They must be equal to each other or opposites of each other. In symbols, we write

$$|a| = |b| \quad \Leftrightarrow \quad a = b \quad \text{or} \quad a = -b$$
$$\uparrow \qquad\qquad\quad \uparrow \qquad\qquad \uparrow$$

Equal in $\qquad\quad$ Equals $\quad$ or $\quad$ Opposites
absolute value

EXAMPLE 5   Solve $|x - 5| = |x - 7|$.

SOLUTION   The quantities $x - 5$ and $x - 7$ must be equal or they must be opposites, because their absolute values are equal:

Equals	*Opposites*
$x - 5 = x - 7$	or $\quad x - 5 = -(x - 7)$
$-5 = -7$	$x - 5 = -x + 7$
No solution here	$2x - 5 = 7$
	$2x = 12$
	$x = 6$

Because the first equation leads to a false statement, it will not give us a solution. (If either of the two equations were to reduce to a true statement, it would mean all real numbers would satisfy the original equation.) In this case, our only solution is $x = 6$.

Problem Set 7.2

Moving Toward Success

"Forget the times of your distress, but never forget what they taught you."

—Herbert Gasser, 1888–1963, American physiologist

1. Do you think you will make mistakes on tests and quizzes, even when you understand the material?

2. If you make a mistake on a test or a quiz, how do you plan on learning from it?

A Use the definition of absolute value to solve each of the following equations. [Examples 1–2]

1. $|x| = 4$
2. $|x| = 7$
3. $2 = |a|$
4. $5 = |a|$
5. $|x| = -3$
6. $|x| = -4$
7. $|a| + 2 = 3$
8. $|a| - 5 = 2$
9. $|y| + 4 = 3$
10. $|y| + 3 = 1$
11. $4 = |x| - 2$
12. $3 = |x| - 5$
13. $|x - 2| = 5$
14. $|x + 1| = 2$
15. $|a - 4| = \frac{5}{3}$
16. $|a + 2| = \frac{7}{5}$
17. $1 = |3 - x|$
18. $2 = |4 - x|$
19. $\left|\frac{3}{5}a + \frac{1}{2}\right| = 1$
20. $\left|\frac{2}{7}a + \frac{3}{4}\right| = 1$
21. $60 = |20x - 40|$
22. $800 = |400x - 200|$
23. $|2x + 1| = -3$
24. $|2x - 5| = -7$
25. $\left|\frac{3}{4}x - 6\right| = 9$
26. $\left|\frac{4}{5}x - 5\right| = 15$
27. $\left|1 - \frac{1}{2}a\right| = 3$
28. $\left|2 - \frac{1}{3}a\right| = 10$

Solve each equation. [Examples 3–5]

29. $|3x + 4| + 1 = 7$
30. $|5x - 3| - 4 = 3$
31. $|3 - 2y| + 4 = 3$
32. $|8 - 7y| + 9 = 1$
33. $3 + |4t - 1| = 8$
34. $2 + |2t - 6| = 10$
35. $\left|9 - \frac{3}{5}x\right| + 6 = 12$
36. $\left|4 - \frac{2}{7}x\right| + 2 = 14$
37. $5 = \left|\frac{2x}{7} + \frac{4}{7}\right| - 3$
38. $7 = \left|\frac{3x}{5} + \frac{1}{5}\right| + 2$
39. $2 = -8 + \left|4 - \frac{1}{2}y\right|$
40. $1 = -3 + \left|2 - \frac{1}{4}y\right|$
41. $|3a + 1| = |2a - 4|$
42. $|5a + 2| = |4a + 7|$
43. $\left|x - \frac{1}{3}\right| = \left|\frac{1}{2}x + \frac{1}{6}\right|$
44. $\left|\frac{1}{10}x - \frac{1}{2}\right| = \left|\frac{1}{5}x + \frac{1}{10}\right|$

45. $|y - 2| = |y + 3|$ **46.** $|y - 5| = |y - 4|$

47. $|3x - 1| = |3x + 1|$ **48.** $|5x - 8| = |5x + 8|$

49. $|3 - m| = |m + 4|$ **50.** $|5 - m| = |m + 8|$

51. $|0.03 - 0.01x| = |0.04 + 0.05x|$

52. $|0.07 - 0.01x| = |0.08 - 0.02x|$

53. $|x - 2| = |2 - x|$

54. $|x - 4| = |4 - x|$

55. $\left|\dfrac{x}{5} - 1\right| = \left|1 - \dfrac{x}{5}\right|$

56. $\left|\dfrac{x}{3} - 1\right| = \left|1 - \dfrac{x}{3}\right|$

57. Work each problem according to the instructions given.

    a. Solve: $4x - 5 = 0$

    b. Solve: $|4x - 5| = 0$

    c. Solve: $4x - 5 = 3$

    d. Solve: $|4x - 5| = 3$

    e. Solve: $|4x - 5| = |2x + 3|$

58. Work each problem according to the instructions given.

    a. Solve: $3x + 6 = 0$

    b. Solve: $|3x + 6| = 0$

    c. Solve: $3x + 6 = 4$

    d. Solve: $|3x + 6| = 4$

    e. Solve: $|3x + 6| = |7x + 4|$

Applying the Concepts

59. Google Earth The Google Earth image shows the western side of The Mall in Washington D.C. The distance between the Lincoln Memorial and the World War II memorial is 840 meters and the distance between the Lincoln Memorial and the Washington Memorial is 1,280 meters. Write an expression that describes the distance between the World War II Memorial and the Washington Memorial using absolute value, and then solve.

60. Skyscrapers The chart shows the heights of the three tallest buildings in the world. Write an expression that finds the difference in height between the Taipei 101 and the Shanghai World Financial Center using absolute values.

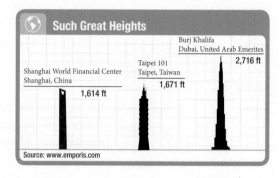

Such Great Heights

Burj Khalifa
Dubai, United Arab Emerites
2,716 ft

Taipei 101
Taipei, Taiwan
1,671 ft

Shanghai World Financial Center
Shanghai, China
1,614 ft

Source: www.emporis.com

61. Amtrak Amtrak's annual passenger revenue for the years 1985–1995 is modeled approximately by the formula

$$R = -60\,|x - 11| + 962$$

where R is the annual revenue in millions of dollars and x is the number of years since January 1, 1980 (Association of American Railroads, Washington, DC, *Railroad Facts, Statistics of Railroads of Class 1,* annual). In what years was the passenger revenue $722 million?

62. Corporate Profits The corporate profits for various U.S. industries vary from year to year. An approximate model for profits of U.S. "communications companies" during a given year between 1990 and 1997 is given by

$$P = -3{,}400\,|x - 5.5| + 36{,}000$$

where P is the annual profits (in millions of dollars) and x is the number of years since January 1, 1990 (U.S. Bureau of Economic Analysis, Income and Product Accounts of the U.S. (1929–1994), *Survey of Current Business,* September 1998). Use the model to determine the years in which profits of "communications companies" were $31.5 billion ($31,500 million).

Maintaining Your Skills

Graph the following inequalities.

63. $x < -2$ **64.** $x < 4$

65. $-2 \le x \le 1$ **66.** $x \le -\dfrac{3}{2}$ or $x \ge 3$

Simplify each expression.

67. $\dfrac{38}{30}$

68. $\dfrac{10}{25}$

69. $\dfrac{240}{6}$

70. $\dfrac{39}{13}$

71. $\dfrac{0+6}{0-3}$

72. $\dfrac{6+6}{6-3}$

73. $\dfrac{4-4}{4-2}$

74. $\dfrac{3+6}{3-3}$

Getting Ready for the Next Section

To understand all of the explanations and examples in the next section, you must be able to work the problems below.

Solve each inequality. Do not graph the solution set.

75. $2x - 5 < 3$

76. $-3 < 2x - 5$

77. $-4 \le 3a + 7$

78. $3a + 7 \le 4$

79. $4t - 3 \le -9$

80. $4t - 3 \ge 9$

Solve each inequality for x.

81. $|x - a| = b$

82. $|x + a| - b = 0$

83. $|ax + b| = c$

84. $|ax - b| - c = 0$

85. $\left| \dfrac{x}{a} + \dfrac{y}{b} \right| = 1$

86. $\left| \dfrac{x}{a} + \dfrac{y}{b} \right| = c$

7.3 Compound Inequalities and Interval Notation

OBJECTIVES

A Solve a linear inequality in one variable and graph the solution set.

B Write solutions to inequalities using interval notation.

C Solve a compound inequality and graph the solution set.

D Solve application problems using inequalities.

TICKET TO SUCCESS

Keep these questions in mind as you read through the section. Then respond in your own words and in complete sentences.

1. What is a linear inequality in one variable?
2. What is the addition property for inequalities?
3. What is the multiplication property for inequalities?
4. What is interval notation?

For Florida, the National Weather Service predicted 3 or more hours of temperature readings between 27 and 32 degrees Fahrenheit during January. Temperatures in this range would cause significant damage to Florida's $9 billion citrus industry. A citrus expert from Spain was consulted about how to manage the damage, but she needed the predictions converted to Celsius in order to render her opinion. In this section, we will begin our work with linear inequalities and their applications like this one.

A *linear inequality in one variable* is any inequality that can be put in the form

$$ax + b < c \qquad (a, b, \text{and } c \text{ are constants}, a \neq 0)$$

where the inequality symbol $<$ can be replaced with any of the other three inequality symbols ($\leq$, $>$, or $\geq$).

Some examples of linear inequalities are

$$3x - 2 \geq 7 \qquad -5y < 25 \qquad 3(x - 4) > 2x$$

A Solving Linear Inequalities

Our first property for inequalities is similar to the addition property we used when solving equations.

Addition Property for Inequalities
For any algebraic expressions, A, B, and C,

$$\text{if} \qquad A < B$$
$$\text{then} \qquad A + C < B + C$$

In words: Adding the same quantity to both sides of an inequality will not change the solution set.

EXAMPLE 1 Solve $3x + 3 < 2x - 1$, and graph the solution.

SOLUTION We use the addition property for inequalities to write all the variable terms on one side and all constant terms on the other side.

$$3x + 3 < 2x - 1$$
$$3x + (-\mathbf{2x}) + 3 < 2x + (-\mathbf{2x}) - 1 \qquad \text{Add } -2x \text{ to each side}$$
$$x + 3 < -1$$
$$x + 3 + (-\mathbf{3}) < -1 + (-\mathbf{3}) \qquad \text{Add } -3 \text{ to each side}$$
$$x < -4$$

The solution set is all real numbers that are less than -4. To show this, we can use set notation and write

$$\{x \mid x < -4\}$$

Or we can graph the solution set on the number line using an open circle at -4 to show that -4 is not part of the solution set. This is the format you used when graphing inequalities previously.

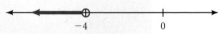

Here is an equivalent graph that uses a parenthesis opening left, instead of an open circle, to represent the end point of the graph.

This graph gives rise to the following notation, called *interval notation,* that is an alternative way to write the solution set.

$$(-\infty, -4)$$

The preceding expression indicates that the solution set is all real numbers from negative infinity up to, but not including, -4.

We have three equivalent representations for the solution set to our original inequality. Here are all three together.

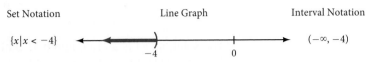

| Set Notation | Line Graph | Interval Notation |

$\{x \mid x < -4\}$ $(-\infty, -4)$

NOTE

The purpose of this discussion is to justify the multiplication property for inequalities. If you are having trouble understanding it, you may want to read the property itself first and then read through the discussion here. What is important is that you understand and can use the multiplication property for inequalities.

Before we state the multiplication property for inequalities, we will take a look at what happens to an inequality statement when we multiply both sides by a positive number and what happens when we multiply by a negative number.

We begin by writing three true inequality statements:

$$3 < 5 \qquad -3 < 5 \qquad -5 < -3$$

We multiply both sides of each inequality by a positive number—say, 4:

$$4(3) < 4(5) \qquad 4(-3) < 4(5) \qquad 4(-5) < 4(-3)$$
$$12 < 20 \qquad\quad -12 < 20 \qquad\quad -20 < -12$$

Notice in each case that the resulting inequality symbol points in the same direction as the original inequality symbol. Multiplying both sides of an inequality by a positive number preserves the *sense* of the inequality.

Let's take the same three original inequalities and multiply both sides by -4.

$$3 < 5 \qquad\qquad -3 < 5 \qquad\qquad -5 < -3$$
$$-4(3) > -4(5) \qquad -4(-3) > -4(5) \qquad -4(-5) > -4(-3)$$
$$-12 > -20 \qquad\quad 12 > -20 \qquad\quad\quad 20 > 12$$

Notice in this case that the resulting inequality symbol always points in the opposite direction from the original one. Multiplying both sides of an inequality by a negative number *reverses* the sense of the inequality. Keeping this in mind, we will now state the multiplication property for inequalities.

NOTE

Because division is defined as multiplication by the reciprocal, we can apply our new property to division as well as to multiplication. We can divide both sides of an inequality by any nonzero number as long as we reverse the direction of the inequality when the number we are dividing by is negative.

> **Multiplication Property for Inequalities**
>
> Let A, B, and C represent algebraic expressions.
>
> $$\begin{array}{lll} \text{If} & A < B & \\ \text{then} & AC < BC & \text{if} \quad C \text{ is positive } (C > 0) \\ \text{or} & AC > BC & \text{if} \quad C \text{ is negative } (C < 0) \end{array}$$
>
> *In words:* Multiplying both sides of an inequality by a positive number always produces an equivalent inequality. Multiplying both sides of an inequality by a negative number reverses the sense of the inequality.

The multiplication property for inequalities does not limit what we can do with inequalities. We are still free to multiply both sides of an inequality by any nonzero number we choose. If the number we multiply by happens to be *negative,* then we *must also reverse* the direction of the inequality.

EXAMPLE 2   Find the solution set for $-2y - 3 \leq 7$.

SOLUTION   We begin by adding 3 to each side of the inequality.

$$-2y - 3 \leq 7$$

$$-2y \leq 10 \qquad \text{Add 3 to both sides}$$

$$-\frac{1}{2}(-2y) \geq -\frac{1}{2}(10) \qquad \text{Multiply by } -\tfrac{1}{2} \text{ and reverse the direction of the inequality symbol}$$

$$y \geq -5$$

The solution set is all real numbers that are greater than or equal to -5. Below are three equivalent ways to represent this solution set.

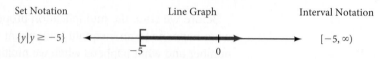

Set Notation	Line Graph	Interval Notation
$\{y \mid y \geq -5\}$		$[-5, \infty)$

Notice how a bracket is used with interval notation to show that -5 is part of the solution set.

When our inequalities become more complicated, we use the same basic steps we used previously to solve equations; that is, we simplify each side of the inequality before we apply the addition property or multiplication property. When we have solved the inequality, we graph the solution on a number line.

EXAMPLE 3   Solve $3(2x - 4) - 7x \leq -3x$.

NOTE
In Examples 2 and 3, notice that each time we multiplied both sides of the inequality by a negative number we also reversed the direction of the inequality symbol. Failure to do so would cause our graph to lie on the wrong side of the endpoint.

SOLUTION   We begin by using the distributive property to separate terms. Next, simplify both sides.

$$3(2x - 4) - 7x \leq -3x \qquad \text{Original inequality}$$

$$6x - 12 - 7x \leq -3x \qquad \text{Distributive property}$$

$$-x - 12 \leq -3x \qquad 6x - 7x = (6 - 7)x = -x$$

$$-12 \leq -2x \qquad \text{Add } x \text{ to both sides}$$

$$-\frac{1}{2}(-12) \geq -\frac{1}{2}(-2x) \qquad \text{Multiply both sides by } -\tfrac{1}{2} \text{ and reverse the direction of the inequality symbol}$$

$$6 \geq x$$

This last line is equivalent to $x \leq 6$. The solution set can be represented in any of the three following ways.

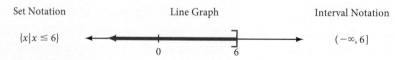

Set Notation	Line Graph	Interval Notation
$\{x \mid x \leq 6\}$		$(-\infty, 6]$

EXAMPLE 4   Solve and graph $-3 \leq 2x - 5 \leq 3$.

SOLUTION   We can extend our properties for addition and multiplication to cover this situation. If we add a number to the middle expression, we must add the same number to the outside expressions. If we multiply the center expression by a number, we must do the same to the outside expressions, remembering to reverse the direction of the inequality symbols if we multiply by a negative number. We begin by adding 5 to all three parts of the inequality.

$$-3 \leq 2x - 5 \leq 3$$

$$2 \leq \quad 2x \leq 8 \qquad \text{Add 5 to all three members}$$

$$1 \leq \quad x \leq 4 \qquad \text{Multiply through by } \tfrac{1}{2}$$

Here are three ways to write this solution set:

Set Notation	Line Graph	Interval Notation
$\{x \mid 1 \leq x \leq 4\}$	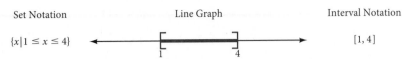	$[1, 4]$

B   Interval Notation and Graphing

The following figure shows the connection between inequalities, interval notation, and number line graphs. We have included the graphs with open and closed circles for those of you who have used this type of graph previously. From this point forward, we will show our graphs using the parentheses/brackets method.

Inequality notation	Interval notation	Graph using parentheses/brackets	Graph using open and closed circles
$x < 2$	$(-\infty, 2)$		
$x \leq 2$	$(-\infty, 2]$		
$x \geq -3$	$[-3, \infty)$		
$x > -3$	$(-3, \infty)$		

This next figure shows the connection between interval notation and number line graphs for a variety of continued inequalities. Again, we have included the graphs with open and closed circles for those of you who have used this type of graph previously.

Inequality notation	Interval notation	Graph using parentheses/brackets	Graph using open and closed circles
$-3 < x < 2$	$(-3, 2)$		
$-3 \leq x \leq 2$	$[-3, 2]$		
$-3 \leq x < 2$	$[-3, 2)$		
$-3 < x \leq 2$	$(-3, 2]$		

C   Interval Notation and Graphing Compound Inequalities

EXAMPLE 5   Solve the compound inequality.

$$3t + 7 \leq -4 \quad \text{or} \quad 3t + 7 \geq 4$$

SOLUTION   We solve each half of the compound inequality separately, and then we graph the solution set.

$$3t + 7 \leq -4 \qquad \text{or } 3t + 7 \geq 4$$

$$3t \leq -11 \qquad \text{or} \qquad 3t \geq -3 \qquad \text{Add } -7$$

$$t \leq -\frac{11}{3} \qquad \text{or} \qquad t \geq -1 \qquad \text{Multiply by } \frac{1}{3}$$

The solution set can be written in any of the following ways:

Set Notation	Line Graph	Interval Notation

$\{t \mid t \leq -\frac{11}{3} \text{ or } t \geq -1\}$ 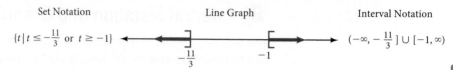 $(-\infty, -\frac{11}{3}] \cup [-1, \infty)$

$-\frac{11}{3}$ -1

D Applications

EXAMPLE 6 A company that manufactures ink cartridges for printers finds that they can sell x cartridges each week at a price of p dollars each, according to the formula $x = 1,300 - 100p$. What price should they charge for each cartridge if they want to sell at least 300 cartridges a week?

SOLUTION Because x is the number of cartridges they sell each week, an inequality that corresponds to selling at least 300 cartridges a week is

$$x \geq 300$$

Substituting $1,300 - 100p$ for x gives us an inequality in the variable p.

$$1,300 - 100p \geq 300$$

$$-100p \geq -1,000 \qquad \text{Add } -1,300 \text{ to each side}$$

$$p \leq 10 \qquad \text{Divide each side by } -100, \text{ and reverse the direction of the inequality symbol}$$

To sell at least 300 cartridges each week, the price per cartridge should be no more than \$10; that is, selling the cartridges for \$10 or less will produce weekly sales of 300 or more cartridges.

EXAMPLE 7 The formula $F = \frac{9}{5}C + 32$ gives the relationship between the Celsius and Fahrenheit temperature scales. If the temperature range on a certain day is 86° to 104° Fahrenheit, what is the temperature range in degrees Celsius?

SOLUTION From the given information we can write $86 \leq F \leq 104$. But, because F is equal to $\frac{9}{5}C + 32$, we can also write

$$86 \leq \frac{9}{5}C + 32 \leq 104$$

$$54 \leq \frac{9}{5}C \leq 72 \qquad \text{Add } -32 \text{ to each member}$$

$$\frac{5}{9}(54) \leq \frac{5}{9}\left(\frac{9}{5}C\right) \leq \frac{5}{9}(72) \qquad \text{Multiply through by } \frac{5}{9}$$

$$30 \leq C \leq 40$$

A temperature range of 86° to 104° Fahrenheit corresponds to a temperature range of 30° to 40° Celsius.

Problem Set 7.3

Moving Toward Success

"Don't find fault, find a remedy."

—Henry Ford, 1863–1947, American industrialist
and founder of Ford Motor Company

1. How do you plan to use your resources list if you have difficulty with a word problem?

2. Should you do all of the assigned homework problems or just a few? Explain.

A Solve each of the following inequalities and graph each solution. [Examples 1–2]

1. $2x \le 3$

2. $5x \ge -115$

3. $\frac{1}{2}x > 2$

4. $\frac{1}{3}x > 4$

5. $-5x \le 25$

6. $-7x \ge 35$

7. $-\frac{3}{2}x > -6$

8. $-\frac{2}{3}x < -8$

9. $-12 \le 2x$

10. $-20 \ge 4x$

11. $-1 \ge -\frac{1}{4}x$

13. $-3x + 1 > 10$

15. $\frac{1}{2} - \frac{m}{12} \le$

17. $\frac{1}{2} \ge -\frac{1}{6} -$

19. $-40 \le 30$

21. $\frac{2}{3}x - 3 <$

23. $10 - \frac{1}{2}y$

A B Si
inequalitie
[Examples

25. $2(3y +$

26. $3(2y$

27. $-(a$

28. $-(a$

29. $\frac{1}{3}t$

30. $\frac{1}{4}$

31. $-$

32.

33. $-\frac{1}{3}(x + 5) \le -\frac{2}{9}(x - 1)$

34. $-\frac{1}{2}(2x + 1) \le -\frac{3}{8}(x + 2)$

Solve each inequality. Write your answer using inequality notation.

35. $20x + 9,300 > 18,000$

36. $20x + 4,800 > 18,000$

B Solve the following continued inequalities. Use both a line graph and interval notation to write each solution [Example 4]

$2 \le m - 5 \le 2$ **38.** $-3 \le m + 1 \le 3$

$-60 < 20a + 20 < 60$

$-60 < 50a - 40 < 60$

$0.5 \le 0.3a - 0.7 \le 1.1$

$0.1 \le 0.4a + 0.1 \le 0.3$

$3 < \frac{1}{2}x + 5 < 6$ **44.** $5 < \frac{1}{4}x + 1 < 9$

$4 < 6 + \frac{2}{3}x < 8$

46. $3 < 7 + \frac{4}{5}x < 15$

C Graph the solution sets for the following compound inequalities. Then write each solution set using interval notation. [Examples 5]

47. $x + 5 \le -2$ or $x + 5 \ge 2$

48. $3x + 2 < -3$ or $3x + 2 > 3$

49. $5y + 1 \le -4$ or $5y + 1 \ge 4$

50. $7y - 5 \le -2$ or $7y - 5 \ge 2$

51. $2x + 5 < 3x - 1$ or $x - 4 > 2x + 6$

52. $3x - 1 > 2x + 4$ or $5x - 2 < 3x + 4$

Translate each of the following phrases into an equivalent inequality statement.

53. x is greater than -2 and at most 4

54. x is less than 9 and at least -3

55. x is less than -4 or at least 1

56. x is at most 1 or more than 6

57. Write each statement using inequality notation.
   a. x is always positive.
   b. x is never negative.
   c. x is greater than or equal to 0.

58. Match each expression on the left with a phrase on the right.
   a. $x^2 \geq 0$ **e.** never true
   b. $x^2 < 0$ **f.** sometimes true
   c. $x^2 \leq 0$ **g.** always true

Solve each inequality by inspection, without showing any work.

59. $x^2 < 0$ **60.** $x^2 \leq 0$

61. $x^2 \geq 0$

62. $\dfrac{1}{x^2} \geq 0$

63. $\dfrac{1}{x^2} < 0$ **64.** $\dfrac{1}{x^2} = 0$

65. Work each problem according to the instructions given.
   a. Evaluate when $x = 0$: $-\dfrac{1}{2}x + 1$
   b. Solve: $-\dfrac{1}{2}x + 1 = -7$
   c. Is 0 a solution to $-\dfrac{1}{2}x + 1 < -7$?
   d. Solve: $-\dfrac{1}{2}x + 1 < -7$

66. Work each problem according to the instructions given.
   a. Evaluate when $x = 0$: $-\dfrac{2}{3}x - 5$
   b. Solve: $-\dfrac{2}{3}x - 5 = 1$
   c. Is 0 a solution to $-\dfrac{2}{3}x - 5 > 1$?
   d. Solve: $-\dfrac{2}{3}x - 5 > 1$

D   Applying the Concepts [Example 6]

Organic Groceries The map shows a range of how many organic grocery stores are found in each state. Use the chart to answer Problems 67 and 68.

67. Find the difference, d, in the number of organic stores between California and the state of Washington.

Source: organic.com 2008

68. If Oregon had 24 organic grocery stores, what is the difference, d, between the number of stores in Oregon and the number of stores found in Nevada.

A store selling art supplies finds that they can sell x sketch pads each week at a price of p dollars each, according to the formula $x = 2{,}000 - 200p$. What price should they charge if they want to sell

69. at least 300 pads each week?

70. more than 600 pads each week?

71. less than 525 pads each week?

72. at most 375 pads each week?

73. Amtrak The average number of passengers carried by Amtrak declined each year for the years 1990–1996 (American Association of Railroads, Washington, DC,

Courtesy of Amtrak

Railroad Facts, Statistics of Railroads of Class 1.) The linear model for the number of passengers carried each year by Amtrak is given by $P = 22{,}419 - 399x$, where P is the number of passengers, in millions, and x is the number of years after January 1, 1990. In what years did Amtrak have more than 20,500 million passengers?

74. Student Loan When considering how much debt to incur in student loans, you learn that it is wise to keep your student loan payment to 8% or less of your starting monthly income. Suppose you anticipate a starting annual salary of $48,000. Set up and solve an inequality that represents the amount of monthly debt for student loans that would be considered manageable.

75. Here is what the United States Geological Survey has to say about the survival rates of the Apapane, one of the endemic birds of Hawaii.

> *Annual survival rates based on 1,584 recaptures of 429 banded individuals 0.72 ± 0.11 for adults and 0.13 ± 0.07 for juveniles.*

Write the survival rates using inequalities. Then give the survival rates in terms of percent.

76. Here is part of a report concerning the survival rates of Western Gulls that appeared on the website of Cornell University.

> *Survival of eggs to hatching is 70% – 80%; of hatched chicks to fledgling 50% – 70%; of fledglings to age of first breeding < 50%.*

Write the survival rates using inequalities without percent.

77. **Temperature** Each of the following temperature ranges are in degrees Fahrenheit. Use the formula $F = \frac{9}{5}C + 32$ to find the corresponding temperature range in degrees Celsius.
   a. 95° to 113°
   b. 68° to 86°
   c. −13° to 14°
   d. −4° to 23°

78. **Temperature** Each of the following temperature ranges are in degrees Celsius. Use the formula $C = \frac{5}{9}(F - 32)$ to find the corresponding temperature range in degrees Fahrenheit.
   a. 5° to 15°
   b. −40° to −25°
   c. 100° to 115°
   d. −10° to 10°

Maintaining Your Skills

The problems that follow review some of the more important skills you have learned in previous sections and chapters. You can consider the time you spend working these problems as time spent studying for exams.

Simplify.

79. $|-3|$

80. $|3|$

81. $-|-3|$

82. $-(-3)$

83. Give a definition for the absolute value of x that involves the number line. (This is the geometric definition.)

84. Give a definition of the absolute value of x that does not involve the number line. (This is the algebraic definition.)

Getting Ready for the Next Section

To understand all of the explanations and examples in the next section, you must be able to work the problems below.

Solve each equation.

85. $2a - 1 = -7$

86. $3x - 6 = 9$

87. $\frac{2}{3}x - 3 = 7$

88. $\frac{2}{3}x - 3 = -7$

89. $x - 5 = x - 7$

90. $x + 3 = x + 8$

91. $x - 5 = -x - 7$

92. $x + 3 = -x + 8$

Extending the Concepts

Assume a, b, and c are positive, and solve each formula for x.

93. $ax + b < c$

94. $\frac{x}{a} + \frac{y}{b} < 1$

95. $-c < ax + b < c$

96. $-1 < \frac{ax + b}{c} < 1$

7.4 Inequalities Involving Absolute Value

OBJECTIVES

A Solve inequalities with absolute value and graph the solution set.

TICKET TO SUCCESS

Keep these questions in mind as you read through the section. Then respond in your own words and in complete sentences.

1. Write an inequality containing absolute value, the solution to which is all the numbers between -5 and 5 on the number line. Why is it important to translate expressions written in symbols into the English language?

2. Translate $|x| < 3$ into words using the definition of absolute value.

3. Explain in words what the inequality $|x - 5| < 2$ means with respect to distance on the number line.

4. Why is there no solution to the inequality $|2x - 3| < 0$?

Image copyright ©hainaultphoto, 2009. Used under license from Shutterstock.com

The chart below shows the heights of the three highest buildings in the world.

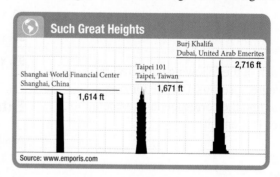

Such Great Heights

Shanghai World Financial Center
Shanghai, China
1,614 ft

Taipei 101
Taipei, Taiwan
1,671 ft

Burj Khalifa
Dubai, United Arab Emerites
2,716 ft

Source: www.emporis.com

The relative size of objects can be compared by extending the use of absolute value to include linear inequalities. In this section, we will again apply the definition of absolute value to solve inequalities involving absolute value. You will see how to compare the heights of these three buildings.

A Solving Inequalities with Absolute Values

The absolute value of x, which is $|x|$, represents the distance that x is from 0 on the number line. We will begin by considering three absolute value equations/inequalities and their English translations.

In Symbols	*In Words*
$|x| = 7$	x is exactly 7 units from 0 on the number line
$|a| < 5$	a is less than 5 units from 0 on the number line
$|y| \geq 4$	y is greater than or equal to 4 units from 0 on the number line

Once we have translated the expression into words, we can use the translation to graph the original equation or inequality. The graph then is used to write a final equation or inequality that does not involve absolute value.

Original Expression	*Graph*	*Final Expression*

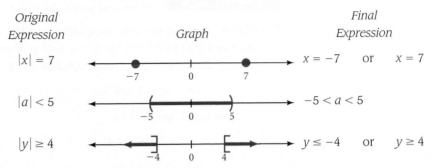

$|x| = 7$		$x = -7$ or $x = 7$
$|a| < 5$		$-5 < a < 5$
$|y| \geq 4$		$y \leq -4$ or $y \geq 4$

Although we will not always write out the English translation of an absolute value inequality, it is important that we understand the translation. Our second expression, $|a| < 5$, means a is within 5 units of 0 on the number line. The graph of this relationship is

which can be written with the following continued inequality:

$$-5 < a < 5$$

We can follow this same kind of reasoning to solve more complicated absolute value inequalities.

EXAMPLE 1   Graph the solution set: $|2x - 5| < 3$

SOLUTION   The absolute value of $2x - 5$ is the distance that $2x - 5$ is from 0 on the number line. We can translate the inequality as "$2x - 5$ is less than 3 units from 0 on the number line"; that is, $2x - 5$ must appear between -3 and 3 on the number line.

A picture of this relationship is

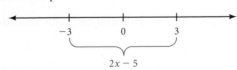

Using the picture, we can write an inequality without absolute value that describes the situation:

$$-3 < 2x - 5 < 3$$

Next, we solve the continued inequality by first adding 5 to all three members and then multiplying all three by $\frac{1}{2}$.

$$-3 < 2x - 5 < 3$$

$$2 < 2x < 8 \qquad \text{Add 5 to all three members}$$

$$1 < x < 4 \qquad \text{Multiply each member by } \frac{1}{2}$$

The graph of the solution set is

We can see from the solution that for the absolute value of $2x - 5$ to be within 3 units of 0 on the number line, x must be between 1 and 4. ■

EXAMPLE 2 Solve and graph $|3a + 7| \leq 4$.

SOLUTION We can read the inequality as, "The distance between $3a + 7$ and 0 is less than or equal to 4." Or, "$3a + 7$ is within 4 units of 0 on the number line." This relationship can be written without absolute value as

$$-4 \leq 3a + 7 \leq 4$$

Solving as usual, we have

$$-4 \leq 3a + 7 \leq 4$$
$$-11 \leq \quad 3a \leq -3 \qquad \text{Add } -7 \text{ to all three members}$$
$$-\frac{11}{3} \leq \quad a \leq -1 \qquad \text{Multiply each member by} \frac{1}{3}$$

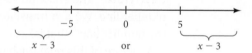

We can see from Examples 1 and 2 that to solve an inequality involving absolute value, we must be able to write an equivalent expression that does not involve absolute value.

EXAMPLE 3 Solve $|x - 3| > 5$ and graph the solution.

SOLUTION We interpret the absolute value inequality to mean that $x - 3$ is more than 5 units from 0 on the number line. The quantity $x - 3$ must be either above 5 or below -5. Here is a picture of the relationship:

An inequality without absolute value that also describes this situation is

$$x - 3 < -5 \quad \text{or} \quad x - 3 > 5$$

Adding 3 to both sides of each inequality we have

$$x < -2 \quad \text{or} \quad x > 8$$

The graph of which is

We can see from Examples 1 and 2 that to solve an inequality involving absolute

EXAMPLE 4 Graph the solution set: $|4t - 3| \geq 9$

SOLUTION The quantity $4t - 3$ is greater than or equal to 9 units from 0. It must be either above 9 or below -9.

$$4t - 3 \leq -9 \quad \text{or} \quad 4t - 3 \geq 9$$
$$4t \leq -6 \quad \text{or} \quad 4t \geq 12 \qquad \text{Add 3}$$

$$t \leq -\frac{6}{4} \quad \text{or} \quad t \geq \frac{12}{4} \quad \text{Multiply by } \frac{1}{4}$$

$$t \leq -\frac{3}{2} \quad \text{or} \quad t \geq 3$$

We can use the results of our first few examples and the material in the previous section to summarize the information we have related to absolute value equations and inequalities.

Rewriting Absolute Value Equations and Inequalities

If c is a positive real number, then each of the following statements on the left is equivalent to the corresponding statement on the right.

With Absolute Value	Without Absolute Value
$\|x\| = c$	$x = -c \quad$ or $\quad x = c$
$\|x\| < c$	$-c < x < c$
$\|x\| > c$	$x < -c \quad$ or $\quad x > c$
$\|ax + b\| = c$	$ax + b = -c \quad$ or $\quad ax + b = c$
$\|ax + b\| < c$	$-c < ax + b < c$
$\|ax + b\| > c$	$ax + b < -c \quad$ or $\quad ax + b > c$

EXAMPLE 5   Solve and graph $|2x + 3| + 4 < 9$.

SOLUTION   Before we can apply the method of solution we used in the previous examples, we must isolate the absolute value on one side of the inequality. To do so, we add -4 to each side.

$$|2x + 3| + 4 < 9$$

$$|2x + 3| + 4 + (-4) < 9 + (-4)$$

$$|2x + 3| < 5$$

From this last line we know that $2x + 3$ must be between -5 and 5.

$$-5 < 2x + 3 < 5$$

$$-8 < 2x < 2 \quad \text{Add } -3 \text{ to each member}$$

$$-4 < x < 1 \quad \text{Multiply each member by } \frac{1}{2}$$

The graph is

EXAMPLE 6   Solve and graph $|4 - 2t| > 2$.

SOLUTION   The inequality indicates that $4 - 2t$ is less than -2 or greater than 2. Writing this without absolute value symbols, we have

$$4 - 2t < -2 \quad \text{or} \quad 4 - 2t > 2$$

To solve these inequalities, we begin by adding -4 to each side.

$$4 + (-4) - 2t < -2 + (-4) \quad \text{or} \quad 4 + (-4) - 2t > 2 + (-4)$$

$$-2t < -6 \quad \text{or} \quad -2t > -2$$

> **NOTE**
> Remember, the multiplication property for inequalities requires that we reverse the direction of the inequality symbol every time we multiply both sides of an inequality by a negative number.

Next we must multiply both sides of each inequality by $-\frac{1}{2}$. When we do so, we must also reverse the direction of each inequality symbol.

$$-2t < -6 \qquad \text{or} \qquad -2t > -2$$

$$-\frac{1}{2}(-2t) > -\frac{1}{2}(-6) \quad \text{or} \quad -\frac{1}{2}(-2t) < -\frac{1}{2}(-2)$$

$$t > 3 \qquad \text{or} \qquad t < 1$$

Although, in situations like this, we are used to seeing the "less than" symbol written first, the meaning of the solution is clear. We want to graph all real numbers that are either greater than 3 or less than 1. Here is the graph.

Because absolute value always results in a nonnegative quantity, we sometimes come across special solution sets when a negative number appears on the right side of an absolute value inequality.

EXAMPLE 7 Solve $|7y - 1| < -2$.

SOLUTION The *left* side is never negative because it is an absolute value. The *right* side is negative. We have a positive quantity (or zero) less than a negative quantity, which is impossible. The solution set is the empty set, $\varnothing$. There is no real number to substitute for y to make this inequality a true statement.

EXAMPLE 8 Solve $|6x + 2| > -5$.

SOLUTION This is the opposite case from that in Example 7. No matter what real number we use for x on the *left* side, the result will always be positive, or zero. The *right* side is negative. We have a positive quantity (or zero) greater than a negative quantity. Every real number we choose for x gives us a true statement. The solution set is the set of all real numbers.

Problem Set 7.4

Moving Toward Success

"If you have made mistakes, even serious ones, there is always another chance for you. What we call failure is not the falling down but the staying down."

—Mary Pickford, 1892–1979, Canadian-born movie actress

1. Does a poor score on a test or difficulties with problems typically derail your focus in this class? Why or why not?

2. How do you intend to prevent setbacks from keeping you from your goals?

A Solve each of the following inequalities using the definition of absolute value. Graph the solution set in each case. [Examples 1–4, 7–8]

1. $|x| < 3$
2. $|x| \le 7$
3. $|x| \ge 2$
4. $|x| > 4$
5. $|x| + 2 < 5$
6. $|x| - 3 < -1$
7. $|t| - 3 > 4$
8. $|t| + 5 > 8$
9. $|y| < -5$
10. $|y| > -3$
11. $|x| \ge -2$
12. $|x| \le -4$
13. $|x - 3| < 7$
14. $|x + 4| < 2$

15. $|a + 5| \geq 4$

16. $|a - 6| \geq 3$

Solve each inequality and graph the solution set.
[Examples 1–5, 7–8]

17. $|a - 1| < -3$

18. $|a + 2| \geq -5$

19. $|2x - 4| < 6$

20. $|2x + 6| < 2$

21. $|3y + 9| \geq 6$

22. $|5y - 1| \geq 4$

23. $|2k + 3| \geq 7$

24. $|2k - 5| \geq 3$

25. $|x - 3| + 2 < 6$

26. $|x + 4| - 3 < -1$

27. $|2a + 1| + 4 \geq 7$

28. $|2a - 6| - 1 \geq 2$

29. $|3x + 5| - 8 < 5$

30. $|6x - 1| - 4 \leq 2$

Solve each inequality, and graph the solution set. Keep in mind that if you multiply or divide both sides of an inequality by a negative number, you must reverse the inequality sign. [Example 6]

31. $|5 - x| > 3$

32. $|7 - x| > 2$

33. $\left|3 - \dfrac{2}{3}x\right| \geq 5$

34. $\left|3 - \dfrac{3}{4}x\right| \geq 9$

35. $\left|2 - \dfrac{1}{2}x\right| > 1$

36. $\left|3 - \dfrac{1}{3}x\right| > 1$

Solve each inequality.

37. $|x - 1| < 0.01$

38. $|x + 1| < 0.01$

39. $|2x + 1| \geq \dfrac{1}{5}$

40. $|2x - 1| \geq \dfrac{1}{8}$

41. $\left|\dfrac{3x - 2}{5}\right| \leq \dfrac{1}{2}$

42. $\left|\dfrac{4x - 3}{2}\right| \leq \dfrac{1}{3}$

43. $\left|2x - \dfrac{1}{5}\right| < 0.3$

44. $\left|3x - \dfrac{3}{5}\right| < 0.2$

45. Write the continued inequality $-4 \leq x \leq 4$ as a single inequality involving absolute value.

46. Write the continued inequality $-8 \leq x \leq 8$ as a single inequality involving absolute value.

47. Write $-1 \leq x - 5 \leq 1$ as a single inequality involving absolute value.

48. Write $-3 \leq x + 2 \leq 3$ as a single inequality involving absolute value.

49. Is 0 a solution to $|5x + 3| > 7$?

50. Is 0 a solution to $|-2x - 5| > 1$?

51. Solve: $|5x + 3| > 7$

52. Solve: $|-2x - 5| > 1$

Maintaining Your Skills

The problems that follow review some of the more important skills you have learned in previous sections and chapters. You can consider the time you spend working these problems as time spent studying for exams.

Simplify each expression as much as possible.

53. $-9 \div \dfrac{3}{2}$

54. $-\dfrac{4}{5} \div (-4)$

55. $3 - 7(-6 - 3)$

56. $(3 - 7)(-6 - 3)$

57. $-4(-2)^3 - 5(-3)^2$

58. $4(2 - 5)^3 - 3(4 - 5)^5$

59. $-2(-3 + 8) - 7(-9 + 6)$

60. $-3 - 6[5 - 2(-3 - 1)]$

61. $\dfrac{2(-3) - 5(-6)}{-1 - 2 - 3}$

62. $\dfrac{4 - 8(3 - 5)}{2 - 4(3 - 5)}$

63. $6(1) - 1(-5) + 1(2)$

64. $-2(-14) + 3(-4) - 1(-10)$

65. $1(0)(1) + 3(1)(4) + (-2)(2)(-1)$

66. $-3(-1 - 1) + 4(-2 + 2) - 5[2 - (-2)]$

67. $4(-1)^2 + 3(-1) - 2$

68. $5 \cdot 2^3 - 3 \cdot 2^2 + 4 \cdot 2 - 5$

Getting Ready for the Next Section

Solve the following equations.

69. $2(3x + 5) - 9x = 25$

70. $3(2x - 9) + 7x = 12$

71. $4x - 3(x - 4) = 21$

72. $7(y + 3) - 3y = 37$

73. $6(2y + 5) - 2y = -10$

74. $3x - 5(2x + 9) = 4$

Extending the Concepts

Solve each inequality for *x*. (Assume *a*, *b*, and *c* are all positive.)

75. $|x - a| < b$

76. $|x - a| > b$

77. $|ax - b| > c$

78. $|ax - b| < c$

79. $|ax + b| \leq c$

80. $|ax + b| \geq c$

7.5 Review of Systems of Equations in Two Variables

OBJECTIVES

A Solve systems of linear equations in two variables by graphing.

B Solve systems of linear equations in two variables by the addition method.

C Solve systems of linear equations in two variables by the substitution method.

TICKET TO SUCCESS

Keep these questions in mind as you read through the section. Then respond in your own words and in complete sentences.

1. How would you define a solution for a linear system of equations?
2. How would you use the addition method to solve a system of linear equations?
3. When would the substitution method be more efficient than the addition method in solving a system of linear equations?
4. Explain what an inconsistent system of linear equations looks like graphically and what would result algebraically when attempting to solve the system.

Image copyright ©Graça Victoria, 2009. Used under license from Shutterstock.com

Suppose you and a friend want to order burritos and tacos for lunch. The restaurant's lighted menu, however, is broken and you can't read it. You order 2 burritos and 3 tacos and are charged a total of $12. Your friend orders 1 burrito and 4 tacos for a total of $11. How much is each burrito? How much is each taco? In this section, we will begin our work with systems of equations in two variables and use these systems to answer your lunch question.

A Solve Systems by Graphing

Previously, we found the graph of an equation of the form $ax + by = c$ to be a straight line. Since the graph is a straight line, the equation is said to be a linear equation. Two linear equations considered together form a *linear system* of equations. For example,

$$3x - 2y = 6$$
$$2x + 4y = 20$$

is a linear system. The solution set to the system is the set of all ordered pairs that satisfy both equations. If we graph each equation on the same set of axes, we can see the solution set (Figure 1).

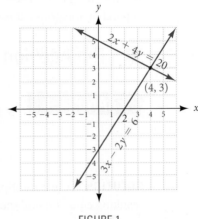

FIGURE 1

The point (4, 3) lies on both lines and therefore must satisfy both equations. It is obvious from the graph that it is the only point that does so. The solution set for the system is {(4, 3)}.

More generally, if $a_1x + b_1y = c_1$ and $a_2x + b_2y = c_2$ are linear equations, then the solution set for the system

$$a_1x + b_1y = c_1$$
$$a_2x + b_2y = c_2$$

can be illustrated through one of the graphs in Figure 2.

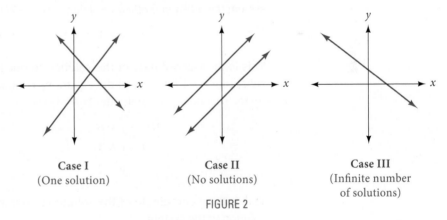

Case I
(One solution)

Case II
(No solutions)

Case III
(Infinite number
of solutions)

FIGURE 2

Case I The two lines intersect at one and only one point. The coordinates of the point give the solution to the system. This is what usually happens.

Case II The lines are parallel and therefore have no points in common. The solution set to the system is the empty set, ∅. In this case, we say the system is *inconsistent*.

Case III The lines coincide; that is, their graphs represent the same line. The solution set consists of all ordered pairs that satisfy either equation. In this case, the equations are said to be *dependent*.

In the beginning of this section we found the solution set for the system

$$3x - 2y = 6$$
$$2x + 4y = 20$$

by graphing each equation and then reading the solution set from the graph. Solving a system of linear equations by graphing is the least accurate method. If the coordinates of the point of intersection are not integers, it can be very difficult to read the solution set from the graph. There is another method of solving a linear system that does not depend on the graph. It is called the *addition method.*

B The Addition Method

EXAMPLE 1 Solve the system.

$$4x + 3y = 10$$
$$2x + y = 4$$

SOLUTION If we multiply the bottom equation by -3, the coefficients of y in the resulting equation and the top equation will be opposites.

$$4x + 3y = 10 \xrightarrow{\text{No change}} 4x + 3y = 10$$
$$2x + y = 4 \xrightarrow[\text{Multiply by } -3]{} -6x - 3y = -12$$

Adding the left and right sides of the resulting equations, we have

$$\begin{array}{rcl} 4x + 3y &=& 10 \\ -6x - 3y &=& -12 \\ \hline -2x &=& -2 \end{array}$$

The result is a linear equation in one variable. We have eliminated the variable y from the equations by addition. (It is for this reason we call this method of solving a linear system the *addition method.*) Solving $-2x = -2$ for x, we have

$$x = 1$$

This is the x-coordinate of the solution to our system. To find the y-coordinate, we substitute $x = 1$ into any of the equations containing both the variables x and y. Let's try the second equation in our original system.

$$2(1) + y = 4$$
$$2 + y = 4$$
$$y = 2$$

This is the y-coordinate of the solution to our system. The ordered pair $(1, 2)$ is the solution to the system.

CHECKING SOLUTIONS We can check our solution by substituting it into both of our equations.

Substituting $x = 1$ and $y = 2$ into $4x + 3y = 10$, we have

$$4(1) + 3(2) \stackrel{?}{=} 10$$
$$4 + 6 \stackrel{?}{=} 10$$
$$10 = 10 \quad \text{A true statement}$$

Substituting $x = 1$ and $y = 2$ into $2x + y = 4$, we have

$$2(1) + 2 \stackrel{?}{=} 4$$
$$2 + 2 \stackrel{?}{=} 4$$
$$4 = 4 \quad \text{A true statement}$$

Our solution satisfies both equations; therefore, it is a solution to our system of equations.

EXAMPLE 2   Solve the system.

$$3x - 5y = -2$$
$$2x - 3y = 1$$

SOLUTION   We can eliminate either variable. Let's decide to eliminate the variable x. We can do so by multiplying the top equation by 2 and the bottom equation by -3, and then adding the left and right sides of the resulting equations.

$$3x - 5y = -2 \xrightarrow{\text{Multiply by 2}} 6x - 10y = -4$$
$$2x - 3y = 1 \xrightarrow[\text{Multiply by } -3]{} \underline{-6x + 9y = -3}$$
$$-y = -7$$
$$y = 7$$

The y-coordinate of the solution to the system is 7. Substituting this value of y into any of the equations with both x- and y-variables gives $x = 11$. The solution to the system is $(11, 7)$. It is the only ordered pair that satisfies both equations.

CHECKING SOLUTIONS   Checking $(11, 7)$ in each equation looks like this:

Substituting $x = 11$ and $y = 7$ into

$3x - 5y = -2$, we have

$3(11) - 5(7) \stackrel{?}{=} -2$

$33 - 35 \stackrel{?}{=} -2$

$-2 = -2$ A true statement

Substituting $x = 11$ and $y = 7$ into

$2x - 3y = 1$, we have

$2(11) - 3(7) \stackrel{?}{=} 1$

$22 - 21 \stackrel{?}{=} 1$

$1 = 1$ A true statement

Our solution satisfies both equations; therefore, $(11, 7)$ is a solution to our system. ▮

EXAMPLE 3   Solve the system.

$$2x - 3y = 4$$
$$4x + 5y = 3$$

SOLUTION   We can eliminate x by multiplying the top equation by -2 and adding it to the bottom equation.

$$2x - 3y = 4 \xrightarrow{\text{Multiply by } -2} -4x + 6y = -8$$
$$4x + 5y = 3 \xrightarrow[\text{No change}]{} \underline{4x + 5y = 3}$$
$$11y = -5$$
$$y = -\frac{5}{11}$$

The y-coordinate of our solution is $-\frac{5}{11}$. If we were to substitute this value of y back into either of our original equations, we would find the arithmetic necessary to solve for x cumbersome. For this reason, it is probably best to go back to the original system and solve it a second time—for x instead of y. Here is how we do that:

$$2x - 3y = 4 \xrightarrow{\text{Multiply by 5}} 10x - 15y = 20$$
$$4x + 5y = 3 \xrightarrow[\text{Multiply by 3}]{} \underline{12x + 15y = 9}$$
$$22x = 29$$
$$x = \frac{29}{22}$$

The solution to our system is $\left(\frac{29}{22}, -\frac{5}{11} \right)$. ▮

The main idea in solving a system of linear equations by the addition method is to use the multiplication property of equality on one or both of the original equations, if necessary, to make the coefficients of either variable opposites. The following box shows some steps to follow when solving a system of linear equations by the addition method.

Strategy Solving a System of Linear Equations by the Addition Method

Step 1: Decide which variable to eliminate. (In some cases, one variable will be easier to eliminate than the other. With some practice, you will notice which one it is.)

Step 2: Use the multiplication property of equality on each equation separately to make the coefficients of the variable that is to be eliminated opposites.

Step 3: Add the respective left and right sides of the system together.

Step 4: Solve for the remaining variable.

Step 5: Substitute the value of the variable from step 4 into an equation containing both variables and solve for the other variable. (Or repeat steps 2–4 to eliminate the other variable.)

Step 6: Check your solution in both equations, if necessary.

EXAMPLE 4 Solve the system.

$$5x - 2y = 5$$
$$-10x + 4y = 15$$

SOLUTION We can eliminate y by multiplying the first equation by 2 and adding the result to the second equation.

$$
\begin{array}{lcl}
5x - 2y = 5 & \xrightarrow{\text{Multiply by 2}} & 10x - 4y = 10 \\
-10x + 4y = 15 & \xrightarrow[\text{No change}]{} & \underline{-10x + 4y = 15} \\
& & 0 = 25
\end{array}
$$

The result is the false statement $0 = 25$, which indicates there is no solution to the system. If we were to graph the two lines, we would find that they are parallel. In a case like this, we say the system is *inconsistent*. Whenever both variables have been eliminated and the resulting statement is false, the solution set for the system will be the empty set, $\varnothing$.

EXAMPLE 5 Solve the system.

$$4x + 3y = 2$$
$$8x + 6y = 4$$

SOLUTION Multiplying the top equation by −2 and adding, we can eliminate the variable x.

$$
\begin{array}{lcl}
4x + 3y = 2 & \xrightarrow{\text{Multiply by } -2} & -8x - 6y = -4 \\
8x + 6y = 4 & \xrightarrow[\text{No change}]{} & \underline{8x + 6y = 4} \\
& & 0 = 0
\end{array}
$$

Both variables have been eliminated and the resulting statement $0 = 0$ is true. In this case, the lines coincide and the equations are said to be *dependent.* The solution set consists of all ordered pairs that satisfy either equation. We can write the solution set as $\{(x, y) \mid 4x + 3y = 2\}$ or $\{(x, y) \mid 8x + 6y = 4\}$. ■

Special Cases

The previous two examples illustrate the two special cases in which the graphs of the equations in the system either coincide or are parallel. In both cases, the left-hand sides of the equations were multiples of each other. In the case of the dependent equations, the right-hand sides were also multiples. We can generalize these observations for the system

$$a_1x + b_1y = c_1$$
$$a_2x + b_2y = c_2$$

Inconsistent System

What Happens	*Geometric Interpretation*	*Algebraic Interpretation*
Both variables are eliminated, and the resulting statement is false.	The lines are parallel, and there is no solution to the system.	$\dfrac{a_1}{a_2} = \dfrac{b_1}{b_2} \neq \dfrac{c_1}{c_2}$

Dependent Equations

What Happens	*Geometric Interpretation*	*Algebraic Interpretation*
Both variables are eliminated, and the resulting statement is true.	The lines coincide, and there are an infinite number of solutions to the system.	$\dfrac{a_1}{a_2} = \dfrac{b_1}{b_2} = \dfrac{c_1}{c_2}$

EXAMPLE 6　Solve the system.

$$\frac{1}{2}x - \frac{1}{3}y = 2$$

$$\frac{1}{4}x + \frac{2}{3}y = 6$$

SOLUTION　Although we could solve this system without clearing the equations of fractions, there is probably less chance for error if we have only integer coefficients to work with. So let's begin by multiplying both sides of the top equation by 6 and both sides of the bottom equation by 12 to clear each equation of fractions.

$$\frac{1}{2}x - \frac{1}{3}y = 2 \xrightarrow{\text{Multiply by 6}} 3x - 2y = 12$$

$$\frac{1}{4}x + \frac{2}{3}y = 6 \xrightarrow[\text{Multiply by 12}]{} 3x + 8y = 72$$

Now we can eliminate x by multiplying the top equation by -1 and leaving the bottom equation unchanged.

$$3x - 2y = 12 \xrightarrow{\text{Multiply by } -1} -3x + 2y = -12$$

$$3x + 8y = 72 \xrightarrow[\text{No change}]{} \underline{3x + 8y = 72}$$

$$10y = 60$$

$$y = 6$$

We can substitute $y = 6$ into any equation that contains both x and y. Let's use $3x - 2y = 12$.

$$3x - 2(6) = 12$$
$$3x - 12 = 12$$
$$3x = 24$$
$$x = 8$$

The solution to the system is $(8, 6)$.

C The Substitution Method

EXAMPLE 7   Solve the system.

$$2x - 3y = -6$$
$$y = 3x - 5$$

SOLUTION   The second equation tells us y is $3x - 5$. Substituting the expression $3x - 5$ for y in the first equation, we have

$$2x - 3(3x - 5) = -6$$

The result of the substitution is the elimination of the variable y. Solving the resulting linear equation in x as usual, we have

$$2x - 9x + 15 = -6$$
$$-7x + 15 = -6$$
$$-7x = -21$$
$$x = 3$$

Putting $x = 3$ into the second equation in the original system, we have

$$y = 3(3) - 5$$
$$= 9 - 5$$
$$= 4$$

The solution to the system is $(3, 4)$.

CHECKING SOLUTIONS Checking $(3, 4)$ in each equation looks like this:

Substituting $x = 3$ and $y = 4$ into $2x - 3y = -6$, we have

$$2(3) - 3(4) \stackrel{?}{=} -6$$
$$6 - 12 \stackrel{?}{=} -6$$
$$-6 = -6 \quad \text{A true statement}$$

Substituting $x = 3$ and $y = 4$ into $y = 3x - 5$, we have

$$4 \stackrel{?}{=} 3(3) - 5$$
$$4 \stackrel{?}{=} 9 - 5$$
$$4 = 4 \quad \text{A true statement}$$

Our solution satisfies both equations; therefore, $(3, 4)$ is a solution to our system.

Strategy   Solving a System of Equations by the Substitution Method

Step 1: Solve either one of the equations for x or y. (This step is not necessary if one of the equations is already in the correct form, as in Example 7.)

Step 2: Substitute the expression for the variable obtained in step 1 into the other equation and solve it.

Step 3: Substitute the solution for step 2 into any equation in the system that contains both variables and solve it.

Step 4: Check your results, if necessary.

EXAMPLE 8   Solve by substitution.

$$2x + 3y = 5$$
$$x - 2y = 6$$

SOLUTION   To use the substitution method, we must solve one of the two equations for x or y. We can solve for x in the second equation by adding $2y$ to both sides.

$$x - 2y = 6$$
$$x = 2y + 6 \qquad \text{Add } 2y \text{ to both sides}$$

Substituting the expression $2y + 6$ for x in the first equation of our system, we have

$$2(2y + 6) + 3y = 5$$
$$4y + 12 + 3y = 5$$
$$7y + 12 = 5$$
$$7y = -7$$
$$y = -1$$

Using $y = -1$ in either equation in the original system, we get $x = 4$. The solution is $(4, -1)$.

> **NOTE**
> Both the substitution method and the addition method can be used to solve any system of linear equations in two variables. Systems like the one in Example 7, however, are easier to solve using the substitution method because one of the variables is already written in terms of the other. A system like the one in Example 2 is easier to solve using the addition method because solving for one of the variables would lead to an expression involving fractions. The system in Example 8 could be solved easily by either method because solving the second equation for x is a one-step process.

USING TECHNOLOGY

Graphing Calculators: Solving a System That Intersects at Exactly One Point

A graphing calculator can be used to solve a system of equations in two variables if the equations intersect at exactly one point. To solve the system shown in Example 3, we first solve each equation for y. Here is the result:

$$2x - 3y = 4 \qquad \text{becomes} \qquad y = \frac{4 - 2x}{-3}$$

$$4x + 5y = 3 \qquad \text{becomes} \qquad y = \frac{3 - 4x}{5}$$

Graphing these two functions on the calculator gives a diagram similar to the one in Figure 3.

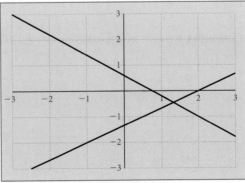

FIGURE 3

Using the Trace and Zoom features, we find that the two lines intersect at $x = 1.32$ and $y = -0.45$, which are the decimal equivalents (accurate to the nearest hundredth) of the fractions found in Example 3.

Problem Set 7.5

Moving Toward Success

"When one has a great deal to put into it, a day has a hundred pockets."

—Friedrich Nietzsche, 1844–1900, German philosopher

1. How would keeping a calendar of your daily schedule help you lay out a successful study plan?

2. Why should you keep a daily to do list and write quiz and exam dates on your calendar?

A Solve each system by graphing both equations on the same set of axes and then reading the solution from the graph.

1. $3x - 2y = 6$
$x - y = 1$

2. $5x - 2y = 10$
$x - y = -1$

3. $y = \dfrac{3}{5}x - 3$
$2x - y = -4$

4. $y = \dfrac{1}{2}x - 2$
$2x - y = -1$

5. $y = \dfrac{1}{2}x$
$y = -\dfrac{3}{4}x + 5$

6. $y = \dfrac{2}{3}x$
$y = -\dfrac{1}{3}x + 6$

7. $3x + 3y = -2$
$y = -x + 4$

8. $2x - 2y = 6$
$y = x - 3$

9. $2x - y = 5$
$y = 2x - 5$

10. $x + 2y = 5$
$y = -\dfrac{1}{2}x + 3$

21. $6x + 3y = -1$
$9x + 5y = 1$

22. $5x + 4y = -1$
$7x + 6y = -2$

23. $4x + 3y = 14$
$9x - 2y = 14$

24. $7x - 6y = 13$
$6x - 5y = 11$

25. $2x - 5y = 3$
$-4x + 10y = 3$

26. $3x - 2y = 1$
$-6x + 4y = -2$

27. $\dfrac{1}{4}x - \dfrac{1}{6}y = -2$
$-\dfrac{1}{6}x + \dfrac{1}{5}y = 4$

28. $-\dfrac{1}{3}x + \dfrac{1}{4}y = 0$
$\dfrac{1}{5}x - \dfrac{1}{10}y = 1$

29. $\dfrac{1}{2}x + \dfrac{1}{3}y = 13$
$\dfrac{2}{5}x + \dfrac{1}{4}y = 10$

30. $\dfrac{1}{2}x + \dfrac{1}{3}y = \dfrac{2}{3}$
$\dfrac{2}{3}x + \dfrac{2}{5}y = \dfrac{14}{15}$

B Solve each of the following systems by the addition method. [Examples 1–6]

11. $x + y = 5$
$3x - y = 3$

12. $x - y = 4$
$-x + 2y = -3$

13. $3x + y = 4$
$4x + y = 5$

14. $6x - 2y = -10$
$6x + 3y = -15$

15. $3x - 2y = 6$
$6x - 4y = 12$

16. $4x + 5y = -3$
$-8x - 10y = 3$

17. $x + 2y = 0$
$2x - 6y = 5$

18. $x + 3y = 3$
$2x - 9y = 1$

19. $2x - 5y = 16$
$4x - 3y = 11$

20. $5x - 3y = -11$
$7x + 6y = -12$

C Solve each of the following systems by the substitution method. [Examples 7–8]

31. $7x - y = 24$
$x = 2y + 9$

32. $3x - y = -8$
$y = 6x + 3$

33. $6x - y = 10$
$y = -\dfrac{3}{4}x - 1$

34. $2x - y = 6$
$y = -\dfrac{4}{3}x + 1$

35. $3y + 4z = 23$
$6y + z = 32$

36. $2x - y = 650$
$3.5x - y = 1{,}400$

37. $y = 3x - 2$
$y = 4x - 4$

38. $y = 5x - 2$
$y = -2x + 5$

39. $2x - y = 5$
$4x - 2y = 10$

40. $-10x + 8y = -6$
$y = \dfrac{5}{4}x$

41. $\dfrac{1}{3}x - \dfrac{1}{2}y = 0$
$x = \dfrac{3}{2}y$

42. $\dfrac{2}{5}x - \dfrac{2}{3}y = 0$
$y = \dfrac{3}{5}x$

59. $\dfrac{3}{4}x - \dfrac{1}{3}y = 1$
$y = \dfrac{1}{4}x$

60. $-\dfrac{2}{3}x - \dfrac{1}{2}y = -1$
$y = -\dfrac{1}{3}x$

61. $\dfrac{1}{4}x - \dfrac{1}{2}y = \dfrac{1}{3}$
$\dfrac{1}{3}x - \dfrac{1}{4}y = \dfrac{2}{3}$

62. $\dfrac{1}{5}x - \dfrac{1}{10}y = -\dfrac{1}{3}$
$\dfrac{2}{3}x - \dfrac{1}{2}y = -\dfrac{1}{6}$

You may want to read Example 3 again before solving the systems that follow.

43. $4x - 7y = 3$
$5x + 2y = -3$

44. $3x - 4y = 7$
$6x - 3y = 5$

45. $9x - 8y = 4$
$2x + 3y = 6$

46. $4x - 7y = 10$
$-3x + 2y = -9$

47. $3x - 5y = 2$
$7x + 2y = 1$

48. $4x - 3y = -1$
$5x + 8y = 2$

Solve each of the following systems by using either the addition or substitution method. Choose the method that is most appropriate for the problem.

49. $x - 3y = 7$
$2x + y = -6$

50. $2x - y = 9$
$x + 2y = -11$

51. $y = \dfrac{1}{2}x + \dfrac{1}{3}$
$y = -\dfrac{1}{3}x + 2$

52. $y = \dfrac{3}{4}x - \dfrac{4}{5}$
$y = \dfrac{1}{2}x - \dfrac{1}{2}$

53. $3x - 4y = 12$
$x = \dfrac{2}{3}y - 4$

54. $-5x + 3y = -15$
$x = \dfrac{4}{5}y - 2$

55. $4x - 3y = -7$
$-8x + 6y = -11$

56. $3x - 4y = 8$
$y = \dfrac{3}{4}x - 2$

57. $3y + z = 17$
$5y + 20z = 65$

58. $x + y = 850$
$1.5x + y = 1,100$

The next two problems are intended to give you practice reading, and paying attention to, the instructions that accompany the problems you are working.

63. Work each problem according to the instructions given.

    a. Simplify: $(3x - 4y) - 3(x - y)$
    b. Find y when x is 0 in $3x - 4y = 8$.
    c. Find the y-intercept: $3x - 4y = 8$
    d. Graph: $3x - 4y = 8$
    e. Find the point where the graphs of $3x - 4y = 8$ and $x - y = 2$ cross.

64. Work each problem according to the instructions given.

    a. Solve: $4x - 5 = 20$
    b. Solve for y: $4x - 5y = 20$
    c. Solve for x: $x - y = 5$
    d. Solve: $4x - 5y = 20$
          $x - y = 5$

65. Multiply both sides of the second equation in the following system by 100, and then solve as usual.

$$x + y = 10,000$$
$$0.06x + 0.05y = 560$$

66. Multiply both sides of the second equation in the following system by 10, and then solve as usual.

$$x + y = 12$$
$$0.20x + 0.50y = 0.30(12)$$

67. What value of c will make the following system a dependent system (one in which the lines coincide)?

$$6x - 9y = 3$$
$$4x - 6y = c$$

68. What value of c will make the following system a dependent system?

$$5x - 7y = c$$
$$-15x + 21y = 9$$

69. Where do the graphs of the lines $x + y = 4$ and $x - 2y = 4$ intersect?

70. Where do the graphs of the line $x = -1$ and $x - 2y = 4$ intersect?

Maintaining Your Skills

71. Find the slope of the line that contains $(-4, -1)$ and $(-2, 5)$.

72. A line has a slope of $\frac{2}{3}$. Find the slope of any line
   a. Parallel to it.
   b. Perpendicular to it.

73. Give the slope and y-intercept of the line $2x - 3y = 6$.

74. Graph the equation of the line with slope -3 and y-intercept 5.

75. Graph the equation of the line with slope $\frac{2}{3}$ that contains the point $(-6, 2)$.

76. Find the slope of the line through $(1, 3)$ and $(-1, -5)$.

77. Find the slope of the line with x-intercept 3 and y-intercept -2.

78. Find the slope of the line whose graph is perpendicular to the graph of $y = 2x + 3$.

Getting Ready for the Next Section

Simplify.

79. $2 - 2(6)$

80. $2(1) - 2 + 3$

81. $(x + 3y) - 1(x - 2z)$

82. $(x + y + z) + (2x - y + z)$

Solve.

83. $-9y = -9$

84. $30x = 38$

85. $3(1) + 2z = 9$

86. $4\left(\dfrac{19}{15}\right) - 2y = 4$

Apply the distributive property, then simplify if possible.

87. $2(5x - z)$

88. $-1(x - 2z)$

89. $3(3x + y - 2z)$

90. $2(2x - y + z)$

Extending the Concepts

91. Find a and b so that the line $ax + by = 7$ passes through the points $(1, -2)$ and $(3, 1)$.

92. Find a and b so that the line $ax + by = 2$ passes through the points $(2, 2)$ and $(6, 7)$.

7.6 Systems of Equations in Three Variables

TICKET TO SUCCESS

Keep these questions in mind as you read through the section. Then respond in your own words and in complete sentences.

1. What is an ordered triple of numbers?
2. Explain what it means for (1, 2, 3) to be a solution to a system of linear equations in three variables.
3. Explain in a general way the procedure you would use to solve a system of three linear equations in three variables.
4. How would you know when a system of linear equations in three variables has no single ordered triple for a solution?

Image copyright ©Alexey Stiop, 2010. Used under license from Shutterstock.com

In the previous section we were working with systems of equations in two variables. However, imagine you are in charge of running a concession stand selling sodas, hot dogs, and chips for a fundraiser. The price for each soda is $.50, each hot dog is $2.00, and each bag of chips is $1.00. You have not kept track of your inventory, but you know you have sold a total of 85 items for a total of $100.00. You also know that you sold twce as many sodas as hot dogs. How many sodas, hot dogs, and chips did you sell? We will now begin to solve systems of equations with three unknowns like this one.

A   Solve Systems in Three Variables

A solution to an equation in three variables such as

$$2x + y - 3z = 6$$

is an ordered triple of numbers (x, y, z). For example, the ordered triples $(0, 0, -2)$, $(2, 2, 0)$, and $(0, 9, 1)$ are solutions to the equation $2x + y - 3z = 6$ since they produce a true statement when their coordinates are substituted for x, y, and z in the equation.

Definition

The **solution set** for a system of three linear equations in three variables is the set of ordered triples that satisfy all three equations.

EXAMPLE 1   Solve the system.

$$x + y + z = 6 \quad (1)$$
$$2x - y + z = 3 \quad (2)$$
$$x + 2y - 3z = -4 \quad (3)$$

SOLUTION   We want to find the ordered triple (x, y, z) that satisfies all three equations. We have numbered the equations so it will be easier to keep track of where they are and what we are doing.

There are many ways to proceed. The main idea is to take two different pairs of equations and eliminate the same variable from each pair. We begin by adding equations (1) and (2) to eliminate the y-variable. The resulting equation is numbered (4):

$$x + y + z = 6 \quad (1)$$
$$\underline{2x - y + z = 3} \quad (2)$$
$$3x + 2z = 9 \quad (4)$$

Adding twice equation (2) to equation (3) will also eliminate the variable y. The resulting equation is numbered (5):

$$
\begin{array}{ll}
4x - 2y + 2z = 6 & \text{Twice (2)} \\
\underline{x + 2y - 3z = -4} & \text{(3)} \\
5x - z = 2 & \text{(5)}
\end{array}
$$

Equations (4) and (5) form a linear system in two variables. By multiplying equation (5) by 2 and adding the result to equation (4), we succeed in eliminating the variable z from the new pair of equations:

$$
\begin{array}{ll}
3x + 2z = 9 & \text{(4)} \\
\underline{10x - 2z = 4} & \text{Twice (5)} \\
13x = 13 & \\
x = 1 &
\end{array}
$$

Substituting $x = 1$ into equation (4), we have

$$3(1) + 2z = 9$$

$$2z = 6$$

$$z = 3$$

Using $x = 1$ and $z = 3$ in equation (1) gives us

$$1 + y + 3 = 6$$

$$y + 4 = 6$$

$$y = 2$$

The solution is the ordered triple $(1, 2, 3)$.

EXAMPLE 2 Solve the system.

$$2x + y - z = 3 \qquad (1)$$

$$3x + 4y + z = 6 \qquad (2)$$

$$2x - 3y + z = 1 \qquad (3)$$

SOLUTION It is easiest to eliminate z from the equations using the addition method. The equation produced by adding (1) and (2) is

$$5x + 5y = 9 \qquad (4)$$

The equation that results from adding (1) and (3) is

$$4x - 2y = 4 \qquad (5)$$

Equations (4) and (5) form a linear system in two variables. We can eliminate the variable y from this system as follows:

$$
\begin{array}{lcl}
5x + 5y = 9 & \xrightarrow{\text{Multiply by 2}} & 10x + 10y = 18 \\
4x - 2y = 4 & \xrightarrow[\text{Multiply by 5}]{} & \underline{20x - 10y = 20} \\
& & 30x \qquad\quad = 38
\end{array}
$$

$$x = \frac{38}{30}$$

$$= \frac{19}{15}$$

Substituting $x = \dfrac{19}{15}$ into equation (5) or equation (4) and solving for y gives

$$y = \frac{8}{15}$$

Using $x = \dfrac{19}{15}$ and $y = \dfrac{8}{15}$ in equation (1), (2), or (3) and solving for z results in

$$z = \frac{1}{15}$$

The ordered triple that satisfies all three equations is $\left(\dfrac{19}{15}, \dfrac{8}{15}, \dfrac{1}{15} \right)$.

EXAMPLE 3   Solve the system.

$$
\begin{aligned}
2x + 3y - z &= 5 \quad &(1) \\
4x + 6y - 2z &= 10 \quad &(2) \\
x - 4y + 3z &= 5 \quad &(3)
\end{aligned}
$$

SOLUTION   Multiplying equation (1) by -2 and adding the result to equation (2) looks like this:

$$
\begin{aligned}
-4x - 6y + 2z &= -10 \quad &-2 \text{ times (1)} \\
4x + 6y - 2z &= 10 \quad &(2) \\
\hline
0 &= 0
\end{aligned}
$$

All three variables have been eliminated, and we are left with a true statement. As was the case in the previous section, this implies that the two equations are dependent. With a system of three equations in three variables, however, a system such as this one can have no solution or an infinite number of solutions. In either case, we have no unique solutions, meaning there is no single ordered triple that is the only solution to the system.

EXAMPLE 4   Solve the system.

$$
\begin{aligned}
x - 5y + 4z &= 8 \quad &(1) \\
3x + y - 2z &= 7 \quad &(2) \\
-9x - 3y + 6z &= 5 \quad &(3)
\end{aligned}
$$

SOLUTION   Multiplying equation (2) by 3 and adding the result to equation (3) produces

$$
\begin{aligned}
9x + 3y - 6z &= 21 \quad &3 \text{ times (2)} \\
-9x - 3y + 6z &= 5 \quad &(3) \\
\hline
0 &= 26
\end{aligned}
$$

In this case, all three variables have been eliminated, and we are left with a false statement. The system is inconsistent: there are no ordered triples that satisfy both equations. The solution set for the system is the empty set, $\varnothing$. If equations (2) and (3) have no ordered triples in common, then certainly (1), (2), and (3) do not either.

EXAMPLE 5   Solve the system.

$$
\begin{aligned}
x + 3y &= 5 \quad &(1) \\
6y + z &= 12 \quad &(2) \\
x - 2z &= -10 \quad &(3)
\end{aligned}
$$

SOLUTION   It may be helpful to rewrite the system as

$$x + 3y \qquad = 5 \qquad (1)$$
$$6y + z = 12 \qquad (2)$$
$$x - \qquad 2z = -10 \qquad (3)$$

Equation (2) does not contain the variable x. If we multiply equation (3) by -1 and add the result to equation (1), we will be left with another equation that does not contain the variable x.

$$x + 3y \qquad = 5 \qquad (1)$$
$$\underline{-x \qquad + 2z = 10} \qquad -1 \text{ times (3)}$$
$$3y + 2z = 15 \qquad (4)$$

Equations (2) and (4) form a linear system in two variables. Multiplying equation (2) by -2 and adding the result to equation (4) eliminates the variable z.

$$6y + z = 12 \xrightarrow{\text{Multiply by } -2} -12y - 2z = -24$$
$$3y + 2z = 15 \xrightarrow[\text{No change}]{} \underline{3y + 2z = 15}$$
$$-9y \qquad = -9$$
$$y = 1$$

Using $y = 1$ in equation (4) and solving for z, we have

$$z = 6$$

Substituting $y = 1$ into equation (1) gives

$$x = 2$$

The ordered triple that satisfies all three equations is $(2, 1, 6)$. ∎

The Geometry Behind Equations in Three Variables

We can graph an ordered triple on a coordinate system with three axes. The graph will be a point in space. The coordinate system is drawn in perspective; you have to imagine that the x-axis comes out of the paper and is perpendicular to both the y-axis and the z-axis. To graph the point $(3, 4, 5)$, we move 3 units in the x-direction, 4 units in the y-direction, and then 5 units in the z-direction, as shown in Figure 1.

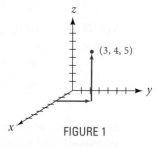

FIGURE 1

Although in actual practice, it is sometimes difficult to graph equations in three variables, if we have to graph a linear equation in three variables, we would find that the graph was a plane in space. A system of three equations in three variables is represented by three planes in space.

There are a number of possible ways in which these three planes can intersect, some of which are shown in the margin on this page. There are still other possibilities that are not among those shown in the margin.

In Example 3, we found that equations 1 and 2 were dependent equations. They represent the same plane; that is, they have all their points in common. But the

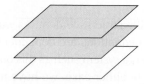

Case 1   The three planes have exactly one point in common. In this case we get one solution to our system, as in Examples 1, 2, and 5.

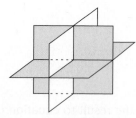

Case 2   The three planes have no points in common because they are all parallel to each other. The system they represent is an inconsistent system.

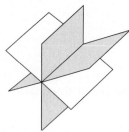

Case 3   The three planes intersect in a line. Any point on the line is a solution to the system.

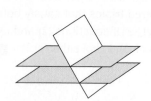

Case 4   In this case the three planes have no points in common. There is no solution to the system; it is an inconsistent system.

system of equations that they came from has either no solution or an infinite number of solutions. It all depends on the third plane. If the third plane coincides with the first two, then the solution to the system is a plane. If the third plane is distinct from but parallel to the first two, then there is no solution to the system. And, finally, if the third plane intersects the first two, but does not coincide with them, then the solution to the system is that line of intersection.

In Example 4, we found that trying to eliminate a variable from the second and third equations resulted in a false statement. This means that the two planes represented by these equations are parallel. It makes no difference where the third plane is; there is no solution to the system in Example 4. (If we were to graph the three planes from Example 4, we would obtain a diagram similar to Case 4 in the margin.)

If, in the process of solving a system of linear equations in three variables, we eliminate all the variables from a pair of equations and are left with a false statement, we will say the system is inconsistent. If we eliminate all the variables and are left with a true statement, then we will say the system has no unique solution.

Problem Set 7.6

Moving Toward Success

"Even a mistake may turn out to be the one thing necessary to a worthwhile achievement."

—Henry Ford, 1863–1947, American industrialist and founder of Ford Motor Company

1. Why is making mistakes important to the process of learning mathematics?

2. What will you do if you notice you are repeatedly making mistakes on certain types of problems?

A Solve the following systems. [Examples 1–5]

1. $x + y + z = 4$
   $x - y + 2z = 1$
   $x - y - 3z = -4$

2. $x - y - 2z = -1$
   $x + y + z = 6$
   $x + y - z = 4$

3. $x + y + z = 6$
   $x - y + 2z = 7$
   $2x - y - 4z = -9$

4. $x + y + z = 0$
   $x + y - z = 6$
   $x - y + 2z = -7$

5. $x + 2y + z = 3$
   $2x - y + 2z = 6$
   $3x + y - z = 5$

6. $2x + y - 3z = -14$
   $x - 3y + 4z = 22$
   $3x + 2y + z = 0$

7. $2x + 3y - 2z = 4$
   $x + 3y - 3z = 4$
   $3x - 6y + z = -3$

8. $4x + y - 2z = 0$
   $2x - 3y + 3z = 9$
   $-6x - 2y + z = 0$

9. $-x + 4y - 3z = 2$
   $2x - 8y + 6z = 1$
   $3x - y + z = 0$

10. $4x + 6y - 8z = 1$
    $-6x - 9y + 12z = 0$
    $x - 2y - 2z = 3$

11. $\frac{1}{2}x - y + z = 0$
    $2x + \frac{1}{3}y + z = 2$
    $x + y + z = -4$

12. $\frac{1}{3}x + \frac{1}{2}y + z = -1$
    $x - y + \frac{1}{5}z = 1$
    $x + y + z = 5$

13. $2x - y - 3z = 1$
    $x + 2y + 4z = 3$
    $4x - 2y - 6z = 2$

14. $3x + 2y + z = 3$
    $x - 3y + z = 4$
    $-6x - 4y - 2z = 1$

15. $2x - y + 3z = 4$
    $x + 2y - z = -3$
    $4x + 3y + 2z = -5$

16. $6x - 2y + z = 5$
    $3x + y + 3z = 7$
    $x + 4y - z = 4$

17. $x + y = 9$
    $y + z = 7$
    $x - z = 2$

18. $x - y = -3$
    $x + z = 2$
    $y - z = 7$

19. $2x + y = 2$
    $y + z = 3$
    $4x - z = 0$

20. $2x + y = 6$
    $3y - 2z = -8$
    $x + z = 5$

21. $2x - 3y = 0$
$6y - 4z = 1$
$x + 2z = 1$

22. $3x + 2y = 3$
$y + 2z = 2$
$6x - 4z = 1$

23. $x + y - z = 2$
$2x + y + 3z = 4$
$x - 2y + 2z = 6$

24. $x + 2y - 2z = 4$
$3x + 4y - z = -2$
$2x + 3y - 3z = -5$

25. $2x + 3y = -\dfrac{1}{2}$
$4x + 8z = 2$
$3y + 2z = -\dfrac{3}{4}$

26. $3x - 5y = 2$
$4x + 6z = \dfrac{1}{3}$
$5y - 7z = \dfrac{1}{6}$

27. $\dfrac{1}{3}x + \dfrac{1}{2}y - \dfrac{1}{6}z = 4$
$\dfrac{1}{4}x - \dfrac{3}{4}y + \dfrac{1}{2}z = \dfrac{3}{2}$
$\dfrac{1}{2}x - \dfrac{2}{3}y - \dfrac{1}{4}z = -\dfrac{16}{3}$

28. $-\dfrac{1}{4}x + \dfrac{3}{8}y + \dfrac{1}{2}z = -1$
$\dfrac{2}{3}x - \dfrac{1}{6}y - \dfrac{1}{2}z = 2$
$\dfrac{3}{4}x - \dfrac{1}{2}y - \dfrac{1}{8}z = 1$

29. $x - \dfrac{1}{2}y - \dfrac{1}{3}z = -\dfrac{4}{3}$
$\dfrac{1}{3}x + y - \dfrac{1}{2}z = 5$
$-\dfrac{1}{4}x + \dfrac{2}{3}y - z = -\dfrac{3}{4}$

30. $x + \dfrac{1}{3}y - \dfrac{1}{2}z = -\dfrac{3}{2}$
$\dfrac{1}{2}x - y + \dfrac{1}{3}z = 8$
$\dfrac{1}{3}x - \dfrac{1}{4}y - z = -\dfrac{5}{6}$

31. $\dfrac{1}{2}x + \dfrac{2}{3}y = \dfrac{5}{2}$
$\dfrac{1}{5}x - \dfrac{1}{2}z = -\dfrac{3}{10}$
$\dfrac{1}{3}y - \dfrac{1}{4}z = \dfrac{3}{4}$

32. $\dfrac{1}{2}x - \dfrac{1}{3}y = \dfrac{1}{6}$
$\dfrac{1}{3}y - \dfrac{1}{3}z = 1$
$\dfrac{1}{5}x - \dfrac{1}{2}z = -\dfrac{4}{5}$

33. $\dfrac{1}{2}x - \dfrac{1}{4}y + \dfrac{1}{2}z = -2$
$\dfrac{1}{4}x - \dfrac{1}{12}y - \dfrac{1}{3}z = \dfrac{1}{4}$
$\dfrac{1}{6}x + \dfrac{1}{3}y - \dfrac{1}{2}z = \dfrac{3}{2}$

34. $\dfrac{1}{2}x + \dfrac{1}{2}y + z = \dfrac{1}{2}$
$\dfrac{1}{2}x - \dfrac{1}{4}y - \dfrac{1}{4}z = 0$
$\dfrac{1}{4}x + \dfrac{1}{12}y + \dfrac{1}{6}z = \dfrac{1}{6}$

Applying the Concepts

35. Electric Current In the following diagram of an electrical circuit, x, y, and z represent the amount of current (in amperes) flowing across the 5-ohm, 20-ohm, and 10-ohm resistors, respectively. (In circuit diagrams resistors are represented by ─⩗⩗⩗─ and potential differences by ─┤├─.)

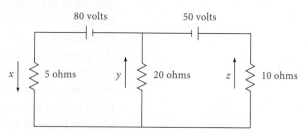

The system of equations used to find the three currents x, y, and z is

$$x - y - z = 0$$
$$5x + 20y = 80$$
$$20y - 10z = 50$$

Solve the system for all variables.

36. Electric Current In the following diagram of an electrical circuit, x, y, and z represent the amount of current (in amperes) flowing across the 10-ohm, 15-ohm, and 5-ohm resistors, respectively.

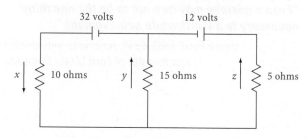

The system of equations used to find the three currents x, y, and z is

$$x - y - z = 0$$
$$10x + 15y = 32$$
$$15y - 5z = 12$$

Solve the system for all variables.

Getting Ready for the Next Section

Simplify.

37. $1(4) - 3(2)$

38. $3(7) - (-2)(5)$

39. $1(1) - 3(-2) + (-2)(-2)$

40. $-4(0)(-2) - (-1)(1)(1) - 1(2)(3)$

41. $-3(-1 - 1) + 4(-2 + 2) - 5[2 - (-2)]$

42. $12 + 4 - (-1) - 6$

Solve.

43. $-5x = 20$

44. $4x - 2x = 8$

Maintaining Your Skills

Solve each system of equations by the elimination method.

45. $2x + y = 3$
$3x - y = 7$

46. $3x - y = -6$
$4x + y = -8$

47. $4x - 5y = 1$
$x - 2y = -2$

48. $6x - 4y = 2$
$2x + y = 10$

Solve by the substitution method.

49. $5x + 2y = 7$
$y = 3x - 2$

50. $-7x - 5y = -1$
$y = x + 5$

Extending the Concepts

Solve each system for the solution (x, y, z, w).

53.
$x + y + z + w = 10$
$x + 2y - z + w = 6$
$x - y - z + 2w = 4$
$x - 2y + z - 3w = -12$

54.
$x + y + z + w = 16$
$x - y + 2z - w = 1$
$x + 3y - z - w = -2$
$x - 3y - 2z + 2w = -4$

7.7 Linear Inequalities in Two Variables

OBJECTIVES

A Graph linear inequalities in two variables.

TICKET TO SUCCESS

Keep these questions in mind as you read through the section. Then respond in your own words and in complete sentences.

1. When graphing a linear inequality in two variables, how do you find the equation of the boundary line?
2. What is the significance of a broken line in the graph of an inequality?
3. When graphing a linear inequality in two variables, how do you know which side of the boundary line to shade?
4. Does the graph of $x + y < 4$ include the boundary line? Explain.

A small movie theater holds 100 people. The owner charges more for adults than for children, so it is important to know the different combinations of adults and children that can be seated at one time. The shaded region in Figure 1 contains all the seating combinations. The line $x + y = 100$ shows the combinations for a full theater: The *y*-intercept corresponds to a theater full of adults, and the *x*-intercept corresponds to a theater full of children. In the shaded region below the line $x + y = 100$ are the combinations that occur if the theater is not full.

Shaded regions like the one shown in Figure 1 are produced by linear inequalities in two variables, which is the topic of this section.

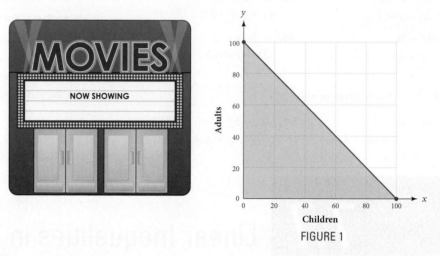

FIGURE 1

A Graphing Inequalities

A *linear inequality in two variables* is any expression that can be put in the form

$$ax + by < c$$

where a, b, and c are real numbers (a and b not both 0). The inequality symbol can be any one of the following four: $<, \leq, >, \geq$.

Some examples of linear inequalities are

$$2x + 3y < 6 \qquad y \geq 2x + 1 \qquad x - y \leq 0$$

Although not all of these examples have the form $ax + by < c$, each one can be put in that form.

The solution set for a linear inequality is a *section of the coordinate plane*. The *boundary* for the section is found by replacing the inequality symbol with an equal sign and graphing the resulting equation. The boundary is included in the solution set (and represented with a *solid line*) if the inequality symbol originally used is $\leq$ or $\geq$. The boundary is not included (and is represented with a *broken line*) if the original symbol is $<$ or $>$.

EXAMPLE 1 Graph the solution set for $x + y \leq 4$.

SOLUTION The boundary for the graph is the graph of $x + y = 4$. The boundary is included in the solution set because the inequality symbol is $\leq$.

Figure 2 is the graph of the boundary:

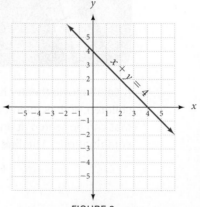

FIGURE 2

The boundary separates the coordinate plane into two regions: the region above the boundary and the region below it. The solution set for $x + y \le 4$ is one of these two regions along with the boundary. To find the correct region, we simply choose any convenient point that is *not* on the boundary. Then we substitute the coordinates of the point into the original inequality, $x + y \le 4$. If the point we choose satisfies the inequality, then it is a member of the solution set, and we can assume that all points on the same side of the boundary as the chosen point are also in the solution set. If the coordinates of our point do not satisfy the original inequality, then the solution set lies on the other side of the boundary.

In this example, a convenient point that is not on the boundary is the origin.

$$\text{Substituting} \rightarrow \quad (0, 0)$$
$$\text{into} \rightarrow \quad x + y \le 4$$
$$\text{gives us} \rightarrow \quad 0 + 0 \le 4$$
$$0 \le 4 \quad \text{A true statement}$$

Because the origin is a solution to the inequality $x + y \le 4$ and the origin is below the boundary, all other points below the boundary are also solutions.

Figure 3 is the graph of $x + y \le 4$.

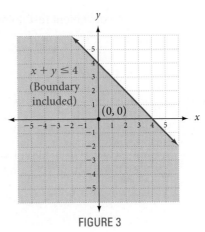

FIGURE 3

The region above the boundary is described by the inequality $x + y > 4$.

Here is a list of steps to follow when graphing the solution set for a linear inequality in two variables.

Strategy   To Graph a Linear Inequalitiy in Two Variables

Step 1: Replace the inequality symbol with an equal sign. The resulting equation represents the boundary for the solution set.

Step 2: Graph the boundary found in step 1 using a solid line if the boundary is included in the solution set (that is, if the original inequality symbol was either $\le$ or $\ge$). Use a broken line to graph the boundary if it is not included in the solution set. (It is not included if the original inequality was either $<$ or $>$.)

Step 3: Choose any convenient point not on the boundary and substitute the coordinates into the original inequality. If the resulting statement is true, the graph lies on the same side of the boundary as the chosen point. If the resulting statement is false, the solution set lies on the opposite side of the boundary.

EXAMPLE 2 Graph the solution set for $y < 2x - 3$.

SOLUTION The boundary is the graph of $y = 2x - 3$, a line with slope 2 and y-intercept -3. The boundary is not included because the original inequality symbol is $<$. We therefore use a broken line to represent the boundary in Figure 4.

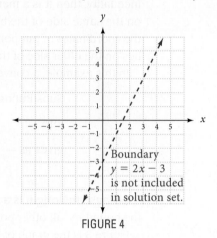

Boundary
$y = 2x - 3$
is not included
in solution set.

FIGURE 4

A convenient test point is again the origin.

Using → (0, 0)

in → $y < 2x - 3$

we have → $0 < 2(0) - 3$

$0 < -3$ A false statement

Because our test point gives us a false statement and it lies above the boundary, the solution set must lie on the other side of the boundary (Figure 5).

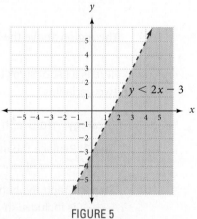

$y < 2x - 3$

FIGURE 5

USING TECHNOLOGY

Graphing Calculators

Most graphing calculators have a Shade command that allows a portion of a graphing screen to be shaded. With this command we can visualize the solution sets to linear inequalities in two variables. Because most graphing calculators cannot draw a dotted line, however, we are not actually "graphing" the solution set, only visualizing it.

Strategy	Visualizing a Linear Inequality in Two Variables on a Graphing Calculator

Step 1: Solve the inequality for y.

Step 2: Replace the inequality symbol with an equal sign. The resulting equation represents the boundary for the solution set.

Step 3: Graph the equation in an appropriate viewing window.

Step 4: Use the Shade command to indicate the solution set:

For inequalities having the $<$ or $\leq$ sign, use Shade(Ymin, Y_1).

For inequalities having the $>$ or $\geq$ sign, use Shade(Y_1, Ymax).

Note: On the TI-83/84, step 4 can be done by manipulating the icons in the left column in the list of Y variables.

Figures 6 and 7 show the graphing calculator screens that help us visualize the solution set to the inequality $y < 2x - 3$ that we graphed in Example 2.

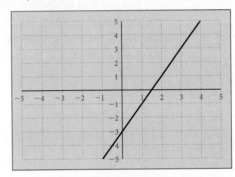

FIGURE 6 $Y_1 = 2X - 3$

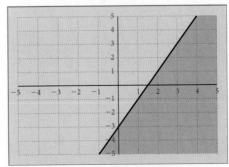

FIGURE 7   Shade (Xmin, Y_1)

Windows: X from -5 to 5, Y from -5 to 5

EXAMPLE 3 Graph the solution set for $x \le 5$.

SOLUTION The boundary is $x = 5$, which is a vertical line. All points in Figure 8 to the left of the boundary have x-coordinates less than 5 and all points to the right have x-coordinates greater than 5.

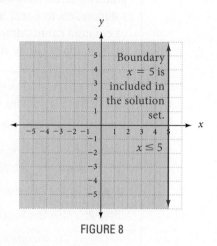

FIGURE 8

EXAMPLE 4 Graph the solution set for $y > \frac{1}{4}x$.

SOLUTION The boundary is the line $y = \frac{1}{4}x$, which has a slope of $\frac{1}{4}$ and passes through the origin. The graph of the boundary line is shown in Figure 9. Since the boundary passes through the origin, we cannot use the origin as our test point. Remember, the test point cannot be on the boundary line. Let's use the point $(0, -4)$ as our test point. It lies below the boundary line. When we substitute the coordinates into our original inequality, the result is the false statement $-4 > 0$. This tells us that the solution set is on the other side of the boundary line. The solution set for our original inequality is shown in Figure 10.

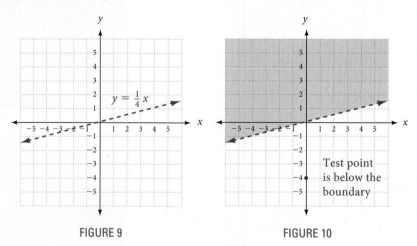

FIGURE 9   **FIGURE 10**

Problem Set 7.7

Moving Toward Success

"If there were dreams to sell, what would you buy?"

—Thomas Lovell Beddoes, 1803–1849, British poet

1. Why are objectives and goals important to set for your study group?

2. Why is it important to come prepared to a study group session?

A Graph the solution set for each of the following. [Examples 1–4]

1. $x + y < 5$
2. $x + y \leq 5$
3. $x - y \geq -3$
4. $x - y > -3$
5. $2x + 3y < 6$
6. $2x - 3y > -6$
7. $-x + 2y > -4$
8. $-x - 2y < 4$
9. $2x + y < 5$
10. $2x + y < -5$
11. $y < 2x - 1$
12. $y \leq 2x - 1$
13. $3x - 4y < 12$
14. $-2x + 3y < 6$
15. $-5x + 2y \leq 10$
16. $4x - 2y \leq 8$
17. $x \geq 3$
18. $x > -2$
19. $y \leq 4$
20. $y > -5$
21. $y < 2x$
22. $y > -3x$
23. $y \geq \frac{1}{2}x$
24. $y \leq \frac{1}{3}x$
25. $y \geq \frac{3}{4}x - 2$
26. $\frac{x}{3} + \frac{y}{2} > 1$
27. $y > -\frac{2}{3}x + 3$
28. $\frac{x}{5} + \frac{y}{4} < 1$
29. $\frac{x}{3} - \frac{y}{2} > 1$
30. $-\frac{x}{4} - \frac{y}{3} > 1$
31. $y \leq -\frac{2}{3}x$
32. $y \geq \frac{1}{4}x$
33. $5x - 3y < 0$
34. $2x + 3y < 0$
35. $\frac{x}{4} + \frac{y}{5} \leq 1$
36. $\frac{x}{2} + \frac{y}{3} < 1$

Fill in the blank with a $<$, $>$, $\leq$, or $\geq$ so that the linear inequality represents the shaded region of the graph.

37. 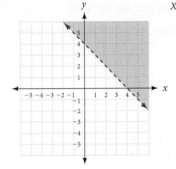 $x + y \_\_ 4$

38. 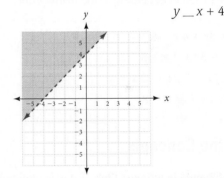 $y \_\_ x + 4$

39. 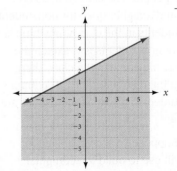 $-x + 2y \_\_ 4$

40.

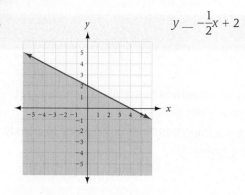

$$y \_ -\frac{1}{2}x + 2$$

The next two problems are intended to give you practice reading, and paying attention to, the instructions that accompany the problems you are working.

41. Work each problem according to the instructions given.

   a. Solve: $\frac{1}{3} + \frac{y}{2} < 1$

   b. Solve: $\frac{1}{3} - \frac{y}{2} < 1$

   c. Solve for y: $\frac{x}{3} + \frac{y}{2} = 1$

   d. Graph: $y < -\frac{4}{3}x + 4$

42. Work each problem according to the instructions given.

   a. Solve: $3x + 4 \geq -8$

   b. Solve: $-3x + 4 \geq -8$

   c. Solve for y: $-3x + 4y = -8$

   d. Graph: $3x + 4y \geq -8$

Applying the Concepts

43. Number of People in a Dance Club A dance club holds a maximum of 200 people. The club charges one price for students and a higher price for nonstudents. If the number of students in the club at any time is x and the number of nonstudents is y, sketch the graph of the number of people in the club. Then, shade the region in the first quadrant that contains all combinations of students and nonstudents that are in the club at any time.

44. Many Perimeters Suppose you have 500 feet of fencing that you will use to build a rectangular livestock pen. Let x represent the length of the pen and y represent the width. Sketch a graph of the situation and shade the region in the first quadrant that contains all possible values of x and y that will give you a rectangle from 500 feet of fencing. (You don't have to use all of the fencing, so the perimeter of the pen could be less than 500 feet.)

45. Gas Mileage You have two cars. The first car travels an average of 12 miles on a gallon of gasoline, and the second averages 22 miles per gallon. Suppose you can afford to buy up to 30 gallons of gasoline this month. If the first car is driven x miles this month, and the second car is driven y miles this month, draw a graph and shade the region in the first quadrant that gives all the possible values of x and y that will keep you from buying more than 30 gallons of gasoline this month.

46. Student Loan Payments When considering how much debt to incur in student loans, it is advisable to keep your student loan payment after graduation to 8% or less of your starting monthly income. Let x represent your starting monthly salary and let y represent your monthly student loan payment. Graph the inequality and shade the region in the first quadrant that is a solution to your inequality.

47. Solar and Wind Energy The chart shows the cost to install either solar panels or a wind turbine. A company that specializes in alternate energy sources is placing an order for more cables for both wind and solar energy solutions. If the total bill is some amount less than $11,440, graph the inequality and shade the region in the first quadrant that contains all the possible combinations of products.

Solar Versus Wind Energy Costs

Solar Energy Equipment Cost:		Wind Energy Equipment Cost:	
Modules	$6200	Turbine	$3300
Fixed Rack	$1570	Tower	$3000
Charge Controller	$971	Cable	$715
Cable	$440		
TOTAL	**$9181**	**TOTAL**	**$7015**

Source: a Limited 2006

48. Gas Prices The chart shows the breakdown of gas prices. CJ has two vehicles, a gasoline car and a diesel truck. Suppose that the price of gas and the price of diesel are both $4.20 per gallon. Shade the region in the first quadrant that contains all the possible combinations of money he paid for just the crude oil if the amount of gasoline and diesel sum to less than 15 gallons. If necessary, round to the nearest cent.

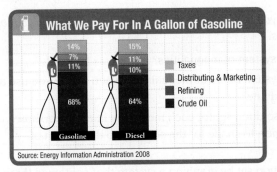

What We Pay For In A Gallon of Gasoline

Gasoline: 14%, 7%, 11%, 68%
Diesel: 15%, 11%, 10%, 64%

Taxes
Distributing & Marketing
Refining
Crude Oil

Source: Energy Information Administration 2008

Maintaining Your Skills

49. $\frac{1}{3} + \frac{y}{5} \leq \frac{26}{15}$

50. $-\frac{1}{3} \geq \frac{1}{6} - \frac{y}{2}$

51. $5t - 4 > 3t - 8$

52. $-3(t - 2) < 6 - 5(t + 1)$

53. $-9 < -4 + 5t < 6$

54. $-3 < 2t + 1 < 3$

Getting Ready for the Next Section

Complete each table using the given equation.

55. $y = 7.5x$

x	y
0	
10	
20	

56. $h = 32t - 16t^2$

t	h
0	
$\frac{1}{2}$	
1	

57. $x = y^2$

x	y
	0
	1
	−1

58. $y = |x|$

x	y
−3	
0	
3	

Extending the Concepts

Graph each inequality.

59. $y < |x + 2|$

60. $y > |x - 2|$

61. $y > |x - 3|$

62. $y < |x + 3|$

63. $y \leq |x - 1|$

64. $y \geq |x + 1|$

65. $y < x^2$

66. $y > x^2 - 2$

67. The Associated Students organization holds a *Night at the Movies* fundraiser. Student tickets are $1.00 each and nonstudent tickets are $2.00 each. The theater holds a maximum of 200 people. The club needs to collect at least $100 to make money. Sketch a graph of the inequality and shade the region in the first quadrant that contains all combinations of students and nonstudents that could attend the movie night so the club makes money.

OBJECTIVES

A Graph the solution to a system of linear inequalities in two variables.

TICKET TO SUCCESS

Keep these questions in mind as you read through the section. Then respond in your own words and in complete sentences.

1. What is the solution set to a system of inequalities?

2. For the boundary lines of a system of linear inequalities, when would you use a dotted line rather than a solid line?

3. Once you have graphed the solution set for each inequality in a system, how would you determine the region to shade for the solution to the system of inequalities?

4. How would you find the solution set by graphing a system that contained three linear inequalities?

Imagine you work at a local coffee house and your boss wants you to create a new house blend using the three most popular coffees. The dark roast coffee you are to use costs $20.00 per pound, the medium roast costs $14.50 per pound, and the flavored coffee costs $16.00 per pound. Your boss wants the cost of the new blend to be at most $17.00 per pound. She will want to see possible combinations of the coffee to keep the price at most $17.00. As discussed previously, words like "at most" imply the use of inequalities. Working through this section will help us visualize solutions to inequalities in two and three variables.

A Graphing Solutions to Linear Inequalities

In the previous chapter, we graphed linear inequalities in two variables. To review, we graph the boundary line using a solid line if the boundary is part of the solution set and a broken line if the boundary is not part of the solution set. Then we test any point that is not on the boundary line in the original inequality. A true statement tells us that the point lies in the solution set; a false statement tells us the solution set is the other region.

Figure 1 shows the graph of the inequality $x + y < 4$. Note that the boundary is not included in the solution set and is therefore drawn with a broken line. Figure 2 shows the graph of $-x + y \leq 3$. Note that the boundary is drawn with a solid line because it is part of the solution set.

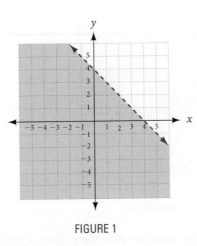

FIGURE 1

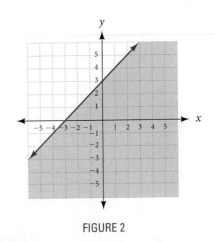

FIGURE 2

If we form a system of inequalities with the two inequalities, the solution set will be all the points common to both solution sets shown in the two figures above; it is the intersection of the two solution sets. Therefore, the solution set for the system of inequalities

$$x + y < 4$$

$$-x + y \leq 3$$

is all the ordered pairs that satisfy both inequalities. It is the set of points that are below the line $x + y = 4$ and also below (and including) the line $-x + y = 3$. The graph of the solution set to this system is shown in Figure 3. We have written the system in Figure 3 with the word *and* just to remind you that the solution set to a system of equations or inequalities is all the points that satisfy both equations or inequalities.

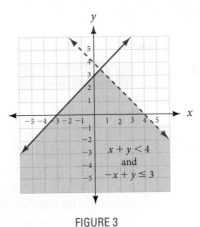

$$x + y < 4$$
$$\text{and}$$
$$-x + y \leq 3$$

FIGURE 3

EXAMPLE 1   Graph the solution to the system of linear inequalities.

$$y < \frac{1}{2}x + 3$$

$$y \geq \frac{1}{2}x - 2$$

SOLUTION   Figures 4 and 5 show the solution set for each of the inequalities separately.

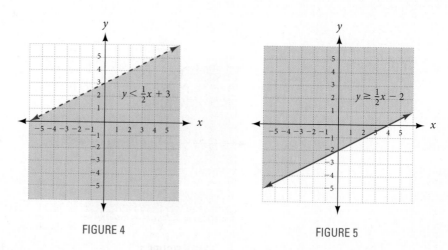

FIGURE 4   FIGURE 5

Figure 6 is the solution set to the system of inequalities. It is the region consisting of points whose coordinates satisfy both inequalities

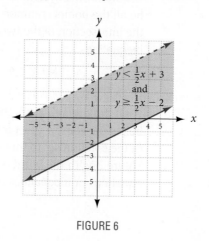

FIGURE 6

EXAMPLE 2   Graph the solution to the system of linear inequalities.

$$x + y < 4$$
$$x \geq 0$$
$$y \geq 0$$

SOLUTION   We graphed the first inequality, $x + y < 4$, in Figure 1 at the beginning of this section. The solution set to the inequality $x \geq 0$, shown in Figure 7, is all the points to the right of the y-axis; that is, all the points with x-coordinates that are greater than or equal to 0. Figure 8 shows the graph of $y \geq 0$. It consists of all points with y-coordinates greater than or equal to 0; that is, all points from the x-axis up.

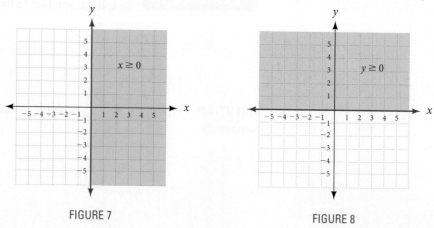

FIGURE 7   FIGURE 8

The regions shown in Figures 7 and 8 overlap in the first quadrant. Therefore, putting all three regions together we have the points in the first quadrant that are below the line $x + y = 4$. This region is shown in Figure 9, and it is the solution to our system of inequalities.

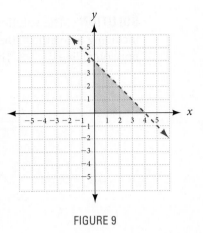

FIGURE 9

Extending the discussion in Example 2, we can name the points in each of the four quadrants using systems of inequalities.

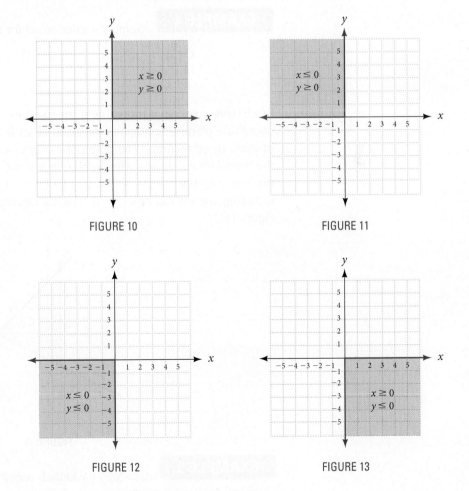

FIGURE 10   FIGURE 11

FIGURE 12   FIGURE 13

EXAMPLE 3 Graph the solution to the system of linear inequalities.

$$x \le 4$$
$$y \ge -3$$

SOLUTION The solution to this system will consist of all points to the left of and including the vertical line $x = 4$ that intersect with all points above and including the horizontal line $y = -3$. The solution set is shown in Figure 14.

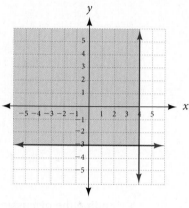

FIGURE 14

EXAMPLE 4 Graph the solution set for the following system.

$$x - 2y \le 4$$
$$x + y \le 4$$
$$x \ge -1$$

SOLUTION We have three linear inequalities, representing three sections of the coordinate plane. The graph of the solution set for this system will be the intersection of these three sections. The graph of $x - 2y \le 4$ is the section above and including the boundary $x - 2y = 4$. The graph of $x + y \le 4$ is the section below and including the boundary line $x + y = 4$. The graph of $x \ge -1$ is all the points to the right of, and including, the vertical line $x = -1$. The intersection of these three graphs is shown in Figure 15.

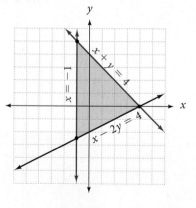

FIGURE 15

EXAMPLE 5 A college basketball arena plans on charging $20 for certain seats and $15 for others. They want to bring in more than $18,000 from all ticket sales and they have reserved at least 500 tickets at the $15 rate. Find a system of inequalities describing all possibilities and sketch the graph. If 620 tickets are sold for $15, at least how many tickets are sold for $20?

SOLUTION   Let x = the number of $20 tickets and y = the number of $15 tickets. We need to write a list of inequalities that describe this situation. That list will form our system of inequalities. First of all, we note that we cannot use negative numbers for either x or y. So, we have our first inequalities:

$$x \geq 0$$
$$y \geq 0$$

Next, we note that they are selling at least 500 tickets for $15, so we can replace our second inequality with $y \geq 500$. Now our system is

$$x \geq 0$$
$$y \geq 500$$

Now, the amount of money brought in by selling $20 tickets is $20x$, and the amount of money brought in by selling $15 tickets is $15y$. If the total income from ticket sales is to be more than $18,000, then $20x + 15y$ must be greater than 18,000. This gives us our last inequality and completes our system.

$$20x + 15y > 18,000$$
$$x \geq 0$$
$$y \geq 500$$

We have used all the information in the problem to arrive at this system of inequalities. The solution set contains all the values of x and y that satisfy all the conditions given in the problem. Here is the graph of the solution set.

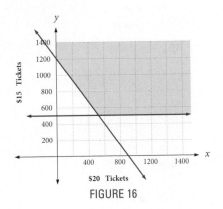

FIGURE 16

If 620 tickets are sold for \$15, then we substitute 620 for y in our first inequality to obtain

$$20x + 15(620) > 18,000 \qquad \text{Substitute 620 for } y$$

$$20x + 9,300 > 18,000 \qquad \text{Multiply}$$

$$20x > 8,700 \qquad \text{Add } -9,300 \text{ to each side}$$

$$x > 435 \qquad \text{Divide each side by 20}$$

If they sell 620 tickets for \$15 each, then they need to sell more than 435 tickets at \$20 each to bring in more than \$18,000.

Problem Set 7.8

Moving Toward Success

"There is no comparison between that which is lost by not succeeding and that which is lost by not trying."

—Francis Bacon, 1561–1626, English philosopher and statesman

1. Do you take notes in class? Why or why not?
2. Research some note taking techniques online. What are some tips that you should employ when taking notes for this class?

A Graph the solution set for each system of linear inequalities. [Examples 1–4]

1. $x + y < 5$
   $2x - y > 4$

2. $x + y < 5$
   $2x - y < 4$

3. $y < \frac{1}{3}x + 4$
   $y \geq \frac{1}{3}x - 3$

4. $y < 2x + 4$
   $y \geq 2x - 3$

5. $x \geq -3$
   $y < 2$

6. $x \leq 4$
   $y \geq -2$

7. $1 \leq x \leq 3$
   $2 \leq y \leq 4$

8. $-4 \leq x \leq -2$
   $1 \leq y \leq 3$

9. $x + y \leq 4$
   $x \geq 0$
   $y \geq 0$

10. $x - y \leq 2$
    $x \geq 0$
    $y \leq 0$

11. $x + y \leq 3$
    $x - 3y \leq 3$
    $x \geq -2$

12. $x - y \leq 4$
    $x + 2y \leq 4$
    $x \geq -1$

13. $x + y \leq 2$
    $-x + y \leq 2$
    $y \geq -2$

14. $x - y \leq 3$
    $-x - y \leq 3$
    $y \leq -1$

15. $x + y < 5$
    $y > x$
    $y \geq 0$

16. $x + y < 5$
    $y > x$
    $x \geq 0$

17. $2x + 3y \leq 6$
    $x \geq 0$
    $y \geq 0$

18. $x + 2y \leq 10$
    $3x + y \leq 12$
    $x \geq 0$
    $y \geq 0$

Fill in the blanks so that the system of inequalities describes the shaded region.

19.

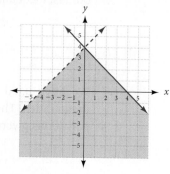

$x + y \underline{} 4$
$-x + y \underline{} 4$

FIGURE 17

20.

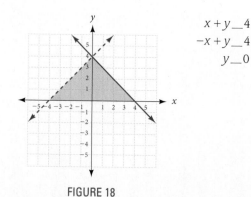

$x + y \underline{} 4$
$-x + y \underline{} 4$
$y \underline{} 0$

FIGURE 18

21.

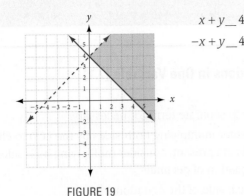

$x + y \_\_ 4$

$-x + y \_\_ 4$

FIGURE 19

22.

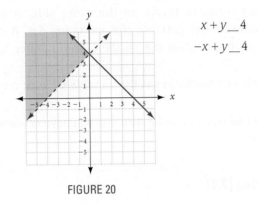

$x + y \_\_ 4$

$-x + y \_\_ 4$

FIGURE 20

Applying the Concepts [Example 5]

23. Office Supplies An office worker wants to purchase some $0.55 postage stamps and also some $0.65 postage stamps totaling no more than $40. He also wants to have at least twice as many $0.55 stamps and more than 15 of the $0.55 stamps.

  a. Find a system of inequalities describing all the possibilities and sketch the graph.

  b. If he purchases 20 of the $0.55 stamps, what is the maximum number of $0.65 stamps he can purchase?

24. Inventory A store sells two brands of DVD players. Customer demand indicates that it is necessary to stock at least twice as many DVD players of brand A as of brand B. At least 30 of brand A and 15 of brand B must be on hand. There is room for not more than 100 DVD players in the store.

  a. Find a system of inequalities describing all possibilities, then sketch the graph.

  b. If there are 35 DVD players of brand A, what is the maximum number of brand B DVD players on hand?

Maintaining Your Skills

For each of the following straight lines, identify the x-intercept, y-intercept, and slope, and sketch the graph.

25. $2x + y = 6$

26. $y = \dfrac{3}{2}x + 4$

27. $x = -2$

28. Graph the equation of the line through $(-1, 3)$ that has slope $m = 2$.

29. Give the slope of the line through $(-3, 2)$ and $(4, -1)$.

30. Line l has a graph parallel to the graph of $2x - 5y = 10$. Find the slope for l.

31. Give the equation of the vertical line through $(4, -7)$.

Simplify.

32. $2x + 3(2x + 2) - 7$

33. $4x - 2(2x - 4) - 9$

34. $4^2 - 3(2 - 3x) - 4x$

35. $2^3 - (5x - 7) - 12$

Solve.

36. If you have 3 times as many dimes than quarters and you have $3.30 in dimes and quarters, how many of each coin do you have?

37. The sum of a number and twice the difference of that number and six is 15. Find the two numbers.

■ Strategy for Solving Linear Equations in One Variable [7.1]

EXAMPLES

1. Solve: $3(2x - 1) = 9$.

$$3(2x - 1) = 9$$
$$6x - 3 = 9$$
$$6x - 3 + \mathbf{3} = 9 + \mathbf{3}$$
$$6x = 12$$
$$\frac{1}{6}(6x) = \frac{1}{6}(12)$$
$$x = 2$$

Step 1:   **a.**   Use the distributive property to separate terms, if necessary.

         b.   If fractions are present, consider multiplying both sides by the LCD to eliminate the fractions. If decimals are present, consider multiplying both sides by a power of 10 to clear the equation of decimals.

         c.   Combine similar terms on each side of the equation.

Step 2:   Use the addition property of equality to get all variable terms on one side of the equation and all constant terms on the other side. A variable term is a term that contains the variable (for example, $5x$). A constant term is a term that does not contain the variable (the number 3, for example).

Step 3:   Use the multiplication property of equality to get the variable by itself on one side of the equation.

Step 4:   Check your solution in the original equation to be sure that you have not made a mistake in the solution process.

■ To Solve an Equation by Factoring [7.1]

2. Solve $x^2 - 5x = -6$.

$$x^2 - 5x + 6 = 0$$
$$(x - 3)(x - 2) = 0$$
$$x - 3 = 0 \quad \text{or} \quad x - 2 = 0$$
$$x = 3 \quad \text{or} \quad x = 2$$

Step 1:   Write the equation in standard form.

Step 2:   Factor the left side.

Step 3:   Use the zero-factor property to set each factor equal to zero.

Step 4:   Solve the resulting linear equations.

■ Absolute Value Equations [7.2]

3. To solve

$$|2x - 1| + 2 = 7$$

we first isolate the absolute value on the left side by adding -2 to each side to obtain

$$|2x - 1| = 5$$
$$2x - 1 = 5 \quad \text{or} \quad 2x - 1 = -5$$
$$2x = 6 \quad \text{or} \quad 2x = -4$$
$$x = 3 \quad \text{or} \quad x = -2$$

To solve an equation that involves absolute value, we isolate the absolute value on one side of the equation and then rewrite the absolute value equation as two separate equations that do not involve absolute value. In general, if b is a positive real number, then

$$|a| = b \quad \text{is equivalent to} \quad a = b \quad \text{or} \quad a = -b$$

■ Addition Property for Inequalities [7.3]

4. Adding 5 to both sides of the inequality $x - 5 < -2$ gives

$$x - 5 + \mathbf{5} < -2 + \mathbf{5}$$
$$x < 3$$

For expressions A, B, and C,

$$\text{if} \qquad A < B$$
$$\text{then} \qquad A + C < B + C$$

Adding the same quantity to both sides of an inequality never changes the solution set.

Multiplication Property for Inequalities [7.3]

5. Multiplying both sides of
$-2x \geq 6$ by $-\frac{1}{2}$ gives

$$-2x \geq 6$$

$$-\frac{1}{2}(-2x) \leq -\frac{1}{2}(6)$$
$$x \leq -3$$

For expressions A, B, and C,

if $\quad A < B$

then $\quad AC < BC \quad$ if $\quad C > 0$ (C is positive)

or $\quad AC > BC \quad$ if $\quad C < 0$ (C is negative)

We can multiply both sides of an inequality by the same nonzero number without changing the solution set as long as each time we multiply by a negative number we also reverse the direction of the inequality symbol.

Absolute Value Inequalities [7.4]

6. To solve

$$|x - 3| + 2 < 6$$

we first add -2 to both sides to obtain

$$|x - 3| < 4$$

which is equivalent to

$$-4 < x - 3 < 4$$
$$-1 < x < 7$$

To solve an inequality that involves absolute value, we first isolate the absolute value on the left side of the inequality symbol. Then we rewrite the absolute value inequality as an equivalent continued or compound inequality that does not contain absolute value symbols. In general, if b is a positive real number, then

$$|a| < b \quad \text{is equivalent to} \quad -b < a < b$$

and

$$|a| > b \quad \text{is equivalent to} \quad a < -b \quad \text{or} \quad a > b$$

Systems of Linear Equations [7.5, 7.6]

7. The solution to the system
$$x + 2y = 4$$
$$x - y = 1$$
is the ordered pair (2, 1). It is the only ordered pair that satisfies both equations.

A system of linear equations consists of two or more linear equations considered simultaneously. The solution set to a linear system in two variables is the set of ordered pairs that satisfy both equations. The solution set to a linear system in three variables consists of all the ordered triples that satisfy each equation in the system.

To Solve a System by the Addition Method [7.5]

8. We can eliminate the y-variable from the system in Example 1 by multiplying both sides of the second equation by 2 and adding the result to the first equation.

$$x + 2y = 4 \xrightarrow{\text{No change}} x + 2y = 4$$
$$x - y = 1 \xrightarrow[\text{Multiply by 2}]{} 2x - 2y = 2$$
$$3x \quad = 6$$
$$x \quad = 2$$

Substituting $x = 2$ into either of the original two equations gives $y = 1$. The solution is (2, 1).

Step 1: Look the system over to decide which variable will be easier to eliminate.

Step 2: Use the multiplication property of equality on each equation separately, if necessary, to ensure that the coefficients of the variable to be eliminated are opposites.

Step 3: Add the left and right sides of the system produced in step 2, and solve the resulting equation.

Step 4: Substitute the solution from step 3 back into any equation with both x- and y-variables, and solve.

Step 5: Check your solution in both equations if necessary.

9. We can apply the substitution method to the system in Example 1 by first solving the second equation for x to get

$$x = y + 1$$

Substituting this expression for x into the first equation we have

$$(y + 1) + 2y = 4$$

$$3y + 1 = 4$$

$$3y = 3$$

$$y = 1$$

Using $y = 1$ in either of the original equations gives $x = 2$.

■ To Solve a System by the Substitution Method [7.5]

Step 1: Solve either of the equations for one of the variables (this step is not necessary if one of the equations has the correct form already).

Step 2: Substitute the results of step 1 into the other equation, and solve.

Step 3: Substitute the results of step 2 into an equation with both x- and y-variables, and solve. (The equation produced in step 1 is usually a good one to use.)

Step 4: Check your solution if necessary.

■ Inconsistent and Dependent Equations [7.5, 7.6]

10. If the two lines are parallel, then the system will be inconsistent and the solution is ∅. If the two lines coincide, then the equations are dependent.

A system of two linear equations that have no solutions in common is said to be an *inconsistent* system, whereas two linear equations that have all their solutions in common are said to be *dependent* equations.

■ Linear Inequalities in Two Variables [7.7]

11. The graph of

$$x - y \leq 3 \text{ is}$$

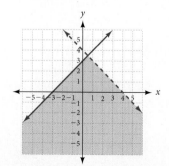

An inequality of the form $ax + by < c$ is a *linear inequality in two variables.* The equation for the boundary of the solution set is given by $ax + by = c$. (This equation is found by simply replacing the inequality symbol with an equal sign.)

To graph a linear inequality, first graph the boundary using a solid line if the boundary is included in the solution set and a broken line if the boundary is not included in the solution set. Next, choose any point not on the boundary and substitute its coordinates into the original inequality. If the resulting statement is true, the graph lies on the same side of the boundary as the test point. A false statement indicates that the solution set lies on the other side of the boundary.

■ Systems of Linear Inequalities [7.8]

12. The solution set for the system

$$x + y < 4$$
$$-x + y \leq 3$$

is shown below.

A system of linear inequalities is two or more linear inequalities considered at the same time. To find the solution set to the system, we graph each of the inequalities on the same coordinate system. The solution set is the region that is common to all the regions graphed.

🚫 **COMMON MISTAKES**

A very common mistake in solving inequalities is to forget to reverse the direction of the inequality symbol when multiplying both sides by a negative number. When this mistake occurs, the graph of the solution set is always drawn on the wrong side of the endpoint.

Solve each equation [7.1]

1. $x - 3 = 7$

2. $5x - 2 = 8$

3. $400 - 100a = 200$

4. $5 - \dfrac{2}{3}a = 7$

5. $4x - 2 = 7x + 7$

6. $\dfrac{3}{2}x - \dfrac{1}{6} = -\dfrac{7}{6}x - \dfrac{1}{6}$

7. $7y - 5 - 2y = 2y - 3$

8. $\dfrac{3y}{4} - \dfrac{1}{2} + \dfrac{3y}{2} = 2 - y$

9. $3(2x + 1) = 18$

10. $-\dfrac{1}{2}(4x - 2) = -x$

11. $8 - 3(2t + 1) = 5(t + 2)$

12. $8 + 4(1 - 3t) = -3(t - 4) + 2$

Solve each equation. [7.1]

13. $\dfrac{5}{6}y^2 = \dfrac{1}{4}y + \dfrac{1}{3}$

14. $9x^2 - 25 = 0$

15. $5x^2 = -10x$

16. $(x + 2)(x - 5) = 8$

17. $x^3 + 4x^2 - 9x - 36 = 0$

Solve each application. In each case be sure to show the equation used. [7.1]

18. Consecutive Numbers The product of two consecutive even integers is 80. Find the two integers.

19. Consecutive Numbers The sum of the squares of two consecutive integers is 41. Find the two integers.

20. Geometry The lengths of the three sides of a right triangle are given by three consecutive integers. Find the three sides.

Solve each equation. [7.2]

21. $|x| = 2$

22. $|a| - 3 = 1$

23. $|x - 3| = 1$

24. $|2y - 3| = 5$

25. $|4x - 3| + 2 = 11$

26. $\left|\dfrac{7}{3} - \dfrac{x}{3}\right| + \dfrac{4}{3} = 2$

27. $|5t - 3| = |3t - 5|$

28. $\left|\dfrac{1}{2} - x\right| = \left|x + \dfrac{1}{2}\right|$

Solve each inequality and graph the solution set. [7.4]

29. $|x| < 5$

30. $|0.01a| \geq 5$

31. $|x| < 0$

32. $|2t + 1| - 3 < 2$

Solve each equation or inequality, if possible. [7.3, 7.2, 7.4]

33. $2x - 3 = 2(x - 3)$

34. $3\left(5x - \dfrac{1}{2}\right) = 15x + 2$

35. $|4y + 8| = -1$

36. $|x| > 0$

37. $|5 - 8t| + 4 \leq 1$

38. $|2x + 1| \geq -4$

Solve each inequality. Write your answer using interval notation. [7.3]

39. $-8a > -4$

40. $6 - a \geq -2$

41. $\dfrac{3}{4}x + 1 \leq 10$

42. $800 - 200x < 1{,}000$

43. $\dfrac{1}{3} \leq \dfrac{1}{6}x \leq 1$

44. $-0.01 \leq 0.02x - 0.01 \leq 0.01$

45. $5t + 1 \leq 3t - 2$ or $-7t \leq -21$

46. $3(x + 1) < 2(x + 2)$ or $2(x - 1) \geq x + 2$

Solve each system using the addition method. [7.5]

47. $x + y = 4$
   $2x - y = 14$

48. $3x + y = 2$
   $2x + y = 0$

49. $2x - 4y = 5$
   $-x + 2y = 3$

50. $5x - 2y = 7$
   $3x + y = 2$

51. $6x - 5y = -5$
   $3x + y = 1$

52. $6x + 4y = 8$
   $9x + 6y = 12$

53. $3x - 7y = 2$
   $-4x + 6y = -6$

54. $6x + 5y = 9$
   $4x + 3y = 6$

55. $-7x + 4y = -1$
   $5x - 3y = 0$

56. $\dfrac{1}{2}x - \dfrac{3}{4}y = -4$
   $\dfrac{1}{4}x + \dfrac{3}{2}y = 13$

57. $\dfrac{2}{3}x - \dfrac{1}{6}y = 0$
   $\dfrac{4}{3}x + \dfrac{5}{6}y = 14$

58. $-\dfrac{1}{2}x + \dfrac{1}{3}y = -\dfrac{13}{6}$
   $\dfrac{4}{5}x + \dfrac{3}{4}y = \dfrac{9}{10}$

Solve each system by the substitution method. [7.5]

59. $x + y = 2$
$y = x - 1$

60. $2x - 3y = 5$
$y = 2x - 7$

61. $x + y = 4$
$2x + 5y = 2$

62. $x + y = 3$
$2x + 5y = -6$

63. $3x + 7y = 6$
$x = -3y + 4$

64. $5x - y = 4$
$y = 5x - 3$

Solve each system. [7.6]

65. $x + y + z = 6$
$x - y - 3z = -8$
$x + y - 2z = -6$

66. $3x + 2y + z = 4$
$2x - 4y + z = -1$
$x + 6y + 3z = -4$

67. $5x + 8y - 4z = -7$
$7x + 4y + 2z = -2$
$3x - 2y + 8z = 8$

68. $5x - 3y - 6z = 5$
$4x - 6y - 3z = 4$
$-x + 9y + 9z = 7$

69. $5x - 2y + z = 6$
$-3x + 4y - z = 2$
$6x - 8y + 2z = -4$

70. $4x - 6y + 8z = 4$
$5x + y - 2z = 4$
$6x - 9y + 12z = 6$

71. $2x - y = 5$
$3x - 2z = -2$
$5y + z = -1$

72. $x - y = 2$
$y - z = -3$
$x - z = -1$

Graph each linear inequality. [7.7]

73. $y \leq 2x - 3$

74. $x \geq -1$

Graph the solution set for each system of linear inequalities. [7.8]

75. $3x + 4y < 12$
$-3x + 2y \leq 6$

76. $3x + 4y < 12$
$-3x + 2y \leq 6$
$y \geq 0$

77. $3x + 4y < 12$
$x \geq 0$
$y \geq 0$

78. $x > -3$
$y > -2$

Simplify each of the following.

1. $4^2 - 8^2$

2. $(4 - 8)^2$

3. $15 - 12 \div 4 - 3 \cdot 2$

4. $6(11 - 13)^3 - 5(8 - 11)^2$

5. $8\left(\dfrac{3}{2}\right) - 24\left(\dfrac{3}{2}\right)^2$

6. $16\left(\dfrac{3}{8}\right) - 32\left(\dfrac{3}{4}\right)^2$

Find the value of each expression when x is -3.

7. $x^2 + 12x - 36$

8. $(x - 6)(x + 6)$

Simplify each of the following expressions.

9. $4(3x - 2) + 3(2x + 5)$

10. $5 - 3[2x - 4(x - 2)]$

11. $-4\left(\dfrac{3}{4}x - \dfrac{3}{2}y\right)$

12. $-6\left(\dfrac{5}{3}x + \dfrac{3}{6}y\right)$

Solve.

13. $-4y - 2 = 6y + 8$

14. $-6 + 2(2x + 3) = 0$

15. $|x + 5| + 3 = 2$

16. $|2x - 3| + 7 = 1$

Solve the following equations for y.

17. $y - 3 = -5(x - 2)$

18. $y - 3 = \dfrac{2}{3}(x - 6)$

Solve the following equations for x.

19. $ax - 5 = bx - 3$

20. $ax + 7 = cx - 2$

Solve each inequality and write your answers using interval notation.

21. $-3t \geq 12$

22. $|2x - 1| \geq 5$

Factor the following expressions.

23. $2x^2 + 5x - 12$

24. $6x^2 - 7x - 20$

For the set $\{-1, 0, \sqrt{2}, 2.35, \sqrt{3}, 4\}$ list all the elements in the following sets.

25. Rational numbers

26. Integers

Solve each of the following systems.

27. $2x - 5y = -7$
$\quad -3x + 4y = 0$

28. $8x + 6y = 4$
$\quad 12x + 9y = 8$

29. $2x + y = 3$
$\quad\quad y = -2x + 3$

30. $2x + 2y = -8z$
$\quad 4y - 2z = -9$
$\quad 3x - 2z = -6$

Graph on a rectangular coordinate system.

31. $0.03x + 0.04y = 0.04$

32. $3x - y < -2$

33. Graph the solution set to the system.
$$x - y < -3$$
$$y < 5$$

34. Find the slope of the line through $\left(\dfrac{2}{3}, -\dfrac{1}{2}\right)$ and $\left(\dfrac{1}{2}, -\dfrac{5}{6}\right)$.

35. Find the slope and y-intercept of $3x - 5y = 15$.

36. Graph the equation of a line with slope $-\dfrac{2}{3}$ and y-intercept $= -3$.

37. Find the slope of the line that is perpendicular to $2x - 5y = 10$ and contains the point $(0, 1)$.

38. Find the slope of the line through $(1, 4)$ and $(-1, -2)$.

Graph the following inequalities.

39. $3x - 7 \geq 5$

40. $4x + 8 < 16$

41. $5 - 3y \leq 8$

42. $7 - 4y < 9$

43. $7 < 2x - 3 < 11$

44. $3x - 4 > 2$ or $2x + 3 \leq -5$

Solve.

45. $2x^2 - x - 15 = 0$

46. $x^3 + x^2 - 12x = 0$

47. $x^2 - 4 = 0$

48. $6x^2 - 8x + 8 = 0$

49. One number is 2 more than 3 times another number. Their sum is 18. Find the two numbers.

Solve the following equations. [7.1]

1. $x - 5 = 7$ **2.** $3y = -4$

3. $5(x - 1) - 2(2x + 3) = 5x - 4$

4. $0.07 - 0.02(3x + 1) = -0.04x + 0.01$

Use the formula $4x - 6y = 12$ to find y when

5. $x = 0$ **6.** $x = 3$

Use the formula $-2x + 4y = -8$ to find x when

7. $y = 0$ **8.** $y = 2$

Solve each equation [7.1]

9. $\frac{1}{5}x^2 = \frac{1}{3}x + \frac{2}{15}$ **10.** $100x^3 = 500x^2$

11. $(x + 1)(x + 2) = 12$

12. $x^3 + 2x^2 - 16x - 32 = 0$

Solve the following equations. [7.2]

13. $\left| \frac{1}{4}x - 1 \right| = \frac{1}{2}$ **14.** $\left| \frac{2}{3}a + 4 \right| = 6$

15. $|3 - 2x| + 5 = 2$ **16.** $5 = |3y + 6| - 4$

Solve the following inequalities. Write the solution set using interval notation, then graph the solution set. [7.3]

17. $-5t \le 30$ **18.** $5 - \frac{3}{2}x > -1$

19. $1.6x - 2 < 0.8x + 2.8$

20. $3(2y + 4) \ge 5(y - 8)$

Solve the following inequalities and graph the solutions. [7.4]

21. $|6x - 1| > 7$

22. $|3x - 5| - 4 \le 3$

23. $|5 - 4x| \ge -7$

24. $|4t - 1| < -3$

Solve the following systems by the addition method. [7.5]

25. $2x - 5y = -8$
$3x + y = 5$

26. $\frac{1}{3}x - \frac{1}{6}y = 3$
$-\frac{1}{5}x + \frac{1}{4}y = 0$

Solve the following systems by the substitution method. [7.5]

27. $2x - 5y = 14$
$y = 3x + 8$

28. $6x - 3y = 0$
$x + 2y = 5$

Solve the system. [7.6]

29. $2x - y + z = 9$
$x + y - 3z = -2$
$3x + y - z = 6$

30. Electric Current In the following diagram of an electrical circuit, x, y, and z represent the amount of current (in amperes) flowing across the 5-ohm, 20-ohm, and 10-ohm resistors, respectively. (In circuit diagrams, resistors are represented by -\/\/\/- and potential differences by —|⊢—.)

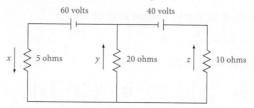

The system of equations used to find the three currents x, y, and z is

$$x - y - z = 0$$
$$5x + 20y = 60$$
$$20y - 10z = 40$$

Solve the system for all variables.

Fill in the blank so that the linear inequality represents by the shaded region. [7.7]

31.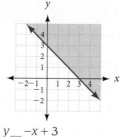

$y \_\_ -x + 3$

32.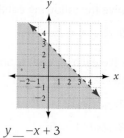

$y \_\_ -x + 3$

Graph the following linear inequalities. [7.7]

33. $3x - 4y < 12$ **34.** $y \le -x + 2$

Graph the solution set for each system of linear inequalities. [7.8]

35. $y < -\frac{1}{2}x + 4$
$x \ge 0$
$y \ge 0$

36. $x + 4y \le 4$
$-3x + 2y > -12$

Chapter 7 Projects

TRANSITIONS

GROUP PROJECT

Break-Even Point

Number of People 2 or 3

Time Needed 10–15 minutes

Equipment Pencil and paper

Background The break-even point for a company occurs when the revenue from sales of a product equals the cost of producing the product. This group project is designed to give you more insight into revenue, cost, and the break-even point.

Procedure A company is planning to open a factory to manufacture calculators.

1. It costs them $120,000 to open the factory, and it will cost $10 for each calculator they make. What is the expression for $C(x)$, the cost of making x calculators?

2. They can sell the calculators for $50 each. What is the expression for $R(x)$, their revenue from selling x calculators? Remember that $R = px$, where p is the price per calculator.

3. Graph both the cost equation $C(x)$ and the revenue equation $R(x)$ on a coordinate system like the one below.

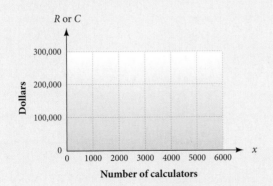

Number of calculators

4. The break-even point is the value of x (the number of calculators) for which the revenue

is equal to the cost. Where is the break-even point on the graph you produced in Part 3? Estimate the break-even point from the graph.

5. Set the cost equal to the revenue and solve for x to find the exact value of the break-even point. How many calculators do they need to make and sell to exactly break even? What will be their revenue and their cost for that many calculators?

6. Write an inequality that gives the values of x that will produce a profit for the company. (A profit occurs when the revenue is larger than the cost.)

7. Write an inequality that gives the values of x that will produce a loss for the company. (A loss occurs when the cost is larger than the revenue.)

8. Profit is the difference between revenue and cost, or $P(x) = R(x) - C(x)$. Write the equation for profit and then graph it on a coordinate system like the one below.

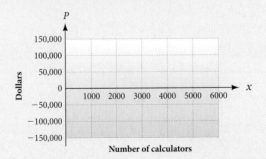

Number of calculators

9. How do you recognize the break-even point and the regions of loss and profit on the graph you produced above?

Zeno's Paradoxes

Zeno of Elea was born at about the same time that Pythagoras (of the Pythagorean theorem) died. He is responsible for three paradoxes that have come to be known as Zeno's paradoxes. One of the three has to do with a race between Achilles and a tortoise. Achilles is much faster than the tortoise, but the tortoise has a head start. According to Zeno's method of reasoning, Achilles can never pass the tortoise because each time he reaches the place where the tortoise was, the tortoise is gone. Research Zeno's paradox concerning Achilles and the tortoise. Put your findings into essay form that begins with a definition for the word "paradox." Then use Zeno's method of reasoning to describe a race between Achilles and the tortoise — if Achilles runs at 10 miles per hour, the tortoise runs at 1 mile per hour, and the tortoise has a 1-mile head start. Next, use the methods shown in this chapter to find the distance and the time at which Achilles reaches the tortoise. Conclude your essay by summarizing what you have done and showing how the two results you have obtained form a paradox.

8 Equations and Functions

Introduction

Maria Gaetana Agnesi is recognized as the first woman to publish a book of mathematics. Maria also went on to become the first female professor of mathematics, at the University of Bologna shown in the above Google Earth photo. She was born in Milan to a wealthy and literate family. The Agnesi home was a gathering place of the most recognized intellectual minds of the day. Maria participated in these gatherings until the death of her mother, when she took over management of the household and the education of her siblings. The texts she wrote for this task were the beginnings of the first published mathematical work written by a woman, *Analytical Institutions*. An illustration of a curve discussed in this book is shown here.

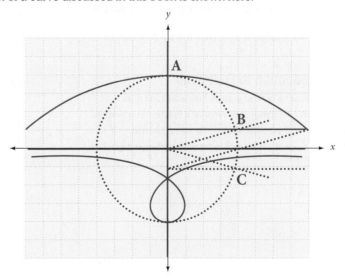

The diagram is drawn on a rectangular coordinate system. The rectangular coordinate system is the foundation upon which we build our study of graphing in this chapter.

1. $16(4.5)^2$

2. $\dfrac{0.2(0.5)^2}{100}$

3. $-\dfrac{1}{3}(-3) + 2$

4. $4(-2) - 1$

5. $\dfrac{4 - (-2)}{3 - 1}$

6. $\dfrac{-3 - (-3)}{2 - (-1)}$

7. $\dfrac{4}{3}(3.14) \cdot 3^3$ (round to the nearest whole number)

8. $3(-2)^2 + 2(-2) - 1$

9. If $2x + 3y = 6$, find x when $y = 0$.

10. If $s = \dfrac{60}{t}$, find s when $t = 15$.

11. Write -0.06 as a fraction in lowest terms.

12. Write -0.6 as a fraction in lowest terms.

13. Solve $2x - 3y = 5$ for y.

14. Solve $y - 4 = -\dfrac{1}{2}(x + 1)$ for y.

15. Which of the following are solutions to $x + y \geq 5$?

(0, 0)   (5, 0)   (4, 3)

16. Which of the following are solutions to $y < \dfrac{1}{2}x + 3$?

(0, 0)   (2, 0)   (−2, 4)

17. Use $h = 64t - 16t^2$ to complete the table.

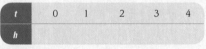

t	0	1	2	3	4
h					

18. Use $s = \dfrac{60}{t}$ to complete the table.

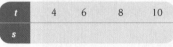

t	4	6	8	10
s				

19. If $y = \dfrac{K}{x}$, find K when $x = 3$ and $y = 45$.

20. If $y = Kxz^2$, find K when $x = 2$, $y = 81$, and $z = 9$.

Chapter Outline

8.1 The Slope of a Line
A   Find the slope of a line from its graph.
B   Find the slope of a line given two points on the line.

8.2 The Equation of a Line
A   Find the equation of a line given its slope and y-intercept.
B   Find the slope and y-intercept from the equation of a line.
C   Find the equation of a line given the slope and a point on the line.
D   Find the equation of a line given two points on the line.

8.3 Introduction to Functions
A   Construct a table or a graph from a function rule.
B   Identify the domain and range of a function or a relation.
C   Determine whether a relation is also a function.
D   Use the vertical line test to determine whether a graph is a function.

8.4 Function Notation
A   Use function notation to find the value of a function for a given value of the variable.
B   Use graphs to visualize the relationship between a function and a variable.
C   Use function notation in formulas.

8.5 Algebra with Composition and Functions
A   Find the sum, difference, product, and quotient of two functions.
B   For two functions f and g, find $f(g(x))$ and $g(f(x))$.

8.6 Variation
A   Set up and solve problems with direct variation.
B   Set up and solve problems with inverse variation.
C   Set up and solve problems with joint variation.

8.1 The Slope of a Line

OBJECTIVES

A Find the slope of a line from its graph.

B Find the slope of a line given two points on the line.

TICKET TO SUCCESS

Keep these questions in mind as you read through the section. Then respond in your own words and in complete sentences.

1. How would you geometrically define the slope of a line?
2. Describe the behavior of a line with a negative slope.
3. Use symbols to define the slope of the line between the points (x_1, y_1) and (x_2, y_2).
4. How are the slopes of two perpendicular lines different than the slopes of two parallel lines?

A highway sign tells us we are approaching a 6% downgrade. As we drive down this hill, each 100 feet we travel horizontally is accompanied by a 6-foot drop in elevation.

In mathematics, we say the slope of the highway is $-0.06 = -\frac{6}{100} = -\frac{3}{50}$. The *slope* is the ratio of the vertical change to the accompanying horizontal change.

In defining the slope of a straight line, we are looking for a number to associate with a straight line that does two things. First, we want the slope of a line to measure the "steepness" of the line; that is, in comparing two lines, the slope of the steeper line should have the larger numerical absolute value. Second, we want a line that *rises* going from left to right to have a *positive* slope. We want a line that *falls* going from left to right to have a *negative* slope. (A line that neither rises nor falls going from left to right must, therefore, have 0 slope.)

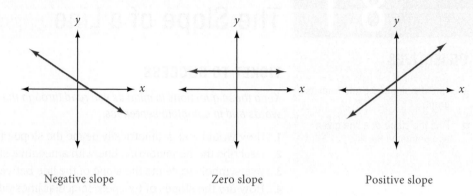

Negative slope   Zero slope   Positive slope

Geometrically, we can define the *slope* of a line as the ratio of the vertical change to the horizontal change encountered when moving from one point to another on the line. The vertical change is sometimes called the *rise*. The horizontal change is called the *run*.

A Finding the Slope From Graphs

EXAMPLE 1   Find the slope of the line $y = 2x - 3$.

SOLUTION   To use our geometric definition, we first graph $y = 2x - 3$ (Figure 1). We then pick any two convenient points and find the ratio of rise to run. By convenient points we mean points with integer coordinates. If we let $x = 2$ in the equation, then $y = 1$. Likewise if we let $x = 4$, then y is 5.

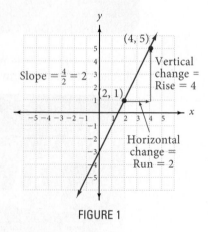

FIGURE 1

Our line has a slope of 2.

Notice that we can measure the vertical change (rise) by subtracting the y-coordinates of the two points shown in Figure 1: $5 - 1 = 4$. The horizontal change (run) is the difference of the x-coordinates: $4 - 2 = 2$. This gives us a second way of defining the slope of a line.

B Finding Slopes Using Two Points

Definition
The **slope** of the line between two points (x_1, y_1) and (x_2, y_2) is given by

$$\text{Slope} = m = \underset{\underset{\text{Geometric form}}{\uparrow}}{\frac{\text{Rise}}{\text{Run}}} = \underset{\underset{\text{Algebraic form}}{\uparrow}}{\frac{y_2 - y_1}{x_2 - x_1}}$$

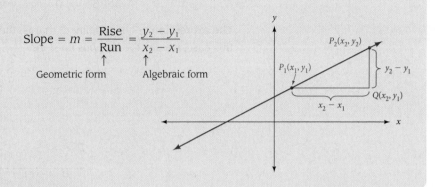

EXAMPLE 2   Find the slope of the line through $(-2, -3)$ and $(-5, 1)$.

SOLUTION $m = \dfrac{y_2 - y_1}{x_2 - x_1} = \dfrac{1 - (-3)}{-5 - (-2)} = \dfrac{4}{-3} = -\dfrac{4}{3}$

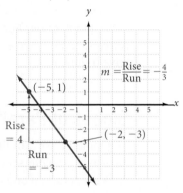

FIGURE 2

Looking at the graph of the line through the two points (Figure 2), we can see that our geometric approach does not conflict with our algebraic approach.

We should note here that it does not matter which ordered pair we call (x_1, y_1) and which we call (x_2, y_2). If we were to reverse the order of subtraction of both the x- and y-coordinates in the preceding example, we would have

$$m = \frac{-3 - 1}{-2 - (-5)} = \frac{-4}{3} = -\frac{4}{3}$$

which is the same as our previous result.

NOTE: The two most common mistakes students make when first working with the formula for the slope of a line are

1. Putting the difference of the x-coordinates over the difference of the y-coordinates.
2. Subtracting in one order in the numerator and then subtracting in the opposite order in the denominator. You would make this mistake in Example 2 if you wrote $1 - (-3)$ in the numerator and then $-2 - (-5)$ in the denominator.

EXAMPLE 3   Find the slope of the line containing $(3, -1)$ and $(3, 4)$.

SOLUTION   Using the definition for slope, we have

$$m = \frac{-1 - 4}{3 - 3} = \frac{-5}{0}$$

The expression $\frac{-5}{0}$ is undefined; that is, there is no real number to associate with it. In this case, we say the line *has no slope*.

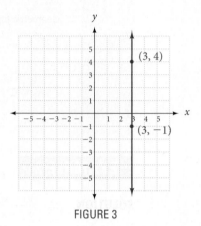

FIGURE 3

The graph of our line is shown in Figure 3. Our line with no slope is a vertical line. All vertical lines have no slope. (And all horizontal lines, as we mentioned earlier, have 0 slope.)

Slopes of Parallel and Perpendicular Lines

In geometry, we call lines in the same plane that never intersect parallel. For two lines to be nonintersecting, they must rise or fall at the same rate. In other words, two lines are *parallel* if and only if they have the *same slope*.

Although it is not as obvious, it is also true that two nonvertical lines are *perpendicular* if and only if the *product of their slopes is* -1. This is the same as saying their slopes are negative reciprocals.

We can state these facts with symbols as follows: If line l_1 has slope m_1 and line l_2 has slope m_2, then

$$l_1 \text{ is parallel to } l_2 \Leftrightarrow m_1 = m_2$$

and

$$l_1 \text{ is perpendicular to } l_2 \Leftrightarrow m_1 \cdot m_2 = -1 \left(\text{or } m_1 = \frac{-1}{m_2} \right)$$

For example, if a line has a slope of $\frac{2}{3}$, then any line parallel to it has a slope of $\frac{2}{3}$. Any line perpendicular to it has a slope of $-\frac{3}{2}$ (the negative reciprocal of $\frac{2}{3}$).

Although we cannot give a formal proof of the relationship between the slopes of perpendicular lines at this level of mathematics, we can offer some justification for the relationship. Figure 4 shows the graphs of two lines. One of the lines has a slope of $\frac{2}{3}$; the other has a slope of $-\frac{3}{2}$. As you can see, the lines are perpendicular.

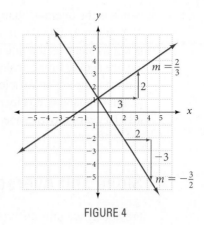

FIGURE 4

Slope and Rate of Change

So far, the slopes we have worked with represent the ratio of the change in y to a corresponding change in x; or, on the graph of the line, the ratio of vertical change to horizontal change in moving from one point on the line to another. However, when our variables represent quantities from the world around us, slope can have additional interpretations.

EXAMPLE 4   On the chart below, find the slope of the line connecting the first point (1955, 0.29) with the highest point (2005, 2.93). Explain the significance of the result.

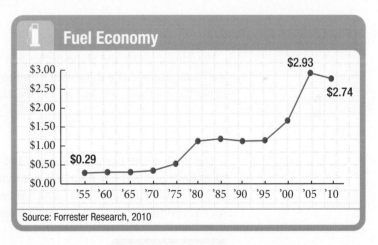

SOLUTION   The slope of the line connecting the first point (1955, 0.29) with the highest point (2005, 2.93), is

$$m = \frac{2.93 - 0.29}{2005 - 1955} = \frac{2.64}{50} = 0.0528$$

The units are dollars/year. If we write this in terms of cents we have

$$m = 5.28 \text{ cents/year}$$

which is the average change in the price of a gallon of gasoline over a 50-year period of time.

Likewise, if we connect the points (1995, 1.10) and (2005, 2.93), the line that results has a slope of

$$m = \frac{2.93 - 1.10}{2005 - 1995} = \frac{1.83}{10} = 0.183 \text{ dollars/year} = 18.3 \text{ cents/year}$$

which is the average change in the price of a gallon of gasoline over a 10-year period. As you can imagine by looking at the chart, the line connecting the first and last point is not as steep as the line connecting the points from 1995 and 2005, and this is what we are seeing numerically with our slope calculations. If we were summarizing this information for an article in the newspaper, we could say, although the price of a gallon of gasoline has increased only 5.28 cents per year over the last 50 years, in the last 10 years the average annual rate of increase was more than triple that, at 18.3 cents per year.

Slope and Average Speed

Previously we introduced the rate equation $d = rt$. Suppose that a boat is traveling at a constant speed of 15 miles per hour in still water. The following table shows the distance the boat will have traveled in the specified number of hours. The graph of this data is shown in Figure 5. Notice that the points all lie along a line.

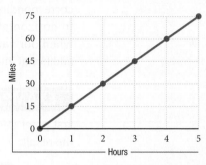

t (hours)	d (miles)
0	0
1	15
2	30
3	45
4	60
5	75

FIGURE 5

We can calculate the slope of this line using any two points from the table. Notice we have graphed the data with t on the horizontal axis and d on the vertical axis. Using the points (2, 30) and (3, 45), the slope will be

$$m = \frac{\text{rise}}{\text{run}} = \frac{45 - 30}{3 - 2} = \frac{15}{1} = 15$$

The units of the rise are miles and the units of the run are hours, so the slope will be in units of miles per hour. We see that the slope is simply the change in distance divided by the change in time, which is how we compute the average speed. Since the speed is constant, the slope of the line represents the speed of 15 miles per hour.

EXAMPLE 5   A car is traveling at a constant speed. A graph (Figure 6) of the distance the car has traveled over time is shown below. Use the graph to find the speed of the car.

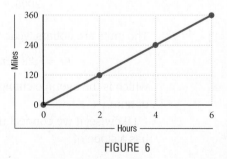

FIGURE 6

SOLUTION   Using the second and third points, we see the rise is $240 - 120 = 120$ miles, and the run is $4 - 2 = 2$ hours. The speed is given by the slope, which is

$$m = \frac{\text{rise}}{\text{run}}$$

$$= \frac{120 \text{ miles}}{2 \text{ hours}}$$

$$= 60 \text{ miles per hour}$$

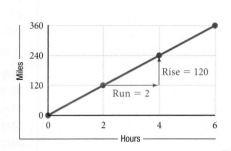

USING TECHNOLOGY

Families of Curves

We can use a graphing calculator to investigate the effects of the numbers a and b on the graph of $y = ax + b$. To see how the number b affects the graph, we can hold a constant and let b vary. Doing so will give us a *family* of curves. Suppose we set $a = 1$ and then let b take on integer values from -3 to 3. The equations we obtain are

$$y = x - 3$$
$$y = x - 2$$
$$y = x - 1$$
$$y = x$$
$$y = x + 1$$
$$y = x + 2$$
$$y = x + 3$$

We will give three methods of graphing this set of equations on a graphing calculator.

Method 1: Y-Variables List

To use the Y-variables list, enter each equation at one of the Y variables, set the graph window, then graph. The calculator will graph the equations in order, starting with Y_1 and ending with Y_7. Following is the Y-variables list, an appropriate window, and a sample of the type of graph obtained (Figure 7).

$Y_1 = X - 3$

$Y_2 = X - 2$

$Y_3 = X - 1$

$Y_4 = X$

$Y_5 = X + 1$

$Y_6 = X + 2$

$Y_7 = X + 3$

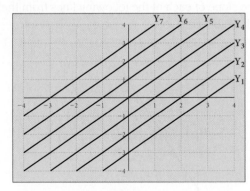

FIGURE 7

Window: X from -4 to 4, Y from -4 to 4

Method 2: Programming

The same result can be obtained by programming your calculator to graph $y = x + b$ for $b = -3, -2, -1, 0, 1, 2$, and 3. Here is an outline of a program that will do this. Check the manual that came with your calculator to find the commands for your calculator.

Step 1: Clear screen
Step 2: Set window for X from -4 to 4 and Y from -4 to 4
Step 3: $-3 \rightarrow$ B
Step 4: Label 1
Step 5: Graph Y = X + B
Step 6: B + 1 $\rightarrow$ B
Step 7: If B < 4, go to 1
Step 8: End

Method 3: Using Lists

On the TI-83/84 you can set Y_1 as follows

$$Y_1 = X + \{-3, -2, -1, 0, 1, 2, 3\}$$

When you press GRAPH, the calculator will graph each line from $y = x + (-3)$ to $y = x + 3$.

Each of the three methods will produce graphs similar to those in Figure 7.

Problem Set 8.1

Moving Toward Success

1. Do you think it is important to take notes as you read this book? Explain.

2. Why is it helpful to put the material you learn in your own words when you take notes or answer homework questions?

A Find the slope of each of the following lines from the given graph. [Example 1]

1.

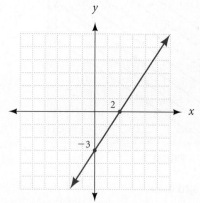

2.

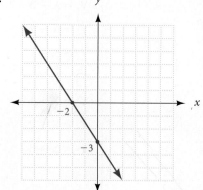

3.

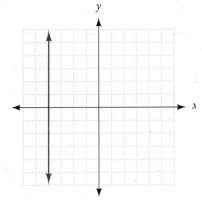

4.

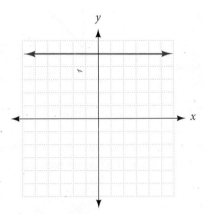

5.

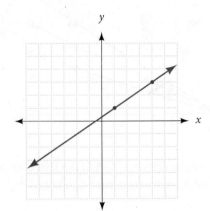

6.

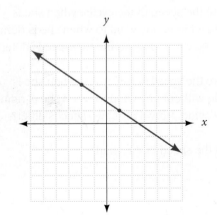

B Find the slope of the line through each of the following pairs of points. Then, plot each pair of points, draw a line through them, and indicate the rise and run in the graph in the manner shown in Example 2. [Examples 2–3]

7. (2, 1), (4, 4)

8. (3, 1), (5, 4)

9. (1, 4), (5, 2)

10. (1, 3), (5, 2)

11. (1, −3), (4, 2)

12. (2, −3), (5, 2)

13. (−3, −2), (1, 3)

14. (−3, −1), (1, 4)

15. (−3, 2), (3, −2)

16. (−3, 3), (3, −1)

17. (2, −5), (3, −2)

18. (2, −4), (3, −1)

Solve for the indicated variable if the line through the two given points has the given slope.

19. $(a, 3)$ and $(2, 6)$, $m = -1$

20. $(a, -2)$ and $(4, -6)$, $m = -3$

21. $(2, b)$ and $(-1, 4b)$, $m = -2$

22. $(-4, y)$ and $(-1, 6y)$, $m = 2$

For each equation below, complete the table, and then use the results to find the slope of the graph of the equation.

23. $2x + 3y = 6$

24. $3x - 2y = 6$

25. $y = \dfrac{2}{3}x - 5$

26. $y = -\dfrac{3}{4}x + 2$

27. Finding Slope from Intercepts Graph the line with x-intercept −4 and y-intercept −2. What is the slope of this line?

28. Parallel Lines Find the slope of a line parallel to the line through (2, 3) and (−8, 1).

29. Parallel Lines Find the slope of a line parallel to the line through (2, 5) and (5, −3).

30. Perpendicular Lines Line l contains the points (5, −6) and (5, 2). Give the slope of any line perpendicular to l.

31. Perpendicular Lines Line l contains the points (3, 4) and (−3, 1). Give the slope of any line perpendicular to l.

32. Parallel Lines Line *l* contains the points (−2, 1) and (4, −5). Find the slope of any line parallel to *l*.

33. Parallel Lines Line *l* contains the points (3, −4) and (−2, −6). Find the slope of any line parallel to *l*.

34. Perpendicular Lines Line *l* contains the points (−2, −5) and (1, −3). Find the slope of any line perpendicular to *l*.

35. Perpendicular Lines Line *l* contains the points (6, −3) and (−2, 7). Find the slope of any line perpendicular to *l*.

36. Graph the line with *x*-intercept 4 and *y*-intercept 2. What is the slope of this line?

37. Determine if each of the following tables could represent ordered pairs from an equation of a line.

a.

x	y
0	5
1	7
2	9
3	11

b.

x	y
−2	−5
0	−2
2	0
4	1

38. The following lines have slope 2, $\frac{1}{2}$, 0, and −1. Match each line to its slope value.

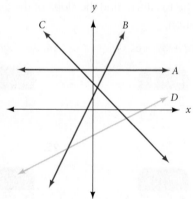

39. Line *l* has a slope of $\frac{2}{3}$. A horizontal change of 12 will always be accompanied by how much of a vertical change?

40. For any line with slope $\frac{4}{5}$, a vertical change of 8 is always accompanied by how much of a horizontal change?

Applying the Concepts [Examples 4–5]

41. An object is traveling at a constant speed. The distance and time data are shown on the given graphs. Use the graphs to find the speed of the object.

a. 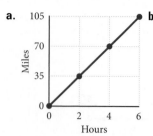 **b.**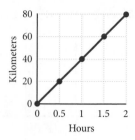

c. 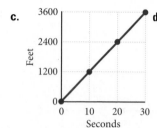 **d.**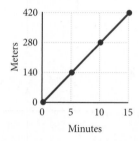

42. Cycling A cyclist is traveling at a constant speed. The graph shows the distance the cyclist travels over time when there is no wind present, when she travels against the wind, and when she travels with the wind.

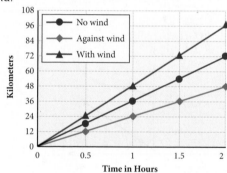

a. Use the concept of slope to find the speed of the cyclist under each of the three conditions.

b. Compare the speed of the cyclist when she is traveling without any wind to when she is riding against the wind. How do the two speeds differ?

c. Compare the speed of the cyclist when she is traveling without any wind to when she is riding with the wind. How do the two speeds differ?

d. What is the speed of the wind?

43. Non-Camera Phone Sales The table and line graph here each show the projected non-camera phone sales each year through 2010. Find the slope of each of the three line segments, A, B, and C.

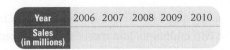

Year	2006	2007	2008	2009	2010
Sales (in millions)					

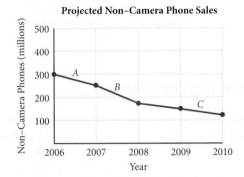

44. Camera Phone Sales The table and line graph shown here show the projected sales of camera phones from 2006 to 2010. Find the slopes of line segments, A, B, and C.

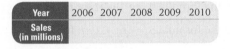

Year	2006	2007	2008	2009	2010
Sales (in millions)					

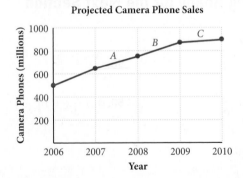

45. Slope of a Highway A sign at the top of the Cuesta Grade, outside of San Luis Obispo, CA, reads "7% downgrade next 3 miles." The diagram shown here is a model of the Cuesta Grade that takes into account the information on that sign.

a. At point B, the graph crosses the y-axis at 1,106 feet. How far is Point A from the origin?

b. What is the slope of the Cuesta Grade?

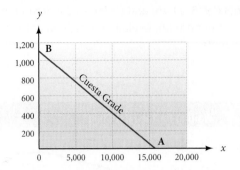

46. Health Care The graph shows the rising cost of health care. What is the slope of the line segment from 2008 to 2011?

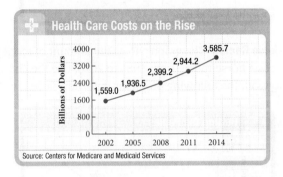

Source: Centers for Medicare and Medicaid Services

47. Solar Energy The graph below shows the annual shipments of solar thermal collectors in the United States. Using the graph below, find the slope of the line connecting the first (1997, 8,000) and last (2008, 17,000) endpoints and then explain in words what the slope represents.

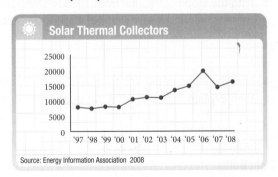

Source: Energy Information Association 2008

48. Age of New Mothers The graph shows the increase in the average age of first time mothers in the U.S. since 1970. Find the slope of the line that connects the points (1975, 21.75) and (1990, 24.25). Round to the nearest hundredth. Explain in words what the slope represents.

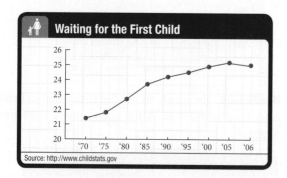

Waiting for the First Child

Source: http://www.childstats.gov

49. Horse Racing The graph shows the amount of money bet on horse racing from 1985 to 2010. Use the chart to work the following problems involving slope.

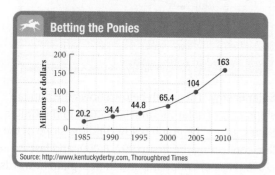

Betting the Ponies

Source: http://www.kentuckyderby.com, Thoroughbred Times

a. Find the slope of the line from 1985 to 1990, and then explain in words what the slope represents.

b. Find the slope of the line from 2000 to 2005, and then explain in words what the slope represents.

50. Light Bulbs The chart shows a comparison of power usage between incandescent and LED light bulbs. Use the chart to work the following problems involving slope.

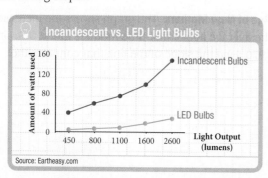

Incandescent vs. LED Light Bulbs

Source: Eartheasy.com

a. Find the slope of the line for the incandescent bulb from the two endpoints and then explain in words what the slope represents.

b. Find the slope of the line for the LED bulb from the two endpoints and then explain in words what the slope represents.

c. Which light bulb is more efficient? Why?

Maintaining Your Skills

The problems that follow review some of the more important skills you have learned in previous sections and chapters.

51. If $3x + 2y = 12$, find y when x is 4.

52. If $y = 3x - 1$, find x when y is 0.

53. Solve the formula $3x + 2y = 12$ for y.

54. Solve the formula $y = 3x - 1$ for x.

55. Solve the formula $A = P + Prt$ for t.

56. Solve the formula $S = \pi r^2 + 2\pi rh$ for h.

Getting Ready for the Next Section

Simplify.

57. $2\left(-\dfrac{1}{2}\right)$

58. $\dfrac{3 - (-1)}{-3 - 3}$

59. $\dfrac{5 - (-3)}{2 - 6}$

60. $3\left(-\dfrac{2}{3}x + 1\right)$

Solve for y.

61. $\dfrac{y - b}{x - 0} = m$

62. $2x + 3y = 6$

63. $y - 3 = -2(x + 4)$

64. $y + 1 = -\dfrac{2}{3}(x - 3)$

65. If $y = -\dfrac{4}{3}x + 5$, find y when x is 0.

66. If $y = -\dfrac{4}{3}x + 5$, find y when x is 3.

8.2 The Equation of a Line

OBJECTIVES

A Find the equation of a line given its slope and *y*-intercept.

B Find the slope and *y*-intercept from the equation of a line.

C Find the equation of a line given the slope and a point on the line.

D Find the equation of a line given two points on the line.

TICKET TO SUCCESS

Keep these questions in mind as you read through the section. Then respond in your own words and in complete sentences.

1. What is the slope-intercept form of the equation of a line?
2. How would you graph the line $y = \frac{1}{2}x + 3$?
3. How would you write the equation of a line using point-slope form?
4. Use symbols to write the equation of a line in standard form.

The table and illustrations show some corresponding temperatures on the Fahrenheit and Celsius temperature scales. For example, water freezes at 32°F and 0°C, and boils at 212°F and 100°C.

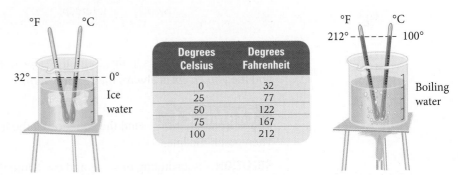

Degrees Celsius	Degrees Fahrenheit
0	32
25	77
50	122
75	167
100	212

If we plot all the points in the table using the *x*-axis for temperatures on the Celsius scale and the *y*-axis for temperatures on the Fahrenheit scale, we see that they line up in a straight line (Figure 1). This means that a linear equation in two variables will give a perfect description of the relationship between the two scales. That equation is

$$F = \frac{9}{5}C + 32$$

The techniques we use to find the equation of a line from a set of points is what this section is all about.

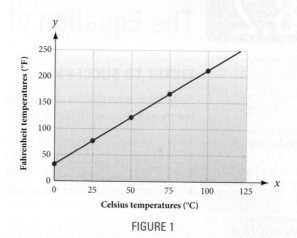

FIGURE 1

A Finding the Line Using the Slope and Intercepts

Suppose line *l* has slope *m* and *y*-intercept *b*. What is the equation of *l*? Because the *y*-intercept is *b*, we know the point $(0, b)$ is on the line. If (x, y) is any other point on *l*, then using the definition for slope, we have

$$\frac{y - b}{x - 0} = m \qquad \text{Definition of slope}$$

$$y - b = mx \qquad \text{Multiply both sides by } x$$

$$y = mx + b \qquad \text{Add } b \text{ to both sides}$$

This last equation is known as the *slope-intercept form* of the equation of a straight line.

Property: Slope-Intercept Form

The equation of any line with slope *m* and *y*-intercept *b* is given by

$$y = mx + b$$

Slope   *y*-intercept

When the equation is in this form, the *slope* of the line is always the *coefficient of x*, and the *y-intercept* is always the *constant term*.

EXAMPLE 1   Find the equation of the line with slope $-\frac{4}{3}$ and *y*-intercept 5. Then graph the line.

SOLUTION   Substituting $m = -\frac{4}{3}$ and $b = 5$ into the equation $y = mx + b$, we have

$$y = -\frac{4}{3}x + 5$$

Finding the equation from the slope and *y*-intercept is just that easy. If the slope is *m* and the *y*-intercept is *b*, then the equation is always $y = mx + b$. Now, let's graph the line.

Because the *y*-intercept is 5, the graph goes through the point $(0, 5)$. To find a second point on the graph, we can use the given slope of $-\frac{4}{3}$. We start at $(0, 5)$ and move 4 units down (that's a rise of −4) and 3 units to the right (a run of 3). The point we end up at is $(3, 1)$. Drawing a line that passes through $(0, 5)$ and $(3, 1)$, we have the graph of our equation. (Note that we could also let the rise = 4 and the run = −3 and obtain the same graph.) The graph is shown in Figure 2.

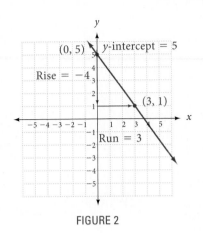

FIGURE 2

B Finding Slopes and Intercepts

EXAMPLE 2
Give the slope and y-intercept for the line $2x - 3y = 5$.

SOLUTION   To use the slope-intercept form, we must solve the equation for y in terms of x.

$$2x - 3y = 5$$

$$-3y = -2x + 5 \qquad \text{Add } -2x \text{ to both sides}$$

$$y = \frac{2}{3}x - \frac{5}{3} \qquad \text{Divide by } -3$$

The last equation has the form $y = mx + b$. The slope must be $m = \frac{2}{3}$, and the y-intercept is $b = -\frac{5}{3}$.

EXAMPLE 3
Graph the equation $2x + 3y = 6$ using the slope and y-intercept.

SOLUTION   Although we could graph this equation by finding ordered pairs that are solutions to the equation and drawing a line through their graphs, it is sometimes easier to graph a line using the slope-intercept form of the equation.

Solving the equation for y, we have

$$2x + 3y = 6$$

$$3y = -2x + 6 \qquad \text{Add } -2x \text{ to both sides}$$

$$y = -\frac{2}{3}x + 2 \qquad \text{Divide by } 3$$

The slope is $m = -\frac{2}{3}$ and the y-intercept is $b = 2$. Therefore, the point $(0, 2)$ is on the graph, and the ratio of rise to run going from $(0, 2)$ to any other point on the line is $-\frac{2}{3}$. If we start at $(0, 2)$ and move 2 units up (that's a rise of 2) and 3 units to the left (a run of -3), we will be at another point on the graph. (We could also go down 2 units and right 3 units and still be assured of ending up at another point on the line because $\frac{2}{-3}$ is the same as $\frac{-2}{3}$.)

NOTE
As we mentioned earlier, the rectangular coordinate system is the tool we use to connect algebra and geometry. Example 3 illustrates this connection, as do the many other examples in this chapter. In Example 3, Descartes' rectangular coordinate system allows us to associate the equation $2x + 3y = 6$ (an algebraic concept) with the straight line (a geometric concept) shown in Figure 3.

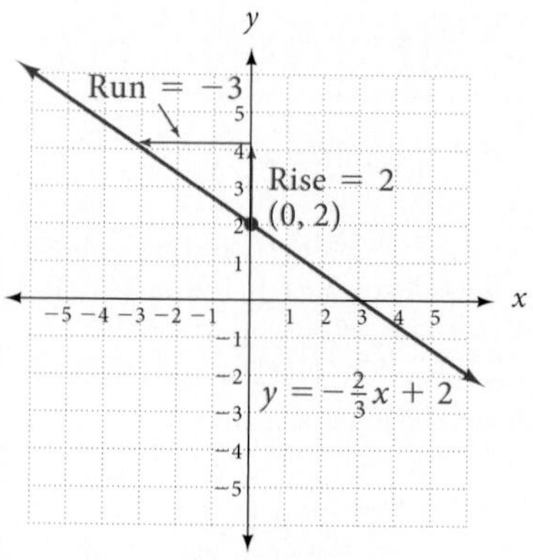

FIGURE 3

C Finding the Line Using the Slope and a Point

A second useful form of the equation of a straight line is the *point-slope form*.

Let line l contain the point (x_1, y_1) and have slope m. If (x, y) is any other point on l, then by the definition of slope we have

$$\frac{y - y_1}{x - x_1} = m$$

Multiplying both sides by $(x - x_1)$ gives us

$$(x - x_1) \cdot \frac{y - y_1}{x - x_1} = m(x - x_1)$$

$$y - y_1 = m(x - x_1)$$

This last equation is known as the *point-slope form* of the equation of a straight line.

Property: Point-Slope Form

The equation of the line through (x_1, y_1) with slope m is given by
$$y - y_1 = m(x - x_1)$$

This form of the equation of a straight line is used to find the equation of a line, either given one point on the line and the slope, or given two points on the line.

EXAMPLE 4 Find the equation of the line with slope -2 that contains the point $(-4, 3)$. Write the answer in slope-intercept form.

SOLUTION

Using $(x_1, y_1) = (-4, 3)$ and $m = -2$

in $y - y_1 = m(x - x_1)$ Point-slope form

gives us $y - 3 = -2(x + 4)$ *Note:* $x - (-4) = x + 4$

$y - 3 = -2x - 8$ Multiply out right side

$y = -2x - 5$ Add 3 to each side

Figure 4 is the graph of the line that contains $(-4, 3)$ and has a slope of -2. Notice that the y-intercept on the graph matches that of the equation we found.

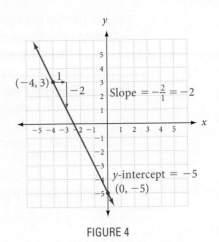

FIGURE 4

D Finding the Line Using Points

EXAMPLE 5 Find the equation of the line that passes through the points $(-3, 3)$ and $(3, -1)$.

SOLUTION We begin by finding the slope of the line

$$m = \frac{3 - (-1)}{-3 - 3} = \frac{4}{-6} = -\frac{2}{3}$$

Using $(x_1, y_1) = (3, -1)$ and $m = -\frac{2}{3}$ in $y - y_1 = m(x - x_1)$ yields

$$y + 1 = -\frac{2}{3}(x - 3)$$

$$y + 1 = -\frac{2}{3}x + 2 \qquad \text{Multiply out right side}$$

$$y = -\frac{2}{3}x + 1 \qquad \text{Add } -1 \text{ to each side}$$

Figure 5 shows the graph of the line that passes through the points $(-3, 3)$ and $(3, -1)$. As you can see, the slope and y-intercept are $-\frac{2}{3}$ and 1, respectively.

NOTE
In Example 5, we could have used the point $(-3, 3)$ instead of $(3, -1)$ and obtained the same equation; that is, using $(x_1, y_1) = (-3, 3)$ and $m = -\frac{2}{3}$ in $y - y_1 = m(x - x_1)$ gives us
$$y - 3 = -\frac{2}{3}(x + 3)$$
$$y - 3 = -\frac{2}{3}x - 2$$
$$y = -\frac{2}{3}x + 1$$
which is the same result we obtained using $(3, -1)$.

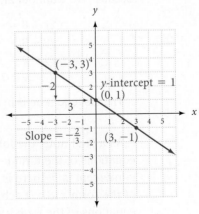

FIGURE 5

The last form of the equation of a line that we will consider in this section is called the *standard form*. It is used mainly to write equations in a form that is free of fractions and is easy to compare with other equations.

> **Property: Standard Form**
>
> If a, b, and c are integers, then the equation of a line is in standard form when it has the form
>
> $$ax + by = c$$

If we were to write the equation

$$y = -\frac{2}{3}x + 1$$

in standard form, we would first multiply both sides by 3 to obtain

$$3y = -2x + 3$$

Then we would add $2x$ to each side, yielding

$$2x + 3y = 3$$

which is a linear equation in standard form.

EXAMPLE 6   Give the equation of the line through $(-1, 4)$ whose graph is perpendicular to the graph of $2x - y = -3$. Write the answer in standard form.

SOLUTION   To find the slope of $2x - y = -3$, we solve for y.

$$2x - y = -3$$
$$y = 2x + 3$$

The slope of this line is 2. The line we are interested in is perpendicular to the line with slope 2 and must, therefore, have a slope of $-\frac{1}{2}$.

Using $(x_1, y_1) = (-1, 4)$ and $m = -\frac{1}{2}$, we have

$$y - y_1 = m(x - x_1)$$
$$y - 4 = -\frac{1}{2}(x + 1)$$

Because we want our answer in standard form, we multiply each side by 2.

$$2y - 8 = -1(x + 1)$$
$$2y - 8 = -x - 1$$
$$x + 2y - 8 = -1$$
$$x + 2y = 7$$

The last equation is in standard form.

As a final note, the following summary reminds us that all horizontal lines have equations of the form $y = b$ and slopes of 0. Because they cross the y-axis at b, the y-intercept is b; there is no x-intercept. Vertical lines have no slope and equations of the form $x = a$. Each will have an x-intercept at a and no y-intercept. Finally, equations of the form $y = mx$ have graphs that pass through the origin. The slope is always m and both the x-intercept and the y-intercept are 0.

FACTS FROM GEOMETRY: Special Equations and Their Graphs, Slopes, and Intercepts

For the equations below, m, a, and b are real numbers.

Equation: $y = mx$
Slope $= m$
x-intercept $= 0$
y-intercept $= 0$

Equation: $x = a$
No slope
x-intercept $= a$
No y-intercept

Equation: $y = b$
Slope $= 0$
No x-intercept
y-intercept $= b$

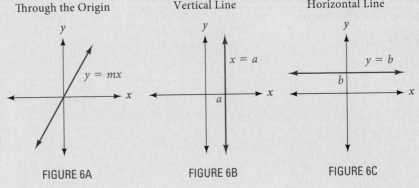

Through the Origin

Vertical Line

Horizontal Line

FIGURE 6A

FIGURE 6B

FIGURE 6C

USING TECHNOLOGY

Graphing Calculators

One advantage of using a graphing calculator to graph lines is that a calculator does not care whether the equation has been simplified or not. To illustrate, in Example 5 we found that the equation of the line with slope $-\frac{2}{3}$ that passes through the point $(3, -1)$ is

$$y + 1 = -\frac{2}{3}(x - 3)$$

Normally, to graph this equation we would simplify it first. With a graphing calculator, we add -1 to each side and enter the equation this way:

$$Y_1 = -(2/3)(X - 3) - 1$$

Problem Set 8.2

Moving Toward Success

"There is no chance, no destiny, no fate, that can hinder or control the firm resolve of a determined soul."

—Ella Wheeler Wilcox, 1850–1919, American author and poet

1. Do you have rules for yourself while you study? Do they yield a quality studying experience? Why or why not?

2. How can you improve your studying routine?

A Write the equation of the line with the given slope and y-intercept. [Example 1]

1. $m = 2, b = 3$

2. $m = -4, b = 2$

3. $m = 1, b = -5$

4. $m = -5, b = -3$

5. $m = \frac{1}{2}, b = \frac{3}{2}$

6. $m = \dfrac{2}{3}, b = \dfrac{5}{6}$

7. $m = 0, b = 4$

8. $m = 0, b = -2$

Give the slope and y-intercept for each of the following equations. Sketch the graph using the slope and y-intercept. Give the slope of any line perpendicular to the given line. [Example 3]

9. $y = 3x - 2$

10. $y = 2x + 3$

11. $2x - 3y = 12$

12. $3x - 2y = 12$

13. $4x + 5y = 20$

14. $5x - 4y = 20$

For each of the following lines, name the slope and y-intercept. Then write the equation of the line in slope-intercept form.

15.

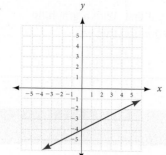

16.

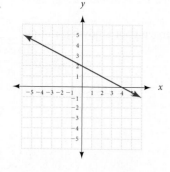

17.

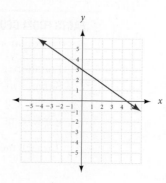

18.

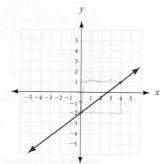

C For each of the following problems, the slope and one point on a line are given. In each case, find the equation of that line. (Write the equation for each line in slope-intercept form.) [Example 4]

19. $(-2, -5); m = 2$

20. $(-1, -5); m = 2$

21. $(-4, 1); m = -\dfrac{1}{2}$

22. $(-2, 1); m = -\dfrac{1}{2}$

23. $\left(-\dfrac{1}{3}, 2\right); m = -3$

24. $\left(-\dfrac{2}{3}, 5\right); m = -3$

25. $(-4, -2), m = \dfrac{2}{3}$

26. $(3, 4), m = -\dfrac{1}{3}$

27. $(-5, 2), m = -\dfrac{1}{4}$

28. $(-4, 3), m = \dfrac{1}{6}$

D Find the equation of the line that passes through each pair of points. Write your answers in standard form. [Example 5]

29. $(-2, -4), (1, -1)$

30. $(2, 4), (-3, -1)$

31. $(-1, -5), (2, 1)$

32. $(-1, 6), (1, 2)$

33. $\left(\frac{1}{3}, -\frac{1}{5}\right), \left(-\frac{1}{3}, -1\right)$

34. $\left(-\frac{1}{2}, -\frac{1}{2}\right), \left(\frac{1}{2}, \frac{1}{10}\right)$

D Find the slope of a line (a) parallel and (b) perpendicular to the given line. [Example 6]

35. $y = 3x - 4$

36. $y = -4x + 1$

37. $3x + y = -2$

38. $2x - y = -4$

39. $2x + 5y = -11$

40. $3x - 5y = -4$

41. The equation $3x - 2y = 10$ is a linear equation in standard form. From this equation, answer the following:

  a. Find the x and y intercepts.

  b. Find a solution to this equation other than the intercepts in part a.

  c. Write this equation in slope-intercept form.

  d. Is the point $(2, 2)$ a solution to the equation?

42. The equation $4x + 3y = 8$ is a linear equation in standard form. From this equation, answer the following:

  a. Find the x and y intercepts.

  b. Find a solution to this equation other than the intercepts in part a.

  c. Write this equation in slope-intercept form.

  d. Is the point $(-3, 2)$ a solution to the equation?

43. The equation $\frac{3x}{4} - \frac{y}{2} = 1$ is a linear equation in standard form. From this equation, answer the following:

  a. Find the x and y intercepts.

  b. Find a solution to this equation other than the intercepts in part a.

  c. Write this equation in slope-intercept form.

  d. Is the point $(1, 2)$ a solution to the equation?

44. The equation $\frac{3x}{5} + \frac{2y}{3} = 1$ is a linear equation in standard form. From this equation, answer the following:

  a. Find the x and y intercepts.

  b. Find a solution to this equation other than the intercepts in part a.

  c. Write this equation in slope-intercept form.

  d. Is the point $(-5, 3)$ a solution to the equation?

The next two problems are intended to give you practice reading, and paying attention to the instructions that accompany the problems you are working. Working these problems is an excellent way to get ready for a test or a quiz.

45. Work each problem according to the instructions given.

  a. Solve: $-2x + 1 = -3$

  b. Find x when y is 0: $-2x + y = -3$

  c. Find y when x is 0: $-2x + y = -3$

  d. Graph: $-2x + y = -3$

  e. Solve for y: $-2x + y = -3$

46. Work each problem according to the instructions given.

  a. Solve: $\frac{x}{3} + \frac{1}{4} = 1$

  b. Find x when y is 0: $\frac{x}{3} + \frac{y}{4} = 1$

  c. Find y when x is 0: $\frac{x}{3} + \frac{y}{4} = 1$

  d. Graph: $\frac{x}{3} + \frac{y}{4} = 1$

  e. Solve for y: $\frac{x}{3} + \frac{y}{4} = 1$

For each of the following lines, name the coordinates of any two points on the line. Then use those two points to find the equation of the line.

47.

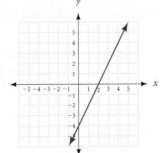

48.

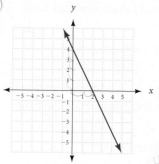

49.

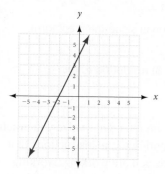

50.

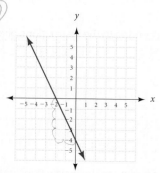

51. Give the slope and y-intercept of $y = -2$. Sketch the graph.

52. For the line $x = -3$, sketch the graph, give the slope, and name any intercepts.

53. Find the equation of the line parallel to the graph of $3x - y = 5$ that contains the point $(-1, 4)$.

54. Find the equation of the line parallel to the graph of $2x - 4y = 5$ that contains the point $(0, 3)$.

55. Line l is perpendicular to the graph of the equation $2x - 5y = 10$ and contains the point $(-4, -3)$. Find the equation for l.

56. Line l is perpendicular to the graph of the equation $-3x - 5y = 2$ and contains the point $(2, -6)$. Find the equation for l.

57. Give the equation of the line perpendicular to the graph of $y = -4x + 2$ that has an x-intercept of -1.

58. Write the equation of the line parallel to the graph of $7x - 2y = 14$ that has an x-intercept of 5.

59. Find the equation of the line with x-intercept 3 and y-intercept 2.

60. Find the equation of the line with x-intercept 2 and y-intercept 3.

Applying the Concepts

61. Horse Racing The graph shows the total amount of money wagered on the Kentucky Derby. Find an equation for the line segment between 2000 and 2005. Write the equation in slope-intercept form. Round the slope and intercept to the nearest tenth.

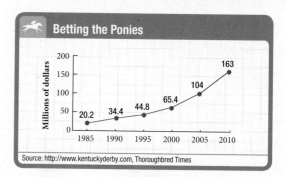

Source: http://www.kentuckyderby.com, Thoroughbred Times

62. Solar Energy The graph shows the annual number of solar thermal collector shipments in the United States. Find an equation for the line segment that connects the points (2004, 13,750) and (2005, 15,000). Write your answer in slope-intercept form.

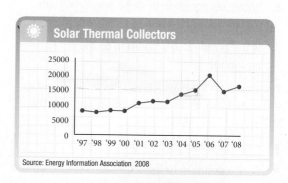

Source: Energy Information Association 2008

63. Deriving the Temperature Equation The table from the introduction to this section is repeated here. The rows of the table give us ordered pairs (C, F).

Degrees Celsius	Degrees Fahrenheit
C	F
0	32
25	77
50	122
75	167
100	212

a. Use any two of the ordered pairs from the table to derive the equation $F = \frac{9}{5}C + 32$.

b. Use the equation from part (a) to find the Fahrenheit temperature that corresponds to a Celsius temperature of 30°.

64. Maximum Heart Rate The table gives the maximum heart rate for adults 30, 40, 50, and 60 years old. Each row of the table gives us an ordered pair (A, M).

Age (years)	Maximum Heart Rate (beats per minute)
A	**M**
30	190
40	180
50	170
60	160

a. Use any two of the ordered pairs from the table to derive the equation $M = 220 - A$, which gives the maximum heart rate M for an adult whose age is A.

b. Use the equation from part (a) to find the maximum heart rate for a 25-year-old adult.

65. Textbook Cost To produce this textbook, suppose the publisher spent $215,000 for typesetting and $9.50 per book for printing and binding. The total cost to produce and print n books can be written as:

Total cost to produce n textbooks	=	Fixed cost + Variable cost
C	=	$215,000 + 9.5n$

a. Suppose the number of books printed in the first printing is 10,000. What is the total cost?

b. If the average cost is the total cost divided by the number of books printed, find the average cost of producing 10,000 textbooks.

c. Find the cost to produce one more textbook when you have already produced 10,000 textbooks.

66. Exercise Heart Rate In an aerobics class, the instructor indicates that her students' exercise heart rate is 60% of their maximum heart rate, where maximum heart rate is 220 minus their age.

a. Determine the equation that gives exercise heart rate E in terms of age A.

b. Use the equation to find the exercise heart rate of a 22-year-old student. Round to the nearest whole number.

67. Daniel starts a new job in which he sells TVs. He earns $1,000 a month plus a certain amount for each TV he sells. The graph below shows the amount per month Daniel will make based on how many TVs he sells.

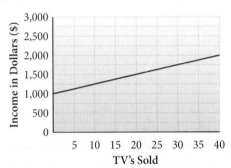

a. How much will he earn if he sells 20 TVs?

b. How much will he earn if he sells 40 TVs?

c. What is the slope of the line?

d. How much does he make for each TV he sells?

e. Find the equation of the line where y is his income and x is TV sales.

68. Mark gets a new job where he sells hats. He will earn $1,250 a month plus a certain amount for each hat he sells. The graph below shows the amount Mark will make based on how many hats he sells.

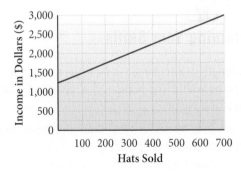

a. How much will he earn if he sells 300 hats?

b. How much does he have to sell to earn $2,500?

c. What is the slope of the line?

d. How much does he make for each hat he sells?

e. Find the equation of the line.

Improving Your Quantitative Literacy

69. Courtney buys a new car. It cost $20,000 and will decrease in value each year. The graph below shows the value of the car after five years.

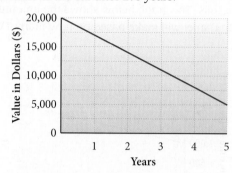

a. How much is the car worth after five years?

b. Find the equation of the line connecting the endpoints.

c. Using your equation, how much will the car be worth in ten years? What's wrong with the answer?

70. Definitions

a. Use an online dictionary to find definitions for the following words: *interpolation* and *extrapolation*

b. Which of the two words describes what you did in problem 69 part (c) above.

Maintaining Your Skills

71. The length of a rectangle is 3 inches more than 4 times the width. The perimeter is 56 inches. Find the length and width.

72. One angle is 10 degrees less than four times another. Find the measure of each angle if:

a. The two angles are complementary.

b. The two angles are supplementary.

73. The cash register in a candy shop contains $66 at the beginning of the day. At the end of the day, it contains $732.50. If the sales tax rate is 7.5%, how much of the total is sales tax?

74. The third angle in an isosceles triangle is 20 degrees less than twice as large as each of the two base angles. Find the measure of each angle.

Getting Ready for the Next Section

Complete each table using the given equation.

75. $y = 7.5x$

x	y
0	
10	
20	

76. $h = 32t - 16t^2$

t	h
0	
$\frac{1}{2}$	
1	

77. $x = y^2$

x	y
	0
	1
	-1

78. $y = |x|$

x	y
-3	
0	
3	

Extending The Concepts

The midpoint M of two points (x_1, y_1) and (x_2, y_2) is defined to be the average of each of their coordinates, so

$$M = \left(\frac{x_1 + x_2}{2}, \frac{y_1 + y_2}{2} \right)$$

For example, the midpoint of $(-2, 3)$ and $(6, 8)$ is given by

$$\left(\frac{-2 + 6}{2}, \frac{3 + 8}{2} \right) = \left(2, \frac{11}{2} \right)$$

For each given pair of points, find the equation of the line that is perpendicular to the line through these points and that passes through their midpoint. Answer using slope-intercept form.

Note: This line is called the perpendicular bisector of the line segment connecting the two points.

79. $(1, 4)$ and $(7, 8)$

80. $(-2, 1)$ and $(6, 7)$

81. $(-5, 1)$ and $(-1, 4)$

82. $(-6, -2)$ and $(2, 1)$

8.3 Introduction to Functions

OBJECTIVES

A Construct a table or a graph from a function rule.

B Identify the domain and range of a function or a relation.

C Determine whether a relation is also a function.

D Use the vertical line test to determine whether a graph is a function.

TICKET TO SUCCESS

Keep these questions in mind as you read through the section. Then respond in your own words and in complete sentences.

1. Define function using the terms domain and range.
2. Which variable is usually associated with the domain of a function?
3. What is a relation?
4. Explain the vertical line test.

Image copyright ©ilker canikligil, 2009. Used under license from Shutterstock.com

International Laws of the Game for soccer state that the length of the playing field can be between 100 and 130 yards. Suppose you are laying out a field at a local park. The width of the field must be 60 yards due to some landscaping. How would you decide what the area of the field could be? In this section, we will learn how to find the area y as a function of the length x to lay out your regulation soccer field.

An Informal Look at Functions

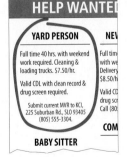

The ad shown in the margin appeared in the help wanted section of the local newspaper the day I was writing this section of the book. We can use the information in the ad to continue an informal discussion of functions.

To begin, suppose you have a job that pays $7.50 per hour and that you work anywhere from 0 to 40 hours per week. The amount of money you make in one week depends on the number of hours you work that week. In mathematics, we say that your weekly earnings are a *function* of the number of hours you work. If we let the variable x represent hours and the variable y represent the money you make, then the relationship between x and y can be written as

$$y = 7.5x \quad \text{for} \quad 0 \le x \le 40$$

A Constructing Tables and Graphs

EXAMPLE 1 Construct a table and graph for the function

$$y = 7.5x \quad \text{for} \quad 0 \le x \le 40$$

SOLUTION Table 1 gives some of the paired data that satisfy the equation $y = 7.5x$. Figure 1 is the graph of the equation with the restriction $0 \leq x \leq 40$.

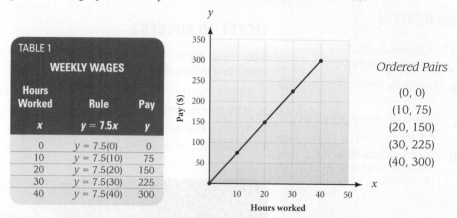

TABLE 1

WEEKLY WAGES

Hours Worked	Rule	Pay
x	$y = 7.5x$	y
0	$y = 7.5(0)$	0
10	$y = 7.5(10)$	75
20	$y = 7.5(20)$	150
30	$y = 7.5(30)$	225
40	$y = 7.5(40)$	300

Ordered Pairs

(0, 0)
(10, 75)
(20, 150)
(30, 225)
(40, 300)

FIGURE 1 **Weekly wages at $7.50 per hour**

The equation $y = 7.5x$ with the restriction $0 \leq x \leq 40$, Table 1, and Figure 1 are three ways to describe the same relationship between the number of hours you work in one week and your gross pay for that week. In all three, we *input* values of x, and then use the function rule to *output* values of y.

B Domain and Range of a Function

We began this discussion by saying that the number of hours worked during the week was from 0 to 40, so these are the values that x can assume. From the line graph in Figure 1, we see that the values of y range from 0 to 300. We call the complete set of values that x can assume the *domain* of the function. The values that are assigned to y are called the *range* of the function.

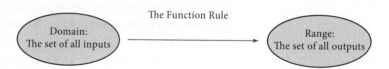

EXAMPLE 2 State the domain and range for the function

$$y = 7.5x, \quad 0 \leq x \leq 40$$

SOLUTION From the previous discussion, we have

$$\text{Domain} = \{x \mid 0 \leq x \leq 40\}$$

$$\text{Range} = \{y \mid 0 \leq y \leq 300\}$$

Function Maps

Another way to visualize the relationship between x and y is with the diagram in Figure 2, which we call a *function map*:

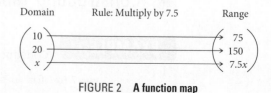

FIGURE 2 **A function map**

Although Figure 2 does not show all the values that x and y can assume, it does give us a visual description of how x and y are related. It shows that values of y in the range come from values of x in the domain according to a specific rule (multiply by 7.5 each time).

A Formal Look at Functions

What is apparent from the preceding discussion is that we are working with paired data. The solutions to the equation $y = 7.5x$ are pairs of numbers; the points on the line graph in Figure 1 come from paired data; and the diagram in Figure 2 pairs numbers in the domain with numbers in the range. We are now ready for the formal definition of a function.

> **Definition**
>
> A **function** is a rule that pairs each element in one set, called the **domain**, with exactly one element from a second set, called the **range**.

In other words, a function is a rule for which each input is paired with exactly one output.

Furthermore, the function rule $y = 7.5x$ from Example 1 produces ordered pairs of numbers (x, y). The same thing happens with all functions: The function rule produces ordered pairs of numbers. We use this result to write an alternative definition for a function.

> **Alternative Definition**
>
> A **function** is a set of ordered pairs in which no two different ordered pairs have the same first coordinate. The set of all first coordinates is called the **domain** of the function. The set of all second coordinates is called the **range** of the function.

The restriction on first coordinates in the alternative definition keeps us from assigning a number in the domain to more than one number in the range.

C   A Relationship That Is Not a Function

You may be wondering if any sets of paired data fail to qualify as functions. The answer is yes, as the next example reveals.

EXAMPLE 3   Table 2 shows the prices of used Ford Mustangs that were listed in the local newspaper. The diagram in Figure 3 is called a *scatter diagram*. It gives a visual representation of the data in Table 2. Why is this data not a function?

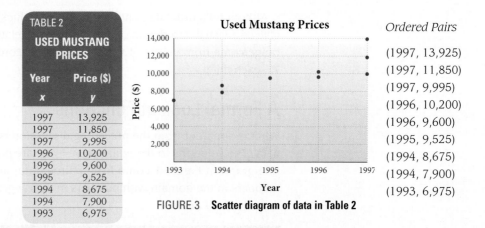

TABLE 2	
USED MUSTANG PRICES	
Year	**Price ($)**
x	**y**
1997	13,925
1997	11,850
1997	9,995
1996	10,200
1996	9,600
1995	9,525
1994	8,675
1994	7,900
1993	6,975

FIGURE 3 Scatter diagram of data in Table 2

Ordered Pairs

(1997, 13,925)
(1997, 11,850)
(1997, 9,995)
(1996, 10,200)
(1996, 9,600)
(1995, 9,525)
(1994, 8,675)
(1994, 7,900)
(1993, 6,975)

SOLUTION In Table 2, the year 1997 is paired with three different prices: $13,925, $11,850, and $9,995. That is enough to disqualify the data from belonging to a function. For a set of paired data to be considered a function, each number in the domain must be paired with exactly one number in the range.

Still, there is a relationship between the first coordinates and second coordinates in the used-car data in Example 3. It is not a function relationship, but it is a relationship. To classify all relationships specified by ordered pairs, whether they are functions or not, we include the following two definitions.

> **Definition**
> A **relation** is a rule that pairs each element in one set, called the **domain**, with one or more elements from a second set, called the **range**.

> **Alternative Definition**
> A **relation** is a set of ordered pairs. The set of all first coordinates is the **domain** of the relation. The set of all second coordinates is the **range** of the relation.

Here are some facts that will help clarify the distinction between relations and functions.

1. Any rule that assigns numbers from one set to numbers in another set is a relation. If that rule makes the assignment so that no input has more than one output, then it is also a function.

2. Any set of ordered pairs is a relation. If none of the first coordinates of those ordered pairs is repeated, the set of ordered pairs is also a function.

3. Every function is a relation.

4. Not every relation is a function.

Graphing Relations and Functions

To give ourselves a wider perspective on functions and relations, we consider some equations whose graphs are not straight lines.

EXAMPLE 4 Kendra is tossing a softball into the air with an underhand motion. The distance of the ball above her hand at any time is given by the function

$$h = 32t - 16t^2 \quad \text{for} \quad 0 \leq t \leq 2$$

where h is the height of the ball in feet and t is the time in seconds. Construct a table that gives the height of the ball at quarter-second intervals, starting with $t = 0$ and ending with $t = 2$. Construct a line graph from the table.

SOLUTION We construct Table 3 using the following values of t: 0, $\frac{1}{4}$, $\frac{1}{2}$, $\frac{3}{4}$, 1, $\frac{5}{4}$, $\frac{3}{2}$, $\frac{7}{4}$, 2. The values of h come from substituting these values of t into the equation $h = 32t - 16t^2$. (This equation comes from physics. If you take a physics class, you will learn how to derive this equation.) Then we construct the graph in Figure 4 from the table. The graph appears only in the first quadrant because neither t nor h can be negative.

TABLE 3		
	TOSSING A SOFTBALL INTO THE AIR	
Time (sec)	**Function Rule**	**Distance (ft)**
t	$h = 32t - 16t^2$	h
0	$h = 32(0) - 16(0)^2 = 0 - 0 = 0$	0
$\frac{1}{4}$	$h = 32\left(\frac{1}{4}\right) - 16\left(\frac{1}{4}\right)^2 = 8 - 1 = 7$	7
$\frac{1}{2}$	$h = 32\left(\frac{1}{2}\right) - 16\left(\frac{1}{2}\right)^2 = 16 - 4 = 12$	12
$\frac{3}{4}$	$h = 32\left(\frac{3}{4}\right) - 16\left(\frac{3}{4}\right)^2 = 24 - 9 = 15$	15
1	$h = 32(1) - 16(1)^2 = 32 - 16 = 16$	16
$\frac{5}{4}$	$h = 32\left(\frac{5}{4}\right) - 16\left(\frac{5}{4}\right)^2 = 40 - 25 = 15$	15
$\frac{3}{2}$	$h = 32\left(\frac{3}{2}\right) - 16\left(\frac{3}{2}\right)^2 = 48 - 36 = 12$	12
$\frac{7}{4}$	$h = 32\left(\frac{7}{4}\right) - 16\left(\frac{7}{4}\right)^2 = 56 - 49 = 7$	7
2	$h = 32(2) - 16(2)^2 = 64 - 64 = 0$	0

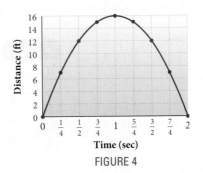

FIGURE 4

Here is a summary of what we know about functions as it applies to this example: We input values of t and output values of h according to the function rule

$$h = 32t - 16t^2 \quad \text{for} \quad 0 \leq t \leq 2$$

The domain is given by the inequality that follows the equation; it is

$$\text{Domain} = \{t \mid 0 \leq t \leq 2\}$$

The range is the set of all outputs that are possible by substituting the values of t from the domain into the equation. From our table and graph, it seems that the range is

$$\text{Range} = \{h \mid 0 \leq h \leq 16\}$$

USING TECHNOLOGY

More About Example 4

Most graphing calculators can easily produce the information in Table 3. Simply set Y_1 equal to $32X - 16X^2$. Then set up the table so it starts at 0 and increases by an increment of 0.25 each time. (On a TI-83/84, use the TBLSET key to set up the table.)

Table Setup

Table minimum = 0
Table increment = .25
Dependent variable: Auto
Independent variable: Auto

Y-Variables Setup

$Y_1 = 32X - 16X^2$

The table will look like this:

X	Y_1
0.00	0
0.25	7
0.50	12
0.75	15
1.00	16
1.25	15
1.50	12

EXAMPLE 5 Sketch the graph of $x = y^2$.

SOLUTION Without going into much detail, we graph the equation $x = y^2$ by finding a number of ordered pairs that satisfy the equation, plotting these points, and then drawing a smooth curve that connects them. A table of values for x and y that satisfy the equation follows, along with the graph of $x = y^2$ shown in Figure 5.

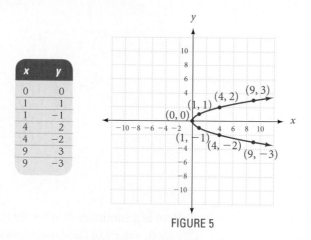

x	y
0	0
1	1
1	−1
4	2
4	−2
9	3
9	−3

FIGURE 5

As you can see from looking at the table and the graph in Figure 5, several ordered pairs whose graphs lie on the curve have repeated first coordinates. For instance, $(1, 1)$ and $(1, -1)$, $(4, 2)$ and $(4, -2)$, and $(9, 3)$ and $(9, -3)$. The graph is therefore not the graph of a function.

D Vertical Line Test

Look back at the scatter diagram for used Mustang prices shown in Figure 3. Notice that some of the points on the diagram lie above and below each other along vertical lines. This is an indication that the data do not constitute a function. Two data points that lie on the same vertical line must have come from two ordered pairs with the same first coordinates.

Now, look at the graph shown in Figure 5. The reason this graph is the graph of a relation, but not of a function, is that some points on the graph have the same first coordinates—for example, the points (4, 2) and (4, −2). Furthermore, any time two points on a graph have the same first coordinates, those points must lie on a vertical line. [To convince yourself, connect the points (4, 2) and (4, −2) with a straight line. You will see that it must be a vertical line.] This allows us to write the following test that uses the graph to determine whether a relation is also a function.

> **Vertical Line Test**
> If a vertical line crosses the graph of a relation in more than one place, the relation cannot be a function. If no vertical line can be found that crosses a graph in more than one place, then the graph is the graph of a function.

If we look back to the graph of $h = 32t - 16t^2$ as shown in Figure 4, we see that no vertical line can be found that crosses this graph in more than one place. The graph shown in Figure 4 is therefore the graph of a function.

EXAMPLE 6   Graph $y = |x|$. Use the graph to determine whether we have the graph of a function. State the domain and range.

SOLUTION   We let x take on values of −4, −3, −2, −1, 0, 1, 2, 3, and 4. The corresponding values of y are shown in the table. The graph is shown in Figure 6.

x	y
−4	4
−3	3
−2	2
−1	1
0	0
1	1
2	2
3	3
4	4

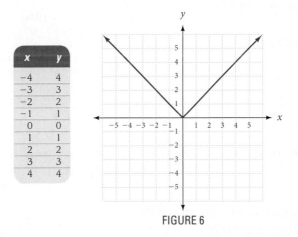

FIGURE 6

Because no vertical line can be found that crosses the graph in more than one place, $y = |x|$ is a function. The domain is all real numbers. The range is $\{y \mid y \geq 0\}$.

Problem Set 8.3

1. Why should you mentally list the items that are known and unknown in a word problem?

2. Why is making a list of difficult problems important?

B For each of the following relations, give the domain and range, and indicate which are also functions. [Example 2]

1. {(1, 3), (2, 5), (4, 1)}

2. {(3, 1), (5, 7), (2, 3)}

3. {(−1, 3), (1, 3), (2, −5)}

4. {(3, −4), (−1, 5), (3, 2)}

5. {(7, −1), (3, −1), (7, 4)}

6. {(5, −2), (3, −2), (5, −1)}

7. {(a, 3), (b, 4), (c, 3), (d, 5)}

8. {(a, 5), (b, 5), (c, 4), (d, 5)}

9. {(a, 1), (a, 2), (a, 3), (a, 4)}

10. {(a, 1), (b, 1), (c, 1), (d, 1)}

D State whether each of the following graphs represents a function. [Example 6]

11.

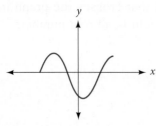

12.

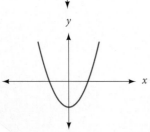

13.

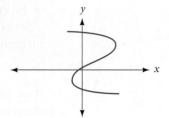

14.

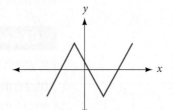

15.

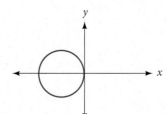

16.

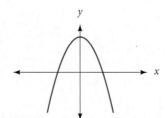

17.

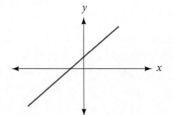

18.

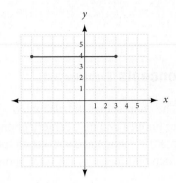

19.

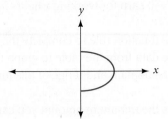

20.

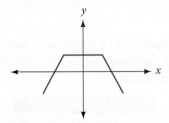

23.

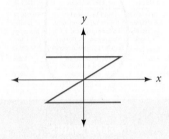

24.

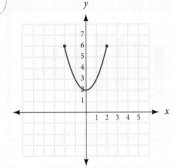

B Determine the domain and range of the following functions. Assume the entire function is shown. [Example 2]

21.

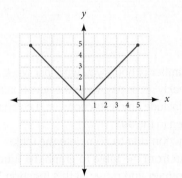

22.

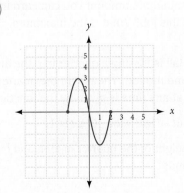

Graph each of the following relations. In each case, use the graph to find the domain and range, and indicate whether the graph is the graph of a function. [Example 5]

25. $y = x^2 - 1$

26. $y = x^2 + 1$

27. $y = x^2 + 4$

28. $y = x^2 - 9$

29. $x = y^2 - 1$

30. $x = y^2 + 1$

31. $x = y^2 + 4$

32. $x = y^2 - 9$

33. $y = |x - 2|$

34. $y = |x + 2|$

35. $y = |x| - 2$

36. $y = |x| + 2$

Applying the Concepts [Examples 3–4]

37. Camera Phones The chart shows the estimated number of camera phones and non-camera phones sold from 2004 to 2010. Using the chart, list all the values in the domain and range for the total phones sales.

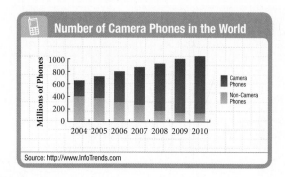

38. Light Bulbs The chart shows a comparison of power usage between incandescent and LED light bulbs. Use the chart to state the domain and range of the function for an LED bulb.

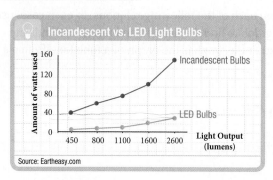

39. Weekly Wages Suppose you have a job that pays $8.50 per hour and you work anywhere from 10 to 40 hours per week.

TABLE 4		
WEEKLY WAGES		
Hours Worked	**Function Rule**	**Gross Pay ($)**
x		y
10		
20		
30		
40		

a. Write an equation, with a restriction on the variable x, that gives the amount of money,

y, you will earn for working x hours in one week.

b. Use the function rule to complete the table.

c. Use the data from the table to graph the function.

d. State the domain and range of this function.

e. What is the minimum amount you can earn in a week with this job? What is the maximum amount?

40. Weekly Wages The ad shown here was in the local newspaper. Suppose you are hired for the job described in the ad.

312 HELP WANTED

TTER	ESPRESSO BAR	NEW
LDREN	OPERATOR	DELIV
16 years old	Must be dependable,	Must be 16
h children	honest, serivce-oriented.	Hardw
l, friendly	Coffee exp desired. 15-30	Morning
hr/week	hrs per wk. $5.25/hour.	10-12 hou
) 444-1237	Start 5/31. Apply in person:	$6.00/ho
5/25	*Espresso Yourself*	*Tribu*
	Central Coast Mall	*San L*
	Deadline 5/23	Dea

TABLE 5		
WEEKLY WAGES		
Hours Worked	**Function Rule**	**Gross Pay ($)**
x		y
15		
20		
25		
30		

a. If x is the number of hours you work per week and y is your weekly gross pay, write the equation for y. (Be sure to include any restrictions on the variable x that are given in the ad.)

b. Use the function rule to complete the table.

c. Use the data from the table to graph the function.

d. State the domain and range of this function.

e. What is the minimum amount you can earn in a week with this job? What is the maximum amount?

41. Tossing a Coin Hali is tossing a quarter into the air with an underhand motion. The height the quarter is above her hand at any time is given by the function

$$h = 16t - 16t^2 \quad \text{for} \quad 0 \leq t \leq 1$$

where h is the height of the quarter in feet, and t is the time in seconds.

a. Fill in the table.

Time (sec)	Function Rule	Height (ft)
t	$h = 16t - 16t^2$	h
0		
0.1		
0.2		
0.3		
0.4		
0.5		
0.6		
0.7		
0.8		
0.9		
1		

b. State the domain and range of this function.

c. Use the data from the table to graph the function.

42. Intensity of Light The formula below gives the intensity of light that falls on a surface at various distances from a 100-watt light bulb:

$$I = \frac{120}{d^2} \quad \text{for} \quad d > 0$$

where I is the intensity of light (in lumens per square foot), and d is the distance (in feet) from the light bulb to the surface.

a. Fill in the table.

Distance (ft)	Function Rule	Intensity
d	$I = \dfrac{120}{d^2}$	I
1		
2		
3		
4		
5		
6		

b. Use the data from the table to graph the function.

43. Area of a Circle The formula for the area A of a circle with radius r is given by $A = \pi r^2$. The formula shows that A is a function of r.

a. Graph the function $A = \pi r^2$ for $0 \le r \le 3$. (On the graph, let the horizontal axis be the r-axis, and let the vertical axis be the A-axis.)

b. State the domain and range of the function $A = \pi r^2$, $0 \le r \le 3$.

44. Area and Perimeter of a Rectangle A rectangle is 2 inches longer than it is wide. Let x = the width, P = the perimeter, and A = the area of the rectangle.

a. Write an equation that will give the perimeter P in terms of the width x of the rectangle. Are there any restrictions on the values that x can assume?

b. Graph the relationship between P and x.

45. Tossing a Ball A ball is thrown straight up into the air from ground level. The relationship between the height h of the ball at any time t is illustrated by the following graph:

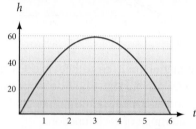

The horizontal axis represents time t, and the vertical axis represents height h.

a. Is this graph the graph of a function?

b. State the domain and range.

c. At what time does the ball reach its maximum height?

d. What is the maximum height of the ball?

e. At what time does the ball hit the ground?

46. Company Profits The amount of profit a company earns is based on the number of items it sells. The relationship between the profit P and number of items it sells x, is illustrated by the following graph:

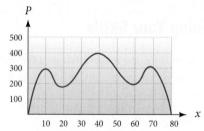

The horizontal axis represents items sold, x, and the vertical axis represents the profit, P.

a. Is this graph the graph of a function?

b. State the domain and range.

c. How many items must the company sell to make their maximum profit?

d. What is their maximum profit?

47. Profits Match each of the following statements to the appropriate graph indicated by labels I-IV.

a. Sarah works 25 hours to earn $250

b. Justin works 35 hours to earn $560

c. Rosemary works 30 hours to earn $360

d. Marcus works 40 hours to earn $320

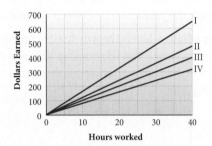

48. Find an equation for each of the functions shown in the graph for problem 47. Show dollars earned, E, as a function of hours worked, t. Then, indicate the domain and range of each function.

a. Graph I: $E =$
Domain = {t | }
Range = {E | }

b. Graph II: $E =$
Domain = {t | }
Range = {E | }

c. Graph III: $E =$
Domain = {t | }
Range = {E | }

d. Graph IV: $E =$
Domain = {t | }
Range = {E | }

Maintaining Your Skills

The problems that follow review some of the more important skills you have learned in previous sections and chapters.

For the equation $y = 3x - 2$:

49. Find y if x is 4.

50. Find y if x is 0.

51. Find y if x is −4.

52. Find y if x is −2.

For the equation $y = x^2 - 3$:

53. Find y if x is 2.

54. Find y if x is −2.

55. Find y if x is 0.

56. Find y if x is −4.

57. If $x - 2y = 4$ and $x = \dfrac{8}{5}$, find y.

58. If $5x - 10y = 15$, find y when x is 3.

59. Let $x = 0$ and $y = 0$ in $y = a(x - 80)^2 + 70$ and solve for a.

60. Find R if $p = 2.5$ and $R = (900 - 300p)p$.

Getting Ready for the Next Section

Simplify. Round to the nearest whole number if necessary.

61. $7.5(20)$

62. $60 \div 7.5$

63. $4(3.14)(9)$

64. $\dfrac{4}{3}(3.14) \cdot 3^3$

65. $4(-2) - 1$

66. $3(3)^2 + 2(3) - 1$

67. If $s = \dfrac{60}{t}$, find s when

a. $t = 10$ **b.** $t = 8$

68. If $y = 3x^2 + 2x - 1$, find y when

a. $x = 0$ **b.** $x = -2$

69. Find the value of $x^2 + 2$ for

a. $x = 5$ **b.** $x = -2$

70. Find the value of $125 \cdot 2^t$ for

a. $t = 0$ **b.** $t = 1$

Extending the Concepts

Graph each of the following relations. In each case, use the graph to find the domain and range, and indicate whether the graph is the graph of a function.

71. $y = 5 - |x|$

72. $y = |x| - 3$

73. $x = |y| + 3$

74. $x = 2 - |y|$

75. $|x| + |y| = 4$

76. $2|x| + |y| = 6$

8.4 Function Notation

OBJECTIVES

A Use function notation to find the value of a function for a given value of the variable.

B Use graphs to visualize the relationship between a function and a variable.

C Use function notation in formulas.

TICKET TO SUCCESS

Keep these questions in mind as you read through the section. Then respond in your own words and in complete sentences.

1. Explain what you are calculating when you find $f(2)$ for a given function f.
2. If $s(t) = \frac{60}{t}$, how do you find $s(10)$?
3. If $f(2) = 3$ for a function f, what is the relationship between the numbers 2 and 3 and the graph of f?
4. If $f(6) = 0$ for a particular function f, then you can immediately graph one of the intercepts. Explain.

Image copyright ©Rob Wilson, 2009. Used under license from Shutterstock.com

Let's return to the discussion that introduced us to functions. If the width of the field 60 yards is multiplied by the length x to get the area y, it can be said that area is a function of length. The exact relationship between x and y is written

$$y = 60x$$

Because the area depends on the length of the touchline, we call y the dependent variable and x the independent variable. If we let f represent all the ordered pairs produced by the equation, then we can write

$$f = \{(x, y) \mid y = 60x \text{ and } 100 \leq x \leq 130\}$$

A Evaluate a Function at a Point

If a job pays $7.50 per hour for working from 0 to 40 hours a week, then the amount of money y earned in one week is a function of the number of hours worked x. The exact relationship between x and y is written

$$y = 7.5x \quad \text{for} \quad 0 \leq x \leq 40$$

Because the amount of money earned, y, depends on the number of hours worked, x, we call y the *dependent variable* and x the *independent variable*. Furthermore, if we let f represent all the ordered pairs produced by the equation, then we can write

$$f = \{(x, y) \mid y = 7.5x \text{ and } 0 \leq x \leq 40\}$$

Once we have named a function with a letter, we can use an alternative notation to represent the dependent variable y. The alternative notation for y is $f(x)$. It is read "f of x" and can be used instead of the variable y when working with functions. The notation y and the notation $f(x)$ are equivalent; that is,

$$y = 7.5x \Leftrightarrow f(x) = 7.5x$$

When we use the notation $f(x)$ we are using *function notation*. The benefit of using function notation is that we can write more information with fewer symbols than we can by using just the variable y. For example, asking how much money a person will make for working 20 hours is simply a matter of asking for $f(20)$. Without function notation, we would have to say, "find the value of y that corresponds to a value of $x = 20$." To illustrate further, using the variable y, we can say, "y is 150 when x is 20." Using the notation $f(x)$, we simply say, "$f(20) = 150$." Each expression indicates that you will earn $150 for working 20 hours.

EXAMPLE 1 If $f(x) = 7.5x$, find $f(0)$, $f(10)$, and $f(20)$.

SOLUTION To find $f(0)$ we substitute 0 for x in the expression $7.5x$ and simplify. We find $f(10)$ and $f(20)$ in a similar manner—by substitution.

$$\text{If} \rightarrow \qquad f(x) = 7.5x$$
$$\text{then} \rightarrow \qquad f(\mathbf{0}) = 7.5(\mathbf{0}) = 0$$
$$f(\mathbf{10}) = 7.5(\mathbf{10}) = 75$$
$$f(\mathbf{20}) = 7.5(\mathbf{20}) = 150$$

If we changed the example in the discussion that opened this section so that the hourly wage was $6.50 per hour, we would have a new equation to work with.

$$y = 6.5x \qquad \text{for} \qquad 0 \le x \le 40$$

Suppose we name this new function with the letter g. Then

$$g = \{(x, y) \mid y = 6.5x \text{ and } 0 \le x \le 40\}$$

and

$$g(x) = 6.5x$$

If we want to talk about both functions in the same discussion, having two different letters, f and g, makes it easy to distinguish between them. For example, because $f(x) = 7.5x$ and $g(x) = 6.5x$, asking how much money a person makes for working 20 hours is simply a matter of asking for $f(20)$ or $g(20)$, avoiding any confusion over which hourly wage we are talking about.

The diagrams shown in Figure 1 further illustrate the similarities and differences between the two functions we have been discussing.

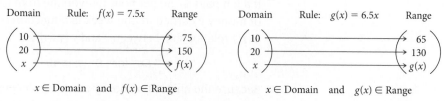

FIGURE 1   **Function maps**

NOTE

Some students like to think of functions as machines. Values of x are put into the machine, which transforms them into values of $f(x)$, which then are output by the machine.

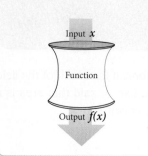

Input x

Function

Output $f(x)$

B Function Notation and Graphs

We can visualize the relationship between x and $f(x)$ or $g(x)$ on the graphs of the two functions. Figure 2 shows the graph of $f(x) = 7.5x$ along with two additional line segments. The horizontal line segment corresponds to $x = 20$, and the vertical line segment corresponds to $f(20)$. Figure 3 shows the graph of $g(x) = 6.5x$ along with the horizontal line segment that corresponds to $x = 20$, and the vertical line segment that corresponds to $g(20)$. (Note that the domain in each case is restricted to $0 \leq x \leq 40$.)

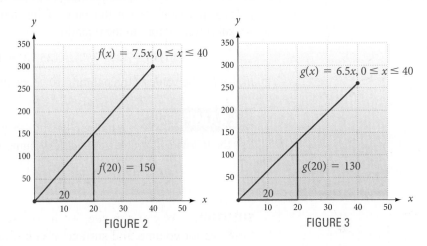

FIGURE 2   FIGURE 3

C Using Function Notation

The remaining examples in this section show a variety of ways to use and interpret function notation.

EXAMPLE 2   If it takes Lorena t minutes to run a mile, then her average speed $s(t)$ in miles per hour is given by the formula

$$s(t) = \frac{60}{t} \quad \text{for} \quad t > 0$$

Find $s(10)$ and $s(8)$, and then explain what they mean.

SOLUTION   To find $s(10)$, we substitute 10 for t in the equation and simplify:

$$s(\mathbf{10}) = \frac{60}{\mathbf{10}} = 6$$

In words: When Lorena runs a mile in 10 minutes, her average speed is 6 miles per hour.

We calculate $s(8)$ by substituting 8 for t in the equation. Doing so gives us

$$s(\mathbf{8}) = \frac{60}{\mathbf{8}} = 7.5$$

In words: Running a mile in 8 minutes is running at a rate of 7.5 miles per hour.

EXAMPLE 3   A painting is purchased as an investment for $125. If its value increases continuously so that it doubles every 5 years, then its value is given by the function

$$V(t) = 125 \cdot 2^{t/5} \quad \text{for} \quad t \geq 0$$

where t is the number of years since the painting was purchased, and $V(t)$ is its value (in dollars) at time t. Find $V(5)$ and $V(10)$, and explain what they mean.

SOLUTION   The expression $V(5)$ is the value of the painting when $t = 5$ (5 years after it is purchased). We calculate $V(5)$ by substituting 5 for t in the equation $V(t) = 125 \cdot 2^{t/5}$. Here is our work:

$$V(\mathbf{5}) = 125 \cdot 2^{5/5} = 125 \cdot 2^1 = 125 \cdot 2 = 250$$

In words: After 5 years, the painting is worth $250.

The expression $V(10)$ is the value of the painting after 10 years. To find this number, we substitute 10 for t in the equation:

$$V(\mathbf{10}) = 125 \cdot 2^{10/5} = 125 \cdot 2^2 = 125 \cdot 4 = 500$$

In words: The value of the painting 10 years after it is purchased is $500.

EXAMPLE 4   A balloon has the shape of a sphere with a radius of 3 inches. Use the following formulas to find the volume and surface area of the balloon.

$$V(r) = \frac{4}{3}\pi r^3 \qquad S(r) = 4\pi r^2$$

SOLUTION   As you can see, we have used function notation to write the two formulas for volume and surface area because each quantity is a function of the radius. To find these quantities when the radius is 3 inches, we evaluate $V(3)$ and $S(3)$.

$V(\mathbf{3}) = \dfrac{4}{3}\pi \mathbf{3}^3 = \dfrac{4}{3}\pi 27 = 36\pi$ cubic inches, or 113 cubic inches   To the nearest whole number

$S(\mathbf{3}) = 4\pi \mathbf{3}^2 = 36\pi$ square inches, or 113 square inches   To the nearest whole number

The fact that $V(3) = 36\pi$ means that the ordered pair $(3, 36\pi)$ belongs to the function V. Likewise, the fact that $S(3) = 36\pi$ tells us that the ordered pair $(3, 36\pi)$ is a member of function S.

We can generalize the discussion at the end of Example 4 this way:

$$(a, b) \in f \qquad \text{if and only if} \qquad f(a) = b$$

USING TECHNOLOGY

More About Example 4

If we look back at Example 4, we see that when the radius of a sphere is 3, the numerical values of the volume and surface area are equal. How unusual is this? Are there other values of r for which $V(r)$ and $S(r)$ are equal? We can answer this question by looking at the graphs of both V and S.

To graph the function $V(r) = \frac{4}{3}\pi r^3$, set $Y_1 = 4\pi X^3/3$. To graph $S(r) = 4\pi r^2$, set $Y_2 = 4\pi X^2$. Graph the two functions in each of the following windows:

> Window 1:   X from -4 to 4, Y from -2 to 10
>
> Window 2:   X from 0 to 4, Y from 0 to 50
>
> Window 3:   X from 0 to 4, Y from 0 to 150

Then use the Trace and Zoom features of your calculator to locate the point in the first quadrant where the two graphs intersect. How do the coordinates of this point compare with the results in Example 4?

EXAMPLE 5   If $f(x) = 3x^2 + 2x - 1$, find $f(0)$, $f(3)$, and $f(-2)$.

SOLUTION   Because $f(x) = 3x^2 + 2x - 1$, we have

$$f(0) = 3(0)^2 + 2(0) - 1 \quad = 0 + 0 - 1 = -1$$

$$f(3) = 3(3)^2 + 2(3) - 1 \quad = 27 + 6 - 1 = 32$$

$$f(-2) = 3(-2)^2 + 2(-2) - 1 = 12 - 4 - 1 = 7$$

In Example 5, the function f is defined by the equation $f(x) = 3x^2 + 2x - 1$. We could just as easily have said $y = 3x^2 + 2x - 1$; that is, $y = f(x)$. Saying $f(-2) = 7$ is exactly the same as saying y is 7 when x is -2.

EXAMPLE 6   If $f(x) = 4x - 1$ and $g(x) = x^2 + 2$, then

$f(5) = 4(5) - 1 = 19$	and	$g(5) = 5^2 + 2 = 27$
$f(-2) = 4(-2) - 1 = -9$	and	$g(-2) = (-2)^2 + 2 = 6$
$f(0) = 4(0) - 1 = -1$	and	$g(0) = 0^2 + 2 = 2$
$f(z) = 4z - 1$	and	$g(z) = z^2 + 2$
$f(a) = 4a - 1$	and	$g(a) = a^2 + 2$

USING TECHNOLOGY

More About Example 6

Most graphing calculators can use tables to evaluate functions. To work Example 6 using a graphing calculator table, set Y_1 equal to $4X - 1$ and Y_2 equal to $X^2 + 2$. Then set the independent variable in the table to Ask instead of Auto. Go to your table and input 5, -2, and 0. Under Y_1 in the table, you will find $f(5)$, $f(-2)$, and $f(0)$. Under Y_2, you will find $g(5)$, $g(-2)$, and $g(0)$.

Table Setup

Table minimum = 0
Table increment = 1
Independent variable: Ask
Dependent variable: Ask

Y Variables Setup

$Y_1 = 4X - 1$
$Y_2 = X^2 + 2$

The table will look like this:

X	Y_1	Y_2
5	19	27
-2	-9	6
0	-1	2

Although the calculator asks us for a table increment, the increment doesn't matter since we are inputting the X-values ourselves.

EXAMPLE 7   If the function f is given by

$$f = \{(-2, 0), (3, -1), (2, 4), (7, 5)\}$$

then $f(-2) = 0$, $f(3) = -1$, $f(2) = 4$, and $f(7) = 5$.

Problem Set 8.4

Moving Toward Success

"Nothing is to be rated higher than the value of the day."

—Johann Wolfgang von Goethe, 1749–1832, German dramatist and novelist

1. When should you review material for the next exam?

   a. Only at the last minute before an exam
   b. After you see it on the exam
   c. Each day
   d. Only after you learn it

2. Should you work problems when you review material? Why or why not?

A Let $f(x) = 2x - 5$ and $g(x) = x^2 + 3x + 4$. Evaluate the following. [Examples 1, 5–6]

1. $f(2)$
2. $f(3)$
3. $f(-3)$
4. $g(-2)$
5. $g(-1)$
6. $f(-4)$
7. $g(-3)$
8. $g(2)$
9. $g(4) + f(4)$
10. $f(2) - g(3)$
11. $f(3) - g(2)$
12. $g(-1) + f(-1)$

A Let $f(x) = 3x^2 - 4x + 1$ and $g(x) = 2x - 1$. Evaluate the following. [Examples 1, 5–6]

13. $f(0)$
14. $g(0)$
15. $g(-4)$
16. $f(1)$
17. $f(-1)$
18. $g(-1)$
19. $g(10)$
20. $f(10)$
21. $f(3)$
22. $g(3)$
23. $g\left(\dfrac{1}{2}\right)$
24. $g\left(\dfrac{1}{4}\right)$
25. $f(a)$
26. $g(b)$

A If $f = \{(1, 4), (-2, 0), \left(3, \dfrac{1}{2}\right), (\pi, 0)\}$ and $g = \{(1, 1), (-2, 2), \left(\dfrac{1}{2}, 0\right)\}$, find each of the following values of f and g. [Example 7]

27. $f(1)$
28. $g(1)$
29. $g\left(\dfrac{1}{2}\right)$
30. $f(3)$
31. $g(-2)$
32. $f(\pi)$

A Let $f(x) = 2x^2 - 8$ and $g(x) = \dfrac{1}{2}x + 1$. Evaluate each of the following. [Examples 1, 5–6]

33. $f(0)$
34. $g(0)$
35. $g(-4)$
36. $f(1)$
37. $f(a)$
38. $g(z)$

39. $f(b)$
40. $g(t)$

41. Graph the function $f(x) = \dfrac{1}{2}x + 2$. Then draw and label the line segments that represent $x = 4$ and $f(4)$.

42. Graph the function $f(x) = -\dfrac{1}{2}x + 6$. Then draw and label the line segments that represent $x = 4$ and $f(4)$.

43. For the function $f(x) = \dfrac{1}{2}x + 2$, find the value of x for which $f(x) = x$.

44. For the function $f(x) = -\dfrac{1}{2}x + 6$, find the value of x for which $f(x) = x$.

45. Graph the function $f(x) = x^2$. Then draw and label the line segments that represent $x = 1$ and $f(1)$, $x = 2$ and $f(2)$, and, finally, $x = 3$ and $f(3)$.

46. Graph the function $f(x) = x^2 - 2$. Then draw and label the line segments that represent $x = 2$ and $f(2)$, and the line segments corresponding to $x = 3$ and $f(3)$.

C Applying the Concepts [Examples 2–4]

47. **Investing in Art** A painting is purchased as an investment for $150. If its value increases continuously so that it doubles every 3 years, then its value is given by the function

$$V(t) = 150 \cdot 2^{t/3} \quad \text{for} \quad t \geq 0$$

where t is the number of years since the painting was purchased, and $V(t)$ is its value (in dollars) at time t. Find $V(3)$ and $V(6)$, and then explain what they mean.

48. **Average Speed** If it takes Minke t minutes to run a mile, then her average speed $s(t)$, in miles per hour, is given by the formula

$$s(t) = \dfrac{60}{t} \quad \text{for} \quad t > 0$$

Find $s(4)$ and $s(5)$, and then explain what they mean.

49. Social Networking The chart shows the percentage of people, by age, who are using various social networking sites. Suppose the following:

f is a function of MySpace users

g is a function of Facebook users

h is a function of Twitter users

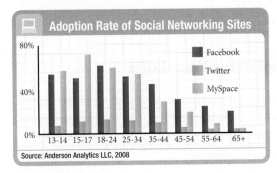

Use this information to determine if each if the following statements are true or false.
a. $g(15\text{–}17) > g(55\text{–}64)$
b. $h(65+) > h(18\text{–}24)$
c. $f(25\text{–}34) < f(15\text{–}17)$

50. Light Bulbs The chart shows a comparison of power usage between incandescent and LED light bulbs. Suppose that

f is the function for an incandescent bulb

g is the function for an LED bulb

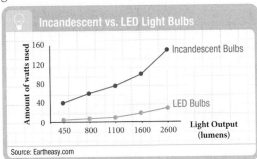

Use this information to approximate the following:
a. $f(2600) - g(2600)$
b. $f(2600) - f(1100)$
c. $g(2600) - f(450)$

51. Social Networking Refer to the table in number 49. Suppose x represents one of the age groups in the chart. Suppose further that we have three functions f, g, and h that do the following:

f is the percentage of people in an age group that use Myspace.

g is the percentage of people in an age group that use Facebook.

h is the percentage of people in an age group that use Twitter.

For each statement below, indicate whether the statement is true or false.
a. $f(55\text{–}64) < f(18\text{–}24)$
b. $f(15\text{–}17) > g(15\text{–}17)$
c. $h(25\text{–}34) > 20\%$
d. $h(18\text{–}24) > h(35\text{–}44) > h(65+)$

52. Mobile Phone Sales Suppose x represents one of the years in the chart. Suppose further that we have three functions f, g, and h that do the following:

f pairs each year with the number of camera phones sold that year.

g pairs each year with the number of non-camera phones sold that year.

h is such that $h(x) = f(x) + g(x)$.

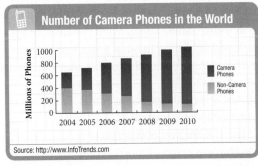

For each statement below, indicate whether the statement is true or false.
a. The domain of f is {2004, 2005, 2006, 2007, 2008, 2009, 2010}
b. $h(2005) = 741{,}000{,}000$
c. $f(2009) > g(2009)$
d. $f(2004) < f(2005)$
e. $h(2010) > h(2007) > h(2004)$

Straight-Line Depreciation Straight-line depreciation is an accounting method used to help spread the cost of new equipment over a number of years. It takes into account both the cost when new and the salvage value, which is the value of the equipment at the time it gets replaced.

53. Value of a Copy Machine The function
$V(t) = -3{,}300t + 18{,}000$, where V is value and t is time in years, can be used to find the value of a large copy machine during the first 5 years of use.
a. What is the value of the copier after 3 years and 9 months?
b. What is the salvage value of this copier if it is replaced after 5 years?
c. State the domain of this function.

d. Sketch the graph of this function.

e. What is the range of this function?

f. After how many years will the copier be worth only $10,000?

Keith Brofsky/Getty Images

54. Value of a Forklift The function

$V = -16,500t + 125,000$, where V is value and t is time in years, can be used to find the value of an electric forklift during the first 6 years of use.

a. What is the value of the forklift after 2 years and 3 months?

b. What is the salvage value of this forklift if it is replaced after 6 years?

c. State the domain of this function.

d. Sketch the graph of this function.

e. What is the range of this function?

f. After how many years will the forklift be worth only $45,000?

Michael Grecco/Getty Images

Getting Ready for the Next Section

Simplify.

55. $(35x - 0.1x^2) - (8x + 500)$

56. $70 + 0.6(M - 70)$

57. $(4x^2 + 3x + 2) + (2x^2 - 5x - 6)$

58. $(4x - 3) + (4x^2 - 7x + 3)$

59. $(4x^2 + 3x + 2) - (2x^2 - 5x - 6)$

60. $(x + 5)^2 - 2(x + 5)$

Multiply.

61. $0.6(M - 70)$　　　　**62.** $x(35 - 0.1x)$

63. $(4x - 3)(x - 1)$

64. $(4x - 3)(4x^2 - 7x + 3)$

Maintaining Your Skills

The problems that follow review some of the more important skills you have learned in previous sections and chapters.

Solve each equation.

65. $|3x - 5| = 7$

66. $|0.04 - 0.03x| = 0.02$

67. $|4y + 2| - 8 = -2$　　　**68.** $4 = |3 - 2y| - 5$

69. $5 + |6t + 2| = 3$　　　**70.** $7 + \left|3 - \frac{3}{4}t\right| = 10$

Extending the Concepts

The graphs of two functions are shown in Figures 4 and 5. Use the graphs to find the following.

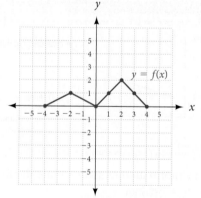

$y = f(x)$

FIGURE 4

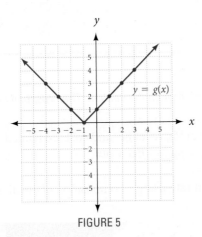

FIGURE 5

71. a. $f(2)$
   b. $f(-4)$
   c. $g(0)$
   d. $g(3)$

72. a. $g(2) - f(2)$
   b. $f(1) + g(1)$
   c. $f[g(3)]$
   d. $g[f(3)]$

73. Step Function Figure 6 shows the graph of the step function C that was used to calculate the first-class postage on a letter weighing x ounces in 2006. Use this graph to answer questions a through d.

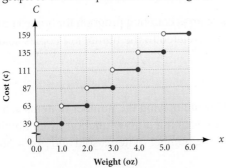

FIGURE 6   **The graph of C(x)**

a. Fill in the following table:

Weight (ounces)	0.6	1.0	1.1	2.5	3.0	4.8	5.0	5.3
Cost (cents)								

b. If a letter costs 87 cents to mail, how much does it weigh? State your answer in words. State your answer as an inequality.

c. If the entire function is shown in Figure 6, state the domain.

d. State the range of the function shown in Figure 6.

74. Step Function A taxi ride in Boston is $2.60 for the first $\frac{1}{7}$ mile, and then $0.40 for each additional $\frac{1}{7}$ of a mile. The following graph shows how much you will pay for a taxi ride of 1 mile or less

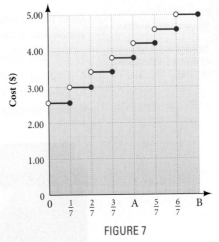

FIGURE 7

a. What is the most you will pay for this taxi ride?

b. How much does it cost to ride the taxi for $\frac{5}{7}$ of a mile?

c. Find the values of A and B on the horizontal axis.

d. If a taxi ride costs $2.50, what distance was the ride?

e. If the complete function is shown in Figure 7, find the domain and range of the function.

OBJECTIVES

A Find the sum, difference, product, and quotient of two functions.

B For two functions f and g, find $f(g(x))$ and $g(f(x))$.

TICKET TO SUCCESS

Keep these questions in mind as you read through the section. Then respond in your own words and in complete sentences.

1. Use function notation to show how profit, revenue, and cost are related.
2. Explain in words the definition for $(f + g)(x) = f(x) + g(x)$.
3. What is the composition of functions?
4. How would you find maximum heart rate using a composition of functions?

A company produces and sells copies of an accounting program for home computers. The price they charge for the program is related to the number of copies sold by the demand function

$$p(x) = 35 - 0.1x$$

We find the revenue for this business by multiplying the number of items sold by the price per item. When we do so, we are forming a new function by combining two existing functions; that is, if $n(x) = x$ is the number of items sold and $p(x) = 35 - 0.1x$ is the price per item, then revenue is

$$R(x) = n(x) \cdot p(x) = x(35 - 0.1x) = 35x - 0.1x^2$$

In this case, the revenue function is the product of two functions. When we combine functions in this manner, we are applying our rules for algebra to functions.

To carry this situation further, we know that the profit function is the difference between two functions. If the cost function for producing x copies of the accounting program is $C(x) = 8x + 500$, then the profit function is

$$P(x) = R(x) - C(x) = (35x - 0.1x^2) - (8x + 500) = -500 + 27x - 0.1x^2$$

The relationship between these last three functions is shown visually in Figure 1.

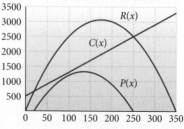

FIGURE 1

Again, when we combine functions in the manner shown, we are applying our rules for algebra to functions. To begin this section, we take a formal look at addition, subtraction, multiplication, and division with functions.

A Algebra with Functions

If we are given two functions f and g with a common domain, we can define four other functions as follows:

Definition

$(f + g)(x) = f(x) + g(x)$ The function $f + g$ is the sum of the functions f and g.

$(f - g)(x) = f(x) - g(x)$ The function $f - g$ is the difference of the functions f and g.

$(fg)(x) = f(x)g(x)$ The function fg is the product of the functions f and g.

$\left(\dfrac{f}{g}\right)(x) = \dfrac{f(x)}{g(x)}$ The function $\dfrac{f}{g}$ is the quotient of the functions f and g, where $g(x) \neq 0$.

EXAMPLE 1   If $f(x) = 4x^2 + 3x$ and $g(x) = -5x - 6$, write the formula for the functions $f + g$ and $f - g$.

SOLUTION   The function $f + g$ is defined by

$$(f + g)(x) = f(x) + g(x)$$
$$= (4x^2 + 3x) + (-5x - 6)$$
$$= 4x^2 - 2x - 6$$

The function $f - g$ is defined by

$$(f - g)(x) = f(x) - g(x)$$
$$= (4x^2 + 3x) - (-5x - 6)$$
$$= 4x^2 + 3x + 5x + 6$$
$$= 4x^2 + 8x + 6$$

EXAMPLE 2   Let $f(x) = 4x - 3$, and $g(x) = 4x^2$. Find $f + g$, fg, and $\dfrac{g}{f}$.

SOLUTION   The function $f + g$, the sum of functions f and g, is defined by

$$(f + g)(x) = f(x) + g(x)$$
$$= (4x - 3) + 4x^2$$
$$= 4x^2 + 4x - 3$$

The product of the functions f and g, fg, is given by

$$(fg)(x) = f(x)g(x)$$
$$= (4x - 3)(4x^2)$$
$$= 16x^3 - 12x^2$$

The quotient of the functions g and f, $\dfrac{g}{f}$, is defined as

$$\left(\frac{g}{f}\right)(x) = \frac{g(x)}{f(x)}$$

$$= \frac{4x^2}{4x - 3}$$

B Composition of Functions

In addition to the four operations used to combine functions shown so far in this section, there is a fifth way to combine two functions to obtain a new function. It is called *composition of functions.* To illustrate the concept, recall the definition of training heart rate: Training heart rate, in beats per minute, is resting heart rate plus 60% of the difference between maximum heart rate and resting heart rate. If your resting heart rate is 70 beats per minute, then your training heart rate is a function of your maximum heart rate M.

$$T(M) = 70 + 0.6(M - 70) = 70 + 0.6M - 42 = 28 + 0.6M$$

But your maximum heart rate is found by subtracting your age in years from 220. So, if x represents your age in years, then your maximum heart rate is

$$M(x) = 220 - x$$

Therefore, if your resting heart rate is 70 beats per minute and your age in years is x, then your training heart rate can be written as a function of x.

$$T(x) = 28 + 0.6(220 - x)$$

This last line is the composition of functions T and M. We input x into function M, which outputs $M(x)$. Then we input $M(x)$ into function T, which outputs $T(M(x))$. This is the training heart rate as a function of age x. Here is a diagram of the situation, also called a function map.

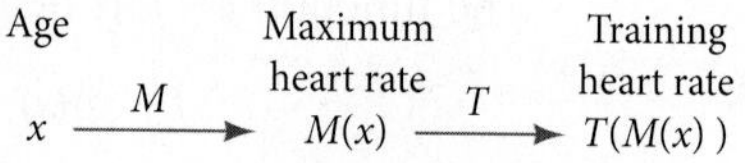

Age		Maximum heart rate		Training heart rate
x	$\xrightarrow{\;\;M\;\;}$	$M(x)$	$\xrightarrow{\;\;T\;\;}$	$T(M(x))$

FIGURE 2

Now let's generalize the preceding ideas into a formal development of composition of functions. To find the composition of two functions f and g, we first require that the range of g have numbers in common with the domain of f. Then the composition of f with g, denoted $f \circ g$, is defined this way:

$$(f \circ g)(x) = f(g(x))$$

To understand this new function, we begin with a number x, and we operate on it with g, giving us $g(x)$. Then we take $g(x)$ and operate on it with f, giving us $f(g(x))$. The only numbers we can use for the domain of the composition of f with g are numbers x in the domain of g, for which $g(x)$ is in the domain of f. The diagrams in Figure 3 illustrate the composition of f with g.

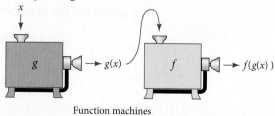

Function machines

FIGURE 3

Composition of functions is not commutative. The composition of f with g, $f \circ g$, may therefore be different from the composition of g with f, $g \circ f$.

$$(g \circ f)(x) = g(f(x))$$

Again, the only numbers we can use for the domain of the composition of g with f are numbers in the domain of f, for which $f(x)$ is in the domain of g. The diagrams in Figure 4 illustrate the composition of g with f.

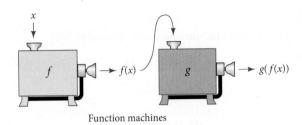

Function machines

FIGURE 4

EXAMPLE 3 If $f(x) = x + 5$ and $g(x) = 2x$, find $(f \circ g)(x)$.

SOLUTION The composition of f with g is

$$(f \circ g)(x) = f(g(x))$$

$$= f(2x)$$

$$= 2x + 5$$

The composition of g with f is

$$(g \circ f)(x) = g(f(x))$$

$$= g(x + 5)$$

$$= 2(x + 5)$$

$$= 2x + 10$$

Problem Set 8.5

Moving Toward Success

"A strong mind always hopes, and has always cause to hope."

—Thomas Carlyle, 1795–1881, Scottish philosopher and author

1. Do you feel anxiety when you are taking a test? Why or why not?

2. What might you do to remain calm and boost your confidence if you feel anxious during a test?

A Let $f(x) = 4x - 3$ and $g(x) = 2x + 5$. Write a formula for each of the following functions. [Example 1]

1. $f + g$
2. $f - g$
3. $g - f$
4. $g + f$

A If the functions f, g, and h are defined by $f(x) = 3x - 5$, $g(x) = x - 2$, and $h(x) = 3x^2$, write a formula for each of the following functions. [Examples 1–2]

5. $g + f$
6. $f + h$
7. $g + h$
8. $f - g$
9. $g - f$
10. $h - g$

11. fh

12. gh

13. $\dfrac{h}{f}$

14. $\dfrac{h}{g}$

15. $\dfrac{f}{h}$

16. $\dfrac{g}{h}$

17. $f + g + h$

18. $h - g + f$

A Let $f(x) = 2x + 1$, $g(x) = 4x + 2$, and $h(x) = 4x^2 + 4x + 1$, and find the following. [Examples 1–2]

19. $(f + g)(2)$

20. $(f - g)(-1)$

21. $(fg)(3)$

22. $\left(\dfrac{f}{g}\right)(-3)$

23. $\left(\dfrac{h}{g}\right)(1)$

24. $(hg)(1)$

25. $(fh)(0)$

26. $(h - g)(-4)$

27. $(f + g + h)(2)$

28. $(h - f + g)(0)$

29. $(h + fg)(3)$

30. $(h - fg)(5)$

B [Example 3]

31. Let $f(x) = x^2$ and $g(x) = x + 4$, and find
   a. $(f \circ g)(5)$
   b. $(g \circ f)(5)$
   c. $(f \circ g)(x)$
   d. $(g \circ f)(x)$

32. Let $f(x) = 3 - x$ and $g(x) = x^3 - 1$, and find
   a. $(f \circ g)(0)$
   b. $(g \circ f)(0)$
   c. $(f \circ g)(x)$

33. Let $f(x) = x^2 + 3x$ and $g(x) = 4x - 1$, and find
   a. $(f \circ g)(0)$
   b. $(g \circ f)(0)$
   c. $(g \circ f)(x)$

34. Let $f(x) = (x - 2)^2$ and $g(x) = x + 1$, and find the following
   a. $(f \circ g)(-1)$
   b. $(g \circ f)(-1)$

For each of the following pairs of functions f and g, show that $(f \circ g)(x) = (g \circ f)(x) = x$.

35. $f(x) = 5x - 4$ and $g(x) = \dfrac{x + 4}{5}$

36. $f(x) = \dfrac{x}{6} - 2$ and $g(x) = 6x + 12$

Use the graph to answer the following problems.

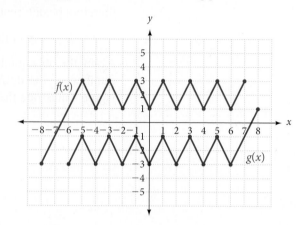

Evaluate.

37. $f(2) + 5$

38. $g(-2) - 5$

39. $f(-3) + g(-3)$

40. $f(5) - g(5)$

41. $(f \circ g)(0)$

42. $(g \circ f)(0)$

43. Find x if $f(x) = -3$

44. Find x if $g(x) = 1$

Use the graph to answer the following problems.

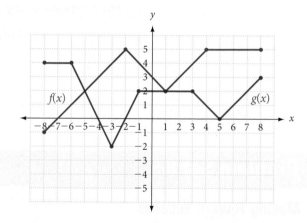

Evaluate.

45. $f(-3) + 2$

46. $g(3) - 3$

47. $f(2) + g(2)$

48. $f(-5) - g(-5)$

49. $(f \circ g)(0)$

50. $(g \circ f)(0)$

51. Find x if $f(x) = 1$

52. Find x if $g(x) = -2$

Applying the Concepts

53. Profit, Revenue, and Cost A company manufactures and sells prerecorded DVDs. Here are the equations they use in connection with their business.

Number of DVDs sold each day: $n(x) = x$

Selling price for each DVD: $p(x) = 11.5 - 0.05x$

Daily fixed costs: $f(x) = 200$

Daily variable costs: $v(x) = 2x$

Find the following functions.
a. Revenue $= R(x) =$ the product of the number of DVDs sold each day and the selling price of each DVD.
b. Cost $= C(x) =$ the sum of the fixed costs and the variable costs.
c. Profit $= P(x) =$ the difference between revenue and cost.
d. Average cost $= \overline{C}(x) =$ the quotient of cost and the number of DVDs sold each day.

54. Profit, Revenue, and Cost A company manufactures and sells flash drives for home computers. Here are the equations they use in connection with their business.

Number of flash drives sold each day: $n(x) = x$

Selling price for each flash drive: $p(x) = 3 - \frac{1}{300}x$

Daily fixed costs: $f(x) = 200$

Daily variable costs: $v(x) = 2x$

Find the following functions.
a. Revenue $= R(x) =$ the product of the number of flash drives sold each day and the selling price of each flash drive.
b. Cost $= C(x) =$ the sum of the fixed costs and the variable costs.
c. Profit $= P(x) =$ the difference between revenue and cost.
d. Average cost $= \overline{C}(x) =$ the quotient of cost and the number of flash drives sold each day.

55. Training Heart Rate Find the training heart rate function, $T(M)$, for a person with a resting heart rate of 62 beats per minute, then find the following.
a. Find the maximum heart rate function, $M(x)$, for a person x years of age.
b. What is the maximum heart rate for a 24-year-old person?
c. What is the training heart rate for a 24-year-old person with a resting heart rate of 62 beats per minute? Round to the nearest whole number.

d. What is the training heart rate for a 36-year-old person with a resting heart rate of 62 beats per minute? Round to the nearest whole number.

e. What is the training heart rate for a 48-year-old person with a resting heart rate of 62 beats per minute? Round to the nearest whole number.

56. Training Heart Rate Find the training heart rate function, $T(M)$, for a person with a resting heart rate of 72 beats per minute, then find the following to the nearest whole number.
a. Find the maximum heart rate function, $M(x)$, for a person x years of age.
b. What is the maximum heart rate for a 20-year-old person?
c. What is the training heart rate for a 20-year-old person with a resting heart rate of 72 beats per minute? Round to the nearest whole number.
d. What is the training heart rate for a 30-year-old person with a resting heart rate of 72 beats per minute? Round to the nearest whole number.
e. What is the training heart rate for a 40-year-old person with a resting heart rate of 72 beats per minute? Round to the nearest whole number.

Maintaining Your Skills

57. $|x - 3| < 1$

58. $|x - 3| > 1$

59. $|6 - x| > 2$

60. $\left|1 - \frac{1}{2}x\right| > 2$

61. $|7x - 1| \leq 6$

62. $|7x - 1| \geq 6$

Getting Ready for the Next Section

63. $16(3.5)^2$

64. $\dfrac{2,400}{100}$

65. $\dfrac{180}{45}$

66. $4(2)(4)^2$

67. $\dfrac{0.0005(200)}{(0.25)^2}$

68. $\dfrac{0.2(0.5)^2}{100}$

69. If $y = Kx$, find K if $x = 5$ and $y = 15$.

70. If $d = Kt^2$, find K if $t = 2$ and $d = 64$.

71. If $V = \dfrac{K}{P}$, find K if $P = 48$ and $V = 50$.

72. If $y = Kxz^2$, find K if $x = 5$, $z = 3$, and $y = 180$.

8.6 Variation

OBJECTIVES

A Set up and solve problems with direct variation.

B Set up and solve problems with inverse variation.

C Set up and solve problems with joint variation.

TICKET TO SUCCESS

Keep these questions in mind as you read through the section. Then respond in your own words and in complete sentences.

1. Give an example of a direct variation statement, and then translate it into symbols.
2. Translate the equation $y = \frac{K}{x}$ into words.
3. For the inverse variation equation $y = \frac{3}{x}$, what happens to the values of y as x gets larger?
4. What do we mean when we say that y varies jointly with x and z?

If you are a runner and you average t minutes for every mile you run during one of your workouts, then your speed s in miles per hour is given by the equation and graph shown here. Figure 1 is shown in the first quadrant only because both t and s are positive.

$$s = \frac{60}{t}$$

Input t	Output s
4	15.0
6	10.0
8	7.5
10	6.0
12	5.0
14	4.3

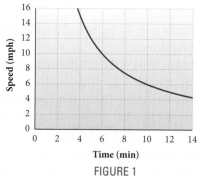

FIGURE 1

You know intuitively that as your average time per mile, t, increases, your speed, s, decreases. Likewise, lowering your time per mile will increase your speed. The equation and Figure 1 also show this to be true: increasing t decreases s, and decreasing t increases s. Quantities that are connected in this way are said to *vary inversely* with each other. Inverse variation is one of the topics we will study in this section.

There are two main types of variation: *direct variation* and *inverse variation.* Variation problems are most common in the sciences, particularly in chemistry and physics.

A   Direct Variation

When we say the variable *y* *varies directly* with the variable *x*, we mean that the relationship can be written in symbols as $y = Kx$, where K is a nonzero constant called the *constant of variation* (or *proportionality constant*). Another way of saying *y* varies directly with *x* is to say *y* is *directly proportional* to *x*.

Study the following list. It gives the mathematical equivalent of some direct variation statements.

English Phrase	Algebraic Equation
y varies directly with *x*	$y = Kx$
s varies directly with the square of *t*	$s = Kt^2$
y is directly proportional to the cube of *z*	$y = Kz^3$
u is directly proportional to the square root of *v*	$u = K\sqrt{v}$

EXAMPLE 1   *y* varies directly with *x*. If *y* is 15 when *x* is 5, find *y* when *x* is 7.

SOLUTION   The first sentence gives us the general relationship between *x* and *y*. The equation equivalent to the statement "*y* varies directly with *x*" is

$$y = Kx$$

The first part of the second sentence in our example gives us the information necessary to evaluate the constant K:

$$\text{When} \rightarrow \quad y = 15$$
$$\text{and} \rightarrow \quad x = 5$$
$$\text{the equation} \rightarrow \quad y = Kx$$
$$\text{becomes} \rightarrow \quad 15 = K \cdot 5$$
$$\text{or} \rightarrow \quad K = 3$$

The equation now can be written specifically as

$$y = 3x$$

Letting $x = 7$, we have

$$y = 3 \cdot 7$$
$$y = 21$$

NOTE
The graph of the direct variation statement $y = 3x$ is a line with slope 3 that passes through the origin.

EXAMPLE 2   A skydiver jumps from a plane. Like any object that falls toward Earth, the distance the skydiver falls is directly proportional to the square of the time he has been falling until he reaches his terminal velocity. If the skydiver falls 64 feet in the first 2 seconds of the jump, then

a. How far will he have fallen after 3.5 seconds?

b. Graph the relationship between distance and time.

c. How long will it take him to fall 256 feet?

SOLUTION We let *t* represent the time the skydiver has been falling, then we can use function notation and let $d(t)$ represent the distance he has fallen.

a. Because $d(t)$ is directly proportional to the square of *t*, we have the general function that describes this situation:

$$d(t) = Kt^2$$

Next, we use the fact that $d(2) = 64$ to find *K*.

$$64 = K(2^2)$$
$$K = 16$$

The specific equation that describes this situation is

$$d(t) = 16t^2$$

To find how far a skydiver will fall after 3.5 seconds, we find $d(3.5)$.

$$d(3.5) = 16(3.5^2)$$
$$d(3.5) = 196$$

A skydiver will have fallen 196 feet after 3.5 seconds.

b. To graph this equation, we use a table.

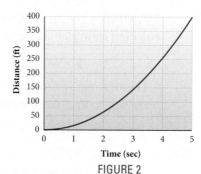

Input *t*	Output $d(t)$
0	0
1	16
2	64
3	144
4	256
5	400

FIGURE 2

c. From the table or the graph (Figure 2), we see that it will take 4 seconds for the skydiver to fall 256 feet.

B Inverse Variation

RUNNING From the introduction to this section, we know that the relationship between the number of minutes (*t*) it takes a person to run a mile and his or her average speed in miles per hour (*s*) can be described with the following equation, table, and figure.

$$s = \frac{60}{t}$$

Input *t*	Output *s*
4	15.0
6	10.0
8	7.5
10	6.0
12	5.0
14	4.3

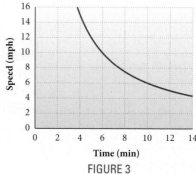

FIGURE 3

If *t* decreases, then *s* will increase, and if *t* increases, then *s* will decrease. The variable *s* is *inversely proportional* to the variable *t*. In this case, the *constant of proportionality* is 60.

PHOTOGRAPHY If you are familiar with the terminology and mechanics associated with photography, you know that the *f*-stop for a particular lens will increase as the aperture (the maximum diameter of the opening of the lens) decreases. In mathematics we say that *f*-stop and aperture vary inversely with each other. The diagram illustrates this relationship.

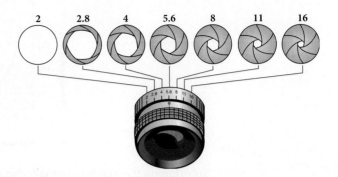

If *f* is the *f*-stop and *d* is the aperture, then their relationship can be written

$$f = \frac{K}{d}$$

In this case, *K* is the constant of proportionality. (Those of you familiar with photography know that *K* is also the focal length of the camera lens.)

We generalize this discussion of inverse variation as follows: If *y* varies inversely with *x*, then

$$y = K \cdot \frac{1}{x} \quad \text{or} \quad y = \frac{K}{x}$$

We can also say *y* is inversely proportional to *x*. The constant *K* is again called the constant of variation or proportionality constant.

English Phrase	Algebraic Equation
y is inversely proportional to *x*	$y = \frac{K}{x}$
s varies inversely with the square of *t*	$s = \frac{K}{t^2}$
y is inversely proportional to x^4	$y = \frac{K}{x^4}$
z varies inversely with the cube root of *t*	$z = \frac{K}{\sqrt[3]{t}}$

EXAMPLE 3 The volume of a gas is inversely proportional to the pressure of the gas on its container. If a pressure of 48 pounds per square inch corresponds to a volume of 50 cubic feet, what pressure is needed to produce a volume of 100 cubic feet?

SOLUTION We can represent volume with *V* and pressure with *P*.

$$V = \frac{K}{P}$$

Using $P = 48$ and $V = 50$, we have

$$50 = \frac{K}{48}$$

$$K = 50(48)$$

$$K = 2,400$$

NOTE
The relationship between pressure and volume as given in this example is known as Boyle's law and applies to situations such as those encountered in a piston-cylinder arrangement. It was Robert Boyle (1627–1691) who, in 1662, published the results of some of his experiments that showed, among other things, that the volume of a gas decreases as the pressure increases. This is an example of inverse variation.

The equation that describes the relationship between P and V is

$$V = \frac{2,400}{P}$$

Here is a graph of this relationship:

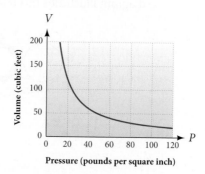

Substituting $V = 100$ into our last equation, we get

$$100 = \frac{2,400}{P}$$

$$100P = 2,400$$

$$P = \frac{2,400}{100}$$

$$P = 24$$

A volume of 100 cubic feet is produced by a pressure of 24 pounds per square inch. ∎

C Joint Variation and Other Variation Combinations

Many times relationships among different quantities are described in terms of more than two variables. If the variable y varies directly with *two* other variables, say x and z, then we say y varies *jointly* with x and z. In addition to joint variation, there are many other combinations of direct and inverse variation involving more than two variables. The following table is a list of some variation statements and their equivalent mathematical forms:

English Phrase	Algebraic Equation
y varies jointly with x and z	$y = Kxz$
z varies jointly with r and the square of s	$z = Krs^2$
V is directly proportional to T and inversely proportional to P	$V = \dfrac{KT}{P}$
F varies jointly with m_1 and m_2 and inversely with the square of r	$F = \dfrac{Km_1m_2}{r^2}$

EXAMPLE 4 y varies jointly with x and the square of z. When x is 5 and z is 3, y is 180. Find y when x is 2 and z is 4.

SOLUTION The general equation is given by

$$y = Kxz^2$$

Substituting $x = 5$, $z = 3$, and $y = 180$, we have

$$180 = K(5)(3)^2$$

$$180 = 45K$$

$$K = 4$$

The specific equation is

$$y = 4xz^2$$

When $x = 2$ and $z = 4$, the last equation becomes

$$y = 4(2)(4)^2$$

$$y = 128$$

EXAMPLE 5 In electricity, the resistance of a cable is directly proportional to its length and inversely proportional to the square of its diameter. If a 100-foot cable 0.5 inch in diameter has a resistance of 0.2 ohm, what will be the resistance of a cable made from the same material if it is 200 feet long with a diameter of 0.25 inch?

SOLUTION Let R = resistance, l = length, and d = diameter. The equation is

$$R = \frac{Kl}{d^2}$$

When $R = 0.2$, $l = 100$, and $d = 0.5$, the equation becomes

$$0.2 = \frac{K(100)}{(0.5)^2}$$

or

$$K = 0.0005$$

Using this value of K in our original equation, the result is

$$R = \frac{0.0005l}{d^2}$$

When $l = 200$ and $d = 0.25$, the equation becomes

$$R = \frac{0.0005(200)}{(0.25)^2}$$

$$R = 1.6 \text{ ohms}$$

Problem Set 8.6

Moving Toward Success

"You cannot make it as a wandering generality. You must become a meaningful specific."

—Zig Ziglar, American sales trainer and motivational speaker

1. Why is your success in this class dependent on actively doing problems?

2. How would the number of classes you may choose to miss relate to your final grade in this class?

Describe each relationship below as direct or inverse.

1. The number of homework problems assigned each night and the time it takes to complete the assignment

2. The uncommonness of a search term and the number of search results

3. The length of the line for an amusement park ride and the number of people in line

4. The age of a home computer and its relative speed

5. The age of a bottle of wine and its price

6. The speed of an internet connection and the time it takes to download a song

7. The length of a song and the time it takes to download

8. The speed of a rollercoaster and your willingness to ride it

9. The diameter of a Ferris wheel and its circumference

10. The measure of angle α to the measure of angle β if $\alpha + \beta = 90°$ ($\alpha \neq 0$)

11. The altitude above sea level and the air temperature

12. The number of years of school and starting salary

A For the following problems, y varies directly with x. [Example 1]

13. If y is 10 when x is 2, find y when x is 6.

14. If y is 20 when x is 5, find y when x is 3.

15. If y is -32 when x is 4, find x when y is -40.

16. If y is -50 when x is 5, find x when y is -70.

B For the following problems, r is inversely proportional to s. [Example 3]

17. If r is -3 when s is 4, find r when s is 2.

18. If r is -10 when s is 6, find r when s is -5.

19. If r is 8 when s is 3, find s when r is 48.

20. If r is 12 when s is 5, find s when r is 30.

A For the following problems, d varies directly with the square of r. [Example 2]

21. If $d = 10$ when $r = 5$, find d when $r = 10$.

22. If $d = 12$ when $r = 6$, find d when $r = 9$.

23. If $d = 100$ when $r = 2$, find d when $r = 3$.

24. If $d = 50$ when $r = 5$, find d when $r = 7$.

A For the following problems, y varies directly with the square root of x. [Example 1]

25. If $y = 6$ when $x = 9$, find y when $x = 25$.

26. If $y = 18$ when $x = 36$, find y when $x = 49$.

27. If $y = 20$ when $x = 25$, find y when $x = 16$.

28. If $y = 35$ when $x = 49$, find y when $x = 36$.

B For the following problems, y varies inversely with the square of x. [Example 3]

29. If $y = 45$ when $x = 3$, find y when x is 5.

30. If $y = 12$ when $x = 2$, find y when x is 6.

31. If $y = 18$ when $x = 3$, find y when x is 2.

32. If $y = 45$ when $x = 4$, find y when x is 5.

C For the following problems, z varies jointly with x and the square of y. [Example 4]

33. If z is 54 when x and y are 3, find z when $x = 2$ and $y = 4$.

34. If z is 80 when x is 5 and y is 2, find z when $x = 2$ and $y = 5$.

35. If z is 64 when $x = 1$ and $y = 4$, find x when $z = 32$ and $y = 1$.

36. If z is 27 when $x = 6$ and $y = 3$, find x when $z = 50$ and $y = 4$.

Applying the Concepts [Examples 2–5]

37. Length of a Spring The length a spring stretches is directly proportional to the force applied. If a force of 5 pounds stretches a spring 3 inches, how much force is necessary to stretch the same spring 10 inches?

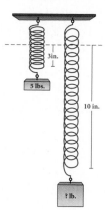

38. Weight and Surface Area The weight of a certain material varies directly with the surface area of that material. If 8 square feet weighs half a pound, how much will 10 square feet weigh?

Cars The chart shows the fastest cars in the world. Use the chart to answer Problems 39 and 40

39. The speed of a car in miles per hour is directly proportional to its speed in feet per second. If the speed of the SSC Ultimate Aero in feet per second is $400\frac{2}{5}$, what is the speed of the Bugatti in feet per second?

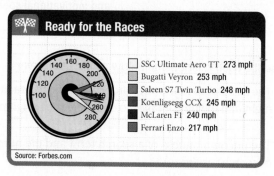

Ready for the Races

- SSC Ultimate Aero TT 273 mph
- Bugatti Veyron 253 mph
- Saleen S7 Twin Turbo 248 mph
- Koenligsegg CCX 245 mph
- McLaren F1 240 mph
- Ferrari Enzo 217 mph

Source: Forbes.com

40. The time that it takes a car to travel a given distance varies inversely with the speed of the car. If a McLaren F1 traveled a certain distance in 2 hours, how long would it take the Saleen S7 Twin Turbo to travel the same distance?

41. **Pressure and Temperature** The temperature of a gas, T, varies directly with its pressure, P. A temperature of 200 K (Kelvin) produces a pressure of 50 pounds per square inch.

   a. Find the equation that relates pressure and temperature.

   b. Graph the equation from part a in the first quadrant only.

   c. What pressure will the gas have at 280° K?

42. **Circumference and Diameter** The circumference of a wheel is directly proportional to its diameter. A wheel has a circumference of 8.5 feet and a diameter of 2.7 feet.

   a. Find the equation that relates circumference and diameter.

   b. Graph the equation from part a in the first quadrant only.

   c. What is the circumference of a wheel that has a diameter of 11.3 feet? Round your answer to the nearest tenth.

43. **Volume and Pressure** The volume of a gas is inversely proportional to the pressure. If a pressure of 36 pounds per square inch corresponds to a volume of 25 cubic feet, what pressure is needed to produce a volume of 75 cubic feet?

44. **Wave Frequency** The frequency of an electromagnetic wave varies inversely with the wavelength. If a wavelength of 200 meters has a frequency of 800 kilocycles per second, what frequency will be associated with a wavelength of 500 meters?

45. **f-Stop and Aperture Diameter** The relative aperture or f-stop for a camera lens is inversely proportional to the diameter of the aperture. An f-stop of 2 corresponds to an aperture diameter of 40 millimeters for the lens on a camera.

   a. Find the equation that relates f-stop and diameter.

   b. Graph the equation from part a in the first quadrant only.

   c. What is the f-stop of this camera when the aperture diameter is 10 millimeters?

46. **f-Stop and Aperture Diameter** The relative aperture or f-stop for a camera lens is inversely proportional to the diameter of the aperture. An f-stop of

2.8 corresponds to an aperture diameter of 75 millimeters for a certain telephoto lens.

   a. Find the equation that relates f-stop and diameter.

   b. Graph the equation from part a in the first quadrant only.

   c. What aperture diameter corresponds to an f-stop of 5.6?

47. **Surface Area of a Cylinder** The surface area of a hollow cylinder varies jointly with the height and radius of the cylinder. If a cylinder with radius 3 inches and height 5 inches has a surface area of 94 square inches, what is the surface area of a cylinder with radius 2 inches and height 8 inches?

48. **Capacity of a Cylinder** The capacity of a cylinder varies jointly with its height and the square of its radius. If a cylinder with a radius of 3 centimeters and a height of 6 centimeters has a capacity of 169.56 cubic centimeters, what will be the capacity of a cylinder with radius 4 centimeters and height 9 centimeters?

49. **Electrical Resistance** The resistance of a wire varies directly with its length and inversely with the square of its diameter. If 100 feet of wire with diameter 0.01 inch has a resistance of 10 ohms, what is the resistance of 60 feet of the same type of wire if its diameter is 0.02 inch?

50. **Volume and Temperature** The volume of a gas varies directly with its temperature and inversely with the pressure. If the volume of a certain gas is 30 cubic feet at a temperature of 300 K and a pressure of 20 pounds per square inch, what is the volume of the same gas at 340 K when the pressure is 30 pounds per square inch?

51. **Music** A musical tone's pitch varies inversely with its wavelength. If one tone has a pitch of 420 vibrations each second and a wavelength of 2.2 meters, find the wavelength of a tone that has a pitch of 720 vibrations each second. Round your answer to the nearest hundredth.

52. **Hooke's Law** Hooke's law states that the stress (force per unit area) placed on a solid object varies directly with the strain (deformation) produced.

   a. Using the variables S_1 for stress and S_2 for strain, state this law in algebraic form.

   b. Find the constant, K, if for one type of material $S_1 = 24$ and $S_2 = 72$.

53. **Gravity** In Book Three of his *Principia*, Isaac Newton (depicted on the postage stamp) states that there is a single force in the universe that holds everything

together, called the force of universal gravity. Newton stated that the force of universal gravity, F, is directly proportional with the product of two masses, m_1 and m_2, and inversely proportional with the square of the distance d between them. Write the equation for Newton's force of universal gravity, using the symbol G as the constant of proportionality.

54. **Boyle's Law and Charles's Law** Boyle's law states that for low pressures, the pressure of an ideal gas kept at a constant temperature varies inversely with the volume of the gas. Charles's law states that for low pressures, the density of an ideal gas kept at a constant pressure varies inversely with the absolute temperature of the gas.

   a. State Boyle's law as an equation using the symbols P, K, and V.

   b. State Charles's law as an equation using the symbols D, K, and T.

Maintaining Your Skills

The problems that follow review some of the more important skills you have learned in previous sections and chapters.

Solve the following equations.

55. $x - 5 = 7$

56. $3y = -4$

57. $5 - \dfrac{4}{7}a = -11$

58. $\dfrac{1}{5}x - \dfrac{1}{2} - \dfrac{1}{10}x + \dfrac{2}{5} = \dfrac{3}{10}x + \dfrac{1}{2}$

59. $5(x - 1) - 2(2x + 3) = 5x - 4$

60. $0.07 - 0.02(3x + 1) = -0.04x + 0.01$

Solve for the indicated variable.

61. $P = 2l + 2w$ for w

62. $A = \dfrac{1}{2}h(b + B)$ for B

Solve the following inequalities. Write the solution set using interval notation, then graph the solution set.

63. $-5t \le 30$

64. $5 - \dfrac{3}{2}x > -1$

65. $1.6x - 2 < 0.8x + 2.8$

66. $3(2y + 4) \ge 5(y - 8)$

Solve the following equations.

67. $\left| \dfrac{1}{4}x - 1 \right| = \dfrac{1}{2}$

68. $\left| \dfrac{2}{3}a + 4 \right| = 6$

69. $|3 - 2x| + 5 = 2$

70. $5 = |3y + 6| - 4$

Solve each inequality and graph the solution set.

71. $\left| \dfrac{x}{5} + 1 \right| \ge \dfrac{4}{5}$

72. $|2 - 6t| < -5$

73. $|3 - 4t| > -5$

74. $|6y - 1| - 4 \le 2$

Extending the Concepts

75. **Human Cannonball** A circus company is deciding where to position the net for the human cannonball so that he will land safely during the act. They do this by firing a 100-pound sack of potatoes out of the cannon at different speeds and then measuring how far from the cannon the sack lands. The results are shown in the table.

Speed in Miles/Hour	Distance in Feet
40	108
50	169
60	243
70	331

   a. Does distance vary directly with the speed, or directly with the square of the speed?

   b. Write the equation that describes the relationship between speed and distance.

   c. If the cannon will fire a human safely at 55 miles/hour, where should they position the net so the cannonball has a safe landing?

   d. How much further will he land if his speed out of the cannon is 56 miles/hour?

The Slope of a Line [8.1]

1. The slope of the line through (6, 9) and (1, −1) is

$$m = \frac{9 - (-1)}{6 - 1} = \frac{10}{5} = 2$$

The *slope* of the line containing points (x_1, y_1) and (x_2, y_2) is given by

$$\text{Slope} = m = \frac{\text{Rise}}{\text{Run}} = \frac{y_2 - y_1}{x_2 - x_1}$$

Horizontal lines have 0 slope, and vertical lines have no slope.
Parallel lines have equal slopes, and perpendicular lines have slopes that are negative reciprocals.

Linear Equations in Two Variables [8.2]

2. The equation $3x + 2y = 6$ is an example of a linear equation in two variables.

A *linear equation in two variables* is any equation that can be put in *standard form* $ax + by = c$. The graph of every linear equation is a straight line.

Relations and Functions [8.3]

3. The relation

$$\{(8, 1), (6, 1), (-3, 0)\}$$

is also a function because no ordered pairs have the same first coordinates. The domain is {8, 6, −3} and the range is {1, 0}.

A *function* is a rule that pairs each element in one set, called the *domain,* with exactly one element from a second set, called the *range.*

A *relation* is any set of ordered pairs. The set of all first coordinates is called the *domain* of the relation, and the set of all second coordinates is the *range* of the relation. A function is a relation in which no two different ordered pairs have the same first coordinates.

Vertical Line Test [8.3]

4. The graph of $x = y^2$ shown in Figure 5 in Section 8.3 fails the vertical line test. It is not the graph of a function.

If a vertical line crosses the graph of a relation in more than one place, the relation cannot be a function. If no vertical line can be found that crosses the graph in more than one place, the relation must be a function.

Function Notation [8.4]

5. If $f(x) = 5x - 3$ then

$$f(0) = 5(0) - 3$$
$$= -3$$
$$f(1) = 5(1) - 3$$
$$= 2$$
$$f(-2) = 5(-2) - 3$$
$$= -13$$
$$f(a) = 5a - 3$$

The alternative notation for y is $f(x)$. It is read "f of x" and can be used instead of the variable y when working with functions. The notation y and the notation $f(x)$ are equivalent; that is, $y = f(x)$.

■ Algebra with Functions [8.5]

If f and g are any two functions with a common domain, then

$(f + g)(x) = f(x) + g(x)$ The function $f + g$ is the sum of the functions f and g.

$(f - g)(x) = f(x) - g(x)$ The function $f - g$ is the difference of the functions f and g.

$(fg)(x) = f(x)g(x)$ The function fg is the product of the functions f and g.

$\dfrac{f}{g}(x) = \dfrac{f(x)}{g(x)}$ The function $\dfrac{f}{g}$ is the quotient of the functions f and g, where $g(x) \neq 0$

■ Composition of Functions [8.5]

If f and g are two functions for which the range of each has numbers in common with the domain of the other, then we have the following definitions:

$$\text{The composition of } f \text{ with } g\text{: } (f \circ g)(x) = f(g(x))$$

$$\text{The composition of } g \text{ with } f\text{: } (g \circ f)(x) = g(f(x))$$

6. If y varies directly with x, then

$$y = Kx$$

Then if y is 18 when x is 6,

$$18 = K \cdot 6$$

or

$$K = 3$$

So the equation can be written more specifically as

$$y = 3x$$

If we want to know what y is when x is 4, we simply substitute:

$$y = 3 \cdot 4$$
$$y = 12$$

■ Variation [8.6]

If y varies *directly* with x (y is directly proportional to x), then

$$y = Kx$$

If y varies *inversely* with x (y is inversely proportional to x), then

$$y = \frac{K}{x}$$

If z varies *jointly* with x and y (z is directly proportional to both x and y), then

$$z = Kxy$$

In each case, K is called the *constant of variation.*

⊘ COMMON MISTAKES

1. When graphing ordered pairs, the most common mistake is to associate the first coordinate with the y-axis and the second with the x-axis. If you make this mistake you would graph (3, 1) by going up 3 and to the right 1, which is just the reverse of what you should do. Remember, the first coordinate is always associated with the horizontal axis, and the second coordinate is always associated with the vertical axis.

2. The two most common mistakes students make when first working with the formula for the slope of a line are the following:
   a. Putting the difference of the x-coordinates over the difference of the y-coordinates.
   b. Subtracting in one order in the numerator and then subtracting in the opposite order in the denominator.

3. When graphing linear inequalities in two variables, remember to graph the boundary with a broken line when the inequality symbol is $<$ or $>$. The only time you use a solid line for the boundary is when the inequality symbol is $\leq$ or $\geq$.

Find the slope of the line through the following pairs of points. [8.1]

1. (5, 2), (3, 6)

2. (−4, 2), (3, 2)

Find x if the line through the two given points has the given slope. [8.1]

3. (4, x), (1, −3); $m = 2$

4. (−4, 7), (2, x); $m = -\dfrac{1}{3}$

5. Find the slope of any line parallel to the line through (3, 8) and (5, −2). [8.2]

6. The line through (5, 3y) and (2, y) is parallel to a line with slope 4. What is the value of y? [8.2]

Give the equation of the line with the following slope and y-intercept. [8.2]

7. $m = 3$, $b = 5$

8. $m = -2$, $b = 0$

Give the slope and y-intercept of each equation. [8.2]

9. $3x - y = 6$

10. $2x - 3y = 9$

Find the equation of the line that contains the given point and has the given slope. [8.2]

11. (2, 4), $m = 2$

12. (−3, 1), $m = -\dfrac{1}{3}$

Find the equation of the line that contains the given pair of points. [8.2]

13. (2, 5), (−3, −5)

14. (−3, 7), (4, 7)

15. (−5, −1), (−3, −4)

16. Find the equation of the line that is parallel to $2x - y = 4$ and contains the point (2, −3). [8.2]

17. Find the equation of the line perpendicular to $y = -3x + 1$ that has an x-intercept of 2. [8.2]

State the domain and range of each relation, and then indicate which relations are also functions. [8.3]

18. {(2, 4), (3, 3), (4, 2)}

19. {(6, 3), (−4, 3), (−2, 0)}

If $f = \{(2, -1), (-3, 0), \left(4, \dfrac{1}{2}\right), (\pi, 2)\}$ and $g = \{(2, 2), (-1, 4), (0, 0)\}$, find the following. [8.4]

20. $f(-3)$

21. $f(2) + g(2)$

Let $f(x) = 2x^2 - 4x + 1$ and $g(x) = 3x + 2$, and evaluate each of the following. [8.5]

22. $f(0)$

23. $g(a)$

24. $f(g(0))$

25. $f(g(1))$

For the following problems, y varies directly with x. [8.6]

26. If y is 6 when x is 2, find y when x is 8.

27. If y is −3 when x is 5, find y when x is −10.

For the following problems, y varies inversely with the square of x. [8.6]

28. If y is 9 when x is 2, find y when x is 3.

29. If y is 4 when x is 5, find y when x is 2.

Solve each application problem. [8.6]

30. **Tension in a Spring** The tension t in a spring varies directly with the distance d the spring is stretched. If the tension is 42 pounds when the spring is stretched 2 inches, find the tension when the spring is stretched twice as far.

31. **Light Intensity** The intensity of a light source, measured in foot-candles, varies inversely with the square of the distance from the source. Four feet from the source the intensity is 9 foot-candles. What is the intensity 3 feet from the source?

Simplify each of the following.

1. -5^2

2. $-|-6|$

3. $5^2 + 7^2$

4. $(5 + 7)^2$

5. $48 \div 8 \cdot 6$

6. $75 \div 15 \cdot 5$

7. $30 - 15 \div 3 + 2$

8. $30 - 15 \div (3 + 2)$

Find the value of each expression when x is 4.

9. $x^2 - 10x + 25$

10. $x^2 - 25$

Reduce the following fractions to lowest terms.

11. $\dfrac{336}{432}$

12. $\dfrac{721}{927}$

Subtract.

13. $\dfrac{3}{15} - \dfrac{2}{20}$

14. $\dfrac{6}{28} - \dfrac{5}{42}$

15. Add $-\dfrac{3}{4}$ to the product of -3 and $\dfrac{5}{12}$.

16. Subtract $\dfrac{2}{5}$ from the product of -3 and $\dfrac{6}{15}$.

Simplify each of the following expressions.

17. $8\left(\dfrac{3}{4}x - \dfrac{1}{2}y\right)$

18. $12\left(\dfrac{5}{6}x + \dfrac{3}{4}y\right)$

Solve the following equations.

19. $\dfrac{2}{3}a - 4 = 6$

20. $\dfrac{3}{4}(8x - 3) + \dfrac{3}{4} = 2$

21. $-4 + 3(3x + 2) = 7$

22. $3x - (x + 2) = -6$

23. $|2x - 3| - 7 = 1$

24. $|2y + 3| = 2y - 7$

Solve the following expressions for y and simplify.

25. $y - 5 = -3(x + 2)$

26. $y + 3 = \dfrac{1}{2}(3x - 4)$

Graph on a rectangular coordinate system.

27. $6x - 5y = 30$

28. $x = 2$

29. $y = \dfrac{1}{2}x - 1$

30. $2x - y \le -3$

31. Find the slope of the line through $(-3, -2)$ and $(-5, 3)$.

32. Find y if the line through $(-1, 2)$ and $(-3, y)$ has a slope of -2.

33. Find the slope of the line $y = 3$.

34. Give the equation of a line with slope $-\dfrac{3}{4}$ and y-intercept = 2.

35. Find the slope and y-intercept of $4x - 5y = 20$.

36. Find the equation of the line with x-intercept -2 and y-intercept -1.

37. Find the equation of the line that is parallel to $5x + 2y = -1$ and contains the point $(-4, -2)$.

38. Find the equation of a line with slope $\dfrac{7}{3}$ that contains the point $(6, 8)$. Write the answer in slope-intercept form.

If $f(x) = x^2 - 3x$, $g(x) = x - 1$ and $h(x) = 3x - 1$, find the following.

39. $h(0)$

40. $f(-1) + g(4)$

41. $(g \circ f)(-2)$

42. $(h - g)(x)$

43. Specify the domain and range for the relation $\{(-1, 3), (2, -1), (3, 3)\}$. Is the relation also a function?

44. y varies inversely with the square of x. If $y = 4$ when $x = 4$; find y when $x = 6$.

45. Zac left a $6 tip for a $30 lunch with his girlfriend. What percent of the cost of lunch was the tip?

46. If the sales tax is 7.25% of the purchase price, how much sales tax will Hailey pay on a $130 dress?

Use the information in the illustration to work Problems 47–48.

Living Expenses Translated

Tokyo, Japan	$143,700
Osaka, Japan	$119,200
Moscow, Russia	$115,400
Geneva, Switzerland	$109,200
Hong Kong, China	$108,700

Source: Mercer Worldwide Cost of Living Survey

47. How much more does it cost, expressed as a percent, to live in Tokyo than in Moscow?

48. How much less does it cost, expressed as a percent, to live in Geneva than in Osaka?

For each of the following straight lines, identify the *x*-intercept, *y*-intercept, and slope, and sketch the graph. [8.1–8.2]

1. $2x + y = 6$

2. $y = -2x - 3$

3. $y = \frac{3}{2}x + 4$

4. $x = -2$

Use slope-intercept form to write the equation of each line below. [8.2]

5.

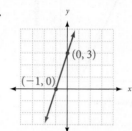

6.

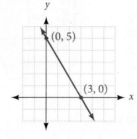

Find the equation for each line. [8.2]

7. The line through $(-1, 3)$ that has slope $m = 2$

8. The line through $(-3, 2)$ and $(4, -1)$

9. The line which contains the point $(5, -3)$ and is parallel to the line of $2x - 5y = 10$

10. The line which contains the point $(-1, -2)$ and is perpendicular to the line of $y = 3x - 2$

State the domain and range for the following relations, and indicate which relations are also functions. [8.3]

11. $\{(-2, 0), (-3, 0), (-2, 1)\}$

12. $y = x^2 - 9$

13. $\{(0, 0), (1, 3), (2, 5)\}$

14. $y = 3x - 5$

Determine the domain and range of the following functions. Assume the *entire* function is shown. [8.3]

15.

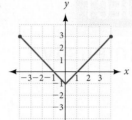

16.

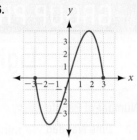

Let $f(x) = x - 2$, $g(x) = 3x + 4$ and $h(x) = 3x^2 - 2x - 8$, and find the following. [8.5]

17. $f(3) + g(2)$

18. $h(0) + g(0)$

19. $f(g(2))$

20. $g(f(2))$

Use the graph to answer questions 21–24.

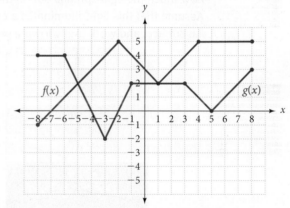

Evaluate. [8.5]

21. $f(-6) + g(-6)$

22. $(f - g)(1)$

23. $(f \circ g)(5)$

24. $(g \circ f)(5)$

Solve the following variation problems. [8.6]

25. Direct Variation Quantity *y* varies directly with the square of *x*. If *y* is 50 when *x* is 5, find *y* when *x* is 3.

26. Joint Variation Quantity *z* varies jointly with *x* and the cube of *y*. If *z* is 15 when *x* is 5 and *y* is 2, find *z* when *x* is 2 and *y* is 3.

Chapter 8 Projects

EQUATIONS AND FUNCTIONS

Light Intensity

Number of People 2–3

Time Needed 15 minutes

Equipment Paper and pencil

Background I found the following diagram while shopping for some track lighting for my home. I was impressed by the diagram because it displays a lot of useful information in a very efficient manner. As the diagram indicates, the amount of light that falls on a surface depends on how far above the surface the light is placed and how much the light spreads out on the surface. Assume that this light illuminates a circle on a flat surface, and work the following problems.

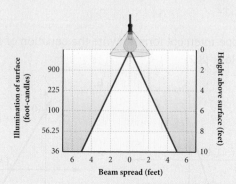

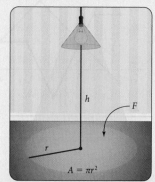

$A = \pi r^2$

Procedure

a. Fill in each table.

Height Above Surface (ft)	Illumination (foot-candles)
2	
4	
6	
8	
10	

Height Above Surface (ft)	Area of Illuminated Region (ft²)
2	
4	
6	
8	
10	

b. Use the templates in Figures 1 and 2 to construct line graphs from the data in the tables.

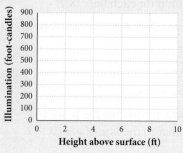

FIGURE 1

FIGURE 2

c. Which of the relationships is direct variation, and which is inverse variation?

d. Let F represent the number of foot-candles that fall on the surface, h the distance the light source is above the surface, and A the area of the illuminated region. Write an equation that shows the relationship between A and h, then write another equation that gives the relationship between F and h.

RESEARCH PROJECT

Descartes and Pascal

René Descartes, the inventor of the rectangular coordinate system, is the person who made the statement, "I think, therefore, I am." Blaise Pascal, another French philosopher, is responsible for the statement, "The heart has its reasons which reason does not know." Although Pascal and Descartes were contemporaries, the philosophies of the two men

René Descartes, 1596–1650

Blaise Pascal, 1623–1662

differed greatly. Research the philosophy of both Descartes and Pascal, and then write an essay that gives the main points of each man's philosophy. In the essay, show how the quotations given here fit in with the philosophy of the man responsible for the quotation.

RESEARCH PROJECT

Descartes and Pascal

9

Rational Exponents and Roots

Introduction

Standing 1482.6 feet over the city of Kuala Lumpur, Malaysia, the Petronas Towers are the tallest twin buildings in the world. Tenants and visitors can pass from one tower to the other by way of a two-story skybridge that stretches between the 41st and 42nd floors. The skybridge is designed to slide in and out of the towers, eliminating possible damage from high winds if it were attached to the main frames. The towers themselves are very complex structures, yet they are built on a simple design involving two overlapping squares, as shown below.

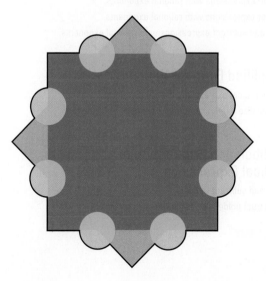

Complex structures often arise out of very simple geometric concepts. In this chapter we discover an interesting geometric figure, the golden rectangle, that is constructed from a square. Although the golden rectangle starts from a simple square, it becomes more interesting and complex the more closely we examine it.

Use rules for exponents to simplify the following.

1. 8^2

2. $(-8)^2$

3. -8^2

4. $-(-8)^2$

5. $(-2)^3$

6. -2^3

7. $12\left(\dfrac{1}{3}\right)$

8. $8\left(\dfrac{1}{4}\right)$

9. $\dfrac{1}{3} + \dfrac{5}{6}$

10. $\dfrac{3}{5} \cdot \dfrac{5}{9}$

11. $-12 - 10$

12. $-7 - (-11)$

13. $2 - \dfrac{3}{2}$

14. $3 + \dfrac{5}{4}$

15. $x^8 \cdot x^{-3}$

16. $(y^2)^3$

17. $(3r^8 s^{20})^5$

18. $\dfrac{x^{-3}}{x^{-4}}$

19. 3^{-3}

20. $\left(\dfrac{3}{2}\right)^{-4}$

Chapter Outline

9.1 Rational Exponents

A Simplify radical expressions using the definition for roots.

B Simplify expressions with rational exponents.

9.2 More Expressions Involving Rational Exponents

A Multiply expressions with rational exponents.

B Divide expressions with rational exponents.

C Factor expressions with rational exponents.

D Add and subtract expressions with rational exponents.

9.3 Simplified Form for Radicals

A Write radical expressions in simplified form.

B Rationalize a denominator that contains only one term.

9.4 Addition and Subtraction of Radical Expressions

A Add and subtract radicals.

B Construct golden rectangles from squares.

9.5 Multiplication and Division of Radical Expressions

A Multiply expressions containing radicals.

B Rationalize a denominator containing two terms.

9.6 Equations with Radicals

A Solve equations containing radicals.

B Graph simple square root and cube root equations in two variables.

9.7 Complex Numbers

A Simplify square roots of negative numbers.

B Simplify powers of i.

C Solve for unknown variables by equating real parts and by equating imaginary parts of two complex numbers.

D Add and subtract complex numbers.

E Multiply complex numbers.

F Divide complex numbers.

9.1 Rational Exponents

TICKET TO SUCCESS

Keep these questions in mind as you read through the section. Then respond in your own words and in complete sentences.

1. What is the definition for the positive square root?
2. What is the rational exponent theorem?
3. For the expression $a^{\frac{m}{n}}$, explain the significance of the numerator m and the significance of the denominator n in the exponent.
4. Briefly explain the concept behind the golden rectangle.

Figure 1 shows a square in which each of the four sides is 1 inch long. To find the square of the length of the diagonal c, we apply the Pythagorean theorem.

$$c^2 = 1^2 + 1^2$$
$$c^2 = 2$$

1 inch

1 inch

FIGURE 1

Because we know that c is positive and that its square is 2, we call c the *positive square root* of 2, and we write $c = \sqrt{2}$. Associating numbers, such as $\sqrt{2}$, with the diagonal of a square or rectangle allows us to analyze some interesting items from geometry. One particularly interesting geometric object, the golden rectangle, that we will study in this section is shown in Figure 2. You may have seen some real-life examples of this object such as the Parthenon, shown on the previous page. It is constructed from a right triangle, and the length of the diagonal is found from the Pythagorean theorem. We will come back to this figure at the end of this section.

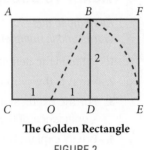

The Golden Rectangle

FIGURE 2

A Radical Expressions

Previously, we developed notation (exponents) to give us the square, cube, or any other power of a number. For instance, if we wanted the square of 3, we wrote $3^2 = 9$. If we wanted the cube of 3, we wrote $3^3 = 27$. In this section, we will develop notation that will take us in the reverse direction; that is, from the square of a number, say 25, back to the original number, 5.

> **Definition**
> If x is a nonnegative real number, then the expression $\sqrt{x}$ is called the **positive square root** of x and is the nonnegative number such that
>
> $$(\sqrt{x})^2 = x$$
>
> In words: $\sqrt{x}$ is the nonnegative number we square to get x.

NOTE
It is a common mistake to assume that an expression like $\sqrt{25}$ indicates both square roots, 5 and -5. The expression $\sqrt{25}$ indicates only the positive square root of 25, which is 5. If we want the negative square root, we must use a negative sign: $-\sqrt{25} = -5$.

The negative square root of x, $-\sqrt{x}$, is defined in a similar manner.

EXAMPLE 1   The positive square root of 64 is 8 because 8 is the positive number with the property $8^2 = 64$. The negative square root of 64 is -8 since -8 is the negative number whose square is 64. We can summarize both of these facts by saying

$$\sqrt{64} = 8 \quad \text{and} \quad -\sqrt{64} = -8$$

The higher roots, cube roots, fourth roots, and so on have definitions similar to that of square roots.

NOTE
We have restricted the even roots in this definition to nonnegative numbers. Even roots of negative numbers exist but are not represented by real numbers; that is, $\sqrt{-4}$ is not a real number since there is no real number whose square is -4.

> **Definition**
> If x is a real number and n is a positive integer, then
>
> Positive square root of x, $\sqrt{x}$, is such that $(\sqrt{x})^2 = x$ $x \geq 0$
>
> Cube root of x, $\sqrt[3]{x}$, is such that $(\sqrt[3]{x})^3 = x$
>
> Positive fourth root of x, $\sqrt[4]{x}$, is such that $(\sqrt[4]{x})^4 = x$ $x \geq 0$
>
> Fifth root of x, $\sqrt[5]{x}$, is such that $(\sqrt[5]{x})^5 = x$
>
> $\vdots$ $\vdots \ \vdots$
>
> The **nth root of x**, $\sqrt[n]{x}$, is such that $(\sqrt[n]{x})^n = x$ $x \geq 0$ if n is even

The following is a table of the most common roots used in this book. Any of the roots that are unfamiliar should be memorized.

Square Roots		Cube Roots	Fourth Roots
$\sqrt{0} = 0$	$\sqrt{49} = 7$	$\sqrt[3]{0} = 0$	$\sqrt[4]{0} = 0$
$\sqrt{1} = 1$	$\sqrt{64} = 8$	$\sqrt[3]{1} = 1$	$\sqrt[4]{1} = 1$
$\sqrt{4} = 2$	$\sqrt{81} = 9$	$\sqrt[3]{8} = 2$	$\sqrt[4]{16} = 2$
$\sqrt{9} = 3$	$\sqrt{100} = 10$	$\sqrt[3]{27} = 3$	$\sqrt[4]{81} = 3$
$\sqrt{16} = 4$	$\sqrt{121} = 11$	$\sqrt[3]{64} = 4$	
$\sqrt{25} = 5$	$\sqrt{144} = 12$	$\sqrt[3]{125} = 5$	
$\sqrt{36} = 6$	$\sqrt{169} = 13$		

NOTATION An expression like $\sqrt[3]{8}$ that involves a root is called a *radical expression*. In the expression $\sqrt[3]{8}$, the 3 is called the *index*, the $\sqrt{}$ is the *radical sign*, and 8 is called the *radicand*. The index of a radical must be a positive integer greater than 1. If no index is written, it is assumed to be 2.

Roots and Negative Numbers

When dealing with negative numbers and radicals, the only restriction concerns negative numbers under even roots. We can have negative signs in front of radicals and negative numbers under odd roots and still obtain real numbers. Here are some examples to help clarify this. In the last section of this chapter, we will see how to deal with even roots of negative numbers.

EXAMPLES   Simplify each expression, if possible.

2. $\sqrt[3]{-8} = -2$ because $(-2)^3 = -8$.

3. $\sqrt{-4}$ is not a real number since there is no real number whose square is -4.

4. $-\sqrt{25} = -5$ is the negative square root of 25.

5. $\sqrt[5]{-32} = -2$ because $(-2)^5 = -32$.

6. $\sqrt[4]{-81}$ is not a real number since there is no real number we can raise to the fourth power and obtain -81.

Variables Under a Radical

From the preceding examples, it is clear that we must be careful that we do not try to take an even root of a negative number. For this reason, we will assume that all variables appearing under a radical sign represent nonnegative numbers.

EXAMPLES   Assume all variables represent nonnegative numbers and simplify each expression as much as possible.

7. $\sqrt{25a^4b^6} = 5a^2b^3$ because $(5a^2b^3)^2 = 25a^4b^6$.

8. $\sqrt[3]{x^6y^{12}} = x^2y^4$ because $(x^2y^4)^3 = x^6y^{12}$.

9. $\sqrt[4]{81r^8s^{20}} = 3r^2s^5$ because $(3r^2s^5)^4 = 81r^8s^{20}$.

B Rational Numbers as Exponents

Next we develop a second kind of notation involving exponents that will allow us to designate square roots, cube roots, and so on in another way.

Consider the equation $x = 8^{1/3}$. Although we have not encountered fractional exponents before, let's assume that all the properties of exponents hold in this case. Cubing both sides of the equation, we have

$$x^3 = (8^{1/3})^3$$

$$x^3 = 8^{(1/3)(3)}$$

$$x^3 = 8^1$$

$$x^3 = 8$$

The last line tells us that x is the number whose cube is 8. It must be true, then, that x is the cube root of 8, $x = \sqrt[3]{8}$. Since we started with $x = 8^{1/3}$, it follows that

$$8^{1/3} = \sqrt[3]{8}$$

It seems reasonable, then, to define fractional exponents as indicating roots. Here is the formal definition.

Definition

If x is a real number and n is a positive integer greater than 1, then

$$x^{1/n} = \sqrt[n]{x} \qquad (x \geq 0 \text{ when } n \text{ is even})$$

In words: The quantity $x^{1/n}$ is the nth root of x.

With this definition we have a way of representing roots with exponents. Here are some examples.

EXAMPLES   Write each expression as a root and then simplify, if possible.

10. $8^{1/3} = \sqrt[3]{8} = 2$

11. $36^{1/2} = \sqrt{36} = 6$

12. $-25^{1/2} = -\sqrt{25} = -5$

13. $(-25)^{1/2} = \sqrt{-25}$, which is not a real number

14. $\left(\dfrac{4}{9}\right)^{1/2} = \sqrt{\dfrac{4}{9}} = \dfrac{2}{3}$

The properties of exponents developed in a previous chapter were applied to integer exponents only. We will now extend these properties to include rational exponents also. We do so without proof.

Properties of Exponents

If a and b are real numbers and r and s are rational numbers, and a and b are nonnegative whenever r and s indicate even roots, then

1. $a^r \cdot a^s = a^{r+s}$ **4.** $a^{-r} = \dfrac{1}{a^r}$ $(a \neq 0)$

2. $(a^r)^s = a^{rs}$ **5.** $\left(\dfrac{a}{b}\right)^r = \dfrac{a^r}{b^r}$ $(b \neq 0)$

3. $(ab)^r = a^r b^r$ **6.** $\dfrac{a^r}{a^s} = a^{r-s}$ $(a \neq 0)$

There are times when rational exponents can simplify our work with radicals. Here are Examples 8 and 9 again, but this time we will work them using rational exponents.

EXAMPLE 15   Write the radical with a rational exponent then simplify.

SOLUTION
$$\sqrt[3]{x^6 y^{12}} = (x^6 y^{12})^{1/3}$$
$$= (x^6)^{1/3}(y^{12})^{1/3}$$
$$= x^2 y^4$$

EXAMPLE 16   Write the radical with a rational exponent then simplify.

SOLUTION
$$\sqrt[4]{81 r^8 s^{20}} = (81 r^8 s^{20})^{1/4}$$
$$= 81^{1/4}(r^8)^{1/4}(s^{20})^{1/4}$$
$$= 3 r^2 s^5$$

> **NOTE**
> Unless we state otherwise, assume all variables throughout this section and problem set represent nonnegative numbers.

So far, the numerators of all the rational exponents we have encountered have been 1. The next theorem extends the work we can do with rational exponents to rational exponents with numerators other than 1.

We can extend our properties of exponents with the following theorem.

> **The Rational Exponent Theorem**
> If a is a nonnegative real number, m is an integer, and n is a positive integer, then
> $$a^{m/n} = (a^{1/n})^m = (a^m)^{1/n}$$

PROOF We can prove this theorem using the properties of exponents. Since $\frac{m}{n} = m\left(\frac{1}{n}\right)$ we have

$$a^{m/n} = a^{m(1/n)} \qquad \bigg| \qquad a^{m/n} = a^{(1/n)(m)}$$
$$= (a^m)^{1/n} \qquad \bigg| \qquad = (a^{1/n})^m$$

Here are some examples that illustrate how we use this theorem.

EXAMPLE 17   Simplify as much as possible.

SOLUTION

$8^{2/3} = (8^{1/3})^2$	Rational exponent theorem
$= 2^2$	Definition of fractional exponents
$= 4$	The square of 2 is 4

> **NOTE**
> On a scientific calculator, Example 17 would look like this:
>
> 8 y^x (2 ÷ 3) =

EXAMPLE 18   Simplify as much as possible.

SOLUTION

$25^{3/2} = (25^{1/2})^3$	Rational exponent theorem
$= 5^3$	Definition of fractional exponents
$= 125$	The cube of 5 is 125

EXAMPLE 19 Simplify as much as possible.

SOLUTION

$$9^{-3/2} = (9^{1/2})^{-3} \qquad \text{Rational exponent theorem}$$

$$= 3^{-3} \qquad \text{Definition of fractional exponents}$$

$$= \frac{1}{3^3} \qquad \text{Property 4 for exponents}$$

$$= \frac{1}{27} \qquad \text{The cube of 3 is 27}$$

EXAMPLE 20 Simplify as much as possible.

SOLUTION

$$\left(\frac{27}{8}\right)^{-4/3} = \left[\left(\frac{27}{8}\right)^{1/3}\right]^{-4} \qquad \text{Rational exponent theorem}$$

$$= \left(\frac{3}{2}\right)^{-4} \qquad \text{Definition of fractional exponents}$$

$$= \left(\frac{2}{3}\right)^{4} \qquad \text{Property 4 for exponents}$$

$$= \frac{16}{81} \qquad \text{The fourth power of } \frac{2}{3} \text{ is } \frac{16}{81}$$

EXAMPLE 21 Assume the variables represent positive quantities and simplify as much as possible.

SOLUTION

$$x^{1/3} \cdot x^{5/6} = x^{1/3+5/6} \qquad \text{Property 1}$$

$$= x^{2/6+5/6} \qquad \text{LCD is 6}$$

$$= x^{7/6} \qquad \text{Add fractions}$$

EXAMPLE 22 Assume the variables represent positive quantities and simplify as much as possible.

SOLUTION

$$(y^{2/3})^{3/4} = y^{(2/3)(3/4)} \qquad \text{Property 2}$$

$$= y^{1/2} \qquad \text{Multiply fractions: } \frac{2}{3} \cdot \frac{3}{4} = \frac{6}{12} = \frac{1}{2}$$

EXAMPLE 23 Assume the variables represent positive quantities and simplify as much as possible.

SOLUTION

$$\frac{z^{1/3}}{z^{1/4}} = z^{1/3-1/4} \qquad \text{Property 6}$$

$$= z^{4/12-3/12} \qquad \text{LCD is 12}$$

$$= z^{1/12} \qquad \text{Subtract fractions}$$

EXAMPLE 24   Assume the variables represent positive quantities and simplify as much as possible.

SOLUTION $\dfrac{(x^{-3}y^{1/2})^4}{x^{10}y^{3/2}} = \dfrac{(x^{-3})^4(y^{1/2})^4}{x^{10}y^{3/2}}$ Property 3

$= \dfrac{x^{-12}y^2}{x^{10}y^{3/2}}$ Property 2

$= x^{-22}y^{1/2}$ Property 6

$= \dfrac{y^{1/2}}{x^{22}}$ Property 4

FACTS FROM GEOMETRY: The Pythagorean Theorem (Again)

Now that we have had some experience working with square roots, we can re-write the Pythagorean theorem using a square root. If triangle ABC is a right tri-angle with $C = 90°$, then the length of the longest side is the *positive square root* of the sum of the squares of the other two sides (see Figure 3).

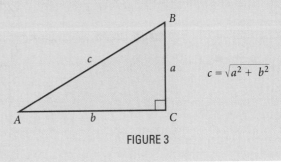

$$c = \sqrt{a^2 + b^2}$$

FIGURE 3

In the introduction to this section, we mentioned the golden rectangle. Its origins can be traced back more than 2,000 years to the Greek civilization that produced Pythagoras, Socrates, Plato, Aristotle, and Euclid. The most important mathematical work to come from that Greek civilization was Euclid's *Elements*, an elegantly written summary of all that was known about geometry at that time in history. Euclid's *Elements*, according to Howard Eves, an authority on the history of mathematics, exercised a greater influence on scientific thinking than any other work. Here is how we construct a golden rectangle from a square of side 2, using the same method that Euclid used in his *Elements*.

FACTS FROM GEOMETRY: Constructing a Golden Rectangle from a Square of Side 2

Step 1: Draw a square with a side of length two. Connect the midpoint of side *CD* to corner *B*. (Note that we have labeled the midpoint of segment *CD* with the letter *O*.)

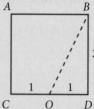

Step 2: Drop the diagonal *OB* from step 1 down so it aligns with side *CD*.

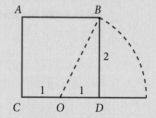

Step 3: Form rectangle *ACEF*. This is a golden rectangle.

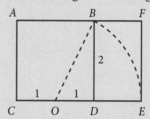

All golden rectangles are constructed from squares. Every golden rectangle, no matter how large or small it is, will have the same shape. To associate a number with the shape of the golden rectangle, we use the ratio of its length to its width. This ratio is called the *golden ratio*. To calculate the golden ratio, we must first find the length of the diagonal we used to construct the golden rectangle. Figure 4 shows the golden rectangle that we constructed from a square of side 2. The length of the diagonal *OB* is found by applying the Pythagorean theorem to triangle *OBD*.

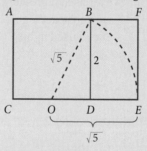

FIGURE 4

The length of segment *OE* is equal to the length of diagonal *OB*; both are $\sqrt{5}$. Since the distance from *C* to *O* is 1, the length *CE* of the golden rectangle is $1 + \sqrt{5}$. Now we can find the golden ratio:

$$\text{Golden ratio} = \frac{\text{length}}{\text{width}} = \frac{CE}{EF} = \frac{1 + \sqrt{5}}{2}$$

USING TECHNOLOGY

Graphing Calculators: A Word of Caution

Some graphing calculators give surprising results when evaluating expressions such as $(-8)^{2/3}$. As you know from reading this section, the expression $(-8)^{2/3}$ simplifies to 4, either by taking the cube root first and then squaring the result, or by squaring the base first and then taking the cube root of the result. Here are three different ways to evaluate this expression on your calculator:

1. $(-8)\wedge(2/3)$ To evaluate $-8^{2/3}$

2. $((-8)\wedge2)\wedge(1/3)$ To evaluate $((-8)^2)^{1/3}$

3. $((-8)\wedge(1/3))\wedge2$ To evaluate $((-8)^{1/3})^2$

Note any difference in the results.

Next, graph each of the following functions, one at a time.

1. $Y_1 = X^{2/3}$ **2.** $Y_2 = (X^2)^{1/3}$ **3.** $Y_3 = (X^{1/3})^2$

The correct graph is shown in Figure 5. Note which of your graphs match the correct graph.

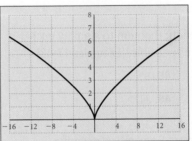

FIGURE 5

Different calculators evaluate exponential expressions in different ways. You should use the method (or methods) that gave you the correct graph.

Problem Set 9.1

Moving Toward Success

"A habit cannot be tossed out the window; it must be coaxed down the stairs a step at a time."

—Mark Twain, 1835–1910, American writer and humorist

1. What is mathematical intuition and how is it important?

2. What does practice and repetition have to do with mathematics?

A Find each of the following roots, if possible.
[Examples 1–6]

1. $\sqrt{144}$ **2.** $-\sqrt{144}$

3. $\sqrt{-144}$ **4.** $\sqrt{-49}$

5. $-\sqrt{49}$ **6.** $\sqrt{49}$

7. $\sqrt[3]{-27}$ **8.** $-\sqrt[3]{27}$

9. $\sqrt[4]{16}$ **10.** $-\sqrt[4]{16}$

11. $\sqrt[4]{-16}$ **12.** $-\sqrt[4]{-16}$

13. $\sqrt{0.04}$ **14.** $\sqrt{0.81}$

15. $\sqrt[3]{0.008}$ **16.** $\sqrt[3]{0.125}$

17. $\sqrt[3]{125}$ **18.** $\sqrt[3]{-125}$

19. $-\sqrt[3]{216}$ **20.** $-\sqrt[3]{-216}$

21. $\sqrt{\dfrac{1}{36}}$ **22.** $\sqrt{\dfrac{9}{25}}$

23. $\sqrt[3]{\dfrac{8}{125}}$ **24.** $\sqrt[3]{-\dfrac{27}{216}}$

A Simplify each expression. Assume all variables represent nonnegative numbers. [Examples 7–9]

25. $\sqrt{36a^8}$

26. $\sqrt{49a^{10}}$

27. $\sqrt[3]{27a^{12}}$

28. $\sqrt[3]{8a^{15}}$

29. $\sqrt[5]{32x^{10}y^5}$

30. $\sqrt[5]{32x^5y^{10}}$

31. $\sqrt[4]{16a^{12}b^{20}}$

32. $\sqrt[4]{81a^{24}b^8}$

B Simplify each expression. [Examples 10–14]

33. $36^{1/2}$

34. $49^{1/2}$

35. $-9^{1/2}$

36. $-16^{1/2}$

37. $8^{1/3}$

38. $-8^{1/3}$

39. $(-8)^{1/3}$

40. $-27^{1/3}$

41. $32^{1/5}$

42. $81^{1/4}$

43. $\left(\dfrac{81}{25}\right)^{1/2}$

44. $\left(\dfrac{64}{125}\right)^{1/3}$

B Use the rational exponent theorem to simplify each of the following as much as possible. [Examples 17–18]

45. $27^{2/3}$

46. $8^{4/3}$

47. $25^{3/2}$

48. $81^{3/4}$

B Simplify each expression. Remember, negative exponents give reciprocals. [Examples 19–20]

49. $27^{-1/3}$

50. $9^{-1/2}$

51. $81^{-3/4}$

52. $4^{-3/2}$

53. $\left(\dfrac{25}{36}\right)^{-1/2}$

54. $\left(\dfrac{16}{49}\right)^{-1/2}$

55. $\left(\dfrac{81}{16}\right)^{-3/4}$

56. $\left(\dfrac{27}{8}\right)^{-2/3}$

57. $16^{1/2} + 27^{1/3}$

58. $25^{1/2} + 100^{1/2}$

59. $8^{-2/3} + 4^{-1/2}$

60. $49^{-1/2} + 25^{-1/2}$

B Use the properties of exponents to simplify each of the following as much as possible. Assume all bases are positive. [Examples 21–24]

61. $x^{3/5} \cdot x^{1/5}$

62. $x^{3/4} \cdot x^{5/4}$

63. $(a^{3/4})^{4/3}$

64. $(a^{2/3})^{3/4}$

65. $\dfrac{x^{1/5}}{x^{3/5}}$

66. $\dfrac{x^{2/7}}{x^{5/7}}$

67. $\dfrac{x^{5/6}}{x^{2/3}}$

68. $\dfrac{x^{7/8}}{x^{8/7}}$

69. $(x^{3/5}y^{5/6}z^{1/3})^{3/5}$

70. $(x^{3/4}y^{1/8}z^{5/6})^{4/5}$

71. $\dfrac{a^{3/4}b^2}{a^{7/8}b^{1/4}}$

72. $\dfrac{a^{1/3}b^4}{a^{3/5}b^{1/3}}$

73. $\dfrac{(y^{2/3})^{3/4}}{(y^{1/3})^{3/5}}$

74. $\dfrac{(y^{5/4})^{2/5}}{(y^{1/4})^{4/3}}$

75. $\left(\dfrac{a^{-1/4}}{b^{1/2}}\right)^8$

76. $\left(\dfrac{a^{-1/5}}{b^{1/3}}\right)^{15}$

77. $\dfrac{(r^{-2}s^{1/3})^6}{r^8s^{3/2}}$

78. $\dfrac{(r^{-5}s^{1/2})^4}{r^{12}s^{5/2}}$

79. $\dfrac{(25a^6b^4)^{1/2}}{(8a^{-9}b^3)^{-1/3}}$

80. $\dfrac{(27a^3b^6)^{1/3}}{(81a^8b^{-4})^{1/4}}$

Applying the Concepts

81. Maximum Speed The maximum speed (v) that an automobile can travel around a curve of radius r without skidding is given by the equation

$$v = \left(\dfrac{5r}{2}\right)^{1/2}$$

where v is in miles per hour and r is measured in feet. What is the maximum speed a car can travel around a curve with a radius of 250 feet without skidding?

82. Golden Ratio The golden ratio is the ratio of the length to the width in any golden rectangle. The exact value of this number is $\dfrac{1 + \sqrt{5}}{2}$. Use a calculator to find a decimal approximation to this number and round it to the nearest thousandth.

83. Chemistry Figure 6 shows part of a model of a magnesium oxide (MgO) crystal. Each corner of the square is at the center of one oxygen ion (O^{2-}), and the center of the middle ion is at the center of the square. The radius for each oxygen ion is 126 picometers (pm), and the radius for each magnesium ion (Mg^{2+}) is 86 picometers.

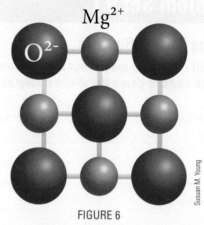

FIGURE 6

Susuan M. Young

a. Find the length of the side of the square to the nearest picometer.

b. Find the length of the diagonal of the square. Write your answer in picometers.

c. If 1 meter is 10^{12} picometers, give the length of the diagonal of the square in meters.

84. Geometry The length of each side of the cube shown in Figure 7 is 1 inch.
  a. Find the length of the diagonal CH.
  b. Find the length of the diagonal CF.

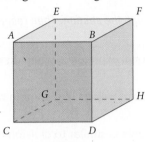

FIGURE 7

85. Comparing Graphs Identify the graph with the correct equation.

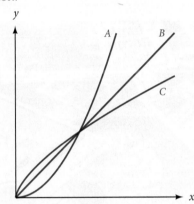

  a. $y = x$
  b. $y = x^2$
  c. $y = x^{2/3}$
  d. What are the two points of intersection of all three graphs?

86. Falling Objects The time t in seconds it takes an object to fall d feet is given by the equation

$$t = \frac{1}{4}\sqrt{d}$$

  a. The Sears Tower in Chicago is 1,450 feet tall. How long would it take a penny to fall to the ground from the top of the Sears Tower? Round to the nearest hundreth of a second.
  b. An object took 30 seconds to fall to the ground. From what distance must it have been dropped?

Maintaining Your Skills

87. $x^2(x^4 - x)$

88. $5x^2(2x^3 - x)$

89. $(x - 3)(x + 5)$

90. $(x - 2)(x + 2)$

91. $(x^2 - 5)^2$

92. $(x^2 + 5)^2$

93. $(x - 3)(x^2 + 3x + 9)$

94. $(x + 3)(x^2 - 3x + 9)$

Getting Ready for the Next Section

Simplify.

95. $x^2(x^4 - x^3)$

96. $(x^2 - 3)(x^2 + 5)$

97. $(3a - 2b)(4a - b)$

98. $(x^2 + 3)^2$

99. $(x^3 - 2)(x^3 + 2)$

100. $(a - b)(a^2 + ab + b^2)$

101. $\dfrac{15x^2y - 20x^4y^2}{5xy}$

102. $\dfrac{12a^3b^2 - 24a^2b^4}{3ab}$

Factor.

103. $x^2 - 3x - 10$

104. $x^2 + x - 12$

105. $6x^2 + 11x - 10$

106. $10x^2 - x - 3$

Use rules of exponents to simplify.

107. $x^{2/3} \cdot x^{4/3}$

108. $x^{1/4} \cdot x^{3/4}$

109. $(t^{1/2})^2$

110. $(x^{3/2})^2$

111. $\dfrac{x^{2/3}}{x^{1/3}}$

112. $\dfrac{x^{1/2}}{x^{1/2}}$

Extending the Concepts

113. Show that the expression $(a^{1/2} + b^{1/2})^2$ is not equal to $a + b$ by replacing a with 9 and b with 4 in both expressions and then simplifying each.

114. Show that the statement $(a^2 + b^2)^{1/2} = a + b$ is not, in general, true by replacing a with 3 and b with 4 and then simplifying both sides.

115. You may have noticed, if you have been using a calculator to find roots, that you can find the fourth root of a number by pressing the square root button twice. Written in symbols, this fact looks like this:

$$\sqrt{\sqrt{a}} = \sqrt[4]{a} \qquad (a \geq 0)$$

Show that this statement is true by rewriting each side with exponents instead of radical notation and then simplifying the left side.

116. Show that the following statement is true by rewriting each side with exponents instead of radical notation and then simplifying the left side.

$$\sqrt[3]{\sqrt{a}} = \sqrt[6]{a} \qquad (a \geq 0)$$

OBJECTIVES

A Multiply expressions with rational exponents.

B Divide expressions with rational exponents.

C Factor expressions with rational exponents.

D Add and subtract expressions with rational exponents.

TICKET TO SUCCESS

Keep these questions in mind as you read through the section. Then respond in your own words and in complete sentences.

1. When multiplying expressions with fractional exponents, when do we add the fractional exponents?

2. When can you use the FOIL method with expressions that contain rational exponents?

3. Explain how division with rational exponents is similar to division of a polynomial by a monomial.

4. What is the first thing you do when adding expressions that contain rational exponents but have different denominators?

© Lana Sundman / Alamy

Suppose you purchased 10 sets of baseball cards for $21 each, for a total investment of $210. Three years later, you find that each set is worth $30, which means that your 10 sets have a total value of $300.

You can calculate the annual rate of return on this investment using a formula that involves rational exponents. The annual rate of return will tell you at what interest rate you would have to invest your original $210 for it to be worth $300 three years later. As you will see at the end of this section, the annual rate of return on this investment is 12.6%, which is a good return on your money.

In this section, we will look at multiplication, division, factoring, and simplification of some expressions that resemble polynomials but contain rational exponents. The problems in this section will be of particular interest to you if you are planning to take either an engineering calculus class or a business calculus class. As was the case in the previous section, we will assume all variables represent nonnegative real numbers. That way, we will not have to worry about the possibility of introducing undefined terms—even roots of negative numbers—into any of our examples. Let's begin this section with a look at multiplication of expressions containing rational exponents.

A Multiplication with Rational Exponents

> **EXAMPLE 1**   Multiply $x^{2/3}(x^{4/3} - x^{1/3})$.

SOLUTION   Applying the distributive property and then simplifying the resulting terms, we have

$$x^{2/3}(x^{4/3} - x^{1/3}) = x^{2/3}x^{4/3} - x^{2/3}x^{1/3} \qquad \text{Distributive property}$$

$$= x^{6/3} - x^{3/3} \qquad\qquad\qquad \text{Add exponents}$$

$$= x^2 - x \qquad\qquad\qquad\quad \text{Simplify}$$

> **EXAMPLE 2**   Multiply $(x^{2/3} - 3)(x^{2/3} + 5)$.

SOLUTION   Applying the FOIL method, we multiply as if we were multiplying two binomials

$$(x^{2/3} - 3)(x^{2/3} + 5) = x^{2/3}x^{2/3} + 5x^{2/3} - 3x^{2/3} - 15$$

$$= x^{4/3} + 2x^{2/3} - 15$$

> **EXAMPLE 3**   Multiply $(3a^{1/3} - 2b^{1/3})(4a^{1/3} - b^{1/3})$.

SOLUTION   Again, we use the FOIL method to multiply

$$(3a^{1/3} - 2b^{1/3})(4a^{1/3} - b^{1/3})$$

$$= 3a^{1/3}4a^{1/3} - 3a^{1/3}b^{1/3} - 2b^{1/3}4a^{1/3} + 2b^{1/3}b^{1/3}$$

$$= 12a^{2/3} - 11a^{1/3}b^{1/3} + 2b^{2/3}$$

> **EXAMPLE 4**   Expand $(t^{1/2} - 5)^2$.

SOLUTION   We can use the definition of exponents and the FOIL method.

$$(t^{1/2} - 5)^2 = (t^{1/2} - 5)(t^{1/2} - 5)$$

$$= t^{1/2}t^{1/2} - 5t^{1/2} - 5t^{1/2} + 25$$

$$= t - 10t^{1/2} + 25$$

We can obtain the same result by using the formula for the square of a binomial, $(a - b)^2 = a^2 - 2ab + b^2$.

$$(t^{1/2} - 5)^2 = (t^{1/2})^2 - 2t^{1/2} \cdot 5 + 5^2$$

$$= t - 10t^{1/2} + 25$$

> **EXAMPLE 5**   Multiply $(x^{3/2} - 2^{3/2})(x^{3/2} + 2^{3/2})$.

SOLUTION   This product has the form $(a - b)(a + b)$, which will result in the difference of two squares, $a^2 - b^2$.

$$(x^{3/2} - 2^{3/2})(x^{3/2} + 2^{3/2}) = (x^{3/2})^2 - (2^{3/2})^2$$

$$= x^3 - 2^3$$

$$= x^3 - 8$$

EXAMPLE 6 Multiply $(a^{1/3} - b^{1/3})(a^{2/3} + a^{1/3}b^{1/3} + b^{2/3})$.

SOLUTION We can find this product by multiplying in columns.

$$\begin{array}{r} a^{2/3} + a^{1/3}b^{1/3} + b^{2/3} \\ a^{1/3} - b^{1/3} \\ \hline a \quad + a^{2/3}b^{1/3} + a^{1/3}b^{2/3} \\ - a^{2/3}b^{1/3} - a^{1/3}b^{2/3} - b \\ \hline a \qquad\qquad\qquad - b \end{array}$$

The product is $a - b$.

B Division with Rational Exponents

Our next example involves division with expressions that contain rational exponents. As you will see, this kind of division is very similar to division of a polynomial by a monomial.

NOTE
For Example 7 assume x and y are positive numbers.

EXAMPLE 7 Divide $\dfrac{15x^{2/3}y^{1/3} - 20x^{4/3}y^{2/3}}{5x^{1/3}y^{1/3}}$.

SOLUTION We can approach this problem in the same way we approached division by a monomial. We simply divide each term in the numerator by the term in the denominator.

$$\frac{15x^{2/3}y^{1/3} - 20x^{4/3}y^{2/3}}{5x^{1/3}y^{1/3}} = \frac{15x^{2/3}y^{1/3}}{5x^{1/3}y^{1/3}} - \frac{20x^{4/3}y^{2/3}}{5x^{1/3}y^{1/3}}$$

$$= 3x^{1/3} - 4xy^{1/3}$$

C Factoring

The next three examples involve factoring. In the first example, we are told what to factor from each term of an expression.

EXAMPLE 8 Factor $3(x - 2)^{1/3}$ from $12(x - 2)^{4/3} - 9(x - 2)^{1/3}$, and then simplify, if possible.

SOLUTION This solution is similar to factoring out the greatest common factor.

$$12(x - 2)^{4/3} - 9(x - 2)^{1/3} = 3(x - 2)^{1/3}[4(x - 2) - 3]$$

$$= 3(x - 2)^{1/3}(4x - 11)$$

Although an expression containing rational exponents is not a polynomial—remember, a polynomial must have exponents that are whole numbers—we are going to treat the expressions that follow as if they were polynomials.

EXAMPLE 9 Factor $x^{2/3} - 3x^{1/3} - 10$ as if it were a trinomial.

SOLUTION We can think of $x^{2/3} - 3x^{1/3} - 10$ as if it is a trinomial in which the variable is $x^{1/3}$. To see this, replace $x^{1/3}$ with y to get

$$y^2 - 3y - 10$$

Since this trinomial in y factors as $(y - 5)(y + 2)$, we can factor our original expression similarly.

$$x^{2/3} - 3x^{1/3} - 10 = (x^{1/3} - 5)(x^{1/3} + 2)$$

Remember, with factoring, we can always multiply our factors to check that we have factored correctly.

EXAMPLE 10 Factor $6x^{2/5} + 11x^{1/5} - 10$ as if it were a trinomial.

SOLUTION We can think of the expression in question as a trinomial in $x^{1/5}$.

$$6x^{2/5} + 11x^{1/5} - 10 = (3x^{1/5} - 2)(2x^{1/5} + 5)$$

D Addition and Subtraction

In our next example, we combine two expressions by applying the methods we used to add and subtract fractions or rational expressions.

EXAMPLE 11 Subtract $(x^2 + 4)^{1/2} - \dfrac{x^2}{(x^2 + 4)^{1/2}}$.

SOLUTION To combine these two expressions, we need to find a least common denominator, change to equivalent fractions, and subtract numerators. The least common denominator is $(x^2 + 4)^{1/2}$.

$$(x^2 + 4)^{1/2} - \frac{x^2}{(x^2 + 4)^{1/2}} = \frac{(x^2 + 4)^{1/2}}{1} \cdot \frac{\boldsymbol{(x^2 + 4)^{1/2}}}{\boldsymbol{(x^2 + 4)^{1/2}}} - \frac{x^2}{(x^2 + 4)^{1/2}}$$

$$= \frac{x^2 + 4 - x^2}{(x^2 + 4)^{1/2}}$$

$$= \frac{4}{(x^2 + 4)^{1/2}}$$

EXAMPLE 12 If you purchase an investment for P dollars and t years later it is worth A dollars, then the annual rate of return r on that investment is given by the formula

$$r = \left(\frac{A}{P}\right)^{1/t} - 1$$

Find the annual rate of return on the baseball card collection in the section opener, purchased for $210 and sold 3 years later for $300.

SOLUTION Using $A = 300$, $P = 210$, and $t = 3$ in the formula, we have

$$r = \left(\frac{300}{210}\right)^{1/3} - 1$$

The easiest way to simplify this expression is with a calculator.

$$\boxed{(}\ 300\ \boxed{\div}\ 210\ \boxed{)}\ \boxed{\wedge}\ \boxed{(}\ \boxed{(}\ 1\ \boxed{\div}\ 3\ \boxed{)}\ \boxed{)}\ \boxed{-}\ 1\ \boxed{=}$$

Allowing three decimal places, the result is 0.126. The annual return on the coin collection is approximately 12.6%. To do as well with a savings account, we would have to invest the original $210 in an account that paid 12.6%, compounded annually.

Problem Set 9.2

Moving Toward Success

"There is nothing in a caterpillar that tells you it's going to be a butterfly."

—R. Buckminster Fuller, 1895–1983, American architect and engineer

1. Go online and research learning styles (e.g., visual, auditory, kinesthetic) to determine how you learn best. You may prefer one style or a combination of styles. How do you prefer to learn?

2. How can you apply your preferred learning style to this class?

A Multiply. (Assume all variables in this problem set represent nonnegative real numbers.) [Examples 1–6]

1. $x^{2/3}(x^{1/3} + x^{4/3})$

2. $x^{2/5}(x^{3/5} - x^{8/5})$

3. $a^{1/2}(a^{3/2} - a^{1/2})$

4. $a^{1/4}(a^{3/4} + a^{7/4})$

5. $2x^{1/3}(3x^{8/3} - 4x^{5/3} + 5x^{2/3})$

6. $5x^{1/2}(4x^{5/2} + 3x^{3/2} + 2x^{1/2})$

7. $4x^{1/2}y^{3/5}(3x^{3/2}y^{-3/5} - 9x^{-1/2}y^{7/5})$

8. $3x^{4/5}y^{1/3}(4x^{6/5}y^{-1/3} - 12x^{-4/5}y^{5/3})$

9. $(x^{2/3} - 4)(x^{2/3} + 2)$

10. $(x^{2/3} - 5)(x^{2/3} + 2)$

11. $(a^{1/2} - 3)(a^{1/2} - 7)$

12. $(a^{1/2} - 6)(a^{1/2} - 2)$

13. $(4y^{1/3} - 3)(5y^{1/3} + 2)$

14. $(5y^{1/3} - 2)(4y^{1/3} + 3)$

15. $(5x^{2/3} + 3y^{1/2})(2x^{2/3} + 3y^{1/2})$

16. $(4x^{2/3} - 2y^{1/2})(5x^{2/3} - 3y^{1/2})$

17. $(t^{1/2} + 5)^2$

18. $(t^{1/2} - 3)^2$

19. $(x^{3/2} + 4)^2$

20. $(x^{3/2} - 6)^2$

21. $(a^{1/2} - b^{1/2})^2$

22. $(a^{1/2} + b^{1/2})^2$

23. $(2x^{1/2} - 3y^{1/2})^2$

24. $(5x^{1/2} + 4y^{1/2})^2$

25. $(a^{1/2} - 3^{1/2})(a^{1/2} + 3^{1/2})$

26. $(a^{1/2} - 5^{1/2})(a^{1/2} + 5^{1/2})$

27. $(x^{3/2} + y^{3/2})(x^{3/2} - y^{3/2})$

28. $(x^{5/2} + y^{5/2})(x^{5/2} - y^{5/2})$

29. $(t^{1/2} - 2^{3/2})(t^{1/2} + 2^{3/2})$

30. $(t^{1/2} - 5^{3/2})(t^{1/2} + 5^{3/2})$

31. $(2x^{3/2} + 3^{1/2})(2x^{3/2} - 3^{1/2})$

32. $(3x^{1/2} + 2^{3/2})(3x^{1/2} - 2^{3/2})$

33. $(x^{1/3} + y^{1/3})(x^{2/3} - x^{1/3}y^{1/3} + y^{2/3})$

34. $(x^{1/3} - y^{1/3})(x^{2/3} + x^{1/3}y^{1/3} + y^{2/3})$

35. $(a^{1/3} - 2)(a^{2/3} + 2a^{1/3} + 4)$

36. $(a^{1/3} + 3)(a^{2/3} - 3a^{1/3} + 9)$

37. $(2x^{1/3} + 1)(4x^{2/3} - 2x^{1/3} + 1)$

38. $(3x^{1/3} - 1)(9x^{2/3} + 3x^{1/3} + 1)$

39. $(t^{1/4} - 1)(t^{1/4} + 1)(t^{1/2} + 1)$

40. $(t^{1/4} - 2)(t^{1/4} + 2)(t^{1/2} + 4)$

B Divide. (Assume all variables represent positive numbers.) [Example 7]

41. $\dfrac{18x^{3/4} + 27x^{1/4}}{9x^{1/4}}$

42. $\dfrac{25x^{1/4} + 30x^{3/4}}{5x^{1/4}}$

43. $\dfrac{12x^{2/3}y^{1/3} - 16x^{1/3}y^{2/3}}{4x^{1/3}y^{1/3}}$

44. $\dfrac{12x^{4/3}y^{1/3} - 18x^{1/3}y^{4/3}}{6x^{1/3}y^{1/3}}$

45. $\dfrac{21a^{7/5}b^{3/5} - 14a^{2/5}b^{8/5}}{7a^{2/5}b^{3/5}}$

46. $\dfrac{24a^{9/5}b^{3/5} - 16a^{4/5}b^{8/5}}{8a^{4/5}b^{3/5}}$

C [Example 8]

47. Factor $3(x - 2)^{1/2}$ from $12(x - 2)^{3/2} - 9(x - 2)^{1/2}$.

48. Factor $4(x + 1)^{1/3}$ from $4(x + 1)^{4/3} + 8(x + 1)^{1/3}$.

49. Factor $5(x - 3)^{7/5}$ from $5(x - 3)^{12/5} - 15(x - 3)^{7/5}$.

50. Factor $6(x + 3)^{8/7}$ from $6(x + 3)^{15/7} - 12(x + 3)^{8/7}$.

51. Factor $3(x + 1)^{1/2}$ from $9x(x + 1)^{3/2} + 6(x + 1)^{1/2}$.

52. Factor $4x(x + 1)^{1/2}$ from $4x^2(x + 1)^{1/2} + 8x(x + 1)^{3/2}$.

C Factor each of the following as if it were a trinomial. [Examples 9–10]

53. $x^{2/3} - 5x^{1/3} + 6$

54. $x^{2/3} - x^{1/3} - 6$

55. $a^{2/5} - 2a^{1/5} - 8$

56. $a^{2/5} + 2a^{1/5} - 8$

57. $2y^{2/3} - 5y^{1/3} - 3$

58. $3y^{2/3} + 5y^{1/3} - 2$

59. $9t^{2/5} - 25$

60. $16t^{2/5} - 49$

61. $4x^{2/7} + 20x^{1/7} + 25$

62. $25x^{2/7} - 20x^{1/7} + 4$

Evaluate the following functions for the given value.

63. If $f(x) = x - 2\sqrt{x} - 8$ find $f(4)$.

64. If $g(x) = x - 2\sqrt{x} - 3$ find $g(9)$.

65. If $f(x) = 2x + 9\sqrt{x} - 5$ find $f(25)$.

66. If $f(t) = t - 2\sqrt{t} - 15$ find $f(9)$.

67. If $g(x) = 2x - \sqrt{x} - 6$ find $g\left(\dfrac{9}{4}\right)$.

68. If $f(x) = 2x + \sqrt{x} - 15$ find $f(9)$.

69. If $f(x) = x^{2/3} - 2x^{1/3} - 8$ find $f(-8)$.

70. If $g(x) = x^{2/3} + 4x^{1/3} - 12$ find $g(8)$.

D Simplify each of the following to a single fraction. (Assume all variables represent positive numbers.) [Example 11]

71. $\dfrac{3}{x^{1/2}} + x^{1/2}$

72. $\dfrac{2}{x^{1/2}} - x^{1/2}$

73. $x^{2/3} + \dfrac{5}{x^{1/3}}$

74. $x^{3/4} - \dfrac{7}{x^{1/4}}$

75. $\dfrac{3x^2}{(x^3 + 1)^{1/2}} + (x^3 + 1)^{1/2}$

76. $\dfrac{x^3}{(x^2 - 1)^{1/2}} + 2x(x^2 - 1)^{1/2}$

77. $\dfrac{x^2}{(x^2 + 4)^{1/2}} - (x^2 + 4)^{1/2}$

78. $\dfrac{x^5}{(x^2 - 2)^{1/2}} + 4x^3(x^2 - 2)^{1/2}$

D **Applying the Concepts** [Example 12]

Round answers to the nearest tenth of a percent.

79. Investing A coin collection is purchased as an investment for $500 and sold 4 years later for $900. Find the annual rate of return on the investment.

80. Investing An investor buys stock in a company for $800. Five years later, the same stock is worth $1,600. Find the annual rate of return on the stocks.

Maintaining Your Skills

Reduce to lowest terms.

81. $\dfrac{x^2 - 9}{x^4 - 81}$

82. $\dfrac{6 - a - a^2}{3 - 2a - a^2}$

Divide.

83. $\dfrac{15x^2y - 20x^4y^2}{5xy}$

84. $\dfrac{12x^3y^2 - 24x^2y^3}{6xy}$

Divide using long division.

85. $\dfrac{10x^2 + 7x - 12}{2x + 3}$

86. $\dfrac{6x^2 - x - 35}{2x - 5}$

87. $\dfrac{x^3 - 125}{x - 5}$

88. $\dfrac{x^3 + 64}{x + 4}$

Getting Ready for the Next Section

89. $\sqrt{25}$

90. $\sqrt{4}$

91. $\sqrt{6^2}$

92. $\sqrt{3^2}$

93. $\sqrt{16x^4y^2}$

94. $\sqrt{4x^6y^8}$

95. $\sqrt{(5y)^2}$

96. $\sqrt{(8x^3)^2}$

97. $\sqrt[3]{27}$

98. $\sqrt[3]{-8}$

99. $\sqrt[3]{2^3}$

100. $\sqrt[3]{(-5)^3}$

101. $\sqrt[3]{8a^3b^3}$

102. $\sqrt[3]{64a^6b^3}$

Fill in the blank.

103. $50 = $ _____ $\cdot 2$

104. $12 = $ _____ $\cdot 3$

105. $48x^4y^3 = $ _____ $\cdot y$

106. $40a^5b^4 = $ _____ $\cdot 5a^2b$

107. $12x^7y^6 = $ _____ $\cdot 3x$

108. $54a^6b^2c^4 = $ _____ $\cdot 2b^2c$

9.3 Simplified Form for Radicals

OBJECTIVES

A Write radical expressions in simplified form.

B Rationalize a denominator that contains only one term.

TICKET TO SUCCESS

Keep these questions in mind as you read through the section. Then respond in your own words and in complete sentences.

1. Explain why property 1 for radicals cannot be applied to the square root of a sum.

2. What is simplified form for an expression that contains a square root?

3. Why is it not necessarily true that $\sqrt{a^2} = a$?

4. What does it mean to rationalize the denominator in an expression?

Earlier in this chapter, we showed how the Pythagorean theorem can be used to construct a golden rectangle. In a similar manner, the Pythagorean theorem can be used to construct the attractive spiral shown here.

The Spiral of Roots

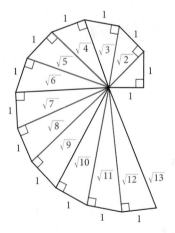

This spiral is called the Spiral of Roots because each of the diagonals is the positive square root of one of the positive integers. In the problem set for this section, we will use the Pythagorean theorem and some of the material in this section to construct this spiral.

In this section, we will use radical notation instead of rational exponents. We will begin by stating the first two properties of radicals. Following this, we will give a definition for simplified form for radical expressions. The examples in this section show how we use the properties of radicals to write radical expressions in simplified form.

A Simplified Form

We will now look at the first two properties of radicals. For these two properties, we will assume a and b are nonnegative real numbers whenever n is an even number.

Property 1 for Radicals

$$\sqrt[n]{ab} = \sqrt[n]{a}\ \sqrt[n]{b}$$

In words: The nth root of a product is the product of the nth roots.

PROOF OF PROPERTY 1

$$\sqrt[n]{ab} = (ab)^{1/n} \qquad \text{Definition of fractional exponents}$$
$$= a^{1/n}b^{1/n} \qquad \text{Exponents distribute over products}$$
$$= \sqrt[n]{a}\ \sqrt[n]{b} \qquad \text{Definition of fractional exponents}$$

Property 2 for Radicals

$$\sqrt[n]{\frac{a}{b}} = \frac{\sqrt[n]{a}}{\sqrt[n]{b}} \qquad (b \neq 0)$$

In words: The nth root of a quotient is the quotient of the nth roots.

The proof of property 2 is similar to the proof of property 1.

The two properties of radicals allow us to change the form of and simplify radical expressions without changing their value.

Simplified Form for Radical Expressions

A radical expression is in *simplified form* if

1. None of the factors of the radicand (the quantity under the radical sign) can be written as powers greater than or equal to the index—that is, no perfect squares can be factors of the quantity under a square root sign, no perfect cubes can be factors of what is under a cube root sign, and so forth;

2. There are no fractions under the radical sign; and

3. There are no radicals in the denominator.

Satisfying the first condition for simplified form actually amounts to taking as much out from under the radical sign as possible. The following examples illustrate the first condition for simplified form.

EXAMPLE 1 Write $\sqrt{50}$ in simplified form.

SOLUTION The largest perfect square that divides 50 is 25. We write 50 as $25 \cdot 2$ and apply property 1 for radicals.

$$\sqrt{50} = \sqrt{25 \cdot 2} \qquad 50 = 25 \cdot 2$$
$$= \sqrt{25}\,\sqrt{2} \qquad \text{Property 1}$$
$$= 5\sqrt{2} \qquad \sqrt{25} = 5$$

We have taken as much as possible out from under the radical sign—in this case, factoring 25 from 50 and then writing $\sqrt{25}$ as 5.

NOTE
Unless we state otherwise, assume all variables throughout this section represent nonnegative numbers. Further, if a variable appears in a denominator, assume the variable cannot be 0.

EXAMPLE 2 Write $\sqrt{48x^4y^3}$, where $x, y \geq 0$ in simplified form.

SOLUTION The largest perfect square that is a factor of the radicand is $16x^4y^2$. Applying property 1 again, we have

$$\sqrt{48x^4y^3} = \sqrt{16x^4y^2 \cdot 3y}$$
$$= \sqrt{16x^4y^2}\,\sqrt{3y}$$
$$= 4x^2y\sqrt{3y}$$

EXAMPLE 3 Write $\sqrt[3]{40a^5b^4}$ in simplified form.

SOLUTION We now want to factor the largest perfect cube from the radicand. We write $40a^5b^4$ as $8a^3b^3 \cdot 5a^2b$ and proceed as we did in Examples 1 and 2.

$$\sqrt[3]{40a^5b^4} = \sqrt[3]{8a^3b^3 \cdot 5a^2b}$$
$$= \sqrt[3]{8a^3b^3}\,\sqrt[3]{5a^2b}$$
$$= 2ab\sqrt[3]{5a^2b}$$

Here are some further examples concerning the first condition for simplified form.

EXAMPLE 4 Write the expression in simplified form.

SOLUTION
$$\sqrt{12x^7y^6} = \sqrt{4x^6y^6 \cdot 3x}$$
$$= \sqrt{4x^6y^6}\,\sqrt{3x}$$
$$= 2x^3y^3\sqrt{3x}$$

EXAMPLE 5 Write the expression in simplified form.

SOLUTION
$$\sqrt[3]{54a^6b^2c^4} = \sqrt[3]{27a^6c^3 \cdot 2b^2c}$$
$$= \sqrt[3]{27a^6c^3}\,\sqrt[3]{2b^2c}$$
$$= 3a^2c\sqrt[3]{2b^2c}$$

B Rationalizing the Denominator

The second property of radicals is used to simplify a radical that contains a fraction.

EXAMPLE 6 Simplify $\sqrt{\dfrac{3}{4}}$.

SOLUTION Applying property 2 for radicals, we have

$$\sqrt{\frac{3}{4}} = \frac{\sqrt{3}}{\sqrt{4}} \qquad \text{Property 2}$$

$$= \frac{\sqrt{3}}{2} \qquad \sqrt{4} = 2$$

The last expression is in simplified form because it satisfies all three conditions for simplified form.

EXAMPLE 7 Write $\sqrt{\dfrac{5}{6}}$ in simplified form.

SOLUTION Proceeding as in Example 6, we have

$$\sqrt{\frac{5}{6}} = \frac{\sqrt{5}}{\sqrt{6}}$$

> **NOTE**
> The idea behind rationalizing the denominator is to produce a perfect square under the square root sign in the denominator. This is accomplished by multiplying both the numerator and denominator by the appropriate radical.

The resulting expression satisfies the second condition for simplified form since neither radical contains a fraction. It does, however, violate condition 3 since it has a radical in the denominator. Getting rid of the radical in the denominator is called *rationalizing the denominator* and is accomplished, in this case, by multiplying the numerator and denominator by $\sqrt{6}$.

$$\frac{\sqrt{5}}{\sqrt{6}} = \frac{\sqrt{5}}{\sqrt{6}} \cdot \frac{\sqrt{6}}{\sqrt{6}}$$
$$= \frac{\sqrt{30}}{\sqrt{6^2}}$$
$$= \frac{\sqrt{30}}{6}$$

EXAMPLE 8 Rationalize the denominator.

SOLUTION
$$\frac{4}{\sqrt{3}} = \frac{4}{\sqrt{3}} \cdot \frac{\sqrt{3}}{\sqrt{3}}$$
$$= \frac{4\sqrt{3}}{\sqrt{3^2}}$$
$$= \frac{4\sqrt{3}}{3}$$

EXAMPLE 9 Rationalize the denominator.

SOLUTION
$$\frac{2\sqrt{3x}}{\sqrt{5y}} = \frac{2\sqrt{3x}}{\sqrt{5y}} \cdot \frac{\sqrt{5y}}{\sqrt{5y}}$$
$$= \frac{2\sqrt{15xy}}{\sqrt{(5y)^2}}$$
$$= \frac{2\sqrt{15xy}}{5y}$$

When the denominator involves a cube root, we must multiply by a radical that will produce a perfect cube under the cube root sign in the denominator, as our next example illustrates.

EXAMPLE 10 Rationalize the denominator in $\dfrac{7}{\sqrt[3]{4}}$.

SOLUTION Since $4 = 2^2$, we can multiply both numerator and denominator by $\sqrt[3]{2}$ and obtain $\sqrt[3]{2^3}$ in the denominator.

$$\frac{7}{\sqrt[3]{4}} = \frac{7}{\sqrt[3]{2^2}}$$

$$= \frac{7}{\sqrt[3]{2^2}} \cdot \frac{\sqrt[3]{2}}{\sqrt[3]{2}}$$

$$= \frac{7\sqrt[3]{2}}{\sqrt[3]{2^3}}$$

$$= \frac{7\sqrt[3]{2}}{2}$$

EXAMPLE 11 Simplify $\sqrt{\dfrac{12x^5y^3}{5z}}$.

SOLUTION We use property 2 to write the numerator and denominator as two separate radicals.

$$\sqrt{\frac{12x^5y^3}{5z}} = \frac{\sqrt{12x^5y^3}}{\sqrt{5z}}$$

Simplifying the numerator, we have

$$\frac{\sqrt{12x^5y^3}}{\sqrt{5z}} = \frac{\sqrt{4x^4y^2}\,\sqrt{3xy}}{\sqrt{5z}}$$

$$= \frac{2x^2y\,\sqrt{3xy}}{\sqrt{5z}}$$

To rationalize the denominator, we multiply the numerator and denominator by $\sqrt{5z}$.

$$\frac{2x^2y\,\sqrt{3xy}}{\sqrt{5z}} \cdot \frac{\sqrt{5z}}{\sqrt{5z}} = \frac{2x^2y\,\sqrt{15xyz}}{\sqrt{(5z)^2}}$$

$$= \frac{2x^2y\,\sqrt{15xyz}}{5z}$$

The Square Root of a Perfect Square

So far in this chapter we have assumed that all our variables are nonnegative when they appear under a square root symbol. There are times, however, when this is not the case.

Consider the following two statements:

$$\sqrt{3^2} = \sqrt{9} = 3 \quad \text{and} \quad \sqrt{(-3)^2} = \sqrt{9} = 3$$

Whether we operate on 3 or −3, the result is the same: both expressions simplify to 3. The other operation we have worked with in the past that produces the same result is absolute value; that is,

$$|3| = 3 \quad \text{and} \quad |-3| = 3$$

This leads us to the next property of radicals.

Property 3 for Radicals
If a is a real number, then $\sqrt{a^2} = |a|$

The result of this discussion and property 3 is simply this:

If we know a is positive, then $\sqrt{a^2} = a$.

If we know a is negative, then $\sqrt{a^2} = |a|$.

If we don't know if a is positive or negative, then $\sqrt{a^2} = |a|$.

EXAMPLES   Simplify each expression. Do *not* assume the variables represent positive numbers.

12. $\sqrt{9x^2} = 3|x|$

13. $\sqrt{x^3} = |x|\sqrt{x}$

14. $\sqrt{x^2 - 6x + 9} = \sqrt{(x-3)^2} = |x-3|$

15. $\sqrt{x^3 - 5x^2} = \sqrt{x^2(x-5)} = |x|\sqrt{x-5}$

As you can see, we must use absolute value symbols when we take a square root of a perfect square, unless we know the base of the perfect square is a positive number. The same idea holds for higher even roots, but not for odd roots. With odd roots, no absolute value symbols are necessary.

EXAMPLES   Simplify each expression.

16. $\sqrt[3]{(-2)^3} = \sqrt[3]{-8} = -2$

17. $\sqrt[3]{(-5)^3} = \sqrt[3]{-125} = -5$

We can extend this discussion to all roots as follows:

Extending Property 3 for Radicals

If a is a real number, then

$$\sqrt[n]{a^n} = |a| \quad \text{if} \quad n \text{ is even}$$

$$\sqrt[n]{a^n} = a \quad \text{if} \quad n \text{ is odd}$$

Problem Set 9.3

Moving Toward Success

"Judge a man by his questions rather than by his answers."

—Voltaire, 1694–1778, French writer and philosopher

1. If your instructor writes something on the board in class, should you write it in your notes? Why or why not?

2. Should you ask questions in class? Why or why not?

Use property 1 for radicals to write each of the following expressions in simplified form. (Assume all variables are nonnegative through Problem 84.) [Examples 1–5]

1. $\sqrt{8}$

2. $\sqrt{32}$

3. $\sqrt{98}$

4. $\sqrt{75}$

5. $\sqrt{288}$

6. $\sqrt{128}$

7. $\sqrt{80}$

8. $\sqrt{200}$

9. $\sqrt{48}$

10. $\sqrt{27}$

11. $\sqrt{675}$

12. $\sqrt{972}$

13. $\sqrt[3]{54}$

14. $\sqrt[3]{24}$

15. $\sqrt[3]{128}$

16. $\sqrt[3]{162}$

17. $\sqrt[3]{432}$

18. $\sqrt[3]{1{,}536}$

19. $\sqrt[5]{64}$

20. $\sqrt[4]{48}$

21. $\sqrt{18x^3}$

22. $\sqrt{27x^5}$

23. $\sqrt[4]{32y^7}$

24. $\sqrt[5]{32y^7}$

25. $\sqrt[3]{40x^4y^7}$

26. $\sqrt[3]{128x^6y^2}$

27. $\sqrt{48a^2b^3c^4}$

28. $\sqrt{72a^4b^3c^2}$

29. $\sqrt[3]{48a^2b^3c^4}$

30. $\sqrt[3]{72a^4b^3c^2}$

31. $\sqrt[5]{64x^8y^{12}}$

32. $\sqrt[4]{32x^9y^{10}}$

33. $\sqrt[5]{243x^7y^{10}z^5}$

34. $\sqrt[5]{64x^8y^4z^{11}}$

Substitute the given numbers into the expression $\sqrt{b^2 - 4ac}$, and then simplify.

35. $a = 2, b = -6, c = 3$

36. $a = 6, b = 7, c = -5$

37. $a = 1, b = 2, c = 6$

38. $a = 2, b = 5, c = 3$

39. $a = \dfrac{1}{2}, b = -\dfrac{1}{2}, c = -\dfrac{5}{4}$

40. $a = \dfrac{7}{4}, b = -\dfrac{3}{4}, c = -2$

B Rationalize the denominator in each of the following expressions. [Examples 6–11]

41. $\dfrac{2}{\sqrt{3}}$

42. $\dfrac{3}{\sqrt{2}}$

43. $\dfrac{5}{\sqrt{6}}$

44. $\dfrac{7}{\sqrt{5}}$

45. $\sqrt{\dfrac{1}{2}}$

46. $\sqrt{\dfrac{1}{3}}$

47. $\sqrt{\dfrac{1}{5}}$

48. $\sqrt{\dfrac{1}{6}}$

49. $\dfrac{4}{\sqrt[3]{2}}$

50. $\dfrac{5}{\sqrt[3]{3}}$

51. $\dfrac{2}{\sqrt[3]{9}}$

52. $\dfrac{3}{\sqrt[3]{4}}$

53. $\sqrt[4]{\dfrac{3}{2x^2}}$

54. $\sqrt[4]{\dfrac{5}{3x^2}}$

55. $\sqrt[4]{\dfrac{8}{y}}$

56. $\sqrt[4]{\dfrac{27}{y}}$

57. $\sqrt[3]{\dfrac{4x}{3y}}$

58. $\sqrt[3]{\dfrac{7x}{6y}}$

59. $\sqrt[3]{\dfrac{2x}{9y}}$

60. $\sqrt[3]{\dfrac{5x}{4y}}$

61. $\sqrt[4]{\dfrac{1}{8x^3}}$

62. $\sqrt[4]{\dfrac{8}{9x^3}}$

Write each of the following in simplified form.

63. $\sqrt{\dfrac{27x^3}{5y}}$

64. $\sqrt{\dfrac{12x^5}{7y}}$

65. $\sqrt{\dfrac{75x^3y^2}{2z}}$

66. $\sqrt{\dfrac{50x^2y^3}{3z}}$

67. $\sqrt[3]{\dfrac{16a^4b^3}{9c}}$

68. $\sqrt[3]{\dfrac{54a^5b^4}{25c^2}}$

69. $\sqrt[3]{\dfrac{8x^3y^6}{9z}}$

70. $\sqrt[3]{\dfrac{27x^6y^3}{2z^2}}$

71. $\sqrt{\sqrt{x^2}}$

72. $\sqrt{\sqrt{2x^3}}$

73. $\sqrt[3]{\sqrt{xy}}$

74. $\sqrt{\sqrt{4x}}$

75. $\sqrt[3]{\sqrt[4]{a}}$

76. $\sqrt[6]{\sqrt[4]{x}}$

77. $\sqrt[3]{\sqrt[3]{6x^{10}}}$

78. $\sqrt[5]{\sqrt{x^{14}y^{11}z}}$

79. $\sqrt[4]{\sqrt[3]{a^{12}b^{24}c^{14}}}$

80. $\sqrt{\sqrt{4a^{17}}}$

81. $\sqrt[3]{\sqrt[5]{3a^{17}b^{16}c^{30}}}$

82. $\left(\sqrt{\sqrt[4]{x^4y^8z^9}}\right)^2$

83. $\left(\sqrt{\sqrt[3]{8ab^6}}\right)^2$

84. $\left(\sqrt[4]{\sqrt[3]{16x^8y^{12}z^3}}\right)^3$

Simplify each expression. Do *not* assume the variables represent positive numbers. [Examples 12–15]

85. $\sqrt{25x^2}$

86. $\sqrt{49x^2}$

87. $\sqrt{27x^3y^2}$

88. $\sqrt{40x^3y^2}$

89. $\sqrt{x^2 - 10x + 25}$

90. $\sqrt{x^2 - 16x + 64}$

91. $\sqrt{4x^2 + 12x + 9}$

92. $\sqrt{16x^2 + 40x + 25}$

93. $\sqrt{4a^4 + 16a^3 + 16a^2}$

94. $\sqrt{9a^4 + 18a^3 + 9a^2}$

95. $\sqrt{4x^3 - 8x^2}$

96. $\sqrt{18x^3 - 9x^2}$

97. Show that the statement $\sqrt{a + b} = \sqrt{a} + \sqrt{b}$ is not true by replacing a with 9 and b with 16 and simplifying both sides.

98. Find a pair of values for a and b that will make the statement $\sqrt{a + b} = \sqrt{a} + \sqrt{b}$ true.

Applying the Concepts

99. Diagonal Distance The distance d between opposite corners of a rectangular room with length l and width w is given by

$$d = \sqrt{l^2 + w^2}$$

How far is it between opposite corners of a living room that measures 10 by 15 feet?

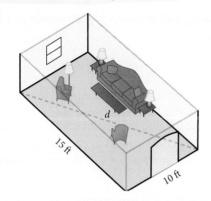

100. Radius of a Sphere The radius r of a sphere with volume V can be found by using the formula

$$r = \sqrt[3]{\frac{3V}{4\pi}}$$

Find the radius of a sphere with volume 9 cubic feet. Write your answer in simplified form. (Use $\frac{22}{7}$ for π.)

Volume

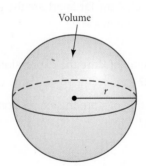

101. Radius of a Cylinder The radius r of a cylinder with volume V and height h can be found by using the formula

$$r = \sqrt{\frac{V}{h\pi}}$$

Find the radius of the cylinder below. Write your answer in simplified form. (Use $\frac{22}{7}$ for π.)

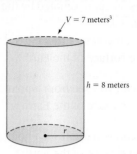

$V = 7$ meters³

$h = 8$ meters

r

102. Radius of a Cone The radius r of a cone with a volume V and height h can be found by using the formula

$$r = \sqrt{\frac{3V}{h\pi}}$$

Find the radius of a cone with a volume of 6 cubic inches and a height of 7 inches. Write your answer in simplified form. (Use $\frac{22}{7}$ for π.)

Volume

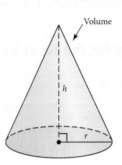

h

r

103. Diagonal of a Box The length of the diagonal of a rectangular box with length l, width w, and height h is given by $d = \sqrt{l^2 + w^2 + h^2}$.
a. Find the length of the diagonal of a rectangular box that is 3 feet wide, 4 feet long, and 12 feet high.
b. Find the length of the diagonal of a rectangular box that is 2 feet wide, 4 feet high, and 6 feet long.

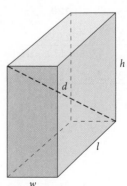

h

d

l

w

104. Distance to the Horizon If you are at a point k miles above the surface of the Earth, the distance you can see, in miles, is approximated by the equation $d = \sqrt{8000k + k^2}$.

   a. How far can you see from a point that is 1 mile above the surface of the Earth?

   b. How far can you see from a point that is 2 miles above the surface of the Earth?

   c. How far can you see from a point that is 3 miles above the surface of the Earth?

105. Spiral of Roots Construct your own spiral of roots by using a ruler. Draw the first triangle by using two 1-inch lines. The first diagonal will have a length of $\sqrt{2}$ inches. Each new triangle will be formed by drawing a 1-inch line segment at the end of the previous diagonal so that the angle formed is 90°.

106. Spiral of Roots Construct a spiral of roots by using line segments of length 2 inches. The length of the first diagonal will be $2\sqrt{2}$ inches. The length of the second diagonal will be $2\sqrt{3}$ inches.

Getting Ready for the Next Section

Simplify the following.

107. $5x - 4x + 6x$ **108.** $12x + 8x - 7x$

109. $35xy^2 - 8xy^2$ **110.** $20a^2b + 33a^2b$

111. $\dfrac{1}{2}x + \dfrac{1}{3}x$ **112.** $\dfrac{2}{3}x + \dfrac{5}{8}x$

Write in simplified form for radicals.

113. $\sqrt{18}$ **114.** $\sqrt{8}$

115. $\sqrt{75xy^3}$ **116.** $\sqrt{12xy}$

117. $\sqrt[3]{8a^4b^2}$ **118.** $\sqrt[3]{27ab^2}$

Maintaining Your Skills

Perform the indicated operations.

119. $\dfrac{8xy^3}{9x^2y} \div \dfrac{16x^2y^2}{18xy^3}$ **120.** $\dfrac{25x^2}{5y^4} \cdot \dfrac{30y^3}{2x^5}$

121. $\dfrac{12a^2 - 4a - 5}{2a + 1} \cdot \dfrac{7a + 3}{42a^2 - 17a - 15}$

122. $\dfrac{20a^2 - 7a - 3}{4a + 1} \cdot \dfrac{25a^2 - 5a - 6}{5a + 2}$

123. $\dfrac{8x^3 + 27}{27x^3 + 1} \div \dfrac{6x^2 + 7x - 3}{9x^2 - 1}$

124. $\dfrac{27x^3 + 8}{8x^3 + 1} \div \dfrac{6x^2 + x - 2}{4x^2 - 1}$

Extending the Concepts

Factor each radicand into the product of prime factors. Then simplify each radical.

125. $\sqrt[3]{8,640}$ **126.** $\sqrt{8,640}$

127. $\sqrt[3]{10,584}$ **128.** $\sqrt{10,584}$

Assume a is a positive number, and rationalize each denominator.

129. $\dfrac{1}{\sqrt[10]{a^3}}$ **130.** $\dfrac{1}{\sqrt[12]{a^7}}$

131. $\dfrac{1}{\sqrt[20]{a^{11}}}$ **132.** $\dfrac{1}{\sqrt[15]{a^{13}}}$

133. Show that the two expressions $\sqrt{x^2 + 1}$ and $x + 1$ are not, in general, equal to each other by graphing $y = \sqrt{x^2 + 1}$ and $y = x + 1$ in the same viewing window.

134. Show that the two expressions $\sqrt{x^2 + 9}$ and $x + 3$ are not, in general, equal to each other by graphing $y = \sqrt{x^2 + 9}$ and $y = x + 3$ in the same viewing window.

135. Approximately how far apart are the graphs in Problem 131 when $x = 2$?

136. Approximately how far apart are the graphs in Problem 132 when $x = 2$?

137. For what value of x are the expressions $\sqrt{x^2 + 1}$ and $x + 1$ equal?

138. For what value of x are the expressions $\sqrt{x^2 + 9}$ and $x + 3$ equal?

OBJECTIVES

A Add and subtract radicals.

B Construct golden rectangles from squares.

TICKET TO SUCCESS

Keep these questions in mind as you read through the section. Then respond in your own words and in complete sentences.

1. What are similar radicals?
2. When can we add two radical expressions?
3. What is the first step when adding or subtracting expressions containing radicals?
4. How would you construct a golden rectangle from a square of side 6?

The yield sign often seen on roadways is an example of an equilateral triangle—a triangle with 3 sides of equal length. The ratio of the height of the sign to one of its sides can be found using addition and subtraction of radical expressions, which is the focus of this section.

We have been able to add and subtract polynomials by combining similar terms. The same idea applies to addition and subtraction of radical expressions.

A Combining Similar Terms

> **Definition**
>
> Two radicals are said to be **similar radicals** if they have the same index and the same radicand.

The expressions $5\sqrt[3]{7}$ and $-8\sqrt[3]{7}$ are similar since the index is 3 in both cases and the radicands are 7. The expressions $3\sqrt[4]{5}$ and $7\sqrt[3]{5}$ are not similar since they have different indices, and the expressions $2\sqrt[5]{8}$ and $3\sqrt[5]{9}$ are not similar because the radicands are not the same.

EXAMPLE 1   Combine $5\sqrt{3} - 4\sqrt{3} + 6\sqrt{3}$.

SOLUTION   All three radicals are similar. We apply the distributive property to get

$$5\sqrt{3} - 4\sqrt{3} + 6\sqrt{3} = (5 - 4 + 6)\sqrt{3}$$
$$= 7\sqrt{3}$$

EXAMPLE 2 Combine $3\sqrt{8} + 5\sqrt{18}$.

SOLUTION The two radicals do not seem to be similar. We must write each in simplified form before applying the distributive property.

$$
\begin{aligned}
3\sqrt{8} + 5\sqrt{18} &= 3\sqrt{4 \cdot 2} + 5\sqrt{9 \cdot 2} \\
&= 3\sqrt{4}\,\sqrt{2} + 5\sqrt{9}\,\sqrt{2} \\
&= 3 \cdot 2\sqrt{2} + 5 \cdot 3\sqrt{2} \\
&= 6\sqrt{2} + 15\sqrt{2} \\
&= (6 + 15)\sqrt{2} \\
&= 21\sqrt{2}
\end{aligned}
$$

The result of Example 2 can be generalized to the following rule for sums and differences of radical expressions.

> **Rule**
> To add or subtract radical expressions, put each in simplified form and apply the distributive property if possible. We can add only similar radicals. We must write each expression in simplified form for radicals before we can tell if the radicals are similar.

EXAMPLE 3 Combine $7\sqrt{75xy^3} - 4y\sqrt{12xy}$, where $x, y \geq 0$.

SOLUTION We write each expression in simplified form and combine similar radicals.

$$
\begin{aligned}
7\sqrt{75xy^3} - 4y\sqrt{12xy} &= 7\sqrt{25y^2}\,\sqrt{3xy} - 4y\sqrt{4}\,\sqrt{3xy} \\
&= 35y\sqrt{3xy} - 8y\sqrt{3xy} \\
&= (35y - 8y)\sqrt{3xy} \\
&= 27y\sqrt{3xy}
\end{aligned}
$$

EXAMPLE 4 Combine $10\sqrt[3]{8a^4b^2} + 11a\sqrt[3]{27ab^2}$.

SOLUTION Writing each radical in simplified form and combining similar terms, we have

$$
\begin{aligned}
10\sqrt[3]{8a^4b^2} + 11a\sqrt[3]{27ab^2} &= 10\sqrt[3]{8a^3}\,\sqrt[3]{ab^2} + 11a\sqrt[3]{27}\,\sqrt[3]{ab^2} \\
&= 20a\sqrt[3]{ab^2} + 33a\sqrt[3]{ab^2} \\
&= 53a\sqrt[3]{ab^2}
\end{aligned}
$$

EXAMPLE 5 Combine $\dfrac{\sqrt{3}}{2} + \dfrac{1}{\sqrt{3}}$.

SOLUTION We begin by writing the second term in simplified form.

$$
\begin{aligned}
\frac{\sqrt{3}}{2} + \frac{1}{\sqrt{3}} &= \frac{\sqrt{3}}{2} + \frac{1}{\sqrt{3}} \cdot \frac{\boldsymbol{\sqrt{3}}}{\boldsymbol{\sqrt{3}}} \\
&= \frac{\sqrt{3}}{2} + \frac{\sqrt{3}}{3} \\
&= \frac{1}{2}\sqrt{3} + \frac{1}{3}\sqrt{3}
\end{aligned}
$$

$$= \left(\frac{1}{2} + \frac{1}{3}\right) \sqrt{3}$$

$$= \frac{5}{6} \sqrt{3} = \frac{5\sqrt{3}}{6}$$

B The Golden Rectangle

Recall our discussion of the golden rectangle from the first section of this chapter. Now that we have practiced adding and subtracting radical expressions, let's put our new skills to use to find a golden rectangle of a different size.

EXAMPLE 6 Construct a golden rectangle from a square of side 4. Then show that the ratio of the length to the width is the golden ratio $\frac{1 + \sqrt{5}}{2}$.

SOLUTION Figure 1 shows the golden rectangle constructed from a square of side 4. The length of the diagonal OB is found from the Pythagorean theorem.

$$OB = \sqrt{2^2 + 4^2} = \sqrt{4 + 16} = \sqrt{20} = 2\sqrt{5}$$

The ratio of the length to the width for the rectangle is the golden ratio.

$$\text{Golden ratio} = \frac{CE}{EF} = \frac{2 + 2\sqrt{5}}{4} = \frac{2(1 + \sqrt{5})}{2 \cdot 2} = \frac{1 + \sqrt{5}}{2}$$

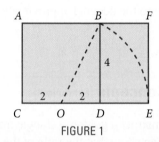

FIGURE 1

Problem Set 9.4

Moving Toward Success

"We make a living by what we get, but we make a life by what we give."

—Winston Churchill, 1874–1965, English politician and Nobel Prize winner

1. Should you read the notes in the side columns of the section pages? Why or why not?

2. How does coming to class prepared help to reduce stress?

A Combine the following expressions. (Assume any variables under an even root are nonnegative.)
[Examples 1–5]

1. $3\sqrt{5} + 4\sqrt{5}$

2. $6\sqrt{3} - 5\sqrt{3}$

3. $3x\sqrt{7} - 4x\sqrt{7}$

4. $6y\sqrt{a} + 7y\sqrt{a}$

5. $5\sqrt[3]{10} - 4\sqrt[3]{10}$

6. $6\sqrt[4]{2} + 9\sqrt[4]{2}$

7. $8\sqrt[5]{6} - 2\sqrt[5]{6} + 3\sqrt[5]{6}$

8. $7\sqrt[6]{7} - \sqrt[6]{7} + 4\sqrt[6]{7}$

9. $3x\sqrt{2} - 4x\sqrt{2} + x\sqrt{2}$

10. $5x\sqrt{6} - 3x\sqrt{6} - 2x\sqrt{6}$

11. $\sqrt{20} - \sqrt{80} + \sqrt{45}$

12. $\sqrt{8} - \sqrt{32} - \sqrt{18}$

13. $4\sqrt{8} - 2\sqrt{50} - 5\sqrt{72}$

14. $\sqrt{48} - 3\sqrt{27} + 2\sqrt{75}$

15. $5x\sqrt{8} + 3\sqrt{32x^2} - 5\sqrt{50x^2}$

16. $2\sqrt{50x^2} - 8x\sqrt{18} - 3\sqrt{72x^2}$

17. $5\sqrt[3]{16} - 4\sqrt[3]{54}$

18. $\sqrt[3]{81} + 3\sqrt[3]{24}$

19. $\sqrt[3]{x^4y^2} + 7x\sqrt[3]{xy^2}$

20. $2\sqrt[3]{x^8y^6} - 3y^2\sqrt[3]{8x^8}$

21. $5a^2\sqrt{27ab^3} - 6b\sqrt{12a^5b}$

22. $9a\sqrt{20a^3b^2} + 7b\sqrt{45a^5}$

23. $b\sqrt[3]{24a^5b} + 3a\sqrt[3]{81a^2b^4}$

24. $7\sqrt[3]{a^4b^3c^2} - 6ab\sqrt[3]{ac^2}$

25. $5x\sqrt[4]{3y^5} + y\sqrt[4]{243x^4y} + \sqrt[4]{48x^4y^5}$

26. $x\sqrt[4]{5xy^8} + y\sqrt[4]{405x^5y^4} + y^2\sqrt[4]{80x^5}$

27. $\frac{\sqrt{2}}{2} + \frac{1}{\sqrt{2}}$

28. $\frac{\sqrt{3}}{3} + \frac{1}{\sqrt{3}}$

29. $\dfrac{\sqrt{5}}{3} + \dfrac{1}{\sqrt{5}}$

30. $\dfrac{\sqrt{6}}{2} + \dfrac{1}{\sqrt{6}}$

31. $\sqrt{x} - \dfrac{1}{\sqrt{x}}$

32. $\sqrt{x} + \dfrac{1}{\sqrt{x}}$

33. $\dfrac{\sqrt{18}}{6} + \sqrt{\dfrac{1}{2}} + \dfrac{\sqrt{2}}{2}$

34. $\dfrac{\sqrt{12}}{6} + \sqrt{\dfrac{1}{3}} + \dfrac{\sqrt{3}}{3}$

35. $\sqrt{6} - \sqrt{\dfrac{2}{3}} + \sqrt{\dfrac{1}{6}}$

36. $\sqrt{15} - \sqrt{\dfrac{3}{5}} + \sqrt{\dfrac{5}{3}}$

37. $\sqrt[3]{25} + \dfrac{3}{\sqrt[3]{5}}$

38. $\sqrt[4]{8} + \dfrac{1}{\sqrt[4]{2}}$

The following problems apply to what you have learned about algebra with functions.

39. Let $f(x) = \sqrt{8x}$ and $g(x) = \sqrt{72x}$, then find

   a. $f(x) + g(x)$

   b. $f(x) - g(x)$

40. Let $f(x) = x + \sqrt{3}$ and $g(x) = x + 2\sqrt{3}$, then find

   a. $f(x) + g(x)$

   b. $f(x) - g(x)$

41. Let $f(x) = 3\sqrt{2x}$ and $g(x) = \sqrt{2x}$, then find

   a. $f(x) + g(x)$

   b. $f(x) - g(x)$

42. Let $f(x) = x\sqrt[3]{64}$ and $g(x) = x\sqrt{81}$, then find

   a. $f(x) + g(x)$

   b. $f(x) - g(x)$

43. Let $f(x) = x\sqrt{2}$ and $g(x) = 2x\sqrt{2}$, then find

   a. $f(x) + g(x)$

   b. $f(x) - g(x)$

44. Let $f(x) = 5 + 2\sqrt{5x}$ and $g(x) = 3\sqrt{5x}$, then find

   a. $f(x) + g(x)$

   b. $f(x) - g(x)$

45. Let $f(x) = \sqrt{2x} - 2$ and $g(x) = 2\sqrt{2x} + 5$ then find

   a. $f(x) + g(x)$

   b. $f(x) - g(x)$

46. Let $f(x) = 1 - 2\sqrt[3]{3x}$ and $g(x) = 2 + 3\sqrt[3]{3x}$, then find

   a. $f(x) + g(x)$

   b. $f(x) - g(x)$

47. Use a calculator to find a decimal approximation for $\sqrt{12}$ and for $2\sqrt{3}$.

48. Use a calculator to find decimal approximations for $\sqrt{50}$ and $5\sqrt{2}$.

49. Use a calculator to find a decimal approximation for $\sqrt{8} + \sqrt{18}$. Is it equal to the decimal approximation for $\sqrt{26}$ or $\sqrt{50}$?

50. Use a calculator to find a decimal approximation for $\sqrt{3} + \sqrt{12}$. Is it equal to the decimal approximation for $\sqrt{15}$ or $\sqrt{27}$?

A Applying the Concepts [Example 6]

51. **Golden Rectangle** Construct a golden rectangle from a square of side 8. Then show that the ratio of the length to the width is the golden ratio $\dfrac{1 + \sqrt{5}}{2}$.

52. **Golden Rectangle** Construct a golden rectangle from a square of side 10. Then show that the ratio of the length to the width is the golden ratio $\dfrac{1 + \sqrt{5}}{2}$.

53. **Golden Rectangle** To show that all golden rectangles have the same ratio of length to width, construct a golden rectangle from a square of side $2x$. Then show that the ratio of the length to the width is the golden ratio.

54. **Golden Rectangle** To show that all golden rectangles have the same ratio of length to width, construct a golden rectangle from a square of side x. Then show that the ratio of the length to the width is the golden ratio.

55. **Equilateral Triangles** A triangle is equilateral if it has three equal sides. The triangle in the figure is equilateral with each side of length $2x$. Find the ratio of the height to a side.

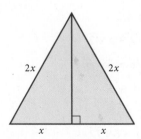

56. Equilateral Triangle The triangle in the figure is equilateral with each side of length $5x$. Find the ratio of the height to a side.

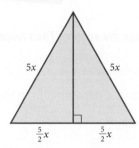

57. Pyramids Refer to the diagram of a square pyramid below. Find the ratio of the height h of the pyramid to the altitude a.

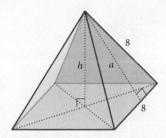

58. Pyramids Refer to the diagram of a square pyramid below. Find the ratio of the height h of the pyramid to the altitude a.

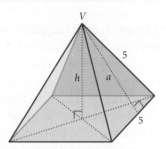

Maintaining Your Skills

Add or subtract as indicated.

59. $\dfrac{2a - 4}{a + 2} - \dfrac{a - 6}{a + 2}$

60. $\dfrac{2a - 3}{a - 2} - \dfrac{a - 1}{a - 2}$

61. $3 + \dfrac{4}{3 - t}$

62. $6 + \dfrac{2}{5 - t}$

63. $\dfrac{3}{2x - 5} - \dfrac{39}{8x^2 - 14x - 15}$

64. $\dfrac{2}{4x - 5} + \dfrac{9}{8x^2 - 38x + 35}$

65. $\dfrac{1}{x - y} - \dfrac{3xy}{x^3 - y^3}$

66. $\dfrac{1}{x + y} + \dfrac{3xy}{x^3 + y^3}$

Getting Ready for the Next Section

Simplify the following.

67. $3 \cdot 2$

68. $5 \cdot 7$

69. $(x + y)(4x - y)$

70. $(2x + y)(x - y)$

71. $(x + 3)^2$

72. $(3x - 2y)^2$

73. $(x - 2)(x + 2)$

74. $(2x + 5)(2x - 5)$

Simplify the following expressions.

75. $2\sqrt{18}$

76. $5\sqrt{36}$

77. $(\sqrt{6})^2$

78. $(\sqrt{2})^2$

79. $(3\sqrt{x})^2$

80. $(2\sqrt{y})^2$

Rationalize the denominator.

81. $\dfrac{\sqrt{3}}{\sqrt{2}}$

82. $\dfrac{\sqrt{5}}{\sqrt{6}}$

Extending the Concepts

Assume all variables represent positive numbers.
Simplify.

83. $\sqrt[5]{32x^5y^5} - y\sqrt[3]{27x^3}$

84. $\sqrt[6]{x^4} + 4\sqrt[3]{8x^2}$

85. $3\sqrt[9]{x^9 y^{18} z^{27}} - 4\sqrt[6]{x^6 y^{12} z^{18}}$

86. $4a\sqrt{b^4 c^6} + 3b\sqrt[3]{a^3 b^3 c^9}$

87. $3c\sqrt[8]{4a^6 b^{18}} + b\sqrt[4]{32a^3 b^5 c^4}$

88. $4x\sqrt[6]{16y^6 z^8} - y\sqrt[3]{32x^3 z^4}$

89. $3\sqrt[9]{8a^{12}b^9} + b\sqrt[3]{16a^4} - 8\sqrt[6]{4a^8 b^6}$

90. $-ac\sqrt{108bc^3} - 4\sqrt[6]{27a^6 b^3 c^{15}} + 3\sqrt[4]{9a^4 b^2 c^{10}}$

OBJECTIVES

A Multiply expressions containing radicals.

B Rationalize a denominator containing two terms.

TICKET TO SUCCESS

Keep these questions in mind as you read through the section. Then respond in your own words and in complete sentences.

1. Explain why $(\sqrt{5} + \sqrt{2})^2 \neq 5 + 2$.

2. Explain in words how you would rationalize the denominator in the expression $\frac{\sqrt{3}}{\sqrt{5} - \sqrt{2}}$.

3. What are conjugates?

4. What result is guaranteed when multiplying radical expressions that are conjugates?

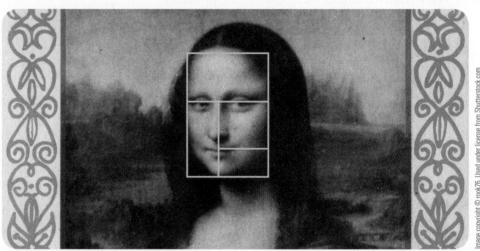

Leonardo da Vinci, as a mathematician and an artist, used the golden ratio in several of his works. He called it the "divine proportion" and considered it as a basis for perfection in the human body. The Mona Lisa, for example, includes the golden ratio in several rectangles. In the above figure, we can see how several of these rectangles can be overlayed on the painting and how the edges of these new figures come to all the important focal points of the woman: her chin, her eye, her nose, and corner of her famously mysterious mouth. This painting seems to be made purposefully to line up with the golden rectangle. In this section we will expand our mathematical work with the golden ratio to include multiplication and division of radical expressions.

We have worked with the golden rectangle more than once in this chapter. Remember the following diagram, the general representation of a golden rectangle:

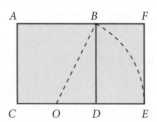

By now you know that in any golden rectangle constructed from a square (of any size) the ratio of the length to the width will be

$$\frac{1 + \sqrt{5}}{2}$$

which we call the golden ratio. What is interesting is that the smaller rectangle on the right of the figure, *BFED*, is also a golden rectangle. We will use the mathematics developed in this section to confirm this fact.

In this section we will look at multiplication and division of expressions that contain radicals. As you will see, multiplication of expressions that contain radicals is very similar to multiplication of polynomials. The division problems in this section are just an extension of the work we did previously when we rationalized denominators.

A   Multiplication of Radical Expressions

EXAMPLE 1   Multiply $(3\sqrt{5})(2\sqrt{7})$.

SOLUTION   We can rearrange the order and grouping of the numbers in this product by applying the commutative and associative properties. Following this, we apply property 1 for radicals and multiply.

$$(3\sqrt{5})(2\sqrt{7}) = (3 \cdot 2)(\sqrt{5} \ \sqrt{7}) \quad \text{Commutative and associative properties}$$
$$= (3 \cdot 2)(\sqrt{5} \cdot \sqrt{7}) \quad \text{Property 1 for radicals}$$
$$= 6\sqrt{35} \qquad\qquad \text{Multiplication}$$

In practice, it is not necessary to show the first two steps.

EXAMPLE 2   Multiply $\sqrt{3}(2\sqrt{6} - 5\sqrt{12})$.

SOLUTION   Applying the distributive property, we have

$$\sqrt{3}(2\sqrt{6} - 5\sqrt{12}) = \sqrt{3} \cdot 2\sqrt{6} - \sqrt{3} \cdot 5\sqrt{12}$$
$$= 2\sqrt{18} - 5\sqrt{36}$$

Writing each radical in simplified form gives

$$2\sqrt{18} - 5\sqrt{36} = 2\sqrt{9} \ \sqrt{2} - 5\sqrt{36}$$
$$= 6\sqrt{2} - 30$$

EXAMPLE 3   Multiply $(\sqrt{3} + \sqrt{5})(4\sqrt{3} - \sqrt{5})$.

SOLUTION   The same principle that applies when multiplying two binomials applies to this product. We must multiply each term in the first expression by each term in the second one. Any convenient method can be used. Let's use the FOIL method.

$$(\sqrt{3} + \sqrt{5})(4\sqrt{3} - \sqrt{5}) = \overset{F}{\sqrt{3} \cdot 4\sqrt{3}} - \overset{O}{\sqrt{3}\sqrt{5}} + \overset{I}{\sqrt{5} \cdot 4\sqrt{3}} - \overset{L}{\sqrt{5} \ \sqrt{5}}$$
$$= 4 \cdot 3 - \sqrt{15} + 4\sqrt{15} - 5$$
$$= 12 + 3\sqrt{15} - 5$$
$$= 7 + 3\sqrt{15}$$

EXAMPLE 4 Expand and simplify $(\sqrt{x} + 3)^2$.

SOLUTION 1 We can write this problem as a multiplication problem and proceed as we did in Example 3:

$$(\sqrt{x} + 3)^2 = (\sqrt{x} + 3)(\sqrt{x} + 3)$$

$$\overset{\text{F}\qquad\text{O}\qquad\text{I}\qquad\text{L}}{= \sqrt{x} \cdot \sqrt{x} + 3\sqrt{x} + 3\sqrt{x} + 3 \cdot 3}$$

$$= x + 3\sqrt{x} + 3\sqrt{x} + 9$$

$$= x + 6\sqrt{x} + 9$$

SOLUTION 2 We can obtain the same result by applying the formula for the square of a sum: $(a + b)^2 = a^2 + 2ab + b^2$.

$$(\sqrt{x} + 3)^2 = (\sqrt{x})^2 + 2(\sqrt{x})(3) + 3^2$$

$$= x + 6\sqrt{x} + 9$$

EXAMPLE 5 Expand $(3\sqrt{x} - 2\sqrt{y})^2$ and simplify the result.

SOLUTION Let's apply the formula for the square of a difference, $(a - b)^2 = a^2 - 2ab + b^2$.

$$(3\sqrt{x} - 2\sqrt{y})^2 = (3\sqrt{x})^2 - 2(3\sqrt{x})(2\sqrt{y}) + (2\sqrt{y})^2$$

$$= 9x - 12\sqrt{xy} + 4y$$

EXAMPLE 6 Expand and simplify $(\sqrt{x + 2} - 1)^2$.

SOLUTION Applying the formula $(a - b)^2 = a^2 - 2ab + b^2$, we have

$$(\sqrt{x + 2} - 1)^2 = (\sqrt{x + 2})^2 - 2\sqrt{x + 2}(1) + 1^2$$

$$= x + 2 - 2\sqrt{x + 2} + 1$$

$$= x + 3 - 2\sqrt{x + 2}$$

EXAMPLE 7 Multiply $(\sqrt{6} + \sqrt{2})(\sqrt{6} - \sqrt{2})$.

SOLUTION We notice the product is of the form $(a + b)(a - b)$, which always gives the difference of two squares, $a^2 - b^2$.

$$(\sqrt{6} + \sqrt{2})(\sqrt{6} - \sqrt{2}) = (\sqrt{6})^2 - (\sqrt{2})^2$$

$$= 6 - 2$$

$$= 4$$

NOTE
We can prove that conjugates always multiply to yield a rational number as follows: If a and b are positive integers, then
$(\sqrt{a} + \sqrt{b})(\sqrt{a} - \sqrt{b})$
$= \sqrt{a}\sqrt{a} - \sqrt{a}\sqrt{b} +$
$\quad \sqrt{a}\sqrt{b} - \sqrt{b}\sqrt{b}$
$= a - \sqrt{ab} + \sqrt{ab} - b$
$= a - b$
which is rational if a and b are rational.

The two expressions $(\sqrt{6} + \sqrt{2})$ and $(\sqrt{6} - \sqrt{2})$ are called *conjugates*. In general, the conjugate of $\sqrt{a} + \sqrt{b}$ is $\sqrt{a} - \sqrt{b}$. If a and b are integers, multiplying conjugates of this form always produces a rational number.

B Division of Radical Expressions

Division with radical expressions is the same as rationalizing the denominator. In a previous section we were able to divide $\sqrt{3}$ by $\sqrt{2}$ by rationalizing the denominator.

$$\frac{\sqrt{3}}{\sqrt{2}} = \frac{\sqrt{3}}{\sqrt{2}} \cdot \frac{\sqrt{2}}{\sqrt{2}} = \frac{\sqrt{6}}{2}$$

We can accomplish the same result with expressions such as

$$\frac{6}{\sqrt{5} - \sqrt{3}}$$

by multiplying the numerator and denominator by the conjugate of the denominator.

EXAMPLE 8   Divide $\dfrac{6}{\sqrt{5} - \sqrt{3}}$. (Rationalize the denominator.)

SOLUTION   Since the product of two conjugates is a rational number, we multiply the numerator and denominator by the conjugate of the denominator.

$$\frac{6}{\sqrt{5} - \sqrt{3}} = \frac{6}{\sqrt{5} - \sqrt{3}} \cdot \frac{(\sqrt{5} + \sqrt{3})}{(\sqrt{5} + \sqrt{3})}$$

$$= \frac{6\sqrt{5} + 6\sqrt{3}}{(\sqrt{5})^2 - (\sqrt{3})^2}$$

$$= \frac{6\sqrt{5} + 6\sqrt{3}}{5 - 3}$$

$$= \frac{6\sqrt{5} + 6\sqrt{3}}{2}$$

The numerator and denominator of this last expression have a factor of 2 in common. We can reduce to lowest terms by factoring 2 from the numerator and then dividing both the numerator and denominator by 2.

$$= \frac{\cancel{2}(3\sqrt{5} + 3\sqrt{3})}{\cancel{2}}$$

$$= 3\sqrt{5} + 3\sqrt{3}$$

EXAMPLE 9   Rationalize the denominator $\dfrac{\sqrt{5} - 2}{\sqrt{5} + 2}$.

SOLUTION   To rationalize the denominator, we multiply the numerator and denominator by the conjugate of the denominator.

$$\frac{\sqrt{5} - 2}{\sqrt{5} + 2} = \frac{\sqrt{5} - 2}{\sqrt{5} + 2} \cdot \frac{(\sqrt{5} - 2)}{(\sqrt{5} - 2)}$$

$$= \frac{5 - 2\sqrt{5} - 2\sqrt{5} + 4}{(\sqrt{5})^2 - 2^2}$$

$$= \frac{9 - 4\sqrt{5}}{5 - 4}$$

$$= \frac{9 - 4\sqrt{5}}{1}$$

$$= 9 - 4\sqrt{5}$$

EXAMPLE 10   A golden rectangle constructed from a square of side 2 is shown in Figure 1. Show that the smaller rectangle *BDEF* is also a golden rectangle by finding the ratio of its length to its width.

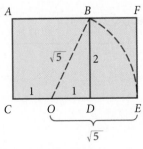

FIGURE 1

SOLUTION First we find expressions for the length and width of the smaller rectangle.

$$\text{Length} = EF = 2$$

$$\text{Width} = DE = \sqrt{5} - 1$$

Next, we find the ratio of length to width.

$$\text{Ratio of length to width} = \frac{EF}{DE} = \frac{2}{\sqrt{5} - 1}$$

To show that the small rectangle is a golden rectangle, we must show that the ratio of length to width is the golden ratio. We do so by rationalizing the denominator.

$$\frac{2}{\sqrt{5} - 1} = \frac{2}{\sqrt{5} - 1} \cdot \frac{\sqrt{5} + 1}{\sqrt{5} + 1}$$

$$= \frac{2(\sqrt{5} + 1)}{5 - 1}$$

$$= \frac{2(\sqrt{5} + 1)}{4}$$

$$= \frac{\sqrt{5} + 1}{2} \qquad \text{Divide out common factor 2}$$

Since addition is commutative, this last expression is the golden ratio. Therefore, the small rectangle in Figure 1 is a golden rectangle.

Problem Set 9.5

Moving Toward Success

"Do not worry if you have built your castles in the air. They are where they should be. Now put the foundations under them."

—Henry David Thoreau, 1817–1862, American author and naturalist

1. Why should you expect to spend more time on challenging material?

2. Each chapter is a building block for the next. How can this fact help you and potentially hurt you if you're not careful?

A Multiply. (Assume all expressions appearing under a square root symbol represent nonnegative numbers throughout this problem set.) [Examples 1–7]

1. $\sqrt{6}\,\sqrt{3}$

2. $\sqrt{6}\,\sqrt{2}$

3. $(2\sqrt{3})(5\sqrt{7})$

4. $(3\sqrt{5})(2\sqrt{7})$

5. $(4\sqrt{6})(2\sqrt{15})(3\sqrt{10})$

6. $(4\sqrt{35})(2\sqrt{21})(5\sqrt{15})$

7. $(3\sqrt[3]{3})(6\sqrt[3]{9})$

8. $(2\sqrt[3]{2})(6\sqrt[3]{4})$

9. $\sqrt{3}(\sqrt{2} - 3\sqrt{3})$

10. $\sqrt{2}(5\sqrt{3} + 4\sqrt{2})$

11. $6\sqrt[3]{4}(2\sqrt[3]{2} + 1)$

12. $7\sqrt[3]{5}(3\sqrt[3]{25} - 2)$

13. $\sqrt[3]{4}(\sqrt[3]{2} + \sqrt[3]{6})$

14. $\sqrt[3]{5}(\sqrt[3]{8} - \sqrt[3]{25})$

15. $\sqrt[3]{x}(\sqrt[3]{x^2y^4} + \sqrt[3]{x^5y})$

16. $\sqrt[3]{x^2y}(2\sqrt[3]{x^4y^2} - \sqrt[3]{x^2y^4})$

17. $\sqrt[4]{2x^3}(\sqrt[4]{8x^6} + \sqrt[4]{16x^9})$

18. $\sqrt[4]{8y^2}(\sqrt[4]{6y^6} - \sqrt[4]{8y^9})$

19. $(\sqrt{3} + \sqrt{2})(3\sqrt{3} - \sqrt{2})$

20. $(\sqrt{5} - \sqrt{2})(3\sqrt{5} + 2\sqrt{2})$

21. $(\sqrt{x} + 5)(\sqrt{x} - 3)$

22. $(\sqrt{x} + 4)(\sqrt{x} + 2)$

23. $(3\sqrt{6} + 4\sqrt{2})(\sqrt{6} + 2\sqrt{2})$

24. $(\sqrt{7} - 3\sqrt{3})(2\sqrt{7} - 4\sqrt{3})$

25. $(\sqrt{3} + 4)^2$

26. $(\sqrt{5} - 2)^2$

27. $(\sqrt{x} - 3)^2$

28. $(\sqrt{x} + 4)^2$

29. $(2\sqrt{a} - 3\sqrt{b})^2$

30. $(5\sqrt{a} - 2\sqrt{b})^2$

31. $(\sqrt{x - 4} + 2)^2$

32. $(\sqrt{x - 3} + 2)^2$

33. $(\sqrt{x - 5} - 3)^2$

34. $(\sqrt{x - 3} - 4)^2$

35. $(\sqrt{3} - \sqrt{2})(\sqrt{3} + \sqrt{2})$

36. $(\sqrt{5} - \sqrt{2})(\sqrt{5} + \sqrt{2})$

37. $(\sqrt{a} + 7)(\sqrt{a} - 7)$

38. $(\sqrt{a} + 5)(\sqrt{a} - 5)$

39. $(5 - \sqrt{x})(5 + \sqrt{x})$

40. $(3 - \sqrt{x})(3 + \sqrt{x})$

41. $(\sqrt{x - 4} + 2)(\sqrt{x - 4} - 2)$

42. $(\sqrt{x + 3} + 5)(\sqrt{x + 3} - 5)$

43. $(\sqrt{3} + 1)^3$

44. $(\sqrt{5} - 2)^3$

45. $(\sqrt[3]{3} + \sqrt[3]{2})(\sqrt[3]{9} + \sqrt[3]{4})$

46. $(\sqrt[3]{5} + \sqrt[3]{3})(\sqrt[3]{9} + \sqrt[3]{25})$

47. $(\sqrt[3]{x^5} + \sqrt[3]{y})(\sqrt[3]{x} + \sqrt[3]{y^2})$

48. $(\sqrt[3]{x^7} - \sqrt[3]{y^4})(\sqrt[3]{x^2} - \sqrt[3]{y^2})$

B Rationalize the denominator in each of the following. [Examples 8–9]

49. $\dfrac{1}{\sqrt{2}}$

50. $\dfrac{x}{\sqrt{2}}$

51. $\dfrac{1}{\sqrt{x}}$

52. $\dfrac{3}{\sqrt{x}}$

53. $\dfrac{4}{\sqrt{3}}$

54. $\dfrac{4}{\sqrt{2x}}$

55. $\dfrac{2x}{\sqrt{6}}$

56. $\dfrac{6x}{\sqrt{3}}$

57. $\dfrac{4}{\sqrt{10x}}$

58. $\sqrt{\dfrac{4x^2}{3x}}$

59. $\dfrac{2}{\sqrt{8}}$

60. $\sqrt{\dfrac{16xy}{4x^2y}}$

61. $\sqrt{\dfrac{32x^3y}{4xy^2}}$

62. $\dfrac{4}{\sqrt{6x^2y^5}}$

63. $\sqrt{\dfrac{12a^3b^3c}{8ab^5c^2}}$

64. $\sqrt{\dfrac{18x^3y^2z^4}{16xy^3z^2}}$

65. $\sqrt[3]{\dfrac{16a^4b^2c^4}{2ab^3c}}$

66. $\sqrt[3]{\dfrac{9x^5y^5z}{3x^3y^7z^2}}$

67. $\dfrac{\sqrt{2}}{\sqrt{6} - \sqrt{2}}$

68. $\dfrac{\sqrt{5}}{\sqrt{5} + \sqrt{3}}$

69. $\dfrac{\sqrt{5}}{\sqrt{5} + 1}$

70. $\dfrac{\sqrt{7}}{\sqrt{7} - 1}$

71. $\dfrac{\sqrt{x}}{\sqrt{x} - 3}$

72. $\dfrac{\sqrt{x}}{\sqrt{x} + 2}$

73. $\dfrac{\sqrt{5}}{2\sqrt{5} - 3}$

74. $\dfrac{\sqrt{7}}{3\sqrt{7} - 2}$

75. $\dfrac{3}{\sqrt{x} - \sqrt{y}}$

76. $\dfrac{2}{\sqrt{x} + \sqrt{y}}$

77. $\dfrac{\sqrt{6} + \sqrt{2}}{\sqrt{6} - \sqrt{2}}$

78. $\dfrac{\sqrt{5} - \sqrt{3}}{\sqrt{5} + \sqrt{3}}$

79. $\dfrac{\sqrt{7} - 2}{\sqrt{7} + 2}$

80. $\dfrac{\sqrt{11} + 3}{\sqrt{11} - 3}$

81. $\dfrac{\sqrt{a} + \sqrt{b}}{\sqrt{a} - \sqrt{b}}$

82. $\dfrac{\sqrt{a} - \sqrt{b}}{\sqrt{a} + \sqrt{b}}$

83. $\dfrac{\sqrt{x} + 2}{\sqrt{x} - 2}$

84. $\dfrac{\sqrt{x} - 3}{\sqrt{x} + 3}$

85. $\dfrac{2\sqrt{3} - \sqrt{7}}{3\sqrt{3} + \sqrt{7}}$

86. $\dfrac{5\sqrt{6} + 2\sqrt{2}}{\sqrt{6} - \sqrt{2}}$

87. $\dfrac{3\sqrt{x} + 2}{1 + \sqrt{x}}$

88. $\dfrac{5\sqrt{x} - 1}{2 + \sqrt{x}}$

89. $\dfrac{2}{\sqrt{3}} + \sqrt{12}$

90. $\sqrt{2} + \dfrac{5}{\sqrt{8}}$

91. $\dfrac{1}{\sqrt{5}} + \sqrt{20}$

92. $\dfrac{4}{\sqrt{2}} + \sqrt{72}$

93. $\dfrac{6}{\sqrt{12}} - \sqrt{75}$

94. $\dfrac{5}{\sqrt{7}} - \sqrt{63} + \sqrt{28}$

95. $\dfrac{1}{\sqrt{3}} + \sqrt{48} + \dfrac{4}{\sqrt{12}}$

96. $\dfrac{4}{\sqrt{2}} + \sqrt{8} + \dfrac{10}{\sqrt{50}}$

97. Show that the product $(\sqrt[3]{2} + \sqrt[3]{3})(\sqrt[3]{4} - \sqrt[3]{6} + \sqrt[3]{9})$ is 5.

98. Show that the product $(\sqrt[3]{x} + 2)(\sqrt[3]{x^2} - 2\sqrt[3]{x} + 4)$ is $x + 8$.

Each statement below is false. Correct the right side of each one.

99. $5(2\sqrt{3}) = 10\sqrt{15}$

100. $3(2\sqrt{x}) = 6\sqrt{3x}$

101. $(\sqrt{x} + 3)^2 = x + 9$

102. $(\sqrt{x} - 7)^2 = x - 49$

103. $(5\sqrt{3})^2 = 15$ **104.** $(3\sqrt{5})^2 = 15$

Applying the Concepts [Example 10]

105. Gravity If an object is dropped from the top of a 100-foot building, the amount of time t (in seconds) that it takes for the object to be h feet from the ground is given by the formula

$$t = \frac{\sqrt{100 - h}}{4}$$

How long does it take before the object is 50 feet from the ground? How long does it take to reach the ground? (When it is on the ground, h is 0.)

106. Gravity Use the formula given in Problem 105 to determine h if t is 1.25 seconds.

107. Golden Rectangle Rectangle *ACEF* in Figure 2 is a golden rectangle. If side *AC* is 6 inches, show that the smaller rectangle *BDEF* is also a golden rectangle.

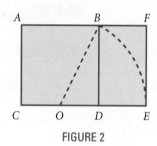

FIGURE 2

108. Golden Rectangle Rectangle *ACEF* in Figure 2 is a golden rectangle. If side *AC* is 1 inch, show that the smaller rectangle *BDEF* is also a golden rectangle.

109. Golden Rectangle If side *AC* in Figure 2 is 2x, show the rectangle *BDEF* is a golden rectangle.

110. Golden Rectangle If side *AC* in Figure 2 is x, show that rectangle *BDEF* is a golden rectangle.

Maintaining Your Skills

Simplify each complex fraction.

111. $\dfrac{\frac{1}{4} - \frac{1}{3}}{\frac{1}{2} + \frac{1}{6}}$ **112.** $\dfrac{\frac{1}{8} - \frac{1}{3}}{\frac{1}{4} - \frac{1}{3}}$

113. $\dfrac{1 - \frac{2}{y}}{1 + \frac{2}{y}}$ **114.** $\dfrac{1 + \frac{3}{y}}{1 - \frac{3}{y}}$

115. $\dfrac{4 + \frac{4}{x} + \frac{1}{x^2}}{4 - \frac{1}{x^2}}$ **116.** $\dfrac{1 - \frac{1}{x} - \frac{6}{x^2}}{1 - \frac{9}{x^2}}$

Getting Ready for the Next Section

Simplify.

117. $(t + 5)^2$ **118.** $(x - 4)^2$

119. $\sqrt{x} \cdot \sqrt{x}$ **120.** $\sqrt{3x} \cdot \sqrt{3x}$

Solve.

121. $3x + 4 = 5^2$ **122.** $4x - 7 = 3^2$

123. $t^2 + 7t + 12 = 0$

124. $x^2 - 3x - 10 = 0$

125. $t^2 + 10t + 25 = t + 7$

126. $x^2 - 4x + 4 = x - 2$

127. $(x + 4)^2 = x + 6$

128. $(x - 6)^2 = x - 4$

Extending the Concepts

Simplify.

129. $\dfrac{x}{\sqrt{x - 2} + 4}$

130. $\dfrac{3}{\sqrt{x - 4} - 7}$

131. $\dfrac{x}{\sqrt{x + 5} - 5}$

132. $\dfrac{2}{\sqrt{2x - 1} + 3}$

133. $\dfrac{3x}{\sqrt{5x} + x}$

134. $\dfrac{4x}{\sqrt{2x^3} + 2x}$

9.6 | Equations with Radicals

OBJECTIVES

A Solve equations containing radicals.

B Graph simple square root and cube root equations in two variables.

TICKET TO SUCCESS

Keep these questions in mind as you read through the section. Then respond in your own words and in complete sentences.

1. What is the squaring property of equality?

2. Under what conditions do we obtain extraneous solutions to equations that contain radical expressions?

3. If we have raised both sides of an equation to a power, when is it not necessary to check for extraneous solutions?

4. Why does the graph of $y = \sqrt{x}$ appear in the first quadrant only?

Image copyright ©George P. Chroma, 2010. Used under license from Shutterstock.com

Imagine you are standing on the roof of a 100-foot-tall building. If you accidentally drop your car keys over the side of the building, how long will it take them to land in the first floor awning, 8 feet from the ground? How long before they hit the ground if they miss the awning? Both answers can be found by solving the following equation:

$$t = \frac{\sqrt{100 - h}}{4}$$

Solving equations like this that involve radicals is our focus in this section.

This section is concerned with solving equations that involve one or more radicals. The first step in solving an equation that contains a radical is to eliminate the radical from the equation. To do so, we need an additional property.

A Equations with Radicals

> **Squaring Property of Equality**
> If both sides of an equation are squared, the solutions to the original equation are solutions to the resulting equation.

We will never lose solutions to our equations by squaring both sides. We may, however, introduce *extraneous solutions*. Extraneous solutions satisfy the equation obtained by squaring both sides of the original equation, but they do not satisfy the original equation.

We know that if two real numbers a and b are equal, then so are their squares.

$$\text{If} \qquad a = b$$
$$\text{then} \qquad a^2 = b^2$$

However, extraneous solutions are introduced when we square opposites; that is, even though opposites are not equal, their squares are. For example,

$$5 = -5 \qquad \text{A false statement}$$
$$(5)^2 = (-5)^2 \qquad \text{Square both sides}$$
$$25 = 25 \qquad \text{A true statement}$$

We are free to square both sides of an equation any time it is convenient. However, we must be aware that doing so may introduce extraneous solutions. We must, therefore, check all our solutions in the original equation if at any time we square both sides of the original equation.

EXAMPLE 1 Solve $\sqrt{3x + 4} = 5$ for x.

SOLUTION We square both sides and proceed as usual.

$$\sqrt{3x + 4} = 5$$
$$(\sqrt{3x + 4})^2 = 5^2$$
$$3x + 4 = 25$$
$$3x = 21$$
$$x = 7$$

Checking $x = 7$ in the original equation, we have

$$\sqrt{3(7) + 4} \overset{?}{=} 5$$
$$\sqrt{21 + 4} \overset{?}{=} 5$$
$$\sqrt{25} \overset{?}{=} 5$$
$$5 = 5$$

The solution $x = 7$ satisfies the original equation.

EXAMPLE 2 Solve $\sqrt{4x - 7} = -3$.

SOLUTION Squaring both sides, we have

$$\sqrt{4x - 7} = -3$$
$$(\sqrt{4x - 7})^2 = (-3)^2$$
$$4x - 7 = 9$$
$$4x = 16$$
$$x = 4$$

NOTE
The fact that there is no solution to the equation in Example 2 was obvious to begin with. Notice that the left side of the equation is the *positive* square root of $4x - 7$, which must be a nonnegative number. The right side of the equation is -3. Since we cannot have a number that is nonnegative equal to a negative number, there is no solution to the equation.

Checking $x = 4$ in the original equation gives

$$\sqrt{4(4) - 7} \overset{?}{=} -3$$
$$\sqrt{16 - 7} \overset{?}{=} -3$$
$$\sqrt{9} \overset{?}{=} -3$$
$$3 = -3 \qquad \text{A false statement}$$

The solution $x = 4$ produces a false statement when checked in the original equation. Since $x = 4$ was the only possible solution, there is no solution to the original equation. The possible solution $x = 4$ is an extraneous solution. It satisfies the equation obtained by squaring both sides of the original equation, but it does not satisfy the original equation.

EXAMPLE 3 Solve $\sqrt{5x - 1} + 3 = 7$.

SOLUTION We must isolate the radical on the left side of the equation. If we attempt to square both sides without doing so, the resulting equation will also contain a radical. Adding -3 to both sides, we have

$$\sqrt{5x - 1} + 3 = 7$$

$$\sqrt{5x - 1} = 4$$

We can now square both sides and proceed as usual.

$$(\sqrt{5x - 1})^2 = 4^2$$

$$5x - 1 = 16$$

$$5x = 17$$

$$x = \frac{17}{5}$$

Checking $x = \frac{17}{5}$, we have

$$\sqrt{5\left(\frac{17}{5}\right) - 1} + 3 \stackrel{?}{=} 7$$

$$\sqrt{17 - 1} + 3 \stackrel{?}{=} 7$$

$$\sqrt{16} + 3 \stackrel{?}{=} 7$$

$$4 + 3 \stackrel{?}{=} 7$$

$$7 = 7$$

Our solution checks.

EXAMPLE 4 Solve $t + 5 = \sqrt{t + 7}$.

SOLUTION This time, squaring both sides of the equation results in a quadratic equation.

$$(t + 5)^2 = (\sqrt{t + 7})^2 \qquad \text{Square both sides}$$

$$t^2 + 10t + 25 = t + 7$$

$$t^2 + 9t + 18 = 0 \qquad \text{Standard form}$$

$$(t + 3)(t + 6) = 0 \qquad \text{Factor the left side}$$

$$t + 3 = 0 \quad \text{or} \quad t + 6 = 0 \qquad \text{Set factors equal to 0}$$

$$t = -3 \quad \text{or} \quad t = -6$$

We must check each solution in the original equation.

Check $t = -3$ Check $t = -6$

$$-3 + 5 \stackrel{?}{=} \sqrt{-3 + 7} \qquad\qquad -6 + 5 \stackrel{?}{=} \sqrt{-6 + 7}$$

$$2 \stackrel{?}{=} \sqrt{4} \qquad\qquad\qquad -1 \stackrel{?}{=} \sqrt{1}$$

$$2 = 2 \qquad\qquad\qquad\qquad -1 = 1$$

A true statement   A false statement

Since $t = -6$ does not check, our only solution is $t = -3$.

EXAMPLE 5 Solve $\sqrt{x-3} = \sqrt{x} - 3$.

SOLUTION We begin by squaring both sides. Note what happens when we square the right side of the equation, and compare the square of the right side with the square of the left side. You must convince yourself that these results are correct. (The note in the margin will help if you are having trouble convincing yourself that what is written below is true.)

$$(\sqrt{x-3})^2 = (\sqrt{x} - 3)^2$$

$$x - 3 = x - 6\sqrt{x} + 9$$

> **NOTE**
> It is very important that you realize that the square of $(\sqrt{x} - 3)$ is not $x + 9$. Remember, when we square a difference with two terms, we use the formula
> $(a - b)^2 = a^2 - 2ab + b^2$
> Applying this formula to $(\sqrt{x} - 3)^2$, we have
> $(\sqrt{x} - 3)^2$
> $= (\sqrt{x})^2 - 2(\sqrt{x})(3) + 3^2$
> $= x - 6\sqrt{x} + 9$

Now we still have a radical in our equation, so we will have to square both sides again. Before we do, though, let's isolate the remaining radical.

$$x - 3 = x - 6\sqrt{x} + 9$$

$-3 = -6\sqrt{x} + 9$	Add $-x$ to each side
$-12 = -6\sqrt{x}$	Add -9 to each side
$2 = \sqrt{x}$	Divide each side by -6
$4 = x$	Square each side

Our only possible solution is $x = 4$, which we check in our original equation as follows:

$$\sqrt{4-3} \stackrel{?}{=} \sqrt{4} - 3$$

$$\sqrt{1} \stackrel{?}{=} 2 - 3$$

$$1 = -1 \qquad \text{A false statement}$$

Substituting 4 for x in the original equation yields a false statement. Since 4 was our only possible solution, there is no solution to our equation.

Here is another example of an equation for which we must apply our squaring property twice before all radicals are eliminated.

EXAMPLE 6 Solve $\sqrt{x+1} = 1 - \sqrt{2x}$.

SOLUTION This equation has two separate terms involving radical signs. Squaring both sides gives

$$x + 1 = 1 - 2\sqrt{2x} + 2x$$

$-x = -2\sqrt{2x}$	Add $-2x$ and -1 to both sides
$x^2 = 4(2x)$	Square both sides
$x^2 - 8x = 0$	Standard form

Our equation is a quadratic equation in standard form. To solve for x, we factor the left side and set each factor equal to 0.

$$x(x - 8) = 0 \qquad \text{Factor left side}$$

$$x = 0 \quad \text{or} \quad x - 8 = 0 \qquad \text{Set factors equal to 0}$$

$$x = 8$$

Since we squared both sides of our equation, we have the possibility that one or both of the solutions are extraneous. We must check each one in the original equation.

$$\text{Check } x = 8 \qquad\qquad \text{Check } x = 0$$

$$\sqrt{8+1} \overset{?}{=} 1 - \sqrt{2\cdot 8} \qquad \sqrt{0+1} \overset{?}{=} 1 - \sqrt{2\cdot 0}$$

$$\sqrt{9} \overset{?}{=} 1 - \sqrt{16} \qquad\qquad \sqrt{1} \overset{?}{=} 1 - \sqrt{0}$$

$$3 \overset{?}{=} 1 - 4 \qquad\qquad 1 \overset{?}{=} 1 - 0$$

$$3 = -3 \qquad\qquad\qquad 1 = 1$$

A false statement   A true statement

Since $x = 8$ does not check, it is an extraneous solution. Our only solution is $x = 0$.

EXAMPLE 7   Solve $\sqrt{x+1} = \sqrt{x+2} - 1$.

SOLUTION   Squaring both sides we have

$$(\sqrt{x+1})^2 = (\sqrt{x+2} - 1)^2$$

$$x + 1 = x + 2 - 2\sqrt{x+2} + 1$$

Once again we are left with a radical in our equation. Before we square each side again, we must isolate the radical on the right side of the equation.

$$x + 1 = x + 3 - 2\sqrt{x+2} \qquad \text{Simplify the right side}$$

$$1 = 3 - 2\sqrt{x+2} \qquad \text{Add } -x \text{ to each side}$$

$$-2 = -2\sqrt{x+2} \qquad \text{Add } -3 \text{ to each side}$$

$$1 = \sqrt{x+2} \qquad\qquad \text{Divide each side by } -2$$

$$1 = x + 2 \qquad\qquad \text{Square both sides}$$

$$-1 = x \qquad\qquad\qquad \text{Add } -2 \text{ to each side}$$

Checking our only possible solution, $x = -1$, in our original equation, we have

$$\sqrt{-1+1} \overset{?}{=} \sqrt{-1+2} - 1$$

$$\sqrt{0} \overset{?}{=} \sqrt{1} - 1$$

$$0 \overset{?}{=} 1 - 1$$

$$0 = 0 \qquad\qquad \text{A true statement}$$

Our solution checks.

It is also possible to raise both sides of an equation to powers greater than 2. We only need to check for extraneous solutions when we raise both sides of an equation to an even power. Raising both sides of an equation to an odd power will not produce extraneous solutions.

EXAMPLE 8   Solve $\sqrt[3]{4x+5} = 3$.

SOLUTION   Cubing both sides we have

$$(\sqrt[3]{4x+5})^3 = 3^3$$

$$4x + 5 = 27$$

$$4x = 22$$

$$x = \frac{22}{4}$$

$$x = \frac{11}{2}$$

We do not need to check $x = \frac{11}{2}$ since we raised both sides to an odd power. ∎

B Graphing

We end this section by looking at graphs of some equations that contain radicals.

EXAMPLE 9   Graph $y = \sqrt{x}$ and $y = \sqrt[3]{x}$.

SOLUTION   The graphs are shown in Figures 1 and 2. Notice that the graph of $y = \sqrt{x}$ appears in the first quadrant only because in the equation $y = \sqrt{x}$, x and y cannot be negative.

The graph of $y = \sqrt[3]{x}$ appears in quadrants 1 and 3 since the cube root of a positive number is also a positive number and the cube root of a negative number is a negative number; that is, when x is positive, y will be positive and when x is negative, y will be negative.

The graphs of both equations will contain the origin since $y = 0$ when $x = 0$ in both equations.

x	y
−4	undefined
−1	undefined
0	0
1	1
4	2
9	3
16	4

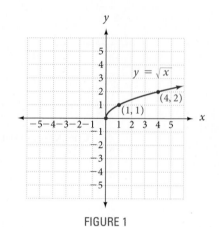

FIGURE 1

x	y
−27	−3
−8	−2
−1	−1
0	0
1	1
8	2
27	3

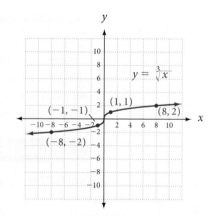

FIGURE 2

Problem Set 9.6

Moving Toward Success

"He that is good for making excuses is seldom good for anything else."

—Benjamin Franklin, 1706–1790, American scientist and diplomat

1. Should you go to your study group to just listen to the other members? Or should you take turns teaching the material? Explain.

2. Why is it important to take quick breaks during your study group sessions?

A Solve each of the following equations.
[Examples 1–4, 8]

1. $\sqrt{2x + 1} = 3$

2. $\sqrt{3x + 1} = 4$

3. $\sqrt{4x + 1} = -5$

4. $\sqrt{6x + 1} = -5$

5. $\sqrt{2y - 1} = 3$

6. $\sqrt{3y - 1} = 2$

7. $\sqrt{5x - 7} = -1$

8. $\sqrt{8x + 3} = -6$

9. $\sqrt{2x - 3} - 2 = 4$

10. $\sqrt{3x + 1} - 4 = 1$

11. $\sqrt{4x + 1} + 3 = 2$

12. $\sqrt{5a - 3} + 6 = 2$

13. $\sqrt[4]{3x + 1} = 2$

14. $\sqrt[4]{4x + 1} = 3$

15. $\sqrt[3]{2x - 5} = 1$

16. $\sqrt[3]{5x + 7} = 2$

17. $\sqrt[3]{3a + 5} = -3$

18. $\sqrt[3]{2a + 7} = -2$

19. $\sqrt{y - 3} = y - 3$

20. $\sqrt{y + 3} = y - 3$

21. $\sqrt{a + 2} = a + 2$

22. $\sqrt{a + 10} = a - 2$

23. $\sqrt{2x + 3} = \dfrac{2x - 7}{3}$

24. $\sqrt{3x - 2} = \dfrac{2x - 3}{3}$

25. $\sqrt{4x - 3} = \dfrac{x + 3}{2}$

26. $\sqrt{4x + 5} = \dfrac{x + 3}{2}$

27. $\sqrt{7x + 2} = \dfrac{2x + 2}{3}$

28. $\sqrt{7x - 3} = \dfrac{4x - 1}{3}$

29. $\sqrt{2x + 4} = \sqrt{1 - x}$

30. $\sqrt{3x + 4} = -\sqrt{2x + 3}$

31. $\sqrt{4a + 7} = -\sqrt{a + 2}$

32. $\sqrt{7a - 1} = \sqrt{2a + 4}$

33. $\sqrt[4]{5x - 8} = \sqrt[4]{4x - 1}$

34. $\sqrt[4]{6x + 7} = \sqrt[4]{x + 2}$

35. $x + 1 = \sqrt{5x + 1}$

36. $x - 1 = \sqrt{6x + 1}$

37. $t + 5 = \sqrt{2t + 9}$

38. $t + 7 = \sqrt{2t + 13}$

39. $\sqrt{y - 8} = \sqrt{8 - y}$

40. $\sqrt{2y + 5} = \sqrt{5y + 2}$

41. $\sqrt[3]{3x + 5} = \sqrt[3]{5 - 2x}$

42. $\sqrt[3]{4x + 9} = \sqrt[3]{3 - 2x}$

A The following equations will require that you square both sides twice before all the radicals are eliminated.
[Examples 5–7]

43. $\sqrt{x - 8} = \sqrt{x} - 2$

44. $\sqrt{x + 3} = \sqrt{x} - 3$

45. $\sqrt{x + 1} = \sqrt{x} + 1$

46. $\sqrt{x - 1} = \sqrt{x} - 1$

47. $\sqrt{x + 8} = \sqrt{x - 4} + 2$

48. $\sqrt{x + 5} = \sqrt{x - 3} + 2$

49. $\sqrt{x - 5} - 3 = \sqrt{x - 8}$

50. $\sqrt{x - 3} - 4 = \sqrt{x - 3}$

51. $\sqrt{x + 4} = 2 - \sqrt{2x}$

52. $\sqrt{5x + 1} = 1 + \sqrt{5x}$

53. $\sqrt{2x + 4} = \sqrt{x + 3} + 1$

54. $\sqrt{2x - 1} = \sqrt{x - 4} + 2$

Let $f(x) = \sqrt{2x - 1}$. Find all values for the variable x that produce the following values of $f(x)$.

55. $f(x) = 0$

56. $f(x) = -10$

57. $f(x) = 2x - 1$

58. $f(x) = \sqrt{5x - 10}$

59. $f(x) = \sqrt{x - 4} + 2$

60. $f(x) = 9$

Let $g(x) = \sqrt{2x + 3}$. Find all values for the variable x that produce the following values of $g(x)$.

61. $g(x) = 0$

62. $g(x) = x$

63. $g(x) = -\sqrt{5x}$

64. $g(x) = \sqrt{x^2 - 5}$

Let $f(x) = \sqrt{2x} - 1$. Find all values for the variable x for which $f(x) = g(x)$.

65. $g(x) = 0$

66. $g(x) = 5$

67. $g(x) = \sqrt{2x + 5}$

68. $g(x) = x - 1$

Let $h(x) = \sqrt[3]{3x + 5}$. Find all values for the variable x for which $h(x) = f(x)$.

69. $f(x) = 2$ **70.** $f(x) = -1$

Let $h(x) = \sqrt[3]{5 - 2x}$. Find all values for the variable x for which $h(x) = f(x)$.

71. $f(x) = 3$ **72.** $f(x) = -1$

B Graph each equation. [Example 9]

73. $y = 2\sqrt{x}$ **74.** $y = -2\sqrt{x}$

75. $y = \sqrt{x} - 2$ **76.** $y = \sqrt{x} + 2$

77. $y = \sqrt{x - 2}$ **78.** $y = \sqrt{x + 2}$

79. $y = 3\sqrt[3]{x}$ **80.** $y = -3\sqrt[3]{x}$

81. $y = \sqrt[3]{x} + 3$ **82.** $y = \sqrt[3]{x} - 3$

83. $y = \sqrt[3]{x + 3}$ **84.** $y = \sqrt[3]{x - 3}$

Applying the Concepts

85. Solving a Formula Solve the following formula for h:

$$t = \frac{\sqrt{100 - h}}{4}$$

86. Solving a Formula Solve the following formula for h:

$$t = \sqrt{\frac{2h - 40t}{g}}$$

87. Pendulum Clock The length of time (T) in seconds it takes the pendulum of a grandfather clock to swing through one complete cycle is given by the formula

$$T = 2\pi \sqrt{\frac{L}{32}}$$

where L is the length, in feet, of the pendulum, and π is approximately $\frac{22}{7}$. How long must the pendulum be if one complete cycle takes 2 seconds?

88. Pendulum Clock Solve the formula in Problem 87 for L.

Pollution A long straight river, 100 meters wide, is flowing at 1 meter per second. A pollutant is entering the river at a constant rate from one of its banks. As the pollutant disperses in the water, it forms a plume that is modeled by the equation $y = \sqrt{x}$. Use this information to answer the following questions.

89. How wide is the plume 25 meters down river from the source of the pollution?

90. How wide is the plume 100 meters down river from the source of the pollution?

91. How far down river from the source of the pollution does the plume reach halfway across the river?

92. How far down river from the source of the pollution does the plume reach the other side of the river?

Maintaining Your Skills

Multiply.

93. $\sqrt{2}(\sqrt{3} - \sqrt{2})$

94. $(\sqrt{x} - 4)(\sqrt{x} + 5)$

95. $(\sqrt{x} + 5)^2$

96. $(\sqrt{5} + \sqrt{3})(\sqrt{5} - \sqrt{3})$

Rationalize the denominator.

97. $\dfrac{\sqrt{x}}{\sqrt{x} + 3}$ **98.** $\dfrac{\sqrt{5} - \sqrt{3}}{\sqrt{5} + \sqrt{3}}$

Getting Ready for the Next Section

Simplify.

99. $\sqrt{25}$ **100.** $\sqrt{49}$

101. $\sqrt{12}$ **102.** $\sqrt{50}$

103. $(-1)^{15}$ **104.** $(-1)^{20}$

105. $(-1)^{50}$ **106.** $(-1)^5$

Solve.

107. $3x = 12$

108. $4 = 8y$

109. $4x - 3 = 5$

110. $7 = 2y - 1$

Perform the indicated operation.

111. $(3 + 4x) + (7 - 6x)$

112. $(2 - 5x) + (-1 + 7x)$

113. $(7 + 3x) - (5 + 6x)$

114. $(5 - 2x) - (9 - 4x)$

115. $(3 - 4x)(2 + 5x)$

116. $(8 + x)(7 - 3x)$

117. $2x(4 - 6x)$

118. $3x(7 + 2x)$

119. $(2 + 3x)^2$

120. $(3 + 5x)^2$

121. $(2 - 3x)(2 + 3x)$

122. $(4 - 5x)(4 + 5x)$

9.7 Complex Numbers

OBJECTIVES

A Simplify square roots of negative numbers.

B Simplify powers of i.

C Solve for unknown variables by equating real parts and by equating imaginary parts of two complex numbers.

D Add and subtract complex numbers.

E Multiply complex numbers.

F Divide complex numbers.

TICKET TO SUCCESS

Keep these questions in mind as you read through the section. Then respond in your own words and in complete sentences.

1. What is the number i?

2. Explain why every real number is a complex number.

3. What kind of number results when we multiply complex conjugates?

4. Explain how to divide complex numbers.

© INTERFOTO/Alamy

Known by some as the Prince of Mathematicians, Carl Friedrich Gauss made profound contributions to many fields of study. In his early twenties, he provided significant proof of the theorem behind complex numbers, greatly clarifying the concept for the academic world. The stamp shown here was issued by Germany in 1977 to commemorate the 200th anniversary of the birth of Carl Gauss. In this section, we will learn about complex numbers and how they are used in algebra.

Working with complex numbers gives us a way to solve a wider variety of equations. For example, the equation $x^2 = -9$ has no real number solutions since the square of a real number is always positive. We have been unable to work with square roots of negative numbers like $\sqrt{-25}$ and $\sqrt{-16}$ for the same reason.

Complex numbers allow us to expand our work with radicals to include square roots of negative numbers and solve equations like $x^2 = -9$ and $x^2 = -64$. Our work with complex numbers is based on the following definition.

A Square Roots of Negative Numbers

Definition
The number i is such that $i = \sqrt{-1}$ (which is the same as saying $i^2 = -1$).

The number i, as we have defined it here, is not a real number. Because of the way we have defined i, we can use it to simplify square roots of negative numbers.

Property
If a is a positive number, then $\sqrt{-a}$ can always be written as $i\sqrt{a}$; that is,
$$\sqrt{-a} = i\sqrt{a} \text{ if } a \text{ is a positive number.}$$

To justify our rule, we simply square the quantity $i\sqrt{a}$ to obtain $-a$. Here is what it looks like when we do so:
$$(i\sqrt{a})^2 = i^2 \cdot (\sqrt{a})^2$$
$$= -1 \cdot a$$
$$= -a$$

Here are some examples that illustrate the use of our new rule.

EXAMPLES Write each square root in terms of the number i.
1. $\sqrt{-25} = i\sqrt{25} = i \cdot 5 = 5i$
2. $\sqrt{-49} = i\sqrt{49} = i \cdot 7 = 7i$
3. $\sqrt{-12} = i\sqrt{12} = i \cdot 2\sqrt{3} = 2i\sqrt{3}$
4. $\sqrt{-17} = i\sqrt{17}$

NOTE
In Examples 3 and 4, we wrote i before the radical simply to avoid confusion. If we were to write the answer to 3 as $2\sqrt{3}i$, some people would think the i was under the radical sign and it is not.

B Powers of i

If we assume all the properties of exponents hold when the base is i, we can write any power of i as i, -1, $-i$, or 1. Using the fact that $i^2 = -1$, we have
$$i^1 = i$$
$$i^2 = -1$$
$$i^3 = i^2 \cdot i = -1(i) = -i$$
$$i^4 = i^2 \cdot i^2 = -1(-1) = 1$$

Since $i^4 = 1$, i^5 will simplify to i, and we will begin repeating the sequence i, -1, $-i$, 1 as we simplify higher powers of i: any power of i simplifies to i, -1, $-i$, or 1. The easiest way to simplify higher powers of i is to write them in terms of i^2. For instance, to simplify i^{21}, we would write it as
$$(i^2)^{10} \cdot i \quad \text{because } 2 \cdot 10 + 1 = 21$$
Then, since $i^2 = -1$, we have
$$(-1)^{10} \cdot i = 1 \cdot i = i$$

| EXAMPLES | Simplify as much as possible. |

5. $i^{30} = (i^2)^{15} = (-1)^{15} = -1$

6. $i^{11} = (i^2)^5 \cdot i = (-1)^5 \cdot i = (-1)i = -i$

7. $i^{40} = (i^2)^{20} = (-1)^{20} = 1$

Definition

A **complex number** is any number that can be put in the form

$$a + bi$$

where a and b are real numbers and $i = \sqrt{-1}$. The form $a + bi$ is called **standard form** for complex numbers. The number a is called the **real part** of the complex number. The number b is called the **imaginary part** of the complex number.

Every real number is a complex number. For example, 8 can be written as $8 + 0i$. Likewise, $-\frac{1}{2}, \pi, \sqrt{3}$, and -9 are complex numbers because they can all be written in the form $a + bi$:

$$-\frac{1}{2} = -\frac{1}{2} + 0i \qquad \pi = \pi + 0i$$
$$\sqrt{3} = \sqrt{3} + 0i \qquad -9 = -9 + 0i$$

The real numbers occur when $b = 0$. When $b \neq 0$, we have complex numbers that contain i, such as $2 + 5i$, $6 - i$, $4i$, and $\frac{1}{2}i$. These numbers are called *imaginary numbers*. The diagram explains this further.

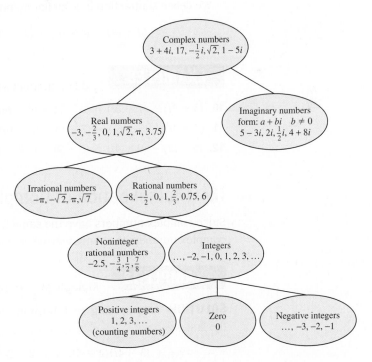

C Equality for Complex Numbers

Two complex numbers are equal if and only if their real parts are equal and their imaginary parts are equal; that is, for real numbers a, b, c, and d,

$$a + bi = c + di \quad \text{if and only if} \quad a = c \quad \text{and} \quad b = d$$

EXAMPLE 8 Find x and y if $3x + 4i = 12 - 8yi$.

SOLUTION Since the two complex numbers are equal, their real parts are equal and their imaginary parts are equal.

$$3x = 12 \quad \text{and} \quad 4 = -8y$$

$$x = 4 \qquad\qquad y = -\frac{1}{2}$$

EXAMPLE 9 Find x and y if $(4x - 3) + 7i = 5 + (2y - 1)i$.

SOLUTION The real parts are $4x - 3$ and 5. The imaginary parts are 7 and $2y - 1$.

$$4x - 3 = 5 \quad \text{and} \quad 7 = 2y - 1$$

$$4x = 8 \qquad\qquad 8 = 2y$$

$$x = 2 \qquad\qquad y = 4$$

D Addition and Subtraction of Complex Numbers

To add two complex numbers, add their real parts and add their imaginary parts; that is, if a, b, c, and d are real numbers, then

$$(a + bi) + (c + di) = (a + c) + (b + d)i$$

If we assume that the commutative, associative, and distributive properties hold for the number i, then the definition of addition is simply an extension of these properties. We define subtraction in a similar manner. If a, b, c, and d are real numbers, then

$$(a + bi) - (c + di) = (a - c) + (b - d)i$$

EXAMPLES Add or subtract as indicated.

10. $(3 + 4i) + (7 - 6i) = (3 + 7) + (4 - 6)i = 10 - 2i$

11. $(7 + 3i) - (5 + 6i) = (7 - 5) + (3 - 6)i = 2 - 3i$

12. $(5 - 2i) - (9 - 4i) = (5 - 9) + (-2 + 4)i = -4 + 2i$

E Multiplication of Complex Numbers

Since complex numbers have the same form as binomials, we find the product of two complex numbers the same way we find the product of two binomials.

EXAMPLE 13 Multiply $(3 - 4i)(2 + 5i)$.

SOLUTION Multiplying each term in the second complex number by each term in the first, we have

$$(3 - 4i)(2 + 5i) = 3 \cdot 2 + 3 \cdot 5i - 2 \cdot 4i - 5i(4i)$$

$$= 6 + 15i - 8i - 20i^2$$

Combining similar terms and using the fact that $i^2 = -1$, we can simplify as follows:

$$6 + 15i - 8i - 20i^2 = 6 + 7i - 20(-1)$$

$$= 6 + 7i + 20$$

$$= 26 + 7i$$

The product of the complex numbers $3 - 4i$ and $2 + 5i$ is the complex number $26 + 7i$.

EXAMPLE 14   Multiply $2i(4 - 6i)$.

SOLUTION   Applying the distributive property gives us

$$2i(4 - 6i) = 2i \cdot 4 - 2i \cdot 6i$$
$$= 8i - 12i^2$$
$$= 12 + 8i$$

EXAMPLE 15   Expand $(3 + 5i)^2$.

SOLUTION   We treat this like the square of a binomial. Remember, $(a + b)^2 = a^2 + 2ab + b^2$.

$$(3 + 5i)^2 = 3^2 + 2(3)(5i) + (5i)^2$$
$$= 9 + 30i + 25i^2$$
$$= 9 + 30i - 25$$
$$= -16 + 30i$$

> **NOTE**
> We can obtain the same result by writing $(3 + 5i)^2$ as $(3 + 5i)$ times $(3 + 5i)$ and applying the FOIL method as we did with the problem in Example 13.

EXAMPLE 16   Multiply $(2 - 3i)(2 + 3i)$.

SOLUTION   This product has the form $(a - b)(a + b)$, which we know results in the difference of two squares, $a^2 - b^2$.

$$(2 - 3i)(2 + 3i) = 2^2 - (3i)^2$$
$$= 4 - 9i^2$$
$$= 4 + 9$$
$$= 13$$

The product of the two complex numbers $2 - 3i$ and $2 + 3i$ is the real number 13. The two complex numbers $2 - 3i$ and $2 + 3i$ are called complex conjugates. The fact that their product is a real number is very useful.

> **Definition**
> The complex numbers $a + bi$ and $a - bi$ are called **complex conjugates.** One important property they have is that their product is the real number $a^2 + b^2$. Here's why:
>
> $$(a + bi)(a - bi) = a^2 - (bi)^2$$
> $$= a^2 - b^2i^2$$
> $$= a^2 - b^2(-1)$$
> $$= a^2 + b^2$$

F Division with Complex Numbers

The fact that the product of two complex conjugates is a real number is the key to division with complex numbers.

EXAMPLE 17 Divide $\dfrac{2 + i}{3 - 2i}$.

SOLUTION We want a complex number in standard form that is equivalent to the quotient $\dfrac{2 + i}{3 - 2i}$. We need to eliminate i from the denominator. Multiplying the numerator and denominator by $3 + 2i$ will give us what we want.

$$\frac{2 + i}{3 - 2i} = \frac{2 + i}{3 - 2i} \cdot \frac{(3 + 2i)}{(3 + 2i)}$$

$$= \frac{6 + 4i + 3i + 2i^2}{9 - 4i^2}$$

$$= \frac{6 + 7i - 2}{9 + 4}$$

$$= \frac{4 + 7i}{13}$$

$$= \frac{4}{13} + \frac{7}{13}i$$

Dividing the complex number $2 + i$ by $3 - 2i$ gives the complex number $\dfrac{4}{13} + \dfrac{7}{13}i$. ∎

EXAMPLE 18 Divide $\dfrac{7 - 4i}{i}$.

SOLUTION The conjugate of the denominator is $-i$. Multiplying the numerator and denominator by this number, we have

$$\frac{7 - 4i}{i} = \frac{7 - 4i}{i} \cdot \frac{-i}{-i}$$

$$= \frac{-7i + 4i^2}{-i^2}$$

$$= \frac{-7i + 4(-1)}{-(-1)}$$

$$= -4 - 7i$$ ∎

Problem Set 9.7

Moving Toward Success

"The most beautiful thing we can experience is the mysterious. It is the source of all true art and science."

—Albert Einstein, 1879–1955, German-born American physicist

1. Why is it important to still get good restful sleep between studying and going to class?

2. How would treating this class as a full-time job help you focus your time?

A Write the following in terms of i, and simplify as much as possible. [Examples 1–4]

1. $\sqrt{-36}$
2. $\sqrt{-49}$
3. $-\sqrt{-25}$
4. $-\sqrt{-81}$
5. $\sqrt{-72}$
6. $\sqrt{-48}$
7. $-\sqrt{-12}$
8. $-\sqrt{-75}$

B Write each of the following as i, -1, $-i$, or 1. [Examples 5–7]

9. i^{28}
10. i^{31}
11. i^{26}
12. i^{37}
13. i^{75}
14. i^{42}

C Find x and y so each of the following equations is true. [Examples 8–9]

15. $2x + 3yi = 6 - 3i$

16. $4x - 2yi = 4 + 8i$

17. $2 - 5i = -x + 10yi$

18. $4 + 7i = 6x - 14yi$

19. $2x + 10i = -16 - 2yi$

20. $4x - 5i = -2 + 3yi$

21. $(2x - 4) - 3i = 10 - 6yi$

22. $(4x - 3) - 2i = 8 + yi$

23. $(7x - 1) + 4i = 2 + (5y + 2)i$

24. $(5x + 2) - 7i = 4 + (2y + 1)i$

D Combine the following complex numbers. [Examples 10–12]

25. $(2 + 3i) + (3 + 6i)$

26. $(4 + i) + (3 + 2i)$

27. $(3 - 5i) + (2 + 4i)$

28. $(7 + 2i) + (3 - 4i)$

29. $(5 + 2i) - (3 + 6i)$

30. $(6 + 7i) - (4 + i)$

31. $(3 - 5i) - (2 + i)$

32. $(7 - 3i) - (4 + 10i)$

33. $[(3 + 2i) - (6 + i)] + (5 + i)$

34. $[(4 - 5i) - (2 + i)] + (2 + 5i)$

35. $[(7 - i) - (2 + 4i)] - (6 + 2i)$

36. $[(3 - i) - (4 + 7i)] - (3 - 4i)$

37. $(3 + 2i) - [(3 - 4i) - (6 + 2i)]$

38. $(7 - 4i) - [(-2 + i) - (3 + 7i)]$

39. $(4 - 9i) + [(2 - 7i) - (4 + 8i)]$

40. $(10 - 2i) - [(2 + i) - (3 - i)]$

E Find the following products. [Examples 13–16]

41. $3i(4 + 5i)$ **42.** $2i(3 + 4i)$

43. $6i(4 - 3i)$ **44.** $11i(2 - i)$

45. $(3 + 2i)(4 + i)$ **46.** $(2 - 4i)(3 + i)$

47. $(4 + 9i)(3 - i)$ **48.** $(5 - 2i)(1 + i)$

49. $(1 + i)^3$ **50.** $(1 - i)^3$

51. $(2 - i)^3$ **52.** $(2 + i)^3$

53. $(2 + 5i)^2$ **54.** $(3 + 2i)^2$

55. $(1 - i)^2$ **56.** $(1 + i)^2$

57. $(3 - 4i)^2$ **58.** $(6 - 5i)^2$

59. $(2 + i)(2 - i)$ **60.** $(3 + i)(3 - i)$

61. $(6 - 2i)(6 + 2i)$ **62.** $(5 + 4i)(5 - 4i)$

63. $(2 + 3i)(2 - 3i)$ **64.** $(2 - 7i)(2 + 7i)$

65. $(10 + 8i)(10 - 8i)$ **66.** $(11 - 7i)(11 + 7i)$

F Find the following quotients. Write all answers in standard form for complex numbers. [Examples 17–18]

67. $\dfrac{2 - 3i}{i}$ **68.** $\dfrac{3 + 4i}{i}$

69. $\dfrac{5 + 2i}{-i}$ **70.** $\dfrac{4 - 3i}{-i}$

71. $\dfrac{4}{2 - 3i}$ **72.** $\dfrac{3}{4 - 5i}$

73. $\dfrac{6}{-3 + 2i}$ **74.** $\dfrac{-1}{-2 - 5i}$

75. $\dfrac{2 + 3i}{2 - 3i}$ **76.** $\dfrac{4 - 7i}{4 + 7i}$

77. $\dfrac{5 + 4i}{3 + 6i}$ **78.** $\dfrac{2 + i}{5 - 6i}$

Applying the Concepts

79. Electric Circuits Complex numbers may be applied to electrical circuits. Electrical engineers use the fact that resistance R to electrical flow of the electrical current I and the voltage V are related by the formula $V = RI$. (Voltage is measured in volts, resistance in ohms, and current in amperes.) Find the resistance to electrical flow in a circuit that has a voltage $V = (80 + 20i)$ volts and current $I = (-6 + 2i)$ amps.

80. Electric Circuits Refer to the information about electrical circuits in Problem 79, and find the current in a circuit that has a resistance of $(4 + 10i)$ ohms and a voltage of $(5 - 7i)$ volts.

Maintaining Your Skills

Solve each equation.

81. $\dfrac{t}{3} - \dfrac{1}{2} = -1$

82. $\dfrac{x}{x - 2} + \dfrac{2}{3} = \dfrac{2}{x - 2}$

83. $2 + \dfrac{5}{y} = \dfrac{3}{y^2}$

84. $1 - \dfrac{1}{y} = \dfrac{12}{y^2}$

Solve each application problem.

85. The sum of a number and its reciprocal is $\dfrac{41}{20}$. Find the number.

86. It takes an inlet pipe 8 hours to fill a tank. The drain can empty the tank in 6 hours. If the tank is full and both the inlet pipe and drain are open, how long will it take to drain the tank?

Extending the Concepts

87. Show that $-i$ and $\dfrac{1}{i}$ (the opposite and the reciprocal of i) are the same number.

88. Show that i^{2n+1} is the same as i for all positive even integers n.

89. Show that $x = 1 + i$ is a solution to the equation $x^2 - 2x + 2 = 0$.

90. Show that $x = 1 - i$ is a solution to the equation $x^2 - 2x + 2 = 0$.

91. Show that $x = 2 + i$ is a solution to the equation $x^3 - 11x + 20 = 0$.

92. Show that $x = 2 - i$ is a solution to the equation $x^3 - 11x + 20 = 0$.

Square Roots [9.1]

1. The number 49 has two square roots, 7 and −7. They are written like this:

$$\sqrt{49} = 7 \qquad -\sqrt{49} = -7$$

Every positive real number x has two square roots. The *positive square root* of x is written $\sqrt{x}$, and the *negative square root* of x is written $-\sqrt{x}$. Both the positive and the negative square roots of x are numbers we square to get x; that is,

$$\left.\begin{array}{c}(\sqrt{x})^2 = x \\ \text{and} \qquad (-\sqrt{x})^2 = x\end{array}\right\} \text{ for } x \geq 0$$

Higher Roots [9.1]

2. $\sqrt[3]{8} = 2$
$\sqrt[3]{-27} = -3$

In the expression $\sqrt[n]{a}$, n is the *index*, a is the *radicand*, and $\sqrt{}$ is the *radical sign*. The expression $\sqrt[n]{a}$ is such that

$$(\sqrt[n]{a})^n = a \qquad a \geq 0 \text{ when } n \text{ is even}$$

Rational Exponents [9.1, 9.2]

3. $25^{1/2} = \sqrt{25} = 5$
$8^{2/3} = (\sqrt[3]{8})^2 = 2^2 = 4$
$9^{3/2} = (\sqrt{9})^3 = 3^3 = 27$

Rational exponents are used to indicate roots. The relationship between rational exponents and roots is as follows:

$$a^{1/n} = \sqrt[n]{a} \qquad \text{and} \qquad a^{m/n} = (a^{1/n})^m = (a^m)^{1/n}$$

$$a \geq 0 \text{ when } n \text{ is even}$$

Properties of Radicals [9.3]

4. $\sqrt{4 \cdot 5} = \sqrt{4}\,\sqrt{5} = 2\sqrt{5}$
$\sqrt{\dfrac{7}{9}} = \dfrac{\sqrt{7}}{\sqrt{9}} = \dfrac{\sqrt{7}}{3}$

If a and b are nonnegative real numbers whenever n is even, then

1. $\sqrt[n]{ab} = \sqrt[n]{a}\,\sqrt[n]{b}$

2. $\sqrt[n]{\dfrac{a}{b}} = \dfrac{\sqrt[n]{a}}{\sqrt[n]{b}} \qquad (b \neq 0)$

Simplified Form for Radicals [9.3]

5. $\sqrt{\dfrac{4}{5}} = \dfrac{\sqrt{4}}{\sqrt{5}}$

$\qquad = \dfrac{2}{\sqrt{5}} \cdot \dfrac{\sqrt{5}}{\sqrt{5}}$

$\qquad = \dfrac{2\sqrt{5}}{5}$

A radical expression is said to be in *simplified form*

1. If there is no factor of the radicand that can be written as a power greater than or equal to the index;
2. If there are no fractions under the radical sign; and
3. If there are no radicals in the denominator.

Addition and Subtraction of Radical Expressions [9.4]

6. $5\sqrt{3} - 7\sqrt{3} = (5 - 7)\sqrt{3}$
$\qquad\qquad = -2\sqrt{3}$
$\sqrt{20} + \sqrt{45} = 2\sqrt{5} + 3\sqrt{5}$
$\qquad\qquad = (2 + 3)\sqrt{5}$
$\qquad\qquad = 5\sqrt{5}$

We add and subtract radical expressions by using the distributive property to combine similar radicals. Similar radicals are radicals with the same index and the same radicand.

Multiplication of Radical Expressions [9.5]

7. $(\sqrt{x} + 2)(\sqrt{x} + 3)$
$= \sqrt{x}\,\sqrt{x} + 3\sqrt{x} + 2\sqrt{x} + 2 \cdot 3$
$= x + 5\sqrt{x} + 6$

We multiply radical expressions in the same way that we multiply polynomials. We can use the distributive property and the FOIL method.

Rationalizing the Denominator [9.3, 9.5]

8. $\dfrac{3}{\sqrt{2}} = \dfrac{3}{\sqrt{2}} \cdot \dfrac{\sqrt{2}}{\sqrt{2}}$

$\quad = \dfrac{3\sqrt{2}}{2}$

$\dfrac{3}{\sqrt{5} - \sqrt{3}}$

$\quad = \dfrac{3}{\sqrt{5} - \sqrt{3}} \cdot \dfrac{\sqrt{5} + \sqrt{3}}{\sqrt{5} + \sqrt{3}}$

$\quad = \dfrac{3\sqrt{5} + 3\sqrt{3}}{5 - 3}$

$\quad = \dfrac{3\sqrt{5} + 3\sqrt{3}}{2}$

When a fraction contains a square root in the denominator, we rationalize the denominator by multiplying numerator and denominator by

1. The square root itself if there is only one term in the denominator, or
2. The conjugate of the denominator if there are two terms in the denominator.

Rationalizing the denominator is also called division of radical expressions.

Squaring Property of Equality [9.6]

9. $\sqrt{2x + 1} = 3$
$(\sqrt{2x + 1})^2 = 3^2$
$\quad 2x + 1 = 9$
$\quad\quad\quad x = 4$

We may square both sides of an equation any time it is convenient to do so, as long as we check all resulting solutions in the original equation.

Complex Numbers [9.7]

10. $3 + 4i$ is a complex number.
Addition
$(3 + 4i) + (2 - 5i) = 5 - i$
Multiplication
$(3 + 4i)(2 - 5i)$
$\quad = 6 - 15i + 8i - 20i^2$
$\quad = 6 - 7i + 20$
$\quad = 26 - 7i$
Division
$\dfrac{2}{3 + 4i} = \dfrac{2}{3 + 4i} \cdot \dfrac{3 - 4i}{3 - 4i}$

$\quad = \dfrac{6 - 8i}{9 + 16}$

$\quad = \dfrac{6}{25} - \dfrac{8}{25}i$

A *complex number* is any number that can be put in the form
$$a + bi$$
where a and b are real numbers and $i = \sqrt{-1}$. The *real part* of the complex number is a, and b is the *imaginary part*.

If a, b, c, and d are real numbers, then we have the following definitions associated with complex numbers:

1. Equality
$$a + bi = c + di \quad \text{if and only if} \quad a = c \text{ and } b = d$$

2. Addition and subtraction
$$(a + bi) + (c + di) = (a + c) + (b + d)i$$
$$(a + bi) - (c + di) = (a - c) + (b - d)i$$

3. Multiplication
$$(a + bi)(c + di) = (ac - bd) + (ad + bc)i$$

4. Division is similar to rationalizing the denominator.

Simplify each expression as much as possible. [9.1]

1. $\sqrt{49}$

2. $(-27)^{1/3}$

3. $16^{1/4}$

4. $9^{3/2}$

5. $\sqrt[5]{32x^{15}y^{10}}$

6. $8^{-4/3}$

Use the properties of exponents to simplify each expression. Assume all bases represent positive numbers. [9.1]

7. $x^{2/3} \cdot x^{4/3}$

8. $(a^{2/3}b^{4/3})^3$

9. $\dfrac{a^{3/5}}{a^{1/4}}$

10. $\dfrac{a^{2/3}b^3}{a^{1/4}b^{1/3}}$

Multiply. [9.2]

11. $(3x^{1/2} + 5y^{1/2})(4x^{1/2} - 3y^{1/2})$

12. $(a^{1/3} - 5)^2$

13. Divide: $\dfrac{28x^{5/6} + 14x^{7/6}}{7x^{1/3}}$. (Assume $x > 0$.) [9.2]

14. Factor $2(x - 3)^{1/4}$ from $8(x - 3)^{5/4} - 2(x - 3)^{1/4}$. [9.2]

15. Simplify $x^{3/4} + \dfrac{5}{x^{1/4}}$ into a single fraction. (Assume $x > 0$.) [9.2]

Write each expression in simplified form for radicals. (Assume all variables represent nonnegative numbers.) [9.3]

16. $\sqrt{12}$

17. $\sqrt{50}$

18. $\sqrt[3]{16}$

19. $\sqrt{18x^2}$

20. $\sqrt{80a^3b^4c^2}$

21. $\sqrt[4]{32a^4b^5c^6}$

Rationalize the denominator in each expression. [9.3]

22. $\dfrac{3}{\sqrt{2}}$

23. $\dfrac{6}{\sqrt[3]{2}}$

Write each expression in simplified form. (Assume all variables represent positive numbers.) [9.3]

24. $\sqrt{\dfrac{48x^3}{7y}}$

25. $\sqrt[3]{\dfrac{40x^2y^3}{3z}}$

Combine the following expressions. (Assume all variables represent positive numbers.) [9.4]

26. $5x\sqrt{6} + 2x\sqrt{6} - 9x\sqrt{6}$

27. $\sqrt{12} + \sqrt{3}$

28. $\dfrac{3}{\sqrt{5}} + \sqrt{5}$

29. $3\sqrt{8} - 4\sqrt{72} + 5\sqrt{50}$

30. $3b\sqrt{27a^5b} + 2a\sqrt{3a^3b^3}$

31. $2x\sqrt[3]{xy^3z^2} - 6y\sqrt[3]{x^4z^2}$

Multiply. [9.5]

32. $\sqrt{2}(\sqrt{3} - 2\sqrt{2})$

33. $(\sqrt{x} - 2)(\sqrt{x} - 3)$

Rationalize the denominator. [9.5]

34. $\dfrac{3}{\sqrt{5} - 2}$

35. $\dfrac{\sqrt{7} + \sqrt{5}}{\sqrt{7} - \sqrt{5}}$

36. $\dfrac{3\sqrt{7}}{3\sqrt{7} - 4}$

Solve each equation. [9.6]

37. $\sqrt{4a + 1} = 1$

38. $\sqrt[3]{3x - 8} = 1$

39. $\sqrt{3x + 1} - 3 = 1$

40. $\sqrt{x + 4} = \sqrt{x} - 2$

Graph each equation. [9.6]

41. $y = 3\sqrt{x}$

42. $y = \sqrt[3]{x} + 2$

Write each of the following as i, -1, $-i$, or 1. [9.7]

43. i^{24}

44. i^{27}

Find x and y so that each of the following equations is true. [9.7]

45. $3 - 4i = -2x + 8yi$

46. $(3x + 2) - 8i = -4 + 2yi$

Combine the following complex numbers. [9.7]

47. $(3 + 5i) + (6 - 2i)$

48. $(2 + 5i) - [(3 + 2i) + (6 - i)]$

Multiply. [9.7]

49. $3i(4 + 2i)$

50. $(2 + 3i)(4 + i)$

51. $(4 + 2i)^2$

52. $(4 + 3i)(4 - 3i)$

Divide. Write all answers in standard form for complex numbers. [9.7]

53. $\dfrac{3+i}{i}$

54. $\dfrac{-3}{2+i}$

55. Construction The roof of the house shown in Figure 1 is to extend up 13.5 feet above the ceiling, which is 36 feet across. Find the length of one side of the roof.

13.5 ft

36 ft

FIGURE 1

56. Surveying A surveyor is attempting to find the distance across a pond. From a point on one side of the pond he walks 25 yards to the end of the pond and then makes a 90-degree turn and walks another 60 yards before coming to a point directly across the pond from the point at which he started. What is the distance across the pond? (See Figure 2.)

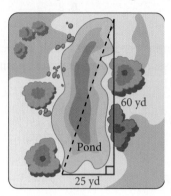

60 yd

Pond

25 yd

FIGURE 2

Simplify.

1. $6 + [2 - 5(7 - 2)]$

2. $3 - 15 \div 5 + 6 \cdot 2$

3. $2 + (5 - 7)^2 - (6 - 5)^3$

4. $\sqrt[4]{48}$

5. $8^{2/3} + 16^{1/4}$

6. $\dfrac{1 - \frac{2}{3}}{1 + \frac{1}{3}}$

7. $\left(-\dfrac{27}{8}\right)^{-1/3}$

8. $12 - [3 - 4 \div 2(7 - 5)]$

Reduce.

9. $\dfrac{105}{189}$

10. $\dfrac{6x^2 - 4xy - 16y^2}{2x - 4y}$

11. $\dfrac{x^2 - x - 6}{x^2 - 2x - 3}$

12. $\dfrac{6a^7b^{-3}}{36a^5b^{-2}}$

Multiply or divide.

13. $(\sqrt{x} - 3)(2\sqrt{x} - 5)$

14. $(7 - i)(3 + i)$

15. $\dfrac{4 + i}{4 + 3i}$

16. $\dfrac{\sqrt{3}}{6 - \sqrt{3}}$

17. $(2x - 5y)(3x - 2y)$

18. $(9x^2 - 25) \cdot \dfrac{x + 5}{3x - 5}$

19. $\dfrac{27x^3y^2}{13\,x^2y^4} \div \dfrac{9xy}{26y}$

20. $\dfrac{\sqrt{x} + \sqrt{y}}{\sqrt{x} - \sqrt{y}}$

Solve.

21. $\dfrac{3}{2} + \dfrac{1}{a - 2} = -\dfrac{1}{a - 2}$

22. $\sqrt{x - 2} = x - 2$

23. $(2x - 5)^2 = 12$

24. $0.03x^2 - 0.02x = 0.08$

25. $\sqrt{x - 3} = 3 - \sqrt{x}$

26. $3x^3 - 28x = -5x^2$

27. $\dfrac{1}{2}x^2 + \dfrac{3x}{8} = \dfrac{5}{16}$

28. $\dfrac{2}{3}(9x - 2) + \dfrac{1}{3} = 4$

29. $|3x - 1| - 2 = 6$

30. $\sqrt[3]{8 - 3x} = -1$

Solve each inequality, and graph the solution.

31. $-3 \le \dfrac{1}{4}x - 5 \le -1$

32. $|2x - 3| > 7$

Solve each system.

33. $2x - 3y = -16$
$3x + y = -2$

34. $\dfrac{3}{4}x + \dfrac{1}{2}y = -9$
$\dfrac{1}{2}x = \dfrac{2}{3}y$

35. $2x + 4y + z = 6$
$3x - y - 2z = 4$
$x + 2y + 3z = -10$

36. $6x + 8y = -9$
$-4x + 6y = -11$

Graph on a rectangular coordinate system.

37. $-2x + 3y = -6$

38. $x + 2y > 2$

Find the slope of the line passing through the points.

39. $(-3, 4)$ and $(-2, -3)$

40. $(-2, -4)$ and $(-3, 6)$

Rationalize the denominator.

41. $\dfrac{6}{\sqrt[4]{8}}$

42. $\sqrt[3]{\dfrac{8y}{4x}}$

Factor completely.

43. $625a^4 - 16b^4$

44. $24a^4 + 10a^2 - 6$

45. Inverse Variation y varies inversely with the square root of x. If $y = \dfrac{1}{2}$ when $x = 16$, find y when $x = 4$.

46. Joint Variation z varies jointly with x and the cube of y. If $z = 80$ when $x = 5$, and $y = 2$, find z when $x = 3$ and $y = -1$.

Let $f(x) = x^2 - 3x$ and $g(x) = x - 1$. Find the following.

47. $f(b)$

48. $(g - f)(3)$

49. $(g \circ f)(3)$

50. $(f \circ g)(-3)$

Use the illustration to answer Questions 51 and 52.

Most Visited Countries

France 79,300,000
United States 58,000,000
Spain 57,300,000
China 53,000,000
Italy 42,700,000   * Annual number of arrivals

Source: TenMojo.com

51. How many more arrivals were there in France than in Spain?

52. Find the difference in the number of people visting the United States and those visiting China.

Simplify each of the following. (Assume all variable bases are positive integers and all variable exponents are positive real numbers throughout this test.) [9.1]

1. $27^{-2/3}$

2. $\left(\dfrac{25}{49}\right)^{-1/2}$

3. $a^{3/4} \cdot a^{-1/3}$

4. $\dfrac{(x^{2/3}y^{-3})^{1/2}}{(x^{3/4}y^{1/2})^{-1}}$

5. $\sqrt{49x^8y^{10}}$

6. $\sqrt[5]{32x^{10}y^{20}}$

7. $\dfrac{(36a^8b^4)^{1/2}}{(27a^9b^6)^{1/3}}$

8. $\dfrac{(x^n y^{1/n})^n}{(x^{1/n}y^n)^{n^2}}$

Multiply. [9.2]

9. $2a^{1/2}(3a^{3/2} - 5a^{1/2})$

10. $(4a^{3/2} - 5)^2$

Factor. [9.2]

11. $3x^{2/3} + 5x^{1/3} - 2$

12. $9x^{2/3} - 49$

Write in simplified form. [9.3]

13. $\sqrt{125x^3y^5}$

14. $\sqrt[3]{40x^7y^8}$

15. $\sqrt{\dfrac{2}{3}}$

16. $\sqrt{\dfrac{12a^4b^3}{5c}}$

Combine. [9.4]

17. $\dfrac{4}{x^{1/2}} + x^{1/2}$

18. $\dfrac{x^2}{(x^2 - 3)^{1/2}} - (x^2 - 3)^{1/2}$

Combine. [9.4]

19. $3\sqrt{12} - 4\sqrt{27}$

20. $2\sqrt[3]{24a^3b^3} - 5a\sqrt[3]{3b^3}$

Multiply. [9.5]

21. $(\sqrt{x} + 7)(\sqrt{x} - 4)$

22. $(3\sqrt{2} - \sqrt{3})^2$

Rationalize the denominator. [9.5]

23. $\dfrac{5}{\sqrt{3} - 1}$

24. $\dfrac{\sqrt{x} - \sqrt{2}}{\sqrt{x} + \sqrt{2}}$

Solve for x. [9.6]

25. $\sqrt{3x + 1} = x - 3$

26. $\sqrt[3]{2x + 7} = -1$

27. $\sqrt{x + 3} = \sqrt{x + 4} - 1$

Graph. [9.6]

28. $y = \sqrt{x - 2}$

29. $y = \sqrt[3]{x} + 3$

30. Find x and y so that the following equation is true. [9.7]

$$(2x + 5) - 4i = 6 - (y - 3)i$$

Perform the indicated operations. [9.7]

31. $(3 + 2i) - [(7 - i) - (4 + 3i)]$

32. $(2 - 3i)(4 + 3i)$

33. $(5 - 4i)^2$

34. $\dfrac{2 - 3i}{2 + 3i}$

35. Show that i^{38} can be written as -1. [9.7]

Match each equation with its graph. [9.6]

36. $y = -\sqrt{x}$

37. $y = \sqrt{x} - 1$

38. $y = \sqrt{x + 1}$

39. $y = \sqrt[3]{x} - 1$

40. $y = \sqrt[3]{x - 1}$

41. $y = \sqrt[3]{x + 1}$

A.

B.

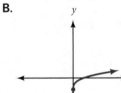

C.

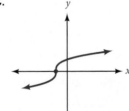

D.

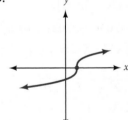

E.

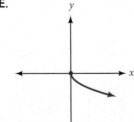

F.

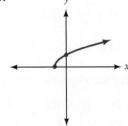

Chapter 9 Projects

RATIONAL EXPONENTS AND ROOTS

GROUP PROJECT

Constructing the Spiral of Roots

Number of People 3

Time Needed 20 minutes

Equipment Two sheets of graph paper (4 or 5 squares per inch) and pencils.

Background The spiral of roots gives us a way to visualize the positive square roots of the counting numbers, and in so doing, we see many line segments whose lengths are irrational numbers.

Procedure You are to construct a spiral of roots from a line segment 1 inch long. The graph paper you have contains either 4 or 5 squares per inch, allowing you to accurately draw 1-inch line segments. Because the lines on the graph paper are perpendicular to one another, if you are careful, you can use the graph paper to connect one line segment to another so that they form a right angle.

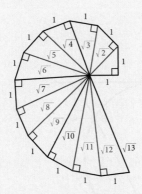

1. Fold one of the pieces of graph paper so it can be used as a ruler.

2. Use the folded paper to draw a line segment 1-inch long, just to the right of the middle of the unfolded paper. On the end of this segment, attach another segment of 1-inch length at a right angle to the first one.

Connect the end points of the segments to form a right triangle. Label each side of this triangle. When you are finished, your work should resemble Figure 1.

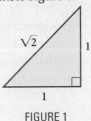

FIGURE 1

3. On the end of the hypotenuse of the triangle, attach a 1-inch line segment so that the two segments form a right angle. (Use the folded paper to do this.) Draw the hypotenuse of this triangle. Label all the sides of this second triangle. Your work should resemble Figure 2.

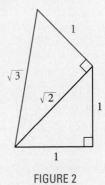

FIGURE 2

4. Continue to draw a new right triangle by attaching 1-inch line segments at right angles to the previous hypotenuse. Label all the sides of each triangle.

5. Stop when you have drawn a hypotenuse $\sqrt{8}$ inches long.

Maria Gaetana Agnesi (1718–1799)

January 9, 1999, was the 200th anniversary of the death of Maria Agnesi, the author of *Instituzioni analitiche ad uso della gioventu italiana* (1748), a calculus textbook considered to be the best book of its time and the first surviving mathematical work written by a woman. Maria Agnesi is also famous for a curve that is named for her. The curve is called the Witch of Agnesi in English, but its actual translation is the Locus of Agnesi. The foundation of the curve is shown in the figure. Research the Witch of Agnesi and then explain, in essay form, how the diagram in the figure is used to produce the Witch of Agnesi. Include a rough sketch of the curve, starting with a circle of diameter *a* as shown in the figure. Then, for comparison, sketch the curve again, starting with a circle of diameter 2*a*.

©Bettmann/CORBIS

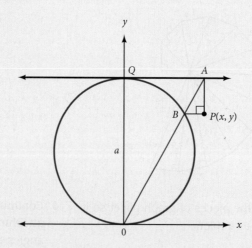

10

Quadratic Functions

Introduction

The city of Khiva, in Uzbekistan, is the birthplace of the mathematician, al-Khwarizmi. Although he was born in Khiva, he spent most of his life in Baghdad, Iraq. The Islamic mathematician wrote on Hindu-Arabic numerals and is credited with being the first to use zero as a place holder. His book *Hisab al-jabr w'al-muqabala* is considered as the first book to be written on algebra. Below is a postage stamp issued by the USSR in 1983, to commemorate the 1200th anniversary of al-Khwarizmi's birth.

"That fondness for science, ... that affability and condescension which God shows to the learned, that promptitude with which he protects and supports them in the elucidation of obscurities and in the removal of difficulties, has encouraged me to compose a short work on calculating by *al-jabr* and *al-muqabala*, confining it to what is easiest and most useful in arithmetic."

—al-Khwarizmi

In his book, al-Khwarizmi solves quadratic equations by a method called completing the square, one of the topics we will study in this chapter.

Simplify.

1. $\dfrac{3 + 5}{2}$

2. $\dfrac{3 - 5}{2}$

3. $2 + \dfrac{25}{4}$

4. $-\dfrac{7}{3} + \dfrac{16}{9}$

Solve.

5. $x^2 = 25$

6. $y^2 = 9$

7. $x^2 = 12$

8. $y^2 = 18$

Find $\left(\dfrac{b}{2}\right)^2$ when

9. $b = 6$

10. $b = 4$

11. $b = 5$

12. $b = 7$

13. $b = \dfrac{8}{3}$

14. $b = \dfrac{2}{5}$

Factor.

15. $x^2 - 6x + 9$

16. $x^2 - 8x + 16$

17. $x^2 + 12x + 36$

18. $125t^3 + 1$

Simplify.

19. $-\dfrac{c}{a} + \left(\dfrac{b}{2a}\right)^2$

20. $(3a - 2)^2 + 2(3a - 2) - 3$

Chapter Outline

10.1 Completing the Square

A Solve quadratic equations by taking the square root of both sides.

B Solve quadratic equations by completing the square.

C Use quadratic equations to solve for missing parts of right triangles.

10.2 The Quadratic Formula

A Solve quadratic equations by the quadratic formula.

B Solve application problems using quadratic equations.

10.3 Additional Items Involving Solutions to Equations

A Find the number and kinds of solutions to a quadratic equation by using the discriminant.

B Find an unknown constant in a quadratic equation so that there is exactly one solution.

C Find an equation from its solutions.

10.4 Equations Quadratic in Form

A Solve equations that are reducible to a quadratic equation.

B Solve application problems using equations quadratic in form.

10.5 Graphing Parabolas

A Graph a parabola.

B Solve application problems using information from a graph.

C Find an equation from its graph.

10.6 Quadratic Inequalities

A Solve quadratic inequalities and graph the solution sets.

B Solve inequalities involving rational expressions.

10.1 Completing the Square

OBJECTIVES

A Solve quadratic equations by taking the square root of both sides.

B Solve quadratic equations by completing the square.

C Use quadratic equations to solve for missing parts of right triangles.

TICKET TO SUCCESS

Keep these questions in mind as you read through the section. Then respond in your own words and in complete sentences.

1. Explain Theorem 1 for square roots.
2. What kind of equation do we solve using the method of completing the square?
3. Explain the strategy for completing the square on $x^2 - 16x = 4$.
4. What is the relationship between the shortest side and the longest side in a 30°-60°-90° triangle?

Table 1 is taken from the trail map given to skiers at the Northstar at Tahoe Ski Resort in Lake Tahoe, California. The table gives the length of each chair lift at Northstar, along with the change in elevation from the beginning of the lift to the end of the lift.

Right triangles are good mathematical models for chair lifts. Can you picture why? In this section, we will use our knowledge of right triangles, along with the new material developed in the section, to solve problems involving chair lifts and a variety of other examples.

TABLE 1

FROM THE TRAIL MAP FOR THE NORTHSTAR AT TAHOE SKI RESORT

Lift Information

Lift	Vertical Rise (feet)	Length (feet)
Big Springs Gondola	480	4,100
Bear Paw Double	120	790
Echo Triple	710	4,890
Aspen Express Quad	900	5,100
Forest Double	1,170	5,750
Lookout Double	960	4,330
Comstock Express Quad	1,250	5,900
Rendezvous Triple	650	2,900
Schaffer Camp Triple	1,860	6,150
Chipmunk Tow Lift	28	280
Bear Cub Tow Lift	120	750

A The Square Root Method

In this section, we will develop the first of our new methods of solving quadratic equations. The new method is called *completing the square*. Completing the square on a quadratic equation allows us to obtain solutions, regardless of whether the equation can be factored. Before we solve equations by completing the square, we need to learn how to solve equations by taking square roots of both sides.

Consider the equation

$$x^2 = 16$$

We could solve it by writing it in standard form, factoring the left side, and proceeding as we have done previously. We can shorten our work considerably, however, if we simply notice that x must be either the positive square root of 16 or the negative square root of 16; that is,

$$\text{If} \rightarrow \quad x^2 = 16$$
$$\text{then} \rightarrow \quad x = \sqrt{16} \quad \text{or} \quad x = -\sqrt{16}$$
$$x = 4 \quad \text{or} \quad x = -4$$

We can generalize this result into a theorem as follows.

Theorem 1 for Square Roots
If $a^2 = b$ where b is a real number, then $a = \sqrt{b}$ or $a = -\sqrt{b}$.

NOTATION The expression $a = \sqrt{b}$ or $a = -\sqrt{b}$ can be written in shorthand form as $a = \pm\sqrt{b}$. The symbol $\pm$ is read "plus or minus."

We can apply Theorem 1 to some fairly complicated quadratic equations.

EXAMPLE 1 Solve $(2x - 3)^2 = 25$.

SOLUTION

$$(2x - 3)^2 = 25$$
$$2x - 3 = \pm\sqrt{25} \qquad \text{Theorem 1}$$
$$2x - 3 = \pm 5 \qquad \sqrt{25} = 5$$
$$2x = 3 \pm 5 \qquad \text{Add 3 to both sides}$$
$$x = \frac{3 \pm 5}{2} \qquad \text{Divide both sides by 2}$$

The last equation can be written as two separate statements:

$$x = \frac{3 + 5}{2} \quad \text{or} \quad x = \frac{3 - 5}{2}$$
$$= \frac{8}{2} \qquad\qquad = \frac{-2}{2}$$
$$= 4 \qquad\qquad \text{or} \qquad = -1$$

The solution set is $\{4, -1\}$.

Notice that we could have solved the equation in Example 1 by expanding the left side, writing the resulting equation in standard form, and then factoring. The problem would look like this:

$$(2x - 3)^2 = 25 \qquad \text{Original equation}$$
$$4x^2 - 12x + 9 = 25 \qquad \text{Expand the left side}$$
$$4x^2 - 12x - 16 = 0 \qquad \text{Add } -25 \text{ to each side}$$
$$4(x^2 - 3x - 4) = 0 \qquad \text{Begin factoring}$$
$$4(x - 4)(x + 1) = 0 \qquad \text{Factor completely}$$
$$x - 4 = 0 \quad \text{or} \quad x + 1 = 0 \qquad \text{Set variable factors equal to 0}$$
$$x = 4 \quad \text{or} \quad x = -1$$

As you can see, solving the equation by factoring leads to the same two solutions.

EXAMPLE 2 Solve $(3x - 1)^2 = -12$ for x.

SOLUTION $(3x - 1)^2 = -12$

$$3x - 1 = \pm\sqrt{-12} \qquad \text{Theorem 1}$$
$$3x - 1 = \pm 2i\sqrt{3} \qquad \sqrt{-12} = 2i\sqrt{3}$$
$$3x = 1 \pm 2i\sqrt{3} \qquad \text{Add 1 to both sides}$$
$$x = \frac{1 \pm 2i\sqrt{3}}{3} \qquad \text{Divide both sides by 3}$$

The solution set is $\left\{ \dfrac{1 + 2i\sqrt{3}}{3}, \dfrac{1 - 2i\sqrt{3}}{3} \right\}$.

Both solutions are complex. Here is a check of the first solution:

$$\text{When} \rightarrow \qquad x = \frac{1 + 2i\sqrt{3}}{3}$$
$$\text{the equation} \rightarrow \qquad (3x - 1)^2 = -12$$
$$\text{becomes} \rightarrow \qquad \left(3 \cdot \frac{1 + 2i\sqrt{3}}{3} - 1 \right)^2 \overset{?}{=} -12$$
$$(1 + 2i\sqrt{3} - 1)^2 \overset{?}{=} -12$$
$$(2i\sqrt{3})^2 \overset{?}{=} -12$$
$$4 \cdot i^2 \cdot 3 \overset{?}{=} -12$$
$$12(-1) \overset{?}{=} -12$$
$$-12 = -12$$

> **NOTE**
> We cannot solve the equation in Example 2 by factoring. If we expand the left side and write the resulting equation in standard form, we are left with a quadratic equation that does not factor:
> $(3x - 1)^2 = -12$
>   Equation from Example 2
> $9x^2 - 6x + 1 = -12$
>   Expand the left side
> $9x^2 - 6x + 13 = 0$
>   Standard form, but not factorable

EXAMPLE 3 Solve $x^2 + 6x + 9 = 12$.

SOLUTION We can solve this equation as we have the equations in Examples 1 and 2 if we first write the left side as $(x + 3)^2$.

$$x^2 + 6x + 9 = 12 \qquad \text{Original equation}$$
$$(x + 3)^2 = 12 \qquad \text{Write } x^2 + 6x + 9 \text{ as } (x + 3)^2$$
$$x + 3 = \pm 2\sqrt{3} \qquad \text{Theorem 1}$$
$$x = -3 \pm 2\sqrt{3} \qquad \text{Add } -3 \text{ to each side}$$

We have two irrational solutions: $-3 + 2\sqrt{3}$ and $-3 - 2\sqrt{3}$. What is important about this problem, however, is the fact that the equation was easy to solve because the left side was a perfect square trinomial.

B Completing the Square

The method of completing the square is simply a way of transforming any quadratic equation into an equation of the form found in the preceding three examples.

The key to understanding the method of completing the square lies in recognizing the relationship between the last two terms of any perfect square trinomial whose leading coefficient is 1.

Consider the following list of perfect square trinomials and their corresponding binomial squares:

$$x^2 - 6x + 9 = (x - 3)^3$$

$$x^2 + 8x + 16 = (x + 4)^2$$

$$x^2 - 10x + 25 = (x - 5)^2$$

$$x^2 + 12x + 36 = (x + 6)^2$$

In each case, the leading coefficient is 1. A more important observation comes from noticing the relationship between the linear and constant terms (middle and last terms) in each trinomial. Observe that the constant term in each case is the square of half the coefficient of x in the middle term. For example, in the last expression, the constant term 36 is the square of half of 12, where 12 is the coefficient of x in the middle term. (Notice also that the second terms in all the binomials on the right side are half the coefficients of the middle terms of the trinomials on the left side.) We can use these observations to build our own perfect square trinomials and, in doing so, solve some quadratic equations.

Consider the following equation:

$$x^2 + 6x = 3$$

We can think of the left side as having the first two terms of a perfect square trinomial. We need only add the correct constant term. If we take half the coefficient of x, we get 3. If we then square this quantity, we have 9. Adding the 9 to both sides, the equation becomes

$$x^2 + 6x + \mathbf{9} = 3 + \mathbf{9}$$

The left side is the perfect square $(x + 3)^2$; the right side is 12.

$$(x + 3)^2 = 12$$

The equation is now in the correct form. We can apply Theorem 1 and finish the solution.

$$(x + 3)^2 = 12$$
$$x + 3 = \pm\sqrt{12} \qquad \text{Theorem 1}$$
$$x + 3 = \pm 2\sqrt{3}$$
$$x = -3 \pm 2\sqrt{3}$$

The solution set is $\{-3 + 2\sqrt{3}, -3 - 2\sqrt{3}\}$. The method just used is called *completing the square* since we complete the square on the left side of the original equation by adding the appropriate constant term.

EXAMPLE 4 Solve $x^2 + 5x - 2 = 0$ by completing the square.

SOLUTION We must begin by adding 2 to both sides. (The left side of the equation, as it is, is not a perfect square because it does not have the correct constant term. We will simply "move" that term to the other side and use our own constant term.)

$$x^2 + 5x = 2 \qquad \text{Add 2 to each side}$$

We complete the square by adding the square of half the coefficient of the linear term to both sides.

$$x^2 + 5x + \frac{\mathbf{25}}{\mathbf{4}} = 2 + \frac{\mathbf{25}}{\mathbf{4}} \qquad \text{Half of 5 is } \frac{5}{2} \text{, the square of which is } \frac{25}{4}$$

$$\left(x + \frac{5}{2}\right)^2 = \frac{33}{4} \qquad 2 + \frac{25}{4} = \frac{8}{4} + \frac{25}{4} = \frac{33}{4}$$

$$x + \frac{5}{2} = \pm\sqrt{\frac{33}{4}} \qquad \text{Theorem 1}$$

$$= \pm\frac{\sqrt{33}}{2} \qquad \text{Simplify the radical}$$

$$x = -\frac{5}{2} \pm \frac{\sqrt{33}}{2} \qquad \text{Add } -\frac{5}{2} \text{ to both sides}$$

$$= \frac{-5 \pm \sqrt{33}}{2}$$

The solution set is $\left\{ \dfrac{-5 + \sqrt{33}}{2}, \dfrac{-5 - \sqrt{33}}{2} \right\}$.

NOTE
We can use a calculator to get decimal approximations to these solutions. If $\sqrt{33} \approx 5.74$, then

$$\frac{-5 + 5.74}{2} = 0.37$$

$$\frac{-5 - 5.74}{2} = -5.37$$

EXAMPLE 5 Solve $3x^2 - 8x + 7 = 0$ for x.

SOLUTION We begin by adding -7 to both sides.

$$3x^2 - 8x + 7 = 0$$

$$3x^2 - 8x = -7 \qquad \text{Add } -7 \text{ to both sides}$$

We cannot complete the square on the left side because the leading coefficient is not 1. We take an extra step and divide both sides by 3:

$$\frac{3x^2}{3} - \frac{8x}{3} = -\frac{7}{3}$$

$$x^2 - \frac{8}{3}x = -\frac{7}{3}$$

Half of $\frac{8}{3}$ is $\frac{4}{3}$, the square of which is $\frac{16}{9}$.

$$x^2 - \frac{8}{3}x + \frac{\mathbf{16}}{\mathbf{9}} = -\frac{7}{3} + \frac{\mathbf{16}}{\mathbf{9}} \qquad \text{Add } \frac{16}{9} \text{ to both sides}$$

$$\left(x - \frac{4}{3}\right)^2 = -\frac{5}{9} \qquad \text{Simplify right side}$$

$$x - \frac{4}{3} = \pm\sqrt{-\frac{5}{9}} \qquad \text{Theorem 1}$$

$$x - \frac{4}{3} = \pm\frac{i\sqrt{5}}{3} \qquad \sqrt{-\frac{5}{9}} = \frac{\sqrt{-5}}{3} = \frac{i\sqrt{5}}{3}$$

$$x = \frac{4}{3} \pm \frac{i\sqrt{5}}{3} \qquad \text{Add } \frac{4}{3} \text{ to both sides}$$

$$x = \frac{4 \pm i\sqrt{5}}{3}$$

The solution set is $\left\{ \dfrac{4 + i\sqrt{5}}{3}, \dfrac{4 - i\sqrt{5}}{3} \right\}$.

Strategy To Solve a Quadratic Equation by Completing the Square

To summarize the method used in the preceding two examples, we list the following steps:

Step 1: Write the equation in the form $ax^2 + bx = c$.

Step 2: If the leading coefficient is not 1, divide both sides by the coefficient so that the resulting equation has a leading coefficient of 1; that is, if $a \neq 1$, then divide both sides by a.

Step 3: Add the square of half the coefficient of the linear term to both sides of the equation.

Step 4: Write the left side of the equation as the square of a binomial, and simplify the right side if possible.

Step 5: Apply Theorem 1, and solve as usual.

C Quadratic Equations and Right Triangles

FACTS FROM GEOMETRY: MORE SPECIAL TRIANGLES

The triangles shown in Figures 1 and 2 occur frequently in mathematics.

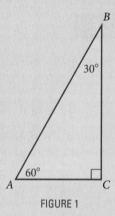

FIGURE 1

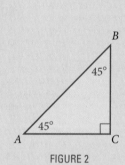

FIGURE 2

Note that both of the triangles are right triangles. We refer to the triangle in Figure 1 as a 30°–60°–90° triangle and the triangle in Figure 2 as a 45°–45°–90° triangle.

EXAMPLE 6 If the shortest side in a 30°–60°–90° triangle is 1 inch, find the lengths of the other two sides.

SOLUTION In Figure 3, triangle ABC is a 30°–60°–90° triangle in which the shortest side AC is 1 inch long. Triangle DBC is also a 30°–60°–90° triangle in which the shortest side DC is 1 inch long.

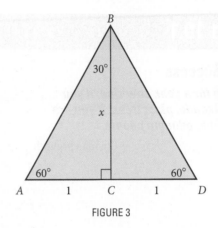

FIGURE 3

Notice that the large triangle ABD is an equilateral triangle because each of its interior angles is 60°. Each side of triangle ABD is 2 inches long. Side AB in triangle ABC is therefore 2 inches. To find the length of side BC, we use the Pythagorean theorem.

$$BC^2 + AC^2 = AB^2$$
$$x^2 + 1^2 = 2^2$$
$$x^2 + 1 = 4$$
$$x^2 = 3$$
$$x = \sqrt{3} \text{ inches}$$

Note that we write only the positive square root because x is the length of a side in a triangle and is therefore a positive number.

EXAMPLE 7   Table 1 in the introduction to this section gives the vertical rise of the Forest Double chair lift as 1,170 feet and the length of the chair lift as 5,750 feet. To the nearest foot, find the horizontal distance covered by a person riding this lift.

SOLUTION   Figure 4 is a model of the Forest Double chair lift. A rider gets on the lift at point A and exits at point B. The length of the lift is AB.

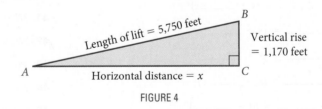

FIGURE 4

To find the horizontal distance covered by a person riding the chair lift, we use the Pythagorean theorem.

$5,750^2 = x^2 + 1,170^2$	Pythagorean theorem
$33,062,500 = x^2 + 1,368,900$	Simplify squares
$x^2 = 33,062,500 - 1,368,900$	Solve for x^2
$x^2 = 31,693,600$	Simplify the right side
$x = \sqrt{31,693,600}$	Theorem 1
$= 5,630 \text{ feet} \quad \text{(to the nearest foot)}$	

A rider getting on the lift at point A and riding to point B will cover a horizontal distance of approximately 5,630 feet.

Problem Set 10.1

Moving Toward Success

"If you are planning for a year, sow rice; if you are planning for a decade, plant trees; if you are planning for a lifetime, educate people."

—Chinese proverb

1. If you focus solely on facts, you may not retain the information as well as if you focus on main ideas and concepts. Why is this true?

2. Make a conscious decision to be interested in what the instructor is teaching or what you're studying. How will this increase your success in this class?

A Solve the following equations. [Examples 1–3]

1. $x^2 = 25$
2. $x^2 = 16$
3. $y^2 = \dfrac{3}{4}$
4. $y^2 = \dfrac{5}{9}$
5. $x^2 + 12 = 0$
6. $x^2 + 8 = 0$
7. $4a^2 - 45 = 0$
8. $9a^2 - 20 = 0$
9. $(2y - 1)^2 = 25$
10. $(3y + 7)^2 = 1$
11. $(2a + 3)^2 = -9$
12. $(3a - 5)^2 = -49$
13. $x^2 + 8x + 16 = -27$
14. $x^2 - 12x + 36 = -8$
15. $4a^2 - 12a + 9 = -4$
16. $9a^2 - 12a + 4 = -9$

B Copy each of the following, and fill in the blanks so that the left side of each is a perfect square trinomial; that is, complete the square.

17. $x^2 + 12x + \underline{\hphantom{xxx}} = (x + \underline{\hphantom{xxx}})^2$
18. $x^2 + 6x + \underline{\hphantom{xxx}} = (x + \underline{\hphantom{xxx}})^2$
19. $x^2 - 4x + \underline{\hphantom{xxx}} = (x - \underline{\hphantom{xxx}})^2$
20. $x^2 - 2x + \underline{\hphantom{xxx}} = (x - \underline{\hphantom{xxx}})^2$
21. $a^2 - 10a + \underline{\hphantom{xxx}} = (a - \underline{\hphantom{xxx}})^2$
22. $a^2 - 8a + \underline{\hphantom{xxx}} = (a - \underline{\hphantom{xxx}})^2$
23. $x^2 + 5x + \underline{\hphantom{xxx}} = (x + \underline{\hphantom{xxx}})^2$
24. $x^2 + 3x + \underline{\hphantom{xxx}} = (x + \underline{\hphantom{xxx}})^2$
25. $y^2 - 7y + \underline{\hphantom{xxx}} = (y - \underline{\hphantom{xxx}})^2$
26. $y^2 - y + \underline{\hphantom{xxx}} = (y - \underline{\hphantom{xxx}})^2$
27. $x^2 + \dfrac{1}{2}x + \underline{\hphantom{xxx}} = (x + \underline{\hphantom{xxx}})^2$
28. $x^2 - \dfrac{3}{4}x + \underline{\hphantom{xxx}} = (x - \underline{\hphantom{xxx}})^2$
29. $x^2 + \dfrac{2}{3}x + \underline{\hphantom{xxx}} = (x + \underline{\hphantom{xxx}})^2$
30. $x^2 - \dfrac{4}{5}x + \underline{\hphantom{xxx}} = (x - \underline{\hphantom{xxx}})^2$

B Solve each of the following quadratic equations by completing the square. [Examples 4–5]

31. $x^2 + 12x = -27$
32. $x^2 - 6x = 16$
33. $a^2 - 2a + 5 = 0$
34. $a^2 + 10a + 22 = 0$
35. $y^2 - 8y + 1 = 0$
36. $y^2 + 6y - 1 = 0$
37. $x^2 - 5x - 3 = 0$
38. $x^2 - 5x - 2 = 0$
39. $2x^2 - 4x - 8 = 0$
40. $3x^2 - 9x - 12 = 0$
41. $3t^2 - 8t + 1 = 0$
42. $5t^2 + 12t - 1 = 0$
43. $4x^2 - 3x + 5 = 0$
44. $7x^2 - 5x + 2 = 0$
45. For the equation $x^2 = -9$:
    a. Can it be solved by factoring?
    b. Solve it.
46. For the equation $x^2 - 10x + 18 = 0$:
    a. Can it be solved by factoring?
    b. Solve it.
47. Solve each equation below by the indicated method.
    a. $x^2 - 6x = 0$ Factoring.
    b. $x^2 - 6x = 0$ Completing the square.

48. Solve each equation below by the indicated method.

    a. $x^2 + ax = 0$ Factoring.

    b. $x^2 + ax = 0$ Completing the square.

49. Solve the equation $x^2 + 2x = 35$

    a. by factoring.

    b. by completing the square.

50. Solve the equation $8x^2 - 10x - 25 = 0$

    a. by factoring.

    b. by completing the square.

51. Is $x = -3 + \sqrt{2}$ a solution to $x^2 - 6x = 7$?

52. Is $x = 2 - \sqrt{5}$ a solution to $x^2 - 4x = 1$?

The next two problems will give you practice solving a variety of equations.

53. Solve each equation.

    a. $5x - 7 = 0$

    b. $5x - 7 = 8$

    c. $(5x - 7)^2 = 8$

    d. $\sqrt{5x - 7} = 0$

    e. $\dfrac{5}{2} - \dfrac{7}{2x} = \dfrac{4}{x}$

54. Solve each equation.

    a. $5x + 11 = 0$

    b. $5x + 11 = 9$

    c. $(5x + 11)^2 = 9$

    d. $\sqrt{5x + 11} = 9$

    e. $\dfrac{5}{3} - \dfrac{11}{3x} = \dfrac{3}{x}$

Simplify the left side of each equation, and then solve for x.

55. $(x + 5)^2 + (x - 5)^2 = 52$

56. $(2x + 1)^2 + (2x - 1)^2 = 10$

57. $(2x + 3)^2 + (2x - 3)^2 = 26$

58. $(3x + 2)^2 + (3x - 2)^2 = 26$

59. $(3x + 4)(3x - 4) - (x + 2)(x - 2) = -4$

60. $(5x + 2)(5x - 2) - (x + 3)(x - 3) = 29$

61. Fill in the table below given the following functions.
$f(x) = (2x - 3)^2, g(x) = 4x^2 - 12x + 9, h(x) = 4x^2 + 9$

x	f(x)	g(x)	h(x)
-2			
-1			
0			
1			
2			

62. Fill in the table below given the following functions.
$f(x) = \left(x + \dfrac{1}{2}\right)^2, g(x) = x^2 + x + \dfrac{1}{4}, h(x) = x^2 + \dfrac{1}{4}$

x	f(x)	g(x)	h(x)
-2			
-1			
0			
1			
2			

63. If $f(x) = (x - 3)^2$ find x if $f(x) = 0$.

64. If $f(x) = (3x - 5)^2$ find x if $f(x) = 0$.

65. If $f(x) = x^2 - 5x - 6$ find

    a. The x-intercepts

    b. The value of x for which $f(x) = 0$

    c. $f(0)$

    d. $f(1)$

66. If $f(x) = 9x^2 - 12x + 4$ find

    a. The x-intercepts

    b. The value of x for which $f(x) = 0$

    c. $f(0)$

    d. $f(1)$

C **Applying the Concepts** [Examples 6–7]

67. Geometry If the shortest side in a 30°–60°–90° triangle is $\dfrac{1}{2}$ inch long, find the lengths of the other two sides.

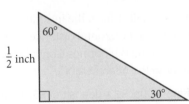

68. Geometry If the length of the longest side of a 30°–60°–90° triangle is x, find the length of the other two sides in terms of x

69. Geometry If the length of the shorter sides of a 45°–45°–90° triangle is 1 inch, find the length of the hypotenuse.

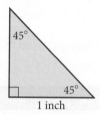

45°

45°

1 inch

70. Geometry If the length of the shorter sides of a 45°–45°–90° triangle is x, find the length of the hypotenuse, in terms of x.

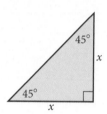

45°

x

45°

x

71. Chair Lift Use Table 1 from the introduction to this section to find the horizontal distance covered by a person riding the Bear Paw Double chair lift. Round your answer to the nearest foot.

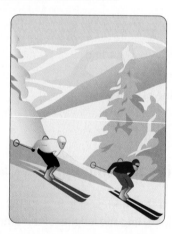

72. Chair Lift Use Table 1 from the introduction to this section to find the horizontal distance covered by a person riding the Big Springs Gondola lift. Round your answer to the nearest foot.

73. Interest Rate Suppose a deposit of $3,000 in a savings account that paid an annual interest rate r (compounded yearly) is worth $3,456 after 2 years. Using the formula $A = P(1 + r)^t$, we have

$$3{,}456 = 3{,}000(1 + r)^2$$

Solve for r to find the annual interest rate.

74. Special Triangles In Figure 5, triangle ABC has angles 45° and 30° and height x. Find the lengths of sides AB, BC, and AC, in terms of x.

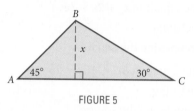

B

x

A 45° 30° C

FIGURE 5

75. Length of an Escalator An escalator in a department store is to carry people a vertical distance of 20 feet between floors. How long is the escalator if it makes an angle of 45° with the ground? (See Figure 6.)

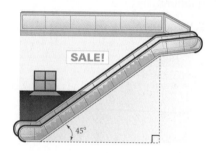

SALE!

45°

FIGURE 6

76. Dimensions of a Tent A two-person tent is to be made so the height at the center is 4 feet. If the sides of the tent are to meet the ground at an angle of 60° and the tent is to be 6 feet in length, how many square feet of material will be needed to make the tent? (Figure 7, assume that the tent has a floor and is closed at both ends.) Give your answer to the nearest tenth of a square foot.

4 ft

60°

6 ft

FIGURE 7

Maintaining Your Skills

Write each of the following in simplified form for radicals.

77. $\sqrt{45}$ **78.** $\sqrt{24}$

79. $\sqrt{27y^5}$ **80.** $\sqrt{8y^3}$

81. $\sqrt[3]{54x^6y^5}$ **82.** $\sqrt[3]{16x^9y^7}$

Rationalize the denominator.

83. $\dfrac{3}{\sqrt{2}}$ **84.** $\dfrac{5}{\sqrt{3}}$

85. $\dfrac{2}{\sqrt[3]{4}}$ **86.** $\dfrac{3}{\sqrt[3]{2}}$

Getting Ready for the Next Section

Simplify.

87. $\sqrt{49 - 4(6)(-5)}$ **88.** $\sqrt{49 - 4(6)2}$

89. $\sqrt{(-27)^2 - 4(0.1)(1{,}700)}$

90. $\sqrt{25 - 4(4)(-10)}$

91. $\dfrac{-7 + \sqrt{169}}{12}$ **92.** $\dfrac{-7 - \sqrt{169}}{12}$

93. Simplify $\sqrt{b^2 - 4ac}$ when $a = 6$, $b = 7$, and $c = -5$.

94. Simplify $\sqrt{b^2 - 4ac}$ when $a = 2$, $b = -6$, and $c = 3$.

Factor.

95. $27t^3 - 8$

96. $125t^3 + 1$

Extending the Concepts

Solve for x.

97. $(x + a)^2 + (x - a)^2 = 10a^2$

98. $(ax + 1)^2 + (ax - 1)^2 = 10$

Assume p and q are positive numbers and solve for x by completing the square on x.

99. $x^2 + px + q = 0$

100. $x^2 - px + q = 0$

101. $3x^2 + px + q = 0$

102. $3x^2 + 2px + q = 0$

Complete the square on x and also on y so that each equation below is written in the form
$$(x - a)^2 + (y - b)^2 = r^2$$
which you will see later in the book as the equation of a circle with center (a, b) and radius r.

103. $x^2 - 10x + y^2 - 6y = -30$

104. $x^2 - 2x + y^2 - 4y = 20$

The Quadratic Formula

OBJECTIVES

A Solve quadratic equations by the quadratic formula.

B Solve application problems using quadratic equations.

TICKET TO SUCCESS

Keep these questions in mind as you read through the section. Then respond in your own words and in complete sentences.

1. What is the quadratic formula?

2. Can you use the quadratic formula to solve an equation that is not quadratic? Explain.

3. When would factoring be a faster method than using the quadratic formula to solve a quadratic equation? When would it not?

4. When will the quadratic formula result in only one solution?

Image copyright ©Voyagerix, 2009. Used under license from Shutterstock.com

Imagine you and a friend are throwing rocks 30 feet down a well to the water below. You throw the rock with an initial velocity of 10 feet per second and your friend throws with an initial velocity of 7 feet per second. The relationships between distance the rocks travel *s* and the time *t* is given by

$$s = 10t + 16t^2 \quad \text{Your rock}$$

$$s = 7t + 16t^2 \quad \text{Your friend's rock}$$

How much sooner will your rock splash than your friend's? Equations such as these require the use of the quadratic formula. In this section, we will use the method of completing the square from the preceding section to derive the quadratic formula. The *quadratic formula* is a very useful tool in mathematics. It allows us to solve all types of quadratic equations.

A The Quadratic Theorem

The Quadratic Theorem

For any quadratic equation in the form $ax^2 + bx + c = 0$, where $a \neq 0$, the two solutions are

$$x = \frac{-b + \sqrt{b^2 - 4ac}}{2a} \quad \text{and} \quad x = \frac{-b - \sqrt{b^2 - 4ac}}{2a}$$

PROOF We will prove the quadratic theorem by completing the square on $ax^2 + bx + c = 0$.

$$ax^2 + bx + c = 0$$

$$ax^2 + bx = -c \qquad \text{Add } -c \text{ to both sides}$$

$$x^2 + \frac{b}{a}x = -\frac{c}{a} \qquad \text{Divide both sides by } a$$

To complete the square on the left side, we add the square of $\frac{1}{2}$ of $\frac{b}{a}$ to both sides $\left(\frac{1}{2} \text{ of } \frac{b}{a} \text{ is } \frac{b}{2a}\right)$.

$$x^2 + \frac{b}{a}x + \left(\frac{b}{2a}\right)^2 = -\frac{c}{a} + \left(\frac{b}{2a}\right)^2$$

We now simplify the right side as a separate step. We square the second term and combine the two terms by writing each with the least common denominator $4a^2$.

$$-\frac{c}{a} + \left(\frac{b}{2a}\right)^2 = -\frac{c}{a} + \frac{b^2}{4a^2} = \frac{\mathbf{4a}}{\mathbf{4a}}\left(\frac{-c}{a}\right) + \frac{b^2}{4a^2} = \frac{-4ac + b^2}{4a^2}$$

It is convenient to write this last expression as

$$\frac{b^2 - 4ac}{4a^2}$$

Continuing with the proof, we have

$$x^2 + \frac{b}{a}x + \left(\frac{b}{2a}\right)^2 = \frac{b^2 - 4ac}{4a^2}$$

$$\left(x + \frac{b}{2a}\right)^2 = \frac{b^2 - 4ac}{4a^2} \qquad \begin{array}{l}\text{Write left side as a} \\ \text{binomial square}\end{array}$$

$$x + \frac{b}{2a} = \pm\frac{\sqrt{b^2 - 4ac}}{2a} \qquad \text{Theorem 1}$$

$$x = -\frac{b}{2a} \pm \frac{\sqrt{b^2 - 4ac}}{2a} \qquad \text{Add } -\frac{b}{2a} \text{ to both sides}$$

$$= \frac{-b \pm \sqrt{b^2 - 4ac}}{2a}$$

Our proof is now complete. What we have is this: if our equation is in the form $ax^2 + bx + c = 0$ (standard form), where $a \neq 0$, the two solutions are always given by the formula

$$x = \frac{-b \pm \sqrt{b^2 - 4ac}}{2a}$$

This formula is known as the *quadratic formula*. If we substitute the coefficients a, b, and c of any quadratic equation in standard form into the formula, we need only perform some basic arithmetic to arrive at the solution set.

EXAMPLE 1   Use the quadratic formula to solve $6x^2 + 7x - 5 = 0$.

SOLUTION   Using $a = 6$, $b = 7$, and $c = -5$ in the formula

$$x = \frac{-b \pm \sqrt{b^2 - 4ac}}{2a}$$

$$\text{we have} \rightarrow x = \frac{-7 \pm \sqrt{49 - 4(6)(-5)}}{2(6)}$$

$$x = \frac{-7 \pm \sqrt{49 + 120}}{12}$$

$$= \frac{-7 \pm \sqrt{169}}{12}$$

$$= \frac{-7 \pm 13}{12}$$

We separate the last equation into the two statements

$$x = \frac{-7 + 13}{12} \quad \text{or} \quad x = \frac{-7 - 13}{12}$$

$$x = \frac{1}{2} \quad \text{or} \quad x = -\frac{5}{3}$$

The solution set is $\left\{ \frac{1}{2}, -\frac{5}{3} \right\}$.

Whenever the solutions to a quadratic equation are rational numbers, as they are in Example 1, it means that the original equation was solvable by factoring. To illustrate, let's solve the equation from Example 1 again, but this time by factoring.

$$6x^2 + 7x - 5 = 0 \qquad \text{Equation in standard form}$$

$$(3x + 5)(2x - 1) = 0 \qquad \text{Factor the left side}$$

$$3x + 5 = 0 \quad \text{or} \quad 2x - 1 = 0 \qquad \text{Set factors equal to 0}$$

$$x = -\frac{5}{3} \quad \text{or} \quad x = \frac{1}{2}$$

When an equation can be solved by factoring, then factoring is usually the faster method of solution. It is best to try to factor first, and then if you have trouble factoring, go to the quadratic formula. It always works.

EXAMPLE 2 Solve $\dfrac{x^2}{3} - x = -\dfrac{1}{2}$.

SOLUTION Multiplying through by 6 and writing the result in standard form, we have

$$2x^2 - 6x + 3 = 0$$

the left side of which is not factorable. Therefore, we use the quadratic formula with $a = 2$, $b = -6$, and $c = 3$. The two solutions are given by

$$x = \frac{-(-6) \pm \sqrt{36 - 4(2)(3)}}{2(2)}$$

$$= \frac{6 \pm \sqrt{12}}{4}$$

$$= \frac{6 \pm 2\sqrt{3}}{4} \qquad \sqrt{12} = \sqrt{4 \cdot 3} = \sqrt{4}\,\sqrt{3} = 2\sqrt{3}$$

We can reduce this last expression to lowest terms by factoring 2 from the numerator and denominator and then dividing the numerator and denominator by 2.

$$x = \frac{2(3 \pm \sqrt{3})}{2 \cdot 2} = \frac{3 \pm \sqrt{3}}{2}$$

EXAMPLE 3 Solve $\dfrac{1}{x + 2} - \dfrac{1}{x} = \dfrac{1}{3}$.

SOLUTION To solve this equation, we must first put it in standard form. To do so, we must clear the equation of fractions by multiplying each side by the LCD for all the denominators, which is $3x(x + 2)$. Multiplying both sides by the LCD, we have

$$3x(x + 2)\left(\frac{1}{x + 2} - \frac{1}{x} \right) = \frac{1}{3} \cdot 3x(x + 2) \qquad \text{Multiply each by the LCD}$$

$$3x(x+2) \cdot \frac{1}{x+2} - 3x(x+2) \cdot \frac{1}{x} = \frac{1}{3} \cdot 3x(x+2)$$

$$3x - 3(x+2) = x(x+2)$$

$$3x - 3x - 6 = x^2 + 2x \qquad \text{Multiplication}$$

$$-6 = x^2 + 2x \qquad \text{Simplify left side}$$

$$0 = x^2 + 2x + 6 \qquad \text{Add 6 to each side}$$

Since the right side of our last equation is not factorable, we use the quadratic formula. From our last equation, we have $a = 1$, $b = 2$, and $c = 6$. Using these numbers for a, b, and c in the quadratic formula gives us

$$x = \frac{-2 \pm \sqrt{4 - 4(1)(6)}}{2(1)}$$

$$= \frac{-2 \pm \sqrt{4 - 24}}{2} \qquad \text{Simplify inside the radical}$$

$$= \frac{-2 \pm \sqrt{-20}}{2} \qquad 4 - 24 = -20$$

$$= \frac{-2 \pm 2i\sqrt{5}}{2} \qquad \sqrt{-20} = i\sqrt{20} = i\sqrt{4}\,\sqrt{5} = 2i\sqrt{5}$$

$$= \frac{2(-1 \pm i\sqrt{5})}{2} \qquad \text{Factor 2 from the numerator}$$

$$= -1 \pm i\sqrt{5} \qquad \text{Divide numerator and denominator by 2}$$

Since neither of the two solutions, $-1 + i\sqrt{5}$ nor $-1 - i\sqrt{5}$, will make any of the denominators in our original equation 0, they are both solutions.

Although the equation in our next example is not a quadratic equation, we solve it by using both factoring and the quadratic formula.

EXAMPLE 4   Solve $27t^3 - 8 = 0$.

SOLUTION   It would be a mistake to add 8 to each side of this equation and then take the cube root of each side because we would lose two of our solutions. Instead, we factor the left side, and then set the factors equal to 0.

$$27t^3 - 8 = 0 \qquad \text{Equation in standard form}$$

$$(3t - 2)(9t^2 + 6t + 4) = 0 \qquad \text{Factor as the difference of two cubes}$$

$$3t - 2 = 0 \quad \text{or} \quad 9t^2 + 6t + 4 = 0 \qquad \text{Set each factor equal to 0}$$

The first equation leads to a solution of $t = \frac{2}{3}$. The second equation does not factor, so we use the quadratic formula with $a = 9$, $b = 6$, and $c = 4$:

$$t = \frac{-6 \pm \sqrt{36 - 4(9)(4)}}{2(9)}$$

$$= \frac{-6 \pm \sqrt{36 - 144}}{18}$$

$$= \frac{-6 \pm \sqrt{-108}}{18}$$

$$= \frac{-6 \pm 6i\sqrt{3}}{18} \qquad \sqrt{-108} = i\sqrt{36 \cdot 3} = 6i\sqrt{3}$$

$$= \frac{6(-1 \pm i\sqrt{3})}{6 \cdot 3} \qquad \text{Factor 6 from the numerator and denominator}$$

$$= \frac{-1 \pm i\sqrt{3}}{3} \qquad \text{Divide out common factor 6}$$

The three solutions to our original equation are

$$\frac{2}{3}, \qquad \frac{-1 + i\sqrt{3}}{3}, \qquad \text{and} \qquad \frac{-1 - i\sqrt{3}}{3}$$

B Applications

EXAMPLE 5 If an object is thrown downward with an initial velocity of 20 feet per second, the distance $s(t)$, in feet, it travels in t seconds is given by the function $s(t) = 20t + 16t^2$. How long does it take the object to fall 40 feet?

SOLUTION We let $s(t) = 40$, and solve for t.

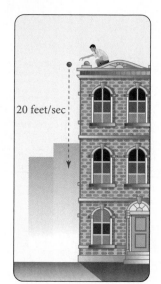

20 feet/sec

$$\text{When} \rightarrow \qquad s(t) = 40$$
$$\text{the function} \rightarrow \quad s(t) = 20t + 16t^2$$
$$\text{becomes} \rightarrow \qquad 40 = 20t + 16t^2$$
$$16t^2 + 20t - 40 = 0$$
$$4t^2 + 5t - 10 = 0 \qquad \text{Divide by 4}$$

Using the quadratic formula, we have

$$t = \frac{-5 \pm \sqrt{25 - 4(4)(-10)}}{2(4)}$$

$$= \frac{-5 \pm \sqrt{185}}{8}$$

$$= \frac{-5 + \sqrt{185}}{8} \qquad \text{or} \qquad \frac{-5 - \sqrt{185}}{8}$$

The second solution is impossible since it is a negative number and time t must be positive. It takes

$$t = \frac{-5 + \sqrt{185}}{8} \qquad \text{or approximately} \qquad \frac{-5 + 13.60}{8} \approx 1.08 \text{ seconds}$$

for the object to fall 40 feet.

Recall that the relationship between profit, revenue, and cost is given by the formula

$$P(x) = R(x) - C(x)$$

where $P(x)$ is the profit, $R(x)$ is the total revenue, and $C(x)$ is the total cost of producing and selling x items.

EXAMPLE 6 A company produces and sells copies of an accounting program for home computers. The total weekly cost (in dollars) to produce x copies of the program is $C(x) = 8x + 500$, and the weekly revenue for selling all x copies of the program is $R(x) = 35x - 0.1x^2$. How many programs must be sold each week for the weekly profit to be $1,200?

SOLUTION Substituting the given expressions for $R(x)$ and $C(x)$ in the equation $P(x) = R(x) - C(x)$, we have a polynomial in x that represents the weekly profit $P(x)$.

$$P(x) = R(x) - C(x)$$
$$= 35x - 0.1x^2 - (8x + 500)$$
$$= 35x - 0.1x^2 - 8x - 500$$
$$= -500 + 27x - 0.1x^2$$

By setting this expression equal to 1,200, we have a quadratic equation to solve that gives us the number of programs x that need to be sold each week to bring in a profit of $1,200.

$$1,200 = -500 + 27x - 0.1x^2$$

We can write this equation in standard form by adding the opposite of each term on the right side of the equation to both sides of the equation. Doing so produces the following equation:

$$0.1x^2 - 27x + 1,700 = 0$$

Applying the quadratic formula to this equation with $a = 0.1$, $b = -27$, and $c = 1,700$, we have

$$x = \frac{27 \pm \sqrt{(-27)^2 - 4(0.1)(1,700)}}{2(0.1)}$$
$$= \frac{27 \pm \sqrt{729 - 680}}{0.2}$$
$$= \frac{27 \pm \sqrt{49}}{0.2}$$
$$= \frac{27 \pm 7}{0.2}$$

Writing this last expression as two separate expressions, we have our two solutions:

$$x = \frac{27 + 7}{0.2} \quad \text{or} \quad x = \frac{27 - 7}{0.2}$$
$$= \frac{34}{0.2} \qquad\qquad = \frac{20}{0.2}$$
$$= 170 \qquad\qquad\quad = 100$$

The weekly profit will be $1,200 if the company produces and sells 100 programs or 170 programs.

What is interesting about the equation we solved in Example 6 is that it has rational solutions, meaning it could have been solved by factoring. But looking back at the equation, factoring does not seem like a reasonable method of solution because the coefficients are either very large or very small. So, there are times when using the quadratic formula is a faster method of solution, even though the equation you are solving is factorable.

USING TECHNOLOGY

Graphing Calculators

More About Example 5

We can solve the problem discussed in Example 5 by graphing the function $Y_1 = 20X + 16X^2$ in a window with X from 0 to 2 (because X is taking the place of t and we know t is a positive quantity) and Y from 0 to 50 (because we are looking for X when Y_1 is 40). Graphing Y_1 gives a graph similar to the graph in Figure 1. Using the Zoom and Trace features at $Y_1 = 40$ gives us X = 1.08 to the nearest hundredth, matching the results we obtained by solving the original equation algebraically.

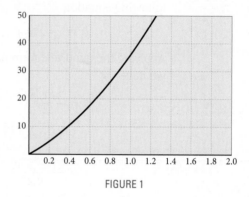

FIGURE 1

More About Example 6

To visualize the functions in Example 6, we set up our calculator this way:

$Y_1 = 35X - 0.1X^2$ Revenue function
$Y_2 = 8X + 500$ Cost function
$Y_3 = Y_1 - Y_2$ Profit function
Window: X from 0 to 350, Y from 0 to 3500

Graphing these functions produces graphs similar to the ones shown in Figure 2. The lower graph is the graph of the profit function. Using the Zoom and Trace features on the lower graph at $Y_3 = 1,200$ produces two corresponding values of X, 170 and 100, which match the results in Example 6.

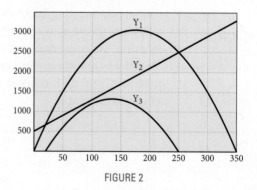

FIGURE 2

We will continue this discussion of the relationship between graphs of functions and solutions to equations in the Using Technology material in the next section.

Problem Set 10.2

Moving Toward Success

"A moment's insight is sometimes worth a lifetime's experience."

—Oliver Wendell Holmes, Jr., 1841–1935, American jurist

1. Why would beginning your study session with the most difficult topics make your session more efficient?

2. When studying, should you take breaks? Why or why not?

A Solve each equation in each problem using the quadratic formula. [Example 1]

1. **a.** $3x^2 + 4x - 2 = 0$

   b. $3x^2 - 4x - 2 = 0$

   c. $3x^2 + 4x + 2 = 0$

   d. $2x^2 + 4x - 3 = 0$

   e. $2x^2 - 4x + 3 = 0$

2. **a.** $3x^2 + 6x - 2 = 0$

   b. $3x^2 - 6x - 2 = 0$

   c. $3x^2 + 6x + 2 = 0$

   d. $2x^2 + 6x + 3 = 0$

   e. $2x^2 + 6x - 3 = 0$

3. **a.** $x^2 - 2x + 2 = 0$
   b. $x^2 - 2x + 5 = 0$
   c. $x^2 + 2x + 2 = 0$

4. **a.** $x^2 - 4x + 5 = 0$
   b. $x^2 + 4x + 5 = 0$
   c. $a^2 + 4a + 1 = 0$

A Solve each equation. Use factoring or the quadratic formula, whichever is appropriate. (Try factoring first. If you have any difficulty factoring, then go right to the quadratic formula.) [Examples 1–2]

5. $\dfrac{1}{6}x^2 - \dfrac{1}{2}x + \dfrac{1}{3} = 0$ 6. $\dfrac{1}{4}x^2 + \dfrac{1}{4}x - \dfrac{1}{2} = 0$

7. $\dfrac{x^2}{2} + 1 = \dfrac{2x}{3}$ 8. $\dfrac{x^2}{2} + \dfrac{2}{3} = -\dfrac{2x}{3}$

9. $y^2 - 5y = 0$ 10. $2y^2 + 10y = 0$

11. $30x^2 + 40x = 0$ 12. $50x^2 - 20x = 0$

13. $\dfrac{2t^2}{3} - t = -\dfrac{1}{6}$ 14. $\dfrac{t^2}{3} - \dfrac{t}{2} = -\dfrac{3}{2}$

15. $0.01x^2 + 0.06x - 0.08 = 0$

16. $0.02x^2 - 0.03x + 0.05 = 0$

17. $2x + 3 = -2x^2$

18. $2x - 3 = 3x^2$

19. $100x^2 - 200x + 100 = 0$

20. $100x^2 - 600x + 900 = 0$

21. $\dfrac{1}{2}r^2 = \dfrac{1}{6}r - \dfrac{2}{3}$

22. $\dfrac{1}{4}r^2 = \dfrac{2}{5}r + \dfrac{1}{10}$

23. $(x - 3)(x - 5) = 1$

24. $(x - 3)(x + 1) = -6$

25. $(x + 3)^2 + (x - 8)(x - 1) = 16$

26. $(x - 4)^2 + (x + 2)(x + 1) = 9$

27. $\dfrac{x^2}{3} - \dfrac{5x}{6} = \dfrac{1}{2}$ 28. $\dfrac{x^2}{6} + \dfrac{5}{6} = -\dfrac{x}{3}$

A Multiply both sides of each equation by its LCD. Then solve the resulting equation. [Example 3]

29. $\dfrac{1}{x + 1} - \dfrac{1}{x} = \dfrac{1}{2}$

30. $\dfrac{1}{x + 1} + \dfrac{1}{x} = \dfrac{1}{3}$

31. $\dfrac{1}{y - 1} + \dfrac{1}{y + 1} = 1$

32. $\dfrac{2}{y + 2} + \dfrac{3}{y - 2} = 1$

33. $\dfrac{1}{x + 2} + \dfrac{1}{x + 3} = 1$

34. $\dfrac{1}{x + 3} + \dfrac{1}{x + 4} = 1$

35. $\dfrac{6}{r^2 - 1} - \dfrac{1}{2} = \dfrac{1}{r + 1}$

36. $2 + \dfrac{5}{r - 1} = \dfrac{12}{(r - 1)^2}$

A Solve each equation. In each case you will have three solutions. [Example 4]

37. $x^3 - 8 = 0$

38. $x^3 - 27 = 0$

39. $8a^3 + 27 = 0$

40. $27a^3 + 8 = 0$

41. $125t^3 - 1 = 0$

42. $64t^3 + 1 = 0$

A Each of the following equations has three solutions. Look for the greatest common factor, then use the quadratic formula to find all solutions.

43. $2x^3 + 2x^2 + 3x = 0$

44. $6x^3 - 4x^2 + 6x = 0$

45. $3y^4 = 6y^3 - 6y^2$

46. $4y^4 = 16y^3 - 20y^2$

47. $6t^5 + 4t^4 = -2t^3$

48. $8t^5 + 2t^4 = -10t^3$

49. Which two of the expressions below are equivalent?

a. $\dfrac{6 + 2\sqrt{3}}{4}$

b. $\dfrac{3 + \sqrt{3}}{2}$

c. $6 + \dfrac{\sqrt{3}}{2}$

50. Which two of the expressions below are equivalent?

a. $\dfrac{8 - 4\sqrt{2}}{4}$

b. $2 - 4\sqrt{2}$

c. $2 - \sqrt{2}$

51. Solve $3x^2 - 5x = 0$

a. by factoring.

b. by the quadratic formula.

52. Solve $3x^2 + 23x - 70 = 0$

a. by factoring.

b. by the quadratic formula.

53. Can the equation $x^2 - 4x + 7 = 0$ be solved by factoring? Solve the equation.

54. Can the equation $x^2 = 5$ be solved by factoring? Solve the equation.

55. Is $x = -1 + i$ a solution to $x^2 + 2x = -2$?

56. Is $x = 2 + 2i$ a solution to $(x - 2)^2 = -4$?

57. Let $f(x) = x^2 - 2x - 3$. Find all values for the variable x, which produce the following values of $f(x)$.

a. $f(x) = 0$

b. $f(x) = -11$

c. $f(x) = -2x + 1$

d. $f(x) = 2x + 1$

58. Let $g(x) = x^2 + 16$. Find all values for the variable x, which produce the following values of $g(x)$.

a. $g(x) = 0$

b. $g(x) = 20$

c. $g(x) = 8x$

d. $g(x) = -8x$

59. Let $f(x) = \dfrac{10}{x^2}$. Find all values for the variable x, for which $f(x) = g(x)$.

a. $g(x) = 3 + \dfrac{1}{x}$

b. $g(x) = 8x - \dfrac{17}{x^2}$

c. $g(x) = 0$

d. $g(x) = 10$

60. Let $h(x) = \dfrac{x + 2}{x}$. Find all values for the variable x, for which $h(x) = f(x)$.

a. $f(x) = 2$

b. $f(x) = x + 2$

c. $f(x) = x - 2$

d. $f(x) = \dfrac{x - 2}{4}$

A Applying the Concepts [Examples 5–6]

61. Falling Object An object is thrown downward with an initial velocity of 5 feet per second. The relationship between the distance s it travels and time t is given by $s = 5t + 16t^2$. How long does it take the object to fall 74 feet?

62. Coin Toss A coin is tossed upward with an initial velocity of 32 feet per second from a height of 16 feet above the ground. The equation giving the object's height h at any time t is $h = 16 + 32t - 16t^2$. Does the object ever reach a height of 32 feet?

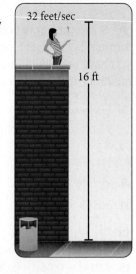

32 feet/sec

16 ft

63. Profit The total cost (in dollars) for a company to manufacture and sell x items per week is $C = 60x + 300$, whereas the revenue brought in by selling all x items is $R = 100x - 0.5x^2$. How many items must be sold to obtain a weekly profit of $300?

64. Profit Suppose a company manufactures and sells x picture frames each month with a total cost of $C = 1,200 + 3.5x$ dollars. If the revenue obtained by selling x frames is $R = 9x - 0.002x^2$, find the number of frames it must sell each month if its monthly profit is to be $2,300.

Kathleen Olson

65. Photograph Cropping The following figure shows a photographic image on a 10.5-centimeter by 8.2-centimeter background. The overall area of the background is to be reduced to 80% of its original area by cutting off (cropping) equal strips on all four sides. What is the width of the strip that is cut from each side?

Shane McKeague/2009

66. Area of a Garden A garden measures 20.3 meters by 16.4 meters. To double the area of the garden, strips of equal width are added to all four sides.
  a. Draw a diagram that illustrates these conditions.
  b. What are the new overall dimensions of the garden?

67. Area and Perimeter A rectangle has a perimeter of 20 yards and an area of 15 square yards.
  a. Write two equations that state these facts in terms of the rectangle's length, l, and its width, w.

  b. Solve the two equations from part a to determine the actual length and width of the rectangle.

  c. Explain why two answers are possible to part b.

68. Population Size Writing in 1829, former President James Madison made some predictions about the growth of the population of the United States. The populations he predicted fit the equation

Library of Congress

$$y = 0.029x^2 - 1.39x + 42$$

where y is the population in millions of people x years from 1829.
  a. Use the equation to determine the approximate year President Madison would have predicted that the U.S. population would reach 100,000,000.
  b. If the U.S. population in 2010 was approximately 300 million, were President Madison's predictions accurate in the long term? Explain why or why not.

Maintaining Your Skills

Divide, using long division.

69. $\dfrac{8y^2 - 26y - 9}{2y - 7}$

70. $\dfrac{6y^2 + 7y - 18}{3y - 4}$

71. $\dfrac{x^3 + 9x^2 + 26x + 24}{x + 2}$

72. $\dfrac{x^3 + 6x^2 + 11x + 6}{x + 3}$

Simplify each expression. (Assume $x, y > 0$.)

73. $25^{1/2}$ **74.** $8^{1/3}$

75. $\left(\dfrac{9}{25}\right)^{3/2}$ **76.** $\left(\dfrac{16}{81}\right)^{3/4}$

77. $8^{-2/3}$ **78.** $4^{-3/2}$

79. $\dfrac{(49x^8y^{-4})^{1/2}}{(27x^{-3}y^9)^{-1/3}}$ **80.** $\dfrac{(x^{-2}y^{1/3})^6}{x^{-10}y^{3/2}}$

Getting Ready for the Next Section

Find the value of $b^2 - 4ac$ when

81. $a = 1, b = -3, c = -40$

82. $a = 2, b = 3, c = 4$

83. $a = 4, b = 12, c = 9$

84. $a = -3, b = 8, c = -1$

Solve.

85. $k^2 - 144 = 0$ **86.** $36 - 20k = 0$

Multiply.

87. $(x - 3)(x + 2)$

88. $(t - 5)(t + 5)$

89. $(x - 3)(x - 3)(x + 2)$

90. $(t - 5)(t + 5)(t - 3)$

Extending the Concepts

So far, all the equations we have solved have had coefficients that were rational numbers. Here are some equations that have irrational coefficients and some that have complex coefficients. Solve each equation. (Remember, $i^2 = -1$.)

91. $x^2 + \sqrt{3}x - 6 = 0$

92. $x^2 - \sqrt{5}x - 5 = 0$

93. $\sqrt{2}x^2 + 2x - \sqrt{2} = 0$

94. $\sqrt{7}x^2 + 2\sqrt{2}x - \sqrt{7} = 0$

95. $x^2 + ix + 2 = 0$ **96.** $x^2 + 3ix - 2 = 0$

10.3 Additional Items Involving Solutions to Equations

OBJECTIVES

A Find the number and kinds of solutions to a quadratic equation by using the discriminant.

B Find an unknown constant in a quadratic equation so that there is exactly one solution.

C Find an equation from its solutions.

TICKET TO SUCCESS

Keep these questions in mind as you read through the section. Then respond in your own words and in complete sentences.

1. What is the discriminant?

2. What kinds of solutions do we get when the discriminant of a quadratic equation is negative?

3. Explain why you would want find an unknown constant of a quadratic equation.

4. What does it mean for a solution to have multiplicity 3?

A shortstop can jump straight up with an initial velocity of 2 feet per second. If he wants to catch a line drive hit over his head, he needs to be able to jump 5 feet off the ground. The equation that models the jump is $h = -16t^2 + 2t$. Can he make the out? Problems like this can be answered without having to name the solutions. We can use an expression called the discriminant to determine if there are any real solutions to this equation. In this section, we will define the discriminant and use it to find the number and types of solutions an equation may have. We will also use the zero-factor property to build equations from their solutions, which may help us decide the initial velocity the shortstop will need to catch the line drive.

A The Discriminant

The quadratic formula

$$x = \frac{-b \pm \sqrt{b^2 - 4ac}}{2a}$$

gives the solutions to any quadratic equation in standard form. When working with quadratic equations, there are times when it is important only to know what kinds of solutions the equation has.

Definition

The expression under the radical in the quadratic formula is called the **discriminant:**

$$\text{Discriminant} = D = b^2 - 4ac$$

The discriminant indicates the number and types of solutions to a quadratic equation, when the original equation has integer coefficients. For example, if we were to use the quadratic formula to solve the equation $2x^2 + 2x + 3 = 0$, we would find the discriminant to be

$$b^2 - 4ac = 2^2 - 4(2)(3) = -20$$

Since the discriminant appears under a square root symbol, we have the square root of a negative number in the quadratic formula. Our solutions therefore would be complex numbers. Similarly, if the discriminant were 0, the quadratic formula would yield

$$x = \frac{-b \pm \sqrt{0}}{2a} = \frac{-b \pm 0}{2a} = \frac{-b}{2a}$$

and the equation would have one rational solution, the number $\dfrac{-b}{2a}$.

The following table gives the relationship between the discriminant and the types of solutions to the equation.

For the equation $ax^2 + bx + c = 0$ where a, b, and c are integers and $a \neq 0$,

If the discriminant $b^2 - 4ac$ is	Then the equation will have
Negative	Two complex solutions containing i
Zero	One rational solution
A positive number that is also a perfect square	Two rational solutions
A positive number that is not a perfect square	Two irrational solutions

In the second and third cases, when the discriminant is 0 or a positive perfect square, the solutions are rational numbers. The quadratic equations in these two cases are the ones that can be factored.

EXAMPLE 1   For $x^2 - 3x - 40 = 0$, give the number and kinds of solutions.

SOLUTION   Using $a = 1$, $b = -3$, and $c = -40$ in $b^2 - 4ac$, we have

$$(-3)^2 - 4(1)(-40) = 9 + 160 = 169.$$

The discriminant is a perfect square. The equation therefore has two rational solutions. ■

EXAMPLE 2 For $2x^2 - 3x + 4 = 0$, give the number and kinds of solutions.

SOLUTION Using $a = 2$, $b = -3$, and $c = 4$, we have

$$b^2 - 4ac = (-3)^2 - 4(2)(4) = 9 - 32 = -23$$

The discriminant is negative, implying the equation has two complex solutions that contain i. ■

EXAMPLE 3 For $4x^2 - 12x + 9 = 0$, give the number and kinds of solutions.

SOLUTION Using $a = 4$, $b = -12$, and $c = 9$, the discriminant is

$$b^2 - 4ac = (-12)^2 - 4(4)(9) = 144 - 144 = 0$$

Since the discriminant is 0, the equation will have one rational solution. ■

EXAMPLE 4 For $x^2 + 6x = 8$, give the number and kinds of solutions.

SOLUTION First, we must put the equation in standard form by adding -8 to each side. If we do so, the resulting equation is

$$x^2 + 6x - 8 = 0$$

Now we identify a, b, and c as 1, 6, and -8, respectively.

$$b^2 - 4ac = 6^2 - 4(1)(-8) = 36 + 32 = 68$$

The discriminant is a positive number but not a perfect square. Therefore, the equation will have two irrational solutions. ■

B Finding an Unknown Constant

EXAMPLE 5 Find an appropriate k so that the equation $4x^2 - kx = -9$ has exactly one rational solution.

SOLUTION We begin by writing the equation in standard form.

$$4x^2 - kx + 9 = 0$$

Using $a = 4$, $b = -k$, and $c = 9$, we have

$$b^2 - 4ac = (-k)^2 - 4(4)(9)$$

$$= k^2 - 144$$

An equation has exactly one rational solution when the discriminant is 0. We set the discriminant equal to 0 and solve.

$$k^2 - 144 = 0$$

$$k^2 = 144$$

$$k = \pm 12$$

Choosing k to be 12 or -12 will result in an equation with one rational solution. ■

C Building Equations From Their Solutions

Suppose we know that the solutions to an equation are $x = 3$ and $x = -2$. We can find equations with these solutions by using the zero-factor property. First, let's write our solutions as equations with 0 on the right side.

If →	$x = 3$	First solution
then →	$x - 3 = 0$	Add -3 to each side
and if →	$x = -2$	Second solution
then →	$x + 2 = 0$	Add 2 to each side

Now, since both $x - 3$ and $x + 2$ are 0, their product must be 0 also. Therefore, we can write

$(x - 3)(x + 2) = 0$	Zero-factor property
$x^2 - x - 6 = 0$	Multiply out the left side

Many other equations have 3 and -2 as solutions. For example, any constant multiple of $x^2 - x - 6 = 0$, such as $5x^2 - 5x - 30 = 0$, also has 3 and -2 as solutions. Similarly, any equation built from positive integer powers of the factors $x - 3$ and $x + 2$ will also have 3 and -2 as solutions. One such equation is

$$(x - 3)^2(x + 2) = 0$$

$$(x^2 - 6x + 9)(x + 2) = 0$$

$$x^3 - 4x^2 - 3x + 18 = 0$$

In mathematics, we distinguish between the solutions to this last equation and those to the equation $x^2 - x - 6 = 0$ by saying $x = 3$ is a solution of *multiplicity 2* in the equation $x^3 - 4x^2 - 3x + 18 = 0$ and a solution of *multiplicity 1* in the equation $x^2 - x - 6 = 0$.

EXAMPLE 6 Find an equation that has solutions $t = 5$, $t = -5$, and $t = 3$.

SOLUTION First, we use the given solutions to write equations that have 0 on their right sides:

If →	$t = 5$	$t = -5$	$t = 3$
then →	$t - 5 = 0$	$t + 5 = 0$	$t - 3 = 0$

Since $t - 5$, $t + 5$, and $t - 3$ are all 0, their product is also 0 by the zero-factor property. An equation with solutions of 5, -5, and 3 is

$(t - 5)(t + 5)(t - 3) = 0$	Zero-factor property
$(t^2 - 25)(t - 3) = 0$	Multiply first two binomials
$t^3 - 3t^2 - 25t + 75 = 0$	Complete the multiplication

The last line gives us an equation with solutions of 5, -5, and 3. Remember, many other equations have these same solutions. ∎

EXAMPLE 7 Find an equation with solutions $x = -\frac{2}{3}$ and $x = \frac{4}{5}$.

SOLUTION The solution $x = -\frac{2}{3}$ can be rewritten as $3x + 2 = 0$ as follows:

$$x = -\frac{2}{3} \qquad \text{The first solution}$$

$$3x = -2 \qquad \text{Multiply each side by 3}$$

$$3x + 2 = 0 \qquad \text{Add 2 to each side}$$

Similarly, the solution $x = \dfrac{4}{5}$ can be rewritten as $5x - 4 = 0$.

$$x = \dfrac{4}{5} \qquad \text{The second solution}$$

$$5x = 4 \qquad \text{Multiply each side by 5}$$

$$5x - 4 = 0 \qquad \text{Add } -4 \text{ to each side}$$

Since both $3x + 2$ and $5x - 4$ are 0, their product is 0 also, giving us the equation we are looking for:

$$(3x + 2)(5x - 4) = 0 \qquad \text{Zero-factor property}$$

$$15x^2 - 2x - 8 = 0 \qquad \text{Multiplication}$$

USING TECHNOLOGY

Graphing Calculators

Solving Equations

Now that we have explored the relationship between equations and their solutions, we can look at how a graphing calculator can be used in the solution process. To begin, let's solve the equation $x^2 = x + 2$ using techniques from algebra: writing it in standard form, factoring, and then setting each factor equal to 0.

$$x^2 - x - 2 = 0 \qquad \text{Standard form}$$

$$(x - 2)(x + 1) = 0 \qquad \text{Factor}$$

$$x - 2 = 0 \quad \text{or} \quad x + 1 = 0 \qquad \text{Set each factor equal to 0}$$

$$x = 2 \quad \text{or} \quad x = -1 \qquad \text{Solve}$$

Our original equation, $x^2 = x + 2$, has two solutions: $x = 2$ and $x = -1$. To solve the equation using a graphing calculator, we need to associate it with an equation (or equations) in two variables. One way to do this is to associate the left side with the equation $y = x^2$ and the right side of the equation with $y = x + 2$. To do so, we set up the functions list in our calculator this way:

$$Y_1 = X^2$$

$$Y_2 = X + 2$$

Window: X from -5 to 5, Y from -5 to 5

Graphing these functions in this window will produce a graph similar to the one shown in Figure 1.

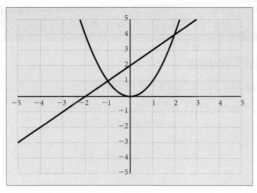

FIGURE 1

If we use the Trace feature to find the coordinates of the points of intersection, we find that the two curves intersect at $(-1, 1)$ and $(2, 4)$. We note that the x-coordinates of these two points match the solutions to the equation $x^2 = x + 2$, which we found using algebraic techniques. This makes sense because if two graphs intersect at a point (x, y), then the coordinates of that point satisfy both equations. If a point (x, y) satisfies both $y = x^2$ and $y = x + 2$, then, for that particular point, $x^2 = x + 2$. From this we conclude that the x-coordinates of the points of intersection are solutions to our original equation. Here is a summary of what we have discovered:

Conclusion I If the graph of two functions $y = f(x)$ and $y = g(x)$ intersect in the coordinate plane, then the x-coordinates of the points of intersection are solutions to the equation $f(x) = g(x)$.

A second method of solving our original equation $x^2 = x + 2$ graphically requires the use of one function instead of two. To begin, we write the equation in standard form as $x^2 - x - 2 = 0$. Next, we graph the function $y = x^2 - x - 2$. The x-intercepts of the graph are the points with y-coordinates of 0. Therefore, they satisfy the equation $0 = x^2 - x - 2$, which is equivalent to our original equation. The graph in Figure 2 shows $Y_1 = X^2 - X - 2$ in a window with X from -5 to 5 and Y from -5 to 5.

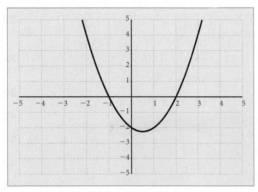

FIGURE 2

Using the Trace feature, we find that the x-intercepts of the graph are $x = -1$ and $x = 2$, which match the solutions to our original equation $x^2 = x + 2$. We can summarize the relationship between solutions to an equation and the intercepts of its associated graph this way:

Conclusion 2 If $y = f(x)$ is a function, then any x-intercept on the graph of $y = f(x)$ is a solution to the equation $f(x) = 0$.

Problem Set 10.3

Moving Toward Success

"The wisest mind has something yet to learn."

—George Santayana, 1863–1952, Spanish-born
American philosopher

1. Should you still read the chapter even if you understand the concepts being taught in class? Why or why not?

2. If you receive a poor score on a quiz or test, do your study habits need to be reevaluated? Why or why not?

A Use the discriminant to find the number and kinds of solutions for each of the following equations. [Examples 1–4]

1. $x^2 - 6x + 5 = 0$

2. $x^2 - x - 12 = 0$

3. $4x^2 - 4x = -1$

4. $9x^2 + 12x = -4$

5. $x^2 + x - 1 = 0$

6. $x^2 - 2x + 3 = 0$

7. $2y^2 = 3y + 1$

8. $3y^2 = 4y - 2$

9. $x^2 - 9 = 0$

10. $4x^2 - 81 = 0$

11. $5a^2 - 4a = 5$

12. $3a = 4a^2 - 5$

B Determine k so that each of the following has exactly one real solution. [Example 5]

13. $x^2 - kx + 25 = 0$ 14. $x^2 + kx + 25 = 0$

15. $x^2 = kx - 36$ 16. $x^2 = kx - 49$

17. $4x^2 - 12x + k = 0$ 18. $9x^2 + 30x + k = 0$

19. $kx^2 - 40x = 25$ 20. $kx^2 - 2x = -1$

21. $3x^2 - kx + 2 = 0$

22. $5x^2 + kx + 1 = 0$

C For each of the following problems, find an equation that has the given solutions. [Examples 6–7]

23. $x = 5, x = 2$

24. $x = -5, x = -2$

25. $t = -3, t = 6$

26. $t = -4, t = 2$

27. $y = 2, y = -2, y = 4$

28. $y = 1, y = -1, y = 3$

29. $x = \dfrac{1}{2}, x = 3$

30. $x = \dfrac{1}{3}, x = 5$

31. $t = -\dfrac{3}{4}, t = 3$

32. $t = -\dfrac{4}{5}, t = 2$

33. $x = 3, x = -3, x = \dfrac{5}{6}$

34. $x = 5, x = -5, x = \dfrac{2}{3}$

35. $a = -\dfrac{1}{2}, a = \dfrac{3}{5}$

36. $a = -\dfrac{1}{3}, a = \dfrac{4}{7}$

37. $x = -\dfrac{2}{3}, x = \dfrac{2}{3}, x = 1$

38. $x = -\dfrac{4}{5}, x = \dfrac{4}{5}, x = -1$

39. $x = 2, x = -2, x = 3, x = -3$

40. $x = 1, x = -1, x = 5, x = -5$

41. Find $f(x)$ if $f(1) = 0$ and $f(-2) = 0$.

42. Find $f(x)$ if $f(4) = 0$ and $f(-3) = 0$.

43. Find $f(x)$ if $f(3 + \sqrt{2}) = 0$ and $f(3 - \sqrt{2}) = 0$.

44. Find $f(x)$ if $f(-3 + \sqrt{15}) = 0$ and $f(-3 - \sqrt{15}) = 0$.

45. Find $f(x)$ if $f(2 + i) = 0$ and $f(2 - i) = 0$.

46. Find $f(x)$ if $f(3 + i\sqrt{5}) = 0$ and $f(3 - i\sqrt{5}) = 0$.

47. Find $f(x)$ if $f\left(\dfrac{5 + \sqrt{7}}{2}\right) = 0$ and $f\left(\dfrac{5 - \sqrt{7}}{2}\right) = 0$.

48. Find $f(x)$ if $f\left(\dfrac{-4 + \sqrt{11}}{3}\right) = 0$ and $f\left(\dfrac{-4 - \sqrt{11}}{3}\right) = 0$.

49. Find $f(x)$ if $f\left(\dfrac{3 + i\sqrt{5}}{2}\right) = 0$ and $f\left(\dfrac{3 - i\sqrt{5}}{2}\right) = 0$.

50. Find $f(x)$ if $f\left(\dfrac{-7 + i\sqrt{14}}{3}\right) = 0$ and $f\left(\dfrac{-7 - i\sqrt{14}}{3}\right) = 0$.

51. Find $f(x)$ if $f(2 + i\sqrt{3}) = 0$ and $f(2 - i\sqrt{3}) = 0$ and $f(0) = 0$.

52. Find $f(x)$ if $f(5 + i\sqrt{7}) = 0$ and $f(5 - i\sqrt{7}) = 0$ and $f(0) = 0$.

53. Indicate which of the graphs represent functions with the following types of solutions.

  a. One real solution

  b. No real solution

  c. Two real solutions

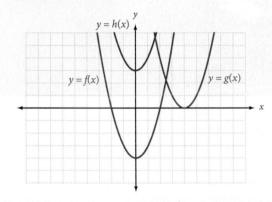

54. Find k, so that the function $9x^2 + kx + 4$ has:

  a. Exactly one real solution

  b. Two distinct real solutions

  c. No real solution

Maintaining Your Skills

Multiply. (Assume all variables represent positive numbers for the rest of the problems in this section.)

55. $a^4(a^{3/2} - a^{1/2})$

56. $(a^{1/2} - 5)(a^{1/2} + 3)$

57. $(x^{3/2} - 3)^2$ **58.** $(x^{1/2} - 8)(x^{1/2} + 8)$

Divide.

59. $\dfrac{30x^{3/4} - 25x^{5/4}}{5x^{1/4}}$

60. $\dfrac{45x^{5/3}y^{7/3} - 36x^{8/3}y^{4/3}}{9x^{2/3}y^{1/3}}$

61. Factor $5(x - 3)^{1/2}$ from $10(x - 3)^{3/2} - 15(x - 3)^{1/2}$.

62. Factor $2(x + 1)^{1/3}$ from $8(x + 1)^{4/3} - 2(x + 1)^{1/3}$.

Factor each of the following.

63. $2x^{2/3} - 11x^{1/3} + 12$

64. $9x^{2/3} + 12x^{1/3} + 4$

Getting Ready for the Next Section

Simplify.

65. $(x + 3)^2 - 2(x + 3) - 8$

66. $(x - 2)^2 - 3(x - 2) - 10$

67. $(2a - 3)^2 - 9(2a - 3) + 20$

68. $(3a - 2)^2 + 2(3a - 2) - 3$

69. $2(4a + 2)^2 - 3(4a + 2) - 20$

70. $6(2a + 4)^2 - (2a + 4) - 2$

Solve.

71. $x^2 = \dfrac{1}{4}$ **72.** $x^2 = -2$

73. $\sqrt{x} = -3$ **74.** $\sqrt{x} = 2$

75. $x + 3 = 4$ **76.** $x + 3 = -2$

77. $y^2 - 2y - 8 = 0$ **78.** $y^2 + y - 6 = 0$

79. $4y^2 + 7y - 2 = 0$ **80.** $6x^2 - 13x - 5 = 0$

Extending the Concepts

Find all solutions to the following equations. Solve using algebra and by graphing. If rounding is necessary, round to the nearest hundredth. A calculator can be used in these problems.

81. $x^2 = 4x + 5$ **82.** $4x^2 = 8x + 5$

83. $x^2 - 1 = 2x$

84. $4x^2 - 1 = 4x$

Find all solutions to each equation. If rounding is necessary, round to the nearest hundredth.

85. $2x^3 - x^2 - 2x + 1 = 0$

86. $3x^3 - 2x^2 - 3x + 2 = 0$

87. $2x^3 + 2 = x^2 + 4x$

88. $3x^3 - 9x = 2x^2 - 6$

10.4 Equations Quadratic in Form

OBJECTIVES

A Solve equations that are reducible to a quadratic equation.

B Solve application problems using equations quadratic in form.

TICKET TO SUCCESS

Keep these questions in mind as you read through the section. Then respond in your own words and in complete sentences.

1. What does it mean for an equation to be quadratic in form?
2. What are all the circumstances in solving equations (that we have studied) in which it is necessary to check for extraneous solutions?
3. How would you start to solve the equation $x + \sqrt{x} - 6 = 0$?
4. What does it mean for a line segment to be divided in "extreme and mean ratio"?

An Olympic platform diver jumps off a platform of height h with an upward velocity of about 8 feet per second. The time it takes her to reach the water is modeled by the equation $16t^2 - 8t - h = 0$. We can use this equation to calculate how long before she passes the platform. We can also use it to determine how long until she reaches the water. Let's put our knowledge of quadratic equations to work to solve a variety of equations like this one.

A Equations Quadratic in Form

> **EXAMPLE 1** Solve $(x + 3)^2 - 2(x + 3) - 8 = 0$.

SOLUTION We can see that this equation is quadratic in form by replacing $x + 3$ with another variable, say y. Replacing $x + 3$ with y we have

$$y^2 - 2y - 8 = 0$$

We can solve this equation by factoring the left side and then setting each factor equal to 0.

$$y^2 - 2y - 8 = 0$$
$$(y - 4)(y + 2) = 0 \qquad \text{Factor}$$
$$y - 4 = 0 \quad \text{or} \quad y + 2 = 0 \qquad \text{Set factors to 0}$$
$$y = 4 \quad \text{or} \qquad y = -2$$

Since our original equation was written in terms of the variable x, we would like our solutions in terms of x also. Replacing y with $x + 3$ and then solving for x we have

$$x + 3 = 4 \quad \text{or} \quad x + 3 = -2$$
$$x = 1 \quad \text{or} \quad x = -5$$

The solutions to our original equation are 1 and -5.

The method we have just shown lends itself well to other types of equations that are quadratic in form, as we will see. In this example, however, there is another method that works just as well. Let's solve our original equation again, but this time, let's begin by expanding $(x + 3)^2$ and $2(x + 3)$.

$$(x + 3)^2 - 2(x + 3) - 8 = 0$$

$$x^2 + 6x + 9 - 2x - 6 - 8 = 0 \qquad \text{Multiply}$$

$$x^2 + 4x - 5 = 0 \qquad \text{Combine similar terms}$$

$$(x - 1)(x + 5) = 0 \qquad \text{Factor}$$

$$x - 1 = 0 \quad \text{or} \quad x + 5 = 0 \qquad \text{Set factors to 0}$$

$$x = 1 \quad \text{or} \quad x = -5$$

As you can see, either method produces the same result.

EXAMPLE 2   Solve $4x^4 + 7x^2 = 2$.

SOLUTION   This equation is quadratic in x^2. We can make it easier to look at by using the substitution $y = x^2$. (The choice of the letter y is arbitrary. We could just as easily use the substitution $m = x^2$.) Making the substitution $y = x^2$ and then solving the resulting equation we have

$$4y^2 + 7y = 2$$

$$4y^2 + 7y - 2 = 0 \qquad \text{Standard form}$$

$$(4y - 1)(y + 2) = 0 \qquad \text{Factor}$$

$$4y - 1 = 0 \quad \text{or} \quad y + 2 = 0 \qquad \text{Set factors to 0}$$

$$y = \frac{1}{4} \quad \text{or} \quad y = -2$$

Now we replace y with x^2 to solve for x:

$$x^2 = \frac{1}{4} \quad \text{or} \quad x^2 = -2$$

$$x = \pm\sqrt{\frac{1}{4}} \quad \text{or} \quad x = \pm\sqrt{-2} \qquad \text{Theorem 1}$$

$$x = \pm\frac{1}{2} \quad \text{or} \quad x = \pm i\sqrt{2}$$

The solution set is $\left\{ \frac{1}{2}, -\frac{1}{2}, i\sqrt{2}, -i\sqrt{2} \right\}$.

EXAMPLE 3   Solve $x + \sqrt{x} - 6 = 0$ for x.

SOLUTION   To see that this equation is quadratic in form, we have to notice that $(\sqrt{x})^2 = x$; that is, the equation can be rewritten as

$$(\sqrt{x})^2 + \sqrt{x} - 6 = 0$$

Replacing $\sqrt{x}$ with y and solving as usual, we have

$$y^2 + y - 6 = 0$$

$$(y + 3)(y - 2) = 0$$

$$y + 3 = 0 \quad \text{or} \quad y - 2 = 0$$

$$y = -3 \quad \text{or} \quad y = 2$$

Again, to find x, we replace y with $\sqrt{x}$ and solve:

$$\sqrt{x} = -3 \quad \text{or} \quad \sqrt{x} = 2$$
$$x = 9 \qquad\qquad x = 4 \qquad \text{Square both sides of each equation}$$

Since we squared both sides of each equation, we have the possibility of obtaining extraneous solutions. We have to check both solutions in our original equation.

When →	$x = 9$	When →	$x = 4$
the equation →$x + \sqrt{x} - 6 = 0$		the equation → $x + \sqrt{x} - 6 = 0$	
becomes → $9 + \sqrt{9} - 6 \overset{?}{=} 0$		becomes → $4 + \sqrt{4} - 6 \overset{?}{=} 0$	
$9 + 3 - 6 \overset{?}{=} 0$		$4 + 2 - 6 \overset{?}{=} 0$	
$6 \neq 0$		$0 = 0$	

This means 9 is   This means 4 is
extraneous.   a solution.

The only solution to the equation $x + \sqrt{x} - 6 = 0$ is $x = 4$.

We should note here that the two possible solutions, 9 and 4, to the equation in Example 3 can be obtained by another method. Instead of substituting for $\sqrt{x}$, we can isolate it on one side of the equation and then square both sides to clear the equation of radicals.

$$x + \sqrt{x} - 6 = 0$$
$$\sqrt{x} = -x + 6 \qquad \text{Isolate } x$$
$$x = x^2 - 12x + 36 \qquad \text{Square both sides}$$
$$0 = x^2 - 13x + 36 \qquad \text{Add } -x \text{ to both sides}$$
$$0 = (x - 4)(x - 9) \qquad \text{Factor}$$
$$x - 4 = 0 \quad \text{or} \quad x - 9 = 0$$
$$x = 4 \qquad\qquad x = 9$$

We obtain the same two possible solutions. Since we squared both sides of the equation to find them, we would have to check each one in the original equation. As was the case in Example 3, only $x = 4$ is a solution; $x = 9$ is extraneous.

B Applications

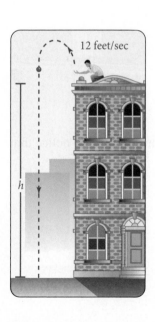
12 feet/sec

EXAMPLE 4 If an object is tossed into the air with an upward velocity of 12 feet per second from the top of a building h feet high, the time it takes for the object to hit the ground below is given by the formula

$$16t^2 - 12t - h = 0$$

Solve this formula for t.

SOLUTION The formula is in standard form and is quadratic in t. The coefficients a, b, and c that we need to apply to the quadratic formula are $a = 16$, $b = -12$, and $c = -h$. Substituting these quantities into the quadratic formula, we have

$$t = \frac{12 \pm \sqrt{144 - 4(16)(-h)}}{2(16)}$$
$$= \frac{12 \pm \sqrt{144 + 64h}}{32}$$

We can factor the perfect square 16 from the two terms under the radical and simplify our radical somewhat.

$$t = \frac{12 \pm \sqrt{16(9 + 4h)}}{32}$$

$$= \frac{12 \pm 4\sqrt{9 + 4h}}{32}$$

Now we can reduce to lowest terms by factoring a 4 from the numerator and denominator.

$$t = \frac{\cancel{4}(3 \pm \sqrt{9 + 4h})}{\cancel{4} \cdot 8}$$

$$= \frac{3 \pm \sqrt{9 + 4h}}{8}$$

If we were given a value of h, we would find that one of the solutions to this last formula would be a negative number. Since time is always measured in positive units, we wouldn't use that solution. ∎

More About the Golden Ratio

Previously, we derived the golden ratio $\frac{1 + \sqrt{5}}{2}$ by finding the ratio of length to width for a golden rectangle. The golden ratio was actually discovered before the golden rectangle by the Greeks who lived before Euclid. The early Greeks found the golden ratio by dividing a line segment into two parts so that the ratio of the shorter part to the longer part was the same as the ratio of the longer part to the whole segment. When they divided a line segment in this manner, they said it was divided in "extreme and mean ratio." Figure 1 illustrates a line segment divided this way.

$$\overset{\bullet}{A} \qquad\qquad \overset{\bullet}{B} \qquad\qquad\qquad \overset{\bullet}{C}$$

FIGURE 1

If point B divides segment AC in "extreme and mean ratio," then

$$\frac{\text{Length of shorter segment}}{\text{Length of longer segment}} = \frac{\text{Length of longer segment}}{\text{Length of whole segment}}$$

$$\frac{AB}{BC} = \frac{BC}{AC}$$

EXAMPLE 5   If the length of segment AB in Figure 1 is 1 inch, find the length of BC so that the whole segment AC is divided in "extreme and mean ratio."

SOLUTION   Using Figure 1 as a guide, if we let x = the length of segment BC, then the length of AC is $x + 1$. If B divides AC into "extreme and mean ratio," then the ratio of AB to BC must equal the ratio of BC to AC. Writing this relationship using the variable x, we have

$$\frac{1}{x} = \frac{x}{x + 1}$$

If we multiply both sides of this equation by the LCD $x(x + 1)$ we have

$$x + 1 = x^2$$

$$0 = x^2 - x - 1 \quad \text{Write equation in standard form}$$

Since this last equation is not factorable, we apply the quadratic formula.

$$x = \frac{1 \pm \sqrt{(-1)^2 - 4(1)(-1)}}{2}$$

$$= \frac{1 \pm \sqrt{5}}{2}$$

Our equation has two solutions, which we approximate using decimals.

$$\frac{1 + \sqrt{5}}{2} \approx 1.618 \qquad \frac{1 - \sqrt{5}}{2} \approx -0.618$$

Since we originally let x equal the length of segment BC, we use only the positive solution to our equation. As you can see, the positive solution is the golden ratio.

USING TECHNOLOGY

Graphing Calculators

More About Example 1

As we mentioned earlier, algebraic expressions entered into a graphing calculator do not have to be simplified to be evaluated. This fact applies to equations as well. We can graph the equation $y = (x + 3)^2 - 2(x + 3) - 8$ to assist us in solving the equation in Example 1. The graph is shown in Figure 2. Using the Zoom and Trace features at the x-intercepts gives us $x = 1$ and $x = -5$ as the solutions to the equation $0 = (x + 3)^2 - 2(x + 3) - 8$.

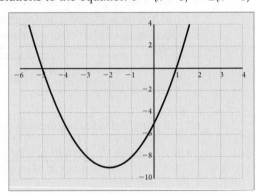

FIGURE 2

More About Example 2

Figure 3 shows the graph of $y = 4x^4 + 7x^2 - 2$. As we expect, the x-intercepts give the real number solutions to the equation $0 = 4x^4 + 7x^2 - 2$. The complex solutions do not appear on the graph.

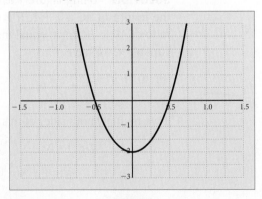

FIGURE 3

More About Example 3

In solving the equation in Example 3, we found that one of the possible solutions was an extraneous solution. If we solve the equation $x + \sqrt{x} - 6 = 0$ by graphing the function $y = x + \sqrt{x} - 6$, we find that the extraneous solution, 9, is not an x-intercept. Figure 4 shows that the only solution to the equation occurs at the x-intercept, 4.

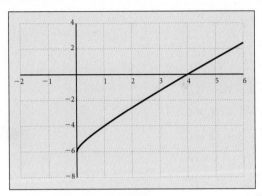

FIGURE 4

Problem Set 10.4

Moving Toward Success

"You've got to do your own growing, no matter how tall your grandfather was."

—Irish proverb

1. Why should you use different methods to study, such as reviewing class notes, working problems, making flashcards, outlining chapters, and studying with a group?

2. Would studying by mixing up the order of topics presented in the chapter better prepare you for an exam? Why or why not?

A Solve each equation. [Examples 1–2]

1. $(x - 3)^2 + 3(x - 3) + 2 = 0$

2. $(x + 4)^2 - (x + 4) - 6 = 0$

3. $2(x + 4)^2 + 5(x + 4) - 12 = 0$

4. $3(x - 5)^2 + 14(x - 5) - 5 = 0$

5. $x^4 - 10x^2 + 9 = 0$

6. $x^4 - 29x^2 + 100 = 0$

7. $x^4 - 7x^2 + 12 = 0$

8. $x^4 - 14x^2 + 45 = 0$

9. $x^4 - 6x^2 - 27 = 0$

10. $x^4 + 2x^2 - 8 = 0$

11. $x^4 + 9x^2 = -20$

12. $x^4 - 11x^2 = -30$

13. $(2a - 3)^2 - 9(2a - 3) = -20$

14. $(3a - 2)^2 + 2(3a - 2) = 3$

15. $2(4a + 2)^2 = 3(4a + 2) + 20$

16. $6(2a + 4)^2 = (2a + 4) + 2$

17. $6t^4 = -t^2 + 5$

18. $3t^4 = -2t^2 + 8$

19. $9x^4 - 49 = 0$

20. $25x^4 - 9 = 0$

A Solve each of the following equations. Remember, if you square both sides of an equation in the process of solving it, you have to check all solutions in the original equation. [Examples 1–3]

21. $x - 7\sqrt{x} + 10 = 0$

22. $x - 6\sqrt{x} + 8 = 0$

23. $t - 2\sqrt{t} - 15 = 0$

24. $t - 3\sqrt{t} - 10 = 0$

25. $6x + 11\sqrt{x} = 35$

26. $2x + \sqrt{x} = 15$

27. $x - 2\sqrt{x} - 8 = 0$

28. $x + 2\sqrt{x} - 3 = 0$

29. $x + 3\sqrt{x} - 18 = 0$

30. $x + 5\sqrt{x} - 14 = 0$

31. $2x + 9\sqrt{x} - 5 = 0$

32. $2x - \sqrt{x} - 6 = 0$

33. $(a - 2) - 11\sqrt{a - 2} + 30 = 0$

34. $(a - 3) - 9\sqrt{a - 3} + 20 = 0$

35. $(2x + 1) - 8\sqrt{2x + 1} + 15 = 0$

36. $(2x - 3) - 7\sqrt{2x - 3} + 12 = 0$

37. $(x^2 + 1)^2 - 2(x^2 + 1) - 15 = 0$

38. $(x^2 - 3)^2 - 4(x^2 - 3) - 12 = 0$

39. $(x^2 + 5)^2 - 6(x^2 + 5) - 27 = 0$

40. $(x^2 - 2)^2 - 6(x^2 - 2) - 7 = 0$

41. $x^{-2} - 3x^{-1} + 2 = 0$

42. $x^{-2} + x^{-1} - 2 = 0$

43. $y^{-4} - 5y^{-2} = 0$

44. $2y^{-4} + 10y^{-2} = 0$

45. $x^{2/3} + 4x^{1/3} - 12 = 0$

46. $x^{2/3} - 2x^{1/3} - 8 = 0$

47. $x^{4/3} + 6x^{2/3} + 9 = 0$

48. $x^{4/3} - 10x^{2/3} + 25 = 0$

49. Solve the formula $16t^2 - vt - h = 0$ for t.

50. Solve the formula $16t^2 + vt + h = 0$ for t.

51. Solve the formula $kx^2 + 8x + 4 = 0$ for x.

52. Solve the formula $k^2x^2 + kx + 4 = 0$ for x.

53. Solve $x^2 + 2xy + y^2 = 0$ for x by using the quadratic formula with $a = 1$, $b = 2y$, and $c = y^2$.

54. Solve $x^2 - 2xy + y^2 = 0$ for x by using the quadratic formula, with $a = 1$, $b = -2y$, and $c = y^2$.

C Applying the Concepts [Examples 4–5]

For Problems 55–56, t is in seconds.

55. Falling Object An object is tossed into the air with an upward velocity of 8 feet per second from the top of a building h feet high. The time it takes for the object to hit the ground below is given by the formula $16t^2 - 8t - h = 0$. Solve this formula for t.

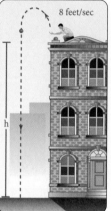

56. Falling Object An object is tossed into the air with an upward velocity of 6 feet per second from the top of a building h feet high. The time it takes for the object to hit the ground below is given by the formula $16t^2 - 6t - h = 0$. Solve this formula for t.

57. Saint Louis Arch The shape of the famous "Gateway to the West" arch in Saint Louis can be modeled by a parabola. The equation for one such parabola is:

$$y = -\frac{1}{150}x^2 + \frac{21}{5}x$$

where x and y are in feet.

a. Sketch the graph of the arch's equation on a coordinate axis.

b. Approximately how far do you have to walk to get from one side of the arch to the other?

58. Area and Perimeter A total of 160 yards of fencing is to be used to enclose part of a lot that borders on a river. This situation is shown in the following diagram.

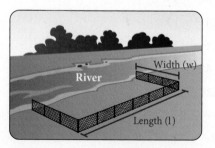

a. Write an equation that gives the relationship between the length and width and the 160 yards of fencing.

b. The formula for the area that is enclosed by the fencing and the river is $A = lw$. Solve the equation in part a for l, and then use the result to write the area in terms of w only.

c. Make a table that gives at least five possible values of w and associated area A.

d. From the pattern in your table shown in part c, what is the largest area that can be enclosed by the 160 yards of fencing? (Try some other table values if necessary.)

Golden Ratio Use Figure 1 from this section as a guide to working Problems 59–60.

59. If AB in Figure 1 is 4 inches, and B divides AC in "extreme and mean ratio," find BC, and then show that BC is 4 times the golden ratio.

60. If AB in Figure 1 is $\frac{1}{2}$ inch, and B divides AC in "extreme and mean ratio," find BC, and then show that the ratio of BC to AB is the golden ratio.

Maintaining Your Skills

Combine, if possible.

61. $5\sqrt{7} - 2\sqrt{7}$

62. $6\sqrt{2} - 9\sqrt{2}$

63. $\sqrt{18} - \sqrt{8} + \sqrt{32}$

64. $\sqrt{50} + \sqrt{72} - \sqrt{8}$

65. $9x\sqrt{20x^3y^2} + 7y\sqrt{45x^5}$

66. $5x^2\sqrt{27xy^3} - 6y\sqrt{12x^5y}$

Multiply.

67. $(\sqrt{5} - 2)(\sqrt{5} + 8)$

68. $(2\sqrt{3} - 7)(2\sqrt{3} + 7)$

69. $(\sqrt{x} + 2)^2$

70. $(3 - \sqrt{x})(3 + \sqrt{x})$

Rationalize the denominator.

71. $\dfrac{\sqrt{7}}{\sqrt{7} - 2}$

72. $\dfrac{\sqrt{5} - \sqrt{2}}{\sqrt{5} + \sqrt{2}}$

Getting Ready for the Next Section

73. Evaluate $y = 3x^2 - 6x + 1$ for $x = 1$.

74. Evaluate $y = -2x^2 + 6x - 5$ for $x = \frac{3}{2}$.

75. Let $P(x) = -0.1x^2 + 27x - 500$ and find $P(135)$.

76. Let $P(x) = -0.1x^2 + 12x - 400$ and find $P(600)$.

Solve.

77. $0 = a(80)^2 + 70$

78. $0 = a(80)^2 + 90$

79. $x^2 - 6x + 5 = 0$

80. $x^2 - 3x - 4 = 0$

81. $-x^2 - 2x + 3 = 0$

82. $-x^2 + 4x + 12 = 0$

83. $2x^2 - 6x + 5 = 0$

84. $x^2 - 4x + 5 = 0$

Fill in the blanks to complete the square.

85. $x^2 - 6x + \square = (x - \square)^2$

86. $x^2 + 10x + \square = (x + \square)^2$

87. $y^2 + 2y + \square = (y + \square)^2$

88. $y^2 - 12y + \square = (y - \square)^2$

Extending the Concepts

Find the x- and y-intercepts.

89. $y = x^3 - 4x$

90. $y = x^4 - 10x^2 + 9$

91. $y = 3x^3 + x^2 - 27x - 9$

92. $y = 2x^3 + x^2 - 8x - 4$

93. The graph of $y = 2x^3 - 7x^2 - 5x + 4$ crosses the x-axis at $x = 4$. Where else does it cross the x-axis?

94. The graph of $y = 6x^3 + x^2 - 12x + 5$ crosses the x-axis at $x = 1$. Where else does it cross the x-axis?

OBJECTIVES

A Graph a parabola.

B Solve application problems using information from a graph.

C Find an equation from its graph.

TICKET TO SUCCESS

Keep these questions in mind as you read through the section. Then respond in your own words and in complete sentences.

1. What is a parabola?

2. What part of the equation of a parabola determines whether the graph is concave up or concave down?

3. Suppose $f(x) = ax^2 + bx + c$ is the equation of a parabola. Explain how $f(4) = 1$ relates to the graph of the parabola.

4. Describe the ways you would find the vertex of a parabola.

Image copyright ©FloridaStock, 2009. Used under license from Shutterstock.com

A company selling beach umbrellas finds it can make a profit of P dollars each week by selling x umbrellas according to the formula $P(x) = -0.06x^2 + 4.5x - 1200$. How many umbrellas must it sell each week to make the maximum profit? What is that profit? Business problems like this one can be found by graphing a geometric figure known as a parabola. As we have seen in our previous work with linear equations, it is always possible to graph figures by making a table of values for P and x that satisfy the equation. As we have also seen, there are other methods that are faster and more accurate. In this section we will begin our work with parabolas and look at several ways to graph them.

The solution set to the equation

$$y = x^2 - 3$$

consists of ordered pairs. One method of graphing the solution set is to find a number of ordered pairs that satisfy the equation and to graph them. We can obtain some ordered pairs that are solutions to $y = x^2 - 3$ by use of a table as follows:

x	$y = x^2 - 3$	y	Solutions
−3	$y = (-3)^2 - 3 = 9 - 3 = 6$	6	(−3, 6)
−2	$y = (-2)^2 - 3 = 4 - 3 = 1$	1	(−2, 1)
−1	$y = (-1)^2 - 3 = 1 - 3 = -2$	−2	(−1, −2)
0	$y = 0^2 - 3 = 0 - 3 = -3$	−3	(0, −3)
1	$y = 1^2 - 3 = 1 - 3 = -2$	−2	(1, −2)
2	$y = 2^2 - 3 = 4 - 3 = 1$	1	(2, 1)
3	$y = 3^2 - 3 = 9 - 3 = 6$	6	(3, 6)

Graphing these solutions and then connecting them with a smooth curve, we have the graph of $y = x^2 - 3$ (Figure 1).

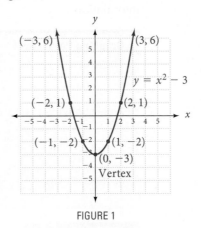

FIGURE 1

This graph is an example of a *parabola*. All equations of the form $y = ax^2 + bx + c$, $a \neq 0$, have parabolas for graphs.

Although it is always possible to graph parabolas by making a table of values of x and y that satisfy the equation, there are other methods that are faster and, in some cases, more accurate.

The important points associated with the graph of a parabola are the highest (or lowest) point on the graph and the x-intercepts. The y-intercepts can also be useful.

A Intercepts for Parabolas

The graph of the equation $y = ax^2 + bx + c$ crosses the y-axis at $y = c$ since substituting $x = 0$ into $y = ax^2 + bx + c$ yields $y = c$.

Since the graph crosses the x-axis when $y = 0$, the x-intercepts are those values of x that are solutions to the quadratic equation $0 = ax^2 + bx + c$.

The Vertex of a Parabola

The highest or lowest point on a parabola is called the *vertex*. The vertex for the graph of $y = ax^2 + bx + c$ will occur when

$$x = \frac{-b}{2a}$$

To see this, we must transform the right side of $y = ax^2 + bx + c$ into an expression that contains x in just one of its terms. This is accomplished by completing the square on the first two terms. Here is what it looks like:

$$y = ax^2 + bx + c$$
$$y = a\left(x^2 + \frac{b}{a}x\right) + c$$
$$y = a\left[x^2 + \frac{b}{a}x + \left(\frac{b}{2a}\right)^2\right] + c - a\left(\frac{b}{2a}\right)^2$$
$$y = a\left(x + \frac{b}{2a}\right)^2 + \frac{4ac - b^2}{4a}$$

It may not look like it, but this last line indicates that the vertex of the graph of $y = ax^2 + bx + c$ has an x-coordinate of $\frac{-b}{2a}$. Since a, b, and c are constants, the only quantity that is varying in the last expression is the x in $\left(x + \frac{b}{2a}\right)^2$. Since the quantity $\left(x + \frac{b}{2a}\right)^2$ is the square of $x + \frac{b}{2a}$, the smallest it will ever be is 0, and that will happen when $x = \frac{-b}{2a}$.

NOTE
What we are doing here is attempting to explain why the vertex of a parabola always has an x-coordinate of $\frac{-b}{2a}$. But the explanation may not be easy to understand the first time you see it. It may be helpful to look over the examples in this section and the notes that accompany these examples and then come back and read over this discussion again.

We can use the vertex point along with the x- and y-intercepts to sketch the graph of any equation of the form $y = ax^2 + bx + c$. Here is a summary of the preceding information.

Graphing Parabolas I

The graph of $y = ax^2 + bx + c$, $a \neq 0$, will have

1. a y-intercept at $y = c$

2. x-intercepts (if they exist) at

$$x = \frac{-b \pm \sqrt{b^2 - 4ac}}{2a}$$

3. a vertex when $x = \dfrac{-b}{2a}$

EXAMPLE 1 Sketch the graph of $y = x^2 - 6x + 5$.

SOLUTION To find the x-intercepts, we let $y = 0$ and solve for x.

$$0 = x^2 - 6x + 5$$

$$0 = (x - 5)(x - 1)$$

$$x = 5 \quad \text{or} \quad x = 1$$

To find the coordinates of the vertex, we first find

$$x = \frac{-b}{2a} = \frac{-(-6)}{2(1)} = 3$$

The x-coordinate of the vertex is 3. To find the y-coordinate, we substitute 3 for x in our original equation.

$$y = 3^2 - 6(3) + 5 = 9 - 18 + 5 = -4$$

The graph crosses the x-axis at 1 and 5 and has its vertex at $(3, -4)$. Plotting these points and connecting them with a smooth curve, we have the graph shown in Figure 2. The graph is a parabola that opens up, so we say the graph is *concave up*. The vertex is the lowest point on the graph. (Note that the graph crosses the y-axis at 5, which is the value of y we obtain when we let $x = 0$.)

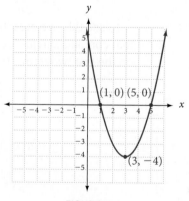

FIGURE 2

Finding the Vertex by Completing the Square

Another way to locate the vertex of the parabola in Example 1 is by completing the square on the first two terms on the right side of the equation $y = x^2 - 6x + 5$. In this case, we would do so by adding 9 to and subtracting 9 from the right side of the equation. This amounts to adding 0 to the equation, so we know we haven't changed its solutions. This is what it looks like:

$$y = (x^2 - 6x \quad) + 5$$

$$y = (x^2 - 6x + \mathbf{9}) + 5 - \mathbf{9}$$

$$y = (x - 3)^2 - 4$$

You may have to look at this last equation a while to see this, but when $x = 3$, then $y = (x - 3)^2 - 4 = 0^2 - 4 = -4$ is the smallest y will ever be. And that is why the vertex is at $(3, -4)$. As a matter of fact, this is the same kind of reasoning we used when we derived the formula $x = \frac{-b}{2a}$ for the x-coordinate of the vertex.

EXAMPLE 2 Graph $y = -x^2 - 2x + 3$.

SOLUTION To find the x-intercepts, we let $y = 0$.

$$0 = -x^2 - 2x + 3$$

$$0 = x^2 + 2x - 3 \qquad \text{Multiply each side by } -1$$

$$0 = (x + 3)(x - 1)$$

$$x = -3 \quad \text{or} \quad x = 1$$

The x-coordinate of the vertex is given by

$$x = \frac{-b}{2a} = \frac{-(-2)}{2(-1)} = \frac{2}{-2} = -1$$

To find the y-coordinate of the vertex, we substitute -1 for x in our original equation to get

$$y = -(-1)^2 - 2(-1) + 3 = -1 + 2 + 3 = 4$$

Our parabola has x-intercepts at -3 and 1 and a vertex at $(-1, 4)$. Figure 3 shows the graph. We say the graph is *concave down* since it opens downward.

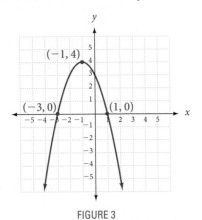

FIGURE 3

Again, we could have obtained the coordinates of the vertex by completing the square on the first two terms on the right side of our equation. To do so, we must first factor -1 from the first two terms. (Remember, the leading coefficient must be 1 to complete the square.) When we complete the square, we add 1 inside the parentheses, which actually decreases the right side of the equation by -1 since everything in the parentheses is multiplied by -1. To make up for it, we add 1 outside the parentheses.

$$y = -1(x^2 + 2x \quad) + 3$$

$$y = -1(x^2 + 2x + \mathbf{1}) + 3 + \mathbf{1}$$

$$y = -1(x + 1)^2 + 4$$

The last line tells us that the *largest* value of y will be 4, and that will occur when $x = -1$.

EXAMPLE 3

Graph $y = 3x^2 - 6x + 1$.

SOLUTION To find the x-intercepts, we let $y = 0$ and solve for x.

$$0 = 3x^2 - 6x + 1$$

Since the right side of this equation does not factor, we can look at the discriminant to see what kind of solutions are possible. The discriminant for this equation is

$$b^2 - 4ac = 36 - 4(3)(1) = 24$$

Since the discriminant is a positive number but not a perfect square, the equation will have irrational solutions. This means that the x-intercepts are irrational numbers and will have to be approximated with decimals using the quadratic formula. Rather than use the quadratic formula, we will find some other points on the graph, but first let's find the vertex.

Here are both methods of finding the vertex:

Using the formula that gives us the x-coordinate of the vertex, we have	To complete the square on the right side of the equation, we factor 3 from the first two terms, add 1 inside the parentheses, and add −3 outside the parentheses (this amounts to adding 0 to the right side).

$$x = \frac{-b}{2a} = \frac{-(-6)}{2(3)} = 1$$

Substituting 1 for x in the equation gives us the y-coordinate of the vertex.

$$y = 3 \cdot 1^2 - 6 \cdot 1 + 1 = -2$$

$$y = 3(x^2 - 2x \qquad) + 1$$

$$y = 3(x^2 - 2x + \mathbf{1}) + 1 - \mathbf{3}$$

$$y = 3(x - 1)^2 - 2$$

In either case, the vertex is $(1, -2)$.

If we can find two points, one on each side of the vertex, we can sketch the graph. Let's let $x = 0$ and $x = 2$ since each of these numbers is the same distance from $x = 1$, and $x = 0$ will give us the y-intercept.

When $x = 0$	When $x = 2$
$y = 3(0)^2 - 6(0) + 1$	$y = 3(2)^2 - 6(2) + 1$
$= 0 - 0 + 1$	$= 12 - 12 + 1$
$= 1$	$= 1$

The two points just found are $(0, 1)$ and $(2, 1)$. Plotting these two points along with the vertex $(1, -2)$, we have the graph shown in Figure 4.

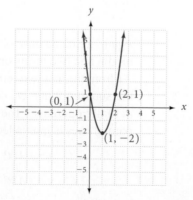

FIGURE 4

EXAMPLE 4   Graph $y = -2x^2 + 6x - 5$.

SOLUTION   Letting $y = 0$, we have

$$0 = -2x^2 + 6x - 5$$

Again, the right side of this equation does not factor. The discriminant is $b^2 - 4ac = 36 - 4(-2)(-5) = -4$, which indicates that the solutions are complex numbers. This means that our original equation does not have x-intercepts. The graph does not cross the x-axis. Let's find the vertex.

Using our formula for the x-coordinate of the vertex, we have

$$x = \frac{-b}{2a} = \frac{-6}{2(-2)} = \frac{6}{4} = \frac{3}{2}$$

To find the y-coordinate, we let $x = \frac{3}{2}$:

$$y = -2\left(\frac{3}{2}\right)^2 + 6\left(\frac{3}{2}\right) - 5$$

$$= \frac{-18}{4} + \frac{18}{2} - 5$$

$$= \frac{-18 + 36 - 20}{4}$$

$$= -\frac{1}{2}$$

Finding the vertex by completing the square is a more complicated matter. To make the coefficient of x^2 a 1, we must factor -2 from the first two terms. To complete the square inside the parentheses, we add $\frac{9}{4}$. Since each term inside the parentheses is multiplied by -2, we add $\frac{9}{2}$ outside the parentheses so that the net result is the same as adding 0 to the right side:

$$y = -2\left(x^2 - 3x \qquad\right) - 5$$

$$y = -2\left(x^2 - 3x + \mathbf{\frac{9}{4}}\right) - 5 + \mathbf{\frac{9}{2}}$$

$$y = -2\left(x - \frac{3}{2}\right)^2 - \frac{1}{2}$$

The vertex is $\left(\frac{3}{2}, -\frac{1}{2}\right)$. Since this is the only point we have so far, we must find two others. Let's let $x = 3$ and $x = 0$ since each point is the same distance from $x = \frac{3}{2}$ and on either side.

When $x = 3$

$$y = -2(3)^2 + 6(3) - 5$$

$$= -18 + 18 - 5$$

$$= -5$$

When $x = 0$

$$y = -2(0)^2 + 6(0) - 5$$

$$= 0 + 0 - 5$$

$$= -5$$

The two additional points on the graph are $(3, -5)$ and $(0, -5)$. Figure 5 shows the graph.

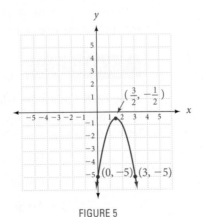

FIGURE 5

The graph is concave down. The vertex is the highest point on the graph.

By looking at the equations and graphs in Examples 1 through 4, we can conclude that the graph of $y = ax^2 + bx + c$ will be concave up when a is positive and concave down when a is negative. Taking this even further, if $a > 0$, then the vertex is the lowest point on the graph, and if $a < 0$, the vertex is the highest point on the graph. Finally, if we complete the square on x in the equation $y = ax^2 + bx + c, a \neq 0$, we can rewrite the equation of our parabola as $y = a(x - h)^2 + k$. When the equation is in this form, the vertex is at the point (h, k).

Here is a summary:

> **Graphing Parabolas II**
>
> The graph of
>
> $$y = a(x - h)^2 + k, a \neq 0$$
>
> will be a parabola with a vertex at (h, k). The vertex will be the highest point on the graph when $a < 0$, and it will be the lowest point on the graph when $a > 0$.

More About Parabolas

Now, we extend the work we did previously by considering parabolas that open left and right instead of up and down.

The graph of the equation $y = x^2 - 6x + 5$ is shown in Figure 6.

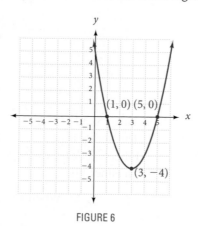

FIGURE 6

Note that the x-intercepts are 1 and 5, and the vertex is at $(3, -4)$.

If we interchange x and y in the equation $y = x^2 - 6x + 5$, we obtain the equation

$$x = y^2 - 6y + 5$$

We can expect the graph of this equation to be similar to the graph shown in Figure 6, except that the role of x and y will be interchanged. Example 5 gives more details.

EXAMPLE 5 Graph $x = y^2 - 6y + 5$.

SOLUTION Since the x-intercepts for $y = x^2 - 6x + 5$ are $x = 1$ and $x = 5$, we expect the y-intercepts for $x = y^2 - 6y + 5$ to be $y = 1$ and $y = 5$. To see that this is true, we let $x = 0$ in $x = y^2 - 6y + 5$, and solve for y:

$$0 = y^2 - 6y + 5$$

$$0 = (y - 1)(y - 5)$$

$$y - 1 = 0 \quad \text{or} \quad y - 5 = 0$$

$$y = 1 \quad \text{or} \quad y = 5$$

The graph of $x = y^2 - 6y + 5$ will cross the y-axis at 1 and 5.

Since the graph in Figure 6 has its vertex at $x = 3, y = -4$, we expect the graph of $x = y^2 - 6y + 5$ to have its vertex at $y = 3, x = -4$. Completing the square on y shows that this is true.

$$x = (y^2 - 6y + 9) + 5 - 9$$

$$x = (y - 3)^2 - 4$$

This last line tells us the smallest value of x is -4, and it occurs when y is 3. Therefore, the vertex must be at the point $(-4, 3)$. Here is that graph:

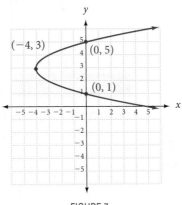

FIGURE 7

A summary of the discussion above is given below.

Parabolas That Open Left and Right

The graph of $x = ay^2 + by + c$ will have

1. an x-intercept at $x = c$

2. y-intercepts (if they exist) at
$$y = \frac{-b \pm \sqrt{b^2 - 4ac}}{2a}$$

3. a vertex when $y = \dfrac{-b}{2a}$

EXAMPLE 6   Graph $x = -y^2 - 2y + 3$.

SOLUTION   The intercepts and vertex are found as shown below. The results are used to sketch the graph in Figure 3.

x-intercept: When $y = 0, x = 3$

y-intercepts: When $x = 0$, we have $-y^2 - 2y + 3 = 0$, which yields solutions $y = -3$ and $y = 1$.

Vertex: $y = \dfrac{-b}{2a} = \dfrac{-(-2)}{2(-1)} = -1.$

Substituting -1 for y in the equation gives us $x = 4$.
The vertex is at $(4, -1)$.

The graph of our equation is shown on the right side of Figure 8. Note the similarities to the graph of $y = -x^2 - 2x + 3$ shown on the left side.

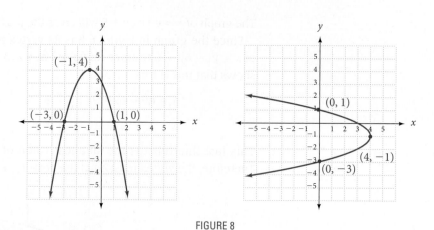

FIGURE 8

EXAMPLE 7 Graph $x = 3y^2 - 6y + 1$.

SOLUTION For this last example, we simply show the graph of our equation alongside the graph of the equation $y = 3x^2 - 6x + 1$. Both are shown in Figure 9.

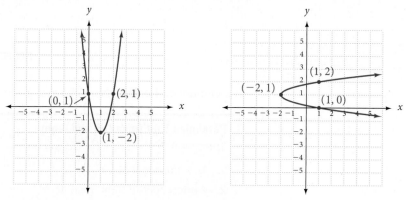

FIGURE 9

B Applications

EXAMPLE 8 A company selling copies of an accounting program for home computers finds that it will make a weekly profit of P dollars from selling x copies of the program, according to the equation

$$P(x) = -0.1x^2 + 27x - 500$$

How many copies of the program should it sell to make the largest possible profit, and what is the largest possible profit?

SOLUTION Since the coefficient of x^2 is negative, we know the graph of this parabola will be concave down, meaning that the vertex is the highest point of the curve. We find the vertex by first finding its x-coordinate.

$$x = \frac{-b}{2a} = \frac{-27}{2(-0.1)} = \frac{27}{0.2} = 135$$

This represents the number of programs the company needs to sell each week to make a maximum profit. To find the maximum profit, we substitute 135 for x in the original equation. (A calculator is helpful for these kinds of calculations.)

$$P(135) = -0.1(135)^2 + 27(135) - 500$$

$$= -0.1(18,225) + 3,645 - 500$$

$$= -1,822.5 + 3,645 - 500$$

$$= 1,322.5$$

The maximum weekly profit is $1,322.50 and is obtained by selling 135 programs a week.

EXAMPLE 9   An art supply store finds that they can sell x sketch pads each week at p dollars each, according to the equation $x = 900 - 300p$. Graph the revenue equation $R = xp$. Then use the graph to find the price p that will bring in the maximum revenue. Finally, find the maximum revenue.

SOLUTION   As it stands, the revenue equation contains three variables. Since we are asked to find the value of p that gives us the maximum value of R, we rewrite the equation using just the variables R and p. Since $x = 900 - 300p$, we have

$$R = xp = (900 - 300p)p$$

The graph of this equation is shown in Figure 10. The graph appears in the first quadrant only since R and p are both positive quantities.

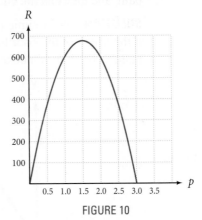

FIGURE 10

From the graph, we see that the maximum value of R occurs when $p = \$1.50$. We can calculate the maximum value of R from the equation.

When → $p = 1.5$

the equation → $R = (900 - 300p)p$

becomes → $R = (900 - 300 \cdot 1.5)1.5$

$$= (900 - 450)1.5$$

$$= 450 \cdot 1.5$$

$$= 675$$

The maximum revenue is $675. It is obtained by setting the price of each sketch pad at $p = \$1.50$.

USING TECHNOLOGY

Graphing Calculators

If you have been using a graphing calculator for some of the material in this course, you are well aware that your calculator can draw all the graphs in this section very easily. It is important, however, that you be able to recognize and sketch the graph of any parabola by hand. It is a skill that all successful intermediate algebra students should possess, even if they are proficient in the use of a graphing calculator. My suggestion is that you work the problems in this section and problem set without your calculator. Then use your calculator to check your results.

C Finding the Equation from the Graph

EXAMPLE 10   At the 1997 Washington County Fair in Oregon, David Smith, Jr., The Bullet, was shot from a cannon. As a human cannonball, he reached a height of 70 feet before landing in a net 160 feet from the cannon. Sketch the graph of his path, and then find the equation of the graph.

SOLUTION   We assume that the path taken by the human cannonball is a parabola. If the origin of the coordinate system is at the opening of the cannon, then the net that catches him will be at 160 on the x-axis. Figure 11 shows the graph:

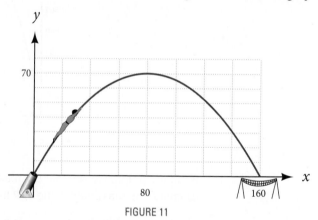

FIGURE 11

Since the curve is a parabola, we know the equation will have the form

$$y = a(x - h)^2 + k$$

Since the vertex of the parabola is at (80, 70), we can fill in two of the three constants in our equation, giving us

$$y = a(x - 80)^2 + 70$$

To find a, we note that the landing point will be (160, 0). Substituting the coordinates of this point into the equation, we solve for a:

$$0 = a(160 - 80)^2 + 70$$

$$0 = a(80)^2 + 70$$

$$0 = 6400a + 70$$

$$a = -\frac{70}{6400} = -\frac{7}{640}$$

The equation that describes the path of the human cannonball is

$$y = -\frac{7}{640}(x - 80)^2 + 70 \text{ for } 0 \le x \le 160$$

USING TECHNOLOGY

Graphing Calculators

Graph the equation found in Example 10 on a graphing calculator using the window shown below. (We will use this graph later in the book to find the angle between the cannon and the horizontal.)

Window: X from 0 to 180, increment 20

Y from 0 to 80, increment 10

On the TI-83/84, an increment of 20 for X means Xscl=20.

Problem Set 10.5

Moving Toward Success

"Write it on your heart that every day is the best day in the year."

—Ralph Waldo Emerson, 1803–1882, American poet and essayist

1. Why should you use all the exam time provided for you, even if you finish early?

2. What are some things you can do with any extra time to ensure the best test grade possible?

A For each of the following equations, give the x-intercepts and the coordinates of the vertex, and sketch the graph. [Examples 1–2]

1. $y = x^2 + 2x - 3$

2. $y = x^2 - 2x - 3$

3. $y = -x^2 - 4x + 5$

4. $y = x^2 + 4x - 5$

5. $y = x^2 - 1$

6. $y = x^2 - 4$

7. $y = -x^2 + 9$

8. $y = -x^2 + 1$

9. $y = 2x^2 - 4x - 6$

10. $y = 2x^2 + 4x - 6$

11. $y = x^2 - 2x - 4$

12. $y = x^2 - 2x - 2$

A Find the vertex and any two convenient points to sketch the graphs of the following. [Examples 3–4]

13. $y = x^2 - 4x - 4$

14. $y = x^2 - 2x + 3$

15. $y = -x^2 + 2x - 5$

16. $y = -x^2 + 4x - 2$

17. $y = x^2 + 1$ 18. $y = x^2 + 4$

19. $y = -x^2 - 3$

20. $y = -x^2 - 2$

A For each of the following equations, find the coordinates of the vertex, and indicate whether the vertex is the highest point on the graph or the lowest point on the graph. (Do not graph.) [Examples 1–4]

21. $y = x^2 - 6x + 5$

22. $y = -x^2 + 6x - 5$

23. $y = -x^2 + 2x + 8$

24. $y = x^2 - 2x - 8$

25. $y = 12 + 4x - x^2$

26. $y = -12 - 4x + x^2$

27. $y = -x^2 - 8x$

28. $y = x^2 + 8x$

Graph each equation. [Examples 5–7]

29. $x = y^2 + 2y - 3$
30. $x = y^2 - 2y - 3$

31. $x = -y^2 - 4y + 5$
32. $x = y^2 + 4y - 5$

33. $x = y^2 - 1$
34. $x = y^2 - 4$

35. $x = -y^2 + 9$
36. $x = -y^2 + 1$

37. $x = 2y^2 - 4y - 6$
38. $x = 2y^2 + 4y - 6$

39. $x = y^2 - 2y - 4$
40. $x = y^2 - 2y + 3$

B Applying the Concepts [Examples 8–9]

41. **Maximum Profit** A company finds that it can make a profit of P dollars each month by selling x patterns, according to the formula $P(x) = -0.002x^2 + 3.5x - 800$. How many patterns must it sell each month to have a maximum profit? What is the maximum profit?

42. **Maximum Profit**
A company selling picture frames finds that it can make a profit of P dollars each month by selling x frames, according to the formula $P(x) = -0.002x^2 + 5.5x - 1,200$. How many frames must it sell each month to have a maximum profit? What is the maximum profit?

Kathleen Olson

43. **Maximum Height** Chaudra is tossing a softball into the air with an underhand motion. The distance of the ball above her hand at any time is given by the function

$$h(t) = 32t - 16t^2$$

for $0 \le t \le 2$

where $h(t)$ is the height of the ball (in feet) and t is the time (in seconds). Find the times at which the ball is in her hand and the maximum height of the ball.

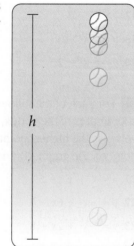

44. **Maximum Area** Justin wants to fence 3 sides of a rectangular exercise yard for his dog. The fourth side of the exercise yard will be a side of the house. He has 80 feet of fencing available. Find the dimensions of the exercise yard that will enclose the maximum area.

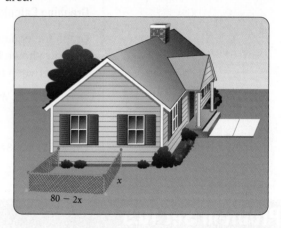

45. **Maximum Revenue** A company that manufactures key chains knows that the number of key chains x it can sell each week is related to the price p of each key chain by the equation $x = 1,200 - 100p$. Graph the revenue equation $R = xp$. Then use the graph to find the price p that will bring in the maximum revenue. Finally, find the maximum revenue.

46. **Maximum Revenue** A company that manufactures CDs for home computers finds that it can sell x CDs each day at p dollars per CD, according to the equation $x = 800 - 100p$. Graph the revenue equation $R = xp$. Then use the graph to find the price p that will bring in the maximum revenue. Finally, find the maximum revenue.

47. **Maximum Revenue** The relationship between the number of calculators x a company sells each day and the price p of each calculator is given by the equation $x = 1,700 - 100p$. Graph the revenue equation $R = xp$, and use the graph to find the price p that will bring in the maximum revenue. Then find the maximum revenue.

48. **Maximum Revenue** The relationship between the number of pencil sharpeners x a company sells each week and the price p of each sharpener is given by the equation $x = 1,800 - 100p$. Graph the revenue equation $R = xp$, and use the graph to find the price p that will bring in the maximum revenue. Then find the maximum revenue.

C [Example 10]

49. Human Cannonball A human cannonball is shot from a cannon at the county fair. He reaches a height of 60 feet before landing in a net 180 feet from the cannon. Sketch the graph of his path, and then find the equation of the graph.

50. Human Cannonball Suppose the equation below gives the height $h(x)$ of the human cannonball after he has traveled x feet horizontally. What is $h(30)$, and what does it represent?

$$h(x) = -\frac{7}{640}(x - 80)^2 + 70$$

Getting Ready for the Next Section

Solve.

51. $x^2 - 2x - 8 = 0$ **52.** $x^2 - x - 12 = 0$

53. $6x^2 - x = 2$ **54.** $3x^2 - 5x = 2$

55. $x^2 - 6x + 9 = 0$ **56.** $x^2 + 8x + 16 = 0$

Maintaining Your Skills

Perform the indicated operations.

57. $(3 - 5i) - (2 - 4i)$ **58.** $2i(5 - 6i)$

59. $(3 + 2i)(7 - 3i)$ **60.** $(4 + 5i)^2$

61. $\dfrac{i}{3 + i}$ **62.** $\dfrac{2 + 3i}{2 - 3i}$

Extending the Concepts

Finding the equation from the graph For each of the next two problems, the graph is a parabola. In each case, find an equation in the form $y = a(x - h)^2 + k$ that describes the graph.

63.

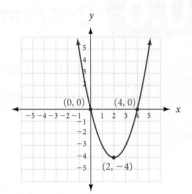

64.

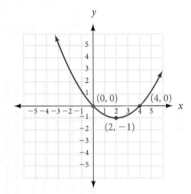

65. Interpreting Graphs The graph below shows the different paths taken by the human cannonball when his velocity out of the cannon is 50 mph and his cannon is inclined at varying angles.

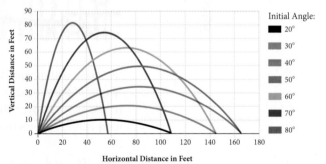

Initial Velocity: 50 miles per hour

a. If his landing net is placed 108 feet from the cannon, at what angle should the cannon be inclined so that he lands in the net?

b. Approximately where do you think he would land if the cannon was inclined at 45°?

c. If the cannon was inclined at 45°, approximately what height do you think he would attain?

d. Do you think there is another angle for which he would travel the same distance he travels at 80°? Give an estimate of that angle.

e. The fact that every landing point can come from two different paths makes us think that the equations that give us the landing points must be what type of equations?

TICKET TO SUCCESS

Keep these questions in mind as you read through the section. Then respond in your own words and in complete sentences.

1. What is the first step in solving a quadratic inequality?

2. Why would you draw the number lines of each factor from a quadratic inequality together?

3. How do you show that the endpoint of a line segment is not part of the graph of a quadratic inequality?

4. How would you use the graph of $y = ax^2 + bx + c$ to help you find the graph of $ax^2 + bx + c < 0$?

Suppose you needed to build a dog run with an area of at least 40 square feet. The length of the run has to be 3 times the width. What are the possibilities for the length? In order to answer this question, you need to know how to work with quadratic inequalities, the focus of this section.

A Quadratic Inequalities

Quadratic inequalities in one variable are inequalities of the form

$$ax^2 + bx + c < 0 \qquad ax^2 + bx + c > 0$$
$$ax^2 + bx + c \le 0 \qquad ax^2 + bx + c \ge 0$$

where a, b, and c are constants, with $a \ne 0$. The technique we will use to solve inequalities of this type involves graphing. Suppose, for example, we wish to find the solution set for the inequality $x^2 - x - 6 > 0$. We begin by factoring the left side to obtain

$$(x - 3)(x + 2) > 0$$

We have two real numbers $x - 3$ and $x + 2$ whose product $(x - 3)(x + 2)$ is greater than zero; that is, their product is positive. The only way the product can be positive is either if both factors, $(x - 3)$ and $(x + 2)$, are positive or if they are both negative. To help visualize where $x - 3$ is positive and where it is negative, we draw a real number line and label it accordingly:

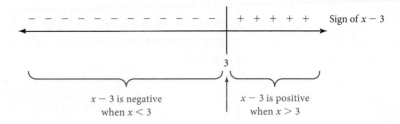

Here is a similar diagram showing where the factor $x + 2$ is positive and where it is negative:

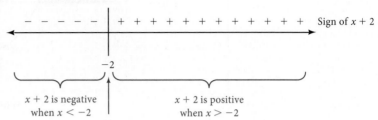

Drawing the two number lines together and eliminating the unnecessary numbers, we have

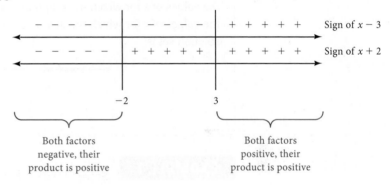

We can see from the preceding diagram that the graph of the solution to $x^2 - x - 6 > 0$ is

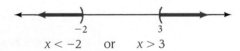

$x < -2$ or $x > 3$

USING TECHNOLOGY

Graphical Solutions to Quadratic Inequalities

We can solve the preceding problem by using a graphing calculator to visualize where the product $(x - 3)(x + 2)$ is positive. First, we graph the function $y = (x - 3)(x + 2)$ as shown in Figure 1.

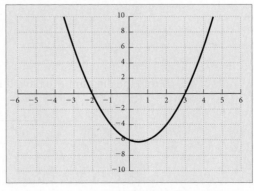

FIGURE 1

Next, we observe where the graph is above the *x*-axis. As you can see, the graph is above the *x*-axis to the right of 3 and to the left of −2, as shown in Figure 2.

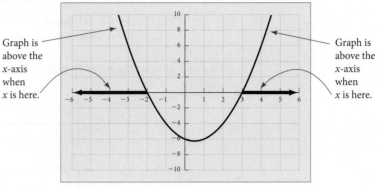

FIGURE 2

When the graph is above the *x*-axis, we have points whose *y*-coordinates are positive. Since these *y*-coordinates are the same as the expression $(x - 3)(x + 2)$, the values of *x* for which the graph of $y = (x - 3)(x + 2)$ is above the *x*-axis are the values of *x* for which the inequality $(x - 3)(x + 2) > 0$ is true. Therefore, our solution set is

$$x < -2 \quad \text{or} \quad x > 3$$

EXAMPLE 1 Solve $x^2 - 2x - 8 \leq 0$ for *x*.

ALGEBRAIC SOLUTION We begin by factoring:

$$x^2 - 2x - 8 \leq 0$$

$$(x - 4)(x + 2) \leq 0$$

The product $(x - 4)(x + 2)$ is negative or zero. The factors must have opposite signs. We draw a diagram showing where each factor is positive and where each factor is negative.

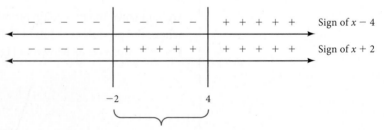

From the diagram, we have the graph of the solution set.

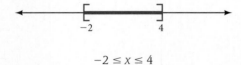

$$-2 \leq x \leq 4$$

When x is here, the graph is on or below the x-axis.

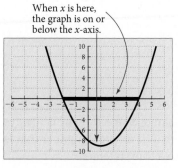

FIGURE 3

GRAPHICAL SOLUTION   To solve this inequality with a graphing calculator, we graph the function $y = (x - 4)(x + 2)$ and observe where the graph is below the x-axis. These points have negative y-coordinates, which means that the product $(x - 4)(x + 2)$ is negative for these points. Figure 3 shows the graph of $y = (x - 4)(x + 2)$, along with the region on the x-axis where the graph contains points with negative y-coordinates.

As you can see, the graph is below the x-axis when x is between -2 and 4. Since our original inequality includes the possibility that $(x - 4)(x + 2)$ is 0, we include the endpoints, -2 and 4, with our solution set.

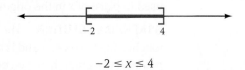

$$-2 \le x \le 4$$

EXAMPLE 2   Solve $6x^2 - x \ge 2$ for x.

ALGEBRAIC SOLUTION

$$6x^2 - x \ge 2$$
$$6x^2 - x - 2 \ge 0 \quad \text{Standard form}$$
$$(3x - 2)(2x + 1) \ge 0$$

The product is positive, so the factors must agree in sign. Here is the diagram showing where that occurs:

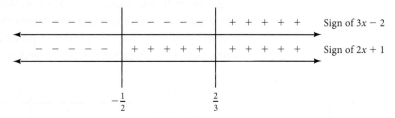

Since the factors agree in sign below $-\frac{1}{2}$ and above $\frac{2}{3}$, the graph of the solution set is

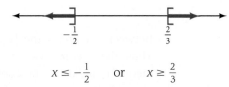

$$x \le -\frac{1}{2} \quad \text{or} \quad x \ge \frac{2}{3}$$

Graph is on or above the x-axis when x is here.   Graph is on or above the x-axis when x is here.

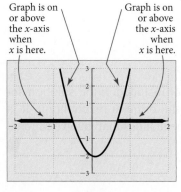

FIGURE 4

GRAPHICAL SOLUTION   To solve this inequality with a graphing calculator, we graph the function $y = (3x - 2)(2x + 1)$ and observe where the graph is above the x-axis. These are the points that have positive y-coordinates, which means that the product $(3x - 2)(2x + 1)$ is positive for these points. Figure 4 shows the graph of $y = (3x - 2)(2x + 1)$, along with the regions on the x-axis where the graph is on or above the x-axis.

To find the points where the graph crosses the x-axis, we need to use either the Trace and Zoom features to zoom in on each point, or the calculator function that finds the intercepts automatically (on the TI-83/84 this is the root/zero function under the CALC key). Whichever method we use, we will obtain the following result:

$$x \le -0.5 \quad \text{or} \quad x \ge 0.67$$

EXAMPLE 3 Solve $x^2 - 6x + 9 \geq 0$.

ALGEBRAIC SOLUTION

$$x^2 - 6x + 9 \geq 0$$

$$(x - 3)^2 \geq 0$$

This is a special case in which both factors are the same. Since $(x - 3)^2$ is always positive or zero, the solution set is all real numbers; that is, any real number that is used in place of x in the original inequality will produce a true statement.

GRAPHICAL SOLUTION The graph of $y = (x - 3)^2$ is shown in Figure 5. Notice that it touches the x-axis at 3 and is above the x-axis everywhere else. This means that every point on the graph has a y-coordinate greater than or equal to 0, no matter what the value of x. The conclusion that we draw from the graph is that the inequality $(x - 3)^2 \geq 0$ is true for all values of x.

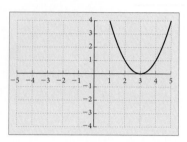

FIGURE 5

B Inequalities Involving Rational Expressions

EXAMPLE 4 Solve $\dfrac{x - 4}{x + 1} \leq 0$.

SOLUTION The inequality indicates that the quotient of $(x - 4)$ and $(x + 1)$ is negative or 0 (less than or equal to 0). We can use the same reasoning we used to solve the first three examples, because quotients are positive or negative under the same conditions that products are positive or negative. Here is the diagram that shows where each factor is positive and where each factor is negative:

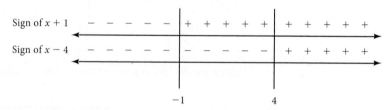

Between -1 and 4 the factors have opposite signs, making the quotient negative. Thus, the region between -1 and 4 is where the solutions lie, because the original inequality indicates the quotient $\dfrac{x - 4}{x + 1}$ is negative. The solution set and its graph are shown here:

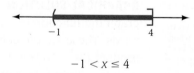

$$-1 < x \leq 4$$

Notice that the left endpoint is open; that is, it is not included in the solution set because $x = -1$ would make the denominator in the original inequality 0. It is important to check all endpoints of solution sets to inequalities that involve rational expressions.

EXAMPLE 5 Solve $\dfrac{3}{x - 2} - \dfrac{2}{x - 3} > 0$.

SOLUTION We begin by adding the two rational expressions on the left side. The common denominator is $(x - 2)(x - 3)$.

$$\frac{3}{x-2} \cdot \frac{(x-3)}{(x-3)} - \frac{2}{x-3} \cdot \frac{(x-2)}{(x-2)} > 0$$

$$\frac{3x-9-2x+4}{(x-2)(x-3)} > 0$$

$$\frac{x-5}{(x-2)(x-3)} > 0$$

This time the quotient involves three factors. Here is the diagram that shows the signs of the three factors:

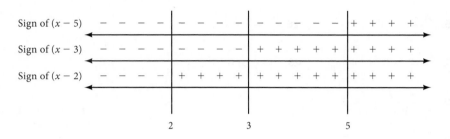

The original inequality indicates that the quotient is positive. For this to happen, either all three factors must be positive, or exactly two factors must be negative. Looking back at the diagram, we see the regions that satisfy these conditions are between 2 and 3 or above 5. Here is our solution set:

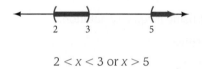

$$2 < x < 3 \text{ or } x > 5$$

Problem Set 10.6

Moving Toward Success

"A life spent making mistakes is not only more honorable, but more useful than a life spent doing nothing."

—George Bernard Shaw, 1856–1950, Irish playwright and Nobel Prize winner

1. Why is it important to review your graded test when you get it back?

2. How will you address any mistakes you made on the test to ensure you don't continue to make the same ones on a future test?

A Solve each of the following inequalities and graph the solution set. [Examples 1–3]

1. $x^2 + x - 6 > 0$

2. $x^2 + x - 6 < 0$

3. $x^2 - x - 12 \leq 0$

4. $x^2 - x - 12 \geq 0$

5. $x^2 + 5x \geq -6$

6. $x^2 - 5x > 6$

7. $6x^2 < 5x - 1$

8. $4x^2 \geq -5x + 6$

9. $x^2 - 9 < 0$

10. $x^2 - 16 \geq 0$

11. $4x^2 - 9 \geq 0$

12. $9x^2 - 4 < 0$

13. $2x^2 - x - 3 < 0$

14. $3x^2 + x - 10 \geq 0$

15. $x^2 - 4x + 4 \geq 0$

16. $x^2 - 4x + 4 < 0$

17. $x^2 - 10x + 25 < 0$

18. $x^2 - 10x + 25 > 0$

19. $(x - 2)(x - 3)(x - 4) > 0$

20. $(x - 2)(x - 3)(x - 4) < 0$

21. $(x + 1)(x + 2)(x + 3) \leq 0$

22. $(x + 1)(x + 2)(x + 3) \geq 0$

B [Examples 4–5]

23. $\dfrac{x - 1}{x + 4} \leq 0$

24. $\dfrac{x + 4}{x - 1} \leq 0$

25. $\dfrac{3x - 8}{x + 6} > 0$

26. $\dfrac{5x - 3}{x + 1} < 0$

27. $\dfrac{x - 2}{x - 6} > 0$

28. $\dfrac{x - 1}{x - 3} \geq 0$

29. $\dfrac{x - 2}{(x + 3)(x - 4)} < 0$

30. $\dfrac{x - 1}{(x + 2)(x - 5)} < 0$

31. $\dfrac{2}{x - 4} - \dfrac{1}{x - 3} > 0$

32. $\dfrac{4}{x + 3} - \dfrac{3}{x + 2} > 0$

33. Write each statement using inequality notation.

    a. $x - 1$ is always positive.

    b. $x - 1$ is never negative.

    c. $x - 1$ is greater than or equal to 0.

34. Match each expression on the left with a phrase on the right.

    a. $(x - 1)^2 \geq 0$ **i.** never true

    b. $(x - 1)^2 < 0$ **ii.** sometimes true

    c. $(x - 1)^2 \leq 0$ **iii.** always true

A Solve each inequality by inspection, without showing any work. [Example 3]

35. $(x - 1)^2 < 0$ **36.** $(x + 2)^2 < 0$

37. $(x - 1)^2 \leq 0$ **38.** $(x + 2)^2 \leq 0$

39. $(x - 1)^2 \geq 0$

40. $(x + 2)^2 \geq 0$

41. $\dfrac{1}{(x - 1)^2} \geq 0$

42. $\dfrac{1}{(x + 2)^2} > 0$

43. $x^2 - 6x + 9 < 0$ **44.** $x^2 - 6x + 9 \leq 0$

45. $x^2 - 6x + 9 > 0$

46. $\dfrac{1}{x^2 - 6x + 9} > 0$

47. The graph of $y = x^2 - 4$ is shown in Figure 6. Use the graph to write the solution set for each of the following:

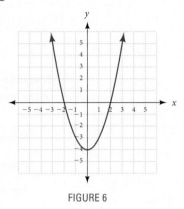

FIGURE 6

    a. $x^2 - 4 < 0$

    b. $x^2 - 4 > 0$

    c. $x^2 - 4 = 0$

48. The graph of $y = 4 - x^2$ is shown in Figure 7. Use the graph to write the solution set for each of the following:

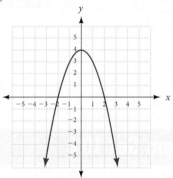

FIGURE 7

    a. $4 - x^2 < 0$

    b. $4 - x^2 > 0$

    c. $4 - x^2 = 0$

49. The graph of $y = x^2 - 3x - 10$ is shown in Figure 8. Use the graph to write the solution set for each of the following:

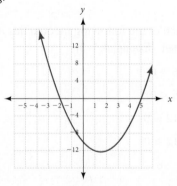

FIGURE 8

    a. $x^2 - 3x - 10 < 0$

    b. $x^2 - 3x - 10 > 0$

    c. $x^2 - 3x - 10 = 0$

50. The graph of $y = x^2 + x - 12$ is shown in Figure 9. Use the graph to write the solution set for each of the following:

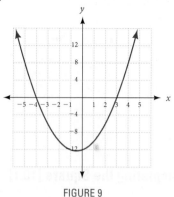

FIGURE 9

a. $x^2 + x - 12 < 0$

b. $x^2 - x - 12 > 0$

c. $x^2 + x - 12 = 0$

51. The graph of $y = x^3 - 3x^2 - x + 3$ is shown in Figure 10. Use the graph to write the solution set for each of the following:

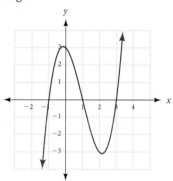

FIGURE 10

a. $x^3 - 3x^2 - x + 3 < 0$

b. $x^3 - 3x^2 - x + 3 > 0$

c. $x^3 - 3x^2 - x + 3 = 0$

52. The graph of $y = x^3 + 4x^2 - 4x - 16$ is shown in Figure 11. Use the graph to write the solution set for each of the following:

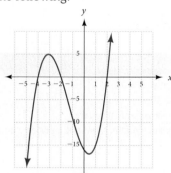

FIGURE 11

a. $x^3 + 4x^2 - 4x - 16 < 0$

b. $x^3 + 4x^2 - 4x - 16 > 0$

c. $x^3 + 4x^2 - 4x - 16 = 0$

Applying the Concepts

53. Dimensions of a Rectangle The length of a rectangle is 3 inches more than twice the width. If the area is to be at least 44 square inches, what are the possibilities for the width?

54. Dimensions of a Rectangle The length of a rectangle is 5 inches less than 3 times the width. If the area is to be less than 12 square inches, what are the possibilities for the width?

55. Revenue A manufacturer of headphones knows that the weekly revenue produced by selling x headphones is given by the equation $R = 1{,}300p - 100p^2$, where p is the price of each set of headphones. What price should she charge for each set of headphones if she wants her weekly revenue to be at least $4,000?

56. Revenue A manufacturer of small calculators knows that the weekly revenue produced by selling x calculators is given by the equation $R = 1{,}700p - 100p^2$, where p is the price of each calculator. What price should be charged for each calculator if the revenue is to be at least $7,000 each week?

Maintaining Your Skills

Use a calculator to evaluate. Give answers to 4 decimal places.

57. $\dfrac{50{,}000}{32{,}000}$

58. $\dfrac{2.4362}{1.9758} - 1$

59. $\dfrac{1}{2}\left(\dfrac{4.5926}{1.3876} - 2\right)$

60. $1 + \dfrac{0.06}{12}$

Solve each equation.

61. $\sqrt{3t - 1} = 2$

62. $\sqrt{4t + 5} + 7 = 3$

63. $\sqrt{x + 3} = x - 3$

64. $\sqrt{x + 3} = \sqrt{x} - 3$

Graph each equation.

65. $y = \sqrt[3]{x - 1}$

66. $y = \sqrt[3]{x} - 1$

Extending the Concepts

Graph the solution set for each inequality.

67. $x^2 - 2x - 1 < 0$

68. $x^2 - 6x + 7 < 0$

69. $x^2 - 8x + 13 > 0$

70. $x^2 - 10x + 18 > 0$

Theorem 1 for Square Roots [10.1]

1. If $(x - 3)^2 = 25$
then $x - 3 = \pm 5$
$x = 3 \pm 5$
$x = 8$ or $x = -2$

If $a^2 = b$, where b is a real number, then

$$a = \sqrt{b} \quad \text{or} \quad a = -\sqrt{b}$$

which can be written as $a = \pm\sqrt{b}$.

To Solve a Quadratic Equation by Completing the Square [10.1]

2. Solve $x^2 - 6x - 6 = 0$
$x^2 - 6x = 6$
$x^2 - 6x + \mathbf{9} = 6 + \mathbf{9}$
$(x - 3)^2 = 15$
$x - 3 = \pm\sqrt{15}$
$x = 3 \pm \sqrt{15}$

Step 1: Write the equation in the form $ax^2 + bx = c$.

Step 2: If $a \neq 1$, divide through by the constant a so the coefficient of x^2 is 1.

Step 3: Complete the square on the left side by adding the square of $\frac{1}{2}$ the coefficient of x to both sides.

Step 4: Write the left side of the equation as the square of a binomial. Simplify the right side if possible.

Step 5: Apply Theorem 1, and solve as usual.

The Quadratic Theorem [10.2]

3. If $2x^2 + 3x - 4 = 0$, then

$$x = \frac{-3 \pm \sqrt{9 - 4(2)(-4)}}{2(2)}$$

$$= \frac{-3 \pm \sqrt{41}}{4}$$

For any quadratic equation in the form $ax^2 + bx + c = 0$, $a \neq 0$, the two solutions are

$$x = \frac{-b \pm \sqrt{b^2 - 4ac}}{2a}$$

This last equation is known as the *quadratic formula.*

The Discriminant [10.3]

4. The discriminant for
$x^2 + 6x + 9 = 0$
is $D = 36 - 4(1)(9) = 0$,
which means the equation
has one rational solution.

The expression $b^2 - 4ac$ that appears under the radical sign in the quadratic formula is known as the *discriminant.*

We can classify the solutions to $ax^2 + bx + c = 0$:

The solutions are	When the discriminant is
Two complex numbers containing i	Negative
One rational number	Zero
Two rational numbers	A positive perfect square
Two irrational numbers	A positive number, but not a perfect square

◼ Equations Quadratic in Form [10.4]

5. The equation
$x^4 - x^2 - 12 = 0$ is quadratic
in x^2. Letting $y = x^2$ we have
$$y^2 - y - 12 = 0$$
$$(y - 4)(y + 3) = 0$$
$$y = 4 \quad \text{or} \quad y = -3$$
Resubstituting x^2 for y, we
have
$$x^2 = 4 \quad \text{or} \quad x^2 = -3$$
$$x = \pm 2 \quad \text{or} \quad x = \pm i\sqrt{3}$$

There are a variety of equations whose form is quadratic. We solve most of them by making a substitution so the equation becomes quadratic, and then solving the equation by factoring or the quadratic formula. For example,

The equation	*is quadratic in*
$(2x - 3)^2 + 5(2x - 3) - 6 = 0$	$2x - 3$
$4x^4 - 7x^2 - 2 = 0$	x^2
$2x - 7\sqrt{x} + 3 = 0$	$\sqrt{x}$

◼ Graphing Parabolas [10.5]

6. The graph of $y = x^2 - 4$ will
be a parabola. It will cross
the x-axis at 2 and -2, and
the vertex will be $(0, -4)$.

The graph of any equation of the form
$$y = ax^2 + bx + c \qquad a \neq 0$$
is a *parabola*. The graph is *concave up* if $a > 0$ and *concave down* if $a < 0$. The highest or lowest point on the graph is called the *vertex* and always has an x-coordinate of $x = \frac{-b}{2a}$.

◼ Quadratic Inequalities [10.6]

7. Solve $x^2 - 2x - 8 > 0$. We
factor and draw the sign
diagram:
$$(x - 4)(x + 2) > 0$$

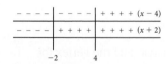

The solution is $x < -2$ or $x > 4$.

We solve quadratic inequalities by manipulating the inequality to get 0 on the right side and then factoring the left side. We then make a diagram that indicates where the factors are positive and where they are negative. From this sign diagram and the original inequality, we graph the appropriate solution set.

Chapter 10 Review

Solve each equation. [10.1]

1. $(2t - 5)^2 = 25$

2. $(3t - 2)^2 = 4$

3. $(3y - 4)^2 = -49$

4. $(2x + 6)^2 = 12$

Solve by completing the square. [10.1]

5. $2x^2 + 6x - 20 = 0$

6. $3x^2 + 15x = -18$

7. $a^2 + 9 = 6a$

8. $a^2 + 4 = 4a$

9. $2y^2 + 6y = -3$

10. $3y^2 + 3 = 9y$

Solve each equation. [10.2]

11. $\frac{1}{6}x^2 + \frac{1}{2}x - \frac{5}{3} = 0$

12. $8x^2 - 18x = 0$

13. $4t^2 - 8t + 19 = 0$

14. $100x^2 - 200x = 100$

15. $0.06a^2 + 0.05a = 0.04$

16. $9 - 6x = -x^2$

17. $(2x + 1)(x - 5) - (x + 3)(x - 2) = -17$

18. $2y^3 + 2y = 10y^2$

19. $5x^2 = -2x + 3$

20. $x^3 - 27 = 0$

21. $3 - \dfrac{2}{x} + \dfrac{1}{x^2} = 0$ **22.** $\dfrac{1}{x - 3} + \dfrac{1}{x + 2} = 1$

23. Profit The total cost (in dollars) for a company to produce x items per week is $C = 7x + 400$. The revenue for selling all x items is $R = 34x - 0.1x^2$. How many items must it produce and sell each week for its weekly profit to be $1,300? [10.2]

24. Profit The total cost (in dollars) for a company to produce x items per week is $C = 70x + 300$. The revenue for selling all x items is $R = 110x - 0.5x^2$. How many items must it produce and sell each week for its weekly profit to be $300? [10.2]

Use the discriminant to find the number and kind of solutions for each equation. [10.3]

25. $2x^2 - 8x = -8$

26. $4x^2 - 8x = -4$

27. $2x^2 + x - 3 = 0$

28. $5x^2 + 11x = 12$

29. $x^2 - x = 1$

30. $x^2 - 5x = -5$

31. $3x^2 + 5x = -4$

32. $4x^2 - 3x = -6$

Determine k so that each equation has exactly one real solution. [10.3]

33. $25x^2 - kx + 4 = 0$

34. $4x^2 + kx + 25 = 0$

35. $kx^2 + 12x + 9 = 0$

36. $kx^2 - 16x + 16 = 0$

37. $9x^2 + 30x + k = 0$

38. $4x^2 + 28x + k = 0$

For each of the following problems, find an equation that has the given solutions. [10.3]

39. $x = 3, x = 5$

40. $x = -2, x = 4$

41. $y = \dfrac{1}{2}, y = -4$

42. $t = 3, t = -3, t = 5$

Find all solutions. [10.4]

43. $(x - 2)^2 - 4(x - 2) - 60 = 0$

44. $6(2y + 1)^2 - (2y + 1) - 2 = 0$

45. $x^4 - x^2 = 12$

46. $x - \sqrt{x} - 2 = 0$

47. $2x - 11\sqrt{x} = -12$

48. $\sqrt{x + 5} = \sqrt{x} + 1$

49. $\sqrt{y + 21} + \sqrt{y} = 7$

50. $\sqrt{y + 9} - \sqrt{y - 6} = 3$

51. Projectile Motion An object is tossed into the air with an upward velocity of 10 feet per second from the top of a building h feet high. The time it takes for the object to hit the ground below is given by the formula $16t^2 - 10t - h = 0$. Solve this formula for t. [10.4]

52. Projectile Motion An object is tossed into the air with an upward velocity of v feet per second from the top of a 10-foot wall. The time it takes for the object to hit the ground below is given by the formula $16t^2 - vt - 10 = 0$. Solve this formula for t. [10.4]

Find the x-intercepts, if they exist, and the vertex for each parabola. Then use them to sketch the graph. [10.5]

53. $y = x^2 - 6x + 8$ **54.** $y = x^2 - 4$

Solve each inequality and graph the solution set. [10.6]

55. $x^2 - x - 2 < 0$

56. $3x^2 - 14x + 8 \leq 0$

57. $2x^2 + 5x - 12 \geq 0$

58. $(x + 2)(x - 3)(x + 4) > 0$

Simplify.

1. $11 + 20 \div 5 - 3 \cdot 5$

2. $4(15 - 19)^2 - 3(17 - 19)^3$

3. $\left(-\dfrac{2}{3}\right)^3$

4. $\left(-\dfrac{5}{4}\right)^2$

5. $4 + 8x - 3(5x - 2)$

6. $3 - 5[2x - 4(x - 2)]$

7. $\left(\dfrac{x^{-5}y^4}{x^{-2}y^{-3}}\right)^{-1}$

8. $\left(\dfrac{a^{-1}b^{-2}}{a^3b^{-3}}\right)^{-2}$

9. $\sqrt[3]{32}$

10. $\sqrt{50x^3}$

11. $8^{-2/3} + 25^{-1/2}$

12. $\left(\dfrac{8}{27}\right)^{-2/3}$

13. $\dfrac{1 - \dfrac{3}{4}}{1 + \dfrac{3}{4}}$

14. $\dfrac{2 + \dfrac{1}{5}}{1 - \dfrac{2}{3}}$

Reduce.

15. $\dfrac{5x^2 - 26xy - 24y^2}{5x + 4y}$

16. $\dfrac{x^2 - x - 6}{x + 2}$

Divide.

17. $\dfrac{3 + i}{i}$

18. $\dfrac{7 - i}{3 - 2i}$

Solve.

19. $\dfrac{7}{5}a - 6 = 15$

20. $\dfrac{3}{5}x - \dfrac{1}{2} = 3$

21. $|a| - 6 = 3$

22. $|a - 2| = -1$

23. $\dfrac{y}{3} + \dfrac{5}{y} = -\dfrac{8}{3}$

24. $\dfrac{a}{2} + \dfrac{3}{a - 3} = \dfrac{a}{a - 3}$

25. $(3x - 4)^2 = 18$

26. $3y^3 - y = 5y^2$

27. $\dfrac{2}{15}x^2 + \dfrac{1}{3}x + \dfrac{1}{5} = 0$

28. $0.06a^2 + 0.01a = -0.02$

29. $\sqrt{y + 3} = y + 3$

30. $\sqrt{x - 2} = 2 - \sqrt{x}$

31. $x - 3\sqrt{x} - 54 = 0$

32. $x^{2/3} - x^{1/3} = 6$

Solve each inequality and graph the solution.

33. $5 \le \dfrac{1}{4}x + 3 \le 8$

34. $|4x - 3| \ge 5$

35. $x^2 + 2x < 8$

36. $\dfrac{x - 1}{x + 3} \ge 0$

Solve each system.

37. $\begin{aligned} 3x - y &= 22 \\ -6x + 2y &= -4 \end{aligned}$

38. $\begin{aligned} 4x - 8y &= 6 \\ 6x - 12y &= 6 \end{aligned}$

Graph on a rectangular coordinate system.

39. $2x - 3y = 12$

40. $y = x^2 - x - 2$

41. Find the equation of the line passing through the points $\left(\dfrac{3}{2}, \dfrac{4}{3}\right)$ and $\left(\dfrac{1}{4}, -\dfrac{1}{3}\right)$.

42. Find the equation of the line perpendicular to $2x - 4y = 8$ and containing the point $(1, 5)$.

Factor completely.

43. $x^2 + 8x + 16 - y^2$

44. $(x - 1)^2 - 5(x - 1) + 4$

Rationalize the denominator.

45. $\dfrac{7}{\sqrt[3]{9}}$

46. $\dfrac{\sqrt{6}}{\sqrt{6} + \sqrt{2}}$

If $A = \{1, 2, 3, 4\}$ and $B = \{0, 2, 4, 6\}$, find the following.

47. $\{x \mid x \in B \text{ and } x \in A\}$

48. $\{x \mid x \notin A \text{ and } x \in B\}$

49. Geometry Find all three angles in a triangle if the smallest angle is one-fourth the largest angle and the remaining angle is 30° more than the smallest angle.

50. Investing A total of $8,000 is invested in two accounts. One account earns 5% per year, and the other earns 4% per year. If the total interest for the first year is $380; how much was invested in each account?

51. Inverse Variation y varies inversely with the square of x. If $y = 4$ when $x = \dfrac{5}{3}$, find y when $x = \dfrac{8}{3}$.

52. Direct Variation w varies directly with the square root of c. If w is 8 when c is 16, find w when c is 9.

Use the illustration to answer the following questions.

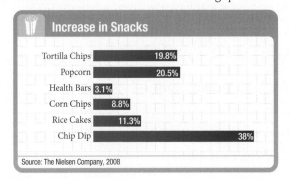

Increase in Snacks

Tortilla Chips 19.8%
Popcorn 20.5%
Health Bars 3.1%
Corn Chips 8.8%
Rice Cakes 11.3%
Chip Dip 38%

Source: The Nielsen Company, 2008

53. If a market sold $2,500 worth of popcorn last year, how much should it sell this year?

54. If a market sold $3,000 worth of tortilla chips last year, how much should it sell this year?

Solve each equation. [10.1, 10.2]

1. $(2x + 4)^2 = 25$

2. $(2x - 6)^2 = -8$

3. $y^2 - 10y + 25 = -4$

4. $(y + 1)(y - 3) = -6$

5. $8t^3 - 125 = 0$

6. $\dfrac{1}{a + 2} - \dfrac{1}{3} = \dfrac{1}{a}$

7. Solve the formula $64(1 + r)^2 = A$ for r. [10.1]

8. Solve $x^2 - 4x = -2$ by completing the square. [10.1]

9. If the length of the longest side of $30°–60°–90°$ triangle is 4 inches, find the lengths of the other two sides. [10.1]

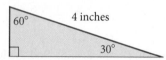

10. Projectile Motion An object projected upward with an initial velocity of 32 feet per second will rise and fall according to the equation $s(t) = 32t - 16t^2$, where s is its distance above the ground at time t. At what times will the object be 12 feet above the ground? [10.2]

11. Revenue The total weekly cost for a company to make x ceramic coffee cups is given by the formula $C(x) = 2x + 100$. If the weekly revenue from selling all x cups is $R(x) = 25x - 0.2x^2$, how many cups must it sell a week to make a profit of $200 a week? [10.2]

12. Find k so that $kx^2 = 12x - 4$ has one rational solution. [10.3]

13. Use the discriminant to identify the number and kind of solutions to $2x^2 - 5x = 7$. [10.3]

Find equations that have the given solutions. [10.3]

14. $x = 5, x = -\dfrac{2}{3}$

15. $x = 2, x = -2, x = 7$

Solve each equation. [10.4]

16. $4x^4 - 7x^2 - 2 = 0$

17. $(2t + 1)^2 - 5(2t + 1) + 6 = 0$

18. $2t - 7\sqrt{t} + 3 = 0$

19. Projectile Motion An object is tossed into the air with an upward velocity of 14 feet per second from the top of a building h feet high. The time it takes for the object to hit the ground below is given by the formula

$16t^2 - 14t - h = 0$. Solve this formula for t. [10.4]

Sketch the graph of each of the following. Give the coordinates of the vertex in each case. [10.5]

20. $y = x^2 - 2x - 3$

21. $y = -x^2 + 2x + 8$

22. Find an equation in the form $y = a(x - h)^2 + k$ that describes the graph of the parabola. [10.5]

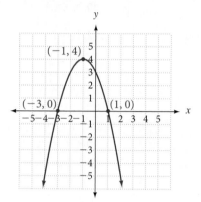

Solve the following inequalities and graph the solution on a number line. [10.6]

23. $x^2 - x - 6 \le 0$

24. $2x^2 + 5x > 3$

25. Profit Find the maximum weekly profit for a company with weekly costs of $C = 5x + 100$ and weekly revenue of $R = 25x - 0.1x^2$. [10.5]

26. The graph of $y = x^3 + 2x^2 - x - 2$ is shown below. Use the graph to write the solution set for each of the following. [10.6]

   a. $x^3 + 2x^2 - x - 2 < 0$

   b. $x^3 + 2x^2 - x - 2 > 0$

   c. $x^3 + 2x^2 - x - 2 = 0$

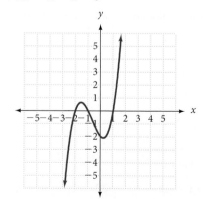

Chapter 10 Projects

QUADRATIC FUNCTIONS

Maximum Volume of a Box

Number of People 5

Time Needed 30 minutes

Equipment Graphing calculator and five pieces of graph paper

Background For many people, having a concrete model to work with allows them to visualize situations that they would have difficulty with if they had only a written description to work with. The purpose of this project is to rework a problem we have worked previously but this time with a concrete model.

Procedure You are going to make boxes of varying dimensions from rectangles that are 11 centimeters wide and 17 centimeters long.

1. Cut a rectangle from your graph paper that is 11 squares for the width and 17 squares for the length. Pretend that each small square is 1 centimeter by 1 centimeter. Do this with five pieces of paper.

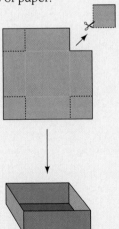

2. One person cuts off one square from each corner of their paper, then folds up the sides to form a box. Write down the length, width, and height of this box. Then calculate its volume.

3. The next person cuts a square that is two units on a side from each corner of their paper, then folds up the sides to form a box. Write down the length, width, and height of this box. Then calculate its volume.

4. The next person follows the same procedure, cutting a still larger square from each corner of their paper. This continues until the squares that are to be cut off are larger than the original piece of paper.

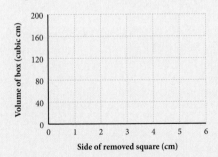

TABLE 1

VOLUME OF A BOX

Side of Square x (cm)	Length of Box L	Width of Box W	Height of Box H	Volume of Box V
1				
2				
3				
4				
5				

5. Enter the data from each box you have created into the table above. Then graph all the points (x, V) from the table.

6. Using what you have learned from filling in the table, write a formula for the volume of the box that is created when a square of side x is cut from each corner of the original piece of paper.

7. Graph the equation you found in Step 6 on a graphing calculator. Use the result to connect the points you plotted in the graph.

8. Use the graphing calculator to find the value of x that will give the maximum volume of the box. Your answer should be accurate to the nearest hundredth.

QUADRATIC FUNCTIONS

GROUP PROJECT

RESEARCH PROJECT

Arlie O. Petters

Arlie O. Petters, Massachusetts Institute of Technology

It can seem at times as if all the mathematicians of note lived 100 or more years ago. However, that is not the case. There are mathematicians doing research today, who are discovering new mathematical ideas and extending what is known about mathematics. One of the current group of research mathematicians is Arlie O. Petters. Dr. Petters earned his Ph.D. from the Massachusetts Institute of Technology in 1991 and currently is working in the mathematics department at Duke University. Use the Internet to find out more about Dr. Petters' education, awards, and research interests. Then use your results to give a profile, in essay form, of a present-day working mathematician.

11 Exponential and Logarithmic Functions

Introduction

The Google Earth image shows St. Andrews University in Scotland. The Scottish mathematician John Napier is the person credited with discovering logarithms. He began his university studies in Scotland at St. Andrews University in 1563, when he was 13 years old. In 1614, at age 64, he published his influential work on logarithms called *A description of the Wonderful Canon of Logarithms*.

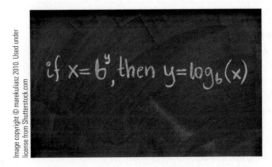

Logarithms, like the one shown above, will be the focus of our work in this chapter.

Simplify the following.

1. 2^0

2. 2^2

3. 2^{-2}

4. $27^{2/3}$

5. $\left(\dfrac{1}{3}\right)^0$

6. $\left(\dfrac{1}{3}\right)^2$

7. $\left(\dfrac{1}{3}\right)^{-2}$

Simplify with a calculator. Round your answers to four decimal places.

8. $2^{-1.25}$

9. $2^{-2.5}$

10. $1,200 \cdot 2^{-1.25}$

11. $1,500 \cdot 2^{-2.5}$

12. $500 \cdot (1.02)^{20}$

13. $500 \cdot (1.02)^{30}$

14. $1 + \dfrac{0.09}{12}$

15. $1 + \dfrac{0.08}{4}$

Solve.

16. $(x + 3)(x) = 2^2$

17. $\dfrac{x + 1}{x - 4} = 25$

Solve each equation for y.

18. $x = 2y - 3$

19. $x = y^2 - 2$

20. $(2y + 4)^2 + y^2 = 4$

Chapter Outline

11.1

Exponential Functions

OBJECTIVES

A Find function values for exponential functions.

B Graph exponential functions.

C Work application problems involving exponential functions.

TICKET TO SUCCESS

Keep these questions in mind as you read through the section. Then respond in your own words and in complete sentences.

1. What is an exponential function?
2. In an exponential function, explain why the base b cannot equal 1.
3. Explain continuously compounded interest.
4. What is the special number e in an exponential function?

To obtain an intuitive idea of how exponential functions behave, we can consider the heights attained by a bouncing ball. When a ball used in the game of racquetball is dropped from any height, the first bounce will reach a height that is $\frac{2}{3}$ of the original height. The second bounce will reach $\frac{2}{3}$ of the height of the first bounce, and so on, as shown in Figure 1.

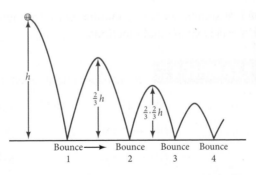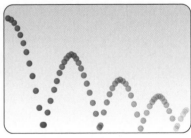

FIGURE 1

If the ball is dropped initially from a height of 1 meter, then during the first bounce it will reach a height of $\frac{2}{3}$ meter. The height of the second bounce will reach $\frac{2}{3}$ of the height reached on the first bounce. The maximum height of any bounce is $\frac{2}{3}$ of the height of the previous bounce.

$$\text{Initial height:} \quad h = 1$$

$$\text{Bounce 1:} \quad h = \frac{2}{3}(1) = \frac{2}{3}$$

$$\text{Bounce 2:} \quad h = \frac{2}{3}\left(\frac{2}{3}\right) = \left(\frac{2}{3}\right)^2$$

$$\text{Bounce 3:} \quad h = \frac{2}{3}\left(\frac{2}{3}\right)^2 = \left(\frac{2}{3}\right)^3$$

$$\text{Bounce 4:} \quad h = \frac{2}{3}\left(\frac{2}{3}\right)^3 = \left(\frac{2}{3}\right)^4$$

$$\vdots \qquad \qquad \vdots$$

$$\text{Bounce } n: \quad h = \frac{2}{3}\left(\frac{2}{3}\right)^n - 1 = \left(\frac{2}{3}\right)^n$$

This last equation is exponential in form. We classify all exponential functions together with the following definition.

> **Definition**
> An **exponential function** is any function that can be written in the form
> $$f(x) = b^x$$
> where b is a positive real number other than 1.

Each of the following is an exponential function:

$$f(x) = 2^x \qquad y = 3^x \qquad f(x) = \left(\frac{1}{4}\right)^x$$

A Evaluating Exponential Functions

The first step in becoming familiar with exponential functions is to find some values for specific exponential functions.

EXAMPLE 1 If the exponential functions f and g are defined by

$$f(x) = 2^x \qquad \text{and} \qquad g(x) = 3^x \qquad \text{then}$$

$$f(0) = 2^0 = 1 \qquad\qquad g(0) = 3^0 = 1$$

$$f(1) = 2^1 = 2 \qquad\qquad g(1) = 3^1 = 3$$

$$f(2) = 2^2 = 4 \qquad\qquad g(2) = 3^2 = 9$$

$$f(3) = 2^3 = 8 \qquad\qquad g(3) = 3^3 = 27$$

$$f(-2) = 2^{-2} = \frac{1}{2^2} = \frac{1}{4} \qquad g(-2) = 3^{-2} = \frac{1}{3^2} = \frac{1}{9}$$

$$f(-3) = 2^{-3} = \frac{1}{2^3} = \frac{1}{8} \qquad g(-3) = 3^{-3} = \frac{1}{3^3} = \frac{1}{27}$$

Half-life is a term used in science and mathematics that quantifies the amount of decay or reduction in an element. The half-life of iodine-131 is 8 days, which means that every 8 days a sample of iodine-131 will decrease to half of its original amount. If we start with A_0 micrograms of iodine-131, then after t days the sample will contain

$$A(t) = A_0 \cdot 2^{-t/8}$$

micrograms of iodine-131.

EXAMPLE 2 A patient is administered a 1,200-microgram dose of iodine-131. How much iodine-131 will be in the patient's system after 10 days and after 16 days?

SOLUTION The initial amount of iodine-131 is $A_0 = 1,200$, so the function that gives the amount left in the patient's system after t days is

$$A(t) = 1,200 \cdot 2^{-t/8}$$

After 10 days, the amount left in the patient's system is

$$A(10) = 1,200 \cdot 2^{-10/8}$$
$$= 1,200 \cdot 2^{-1.25}$$
$$\approx 504.5 \text{ micrograms}$$

After 16 days, the amount left in the patient's system is

$$A(16) = 1,200 \cdot 2^{-16/8}$$
$$= 1,200 \cdot 2^{-2}$$
$$= 300 \text{ micrograms}$$

> **NOTE**
> Recall that the symbol $\approx$ is read "is approximately equal to."

B Graphing Exponential Functions

We will now turn our attention to the graphs of exponential functions. Since the notation y is easier to use when graphing, and $y = f(x)$, for convenience we will write the exponential functions as

$$y = b^x$$

EXAMPLE 3 Sketch the graph of the exponential function $y = 2^x$.

SOLUTION Using the results of Example 1, we have the following table. Graphing the ordered pairs given in the table and connecting them with a smooth curve, we have the graph of $y = 2^x$ shown in Figure 2.

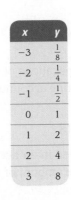

x	y
-3	$\frac{1}{8}$
-2	$\frac{1}{4}$
-1	$\frac{1}{2}$
0	1
1	2
2	4
3	8

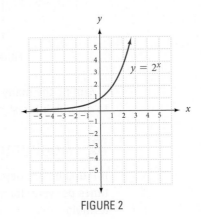

FIGURE 2

Notice that the graph does not cross the x-axis. It *approaches* the x-axis—in fact, we can get it as close to the x-axis as we want without it actually intersecting the x-axis. For the graph of $y = 2^x$ to intersect the x-axis, we would have to find a value of x that would make $2^x = 0$. Because no such value of x exists, the graph of $y = 2^x$ cannot intersect the x-axis.

EXAMPLE 4 Sketch the graph of $y = \left(\dfrac{1}{3}\right)^x$.

SOLUTION The table shown here gives some ordered pairs that satisfy the equation. Using the ordered pairs from the table, we have the graph shown in Figure 3.

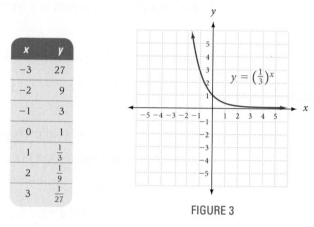

x	y
−3	27
−2	9
−1	3
0	1
1	$\frac{1}{3}$
2	$\frac{1}{9}$
3	$\frac{1}{27}$

FIGURE 3

The graphs of all exponential functions have two things in common: each crosses the y-axis at $(0, 1)$ since $b^0 = 1$; and none can cross the x-axis since $b^x = 0$ is impossible because of the restrictions on b.

Figures 4 and 5 show some families of exponential curves to help you become more familiar with them on an intuitive level.

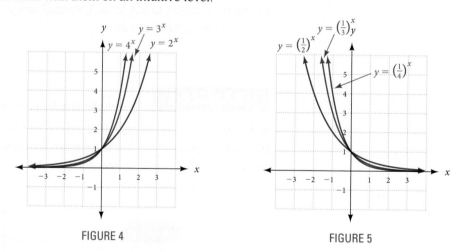

FIGURE 4   **FIGURE 5**

Among the many applications of exponential functions are the applications having to do with interest-bearing accounts. Here are the details.

C Applications

If P dollars are deposited in an account with annual interest rate r, compounded n times per year, then the amount of money in the account after t years is given by the formula

$$A(t) = P\left(1 + \frac{r}{n}\right)^{nt}$$

EXAMPLE 5   Suppose you deposit $500 in an account with an annual interest rate of 8% compounded quarterly. Find an equation that gives the amount of money in the account after t years. Then find

a. the amount of money in the account after 5 years.
b. the number of years it will take for the account to contain $1,000.

SOLUTION   First, we note that $P = 500$ and $r = 0.08$. Interest that is compounded quarterly is compounded 4 times a year, giving us $n = 4$. Substituting these numbers into the preceding formula, we have our function

$$A(t) = 500\left(1 + \frac{0.08}{4}\right)^{4t} = 500(1.02)^{4t}$$

a. To find the amount after 5 years, we let $t = 5$:

$$A(5) = 500(1.02)^{4 \cdot 5}$$

$$= 500(1.02)^{20}$$

$$\approx \$742.97$$

Our answer is found on a calculator and then rounded to the nearest cent.

b. To see how long it will take for this account to total $1,000, we graph the equation $Y_1 = 500(1.02)^{4X}$ on a graphing calculator and then look to see where it intersects the line $Y_2 = 1,000$. The two graphs are shown in Figure 6.

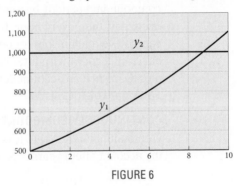

FIGURE 6

Using Zoom and Trace, or the Intersect function on the graphing calculator, we find that the two curves intersect at $X \approx 8.75$ and $Y = 1,000$. This means that our account will contain $1,000 after the money has been on deposit for 8.75 years.

The Natural Exponential Function

A very commonly occurring exponential function is based on a special number we denote with the letter e. The number e is a number like π. It is irrational and occurs in many formulas that describe the world around us. Like π, it can be approximated with a decimal number. Whereas π is approximately 3.1416, e is approximately 2.7183. (If you have a calculator with a key labeled $\boxed{e^x}$, you can use it to find e^1 to find a more accurate approximation to e.) We cannot give a more precise definition of the number e without using some of the topics taught in calculus. For the work we are going to do with the number e, we only need to know that it is an irrational number that is approximately 2.7183.

Here are a table and graph for the natural exponential function.

$$y = f(x) = e^x$$

x	$f(x) = e^x$
−2	$f(-2) = e^{-2} = \frac{1}{e^2} = 0.135$
−1	$f(-1) = e^{-1} = \frac{1}{e} \approx 0.368$
0	$f(0) = e^0 = 1$
1	$f(1) = e^1 = e \approx 2.72$
2	$f(2) = e^2 \approx 7.39$
3	$f(3) = e^3 \approx 20.09$

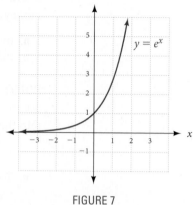

FIGURE 7

One common application of natural exponential functions is with interest-bearing accounts. In Example 5, we worked with the formula

$$A = P\left(1 + \frac{r}{n}\right)^{nt}$$

which gives the amount of money in an account if P dollars are deposited for t years at annual interest rate r, compounded n times per year. In Example 5, the number of compounding periods was 4. What would happen if we let the number of compounding periods become larger and larger, so that we compounded the interest every day, then every hour, then every second, and so on? If we take this as far as it can go, we end up compounding the interest every moment. When this happens, we have an account with interest that is compounded continuously, and the amount of money in such an account depends on the number e.

Here are the details.

CONTINUOUSLY COMPOUNDED INTEREST If P dollars are deposited in an account with annual interest rate r, compounded continuously, then the amount of money in the account after t years is given by the formula

$$A(t) = Pe^{rt}$$

EXAMPLE 6 Suppose you deposit $500 in an account with an annual interest rate of 8% compounded continuously. Find an equation that gives the amount of money in the account after t years. Then find the amount of money in the account after 5 years.

SOLUTION Since the interest is compounded continuously, we use the formula $A(t) = Pe^{rt}$. Substituting $P = 500$ and $r = 0.08$ into this formula we have

$$A(t) = 500e^{0.08t}$$

After 5 years, this account will contain

$$A(5) = 500e^{0.08 \cdot 5}$$

$$= 500e^{0.4}$$

$$\approx \$745.91$$

to the nearest cent. Compare this result with the answer to Example 5a.

Problem Set 11.1

Moving Toward Success

"The few who do are the envy of the many who only watch."

—Jim Rohn, 1930–2009, American businessman and author

1. How many hours a day outside of class are you setting aside for this course? Is that enough?

2. To where do the hours in your day go? How do you plan on fitting study time for this class into each day?

A Let $f(x) = 3^x$ and $g(x) = \left(\frac{1}{2}\right)^x$, and evaluate each of the following. [Examples 1–2]

1. $g(0)$

2. $f(0)$

3. $g(-1)$

4. $g(-4)$

5. $f(-3)$

6. $f(-1)$

7. $f(2) + g(-2)$

8. $f(2) - g(-2)$

B Graph each of the following functions. [Examples 3–4]

9. $y = 4^x$

10. $y = 2^{-x}$

11. $y = 3^{-x}$

12. $y = \left(\frac{1}{3}\right)^{-x}$

13. $y = 2^{x+1}$

14. $y = 2^{x-3}$

15. $y = e^x$

16. $y = e^{-x}$

Graph each of the following functions on the same coordinate system for positive values of x only.

17. $y = 2x, y = x^2, y = 2^x$

18. $y = 3x, y = x^3, y = 3^x$

19. On a graphing calculator, graph the family of curves $y = b^x$, $b = 2, 4, 6, 8$.

20. On a graphing calculator, graph the family of curves $y = b^x$, $b = \frac{1}{2}, \frac{1}{4}, \frac{1}{6}, \frac{1}{8}$.

C **Applying the Concepts** [Examples 5–6]

21. Bouncing Ball Suppose the ball mentioned in the introduction to this section is dropped from a height of 6 feet above the ground. Find an exponential

equation that gives the height h the ball will attain during the nth bounce. How high will it bounce on the fifth bounce?

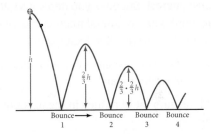

22. Bouncing Ball A golf ball is manufactured so that if it is dropped from A feet above the ground onto a hard surface, the maximum height of each bounce will be $\frac{1}{2}$ of the height of the previous bounce. Find an exponential equation that gives the height h the ball will attain during the nth bounce. If the ball is dropped from 10 feet above the ground onto a hard surface, how high will it bounce on the 8th bounce?

23. Exponential Decay The half-life of iodine-131 is 8 days. If a patient is administered a 1,400-microgram dose of iodine-131, how much iodine-131 will be in the patient's system after 8 days and after 11 days? (See Example 2.)

24. Exponential Decay The half-life of caffeine is about 6 hours. The amount of caffeine remaining in a person's system can be given by the formula, $C = C_0 \cdot 2^{-t/6}$, where C_0 is the amount of caffeine in the system initially and t is time in hours since ingestion. If a person drinks a cup of coffee containing 150 mg of caffeine at 8:00 AM, how much caffeine is in their body after 12 hours? At midnight?

25. Exponential Decay The half-life of ibuprofen is about 2 hours. The amount of ibuprofen remaining in a person's system can be given by the formula, $I = I_0 \cdot 2^{-t/2}$, where I_0 is the amount of ibuprofen in the system initially and t is time in hours since ingestion. If a person takes 600 mg of ibuprofen, how much ibuprofen is still in their body after 4 hours? After 8 hours?

26. Exponential Growth Automobiles built before 1993 use Freon in their air conditioners. The federal government now prohibits the manufacture of Freon. Because the supply of Freon is decreasing, the price per pound is increasing exponentially. Current estimates put the formula for the price per pound of Freon at $p(t) = 1.89(1.25)^t$, where t is the number of years since 1990. Find the price of Freon in 1995 and 1990. How much will Freon cost in the year 2014?

27. Compound Interest Suppose you deposit $1,200 in an account with an annual interest rate of 6% compounded quarterly. (See Example 5.)

a. Find an equation that gives the amount of money in the account after t years.

b. Find the amount of money in the account after 8 years.

c. How many years will it take for the account to contain $2,400?

d. If the interest were compounded continuously, how much money would the account contain after 8 years?

28. Compound Interest Suppose you deposit $500 in an account with an annual interest rate of 8% compounded monthly.

a. Find an equation that gives the amount of money in the account after t years.

b. Find the amount of money in the account after 5 years.

c. How many years will it take for the account to contain $1,000?

d. If the interest were compounded continuously, how much money would the account contain after 5 years?

29. Health Care In 1990, $699 billion were spent on health care expenditures. The amount of money, E, in billions spent on health care expenditures can be estimated using the function $E(t) = 78.16(1.11)^t$, where t is time in years since 1970 (U.S. Census Bureau).

a. How close was the estimate determined by the function in estimating the actual amount of money spent on health care expenditures in 1990?

b. What were the expected health care expenditures for 2008, 2009, and 2010? Round to the nearest billion.

30. Exponential Growth The cost of a can of Coca-Cola on January 1, 1960, was 10 cents. The function below gives the cost of a can of Coca-Cola t years after that.

$$C(t) = 0.10e^{0.0576t}$$

a. Use the function to fill in the table below. (Round to the nearest cent.)

Years Since 1960 t	Cost $C(t)$
0	$0.10
15	
40	
50	
90	

b. Use the table to find the cost of a can of Coca-Cola at the beginning of the year 2000.

c. In what year will a can of Coca-Cola cost $17.84?

31. Value of a Painting A painting is purchased as an investment for $150. If the painting's value doubles every 3 years, then its value is given by the function

$$V(t) = 150 \cdot 2^{t/3} \quad \text{for } t \geq 0$$

where t is the number of years since it was purchased, and $V(t)$ is its value (in dollars) at that time. Graph this function.

32. Value of a Painting A painting is purchased as an investment for $125. If the painting's value doubles every 5 years, then its value is given by the function

$$V(t) = 125 \cdot 2^{t/5} \quad \text{for } t \geq 0$$

where t is the number of years since it was purchased, and $V(t)$ is its value (in dollars) at that time. Graph this function.

33. Value of a Painting When will the painting mentioned in Problem 29 be worth $600?

34. Value of a Painting When will the painting mentioned in Problem 30 be worth $250?

35. Value of a Crane The function

$$V(t) = 450,000(1 - 0.30)^t$$

where V is value and t is time in years, can be used to find the value of a crane for the first 6 years of use.

Sandy Felsenthal/Corbis

a. What is the value of the crane after 3 years and 6 months?

b. State the domain of this function.

c. Sketch the graph of this function.

d. State the range of this function.

e. After how many years will the crane be worth only $85,000?

36. Value of a Printing Press The function $V(t) = 375,000(1 - 0.25)^t$, where V is value and t is time in years, can be used to find the value of a printing press during the first 7 years of use.

Lester Lefkowitz/Taxi/Getty Images

a. What is the value of the printing press after 4 years and 9 months?

b. State the domain of this function.

c. Sketch the graph of this function.

d. State the range of this function.

e. After how many years will the printing press be worth only $65,000?

Maintaining Your Skills

For each of the following relations, specify the domain and range, then indicate which are also functions.

37. $\{(1, 2), (3, 4), (4, 1)\}$

38. $\{(-2, 6), (-2, 8), (2, 3)\}$

State the domain for each of the following functions.

39. $y = \sqrt{3x + 1}$

40. $y = \dfrac{-4}{x^2 + 2x - 35}$

If $f(x) = 2x^2 - 18$ and $g(x) = 2x - 6$, find

41. $f(0)$ **42.** $g(f(0))$

43. $\dfrac{g(x + h) - g(x)}{h}$ **44.** $\dfrac{g(x)}{f(x)}$

Factor completely.

45. $2x^2 + 5x - 12$

46. $9x^2 - 33x - 12$

47. $5x^3 - 22x^2 + 8x$

48. $5x^3 - 3x^2 - 10x + 6$

49. $x^4 - 64x$

50. $5x^2 + 50x + 105$

Getting Ready for the Next Section

Solve each equation for y.

51. $x = 2y - 3$ **52.** $x = \dfrac{y + 7}{5}$

53. $x = y^2 - 2$ **54.** $x = (y + 4)^3$

55. $x = \dfrac{y - 4}{y - 2}$ **56.** $x = \dfrac{y + 5}{y - 3}$

57. $x = \sqrt{y - 3}$ **58.** $x = \sqrt{y} + 5$

Fill in the table of ordered pairs.

59. a. $y = 2x - 6$

x	y
0	
	0
5	
	10

b. $y = \dfrac{1}{2}x + 3$

x	y
0	
	5
10	
	0

60. a. $y = \frac{2}{3}x - 4$

x	y
−6	
	0
3	
9	

b. $y = \frac{3}{2}x + 6$

x	y
−8	
	6
2	
	3

Extending the Concepts

61. Drag Racing Previously we mentioned the dragster equipped with a computer. The table gives the speed of the dragster every second during one race at the 1993 Winternationals. Figure 8 is a line graph constructed from the data in the table.

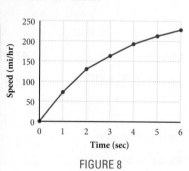

SPEED OF A DRAGSTER	
Elapsed Time (sec)	**Speed (mph)**
0	0.0
1	72.7
2	129.9
3	162.8
4	192.2
5	212.4
6	228.1

FIGURE 8

The graph of the following function contains the first point and the last point shown in Figure 8; that is, both (0, 0) and (6, 228.1) satisfy the function. Graph the function to see how close it comes to the other points in Figure 8.

$$s(t) = 250(1 - 1.5^{-t})$$

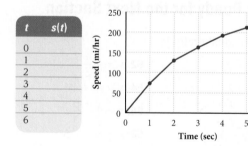

t	s(t)
0	
1	
2	
3	
4	
5	
6	

62. The graphs of two exponential functions are given in Figures 9 and 10. Use the graph to find the following:

a. $f(0)$ **b.** $f(-1)$

c. $f(1)$ **d.** $g(0)$

e. $g(1)$ **f.** $g(-1)$

g. $f(g(0))$ **h.** $g(f(0))$

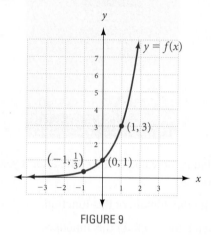

FIGURE 9

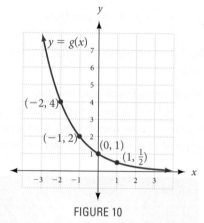

FIGURE 10

<table>
<tr><td>

11.2

The Inverse of a Function

</td></tr>
</table>

OBJECTIVES

A Find the equation of the inverse of a function.

B Sketch a graph of a function and its inverse.

TICKET TO SUCCESS

Keep these questions in mind as you read through the section. Then respond in your own words and in complete sentences.

1. What is the inverse of a function?
2. What is the relationship between the graph of a function and the graph of its inverse?
3. What is a one-to-one function?
4. In words, explain inverse function notation.

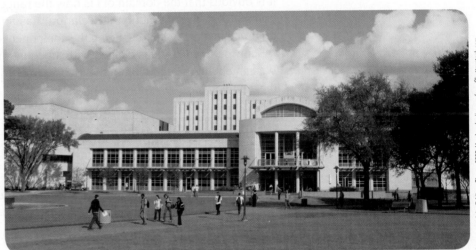

The following diagram (Figure 1) shows the route Justin takes to his college. He leaves his home and drives 3 miles east and then turns left and drives 2 miles north. When he leaves college to drive home, he drives the same two segments but in the reverse order and the opposite direction; that is, he drives 2 miles south, turns right, and drives 3 miles west. When he arrives home, he is right where he started. His route home "undoes" his route to college, leaving him where he began.

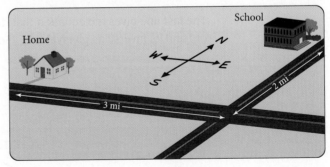

FIGURE 1

As you will see, the relationship between a function and its inverse function is similar to the relationship between Justin's route from home to college and his route from college to home.

A Finding the Inverse Relation

Suppose the function f is given by

$$f = \{(1, 4), (2, 5), (3, 6), (4, 7)\}$$

The inverse of f is obtained by reversing the order of the coordinates in each ordered pair in f. The inverse of f is the relation given by

$$g = \{(4, 1), (5, 2), (6, 3), (7, 4)\}$$

It is obvious that the domain of f is now the range of g, and the range of f is now the domain of g. Every function (or relation) has an inverse that is obtained from the original function by interchanging the components of each ordered pair.

Suppose a function f is defined with an equation instead of a list of ordered pairs. We can obtain the equation of the inverse of f by interchanging the role of x and y in the equation for f.

EXAMPLE 1   If the function f is defined by $f(x) = 2x - 3$, find the equation that represents the inverse of f.

SOLUTION   Since the inverse of f is obtained by interchanging the components of all the ordered pairs belonging to f, and each ordered pair in f satisfies the equation $y = 2x - 3$, we simply exchange x and y in the equation $y = 2x - 3$ to get the formula for the inverse of f.

$$x = 2y - 3$$

We now solve this equation for y in terms of x.

$$x + 3 = 2y$$

$$\frac{x + 3}{2} = y$$

$$y = \frac{x + 3}{2}$$

The last line gives the equation that defines the inverse of f. Let's compare the graphs of f and its inverse as given above. (See Figure 2.)

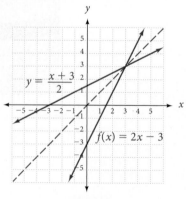

FIGURE 2

The graphs of f and its inverse have symmetry about the line $y = x$. This is a reasonable result since the one function was obtained from the other by interchanging x and y in the equation. The ordered pairs (a, b) and (b, a) always have symmetry about the line $y = x$. ■

B   Graph a Function with its Inverse

EXAMPLE 2   Graph the function $y = x^2 - 2$ and its inverse. Give the equation for the inverse.

SOLUTION   We can obtain the graph of the inverse of $y = x^2 - 2$ by graphing $y = x^2 - 2$ by the usual methods and then reflecting the graph about the line $y = x$.

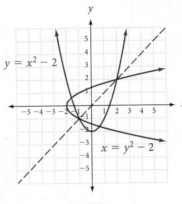

FIGURE 3

The equation that corresponds to the inverse of $y = x^2 - 2$ is obtained by interchanging x and y to get $x = y^2 - 2$.

We can solve the equation $x = y^2 - 2$ for y in terms of x as follows:

$$x = y^2 - 2$$

$$x + 2 = y^2$$

$$y = \pm\sqrt{x + 2}$$

■

USING TECHNOLOGY

Graphing an Inverse

One way to graph a function and its inverse is to use parametric equations. To graph the function $y = x^2 - 2$ and its inverse from Example 2, first set your graphing calculator to parametric mode. Then define the following set of parametric equations (Figure 4).

$$X1 = t; \ Y1 = t^2 - 2$$

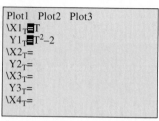

FIGURE 4

Set the window variables so that

$$-3 \le t \le 3, \ \text{step} = 0.05; \ 4 \le x \le 4; \ 4 \le y \le 4$$

Graph the function using the zoom-square command. Your graph should look similar to Figure 5.

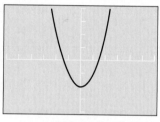

FIGURE 5

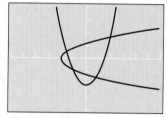

FIGURE 6

To graph the inverse, we need to interchange the roles of x and y for the original function. This is easily done by defining a new set of parametric equations that is just the reverse of the pair given above:

$$X2 = t^2 - 2, \ Y2 = t$$

Press GRAPH again, and you should now see the graphs of the original function and its inverse (Figure 6). If you trace to any point on either graph, you can alternate between the two curves to see how the coordinates of the corresponding ordered pairs compare. As Figure 7 illustrates, the coordinates of a point on one graph are reversed for the other graph.

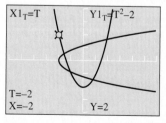

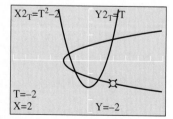

FIGURE 7

Comparing the graphs from Examples 1 and 2, we observe that the inverse of a function is not always a function. In Example 1, both f and its inverse have graphs that are nonvertical straight lines and therefore both represent functions. In Example 2, the inverse of function f is not a function since a vertical line crosses it in more than one place.

One-to-One Functions

We can distinguish between those functions with inverses that are also functions and those functions with inverses that are not functions with the following definition.

> **Definition**
> A function is a **one-to-one function** if every element in the range comes from exactly one element in the domain.

This definition indicates that a one-to-one function will yield a set of ordered pairs in which no two different ordered pairs have the same second coordinates. For example, the function

$$f = \{(2, 3), (-1, 3), (5, 8)\}$$

is not one-to-one because the element 3 in the range comes from both 2 and -1 in the domain. On the other hand, the function

$$g = \{(5, 7), (3, -1), (4, 2)\}$$

is a one-to-one function because every element in the range comes from only one element in the domain.

Horizontal Line Test

If we have the graph of a function, then we can determine if the function is one-to-one with the following test. If a horizontal line crosses the graph of a function in more than one place, then the function is not a one-to-one function because the points at which the horizontal line crosses the graph will be points with the same y-coordinates but different x-coordinates. Therefore, the function will have an element in the range (the y-coordinate) that comes from more than one element in the domain (the x-coordinates).

Of the functions we have covered previously, all the linear functions and exponential functions are one-to-one functions because no horizontal lines can be found that will cross their graphs in more than one place.

Functions Whose Inverses Are Also Functions

Because one-to-one functions do not repeat second coordinates, when we reverse the order of the ordered pairs in a one-to-one function, we obtain a relation in which no two ordered pairs have the same first coordinate—by definition, this relation must be a function. In other words, every one-to-one function has an inverse that is itself a function. Because of this, we can use function notation to represent that inverse.

> **Inverse Function Notation**
> If $y = f(x)$ is a one-to-one function, then the inverse of f is also a function and can be denoted by $y = f^{-1}(x)$.

NOTE

The notation f^{-1} does not represent the reciprocal of f; that is, the -1 in this notation is not an exponent. The notation f^{-1} is defined as representing the inverse function for a one-to-one function.

To illustrate, in Example 1 we found the inverse of $f(x) = 2x - 3$ was the function $y = \frac{x + 3}{2}$. We can write this inverse function with inverse function notation as

$$f^{-1}(x) = \frac{x + 3}{2}$$

However, the inverse of the function in Example 2 is not itself a function, so we do not use the notation $f^{-1}(x)$ to represent it.

EXAMPLE 3 Find the inverse of $g(x) = \frac{x - 4}{x - 2}$.

SOLUTION To find the inverse for g, we begin by replacing $g(x)$ with y to obtain

$$y = \frac{x - 4}{x - 2} \qquad \text{The original function}$$

To find an equation for the inverse, we exchange x and y.

$$x = \frac{y - 4}{y - 2} \qquad \text{The inverse of the original function}$$

To solve for y, we first multiply each side by $y - 2$ to obtain

$$x(y - 2) = y - 4$$

$$xy - 2x = y - 4 \qquad \text{Distributive property}$$

$$xy - y = 2x - 4 \qquad \text{Collect all terms containing } y \text{ on the left side}$$

$$y(x - 1) = 2x - 4 \qquad \text{Factor } y \text{ from each term on the left side}$$

$$y = \frac{2x - 4}{x - 1} \qquad \text{Divide each side by } x - 1$$

NOTE

Asymptotes will be covered in the next chapter.

Because our original function is one-to-one, as verified by the graph in Figure 8, its inverse is also a function. Therefore, we can use inverse function notation to write

$$g^{-1}(x) = \frac{2x - 4}{x - 1}$$

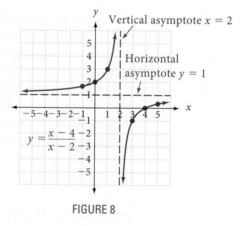

FIGURE 8

EXAMPLE 4 Graph the function $y = 2^x$ and its inverse $x = 2^y$.

SOLUTION We graphed $y = 2^x$ in the preceding section. We simply reflect its graph about the line $y = x$ to obtain the graph of its inverse $x = 2^y$. (See Figure 9.)

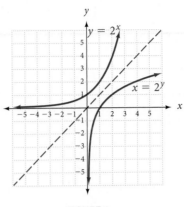

FIGURE 9

As you can see from the graph, $x = 2^y$ is a function. However, we do not have the mathematical tools to solve this equation for y. Therefore, we are unable to use the inverse function notation to represent this function. In the next section, we will give a definition that solves this problem. For now, we simply leave the equation as $x = 2^y$.

Functions, Relations, and Inverses—A Summary

Here is a summary of some of the things we know about functions, relations, and their inverses:

1. Every function is a relation, but not every relation is a function.

2. Every function has an inverse, but only one-to-one functions have inverses that are also functions.

3. The domain of a function is the range of its inverse, and the range of a function is the domain of its inverse.

4. If $y = f(x)$ is a one-to-one function, then we can use the notation $y = f^{-1}(x)$ to represent its inverse function.

5. The graph of a function and its inverse have symmetry about the line $y = x$.

6. If (a, b) belongs to the function f, then the point (b, a) belongs to its inverse.

Problem Set 11.2

Moving Toward Success

"The beautiful thing about learning is nobody can take it away from you."

—B.B. King, 1925–present, American Blues guitarist and songwriter

1. Why will increasing the effectiveness of the time you spend learning help you?

2. How will keeping the same studying schedule you had at the beginning of this class benefit you?

A For each of the following one-to-one functions, find the equation of the inverse. Write the inverse using the notation $f^{-1}(x)$. [Examples 1,3]

1. $f(x) = 3x - 1$

2. $f(x) = 2x - 5$

3. $f(x) = x^3$

4. $f(x) = x^3 - 2$

5. $f(x) = \dfrac{x - 3}{x - 1}$

6. $f(x) = \dfrac{x - 2}{x - 3}$

7. $f(x) = \dfrac{x - 3}{4}$

8. $f(x) = \dfrac{x + 7}{2}$

9. $f(x) = \dfrac{1}{2}x - 3$

10. $f(x) = \dfrac{1}{3}x + 1$

11. $f(x) = \dfrac{2x + 1}{3x + 1}$

12. $f(x) = \dfrac{3x + 2}{5x + 1}$

B For each of the following relations, sketch the graph of the relation and its inverse, and write an equation for the inverse. [Examples 2, 4]

13. $y = 2x - 1$ **14.** $y = 3x + 1$

15. $y = x^2 - 3$

16. $y = x^2 + 1$

17. $y = x^2 - 2x - 3$

18. $y = x^2 + 2x - 3$

19. $y = 3^x$ **20.** $y = \left(\dfrac{1}{2}\right)^x$

21. $y = 4$ **22.** $y = -2$

23. $y = \dfrac{1}{2}x^3$ **24.** $y = x^3 - 2$

25. $y = \dfrac{1}{2}x + 2$ **26.** $y = \dfrac{1}{3}x - 1$

27. $y = \sqrt{x + 2}$ **28.** $y = \sqrt{x} + 2$

29. Determine if the following functions are one-to-one.

a.

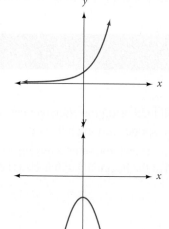

b.

c.

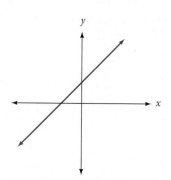

30. Could the following tables of values represent ordered pairs from one-to-one functions?

a.

x	y
-2	5
-1	4
0	3
1	4
2	5

b.

x	y
1.5	0.1
2.0	0.2
2.5	0.3
3.0	0.4
3.5	0.5

31. If $f(x) = 3x - 2$, then $f^{-1}(x) = \dfrac{x + 2}{3}$. Use these two functions to find

    a. $f(2)$ **b.** $f^{-1}(2)$

    c. $f(f^{-1}(2))$ **d.** $f^{-1}(f(2))$

32. If $f(x) = \dfrac{1}{2}x + 5$, then $f^{-1}(x) = 2x - 10$. Use these two functions to find

    a. $f(-4)$ **b.** $f^{-1}(-4)$

    c. $f(f^{-1}(-4))$ **d.** $f^{-1}(f(-4))$

33. Let $f(x) = \dfrac{1}{x}$, and find $f^{-1}(x)$.

34. Let $f(x) = \dfrac{a}{x}$, and find $f^{-1}(x)$. (*a* is a real number constant.)

Applying the Concepts

35. Reading Tables Evaluate each of the following functions using the functions defined by Tables 1 and 2.

    a. $f(g(-3))$ **b.** $g(f(-6))$

    c. $g(f(2))$ **d.** $f(g(3))$

    e. $f(g(-2))$ **f.** $g(f(3))$

    g. What can you conclude about the relationship between functions *f* and *g*?

TABLE 1	
x	f(x)
-6	3
2	-3
3	-2
6	4

TABLE 2	
x	g(x)
-3	2
-2	3
3	-6
4	6

36. Reading Tables Use the functions defined in Tables 1 and 2 in Problem 35 to answer the following questions.

   a. What are the domain and range of f?

   b. What are the domain and range of g?

   c. How are the domain and range of f related to the domain and range of g?

   d. Is f a one-to-one function?

   e. Is g a one-to-one function?

Maintaining Your Skills

Solve each equation.

37. $(2x - 1)^2 = 25$

38. $(3x + 5)^2 = -12$

39. What number would you add to $x^2 - 10x$ to make it a perfect square trinomial?

40. What number would you add to $x^2 - 5x$ to make it a perfect square trinomial?

Solve by completing the square.

41. $x^2 - 10x + 8 = 0$ **42.** $x^2 - 5x + 4 = 0$

43. $3x^2 - 6x + 6 = 0$ **44.** $4x^2 - 16x - 8 = 0$

Getting Ready for the Next Section

Simplify.

45. 3^{-2} **46.** 2^3

Solve.

47. $2 = 3x$ **48.** $3 = 5x$

49. $4 = x^3$ **50.** $12 = x^2$

Fill in the blanks to make each statement true.

51. $8 = 2^{\square}$ **52.** $27 = 3^{\square}$

53. $10,000 = 10^{\square}$ **54.** $1,000 = 10^{\square}$

55. $81 = 3^{\square}$ **56.** $81 = 9^{\square}$

57. $6 = 6^{\square}$ **58.** $1 = 5^{\square}$

Extending the Concepts

For each of the following functions, find $f^{-1}(x)$. Then show that $f(f^{-1}(x)) = x$.

59. $f(x) = 3x + 5$

60. $f(x) = 6 - 8x$

61. $f(x) = x^3 + 1$

62. $f(x) = x^3 - 8$

63. $f(x) = \dfrac{x - 4}{x - 2}$

64. $f(x) = \dfrac{x - 3}{x - 1}$

65. The graphs of a function and its inverse are shown in the figure. Use the graphs to find the following:

   a. $f(0)$ **b.** $f(1)$

   c. $f(2)$ **d.** $f^{-1}(1)$

   e. $f^{-1}(2)$ **f.** $f^{-1}(5)$

   g. $f^{-1}(f(2))$ **h.** $f(f^{-1}(5))$

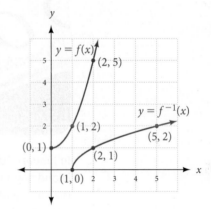

TICKET TO SUCCESS

Keep these questions in mind as you read through the section. Then respond in your own words and in complete sentences.

1. What is a logarithm?
2. What is the relationship between $y = 2^x$ and $y = \log_2 x$? How are their graphs related?
3. Will the graph of $y = \log_b x$ ever appear in the second or third quadrants? Explain.
4. How would you use the identity $b^{\log_b x} = x$ to simplify an equation involving a logarithm?

In January 2010, news organizations reported that an earthquake had rocked the tiny island nation of Haiti and caused massive destruction. They reported the strength of the earthquake as having measured 7.0 on the Richter scale. For comparison, Table 1 gives the Richter magnitude of a number of other earthquakes.

TABLE 1		
EARTHQUAKES		
Year	Earthquake	Richter Magnitude
1971	Los Angeles	6.6
1985	Mexico City	8.1
1989	San Francisco	7.1
1992	Kobe, Japan	7.2
1994	Northridge	6.6
1999	Armenia, Colombia	6.0
2004	Indian Ocean	9.1
2010	Haiti	7.0

Although the sizes of the numbers in the table do not seem to be very different, the intensity of the earthquakes they measure can be very different. For example, the 1989 San Francisco earthquake was more than 10 times stronger than the 1999 earthquake in Colombia. The reason behind this is that the Richter scale is a *logarithmic scale*. In this section, we start our work with logarithms, which will give you an understanding of the Richter scale. Let's begin.

A Logarithmic Functions

As you know from your work in the previous sections, equations of the form

$$y = b^x \qquad b > 0, b \neq 1$$

are called exponential functions. Because the equation of the inverse of a function can be obtained by exchanging x and y in the equation of the original function, the inverse of an exponential function must have the form

$$x = b^y \qquad b > 0, b \neq 1$$

Now, this last equation is actually the equation of a logarithmic function, as the following definition indicates:

> **Definition**
>
> The equation $y = \log_b x$ is read "y is the logarithm to the base b of x" and is equivalent to the equation
>
> $$x = b^y \qquad b > 0, b \neq 1$$
>
> *In words*: y is the number we raise b to in order to get x.

NOTATION When an equation is in the form $x = b^y$, it is said to be in exponential form. On the other hand, if an equation is in the form $y = \log_b x$, it is said to be in logarithmic form.

Here are some equivalent statements written in both forms.

NOTE
The ability to change from logarithmic form to exponential form is the first important thing to know about logarithms.

Exponential Form		Logarithmic Form
$8 = 2^3$	$\Leftrightarrow$	$\log_2 8 = 3$
$25 = 5^2$	$\Leftrightarrow$	$\log_5 25 = 2$
$0.1 = 10^{-1}$	$\Leftrightarrow$	$\log_{10} 0.1 = -1$
$\frac{1}{8} = 2^{-3}$	$\Leftrightarrow$	$\log_2 \frac{1}{8} = -3$
$r = z^s$	$\Leftrightarrow$	$\log_z r = s$

B Logarithmic Equations

EXAMPLE 1   Solve $\log_3 x = -2$ for x.

SOLUTION   In exponential form the equation looks like this:

$$x = 3^{-2}$$

$$\text{or} \quad x = \frac{1}{9}$$

The solution is $\frac{1}{9}$.

EXAMPLE 2   Solve $\log_x 4 = 3$.

SOLUTION   Again, we use the definition of logarithms to write the equation in exponential form.

$$4 = x^3$$

Taking the cube root of both sides, we have

$$\sqrt[3]{4} = \sqrt[3]{x^3}$$

$$x = \sqrt[3]{4}$$

The solution set is $\{\sqrt[3]{4}\}$.

EXAMPLE 3 Solve $\log_8 4 = x$.

SOLUTION We write the equation again in exponential form.

$$4 = 8^x$$

Since both 4 and 8 can be written as powers of 2, we write them in terms of powers of 2.

$$2^2 = (2^3)^x$$

$$2^2 = 2^{3x}$$

The only way the left and right sides of this last line can be equal is if the exponents are equal—that is, if

$$2 = 3x$$

$$\text{or} \quad x = \frac{2}{3}$$

The solution is $\frac{2}{3}$. We check as follows:

$$\log_8 4 = \frac{2}{3} \quad \Leftrightarrow \quad 4 = 8^{2/3}$$

$$4 = (\sqrt[3]{8})^2$$

$$4 = 2^2$$

$$4 = 4$$

The solution checks when used in the original equation.

> **NOTE**
> The first step in each of these first three examples is the same. In each case the first step in solving the equation is to put the equation in exponential form.

C Graphing Logarithmic Functions

Graphing logarithmic functions can be done using the graphs of exponential functions and the fact that the graphs of inverse functions have symmetry about the line $y = x$. Here's an example to illustrate.

EXAMPLE 4 Graph the equation $y = \log_2 x$.

SOLUTION The equation $y = \log_2 x$ is, by definition, equivalent to the exponential equation

$$x = 2^y$$

which is the equation of the inverse of the function

$$y = 2^x$$

We simply reflect the graph of $y = 2^x$ about the line $y = x$ to get the graph of $x = 2^y$, which is also the graph of $y = \log_2 x$. (See Figure 1.)

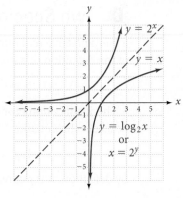

FIGURE 1

NOTE

From the graph of $y = \log_2 x$ it is apparent that $x > 0$ since the graph does not appear to the left of the y-axis. In general, the only variable in the expression $y = \log_b x$ that can be negative is y.

It is apparent from the graph that $y = \log_2 x$ is a function since no vertical line will cross its graph in more than one place. The same is true for all logarithmic equations of the form $y = \log_b x$, where b is a positive number other than 1. Note also that the graph of $y = \log_b x$ will always appear to the right of the y-axis, meaning that x will always be positive in the equation $y = \log_b x$.

USING TECHNOLOGY

Graphing Logarithmic Functions

As demonstrated in Example 4, we can graph the logarithmic function $y = \log_2 x$ as the inverse of the exponential function $y = 2^x$. Your graphing calculator most likely has a command to do this. First, define the exponential function as $Y_1 = 2^x$. To see the line of symmetry, define a second function $Y_2 = x$. Set the window variables so that

$$-6 \leq x \leq 6; \; -6 \leq y \leq 6$$

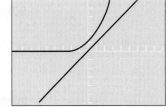

FIGURE 2

and use your zoom-square command to graph both functions. Your graph should look similar to the one shown in Figure 2. Now use the appropriate command to graph the inverse of the exponential function defined as Y_1 (Figure 3).

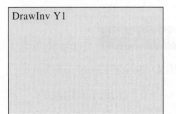

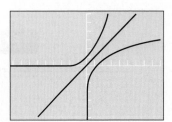

FIGURE 3

D Two Special Identities

If b is a positive real number other than 1, then each of the following is a consequence of the definition of a logarithm:

$$(1)\ b^{\log_b x} = x \quad \text{and} \quad (2)\ \log_b b^x = x$$

The justifications for these identities are similar. Let's consider only the first one. Consider the equation

$$y = \log_b x$$

By definition, it is equivalent to

$$x = b^y$$

Substituting $\log_b x$ for y in the last line gives us

$$x = b^{\log_b x}$$

The next examples in this section show how these two special properties can be used to simplify equations involving logarithms.

EXAMPLE 5 Simplify $\log_2 8$.

SOLUTION Substitute 2^3 for 8.

$$\log_2 8 = \log_2 2^3$$

$$= 3$$

EXAMPLE 6 Simplify $\log_{10} 10,000$.

SOLUTION 10,000 can be written as 10^4.

$$\log_{10} 10,000 = \log_{10} 10^4$$

$$= 4$$

EXAMPLE 7 Simplify $\log_b b$ ($b > 0$, $b \neq 1$).

SOLUTION Since $b^1 = b$, we have

$$\log_b b = \log_b b^1$$

$$= 1$$

EXAMPLE 8 Simplify $\log_b 1$ ($b > 0$, $b \neq 1$).

SOLUTION Since $1 = b^0$, we have

$$\log_b 1 = \log_b b^0$$

$$= 0$$

EXAMPLE 9 Simplify $\log_4(\log_5 5)$.

SOLUTION Since $\log_5 5 = 1$,

$$\log_4(\log_5 5) = \log_4 1$$

$$= 0$$

Application

As we mentioned in the introduction to this section, one application of logarithms is in measuring the magnitude of an earthquake. If an earthquake has a shock wave T times greater than the smallest shock wave that can be measured on a seismograph, then the magnitude M of the earthquake, as measured on the Richter scale, is given by the formula

$$M = \log_{10} T$$

(When we talk about the size of a shock wave, we are talking about its amplitude. The amplitude of a wave is half the difference between its highest point and its lowest point, as shown in Figure 4.)

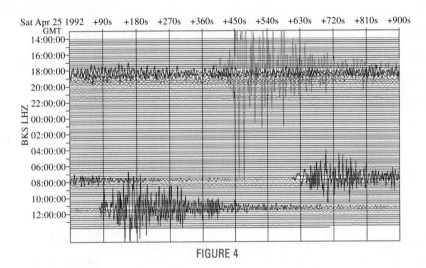

FIGURE 4

To illustrate the discussion, an earthquake that produces a shock wave that is 10,000 times greater than the smallest shock wave measurable on a seismograph will have a magnitude M on the Richter scale of

$$M = \log_{10} 10{,}000 = 4$$

EXAMPLE 10   If an earthquake has a magnitude of $M = 5$ on the Richter scale, what can you say about the size of its shock wave?

SOLUTION   To answer this question, we put $M = 5$ into the formula $M = \log_{10} T$ to obtain

$$5 = \log_{10} T$$

Writing this equation in exponential form, we have

$$T = 10^5 = 100{,}000$$

We can say that an earthquake that measures 5 on the Richter scale has a shock wave 100,000 times greater than the smallest shock wave measurable on a seismograph.

From Example 10 and the discussion that preceded it, we find that an earthquake of magnitude 5 has a shock wave that is 10 times greater than an earthquake of magnitude 4 because 100,000 is 10 times 10,000.

Problem Set 11.3

Moving Toward Success

"Just as iron rusts from disuse, even so does inaction spoil the intellect."

—Leonardo da Vinci, 1452–1519, Italian painter and sculptor

1. Why is it a poor choice to just come to class and not read the textbook?
2. Why is it a poor choice to rarely come to class and only read the textbook?

A Write each of the following equations in logarithmic form.

1. $2^4 = 16$
2. $3^2 = 9$
3. $125 = 5^3$
4. $16 = 4^2$
5. $0.01 = 10^{-2}$
6. $0.001 = 10^{-3}$
7. $2^{-5} = \dfrac{1}{32}$
8. $4^{-2} = \dfrac{1}{16}$
9. $\left(\dfrac{1}{2}\right)^{-3} = 8$
10. $\left(\dfrac{1}{3}\right)^{-2} = 9$
11. $27 = 3^3$
12. $81 = 3^4$

A Write each of the following equations in exponential form.

13. $\log_{10} 100 = 2$
14. $\log_2 8 = 3$
15. $\log_2 64 = 6$
16. $\log_2 32 = 5$
17. $\log_8 1 = 0$
18. $\log_9 9 = 1$
19. $\log_{10} 0.001 = -3$
20. $\log_{10} 0.0001 = -4$
21. $\log_6 36 = 2$
22. $\log_7 49 = 2$
23. $\log_5 \dfrac{1}{25} = -2$
24. $\log_3 \dfrac{1}{81} = -4$

B Solve each of the following equations for x. [Examples 1–3]

25. $\log_3 x = 2$
26. $\log_4 x = 3$
27. $\log_5 x = -3$
28. $\log_2 x = -4$
29. $\log_2 16 = x$
30. $\log_3 27 = x$
31. $\log_8 2 = x$
32. $\log_{25} 5 = x$
33. $\log_x 4 = 2$
34. $\log_x 16 = 4$
35. $\log_x 5 = 3$
36. $\log_x 8 = 2$

C Sketch the graph of each of the following logarithmic equations. [Example 4]

37. $y = \log_3 x$
38. $y = \log_{1/2} x$
39. $y = \log_{1/3} x$
40. $y = \log_4 x$
41. $y = \log_5 x$
42. $y = \log_{1/5} x$
43. $y = \log_{10} x$
44. $y = \log_{1/4} x$

D Simplify each of the following. [Example 5–9]

45. $\log_2 16$
46. $\log_3 9$
47. $\log_{25} 125$
48. $\log_9 27$
49. $\log_{10} 1{,}000$
50. $\log_{10} 10{,}000$
51. $\log_3 3$
52. $\log_4 4$
53. $\log_5 1$
54. $\log_{10} 1$
55. $\log_3(\log_6 6)$
56. $\log_5(\log_3 3)$
57. $\log_4[\log_2(\log_2 16)]$
58. $\log_4[\log_3(\log_2 8)]$

Applying the Concepts [Example 10]

$pH = -\log_{10}[H+]$, where $[H+]$ is the concentration of the hydrogen ions in solution. An acid solution has a pH below 7, and a basic solution has a pH higher than 7.

59. In distilled water, the concentration of hydrogen ions is $[H^+] = 10^{-7}$. What is the pH?

60. Find the pH of a bottle of vinegar if the concentration of hydrogen ions is $[H^+] = 10^{-3}$.

61. A hair conditioner has a pH of 6. Find the concentration of hydrogen ions, $[H^+]$, in the conditioner.

62. If a glass of orange juice has a pH of 4, what is the concentration of hydrogen ions, $[H^+]$, in the juice?

63. **Magnitude of an Earthquake** Find the magnitude M of an earthquake with a shock wave that measures $T = 100$ on a seismograph.

64. Magnitude of an Earthquake Find the magnitude M of an earthquake with a shock wave that measures $T = 100{,}000$ on a seismograph.

65. Shock Wave If an earthquake has a magnitude of 8 on the Richter scale, how many times greater is its shock wave than the smallest shock wave measurable on a seismograph?

66. Shock Wave If an earthquake has a magnitude of 6 on the Richter scale, how many times greater is its shock wave than the smallest shock wave measurable on a seismograph?

Maintaining Your Skills

Fill in the blanks to complete the square.

67. $x^2 + 10x + \boxed{} = (x + \boxed{})^2$

68. $x^2 + 4x + \boxed{} = (x + \boxed{})^2$

69. $y^2 - 2y + \boxed{} = (y - \boxed{})^2$

70. $y^2 + 3y + \boxed{} = (y + \boxed{})^2$

Solve.

71. $-y^2 = 9$

72. $7 + y^2 = 11$

73. $-x^2 - 8 = -4$

74. $10x^2 = 100$

75. $2x^2 + 4x - 3 = 0$

76. $3x^2 + 4x - 2 = 0$

77. $(2y - 3)(2y - 1) = -4$

78. $(y - 1)(3y - 3) = 10$

79. $t^3 - 125 = 0$

80. $8t^3 + 1 = 0$

81. $4x^5 - 16x^4 = 20x^3$

82. $3x^4 + 6x^2 = 6x^3$

83. $\dfrac{1}{x - 3} + \dfrac{1}{x + 2} = 1$

84. $\dfrac{1}{x + 3} + \dfrac{1}{x - 2} = 1$

Getting Ready for the Next Section

Simplify.

85. $8^{2/3}$

86. $27^{2/3}$

Solve.

87. $(x + 2)(x) = 2^3$

88. $(x + 3)(x) = 2^2$

89. $\dfrac{x - 2}{x + 1} = 9$

90. $\dfrac{x + 1}{x - 4} = 25$

Write in exponential form.

91. $\log_2 [(x + 2)(x)] = 3$

92. $\log_4 [x(x - 6)] = 2$

93. $\log_3 \left(\dfrac{x - 2}{x + 1} \right) = 4$

94. $\log_5 \left(\dfrac{x - 1}{x - 4} \right) = 2$

Extending the Concepts

95. The graph of the exponential function $y = f(x) = b^x$ is shown here. Use the graph to complete parts a through d.

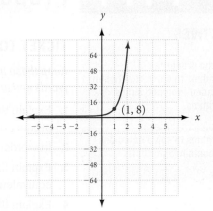

a. Fill in the table.

x	f(x)
−1	
0	
1	
2	

b. Fill in the table.

x	f⁻¹(x)
−1	
0	
1	
2	

c. Find the equation for $f(x)$.

d. Find the equation for $f^{-1}(x)$.

96. The graph of the exponential function $y = f(x) = b^x$ is shown here. Use the graph to complete parts a through d.

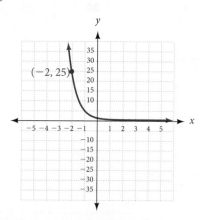

$(-2, 25)$

a. Fill in the table.

x	f(x)
−1	
0	
1	
2	

b. Fill in the table.

x	f⁻¹(x)
−1	
0	
1	
2	

c. Find the equation for $f(x)$.

d. Find the equation for $f^{-1}(x)$.

11.4 Properties of Logarithms

OBJECTIVES

A Use the properties of logarithms to convert between expanded form and single logarithms.

B Use the properties of logarithms to solve equations that contain logarithms.

TICKET TO SUCCESS

Keep these questions in mind as you read through the section. Then respond in your own words and in complete sentences.

1. Explain why the following statement is false: "The logarithm of a product is the product of the logarithms."

2. In words, explain property 2 of logarithms.

3. Explain the difference between $\log_b m + \log_b n$ and $\log_b(m + n)$. Are they equivalent?

4. Explain the difference between $\log_b(mn)$ and $(\log_b m)(\log_b n)$. Are they equivalent?

A *decibel* is a measurement of sound intensity, more specifically, the difference in intensity between two sounds. This difference is equal to ten times the logarithm of the ratio of the two sounds.

The following table and bar chart compare the intensity of sounds we are familiar with.

Decibels	Comparable to
10	A light whisper
20	Quiet conversation
30	Normal conversation
40	Light traffic
50	Typewriter, loud conversation
60	Noisy office
70	Normal traffic, quiet train
80	Rock music, subway
90	Heavy traffic, thunder
100	Jet plane at takeoff

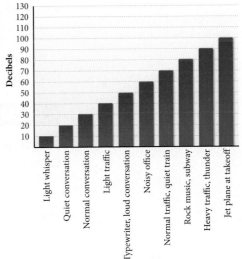

The precise definition for a *decibel* is

$$D = 10 \log_{10}\left(\frac{I}{I_0}\right)$$

where I is the intensity of the sound being measured, and I_0 is the intensity of the least audible sound. In this section, we will see that the preceding formula can also be written as

$$D = 10(\log_{10} I - \log_{10} I_0)$$

The rules we use to rewrite expressions containing logarithms are called the *properties of logarithms.* There are three of them.

A   Properties of Logarithms

For the following three properties, x, y, and b are all positive real numbers, $b \neq 1$, and r is any real number.

Property 1

$$\log_b (xy) = \log_b x + \log_b y$$

In words: The logarithm of a *product* is the *sum* of the logarithms.

Property 2

$$\log_b\left(\frac{x}{y}\right) = \log_b x - \log_b y$$

In words: The logarithm of a *quotient* is the *difference* of the logarithms.

Property 3

$$\log_b x^r = r \log_b x$$

In words: The logarithm of a number raised to a *power* is the *product* of the power and the logarithm of the number.

PROOF OF PROPERTY 1 To prove property 1, we simply apply the first identity for logarithms given in the preceding section.

$$b^{\log_b xy} = xy = (b^{\log_b x})(b^{\log_b y}) = b^{\log_b x + \log_b y}$$

Since the first and last expressions are equal and the bases are the same, the exponents $\log_b xy$ and $\log_b x + \log_b y$ must be equal. Therefore,

$$\log_b xy = \log_b x + \log_b y$$

The proofs of properties 2 and 3 proceed in much the same manner, so we will omit them here. The examples that follow show how the three properties can be used.

EXAMPLE 1 Expand, using the properties of logarithms: $\log_5 \dfrac{3xy}{z}$

SOLUTION Applying property 2, we can write the quotient of $3xy$ and z in terms of a difference.

$$\log_5 \frac{3xy}{z} = \log_5 3xy - \log_5 z$$

Applying property 1 to the product $3xy$, we write it in terms of addition.

$$\log_5 \frac{3xy}{z} = \log_5 3 + \log_5 x + \log_5 y - \log_5 z$$

EXAMPLE 2 Expand, using the properties of logarithms:

$$\log_2 \frac{x^4}{\sqrt{y} \cdot z^3}$$

SOLUTION We write $\sqrt{y}$ as $y^{1/2}$ and apply the properties.

$$\log_2 \frac{x^4}{\sqrt{y} \cdot z^3} = \log_2 \frac{x^4}{y^{1/2} z^3} \qquad \sqrt{y} = y^{1/2}$$

$$= \log_2 x^4 - \log_2(y^{1/2} \cdot z^3) \qquad \text{Property 2}$$

$$= \log_2 x^4 - (\log_2 y^{1/2} + \log_2 z^3) \qquad \text{Property 1}$$

$$= \log_2 x^4 - \log_2 y^{1/2} - \log_2 z^3 \qquad \text{Remove parentheses}$$

$$= 4 \log_2 x - \frac{1}{2} \log_2 y - 3 \log_2 z \qquad \text{Property 3}$$

We can also use the three properties to write an expression in expanded form as just one logarithm.

EXAMPLE 3 Write the following as a single logarithm:

$$2 \log_{10} a + 3 \log_{10} b - \frac{1}{3} \log_{10} c$$

SOLUTION We begin by applying property 3.

$$2 \log_{10} a + 3 \log_{10} b - \frac{1}{3} \log_{10} c$$

$$= \log_{10} a^2 + \log_{10} b^3 - \log_{10} c^{1/3} \qquad \text{Property 3}$$

$$= \log_{10}(a^2 \cdot b^3) - \log_{10} c^{1/3} \qquad \text{Property 1}$$

$$= \log_{10} \frac{a^2 b^3}{c^{1/3}} \qquad \text{Property 2}$$

$$= \log_{10} \frac{a^2 b^3}{\sqrt[3]{c}} \qquad c^{1/3} = \sqrt[3]{c}$$

B   Equations with Logarithms

The properties of logarithms along with the definition of logarithms are useful in solving equations that involve logarithms.

> **EXAMPLE 4**   Solve $\log_2(x + 2) + \log_2 x = 3$ for x.

SOLUTION   Applying property 1 to the left side of the equation allows us to write it as a single logarithm.

$$\log_2(x + 2) + \log_2 x = 3$$

$$\log_2[(x + 2)(x)] = 3$$

The last line can be written in exponential form using the definition of logarithms.

$$(x + 2)(x) = 2^3$$

Solve as usual.

$$x^2 + 2x = 8$$

$$x^2 + 2x - 8 = 0$$

$$(x + 4)(x - 2) = 0$$

$$x + 4 = 0 \quad \text{or} \quad x - 2 = 0$$

$$x = -4 \quad \text{or} \quad x = 2$$

In the previous section, we noted the fact that x in the expression $y = \log_b x$ cannot be a negative number. Since substitution of $x = -4$ into the original equation gives

$$\log_2(-2) + \log_2(-4) = 3$$

which contains logarithms of negative numbers, we cannot use -4 as a solution. The solution set is {2}.

Problem Set 11.4

Moving Toward Success

"Our greatest weakness lies in giving up. The most certain way to succeed is always to try just one more time."

—Thomas Edison, 1847–1931, American inventor and entrepreneur

1. What can you do to stay focused in this course if things are not going well for you?

2. How is patience important while taking this class?

A Use the three properties of logarithms given in this section to expand each expression as much as possible. [Examples 1–2]

1. $\log_3 4x$

2. $\log_2 5x$

3. $\log_6 \dfrac{5}{x}$

4. $\log_3 \dfrac{x}{5}$

5. $\log_2 y^5$

6. $\log_7 y^3$

7. $\log_9 \sqrt[3]{z}$

8. $\log_8 \sqrt{z}$

9. $\log_6 x^2 y^4$

10. $\log_{10} x^2 y^4$

11. $\log_5 \sqrt{x} \cdot y^4$

12. $\log_8 \sqrt[3]{xy^6}$

13. $\log_b \dfrac{xy}{z}$

14. $\log_b \dfrac{3x}{y}$

15. $\log_{10} \dfrac{4}{xy}$

16. $\log_{10} \dfrac{5}{4y}$

17. $\log_{10} \dfrac{x^2 y}{\sqrt{z}}$

18. $\log_{10} \dfrac{\sqrt{x} \cdot y}{z^3}$

19. $\log_{10} \dfrac{x^3 \sqrt{y}}{z^4}$

20. $\log_{10} \dfrac{x^4 \sqrt[3]{y}}{\sqrt{z}}$

21. $\log_b \sqrt[3]{\dfrac{x^2 y}{z^4}}$

22. $\log_b \sqrt[4]{\dfrac{x^4 y^3}{z^5}}$

A Write each expression as a single logarithm.
[Example 3]

23. $\log_b x + \log_b z$ **24.** $\log_b x - \log_b z$

25. $2 \log_3 x - 3 \log_3 y$

26. $4 \log_2 x + 5 \log_2 y$

27. $\dfrac{1}{2} \log_{10} x + \dfrac{1}{3} \log_{10} y$

28. $\dfrac{1}{3} \log_{10} x - \dfrac{1}{4} \log_{10} y$

29. $3 \log_2 x + \dfrac{1}{2} \log_2 y - \log_2 z$

30. $2 \log_3 x + 3 \log_3 y - \log_3 z$

31. $\dfrac{1}{2} \log_2 x - 3 \log_2 y - 4 \log_2 z$

32. $3 \log_{10} x - \log_{10} y - \log_{10} z$

33. $\dfrac{3}{2} \log_{10} x - \dfrac{3}{4} \log_{10} y - \dfrac{4}{5} \log_{10} z$

34. $3 \log_{10} x - \dfrac{4}{3} \log_{10} y - 5 \log_{10} z$

B Solve each of the following equations. [Example 4]

35. $\log_2 x + \log_2 3 = 1$ **36.** $\log_3 x + \log_3 3 = 1$

37. $\log_3 x - \log_3 2 = 2$ **38.** $\log_3 x + \log_3 2 = 2$

39. $\log_3 x + \log_3 (x - 2) = 1$

40. $\log_6 x + \log_6 (x - 1) = 1$

41. $\log_3 (x + 3) - \log_3 (x - 1) = 1$

42. $\log_4 (x - 2) - \log_4 (x + 1) = 1$

43. $\log_2 x + \log_2 (x - 2) = 3$

44. $\log_4 x + \log_4 (x + 6) = 2$

45. $\log_8 x + \log_8 (x - 3) = \dfrac{2}{3}$

46. $\log_{27} x + \log_{27} (x + 8) = \dfrac{2}{3}$

47. $\log_5 \sqrt{x} + \log_5 \sqrt{6x + 5} = 1$

48. $\log_2 \sqrt{x} + \log_2 \sqrt{6x + 5} = 1$

Applying the Concepts

49. Decibel Formula Use the properties of logarithms to rewrite the decibel formula $D = 10 \log_{10} \left(\dfrac{I}{I_0} \right)$ as

$$D = 10(\log_{10} I - \log_{10} I_0).$$

50. Decibel Formula In the decibel formula $D = 10 \log_{10} \left(\dfrac{I}{I_0} \right)$, the threshold of hearing, I_0, is

$$I_0 = 10^{-12} \text{ watts/meter}^2$$

Substitute 10^{-12} for I_0 in the decibel formula, and then show that it simplifies to

$$D = 10(\log_{10} I + 12)$$

51. Acoustic Powers The formula $N = \log_{10} \dfrac{P_1}{P_2}$ is used in radio electronics to find the ratio of the acoustic powers of two electric circuits in terms of their electric powers. Find N if P_1 is 100 and P_2 is 1. Then use the same two values of P_1 and P_2 to find N in the formula $N = \log_{10} P_1 - \log_{10} P_2$.

52. Henderson-Hasselbalch Formula Doctors use the Henderson-Hasselbalch formula to calculate the pH of a person's blood. pH is a measure of the acidity and/or the alkalinity of a solution. This formula is represented as

$$pH = 6.1 + \log_{10} \left(\dfrac{x}{y} \right)$$

where x is the base concentration and y is the acidic concentration. Rewrite the Henderson–Hasselbalch formula so that the logarithm of a quotient is not involved.

Maintaining Your Skills

Divide.

53. $\dfrac{12x^2 + y^2}{36}$ **54.** $\dfrac{x^2 + 4y^2}{16}$

55. Divide $25x^2 + 4y^2$ by 100

56. Divide $4x^2 + 9y^2$ by 36

Use the discriminant to find the number and kind of solutions to the following equations.

57. $2x^2 - 5x + 4 = 0$

58. $4x^2 - 12x = -9$

For each of the following problems, find an equation with the given solutions.

59. $x = -3, x = 5$

60. $x = 2, x = -2, x = 1$

61. $y = \dfrac{2}{3}, y = 3$

62. $y = -\dfrac{3}{5}, y = 2$

Getting Ready for the Next Section

Simplify.

63. 5^0

64. 4^1

65. $\log_3 3$

66. $\log_5 5$

67. $\log_b b^4$

68. $\log_a a^k$

Use a calculator to find each of the following. Write your answer in scientific notation with the first number in each answer rounded to the nearest tenth.

69. $10^{-5.6}$

70. $10^{-4.1}$

Divide and round to the nearest whole number.

71. $\dfrac{2.00 \times 10^8}{3.96 \times 10^6}$

72. $\dfrac{3.25 \times 10^{12}}{1.72 \times 10^{10}}$

11.5 Common Logarithms and Natural Logarithms

OBJECTIVES

A Use a calculator to find common logarithms.

B Use a calculator to find a number given its common logarithm.

C Simplify expressions containing natural logarithms.

TICKET TO SUCCESS

Keep these questions in mind as you read through the section. Then respond in your own words and in complete sentences.

1. What is a common logarithm?
2. What is a natural logarithm?
3. What is an antilogarithm?
4. Find ln e, and explain how you arrived at your answer.

Image copyright ©irin-k, 2009. Used under license from Shutterstock.com

Acid rain was first discovered in the 1960s by Gene Likens and his research team, who studied the damage caused by acid rain to Hubbard Brook in New Hampshire. Acid rain is rain with a pH of 5.6 and below. As you work your way through this section, you will see that pH is defined in terms of common logarithms. When you are finished with this section, you will have a more detailed knowledge of pH and acid rain.

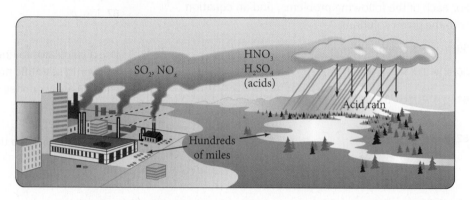

There are two kinds of logarithms that occur more frequently than other logarithms. They are logarithms with base 10 and natural logarithms, or logarithms with base e. Logarithms with a base of 10 are very common because our number system is a base-10 number system. For this reason, we call base-10 logarithms *common logarithms.*

A Common Logarithms

Definition

A **common logarithm** is a logarithm with a base of 10. Because common logarithms are used so frequently, it is customary, in order to save time, to omit notating the base; that is,

$$\log_{10} x = \log x$$

When the base is not shown, it is assumed to be 10.

> **NOTE**
> Remember, when the base is not written it is assumed to be 10.

Common logarithms of powers of 10 are simple to evaluate. We need only recognize that $\log 10 = \log_{10} 10 = 1$ and apply the third property of logarithms: $\log_b x^r = r \log_b x$.

$$\log 1,000 = \log 10^3 \ = 3 \log 10 \ \ = 3(1) \ \ = 3$$

$$\log 100 \ \ = \log 10^2 \ = 2 \log 10 \ \ = 2(1) \ \ = 2$$

$$\log 10 \ \ \ = \log 10^1 \ = 1 \log 10 \ \ = 1(1) \ \ = 1$$

$$\log 1 \ \ \ \ = \log 10^0 \ = 0 \log 10 \ \ = 0(1) \ \ = 0$$

$$\log 0.1 \ \ \ = \log 10^{-1} = -1 \log 10 = -1(1) = -1$$

$$\log 0.01 \ \ = \log 10^{-2} = -2 \log 10 = -2(1) = -2$$

$$\log 0.001 = \log 10^{-3} = -3 \log 10 = -3(1) = -3$$

To find common logarithms of numbers that are not powers of 10, we use a calculator with a **log** key.

Check the following logarithms to be sure you know how to use your calculator. (These answers have been rounded to the nearest ten-thousandth.)

$$\log 7.02 \approx 0.8463$$

$$\log 1.39 \approx 0.1430$$

$$\log 6.00 \approx 0.7782$$

$$\log 9.99 \approx 0.9996$$

EXAMPLE 1   Use a calculator to find log 2,760.

SOLUTION $\log 2{,}760 \approx 3.4409$

To work this problem on a scientific calculator, we simply enter the number 2,760 and press the key labeled $\boxed{\textbf{log}}$. On a graphing calculator we press the $\boxed{\textbf{log}}$ key first, then 2,760.

The 3 in the answer is called the *characteristic,* and the decimal part of the logarithm is called the *mantissa.*   ∎

EXAMPLE 2   Find log 0.0391.

SOLUTION $\log 0.0391 \approx -1.4078$ ∎

EXAMPLE 3   Find log 0.00523.

SOLUTION $\log 0.00523 \approx -2.2815$ ∎

B   Antilogarithms

EXAMPLE 4   Find x if log $x = 3.8774$.

SOLUTION   We are looking for the number whose logarithm is 3.8774. On a scientific calculator, we enter 3.8774 and press the key labeled $\boxed{\textbf{10}^x}$. On a graphing calculator we press $\boxed{\textbf{10}^x}$ first, then 3.8774. The result is 7,540 to four significant digits. Here's why:

$$\text{If} \qquad \log x = 3.8774$$

$$\text{then} \qquad x = 10^{3.8774}$$

$$\approx 7{,}540$$

The number 7,540 is called the *antilogarithm* or just *antilog* of 3.8774; that is, 7,540 is the number whose logarithm is 3.8774.   ∎

EXAMPLE 5   Find x if log $x = -2.4179$.

SOLUTION   Using the $\boxed{\textbf{10}^x}$ key, the result is 0.00382.

$$\text{If} \qquad \log x = -2.4179$$

$$\text{then} \qquad x = 10^{-2.4179}$$

$$\approx 0.00382$$

The antilog of -2.4179 is 0.00382; that is, the logarithm of 0.00382 is -2.4179.   ∎

Applications

Previously, we found that the magnitude M of an earthquake that produces a shock wave T times larger than the smallest shock wave that can be measured on a seismograph is given by the formula

$$M = \log_{10} T$$

We can rewrite this formula using our shorthand notation for common logarithms as

$$M = \log T$$

EXAMPLE 6   The San Francisco earthquake of 1906 is estimated to have measured 8.3 on the Richter scale. The San Fernando earthquake of 1971 measured 6.6 on the Richter scale. Find T for each earthquake, and then give some indication of how much stronger the 1906 earthquake was than the 1971 earthquake.

SOLUTION   For the 1906 earthquake:

If $\log T = 8.3$, then $T = 2.00 \times 10^8$.

For the 1971 earthquake:

If $\log T = 6.6$, then $T = 3.98 \times 10^6$.

Dividing the two values of T and rounding our answer to the nearest whole number, we have

$$\frac{2.00 \times 10^8}{3.98 \times 10^6} \approx 50$$

The shock wave for the 1906 earthquake was approximately 50 times stronger than the shock wave for the 1971 earthquake.

In chemistry, the pH of a solution is the measure of the acidity of the solution. The definition for pH involves common logarithms. Here it is:

$$pH = -\log[H^+]$$

where $[H^+]$ is the concentration of hydrogen ions in moles per liter. The range for pH is from 0 to 14. Pure water, a neutral solution, has a pH of 7. An acidic solution, such as vinegar, will have a pH less than 7, and an alkaline solution, such as ammonia, has a pH above 7.

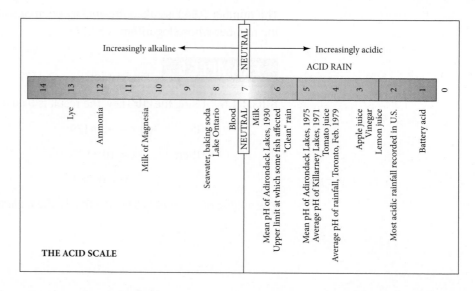

EXAMPLE 7   Normal rainwater has a pH of 5.6. What is the concentration of hydrogen ions in normal rainwater?

SOLUTION   Substituting 5.6 for pH in the formula pH = −log[H⁺], we have

$5.6 = -\log[H^+]$	Substitution
$\log[H^+] = -5.6$	Isolate the logarithm
$[H^+] = 10^{-5.6}$	Write in exponential form
$\approx 2.5 \times 10^{-6}$ mole per liter	Answer in scientific notation

EXAMPLE 8   The concentration of hydrogen ions in a sample of acid rain known to kill fish is 3.2×10^{-5} mole per liter. Find the pH of this acid rain to the nearest tenth.

SOLUTION   Substituting 3.2×10^{-5} for [H⁺] in the formula pH = −log[H⁺], we have

$pH = -\log[3.2 \times 10^{-5}]$	Substitution
$\approx -(-4.5)$	Evaluate the logarithm
$= 4.5$	Simplify

C Natural Logarithms

Suppose you pour yourself a cup of hot coffee when you begin to make breakfast. Newton's Law of Cooling can be used to model the temperature change of the coffee while sitting at room temperature. The equation used to determine the change in heat is

$$T = e^{kt + C} + R$$

where T is the temperature of the coffee at time t, R is the temperature of the room and k and C are constants. The e is used to denote the base of a natural logarithm, which leads us to our next definition.

Definition

A **natural logarithm** is a logarithm with a base of e. The natural logarithm of x is denoted by ln x; that is,

$$\ln x = \log_e x$$

We can assume that all our properties of exponents and logarithms hold for expressions with a base of e since e is a real number. Here are some examples intended to make you more familiar with the number e and natural logarithms.

EXAMPLE 9   Simplify each of the following expressions.

a. $e^0 = 1$

b. $e^1 = e$

c. $\ln e = 1$ In exponential form, $e^1 = e$

d. $\ln 1 = 0$ In exponential form, $e^0 = 1$

e. $\ln e^3 = 3$

f. $\ln e^{-4} = -4$

g. $\ln e^t = t$

EXAMPLE 10 Use the properties of logarithms to expand the expression $\ln Ae^{5t}$.

SOLUTION Since the properties of logarithms hold for natural logarithms, we have

$$\ln Ae^{5t} = \ln A + \ln e^{5t}$$

$$= \ln A + 5t \ln e$$

$$= \ln A + 5t \qquad \text{Because } \ln e = 1$$

EXAMPLE 11 If $\ln 2 = 0.6931$ and $\ln 3 = 1.0986$, find

a. $\ln 6$ **b.** $\ln 0.5$ **c.** $\ln 8$

SOLUTION

a. Since $6 = 2 \cdot 3$, we have

$$\ln 6 = \ln (2 \cdot 3)$$

$$= \ln 2 + \ln 3$$

$$= 0.6931 + 1.0986$$

$$= 1.7917$$

b. Writing 0.5 as $\frac{1}{2}$ and applying property 2 for logarithms gives us

$$\ln 0.5 = \ln \frac{1}{2}$$

$$= \ln 1 - \ln 2$$

$$= 0 - 0.6931$$

$$= -0.6931$$

c. Writing 8 as 2^3 and applying property 3 for logarithms, we have

$$\ln 8 = \ln 2^3$$

$$= 3 \ln 2$$

$$= 3(0.6931)$$

$$= 2.0793$$

Problem Set 11.5

Moving Toward Success

"We don't stop playing because we grow old;
we grow old because we stop playing."

—George Bernard Shaw, 1856–1950, Irish playwright
and Nobel Prize winner

1. Think of some fun things you may rather be

doing than studying. How can you work those things into your day and still meet your study schedule?

2. How might you reward yourself for doing something well in this class?

A Find the following logarithms. Round to four decimal places. [Examples 1–3]

1. log 378
2. log 426
3. log 37.8
4. log 42,600
5. log 3,780
6. log 0.4260
7. log 0.0378
8. log 0.0426
9. log 37,800
10. log 4,900
11. log 600
12. log 900
13. log 2,010
14. log 10,200
15. log 0.00971
16. log 0.0312
17. log 0.0314
18. log 0.00052
19. log 0.399
20. log 0.111

B Find x in the following equations. [Examples 4–5]

21. $\log x = 2.8802$
22. $\log x = 4.8802$
23. $\log x = -2.1198$
24. $\log x = -3.1198$
25. $\log x = 3.1553$
26. $\log x = 5.5911$
27. $\log x = -5.3497$
28. $\log x = -1.5670$
29. $\log x = -7.0372$
30. $\log x = -4.2000$
31. $\log x = 10$
32. $\log x = -1$
33. $\log x = -10$
34. $\log x = 1$
35. $\log x = 20$
36. $\log x = -20$
37. $\log x = -2$
38. $\log x = 4$
39. $\log x = \log_2 8$
40. $\log x = \log_3 9$

C Simplify each of the following expressions. [Example 9]

41. $\ln e$
42. $\ln 1$
43. $\ln e^5$
44. $\ln e^{-3}$
45. $\ln e^x$
46. $\ln e^y$

C Use the properties of logarithms to expand each of the following expressions. [Example 10]

47. $\ln 10e^{3t}$
48. $\ln 10e^{4t}$
49. $\ln Ae^{-2t}$
50. $\ln Ae^{-3t}$

C If $\ln 2 = 0.6931$, $\ln 3 = 1.0986$, and $\ln 5 = 1.6094$, find each of the following. [Example 11]

51. $\ln 15$
52. $\ln 10$
53. $\ln \dfrac{1}{3}$
54. $\ln \dfrac{1}{5}$
55. $\ln 9$
56. $\ln 25$
57. $\ln 16$
58. $\ln 81$

Use a calculator to evaluate each expression. Round your answers four decimal places, if necessary.

59. a. $\dfrac{\log 25}{\log 15}$ b. $\log \dfrac{25}{15}$

60. a. $\dfrac{\log 12}{\log 7}$ b. $\log \dfrac{12}{7}$

61. a. $\dfrac{\log 4}{\log 8}$ b. $\log \dfrac{4}{8}$

62. a. $\dfrac{\log 3}{\log 9}$ b. $\log \dfrac{3}{9}$

Applying the Concepts [Examples 6–8]

Measuring Acidity Previously we indicated that the pH of a solution is defined in terms of logarithms as

$$pH = -\log[H^+]$$

where $[H^+]$ is the concentration of hydrogen ions in that solution. Round to the nearest hundredth.

63. Find the pH of orange juice if the concentration of hydrogen ions in the juice is $[H^+] = 6.50 \times 10^{-4}$.

64. Find the pH of milk if the concentration of hydrogen ions in milk is $[H^+] = 1.88 \times 10^{-6}$.

65. Find the concentration of hydrogen ions in a glass of wine if the pH is 4.75.

66. Find the concentration of hydrogen ions in a bottle of vinegar if the pH is 5.75.

The Richter Scale Find the relative size T of the shock wave of earthquakes with the following magnitudes, as measured on the Richter scale. Round to the nearest hundredth.

67. 5.5

68. 6.6

69. 8.3

70. 8.7

71. Shock Wave How much larger is the shock wave of an earthquake that measures 6.5 on the Richter scale than one that measures 5.5 on the same scale?

72. Shock Wave How much larger is the shock wave of an earthquake that measures 8.5 on the Richter scale than one that measures 5.5 on the same scale?

73. Earthquake The chart below is a partial listing of earthquakes that were recorded in Canada in 2000. Complete the chart by computing the magnitude on the Richter scale, M, or the number of times the associated shock wave is larger than the smallest measurable shock wave, T.

Location	Date	Magnitude, M	Shock Wave, T
Moresby Island	January 23	4.0	
Vancouver Island	April 30		1.99×10^5
Quebec City	June 29	3.2	
Mould Bay	November 13	5.2	
St. Lawrence	December 14		5.01×10^3

Source: National Resources Canada, National Earthquake Hazards Program.

74. Earthquake On January 6, 2001, an earthquake with a magnitude of 7.7 on the Richter scale hit southern India (*National Earthquake Information Center*). By what factor was this earthquake's shock wave greater than the smallest measurable shock wave?

Depreciation The annual rate of depreciation r on a used car that is purchased for P dollars and is worth W dollars t years later can be found from the formula

$$\log(1 - r) = \frac{1}{t} \log \frac{W}{P}$$

75. Find the annual rate of depreciation on a car that is purchased for $9,000 and sold 5 years later for $4,500.

76. Find the annual rate of depreciation on a car that is purchased for $9,000 and sold 4 years later for $3,000. Round to the nearest tenth of a percent.

Two cars depreciate in value according to the following depreciation tables. In each case, find the annual rate of depreciation.

77.

Age in Years	Value in Dollars
new	7,550
5	5,750

78.

Age in Years	Value in Dollars
new	7,550
3	5,750

79. Getting Close to *e* Use a calculator to complete the following table.

x	$(1 + x)^{1/x}$
1	
0.5	
0.1	
0.01	
0.001	
0.0001	
0.00001	

What number does the expression $(1 + x)^{1/x}$ seem to approach as x gets closer and closer to zero?

80. Getting Close to *e* Use a calculator to complete the following table.

x	$\left(1 + \frac{1}{x}\right)^x$
1	
10	
50	
100	
500	
1,000	
10,000	
1,000,000	

What number does the expression $\left(1 + \frac{1}{x}\right)^x$ seem to approach as x gets larger and larger?

Maintaining Your Skills

Solve each equation.

81. $(y + 3)^2 + y^2 = 9$

82. $(2y + 4)^2 + y^2 = 4$

83. $(x + 3)^2 + 1^2 = 2$

84. $(x - 3)^2 + (-1)^2 = 10$

Solve each equation.

85. $x^4 - 2x^2 - 8 = 0$

86. $x^{2/3} - 5x^{1/3} + 6 = 0$

87. $2x - 5\sqrt{x} + 3 = 0$

88. $(3x + 1) - 6\sqrt{3x + 1} + 8 = 0$

Getting Ready for the Next Section

Solve.

89. $5(2x + 1) = 12$

90. $4(3x - 2) = 21$

Use a calculator to evaluate. Give answers to four decimal places.

91. $\dfrac{100{,}000}{32{,}000}$

92. $\dfrac{1.4982}{3.5681} + 3$

93. $\dfrac{1}{2}\left(\dfrac{-0.6931}{1.4289} + 3\right)$

94. $1 + \dfrac{0.04}{52}$

Use the power rule to rewrite the following logarithms.

95. $\log 1.05^t$

96. $\log 1.033^t$

Use identities to simplify.

97. $\ln e^{0.05t}$

98. $\ln e^{-0.000121t}$

11.6 Exponential Equations and Change of Base

OBJECTIVES

A Solve exponential equations.

B Use the change-of-base property.

C Solve application problems involving logarithmic or exponential equations.

TICKET TO SUCCESS

Keep these questions in mind as you read through the section. Then respond in your own words and in complete sentences.

1. What is doubling time?
2. What is an exponential equation and how would logarithms help you solve it?
3. What is the change-of-base property?
4. Write an application modeled by the equation $A = 10{,}000\left(1 + \dfrac{0.08}{2}\right)^{2(5)}$.

For items involved in exponential growth, the time it takes for a quantity to double is called the *doubling time.* For example, if you invest $5,000 in an account that pays 5% annual interest, compounded quarterly, you may want to know how long it will take for your money to double in value. You can find this doubling time if you can solve the equation

$$10,000 = 5,000 \, (1.0125)^{4t}$$

As you will see as you progress through this section, logarithms are the key to solving equations of this type.

A   Exponential Equations

Logarithms are very important in solving equations in which the variable appears as an exponent. The equation

$$5^x = 12$$

is an example of one such equation. Equations of this form are called *exponential equations.* Since the quantities 5^x and 12 are equal, so are their common logarithms. We begin our solution by taking the logarithm of both sides.

$$\log 5^x = \log 12$$

We now apply property 3 for logarithms, $\log x^r = r \log x$, to turn x from an exponent into a coefficient.

$$x \log 5 = \log 12$$

Dividing both sides by log 5 gives us

$$x = \frac{\log 12}{\log 5}$$

If we want a decimal approximation to the solution, we can find log 12 and log 5 on a calculator and divide.

$$x \approx \frac{1.0792}{0.6990}$$

$$\approx 1.5439$$

The complete problem looks like this:

$$5^x = 12$$

$$\log 5^x = \log 12$$

$$x \log 5 = \log 12$$

$$x = \frac{\log 12}{\log 5}$$

$$\approx \frac{1.0792}{0.6990}$$

$$\approx 1.5439$$

Here is another example of solving an exponential equation using logarithms.

> **NOTE**
> Your answers to the problems in the problem sets may differ from those in the book slightly, depending on whether you round off intermediate answers or carry out the complete problem on a calculator. The differences will occur only in the right-most digit of your answers.

EXAMPLE 1   Solve $25^{2x+1} = 15$ for x.

SOLUTION   Taking the logarithm of both sides and then writing the exponent $(2x + 1)$ as a coefficient, we proceed as follows:

$$25^{2x+1} = 15$$

$$\log 25^{2x+1} = \log 15 \qquad \text{Take the log of both sides}$$

$$(2x + 1) \log 25 = \log 15 \qquad \text{Property 3}$$

$$2x + 1 = \frac{\log 15}{\log 25} \qquad \text{Divide by log 25}$$

$$2x = \frac{\log 15}{\log 25} - 1 \qquad \text{Add } -1 \text{ to both sides}$$

$$x = \frac{1}{2}\left(\frac{\log 15}{\log 25} - 1\right) \qquad \text{Multiply both sides by } \tfrac{1}{2}$$

Using a calculator, we can write a decimal approximation to the answer.

$$x \approx \frac{1}{2}\left(\frac{1.1761}{1.3979} - 1\right)$$

$$\approx \frac{1}{2}(0.8413 - 1)$$

$$\approx \frac{1}{2}(-0.1587)$$

$$\approx -0.0793$$

If you invest P dollars in an account with an annual interest rate r that is compounded n times a year, then t years later the amount of money in that account will be

$$A = P\left(1 + \frac{r}{n}\right)^{nt}$$

EXAMPLE 2   How long does it take for $5,000 to double if it is deposited in an account that yields 5% interest compounded once a year?

SOLUTION   Substituting $P = 5{,}000$, $r = 0.05$, $n = 1$, and $A = 10{,}000$ into our formula, we have

$$10{,}000 = 5{,}000(1 + 0.05)^t$$

$$10{,}000 = 5{,}000(1.05)^t$$

$$2 = (1.05)^t \qquad \text{Divide by 5,000}$$

This is an exponential equation. We solve by taking the logarithm of both sides.

$$\log 2 = \log(1.05)^t$$

$$\log 2 = t \log 1.05$$

Dividing both sides by log 1.05, we have

$$t = \frac{\log 2}{\log 1.05}$$

$$\approx 14.2$$

It takes a little over 14 years for $5,000 to double if it earns 5% interest per year, compounded once a year.

B Change of Base

There is a fourth property of logarithms we have not yet considered. This last property allows us to change from one base to another and is therefore called the *change-of-base property*.

Property 4

If a and b are both positive numbers other than 1, and if $x > 0$, then

$$\log_a x = \frac{\log_b x}{\log_b a}$$

$$\uparrow \qquad \uparrow$$

Base a Base b

The logarithm on the left side has a base of a, and both logarithms on the right side have a base of b. This allows us to change from base a to any other base b that is a positive number other than 1. Here is a proof of property 4 for logarithms.

PROOF We begin by writing the identity

$$a^{\log_a x} = x$$

Taking the logarithm base b of both sides and writing the exponent $\log_a x$ as a coefficient, we have

$$\log_b a^{\log_a x} = \log_b x$$

$$\log_a x \log_b a = \log_b x$$

Dividing both sides by $\log_b a$, we have the desired result:

$$\frac{\log_a x \log_b a}{\log_b a} = \frac{\log_b x}{\log_b a}$$

$$\log_a x = \frac{\log_b x}{\log_b a}$$

We can use this property to find logarithms we could not otherwise compute on our calculators—that is, logarithms with bases other than 10 or e. The next example illustrates the use of this property.

EXAMPLE 3 Find $\log_8 24$.

SOLUTION Since we do not have base-8 logarithms on our calculators, we can change this expression to an equivalent expression that contains only base-10 logarithms:

$$\log_8 24 = \frac{\log 24}{\log 8} \qquad \text{Property 4}$$

Don't be confused. We did not just drop the base, we changed to base 10. We could have written the last line like this:

$$\log_8 24 = \frac{\log_{10} 24}{\log_{10} 8}$$

From our calculators, we write

$$\log_8 24 \approx \frac{1.3802}{0.9031}$$

$$\approx 1.5283$$

C Applications

EXAMPLE 4 Suppose that the population of a small city is 32,000 in the beginning of 2010. The city council assumes that the population size t years later can be estimated by the equation

$$P = 32,000e^{0.05t}$$

Approximately when will the city have a population of 50,000?

SOLUTION   We substitute 50,000 for P in the equation and solve for t.

$$50{,}000 = 32{,}000e^{0.05t}$$

$$1.5625 = e^{0.05t} \qquad\qquad \frac{50{,}000}{32{,}000} = 1.5625$$

To solve this equation for t, we can take the natural logarithm of each side.

$$\ln 1.5625 = \ln e^{0.05t}$$

$$\ln 1.5625 = 0.05t \ln e \qquad \text{Property 3 for logarithms}$$

$$\ln 1.5625 = 0.05t \qquad\qquad \text{Because } \ln e = 1$$

$$t = \frac{\ln 1.5625}{0.05} \qquad\qquad \text{Divide each side by 0.05}$$

$$\approx 8.93 \text{ years}$$

We can estimate that the population will reach 50,000 toward the end of 2018.

USING TECHNOLOGY

Solving Exponential Equations

We can solve the equation $50{,}000 = 32{,}000e^{0.05t}$ from Example 4 with a graphing calculator by defining the expression on each side of the equation as a function. The solution to the equation will be the x-value of the point where the two graphs intersect.

First, define functions $Y_1 = 50{,}000$ and $Y_2 = 32{,}000e^{(0.05x)}$ as shown in Figure 1. Set your window variables so that

$$0 \le x \le 15;\ 0 \le y \le 70{,}000,\ \text{scale} = 10{,}000$$

Graph both functions, then use the appropriate command on your calculator to find the coordinates of the intersection point. From Figure 2 we see that the x-coordinate of this point is $x \approx 8.93$.

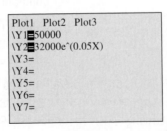

FIGURE 1

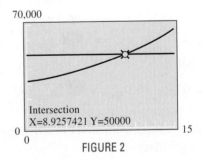

FIGURE 2

Problem Set 11.6

Moving Toward Success

"Wonder is the beginning of wisdom."

— Greek proverb

1. Can you study too much? Why or why not?
2. What do you plan to do to help yourself if you feel burned out during this class?

A Solve each exponential equation. Use a calculator to write the answer to four decimal places. [Example 1]

1. $3^x = 5$
2. $4^x = 3$
3. $5^x = 3$
4. $3^x = 4$
5. $5^{-x} = 12$
6. $7^{-x} = 8$
7. $12^{-x} = 5$
8. $8^{-x} = 7$
9. $8^{x+1} = 4$
10. $9^{x+1} = 3$
11. $4^{x-1} = 4$
12. $3^{x-1} = 9$
13. $3^{2x+1} = 2$
14. $2^{2x+1} = 3$
15. $3^{1-2x} = 2$
16. $2^{1-2x} = 3$
17. $15^{3x-4} = 10$
18. $10^{3x-4} = 15$
19. $6^{5-2x} = 4$
20. $9^{7-3x} = 5$

B Use the change-of-base property and a calculator to find a decimal approximation to each of the following logarithms. [Example 3]

21. $\log_8 16$
22. $\log_9 27$
23. $\log_{16} 8$
24. $\log_{27} 9$
25. $\log_7 15$
26. $\log_3 12$
27. $\log_{15} 7$
28. $\log_{12} 3$
29. $\log_8 240$
30. $\log_6 180$
31. $\log_4 321$
32. $\log_5 462$

Find a decimal approximation to each of the following natural logarithms.

33. $\ln 345$
34. $\ln 3,450$
35. $\ln 0.345$
36. $\ln 0.0345$
37. $\ln 10$
38. $\ln 100$
39. $\ln 45,000$
40. $\ln 450,000$

C **Applying the Concepts** [Examples 2, 4]

Find estimates for the following.

41. **Compound Interest** How long will it take for $500 to double if it is invested at 6% annual interest compounded twice a year?

42. **Compounded Interest** How long will it take for $500 to double if it is invested at 6% annual interest compounded 12 times a year?

43. **Compound Interest** How long will it take for $1,000 to triple if it is invested at 12% annual interest compounded 6 times a year?

44. **Compound Interest** How long will it take for $1,000 to become $4,000 if it is invested at 12% annual interest compounded 6 times a year?

45. **Doubling Time** How long does it take for an amount of money P to double itself if it is invested at 8% interest compounded 4 times a year?

46. **Tripling Time** How long does it take for an amount of money P to triple itself if it is invested at 8% interest compounded 4 times a year?

47. **Tripling Time** If a $25 investment is worth $75 today, how long ago must that $25 have been invested at 6% interest compounded twice a year?

48. **Doubling Time** If a $25 investment is worth $50 today, how long ago must that $25 have been invested at 6% interest compounded twice a year?

Recall that if P dollars are invested in an account with annual interest rate r, compounded continuously, then the amount of money in the account after t years is given by the formula $A(t) = Pe^{rt}$.

49. **Continuously Compounded Interest** Repeat Problem 41 if the interest is compounded continuously.

50. **Continuously Compounded Interest** Repeat Problem 44 if the interest is compounded continuously.

51. Continuously Compounded Interest How long will it take $500 to triple if it is invested at 6% annual interest, compounded continuously?

52. Continuously Compounded Interest How long will it take $500 to triple if it is invested at 12% annual interest, compounded continuously?

Maintaining Your Skills

Find the vertex for each parabola. Is the vertex the lowest or the highest point of the parabola?

53. $y = 2x^2 + 8x - 15$

54. $y = 3x^2 - 9x - 10$

55. $y = 12x - 4x^2$

56. $y = 18x - 6x^2$

57. Maximum Height An object is projected into the air with an initial upward velocity of 64 feet per second. Its height h at any time t is given by the formula $h = 64t - 16t^2$. Find the time at which the object reaches its maximum height. Then, find the maximum height.

58. Maximum Height An object is projected into the air with an initial upward velocity of 64 feet per second from the top of a building 40 feet high. If the height h of the object t seconds after it is projected into the air is $h = 40 + 64t - 16t^2$, find the time at which the object reaches its maximum height. Then, find the maximum height it attains.

Extending the Concepts

59. Exponential Growth Suppose that the population in a small city is 32,000 at the beginning of 2010 and that the city council assumes that the population size t years later can be estimated by the equation

$$P(t) = 32,000e^{0.05t}$$

Approximately when will the city have a population of 64,000?

60. Exponential Growth Suppose the population of a city is given by the equation

$$P(t) = 100,000e^{0.05t}$$

where t is the number of years from the present time. How large is the population now? (Now corresponds to a certain value of t. Once you realize what that value of t is, the problem becomes very simple.)

61. Exponential Growth Suppose the population of a city is given by the equation

$$P(t) = 15,000e^{0.04t}$$

where t is the number of years from the present time. About how long will it take for the population to reach 45,000?

62. Exponential Growth Suppose the population of a city is given by the equation

$$P(t) = 15,000e^{0.08t}$$

where t is the number of years from the present time. About how long will it take for the population to reach 45,000?

63. Solve the formula $A = Pe^{rt}$ for t.

64. Solve the formula $A = Pe^{-rt}$ for t.

65. Solve the formula $A = P \cdot 2^{-kt}$ for t.

66. Solve the formula $A = P \cdot 2^{kt}$ for t.

67. Solve the formula $A = P(1 - r)^t$ for t.

68. Solve the formula $A = P(1 + r)^t$ for t.

Chapter 11 Summary

EXAMPLES

Exponential Functions [11.1]

1. For the exponential function
$f(x) = 2^x$,

$f(0) = 2^0 = 1$
$f(1) = 2^1 = 2$
$f(2) = 2^2 = 4$
$f(3) = 2^3 = 8$

Any function of the form

$$f(x) = b^x$$

where $b > 0$ and $b \neq 1$, is an *exponential function*.

One-to-One Functions [11.2]

2. The function $f(x) = x^2$ is not one-to-one because 9, which is in the range, comes from both 3 and -3 in the domain.

A function is a *one-to-one function* if every element in the range comes from exactly one element in the domain.

Inverse Functions [11.2]

3. The inverse of $f(x) = 2x - 3$ is

$$f^{-1}(x) = \frac{x + 3}{2}$$

The *inverse* of a function is obtained by reversing the order of the coordinates of the ordered pairs belonging to the function. Only one-to-one functions have inverses that are also functions.

Definition of Logarithms [11.3]

4. The definition allows us to write expressions like
$y = \log_3 27$
equivalently in exponential form as
$3^y = 27$
which makes it apparent that y is 3.

If b is a positive number not equal to 1, then the expression

$$y = \log_b x$$

is equivalent to $x = b^y$; that is, in the expression $y = \log_b x$, y is the number to which we raise b in order to get x. Expressions written in the form $y = \log_b x$ are said to be in *logarithmic form*. Expressions like $x = b^y$ are in *exponential form*.

Two Special Identities [11.3]

5. Examples of the two special identities are
$5^{\log_5 12} = 12$
and
$\log_8 8^3 = 3$

For $b > 0$, $b \neq 1$, the following two expressions hold for all positive real numbers x:

(1) $b^{\log_b x} = x$

(2) $\log_b b^x = x$

Properties of Logarithms [11.4]

6. We can rewrite the expression

$$\log_{10} \frac{45^6}{273}$$

using the properties of logarithms, as
$6 \log_{10} 45 - \log_{10} 273$

If x, y, and b are positive real numbers, $b \neq 1$, and r is any real number, then:

1. $\log_b (xy) = \log_b x + \log_b y$

2. $\log_b \left(\dfrac{x}{y} \right) = \log_b x - \log_b y$

3. $\log_b x^r = r \log_b x$

Common Logarithms [11.5]

7. $\log_{10} 10{,}000 = \log 10{,}000$
$= \log 10^4$
$= 4$

Common logarithms are logarithms with a base of 10. To save time in writing, we omit the base when working with common logarithms; that is,

$$\log x = \log_{10} x$$

Natural Logarithms [11.5]

8. $\ln e = 1$
$\ln 1 = 0$

Natural logarithms, written *ln x,* are logarithms with a base of *e,* where the number *e* is an irrational number (like the number π). A decimal approximation for *e* is 2.7183. All the properties of exponents and logarithms hold when the base is *e.*

Change of Base [11.6]

9. $\log_6 475 = \dfrac{\log 475}{\log 6}$

$\approx \dfrac{2.6767}{0.7782}$

≈ 3.44

If *x, a,* and *b* are positive real numbers, $a \neq 1$ and $b \neq 1$, then

$$\log_a x = \frac{\log_b x}{\log_b a}$$

> ## 🚫 COMMON MISTAKES
>
> The most common mistakes that occur with logarithms come from trying to apply the three properties of logarithms to situations in which they don't apply. For example, a very common mistake looks like this:
>
> $$\frac{\log 3}{\log 2} = \log 3 - \log 2 \quad \text{Mistake}$$
>
> This is not a property of logarithms. To write the equation $\log 3 - \log 2$, we would have to start with
>
> $$\log \frac{3}{2} \quad NOT \quad \frac{\log 3}{\log 2}$$
>
> There is a difference.

Chapter 11 Review

Let $f(x) = 2^x$ and $g(x) = \left(\dfrac{1}{3}\right)^x$, and find the following. [11.1]

1. $f(4)$

2. $f(-1)$

3. $g(2)$

4. $f(2) - g(-2)$

5. $f(-1) + g(1)$

6. $g(-1) + f(2)$

7. The graph of $y = f(x)$

8. The graph of $y = g(x)$

For each relation that follows, sketch the graph of the relation and its inverse, and write an equation for the inverse. [11.2]

9. $y = 2x + 1$

10. $y = x^2 - 4$

For each of the following functions, find the equation of the inverse. Write the inverse using the notation $f^{-1}(x)$ if the inverse is itself a function. [11.2]

11. $f(x) = 2x + 3$

12. $f(x) = x^2 - 1$

13. $f(x) = \frac{1}{2}x + 2$

14. $f(x) = 4 - 2x^2$

Write each equation in logarithmic form. [11.3]

15. $3^4 = 81$

16. $7^2 = 49$

17. $0.01 = 10^{-2}$

18. $2^{-3} = \frac{1}{8}$

Write each equation in exponential form. [11.3]

19. $\log_2 8 = 3$

20. $\log_3 9 = 2$

21. $\log_4 2 = \frac{1}{2}$

22. $\log_4 4 = 1$

Solve for x. [11.3]

23. $\log_5 x = 2$

24. $\log_{16} 8 = x$

25. $\log_x 0.01 = -2$

Graph each equation. [11.3]

26. $y = \log_2 x$

27. $y = \log_{1/2} x$

Simplify each expression. [11.3]

28. $\log_4 16$

29. $\log_{27} 9$

30. $\log_4(\log_3 3)$

Use the properties of logarithms to expand each expression. [11.4]

31. $\log_2 5x$

32. $\log_{10} \dfrac{2x}{y}$

33. $\log_a \dfrac{y^3\sqrt{x}}{z}$

34. $\log_{10} \dfrac{x^2}{y^3 z^4}$

Write each expression as a single logarithm. [11.4]

35. $\log_2 x + \log_2 y$

36. $\log_3 x - \log_3 4$

37. $2 \log_a 5 - \dfrac{1}{2} \log_a 9$

38. $3 \log_2 x + 2 \log_2 y - 4 \log_2 z$

Solve each equation. [11.4]

39. $\log_2 x + \log_2 4 = 3$

40. $\log_2 x - \log_2 3 = 1$

41. $\log_3 x + \log_3(x - 2) = 1$

42. $\log_4(x + 1) - \log_4(x - 2) = 1$

43. $\log_6(x - 1) + \log_6 x = 1$

44. $\log_4(x - 3) + \log_4 x = 1$

Evaluate each expression. Round to four decimal places. [11.5]

45. $\log 346$

46. $\log 0.713$

Find x. [11.5]

47. $\log x = 3.9652$

48. $\log x = -1.6003$

Simplify. [11.5]

49. $\ln e$

50. $\ln 1$

51. $\ln e^2$

52. $\ln e^{-4}$

Use the formula $pH = -\log[H^+]$ to find the pH of a solution with the given hydrogen ion concentration. Round to the nearest tenth. [11.5]

53. $[H^+] = 7.9 \times 10^{-3}$

54. $[H^+] = 8.1 \times 10^{-6}$

Find $[H^+]$ for a solution with the given pH. [11.5]

55. $pH = 2.7$

56. $pH = 7.5$

Solve each equation. [11.6]

57. $4^x = 8$

58. $4^{3x+2} = 5$

Use the change-of-base property and a calculator to evaluate each expression. Round your answers to the nearest hundredth. [11.6]

59. $\log_{16} 8$

60. $\log_{12} 421$

Use the formula $A = P\left(1 + \dfrac{r}{n}\right)^{nt}$ to solve each of the following problems. [11.6]

61. Investing How long does it take $5,000 to double if it is deposited in an account that pays 16% annual interest compounded once a year?

62. Investing How long does it take $10,000 to triple if it is deposited in an account that pays 12% annual interest compounded 6 times a year?

Simplify.

1. $-8 + 2[5 - 3(-2 - 3)]$

2. $6(2x - 3) + 4(3x - 2)$

3. $\dfrac{3}{4} \div \dfrac{1}{8} \cdot \dfrac{3}{2}$

4. $\dfrac{3}{4} - \dfrac{1}{8} + \dfrac{3}{2}$

5. $3\sqrt{48} - 3\sqrt{75} + 2\sqrt{27}$

6. $(\sqrt{6} + 3\sqrt{2})(2\sqrt{6} + \sqrt{2})$

7. $[(6 + 2i) - (3 - 4i)] - (5 - i)$

8. $(1 + i)^2$

9. $\log_6(\log_5 5)$

10. $\log_5[\log_2(\log_3 9)]$

11. $\dfrac{1 - \dfrac{1}{x - 3}}{1 + \dfrac{1}{x - 3}}$

12. $1 + \dfrac{x}{1 + \dfrac{1}{x}}$

Reduce the following to lowest terms.

13. $\dfrac{452}{791}$

14. $\dfrac{468}{585}$

Multiply.

15. $\left(3t^2 + \dfrac{1}{4}\right)\left(4t^2 - \dfrac{1}{3}\right)$

16. $(3x - 2)(x^2 - 3x - 2)$

Divide.

17. $\dfrac{y^2 + 7y + 7}{y + 2}$

18. $\dfrac{9x^2 + 9x - 18}{3x - 4}$

Add or subtract.

19. $\dfrac{-3}{x^2 - 2x - 8} + \dfrac{4}{x^2 - 16}$

20. $\dfrac{7}{4x^2 - x - 3} - \dfrac{1}{4x^2 - 7x + 3}$

Solve.

21. $6 - 3(2x - 4) = 2$

22. $\dfrac{2}{3}(6x - 5) + \dfrac{1}{3} = 13$

23. $28x^2 = 3x + 1$

24. $2x - 1 = x^2$

25. $\dfrac{1}{x + 3} + \dfrac{1}{x - 2} = 1$

26. $\dfrac{1}{x^2 + 3x - 4} + \dfrac{3}{x^2 - 1} = \dfrac{-1}{x^2 + 5x + 4}$

27. $x - 3\sqrt{x} + 2 = 0$

28. $3(4y - 1)^2 + (4y - 1) - 10 = 0$

29. $\log_3 x = 3$

30. $\log_x 0.1 = -1$

31. $\log_3(x - 3) - \log_3(x + 2) = 1$

32. $\log_4(x - 5) + \log_4(x + 1) = 2$

Solve each system.

33. $-9x + 3y = -1$
$\quad\;\; 5x - 2y = -2$

34. $4x + 7y = -3$
$\quad\;\; x = -2y - 2$

35. $x + 2y = 4$
$\quad\; x + 2z = 1$
$\quad\; y - 2z = 5$

36. $2x + 3y - 8z = 2$
$\quad\; 3x - y + 2z = 10$
$\quad\; 4x + y + 8z = 16$

Graph each line.

37. $5x + 4y = 20$

38. $x = 3$

Write in symbols.

39. The difference of 7 and y is $-x$.

40. The difference of $9a$ and $4b$ is less than their sum.

Write in scientific notation.

41. 0.0000972

42. 47,000,000

Simplify.

43. $(2.6 \times 10^{-6})(1.4 \times 10^{19})$

44. $\dfrac{(5.6 \times 10^3)(1.5 \times 10^{-6})}{(9.8 \times 10^{11})}$

Factor completely.

45. $8a^3 - 125$

46. $50a^4 + 10a^2 - 4$

Specify the domain and range for the relation and state whether the relation is a function.

47. $\{(2, -3), (2, -1), (-3, 3)\}$

48. $\{(3, 5), (4, 5), (6, 4)\}$

Use the illustration to work Problems 49 and 50.

Source: U.S. Census Bureau, June 2010

49. Estimate the population of the United States in millions.

50. Estimate the world population in millions.

Graph each exponential function. [11.1]

1. $f(x) = 2^x$

2. $g(x) = 3^{-x}$

Sketch the graph of each function and its inverse. Find $f^{-1}(x)$ for Problem 3. [11.2]

3. $f(x) = 2x - 3$

4. $f(x) = x^2 - 4$

Solve for x. [11.3]

5. $\log_4 x = 3$

6. $\log_x 5 = 2$

Graph each of the following. [11.3]

7. $y = \log_2 x$

8. $y = \log_{\frac{1}{2}} x$

Evaluate each of the following. Round to the nearest hundredth if necessary. [11.3, 11.4, 11.5]

9. $\log_8 4$

10. $\log_7 21$

11. $\log 23{,}400$

12. $\log 0.0123$

13. $\ln 46.2$

14. $\ln 0.0462$

Use the properties of logarithms to expand each expression. [11.4]

15. $\log_2 \dfrac{6x^2}{y}$

16. $\log \dfrac{\sqrt{x}}{(y^4)\sqrt[5]{z}}$

Write each expression as a single logarithm. [11.4]

17. $2 \log_3 x - \dfrac{1}{2} \log_3 y$

18. $\dfrac{1}{3} \log x - \log y - 2 \log z$

Use a calculator to find x. Round to the nearest thousandth. [11.5]

19. $\log x = 4.8476$

20. $\log x = -2.6478$

Solve for x. Round to the nearest thousandth if necessary. [11.4, 11.6]

21. $5 = 3^x$

22. $4^{2x-1} = 8$

23. $\log_5 x - \log_5 3 = 1$

24. $\log_2 x + \log_2(x - 7) = 3$

25. Use the formula pH $= -\log[\text{H}^+]$ to find the pH of a solution in which $[\text{H}^+] = 6.6 \times 10^{-7}$. [11.5]

26. Compound Interest If $400 is deposited in an account that earns 10% annual interest compounded twice a year, how much money will be in the account after 5 years? [11.1]

27. Compound Interest How long will it take $600 to become $1,800 if the $600 is deposited in an account that earns 8% annual interest compounded four times a year? [11.6]

Determine if the function is one-to-one. [11.2]

28. $f = \{(1, 2), (2, 3), (3, 4), (4, 5)\}$

29. $g = \{(0, 1), (1, 0), (-1, 0), (-2, 1)\}$

30.

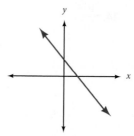

31.

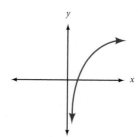

32.

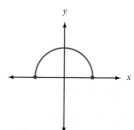

33.

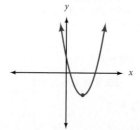

Chapter 11 Projects

EXPONENTIAL AND LOGARITHMIC FUNCTIONS

GROUP PROJECT

Two Depreciation Models

Number of People 3

Time Needed 20 minutes

Equipment Paper, pencil, and graphing calculator

Background Recently, a consumer magazine contained an article on leasing a computer. The original price of the computer was $2,579. The term of the lease was 24 months. At the end of the lease, the computer could be purchased for its residual value, $188. This is enough information to find an equation that will give us the value of the computer t months after it has been purchased.

Procedure We will find models for two types of depreciation: linear depreciation and exponential depreciation. Here are the general equations:

Linear Depreciation   *Exponential Depreciation*

$$V = mt + b \qquad\qquad V = V_0 e^{-kt}$$

1. Let t represent time in months and V represent the value of the computer at time t; then find two ordered pairs (t, V) that correspond to the initial price of the computer and another ordered pair that corresponds to the residual value of $188 when the computer is 2 years old.

2. Use the two ordered pairs to find m and b in the linear depreciation model; then write the equation that gives us linear depreciation.

3. Use the two ordered pairs to find V_0 and k in the exponential depreciation model; then write the equation that gives us exponential depreciation.

4. Graph each of the equations on your graphing calculator; then sketch the graphs on the following templates.

5. Find the value of the computer after 1 year, using both models.

6. Which of the two models do you think best describes the depreciation of a computer?

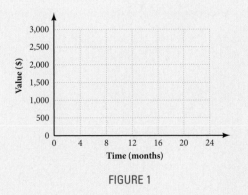

FIGURE 1

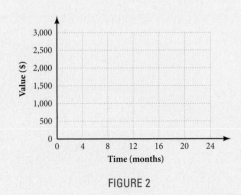

FIGURE 2

RESEARCH PROJECT
Drag Racing

The movie *Heart Like a Wheel* is based on the racing career of drag racer Shirley Muldowney. The movie includes a number of races. Choose four races as the basis for your report. For each race you choose, give a written description of the events leading to the race, along with the details of the race itself. Then draw a graph that shows the speed of the dragster as a function of time. Do

© picturesbyrob / Alamy

this for each race. One such graph from earlier in the chapter is shown here as an example. (For each graph, label the horizontal axis from 0 to 12 seconds, and label the vertical axis from 0 to 200 miles per hour.) Follow each graph with a description of any significant events that happen during the race (such as a motor malfunction or a crash) and their correlation to the graph.

Elapsed Time (sec)	Speed (mph)
0	0.0
1	72.7
2	129.9
3	162.8
4	192.2
5	212.4
6	228.1

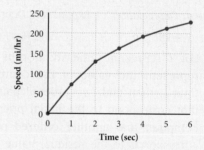

12

Conic Sections

Because of their perfect symmetry, circles have been used for thousands of years in many disciplines, including art, science, and religion. The Google Earth photograph above shows Stonehenge, a 4,500-year-old site in England. The arrangement of the stones is based on a circular plan that is thought to have both religious and astronomical significance.

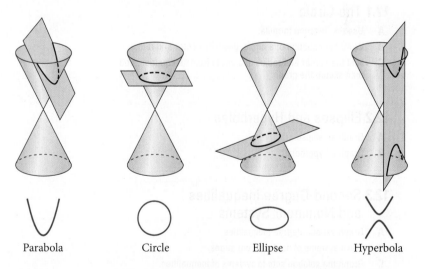

| Parabola | Circle | Ellipse | Hyperbola |

Curved shapes such as circles, ellipses, hyperbolas, and parabolas can be created by the intersection of the planes and cones shown above. Take a moment to study how each plane slices a cone to yield the specific curve displayed below it. These curves are called conic sections, and are the focus of this chapter.

Evaluate.

1. $\sqrt{4 + 25}$ **2.** $\sqrt{16 + 4}$ **3.** $\sqrt{(-2)^2 + 3^2}$ **4.** $\sqrt{1^2 + (-5)^2}$

Solve.

5. $x^2 = 16$ **6.** $-y^2 = 25$ **7.** $0 = x^2 - 6x + 8$

8. $0 = y^2 + 2y - 8$ **9.** $2 = (x - 3)^2 + 1$ **10.** $10 = (-1)^2 + (y + 1)^2$

Fill in the blanks to complete the square.

11. $x^2 + 6x + \boxed{} = (x + \boxed{})^2$ **12.** $x^2 - 4x + \boxed{} = (x - \boxed{})^2$

13. $y^2 + 2y + \boxed{} = (y + \boxed{})^2$ **14.** $y^2 + 1y + \boxed{} = (y + \boxed{})^2$

Factor.

15. $5y^2 + 16y + 12$

16. $3x^2 + 17x - 28$

17. Divide $12x^2 + y^2$ by 36.   **18.** Divide $x^2 + 4y^2$ by 16.   **19.** Solve $x + 3y = 3$ for y.

20. Solve $2x + y = 3$ for y.

Chapter Outline

12.1 The Circle
  A   Use the distance formula.
  B   Write the equation of a circle, given its center and radius.
  C   Find the center and radius of a circle from its equation, and then sketch the graph.

12.2 Ellipses and Hyperbolas
  A   Graph an ellipse.
  B   Graph a hyperbola.

12.3 Second-Degree Inequalities and Nonlinear Systems
  A   Graph second-degree inequalities.
  B   Solve systems of nonlinear equations.
  C   Graph the solution sets to systems of inequalities.

12.1 The Circle

TICKET TO SUCCESS

Keep these questions in mind as you read through the section. Then respond in your own words and in complete sentences.

1. What is the distance formula?
2. What is the mathematical definition of a circle?
3. How are the distance formula and the equation of a circle related?
4. How would you find the center and radius of a circle that can be identified by the equation $(x - a)^2 + (y - b)^2 = r^2$?

We begin our work with conic sections by studying circles. Before we find the general equation of a circle, we must first derive what is known as the *distance formula*.

A The Distance Formula

Suppose (x_1, y_1) and (x_2, y_2) are any two points in the first quadrant. (Actually, we could choose the two points to be anywhere on the coordinate plane. It is just more convenient to have them in the first quadrant.) We can name the points P_1 and P_2, respectively, and draw the diagram shown in Figure 1.

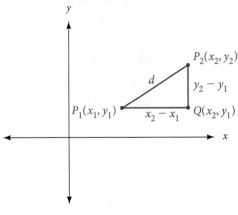

FIGURE 1

Notice the coordinates of point Q. The x-coordinate is x_2 since Q is directly below point P_2. The y-coordinate of Q is y_1 since Q is directly across from point P_1. It is evident from the diagram that the length of P_2Q is $y_2 - y_1$ and the length of P_1Q is $x_2 - x_1$. Using the Pythagorean theorem, we have

$$(P_1P_2)^2 = (P_1Q)^2 + (P_2Q)^2$$

or

$$d^2 = (x_2 - x_1)^2 + (y_2 - y_1)^2$$

Taking the square root of both sides, we have

$$d = \sqrt{(x_2 - x_1)^2 + (y_2 - y_1)^2}$$

We know this is the positive square root since d is the distance from P_1 to P_2 and therefore must be positive. This formula is called the *distance formula.*

EXAMPLE 1   Find the distance between $(3, 5)$ and $(2, -1)$.

SOLUTION   If we let $(3, 5)$ be (x_1, y_1) and $(2, -1)$ be (x_2, y_2) and apply the distance formula, we have

$$d = \sqrt{(2 - 3)^2 + (-1 - 5)^2}$$
$$= \sqrt{(-1)^2 + (-6)^2}$$
$$= \sqrt{1 + 36}$$
$$= \sqrt{37}$$

EXAMPLE 2   Find x if the distance from $(x, 5)$ to $(3, 4)$ is $\sqrt{2}$.

SOLUTION   Using the distance formula, we have

$$\sqrt{2} = \sqrt{(x - 3)^2 + (5 - 4)^2}$$
$$2 = (x - 3)^2 + 1^2$$
$$2 = x^2 - 6x + 9 + 1$$
$$0 = x^2 - 6x + 8$$
$$0 = (x - 4)(x - 2)$$
$$x = 4 \quad \text{or} \quad x = 2$$

The two solutions are 4 and 2, which indicates there are two points, $(4, 5)$ and $(2, 5)$, which are $\sqrt{2}$ units from $(3, 4)$.

B   Circles

In the 1990s, designs like the one shown in Figure 2 began appearing in England's agricultural fields. Whoever made these designs chose the circle as their basic shape.

FIGURE 2

We can model circles very easily in algebra by using equations that are based on the distance formula.

> **Circle Theorem**
> The equation of the circle with center at (a, b) and radius r is given by
> $$(x - a)^2 + (y - b)^2 = r^2$$

PROOF By definition, all points on the circle are a distance r from the center (a, b). If we let (x, y) represent any point on the circle, then (x, y) is r units from (a, b). Applying the distance formula, we have

$$r = \sqrt{(x - a)^2 + (y - b)^2}$$

Squaring both sides of this equation gives the equation of the circle.

$$(x - a)^2 + (y - b)^2 = r^2$$

We can use the circle theorem to find the equation of a circle, given its center and radius, or to find its center and radius, given the equation.

EXAMPLE 3   Find the equation of the circle with center at $(-3, 2)$ having a radius of 5.

SOLUTION   We have $(a, b) = (-3, 2)$ and $r = 5$. Applying the circle theorem yields

$$[x - (-3)]^2 + (y - 2)^2 = 5^2$$
$$(x + 3)^2 + (y - 2)^2 = 25$$

EXAMPLE 4   Give the equation of the circle with radius 3 whose center is at the origin.

SOLUTION   The coordinates of the center are $(0, 0)$, and the radius is 3. The equation must be

$$(x - 0)^2 + (y - 0)^2 = 3^2$$
$$x^2 + y^2 = 9$$

We can see from Example 4 that the equation of any circle with its center at the origin and radius r will be

$$x^2 + y^2 = r^2$$

C Graphing Circles

EXAMPLE 5 Find the center and radius, and sketch the graph, of the circle with an equation of

$$(x - 1)^2 + (y + 3)^2 = 4$$

SOLUTION Writing the equation in the form

$$(x - a)^2 + (y - b)^2 = r^2$$

we have

$$(x - 1)^2 + [y - (-3)]^2 = 2^2$$

The center is at $(1, -3)$ and the radius is 2. The graph is shown in Figure 3.

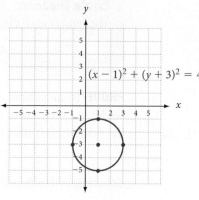

FIGURE 3

EXAMPLE 6 Sketch the graph of $x^2 + y^2 = 9$.

SOLUTION Since the equation can be written in the form

$$(x - 0)^2 + (y - 0)^2 = 3^2$$

it must have its center at $(0, 0)$ and a radius of 3. The graph is shown in Figure 4.

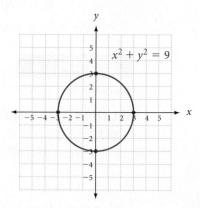

FIGURE 4

EXAMPLE 7   Sketch the graph of $x^2 + y^2 + 6x - 4y - 12 = 0$.

SOLUTION   To sketch the graph, we must find the center and radius. The center and radius can be identified if the equation has the form

$$(x - a)^2 + (y - b)^2 = r^2$$

The original equation can be written in this form by completing the squares on x and y.

$$x^2 + y^2 + 6x - 4y - 12 = 0$$

$$x^2 + 6x + y^2 - 4y = 12$$

$$x^2 + 6x + \mathbf{9} + y^2 - 4y + \mathbf{4} = 12 + \mathbf{9} + \mathbf{4}$$

$$(x + 3)^2 + (y - 2)^2 = 25$$

$$(x + 3)^2 + (y - 2)^2 = 5^2$$

From the last line, it is apparent that the center is at $(-3, 2)$ and the radius is 5. The graph is shown in Figure 5.

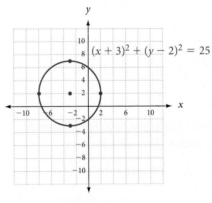

FIGURE 5

Problem Set 12.1

Moving Toward Success

"Spread love everywhere you go. Let no one ever come to you without leaving happier."

—Mother Teresa, 1910–1997, Albanian-born humanitarian and missionary

1. How will communicating your feelings about this class in an appropriate manner aid your success in it?

2. Why is maintaining motivation in this class more important now than ever?

A Find the distance between the following points. [Examples 1–2]

1. $(3, 7)$ and $(6, 3)$

2. $(4, 7)$ and $(8, 1)$

3. $(0, 9)$ and $(5, 0)$

4. $(-3, 0)$ and $(0, 4)$

5. $(3, -5)$ and $(-2, 1)$

6. $(-8, 9)$ and $(-3, -2)$

7. $(-1, -2)$ and $(-10, 5)$

8. $(-3, -8)$ $(-1, 6)$

9. Find x so the distance between $(x, 2)$ and $(1, 5)$ is $\sqrt{13}$.

10. Find x so the distance between $(-2, 3)$ and $(x, 1)$ is 3.

11. Find y so the distance between $(7, y)$ and $(8, 3)$ is 1.

12. Find y so the distance between $(3, -5)$ and $(3, y)$ is 9.

B   Write the equation of the circle with the given center and radius. [Examples 3–4]

13. Center $(2, 3)$; $r = 4$

14. Center $(3, -1)$; $r = 5$

15. Center $(3, -2)$; $r = 3$

16. Center $(-2, 4)$; $r = 1$

17. Center $(-5, -1)$; $r = \sqrt{5}$

18. Center $(-7, -6)$; $r = \sqrt{3}$

19. Center $(0, -5)$; $r = 1$

20. Center $(0, -1)$; $r = 7$

21. Center $(0, 0)$; $r = 2$

22. Center $(0, 0)$; $r = 5$

C   Give the center and radius, and sketch the graph of each of the following circles. [Examples 5–7]

23. $x^2 + y^2 = 4$

24. $x^2 + y^2 = 16$

25. $(x - 1)^2 + (y - 3)^2 = 25$

26. $(x - 4)^2 + (y - 1)^2 = 36$

27. $(x + 2)^2 + (y - 4)^2 = 8$

28. $(x - 3)^2 + (y + 1)^2 = 12$

29. $(x + 1)^2 + (y + 1)^2 = 1$

30. $(x + 3)^2 + (y + 2)^2 = 9$

31. $x^2 + y^2 - 6y = 7$

32. $x^2 + y^2 + 10x = 0$

33. $x^2 + y^2 - 4x - 6y = -4$

34. $x^2 + y^2 - 4x + 2y = 4$

35. $x^2 + y^2 + 2x + y = \dfrac{11}{4}$

36. $x^2 + y^2 - 6x - y = -\dfrac{1}{4}$

Both of the following circles pass through the origin. In each case, find the equation.

37.

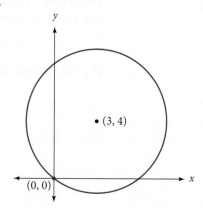

38.

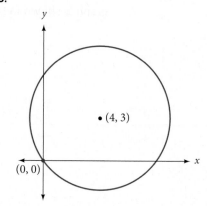

39. Find the equations of circles A, B, and C in the following diagram. The three points are the centers of the three circles.

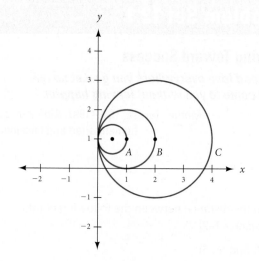

40. Each of the following circles passes through the origin. The centers are as shown. Find the equation of each circle.

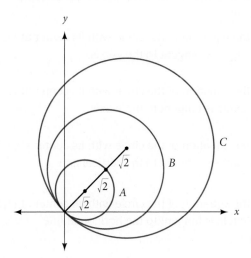

41. Find the equation of each of the three circles shown here.

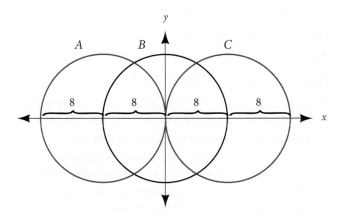

42. Find the equation of each of the three circles shown here.

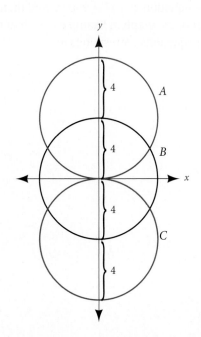

43. Find the equation of each of the three circles shown here.

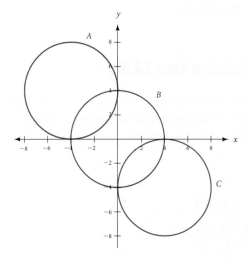

44. A parabola and a circle each contain the points $(-4, 0)$, $(0, 4)$, and $(4, 0)$. Sketch the graph of each curve on the same coordinate system, then write an equation for each curve.

45. Ferris Wheel A giant Ferris wheel has a diameter of 240 feet and sits 12 feet above the ground. As shown in the diagram below, the wheel is 500 feet from the entrance to the park. The xy-coordinate system containing the wheel has its origin on the ground at the center of the entrance. Write an equation that models the shape of the wheel.

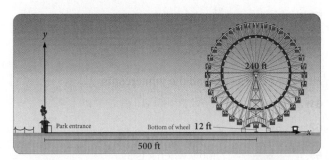

46. Magic Rings A magician is holding two rings that seem to lie in the same plane and intersect in two points. Each ring is 10 inches in diameter.

   a. Find the equation of each ring if a coordinate system is placed with its origin at the center of the first ring and the x-axis contains the center of the second ring.

b. Find the equation of each ring if a coordinate system is placed with its origin at the center of the second ring and the x-axis contains the center of the first ring.

Maintaining Your Skills

Find the equation of the inverse of each of the following functions. Write the inverse using the notation $f^{-1}(x)$, if the inverse is itself a function.

47. $f(x) = 3^x$

48. $f(x) = 5^x$

49. $f(x) = 2x + 3$

50. $f(x) = 3x - 2$

51. $f(x) = \dfrac{x + 3}{5}$

52. $f(x) = \dfrac{x - 2}{6}$

Getting Ready for the Next Section

Solve.

53. $y^2 = 9$

54. $x^2 = 25$

55. $-y^2 = 4$

56. $-x^2 = 16$

57. $\dfrac{-x^2}{9} = 1$

58. $\dfrac{y^2}{100} = 1$

59. Divide $4x^2 + 9y^2$ by 36.

60. Divide $25x^2 + 4y^2$ by 100.

Find the x-intercepts and the y-intercepts.

61. $3x - 4y = 12$

62. $y = 3x^2 + 5x - 2$

63. If $\dfrac{x^2}{25} + \dfrac{y^2}{9} = 1$, find y when x is 3.

64. If $\dfrac{x^2}{25} + \dfrac{y^2}{9} = 1$, find y when x is -4.

Extending the Concepts

A circle is *tangent to* a line if it touches, but does not cross, the line.

65. Find the equation of the circle with its center at $(2, 3)$ if the circle is tangent to the y-axis.

66. Find the equation of the circle with its center at $(3, 2)$ if the circle is tangent to the x-axis.

67. Find the equation of the circle with its center at $(2, 3)$ if the circle is tangent to the vertical line $x = 4$.

68. Find the equation of the circle with its center at $(3, 2)$ if the circle is tangent to the horizontal line $y = 6$.

Find the distance from the origin to the center of each of the following circles.

69. $x^2 + y^2 - 6x + 8y = 144$

70. $x^2 + y^2 - 8x + 6y = 144$

71. $x^2 + y^2 - 6x - 8y = 144$

72. $x^2 + y^2 + 8x + 6y = 144$

73. If we were to solve the equation $x^2 + y^2 = 9$ for y, we would obtain the equation $y = \pm\sqrt{9 - x^2}$. This last equation is equivalent to the two equations $y = \sqrt{9 - x^2}$, in which y is always positive, and $y = -\sqrt{9 - x^2}$, in which y is always negative. Look at the graph of $x^2 + y^2 = 9$ in Example 6 of this section and indicate what part of the graph each of the two equations corresponds to.

74. Solve the equation $x^2 + y^2 = 9$ for x, and then indicate what part of the graph in Example 6 each of the two resulting equations corresponds to.

OBJECTIVES

A   Graph an ellipse.

B   Graph a hyperbola.

TICKET TO SUCCESS

Keep these questions in mind as you read through the section. Then respond in your own words and in complete sentences.

1. What is an ellipse?
2. How would you find the *x*- and *y*-intercepts of an ellipse by looking at its equation in standard form?
3. Explain how to find the asymptotes of a graph of a hyperbola.
4. How would you find the vertices of a hyperbola with a center at (h, k)?

The photograph below shows Halley's comet as it passed close to Earth in 1986. Like the planets in our solar system, Halley's comet orbits the sun in an elliptical path (Figure 1). While it takes the earth one year to complete one orbit around the sun, it takes Halley's comet 76 years. The first known sighting of Halley's comet was in 239 B.C. Its most famous appearance occurred in 1066 A.D., when it was seen a few months before the Battle of Hastings and believed to be an omen.

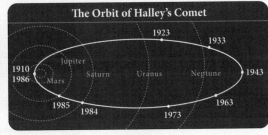

FIGURE 1

A   Ellipses

We begin this section with an introductory look at equations that produce ellipses. To simplify our work, we consider only ellipses that are centered about the origin.

Suppose we want to graph the equation

$$\frac{x^2}{25} + \frac{y^2}{9} = 1$$

We can find other ordered pairs on the graph by substituting values for x (or y) and then solving for y (or x). For example, if we let $x = 3$, then

$$\frac{3^2}{25} + \frac{y^2}{9} = 1$$

$$\frac{9}{25} + \frac{y^2}{9} = 1$$

$$0.36 + \frac{y^2}{9} = 1$$

$$\frac{y^2}{9} = 0.64$$

$$y^2 = 5.76$$

$$y = \pm 2.4$$

This would give us the two ordered pairs $(3, -2.4)$ and $(3, 2.4)$.

We can find the y-intercepts by letting $x = 0$, and we can find the x-intercepts by letting $y = 0$:

When $x = 0$

$$\frac{0^2}{25} + \frac{y^2}{9} = 1$$

$$y^2 = 9$$

$$y = \pm 3$$

When $y = 0$

$$\frac{x^2}{25} + \frac{0^2}{9} = 1$$

$$x^2 = 25$$

$$x = \pm 5$$

The graph crosses the y-axis at $(0, 3)$ and $(0, -3)$ and the x-axis at $(5, 0)$ and $(-5, 0)$. Graphing these points and then connecting them with a smooth curve gives the graph shown in Figure 2.

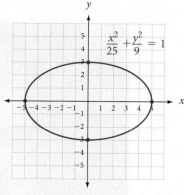

FIGURE 2

A graph of the type shown in Figure 2 is called an *ellipse*. If we were to find some other ordered pairs that satisfy our original equation, we would find that their graphs lie on the ellipse. Also, the coordinates of any point on the ellipse will satisfy the equation. We can generalize these results as follows.

The Ellipse

The graph of any equation of the form

$$\frac{x^2}{a^2} + \frac{y^2}{b^2} = 1 \qquad \text{Standard form}$$

will be an **ellipse** centered at the origin. The ellipse will cross the x-axis at $(a, 0)$ and $(-a, 0)$. It will cross the y-axis at $(0, b)$ and $(0, -b)$. When a and b are equal, the ellipse will be a circle.

The most convenient way to graph an ellipse is to locate the intercepts.

EXAMPLE 1 Sketch the graph of $4x^2 + 9y^2 = 36$.

SOLUTION To write the equation in the form

$$\frac{x^2}{a^2} + \frac{y^2}{b^2} = 1$$

we must divide both sides by 36.

$$\frac{4x^2}{36} + \frac{9y^2}{36} = \frac{36}{36}$$

$$\frac{x^2}{9} + \frac{y^2}{4} = 1$$

The graph crosses the *x*-axis at (3, 0), (−3, 0) and the *y*-axis at (0, 2), (0, −2), as shown in Figure 3.

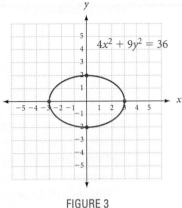

FIGURE 3

B Hyperbolas

Figure 4 shows Europa, one of Jupiter's moons, as it was photographed by the Galileo space probe in the late 1990s. To speed up the trip from Earth to Jupiter—nearly a billion miles—Galileo made use of the *slingshot effect.* This involves flying a hyperbolic path very close to a planet, so that gravity can be used to gain velocity as the space probe hooks around the planet (Figure 5).

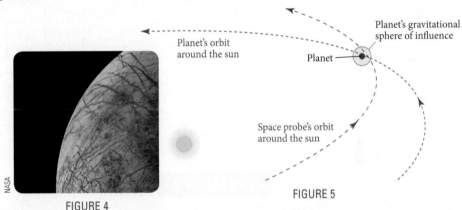

FIGURE 4

FIGURE 5

We use the next part of this section to consider equations that produce hyperbolas.
Consider the equation

$$\frac{x^2}{9} - \frac{y^2}{4} = 1$$

If we were to find a number of ordered pairs that are solutions to the equation and connect their graphs with a smooth curve, we would have Figure 6.

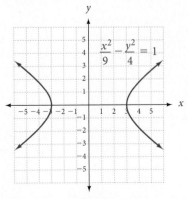

FIGURE 6

This graph is an example of a *hyperbola*. Notice that the graph has x-intercepts at $(3, 0)$ and $(-3, 0)$. The graph has no y-intercepts and hence does not cross the y-axis since substituting $x = 0$ into the equation yields

$$\frac{0^2}{9} - \frac{y^2}{4} = 1$$

$$-y^2 = 4$$

$$y^2 = -4$$

for which there is no real solution. We can, however, use the number below y^2 to help sketch the graph. If we draw a rectangle that has its sides parallel to the x- and y-axes and that passes through the x-intercepts and the points on the y-axis corresponding to the square roots of the number below y^2, 2 and -2, it looks like the rectangle in Figure 7. The lines that connect opposite corners of the rectangle are called *asymptotes*. The graph of the hyperbola

$$\frac{x^2}{9} - \frac{y^2}{4} = 1$$

will approach these lines. Figure 7 is the graph.

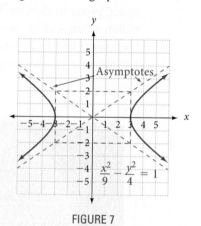

FIGURE 7

EXAMPLE 2 Graph the equation $\dfrac{y^2}{9} - \dfrac{x^2}{16} = 1$.

SOLUTION In this case, the y-intercepts are 3 and -3, and the x-intercepts do not exist. We can use the square roots of the number below x^2, however, to find the asymptotes associated with the graph. The sides of the rectangle used to draw the asymptotes must pass through 3 and -3 on the y-axis and 4 and -4 on the x-axis. (See Figure 8.)

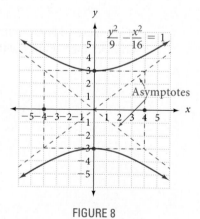

FIGURE 8

Here is a summary of what we have for hyperbolas.

The Hyperbola

The graph of the equation

$$\frac{x^2}{a^2} - \frac{y^2}{b^2} = 1$$

will be a hyperbola centered at the origin. The graph will have x-intercepts at $-a$ and a.

The graph of the equation

$$\frac{y^2}{a^2} - \frac{x^2}{b^2} = 1$$

will be a hyperbola centered at the origin. The graph will have y-intercepts at $-a$ and a.

As an aid in sketching either of the preceding equations, the asymptotes can be found by drawing lines through opposite corners of the rectangle whose sides pass through $-a$, a, $-b$, and b on the axes.

Ellipses and Hyperbolas Not Centered at the Origin

The following equation is that of an ellipse with its center at the point (4, 1):

$$\frac{(x-4)^2}{9} + \frac{(y-1)^2}{4} = 1$$

To see why the center is at (4, 1) we substitute x' (read "x prime") for $x - 4$ and y' for $y - 1$ in the equation. That is,

If → $x' = x - 4$

and → $y' = y - 1$

the equation → $\dfrac{(x-4)^2}{9} + \dfrac{(y-1)^2}{4} = 1$

becomes → $\dfrac{(x')^2}{9} + \dfrac{(y')^2}{4} = 1$

This is the equation of an ellipse in a coordinate system with an x'-axis and a y'-axis. We call this new coordinate system the $x'y'$-coordinate system. The center of our ellipse is at the origin in the $x'y'$-coordinate system. The question is this: What are the coordinates of the center of this ellipse in the original xy-coordinate system? To answer this question, we go back to our original substitutions:

$$x' = x - 4$$
$$y' = y - 1$$

In the $x'y'$-coordinate system, the center of our ellipse is at $x' = 0, y' = 0$ (the origin of the $x'y'$ system). Substituting these numbers for x' and y', we have

$$0 = x - 4$$
$$0 = y - 1$$

Solving these equations for x and y will give us the coordinates of the center of our ellipse in the xy-coordinate system. As you can see, the solutions are $x = 4$ and $y = 1$. Therefore, in the xy-coordinate system, the center of our ellipse is at the point (4, 1). Figure 9 shows the graph.

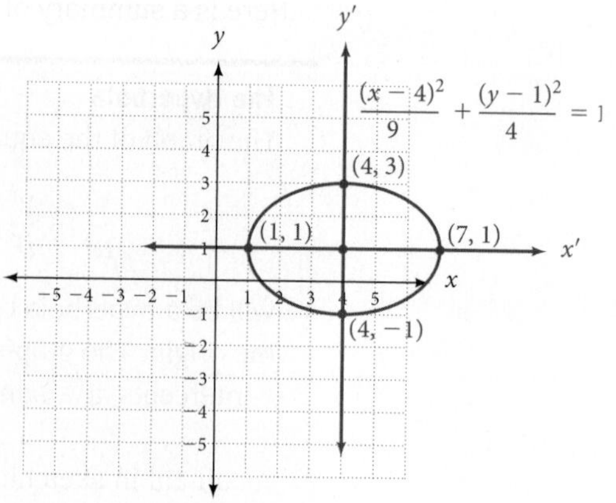

FIGURE 9

The coordinates of all points labeled in Figure 9 are given with respect to the *xy*-coordinate system. The *x'* and *y'* are shown simply for reference in our discussion. Note that the horizontal distance from the center to the vertices is 3—the square root of the denominator of the $(x - 4)^2$ term. Likewise, the vertical distance from the center to the other vertices is 2—the square root of the denominator of the $(y - 1)^2$ term.

We summarize the information above with the following:

An Ellipse with Center at (*h, k*)
The graph of the equation

$$\frac{(x - h)^2}{a^2} + \frac{(y - k)^2}{b^2} = 1$$

will be an ellipse with center at (h, k). The vertices of the ellipse will be at the points, $(h + a, k)$, $(h - a, k)$, $(h, k + b)$, and $(h, k - b)$.

EXAMPLE 3   Graph the ellipse: $x^2 + 9y^2 + 4x - 54y + 76 = 0$.

SOLUTION   To identify the coordinates of the center, we must complete the square on *x* and also on *y*. To begin, we rearrange the terms so that those containing *x* are together, those containing *y* are together, and the constant term is on the other side of the equal sign. Doing so gives us the following equation:

$$x^2 + 4x \qquad + 9y^2 - 54y \qquad = -76$$

Before we can complete the square on y, we must factor 9 from each term containing y:

$$x^2 + 4x \qquad + 9(y^2 - 6y \quad) = -76$$

To complete the square on *x*, we add 4 to each side of the equation. To complete the square on *y*, we add 9 inside the parentheses. This increases the left side of the equation by 81 since each term within the parentheses is multiplied by 9. Therefore, we must also add 81 to the right side of the equation.

$$x^2 + 4x + \mathbf{4} + 9(y^2 - 6y + \mathbf{9}) = -76 + \mathbf{4} + \mathbf{81}$$
$$(x + 2)^2 + 9(y - 3)^2 = 9$$

To identify the distances to the vertices, we divide each term on both sides by 9.

$$\frac{(x + 2)^2}{9} + \frac{9(y - 3)^2}{9} = \frac{9}{9}$$

$$\frac{(x + 2)^2}{9} + \frac{(y - 3)^2}{1} = 1$$

The graph is an ellipse with center at $(-2, 3)$, as shown in Figure 10.

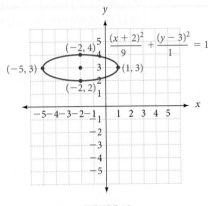

FIGURE 10

The ideas associated with graphing hyperbolas whose centers are not at the origin parallel the ideas just presented about graphing ellipses whose centers have been moved off the origin. Without showing the justification for doing so, we state the following guidelines for graphing hyperbolas:

Hyperbolas with Centers at (h, k)

The graphs of the equations

$$\frac{(x - h)^2}{a^2} - \frac{(y - k)^2}{b^2} = 1 \text{ and } \frac{(y - k)^2}{b^2} - \frac{(x - h)^2}{a^2} = 1$$

will be hyperbolas with their centers at (h, k). The vertices of the graph of the first equation will be at the points $(h + a, k)$, and $(h - a, k)$, and the vertices for the graph of the second equation will be at $(h, k + b)$, $(h, k - b)$. In either case, the asymptotes can be found by connecting opposite corners of the rectangle that contains the four points $(h + a, k)$, $(h - a, k)$, $(h, k + b)$, and $(h, k - b)$.

EXAMPLE 4   Graph the hyperbola: $4x^2 - y^2 + 4y - 20 = 0$.

SOLUTION   To identify the coordinates of the center of the hyperbola, we need to complete the square on y. (Because there is no linear term in x, we do not need to complete the square on x. The x-coordinate of the center will be $x = 0$.)

$$4x^2 - y^2 + 4y - 20 = 0$$

$$4x^2 - y^2 + 4y = 20 \quad \text{Add 20 to each side}$$

$$4x^2 - 1(y^2 - 4y) = 20 \quad \text{Factor } -1 \text{ from each term containing } y$$

To complete the square on y, we add 4 to the terms inside the parentheses. Doing so adds -4 to the left side of the equation because everything inside the parentheses is multiplied by -1. To keep from changing the equation we must also add -4 to the right side.

$$4x^2 - 1(y^2 - 4y + \mathbf{4}) = 20 - \mathbf{4} \quad \text{Add } -4 \text{ to each side}$$

$$4x^2 - 1(y - 2)^2 = 16 \quad y^2 - 4y + 4 = (y - 2)^2$$

$$\frac{4x^2}{16} - \frac{(y - 2)^2}{16} = \frac{16}{16} \quad \text{Divide each side by 16}$$

$$\frac{x^2}{4} - \frac{(y-2)^2}{16} = 1 \qquad \text{Simplify each term}$$

This is the equation of a hyperbola with center at (0, 2). The graph opens to the right and left as shown in Figure 11.

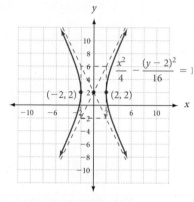

FIGURE 11

Problem Set 12.2

Moving Toward Success

"Our attitude toward life determines life's attitude towards us."

—Earl Nightingale, 1921–1989, American motivational author and radio broadcaster

1. Explain why cramming for a test would increase your anxiety during the test.

2. Why will staying positive before and during a test reduce your stress level?

A Graph each of the following. Be sure to label both the x- and y-intercepts. [Example 1]

1. $\frac{x^2}{9} + \frac{y^2}{16} = 1$

2. $\frac{x^2}{25} + \frac{y^2}{4} = 1$

3. $\frac{x^2}{16} + \frac{y^2}{9} = 1$

4. $\frac{x^2}{4} + \frac{y^2}{25} = 1$

5. $\frac{x^2}{3} + \frac{y^2}{4} = 1$

6. $\frac{x^2}{4} + \frac{y^2}{3} = 1$

7. $4x^2 + 25y^2 = 100$

8. $4x^2 + 9y^2 = 36$

9. $x^2 + 8y^2 = 16$

10. $12x^2 + y^2 = 36$

B Graph each of the following. Show the intercepts and the asymptotes in each case. [Example 2]

11. $\frac{x^2}{9} - \frac{y^2}{16} = 1$

12. $\frac{x^2}{25} - \frac{y^2}{4} = 1$

13. $\frac{x^2}{16} - \frac{y^2}{9} = 1$

14. $\frac{x^2}{4} - \frac{y^2}{25} = 1$

15. $\frac{y^2}{9} - \frac{x^2}{16} = 1$

16. $\frac{y^2}{25} - \frac{x^2}{4} = 1$

17. $\frac{y^2}{36} - \frac{x^2}{4} = 1$

18. $\frac{y^2}{4} - \frac{x^2}{36} = 1$

19. $x^2 - 4y^2 = 4$

20. $y^2 - 4x^2 = 4$

21. $16y^2 - 9x^2 = 144$

22. $4y^2 - 25x^2 = 100$

Find the x- and y-intercepts, if they exist, for each of the following. Do not graph.

23. $0.4x^2 + 0.9y^2 = 3.6$

24. $1.6x^2 + 0.9y^2 = 14.4$

25. $\frac{x^2}{0.04} - \frac{y^2}{0.09} = 1$

26. $\frac{y^2}{0.16} - \frac{x^2}{0.25} = 1$

27. $\frac{25x^2}{9} + \frac{25y^2}{4} = 1$

28. $\frac{16x^2}{9} + \frac{16y^2}{25} = 1$

A Graph each of the following ellipses. In each case, label the coordinates of the center and the vertices. [Example 3]

29. $\dfrac{(x-4)^2}{4} + \dfrac{(y-2)^2}{9} = 1$

30. $\dfrac{(x-2)^2}{4} + \dfrac{(y-4)^2}{9} = 1$

31. $4x^2 + y^2 - 4y - 12 = 0$

32. $4x^2 + y^2 - 24x - 4y + 36 = 0$

33. $x^2 + 9y^2 + 4x - 54y + 76 = 0$

34. $4x^2 + y^2 - 16x + 2y + 13 = 0$

B Graph each of the following hyperbolas. In each case, label the coordinates of the center and the vertices and show the asymptotes. [Example 4]

35. $\dfrac{(x-2)^2}{16} - \dfrac{y^2}{4} = 1$

36. $\dfrac{(y-2)^2}{16} - \dfrac{x^2}{4} = 1$

37. $9y^2 - x^2 - 4x + 54y + 68 = 0$

38. $4x^2 - y^2 - 24x + 4y + 28 = 0$

39. $4y^2 - 9x^2 - 16y + 72x - 164 = 0$

40. $4x^2 - y^2 - 16x - 2y + 11 = 0$

Solve.

41. Find y when x is 4 in the equation $\dfrac{x^2}{25} + \dfrac{y^2}{9} = 1$.

42. Find x when y is 3 in the equation $\dfrac{x^2}{4} + \dfrac{y^2}{25} = 1$.

43. Find y when x is 1.8 in $16x^2 + 9y^2 = 144$.

44. Find y when x is 1.6 in $49x^2 + 4y^2 = 196$.

45. Give the equations of the two asymptotes in the graph you found in Problem 15.

46. Give the equations of the two asymptotes in the graph you found in Problem 16.

Find an equation for each graph.

47.

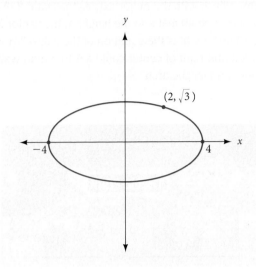

48.

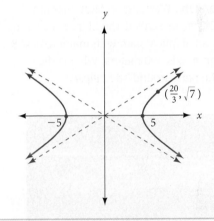

Applying the Concepts

The diagram shows the minor axis and the major axis for an ellipse.

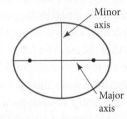

In any ellipse, the length of the major axis is $2a$, and the length of the minor axis is $2b$ (these are the same a and b that appear in the general equation of an ellipse). Each of the two points shown on the major axis is a focus of the ellipse. If the distance from the center of the ellipse to each focus is c, then it is always true that $a^2 = b^2 + c^2$. You will need this information for some of the problems that follow.

49. Archway A new theme park is planning an archway at its main entrance. The arch is to be in the form of a semi-ellipse with the major axis as the span. If the span is to be 40 feet and the height at the center is to be 10 feet, what is the equation of the ellipse? How far left and right of center could a 6-foot man walk upright under the arch?

50. The Ellipse President's Park, located between the White House and the Washington Monument in Washington, DC, is also called The Ellipse. The park is enclosed by an elliptical path with major axis 458 meters and minor axis 390 meters. What is the equation for the path around The Ellipse?

51. Elliptical Pool Table A children's science museum plans to build an elliptical pool table to demonstrate that a ball rolled from a particular point (focus) will always go into a hole located at another particular point (the other focus). The focus needs to be 1 foot from the vertex of the ellipse. If the table is to be 8 feet long, how wide should it be? *Hint:* The distance from the center to each focus point is represented by c and is found by using the equation $a^2 = b^2 + c^2$.

52. Lithotripter A lithotripter, used to break up kidney stones, is based on the ellipse $\frac{x^2}{36} + \frac{y^2}{25} = 1$. Determine how many units the kidney stone and the wave source (focus points) must be placed from the center of the ellipse. *Hint:* The distance from the center to each focus point is represented by c and is found by using the equation $a^2 = b^2 + c^2$.

Maintaining Your Skills

Let $f(x) = \frac{2}{x-2}$ and $g(x) = \frac{2}{x+2}$. Find the following:

53. $f(4)$ **54.** $g(4)$

55. $f(g(0))$ **56.** $g(f(0))$

57. $f(x) + g(x)$ **58.** $f(x) - g(x)$

Getting Ready for the Next Section

59. Which of the following are solutions to $x^2 + y^2 < 16$?
$(0, 0)$ $(4, 0)$ $(0, 5)$

60. Which of the following are solutions to $y \leq x^2 - 2$?
$(0, 0)$ $(-2, 0)$ $(0, -2)$

Expand and multiply.

61. $(2y + 4)^2$ **62.** $(y + 3)^2$

63. Solve $x - 2y = 4$ for x.

64. Solve $2x + 3y = 6$ for y.

Simplify.

65. $x^2 - 2(x^2 - 3)$ **66.** $x^2 + (x^2 - 4)$

Factor.

67. $5y^2 + 16y + 12$

68. $3x^2 + 17x - 28$

Solve.

69. $y^2 = 4$ **70.** $x^2 = 25$

71. $-x^2 + 6 = 2$

72. $5y^2 + 16y + 12 = 0$

12.3 Second-Degree Inequalities and Nonlinear Systems

OBJECTIVES

A Graph second-degree inequalities.

B Solve systems of nonlinear equations.

C Graph the solution sets to systems of inequalities.

TICKET TO SUCCESS

Keep these questions in mind as you read through the section. Then respond in your own words and in complete sentences.

1. What is a second-degree inequality and how might its graph differ from a linear inequality?
2. What is the significance of a broken line when graphing second-degree inequalities?
3. Is solving a system of nonlinear equations any different from solving a system of linear equations? Explain.
4. When solving a nonlinear system whose graphs are both circles, how many solutions sets can you expect?

Image copyright ©walex101, 2010. Used under license from Shutterstock.com

As discussed in the last section, the most recent sighting of Halley's comet was in 1986. Comets and planets both orbit the sun in an elliptical path. If the equation for the Earth and the comet are known, can you predict if the two objects will collide? In this section we will work with systems of nonlinear equations like these to predict whether ellipses and other nonlinear pathways will intersect.

Previously we graphed linear inequalities by first graphing the boundary and then choosing a test point not on the boundary to indicate the region used for the solution set. The problems in this section are very similar. We will use the same general methods for graphing second-degree inequalities in two variables that we used when we graphed linear inequalities in two variables.

A Second-Degree Inequalities

A second-degree inequality is an inequality that contains at least one squared variable.

EXAMPLE 1 Graph $x^2 + y^2 < 16$.

SOLUTION The boundary is $x^2 + y^2 = 16$, which is a circle with center at the origin and a radius of 4. Since the inequality sign is $<$, the boundary is not included in the solution set and therefore must be represented with a broken line. The graph of the boundary is shown in Figure 1.

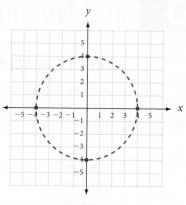

FIGURE 1

The solution set for $x^2 + y^2 < 16$ is either the region inside the circle or the region outside the circle. To see which region represents the solution set, we choose a convenient point not on the boundary and test it in the original inequality. The origin (0, 0) is a convenient point. Since the origin satisfies the inequality $x^2 + y^2 < 16$, all points in the same region will also satisfy the inequality. The graph of the solution set is shown in Figure 2.

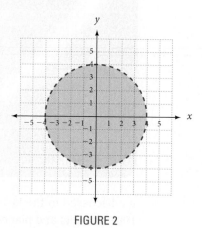

FIGURE 2

EXAMPLE 2 Graph the inequality $y \leq x^2 - 2$.

SOLUTION The parabola $y = x^2 - 2$ is the boundary and is included in the solution set. Using (0, 0) as the test point, we see that $0 \leq 0^2 - 2$ is a false statement, which means that the region containing (0, 0) is not in the solution set. (See Figure 3.)

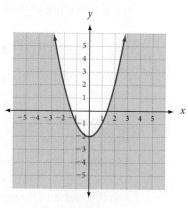

FIGURE 3

EXAMPLE 3   Graph $4y^2 - 9x^2 < 36$.

SOLUTION   The boundary is the hyperbola $4y^2 - 9x^2 = 36$ and is not included in the solution set. Testing $(0, 0)$ in the original inequality yields a true statement, which means that the region containing the origin is the solution set. (See Figure 4.)

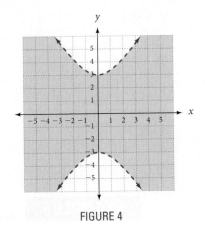

FIGURE 4

B   Nonlinear Systems

In an earlier chapter, we worked with equations of linear systems. Now we will practice solving nonlinear systems, which include at least one nonlinear equation. Remember, nonlinear equations can produce graphs in the shapes of circles, ellipses, hyperbolas, and parabolas.

EXAMPLE 4   Solve the system.

$$x^2 + y^2 = 4$$
$$x - 2y = 4$$

SOLUTION   In this case the substitution method is the most convenient. Solving the second equation for x in terms of y, we have

$$x - 2y = 4$$
$$x = 2y + 4$$

We now substitute $2y + 4$ for x in the first equation in our original system and proceed to solve for y.

$$(2y + 4)^2 + y^2 = 4$$

$$4y^2 + 16y + 16 + y^2 = 4 \qquad \text{Expand } (2y + 4)^2$$

$$5y^2 + 16y + 16 = 4 \qquad \text{Simplify left side}$$

$$5y^2 + 16y + 12 = 0 \qquad \text{Add } -4 \text{ to each side}$$

$$(5y + 6)(y + 2) = 0 \qquad \text{Factor}$$

$$5y + 6 = 0 \quad \text{or} \quad y + 2 = 0 \qquad \text{Set factors equal to 0}$$

$$y = -\frac{6}{5} \quad \text{or} \quad y = -2 \qquad \text{Solve}$$

These are the y-coordinates of the two solutions to the system. Substituting $y = -\frac{6}{5}$ into $x - 2y = 4$ and solving for x gives us $x = \frac{8}{5}$. Using $y = -2$ in the same equation yields $x = 0$. The two solutions to our system are $\left(\frac{8}{5}, -\frac{6}{5}\right)$ and $(0, -2)$. Although graphing the system is not necessary, it does help us visualize the situation. (See Figure 5.)

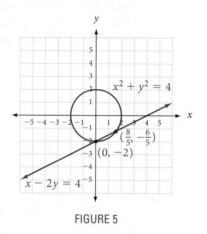

FIGURE 5

EXAMPLE 5 Solve the system.

$$16x^2 - 4y^2 = 64$$

$$x^2 + y^2 = 9$$

SOLUTION Since each equation is of the second degree in both x and y, it is easier to solve this system by eliminating one of the variables by addition. To eliminate y, we multiply the bottom equation by 4 and add the result to the top equation.

$$16x^2 - 4y^2 = 64$$

$$\underline{4x^2 + 4y^2 = 36}$$

$$20x^2 = 100$$

$$x^2 = 5$$

$$x = \pm\sqrt{5}$$

The x-coordinates of the points of intersection are $\sqrt{5}$ and $-\sqrt{5}$. We substitute each back into the second equation in the original system and solve for y.

When $\rightarrow$ $x = \sqrt{5}$

$$(\sqrt{5})^2 + y^2 = 9$$

$$5 + y^2 = 9$$

$$y^2 = 4$$

$$y = \pm 2$$

When $\rightarrow$ $x = -\sqrt{5}$

$$(-\sqrt{5})^2 + y^2 = 9$$

$$5 + y^2 = 9$$

$$y^2 = 4$$

$$y = \pm 2$$

The four points of intersection are $(\sqrt{5}, 2)$, $(\sqrt{5}, -2)$, $(-\sqrt{5}, 2)$, and $(-\sqrt{5}, -2)$. Graphically, the situation is as shown in Figure 6.

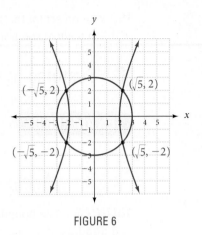

FIGURE 6

EXAMPLE 6   Solve the system.

$$x^2 - 2y = 2$$

$$y = x^2 - 3$$

SOLUTION   We can solve this system using the substitution method. Replacing y in the first equation with $x^2 - 3$ from the second equation, we have

$$x^2 - 2(x^2 - 3) = 2$$

$$-x^2 + 6 = 2$$

$$x^2 = 4$$

$$x = \pm 2$$

Using either 2 or -2 in the equation $y = x^2 - 3$ gives us $y = 1$. The system has two solutions: $(2, 1)$ and $(-2, 1)$.

EXAMPLE 7   The sum of the squares of two numbers is 34. The difference of their squares is 16. Find the two numbers.

SOLUTION   Let x and y be the two numbers. The sum of their squares is $x^2 + y^2$, and the difference of their squares is $x^2 - y^2$. (We can assume here that x^2 is the larger number.) The system of equations that describes the situation is

$$x^2 + y^2 = 34$$

$$x^2 - y^2 = 16$$

We can eliminate y by simply adding the two equations. The result of doing so is

$$2x^2 = 50$$

$$x^2 = 25$$

$$x = \pm 5$$

Substituting $x = 5$ into either equation in the system gives $y = \pm 3$. Using $x = -5$ gives the same results, $y = \pm 3$. The four pairs of numbers that are solutions to the original problem are 5 and 3, -5 and 3, 5 and -3, -5 and -3.

We now turn our attention to systems of inequalities. To solve a system of inequalities by graphing, we simply graph each inequality on the same set of axes.

The solution set for the system is the region common to both graphs—the intersection of the individual solution sets.

C Systems of Inequalities

EXAMPLE 8 Graph the solution set for the system.

$$x^2 + y^2 \leq 9$$
$$\frac{x^2}{4} + \frac{y^2}{25} \geq 1$$

SOLUTION The boundary for the top inequality is a circle with center at the origin and a radius of 3. The solution set lies inside the boundary. The boundary for the second inequality is an ellipse. In this case, the solution set lies outside the boundary. (See Figure 7.)

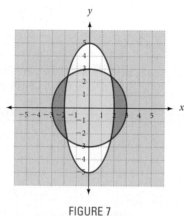

FIGURE 7

The solution set is the intersection of the two individual solution sets.

Problem Set 12.3

Moving Toward Success

"He who would learn to fly one day must first learn to stand and walk and run and climb and dance; one cannot fly into flying."

—Friedrich Nietzsche, 1844–1900, German philosopher

1. Why is it important to go into an exam knowing you can work any problem on your list of difficult problems?

2. How would visualizing yourself working through possible problems increase your potential for success during a test?

A Graph each of the following inequalities.
[Examples 1–3]

1. $x^2 + y^2 \leq 49$

2. $x^2 + y^2 < 49$

3. $(x - 2)^2 + (y + 3)^2 < 16$

4. $(x + 3)^2 + (y - 2)^2 \geq 25$

5. $y < x^2 - 6x + 7$

6. $y \geq x^2 + 2x - 8$

7. $4x^2 + 25y^2 \leq 100$

8. $25x^2 - 4y^2 > 100$

B Solve each of the following systems of equations.
[Examples 4–6]

9. $x^2 + y^2 = 9$
   $2x + y = 3$

10. $x^2 + y^2 = 9$
    $x + 2y = 3$

11. $x^2 + y^2 = 16$
$x + 2y = 8$

12. $x^2 + y^2 = 16$
$x - 2y = 8$

13. $x^2 + y^2 = 25$
$x^2 - y^2 = 25$

14. $x^2 + y^2 = 4$
$2x^2 - y^2 = 5$

15. $x^2 + y^2 = 9$
$y = x^2 - 3$

16. $x^2 + y^2 = 4$
$y = x^2 - 2$

17. $x^2 + y^2 = 16$
$y = x^2 - 4$

18. $x^2 + y^2 = 1$
$y = x^2 - 1$

19. $3x + 2y = 10$
$y = x^2 - 5$

20. $4x + 2y = 10$
$y = x^2 - 10$

21. $y = x^2 + 2x - 3$
$y = -x + 1$

22. $y = -x^2 - 2x + 3$
$y = -x + 1$

23. $y = x^2 - 6x + 5$
$y = x - 5$

24. $y = x^2 - 2x - 4$
$y = x - 4$

25. $4x^2 - 9y^2 = 36$
$4x^2 + 9y^2 = 36$

26. $4x^2 + 25y^2 = 100$
$4x^2 - 25y^2 = 100$

27. $x - y = 4$
$x^2 + y^2 = 16$

28. $x + y = 2$
$x^2 - y^2 = 4$

29. In a previous problem set, you found the equations for the three circles below. The equations are

Circle A	Circle B	Circle C
$(x + 8)^2 + y^2 = 64$	$x^2 + y^2 = 64$	$(x - 8)^2 + y^2 = 64$

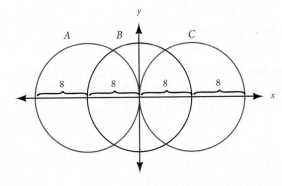

a. Find the points of intersection of circles A and B.

b. Find the points of intersection of circles B and C.

30. A magician is holding two rings that seem to lie in the same plane and intersect in two points. Each ring is 10 inches in diameter. If a coordinate system is placed with its origin at the center of the first ring and the x-axis contains the center of the second ring, then the equations are as follows:

First Ring   Second Ring
$x^2 + y^2 = 25$ $(x - 5)^2 + y^2 = 25$

Find the points of intersection of the two rings.

C Graph the solution sets to the following systems. [Example 8]

31. $x^2 + y^2 < 9$
$y \geq x^2 - 1$

32. $x^2 + y^2 \leq 16$
$y < x^2 + 2$

33. $\dfrac{x^2}{9} + \dfrac{y^2}{25} \leq 1$

$\dfrac{x^2}{4} - \dfrac{y^2}{9} > 1$

34. $\dfrac{x^2}{4} + \dfrac{y^2}{16} \geq 1$

$\dfrac{x^2}{9} - \dfrac{y^2}{25} < 1$

35. $4x^2 + 9y^2 \leq 36$

$y > x^2 + 2$

36. $9x^2 + 4y^2 \geq 36$

$y < x^2 + 1$

37. A parabola and a circle each contain the points $(-4, 0), (0, 4)$, and $(4, 0)$, as shown below. The equations for each of the curves are

Circle	Parabola
$x^2 + y^2 = 16$	$y = 4 - \dfrac{1}{4}x^2$

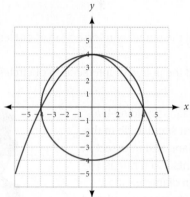

Write a system of inequalities that describes the regions in quadrants I and II that are between the two curves, if the boundaries are not included with the shaded region.

38. Find a system of inequalities that describes the shaded region in the figure below if the boundaries of the shaded region are included and equations of the circles are given by

Circle A	Circle B	Circle C
$(x + 8)^2 + y^2 = 64$	$x^2 + y^2 = 64$	$(x - 8)^2 + y^2 = 64$

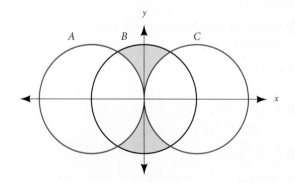

Applying the Concepts [Example 7]

39. Number Problems The sum of the squares of two numbers is 89. The difference of their squares is 39. Find the numbers.

40. Number Problems The difference of the squares of two numbers is 35. The sum of their squares is 37. Find the numbers.

41. Number Problems One number is 3 less than the square of another. Their sum is 9. Find the numbers.

42. Number Problems The square of one number is 2 less than twice the square of another. The sum of the squares of the two numbers is 25. Find the numbers.

Maintaining Your Skills

43. Let $g(x) = \left(x + \dfrac{2}{5}\right)^2$. Find all values for the variable x for which $g(x) = 0$.

44. Let $f(x) = (x + 2)^2 - 25$. Find all values for the variable x for which $f(x) = 0$.

For the problems below, let $f(x) = x^2 + 4x - 4$. Find all values for the variable x for which $f(x) = g(x)$.

45. $g(x) = 1$ **46.** $g(x) = -7$

47. $g(x) = x - 6$ **48.** $g(x) = x + 6$

Distance Formula [12.1]

1. The distance between (5, 2) and (−1, 1) is

$$d = \sqrt{(5 + 1)^2 + (2 - 1)^2}$$
$$= \sqrt{37}$$

The distance between the two points (x_1, y_1) and (x_2, y_2) is given by the formula

$$d = \sqrt{(x_2 - x_1)^2 + (y_2 - y_1)^2}$$

The Circle [12.1]

2. The graph of the circle $(x - 3)^2 + (y + 2)^2 = 25$ will have its center at (3, −2) and the radius will be 5.

The graph of any equation of the form

$$(x - a)^2 + (y - b)^2 = r^2$$

will be a circle having its center at (a, b) and a radius of r.

The Ellipse [12.2]

3. The ellipse $\frac{x^2}{9} + \frac{y^2}{4} = 1$ will cross the x-axis at 3 and −3 and will cross the y-axis at 2 and −2.

Any equation that can be put in the form

$$\frac{x^2}{a^2} + \frac{y^2}{b^2} = 1$$

will have an ellipse for its graph. The x-intercepts will be at a and $-a$, and the y-intercepts will be at b and $-b$.

The Hyperbola [12.2]

4. The hyperbola $\frac{x^2}{4} - \frac{y^2}{9} = 1$ will cross the x-axis at 2 and −2. It will not cross the y-axis.

The graph of an equation that can be put in either of the forms

$$\frac{x^2}{a^2} - \frac{y^2}{b^2} = 1 \quad \text{or} \quad \frac{y^2}{a^2} - \frac{x^2}{b^2} = 1$$

will be a hyperbola. The x-intercepts, for the first equation, will be at a and $-a$. The y-intercepts, for the second equation, will be at a and $-a$. Two straight lines, called *asymptotes*, are associated with the graph of every hyperbola. Although the asymptotes are not part of the hyperbola, they are useful in sketching the graph.

Second-Degree Inequalities in Two Variables [12.3]

5. The graph of the inequality
$$x^2 + y^2 < 9$$
is all points inside the circle with center at the origin and radius 3. The circle itself is not part of the solution and therefore is shown with a broken curve.

We graph second-degree inequalities in two variables in much the same way that we graphed linear inequalities; that is, we begin by graphing the boundary, using a solid curve if the boundary is included in the solution (this happens when the inequality symbol is $\geq$ or $\leq$) or a broken curve if the boundary is not included in the solution (when the inequality symbol is $>$ or $<$). After we have graphed the boundary, we choose a test point that is not on the boundary and try it in the original inequality. A true statement indicates we are in the region of the solution. A false statement indicates we are not in the region of the solution.

6. We can solve the system

$$x^2 + y^2 = 4$$
$$x = 2y + 4$$

by substituting $2y + 4$ from the second equation for x in the first equation:

$$(2y + 4)^2 + y^2 = 4$$
$$4y^2 + 16y + 16 + y^2 = 4$$
$$5y^2 + 16y + 12 = 0$$
$$(5y + 6)(y + 2) = 0$$
$$y = -\frac{6}{5} \quad \text{or} \quad y = -2$$

Substituting these values of y into the second equation in our system gives $x = \frac{8}{5}$ and $x = 0$. The solutions are $\left(\frac{8}{5}, -\frac{6}{5}\right)$ and $(0, -2)$.

Systems of Nonlinear Equations [12.3]

A system of nonlinear equations is two equations, at least one of which is not linear, considered at the same time. The solution set for the system consists of all ordered pairs that satisfy both equations. In most cases we use the substitution method to solve these systems; however, the addition method can be used if like variables are raised to the same power in both equations. It is sometimes helpful to graph each equation in the system on the same set of axes to anticipate the number and approximate positions of the solutions.

Chapter 12 Review

Find the distance between the following points. [12.1]

1. $(2, 6), (-1, 5)$

2. $(3, -4), (1, -1)$

3. $(0, 3), (-4, 0)$

4. $(-3, 7), (-3, -2)$

5. Find x so that the distance between $(x, -1)$ and $(2, -4)$ is 5. [12.1]

6. Find y so that the distance between $(3, -4)$ and $(-3, y)$ is 10. [12.1]

Find the equation of each circle. [12.1]

7. Center at the origin, x-intercepts ± 5

8. Center at the origin, y-intercepts ± 3

9. Center at $(-2, 3)$ and passing through the point $(2, 0)$

10. Center at $(-6, 8)$ and passing through the origin

Give the center and radius of each circle, and then sketch the graph. [12.1]

11. $x^2 + y^2 = 4$

12. $(x - 3)^2 + (y + 1)^2 = 16$

13. $x^2 + y^2 - 6x + 4y = -4$

14. $x^2 + y^2 + 4x - 2y = 4$

Graph each of the following. Label the x- and y-intercepts. [12.2]

15. $\dfrac{x^2}{4} + \dfrac{y^2}{9} = 1$

16. $4x^2 + y^2 = 16$

Graph the following. Show the asymptotes. [12.2]

17. $\dfrac{x^2}{4} - \dfrac{y^2}{9} = 1$

18. $4x^2 - y^2 = 16$

Graph each of the following inequalities. [12.3]

19. $x^2 + y^2 < 9$

20. $(x + 2)^2 + (y - 1)^2 \leq 4$

21. $y \geq x^2 - 1$

22. $9x^2 + 4y^2 \leq 36$

Graph the solution set for each system. [12.3]

23. $x^2 + y^2 < 16$
$y > x^2 - 4$

24. $x + y \leq 2$
$-x + y \leq 2$
$y \geq -2$

Solve each system of equations. [12.3]

25. $x^2 + y^2 = 16$
$2x + y = 4$

26. $x^2 + y^2 = 4$
$y = x^2 - 2$

27. $9x^2 - 4y^2 = 36$
$9x^2 + 4y^2 = 36$

Simplify.

1. $\dfrac{-9 - 6}{4 - 7}$

2. $\dfrac{5(-6) + 3(-2)}{4(-3) + 3}$

3. $\dfrac{18a^7b^{-4}}{36a^2b^{-8}}$

4. $\dfrac{x^{3/4}y^2}{x^{1/2}y^{1/4}}$

5. $\dfrac{12x^2y^3 - 16x^2y + 8xy^3}{4xy}$

6. $\dfrac{y^2 - y - 6}{y^2 - 4}$

7. $\log_4 16$

8. $\log_3 27$

Factor completely.

9. $ab^3 + b^3 + 6a + 6$

10. $8x^2 - 5x - 3$

Solve.

11. $6 - 2(5x - 1) + 4x = 20$

12. $3 - 2(3x - 4) = -1$

13. $(x + 1)(x + 2) = 12$

14. $1 - \dfrac{2}{x} = \dfrac{8}{x^2}$

15. $t - 6 = \sqrt{t - 4}$

16. $\sqrt{7x - 4} = -2$

17. $(4x - 3)^2 = -50$

18. $(x - 3)^2 = -3$

Solve and graph the solution on the number line.

19. $|4x + 3| - 6 > 5$

20. $|2x + 5| - 2 < 9$

Graph on a rectangular coordinate system.

21. $y < \dfrac{1}{2}x + 3$

22. $3x - y < -2$

23. $y = (x - 2)^2 - 3$

24. $y = x^2 - 2x - 3$

Multiply.

25. $\dfrac{3y^2 - 3y}{3y - 12} \cdot \dfrac{y^2 - 2y - 8}{y^2 + 3y + 2}$

26. $\dfrac{x^2 - 16}{x^2 + 5x + 6} \cdot \dfrac{x^2 + 6x + 9}{x^3 + 4x^2}$

27. $(2 + 3i)(1 - 4i)$

28. $(2 + 3i)(1 + 4i)$

Rationalize the denominator.

29. $\dfrac{3}{\sqrt{7} - \sqrt{3}}$

30. $\dfrac{5\sqrt{6}}{2\sqrt{6} + 7}$

Find the inverse, $f^{-1}(x)$.

31. $f(x) = 4x + 1$

32. $f(x) = \dfrac{1}{2}x + 3$

33. Find x if $\log x = 3.9786$ to the nearest hundreth.

34. Find $\log_6 14$ to the nearest hundredth.

35. Solve $mx + 2 = nx - 3$ for x.

36. Solve $S = 2x^2 + 4xy$ for y.

37. Find the slope of the line through $(2, -3)$ and $(-1, -3)$.

38. Find the slope and y-intercept for $2x - 3y = 12$.

39. If $f(x) = -\dfrac{3}{2}x + 1$, find $f(4)$.

40. If $C(t) = 80\left(\dfrac{1}{2}\right)^{t/5}$, find $C(5)$ and $C(10)$.

41. Find an equation that has solutions $x = 1$ and $x = \dfrac{2}{3}$.

42. Find an equation with solutions $t = -\dfrac{3}{4}$ and $t = \dfrac{1}{5}$.

43. If $f(x) = 3x - 7$ and $g(x) = 4 - x$, find $(f \circ g)(x)$.

44. If $f(x) = 4 - x^2$ and $g(x) = 5x - 1$, find $(g \circ f)(x)$.

45. Add -8 to the difference of -7 and 4.

46. Subtract 9 from the sum of -2 and 5.

47. Variation y varies directly with x. If y is 24 when x is 8, find y when x is 2.

48. Variation y varies inversely with the square of x. If y is 3 when x is 3, find y when x is 6.

49. Mixture How much 30% alcohol solution and 70% alcohol solution must be mixed to get 16 gallons of 60% alcohol solution?

50. Mixture How many gallons of 20% alcohol solution and 60% alcohol solution must be mixed to get 16 gallons of 30% alcohol solution?

Use the illustration to answer Questions 51 and 52. Round your answers to the nearest minute.

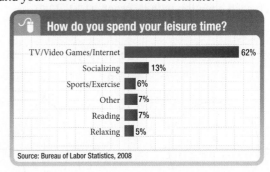

How do you spend your leisure time?

TV/Video Games/Internet	62%
Socializing	13%
Sports/Exercise	6%
Other	7%
Reading	7%
Relaxing	5%

Source: Bureau of Labor Statistics, 2008

51. If 150 people were surveyed, how many people spend their leisure time playing video games, watching TV or surfing the internet?

52. If 150 people were surveyed, how many people spend their leisure time reading or relaxing?

Find the distance between the points. [12.1]

1. $(-9, 4)$ and $(3, -1)$ **2.** $(4, -6)$ and $(0, -4)$

3. Find x so that $(x, 2)$ is $2\sqrt{5}$ units from $(-1, 4)$. [12.1]

4. Give the equation of the circle with center at $(-2, 4)$ and radius 3. [12.1]

5. Give the equation of the circle with center at the origin that contains the point $(-3, -4)$. [12.1]

Find the center and radius of the circle. [12.1]

6. $x^2 + (y + 2)^2 = 64$

7. $x^2 + y^2 - 10x + 6y = 5$

Graph each of the following. [12.1, 12.2, 12.3]

8. $(x - 1)^2 + (y + 2)^2 = 9$ **9.** $\dfrac{x^2}{25} + \dfrac{y^2}{4} = 1$

10. $4x^2 - y^2 = 16$

11. $25x^2 + 150x + 4y^2 + 125 = 0$

12. $(x - 2)^2 + (y + 1)^2 \le 9$ **13.** $y > x^2 - 4$

Solve the following systems. [12.3]

14. $\begin{aligned} x^2 + y^2 &= 25 \\ 2x + y &= 5 \end{aligned}$

15. $\begin{aligned} x^2 + y^2 &= 16 \\ y &= x^2 - 4 \end{aligned}$

16. Graph the solution set to the system. [12.3]
$$\begin{aligned} x^2 + y^2 &\ge 1 \\ y &< -x^2 + 1 \end{aligned}$$

Match each equation with its graph. [12.1, 12.2]

17. $x^2 + 4y^2 = 16$ **18.** $x^2 + y^2 = 16$

19. $x^2 - 4y^2 = 16$ **20.** $4x^2 + y^2 = 16$

A.

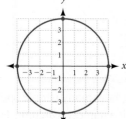

B.

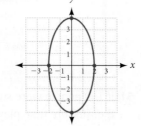

C.

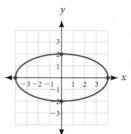

D.

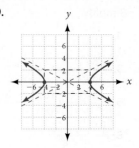

Find an equation for each graph. [12.1, 12.2]

21.

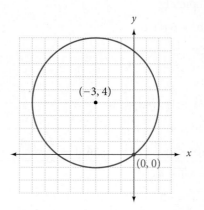

22.

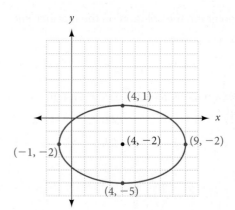

23.

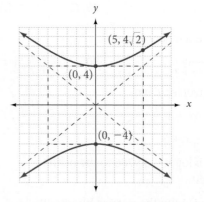

Chapter 12 Projects

CONIC SECTIONS

GROUP PROJECT

Constructing Ellipses

Number of People 4

Time Needed 20 minutes

Equipment Graph paper, pencils, string, and thumbtacks

Background The geometric definition for an ellipse is the set of points the sum of whose distances from two fixed points (called foci) is a constant. We can use this definition to draw an ellipse using thumbtacks, string, and a pencil.

Procedure 1. Start with a piece of string 7 inches long. Place thumbtacks through the string $\frac{1}{2}$ inch from each end, then tack the string to a pad of graph paper so that the tacks are 4 inches apart. Pull the string tight with the tip of a pencil, then trace all the way around the two tacks. (See Figure 1.) The resulting diagram will be an ellipse.

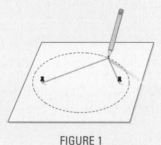

FIGURE 1

2. The line segment that passes through the tacks (these are the foci) and connects opposite ends of the ellipse is called the major axis. The line segment perpendicular to the major axis that passes through the center of the ellipse and connects opposite ends of the ellipse is called the minor axis. (See Figure 2.) Measure the length of the major axis and the length of the minor axis. Record your results in Table 1.

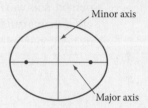

FIGURE 2

3. Explain how drawing the ellipse as you have in step 2 shows that the geometric definition of an ellipse given at the beginning of this project is, in fact, correct.

4. Next, move the tacks so that they are 3 inches apart. Trace out that ellipse. Measure the length of the major axis and the length of the minor axis, and record your results in Table 1.

5. Repeat step 4 with the tacks 2 inches apart.

6. If the length of the string between the tacks stays at 6 inches, and the tacks were placed 6 inches apart, then the resulting ellipse would be a _____. If the tacks were placed 0 inches apart, then the resulting ellipse would be a _____.

TABLE 1			
ELLIPSES (ALL LENGTHS ARE INCHES)			
Length of String	**Distance Between Foci**	**Length of Major Axis**	**Length of Minor Axis**
6	4		
6	3		
6	2		

Hypatia of Alexandria

As previously discussed, the first woman mentioned in the history of mathematics is Hypatia of Alexandria. Research the life of Hypatia, and then write an essay that begins with a description of the time and place in which she lived. Continue with an indication of the type of person she was, her accomplishments in areas other than mathematics, and how she was viewed by her contemporaries.

Bettmann/Corbis

13

Sequences and Series

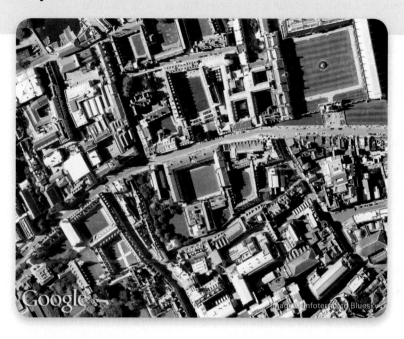

Introduction

The Google image above shows The University of Cambrige, the second oldest university in England. Isaac Newton graduated from Cambrige in 1665, with no great distinction. With the plague raging in Europe, Newton secluded himself in his family's English country home. He spent much of the next two years intensely studying science and mathematics. Though other mathematicians had described a formula for binomial expansion, or multiplying any expression in the form $(x + y)^n$, Newton was the first to apply it to negative and fractional exponents. His consideration of these aspects of the formula led directly to his development of calculus. In this chapter, we will work with series and sequences, including binomial expansion.

Evaluate each expression for the given values of the variable.

1. $2x + 3$ for $x = 1, 2, 3, 4$ **2.** $4x - 1$ for $x = 1, 2, 3, 4$ **3.** $\dfrac{x}{x + 3}$ for $x = 1, 2, 3, 4$ **4.** $\dfrac{x + 1}{x + 2}$ for $x = 1, 2, 3, 4$

5. $(-2)^x$ for $x = 1, 2, 3, 4$ **6.** $(-3)^x$ for $x = 1, 2, 3, 4$

Simplify the following.

7. $\dfrac{2\sqrt{2}}{2}$ **8.** $\dfrac{2}{\sqrt{2}}$ **9.** $3 \cdot 3^{x-1}$ **10.** $5 \cdot 5^{x-1}$

11. $\dfrac{xy^7}{xy^3}$ **12.** $\dfrac{ab^5}{ab^3}$ **13.** $\dfrac{xy^{-6}}{xy^{-3}}$ **14.** $\dfrac{1}{b^9}$

Expand and multiply.

15. $(a + b)^2$ **16.** $(a - b)^2$ **17.** $(a + b)^3$ **18.** $(a - b)^3$

Chapter Outline

13.1 Sequences
A Use a formula to find the terms of a sequence.
B Use a recursive formula.
C Find the general term of a sequence.

13.2 Series
A Expand and simplify an expression written with summation notation.
B Write a series using summation notation.

13.3 Arithmetic Sequences
A Find the common difference for an arithmetic sequence.
B Find the general term for an arithmetic sequence.
C Find the sum of the first n terms of an arithmetic sequence.

13.4 Geometric Sequences
A Find the common ratio, general term, and the nth term for a geometric sequence.
B Find the sum of the first n terms of a geometric sequence.
C Find the sum of all terms of an infinite geometric sequence.

13.5 The Binomial Expansion
A Calculate binomial coefficients.
B Use the binomial formula to expand binomial powers.
C Find a particular term of a binomial expansion.

OBJECTIVES

A Use a formula to find the terms of a sequence.

B Use a recursive formula.

C Find the general term of a sequence.

TICKET TO SUCCESS

Keep these questions in mind as you read through the section. Then respond in your own words and in complete sentences.

1. How are subscripts used to denote the terms of a sequence?
2. What is the general term of a sequence?
3. What is a decreasing sequence?
4. What is meant by a recursion formula for a sequence?

Suppose you are helping a friend stack wood to get his mountain cabin ready for winter. The first row in the stack has 50 logs, the second has 48, the third 46, and so on. If the stack is 10 rows high, how many logs are in the stack? In order to answer this question, you will need a working knowledge of sequences, the topic of this section.

A Terms of a Sequence

Many of the sequences in this chapter will be familiar to you on an intuitive level because you have worked with them for some time now. Here are some of those sequences:

The sequence of odd numbers

$$1, 3, 5, 7, \ldots$$

The sequence of even numbers

$$2, 4, 6, 8, \ldots$$

The sequence of squares

$$1^2, 2^2, 3^2, 4^2, \ldots = 1, 4, 9, 16, \ldots$$

The numbers in each of these sequences can be found from the formulas that define functions. For example, the sequence of even numbers can be found from the function

$$f(x) = 2x$$

by finding $f(1), f(2), f(3), f(4)$, and so forth. This gives us justification for the formal definition of a sequence.

Definition

A **sequence** is a function whose domain is the set of positive integers $\{1, 2, 3, 4, \dots \}$.

As you can see, sequences are simply functions with a specific domain. If we want to form a sequence from the function $f(x) = 3x + 5$, we simply find $f(1)$, $f(2), f(3)$, and so on. Doing so gives us the sequence

$$8, 11, 14, 17, \dots$$

because $f(1) = 3(1) + 5 = 8$, $f(2) = 3(2) + 5 = 11$, $f(3) = 3(3) + 5 = 14$, and $f(4) = 3(4) + 5 = 17$.

Notation Because the domain for a sequence is always the set $\{1, 2, 3, \dots \}$, we can simplify the notation we use to represent the terms of a sequence. Using the letter a instead of f, and subscripts instead of numbers enclosed by parentheses, we can represent the sequence from the previous discussion as follows:

$$a_n = 3n + 5$$

Instead of $f(1)$, we write a_1 for the *first term* of the sequence.
Instead of $f(2)$, we write a_2 for the *second term* of the sequence.
Instead of $f(3)$, we write a_3 for the *third term* of the sequence.
Instead of $f(4)$, we write a_4 for the *fourth term* of the sequence.
Instead of $f(n)$, we write a_n for the *nth term* of the sequence.

The *n*th term is also called the *general term* of the sequence. The general term is used to define the other terms of the sequence. That is, if we are given the formula for the general term a_n, we can find any other term in the sequence. The following examples illustrate.

EXAMPLE 1 Find the first four terms of the sequence whose general term is given by $a_n = 2n - 1$.

SOLUTION The subscript notation a_n works the same way function notation works. To find the first, second, third, and fourth terms of this sequence, we simply substitute 1, 2, 3, and 4 for n in the formula $2n - 1$.

$$\text{If} \rightarrow \quad \text{the general term is } a_n = 2n - 1$$
$$\text{then} \rightarrow \quad \text{the first term is } a_1 = 2(1) - 1 = 1$$
$$\text{the second term is } a_2 = 2(2) - 1 = 3$$
$$\text{the third term is } a_3 = 2(3) - 1 = 5$$
$$\text{the fourth term is } a_4 = 2(4) - 1 = 7$$

The first four terms of the sequence in Example 1 are the odd numbers 1, 3, 5, and 7. The whole sequence can be written as

$$1, 3, 5, \dots, 2n - 1, \dots$$

Because each term in this sequence is larger than the preceding term, we say the sequence is an *increasing sequence*.

EXAMPLE 2 Write the first four terms of the sequence defined by
$$a_n = \frac{1}{n+1}.$$

SOLUTION Replacing n with 1, 2, 3, and 4, we have, respectively, the first four terms:

$$\text{First term} = a_1 = \frac{1}{1+1} = \frac{1}{2}$$

$$\text{Second term} = a_2 = \frac{1}{2+1} = \frac{1}{3}$$

$$\text{Third term} = a_3 = \frac{1}{3+1} = \frac{1}{4}$$

$$\text{Fourth term} = a_4 = \frac{1}{4+1} = \frac{1}{5}$$

The sequence in Example 2 defined by
$$a_n = \frac{1}{n+1}$$
can be written as
$$\frac{1}{2}, \frac{1}{3}, \frac{1}{4}, \ldots, \frac{1}{n+1}, \ldots$$

Because each term in the sequence is smaller than the term preceding it, the sequence is said to be a *decreasing sequence*.

EXAMPLE 3 Find the fifth and sixth terms of the sequence whose general term is given by $a_n = \frac{(-1)^n}{n^2}$.

SOLUTION For the fifth term, we replace n with 5. For the sixth term, we replace n with 6.

$$\text{Fifth term} = a_5 = \frac{(-1)^5}{5^2} = \frac{-1}{25}$$

$$\text{Sixth term} = a_6 = \frac{(-1)^6}{6^2} = \frac{1}{36}$$

The sequence in Example 3 can be written as
$$-1, \frac{1}{4}, -\frac{1}{9}, \frac{1}{16}, \ldots, \frac{(-1)^n}{n^2}, \ldots$$

Because the terms alternate in sign—if one term is positive, then the next term is negative—we call this an *alternating sequence*. The first three examples all illustrate how we work with a sequence in which we are given a formula for the general term.

USING TECHNOLOGY

Finding Sequences on a Graphing Calculator

Method 1: Using a Table

We can use the table function on a graphing calculator to view the terms of a sequence. To view the terms of the sequence $a_n = 3n + 5$, we set $Y_1 = 3X + 5$. Then we use the table setup feature on the calculator to set the table minimum to 1, and the table increment to 1 also. Here is the setup and result for a TI-83/84.

Table Setup   *Y Variables Setup*

Table minimum = 1 $Y_1 = 3x + 5$
Table increment = 1
Independent variable: Auto
Dependent variable: Auto
The table will look like this:

X	Y
1	8
2	11
3	14
4	17
5	20

To find any particular term of a sequence, we change the independent variable setting to Ask, and then input the number of the term of the sequence we want to find. For example, if we want term a_{100}, then we input 100 for the independent variable, and the table gives us the value of 305 for that term.

Method 2: Using the Built-in seq(Function

Using this method, first find the seq(function. On a TI-83/84 it is found in the LIST OPS menu. To find terms a_1 through a_7 for $a_n = 3n + 5$, we first bring up the seq(function on our calculator, then we input the following four items, in order, separated by commas: 3X + 5, X, 1, 7. Then we close the parentheses. Our screen will look like this:

$$\text{seq}(3X + 5, X, 1, 7)$$

Pressing ‍ENTER‍ displays the first five terms of the sequence. Pressing the right arrow key repeatedly brings the remaining members of the sequence into view.

Method 3: Using the Built-in Seq Mode

Press the ‍MODE‍ key on your TI-83/84 and then select Seq (it's next to Func Par and Pol). Go to the Y variables list and set nMin = 1 and $u(n) = 3n + 5$. Then go to the TBLSET key to set up your table like the one shown in Method 1. Pressing ‍TABLE‍ will display the sequence you have defined.

B Recursion Formulas

Let's go back to one of the first sequences we looked at in this section:

$$8, 11, 14, 17, \ldots$$

Each term in the sequence can be found by simply substituting positive integers for n in the formula $a_n = 3n + 5$. Another way to look at this sequence, however, is to notice that each term can be found by adding 3 to the preceding term—a process known as recursion. We could give all the terms of this sequence by simply saying

Start with 8, and then add 3 to each term to get the next term.

The same idea, expressed in symbols, looks like this:

$$a_1 = 8 \quad \text{and} \quad a_n = a_{n-1} + 3 \quad \text{for } n > 1$$

This formula is called a *recursion formula* because each term is written *recursively* using the term or terms that precede it.

EXAMPLE 4 Write the first four terms of the sequence given recursively by

$$a_1 = 4 \quad \text{and} \quad a_n = 5a_{n-1} \quad \text{for } n > 1$$

SOLUTION The formula tells us to start the sequence with the number 4, and then multiply each term by 5 to get the next term. Therefore,

$$a_1 = 4$$
$$a_2 = 5a_1 = 5(4) = 20$$
$$a_3 = 5a_2 = 5(20) = 100$$
$$a_4 = 5a_3 = 5(100) = 500$$

The sequence is 4, 20, 100, 500, ...

USING TECHNOLOGY

Recursion Formulas on a Graphing Calculator

We can use a TI-83/84 graphing calculator to view the sequence defined recursively as

$$a_1 = 8, a_n = a_{n-1} + 3$$

First, put your TI-83/84 calculator in sequence mode by pressing the MODE key, and then selecting Seq (it's next to Func Par and Pol). Go to the Y variables list and set nMin = 1 and $u(n) = u(n - 1) + 3$. (The u is above the 7, and the n is on the X, T, θ, n and is automatically displayed if that key is pressed when the calculator is in the Seq mode.) Pressing TABLE will display the sequence you have defined.

C Finding the General Term

In the first four examples, we found some terms of a sequence after being given the general term. In the next two examples, we will do the reverse. That is, given some terms of a sequence, we will find the formula for the general term.

EXAMPLE 5 Find a formula for the nth term of the sequence 2, 8, 18, 32, ...

SOLUTION Solving a problem like this involves some guessing. Looking over the first four terms, we see each is twice a perfect square.

$$2 = 2(1)$$
$$8 = 2(4)$$

$$18 = 2(9)$$

$$32 = 2(16)$$

If we write each square with an exponent of 2, the formula for the nth term becomes obvious.

$$a_1 = 2 = 2(1)^2$$

$$a_2 = 8 = 2(2)^2$$

$$a_3 = 18 = 2(3)^2$$

$$a_4 = 32 = 2(4)^2$$

$$\vdots$$

$$a_n = \quad 2(n)^2 = 2n^2$$

The general term of the sequence 2, 8, 18, 32, . . . is $a_n = 2n^2$. ■

EXAMPLE 6 Find the general term for the sequence $2, \dfrac{3}{8}, \dfrac{4}{27}, \dfrac{5}{64}, \ldots$

SOLUTION The first term can be written as $\dfrac{2}{1}$. The denominators are all perfect cubes. The numerators are all 1 more than the base of the cubes in the denominators.

$$a_1 = \frac{2}{1} = \frac{1+1}{1^3}$$

$$a_2 = \frac{3}{8} = \frac{2+1}{2^3}$$

$$a_3 = \frac{4}{27} = \frac{3+1}{3^3}$$

$$a_4 = \frac{5}{64} = \frac{4+1}{4^3}$$

Observing this pattern, we recognize the general term to be

$$a_n = \frac{n+1}{n^3}$$

■

NOTE
Finding the nth term of a sequence from the first few terms is not always automatic. That is, it sometimes takes a while to recognize the pattern. Don't be afraid to guess at the formula for the general term. Many times, an incorrect guess leads to the correct formula.

Problem Set 13.1

Moving Toward Success

"Life consists not in holding good cards but in playing those you hold well."

—Josh Billings, 1818–1885, American humorist and lecturer

1. What has been your most useful study tool for this class?

2. Do you prefer to study for your final exam with a study partner or group, or by yourself? Explain.

A Write the first five terms of the sequences with the following general terms. [Examples 1–3]

1. $a_n = 3n + 1$

2. $a_n = 2n + 3$

3. $a_n = 4n - 1$

4. $a_n = n + 4$

5. $a_n = n$

6. $a_n = -n$

7. $a_n = n^2 + 3$

8. $a_n = n^3 + 1$

9. $a_n = \dfrac{n}{n+3}$

10. $a_n = \dfrac{n+1}{n+2}$

11. $a_n = \dfrac{1}{n^2}$

12. $a_n = \dfrac{1}{n^3}$

13. $a_n = 2^n$

14. $a_n = 3^{-n}$

15. $a_n = 1 + \dfrac{1}{n}$

16. $a_n = n - \dfrac{1}{n}$

17. $a_n = (-2)^n$

18. $a_n = (-3)^n$

19. $a_n = 4 + (-1)^n$

20. $a_n = 10 + (-2)^n$

21. $a_n = (-1)^{n+1} \cdot \dfrac{n}{2n - 1}$

22. $a_n = (-1)^n \cdot \dfrac{2n + 1}{2n - 1}$

23. $a_n = n^2 \cdot 2^{-n}$ **24.** $a_n = n^n$

B Write the first five terms of the sequences defined by the following recursion formulas. [Example 4]

25. $a_1 = 3 \qquad a_n = -3a_{n-1} \qquad n > 1$

26. $a_1 = 3 \qquad a_n = a_{n-1} - 3 \qquad n > 1$

27. $a_1 = 1 \qquad a_n = 2a_{n-1} + 3 \qquad n > 1$

28. $a_1 = 1 \qquad a_n = a_{n-1} + n \qquad n > 1$

29. $a_1 = 2 \qquad a_n = 2a_{n-1} - 1 \qquad n > 1$

30. $a_1 = -4 \qquad a_n = -2a_{n-1} \qquad n > 1$

31. $a_1 = 5 \qquad a_n = 3a_{n-1} - 4 \qquad n > 1$

32. $a_1 = -3 \qquad a_n = -2a_{n-1} + 5 \qquad n > 1$

33. $a_1 = 4 \qquad a_n = 2a_{n-1} - a_1 \qquad n > 1$

34. $a_1 = -3 \qquad a_n = -2a_{n-1} - n \qquad n > 1$

C Determine the general term for each of the following sequences. [Examples 5–6]

35. $4, 8, 12, 16, 20, \ldots$

36. $7, 10, 13, 16, \ldots$

37. $1, 4, 9, 16, \ldots$

38. $3, 12, 27, 48, \ldots$

39. $4, 8, 16, 32, \ldots$

40. $-2, 4, -8, 16, \ldots$

41. $\dfrac{1}{4}, \dfrac{1}{8}, \dfrac{1}{16}, \dfrac{1}{32}, \ldots$

42. $\dfrac{1}{4}, \dfrac{2}{9}, \dfrac{3}{16}, \dfrac{4}{25}, \ldots$

43. $5, 8, 11, 14, \ldots$

44. $7, 5, 3, 1, \ldots$

45. $-2, -6, -10, -14, \ldots$

46. $-2, 2, -2, 2, \ldots$

47. $1, -2, 4, -8, \ldots$

48. $-1, 3, -9, 27, \ldots$

49. $\log_2 3, \log_3 4, \log_4 5, \log_5 6, \ldots$

50. $0, \dfrac{3}{5}, \dfrac{8}{10}, \dfrac{15}{17}, \ldots$

Applying the Concepts

51. Salary Increase The entry level salary for a teacher is $28,000 with 4% increases after every year of service.
  a. Write a sequence for this teacher's salary for the first 5 years.

  b. Find the general term of the sequence in part a.

52. Holiday Account To save money for holiday presents, a person deposits $5 in a savings account on January 1, and then deposits an additional $5 every week thereafter until Christmas.
  a. Write a sequence for the money in that savings account for the first 10 weeks of the year.

  b. Write the general term of the sequence in part a.

  c. If there are 50 weeks from January 1 to Christmas, how much money will be available for spending on Christmas presents?

53. Akaka Falls If a boulder fell from the top of Akaka Falls in Hawaii, the distance, in feet, the boulder would fall in each consecutive second would be modeled by a sequence whose general term is $a_n = 32n - 16$, where n represents the number of seconds.

George H. H. Huey/Corbis

  a. Write a sequence for the first 5 seconds the boulder falls.
  b. What is the total distance the boulder fell in 5 seconds?
  c. If Akaka Falls is approximately 420 feet high, will the boulder hit the ground within 5 seconds?

54. Health Expenditures The total national health expenditures, in billions of dollars, increased by approximately $50 billion from 1995 to 1996 and from 1996 to 1997. The total national health expenditures for 1997 was approximately $1,125 billion (U.S. Census Bureau, National Health Expenditures).
  a. If the same approximate increase is assumed to continue, what would have been the estimated total expenditures for 1998, 1999, and 2000?

b. Write the general term for the sequence, where n represents the number of years since 1997.

c. If health expenditures continue to increase at this approximate rate, what are the expected expenditures in 2015?

55. Polygons The formula for the sum of the interior angles of a polygon with n sides is $a_n = 180°(n - 2)$.

a. Write a sequence to represent the sum of the interior angles of a polygon with 3, 4, 5, and 6 sides.

b. What would be the sum of the interior angles of a polygon with 20 sides?

c. What happens when $n = 2$ to indicate that a polygon cannot be formed with only two sides?

56. Pendulum A pendulum swings 10 feet left to right on its first swing. On each swing following the first, the pendulum swings $\frac{4}{5}$ of the previous swing.

a. Write a sequence for the distance traveled by the pendulum on the first, second, and third swing.

b. Write a general term for the sequence, where n represents the number of the swing.

c. How far will the pendulum swing on its tenth swing? (Round to the nearest hundredth.)

Maintaining Your Skills

Find x in each of the following.

57. $\log_9 x = \dfrac{3}{2}$

58. $\log_x \dfrac{1}{4} = -2$

Simplify each expression.

59. $\log_2 32$

60. $\log_{10} 10,000$

61. $\log_3[\log_2 8]$

62. $\log_5[\log_6 6]$

Getting Ready for the Next Section

Simplify.

63. $-2 + 6 + 4 + 22$

64. $9 - 27 + 81 - 243$

65. $-8 + 16 - 32 + 64$

66. $-4 + 8 - 16 + 32 - 64$

67. $(1 - 3) + (4 - 3) + (9 - 3) + (16 - 3)$

68. $(1 - 3) + (9 + 1) + (16 + 1) + (25 + 1) + (36 + 1)$

69. $-\dfrac{1}{3} + \dfrac{1}{9} - \dfrac{1}{27} + \dfrac{1}{81}$

70. $\dfrac{1}{2} + \dfrac{2}{3} + \dfrac{3}{4} + \dfrac{4}{5}$

71. $\dfrac{1}{3} + \dfrac{1}{2} + \dfrac{3}{5} + \dfrac{2}{3}$

72. $\dfrac{1}{16} + \dfrac{1}{32} + \dfrac{1}{64}$

Extending the Concepts

73. As n increases, the terms in the sequence

$$a_n = \left(1 + \frac{1}{n}\right)^n$$

get closer and closer to the number e (that's the same e we used in defining natural logarithms). It takes some fairly large values of n, however, before we can see this happening. Use a calculator to find a_{100}, $a_{1,000}$, $a_{10,000}$, and $a_{100,000}$, and compare them to the decimal approximation we gave for the number e.

74. The sequence

$$a_n = \left(1 + \frac{1}{n}\right)^{-n}$$

gets close to the number $\dfrac{1}{e}$ as n becomes large. Use a calculator to find approximations for a_{100} and $a_{1,000}$, and then compare them to $\dfrac{1}{2.7183}$.

75. Write the first ten terms of the sequence defined by the recursion formula

$$a_1 = 1, a_2 = 1, a_n = a_{n-1} + a_{n-2} \quad n > 2$$

76. Write the first ten terms of the sequence defined by the recursion formula

$$a_1 = 2, a_2 = 2, a_n = a_{n-1} + a_{n-2} \quad n > 2$$

13.2 Series

OBJECTIVES

A Expand and simplify an expression written with summation notation.

B Write a series using summation notation.

TICKET TO SUCCESS

Keep these questions in mind as you read through the section. Then respond in your own words and in complete sentences.

1. What is the difference between a sequence and a series?
2. In words, explain summation notation.
3. What is the index of summation?
4. Explain the summation notation $\sum\limits_{i=1}^{4}$ in the series $\sum\limits_{i=1}^{4}(2i + 1)$.

Imagine you are studying bacterial growth in a bacteriology class. Your study begins with a colony of 50 bacteria in a petri dish. After 1 day, they have reproduced to become 200 bacteria, and 800 after 2 days. How many bacteria will be present after 4 days? To solve this problem, we must construct a series. In this section we will work with series and the notation associated with them. You will also solve this problem in the problem set at the end of the section.

A Finite Series

There is an interesting relationship between the sequence of odd numbers and the sequence of squares that is found by adding the terms in the sequence of odd numbers.

$$1 \qquad\qquad = 1$$
$$1 + 3 \qquad\; = 4$$
$$1 + 3 + 5 \quad\; = 9$$
$$1 + 3 + 5 + 7 = 16$$

When we add the terms of a sequence the result is called a series.

Definition

The sum of a number of terms in a sequence is called a **series**.

A sequence can be finite or infinite, depending on whether the sequence ends at the nth term. For example,

$$1, 3, 5, 7, 9$$

is a finite sequence, but

$$1, 3, 5, \ldots$$

is an infinite sequence. Associated with each of the preceding sequences is a series found by adding the terms of the sequence.

$$1 + 3 + 5 + 7 + 9 \quad \text{Finite series}$$

$$1 + 3 + 5 + \ldots \quad \text{Infinite series}$$

In this section, we will consider only finite series. We can introduce a new kind of notation here that is a compact way of indicating a finite series. The notation is called *summation notation,* or *sigma notation* because it is written using the Greek letter sigma. The expression

$$\sum_{i=1}^{4} (8i - 10)$$

is an example of an expression that uses summation notation. The summation notation in this expression is used to indicate the sum of all the expressions $8i - 10$ from $i = 1$ up to and including $i = 4$. That is,

$$\sum_{i=1}^{4} (8i - 10) = (8 \cdot 1 - 10) + (8 \cdot 2 - 10) + (8 \cdot 3 - 10) + (8 \cdot 4 - 10)$$

$$= -2 + 6 + 14 + 22$$

$$= 40$$

The letter i as used here is called the *index of summation,* or just *index* for short. Here are some examples illustrating the use of summation notation.

EXAMPLE 1 Expand and simplify $\sum_{i=1}^{5} (i^2 - 1)$.

SOLUTION We replace i in the expression $i^2 - 1$ with all consecutive integers from 1 up to 5, including 1 and 5.

$$\sum_{i=1}^{5} (i^2 - 1) = (1^2 - 1) + (2^2 - 1) + (3^2 - 1) + (4^2 - 1) + (5^2 - 1)$$

$$= 0 + 3 + 8 + 15 + 24$$

$$= 50$$

EXAMPLE 2 Expand and simplify $\sum_{i=3}^{6} (-2)^i$.

SOLUTION We replace i in the expression $(-2)^i$ with the consecutive integers beginning at 3 and ending at 6.

$$\sum_{i=3}^{6} (-2)^i = (-2)^3 + (-2)^4 + (-2)^5 + (-2)^6$$

$$= -8 + 16 + (-32) + 64$$

$$= 40$$

USING TECHNOLOGY

Summing Series on a Graphing Calculator

A TI-83/84 graphing calculator has a built-in sum(function that, when used with the seq(function, allows us to add the terms of a series. Let's repeat Example 1 using our graphing calculator. First, we go to LIST and select MATH. The fifth option in that list is sum(, which we select. Then we go to LIST again and select OPS. From that list we select seq(. Next we enter X^2 − 1, X, 1, 5, and then we close both sets of parentheses. Our screen shows the following:

$$\text{sum(seq(X\textasciicircum2} - 1, X, 1, 5)) \quad \text{which will give us} \quad \sum_{i=1}^{5} (i^2 - 1)$$

When we press ⬛ **ENTER** ⬛ the calculator displays 50, which is the same result we obtained in Example 1.

EXAMPLE 3   Expand $\displaystyle\sum_{i=2}^{5} (x^i - 3)$.

SOLUTION We must be careful not to confuse the letter x with i. The index i is the quantity we replace by the consecutive integers from 2 to 5, not x.

$$\sum_{i=2}^{5} (x^i - 3) = (x^2 - 3) + (x^3 - 3) + (x^4 - 3) + (x^5 - 3)$$

B Writing with Summation Notation

In the first three examples, we were given an expression with summation notation and asked to expand it. The next examples in this section illustrate how we can write an expression in expanded form as an expression involving summation notation.

EXAMPLE 4   Write with summation notation: $1 + 3 + 5 + 7 + 9$.

SOLUTION A formula that gives us the terms of this sum is

$$a_i = 2i - 1$$

where i ranges from 1 up to and including 5. Notice we are using the subscript i in exactly the same way we used the subscript n in the previous section—to indicate the general term. Writing the sum

$$1 + 3 + 5 + 7 + 9$$

with summation notation looks like this:

$$\sum_{i=1}^{5} (2i - 1)$$

EXAMPLE 5   Write with summation notation: $3 + 12 + 27 + 48$.

SOLUTION We need a formula, in terms of i, that will give each term in the sum. Writing the sum as

$$3 \cdot 1^2 + 3 \cdot 2^2 + 3 \cdot 3^2 + 3 \cdot 4^2$$

we see the formula

$$a_i = 3 \cdot i^2$$

where i ranges from 1 up to and including 4. Using this formula and summation notation, we can represent the sum

$$3 + 12 + 27 + 48$$

as

$$\sum_{i=1}^{4} 3i^2$$

EXAMPLE 6 Write with summation notation:

$$\frac{x+3}{x^3} + \frac{x+4}{x^4} + \frac{x+5}{x^5} + \frac{x+6}{x^6}$$

SOLUTION A formula that gives each of these terms is

$$a_i = \frac{x+i}{x^i}$$

where i assumes all integer values between 3 and 6, including 3 and 6. The sum can be written as

$$\sum_{i=3}^{6} \frac{x+i}{x^i}$$

Problem Set 13.2

Moving Toward Success

"If I had permitted my failures, or what seemed to me at the time a lack of success, to discourage me, I cannot see any way in which I would ever have made progress."

—Calvin Coolidge, 1872–1933,
30th President of the United States

1. How have you found the most success when stuck on a problem?

2. Which study strategies have worked best for you during this class?

A Expand and simplify each of the following.
[Examples 1–2]

1. $\sum_{i=1}^{4} (2i + 4)$

2. $\sum_{i=1}^{5} (3i - 1)$

3. $\sum_{i=2}^{3} (i^2 - 1)$

4. $\sum_{i=3}^{6} (i^2 + 1)$

5. $\sum_{i=1}^{4} (i^2 - 3)$

6. $\sum_{i=2}^{6} (2i^2 + 1)$

7. $\sum_{i=1}^{4} \frac{i}{1+i}$

8. $\sum_{i=1}^{3} \frac{i^2}{2i-1}$

9. $\sum_{i=1}^{4} (-3)^i$

10. $\sum_{i=1}^{4} \left(-\frac{1}{3}\right)^i$

11. $\sum_{i=3}^{6} (-2)^i$

12. $\sum_{i=4}^{6} \left(-\frac{1}{2}\right)^i$

13. $\sum_{i=2}^{6} (-2)^i$

14. $\sum_{i=2}^{5} (-3)^i$

15. $\sum_{i=1}^{5} \left(-\frac{1}{2}\right)^i$

16. $\sum_{i=3}^{6} \left(-\frac{1}{3}\right)^i$

17. $\sum_{i=2}^{5} \frac{i-1}{i+1}$

18. $\sum_{i=2}^{4} \frac{i^2-1}{i^2+1}$

A Expand the following. [Example 3]

19. $\sum_{i=1}^{5} (x + i)$

20. $\sum_{i=2}^{7} (x + 1)^i$

21. $\sum_{i=1}^{4} (x - 2)^i$

22. $\sum_{i=2}^{5} \left(x + \frac{1}{i}\right)^2$

23. $\sum_{i=1}^{5} \frac{x+i}{x-1}$

24. $\displaystyle\sum_{i=1}^{6} \frac{x - 3i}{x + 3i}$

25. $\displaystyle\sum_{i=3}^{8} (x + i)^{i}$

26. $\displaystyle\sum_{i=1}^{5} (x + i)^{i+1}$

27. $\displaystyle\sum_{i=3}^{6} (x - 2i)^{i+3}$

28. $\displaystyle\sum_{i=5}^{8} \left(\frac{x - i}{x + i} \right)^{2i}$

B Write each of the following sums with summation notation. Do not calculate the sum. [Example 4]

29. $2 + 4 + 8 + 16$

30. $3 + 5 + 7 + 9 + 11$

31. $4 + 8 + 16 + 32 + 64$

32. $3 + 8 + 15 + 24$

B Write each of the following sums with summation notation. Do not calculate the sum. *Note:* More than one answer is possible. [Examples 4–6]

33. $5 + 9 + 13 + 17 + 21$

34. $3 - 6 + 12 - 24 + 48$

35. $-4 + 8 - 16 + 32$

36. $15 + 24 + 35 + 48 + 63$

37. $\dfrac{3}{4} + \dfrac{4}{5} + \dfrac{5}{6} + \dfrac{6}{7} + \dfrac{7}{8}$

38. $\dfrac{1}{2} + \dfrac{2}{3} + \dfrac{3}{4} + \dfrac{4}{5}$

39. $\dfrac{1}{3} + \dfrac{2}{5} + \dfrac{3}{7} + \dfrac{4}{9}$

40. $\dfrac{3}{1} + \dfrac{5}{3} + \dfrac{7}{5} + \dfrac{9}{7}$

41. $(x - 2)^{6} + (x - 2)^{7} + (x - 2)^{8} + (x - 2)^{9}$

42. $(x + 1)^{3} + (x + 2)^{4} + (x + 3)^{5} + (x + 4)^{6} + (x + 5)^{7}$

43. $\left(1 + \dfrac{1}{x}\right)^{2} + \left(1 + \dfrac{2}{x}\right)^{3} + \left(1 + \dfrac{3}{x}\right)^{4} + \left(1 + \dfrac{4}{x}\right)^{5}$

44. $\dfrac{x - 1}{x + 2} + \dfrac{x - 2}{x + 4} + \dfrac{x - 3}{x + 6} + \dfrac{x - 4}{x + 8} + \dfrac{x - 5}{x + 10}$

45. $\dfrac{x}{x + 3} + \dfrac{x}{x + 4} + \dfrac{x}{x + 5}$

46. $\dfrac{x - 3}{x^{3}} + \dfrac{x - 4}{x^{4}} + \dfrac{x - 5}{x^{5}} + \dfrac{x - 6}{x^{6}}$

47. $x^{2}(x + 2) + x^{3}(x + 3) + x^{4}(x + 4)$

48. $x(x + 2)^{2} + x(x + 3)^{3} + x(x + 4)^{4}$

49. Repeating Decimals Any repeating, nonterminating decimal may be viewed as a series. For instance, $\dfrac{2}{3} = 0.6 + 0.06 + 0.006 + 0.0006 + \cdots$. Write the following fractions as series.

a. $\dfrac{1}{3}$

b. $\dfrac{2}{9}$

c. $\dfrac{3}{11}$

50. Repeating Decimals Refer to the previous exercise, and express the following repeating decimals as fractions.

a. $0.55555\cdots$ b. $1.33333\cdots$

c. $0.29292929\cdots$

Applying the Concepts

51. Skydiving A skydiver jumps from a plane and falls 16 feet the first second, 48 feet the second second, and 80 feet the third second. If he continues to fall in the same manner, how far will he fall the seventh second? What is the distance he falls in 7 seconds?

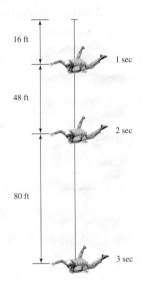

16 ft   1 sec

48 ft   2 sec

80 ft   3 sec

52. Bacterial Growth After 1 day, a colony of 50 bacteria reproduces to become 200 bacteria. After 2 days, they reproduce to become 800 bacteria. If they continue to reproduce at this rate, how many bacteria will be present after 4 days?

Start   1 day   2 days

53. Akaka Falls In Section 13.1, when a boulder fell from the top of Akaka Falls in Hawaii, the sequence generated during the first 5 seconds the boulder fell was 16, 48, 80, 112, 144.

  a. Write a finite series that represents the sum of this sequence.

  b. The general term of this sequence was given as $a_n = 32n - 16$. Write the series produced in part a in summation notation.

54. Pendulum A pendulum swings 12 feet left to right on its first swing, and on each swing following the first, swings $\frac{3}{4}$ of the previous swing. The distance the pendulum traveled in 5 seconds can be expressed in summation notation $\sum_{i=1}^{5} 12\left(\frac{3}{4}\right)^{i-1}$. Expand the summation notation and simplify. Round your final answer to the nearest tenth.

Getting Ready for the Next Section

Simplify.

67. $2 + 9(8)$

68. $\frac{1}{2} + 9\left(\frac{1}{2}\right)$

69. $\frac{10}{2}\left(\frac{1}{2} + 5\right)$

70. $\frac{10}{2}(2 + 74)$

71. $3 + (n - 1)2$

72. $7 + (n - 1)3$

Extending the Concepts

Solve each of the following equations for x.

73. $\sum_{i=1}^{4} (x - i) = 16$

74. $\sum_{i=3}^{8} (2x + i) = 30$

Maintaining Your Skills

Use the properties of logarithms to expand each of the following expressions.

55. $\log_2 x^3 y$

56. $\log_7 \dfrac{x^2}{y^4}$

57. $\log_{10} \dfrac{\sqrt[3]{x}}{y^2}$

58. $\log_{10} \sqrt[3]{\dfrac{x}{y^2}}$

Write each expression as a single logarithm.

59. $\log_{10} x - \log_{10} y^2$

60. $\log_{10} x^2 + \log_{10} y^2$

61. $2 \log_3 x - 3 \log_3 y - 4 \log_3 z$

62. $\dfrac{1}{2} \log_6 x + \dfrac{1}{3} \log_6 y + \dfrac{1}{4} \log_6 z$

Solve each equation.

63. $\log_4 x - \log_4 5 = 2$

64. $\log_3 6 + \log_3 x = 4$

65. $\log_2 x + \log_2 (x - 7) = 3$

66. $\log_5 (x + 1) + \log_5 (x - 3) = 1$

13.3 Arithmetic Sequences

OBJECTIVES

A Find the common difference for an arithmetic sequence.

B Find the general term for an arithmetic sequence.

C Find the sum of the first n terms of an arithmetic sequence.

TICKET TO SUCCESS

Keep these questions in mind as you read through the section. Then respond in your own words and in complete sentences.

1. Explain how to determine if a sequence is arithmetic.

2. What is a common difference?

3. Explain the formula $a_n = a_1 + (n - 1)d$ in order to find the nth term of an arithmetic sequence.

4. Explain the theorem that gives the formula for finding the sum of the first n terms of an arithmetic sequence.

Image copyright ©Phase4Photography, 2010. Used under license from Shutterstock.com

After an injury, your trainer tells you to return to jogging slowly. She suggests that you jog 10 minutes per day the first week, and increase by 5 minutes per day each subsequent week. How long will it be before you are up to 45 minutes per day? Problems like this require that you know how to solve arithmetic sequences, the topic of this section.

A Common Difference of an Arithmetic Sequence

> **Definition**
> An **arithmetic sequence** is a sequence of numbers in which each term is obtained from the preceding term by adding the same amount each time. An arithmetic sequence is also called an **arithmetic progression**.

The sequence

$$2, 6, 10, 14, \ldots$$

is an example of an arithmetic sequence, because each term is obtained from the preceding term by adding 4 each time. The amount we add each time—in this case, 4—is called the *common difference*, because it can be obtained by subtracting any two consecutive terms. (The term with the larger subscript must be written first.) The common difference is denoted by d.

EXAMPLE 1 Give the common difference d for the arithmetic sequence 4, 10, 16, 22, ...

SOLUTION Because each term can be obtained from the preceding term by adding 6, the common difference is 6. That is, $d = 6$.

EXAMPLE 2 Give the common difference for 100, 93, 86, 79, ...

SOLUTION The common difference in this case is $d = -7$, since adding -7 to any term always produces the next consecutive term.

EXAMPLE 3 Give the common difference for $\frac{1}{2}, 1, \frac{3}{2}, 2, \ldots$.

SOLUTION The common difference is $d = \frac{1}{2}$.

B The General Term

The general term a_n of an arithmetic progression can always be written in terms of the first term a_1 and the common difference d. Consider the sequence from Example 1:

$$4, 10, 16, 22, \ldots$$

We can write each term using the first term 4 and the common difference 6.

$$4, \quad 4 + (1 \cdot 6), \quad 4 + (2 \cdot 6), \quad 4 + (3 \cdot 6), \ldots$$
$$a_1, \quad a_2, \quad a_3, \quad a_4, \ldots$$

Observing the relationship between the subscript on the terms in the second line and the coefficients of the 6s in the first line, we write the general term for the sequence as

$$a_n = 4 + (n - 1)6$$

We generalize this result to include the general term of any arithmetic sequence.

Arithmetic Sequence

The **general term** of an arithmetic progression with first term a_1 and common difference d is given by
$$a_n = a_1 + (n - 1)d$$

EXAMPLE 4 Find the general term for the sequence
$$7, 10, 13, 16, \ldots$$

SOLUTION The first term is $a_1 = 7$, and the common difference is $d = 3$. Substituting these numbers into the formula for the general term, we have

$$a_n = 7 + (n - 1)3$$

which we can simplify, if we choose, to

$$a_n = 7 + 3n - 3$$
$$= 3n + 4$$

EXAMPLE 5 Find the general term of the arithmetic progression whose third term a_3 is 7 and whose eighth term a_8 is 17.

SOLUTION According to the formula for the general term, the third term can be written as $a_3 = a_1 + 2d$, and the eighth term can be written as $a_8 = a_1 + 7d$. Because these terms are also equal to 7 and 17, respectively, we can write

$$a_3 = a_1 + 2d = 7$$

$$a_8 = a_1 + 7d = 17$$

To find a_1 and d, we simply solve the system:

$$a_1 + 2d = 7$$

$$a_1 + 7d = 17$$

We add the opposite of the top equation to the bottom equation. The result is

$$5d = 10$$

$$d = 2$$

To find a_1, we simply substitute 2 for d in either of the original equations and get

$$a_1 = 3$$

The general term for this progression is

$$a_n = 3 + (n - 1)2$$

which we can simplify to

$$a_n = 2n + 1$$

C Sum of the First n Terms

The sum of the first n terms of an arithmetic sequence is denoted by S_n. The following theorem gives the formula for finding S_n, which is sometimes called the nth *partial sum*.

Theorem 1

The sum of the first n terms of an arithmetic sequence whose first term is a_1 and whose nth term is a_n is given by

$$S_n = \frac{n}{2}(a_1 + a_n)$$

PROOF We can write S_n in expanded form as

$$S_n = a_1 + [a_1 + d] + [a_1 + 2d] + \cdots + [a_1 + (n - 1)d]$$

We can arrive at this same series by starting with the last term a_n and subtracting d each time. Writing S_n this way, we have

$$S_n = a_n + [a_n - d] + [a_n - 2d] + \cdots + [a_n - (n - 1)d]$$

If we add the preceding two expressions term by term, we have

$$2S_n = (a_1 + a_n) + (a_1 + a_n) + (a_1 + a_n) + \cdots + (a_1 + a_n)$$

$$2S_n = n(a_1 + a_n)$$

$$S_n = \frac{n}{2}(a_1 + a_n)$$

EXAMPLE 6 Find the sum of the first 10 terms of the arithmetic progression 2, 10, 18, 26, ...

SOLUTION The first term is 2, and the common difference is 8. The tenth term is

$$a_{10} = 2 + 9(8)$$
$$= 2 + 72$$
$$= 74$$

Substituting $n = 10$, $a_1 = 2$, and $a_{10} = 74$ into the formula

$$S_n = \frac{n}{2}(a_1 + a_n)$$

we have

$$S_{10} = \frac{10}{2}(2 + 74)$$
$$= 5(76)$$
$$= 380$$

The sum of the first 10 terms is 380.

Problem Set 13.3

Moving Toward Success

"Your worth consists in what you are and not in what you have."

—Thomas Edison, 1847–1931, American inventor and entrepreneur

1. How does mathematics influence your critical thinking skills?
2. How can you use critical thinking skills outside of this class?

A Determine which of the following sequences are arithmetic progressions. For those that are arithmetic progressions, identify the common difference d. [Examples 1–3]

1. 1, 2, 3, 4, . . .

2. 4, 6, 8, 10, . . .

3. 1, 2, 4, 7, . . .

4. 1, 2, 4, 8, . . .

5. 50, 45, 40, . . .

6. $1, \frac{1}{2}, \frac{1}{4}, \frac{1}{8}, \ldots$

7. 1, 4, 9, 16, . . .

8. 5, 7, 9, 11, . . .

9. $\frac{1}{3}, 1, \frac{5}{3}, \frac{7}{3}, \ldots$

10. 5, 11, 17, . . .

A B C Each of the following problems refers to arithmetic sequences. [Examples 1–6]

11. If $a_1 = 3$ and $d = 4$, find a_n and a_{24}.

12. If $a_1 = 5$ and $d = 10$, find a_n and a_{100}.

13. If $a_1 = 6$ and $d = -2$, find a_{10} and S_{10}.

14. If $a_1 = 7$ and $d = -1$, find a_{24} and S_{24}.

15. If $a_6 = 17$ and $a_{12} = 29$, find the term a_1, the common difference d, and then find a_{30}.

16. If $a_5 = 23$ and $a_{10} = 48$, find the first term a_1, the common difference d, and then find a_{40}.

17. If the third term is 16 and the eighth term is 26, find the first term, the common difference, and then find a_{20} and S_{20}.

18. If the third term is 16 and the eighth term is 51, find the first term, the common difference, and then find a_{50} and S_{50}.

19. If $a_1 = 3$ and $d = 4$, find a_{20} and S_{20}.

20. If $a_1 = 40$ and $d = -5$, find a_{25} and S_{25}.

21. If $a_4 = 14$ and $a_{10} = 32$, find a_{40} and S_{40}.

22. If $a_7 = 0$ and $a_{11} = -\frac{8}{3}$, find a_{61} and S_{61}.

23. If $a_6 = -17$ and $S_6 = -12$, find a_1 and d.

24. If $a_{10} = 12$ and $S_{10} = 40$, find a_1 and d.

25. Find a_{85} for the sequence 14, 11, 8, 5, . . .

26. Find S_{100} for the sequence $-32, -25, -18,$ $-11, . . .$

27. If $S_{20} = 80$ and $a_1 = -4$, find d and a_{39}.

28. If $S_{24} = 60$ and $a_1 = 4$, find d and a_{116}.

29. Find the sum of the first 100 terms of the sequence 5, 9, 13, 17, ...

30. Find the sum of the first 50 terms of the sequence 8, 11, 14, 17, ...

31. Find a_{35} for the sequence 12, 7, 2, -3, ...

32. Find a_{45} for the sequence 25, 20, 15, 10, ...

33. Find the tenth term and the sum of the first 10 terms of the sequence $\frac{1}{2}$, 1, $\frac{3}{2}$, 2, ...

34. Find the 15th term and the sum of the first 15 terms of the sequence $-\frac{1}{3}$, 0, $\frac{1}{3}$, $\frac{2}{3}$, ...

Applying the Concepts

Straight-Line Depreciation Straight-line depreciation is an accounting method used to help spread the cost of new equipment over a number of years. The value at any time during the life of the machine can be found with a linear equation in two variables. For income tax purposes, however, it is the value at the end of the year that is most important, and for this reason sequences can be used.

35. Value of a Copy Machine

A large copy machine sells for $18,000 when it is new. Its value decreases $3,300 each year after that. We can use an arithmetic sequence to find the value of the machine at the end  of each year. If we let a_0 represent the value when it is

purchased, then a_1 is the value after 1 year, a_2 is the value after 2 years, and so on.

   a. Write the first 5 terms of the sequence.

   b. What is the common difference?

   c. Construct a line graph for the first 5 terms of the sequence.

   d. Use the line graph to estimate the value of the copy machine 2.5 years after it is purchased.

   e. Write the sequence from part a using a recursive formula.

36. Value of a Forklift An electric forklift sells for $125,000 when new. Each year after that, it decreases $16,500 in value.

   a. Write an arithmetic sequence that gives the value of the forklift at the end of each of the first 5 years after it is purchased.

   b. What is the common difference for this sequence?

   c. Construct a line graph for this sequence.

   d. Use the line graph to estimate the value of the forklift 3.5 years after it is purchased.

   e. Write the sequence from part a using a recursive formula.

37. Distance A rocket travels vertically 1,500 feet in its first second of flight, and then about 40 feet less each succeeding second. Use these estimates to answer the following questions.

   a. Write a sequence of the vertical distance traveled by a rocket in each of its first 6 seconds.

   b. Is the sequence in part a an arithmetic sequence? Explain why or why not.

   c. What is the general term of the sequence in part a?

38. Depreciation Suppose an automobile sells for N dollars new, and then depreciates 40% each year.

   a. Write a sequence for the value of this automobile (in terms of N) for each year.

b. What is the general term of the sequence in part a?

c. Is the sequence in part a an arithmetic sequence? Explain why it is or is not.

39. Triangular Numbers The first four triangular numbers are {1, 3, 6, 10, . . . }, and are illustrated in the following diagram.

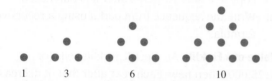

a. Write a sequence of the first 15 triangular numbers.

b. Write the recursive general term for the sequence of triangular numbers.

c. Is the sequence of triangular numbers an arithmetic sequence? Explain why it is or is not.

40. Arithmetic Means Three (or more) arithmetic means between two numbers may be found by forming an arithmetic sequence using the original two numbers and the arithmetic means. For example, three arithmetic means between 10 and 34 may be found by examining the sequence {10, a, b, c, 34}. For the sequence to be arithmetic, the common difference must be 6; therefore, $a = 16$, $b = 22$, and $c = 28$. Use this idea to answer the following questions.

a. Find four arithmetic means between 10 and 35.

b. Find three arithmetic means between 2 and 62.

c. Find five arithmetic means between 4 and 28.

Maintaining Your Skills

Find the following logarithms. Round to four decimal places.

41. log 576

42. log 57,600

43. log 0.0576

44. log 0.000576

Find x.

45. $\log x = 2.6484$

46. $\log x = 7.9832$

47. $\log x = -7.3516$

48. $\log x = -2.0168$

Getting Ready for the Next Section

Simplify.

49. $\dfrac{1}{8}\left(\dfrac{1}{2}\right)$

50. $\dfrac{1}{4}\left(\dfrac{1}{2}\right)$

51. $\dfrac{3\sqrt{3}}{3}$

52. $\dfrac{3}{\sqrt{3}}$

53. $2 \cdot 2^{n-1}$

54. $3 \cdot 3^{n-1}$

55. $\dfrac{ar^6}{ar^3}$

56. $\dfrac{ar^7}{ar^4}$

57. $\dfrac{\dfrac{1}{5}}{1 - \dfrac{1}{2}}$

58. $\dfrac{\dfrac{9}{10}}{1 - \dfrac{1}{10}}$

59. $-\dfrac{3[(-2)^8 - 1]}{-2 - 1}$

60. $\dfrac{4\left[\left(\dfrac{1}{2}\right)^6 - 1\right]}{\dfrac{1}{2} - 1}$

Extending the Concepts

61. Find a_1 and d for an arithmetic sequence with $S_7 = 147$ and $S_{13} = 429$.

62. Find a_1 and d for an arithmetic sequence with $S_{12} = 282$ and $S_{19} = 646$.

63. Find d for an arithmetic sequence in which $a_{15} - a_7 = -24$.

64. Find d for an arithmetic sequence in which $a_{25} - a_{15} = 120$.

65. Find the sum of the first 50 terms of the sequence $1, 1 + \sqrt{2}, 1 + 2\sqrt{2}, \ldots$.

66. Given $a_1 = -3$ and $d = 5$ in an arithmetic sequence, and $S_n = 890$, find n.

13.4 Geometric Sequences

OBJECTIVES

A Find the common ratio, general term, and the *n*th term for a geometric sequence.

B Find the sum of the first *n* terms of a geometric sequence.

C Find the sum of all terms of an infinite geometric sequence.

TICKET TO SUCCESS

Keep these questions in mind as you read through the section. Then respond in your own words and in complete sentences.

1. What is a geometric sequence?
2. What is a common ratio?
3. Explain the formula $a_n = a_1 r^{n-1}$ in order to find the *n*th term of a geometric sequence.
4. When is the sum of an infinite geometric series a finite number?
5. Explain how a repeating decimal can be represented as an infinite geometric series.

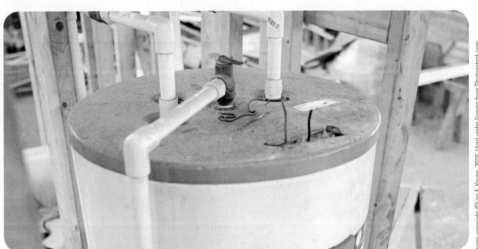

Image copyright ©Lisa F. Young, 2010. Used under license from Shutterstock.com

Suppose your landlord buys a new water heater for your apartment. Once it is fully installed, the technician tells you that the water temperature is currently 75 degrees and will increase by 5% every hour. If the optimal temperature in the water heater is 130 degrees, how long before you can take a hot shower? After working through this section, you will be able to use geometric sequences to answer this question. The problems in this section are similar to the problems in the preceding section.

A Terms of Geometric Sequences

> **Definition**
> A sequence of numbers in which each term is obtained from the previous term by multiplying by the same amount each time is called a **geometric sequence**. Geometric sequences are also called **geometric progressions**.

The sequence

$$3, 6, 12, 24, \ldots$$

is an example of a geometric progression. Each term is obtained from the previous term by multiplying by 2. The amount by which we multiply each time—in this case, 2—is called the *common ratio*. The common ratio is denoted by *r* and can be found by taking the ratio of any two consecutive terms. (The term with the larger subscript must be in the numerator.)

EXAMPLE 1 Find the common ratio for the geometric progression.

$$\frac{1}{2}, \frac{1}{4}, \frac{1}{8}, \frac{1}{16}, \ldots$$

SOLUTION Because each term can be obtained from the term before it by multiplying by $\frac{1}{2}$, the common ratio is $\frac{1}{2}$. That is, $r = \frac{1}{2}$.

EXAMPLE 2 Find the common ratio for $\sqrt{3}, 3, 3\sqrt{3}, 9, \ldots$

SOLUTION If we take the ratio of the third term to the second term, we have

$$\frac{3\sqrt{3}}{3} = \sqrt{3}$$

The common ratio is $r = \sqrt{3}$.

Geometric Sequences

The **general term** a_n of a geometric sequence with first term a_1 and common ratio r is given by

$$a_n = a_1 r^{n-1}$$

To see how we arrive at this formula, consider the following geometric progression whose common ratio is 3:

$$2, 6, 18, 54, \ldots$$

We can write each term of the sequence in terms of the first term 2 and the common ratio 3.

$$2 \cdot 3^0, \quad 2 \cdot 3^1, \quad 2 \cdot 3^2, \quad 2 \cdot 3^3, \ldots$$
$$a_1, \quad\quad a_2, \quad\quad a_3, \quad\quad a_4, \ldots$$

Observing the relationship between the two preceding lines, we find we can write the general term of this progression as

$$a_n = 2 \cdot 3^{n-1}$$

Because the first term can be designated by a_1 and the common ratio by r, the formula $a_n = 2 \cdot 3^{n-1}$ coincides with the formula $a_n = a_1 r^{n-1}$

EXAMPLE 3 Find the general term for the geometric progression

$$5, 10, 20, \ldots$$

SOLUTION The first term is $a_1 = 5$, and the common ratio is $r = 2$. Using these values in the formula

$$a_n = a_1 r^{n-1}$$

we have

$$a_n = 5 \cdot 2^{n-1}$$

EXAMPLE 4   Find the tenth term of the sequence $3, \frac{3}{2}, \frac{3}{4}, \frac{3}{8}, \ldots$.

SOLUTION The sequence is a geometric progression with first term $a_1 = 3$ and common ratio $r = \frac{1}{2}$. The tenth term is

$$a_{10} = 3\left(\frac{1}{2}\right)^9 = \frac{3}{512}$$

■

EXAMPLE 5   Find the general term for the geometric progression whose fourth term is 16 and whose seventh term is 128.

SOLUTION The fourth term can be written as $a_4 = a_1 r^3$, and the seventh term can be written as $a_7 = a_1 r^6$.

$$a_4 = a_1 r^3 = 16$$

$$a_7 = a_1 r^6 = 128$$

We can solve for r by using the ratio $\dfrac{a_7}{a_4}$.

$$\frac{a_7}{a_4} = \frac{a_1 r^6}{a_1 r^3} = \frac{128}{16}$$

$$r^3 = 8$$

$$r = 2$$

The common ratio is 2. To find the first term, we substitute $r = 2$ into either of the original two equations. The result is

$$a_1 = 2$$

The general term for this progression is

$$a_n = 2 \cdot 2^{n-1}$$

which we can simplify by adding exponents, because the bases are equal:

$$a_n = 2^n$$

■

B Sum of the First n Terms

As was the case in the preceding section, the sum of the first n terms of a geometric progression is denoted by S_n, which is called the nth *partial sum* of the progression.

> **Theorem 2**
> The sum of the first n terms of a geometric progression with first term a_1 and common ratio r is given by the formula
> $$S_n = \frac{a_1(r^n - 1)}{r - 1}$$

PROOF We can write the sum of the first n terms in expanded form.

$$S_n = a_1 + a_1 r + a_1 r^2 + \cdots + a_1 r^{n-1} \qquad (1)$$

Then multiplying both sides by r, we have

$$rS_n = a_1 r + a_1 r^2 + a_1 r^3 + \cdots + a_1 r^n \qquad (2)$$

If we subtract the left side of equation (1) from the left side of equation (2) and do the same for the right sides, we end up with

$$rS_n - S_n = a_1 r^n - a_1$$

We factor S_n from both terms on the left side and a_1 from both terms on the right side of this equation:

$$S_n(r - 1) = a_1(r^n - 1)$$

Dividing both sides by $r - 1$ gives the desired result:

$$S_n = \frac{a_1(r^n - 1)}{r - 1}$$

EXAMPLE 6 Find the sum of the first 10 terms of the geometric progression 5, 15, 45, 135, ...

SOLUTION The first term is $a_1 = 5$, and the common ratio is $r = 3$. Substituting these values into the formula for S_{10}, we have the sum of the first 10 terms of the sequence:

$$S_{10} = \frac{5(3^{10} - 1)}{3 - 1}$$

$$= \frac{5(3^{10} - 1)}{2}$$

The answer can be left in this form. A calculator will give the result as 147,620. ■

C Infinite Geometric Series

Suppose the common ratio for a geometric sequence is a number whose absolute value is less than 1—for instance, $\frac{1}{2}$. The sum of the first n terms is given by the formula

$$S_n = \frac{a_1\left[\left(\frac{1}{2}\right)^n - 1\right]}{\frac{1}{2} - 1}$$

As n becomes larger and larger, the term $\left(\frac{1}{2}\right)^n$ will become closer and closer to 0. That is, for $n = 10$, 20, and 30, we have the following approximations:

$$\left(\frac{1}{2}\right)^{10} \approx 0.001$$

$$\left(\frac{1}{2}\right)^{20} \approx 0.000001$$

$$\left(\frac{1}{2}\right)^{30} \approx 0.000000001$$

so that for large values of n, there is little difference between the expression

$$\frac{a_1(r^n - 1)}{r - 1}$$

and the expression

$$\frac{a_1(0 - 1)}{r - 1} = \frac{-a_1}{r - 1} = \frac{a_1}{1 - r} \qquad \text{if} \qquad |r| < 1$$

In fact, the sum of the terms of a geometric sequence in which $|r| < 1$ actually becomes the expression

$$\frac{a_1}{1 - r}$$

as n approaches infinity. To summarize, we have the following:

The Sum of an Infinite Geometric Series

If a geometric sequence has first term a_1 and common ratio r such that $|r| < 1$, then the following is called an **infinite geometric series**:

$$S = \sum_{i=0}^{\infty} a_1 r^i = a_1 + a_1 r + a_1 r^2 + a_1 r^3 + \cdots$$

Its sum is given by the formula

$$S = \frac{a_1}{1 - r}$$

EXAMPLE 7   Find the sum of the infinite geometric series

$$\frac{1}{5} + \frac{1}{10} + \frac{1}{20} + \frac{1}{40} + \cdots$$

SOLUTION The first term is $a_1 = \frac{1}{5}$, and the common ratio is $r = \frac{1}{2}$, which has an absolute value less than 1. Therefore, the sum of this series is

$$S = \frac{a_1}{1 - r} = \frac{\dfrac{1}{5}}{1 - \dfrac{1}{2}} = \frac{\dfrac{1}{5}}{\dfrac{1}{2}} = \frac{2}{5}$$

EXAMPLE 8   Show that 0.999 . . . is equal to 1.

SOLUTION We begin by writing 0.999 . . . as an infinite geometric series.

$$0.999\ldots = 0.9 + 0.09 + 0.009 + 0.0009 + \cdots$$

$$= \frac{9}{10} + \frac{9}{100} + \frac{9}{1{,}000} + \frac{9}{10{,}000} + \cdots$$

$$= \frac{9}{10} + \frac{9}{10}\left(\frac{1}{10}\right) + \frac{9}{10}\left(\frac{1}{10}\right)^2 + \frac{9}{10}\left(\frac{1}{10}\right)^3 + \cdots$$

As the last line indicates, we have an infinite geometric series with $a_1 = \frac{9}{10}$ and $r = \frac{1}{10}$. The sum of this series is given by

$$S = \frac{a_1}{1 - r} = \frac{\dfrac{9}{10}}{1 - \dfrac{1}{10}} = \frac{\dfrac{9}{10}}{\dfrac{9}{10}} = 1$$

Problem Set 13.4

Moving Toward Success

"Ordinary riches can be stolen, real riches cannot. In your soul are infinitely precious things that cannot be taken from you."

—Oscar Wilde, 1854–1900, Irish novelist and poet

1. How has taking this class benefited you?
2. What was your favorite thing about this class?

A Identify those sequences that are geometric progressions. For those that are geometric, give the common ratio r. [Examples 1–2]

1. $1, 5, 25, 125, \ldots$

2. $6, 12, 24, 48, \ldots$

3. $\dfrac{1}{2}, \dfrac{1}{6}, \dfrac{1}{18}, \dfrac{1}{54}, \ldots$

4. $5, 10, 15, 20, \ldots$

5. $4, 9, 16, 25, \ldots$

6. $-1, \dfrac{1}{3}, -\dfrac{1}{9}, \dfrac{1}{27}, \ldots$

7. $-2, 4, -8, 16, \ldots$

8. $1, 8, 27, 64, \ldots$

9. $4, 6, 8, 10, \ldots$

10. $1, -3, 9, -27, \ldots$

A **B** Each of the following problems gives some information about a specific geometric progression. [Examples 3–6]

11. If $a_1 = 4$ and $r = 3$, find a_n.

12. If $a_1 = 5$ and $r = 2$, find a_n.

13. If $a_1 = -2$ and $r = -\dfrac{1}{2}$, find a_6.

14. If $a_1 = 25$ and $r = -\dfrac{1}{5}$, find a_6.

15. If $a_1 = 3$ and $r = -1$, find a_{20}.

16. If $a_1 = -3$ and $r = -1$, find a_{20}.

17. If $a_1 = 10$ and $r = 2$, find S_{10}.

18. If $a_1 = 8$ and $r = 3$, find S_5.

19. If $a_1 = 1$ and $r = -1$, find S_{20}.

20. If $a_1 = 1$ and $r = -1$, find S_{21}.

21. Find a_8 for $\dfrac{1}{5}, \dfrac{1}{10}, \dfrac{1}{20}, \ldots$

22. Find a_8 for $\dfrac{1}{2}, \dfrac{1}{10}, \dfrac{1}{50}, \ldots$

23. Find S_5 for $-\dfrac{1}{2}, -\dfrac{1}{4}, -\dfrac{1}{8}, \ldots$

24. Find S_6 for $-\dfrac{1}{2}, 1, -2, \ldots$

25. Find a_{10} and S_{10} for $\sqrt{2}, 2, 2\sqrt{2}, \ldots$

26. Find a_8 and S_8 for $\sqrt{3}, 3, 3\sqrt{3}, \ldots$

27. Find a_6 and S_6 for $100, 10, 1, \ldots$

28. Find a_6 and S_6 for $100, -10, 1, \ldots$

29. If $a_4 = 40$ and $a_6 = 160$, find r.

30. If $a_5 = \dfrac{1}{8}$ and $a_8 = \dfrac{1}{64}$, find r.

31. Given the sequence $-3, 6, -12, 24, \ldots$, find a_8 and S_8.

32. Given the sequence $4, 2, 1, \dfrac{1}{2}, \ldots$, find a_9 and S_9.

33. Given $a_7 = 13$ and $a_{10} = 104$, find r.

34. Given $a_5 = -12$ and $a_8 = 324$, find r.

C Find the sum of each geometric series. [Examples 7–8]

35. $\dfrac{1}{2} + \dfrac{1}{4} + \dfrac{1}{8} + \cdots$

36. $\dfrac{1}{3} + \dfrac{1}{9} + \dfrac{1}{27} + \cdots$

37. $4 + 2 + 1 + \cdots$

38. $8 + 4 + 2 + \cdots$

39. $2 + 1 + \dfrac{1}{2} + \cdots$

40. $3 + 1 + \dfrac{1}{3} + \cdots$

41. $\dfrac{4}{3} - \dfrac{2}{3} + \dfrac{1}{3} + \cdots$

42. $6 - 4 + \dfrac{8}{3} + \cdots$

43. $\dfrac{2}{5} + \dfrac{4}{25} + \dfrac{8}{125} + \cdots$

44. $\dfrac{3}{4} + \dfrac{9}{16} + \dfrac{27}{64} + \cdots$

45. $\dfrac{3}{4} + \dfrac{1}{4} + \dfrac{1}{12} + \cdots$

46. $\dfrac{5}{3} + \dfrac{1}{3} + \dfrac{1}{15} + \cdots$

47. Show that $0.444 \ldots$ is the same as $\dfrac{4}{9}$.

48. Show that $0.333 \ldots$ is the same as $\dfrac{1}{3}$.

49. Show that $0.272727 \ldots$ is the same as $\dfrac{3}{11}$.

50. Show that $0.545454 \ldots$ is the same as $\dfrac{6}{11}$.

Applying the Concepts

Declining-Balance Depreciation The declining-balance method of depreciation is an accounting method businesses use to deduct most of the cost of new equipment during the first few years of purchase. The value at any time during the life of the machine can be found with a linear equation in two variables. For income tax purposes, however, it is the value at the end of the year that is most important, and for this reason sequences can be used.

51. Value of a Crane A construction crane sells for $450,000 if purchased new. After that, the value decreases by 30% each year. We can use a geometric sequence to find the value of the crane at the end of each year. If we let a_0 represent the value when it is purchased, then a_1 is the value after 1 year, a_2 is the value after 2 years, and so on.
a. Write the first five terms of the sequence.

b. What is the common ratio?
c. Construct a line graph for the first five terms of the sequence.
d. Use the line graph to estimate the value of the crane 4.5 years after it is purchased.

e. Write the sequence from part a using a recursive formula.

Image copyright © Mack7777.
Used under license from Shutterstock.com.

52. Value of a Printing Press A large printing press sells for $375,000 when it is new. After that, its value decreases 25% each year.
a. Write a geometric sequence that gives the value of the press at the end of each of the first 5 years after it is purchased.

b. What is the common ratio for this sequence?
c. Construct a line graph for this sequence.
d. Use the line graph to estimate the value of the printing press 1.5 years after it is purchased.

e. Write the sequence from part a using a recursive formula.

Lester Lefkowitz/Getty Images

53. Adding Terms Given the geometric series
$$\frac{1}{3} + \frac{1}{9} + \frac{1}{27} + \cdots,$$
a. Find the sum of all the terms.
b. Find the sum of the first six terms.
c. Find the sum of all but the first six terms.

54. Perimeter Triangle ABC has a perimeter of 40 inches. A new triangle XYZ is formed by connecting the midpoints of the sides of the first triangle, as shown in the following figure. Because midpoints are joined, the perimeter of triangle XYZ will be one-half the perimeter of triangle ABC, or 20 inches.

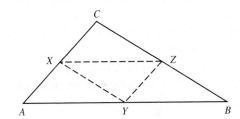

Midpoints of the sides of triangle XYZ are used to form a new triangle RST. If this pattern of using midpoints to draw triangles is continued seven more times, so there are a total of 10 triangles drawn, what will be the sum of the perimeters of these ten triangles?

55. Bouncing Ball A ball is dropped from a height of 20 feet. Each time it bounces it returns to $\frac{7}{8}$ of the height it fell from. If the ball is allowed to bounce an infinite number of times, find the total vertical distance that the ball travels.

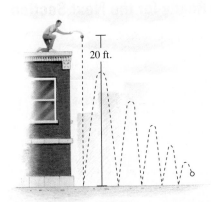

20 ft.

56. Stacking Paper Assume that a thin sheet of paper is 0.002 inch thick. The paper is torn in half, and the two halves placed together.

   a. How thick is the pile of torn paper?

   b. The pile of paper is torn in half again, and then the two halves placed together and torn in half again. The paper is large enough so this process may be performed a total of 5 times. How thick is the pile of torn paper?

   c. Refer to the tearing and piling process described in part b. Assuming that the original paper is large enough, how thick is the pile of torn paper if 25 tears are made?

57. Pendulum A pendulum swings 15 feet left to right on its first swing. On each swing following the first, the pendulum swings $\frac{4}{5}$ of the previous swing.

   a. Write the general term for this geometric sequence.

   b. If the pendulum is allowed to swing an infinite number of times, what is the total distance the pendulum will travel?

58. Bacteria Growth A bacteria colony of 50 bacteria doubles each day. The number of bacteria in the colony each day can be modeled as a geometric sequence.

   a. What is the first term of the sequence? What is the common ratio?

   b. Write the general term for this sequence.

   c. What would be the total number of bacteria in the colony after 10 days?

Maintaining Your Skills

Find each of the following to the nearest hundredth.

59. $\log_4 20$

60. $\log_7 21$

61. $\ln 576$

62. $\ln 5{,}760$

63. Solve the formula $A = 10e^{5t}$ for t.

64. Solve the formula $A = P \cdot 2^{-5t}$ for t.

Getting Ready for the Next Section

Expand and multiply.

65. $(x + y)^0$

66. $(x + y)^1$

67. $(x + y)^2$

68. $(x + y)^3$

Simplify.

69. $\dfrac{6 \cdot 5 \cdot 4 \cdot 3 \cdot 2 \cdot 1}{(2 \cdot 1)(4 \cdot 3 \cdot 2 \cdot 1)}$

70. $\dfrac{7 \cdot 6 \cdot 5 \cdot 4 \cdot 3 \cdot 2 \cdot 1}{(5 \cdot 4 \cdot 3 \cdot 2 \cdot 1)(2 \cdot 1)}$

Extending the Concepts

71. Find the fraction form of 0.636363 . . . .

72. Find the fraction form of 0.123123123 . . . .

73. The sum of an infinite geometric series is 6 and the first term is 4. Find the common ratio.

74. The sum of an infinite geometric series is $\frac{5}{2}$ and the first term is 3. Find the common ratio.

75. Sierpinski Triangle In the sequence that follows, the figures are moving toward what is known as the Sierpinski triangle. To construct the figure in stage 2, we remove the triangle formed from the midpoints of the sides of the shaded region in stage 1. Likewise, the figure in stage 3 is found by removing the triangles formed by connecting the midponts of the sides of the shaded regions in stage 2. If we repeat this process an infinite number of times, we arrive at the Sierpinski triangle.

   a. If the shaded region in stage 1 has an area of 1, find the area of the shaded regions in stages 2 through 4.

   b. Do the areas you found in part a form an arithmetic sequence or a geometric sequence?

Stage 1   Stage 2   Stage 3   Stage 4

   c. The Sierpinski triangle is the triangle that is formed after the process of forming the stages shown in the figure is repeated an infinite number of times. What do you think the area of the shaded region of the Sierpinski triangle will be?

   d. Suppose the perimeter of the shaded region of the triangle in stage 1 is 1. If we were to find the perimeters of the shaded regions in the other stages, would we have an increasing sequence or a decreasing sequence?

The Binomial Expansion

OBJECTIVES

A Calculate binomial coefficients.

B Use the binomial formula to expand binomial powers.

C Find a particular term of a binomial expansion.

TICKET TO SUCCESS

Keep these questions in mind as you read through the section. Then respond in your own words and in complete sentences.

1. What is Pascal's triangle?
2. What is factorial notation and how does it relate to binomial coefficients?
3. State the binomial formula.
4. When is it more efficient to use the binomial formula rather than Pascal's triangle to expand a binomial raised to a whole-number exponent?

Image copyright ©Tischenko Irina, 2010. Used under license from Shutterstock.com

Previously, you were introduced to Pascal's triangle, named for the French philosopher Blaise Pascal. Pascal's triangle is a visual representation of binomial expansion. In this section, you will work with Pascal's triangle and see how patterns such as binomial expansion emerge within it.

Now we will learn how to write and apply the formula for the expansion of expressions of the form $(x + y)^n$, where n is any positive integer. To write the formula, we must generalize the information in the following chart:

> **NOTE**
> The polynomials to the right have been found by expanding the binomials on the left—we just haven't shown the work.

$$(x + y)^0 = 1$$

$$(x + y)^1 = x + y$$

$$(x + y)^2 = x^2 + 2xy + y^2$$

$$(x + y)^3 = x^3 + 3x^2y + 3xy^2 + y^3$$

$$(x + y)^4 = x^4 + 4x^3y + 6x^2y^2 + 4xy^3 + y^4$$

$$(x + y)^5 = x^5 + 5x^4y + 10x^3y^2 + 10x^2y^3 + 5xy^4 + y^5$$

There are a number of similarities to notice among the polynomials on the right. Here is a list:

1. In each polynomial, the sequence of exponents on the variable x decreases to 0 from the exponent on the binomial at the left. (The exponent 0 is not shown, since $x^0 = 1$.)

2. In each polynomial, the exponents on the variable y increase from 0 to the exponent on the binomial at the left. (Because $y^0 = 1$, it is not shown in the first term.)

3. The sum of the exponents on the variables in any single term is equal to the exponent on the binomial at the left.

The pattern in the coefficients of the polynomials on the right can best be seen by writing the right side again without the variables. It looks like this:

Row 0						1				
Row 1					1		1			
Row 2				1		2		1		
Row 3			1		3		3		1	
Row 4		1		4		6		4		1
Row 5	1		5		10		10		5		1

This triangle-shaped array of coefficients is called *Pascal's triangle*. Notice that each entry in the triangular array is obtained by adding the two numbers on the line directly above it. Each row also begins and ends with the number 1.

If we were to continue Pascal's triangle, the next two rows would be

Row 6		1	6	15	20	15	6	1	
Row 7	1	7	21	35	35	21	7	1	

Briefly, let's revisit Row 6 in terms of variables. Row 6 represents the expansion of the expression $(x + y)^6$, which is

$$x^6 + 6x^5y + 15x^4y^2 + 20x^3y^3 + 15x^2y^4 + 6xy^5 + y^6$$

Take a moment to practice writing the expansion of $(x + y)^7$ represented by the coefficients in Row 7.

A Binomials Coefficients

The coefficients for the terms in the expansion of $(x + y)^n$ are given in the nth row of Pascal's triangle. There is an alternative method of finding these coefficients that does not involve Pascal's triangle. The alternative method involves *factorial notation*.

Definition

The expression $n!$ is read "n factorial" and is the product of all the consecutive positive integers from n down to 1. For example,

$$1! = 1$$

$$2! = 2 \cdot 1 = 2$$

$$3! = 3 \cdot 2 \cdot 1 = 6$$

$$4! = 4 \cdot 3 \cdot 2 \cdot 1 = 24$$

$$5! = 5 \cdot 4 \cdot 3 \cdot 2 \cdot 1 = 120$$

The expression $0!$ is defined to be 1.

We use factorial notation to define binomial coefficients as follows:

Definition

The expression $\binom{n}{r}$ is read as **n choose r** and called a **binomial coefficient.** It is defined by

$$\binom{n}{r} = \frac{n!}{r!(n-r)!}$$

where **n** is a number of ways we can **choose** from a set of things **r**.

Let's practice finding binomial coefficients using factorial notation before we move on to expansion.

EXAMPLE 1 Calculate the following binomial coefficients.

$$\binom{7}{5}, \binom{6}{2}, \binom{3}{0}$$

SOLUTION We simply apply the definition for binomial coefficients:

$$\binom{7}{5} = \frac{7!}{5!(7-5)!}$$

$$= \frac{7!}{5! \cdot 2!}$$

$$= \frac{7 \cdot 6 \cdot 5 \cdot 4 \cdot 3 \cdot 2 \cdot 1}{(5 \cdot 4 \cdot 3 \cdot 2 \cdot 1)(2 \cdot 1)}$$

$$= \frac{42}{2}$$

$$= 21$$

$$\binom{6}{2} = \frac{6!}{2!(6-2)!}$$

$$= \frac{6!}{2! \cdot 4!}$$

$$= \frac{6 \cdot 5 \cdot 4 \cdot 3 \cdot 2 \cdot 1}{(2 \cdot 1)(4 \cdot 3 \cdot 2 \cdot 1)}$$

$$= \frac{30}{2}$$

$$= 15$$

$$\binom{3}{0} = \frac{3!}{0!(3-0)!}$$

$$= \frac{3!}{0! \cdot 3!}$$

$$= \frac{3 \cdot 2 \cdot 1}{(1)(3 \cdot 2 \cdot 1)}$$

$$= 1$$

If we were to calculate all the binomial coefficients in the following array, we would find they match exactly with the numbers in Pascal's triangle. That is why they are called binomial coefficients—because they are the coefficients of the expansion of $(x + y)^n$.

$$\binom{0}{0}$$

$$\binom{1}{0} \qquad \binom{1}{1}$$

$$\binom{2}{0} \qquad \binom{2}{1} \qquad \binom{2}{2}$$

$$\binom{3}{0} \qquad \binom{3}{1} \qquad \binom{3}{2} \qquad \binom{3}{3}$$

$$\binom{4}{0} \qquad \binom{4}{1} \qquad \binom{4}{2} \qquad \binom{4}{3} \qquad \binom{4}{4}$$

$$\binom{5}{0} \qquad \binom{5}{1} \qquad \binom{5}{2} \qquad \binom{5}{3} \qquad \binom{5}{4} \qquad \binom{5}{5}$$

Using the new notation to represent the entries in Pascal's triangle, we can summarize everything we have noticed about the expansion of binomial powers of the form $(x + y)^n$.

B The Binomial Expansion

Binomial Expansion Formula

If x and y represent real numbers and n is a positive integer, then the following formula is known as the **binomial expansion,** or **binomial formula**:

$$(x + y)^n = \binom{n}{0} x^n y^0 + \binom{n}{1} x^{n-1} y^1 + \binom{n}{2} x^{n-2} y^2 + \cdots + \binom{n}{n} x^0 y^n$$

Notice the exponents on the variables in the formula above are calculated using the similarities we observed at the beginning of the section.

It does not make any difference, when expanding binomial powers of the form $(x + y)^n$, whether we use Pascal's triangle or the formula

$$\binom{n}{r} = \frac{n!}{r!(n-r)!}$$

to calculate the coefficients. We will now show examples of both methods.

EXAMPLE 2 Expand $(x - 2)^3$.

SOLUTION Applying the binomial formula, we have

$$(x - 2)^3 = \binom{3}{0} x^3 (-2)^0 + \binom{3}{1} x^2 (-2)^1 + \binom{3}{2} x^1 (-2)^2 + \binom{3}{3} x^0 (-2)^3$$

The following coefficients we found using the binomial formula

$$\binom{3}{0}, \binom{3}{1}, \binom{3}{2}, \text{ and } \binom{3}{3}$$

can also be found in the third row of Pascal's triangle based on the exponents of the binomial we were given to expand. They are 1, 3, 3, and 1.

$$(x - 2)^3 = 1x^3(-2)^0 + 3x^2(-2)^1 + 3x^1(-2)^2 + 1x^0(-2)^3$$

$$= x^3 - 6x^2 + 12x - 8$$

EXAMPLE 3 Expand $(3x + 2y)^4$.

SOLUTION The coefficients can be found in the fourth row of Pascal's triangle.

$$1, 4, 6, 4, 1$$

Here is the expansion of $(3x + 2y)^4$:

$$(3x + 2y)^4 = 1(3x)^4 + 4(3x)^3(2y) + 6(3x)^2(2y)^2 + 4(3x)(2y)^3 + 1(2y)^4$$

$$= 81x^4 + 216x^3y + 216x^2y^2 + 96xy^3 + 16y^4$$

EXAMPLE 4   Write the first three terms in the expansion of $(x + 5)^9$.

SOLUTION The coefficients of the first three terms are

$$\binom{9}{0}, \binom{9}{1}, \text{ and } \binom{9}{2}$$

which we calculate using factorial notation as follows:

$$\binom{9}{0} = \left(\frac{9!}{0! \cdot 9!}\right) = \frac{9 \cdot 8 \cdot 7 \cdot 6 \cdot 5 \cdot 4 \cdot 3 \cdot 2 \cdot 1}{(1)(9 \cdot 8 \cdot 7 \cdot 6 \cdot 5 \cdot 4 \cdot 3 \cdot 2 \cdot 1)} = \frac{1}{1} = 1$$

$$\binom{9}{1} = \frac{9!}{1! \cdot 8!} = \frac{9 \cdot 8 \cdot 7 \cdot 6 \cdot 5 \cdot 4 \cdot 3 \cdot 2 \cdot 1}{(1)(8 \cdot 7 \cdot 6 \cdot 5 \cdot 4 \cdot 3 \cdot 2 \cdot 1)} = \frac{9}{1} = 9$$

$$\binom{9}{2} = \frac{9!}{2! \cdot 7!} = \frac{9 \cdot 8 \cdot 7 \cdot 6 \cdot 5 \cdot 4 \cdot 3 \cdot 2 \cdot 1}{(2 \cdot 1)(7 \cdot 6 \cdot 5 \cdot 4 \cdot 3 \cdot 2 \cdot 1)} = \frac{72}{2} = 36$$

From the binomial formula, we write the first three terms.

$$(x + 5)^9 = 1 \cdot x^9 + 9 \cdot x^8(5) + 36x^7(5)^2 + \cdots$$

$$= x^9 + 45x^8 + 900x^7 + \cdots$$

C The *k*th Term of a Binomial Expansion

If we look at each term in the expansion of $(x + y)^n$ as a term in a sequence, $a_1, a_2, a_3, \ldots$, we can write

$$a_1 = \binom{n}{0} x^n y^0$$

$$a_2 = \binom{n}{1} x^{n-1} y^1$$

$$a_3 = \binom{n}{2} x^{n-2} y^2$$

$$a_4 = \binom{n}{3} x^{n-3} y^3 \qquad \text{and so on}$$

To write the formula for the general term k, we simply notice that the exponent on y and the number below n in the coefficient are both 1 less than the term number in the sequence. This observation allows us to write the following:

The General Term of a Binomial Expansion
The *k*th term in the expansion of $(x + y)^n$ is

$$a_k = \binom{n}{k - 1} x^{n-(k-1)} y^{k-1}$$

EXAMPLE 5   Find the fifth term in the expansion of $(2x + 3y)^{12}$.

SOLUTION Applying the preceding formula, we have

$$a_5 = \binom{12}{4} (2x)^8 (3y)^4 \qquad k = 5, 5 - 1 = 4 \text{ to get the number below } n$$

$$= \frac{12!}{4! \cdot 8!}\, (2x)^8 (3y)^4$$

Notice that once we have one of the exponents, the other exponent and the denominator of the coefficient are determined: the two exponents add to 12 and match the numbers in the denominator of the coefficient.

Making the calculations from the preceding formula, we have

$$a_5 = 495(256x^8)(81y^4)$$

$$= 10,264,320x^8 y^4$$

Problem Set 13.5

Moving Toward Success

"In three words I can sum up everything I've learned about life: it goes on."

—Robert Frost, 1874–1963, American poet

1. What is one thing you wish you had done differently in this class? Explain.

2. Will you apply the study skills you have learned in this class to other classes? Why or why not?

A **B** Use the binomial formula to expand each of the following. [Examples 2–3]

1. $(x + 2)^4$

2. $(x - 2)^5$

3. $(x + y)^6$

4. $(x - 1)^6$

5. $(2x + 1)^5$

6. $(2x - 1)^4$

7. $(x - 2y)^5$

8. $(2x + y)^5$

9. $(3x - 2)^4$

10. $(2x - 3)^4$

11. $(4x - 3y)^3$

12. $(3x - 4y)^3$

13. $(x^2 + 2)^4$

14. $(x^2 - 3)^3$

15. $(x^2 + y^2)^3$

16. $(x^2 - 3y)^4$

17. $(2x + 3y)^4$

18. $(2x - 1)^5$

19. $\left(\dfrac{x}{2} + \dfrac{y}{3}\right)^3$

20. $\left(\dfrac{x}{3} - \dfrac{y}{2}\right)^4$

21. $\left(\dfrac{x}{2} - 4\right)^3$

22. $\left(\dfrac{x}{3} + 6\right)^3$

23. $\left(\dfrac{x}{3} + \dfrac{y}{2}\right)^4$

24. $\left(\dfrac{x}{2} - \dfrac{y}{3}\right)^4$

B Write the first four terms in the expansion of the following. [Example 4]

25. $(x + 2)^9$

26. $(x - 2)^9$

27. $(x - y)^{10}$

28. $(x + 2y)^{10}$

29. $(x + 3)^{25}$

30. $(x - 1)^{40}$

31. $(x - 2)^{60}$

32. $\left(x + \dfrac{1}{2}\right)^{30}$

33. $(x - y)^{18}$

34. $(x - 2y)^{65}$

B Write the first three terms in the expansion of each of the following. [Example 4]

35. $(x + 1)^{15}$

36. $(x - 1)^{15}$

37. $(x - y)^{12}$

38. $(x + y)^{12}$

39. $(x + 2)^{20}$

40. $(x - 2)^{20}$

B C Write the first two terms in the expansion of each of the following. [Examples 4–5]

41. $(x + 2)^{100}$

42. $(x - 2)^{50}$

43. $(x + y)^{50}$

44. $(x - y)^{100}$

45. Find the ninth term in the expansion of $(2x + 3y)^{12}$.

46. Find the sixth term in the expansion of $(2x + 3y)^{12}$.

47. Find the fifth term of $(x - 2)^{10}$.

48. Find the fifth term of $(2x - 1)^{10}$.

49. Find the sixth term in the expansion of $(x - 2)^{12}$.

50. Find the ninth term in the expansion of $(7x - 1)^{10}$.

51. Find the third term in the expansion of $(x - 3y)^{25}$.

52. Find the 24th term in the expansion of $(2x - y)^{26}$.

53. Write the formula for the 12th term of $(2x + 5y)^{20}$. Do not simplify.

54. Write the formula for the eighth term of $(2x + 5y)^{20}$. Do not simplify.

55. Write the first three terms of the expansion of $(x^2y - 3)^{10}$.

56. Write the first three terms of the expansion of $\left(x - \frac{1}{x}\right)^{50}$.

Applying the Concepts

57. Probability The third term in the expansion of $\left(\frac{1}{2} + \frac{1}{2}\right)^{7}$ will give the probability that in a family with 7 children, 5 will be boys and 2 will be girls. Find the third term.

58. Probability The fourth term in the expansion of $\left(\frac{1}{2} + \frac{1}{2}\right)^{8}$ will give the probability that in a family with 8 children, 3 will be boys and 5 will be girls. Find the fourth term.

Maintaining Your Skills

Solve each equation. Write your answers to the nearest hundreth.

59. $5^x = 7$

60. $10^x = 15$

61. $8^{2x+1} = 16$

62. $9^{3x-1} = 27$

63. Compound Interest How long will it take $400 to double if it is invested in an account with an annual interest rate of 10% compounded four times a year?

64. Compound Interest How long will it take $200 to become $800 if it is invested in an account with an annual interest rate of 8% compounded four times a year?

Extending the Concepts

65. Calculate both $\binom{8}{5}$ and $\binom{8}{3}$ to show that they are equal.

66. Calculate both $\binom{10}{8}$ and $\binom{10}{2}$ to show that they are equal.

67. Simplify $\binom{20}{12}$ and $\binom{20}{8}$.

68. Simplify $\binom{15}{10}$ and $\binom{15}{5}$.

69. Pascal's Triangle Copy the first eight rows of Pascal's triangle into the eight rows of the triangular array below. (Each number in Pascal's triangle will go into one of the hexagons in the array.) Next, color in each hexagon that contains an odd number. What pattern begins to emerge from this coloring process?

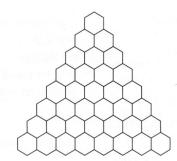

Sequences [13.1]

EXAMPLES

1. In the sequence
$$1, 3, 5, \ldots, 2n - 1, \ldots,$$
$a_1 = 1, a_2 = 3, a_3 = 5$, and
$a_n = 2n - 1$.

A *sequence* is a function whose domain is the set of positive integers. The terms of a sequence are denoted by

$$a_1, a_2, a_3, \ldots, a_n, \ldots$$

where a_1 (read "a sub 1") is the first term, a_2 the second term, and a_n the nth or *general term*.

Summation Notation [13.2]

2. $\displaystyle\sum_{i=3}^{6} (-2)^i$
$= (-2)^3 + (-2)^4 + (-2)^5$
$\quad + (-2)^6$
$= -8 + 16 + (-32) + 64$
$= 40$

The notation

$$\sum_{i=1}^{n} a_i = a_1 + a_2 + a_3 + \cdots + a_n$$

is called *summation notation* or *sigma notation*. The letter i as used here is called the *index of summation* or just *index*.

Arithmetic Sequences [13.3]

3. For the sequence
$$3, 7, 11, 15, \ldots,$$
$a_1 = 3$ and $d = 4$. The general term is
$$a_n = 3 + (n - 1)4$$
$$\quad = 4n - 1$$
Using this formula to find the tenth term, we have
$$a_{10} = 4(10) - 1 = 39$$
The sum of the first 10 terms is
$$S_{10} = \frac{10}{2}(3 + 39) = 210$$

An *arithmetic sequence* is a sequence in which each term comes from the preceding term by adding a constant amount each time. If the first term of an arithmetic sequence is a_1 and the amount we add each time (called the *common difference*) is d, then the nth term of the progression is given by

$$a_n = a_1 + (n - 1)d$$

The sum of the first n terms of an arithmetic sequence is

$$S_n = \frac{n}{2}(a_1 + a_n)$$

S_n is called the nth *partial sum*.

Geometric Sequences [13.4]

4. For the geometric progression
$$3, 6, 12, 24, \ldots,$$
$a_1 = 3$ and $r = 2$. The general term is
$$a_n = 3 \cdot 2^{n-1}$$
The sum of the first 10 terms is
$$S_{10} = \frac{3(2^{10} - 1)}{2 - 1} = 3{,}069$$

A *geometric sequence* is a sequence of numbers in which each term comes from the previous term by multiplying by a constant amount each time. The constant by which we multiply each term to get the next term is called the *common ratio*. If the first term of a geometric sequence is a_1 and the common ratio is r, then the formula that gives the general term a_n is

$$a_n = a_1 r^{n-1}$$

The sum of the first n terms of a geometric sequence is given by the formula

$$S_n = \frac{a_1(r^n - 1)}{r - 1}$$

The Sum of an Infinite Geometric Series [13.4]

5. The sum of the series

$$\frac{1}{3} + \frac{1}{6} + \frac{1}{12} + \cdots$$

is

$$S = \frac{\frac{1}{3}}{1 - \frac{1}{2}} = \frac{\frac{1}{3}}{\frac{1}{2}} = \frac{2}{3}$$

If a geometric sequence has first term a_1 and common ratio r such that $|r| < 1$, then the following is called an *infinite geometric series:*

$$S = \sum_{i=0}^{\infty} a_1 r^i = a_1 + a_1 r + a_1 r^2 + a_1 r^3 + \cdots$$

Its sum is given by the formula

$$S = \frac{a_1}{1 - r}$$

Factorials [13.5]

The notation $n!$ is called n *factorial* and is defined to be the product of each consecutive integer from n down to 1. That is,

$$0! = 1 \qquad \text{(By definition)}$$
$$1! = 1$$
$$2! = 2 \cdot 1$$
$$3! = 3 \cdot 2 \cdot 1$$
$$4! = 4 \cdot 3 \cdot 2 \cdot 1$$

and so on.

Binomial Coefficients [13.5]

6. $\dbinom{7}{3} = \dfrac{7!}{3!(7-3)!}$

$= \dfrac{7!}{3! \cdot 4!}$

$= \dfrac{7 \cdot 6 \cdot 5 \cdot 4 \cdot 3 \cdot 2 \cdot 1}{(3 \cdot 2 \cdot 1)(4 \cdot 3 \cdot 2 \cdot 1)}$

$= 35$

The notation $\dbinom{n}{r}$ is called a *binomial coefficient* and is defined by

$$\binom{n}{r} = \frac{n!}{r!(n-r)!}$$

Binomial coefficients can be found by using the formula above or by *Pascal's triangle,* which is

```
            1
          1   1
        1   2   1
      1   3   3   1
    1   4   6   4   1
  1   5   10   10   5   1
```

and so on.

Binomial Expansion [13.5]

7. $(x+2)^4$

$= x^4 + 4x^3 \cdot 2 + 6x^2 \cdot 2^2$
$\quad + 4x \cdot 2^3 + 2^4$

$= x^4 + 8x^3 + 24x^2 + 32x + 16$

If n is a positive integer, then the formula for expanding $(x+y)^n$ is given by

$$(x+y)^n = \binom{n}{0}x^n y^0 + \binom{n}{1}x^{n-1}y^1 + \binom{n}{2}x^{n-2}y^2 + \cdots + \binom{n}{n}x^0 y^n$$

Write the first four terms of the sequence with the following general terms. [13.1]

1. $a_n = 2n + 5$

2. $a_n = 3n - 2$

3. $a_n = n^2 - 1$

4. $a_n = \dfrac{n + 3}{n + 2}$

5. $a_1 = 4, a_n = 4a_{n-1}, n > 1$

6. $a_1 = \dfrac{1}{4}, a_n = \dfrac{1}{4}a_{n-1}, n > 1$

Determine the general term for each of the following sequences. [13.1]

7. $2, 5, 8, 11, \ldots$

8. $-3, -1, 1, 3, 5, \ldots$

9. $1, 16, 81, 256, \ldots$

10. $2, 5, 10, 17, \ldots$

11. $\dfrac{1}{2}, \dfrac{1}{4}, \dfrac{1}{8}, \dfrac{1}{16}, \ldots$

12. $2, \dfrac{3}{4}, \dfrac{4}{9}, \dfrac{5}{16}, \dfrac{6}{25}, \ldots$

Expand and simplify each of the following. [13.2]

13. $\displaystyle\sum_{i=1}^{4} (2i + 3)$

14. $\displaystyle\sum_{i=1}^{3} (2i^2 - 1)$

15. $\displaystyle\sum_{i=2}^{3} \dfrac{i^2}{i + 2}$

16. $\displaystyle\sum_{i=1}^{4} (-2)^{i-1}$

17. $\displaystyle\sum_{i=3}^{5} (4i + i^2)$

18. $\displaystyle\sum_{i=4}^{6} \dfrac{i + 2}{i}$

Write each of the following sums with summation notation. [13.2]

19. $3 + 6 + 9 + 12$

20. $3 + 7 + 11 + 15$

21. $5 + 7 + 9 + 11 + 13$

22. $4 + 9 + 16$

23. $\dfrac{1}{3} + \dfrac{1}{4} + \dfrac{1}{5} + \dfrac{1}{6}$

24. $\dfrac{1}{3} + \dfrac{2}{9} + \dfrac{3}{27} + \dfrac{4}{81} + \dfrac{5}{243}$

25. $(x - 2) + (x - 4) + (x - 6)$

26. $\dfrac{x}{x + 1} + \dfrac{x}{x + 2} + \dfrac{x}{x + 3} + \dfrac{x}{x + 4}$

Determine which of the following sequences are arithmetic progressions, geometric progressions, or neither. [13.3, 13.4]

27. $1, -3, 9, -27, \ldots$

28. $7, 9, 11, 13, \ldots$

29. $5, 11, 17, 23, \ldots$

30. $\dfrac{1}{2}, \dfrac{1}{3}, \dfrac{1}{4}, \dfrac{1}{5}, \ldots$

31. $4, 8, 16, 32, \ldots$

32. $\dfrac{1}{2}, \dfrac{1}{4}, \dfrac{1}{8}, \dfrac{1}{16}, \ldots$

33. $12, 9, 6, 3, \ldots$

34. $2, 5, 9, 14, \ldots$

Each of the following problems refers to arithmetic progressions. [13.3]

35. If $a_1 = 2$ and $d = 3$, find a_n and a_{20}.

36. If $a_1 = 5$ and $d = -3$, find a_n and a_{16}.

37. If $a_1 = -2$ and $d = 4$, find a_{10} and S_{10}.

38. If $a_1 = 3$ and $d = 5$, find a_{16} and S_{16}.

39. If $a_5 = 21$ and $a_8 = 33$, find the first term a_1, the common difference d, and then find a_{10}.

40. If $a_3 = 14$ and $a_7 = 26$, find the first term a_1, the common difference d, and then find a_9 and S_9.

41. If $a_4 = -10$ and $a_8 = -18$, find the first term a_1, the common difference d, and then find a_{20} and S_{20}.

42. Find the sum of the first 100 terms of the sequence $3, 7, 11, 15, 19, \ldots$

43. Find a_{40} for the sequence $100, 95, 90, 85, 80, \ldots$

Each of the following problems refers to infinite geometric progressions. [13.4]

44. If $a_1 = 3$ and $r = 2$, find a_n and a_{20}.

45. If $a_1 = 5$ and $r = -2$, find a_n and a_{16}.

46. If $a_1 = 4$ and $r = \frac{1}{2}$, find a_n and a_{10}.

47. If $a_1 = -2$ and $r = \frac{1}{3}$, find the sum.

48. If $a_1 = 4$ and $r = \frac{1}{2}$, find the sum.

49. If $a_3 = 12$ and $a_4 = 24$, find the first term a_1, the common ratio r, and then find a_6.

50. Find the tenth term of the sequence $3, 3\sqrt{3}, 9, 9\sqrt{3}, \ldots$

Use the binomial formula to expand each of the following. [13.5]

51. $(x - 2)^4$

52. $(2x + 3)^4$

53. $(3x + 2y)^3$

54. $(x^2 - 2)^5$

55. $\left(\frac{x}{2} + 3\right)^4$

56. $\left(\frac{x}{3} - \frac{y}{2}\right)^3$

Use the binomial formula to write the first three terms in the expansion of the following. [13.5]

57. $(x + 3y)^{10}$

58. $(x - 3y)^9$

59. $(x + y)^{11}$

60. $(x - 2y)^{12}$

Use the binomial formula to write the first two terms in the expansion of the following. [13.5]

61. $(x - 2y)^{16}$

62. $(x + 2y)^{32}$

63. $(x - 1)^{50}$

64. $(x + y)^{150}$

65. Find the sixth term in $(x - 3)^{10}$.

66. Find the fourth term in $(2x + 1)^9$.

Chapter 13 Cumulative Review

Simplify.

1. $|-2|$

2. $-|-5|$

3. $2^3 + 3(2 + 20 \div 4)$

4. $4 + 2(3 - 2^2 + 5)$

5. $-5(2x + 3) + 8x$

6. $4 - 2[3x - 4(x + 2)]$

7. $(3y + 2)^2 - (3y - 2)^2$

8. $(x + 2)^2 - 3(x + 2) + 4$

9. $\dfrac{\frac{1}{5} - \frac{1}{4}}{\frac{1}{2} + \frac{3}{4}}$

10. $\dfrac{\frac{1}{x} - \frac{3}{y}}{\frac{3}{x} - \frac{1}{y}}$

11. $x^{2/3} \cdot x^{1/5}$

12. $a^{-2/5} \cdot a^3$

13. $\sqrt{48x^5y^3}$ (Assume x and y are positive.)

14. $\sqrt[3]{27x^4y^3}$

Solve.

15. $5y - 2 = -3y + 6$

16. $4x - 3 + 2 = 3$

17. $|3x - 1| - 2 = 6$

18. $|3x - 5| + 6 = 2$

19. $2x^2 = 5x + 3$

20. $x^3 - 3x^2 - 4x + 12 = 0$

21. $\dfrac{2}{x + 1} = \dfrac{4}{5}$

22. $\dfrac{6}{a + 2} = \dfrac{5}{a - 3}$

23. $4x^2 + 6x = -5$

24. $2x^2 - 3x + 4 = 0$

25. $\log_2 x = 3$

26. $\log_2 x + \log_2 5 = 1$

27. $8^{x + 3} = 4$

28. $22^{x - 1} = 10$

Solve each of the following sytems of equations.

29. $4x + 2y = 4$
$y = -3x + 1$

30. $2x - 5y = -3$
$3x + 2y = -5$

31. $-x + 2y - 3z = -4$
$2x - 2y - 3z = -1$
$-x + 2y + 2z = -3$

32. $x + 2y = 0$
$3y + z = -3$
$2x - z = 5$

Multiply.

33. $(x^{1/5} + 3)(x^{1/5} - 3)$

34. $(x^{3/5} + 2)(x^{3/5} - 2)$

Graph.

35. $2x - 3y = 12$

36. $2x - 3y = 6$

37. $y = 2^x$

38. $y = \left(\dfrac{1}{2}\right)^x$

39. $y = \log_2 x$

40. $y = \log_{1/2} x$

41. $x^2 + 4x + y^2 - 6y = 12$

42. $9x^2 - 4y^2 = 36$

Find the general term for each sequence.

43. $3, 14, 25, \ldots$

44. $16, 8, 4, \ldots$

Find the equation of the line through the given points. Write your answer in slope-intercept form.

45. $(-6, -1)$ and $(-3, -5)$

46. $(2, 5)$ and $(6, -3)$

Find the distance between the given points.

47. $(-3, 1)$ and $(4, 5)$

48. $(1, -2)$ and $(3, -4)$

Projectile Motion An object projected upward with an initial velocity of 48 feet per second will rise and fall according to the equation $s = 48t - 16t^2$, where s is its distance above the ground at time t.

49. At what times will the object be 20 feet above the ground?

50. At what times will the object be 11 feet above the ground?

Expand and simplify.

51. $\displaystyle\sum_{i=1}^{5} (2i - 1)$

52. $\displaystyle\sum_{i=1}^{4} (x - i)$

Use the information in the illustration to work Problems 53 and 54.

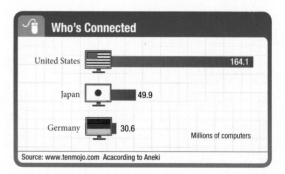

Who's Connected

United States — 164.1

Japan — 49.9

Germany — 30.6

Millions of computers

Source: www.tenmojo.com Acacording to Aneki

53. Write the number of computers in the United States in scientific notation.

54. Write the number of computers in Japan in scientific notation.

Write the first five terms of the sequences with the following general terms. [13.1]

1. $a_n = 3n - 5$

2. $a_n = n^2 + 1$

3. $a_n = 2n^3$

4. $a_n = \dfrac{n + 1}{n^2}$

5. $a_1 = 3, a_n = a_{n-1} + 4, n > 1$

6. $a_1 = 4, a_n = -2a_{n-1}, n > 1$

Give the general term for each sequence. [13.1]

7. $6, 10, 14, 18, \ldots$

8. $1, 2, 4, 8, \ldots$

9. $\dfrac{1}{2}, \dfrac{1}{4}, \dfrac{1}{8}, \dfrac{1}{16}, \ldots$

10. $-3, 9, -27, 81, \ldots$

Expand and simplify each of the following. [13.2]

11. $\displaystyle\sum_{i=1}^{5}(5i + 3)$

12. $\displaystyle\sum_{i=3}^{5} 2^{i-1}$

13. $\displaystyle\sum_{i=2}^{6}(i^2 + 2i)$

14. $\displaystyle\sum_{i=1}^{3}(x + 2i)$

Write each sum using summation notation. [13.2]

15. $4 + 7 + 10 + 13 + 16$

16. $\dfrac{2}{4} + \dfrac{3}{5} + \dfrac{4}{6} + \dfrac{5}{7}$

17. $(x + 4)^5 + (x + 4)^6 + (x + 4)^7 + (x + 4)^8$

18. $\dfrac{x}{x - 1} + \dfrac{x^2}{x - 2} + \dfrac{x^3}{x - 3}$

Determine whether each sequence is arithmetic, geometric, or neither. If the sequence is arithmetic, give the common difference d. If the sequence is geometric, give the common ratio r. [13.3, 13.4]

19. $-1, 5, -25, 125, \ldots$

20. $-2, -3, 2, 1, \ldots$

21. $1, \dfrac{1}{2}, 0, -\dfrac{1}{2}, \ldots$

22. $-2, 3, 8, 13, \ldots$

23. Find the first term of an arithmetic progression if $a_5 = 11$ and $a_9 = 19$. [13.3]

24. Find the second term of a geometric progression for which $a_3 = 18$ and $a_5 = 162$. [13.4]

Find the sum of the first 10 terms of the following arithmetic progressions. [13.3]

25. $5, 11, 17, \ldots$

26. $25, 20, 15, \ldots$

27. Write a formula for the sum of the first 50 terms of the geometric progression $3, 6, 12, \ldots$ [13.4]

28. Find the sum of $\dfrac{1}{2} + \dfrac{1}{6} + \dfrac{1}{18} + \dfrac{1}{54} + \ldots$ [13.4]

Use the binomial formula to expand each of the following. [13.5]

29. $(x - 3)^4$

30. $(2x - 1)^5$

31. Find the first 3 terms in the expansion of $(x - 1)^{20}$. [13.5]

32. Find the sixth term in $(2x - 3y)^8$. [13.5]

33. Write down the first eight rows of Pascal's triangle in the triangular array below. [13.5]

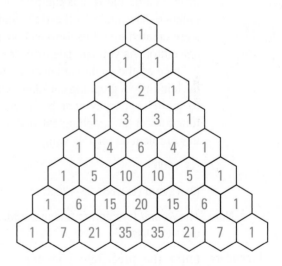

Chapter 13 Projects

SEQUENCES AND SERIES

GROUP PROJECT

Credit Card Payments

Number of People 2

Time Needed 20–30 minutes

Equipment Paper, pencil, and graphing calculator

Background Suppose you run up a balance of $1,000 on a credit card that charges 1.65% interest each month (i.e., an annual rate of 19.8%). If you stop using the card and make the minimum payment of $20 each month, how long will it take you to pay off the balance on the card? The answer can be found by using the formula

$$U_n = (1.0165)U_{n-1} - 20$$

where U_n stands for the current unpaid balance on the card, and U_{n-1} is the previous month's balance. The table to the right and Figure 1 were created from this formula and a graphing calculator. As you can see from the table, the balance on the credit card decreases very little each month. A graphing calculator can be used to model each payment by using the recall function on the graphing calculator. To set up this problem do the following:

(1) 1000   **ENTER**

(2) Round (1.0165ANS—20,2)   **ENTER**

Note The *Round* function is under the MATH key in the NUM list.

Procedure Enter the preceding commands into your calculator. While one person in the group hits the ENTER key, another person counts, by keeping a tally of the "payments" made. The credit card is paid off when the calculator displays a negative number.

MONTHLY CREDIT CARD BALANCES				
Previous Balance $U_{(n-1)}$	Monthly Interest Rate	Payment Number n	Monthly Payment	New Balance $U_{(n)}$
$1,000.00	1.65%	1	$20	$996.50
$996.50	1.65%	2	$20	$992.94
$992.94	1.65%	3	$20	$989.32
$989.32	1.65%	4	$20	$985.64
$985.64	1.65%	5	$20	$981.90
.	.	.	.	.
.	.	.	.	.
.	.	.	.	.

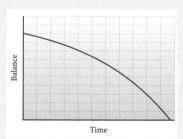

FIGURE 1

1. How many months did it take to pay off the credit card?

2. What was the amount of the last payment?

3. What was the total interest paid to the credit card company?

4. How much would you save if you paid $25 per month instead of $20?

5. If the credit card company raises the interest rate to 21.5% annual interest, how long will it take to pay off the balance? How much more would it cost in interest?

6. On larger balances, many credit card companies require a minimum payment of 2% of the outstanding balance. What is the recursion formula for this? How much would this save or cost you in interest?

Building Squares from Odd Numbers

A relationship exists between the sequence of squares and the sequence of odd numbers. In *The Book of Squares*, written in 1225, Leonardo Fibonacci has this to say about that relationship:

Corbis/Bettmann

> *I thought about the origin of all square numbers and discovered that they arise out of the increasing sequence of odd numbers.*

Work with the sequence of odd numbers until you discover how it can be used to produce the sequence of squares. Then write an essay in which you give a written description of the relationship between the two sequences, along with a diagram that illustrates the relationship. See if you can use summation notation to write an equation that summarizes the whole relationship. Your essay should be clear and concise and written so that any of your classmates can read it and understand the relationship you are describing.

Answers to Odd-Numbered Problems

Chapter 1

Getting Ready for Chapter 1

1. 125 **2.** 3 **3.** 0.5 **4.** 0.5 **5.** 1.4 **6.** 1.8 **7.** 5 **8.** 3 **9.** 45 **10.** 32 **11.** 29 **12.** 4 **13.** 13 **14.** 44
15. 10 **16.** 21 **17.** 210 **18.** 231 **19.** 630 **20.** 3

Problem Set 1.1

1. $x + 5 = 14$ **3.** $5y < 30$ **5.** $3y \le y + 6$ **7.** $\frac{x}{3} = x + 2$ **9.** 9 **11.** 49 **13.** 8 **15.** 64 **17.** 16 **19.** 100 **21.** 121
23. 100 **25.** 9 **27.** 18 **29.** 8 **31.** 12 **33.** 18 **35.** 42 **37.** 53 **39.** 11 **41.** 16 **43.** 17 **45.** 42 **47.** 30
49. 30 **51.** 24 **53.** 80 **55.** 27 **57.** 35 **59.** 13 **61.** 4 **63.** 37 **65.** 37 **67.** 16 **69.** 16 **71.** 81 **73.** 41
75. 345 **77.** 2,345 **79.** 2 **81.** 148 **83.** 36 **85.** 36 **87.** 58 **89.** 62 **91.** 5 **93.** 10 **95.** 25 **97.** 10 **99.** 10
101. 420 calories **103.** $404,495 **105.**

Activity	Calories Burned in 1 Hour by a 150-Pound Person
Bicycling	374
Bowling	265
Handball	680
Jogging	680
Skiing	544

Problem Set 1.2

1–7.

9. $\frac{18}{24}$ **11.** $\frac{12}{24}$ **13.** $\frac{15}{24}$ **15.** $\frac{36}{60}$ **17.** $\frac{22}{60}$ **19.** 2 **21.** 25 **23.** 6 **25.** $-10, \frac{1}{10}, 10$ **27.** $-\frac{3}{4}, \frac{4}{3}, \frac{3}{4}$ **29.** $-\frac{11}{2}, \frac{2}{11}, \frac{11}{2}$
31. $3, -\frac{1}{3}, 3$ **33.** $\frac{2}{5}, -\frac{5}{2}, \frac{2}{5}$ **35.** $-x, \frac{1}{x}, |x|$ **37.** $<$ **39.** $>$ **41.** $>$ **43.** $>$ **45.** $<$ **47.** $<$ **49.** 6 **51.** 22
53. 3 **55.** 7 **57.** 3 **59.** $\frac{8}{15}$ **61.** $\frac{3}{2}$ **63.** $\frac{5}{4}$ **65.** 1 **67.** 1 **69.** 1 **71.** $\frac{1}{9}$ **73.** $\frac{1}{25}$ **75. a.** 4 **b.** 14 **c.** 24 **d.** 34
77. a. 12 **b.** 20 **c.** 100 **d.** 1,024 **79. a.** 17 **b.** 25 **c.** 25 **81.** 4 inches; 1 in² **83.** 4.5 inches; 1.125 in²
85. 10.25 centimeters; 5 cm² **87.** $-8, -2$ **89.** $-64°$ F; $-54°$ F **91.** 1,150 ft **93.** $-15°$F **95.** $-25°$F

Problem Set 1.3

1. 3 **3.** -14 **5.** -3 **7.** -25 **9.** -12 **11.** -25 **13.** 11 **15.** 10.8
17. $3 + 5 = 8; 3 + (-5) = -2; -3 + 5 = 2; -3 + (-5) = -8$
19. $15 + 20 = 35; 15 + (-20) = -5; -15 + 20 = 5; -15 + (-20) = -35$ **21.** 23, 28 **23.** 30, 35 **25.** $-12, -18$
27. Yes **29.** -3 **31.** 0 **33.** -10 **35.** -16 **37.** 35 **39.** 0 **41.** -11.3 **43.** -3.6 **45.** -8 **47.** 6 **49.** 6
51. -4 **53.** -10 **55.** -21 **57.** -12 **59.** -2 **61.** 4 **63.** 0 **65.** 7 **67.** 5 **69.** 12 **71.** 25 **73.** -40
75. 1 **77.** 19 **79.** -198 **81.** 45° **83.** 150° **85.** 577.7 million **87.** $73 + 10 - 8$, 75°F **89.** $32 **91.** 949 strikeouts
93. $-35 + 15 - 20 = -\$40$ **95.** $8 + (-5) = 3$ **97.** $4,500; $3,950; $3,400; $2,850; $2,300; yes **99.** 769 feet **101.** 439 feet
103. 2 seconds

Problem Set 1.4

1. -42 **3.** -16 **5.** 3 **7.** 121 **9.** 6 **11.** -60 **13.** 24 **15.** 49 **17.** -27 **19.** 6 **21.** 10 **23.** 9 **25.** 45
27. 14 **29.** -2 **31.** 216 **33.** -2 **35.** -18 **37.** 29 **39.** 38 **41.** -5 **43.** 37 **45. a.** 60 **b.** -23 **c.** 17 **d.** -2
47. $-\frac{10}{21}$ **49.** -4 **51.** $\frac{9}{16}$ **53. a.** 27 **b.** 3 **c.** -27 **d.** -27 **55.** 9 **57.** $\frac{25}{4}$ **59.** 4 **61.** $\frac{9}{4}$ **63.** 0 **65.** 0
67. 0 **69.** 8 **71.** -80 **73.** $\frac{1}{8}$ **75.** -24 **77.** 4 **79.** 24 **81.** 81 **83.** $\frac{9}{4}$ **85.** $2.50 **87.** $60 **89.** 1° F
91. 100 mg

Problem Set 1.5

1. -2 **3.** -3 **5.** $-\frac{1}{3}$ **7.** 3 **9.** $\frac{1}{7}$ **11.** 0 **13.** 9 **15.** -15 **17.** -36 **19.** $-\frac{1}{4}$ **21.** $\frac{16}{15}$ **23.** $\frac{4}{3}$ **25.** $-\frac{8}{13}$

27. -1 **29.** 1 **31.** $\frac{3}{5}$ **33.** $-\frac{5}{3}$ **35.** -2 **37.** -3 **39.** Undefined **41.** Undefined **43.** 5 **45.** $-\frac{7}{3}$ **47.** -1

49. -7 **51.** $\frac{15}{17}$ **53.** $-\frac{32}{17}$ **55.** $\frac{1}{3}$ **57.** 1 **59.** 1 **61.** -2 **63.** $\frac{9}{7}$ **65.** $\frac{16}{11}$ **67.** -1 **69.** $\frac{7}{20}$ **71.** -1

73. a. $\frac{3}{2}$ **b.** $\frac{3}{2}$ **75. a.** $-\frac{5}{7}$ **b.** $-\frac{5}{7}$ **77. a.** 2.618 **b.** 0.382 **c.** 3 **79. a.** 25 **b.** -25 **c.** -25 **d.** -25 **e.** 25

81. a. 10 **b.** 0 **c.** -100 **d.** -20 **83.** 3 **85.** -10 **87.** -3 **89.** -8 **91.** 425 yards **93.** \$350 **95.** Drops 3.5° F each hour

97. 3 capsules **99.** 0.3 mg **101. a.** \$20,000 **b.** \$50,000 **c.** Yes

Problem Set 1.6

1. $(4 + 2) + x = 6 + x$ **3.** $x + (2 + 7) = x + 9$ **5.** $(3 \cdot 5)x = 15x$ **7.** $(9 \cdot 6)y = 54y$ **9.** $\left(\frac{1}{2} \cdot 3\right)a = \frac{3}{2}a$ **11.** $\left(\frac{1}{3} \cdot 3\right)x = x$

13. $\left(\frac{1}{2} \cdot 2\right)y = y$ **15.** $\left(\frac{3}{4} \cdot \frac{4}{3}\right)x = x$ **17.** $\left(\frac{6}{5} \cdot \frac{5}{6}\right)a = a$ **19.** $8x + 16$ **21.** $8x - 16$ **23.** $4y + 4$ **25.** $18x + 15$ **27.** $6a + 14$

29. $54y - 72$ **31.** $\frac{3}{2}x - 3$ **33.** $x + 2$ **35.** $3x + 3y$ **37.** $8a - 8b$ **39.** $12x + 18y$ **41.** $12a - 8b$ **43.** $3x + 2y$ **45.** $4a + 25$

47. $6x + 12$ **49.** $14x + 38$ **51.** $0.09x + 180$ **53.** $0.15x + 75$ **55.** $3x - 2y$ **57.** $4x + 3y$ **59.** $2x + 1$ **61.** $6x - 3$

63. $5x + 10$ **65.** $6x + 5$ **67.** $5x + 6$ **69.** $6m - 5$ **71.** $7 + 3x$ **73.** $6x - 4y$ **75.** $0.15x + 90$ **77.** $0.12x + 60$

79. $a + 1$ **81.** $1 - a$ **83.** Commutative **85.** Multiplicative inverse **87.** Commutative **89.** Distributive

91. Commutative, associative **93.** Commutative, associative **95.** Commutative **97.** Commutative, associative

99. Commutative **101.** Additive inverse **103.** $3x + 6$ **105.** $9a + 9b$ **107.** 0 **109.** 0 **111.** 10 **113.** No

115. No **117.** $3(6,200 - 100) = 18,600 - 300 = \$18,300$ **119.** $8 \div 4 \neq 4 \div 8$

121. $12(2,400 - 480) = \$23,040$; $12(2,400) - 12(480) = \$23,040$

Problem Set 1.7

1. 0, 1 **3.** $-3, -2.5, 0, 1, \frac{3}{2}$ **5.** All **7.** $-10, -8, -2, 9$ **9.** π **11.** T **13.** F **15.** F **17.** T **19.** Composite, $2^4 \cdot 3$

21. Prime **23.** Composite, $3 \cdot 11 \cdot 31$ **25.** $2^4 \cdot 3^2$ **27.** $2 \cdot 19$ **29.** $3 \cdot 5 \cdot 7$ **31.** $2^2 \cdot 3^2 \cdot 5$ **33.** $5 \cdot 7 \cdot 11$ **35.** 11^2

37. $2^2 \cdot 3 \cdot 5 \cdot 7$ **39.** $2^2 \cdot 5 \cdot 31$ **41.** $\frac{7}{11}$ **43.** $\frac{5}{7}$ **45.** $\frac{11}{13}$ **47.** $\frac{14}{15}$ **49.** $\frac{5}{9}$ **51.** $\frac{5}{8}$ **53. a.** -30 **b.** 130 **c.** $-4,000$ **d.** $-\frac{5}{8}$

55. $\frac{2}{9}$ **57.** $\frac{3}{2}$ **59.** 64 **61.** 509 **63. a.** 0.35 **b.** 0.28 **c.** 0.25 **65.** $6^3 = (2 \cdot 3)^3 = 2^3 \cdot 3^3$ **67.** $9^4 \cdot 16^2 = (3^2)^4(2^4)^2 = 2^8 \cdot 3^8$

69. $3 \cdot 8 + 3 \cdot 7 + 3 \cdot 5 = 24 + 21 + 15 = 60 = 2^2 \cdot 3 \cdot 5$ **71.** $3 \cdot 7 \cdot 13$ **73.** Irrational numbers **75.** 8, 21, 34

Problem Set 1.8

1. $\frac{2}{3}$ **3.** $-\frac{1}{4}$ **5.** $\frac{1}{2}$ **7.** $\frac{x - 1}{3}$ **9.** $\frac{3}{2}$ **11.** $\frac{x + 6}{2}$ **13.** $-\frac{3}{5}$ **15.** $\frac{10}{a}$ **17.** $\frac{7}{8}$ **19.** $\frac{1}{10}$ **21.** $\frac{7}{9}$ **23.** $\frac{7}{3}$

25. $\frac{1}{4}$ **27.** $\frac{7}{6}$ **29.** $\frac{19}{24}$ **31.** $\frac{13}{60}$ **33.** $\frac{29}{35}$ **35.** $\frac{949}{1,260}$ **37.** $\frac{13}{420}$ **39.** $\frac{41}{24}$ **41.** $\frac{5}{4}$ **43.** $-\frac{3}{2}$ **45.** $\frac{3}{2}$ **47.** $-\frac{2}{3}$

49. $\frac{7}{3}$ **51.** $\frac{1}{125}$ **53.** 4 **55.** -2 **57.** -18 **59.** -16 **61.** -2 **63.** 2 **65.** -144 **67.** -162 **69.** $\frac{5x + 4}{20}$ **71.** $\frac{4 + a}{12}$

73. $\frac{9x + 4}{12}$ **75.** $\frac{10 + 3x}{5x}$ **77.** $\frac{3y + 28}{7y}$ **79.** $\frac{60 + 19a}{20a}$ **81.** $\frac{2}{3}x$ **83.** $-\frac{1}{4}x$ **85.** $\frac{14}{15}x$ **87.** $\frac{11}{12}x$ **89.** $\frac{41}{40}x$ **91.** $\frac{x - 1}{x}$

93. a. $\frac{1}{4}$ **b.** $\frac{5}{4}$ **c.** $-\frac{3}{8}$ **d.** $-\frac{3}{2}$ **95.** $\frac{1}{3}$ **97.** $\frac{3}{4}$ **99. a.** $\frac{3}{2}$ **b.** $\frac{4}{3}$ **c.** $\frac{5}{4}$ **101. a.** 8 **b.** 7 **c.** 8

Chapter 1 Review

1. $-7 + (-10) = -17$ **2.** $(-7 + 4) + 5 = 2$ **3.** $(-3 + 12) + 5 = 14$ **4.** $4 - 9 = -5$ **5.** $9 - (-3) = 12$ **6.** $-7 - (-9) = 2$

7. $(-3)(-7) - 6 = 15$ **8.** $5(-6) + 10 = -20$ **9.** $2[(-8)(3x)] = -48x$ **10.** $\frac{-25}{-5} = 5$ **11.** 1.8 **12.** -10 **13.** $-6, \frac{1}{6}$

14. $\frac{12}{5}, -\frac{5}{12}$ **15.** -5 **16.** $-\frac{5}{4}$ **17.** 3 **18.** -38 **19.** -25 **20.** 1 **21.** -3 **22.** 22 **23.** -4 **24.** -20 **25.** -30

26. -12 **27.** -24 **28.** 12 **29.** -4 **30.** $-\frac{2}{3}$ **31.** 23 **32.** 47 **33.** -3 **34.** -35 **35.** 32 **36.** 2 **37.** 20 **38.** -3

39. -98 **40.** 70 **41.** 2 **42.** $\frac{17}{2}$ **43.** Undefined **44.** 3 **45.** Associative **46.** Multiplicative identity

47. Commutative **48.** Additive inverse **49.** $12 + x$ **50.** $28a$ **51.** x **52.** y **53.** $14x + 21$ **54.** $6a - 12$ **55.** $\frac{5}{2}x - 3$

56. $-\frac{3}{2}x + 3$ **57.** $-\frac{1}{3}, 0, 5, -4.5, \frac{2}{5}, -3$ **58.** 0, 5 **59.** $\sqrt{7}, \pi$ **60.** 0, 5, -3 **61.** $2 \cdot 3^2 \cdot 5$ **62.** $2^3 \cdot 3 \cdot 5 \cdot 7$ **63.** $\frac{173}{210}$

64. $\frac{2x + 7}{12}$

Chapter 1 Test

1. $x + 3 = 8$ **2.** $5y = 15$ **3.** 40   **4.** 16   **5.** $4, -\frac{1}{4}, 4$ **6.** $-\frac{3}{4}, \frac{4}{3}, \frac{3}{4}$ **7.** -4 **8.** 17   **9.** -12 **10.** 0   **11.** c   **12.** e

13. d   **14.** a   **15.** -21 **16.** 64   **17.** -2 **18.** $-\frac{8}{27}$ **19.** 4   **20.** 204   **21.** 25   **22.** 52   **23.** 2   **24.** 2   **25.** $8 + 2x$

26. $10x$ **27.** $6x + 10$ **28.** $-2x + 1$ **29.** $-8, 1$ **30.** $-8, \frac{3}{4}, 1, 1.5$ **31.** $\sqrt{2}$ **32.** $-8, \frac{3}{4}, 1, \sqrt{2}, 1.5$ **33.** $2^4 \cdot 37$

34. $2^2 \cdot 5 \cdot 67$ **35.** $\frac{25}{42}$ **36.** $\frac{8}{x}$ **37.** $8 + (-3) = 5$ **38.** $-24 - 2 = -26$ **39.** $(-5)(-4) = 20$ **40.** $\frac{-24}{-2} = 12$ **41.** 12

42. $\frac{1}{2}$ **43.** 425 million   **44.** 1.35 billion   **45.** 18 million

Chapter 2

Getting Ready for Chapter 2

1. 4   **2.** -14 **3.** 33   **4.** -35 **5.** $-6y - 3$ **6.** 10.8   **7.** $-\frac{5}{4}$ **8.** $-\frac{6}{5}$ **9.** x **10.** x **11.** x **12.** x **13.** $-3x - 4$

14. -9.7 **15.** $0.04x + 280$ **16.** $0.09x + 180$ **17.** $3x - 11$ **18.** $5x - 13$ **19.** $10x - 43$ **20.** $12x - 22$

Problem Set 2.1

1. $-3x$ **3.** $-a$ **5.** $12x$ **7.** $6a$ **9.** $6x - 3$ **11.** $7a + 5$ **13.** $5x - 5$ **15.** $4a + 2$ **17.** $-9x - 2$ **19.** $12a + 3$ **21.** $10x - 1$
23. $21y + 6$ **25.** $-6x + 8$ **27.** $-2a + 3$ **29.** $-4x + 26$ **31.** $4y - 16$ **33.** $-6x - 1$ **35.** $2x - 12$ **37.** $10a + 33$
39. $4x - 9$ **41.** $7y - 39$ **43.** $-19x - 14$ **45.** 5   **47.** -9 **49.** 4   **51.** 4   **53.** -37 **55.** -41 **57.** 64   **59.** 64
61. 144   **63.** 144   **65.** 3   **67.** 0   **69.** 15   **71.** 6

73. a.

n	1	2	3	4
3n	3	6	9	12

b.

n	1	2	3	4
n³	1	8	27	64

75. $1, 4, 7, 10, \ldots$ **77.** $0, 1, 4, 9, \ldots$ **79.** $-6y + 4$ **81.** $0.17x$ **83.** $2x$ **85.** $5x - 4$ **87.** $7x - 5$ **89.** $-2x - 9$
91. $7x + 2$ **93.** $-7x + 6$ **95.** $7x$ **97.** $-y$ **99.** $10y$ **101.** $0.17x + 180$ **103.** $0.22x + 60$ **105.** $a = 4; 13x$ **107.** 49
109. 40   **111. a.** $42° F$ **b.** $28° F$ **c.** $-14° F$ **113. a.** \$37.50 **b.** \$40.00 **c.** \$42.50   **115.** $0.71G; \$887.50$ **117.** 68 mph   **119.** 12
121. -3 **123.** -9.7 **125.** $-\frac{5}{4}$ **127.** -3 **129.** $a - 12$ **131.** $-\frac{7}{2}$ **133.** $\frac{51}{40}$ **135.** 7   **137.** $-\frac{1}{24}$ **139.** $-\frac{7}{9}$ **141.** $-\frac{23}{420}$

Problem Set 2.2

1. 11   **3.** 4   **5.** $-\frac{3}{4}$ **7.** -5.8 **9.** -17 **11.** $-\frac{1}{8}$ **13.** -6 **15.** -3.6 **17.** -7 **19.** $-\frac{7}{45}$ **21.** 3   **23.** $\frac{11}{8}$ **25.** 21
27. 7   **29.** 3.5   **31.** 22   **33.** -2 **35.** -16 **37.** -3 **39.** 10   **41.** -12 **43.** 4   **45.** 2   **47.** -5 **49.** -1 **51.** -3
53. 8   **55.** -8 **57.** 2   **59.** 11   **61.** 10   **63. a.** 6% **b.** 5% **c.** 2% **d.** 75%   **65.** $x + 55 + 55 = 180; 70°$
67. a. 65.9% **b.** 71.5% **c.** 34.1% **d.** 28.5%   **69.** y **71.** x **73.** 6   **75.** 6   **77.** -9 **79.** $-\frac{15}{8}$ **81.** -18 **83.** $-\frac{5}{4}$
85. $3x$ **87.** $18x$ **89.** x **91.** y **93.** x **95.** a

Problem Set 2.3

1. 2   **3.** 4   **5.** $-\frac{1}{2}$ **7.** -2 **9.** 3   **11.** 4   **13.** 0   **15.** 0   **17.** 6   **19.** -50 **21.** $\frac{3}{2}$ **23.** 12   **25.** -3 **27.** 32
29. -8 **31.** $\frac{1}{2}$ **33.** 4   **35.** 8   **37.** -4 **39.** 4   **41.** -15 **43.** $-\frac{1}{2}$ **45.** 3   **47.** 1   **49.** $\frac{1}{4}$ **51.** -3 **53.** 3   **55.** 2
57. $-\frac{3}{2}$ **59.** $-\frac{3}{2}$ **61.** 1   **63.** 1   **65.** -2 **67.** -2 **69.** $-\frac{4}{5}$ **71.** 1   **73.** $\frac{7}{2}$ **75.** $\frac{5}{4}$ **77.** 40   **79.** $\frac{7}{3}$ **81.** 200 tickets
83. \$1,391 per month   **85.** 725 ft.   **87.** 2   **89.** 6   **91.** 3,000   **93.** $3x - 11$ **95.** $0.09x + 180$ **97.** $-6y + 3$
99. $4x - 11$ **101.** $5x$ **103.** $0.17x$ **105.** $6x - 10$ **107.** $\frac{3}{2}x + 3$ **109.** $-x + 2$ **111.** $10x - 43$ **113.** $-3x - 13$
115. $-6y + 4$

Problem Set 2.4

1. 3   **3.** −2   **5.** −1   **7.** 2   **9.** −4   **11.** −2   **13.** 0   **15.** 1   **17.** $\frac{1}{2}$ **19.** 7   **21.** 8   **23.** $-\frac{1}{3}$ **25.** $\frac{3}{4}$ **27.** 75

29. 2   **31.** 6   **33.** 8   **35.** 0   **37.** $\frac{3}{7}$ **39.** 1   **41.** 1   **43.** −1   **45.** 6   **47.** $\frac{3}{4}$ **49.** 3   **51.** $\frac{3}{4}$ **53.** 8   **55.** 6

57. −2   **59.** −2   **61.** 2   **63.** −6   **65.** 2   **67.** 20   **69.** 4,000   **71.** 700   **73.** 11   **75.** 7

77. a. $\frac{5}{4}$ **b.** $\frac{15}{2}$ **c.** $6x + 20$ **d.** 15 **e.** $4x − 20$ **f.** $\frac{45}{2}$ **79.** 2004   **81.** 14   **83.** −3   **85.** $\frac{1}{4}$ **87.** $\frac{1}{3}$ **89.** $-\frac{3}{2}x + 3$ **91.** $\frac{3}{2}$

93. 4   **95.** 1   **97.**

x	3(x + 2)	3x + 2	3x + 6
0	6	2	6
1	9	5	9
2	12	8	12
3	15	11	15

99.

a	(2a + 1)²	4a² + 4a + 1
1	9	9
2	25	25
3	49	49

Problem Set 2.5

1. 100 feet   **3. a.** 2 **b.** $\frac{3}{2}$ **c.** 0   **5. a.** 2 **b.** −3 **c.** 1   **7.** −1   **9.** 2   **11.** 12   **13.** 10   **15.** −2   **17.** 1   **19. a.** 2 **b.** 4

21. a. 5 **b.** 18   **23. a.** 20 **b.** 75   **25.** $l = \frac{A}{w}$ **27.** $h = \frac{V}{lw}$ **29.** $a = P − b − c$ **31.** $x = 3y − 1$ **33.** $y = 3x + 6$

35. $y = -\frac{2}{3}x + 2$ **37.** $y = −2x − 5$ **39.** $y = -\frac{2}{3}x + 1$ **41.** $w = \frac{P − 2l}{2}$ **43.** $v = \frac{h − 16t^2}{t}$ **45.** $h = \frac{A − \pi r^2}{2\pi r}$

47. a. $y = −2x − 5$ **b.** $y = 4x − 7$ **49. a.** $y = \frac{3}{4}x + \frac{1}{4}$ **b.** $y = \frac{3}{4}x − 5$ **51. a.** $y = \frac{3}{5}x + 1$ **b.** $y = \frac{1}{2}x + 2$

53. $y = \frac{3}{7}x − 3$ **55.** $y = 2x + 8$ **57. a.** $\frac{15}{4}$ **b.** 17 **c.** $y = -\frac{4}{5}x + 4$ **d.** $x = -\frac{5}{4}y + 5$ **59.** 60°; 150°   **61.** 45°; 135°   **63.** 10

65. 240   **67.** 25%   **69.** 35%   **71.** 64   **73.** 2,000   **75.** 100° C; yes   **77.** 20° C; yes   **79.** $C = \frac{5}{9}(F − 32)$ **81.** 13.9%

83. 60%   **85.** 26.5%   **87.** 7 meters   **89.** 1.5 inches   **91.** 132 feet   **93.** 0.22 centimeters   **95.** The sum of 4 and 1 is 5.

97. The difference of 6 and 2 is 4.   **99.** The difference of a number and 5 is −12.

101. The sum of a number and 3 is four times the difference of that number and 3.   **103.** $2(6 + 3) = 18$ **105.** $2(5) + 3 = 13$

107. $x + 5 = 13$ **109.** $5(x + 7) = 30$ **111. a.** 95 **b.** −41 **c.** −95 **d.** 41   **113. a.** 9 **b.** −73 **c.** 9 **d.** −73

Problem Set 2.6

Along with the answers to the odd-numbered problems in this problem set and the next, we are including some of the equations used to solve the problems. Be sure that you try the problems on your own before looking to see the correct equations.

1. 8   **3.** $2x + 4 = 14$; 5   **5.** −1   **7.** The two numbers are x and $x + 2$; $x + (x + 2) = 8$; 3 and 5   **9.** 6 and 14

11. Shelly is 39, and Michele is 36   **13.** Evan is 11, and Cody is 22   **15.** Barney is 27; Fred is 31   **17.** Lacy is 16; Jack is 32

19. Patrick is 18; Pat is 38   **21.** 9 inches   **23.** 15 feet   **25.** 11 feet, 18 feet, 33 feet   **27.** 26 feet, 13 feet, 14 feet

29. Length = 11 inches; width = 6 inches   **31.** Length = 25 inches; width = 9 inches   **33.** Length = 15 feet; width = 3 feet

35. 9 dimes; 14 quarters   **37.** 12 quarters; 27 nickels   **39.** 8 nickels; 17 dimes   **41.** 7 nickels; 10 dimes; 12 quarters

43. 3 nickels; 9 dimes; 6 quarters   **45.** $5x$ **47.** $1.075x$ **49.** $0.09x + 180$ **51.** 6,000   **53.** 30   **55.** 4 is less than 10.

57. 9 is greater than or equal to −5   **59.** $<$ **61.** $<$ **63.** 2   **65.** 12

67. a. e **b.** i **c.** x **d.** The larger the area of space used to store the letter in a typecase, the more often the letter is printed.

Problem Set 2.7

1. 5 and 6   **3.** −4 and −5   **5.** 13 and 15   **7.** 52 and 54   **9.** −14 and −16   **11.** 17, 19, and 21   **13.** 42, 44, and 46

15. $4,000 invested at 8%, $6,000 invested at 9%   **17.** $700 invested at 10%, $1,200 invested at 12%

19. $500 at 8%, $1,000 at 9%, $1,500 at 10%   **21.** 45°, 45°, 90°   **23.** 22.5°, 45°, 112.5°   **25.** 53°, 90°   **27.** 80°, 60°, 40°

29. 16 adult and 22 children's tickets   **31.** 16 minutes   **33.** 39 hours   **35.** They are in offices 7329 and 7331.

37. Kendra is 8 years old and Marissa is 10 years old.   **39.** Jeff   **41.** $10.38   **43.** Length = 12 meters; width = 10 meters

45. 59°, 60°, 61°   **47.** $54.00   **49.** Yes   **51. a.** 9 **b.** 3 **c.** −9 **d.** −3   **53. a.** −8 **b.** 8 **c.** 8 **d.** −8   **55.** −2.3125

57. $\frac{10}{3}$ **59.** −3   **61.** −8   **63.** 0   **65.** 8   **67.** −24   **69.** 28   **71.** $-\frac{1}{2}$ **73.** −16   **75.** −25   **77.** 24

Problem Set 2.8

1. $x < 12$
0   12

3. $a \le 12$
0   12

5. $x > 13$
0   13

7. $y \ge 4$
0   4

9. $x > 9$
0   9

11. $x < 2$
0   2

13. $a \le 5$
0   5

15. $x > 15$
0   15

17. $x < -3$
-3   0

19. $x \le 6$
0   6

21. $x \ge -50$
-50   0

23. $y < -6$
-6   0

25. $x < 6$
0   6

27. $y \ge -5$
-5   0

29. $x < 3$
0   3

31. $x \le 18$
0   18

33. $a < -20$
-20   0

35. $y < 25$
0   25

37. $a \le 3$
0   3

39. $x \ge \frac{15}{2}$
0    $\frac{15}{2}$

41. $x < -1$
-1   0

43. $y \ge -2$
-2   0

45. $x < -1$
-1   0

47. $m \le -6$
-6   0

49. $x \le -5$
-5   0

51. $y < -\frac{3}{2}x + 3$ **53.** $y < \frac{2}{5}x - 2$ **55.** $y \le \frac{3}{7}x + 3$

57. $y \le \frac{1}{2}x + 1$ **59. a.** 3 **b.** 2 **c.** No **d.** $x > 2$ **61.** $x < 3$ **63.** $x \ge 3$ **65.** $x \le 50$ **67.** At least 291

69. $2x + 6 < 10; x < 2$ **71.** $4x > x - 8; x > -\frac{8}{3}$ **73.** $2x + 2(3x) \ge 48; x \ge 6$; the width is at least 6 meters.

75. $x + (x + 2) + (x + 4) > 24; x > 6$; the shortest side is even and greater than 6 inches.   **77.** $t \ge 100$

79. Lose money if they sell less than 200 tickets; make a profit if they sell more than 200 tickets   **81.** $2x + 1$ **83.** $-2x - 4$

85. $\frac{5}{2}a + \frac{4}{3}$ **87.** $-2a - \frac{5}{2}$ **89.** $\frac{2}{3}x$ **91.** $-\frac{1}{6}x$ **93.** $\frac{11}{12}x$ **95.** $\frac{41}{40}x$

Chapter 2 Review

1. $-3x$ **2.** $-2x - 3$ **3.** $4a - 7$ **4.** $-2a + 3$ **5.** $-6y$ **6.** $-6x + 1$ **7.** 19   **8.** -11 **9.** -18 **10.** -13 **11.** 8

12. 6   **13.** -8 **14.** $\frac{15}{14}$ **15.** 2   **16.** -5 **17.** -5 **18.** 0   **19.** 12   **20.** -8 **21.** -1 **22.** 1   **23.** 5   **24.** 2   **25.** 0

26. 0   **27.** 10   **28.** 2   **29.** 2   **30.** $-\frac{8}{3}$ **31.** 0   **32.** -5 **33.** 0   **34.** -4 **35.** -8 **36.** 4   **37.** $y = \frac{2}{5}x - 2$

38. $y = \frac{5}{2}x - 5$ **39.** $h = \frac{V}{\pi r^2}$ **40.** $w = \frac{P - 2l}{2}$ **41.** 206.4   **42.** 9%   **43.** 11   **44.** 5 meters; 25 meters

45. \$500 at 9%; \$800 at 10%   **46.** 10 nickels; 5 dimes   **47.** $x > -2$ **48.** $x < 2$ **49.** $a \ge 6$ **50.** $a < -15$

51.
-8   0

52.
0   7

53.
-3   0

Chapter 2 Cumulative Review

1. -14 **2.** -7 **3.** -12 **4.** 100 **5.** 30 **6.** 8 **7.** $-\frac{8}{27}$ **8.** 16 **9.** $-\frac{1}{4}$ **10.** $\frac{3}{4}$ **11.** $-\frac{4}{5}$ **12.** $\frac{8}{35}$ **13.** $-\frac{8}{3}$

14. 0 **15.** $\frac{8}{9}$ **16.** $-\frac{1}{15}$ **17.** $2x$ **18.** $-72x$ **19.** $2x-1$ **20.** $11x+7$ **21.** 4 **22.** -16 **23.** -50 **24.** $-\frac{3}{2}$ **25.** 5

26. $\frac{3}{2}$ **27.** 1 **28.** $\frac{1}{3}$ **29.** 12 **30.** 48 **31.** $c = P - a - b$ **32.** $y = -\frac{3}{4}x + 3$

33. $x \le 6$ ←——————●——————→
                              6

34. $x > 3$ ←————⊕————————→
                        3

35. $x > 3$ ←————⊕————————→
                      3

36. $x < 4$ ←———————⊕————→
                            4

37. $x - 5 = 12$ **38.** $x + 7 = 4$

39. $-5, -3$ **40.** $-5, -3, -1.7, 2.3, \frac{12}{7}$ **41.** 11 **42.** 25 **43.** 15 **44.** 75% **45.** 14 **46.** 12 **47.** $48°, 90°$ **48.** $65°$

49. 18 hours   **50.** 3 oz   **51.** 32.4 million copies   **52.** 2.6 million copies

Chapter 2 Test

1. $-4x + 5$ **2.** $3a - 4$ **3.** $-3y - 12$ **4.** $11x + 28$ **5.** 22 **6.** 25

7. a.

n	$(n+1)^2$
1	4
2	9
3	16
4	25

b.

n	$n^2 + 1$
1	2
2	5
3	10
4	17

8. 6 **9.** $\frac{4}{3}$ **10.** 2 **11.** $-\frac{1}{2}$ **12.** -3 **13.** 70 **14.** -3 **15.** -1

16. 5.7 **17.** $2{,}000$ **18.** 3 **19.** -8 **20.** $y = -\frac{2}{5}x + 4$ **21.** $v = \frac{h - x - 16t^2}{t}$ **22.** Rick is 20, Dave is 40

23. $\ell = 20$ in., $w = 10$ in.   **24.** 8 quarters, 15 dimes   **25.** \$800 at 7% and \$1,400 at 9%

26. $x < 1$ ←————⊕————→
                      1

27. $a < -4$ ←————————⊕————→
                              -4

28. $x \le -3$ ←————————●————→
                              -3

29. $m \ge -2$ ←———————●————————→
                              -2

30. 4.1% **31.** 3.4%

Chapter 3

Getting Ready for Chapter 3

1. 3 **2.** 9 **3.** 1 **4.** 0 **5.** $\frac{3}{2}x - 3$ **6.** $-\frac{2}{3}x + 2$ **7.** 0 **8.** 15 **9.** 4 **10.** 2 **11.** -2 **12.** 9 **13.** 1 **14.** -2

15. 1 **16.** 0 **17.** $y = \frac{7}{2}x - 7$ **18.** $y = -\frac{2}{5}x - \frac{4}{5}$ **19.** $y = -2x - 5$ **20.** $y = -\frac{2}{3}x + 1$

Problem Set 3.1

1–17.

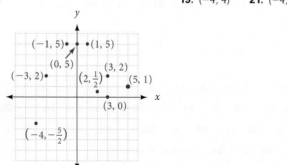

19. $(-4, 4)$ **21.** $(-4, 2)$ **23.** $(-3, 0)$ **25.** $(2, -2)$ **27.** $(-5, -5)$

29. Yes   **31.** No   **33.** Yes   **35.** No

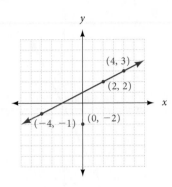

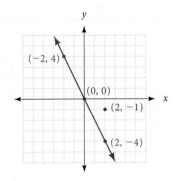

37. Yes   **39.** No   **41.** No   **43.** No

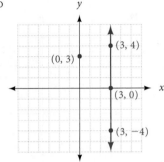

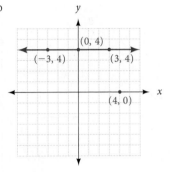

45. a. 60 watts   **b.** 20 watts   **c.** 100 watts

47. a. (5, 40), (10, 80), (20, 160), answers may vary   **b.** $320   **c.** 30 hours   **d.** No, she should make $280 for working 35 hours.

49. $A = (1, 2)$, $B = (6, 7)$ **51.** $A = (2, 2)$, $B = (2, 5)$, $C = (7, 5)$ **53. a.** -3 **b.** 6   **c.** 0   **d.** -4 **55. a.** 4   **b.** 2   **c.** -1 **d.** 9

57. $\frac{x+3}{5}$ **59.** $\frac{2-a}{7}$ **61.** $\frac{1-2y}{14}$ **63.** $\frac{x+6}{2x}$ **65.** $\frac{x}{6}$ **67.** $\frac{8+x}{2x}$

Problem Set 3.2

1. $(0, 6), (3, 0), (6, -6)$ **3.** $(0, 3), (4, 0), (-4, 6)$ **5.** $(1, 1), \left(\frac{3}{4}, 0\right), (5, 17)$ **7.** $(2, 13), (1, 6), (0, -1)$ **9.** $(-5, 4), (-5, -3), (-5, 0)$

11.

x	y
1	3
−3	−9
4	12
6	18

13.

x	y
0	0
$-\frac{1}{2}$	−2
−3	−12
3	12

15.

x	y
2	3
3	2
5	0
9	−4

17.

x	y
2	0
3	2
1	−2
−3	−10

19.

x	y
0	−1
−1	−7
−3	−19
$\frac{3}{2}$	8

21. $(0, -2)$ **23.** $(1, 5), (0, -2), (-2, -16)$ **25.** $(1, 6), (-2, -12), (0, 0)$ **27.** $(2, -2)$ **29.** $(3, 0), (3, -3)$

31. (2008, 120,000), (2009, 160,000), (2010, 200,000)   **33.** 12 inches

35. a. Yes   **b.** No, she should earn $108 for working 9 hours.   **c.** No, she should earn $84 for working 7 hours.   **d.** Yes

37. a. $375,000   **b.** At the end of six years   **c.** No, the crane will be worth $195,000 after nine years.   **d.** $600,000

39. -3 **41.** 2   **43.** 0   **45.** $y = -5x + 4$ **47.** $y = \frac{3}{2}x - 3$ **49.** 6   **51.** 3   **53.** 6   **55.** 4

Problem Set 3.3

1.
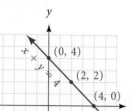

3.
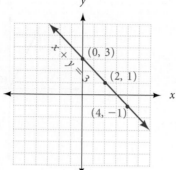

5.
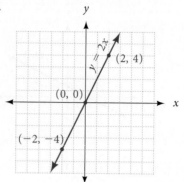

7.
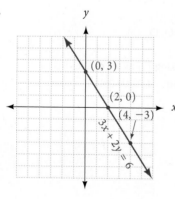

9.
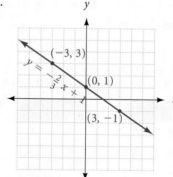

11.
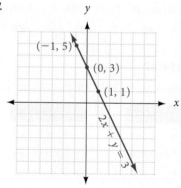

13.
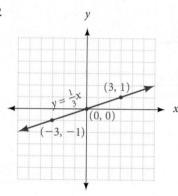

15.
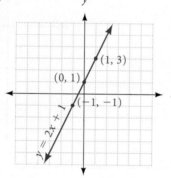

17.
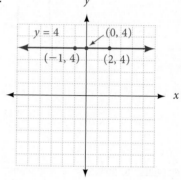

19.
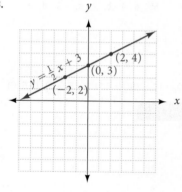

21.
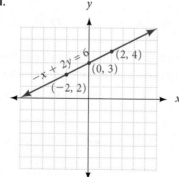

23.
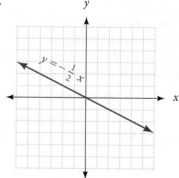

25.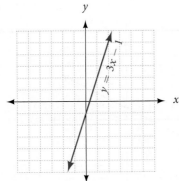

27.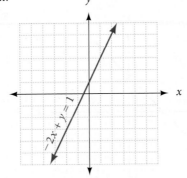

29.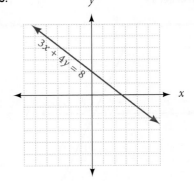

31.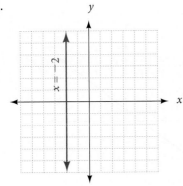

33.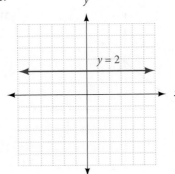

35.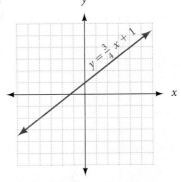

37.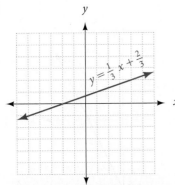

39.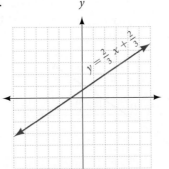

41.

Equation	H, V, and/or O
$x = 3$	V
$y = 3$	H
$y = 3x$	O
$y = 0$	O, H

43.

Equation	H, V, and/or O
$x = -\frac{3}{5}$	V
$y = -\frac{3}{5}$	H
$y = -\frac{3}{5}x$	O
$x = 0$	O, V

45. a. $\frac{5}{2}$ **b.** 5 **c.** 2 **d.** 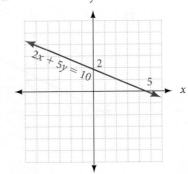 **e.** $y = -\frac{2}{5}x + 2$

47. a. Yes **b.** No **c.** Yes   **49. a.** 2 **b.** -3 **51. a.** -4 **b.** 2   **53. a.** 6 **b.** 2   **55.** $2x + 5$ **57.** $2x - 6$ **59.** $3x + \frac{15}{2}$

61. $6x + 9$ **63.** $2x + 50$ **65.** $6x - 28$ **67.** $\frac{3}{2}x + 9y$ **69.** $\frac{5}{2}x - 5y + 6$

Problem Set 3.4

1.

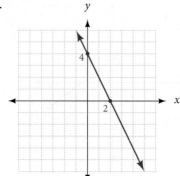

3.

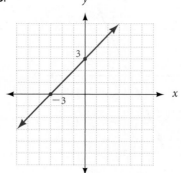

5.

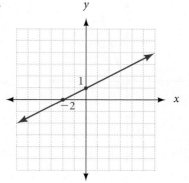

7.

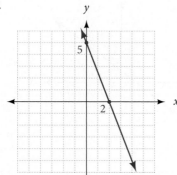

9.

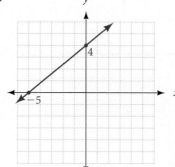

11.

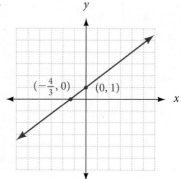

13.

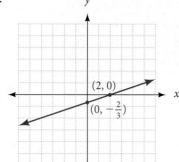

15.

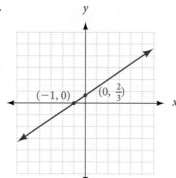

17.

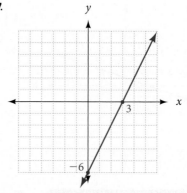

19.

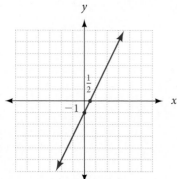

21.

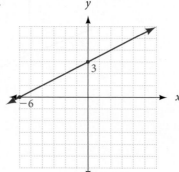

23.

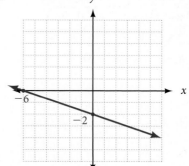

25.

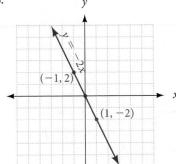

27.

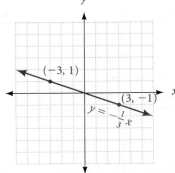

29.

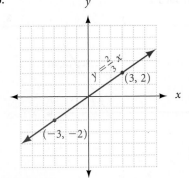

31.

Equation	x-intercept	y-intercept
$3x + 4y = 12$	4	3
$3x + 4y = 4$	$\frac{4}{3}$	1
$3x + 4y = 3$	1	$\frac{3}{4}$
$3x + 4y = 2$	$\frac{2}{3}$	$\frac{1}{2}$

33.

Equation	x-intercept	y-intercept
$x - 3y = 2$	2	$-\frac{2}{3}$
$y = \frac{1}{3}x - \frac{2}{3}$	2	$-\frac{2}{3}$
$x - 3y = 0$	0	0
$y = \frac{1}{3}x$	0	0

35. a. 0 **b.** $-\frac{3}{2}$ **c.** 1 **d.**

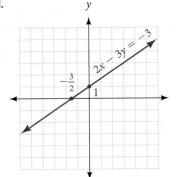

e. $y = \frac{2}{3}x + 1$ **37.**

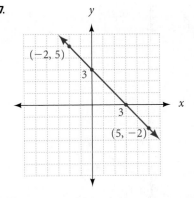

39. x-intercept $= -1$; y-intercept $= -3$ **41.**

x	y
−2	1
0	−1
−1	0
1	−2

43. x-intercept: -425; y-intercept: $\frac{17}{7}$

45. a. $10x + 12y = 480$ **b.** x-intercept $= 48$; y-intercept $= 40$ **c.**

d. 10 hours   **e.** 18 hours

47. a. $\frac{3}{2}$ **b.** $\frac{3}{2}$ **49. a.** $\frac{3}{2}$ **b.** $\frac{3}{2}$ **51.** 12   **53.** -11

55. -6 **57.** 7   **59.** -2 **61.** -1 **63.** 14   **65.** -6

Problem Set 3.5

1.
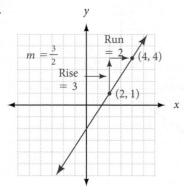

3.
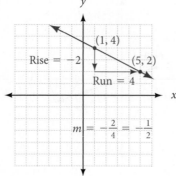

5.
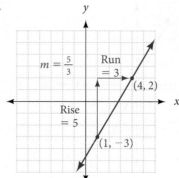

7.
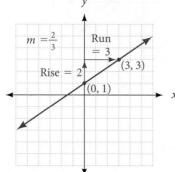

9.
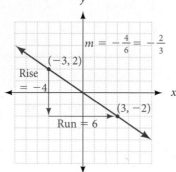

11.
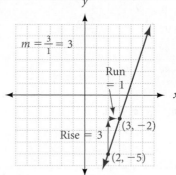

13.
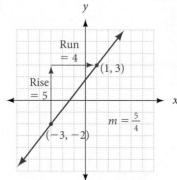

15.
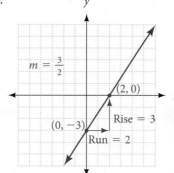

17.
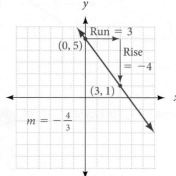

19.
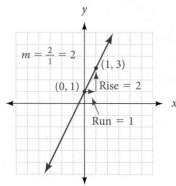

21.
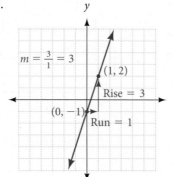

23. Slope = 3, y-intercept = 2

25. Slope = 2, y-intercept = -2 **27.**

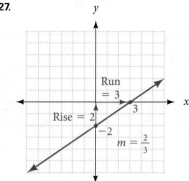

29.

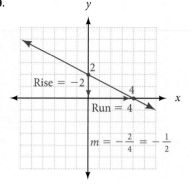

31.

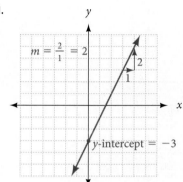

33.

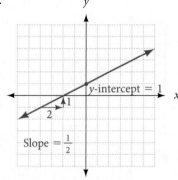

35. 6   **37.**

Equation	Slope
$x = 3$	Undefined
$y = 3$	0
$y = 3x$	3

39.

Equation	Slope
$y = -\frac{2}{3}$	0
$x = -\frac{2}{3}$	Undefined
$y = -\frac{2}{3}x$	$-\frac{2}{3}$

41. 154.23   **43.** Slopes: A, 3.3; B, 3.1; C, 5.3; D, 1.9   **45.** $y = 2x + 4$ **47.** $y = -3x + 3$ **49.** $y = \frac{4}{5}x - 4$ **51.** $y = -2x - 5$

53. $y = -\frac{2}{3}x + 1$ **55.** $y = \frac{3}{2}x + 1$ **57.** 3   **59.** 15   **61.** -6 **63.** 6   **65.** -39 **67.** 4   **69.** 10   **71.** -20

Problem Set 3.6

1.

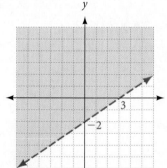

3.

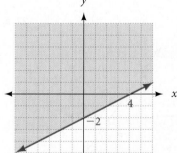

5.

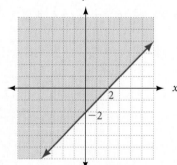

7.

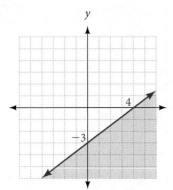

9.

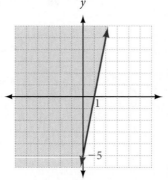

11.

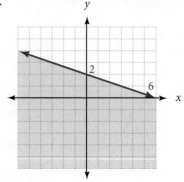

13.

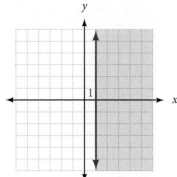

15.

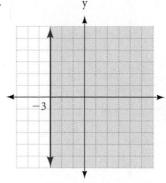

17.

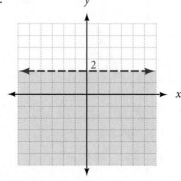

19.

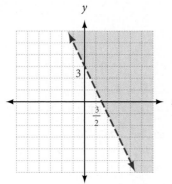

21.

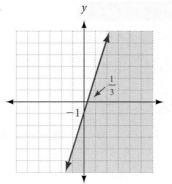

23.

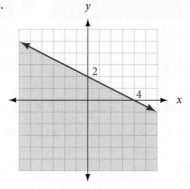

25. a. $y < \frac{8}{3}$ **b.** $y > -\frac{8}{3}$ **c.** $y = -\frac{4}{3}x + 4$ **d.**

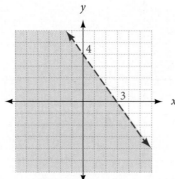

27. a. $y < \frac{2}{5}x + 2$ **b.** $y > \frac{2}{5}x + 2$

29.

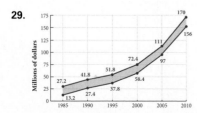

31. $y = -2x + 5$ **33.** $y = -\frac{1}{3}x + 2$

35.

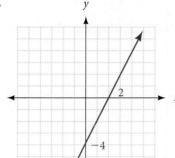

37.

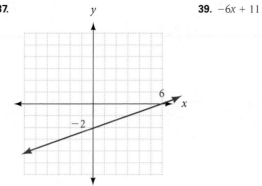

39. $-6x + 11$ **41.** -8 **43.** -4

45. $w = \dfrac{P - 2l}{2}$ **47.**

49. $y \geq \dfrac{3}{2}x - 6$ **51.** Width 2 inches, length 11 inches

Problem Set 3.7

1.

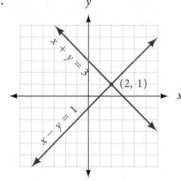

3.

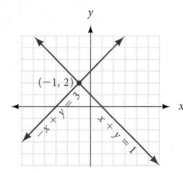

5.

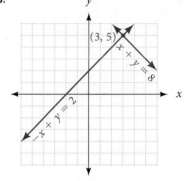

7.

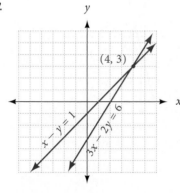

9.

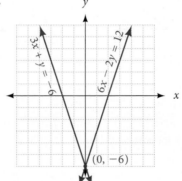

11.

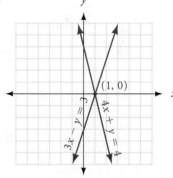

13.

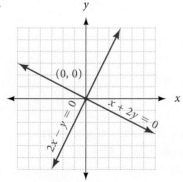

15.

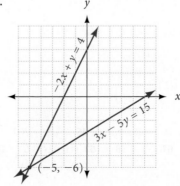

17.

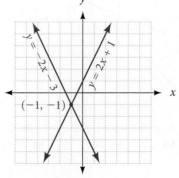

19.

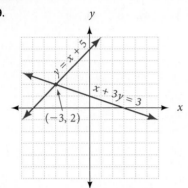

21.

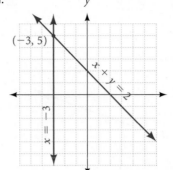

23.

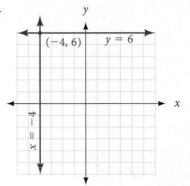

25. ∅

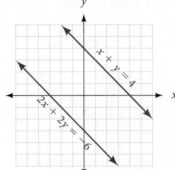

27. Any point on the line

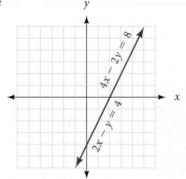

29. a. $4x - 5y$ **b.** 1 **c.** -2 **d.**

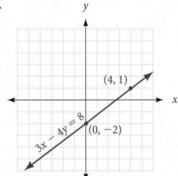

e. $(0, -2)$ **31.**

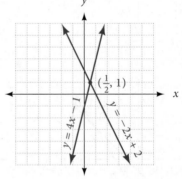

33.

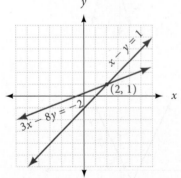

35. No, not for positive values

37. a. 25 hours **b.** Gigi's **c.** Marcy's   **39.** $2x$ **41.** $7x$ **43.** $13x$ **45.** $-12x - 20y$ **47.** $3x + 8y$ **49.** $-4x + 2y$ **51.** 1

53. 0   **55.** -5 **57.** $10x + 75$ **59.** $11x - 35$ **61.** $-x + 58$ **63.** $-0.02x + 6$ **65.** $-0.003x - 28$ **67.** $1.31x - 125$

Problem Set 3.8

1. $(2, 1)$ **3.** $(3, 7)$ **5.** $(2, -5)$ **7.** $(-1, 0)$ **9.** Lines coincide   **11.** $(4, 8)$ **13.** $\left(\frac{1}{5}, 1\right)$ **15.** $(1, 0)$ **17.** $(-1, -2)$

19. $\left(-5, \frac{3}{4}\right)$ **21.** $(-4, 5)$ **23.** $(-3, -10)$ **25.** $(3, 2)$ **27.** $\left(5, \frac{1}{3}\right)$ **29.** $\left(-2, \frac{2}{3}\right)$ **31.** $(2, 2)$ **33.** Lines are parallel; $\varnothing$

35. $(1, 1)$ **37.** Lines are parallel; $\varnothing$ **39.** $(10, 12)$ **41.** $(6, 8)$ **43.** $(7{,}000, 8{,}000)$ **45.** $(3{,}000, 8{,}000)$ **47.** $(11, 12)$

49. $(10, 12)$ **51.** 2001   **53.** 1   **55.** 2   **57.** 1   **59.** $x = 3y - 1$ **61.** 1   **63.** 5   **65.** 34.5   **67.** 33.95

69. slope $= 3$; y-int $= -3$ **71.** slope $= \frac{2}{5}$; y-int $= -5$ **73.** slope $= -1$ **75.** slope $= 2$ **77.** slope $= 1$ **79.** 11   **81.** 9

Problem Set 3.9

1. $(4, 7)$ **3.** $(3, 17)$ **5.** $\left(\frac{3}{2}, 2\right)$ **7.** $(2, 4)$ **9.** $(0, 4)$ **11.** $(-1, 3)$ **13.** $(1, 1)$ **15.** $(2, -3)$ **17.** $\left(-2, \frac{3}{5}\right)$ **19.** $(-3, 5)$

21. Lines are parallel; $\varnothing$ **23.** $(3, 1)$ **25.** $\left(\frac{1}{2}, \frac{3}{4}\right)$ **27.** $(2, 6)$ **29.** $(4, 4)$ **31.** $(5, -2)$ **33.** $(18, 10)$ **35.** Lines coincide

37. $(8, 4)$ **39.** $(6, 12)$ **41.** $(16, 32)$ **43.** $(10, 12)$ **45. a.** $\frac{25}{4}$ **b.** $y = \frac{4}{5}x - 4$ **c.** $x = y + 5$ **d.** $(5, 0)$

47. 2001   **49. a.** 560 miles **b.** Car **c.** Truck **d.** Miles ≥ 0 **51.** 47   **53.** 14   **55.** 70   **57.** 35   **59.** 5   **61.** 6,540

63. 1,760   **65.** 20   **67.** 63   **69.** 53

Chapter 3 Review

1. $(4, -6)$, $(0, 6)$, $(1, 3)$, $(2, 0)$ **2.** $(5, -2)$, $(0, -4)$, $(15, 2)$, $(10, 0)$ **3.** $(4, 2)$, $(2, -2)$, $\left(\frac{9}{2}, 3\right)$ **4.** $(2, 13)$, $\left(-\frac{3}{5}, 0\right)$, $\left(-\frac{6}{5}, -3\right)$

5. $(2, -3)$, $(-1, -3)$, $(-3, -3)$ **6.** $(6, 5)$, $(6, 0)$, $(6, -1)$ **7.** $\left(2, -\frac{3}{2}\right)$ **8.** $\left(-\frac{8}{3}, -1\right)$, $(-3, -2)$

9.–14.

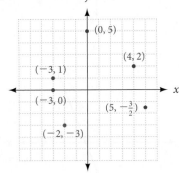

15. $(-2, 0)$, $(0, -2)$, $(1, -3)$

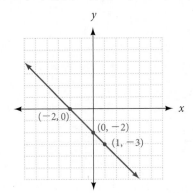

16. $(-1, -3)$, $(1, 3)$, $(0, 0)$

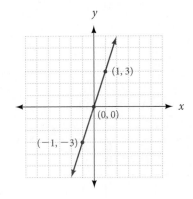

17. $(1, 1)$, $(0, -1)$, $(-1, -3)$

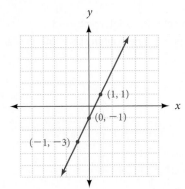

18. $(-3, 0)$, $(-3, 5)$, $(-3, -5)$

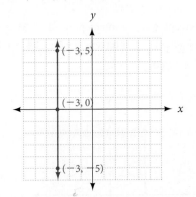

19.

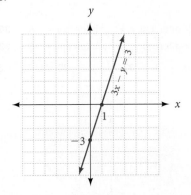

20.

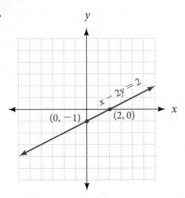

21.

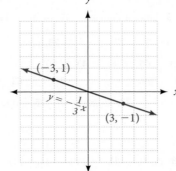

22.

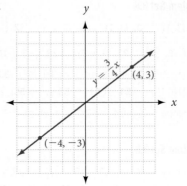

23.

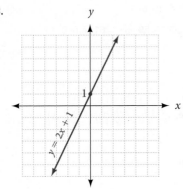

24.

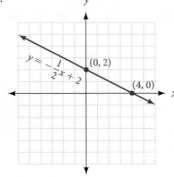

25.

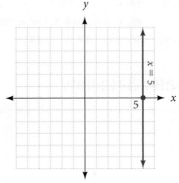

26.

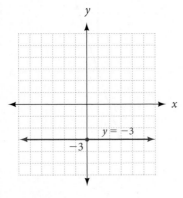

27.

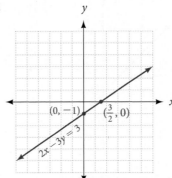

28.

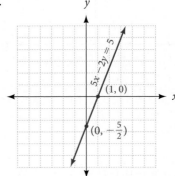

29. x-intercept = 2; y-intercept = -6 **30.** x-intercept = 12; y-intercept = -4 **31.** x-intercept = 3; y-intercept = -3

32. x-intercept = 2; y-intercept = -6 **33.** no x-intercept; y-intercept = -5 **34.** x-intercept = 4; no y-intercept   **35.** $m = 2$

36. $m = -1$ **37.** $m = 2$ **38.** $m = 2$ **39.** $x = 6$ **40.** $y = -25$

41.

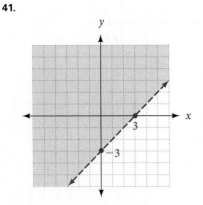

42.

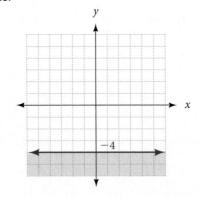

43.

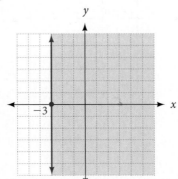

44.

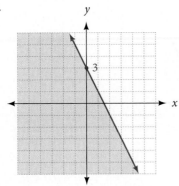

45. $(4, -2)$ **46.** $(-3, 2)$ **47.** $(3, -2)$ **48.** $(2, 0)$ **49.** $(2, 1)$ **50.** $(-1, -2)$ **51.** $(1, -3)$ **52.** $(-2, 5)$ **53.** Lines coincide
54. $(2, -2)$ **55.** $(1, 1)$ **56.** Lines are parallel; $\varnothing$ **57.** $(-2, -3)$ **58.** $(13, 5)$ **59.** $(-2, 7)$ **60.** $(-4, -2)$ **61.** $(-1, 4)$
62. $(2, -6)$ **63.** Lines are parallel; $\varnothing$ **64.** $(-3, -10)$ **65.** $(1, -2)$ **66.** Lines coincide

Chapter 3 Cumulative Review

1. 17 **2.** 32 **3.** 48 **4.** 36 **5.** Undefined   **6.** 16 **7.** $\frac{1}{35}$ **8.** $\frac{5}{4}$ **9.** $2x - 2$ **10.** $-10a + 29$ **11.** -8 **12.** 0
13. 7 **14.** -6 **15.** $x \le -9$ ⟵━━━●━━━⟶   **16.** $x \ge 12$ ⟵━━━●━━━⟶
$\qquad\qquad\qquad\qquad\qquad\qquad\qquad -9 \qquad\qquad\qquad\qquad\qquad\qquad\qquad\quad 12$

17.

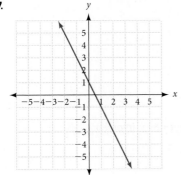

18.

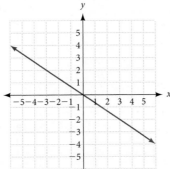

19.

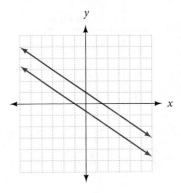

20. $(1, 0)$

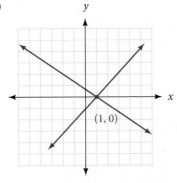

21. Lines coincide.   **22.** $(1, -3)$ **23.** $(2, 3)$ **24.** $(6, 7)$ **25.** $(9, 3)$

26. $(4, -1)$ **27.** $(4, 0)$ **28.** $(3, -1)$ **29.** $(-1, -3)$ **30.** Lines coincide.   **31.** -8 **32.** 32 **33.** -5 **34.** -17
35. $2^2 \cdot 3^2 \cdot 5$ **36.** $2^2 \cdot 3 \cdot 5^2$ **37.** 13 **38.** -21 **39.** $(-2, -4)$ **40.** $\left(2, -\frac{4}{5}\right), \left(\frac{5}{3}, -1\right)$
41. x-intercept $= 4$; y-intercept $= -3$ **42.** $m = \frac{1}{3}$ **43.** $m = \frac{5}{4}$ **44.** $m = -\frac{1}{12}$

45. $y = \frac{2}{3}x + 3$

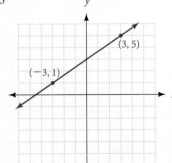

46. $y = 2x - 7$

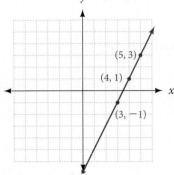

47. $m = \frac{3}{2}$ **48.** x-intercept $= 4$; y-intercept $= 2$ **49.** 7 Dimes, 8 Nickels   **50.** \$400 at 5%, \$600 at 6%   **51.** 156 people

52. 250 people

Chapter 3 Test

1-4.

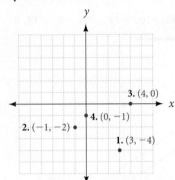

5. $(0, -2), (5, 0), (10, 2), \left(-\frac{5}{2}, -3\right)$ **6.** $(2, 5), (0, -3)$

7.

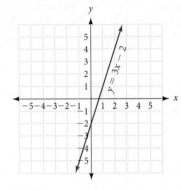

8.

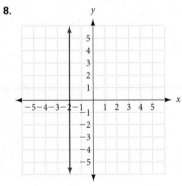

9. x-intercept $= 5$; y-intercept $= -3$

10. x-intercept $= -\frac{2}{3}$; y-intercept $= 1$ **11.** x-intercept $= 3$; y-intercept $= -2$ **12.** -2 **13.** $-\frac{13}{5}$ **14.** $-\frac{1}{4}$ **15.** 0

16. No slope   **17.**

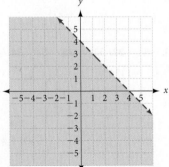

18.

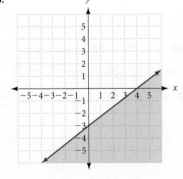

19. $(-4, 2)$ **20.** $(1, 2)$ **21.** $(3, -2)$ **22.** Lines are parallel; $\varnothing$ **23.** $(-3, -4)$ **24.** $(5, -3)$ **25.** $(2, -2)$ **26.** Lines coincide.
27. $(2, 7)$ **28.** $(4, 3)$ **29.** $(2, 9)$ **30.** $(-5, -1)$

Chapter 4

Getting Ready for Chapter 4

1. $\frac{5}{14}$ **2.** 3.2 **3.** -11 **4.** -7 **5.** -14 **6.** 14 **7.** 16 **8.** 26 **9.** $-4x^2 + 2x + 6$ **10.** $9x^2$ **11.** $-12x$ **12.** $6x^2$
13. $-13x$ **14.** $-x$ **15.** $10x^2$ **16.** -4 **17.** $-4x^2 - 20x$ **18.** $-4x^2 - 14x$ **19.** $2x^3 - 10x^2$ **20.** $10x^2 - 50x$

Problem Set 4.1

1. Base 4; exponent 2; 16   **3.** Base 0.3; exponent 2; 0.09   **5.** Base 4; exponent 3; 64   **7.** Base -5; exponent 2; 25
9. Base 2; exponent 3; -8 **11.** Base 3; exponent 4; 81   **13.** Base $\frac{2}{3}$; exponent 2; $\frac{4}{9}$ **15.** Base $\frac{1}{2}$; exponent 4; $\frac{1}{16}$

17. a.

Number x	Square x^2
1	1
2	4
3	9
4	16
5	25
6	36
7	49

b. Either *larger* or *greater* will work.
19. x^9 **21.** y^{30} **23.** 2^{12} **25.** x^{28} **27.** x^{10} **29.** 5^{12} **31.** y^9 **33.** 2^{50} **35.** a^{3x}
37. b^{xy} **39.** $16x^2$ **41.** $32y^5$ **43.** $81x^4$ **45.** $0.25a^2b^2$ **47.** $64x^3y^3z^3$ **49.** $8x^{12}$
51. $16a^6$ **53.** x^{14} **55.** a^{11} **57.** $128x^7$ **59.** $432x^{10}$ **61.** $16x^4y^6$ **63.** $\frac{8}{27}a^{12}b^{15}$
65. x^2 **67.** $4x$ **69.** 2 **71.** $4x$ **73. a.** 32 **b.** 64 **c.** 32 **d.** 64

75.

Number x	Square x^2
-3	9
-2	4
-1	1
0	0
1	1
2	4
3	9

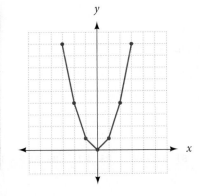

77.

Number x	Square x^2
-2.5	6.25
-1.5	2.25
-0.5	0.25
0	0
0.5	0.25
1.5	2.25
2.5	6.25

79. 4.32×10^4 **81.** 5.7×10^2 **83.** 2.38×10^5 **85.** 2,490   **87.** 352   **89.** 28,000

91. 275,625 sq ft   **93.** 27 inches³   **95.** 15.6 inches³   **97.** 36 inches³   **99.** 6.5×10^8 seconds   **101.** \$740,000
103. \$180,000   **105.** 219 inches³   **107.** 182 inches³   **109.** -3 **111.** 11   **113.** -5 **115.** 5   **117.** 2   **119.** 6   **121.** 4
123. 3   **125.** 2^7 **127.** $2 \cdot 5^3$ **129.** $2^4 \cdot 3^2 \cdot 5$ **131.** $2^2 \cdot 5 \cdot 41$ **133.** $2^3 \cdot 3^3$ **135.** $2^3 \cdot 3^3 \cdot 5^3$ **137.** 5^6 **139.** $2^6 \cdot 3^3$

Problem Set 4.2

1. $\frac{1}{9}$ **3.** $\frac{1}{36}$ **5.** $\frac{1}{64}$ **7.** $\frac{1}{125}$ **9.** $\frac{2}{x^3}$ **11.** $\frac{1}{8x^3}$ **13.** $\frac{1}{25y^2}$ **15.** $\frac{1}{100}$

17.

Number x	Square x^2	Power of 2 2^x
-3	9	$\frac{1}{8}$
-2	4	$\frac{1}{4}$
-1	1	$\frac{1}{2}$
0	0	1
1	1	2
2	4	4
3	9	8

19. $\frac{1}{25}$ **21.** x^6 **23.** 64   **25.** $8x^3$ **27.** 6^{10} **29.** $\frac{1}{6^{10}}$ **31.** $\frac{1}{2^8}$ **33.** 2^8

35. $27x^3$ **37.** $81x^4y^4$ **39.** 1   **41.** $2a^2b$ **43.** $\frac{1}{49y^6}$ **45.** $\frac{1}{x^8}$ **47.** $\frac{1}{y^3}$ **49.** x^2 **51.** a^6 **53.** $\frac{1}{y^9}$ **55.** y^{40}

57. $\frac{1}{x}$ **59.** x^9 **61.** a^{16} **63.** $\frac{1}{a^4}$ **65. a.** 32 **b.** 32 **c.** $\frac{1}{32}$ **d.** $\frac{1}{32}$ **67. a.** $\frac{1}{5}$ **b.** $\frac{1}{8}$ **c.** $\frac{1}{x}$ **d.** $\frac{1}{x^2}$

69.

Number x	Power of 2 2^x
-3	$\frac{1}{8}$
-2	$\frac{1}{4}$
-1	$\frac{1}{2}$
0	1
1	2
2	4
3	8

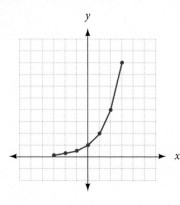

71. 4.8×10^{-3} **73.** 2.5×10^1 **75.** 9×10^{-6}

77.

Expanded Form	Scientific Notation $n \times 10^r$
0.000357	3.57×10^{-4}
0.00357	3.57×10^{-3}
0.0357	3.57×10^{-2}
0.357	3.57×10^{-1}
3.57	3.57×10^{0}
35.7	3.57×10^{1}
357	3.57×10^{2}
3,570	3.57×10^{3}
35,700	3.57×10^{4}

79. 0.00423   **81.** 0.00008   **83.** 4.2   **85.** $\$5.2 \times 10^7$ **87.** 0.002

89. 2.5×10^4 **91.** 2.35×10^5 **93.** 8.2×10^{-4} **95.** 100 inches^2; 400 inches^2; 4   **97.** x^2; $4x^2$; 4   **99.** 216 inches^3; 1,728 inches^3; 8

101. x^3; $8x^3$; 8   **103.** 13.5   **105.** 8   **107.** 26.52   **109.** 12   **111.** x^8 **113.** x **115.** $\frac{1}{y^2}$ **117.** 340   **119.** $7x$ **121.** $2a$

123. $10y$

Problem Set 4.3

1. $12x^7$ **3.** $-16y^{11}$ **5.** $32x^2$ **7.** $200a^6$ **9.** $-24a^3b^3$ **11.** $24x^6y^8$ **13.** $3x$ **15.** $\dfrac{6}{y^3}$ **17.** $\dfrac{1}{2a}$ **19.** $-\dfrac{3a}{b^2}$ **21.** $\dfrac{x^2}{9z^2}$

23.

a	b	ab	$\dfrac{a}{b}$	$\dfrac{b}{a}$
10	$5x$	$50x$	$\dfrac{2}{x}$	$\dfrac{x}{2}$
$20x^3$	$6x^2$	$120x^5$	$\dfrac{10x}{3}$	$\dfrac{3}{10x}$
$25x^5$	$5x^4$	$125x^9$	$5x$	$\dfrac{1}{5x}$
$3x^{-2}$	$3x^2$	9	$\dfrac{1}{x^4}$	x^4
$-2y^4$	$8y^7$	$-16y^{11}$	$-\dfrac{1}{4y^3}$	$-4y^3$

25. 6×10^8 **27.** 1.75×10^{-1} **29.** 1.21×10^{-6} **31.** 4.2×10^3

33. 3×10^{10} **35.** 5×10^{-3} **37.** $8x^2$ **39.** $-11x^5$ **41.** 0 **43.** $4x^3$ **45.** $31ab^2$ **47.**

a	b	ab	$a + b$
$5x$	$3x$	$15x^2$	$8x$
$4x^2$	$2x^2$	$8x^4$	$6x^2$
$3x^3$	$6x^3$	$18x^6$	$9x^3$
$2x^4$	$-3x^4$	$-6x^8$	$-x^4$
x^5	$7x^5$	$7x^{10}$	$8x^5$

49. $4x^3$ **51.** $\dfrac{1}{b^2}$ **53.** $\dfrac{6y^{10}}{x^4}$ **55.** 2×10^6 **57.** 1×10^1 **59.** 4.2×10^{-6} **61.** $9x^3$ **63.** $-20a^2$
65. $6x^5y^2$ **67.** $x^2y + x$ **69.** $x + y$ **71.** $x^2 - 4$ **73.** $x^2 - x - 6$ **75.** $x^2 - 5x$ **77.** $x^2 - 8x$
79. a. 2 **b.** $-3b$ **c.** $5b^2$ **81. a.** $3xy$ **b.** $2y$ **c.** $-\dfrac{y^2}{2}$ **83.** -5 **85.** 6 **87.** 76 **89.** $6x^2$
91. $2x$ **93.** $-2x - 9$ **95.** 11 **97.** -8 **99.** 9 **101.** 0

103.

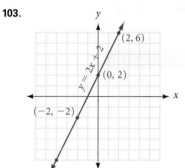

105.

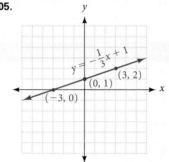

Problem Set 4.4

1. Trinomial, 3 **3.** Trinomial, 3 **5.** Binomial, 1 **7.** Binomial, 2 **9.** Monomial, 2 **11.** Monomial, 0 **13.** $5x^2 + 5x + 9$
15. $5a^2 - 9a + 7$ **17.** $x^2 + 6x + 8$ **19.** $6x^2 - 13x + 5$ **21.** $x^2 - 9$ **23.** $3y^2 - 11y + 10$ **25.** $6x^3 + 5x^2 - 4x + 3$
27. $2x^2 - x + 1$ **29.** $2a^2 - 2a - 2$ **31.** $-\dfrac{1}{9}x^3 - \dfrac{2}{3}x^2 - \dfrac{5}{2}x + \dfrac{7}{4}$ **33.** $-4y^2 + 15y - 22$ **35.** $-2x$ **37.** $4x$ **39.** $x^3 - 27$
41. $x^3 + 6x^2 + 12x + 8$ **43.** $4x - 8$ **45.** $x^2 - 33x + 63$ **47.** $8y^2 + 4y + 26$ **49.** $75x^2 - 150x - 75$ **51.** $12x + 2$ **53.** 4
55. a. 0 **b.** 0 **57. a.** $2,200$ **b.** $-1,800$ **59.** 1.44×10^{12} in^2 **61.** 18π inches^3
63. First year $53{,}160x + 63{,}935y$; Second year $53{,}160(2x) + 63{,}935y$; $106{,}320x + 63{,}935y$ **65.** 5 **67.** -6 **69.** $-20x^2$ **71.** $-21x$
73. $2x$ **75.** $6x - 18$ **77.** $-15x^2$ **79.** $6x^3$ **81.** $6x^4$

Problem Set 4.5

1. $6x^2 + 2x$ **3.** $6x^4 - 4x^3 + 2x^2$ **5.** $2a^3b - 2a^2b^2 + 2ab$ **7.** $3y^4 + 9y^3 + 12y^2$ **9.** $8x^5y^2 + 12x^4y^3 + 32x^2y^4$ **11.** $x^2 + 7x + 12$
13. $x^2 + 7x + 6$ **15.** $x^2 + 2x + \dfrac{3}{4}$ **17.** $a^2 + 2a - 15$ **19.** $xy + bx - ay - ab$ **21.** $x^2 - 36$ **23.** $y^2 - \dfrac{25}{36}$ **25.** $2x^2 - 11x + 12$
27. $2a^2 + 3a - 2$ **29.** $6x^2 - 19x + 10$ **31.** $2ax + 8x + 3a + 12$ **33.** $25x^2 - 16$ **35.** $2x^2 + \dfrac{5}{2}x - \dfrac{3}{4}$ **37.** $3 - 10a + 8a^2$

39.

	x	3
x	x^2	$3x$
2	$2x$	6

$(x + 2)(x + 3) = x^2 + 2x + 3x + 6$
$= x^2 + 5x + 6$

41.

	x	x	2
x	x^2	x^2	$2x$
1	x	x	2

$(x + 1)(2x + 2) = 2x^2 + 4x + 2$

43. $a^3 - 6a^2 + 11a - 6$ **45.** $x^3 + 8$ **47.** $2x^3 + 17x^2 + 26x + 9$ **49.** $5x^4 - 13x^3 + 20x^2 + 7x + 5$ **51.** $2x^4 + x^2 - 15$
53. $6a^6 + 15a^4 + 4a^2 + 10$ **55.** $x^3 + 12x^2 + 47x + 60$ **57.** $x^2 - 5x + 8$ **59.** $8x^2 - 6x - 5$ **61.** $x^2 - x - 30$ **63.** $x^2 + 4x - 6$

65. $x^2 + 13x$ **67.** $x^2 + 2x - 3$ **69.** $a^2 - 3a + 6$ **71. a.** $x^2 - 1$ **b.** $x^2 + 2x + 1$ **c.** $x^3 + 3x^2 + 3x + 1$ **d.** $x^4 + 4x^3 + 6x^2 + 4x + 1$

73. a. $x^2 - 1$ **b.** $x^2 - x - 2$ **c.** $x^2 - 2x - 3$ **d.** $x^2 - 3x - 4$ **75.** 0 **77.** -5 **79.** $-2, 3$ **81.** Never 0

83. $x(6{,}200 - 25x) = 6{,}200x - 25x^2$; $18,375 **85.** $A = x(2x + 5) = 2x^2 + 5x$ **87.** $A = x(x + 1) = x^2 + x$

89. $R = (1{,}200 - 100p)p = 1{,}200p - 100p^2$ **91.** 169 **93.** $-10x$ **95.** 0 **97.** 0 **99.** $-12x + 16$ **101.** $x^2 + x - 2$

103. $x^2 + 6x + 9$ **105.** **107.** $(1, -1)$ **109.** $\left(-\frac{1}{2}, -\frac{1}{2}\right)$ **111.** 6 dimes, 5 quarters

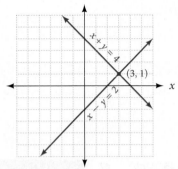

Problem Set 4.6

1. $x^2 - 4x + 4$ **3.** $a^2 + 6a + 9$ **5.** $x^2 - 10x + 25$ **7.** $a^2 - a + \frac{1}{4}$ **9.** $x^2 + 20x + 100$ **11.** $a^2 + 1.6a + 0.64$

13. $4x^2 - 4x + 1$ **15.** $16a^2 + 40a + 25$ **17.** $9x^2 - 12x + 4$ **19.** $9a^2 + 30ab + 25b^2$ **21.** $16x^2 - 40xy + 25y^2$

23. $49m^2 + 28mn + 4n^2$ **25.** $36x^2 - 120xy + 100y^2$ **27.** $x^4 + 10x^2 + 25$ **29.** $a^4 + 2a^2 + 1$ **31.** $y^2 + 3y + \frac{9}{4}$

33. $a^2 + a + \frac{1}{4}$ **35.** $x^2 + \frac{3}{2}x + \frac{9}{16}$ **37.** $t^2 + \frac{2}{5}t + \frac{1}{25}$

39.

x	(x + 3)²	x² + 9	x² + 6x + 9
1	16	10	16
2	25	13	25
3	36	18	36
4	49	25	49

41.

a	b	(a + b)²	a² + b²	a² + ab + b²	a² + 2ab + b²
1	1	4	2	3	4
3	5	64	34	49	64
3	4	49	25	37	49
4	5	81	41	61	81

43. $a^2 - 25$ **45.** $y^2 - 1$ **47.** $81 - x^2$ **49.** $4x^2 - 25$ **51.** $16x^2 - \frac{1}{9}$ **53.** $4a^2 - 49$ **55.** $36 - 49x^2$ **57.** $x^4 - 9$ **59.** $a^4 - 16$

61. $25y^8 - 64$ **63.** $2x^2 - 34$ **65.** $-12x^2 + 20x + 8$ **67.** $a^2 + 4a + 6$ **69.** $8x^3 + 36x^2 + 54x + 27$ **71. a.** 11 **b.** 1 **c.** 11

73. a. 1 **b.** 13 **c.** 1 **75.** Both equal 25. **77.** $x^2 + (x + 1)^2 = 2x^2 + 2x + 1$ **79.** $x^2 + (x + 1)^2 + (x + 2)^2 = 3x^2 + 6x + 5$

81. $a^2 + ab + ba + b^2 = a^2 + 2ab + b^2$ **83.** $2x^2$ **85.** x^2 **87.** $3x$ **89.** $3xy$ **91.** $5x$ **93.** $\frac{a^4}{2b^2}$ **95.** $\frac{12}{20} = \frac{3}{5}$

97.

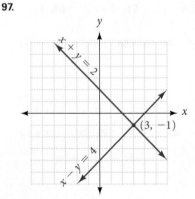

99.

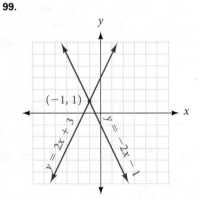

Problem Set 4.7

1. $x - 2$ **3.** $3 - 2x^2$ **5.** $5xy - 2y$ **7.** $7x^4 - 6x^3 + 5x^2$ **9.** $10x^4 - 5x^2 + 1$ **11.** $-4a + 2$ **13.** $-8a^4 - 12a^3$ **15.** $-4b - 5a$

17. $-6a^2b + 3ab^2 - 7b^3$ **19.** $-\frac{a}{2} - b - \frac{b^2}{2a}$ **21.** $3x + 4y$ **23.** $-y + 3$ **25.** $x^2 + 8x - 9$ **27.** $y^2 - 3y + 1$ **29.** $5y - 4$

31. $xy - x^2y^2$ **33.** $-1 + xy$ **35.** $-a + 1$ **37.** $x^2 - 3xy + y^2$ **39.** $2 - 3b + 5b^2$ **41.** $-2xy + 1$ **43.** $xy - \frac{1}{2}$ **45.** $\frac{1}{4x} - \frac{1}{2a} + \frac{3}{4}$

47. $\frac{4x^2}{3} + \frac{2}{3x} + \frac{1}{x^2}$ **49.** $3a^{3m} - 9a^m$ **51.** $2x^{4m} - 5x^{2m} + 7$ **53.** $3x^2 - x + 6$ **55.** 4   **57.** $x + 5$ **59. a.** 7 **b.** 7 **c.** 23

61. a. 19 **b.** 34 **c.** 19   **63. a.** 2 **b.** 2   **65.** Both equal 7.   **67.** $\frac{3(10)+8}{2} = 19$; $3(10) + 4 = 34$ **69.** $146\frac{20}{27}$ **71.** $2x + 5$

73. $x^2 - 3x$ **75.** $2x^3 - 10x^2$ **77.** $-2x$ **79.** 2   **81. a.** 2,160 million **b.** 533.5 million   **83.** $(7, -1)$ **85.** $(2, 3)$

87. $(1, 1)$ **89.** Lines coincide.

Problem Set 4.8

1. $x - 2$ **3.** $a + 4$ **5.** $x - 3$ **7.** $x + 3$ **9.** $a - 5$ **11.** $x + 2 + \frac{2}{x+3}$ **13.** $a - 2 + \frac{12}{a+5}$ **15.** $x + 4 + \frac{9}{x-2}$

17. $x + 4 + \frac{-10}{x+1}$ **19.** $a + 1 + \frac{-1}{a+2}$ **21.** $x - 3 + \frac{17}{2x+4}$ **23.** $3a - 2 + \frac{7}{2a+3}$ **25.** $2a^2 - a - 3$ **27.** $x^2 - x + 5$

29. $x^2 + x + 1$ **31.** $x^2 + 2x + 4$ **33. a.** 19 **b.** 19 **c.** 5   **35. a.** 7 **b.** $\frac{25}{7}$ **c.** $\frac{25}{7}$ **37.** 404 yards   **39.** 5, 20

41. \$800 at 8%, \$400 at 9%   **43.** Eight \$5 bills, twelve \$10 bills

Chapter 4 Review

1. -1 **2.** -64 **3.** $\frac{9}{49}$ **4.** y^{12} **5.** x^{30} **6.** x^{35} **7.** 2^{24} **8.** $27y^3$ **9.** $-8x^3y^3z^3$ **10.** $\frac{1}{49}$ **11.** $\frac{4}{x^5}$ **12.** $\frac{1}{27y^3}$

13. a^6 **14.** $\frac{1}{x^4}$ **15.** x^{15} **16.** $\frac{1}{x^5}$ **17.** 1   **18.** -3 **19.** $9x^6y^4$ **20.** $64a^{22}b^{20}$ **21.** $-\frac{1}{27x^3y^6}$ **22.** b **23.** $\frac{1}{x^{35}}$ **24.** $5x^7$

25. $\frac{6y^7}{x^5}$ **26.** $4a^6$ **27.** $2x^5y^2$ **28.** 6.4×10^7 **29.** 2.3×10^8 **30.** 8×10^3 **31.** $8a^2 - 12a - 3$ **32.** $-12x^2 + 5x - 14$

33. $-4x^2 - 6x$ **34.** 14   **35.** $12x^2 - 21x$ **36.** $24x^5y^2 - 40x^4y^3 + 32x^3y^4$ **37.** $a^3 + 6a^2 + a - 4$ **38.** $x^3 + 125$

39. $6x^2 - 29x + 35$ **40.** $25y^2 - \frac{1}{25}$ **41.** $a^4 - 9$ **42.** $a^2 - 10a + 25$ **43.** $9x^2 + 24x + 16$ **44.** $y^4 + 6y^2 + 9$ **45.** $-2b - 4a$

46. $-5x^4y^3 + 4x^2y^2 + 2x$ **47.** $x + 9$ **48.** $2x + 5$ **49.** $x^2 - 4x + 16$ **50.** $x - 3 + \frac{16}{3x+2}$ **51.** $x^2 - 4x + 5 + \frac{5}{2x+1}$

Chapter 4 Cumulative Review

1. $\frac{3}{4}$ **2.** -9 **3.** 24   **4.** 105   **5.** -5 **6.** -12 **7.** 1   **8.** 63   **9.** $-3x + y$ **10.** $9a + 15$ **11.** y^{16} **12.** $16y^4$

13. $108a^{18}b^{29}$ **14.** $\frac{1152y^{13}}{x^3}$ **15.** 2.8×10^{-2} **16.** 5×10^{-5} **17.** $6a^2b$ **18.** $-13x + 10$ **19.** $25x^2 - 10x + 1$ **20.** $x^3 - 1$

21. 4   **22.** $\frac{5}{2}$ **23.** 4   **24.** -2 **25.** $a < -24$ ⟵————————⊕——————
$\qquad\qquad\qquad\qquad\qquad\qquad\qquad\quad -24$

26. $t \geq -\frac{32}{9}$ ⟵————●——————————⟶
$\qquad\qquad -\frac{32}{9}$ **27.** $x < 1$ ⟵————————⊕——————
$\qquad\qquad\qquad\qquad\qquad\qquad 1$

28. $x \geq -2$ ⟵————●——————————⟶
$\qquad\quad -2$ **29.** $2x + 7 + \frac{11}{2x-3}$ **30.** $3 - \frac{2}{x^3} + \frac{4}{x^4}$

31.

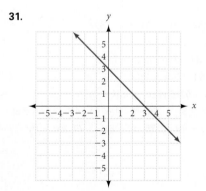

32.

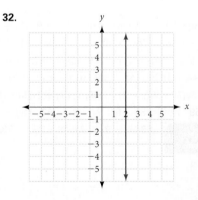

33. Lines are parallel; $\varnothing$ **34.** $(0, 2)$

35. $(3, 1)$ **36.** Lines are parallel; $\varnothing$ **37.** $(-2, 1)$ **38.** $(3, 2)$ **39.** $(4, 5)$ **40.** $(5, 1)$ **41.** 21   **42.** 3   **43.** $h = \frac{2A}{b}$

44. $y = -2x + 1$ **45.** 0   **46.** $-\sqrt{2}, \pi$ **47.** Additive inverse property   **48.** Commutative property of addition

49. x-intercept $= -2$; y-intercept $= 4$ **50.** $m = \frac{2}{5}$ **51.** \$10,775 more   **52.** \$21,275 less

Chapter 4 Test

1. 81   **2.** $\frac{9}{16}$ **3.** $72x^{18}$ **4.** $\frac{1}{9}$ **5.** 1   **6.** a^2 **7.** x **8.** 2.78×10^{-2} **9.** 243,000   **10.** $\frac{y^2}{2x^4}$ **11.** $3a$ **12.** $10x^5$

13. 9×10^9 **14.** $8x^2 + 2x + 2$ **15.** $3x^2 + 4x + 6$ **16.** $3x - 4$ **17.** 10   **18.** $6a^4 - 10a^3 + 8a^2$ **19.** $x^2 + \frac{5}{6}x + \frac{1}{6}$

20. $8x^2 + 2x - 15$ **21.** $x^3 - 27$ **22.** $x^2 + 10x + 25$ **23.** $9a^2 - 12ab + 4b^2$ **24.** $9x^2 - 16y^2$ **25.** $a^4 - 9$ **26.** $2x^2 + 3x - 1$

27. $4x + 3 + \frac{4}{2x - 3}$ **28.** $3x^2 + 9x + 25 + \frac{76}{x - 3}$ **29.** 15.625cm^3 **30.** $V = w^3$ **31.** 6.37×10^7 **32.** 5.11×10^7

33. 5×10^{-3} **34.** 2×10^{-3}

Chapter 5

Getting Ready for Chapter 5

1. $y^2 - 16y + 64$ **2.** $x^2 + 7x + 12$ **3.** $x^2 - 4x - 12$ **4.** $4x^2 + 5x + 15$ **5.** $x^2 - 16$ **6.** $x^2 - 36$ **7.** $x^2 - 64$ **8.** $x^4 - 16$

9. $x^3 + 2x^2 + 4x$ **10.** $8x^3 - 20x^2 + 50x$ **11.** $-2x^2 - 4x - 8$ **12.** $20x^2 - 50x + 125$ **13.** $x^3 - 8$ **14.** $8x^3 + 125$

15. $3x^4 - 27x^2$ **16.** $2x^5 + 20x^4 + 50x^3$ **17.** $y^5 + 36y^3$ **18.** $6a^2 - 11a + 4$ **19.** $6x^3 - 12x^2 - 48x$ **20.** $2a^5b + 6a^4b + 2a^3b$

Problem Set 5.1

1. $5(3x + 5)$ **3.** $3(2a + 3)$ **5.** $4(x - 2y)$ **7.** $3(x^2 - 2x - 3)$ **9.** $3(a^2 - a - 20)$ **11.** $4(6y^2 - 13y + 6)$ **13.** $x^2(9 - 8x)$

15. $13a^2(1 - 2a)$ **17.** $7xy(3x - 4y)$ **19.** $11ab^2(2a - 1)$ **21.** $7x(x^2 + 3x - 4)$ **23.** $11(11y^4 - x^4)$ **25.** $25x^2(4x^2 - 2x + 1)$

27. $8(a^2 + 2b^2 + 4c^2)$ **29.** $4ab(a - 4b + 8ab)$ **31.** $11a^2b^2(11a - 2b + 3ab)$ **33.** $12x^2y^3(1 - 6x^3 - 3x^2y)$ **35.** $(x + 3)(y + 5)$

37. $(x + 2)(y + 6)$ **39.** $(a - 3)(b + 7)$ **41.** $(a - b)(x + y)$ **43.** $(2x - 5)(a + 3)$ **45.** $(b - 2)(3x - 4)$ **47.** $(x + 2)(x + a)$

49. $(x - b)(x - a)$ **51.** $(x + y)(a + b + c)$ **53.** $(3x + 2)(2x + 3)$ **55.** $(10x - 1)(2x + 5)$ **57.** $(4x + 5)(5x + 1)$ **59.** $(x + 2)(x^2 + 3)$

61. $(3x - 2)(2x^2 + 5)$ **63.** 6   **65.** $3(4x^2 + 2x + 1)$ **67.** $x^2 - 5x - 14$ **69.** $x^2 - x - 6$ **71.** $x^3 + 27$ **73.** $2x^3 + 9x^2 - 2x - 3$

75. $18x^7 - 12x^6 + 6x^5$ **77.** $x^2 + x + \frac{2}{9}$ **79.** $12x^2 - 10xy - 12y^2$ **81.** $81a^2 - 1$ **83.** $x^2 - 18x + 81$ **85.** $x^3 + 8$

87. $y^2 - 16y + 64$ **89.** $x^2 - 4x - 12$ **91.** $2y^2 - 7y + 12$ **93.** $2x^2 + 7x - 11$ **95.** $3x + 8$ **97.** $3x^2 + 4x - 5$

Problem Set 5.2

1. $(x + 3)(x + 4)$ **3.** $(x + 1)(x + 2)$ **5.** $(a + 3)(a + 7)$ **7.** $(x - 2)(x - 5)$ **9.** $(y - 3)(y - 7)$ **11.** $(x - 4)(x + 3)$

13. $(y + 4)(y - 3)$ **15.** $(x + 7)(x - 2)$ **17.** $(r - 9)(r + 1)$ **19.** $(x - 6)(x + 5)$ **21.** $(a + 7)(a + 8)$ **23.** $(y + 6)(y - 7)$

25. $(x + 6)(x + 7)$ **27.** $2(x + 1)(x + 2)$ **29.** $3(a + 4)(a - 5)$ **31.** $100(x - 2)(x - 3)$ **33.** $100(p - 5)(p - 8)$ **35.** $x^2(x + 3)(x - 4)$

37. $2r(r + 5)(r - 3)$ **39.** $2y^2(y + 1)(y - 4)$ **41.** $x^3(x + 2)^2$ **43.** $3y^2(y + 1)(y - 5)$ **45.** $4x^2(x - 4)(x - 9)$

47. $(x + 2y)(x + 3y)$ **49.** $(x - 4y)(x - 5y)$ **51.** $(a + 4b)(a - 2b)$ **53.** $(a - 5b)^2$ **55.** $(a + 5b)^2$ **57.** $(x - 6a)(x + 8a)$

59. $(x + 4b)(x - 9b)$ **61.** $(x^2 - 3)(x^2 - 2)$ **63.** $(x - 100)(x + 20)$ **65.** $\left(x - \frac{1}{2}\right)\left(x - \frac{1}{2}\right)$ **67.** $(x + 0.2)(x + 0.4)$ **69.** $x + 16$

71. $4x^2 - x - 3$ **73.** $6a^2 + 13a + 2$ **75.** $6a^2 + 7a + 2$ **77.** $6a^2 + 8a + 2$ **79.** $\frac{4}{25}$ **81.** $72a^{12}$ **83.** $\frac{1}{16x^2}$ **85.** $\frac{a^3}{6b^2}$

87. $\frac{1}{3x^7y^2}$ **89.** $-21x^4y^5$ **91.** $-20ab^3$ **93.** $-27a^5b^8$

Problem Set 5.3

1. $(2x + 1)(x + 3)$ **3.** $(2a - 3)(a + 1)$ **5.** $(3x + 5)(x - 1)$ **7.** $(3y + 1)(y - 5)$ **9.** $(2x + 3)(3x + 2)$ **11.** $(2x - 3y)^2$

13. $(4y + 1)(y - 3)$ **15.** $(4x - 5)(5x - 4)$ **17.** $(10a - b)(2a + 5b)$ **19.** $(4x - 5)(5x + 1)$ **21.** $(6m - 1)(2m + 3)$

23. $(4x + 5)(5x + 3)$ **25.** $(3a - 4b)(4a - 3b)$ **27.** $(3x - 7y)(x + 2y)$ **29.** $(2x + 5)(7x - 3)$ **31.** $(3x - 5)(2x - 11)$

33. $(5t - 19)(3t - 2)$ **35.** $2(2x + 3)(x - 1)$ **37.** $2(4a - 3)(3a - 4)$ **39.** $x(5x - 4)(2x - 3)$ **41.** $x^2(3x + 2)(2x - 5)$

43. $2a(5a + 2)(a - 1)$ **45.** $3x(5x + 1)(x - 7)$ **47.** $5y(7y + 2)(y - 2)$ **49.** $a^2(5a + 1)(3a - 1)$ **51.** $2y(2x - y)(3x - 7y)$

53. Both equal 25.   **55.** $4x^2 - 9$ **57.** $x^4 - 81$ **59.** $12x - 35$ **61.** $9x + 8$ **63.** $16(t - 1)(t - 3)$ **65.** $h = 4(679 - 4t^2)$, 2,140 feet

67. a. $h = 2(4 - t)(1 + 8t)$ **b.**

Time t (seconds)	Height h (feet)
0	8
1	54
2	68
3	50
4	0

69. $x^2 - 9$ **71.** $x^2 - 25$ **73.** $x^2 - 49$ **75.** $x^2 - 81$ **77.** $4x^2 - 9y^2$

79. $x^4 - 16$ **81.** $x^2 + 6x + 9$ **83.** $x^2 + 10x + 25$ **85.** $x^2 + 14x + 49$ **87.** $x^2 + 18x + 81$ **89.** $4x^2 + 12x + 9$
91. $16x^2 - 16xy + 4y^2$ **93.** $6x^3 + 5x^2 - 4x + 3$ **95.** $x^4 + 11x^3 - 6x - 9$ **97.** $2x^5 + 8x^3 - 2x - 7$ **99.** $-8x^5 + 2x^4 + 2x^2 + 12$
101. $7x^5y^3$ **103.** $3a^2b^3$

Problem Set 5.4

1. $(x + 3)(x - 3)$ **3.** $(a + 6)(a - 6)$ **5.** $(x + 7)(x - 7)$ **7.** $4(a + 2)(a - 2)$ **9.** Cannot be factored   **11.** $(5x + 13)(5x - 13)$
13. $(3a + 4b)(3a - 4b)$ **15.** $(3 + m)(3 - m)$ **17.** $(5 + 2x)(5 - 2x)$ **19.** $2(x + 3)(x - 3)$ **21.** $32(a + 2)(a - 2)$
23. $2y(2x + 3)(2x - 3)$ **25.** $(a^2 + b^2)(a + b)(a - b)$ **27.** $(4m^2 + 9)(2m + 3)(2m - 3)$ **29.** $3xy(x + 5y)(x - 5y)$ **31.** $(x - 1)^2$
33. $(x + 1)^2$ **35.** $(a - 5)^2$ **37.** $(y + 2)^2$ **39.** $(x - 2)^2$ **41.** $(m - 6)^2$ **43.** $(2a + 3)^2$ **45.** $(7x - 1)^2$ **47.** $(3y - 5)^2$
49. $(x + 5y)^2$ **51.** $(3a + b)^2$ **53.** $\left(y - \frac{3}{2}\right)^2$ **55.** $\left(a + \frac{1}{2}\right)^2$ **57.** $\left(x - \frac{7}{2}\right)^2$ **59.** $\left(x - \frac{3}{8}\right)^2$ **61.** $3(a + 3)^2$ **63.** $2(x + 5y)^2$
65. $x(x + 2)^2$ **67.** $y^2(y - 4)^2$ **69.** $5x(x + 3y)^2$ **71.** $3y^2(2y - 5)^2$ **73.** $(x + 3 + y)(x + 3 - y)$ **75.** $(x + y + 3)(x + y - 3)$ **77.** 14
79. 25   **81. a.** 1   **b.** 8   **c.** 27   **d.** 64   **e.** 125   **83. a.** $x^3 - x^2 + x$ **b.** $x^2 - x + 1$ **c.** $x^3 + 1$
85. a. $x^3 - 2x^2 + 4x$ **b.** $2x^2 - 4x + 8$ **c.** $x^3 + 8$ **87. a.** $x^3 - 3x^2 + 9x$ **b.** $3x^2 - 9x + 27$ **c.** $x^3 + 27$
89. a. $x^3 - 4x^2 + 16x$ **b.** $4x^2 - 16x + 64$ **c.** $x^3 + 64$ **91. a.** $x^3 - 5x^2 + 25x$ **b.** $5x^2 - 25x + 125$ **c.** $x^3 + 125$ **93.** $0.19
95. $4y^2 - 6y - 3$ **97.** $12x^5 - 9x^3 + 3$ **99.** $-6x^3 - 4x^2 + 2x$ **101.** $-7x^3 + 4x^2 - 11$ **103.** $x - 2 + \frac{2}{x - 3}$
105. $3x - 2 + \frac{9}{2x + 3}$ **107.** $\left(t - \frac{1}{5}\right)^2$

Problem Set 5.5

1. $(x - y)(x^2 + xy + y^2)$ **3.** $(a + 2)(a^2 - 2a + 4)$ **5.** $(3 + x)(9 - 3x + x^2)$ **7.** $(y - 1)(y^2 + y + 1)$ **9.** $(y - 4)(y^2 + 4y + 16)$
11. $(5h - t)(25h^2 + 5ht + t^2)$ **13.** $(x - 6)(x^2 + 6x + 36)$ **15.** $2(y - 3)(y^2 + 3y + 9)$ **17.** $2(a - 4b)(a^2 + 4ab + 16b^2)$
19. $2(x + 6y)(x^2 - 6xy + 36y^2)$ **21.** $10(a - 4b)(a^2 + 4ab + 16b^2)$ **23.** $10(r - 5)(r^2 + 5r + 25)$ **25.** $(4 + 3a)(16 - 12a + 9a^2)$
27. $(2x - 3y)(4x^2 + 6xy + 9y^2)$ **29.** $\left(t + \frac{1}{3}\right)\left(t^2 - \frac{1}{3}t + \frac{1}{9}\right)$ **31.** $\left(3x - \frac{1}{3}\right)\left(9x^2 + x + \frac{1}{9}\right)$ **33.** $(4a + 5b)(16a^2 - 20ab + 25b^2)$
35. $\left(\frac{1}{2}x - \frac{1}{3}y\right)\left(\frac{1}{4}x^2 + \frac{1}{6}xy + \frac{1}{9}y^2\right)$ **37.** $(a - b)(a^2 + ab + b^2)(a + b)(a^2 - ab + b^2)$ **39.** $(2x - y)(4x^2 + 2xy + y^2)(2x + y)(4x^2 - 2xy + y^2)$
41. $(x - 5y)(x^2 + 5xy + 25y^2)(x + 5y)(x^2 - 5xy + 25y^2)$ **43.** $2x^5 - 8x^3$ **45.** $3x^4 - 18x^3 + 27x^2$ **47.** $y^3 + 25y$ **49.** $15a^2 - a - 2$
51. $4x^4 - 12x^3 - 40x^2$ **53.** $2ab^5 - 8ab^4 + 2ab^3$ **55.** 37.59 quadrillion Btu   **57.** $x = 3y + 4$ **59.** $x = \frac{3}{2}y + 2$ **61.** $y = \frac{1}{2}x + 3$
63. $y = \frac{2}{3}x - 4$

Problem Set 5.6

1. $(x + 9)(x - 9)$ **3.** $(x + 5)(x - 3)$ **5.** $(x + 3)^2$ **7.** $(y - 5)^2$ **9.** $2ab(a^2 + 3a + 1)$ **11.** Cannot be factored
13. $3(2a + 5)(2a - 5)$ **15.** $(3x - 2y)^2$ **17.** $4x(x^2 + 4y^2)$ **19.** $2y(y + 5)^2$ **21.** $a^4(a^2 + 4b^2)$ **23.** $(x + 4)(y + 3)$
25. $(x - 3)(x^2 + 3x + 9)$ **27.** $(x + 2)(y - 5)$ **29.** $5(a + b)^2$ **31.** Cannot be factored   **33.** $3(x + 2y)(x + 3y)$ **35.** $(2x + 19)(x - 2)$
37. $100(x - 2)(x - 1)$ **39.** $(x + 8)(x - 8)$ **41.** $(x + a)(x + 3)$ **43.** $a^5(7a + 3)(7a - 3)$ **45.** Cannot be factored
47. $a(5a + 1)(5a + 3)$ **49.** $(x + y)(a - b)$ **51.** $3a^2b(4a + 1)(4a - 1)$ **53.** $5x^2(2x + 3)(2x - 3)$ **55.** $(3x + 41y)(x - 2y)$
57. $2x^3(2x - 3)(4x - 5)$ **59.** $(2x + 3)(x + a)$ **61.** $(y^2 + 1)(y + 1)(y - 1)$ **63.** $3x^2y^2(2x + 3y)^2$ **65.** $16(t - 1)(t - 3)$
67. $(9x + 8)(6x + 7)$ **69.** 5   **71.** $-\frac{3}{2}$ **73.** $-\frac{3}{4}$ **75.** 1   **77.** 30   **79.** -6 **81.** 3   **83.** 2

Problem Set 5.7

1. $-2, 1$ **3.** 4, 5   **5.** $0, -1, 3$ **7.** $-\frac{2}{3}, -\frac{3}{2}$ **9.** $0, -\frac{4}{3}, \frac{4}{3}$ **11.** $0, -\frac{1}{3}, -\frac{3}{5}$ **13.** $-1, -2$ **15.** 4, 5   **17.** $6, -4$ **19.** 2, 3
21. -3 **23.** $4, -4$ **25.** $\frac{3}{2}, -4$ **27.** $-\frac{2}{3}$ **29.** 5   **31.** $4, -\frac{5}{2}$ **33.** $\frac{5}{3}, -4$ **35.** $\frac{7}{2}, -\frac{7}{2}$ **37.** $0, -6$ **39.** 0, 3   **41.** 0, 4
43. 0, 5   **45.** 2, 5   **47.** $\frac{1}{2}, -\frac{4}{3}$ **49.** $4, -\frac{5}{2}$ **51.** $8, -10$ **53.** 5, 8   **55.** 6, 8   **57.** -4 **59.** 5, 8   **61.** $6, -8$ **63.** $0, -\frac{3}{2}, -4$
65. $0, 3, -\frac{5}{2}$ **67.** $0, \frac{1}{2}, -\frac{5}{2}$ **69.** $0, \frac{3}{5}, -\frac{3}{2}$ **71.** $\frac{1}{2}, \frac{3}{2}$ **73.** $-5, 4$ **75.** $-7, -6$ **77.** $-3, -1$ **79.** 2, 3   **81.** $-15, 10$
83. $-5, 3$ **85.** $-3, -2, 2$ **87.** $-4, -1, 4$ **89.** $x^2 - 8x + 15 = 0$ **91. a.** $x^2 - 5x + 6 = 0$ **b.** $x^2 - 7x + 6 = 0$ **c.** $x^2 - x - 6 = 0$
93. $x(x + 1) = 72$ **95.** $x(x + 2) = 99$ **97.** $x(x + 2) = 5[x + (x + 2)] - 10$ **99.** Bicycle $75, suit $15   **101.** House $2,400, lot $600
103. $\frac{1}{8}$ **105.** x^8 **107.** x^{18} **109.** 5.6×10^{-3} **111.** 5.67×10^9

Problem Set 5.8

1. Two consecutive even integers are x and $x + 2$; $x(x + 2) = 80$; 8, 10 and $-10, -8$ **3.** 9, 11 and $-11, -9$
5. $x(x + 2) = 5(x + x + 2) - 10$; 8, 10 and 0, 2   **7.** 8, 6   **9.** The numbers are x and $5x + 2$; $x(5x + 2) = 24$; 2, 12 and $-\frac{12}{5}, -10$

11. 5, 20 and 0, 0   **13.** Let x = the width; $x(x + 1) = 12$; width 3 inches, length 4 inches

15. Let x = the base; $\frac{1}{2}(x)(2x) = 9$; base 3 inches   **17.** $x^2 + (x + 2)^2 = 10^2$; 6 inches and 8 inches   **19.** 12 meters

21. $1{,}400 = 400 + 700x - 100x^2$; 200 or 500 items   **23.** 200 DVD's or 800 DVD's

25. $R = xp = (1{,}200 - 100p)p = 3{,}200$; \$4 or \$8   **27.** \$7 or \$10   **29. a.** 5 feet **b.** 12 feet

31. a. 25 seconds later **b.**

t (sec)	h (feet)
0	100
5	1,680
10	2,460
15	2,440
20	1,620
25	0

33. a. 2 seconds **b.** 256 feet   **35.** $200x^{24}$ **37.** x^7 **39.** 8×10^1 **41.** $10ab^2$ **43.** $6x^4 + 6x^3 - 2x^2$ **45.** $9y^2 - 30y + 25$

47. $4a^4 - 49$

Chapter 5 Review

1. $10(x - 2)$ **2.** $x^2(4x - 9)$ **3.** $5(x - y)$ **4.** $x(7x^2 + 2)$ **5.** $7(7a^3 - 2b^3)$ **6.** $6ab(b + 3a^2b^2 - 4a)$ **7.** $(x + a)(y + b)$

8. $(x - 5)(y + 4)$ **9.** $(2x - 3)(y + 5)$ **10.** $(5x - 4a)(x - 2b)$ **11.** $(y + 7)(y + 2)$ **12.** $(w + 5)(w + 10)$ **13.** $(y + 9)(y + 11)$

14. $(y + 6)(y + 2)$ **15.** $(2x + 3)(x + 5)$ **16.** $(2y - 5)(2y - 1)$ **17.** $(2r + 3t)(3r - 2t)$ **18.** $(2x - 7)(5x + 3)$ **19.** $(n + 9)(n - 9)$

20. $(2y + 3)(2y - 3)$ **21** Cannot be factored.   **22.** $(6y + 11x)(6y - 11x)$ **23.** $(8t + 1)^2$ **24.** $(4n - 3)^2$ **25.** $(2r - 3t)^2$

26. $(3m + 5n)^2$ **27.** $2(x + 4)(x + 6)$ **28.** $a(a - 3)(a - 7)$ **29.** $3m(m + 1)(m - 7)$ **30.** $5y^2(y - 2)(y + 4)$ **31.** $2(2x + 1)(2x + 3)$

32. $a(3a + 1)(a - 5)$ **33.** $2m(2m - 3)(5m - 1)$ **34.** $5y(2x - 3y)(3x - y)$ **35.** $4(x + 5)^2$ **36.** $x(2x + 3)^2$ **37.** $5(x + 3)(x - 3)$

38. $3x(2x + 3y)(2x - 3y)$ **39.** $(3x + 2y)(9x^2 - 6xy + 4y^2)$ **40.** $(5x - 4y)(25x^2 + 20xy + 16y^2)$ **41.** $3ab(2a + b)(a + 5b)$

42. $x^3(x + 1)(x - 1)$ **43.** $y^4(4y^2 + 9)$ **44.** $3ab^2(2a - b)(3a + 2b)$ **45.** $-2, 5$ **46.** $7, -7$ **47.** $-\frac{3}{2}, -\frac{2}{3}$ **48.** $0, -\frac{5}{3}, \frac{2}{3}$

49. 10, 12 and $-12, -10$ **50.** 10, 11 and $-11, -10$ **51.** 5, 15   **52.** 2 inches

Chapter 5 Cumulative Review

1. -6 **2.** 29   **3.** 9   **4.** 9   **5.** -64 **6.** $\frac{1}{81}$ **7.** 2   **8.** -3 **9.** $5a - 5$ **10.** $-8a - 3$ **11.** x^{40} **12.** 1

13. $15x^2 + 14x - 8$ **14.** $a^4 - 49$ **15.** -6 **16.** 10   **17.** -21 **18.** -2 **19.** $0, \frac{7}{2}, 7$ **20.** $-\frac{9}{4}, \frac{9}{4}$ **21.** $x < 4$

22. $x \geq 7$ **23.**

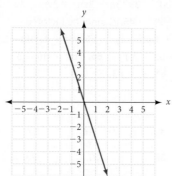

24.

25.

26.

27. $(4, 0)$ and $\left(\frac{16}{3}, 1\right)$

28. x-intercept = 3; y-intercept = -10 **29.** $m = \frac{7}{9}$ **30.** $y = \frac{2}{3}x + 8$ **31.** $y = -\frac{2}{5}x - \frac{2}{3}$ **32.** slope = $\frac{3}{4}$; y-intercept = 4

33. $(-2, 1)$ **34.** Lines are parallel; $\varnothing$ **35.** $(2, 5)$

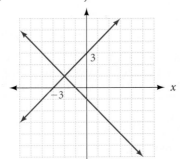

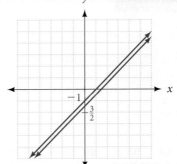

36. $(2, -4)$ **37.** $(1, 2)$ **38.** $(1500, 3500)$ **39.** $(n - 9)(n + 4)$ **40.** $(2x + 5y)(7x - 2y)$ **41.** $(4 - a)(4 + a)$ **42.** $(7x - 1)^2$

43. $5y(3x - y)^2$ **44.** $3x(3x + y)(2x - y)$ **45.** $(3x - 2)(y + 5)$ **46.** $(a + 4)(a^2 - 4a + 16)$ **47.** $2, -\frac{1}{2}, 2$ **48.** $-\frac{1}{5}, 5, \frac{1}{5}$

49. $4x + 5 + \frac{2}{x - 3}$ **50.** $x^2 + 3x + 9$ **51.** 34 inches, 38 inches   **52.** 6 burgers, 8 fries   **53.** 5,400   **54.** 2,100

Chapter 5 Test

1. $5(x - 2)$ **2.** $9xy(2x - 1 - 4y)$ **3.** $(x - 3b)(x + 2a)$ **4.** $(y + 4)(x - 7)$ **5.** $(x - 3)(x - 2)$ **6.** $(x - 3)(x + 2)$

7. $(a - 4)(a + 4)$ **8.** Does not factor   **9.** $(x^2 + 9)(x - 3)(x + 3)$ **10.** $3(3x - 5y)(3x + 5y)$ **11.** $(x - 3)(x + 3)(x + 5)$

12. $(x + 5)(x - b)$ **13.** $2(2a + 1)(a + 5)$ **14.** $3(m + 2)(m - 3)$ **15.** $(2y - 1)(3y + 5)$ **16.** $2x(2x + 1)(3x - 5)$

17. $(a + 4b)(a^2 - 4ab + 16b^2)$ **18.** $2(3x - 2)(9x^2 + 6x + 4)$ **19.** $-3, -4$ **20.** 2   **21.** $-6, 6$ **22.** $-4, 5$ **23.** 5, 6

24. $-4, 0, 4$ **25.** $-\frac{5}{2}, 3$ **26.** $-\frac{1}{3}, 0, 1$ **27.** 4 and 16   **28.** $-3, -1$ and 3, 5   **29.** 3 ft by 14 ft   **30.** 5 m, 12 m

31. 200 or 300   **32.** $3 or $6   **33. a.** 8 ft   **b.** 15 ft   **34. a.** $t = \frac{1}{2}, 2$ sec   **b.** $t = 3$ sec

Chapter 6

Getting Ready for Chapter 6

1. $\frac{5}{33}$ **2.** $\frac{7}{25}$ **3.** $-\frac{2}{3}$ **4.** $\frac{8}{5}$ **5.** Undefined   **6.** $\frac{11}{35}$ **7.** $\frac{11}{35}$ **8.** $x^2 + 2x$ **9.** $x^2 - x - 12$ **10.** $(x + 5)(x - 5)$

11. $2(x - 2)$ **12.** $(x + 3)(x + 4)$ **13.** $a(a - 4)$ **14.** $y(xy + 1)$ **15.** 36   **16.** 6   **17.** $-5, 4$ **18.** $-4, 2$ **19.** 2   **20.** $-6, 6$

Problem Set 6.1

1. a. $\frac{1}{4}$ **b.** $\frac{1}{x - 1}$; $x \neq \pm 1$ **c.** $\frac{x}{x + 1}$; $x \neq \pm 1$ **d.** $\frac{x^2 + x + 1}{x + 1}$; $x \neq \pm 1$ **e.** $\frac{x^2 + x + 1}{x^2}$; $x \neq 0, 1$ **3.** $\frac{1}{x - 2}$, $x \neq 2$ **5.** $\frac{1}{a + 3}$, $a \neq -3, 3$

7. $\frac{1}{x - 5}$, $x \neq -5, 5$ **9.** $\frac{(x + 2)(x - 2)}{2}$ **11.** $\frac{2(x - 5)}{3(x - 2)}$, $x \neq 2$ **13.** 2   **15.** $\frac{5(x - 1)}{4}$ **17.** $\frac{1}{x - 3}$ **19.** $\frac{1}{x + 3}$ **21.** $\frac{1}{a - 5}$ **23.** $\frac{1}{3x + 2}$

25. $\frac{x + 5}{x + 2}$ **27.** $\frac{2m(m + 2)}{m - 2}$ **29.** $\frac{x - 1}{x - 4}$ **31.** $\frac{2(2x - 3)}{2x + 3}$ **33.** $\frac{2x - 3}{2x + 3}$ **35.** $\frac{1}{(x^2 + 9)(x - 3)}$ **37.** $\frac{3x - 5}{(x^2 + 4)(x - 2)}$ **39.** $\frac{2x(7x + 6)}{2x + 1}$ **41.** $\frac{x^2 + xy + y^2}{x + y}$

43. $\frac{x^2 - 2x + 4}{x - 2}$ **45.** $\frac{x^2 - 2x + 4}{x - 1}$ **47.** $\frac{x + 2}{x + 5}$ **49.** $\frac{x + a}{x + b}$ **51.** $\frac{x + a}{x + 3}$ **53. a.** $x^2 - 16$ **b.** $x^2 - 8x + 16$ **c.** $4x^3 - 32x^2 + 64x$ **d.** $\frac{x}{4}$

55. $\frac{4}{3}$ **57.** $\frac{4}{5}$ **59.** $\frac{8}{1}$ **61.** Bugatti Veyron and the SSC Ultimate Aero   **63.** 40.7 miles/hour   **65.** 39.25 feet/minute

67. 6.3 feet/second   **69.** 48 miles/gallon   **71.** $5 + 4 = 9$ **73.**

x	$\frac{x - 3}{3 - x}$
-2	-1
-1	-1
0	-1
1	-1
2	-1

The entries are all -1 because the numerator and denominator are opposites. Or, $\frac{x - 3}{3 - x} = \frac{-1(x - 3)}{3 - x} = -1$.

75.

x	$\dfrac{x-5}{x^2-25}$	$\dfrac{1}{x+5}$
0	$\dfrac{1}{5}$	$\dfrac{1}{5}$
2	$\dfrac{1}{7}$	$\dfrac{1}{7}$
−2	$\dfrac{1}{3}$	$\dfrac{1}{3}$
5	Undefined	$\dfrac{1}{10}$
−5	Undefined	Undefined

77. $\dfrac{5}{14}$ **79.** $\dfrac{9}{10}$ **81.** $(x+3)(x-3)$ **83.** $3(x-3)$ **85.** $(x-5)(x+4)$ **87.** $a(a+5)$

89. $(a-5)(a+4)$ **91.** 5.8   **93.** 0   **95.** $16a^2b^2$ **97.** $19x^4+21x^2-42$ **99.** $7a^4b^4+9b^3-11a^3$

101. Disney, Nike

Company	P/E
Yahoo	35.33
Google	84.53
Disney	18.63
Nike	17.51
eBay	56.11

Problem Set 6.2

1. 2   **3.** $\dfrac{x}{2}$ **5.** $\dfrac{3}{2}$ **7.** $\dfrac{1}{2(x-3)}$ **9.** $\dfrac{4a(a+5)}{7(a+4)}$ **11.** $\dfrac{y-2}{4}$ **13.** $\dfrac{2(x+4)}{x-2}$ **15.** $\dfrac{x+3}{(x-3)(x+1)}$ **17.** 1   **19.** $\dfrac{y-5}{(y+2)(y-2)}$

21. $\dfrac{x^2+5x+25}{x-5}$ **23.** 1   **25.** $\dfrac{a+3}{a+4}$ **27.** $\dfrac{2y-3}{y-6}$ **29.** $\dfrac{x-1}{(x-2)(x^2-x+1)}$ **31.** $\dfrac{3}{2}$ **33. a.** $\dfrac{4}{13}$ **b.** $\dfrac{x+1}{x^2+x+1}$ **c.** $\dfrac{x-1}{x^2+x+1}$ **d.** $\dfrac{x+1}{x-1}$

35. $2(x-3)$ **37.** $2a$ **39.** $(x+2)(x+1)$ **41.** $-2x(x-5)$ **43.** $\dfrac{2(x+5)}{x(y+5)}$ **45.** $\dfrac{2x}{x-y}$ **47.** $\dfrac{1}{x-2}$ **49.** $\dfrac{1}{5}$ **51.** $\dfrac{1}{100}$ **53.** 359.3 ft/s

55. 2.7 miles   **57.** 742 miles/hour   **59.** 0.45 mile/hour   **61.** 8.8 miles/hour   **63.** $\dfrac{4}{5}$ **65.** $\dfrac{11}{35}$ **67.** $-\dfrac{4}{35}$ **69.** $2x-6$

71. x^2-x-20 **73.** $\dfrac{1}{x-3}$ **75.** $\dfrac{x-6}{2(x-5)}$ **77.** x^2-x-30 **79.** 3   **81.** $\dfrac{11}{4}$ **83.** $9x^2$ **85.** $6ab^2$

Problem Set 6.3

1. $\dfrac{7}{x}$ **3.** $\dfrac{4}{a}$ **5.** 1   **7.** $y+1$ **9.** $x+2$ **11.** $x-2$ **13.** $\dfrac{6}{x+6}$ **15.** $\dfrac{(y+2)(y-2)}{2y}$ **17.** $\dfrac{3+2a}{6}$ **19.** $\dfrac{7x+3}{4(x+1)}$ **21.** $\dfrac{1}{5}$

23. $\dfrac{1}{3}$ **25.** $\dfrac{3}{x-2}$ **27.** $\dfrac{4}{a+3}$ **29.** $\dfrac{2(x-10)}{(x+5)(x-5)}$ **31.** $\dfrac{x+2}{x+3}$ **33.** $\dfrac{a+1}{a+2}$ **35.** $\dfrac{1}{(x+3)(x+4)}$ **37.** $\dfrac{y}{(y+5)(y+4)}$ **39.** $\dfrac{3(x-1)}{(x+4)(x+1)}$

41. $\dfrac{1}{3}$ **43. a.** $\dfrac{2}{27}$ **b.** $\dfrac{8}{3}$ **c.** $\dfrac{11}{18}$ **d.** $\dfrac{3x+10}{(x-2)^2}$ **e.** $\dfrac{(x+2)^2}{3x+10}$ **f.** $\dfrac{x+3}{x+2}$

45.

Number	Reciprocal	Sum	Sum
x	$\dfrac{1}{x}$	$1+\dfrac{1}{x}$	$\dfrac{x+1}{x}$
1	1	2	2
2	$\dfrac{1}{2}$	$\dfrac{3}{2}$	$\dfrac{3}{2}$
3	$\dfrac{1}{3}$	$\dfrac{4}{3}$	$\dfrac{4}{3}$
4	$\dfrac{1}{4}$	$\dfrac{5}{4}$	$\dfrac{5}{4}$

47.

x	$x+\dfrac{4}{x}$	$\dfrac{x^2+4}{x}$	$x+4$
1	5	5	5
2	4	4	6
3	$\dfrac{13}{3}$	$\dfrac{13}{3}$	7
4	5	5	8

49. $\dfrac{x+3}{x+2}$ **51.** $\dfrac{x+2}{x+3}$ **53.** 3   **55.** 0   **57.** Undefined   **59.** $-\dfrac{3}{2}$ **61.** $2x+15$ **63.** x^2-5x **65.** -6 **67.** -2 **69.** 3

71. -2 **73.** $\dfrac{1}{5}$ **75.** $-2, -3$ **77.** $3, -2$ **79.** $0, 5$ **81.** 58° C

Problem Set 6.4

1. -3 **3.** 20   **5.** -1 **7.** 5   **9.** -2 **11.** 4   **13.** 3, 5   **15.** $-8, 1$ **17.** 2   **19.** 1   **21.** 8   **23.** 5

25. $\varnothing$; 2 does not check   **27.** 3   **29.** $\varnothing$; -3 does not check   **31.** 0   **33.** 2; 3 does not check   **35.** -4 **37.** -1 **39.** $-6, -7$

41. -1; -3 does not check   **43. a.** $\dfrac{1}{5}$ **b.** 5 **c.** $\dfrac{25}{3}$ **d.** 3 **e.** $\dfrac{1}{5}, 5$ **45. a.** $\dfrac{7}{(a-6)(a+2)}$ **b.** $\dfrac{a-5}{a-6}$ **c.** 7; -1 does not check   **47.** $\dfrac{1}{3}$

49. 30   **51.** 1   **53.** -3 **55.** 26 centimeters per second   **57.** 7   **59.** Length 13 inches, width 4 inches   **61.** 6, 8, or $-8, -6$

63. 6 inches, 8 inches

Problem Set 6.5

1. $\frac{1}{4}, \frac{3}{4}$ **3.** $x + \frac{1}{x} = \frac{13}{6}; \frac{2}{3}$ and $\frac{3}{2}$ **5.** -2 **7.** $\frac{1}{x} + \frac{1}{x+2} = \frac{5}{12}$; 4 and 6

9. Let x = the speed of the boat in still water.

	d	r	t
Upstream	26	$x-3$	$\frac{26}{x-3}$
Downstream	38	$x+3$	$\frac{38}{x+3}$

The equation is $\frac{26}{x-3} = \frac{38}{x+3}$; $x = 16$ miles per hour.

11. 300 miles/hour **13.** 170 miles/hour, 190 miles/hour **15.** 9 miles/hour **17.** 8 miles/hour

19. $\frac{1}{12} - \frac{1}{15} = \frac{1}{x}$; 60 hours **21.** $5\frac{5}{11}$ minutes **23.** 12 minutes

25.

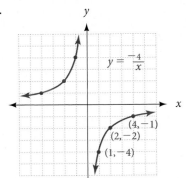

$y = \frac{-4}{x}$

$(4,-1)$
$(2,-2)$
$(1,-4)$

27.

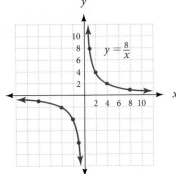

$y = \frac{8}{x}$

29.

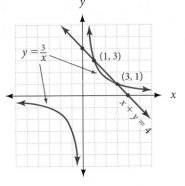

$y = \frac{3}{x}$
$(1, 3)$
$(3, 1)$
$x + y = 4$

31. $\frac{3}{4}$ **33.** $\frac{3}{2}$ **35.** $2x^3y^3$ **37.** $\frac{x^2y^3}{2}$ **39.** $x(xy + 1)$ **41.** $\frac{x^2y^2}{2}$ **43.** $\frac{x-2}{x-3}$ **45.** About 19.9 million

47. $5ab(3a^2b^2 - 4a - 7b)$ **49.** $(x-6)(x+2)$ **51.** $(x^2+4)(x+2)(x-2)$ **53.** $5x(x-6)(x+1)$

Problem Set 6.6

1. 6 **3.** $\frac{1}{6}$ **5.** xy^2 **7.** $\frac{xy}{2}$ **9.** $\frac{y}{x}$ **11.** $\frac{a+1}{a-1}$ **13.** $\frac{x+1}{2(x-3)}$ **15.** $a-3$ **17.** $\frac{y+3}{y+2}$ **19.** $x+y$ **21.** $\frac{a+1}{a}$ **23.** $\frac{1}{x}$

25. $\frac{2a+3}{3a+4}$ **27.** $\frac{x-1}{x+2}$ **29.** $\frac{x+4}{x+1}$ **31.** $\frac{7}{3}, \frac{17}{7}, \frac{41}{17}$

33.

Number x	Reciprocal $\frac{1}{x}$	Quotient $\frac{x}{\frac{1}{x}}$	Square x^2
1	1	1	1
2	$\frac{1}{2}$	4	4
3	$\frac{1}{3}$	9	9
4	$\frac{1}{4}$	16	16

35.

Number x	Reciprocal $\frac{1}{x}$	Sum $1 + \frac{1}{x}$	Quotient $\frac{1 + \frac{1}{x}}{\frac{1}{x}}$
1	1	2	2
2	$\frac{1}{2}$	$\frac{3}{2}$	3
3	$\frac{1}{3}$	$\frac{4}{3}$	4
4	$\frac{1}{4}$	$\frac{5}{4}$	5

37. 407 kilometers per hour **39.** $\frac{7}{2}$ **41.** $-3, 2$ **43.** $x < 1$ **45.** $x \geq -7$ **47.** $x < 6$ **49.** $x \leq 2$

Problem Set 6.7

1. 1 **3.** 10 **5.** 40 **7.** $\frac{5}{4}$ **9.** $\frac{7}{3}$ **11.** 3, -4 **13.** 4, -4 **15.** 2, -4 **17.** 6, -1 **19.** 12.6 million metric tons

21. 15 hits **23.** 21 milliliters **25.** 45.5 grams **27.** 14.7 inches **29.** 12.5 miles **31.** 17 minutes **33.** 343 miles **35.** $h = 9$

37. $y = 14$ **39.** $\frac{x+2}{x+3}$ **41.** $\frac{2(x+5)}{x-4}$ **43.** $\frac{1}{x-4}$

Chapter 6 Review

1. $\frac{1}{2(x-2)}, x \neq 2$ **2.** $\frac{1}{a-6}, a \neq -6, 6$ **3.** $\frac{2x-1}{x+3}, x \neq -3$ **4.** $\frac{1}{x+4}, x \neq -4$ **5.** $\frac{3x-2}{2x-3}, x \neq \frac{3}{2}, -6, 0$ **6.** $\frac{1}{(x-2)(x^2+4)}, x \neq -2, 2$

7. $x - 2, x \neq -7$ **8.** $a + 8, a \neq -8$ **9.** $\frac{y+b}{y+5}, x \neq -a, y \neq -5$ **10.** $\frac{x}{2}$ **11.** $\frac{x+2}{x-4}$ **12.** $(a-6)^2$ **13.** $\frac{x-1}{x+2}$ **14.** 1 **15.** $x - 9$

16. $\frac{13}{a+8}$ **17.** $\frac{x^2+5x+45}{x(x+9)}$ **18.** $\frac{4x+5}{4(x+5)}$ **19.** $\frac{1}{(x+6)(x+2)}$ **20.** $\frac{a-6}{(a+5)(a+3)}$ **21.** 4 **22.** 9 **23.** 1, 6

24. $\varnothing$; 2 does not check **25.** 5 **26.** 3 and $\frac{7}{3}$ **27.** 15 miles/hour **28.** 84 hours **29.** $\frac{1}{2}$

30. $\frac{y+3}{y+7}$ **31.** $\frac{4a-7}{a-1}$ **32.** $\frac{2}{5}$ **33.** $\frac{2}{9}$ **34.** 12 **35.** $-6, 6$ **36.** $-6, 8$

Chapter 6 Cumulative Review

1. -3 **2.** -6 **3.** -4 **4.** -3 **5.** $-4x - 4$ **6.** $4 - x$ **7.** x^4 **8.** $\frac{1}{x^4}$ **9.** 6 **10.** 1 **11.** $-3x$

12. $-2a^3 - 10a^2 - 5a + 13$ **13.** $x - 7$ **14.** $\frac{5x+3}{5(x+3)}$ **15.** -2 **16.** $\frac{11}{4}$ **17.** $-\frac{3}{7}, \frac{3}{7}$ **18.** $0, 2, \frac{7}{2}$ **19.** 6 **20.** No solution

21. $(3, -1)$ **22.** $\left(\frac{1}{3}, -\frac{1}{2}\right)$ **23.** $(17, 5)$ **24.** $(6, -12)$

25.

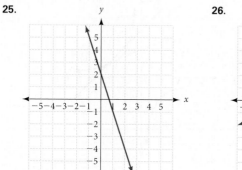

26.

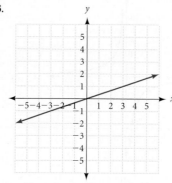

27. $a \geq 6$ **28.** $x > -3$

29. $(x+a)(y+5)$ **30.** $(a-5)(a+7)$ **31.** $(4y-3)(5y-3)$ **32.** $(2r-3t)(2r+3t)$ **33.** $3(x^2+4y^2)$

34. $(4x+9y)^2$ **35.** $(2, 0)$

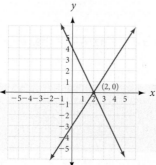

36. $(-2, 1)$

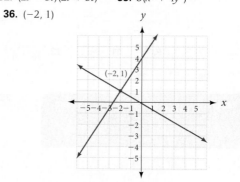

37. x-intercept = 5; y-intercept = 2 **38.** $m = 3$; y-intercept = -4 **39.** $y = -5x - 1$ **40.**

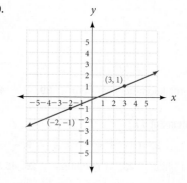

41. 8 **42.** -14 **43.** Associative property of addition

44. Commutative property of addition **45.** $x - 7$

46. $\frac{x+1}{4(x+2)}$ **47.** $\frac{1}{2}$ **48.** $\frac{2x+3}{3x+2}$ **49.** $\frac{2x-3}{x+4}$ **50.** $\frac{x-3}{2(x-4)}$ **51.** 1,003,861 sq mi

52. 401,544 sq mi

Chapter 6 Test

1. $\frac{x+4}{x-4}$ **2.** $\frac{2}{a+2}$ **3.** $\frac{y+7}{x+a}$ **4.** 3   **5.** $(x-7)(x-1)$ **6.** $\frac{x+4}{3x-4}$ **7.** $(x-3)(x+2)$ **8.** $-\frac{3}{x-2}$ **9.** $\frac{2x+3}{(x-3)(x+3)}$

10. $\frac{3x}{(x-2)(x+1)}$ **11.** $\frac{11}{5}$ **12.** 1   **13.** 6   **14.** 15 miles/hour   **15.** 60 hours   **16.** $\frac{x+1}{x-1}$ **17.** $\frac{x+4}{x+2}$ **18.** $\frac{1}{2}, \frac{1}{3}$

19. 132 parts   **20.** 3.1

21. Koenigsegg CCX: 359.3 ft/sec.; Saleen S7: 363.7 ft/sec.; Bugatti Veyron: 371.1 ft/sec.; SSC Ultimate Aero: 376.9 ft/sec

Chapter 7

Getting Ready for Chapter 7

1. 540   **2.** 750   **3.** -2 **4.** -6 **5.** 4   **6.** 29   **7.** -4 **8.** 18   **9.** $-6x+9y$ **10.** $12x+15y$ **11.** $3x-2y$

12. $3x+8y$ **13.** $3x+7y$ **14.** $6x+7y$ **15.** 2   **16.** 3   **17.** -1 **18.** $\frac{8}{15}$ **19.** 4   **20.** $x>435$

Problem Set 7.1

1. 8   **3.** 5   **5.** 2   **7.** -7 **9.** $-\frac{9}{2}$ **11.** -4 **13.** $-\frac{4}{3}$ **15.** -4 **17.** -2 **19.** $\frac{3}{4}$ **21.** 12   **23.** -10 **25.** 7

27. -7 **29.** 3   **31.** $\frac{4}{5}$ **33.** 1   **35.** 4   **37.** 17   **39.** 2   **41.** 6   **43.** $-\frac{4}{3}$ **45.** $-\frac{3}{2}$ **47.** $\frac{5}{3}$ **49.** $\frac{3}{2}$ **51.** 1

53. No solution   **55.** All real numbers are solutions.   **57.** No solution   **59.** $6, -1$ **61.** $0, 2, 3$ **63.** $\frac{1}{3}, -4$ **65.** $\frac{2}{3}, \frac{3}{2}$

67. $5, -5$ **69.** $0, -3, 7$ **71.** $-4, \frac{5}{2}$ **73.** $0, \frac{4}{3}$ **75.** $-\frac{1}{5}, \frac{1}{3}$ **77.** $0, -\frac{4}{3}, \frac{4}{3}$ **79.** $-10, 0$ **81.** $-5, 1$ **83.** $1, 2$

85. $-2, 3$ **87.** $-2, \frac{1}{4}$ **89.** $-3, -2, 2$ **91.** $-2, -5, 5$ **93.** $-\frac{3}{2}, -2, 2$ **95.** $-3, -\frac{3}{2}, \frac{3}{2}$ **97.** $-2, \frac{5}{3}$ **99.** $-1, 9$

101. $0, -3$ **103.** $-4, -2$ **105.** $-4, 2$ **107.** 3 hours   **109.** $-5, -4$ or $4, 5$ **111.** 24 feet   **113.** $6, 8, 10$

115. 2 feet, 8 feet   **117.** 18 inches, 4 inches   **119.** 0 and 2 seconds   **121.** 1 and 2 seconds   **123.** 0 and $\frac{3}{2}$ seconds

125. 2 and 3 seconds   **127.** \$4 or \$8   **129.** \$7 or \$10   **131.** $(1, 2)$ **133.** $(15, 12)$ **135.** $(3, -2, 1)$

137. \$4,000 at 5%, \$8,000 at 6%   **139.** 4   **141.** 5   **143.** 3

Problem Set 7.2

1. $-4, 4$ **3.** $-2, 2$ **5.** $\varnothing$ **7.** $-1, 1$ **9.** $\varnothing$ **11.** $-6, 6$ **13.** $-3, 7$ **15.** $\frac{17}{3}, \frac{7}{3}$ **17.** $2, 4$ **19.** $-\frac{5}{2}, \frac{5}{6}$ **21.** $-1, 5$

23. $\varnothing$ **25.** $-4, 20$ **27.** $-4, 8$ **29.** $-\frac{10}{3}, \frac{2}{3}$ **31.** $\varnothing$ **33.** $-1, \frac{3}{2}$ **35.** $5, 25$ **37.** $-30, 26$ **39.** $-12, 28$ **41.** $-5, \frac{3}{5}$

43. $1, \frac{1}{9}$ **45.** $-\frac{1}{2}$ **47.** 0   **49.** $-\frac{1}{2}$ **51.** $-\frac{1}{6}, -\frac{7}{4}$ **53.** All real numbers   **55.** All real numbers

57. a. $\frac{5}{4}$ **b.** $\frac{5}{4}$ **c.** 2   **d.** $\frac{1}{2}, 2$ **e.** $\frac{1}{3}, 4$ **59.** $|840 - 1{,}280| = 440$ **61.** 1987 and 1995

63. 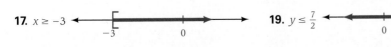   **65.**    **67.** $\frac{19}{15}$ **69.** 40   **71.** -2 **73.** 0

75. $x < 4$ **77.** $-\frac{11}{3} \le a$ **79.** $t \le -\frac{3}{2}$ **81.** $x = a \pm b$ **83.** $x = \frac{-b \pm c}{a}$ **85.** $x = -\frac{a}{b}y \pm a$

Problem Set 7.3

1. $x \le \frac{3}{2}$

3. $x > 4$

5. $x \ge -5$

7. $x < 4$

9. $x \ge -6$

11. $x \ge 4$

13. $x < -3$

15. $m \ge -1$

17. $x \ge -3$

19. $y \le \frac{7}{2}$

21. $x < 6$

23. $y \ge -52$

25. $(-\infty, -2]$ **27.** $[1, \infty)$

29. $(-\infty, 3)$ **31.** $(-\infty, -1]$ **33.** $[-17, \infty)$ **35.** $x > 435$ **37.** $[3, 7]$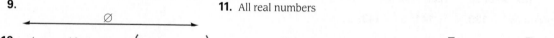

39. $(-4, 2)$ **41.** $[4, 6]$

43. $(-4, 2)$ **45.** $(-3, 3)$

47. $(-\infty, -7] \cup [-3, \infty)$ **49.** $(-\infty, -1] \cup [\frac{3}{5}, \infty)$

51. $(-\infty, -10) \cup (6, \infty)$ **53.** $-2 < x \le 4$ **55.** $x < -4$ or $x \ge 1$

57. a. $x > 0$ **b.** $x \ge 0$ **c.** $x \ge 0$ **59.** $\varnothing$ **61.** All real numbers   **63.** $\varnothing$ **65. a.** 1   **b.** 16   **c.** no   **d.** $x > 16$

67. $31 < d < 49$ **69.** $p \le 8.5$; set the price at \$8.50 or less per pad   **71.** $p > 7.25$; charge more than \$7.25 per pad

73. During the years 1990 through 1994   **75.** Adults: $0.61 \le r \le 0.83$; 61% – 83%   Juveniles: $0.06 \le r \le 0.20$; 6% – 20%

77. a. $35° \le C \le 45°$ **b.** $20° \le C \le 30°$ **c.** $-25° \le C \le -10°$ **d.** $-20° \le C \le -5°$ **79.** 3   **81.** -3

83. The distance between x and 0 on the number line   **85.** -3 **87.** 15   **89.** No solution   **91.** -1 **93.** $x < \frac{c-b}{a}$

95. $\frac{-c-b}{a} < x < \frac{c-b}{a}$

Problem Set 7.4

1. $-3 < x < 3$ **3.** $x \le -2$ or $x \ge 2$

5. $-3 < x < 3$ **7.** $t < -7$ or $t > 7$

9. $\varnothing$ **11.** All real numbers

13. $-4 < x < 10$ **15.** $a \le -9$ or $a \ge -1$

17. $\varnothing$ **19.** $-1 < x < 5$

21. $y \le -5$ or $y \ge -1$ **23.** $k \le -5$ or $k \ge 2$

25. $-1 < x < 7$ **27.** $a \le -2$ or $a \ge 1$

29. $-6 < x < \frac{8}{3}$ **31.** $x < 2$ or $x > 8$

33. $x \le -3$ or $x \ge 12$ **35.** $x < 2$ or $x > 6$

37. $0.99 < x < 1.01$ **39.** $x \le -\frac{3}{5}$ or $x \ge -\frac{2}{5}$ **41.** $-\frac{1}{6} \le x \le \frac{3}{2}$ **43.** $-0.05 < x < 0.25$ **45.** $|x| \le 4$ **47.** $|x - 5| \le 1$
49. no   **51.** $x < -2$ or $x > \frac{4}{5}$ **53.** -6 **55.** 66   **57.** -13 **59.** 11   **61.** -4 **63.** 13   **65.** 16   **67.** -1
69. -5 **71.** 9   **73.** -4 **75.** $a - b < x < a + b$ **77.** $x < \frac{b-c}{a}$ or $x > \frac{b+c}{a}$ **79.** $\frac{-b-c}{a} \le x \le \frac{c-b}{a}$

Problem Set 7.5

1. $(4, 3)$

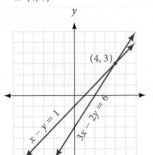

3. $(-5, -6)$

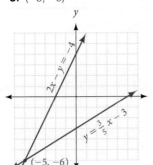

5. $(4, 2)$

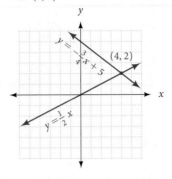

7. Lines are parallel; no solution. **9.** Lines coincide; any solution to one of the equations is a solution to the other.

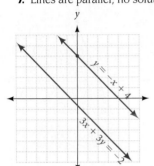

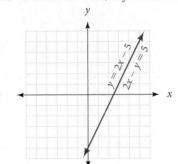

11. $(2, 3)$ **13.** $(1, 1)$ **15.** Lines coincide: $\{(x, y) \mid 3x - 2y = 6\}$ **17.** $\left(1, -\frac{1}{2}\right)$ **19.** $\left(\frac{1}{2}, -3\right)$ **21.** $\left(-\frac{8}{3}, 5\right)$ **23.** $(2, 2)$

25. Parallel lines; $\varnothing$ **27.** $(12, 30)$ **29.** $(10, 24)$ **31.** $(3, -3)$ **33.** $\left(\frac{4}{3}, -2\right)$ **35.** $y = 5, z = 2$ **37.** $(2, 4)$

39. Lines coincide: $\{(x, y) \mid 2x - y = 5\}$ **41.** Lines coincide: $\left\{(x, y) \mid x = \frac{3}{2}y\right\}$ **43.** $\left(-\frac{15}{43}, -\frac{27}{43}\right)$ **45.** $\left(\frac{60}{43}, \frac{46}{43}\right)$ **47.** $\left(\frac{9}{41}, -\frac{11}{41}\right)$

49. $\left(-\frac{11}{7}, -\frac{20}{7}\right)$ **51.** $\left(2, \frac{4}{3}\right)$ **53.** $(-12, -12)$ **55.** Lines are parallel; $\varnothing$ **57.** $y = 5, z = 2$ **59.** $\left(\frac{3}{2}, \frac{3}{8}\right)$ **61.** $\left(\frac{12}{5}, \frac{8}{15}\right)$

63. a. $-y$ **b.** -2 **c.** -2 **d.** graph below **e.** $(0, -2)$ **65.** $(6{,}000, 4{,}000)$ **67.** 2   **69.** $(4, 0)$ **71.** 3   **73.** $m = \frac{2}{3}, b = -2$

75. $y = \frac{2}{3}x + 6$

77. $m = \frac{2}{3}$ **79.** -10 **81.** $3y + 2z$

83. 1   **85.** 3   **87.** $10x - 2z$

89. $9x + 3y - 6z$ **91.** $a = 3, b = -2$

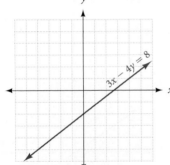

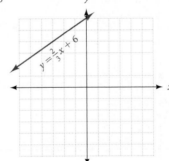

Problem Set 7.6

1. $(1, 2, 1)$ **3.** $(2, 1, 3)$ **5.** $(2, 0, 1)$ **7.** $\left(\frac{1}{2}, \frac{2}{3}, -\frac{1}{2}\right)$ **9.** No solution, inconsistent system   **11.** $(4, -3, -5)$

13. No unique solution   **15.** $(4, -5, -3)$ **17.** No unique solution   **19.** $\left(\frac{1}{2}, 1, 2\right)$ **21.** $\left(\frac{1}{2}, \frac{1}{3}, \frac{1}{4}\right)$ **23.** $\left(\frac{10}{3}, -\frac{5}{3}, -\frac{1}{3}\right)$

25. $\left(\frac{1}{4}, -\frac{1}{3}, \frac{1}{8}\right)$ **27.** $(6, 8, 12)$ **29.** $(3, 6, 4)$ **31.** $(1, 3, 1)$ **33.** $(-1, 2, -2)$ **35.** 4 amp, 3 amp, 1 amp   **37.** -2 **39.** 11

41. -14 **43.** -4 **45.** $(2, -1)$ **47.** $(4, 3)$ **49.** $(1, 1)$ **51.** $(5, 2)$ **53.** $(1, 2, 3, 4)$

Problem Set 7.7

1.
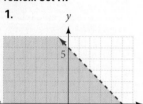

3.
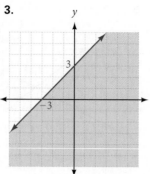

5.
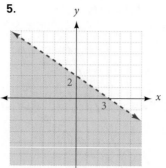

7.
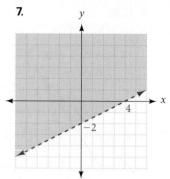

9.
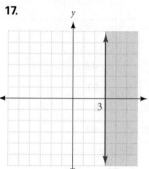

11.
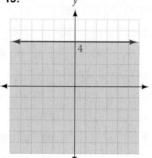

13.
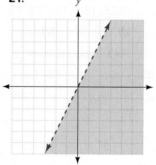

15.
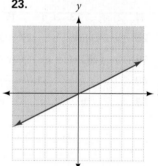

17.
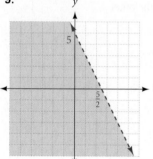

19.
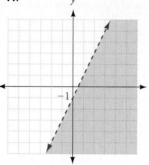

21.
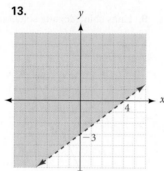

23.
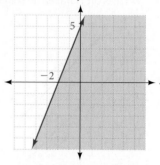

25.
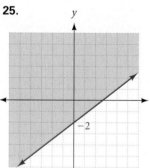

27.
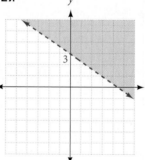

29.
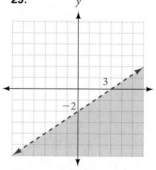

31.
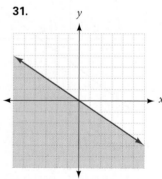

33.
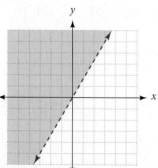

35.
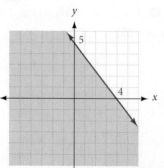

37. >   **39.** ≤

41. a. $y < \frac{4}{3}$ **b.** $y > -\frac{4}{3}$ **c.** $y = -\frac{2}{3}x + 2$ **d.** (graph below)

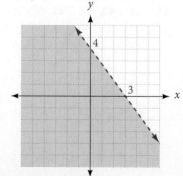

43.

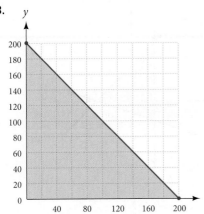

45.

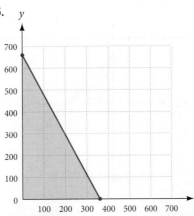

47.

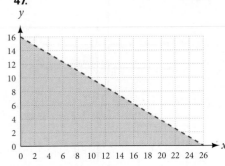

49. $y \leq 7$ **51.** $t > -2$ **53.** $-1 < t < 2$

55.

x	y
0	0
10	75
20	150

57.

x	y
0	0
1	1
1	-1

59.

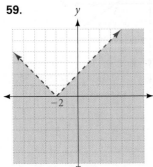

61.

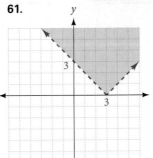

63.

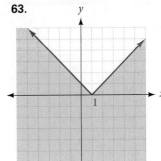

65.

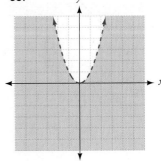

67.

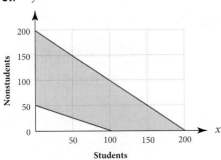

Problem Set 7.8

1.

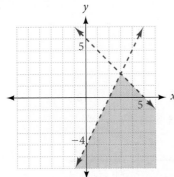

3.

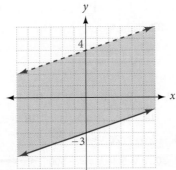

5.

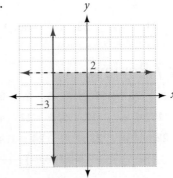

7.

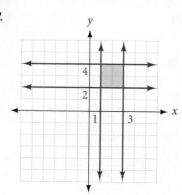

9.

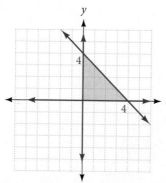

11.

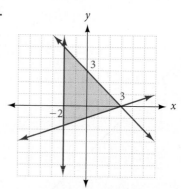

13.

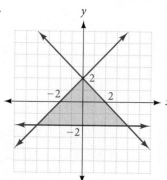

15.

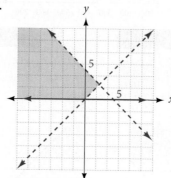

17.

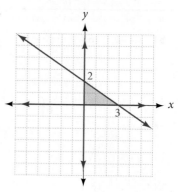

19. $\leq, <$ **21.** $\geq, <$

23. a. $0.55x + 0.65y \leq 40$
$x \geq 2y$
$x > 15$
$y \geq 0$
b. 10 65-cent stamps

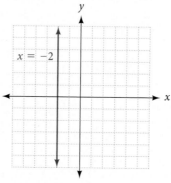

25. x-intercept $= 3$, y-intercept $= 6$, slope $= -2$

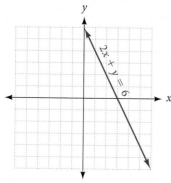

27. x-intercept $= -2$, no y-intercept, no slope

$x = -2$

29. $m = -\frac{3}{7}$ **31.** $x = 4$ **33.** -1 **35.** $3 - 5x$ **37.** 6 and 9

Chapter 7 Review

1. 10   **2.** 2   **3.** 2   **4.** -3 **5.** -3 **6.** 0   **7.** $\frac{2}{3}$ **8.** $\frac{10}{13}$ **9.** $\frac{5}{2}$ **10.** 1   **11.** $-\frac{5}{11}$ **12.** $-\frac{2}{9}$ **13.** $-\frac{1}{2}, \frac{4}{5}$

14. $-\frac{5}{3}, \frac{5}{3}$ **15.** $0, -2$ **16.** $-3, 6$ **17.** $-4, 3, -3$ **18.** $-10, -8$ or $8, 10$ **19.** $-5, -4$ or $4, 5$ **20.** 3, 4, 5

21. $-2, 2$ **22.** $-4, 4$ **23.** 2, 4   **24.** $-1, 4$ **25.** $-\frac{3}{2}, 5$ **26.** 5, 9   **27.** $-1, 1$ **28.** 0

29. $-5 < x < 5$ **30.** $a \geq 500$ or $a \leq -500$

31. $\varnothing$ **32.** $-3 < x < 2$ **33.** $\varnothing$ **34.** $\varnothing$ **35.** $\varnothing$ **36.** All real numbers except 0.   **37.** $\varnothing$ **38.** All real numbers

39. $\left(-\infty, \frac{1}{2}\right)$ **40.** $(-\infty, 8]$ **41.** $(-\infty, 12]$ **42.** $(-1, \infty)$ **43.** $[2, 6]$ **44.** $[0, 1]$ **45.** $\left(-\infty, -\frac{3}{2}\right] \cup [3, \infty)$ **46.** $(-\infty, 1) \cup [4, \infty)$

47. $(6, -2)$ **48.** $(2, -4)$ **49.** No solution (parallel lines)   **50.** $(1, -1)$ **51.** $(0, 1)$

52. No unique solution (lines coincide)   **53.** $(3, 1)$ **54.** $\left(\frac{3}{2}, 0\right)$ **55.** $(3, 5)$ **56.** $(4, 8)$ **57.** $(3, 12)$ **58.** $(3, -2)$

59. $\left(\frac{3}{2}, \frac{1}{2}\right)$ **60.** $(4, 1)$ **61.** $(6, -2)$ **62.** $(7, -4)$ **63.** $(-5, 3)$ **64.** No solution (parallel lines)   **65.** $(3, -1, 4)$

66. $\left(2, \frac{1}{2}, -3\right)$ **67.** $\left(-1, \frac{1}{2}, \frac{3}{2}\right)$ **68.** $\left(2, \frac{1}{3}, \frac{2}{3}\right)$ **69.** No unique solution (dependent system)

70. No unique solution (dependent system)   **71.** $(2, -1, 4)$

72. No unique solution (dependent system)   **73.**   **74.**

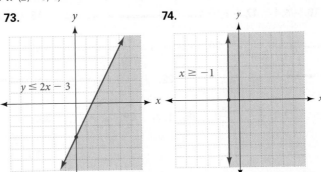

75.   **76.**   **77.**   **78.**

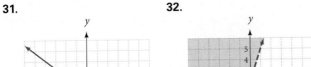

Chapter 7 Cumulative Review

1. -48 **2.** 16 **3.** 6 **4.** -93 **5.** -42 **6.** -12 **7.** -63 **8.** -27 **9.** $18x + 7$ **10.** $6x - 19$ **11.** $-3x + 6y$
12. $-10x - 3y$ **13.** -1 **14.** 0 **15.** $\varnothing$ **16.** $\varnothing$ **17.** $y = -5x + 13$ **18.** $y = \frac{2}{3}x - 1$ **19.** $x = \frac{2}{a - b}$
20. $x = \frac{9}{c - a}$ **21.** $(-\infty, -4]$ **22.** $(-\infty, -2] \cup [3, \infty)$ **23.** $(2x - 3)(x + 4)$ **24.** $(3x + 4)(2x - 5)$ **25.** $0, -1, 2.35, 4$ **26.** $-1, 0, 4$
27. $(4, 3)$ **28.** Lines are parallel; $\varnothing$ **29.** Lines coincide   **30.** $\left(-\frac{45}{31}, -\frac{57}{31}, \frac{51}{62}\right)$

34. 2 **35.** $m = \frac{3}{5}, b = -3$

31.   **32.**   **33.**

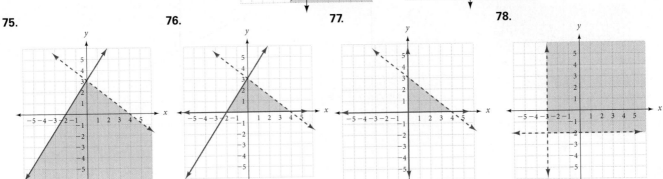

36. $y = -\frac{2}{3}x - 3$ **37.** $m = \frac{5}{2}$ **38.** $m = 3$ **39.**

40.

41.

42.

43.

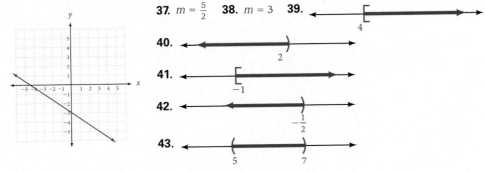

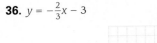

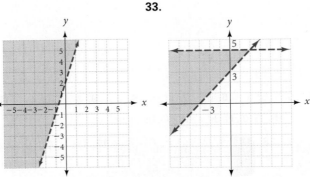

44.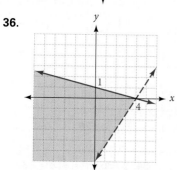

45. $-\frac{5}{2}, 3$ **46.** $0, -4, 3$ **47.** $-2, 2$ **48.** $-\frac{2}{3}, 2$ **49.** $14, 4$

Chapter 7 Test

1. 12 **2.** $-\frac{4}{3}$ **3.** $-\frac{7}{4}$ **4.** 2 **5.** -2 **6.** 0 **7.** 4 **8.** 8 **9.** $-\frac{1}{3}, 2$ **10.** $0, 5$ **11.** $-5, 2$ **12.** $-4, -2, 4$ **13.** $2, 6$

14. $-15, 3$ **15.** $\varnothing$ **16.** $-5, 1$ **17.** $[-6, \infty)$ 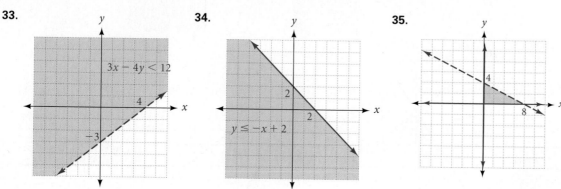   **18.** $(-\infty, 4)$

19. $(-\infty, 6)$ **20.** $[-52, \infty)$

21. $x < -1$ or $x > \frac{4}{3}$ **22.** $-\frac{2}{3} \le x \le 4$ **23.** All real numbers   **24.** $\varnothing$

25. $(1, 2)$ **26.** $(15, 12)$ **27.** $\left(-\frac{54}{13}, -\frac{58}{13}\right)$ **28.** $(1, 2)$ **29.** $(3, -2, 1)$ **30.** $\left(\frac{20}{7}, \frac{16}{7}, \frac{4}{7}\right)$ **31.** $\ge$ **32.** $<$

33.

$3x - 4y < 12$

34.

$y \le -x + 2$

35.

36.

Chapter 8

Getting Ready for Chapter 8

1. 324 **2.** 0.0005 **3.** 3 **4.** -9 **5.** 3 **6.** 0 **7.** 113 **8.** 7 **9.** 3 **10.** 4 **11.** $-\frac{3}{50}$ **12.** $-\frac{3}{5}$

13. $y = \frac{2}{3}x - \frac{5}{3}$ **14.** $y = -\frac{1}{2}x + \frac{7}{2}$ **15.** $(5, 0), (4, 3)$ **16.** $(0, 0), (2, 0)$

17.

t	0	1	2	3	4
h	0	48	64	48	0

18.

t	4	6	8	10
s	15	10	7.5	6

19. 135 **20.** $\frac{1}{2}$

Problem Set 8.1

1. $\frac{3}{2}$ **3.** No slope   **5.** $\frac{2}{3}$

7.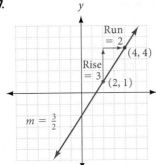

9.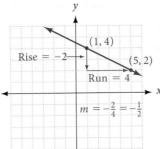

11.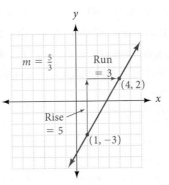

13.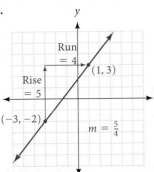

15.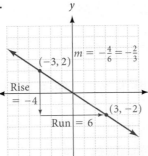

17.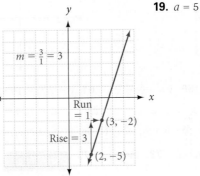

19. $a = 5$ **21.** $b = 2$

23.

x	y
0	2
3	0

Slope $= -\frac{2}{3}$

25.

x	y
0	−5
3	−3

Slope $= \frac{2}{3}$

27. $-\frac{1}{2}$ **29.** $-\frac{8}{3}$ **31.** -2 **33.** $\frac{2}{5}$ **35.** $\frac{4}{5}$ **37. a.** yes   **b.** no   **39.** 8

41. a. 17.5 mph   **b.** 40 km/h   **c.** 120 ft/sec   **d.** 28 m/min

43. Slopes: *A:* -50; *B:* -75; *C:* -25

Year	2006	2007	2008	2009	2010
Sales (in millions)	300	250	175	150	125

45. a. 15,800 ft.   **b.** $-\frac{7}{100}$

47. 818.18 solar thermal collector shipments/year. Between 1997 and 2008 the number of solar thermal collector shipments increased an average of 818.18 shipments per year.

49. a. $2.84 million/year. From the years 1985 to 1990, the amount of money bet on horse races increased at an average of $2.84 million a year.

b. $7.72 million/year. From the years 2000 to 2005, the amount of money bet on horse races increased at an average of $7.72 million a year.

51. 0   **53.** $y = -\frac{3}{2}x + 6$ **55.** $t = \frac{A - P}{Pr}$ **57.** -1 **59.** -2 **61.** $y = mx + b$ **63.** $y = -2x - 5$ **65.** 5

Problem Set 8.2

1. $y = 2x + 3$ **3.** $y = x - 5$ **5.** $y = \frac{1}{2}x + \frac{3}{2}$ **7.** $y = 4$

9. Slope $= 3$,

y-intercept $= -2$,

perpendicular slope $= -\frac{1}{3}$

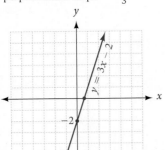

11. Slope $= \frac{2}{3}$,

y-intercept $= -4$,

perpendicular slope $= -\frac{3}{2}$

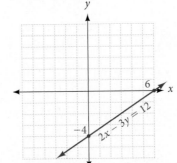

13. Slope $= -\frac{4}{5}$,

y-intercept $= 4$,

perpendicular slope $= \frac{5}{4}$

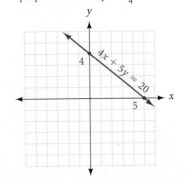

15. Slope = $\frac{1}{2}$, y-intercept = -4, $y = \frac{1}{2}x - 4$ **17.** Slope = $-\frac{2}{3}$, y-intercept = 3, $y = -\frac{2}{3}x + 3$ **19.** $y = 2x - 1$ **21.** $y = -\frac{1}{2}x - 1$

23. $y = -3x + 1$ **25.** $y = \frac{2}{3}x + \frac{2}{3}$ **27.** $y = -\frac{1}{4}x + \frac{3}{4}$ **29.** $x - y = 2$ **31.** $2x - y = 3$ **33.** $6x - 5y = 3$ **35. a.** 3 **b.** $-\frac{1}{3}$

37. a. -3 **b.** $\frac{1}{3}$ **39. a.** $-\frac{2}{5}$ **b.** $\frac{5}{2}$ **41. a.** $x: \frac{10}{3}$ $y: -5$ **b.** $(4, 1)$; answers will vary **c.** $y = \frac{3}{2}x - 5$ **d.** no

43. a. $x: \frac{4}{3}$ $y: -2$ **b.** $(2, 1)$; answers will vary **c.** $y = \frac{3}{2}x - 2$ **d.** no

45. a. 2 **b.** $\frac{3}{2}$ **c.** -3 **d.** (graph below) **e.** $y = 2x - 3$ **47.** $(0, -4), (2, 0); y = 2x - 4$ **49.** $(-2, 0)$ $(0, 4); y = 2x + 4$

51. Slope = 0, y-intercept = -2 (graph below)   **53.** $y = 3x + 7$ **55.** $y = -\frac{5}{2}x - 13$

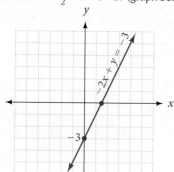

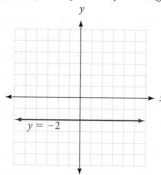

57. $y = \frac{1}{4}x + \frac{1}{4}$ **59.** $y = -\frac{2}{3}x + 2$ **61.** $y = 7.7x - 15{,}374.6$ **63. b.** $86°$ F   **65. a.** \$310,000 **b.** \$31 **c.** \$9.50

67. a. \$1,500 **b.** \$2,000 **c.** $m = 25$ **d.** \$25 **e.** $y = 25x + 1000$ **69. a.** \$5,000 **b.** $y = -3000x + 20000$ **c.** $-\$10{,}000$. When we try to estimate the cars worth, we get a negative number, which can't happen.

71. 23 inches, 5 inches   **73.** \$46.50   **75.**

x	y
0	0
10	75
20	150

77.

x	y
0	0
1	1
1	−1

79. $y = -\frac{3}{2}x + 12$ **81.** $y = -\frac{4}{3}x - \frac{3}{2}$

Problem Set 8.3

1. Domain = {1, 2, 4}; Range = {3, 5, 1}; a function   **3.** Domain = {−1, 1, 2}; Range = {3, −5}; a function

5. Domain = {7, 3}; Range = {−1, 4}; not a function   **7.** Domain = {a, b, c, d}; Range = {3, 4, 5}; a function

9. Domain = {a}; Range = {1, 2, 3, 4}; not a function   **11.** Yes   **13.** No   **15.** No   **17.** Yes   **19.** Yes

21. Domain = $\{x \mid -5 \le x \le 5\}$; Range = $\{y \mid 0 \le y \le 5\}$ **23.** Domain = $\{x \mid -5 \le x \le 3\}$; Range = $\{y \mid y = 4\}$

25. Domain = All real numbers; Range = $\{y \mid y \ge -1\}$; a function   **27.** Domain = All real numbers; Range = $\{y \mid y \ge 4\}$; a function   **29.** Domain = $\{x \mid x \ge -1\}$; Range = All real numbers; not a function   **31.** Domain = $\{x \mid x \ge 4\}$; Range = All real numbers; not a function

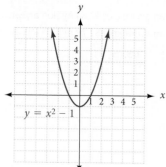

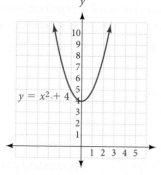

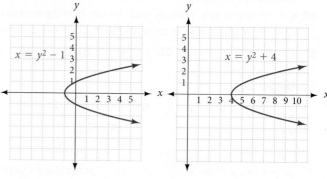

33. Domain = All real numbers;
Range = $\{y \mid y \geq 0\}$;
a function

35. Domain = All real numbers;
Range = $\{y \mid y \geq -2\}$;
a function

37. Domain = $\{2004, 2005, 2006, 2007, 2008, 2009, 2010\}$;
Range = $\{680, 730, 800, 900, 920, 990, 1030\}$;

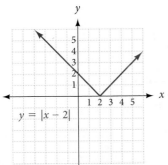

$y = |x - 2|$

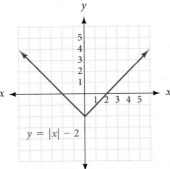

$y = |x| - 2$

39. a. $y = 8.5x$ for $10 \leq x \leq 40$ **b.** (table below)   **c.**

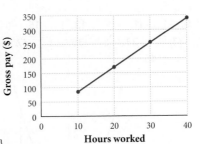

Hours Worked	Function Rule	Gross Pay ($)
x	$y = 8.5x$	y
10	$y = 8.5(10) = 85$	85
20	$y = 8.5(20) = 170$	170
30	$y = 8.5(30) = 255$	255
40	$y = 8.5(40) = 340$	340

d. Domain = $\{x \mid 10 \leq x \leq 40\}$; Range = $\{y \mid 85 \leq y \leq 340\}$
e. Minimum = \$85; Maximum = \$340

41. a.

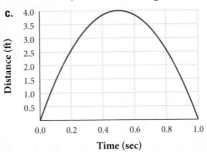

Time (sec)	Function Rule	Distance (ft)
t	$h = 16t - 16t^2$	h
0	$h = 16(0) - 16(0)^2$	0
0.1	$h = 16(0.1) - 16(0.1)^2$	1.44
0.2	$h = 16(0.2) - 16(0.2)^2$	2.56
0.3	$h = 16(0.3) - 16(0.3)^2$	3.36
0.4	$h = 16(0.4) - 16(0.4)^2$	3.84
0.5	$h = 16(0.5) - 16(0.5)^2$	4
0.6	$h = 16(0.6) - 16(0.6)^2$	3.84
0.7	$h = 16(0.7) - 16(0.7)^2$	3.36
0.8	$h = 16(0.8) - 16(0.8)^2$	2.56
0.9	$h = 16(0.9) - 16(0.9)^2$	1.44
1	$h = 16(1) - 16(1)^2$	0

b. Domain = $\{t \mid 0 \leq t \leq 1\}$; Range = $\{h \mid 0 \leq h \leq 4\}$
c.

43. a.

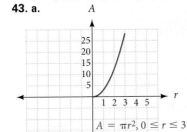

$A = \pi r^2, 0 \leq r \leq 3$

b. Domain = $\{r \mid 0 \leq r \leq 3\}$
Range = $\{A \mid 0 \leq A \leq 9\pi\}$

45. a. Yes
b. Domain = $\{t \mid 0 \leq t \leq 6\}$; Range = $\{h \mid 0 \leq h \leq 60\}$
c. $t = 3$
d. $h = 60$
e. $t = 6$

47. a. III **b.** I **c.** II **d.** IV   **49.** 10   **51.** -14 **53.** 1   **55.** -3 **57.** $-\dfrac{6}{5}$

59. $-\dfrac{7}{640}$ **61.** 150   **63.** 113   **65.** -9 **67. a.** 6 **b.** 7.5   **69. a.** 27 **b.** 6

71. Domain = All real numbers
Range = $\{y \mid y \le 5\}$
A Function

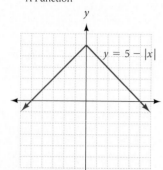

73. Domain = $\{x \mid x \ge 3\}$
Range = All real numbers
Not a Function

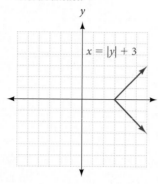

75. Domain = $\{x \mid -4 \le x \le 4\}$
Range = $\{y \mid -4 \le y \le 4\}$
Not a Function

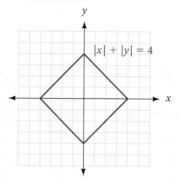

Problem Set 8.4

1. -1 **3.** -11 **5.** 2 **7.** 4 **9.** 35 **11.** -13 **13.** 1 **15.** -9 **17.** 8 **19.** 19 **21.** 16 **23.** 0

25. $3a^2 - 4a + 1$ **27.** 4 **29.** 0 **31.** 2 **33.** -8 **35.** -1 **37.** $2a^2 - 8$ **39.** $2b^2 - 8$

41.

43. $x = 4$ **45.**

47. $V(3) = 300$, the painting is worth \$300 in 3 years; $V(6) = 600$, the painting is worth \$600 in 6 years.

49. a. True **b.** False **c.** True
51. a. True **b.** True **c.** False **d.** True
53. a. \$5,625 **b.** \$1,500 **c.** $\{t \mid 0 \le t \le 5\}$
d.

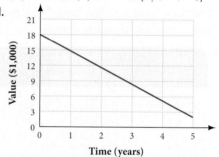

e. $\{V(t) \mid 1,500 \le V(t) \le 18,000\}$
f. about 2.42 years

55. $-0.1x^2 + 27x - 500$ **57.** $6x^2 - 2x - 4$ **59.** $2x^2 + 8x + 8$ **61.** $0.6M - 42$ **63.** $4x^2 - 7x + 3$ **65.** $-\frac{2}{3}, 4$ **67.** $-2, 1$

69. $\varnothing$ **71. a.** 2 **b.** 0 **c.** 1 **d.** 4

73. a.

Weight (ounces)	0.6	1.0	1.1	2.5	3.0	4.8	5.0	5.3
Cost (cents)	39	39	63	87	87	135	135	159

b. More than 2 ounces, but not more than 3 ounces; $2 < x \le 3$
c. Domain: $\{x \mid 0 < x \le 6\}$
d. Range: $\{C \mid C = 39, 63, 87, 111, 135, 159\}$

Problem Set 8.5

1. $6x + 2$ **3.** $-2x + 8$ **5.** $4x - 7$ **7.** $3x^2 + x - 2$ **9.** $-2x + 3$ **11.** $9x^3 - 15x^2$ **13.** $\frac{3x^2}{3x - 5}$ **15.** $\frac{3x - 5}{3x^2}$

17. $3x^2 + 4x - 7$ **19.** 15 **21.** 98 **23.** $\frac{3}{2}$ **25.** 1 **27.** 40 **29.** 147 **31. a.** 81 **b.** 29 **c.** $(x + 4)^2$ **d.** $x^2 + 4$

33. a. -2 **b.** -1 **c.** $4x^2 + 12x - 1$ **35.** $(f \circ g)(x) = 5\left(\frac{x + 4}{5}\right) - 4 = x + 4 - 4 = x; (g \circ f)(x) = \frac{(5x - 4) + 4}{5} = \frac{5x}{5} = x$ **37.** 6 **39.** 2

41. 3 **43.** -8 **45.** 6 **47.** 5 **49.** 3 **51.** -6

53. a. $R(x) = 11.5x - 0.05x^2$ **b.** $C(x) = 2x + 200$ **c.** $P(x) = -0.05x^2 + 9.5x - 200$ **d.** $\overline{C}(x) = 2 + \frac{200}{x}$

55. $T(M) = 24.8 + 0.6M$ **a.** $M(x) = 220 - x$ **b.** $M(24) = 196$ **c.** 142 **d.** 135 **e.** 128 **57.** $2 < x < 4$ **59.** $x < 4$ or $x > 8$

61. $-\frac{5}{7} \le x \le 1$ **63.** 196 **65.** 4 **67.** 1.6 **69.** 3 **71.** 2,400

Problem Set 8.6

1. Direct **3.** Direct **5.** Direct **7.** Direct **9.** Direct **11.** Inverse **13.** 30 **15.** 5 **17.** -6 **19.** $\frac{1}{2}$ **21.** 40

23. 225 **25.** 10 **27.** 16 **29.** $\frac{81}{5}$ **31.** $\frac{81}{2}$ **33.** 64 **35.** 8 **37.** $\frac{50}{3}$ pounds **39.** $371 \frac{3}{5}$ feet/second

41. a. $T = 4P$ **c.** 70 pounds per square inch   **43.** 12 pounds per square inch   **45. a.** $f = \frac{80}{d}$ **c.** An f-stop of 8

b.

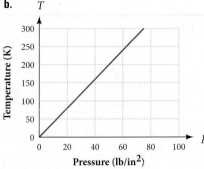

b.

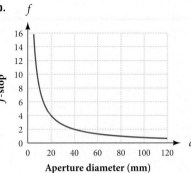

47. $\frac{1,504}{15}$ square inches   **49.** 1.5 ohms   **51.** 1.28 meters   **53.** $F = G\,\frac{m_1 m_2}{d^2}$ **55.** 12   **57.** 28   **59.** $-\frac{7}{4}$ **61.** $w = \frac{P - 2l}{2}$

63. $t \geq -6$ **65.** $x < 6$ **67.** 6, 2   **69.** $\varnothing$ **71.** $x \leq -9$ or $x \geq -1$

73. All real numbers   **75. a.** Square of speed **b.** $d = 0.0675 s^2$ **c.** About 204 feet from the cannon **d.** About 7.5 feet further away

Chapter 8 Review

1. -2 **2.** 0   **3.** 3   **4.** 5   **5.** -5 **6.** 6   **7.** $y = 3x + 5$ **8.** $y = -2x$ **9.** $m = 3, b = -6$ **10.** $m = \frac{2}{3}, b = -3$

11. $y = 2x$ **12.** $y = -\frac{1}{3}x$ **13.** $y = 2x + 1$ **14.** $y = 7$ **15.** $y = -\frac{3}{2}x - \frac{17}{2}$ **16.** $y = 2x - 7$ **17.** $y = \frac{1}{3}x - \frac{2}{3}$

18. Domain = {2, 3, 4}; Range = {4, 3, 2}; a function   **19.** Domain = {6, −4, −2}; Range = {3, 0}; a function

20. 0   **21.** 1   **22.** 1   **23.** $3a + 2$ **24.** 1   **25.** 31   **26.** 24   **27.** 6   **28.** 4   **29.** 25   **30.** 84 pounds

31. 16 foot-candles

Chapter 8 Cumulative Review

1. -25 **2.** -6 **3.** 74   **4.** 144   **5.** 36   **6.** 25   **7.** 27   **8.** 27   **9.** 1   **10.** -9 **11.** $\frac{7}{9}$ **12.** $\frac{7}{9}$ **13.** $\frac{1}{10}$ **14.** $\frac{2}{21}$ **15.** -2

16. $-\frac{8}{5}$ **17.** $6x - 4y$ **18.** $10x + 9y$ **19.** 15   **20.** $\frac{7}{12}$ **21.** $\frac{5}{9}$ **22.** -2 **23.** $\frac{11}{2}, -\frac{5}{2}$ **24.** $\varnothing$ **25.** $y = -3x - 1$

26. $y = \frac{3}{2}x - 5$

27.

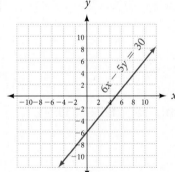

28.

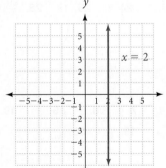

29.

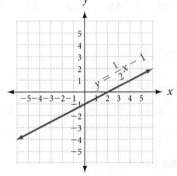

30.

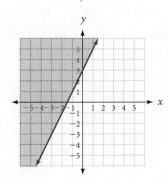

31. $-\frac{5}{2}$ **32.** 6   **33.** 0   **34.** $y = -\frac{3}{4}x + 2$ **35.** $m = \frac{4}{5}; b = -4$ **36.** $y = -\frac{1}{2}x - 1$

37. $y = -\frac{5}{2}x - 12$ **38.** $y = \frac{7}{3}x - 6$ **39.** -1 **40.** 7   **41.** 9   **42.** $2x$

43. Domain = {−1, 2, 3}; Range = {−1, 3}; a function   **44.** $\frac{16}{9}$ **45.** 20%   **46.** $9.43

47. 25% more   **48.** 8% less

Chapter 8 Test

1. x-intercept = 3;
y-intercept = 6; slope = -2

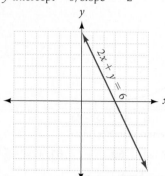

2. x-intercept = $-\frac{3}{2}$;
y-intercept = -3; slope = -2

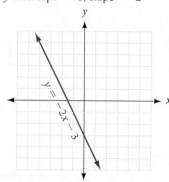

3. x-intercept = $-\frac{8}{3}$;
y-intercept = 4; slope = $\frac{3}{2}$

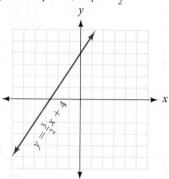

4. x-intercept = -2;
no y-intercept; undefined slope

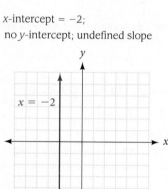

5. $y = 3x + 3$　**6.** $y = -\frac{5}{3}x + 5$　**7.** $y = 2x + 5$　**8.** $y = -\frac{3}{7}x + \frac{5}{7}$

9. $y = \frac{2}{5}x - 5$　**10.** $y = -\frac{1}{3}x - \frac{7}{3}$

11. Domain = $\{-2, -3\}$; Range = $\{0, 1\}$; not a function　**12.** Domain = All real numbers; Range= $\{y \mid y \geq -9\}$; a function

13. Domain = $\{0, 1, 2\}$; Range = $\{0, 3, 5\}$; a function　**14.** Domain = All real numbers; Range = All real numbers; a function

15. Domain = $\{x \mid -4 \leq x \leq 4\}$; Range = $\{y \mid -1 \leq y \leq 3\}$　**16.** Domain = $\{x \mid -3 \leq x \leq 3\}$; Range = $\{y \mid -4 \leq y \leq 4\}$

17. 11　**18.** -4　**19.** 8　**20.** 4　**21.** 5　**22.** 0　**23.** 3　**24.** 0　**25.** 18　**26.** $\frac{81}{4}$

Chapter 9

Getting Ready for Chapter 9

1. 64　**2.** 64　**3.** -64　**4.** -64　**5.** -8　**6.** -8　**7.** 4　**8.** 2　**9.** $\frac{7}{6}$　**10.** $\frac{1}{3}$　**11.** -22　**12.** 4　**13.** $\frac{1}{2}$

14. $\frac{17}{4} = 4\frac{1}{4}$　**15.** x^5　**16.** y^6　**17.** $243r^{40}s^{100}$　**18.** x　**19.** $\frac{1}{27}$　**20.** $\frac{16}{81}$

Problem Set 9.1

1. 12　**3.** Not a real number　**5.** -7　**7.** -3　**9.** 2　**11.** Not a real number　**13.** 0.2　**15.** 0.2　**17.** 5　**19.** -6

21. $\frac{1}{6}$　**23.** $\frac{2}{5}$　**25.** $6a^4$　**27.** $3a^4$　**29.** $2x^2y$　**31.** $2a^3b^5$　**33.** 6　**35.** -3　**37.** 2　**39.** -2　**41.** 2　**43.** $\frac{9}{5}$　**45.** 9

47. 125　**49.** $\frac{1}{3}$　**51.** $\frac{1}{27}$　**53.** $\frac{6}{5}$　**55.** $\frac{8}{27}$　**57.** 7　**59.** $\frac{3}{4}$　**61.** $x^{4/5}$　**63.** a　**65.** $\frac{1}{x^{2/5}}$　**67.** $x^{1/6}$　**69.** $x^{9/25}y^{1/2}z^{1/5}$

71. $\frac{b^{7/4}}{a^{1/8}}$　**73.** $y^{3/10}$　**75.** $\frac{1}{a^2b^4}$　**77.** $\frac{s^{1/2}}{r^{20}}$　**79.** $10b^3$　**81.** 25 mph

83. a. 424 picometers **b.** 600 picometers **c.** 6×10^{-10} meters　**85. a.** B **b.** A **c.** C **d.** $(0, 0)$ and $(1, 1)$　**87.** $x^6 - x^3$　**89.** $x^2 + 2x - 15$

91. $x^4 - 10x^2 + 25$　**93.** $x^3 - 27$　**95.** $x^6 - x^5$　**97.** $12a^2 - 11ab + 2b^2$　**99.** $x^6 - 4$　**101.** $3x - 4x^3y$　**103.** $(x - 5)(x + 2)$

105. $(3x - 2)(2x + 5)$　**107.** x^2　**109.** t　**111.** $x^{1/3}$　**113.** $(9^{1/2} + 4^{1/2})^2 = (3 + 2)^2 = 5^2 = 25 \neq 9 + 4$　**115.** $\sqrt{\sqrt{a}} = (a^{1/2})^{1/2} = a^{1/4} = \sqrt[4]{a}$

Problem Set 9.2

1. $x + x^2$ **3.** $a^2 - a$ **5.** $6x^3 - 8x^2 + 10x$ **7.** $12x^2 - 36y^2$ **9.** $x^{4/3} - 2x^{2/3} - 8$ **11.** $a - 10a^{1/2} + 21$ **13.** $20y^{2/3} - 7y^{1/3} - 6$

15. $10x^{4/3} + 21x^{2/3}y^{1/2} + 9y$ **17.** $t + 10t^{1/2} + 25$ **19.** $x^3 + 8x^{3/2} + 16$ **21.** $a - 2a^{1/2}b^{1/2} + b$ **23.** $4x - 12x^{1/2}y^{1/2} + 9y$ **25.** $a - 3$

27. $x^3 - y^3$ **29.** $t - 8$ **31.** $4x^3 - 3$ **33.** $x + y$ **35.** $a - 8$ **37.** $8x + 1$ **39.** $t - 1$ **41.** $2x^{1/2} + 3$ **43.** $3x^{1/3} - 4y^{1/3}$

45. $3a - 2b$ **47.** $3(x - 2)^{1/2}(4x - 11)$ **49.** $5(x - 3)^{7/5}(x - 6)$ **51.** $3(x + 1)^{1/2}(3x^2 + 3x + 2)$ **53.** $(x^{1/3} - 2)(x^{1/3} - 3)$

55. $(a^{1/5} - 4)(a^{1/5} + 2)$ **57.** $(2y^{1/3} + 1)(y^{1/3} - 3)$ **59.** $(3t^{1/5} + 5)(3t^{1/5} - 5)$ **61.** $(2x^{1/7} + 5)^2$ **63.** -8 **65.** 90 **67.** -3

69. 0 **71.** $\frac{3 + x}{x^{1/2}}$ **73.** $\frac{x + 5}{x^{1/3}}$ **75.** $\frac{x^3 + 3x^2 + 1}{(x^3 + 1)^{1/2}}$ **77.** $\frac{-4}{(x^2 + 4)^{1/2}}$ **79.** 15.8% **81.** $\frac{1}{x^2 + 9}$ **83.** $3x - 4x^3y$ **85.** $5x - 4$

87. $x^2 + 5x + 25$ **89.** 5 **91.** 6 **93.** $4x^2y$ **95.** $5y$ **97.** 3 **99.** 2 **101.** $2ab$ **103.** 25 **105.** $48x^4y^2$ **107.** $4x^6y^6$

Problem Set 9.3

1. $2\sqrt{2}$ **3.** $7\sqrt{2}$ **5.** $12\sqrt{2}$ **7.** $4\sqrt{5}$ **9.** $4\sqrt{3}$ **11.** $15\sqrt{3}$ **13.** $3\sqrt[3]{2}$ **15.** $4\sqrt[3]{2}$ **17.** $6\sqrt[3]{2}$ **19.** $2\sqrt[5]{2}$ **21.** $3x\sqrt{2x}$

23. $2y\sqrt[4]{2y^3}$ **25.** $2xy^2\sqrt[3]{5xy}$ **27.** $4abc^2\sqrt[3]{3b}$ **29.** $2bc\sqrt[6]{6a^2c}$ **31.** $2xy^2\sqrt[5]{2x^3y^2}$ **33.** $3xy^2z\sqrt[5]{x^2}$ **35.** $2\sqrt{3}$

37. $\sqrt{-20}$, which is not a real number   **39.** $\frac{\sqrt{11}}{2}$ **41.** $\frac{2\sqrt{3}}{3}$ **43.** $\frac{5\sqrt{6}}{6}$ **45.** $\frac{\sqrt{2}}{2}$ **47.** $\frac{\sqrt{5}}{5}$ **49.** $2\sqrt[3]{4}$ **51.** $\frac{2\sqrt{3}}{3}$

53. $\frac{\sqrt{24x^2}}{2x}$ **55.** $\frac{\sqrt{8y^3}}{y}$ **57.** $\frac{\sqrt{36xy}}{3y}$ **59.** $\frac{\sqrt{6xy^2}}{3y}$ **61.** $\frac{\sqrt[4]{2x}}{2x}$ **63.** $\frac{3x\sqrt{15xy}}{5y}$ **65.** $\frac{5xy\sqrt{6xz}}{2z}$ **67.** $\frac{2ab\sqrt[3]{6ac^2}}{3c}$

69. $\frac{2xy^2\sqrt[3]{3z^2}}{3z}$ **71.** $\sqrt{x}$ **73.** $\sqrt[6]{xy}$ **75.** $\sqrt[12]{a}$ **77.** $x\sqrt[9]{6x}$ **79.** $ab^2c\sqrt[6]{c}$ **81.** $abc^2\sqrt[15]{3a^2b}$ **83.** $2b^2\sqrt[3]{a}$

85. $5|x|$ **87.** $3|xy|\sqrt{3x}$ **89.** $|x - 5|$ **91.** $|2x + 3|$ **93.** $2|a(a + 2)|$ **95.** $2|x|\sqrt{x - 2}$

97. $\sqrt{9 + 16} = \sqrt{25} = 5; \sqrt{9} + \sqrt{16} = 3 + 4 = 7$ **99.** $5\sqrt{13}$ feet   **101.** $r = \frac{7\sqrt{11}}{44}$ meters   **103. a.** 13 feet **b.** $2\sqrt{14} \approx 7.5$ feet

107. $7x$ **109.** $27xy^2$ **111.** $\frac{5}{6}x$ **113.** $3\sqrt{2}$ **115.** $5y\sqrt{3xy}$ **117.** $2a\sqrt[3]{ab^2}$ **119.** $\frac{y^3}{x^2}$ **121.** 1 **123.** $\frac{4x^2 - 6x + 9}{9x^2 - 3x + 1}$ **125.** $12\sqrt[3]{5}$

127. $6\sqrt[3]{49}$ **129.** $\frac{\sqrt[10]{a^7}}{a}$ **131.** $\frac{\sqrt[20]{a^9}}{a}$ **133.**

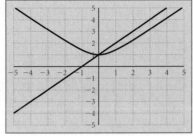

135. About $\frac{3}{4}$ of a unit   **137.** $x = 0$

Problem Set 9.4

1. $7\sqrt{5}$ **3.** $-x\sqrt{7}$ **5.** $\sqrt[3]{10}$ **7.** $9\sqrt[5]{6}$ **9.** 0 **11.** $\sqrt{5}$ **13.** $-32\sqrt{2}$ **15.** $-3x\sqrt{2}$ **17.** $-2\sqrt[3]{2}$ **19.** $8x\sqrt[3]{xy^2}$

21. $3a^2b\sqrt{3ab}$ **23.** $11ab\sqrt[3]{3a^2b}$ **25.** $10xy\sqrt[4]{3y}$ **27.** $\sqrt{2}$ **29.** $\frac{8\sqrt{5}}{15}$ **31.** $\frac{(x - 1)\sqrt{x}}{x}$ **33.** $\frac{3\sqrt{2}}{2}$ **35.** $\frac{5\sqrt{6}}{6}$ **37.** $\frac{8\sqrt{25}}{5}$

39. a. $8\sqrt{2x}$ **b.** $-4\sqrt{2x}$ **41. a.** $4\sqrt{2x}$ **b.** $2\sqrt{2x}$ **43. a.** $3x\sqrt{2}$ **b.** $-x\sqrt{2}$ **45. a.** $3\sqrt{2x} + 3$ **b.** $-\sqrt{2x} - 7$

47. $\sqrt{12} \approx 3.464; 2\sqrt{3} \approx 2(1.732) = 3.464$ **49.** $\sqrt{8} + \sqrt{18} \approx 2.828 + 4.243 = 7.071; \sqrt{50} \approx 7.071; \sqrt{26} \approx 5.099$ **55.** $\frac{\sqrt{3}}{2}$ **57.** $\frac{\sqrt{6}}{3}$

59. 1 **61.** $\frac{13 - 3t}{3 - t}$ **63.** $\frac{6}{4x + 3}$ **65.** $\frac{x - y}{x^2 + xy + y^2}$ **67.** 6 **69.** $4x^2 + 3xy - y^2$ **71.** $x^2 + 6x + 9$ **73.** $x^2 - 4$ **75.** $6\sqrt{2}$

77. 6 **79.** $9x$ **81.** $\frac{\sqrt{6}}{2}$ **83.** $-xy$ **85.** $-xy^2z^3$ **87.** $5b^2c\sqrt[4]{2a^3b}$ **89.** $-3ab\sqrt[3]{2a}$

Problem Set 9.5

1. $3\sqrt{2}$ **3.** $10\sqrt{21}$ **5.** 720 **7.** 54 **9.** $\sqrt{6} - 9$ **11.** $24 + 6\sqrt[3]{4}$ **13.** $2 + 2\sqrt[3]{3}$ **15.** $xy\sqrt[3]{y} + x^2\sqrt[3]{y}$

17. $2x^2\sqrt[4]{x} + 2x^3\sqrt[4]{2}$ **19.** $7 + 2\sqrt{6}$ **21.** $x + 2\sqrt{x} - 15$ **23.** $34 + 20\sqrt{3}$ **25.** $19 + 8\sqrt{3}$ **27.** $x - 6\sqrt{x} + 9$

29. $4a - 12\sqrt{ab} + 9b$ **31.** $x + 4\sqrt{x - 4}$ **33.** $x + 4 - 6\sqrt{x - 5}$ **35.** 1 **37.** $a - 49$ **39.** $25 - x$ **41.** $x - 8$

43. $10 + 6\sqrt{3}$ **45.** $5 + \sqrt[3]{12} + \sqrt[3]{18}$ **47.** $x^2 + x\sqrt[3]{x^2y^2} + \sqrt[3]{xy} + y$ **49.** $\frac{\sqrt{2}}{2}$ **51.** $\frac{\sqrt{x}}{x}$ **53.** $\frac{4\sqrt{3}}{3}$ **55.** $\frac{x\sqrt{6}}{3}$ **57.** $\frac{2\sqrt{10x}}{5x}$

59. $\frac{\sqrt{2}}{2}$ **61.** $\frac{2x\sqrt{2y}}{y}$ **63.** $\frac{a\sqrt{6c}}{2bc}$ **65.** $\frac{2ac\sqrt[3]{b^2}}{b}$ **67.** $\frac{1 + \sqrt{3}}{2}$ **69.** $\frac{5 - \sqrt{5}}{4}$ **71.** $\frac{x + 3\sqrt{x}}{x - 9}$ **73.** $\frac{10 + 3\sqrt{5}}{11}$ **75.** $\frac{3\sqrt{x} + 3\sqrt{y}}{x - y}$

77. $2 + \sqrt{3}$ **79.** $\frac{11 - 4\sqrt{7}}{3}$ **81.** $\frac{a + 2\sqrt{ab} + b}{a - b}$ **83.** $\frac{x + 4\sqrt{x} + 4}{x - 4}$ **85.** $\frac{5 - \sqrt{21}}{4}$ **87.** $\frac{\sqrt{x} - 3x + 2}{1 - x}$ **89.** $\frac{8\sqrt{3}}{3}$ **91.** $\frac{11\sqrt{5}}{5}$

93. $-4\sqrt{3}$ **95.** $5\sqrt{3}$ **99.** $10\sqrt{3}$ **101.** $x + 6\sqrt{x} + 9$ **103.** 75 **105.** $\frac{5\sqrt{2}}{4}$ seconds; $\frac{5}{2}$ seconds   **111.** $-\frac{1}{8}$ **113.** $\frac{y - 2}{y + 2}$

115. $\frac{2x + 1}{2x - 1}$ **117.** $t^2 + 10t + 25$ **119.** x **121.** 7 **123.** $-4, -3$ **125.** $-6, -3$ **127.** $-5, -2$ **129.** $\frac{x(\sqrt{x - 2} - 4)}{x - 18}$

131. $\frac{x(\sqrt{x + 5} + 5)}{x - 20}$ **133.** $\frac{3(\sqrt{5x} - x)}{5 - x}$

Problem Set 9.6

1. 4   **3.** ∅   **5.** 5   **7.** ∅   **9.** $\frac{39}{2}$ **11.** ∅   **13.** 5   **15.** 3   **17.** $-\frac{32}{3}$ **19.** 3, 4   **21.** −1, −2   **23.** 11   **25.** 3, 7
27. $-\frac{1}{4}$, 14   **29.** −1   **31.** ∅   **33.** 7   **35.** 0, 3   **37.** −4   **39.** 8   **41.** 0   **43.** 9   **45.** 0   **47.** 8
49. ∅; 9 does not check   **51.** 0; 32 does not check   **53.** 6, −2 does not check   **55.** $\frac{1}{2}$ **57.** $\frac{1}{2}$, 1   **59.** 5, 13   **61.** $-\frac{3}{2}$ **63.** ∅
65. $\frac{1}{2}$ **67.** ∅   **69.** 1   **71.** −11

73.

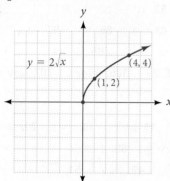

75.

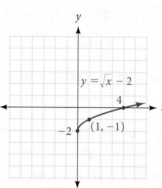

77.

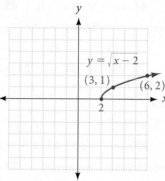

79.

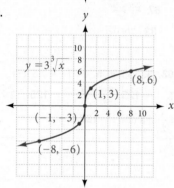

81.

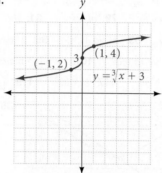

83.

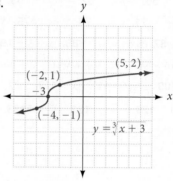

85. $h = 100 - 16t^2$ **87.** $\frac{392}{121} \approx 3.24$ feet   **89.** 5 meters   **91.** 2,500 meters   **93.** $\sqrt{6} - 2$ **95.** $x + 10\sqrt{x} + 25$
97. $\frac{x - 3\sqrt{x}}{x - 9}$ **99.** 5   **101.** $2\sqrt{3}$ **103.** −1   **105.** 1   **107.** 4   **109.** 2   **111.** $10 - 2x$ **113.** $2 - 3x$
115. $6 + 7x - 20x^2$ **117.** $8x - 12x^2$ **119.** $4 + 12x + 9x^2$ **121.** $4 - 9x^2$

Problem Set 9.7

1. $6i$ **3.** $-5i$ **5.** $6i\sqrt{2}$ **7.** $-2i\sqrt{3}$ **9.** 1   **11.** −1   **13.** $-i$ **15.** $x = 3, y = -1$ **17.** $x = -2, y = -\frac{1}{2}$ **19.** $x = -8, y = -5$
21. $x = 7, y = \frac{1}{2}$ **23.** $x = \frac{3}{7}, y = \frac{2}{5}$ **25.** $5 + 9i$ **27.** $5 - i$ **29.** $2 - 4i$ **31.** $1 - 6i$ **33.** $2 + 2i$ **35.** $-1 - 7i$ **37.** $6 + 8i$
39. $2 - 24i$ **41.** $-15 + 12i$ **43.** $18 + 24i$ **45.** $10 + 11i$ **47.** $21 + 23i$ **49.** $-2 + 2i$ **51.** $2 - 11i$ **53.** $-21 + 20i$ **55.** $-2i$
57. $-7 - 24i$ **59.** 5   **61.** 40   **63.** 13   **65.** 164   **67.** $-3 - 2i$ **69.** $-2 + 5i$ **71.** $\frac{8}{13} + \frac{12}{13}i$ **73.** $-\frac{18}{13} - \frac{12}{13}i$
75. $-\frac{5}{13} + \frac{12}{13}i$ **77.** $\frac{13}{15} - \frac{2}{5}i$ **79.** $R = -11 - 7i$ ohms   **81.** $-\frac{3}{2}$ **83.** $-3, \frac{1}{2}$ **85.** $\frac{5}{4}$ or $\frac{4}{5}$ **87.** $\frac{1}{i} \cdot \frac{i}{i} = \frac{i}{i^2} = \frac{i}{-1} = -i$
89.

$$(1 + i)^2 - 2(1 + i) + 2 \stackrel{?}{=} 0$$
$$1 + 2i + i^2 - 2 - 2i + 2 \stackrel{?}{=} 0$$
$$1 - 1 - 2 + 2 + 2i - 2i \stackrel{?}{=} 0$$
$$0 = 0$$

91.

$$(2 + i)^3 - 11(2 + i) + 20 \stackrel{?}{=} 0$$
$$8 + 12i + 6i^2 + i^3 - 22 - 11i + 20 \stackrel{?}{=} 0$$
$$8 - 6 - 22 + 20 + 12i - i - 11i \stackrel{?}{=} 0$$
$$0 = 0$$

Chapter 9 Review

1. 7   **2.** −3   **3.** 2   **4.** 27   **5.** $2x^3y^2$ **6.** $\frac{1}{16}$ **7.** x^2 **8.** a^2b^4 **9.** $a^{7/20}$ **10.** $a^{5/12}b^{8/3}$ **11.** $12x + 11x^{1/2}y^{1/2} - 15y$
12. $a^{2/3} - 10a^{1/3} + 25$ **13.** $4x^{1/2} + 2x^{5/6}$ **14.** $2(x - 3)^{1/4}(4x - 13)$ **15.** $\frac{x + 5}{x^{1/4}}$ **16.** $2\sqrt{3}$ **17.** $5\sqrt{2}$ **18.** $2\sqrt[3]{2}$ **19.** $3x\sqrt{2}$
20. $4ab^2c\sqrt{5a}$ **21.** $2abc\sqrt[4]{2bc^2}$ **22.** $\frac{3\sqrt{2}}{2}$ **23.** $3\sqrt[3]{4}$ **24.** $\frac{4x\sqrt{21xy}}{7y}$ **25.** $\frac{2y\sqrt[3]{45x^2z^2}}{3z}$ **26.** $-2x\sqrt{6}$ **27.** $3\sqrt{3}$
28. $\frac{8\sqrt{5}}{5}$ **29.** $7\sqrt{2}$ **30.** $11a^2b\sqrt{3ab}$ **31.** $-4xy\sqrt[3]{xz^2}$ **32.** $\sqrt{6} - 4$ **33.** $x - 5\sqrt{x} + 6$ **34.** $3\sqrt{5} + 6$ **35.** $6 + \sqrt{35}$

36. $\frac{63 + 12\sqrt{7}}{47}$ **37.** 0 **38.** 3 **39.** 5 **40.** $\varnothing$ **41.**

42.

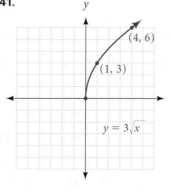

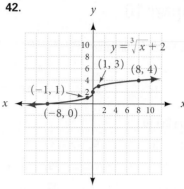

43. 1 **44.** $-i$ **45.** $x = -\frac{3}{2}, y = -\frac{1}{2}$

46. $x = -2, y = -4$ **47.** $9 + 3i$ **48.** $-7 + 4i$

49. $-6 + 12i$ **50.** $5 + 14i$ **51.** $12 + 16i$

52. 25 **53.** $1 - 3i$ **54.** $-\frac{6}{5} + \frac{3}{5}i$ **55.** 22.5 feet

56. 65 yards

Chapter 9 Cumulative Review

1. -17 **2.** 12 **3.** 5 **4.** $2\sqrt[4]{3}$ **5.** 6 **6.** $\frac{1}{4}$ **7.** $-\frac{2}{3}$ **8.** 13 **9.** $\frac{5}{9}$ **10.** $3x + 4y$ **11.** $\frac{x + 2}{x + 1}$ **12.** $\frac{a^2}{6b}$

13. $2x - 11\sqrt{x} + 15$ **14.** $22 + 4i$ **15.** $\frac{19}{25} - \frac{8}{25}i$ **16.** $\frac{2\sqrt{3} + 1}{11}$ **17.** $6x^2 - 19xy + 10y^2$ **18.** $3x^2 + 20x + 25$ **19.** $\frac{6}{y^2}$

20. $\frac{x + 2\sqrt{xy} + y}{x - y}$ **21.** $\frac{2}{3}$ **22.** 2, 3 **23.** $\frac{5 \pm 2\sqrt{3}}{2}$ **24.** $-\frac{4}{3}, 2$ **25.** 4 **26.** $-4, 0, \frac{7}{3}$ **27.** $-\frac{5}{4}, \frac{1}{2}$ **28.** $\frac{5}{6}$ **29.** $-\frac{7}{3}, 3$

30. 3 **31.** $8 \le x \le 16$ **32.** $x < -2$ or $x > 5$ **33.** $(-2, 4)$

34. $(-8, -6)$ **35.** $\left(-\frac{36}{35}, \frac{116}{35}, -\frac{26}{5}\right)$ **36.** $\left(\frac{1}{2}, -\frac{3}{2}\right)$

37.

38.

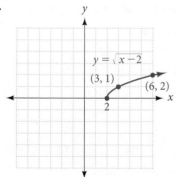

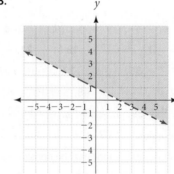

39. -7 **40.** -10 **41.** $3\sqrt[4]{2}$ **42.** $\frac{\sqrt[3]{2x^2y}}{x}$

43. $(5a - 2b)(5a + 2b)(25a^2 + 4b^2)$

44. $2(3a^2 - 1)(4a^2 + 3)$ **45.** $y = 1$ **46.** $z = -6$

47. $b^2 - 3b$ **48.** 2 **49.** -1 **50.** 28

51. 22,000,000 arrivals **52.** 5,000,000 people

Chapter 9 Test

1. $\frac{1}{9}$ **2.** $\frac{7}{5}$ **3.** $a^{5/12}$ **4.** $\frac{x^{13/12}}{y}$ **5.** $7x^4y^5$ **6.** $2x^2y^4$ **7.** $2a$ **8.** $x^{n^2 - n}y^{1 - n^3}$ **9.** $6a^2 - 10a$ **10.** $16a^3 - 40a^{3/2} + 25$

11. $(3x^{1/3} - 1)(x^{1/3} + 2)$ **12.** $(3x^{1/3} - 7)(3x^{1/3} + 7)$ **13.** $5xy^2\sqrt{5xy}$ **14.** $2x^2y^2\sqrt[3]{5xy^2}$ **15.** $\frac{\sqrt{6}}{3}$ **16.** $\frac{2a^2b\sqrt{15bc}}{5c}$

17. $\frac{4 + x}{x^{1/2}}$ **18.** $\frac{3}{(x^2 - 3)^{1/2}}$ **19.** $-6\sqrt{3}$ **20.** $-ab\sqrt[3]{3}$ **21.** $x + 3\sqrt{x} - 28$ **22.** $21 - 6\sqrt{6}$ **23.** $\frac{5\sqrt{3} + 5}{2}$ **24.** $\frac{x - 2\sqrt{2x} + 2}{x - 2}$

25. 8; 1 does not check **26.** -4 **27.** -3

28.

29.

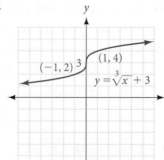

30. $x = \frac{1}{2}; y = 7$ **31.** $6i$ **32.** $17 - 6i$

33. $9 - 40i$ **34.** $-\frac{5}{13} - \frac{12}{13}i$

35. $i^{38} = (i^2)^{19} = (-1)^{19} = -1$

36. E **37.** A **38.** F **39.** B **40.** D **41.** C

Chapter 10

Getting Ready for Chapter 10

1. 4   **2.** -1 **3.** $\frac{33}{4} = 8\frac{1}{4}$ **4.** $-\frac{5}{9}$ **5.** ± 5 **6.** ± 3 **7.** $\pm 2\sqrt{3}$ **8.** $\pm 3\sqrt{2}$ **9.** 9   **10.** 4   **11.** $\frac{25}{4}$ **12.** $\frac{49}{4}$
13. $\frac{16}{9}$ **14.** $\frac{1}{25}$ **15.** $(x-3)^2$ **16.** $(x-4)^2$ **17.** $(x+6)^2$ **18.** $(5t+1)(25t^2-5t+1)$ **19.** $\frac{b^2-4ac}{4a^2}$ **20.** $9a^2-6a-3$

Problem Set 10.1

1. ± 5 **3.** $\pm\frac{\sqrt{3}}{2}$ **5.** $\pm 2i\sqrt{3}$ **7.** $\pm\frac{3\sqrt{5}}{2}$ **9.** $-2, 3$ **11.** $\frac{-3\pm 3i}{2}$ **13.** $-4\pm 3i\sqrt{3}$ **15.** $\frac{3\pm 2i}{2}$
17. $x^2+12x+36=(x+6)^2$ **19.** $x^2-4x+4=(x-2)^2$ **21.** $a^2-10a+25=(a-5)^2$ **23.** $x^2+5x+\frac{25}{4}=\left(x+\frac{5}{2}\right)^2$
25. $y^2-7y+\frac{49}{4}=\left(y-\frac{7}{2}\right)^2$ **27.** $x^2+\frac{1}{2}x+\frac{1}{16}=\left(x+\frac{1}{4}\right)^2$ **29.** $x^2+\frac{2}{3}x+\frac{1}{9}=\left(x+\frac{1}{3}\right)^2$ **31.** $-3, -9$ **33.** $1\pm 2i$ **35.** $4\pm\sqrt{15}$
37. $\frac{5\pm\sqrt{37}}{2}$ **39.** $1\pm\sqrt{5}$ **41.** $\frac{4\pm\sqrt{13}}{3}$ **43.** $\frac{3\pm i\sqrt{71}}{8}$ **45. a.** no   **b.** $\pm 3i$ **47. a.** $0, 6$ **b.** $0, 6$ **49. a.** $-7, 5$ **b.** $-7, 5$
51. no   **53. a.** $\frac{7}{5}$ **b.** 3   **c.** $\frac{7\pm 2\sqrt{2}}{5}$ **d.** $\frac{7}{5}$ **e.** 3   **55.** ± 1 **57.** ± 1 **59.** ± 1

61.

x	$f(x)$	$g(x)$	$h(x)$
-2	49	49	25
-1	25	25	13
0	9	9	9
1	1	1	13
2	1	1	25

63. 3   **65. a.** $-1, 6$ **b.** $-1, 6$ **c.** -6 **d.** -10 **67.** $\frac{\sqrt{3}}{2}$ inch, 1 inch   **69.** $\sqrt{2}$ inches
71. 781 feet   **73.** 7.3% to the nearest tenth   **75.** $20\sqrt{2}\approx 28$ feet   **77.** $3\sqrt{5}$
79. $3y^2\sqrt{3y}$ **81.** $3x^2y\sqrt[3]{2y^2}$ **83.** $\frac{3\sqrt{2}}{2}$ **85.** $\sqrt[3]{2}$ **87.** 13   **89.** 7   **91.** $\frac{1}{2}$ **93.** 13
95. $(3t-2)(9t^2+6t+4)$ **97.** $x=\pm 2a$ **99.** $x=\frac{-p\pm\sqrt{p^2-4q}}{2}$ **101.** $x=\frac{-p\pm\sqrt{p^2-12q}}{6}$
103. $(x-5)^2+(y-3)^2=2^2$

Problem Set 10.2

1. a. $\frac{-2\pm\sqrt{10}}{3}$ **b.** $\frac{2\pm\sqrt{10}}{3}$ **c.** $\frac{-2\pm i\sqrt{2}}{3}$ **d.** $\frac{-2\pm\sqrt{10}}{2}$ **e.** $\frac{2\pm i\sqrt{2}}{2}$ **3. a.** $1\pm i$ **b.** $1\pm 2i$ **c.** $-1\pm i$ **5.** $1, 2$ **7.** $\frac{2\pm i\sqrt{14}}{3}$
9. $0, 5$ **11.** $0, -\frac{4}{3}$ **13.** $\frac{3\pm\sqrt{5}}{4}$ **15.** $-3\pm\sqrt{17}$ **17.** $\frac{-1\pm i\sqrt{5}}{2}$ **19.** 1   **21.** $\frac{1\pm i\sqrt{47}}{6}$ **23.** $4\pm\sqrt{2}$ **25.** $\frac{1}{2}, 1$
27. $-\frac{1}{2}, 3$ **29.** $\frac{-1\pm i\sqrt{7}}{2}$ **31.** $1\pm\sqrt{2}$ **33.** $\frac{-3\pm\sqrt{5}}{2}$ **35.** $3, -5$ **37.** $2, -1\pm i\sqrt{3}$ **39.** $-\frac{3}{2}, \frac{3\pm 3i\sqrt{3}}{4}$
41. $\frac{1}{5}, \frac{-1\pm i\sqrt{3}}{10}$ **43.** $0, \frac{-1\pm i\sqrt{5}}{2}$ **45.** $0, 1\pm i$ **47.** $0, \frac{-1\pm i\sqrt{2}}{3}$ **49.** a and b   **51. a.** $\frac{5}{3}, 0$ **b.** $\frac{5}{3}, 0$ **53.** No, $2\pm i\sqrt{3}$
55. Yes   **57. a.** $-1, 3$ **b.** $1\pm i\sqrt{7}$ **c.** ± 2 **d.** $2\pm 2\sqrt{2}$ **59. a.** $-2, \frac{5}{3}$ **b.** $\frac{3}{2}, \frac{-3\pm 3i\sqrt{3}}{4}$ **c.** $\varnothing$ **d.** ± 1 **61.** 2 seconds
63. 20 or 60 items   **65.** 0.49 centimeter (8.86 cm is impossible)   **67. a.** $\ell + w = 10, \ell w = 15$ **b.** 8.16 yards, 1.84 yards
  c. Two answers are possible because either dimension (long or short) may be considered the length.
69. $4y+1-\frac{2}{2y-7}$ **71.** $x^2+7x+12$ **73.** 5   **75.** $\frac{27}{125}$ **77.** $\frac{1}{4}$ **79.** $21x^3y$ **81.** 169   **83.** 0   **85.** ± 12 **87.** x^2-x-6
89. $x^3-4x^2-3x+18$ **91.** $-2\sqrt{3}, \sqrt{3}$ **93.** $\frac{-\sqrt{2}\pm\sqrt{6}}{2}$ **95.** $-2i, i$

Problem Set 10.3

1. $D=16$, two rational   **3.** $D=0$, one rational   **5.** $D=5$, two irrational   **7.** $D=17$, two irrational   **9.** $D=36$, two rational
11. $D=116$, two irrational   **13.** ± 10 **15.** ± 12 **17.** 9   **19.** -16 **21.** $\pm 2\sqrt{6}$ **23.** $x^2-7x+10=0$ **25.** $t^2-3t-18=0$
27. $y^3-4y^2-4y+16=0$ **29.** $2x^2-7x+3=0$ **31.** $4t^2-9t-9=0$ **33.** $6x^3-5x^2-54x+45=0$ **35.** $10a^2-a-3=0$
37. $9x^3-9x^2-4x+4=0$ **39.** $x^4-13x^2+36=0$ **41.** $f(x)=x^2+x-2$ **43.** $f(x)=x^2-6x+7$ **45.** $f(x)=x^2-4x+5$
47. $f(x)=2x^2-10x+9$ **49.** $f(x)=2x^2-6x+7$ **51.** $f(x)=x^3-4x^2+7x$ **53. a.** $g(x)$ **b.** $h(x)$ **c.** $f(x)$
55. $a^{11/2}-a^{9/2}$ **57.** $x^3-6x^{3/2}+9$ **59.** $6x^{1/2}-5x$ **61.** $5(x-3)^{1/2}(2x-9)$ **63.** $(2x^{1/3}-3)(x^{1/3}-4)$ **65.** x^2+4x-5
67. $4a^2-30a+56$ **69.** $32a^2+20a-18$ **71.** $\pm\frac{1}{2}$ **73.** no solution   **75.** 1   **77.** $-2, 4$ **79.** $-2, \frac{1}{4}$
81. $-1, 5$ **83.** $1+\sqrt{2}\approx 2.41; 1-\sqrt{2}\approx -0.41$ **85.** $-1, \frac{1}{2}, 1$ **87.** $\frac{1}{2}, \sqrt{2}\approx 1.41, -\sqrt{2}\approx -1.41$

Problem Set 10.4

1. $1, 2$ **3.** $-8, -\frac{5}{2}$ **5.** $\pm 3, \pm 1$ **7.** $\pm 2, \pm\sqrt{3}$ **9.** $\pm 3, \pm i\sqrt{3}$ **11.** $\pm 2i, \pm i\sqrt{5}$ **13.** $\frac{7}{2}, 4$ **15.** $-\frac{9}{8}, \frac{1}{2}$ **17.** $\pm\frac{\sqrt{30}}{6}, \pm i$
19. $\pm\frac{\sqrt{21}}{3}, \pm\frac{i\sqrt{21}}{3}$ **21.** $4, 25$ **23.** 25; 9 does not check   **25.** $\frac{25}{9}; \frac{49}{4}$ does not check   **27.** 16; 4 does not check
29. 9; 36 does not check   **31.** $\frac{1}{4}$; 25 does not check   **33.** 27, 38   **35.** 4, 12   **37.** $\pm 2, \pm 2i$ **39.** $\pm 2, \pm 2i\sqrt{2}$ **41.** $\frac{1}{2}, 1$
43. $\pm\frac{\sqrt{5}}{5}$ **45.** $-216, 8$ **47.** $\pm 3i\sqrt{3}$ **49.** $t=\frac{v\pm\sqrt{v^2+64h}}{32}$ **51.** $x=\frac{-4\pm 2\sqrt{4-k}}{k}$ **53.** $x=-y$ **55.** $t=\frac{1\pm\sqrt{1+h}}{4}$

57. a.

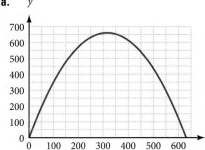

b. 630 feet   **61.** $3\sqrt{7}$ **63.** $5\sqrt{2}$ **65.** $39x^2y\sqrt{5x}$ **67.** $-11 + 6\sqrt{5}$

69. $x + 4\sqrt{x} + 4$ **71.** $\frac{7 + 2\sqrt{7}}{3}$ **73.** -2 **75.** $1{,}322.5$ **77.** $-\frac{7}{640}$

79. $1, 5$ **81.** $-3, 1$ **83.** $\frac{3}{2} \pm \frac{1}{2}i$ **85.** $9; 3$ **87.** $1; 1$

89. x-intercepts $= -2, 0, 2$; y-intercept $= 0$

91. x-intercepts $= -3, -\frac{1}{3}, 3$; y-intercept $= -9$ **93.** $\frac{1}{2}$ and -1

Problem Set 10.5

1. x-intercepts $= -3, 1$;
Vertex $= (-1, -4)$

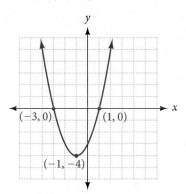

3. x-intercepts $= -5, 1$;
Vertex $= (-2, 9)$

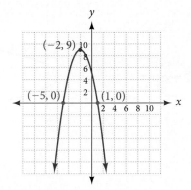

5. x-intercepts $= -1, 1$;
Vertex $= (0, -1)$

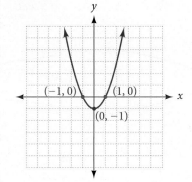

7. x-intercepts $= 3, -3$;
Vertex $= (0, 9)$

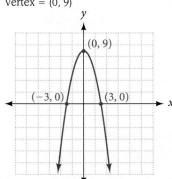

9. x-intercepts $= -1, 3$;
Vertex $= (1, -8)$

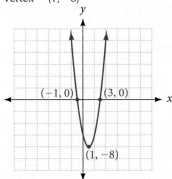

11. x-intercepts $= 1 + \sqrt{5}, 1 - \sqrt{5}$;
Vertex $= (1, -5)$

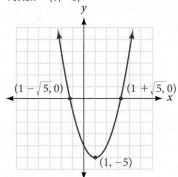

13. Vertex $= (2, -8)$

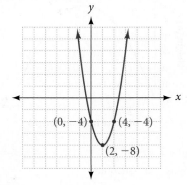

15. Vertex $= (1, -4)$

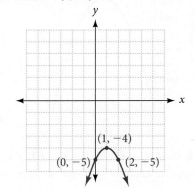

17. Vertex $= (0, 1)$

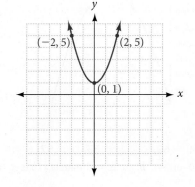

19. Vertex = $(0, -3)$

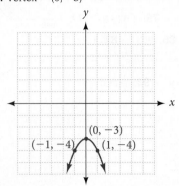

$(0, -3)$
$(-1, -4)$ $(1, -4)$

21. $(3, -4)$ lowest   **23.** $(1, 9)$ highest   **25.** $(2, 16)$ highest   **27.** $(-4, 16)$ highest

29.

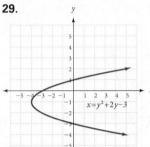

$x = y^2 + 2y - 3$

31.

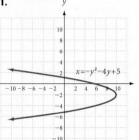

$x = -y^2 - 4y + 5$

33.

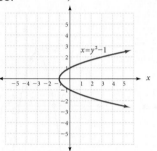

$x = y^2 - 1$

35.

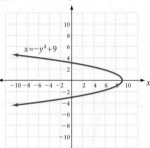

$x = -y^2 + 9$

37.

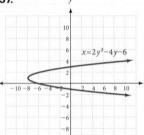

$x = 2y^2 - 4y - 6$

39.

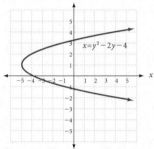

$x = y^2 - 2y - 4$

41. 875 patterns; maximum profit $731.25

43. The ball is in her hand when $h(t) = 0$, which means $t = 0$ or $t = 2$ seconds. Maximum height is $h(1) = 16$ feet.

45. Maximum $R = \$3,600$ when $p = \$6.00$ **47.** Maximum $R = \$7,225$ when $p = \$8.50$ **49.** $y = -\frac{1}{135}(x - 90)^2 + 60$

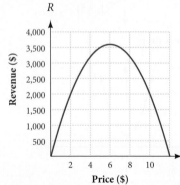

Price ($)

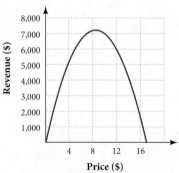

Price ($)

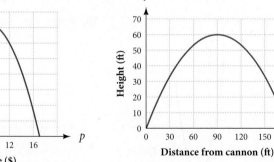

Distance from cannon (ft)

51. $-2, 4$ **53.** $-\frac{1}{2}, \frac{2}{3}$ **55.** 3   **57.** $1 - i$ **59.** $27 + 5i$ **61.** $\frac{1}{10} + \frac{3}{10}i$ **63.** $y = (x - 2)^2 - 4$

65. **a.** 20° or 70°   **b.** At approximately 170 feet   **c.** Approximately 42 or 43 feet   **d.** Approximately 10°   **e.** Quadratic

Problem Set 10.6

1.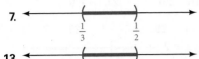

3.

5.

7.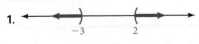

9.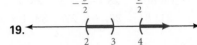

11.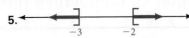

13.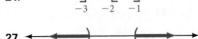

15. All real numbers   **17.** No solution; $\varnothing$ **19.**

21.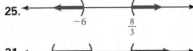

23.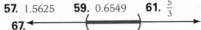

25.

27.

29.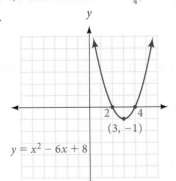

31.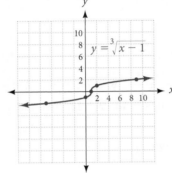

33. a. $x - 1 > 0$ **b.** $x - 1 \geq 0$ **c.** $x - 1 \geq 0$ **35.** $\varnothing$ **37.** $x = 1$ **39.** all real numbers   **41.** $x \neq 1$ **43.** $\varnothing$ **45.** $x \neq 3$

47. a. $-2 < x < 2$ **b.** $x < -2$ or $x > 2$ **c.** $x = -2$ or $x = 2$ **49. a.** $-2 < x < 5$ **b.** $x < -2$ or $x > 5$ **c.** $x = -2$ or $x = 5$

51. a. $x < -1$ or $1 < x < 3$ **b.** $-1 < x < 1$ or $x > 3$ **c.** $x = -1$ or $x = 1$ or $x = 3$ **53.** $x \geq 4$; the width is at least 4 inches

55. $5 \leq p \leq 8$; charge at least \$5 but no more than \$8 for each set   **57.** 1.5625   **59.** 0.6549   **61.** $\frac{5}{3}$

63. 6; 1 does not check   **65.**

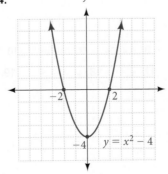

67.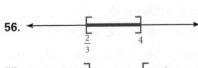

69.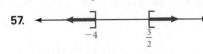

Chapter 10 Review

1. $0, 5$ **2.** $0, \frac{4}{3}$ **3.** $\frac{4 \pm 7i}{3}$ **4.** $-3 \pm \sqrt{3}$ **5.** $-5, 2$ **6.** $-3, -2$ **7.** 3   **8.** 2   **9.** $\frac{-3 \pm \sqrt{3}}{2}$ **10.** $\frac{3 \pm \sqrt{5}}{2}$ **11.** $-5, 2$

12. $0, \frac{9}{4}$ **13.** $\frac{2 \pm i\sqrt{15}}{2}$ **14.** $1 \pm \sqrt{2}$ **15.** $-\frac{4}{3}, \frac{1}{2}$ **16.** 3   **17.** $5 \pm \sqrt{7}$ **18.** $0, \frac{5 \pm \sqrt{21}}{2}$ **19.** $-1, \frac{3}{5}$ **20.** $3, \frac{-3 \pm 3i\sqrt{3}}{2}$

21. $\frac{1 \pm i\sqrt{2}}{3}$ **22.** $\frac{3 \pm \sqrt{29}}{2}$ **23.** 100 or 170 items   **24.** 20 or 60 items   **25.** $D = 0$; 1 rational   **26.** $D = 0$; 1 rational

27. $D = 25$; 2 rational   **28.** $D = 361$; 2 rational   **29.** $D = 5$; 2 irrational   **30.** $D = 5$; 2 irrational   **31.** $D = -23$; 2 complex

32. $D = -87$; 2 complex   **33.** ± 20 **34.** ± 20 **35.** 4   **36.** 4   **37.** 25   **38.** 49   **39.** $x^2 - 8x + 15 = 0$

40. $x^2 - 2x - 8 = 0$ **41.** $2y^2 + 7y - 4 = 0$ **42.** $t^3 - 5t^2 - 9t + 45 = 0$ **43.** $-4, 12$ **44.** $-\frac{3}{4}, -\frac{1}{6}$ **45.** $\pm 2, \pm i\sqrt{3}$

46. 4; 1 does not check   **47.** $\frac{9}{4}, 16$ **48.** 4   **49.** 4   **50.** 7   **51.** $t = \frac{5 \pm \sqrt{25 + 16h}}{16}$ **52.** $t = \frac{v \pm \sqrt{v^2 + 640}}{32}$

53.

$y = x^2 - 6x + 8$
$(3, -1)$

54.

$y = x^2 - 4$

55.

56.

57.

58.

Chapter 10 Cumulative Review

1. 0   **2.** 88   **3.** $-\frac{8}{27}$ **4.** $\frac{25}{16}$ **5.** $-7x + 10$ **6.** $10x - 37$ **7.** $\frac{x^3}{y^7}$ **8.** $\frac{a^8}{b^2}$ **9.** $2\sqrt[3]{4}$ **10.** $5x\sqrt{2x}$ **11.** $\frac{9}{20}$

12. $\frac{9}{4}$ **13.** $\frac{1}{7}$ **14.** $\frac{33}{5}$ **15.** $x - 6y$ **16.** $x - 3$ **17.** $1 - 3i$ **18.** $\frac{23}{13} + \frac{11}{13i}$ **19.** 15   **20.** $\frac{35}{6}$ **21.** $-9, 9$

22. No solution   **23.** $-5, -3$ **24** $a = 2$ **25.** $\frac{4 \pm 3\sqrt{2}}{3}$ **26.** $0, \frac{5 \pm \sqrt{37}}{6}$ **27.** $-\frac{3}{2}, -1$ **28.** $\frac{-1 \pm i\sqrt{47}}{12}$ **29.** $-2, -3$

30. $\frac{9}{4}$ **31.** 81   **32.** $-8, 27$ **33.** $8 \le x \le 20$    **34.** $x \le -\frac{1}{2}, x \ge 2$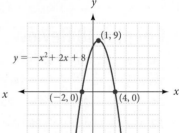

35. $-4 < x < 2$ 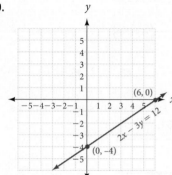   **36.** $x < -3$ or $x \ge 1$ 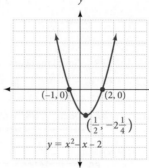   **37.** No solution   **38.** No solution

39. 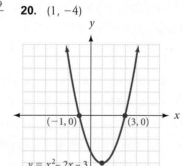   **40.**

41. $y = \frac{4}{3}x - \frac{2}{3}$ **42.** $y = -2x + 7$

43. $(x + 4 - y)(x + 4 + y)$ **44.** $(x - 5)(x - 2)$

45. $\frac{7\sqrt[3]{3}}{3}$ **46.** $\frac{3 - \sqrt{3}}{2}$ **47.** $\{2, 4\}$ **48.** $\{0, 6\}$

49. $25°, 55°, 100°$ **50.** $6,000 at 5%; $2,000 at 4%

51. $\frac{25}{16}$ **52.** 6   **53.** $3,012.50   **54.** $3,594

Chapter 10 Test

1. $-\frac{9}{2}, \frac{1}{2}$ **2.** $3 \pm i\sqrt{2}$ **3.** $5 \pm 2i$ **4.** $1 \pm i\sqrt{2}$ **5.** $\frac{5}{2}, \frac{-5 \pm 5i\sqrt{3}}{4}$ **6.** $-1 \pm i\sqrt{5}$ **7.** $r = -1 \pm \frac{\sqrt{A}}{8}$ **8.** $2 \pm \sqrt{2}$

9. 2 in., $2\sqrt{3}$ in.   **10.** $\frac{1}{2}$ second, $\frac{3}{2}$ seconds   **11.** 15 or 100 cups   **12.** 9   **13.** Two rational solutions

14. $3x^2 - 13x - 10 = 0$ **15.** $x^3 - 7x^2 - 4x + 28 = 0$

16. $\pm\frac{1}{2}i, \pm\sqrt{2}$ **17.** $\frac{1}{2}, 1$ **18.** $\frac{1}{4}, 9$ **19.** $t = \frac{7 \pm \sqrt{16h + 49}}{16}$ **20.** $(1, -4)$ **21.** $(1, 9)$

22. $y = -(x + 1)^2 + 4$

23.

24.   **25.** $900

26. a. $x < -2$ or $-1 < x < 1$
  b. $-2 < x < -1$ or $x > 1$
  c. $x = -2$ or $x = -1$ or $x = 1$

Chapter 11

Getting Ready for Chapter 11

1. 1   **2.** 4   **3.** $\frac{1}{4}$ **4.** 9   **5.** 1   **6.** $\frac{1}{9}$ **7.** 9   **8.** 0.4204   **9.** 0.1768   **10.** 504.5378   **11.** 265.1650   **12.** 742.9737

13. 905.6808   **14.** 1.0075   **15.** 1.02   **16.** $-4, 1$ **17.** $\frac{101}{24}$ **18.** $y = \frac{x + 3}{2}$ **19.** $y = \pm\sqrt{x + 2}$ **20.** $-\frac{6}{5}, -2$

Problem Set 11.1

1. 1   **3.** 2   **5.** $\frac{1}{27}$ **7.** 13

9.

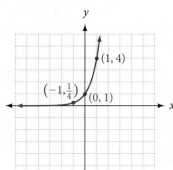

11.

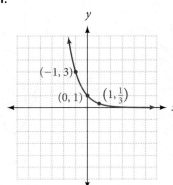

13.

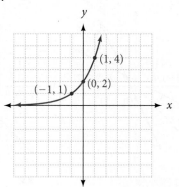

15.

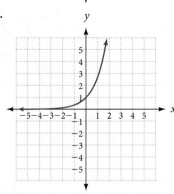

17.

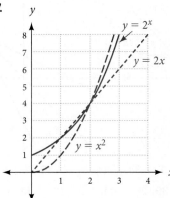

19.

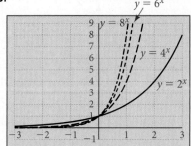

21. $h = 6 \cdot \left(\frac{2}{3}\right)^n$; 5th bounce: $6\left(\frac{2}{3}\right)^5 \approx 0.79$ feet   **23.** After 8 days, 700 micrograms; After 11 days, $1{,}400 \cdot 2^{-11/8} \approx 539.8$ micrograms

25. After 4 hours, 150 mg; after 8 hours, 37.5 mg   **27. a.** $A(t) = 1{,}200\left(1 + \frac{.06}{4}\right)^{4t}$ **b.** \$1,932.39 **c.** About 11.64 years **d.** \$1,939.29

29. a. The function underestimated the expenditures by \$69 billion.   **b.** \$4,123 billion; \$4,577 billion; \$5,080 billion

31.

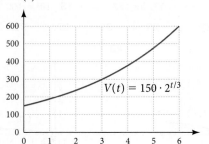

33. After 6 years   **35. a.** \$129,138.48   **b.** $\{t \mid 0 \le t \le 6\}$

c.

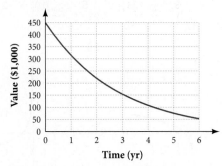

d. $\{V(t) \mid 52{,}942.05 \le V(t) \le 450{,}000\}$

e. After approximately 4 years 8 months

37. $D = \{1, 3, 4\}$, $R = \{2, 4, 1\}$, a function   **39.** $\left\{x \mid x \ge -\frac{1}{3}\right\}$ **41.** -18 **43.** 2   **45.** $(2x - 3)(x + 4)$ **47.** $x(5x - 2)(x - 4)$

49. $x(x - 4)(x^2 + 4x + 16)$ **51.** $y = \frac{x + 3}{2}$ **53.** $y = \pm\sqrt{x + 2}$ **55.** $y = \frac{2x - 4}{x - 1}$ **57.** $y = x^2 + 3$

59. a.

x	y
0	-6
3	0
5	4
8	10

b.

x	y
0	3
4	5
10	8
-6	0

61.

t	s(t)
0	0
1	83.3
2	138.9
3	175.9
4	200.6
5	217.1
6	228.1

Problem Set 11.2

1. $f^{-1}(x) = \frac{x+1}{3}$ **3.** $f^{-1}(x) = \sqrt[3]{x}$ **5.** $f^{-1}(x) = \frac{x-3}{x-1}$ **7.** $f^{-1}(x) = 4x + 3$ **9.** $f^{-1}(x) = 2(x+3) = 2x + 6$ **11.** $f^{-1}(x) = \frac{1-x}{3x-2}$

13. 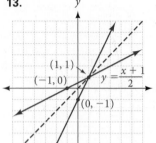   **15.** 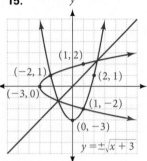   **17.** 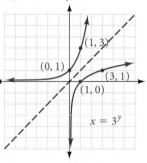   **19.**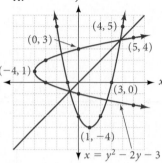

21. 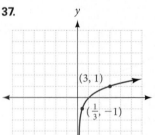   **23.** 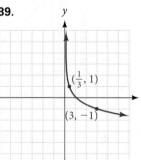   **25.** 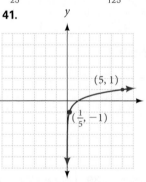   **27.**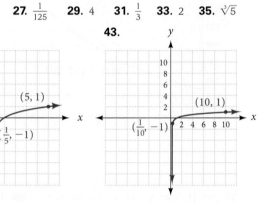

29. a. Yes **b.** No **c.** Yes   **31. a.** 4 **b.** $\frac{4}{3}$ **c.** 2 **d.** 2   **33.** $f^{-1}(x) = \frac{1}{x}$ **35. a.** -3 **b.** -6 **c.** 2 **d.** 3 **e.** -2 **f.** 3

g. They are inverses of each other.   **37.** $-2, 3$ **39.** 25   **41.** $5 \pm \sqrt{17}$ **43.** $1 \pm i$ **45.** $\frac{1}{9}$ **47.** $\frac{2}{3}$ **49.** $\sqrt[3]{4}$ **51.** 3

53. 4   **55.** 4   **57.** 1   **59.** $f^{-1}(x) = \frac{x-5}{3}$ **61.** $f^{-1}(x) = \sqrt[3]{x-1}$ **63.** $f^{-1}(x) = \frac{2x-4}{x-1}$

65. a. 1 **b.** 2 **c.** 5 **d.** 0 **e.** 1 **f.** 2 **g.** 2 **h.** 5

Problem Set 11.3

1. $\log_2 16 = 4$ **3.** $\log_5 125 = 3$ **5.** $\log_{10} 0.01 = -2$ **7.** $\log_2 \frac{1}{32} = -5$ **9.** $\log_{1/2} 8 = -3$ **11.** $\log_3 27 = 3$ **13.** $10^2 = 100$

15. $2^6 = 64$ **17.** $8^0 = 1$ **19.** $10^{-3} = 0.001$ **21.** $6^2 = 36$ **23.** $5^{-2} = \frac{1}{25}$ **25.** 9   **27.** $\frac{1}{125}$ **29.** 4   **31.** $\frac{1}{3}$ **33.** 2   **35.** $\sqrt[5]{5}$

37. 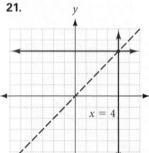   **39.** 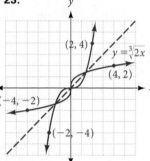   **41.** 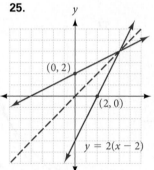   **43.**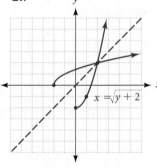

45. 4   **47.** $\frac{3}{2}$ **49.** 3   **51.** 1   **53.** 0   **55.** 0   **57.** $\frac{1}{2}$ **59.** 7   **61.** 10^{-6} **63.** 2   **65.** 10^8 times as large   **67.** 25; 5

69. 1; 1   **71.** $\pm 3i$ **73.** $\pm 2i$ **75.** $\frac{-2 \pm \sqrt{10}}{2}$ **77.** $\frac{2 \pm i\sqrt{3}}{2}$ **79.** $5, \frac{-5 \pm 5i\sqrt{3}}{2}$ **81.** $0, 5, -1$ **83.** $\frac{3 \pm \sqrt{29}}{2}$ **85.** 4   **87.** $-4, 2$

89. $-\frac{11}{8}$ **91.** $2^3 = (x+2)(x)$ **93.** $3^4 = \frac{x-2}{x+1}$ **95. a.**

x	$f(x)$
-1	$\frac{1}{8}$
0	1
1	8
2	64

b.

x	$f^{-1}(x)$
$\frac{1}{8}$	-1
1	0
8	1
64	2

c. $f(x) = 8^x$ **d.** $f^{-1}(x) = \log_8 x$

Problem Set 11.4

1. $\log_3 4 + \log_3 x$ **3.** $\log_6 5 - \log_6 x$ **5.** $5\log_2 y$ **7.** $\frac{1}{3}\log_9 z$ **9.** $2\log_6 x + 4\log_6 y$ **11.** $\frac{1}{2}\log_5 x + 4\log_5 y$

13. $\log_b x + \log_b y - \log_b z$ **15.** $\log_{10} 4 - \log_{10} x - \log_{10} y$ **17.** $2\log_{10} x + \log_{10} y - \frac{1}{2}\log_{10} z$ **19.** $3\log_{10} x + \frac{1}{2}\log_{10} y - 4\log_{10} z$

21. $\frac{2}{3}\log_b x + \frac{1}{3}\log_b y - \frac{4}{3}\log_b z$ **23.** $\log_b xz$ **25.** $\log_3 \frac{x^2}{y^3}$ **27.** $\log_{10} \sqrt{x}\,\sqrt[3]{y}$ **29.** $\log_2 \frac{x^3\sqrt{y}}{z}$ **31.** $\log_2 \frac{\sqrt{x}}{y^3 z^4}$

33. $\log_{10} \frac{x^{3/2}}{y^{3/4} z^{4/5}}$ **35.** $\frac{2}{3}$ **37.** 18 **39.** 3; -1 does not check **41.** 3 **43.** 4; -2 does not check **45.** 4; -1 does not check

47. $\frac{5}{3}$; $-\frac{5}{2}$ does not check **49.** $D = 10(\log_{10} I - \log_{10} I_0)$ **51.** $N = 2$ for both **53.** $\frac{x^2}{3} + \frac{y^2}{36}$ **55.** $\frac{x^2}{4} + \frac{y^2}{25}$

57. $D = -7$; two complex **59.** $x^2 - 2x - 15 = 0$ **61.** $3y^2 - 11y + 6 = 0$ **63.** 1 **65.** 1 **67.** 4 **69.** 2.5×10^{-6} **71.** 51

Problem Set 11.5

1. 2.5775 **3.** 1.5775 **5.** 3.5775 **7.** -1.4225 **9.** 4.5775 **11.** 2.7782 **13.** 3.3032 **15.** -2.0128 **17.** -1.5031

19. -0.3990 **21.** 759 **23.** 0.00759 **25.** 1,430 **27.** 0.00000447 **29.** 0.0000000918 **31.** 10^{10} **33.** 10^{-10} **35.** 10^{20}

37. $\frac{1}{100}$ **39.** 1,000 **41.** 1 **43.** 5 **45.** x **47.** $\ln 10 + 3t$ **49.** $\ln A - 2t$ **51.** 2.7080 **53.** -1.0986 **55.** 2.1972

57. 2.7724 **59. a.** 1.1886 **b.** 0.2218 **61. a.** 0.6667 **b.** -0.3010 **63.** 3.19 **65.** 1.78×10^{-5} **67.** 3.16×10^5

69. 2.00×10^8 **71.** 10 times larger

73.

Location	Date	Magnitude, M	Shock Wave, T
Moresby Island	January 23	4.0	1.00×10^4
Vancouver Island	April 30	5.3	1.99×10^5
Quebec City	June 29	3.2	1.58×10^3
Mould Bay	November 13	5.2	1.58×10^5
St. Lawrence	December 14	3.7	5.01×10^3

75. 12.9% **77.** 5.3% **79.**

x	$(1 + x)^{1/x}$
1	2
0.5	2.25
0.1	2.5937
0.01	2.7048
0.001	2.7169
0.0001	2.7181
0.00001	2.7183

e

81. $0, -3$ **83.** $-4, -2$ **85.** $\pm 2, \pm i\sqrt{2}$ **87.** $1, \frac{9}{4}$ **89.** 0.7 **91.** 3.125 **93.** 1.2575

95. $t \cdot \log(1.05)$ **97.** $0.05t$

Problem Set 11.6

1. 1.4650 **3.** 0.6826 **5.** -1.5440 **7.** -0.6477 **9.** -0.3333 **11.** 2 **13.** -0.1845 **15.** 0.1845 **17.** 1.6168

19. 2.1131 **21.** 1.3333 **23.** 0.75 **25.** 1.3917 **27.** 0.7186 **29.** 2.6356 **31.** 4.1632 **33.** 5.8435 **35.** -1.0642

37. 2.3026 **39.** 10.7144 **41.** 11.7 years **43.** 9.25 years **45.** 8.75 years **47.** 18.58 years **49.** 11.55 years

51. 18.3 years **53.** lowest point: $(-2, -23)$ **55.** highest point: $\left(\frac{3}{2}, 9\right)$ **57.** 2 seconds, 64 feet

59. 13.9 years later or toward the end of 2023 **61.** 27.5 years **63.** $t = \frac{1}{r}\ln\frac{A}{P}$ **65.** $t = \frac{1}{k}\frac{\log P - \log A}{\log 2}$ **67.** $t = \frac{\log A - \log P}{\log(1 - r)}$

Chapter 11 Review

1. 16 **2.** $\frac{1}{2}$ **3.** $\frac{1}{9}$ **4.** -5 **5.** $\frac{5}{6}$ **6.** 7

7.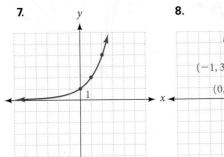

8.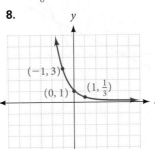

9.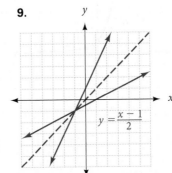

10.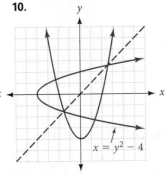

11. $f^{-1}(x) = \frac{x - 3}{2}$ **12.** $y = \pm\sqrt{x + 1}$ **13.** $f^{-1}(x) = 2x - 4$ **14.** $y = \pm\sqrt{\frac{4 - x}{2}}$ **15.** $\log_3 81 = 4$ **16.** $\log_7 49 = 2$

17. $\log_{10} 0.01 = -2$ **18.** $\log_2 \frac{1}{8} = -3$ **19.** $2^3 = 8$ **20.** $3^2 = 9$ **21.** $4^{1/2} = 2$ **22.** $4^1 = 4$ **23.** 25 **24.** $\frac{3}{4}$ **25.** 10

26.

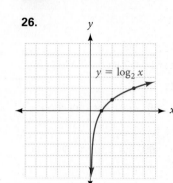

$y = \log_2 x$

27.

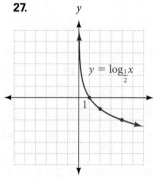

$y = \log_{\frac{1}{2}} x$

28. 2   **29.** $\frac{2}{3}$ **30.** 0   **31.** $\log_2 5 + \log_2 x$

32. $\log_{10} 2 + \log_{10} x - \log_{10} y$ **33.** $\frac{1}{2}\log_a x + 3\log_a y - \log_a z$

34. $2\log_{10} x - 3\log_{10} y - 4\log_{10} z$ **35.** $\log_2 xy$ **36.** $\log_3 \frac{x}{4}$

37. $\log_a \frac{25}{3}$ **38.** $\log_2 \frac{x^3 y^2}{z^4}$ **39.** 2   **40.** 6

41. Possible solutions -1 and 3; only 3 checks; 3   **42.** 3

43. Possible solutions -2 and 3; only 3 checks; 3

44. Possible solutions -1 and 4; only 4 checks; 4   **45.** 2.5391

46. -0.1469 **47.** 9,230   **48.** 0.0251   **49.** 1   **50.** 0   **51.** 2   **52.** -4 **53.** 2.1   **54.** 5.1   **55.** 2.0×10^{-3}

56. 3.2×10^{-8} **57.** $\frac{3}{2}$ **58.** $x = \frac{1}{3}\left(\frac{\log 5}{\log 4} - 2\right) \approx -0.28$ **59.** 0.75   **60.** 2.43   **61.** About 4.67 years   **62.** About 9.25 years

Chapter 11 Cumulative Review

1. 32   **2.** $24x - 26$ **3.** 9   **4.** $\frac{17}{8}$ **5.** $3\sqrt{3}$ **6.** $18 + 14\sqrt{3}$ **7.** $-2 + 7i$ **8.** $2i$ **9.** 0   **10.** 0   **11.** $\frac{x-4}{x-2}$ **12.** $\frac{x^2+x+1}{x+1}$

13. $\frac{4}{7}$ **14.** $\frac{4}{5}$ **15.** $12t^4 - \frac{1}{12}$ **16.** $3x^3 - 11x^2 + 4$ **17.** $y + 5 - \frac{3}{y+2}$ **18.** $3x + 7 + \frac{10}{3x-4}$ **19.** $\frac{1}{(x+4)(x+2)}$ **20.** $\frac{24}{(4x+3)(4x-3)}$

21. $\frac{8}{3}$ **22.** 4   **23.** $-\frac{1}{7}, \frac{1}{4}$ **24.** 1   **25.** $\frac{1 \pm \sqrt{29}}{2}$ **26.** $-\frac{12}{5}$ **27.** 1, 4   **28.** $-\frac{1}{4}, \frac{2}{3}$ **29.** 27   **30.** 10   **31.** $\varnothing$ **32.** 7

33. $\left(\frac{8}{3}, \frac{23}{3}\right)$ **34.** $(8, -5)$ **35.** $\left(8, -2, -\frac{7}{2}\right)$ **36.** $\left(3, 0, \frac{1}{2}\right)$

37.

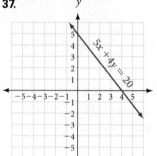

$5x + 4y = 20$

38.

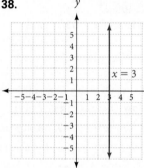

$x = 3$

39. $7 - y = -x$ **40.** $9a - 4b < 9a + 4b$ **41.** 9.72×10^{-5}

42. 4.7×10^7 **43.** 3.6×10^{13} **44.** 8.6×10^{-15}

45. $(2a - 5)(4a^2 + 10a + 25)$ **46.** $2(5a^2 + 2)(5a^2 - 1)$

47. Domain = $\{2, -3\}$; Range = $\{-3, -1, 3\}$; not a function

48. Domain = $\{3, 4, 6\}$; Range = $\{4, 5\}$; a function

49. 300 million   **50.** 6,550 million

Chapter 11 Test

1.

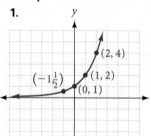

$(2, 4)$, $(-1\frac{1}{2})$, $(1, 2)$, $(0, 1)$

2.

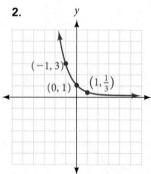

$(-1, 3)$, $(0, 1)$, $(1, \frac{1}{3})$

3.

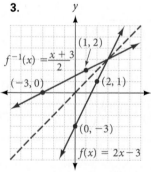

$f^{-1}(x) = \frac{x+3}{2}$, $(1, 2)$, $(-3, 0)$, $(2, 1)$, $(0, -3)$, $f(x) = 2x - 3$

4.

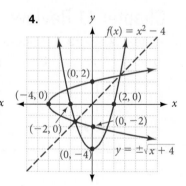

$f(x) = x^2 - 4$, $(0, 2)$, $(-4, 0)$, $(2, 0)$, $(-2, 0)$, $(0, -2)$, $(0, -4)$, $y = \pm\sqrt{x+4}$

5. 64   **6.** $\sqrt{5}$ **7.**

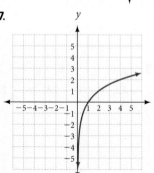

8.

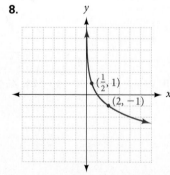

$\left(\frac{1}{2}, 1\right)$, $(2, -1)$

9. .67   **10.** 1.56   **11.** 4.37   **12.** -1.91 **13.** 3.83   **14.** -3.07 **15.** $\log_2 6 + 2\log_2 x - \log_2 y$

16. $\frac{1}{2}\log x - 4\log y - \frac{1}{5}\log z$ **17.** $\log_3 \frac{x^2}{\sqrt{y}}$ **18.** $\log \frac{\sqrt[3]{x}}{yz^2}$ **19.** 70,404.432   **20.** 0.002   **21.** 1.465   **22.** 1.25   **23.** 15
24. 8   **25.** 6.18   **26.** $651.56   **27.** 13.87 years   **28.** yes   **29.** no   **30.** yes   **31.** yes   **32.** no   **33.** no

Chapter 12

Getting Ready for Chapter 12

1. $\sqrt{29}$ **2.** $2\sqrt{5}$ **3.** $\sqrt{13}$ **4.** $\sqrt{26}$ **5.** ± 4 **6.** $\pm 5i$ **7.** 4, 2   **8.** -4, 2   **9.** 4, 2   **10.** -4, 2   **11.** 9; 3   **12.** 4; 2
13. 1; 1   **14.** $\frac{1}{4}$; $\frac{1}{2}$ **15.** $(5y + 6)(y + 2)$ **16.** $(3x - 4)(x + 7)$ **17.** $\frac{x^2}{3} + \frac{y^2}{36}$ **18.** $\frac{x^2}{16} + \frac{y^2}{4}$ **19.** $y = -\frac{1}{3}x + 1$ **20.** $y = -2x + 3$

Problem Set 12.1

1. 5   **3.** $\sqrt{106}$ **5.** $\sqrt{61}$ **7.** $\sqrt{130}$ **9.** 3 or -1 **11.** 3   **13.** $(x - 2)^2 + (y - 3)^2 = 16$ **15.** $(x - 3)^2 + (y + 2)^2 = 9$
17. $(x + 5)^2 + (y + 1)^2 = 5$ **19.** $x^2 + (y + 5)^2 = 1$ **21.** $x^2 + y^2 = 4$

23. Center = (0, 0),   **25.** Center = (1, 3),   **27.** Center = $(-2, 4)$,   **29.** Center = $(-1, -1)$
Radius = 2   Radius = 5   Radius = $2\sqrt{2}$ Radius = 1

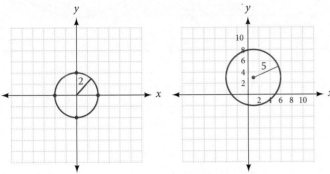
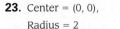
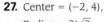
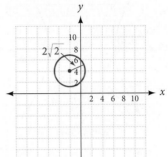
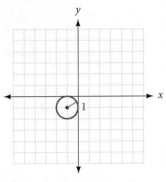

31. Center = (0, 3),   **33.** Center = (2, 3),   **35.** Center = $\left(-1, -\frac{1}{2}\right)$,   **37.** $(x - 3)^2 + (y - 4)^2 = 25$
Radius = 4   Radius = 3   Radius = 2

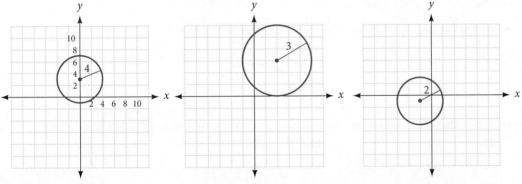
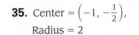

39. A: $\left(x - \frac{1}{2}\right)^2 + (y - 1)^2 = \frac{1}{4}$ **41.** A: $(x + 8)^2 + y^2 = 64$ **43.** A: $(x + 4)^2 + (y - 4)^2 = 16$
B: $(x - 1)^2 + (y - 1)^2 = 1$ B: $x^2 + y^2 = 64$ B: $x^2 + y^2 = 16$
C: $(x - 2)^2 + (y - 1)^2 = 4$ C: $(x - 8)^2 + y^2 = 64$ C: $(x - 4)^2 + (y + 4)^2 = 16$

45. $(x - 500)^2 + (y - 132)^2 = 120^2$ **47.** $f^{-1}(x) = \log_3 x$ **49.** $f^{-1}(x) = \frac{x - 3}{2}$ **51.** $f^{-1}(x) = 5x - 3$ **53.** $y = \pm 3$ **55.** $y = \pm 2i$

57. $x = \pm 3i$ **59.** $\frac{x^2}{9} + \frac{y^2}{4}$ **61.** x-intercept 4, y-intercept -3 **63.** ± 2.4 **65.** $(x - 2)^2 + (y - 3)^2 = 4$ **67.** $(x - 2)^2 + (y - 3)^2 = 4$

69. 5   **71.** 5   **73.** $y = \sqrt{9 - x^2}$ corresponds to the top half; $y = -\sqrt{9 - x^2}$ to the bottom half.

Problem Set 12.2

1.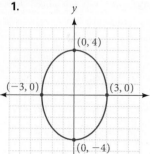

3.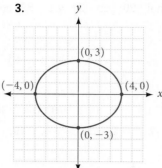

5.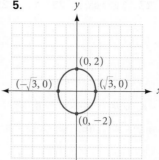

7.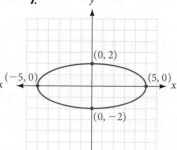

9.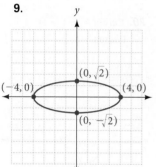

11.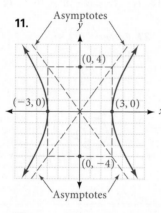

13.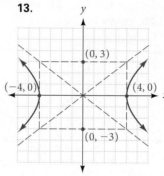

15.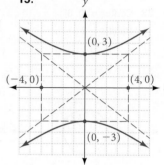

17.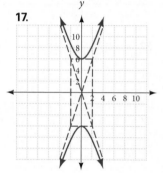

19.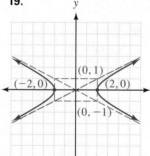

21.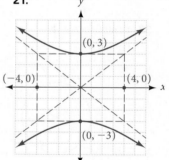

23. x-intercepts $= \pm 3$, y-intercepts $= \pm 2$ **25.** x-intercepts $= \pm 0.2$, no y-intercepts   **27.** x-intercepts $= \pm\frac{3}{5}$, y-intercepts $= \pm\frac{2}{5}$

29.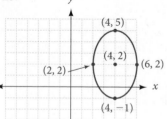

31.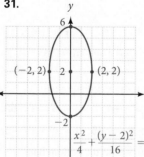

33.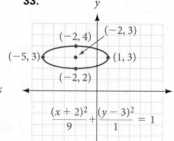

35.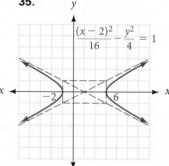

37.

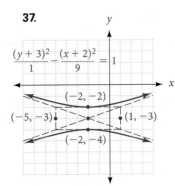

$$\frac{(y+3)^2}{1} - \frac{(x+2)^2}{9} = 1$$

39.

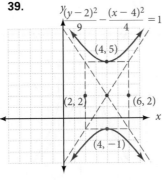

$$\frac{(y-2)^2}{9} - \frac{(x-4)^2}{4} = 1$$

41. $\pm\frac{9}{5}$ **43.** ±3.2 **45.** $y = \frac{3}{4}x, y = -\frac{3}{4}x$ **47.** $\frac{x^2}{16} + \frac{y^2}{4} = 1$

49. The equation is $\frac{x^2}{20^2} + \frac{y^2}{10^2} = 1$. A 6-foot man could walk upright under the arch anywhere between 16 feet to the left and 16 feet the right of the center.

51. About 5.3 feet wide   **53.** 1   **55.** -2 **57.** $\frac{4x}{(x-2)(x+2)}$

59. $(0, 0)$ **61.** $4y^2 + 16y + 16$ **63.** $x = 2y + 4$

65. $-x^2 + 6$ **67.** $(5y+6)(y+2)$ **69.** ±2 **71.** ±2

Problem Set 12.3

1.

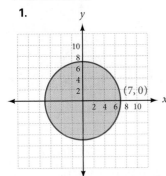

3.

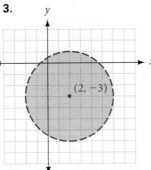

5.

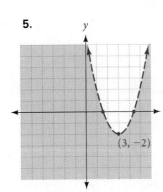

7.

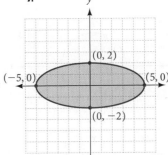

9. $(0, 3), \left(\frac{12}{5}, -\frac{9}{5}\right)$ **11.** $(0, 4), \left(\frac{16}{5}, \frac{12}{5}\right)$ **13.** $(5, 0), (-5, 0)$ **15.** $(0, -3), (\sqrt{5}, 2), (-\sqrt{5}, 2)$ **17.** $(0, -4), (\sqrt{7}, 3), (-\sqrt{7}, 3)$

19. $(-4, 11), \left(\frac{5}{2}, \frac{5}{4}\right)$ **21.** $(-4, 5), (1, 0)$ **23.** $(2, -3), (5, 0)$ **25.** $(3, 0), (-3, 0)$ **27.** $(4, 0), (0, -4)$

29. a. $(-4, 4\sqrt{3})$ and $(-4, -4\sqrt{3})$ **b.** $(4, 4\sqrt{3})$ and $(4, -4\sqrt{3})$

31.

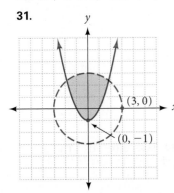

33.

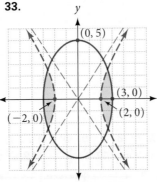

35.   No intersection

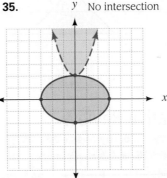

37. $x^2 + y^2 < 16, y > 4 - \frac{1}{4}x^2$ **39.** 8, 5 or -8, -5 or 8, -5 or -8, 5   **41.** 6, 3 or 13, -4 **43.** $-\frac{2}{5}$ **45.** $-5, 1$ **47.** $-2, -1$

Chapter 12 Review

1. $\sqrt{10}$ **2.** $\sqrt{13}$ **3.** 5   **4.** 9   **5.** $-2, 6$ **6.** $-12, 4$ **7.** $x^2 + y^2 = 25$ **8.** $x^2 + y^2 = 9$ **9.** $(x + 2)^2 + (y - 3)^2 = 25$

10. $(x + 6)^2 + (y - 8)^2 = 100$

11. $(0, 0); r = 2$

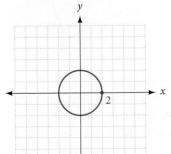

12. $(3, -1); r = 4$

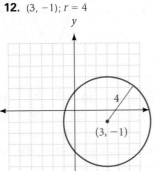

13. $(3, -2); r = 3$

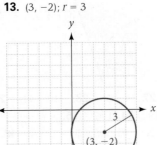

14. $(-2, 1); r = 3$

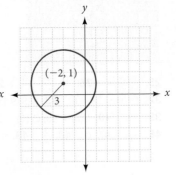

15.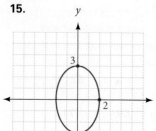

16.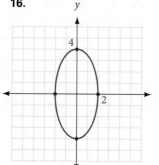

17.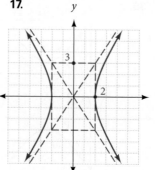

18.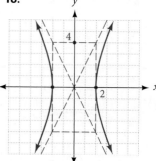

19.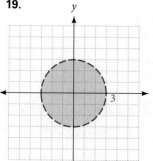

20.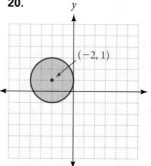

21.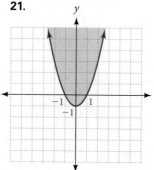

22.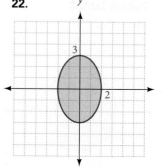

23.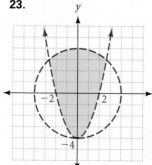

24.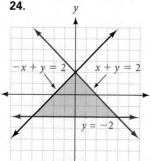

25. $(0, 4), \left(\frac{16}{5}, -\frac{12}{5}\right)$ **26.** $(0, -2), (\sqrt{3}, 1), (-\sqrt{3}, 1)$

27. $(-2, 0), (2, 0)$

23.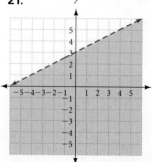

24.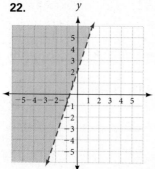

Chapter 12 Cumulative Review

1. 5 **2.** 4 **3.** $\frac{a^5b^4}{2}$ **4.** $x^{1/4}y^{7/4}$ **5.** $3xy^2 - 4x + 2y^2$ **6.** $\frac{y-3}{y-2}$ **7.** 2 **8.** 3 **9.** $(b^3 + 6)(a + 1)$ **10.** $(8x + 3)(x - 1)$ **11.** -2

12. 2 **13.** $-5, 2$ **14.** $-2, 4$ **15.** 8; 5 does not check **16.** No solution **17.** $\frac{3 \pm 5i\sqrt{2}}{4}$ **18.** $3 \pm i\sqrt{3}$

19. $x < -\frac{7}{2}$ or $x > 2$ 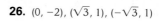   **20.** $-8 < x < 3$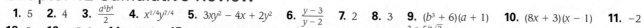

21.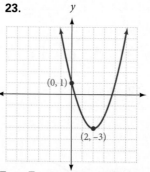

22.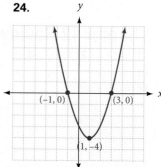

23.

24.

25. $\frac{y(y-1)}{y+1}$ **26.** $\frac{(x-4)(x+3)}{x^2(x+2)}$ **27.** $14 - 5i$ **28.** $-10 + 11i$ **29.** $\frac{3\sqrt{7} + 3\sqrt{3}}{4}$ **30.** $\frac{-12 + 7\sqrt{6}}{5}$ **31.** $f^{-1}(x) = \frac{1}{4}x - \frac{1}{4}$

32. $f^{-1}(x) = 2x - 6$ **33.** 9,519.19 **34.** 1.47 **35.** $x = -\frac{5}{m-n}$ **36.** $y = \frac{S - 2x^2}{4x}$ **37.** 0 **38.** slope $= \frac{2}{3}$; y-intercept $= -4$

39. -5 **40.** $C(5) = 40$; $C(10) = 20$ **41.** $3x^2 - 5x + 2 = 0$ **42.** $20t^2 + 11t - 3 = 0$ **43.** $-3x + 5$ **44.** $-5x^2 + 19$ **45.** -19

46. -6 **47.** 6 **48.** $\frac{3}{4}$ **49.** 4 gal of 30%; 12 gal of 70% **50.** 12 gal of 20%; 4 gal of 60% **51.** 93 people **52.** 18 people

Chapter 12 Test

1. 13 **2.** $2\sqrt{5}$ **3.** $-5, 3$ **4.** $(x + 2)^2 + (y - 4)^2 = 9$ **5.** $x^2 + y^2 = 25$ **6.** Center = $(0, -2)$; Radius = 8

7. Center = $(5, -3)$; Radius = $\sqrt{39}$

8.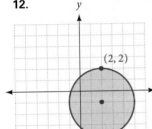

9.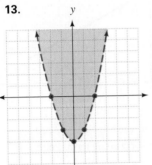

10.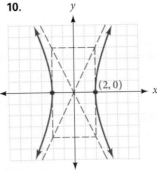

11.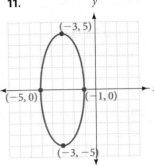

12.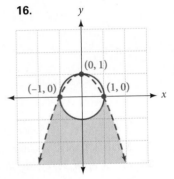

13.

14. $(0, 5), (4, -3)$

15. $(\pm\sqrt{7}, 3), (0, -4)$

16.

17. C **18.** A **19.** D **20.** B **21.** $(x + 3)^2 + (y - 4)^2 = 25$ **22.** $\frac{(x - 4)^2}{25} + \frac{(y + 2)^2}{9} = 1$ **23.** $\frac{y^2}{16} - \frac{x^2}{25} = 1$

Chapter 13

Getting Ready for Chapter 13

1. 5, 7, 9, 11 **2.** 3, 7, 11, 15 **3.** $\frac{1}{4}, \frac{2}{5}, \frac{1}{2}, \frac{4}{7}$ **4.** $\frac{2}{3}, \frac{3}{4}, \frac{4}{5}, \frac{5}{6}$ **5.** $-2, 4, -8, 16$ **6.** $-3, 9, -27, 81$ **7.** $\sqrt{2}$ **8.** $\sqrt{2}$

9. 3^x **10.** 5^x **11.** y^4 **12.** b^2 **13.** $\frac{1}{y^3}$ **14.** $\frac{1}{b^9}$ **15.** $a^2 + 2ab + b^2$ **16.** $a^2 - 2ab + b^2$ **17.** $a^3 + 3a^2b + 3ab^2 + b^2$

18. $a^3 - 3a^2b + 3ab^2 - b^3$

Problem Set 13.1

1. 4, 7, 10, 13, 16 **3.** 3, 7, 11, 15, 19 **5.** 1, 2, 3, 4, 5 **7.** 4, 7, 12, 19, 28 **9.** $\frac{1}{4}, \frac{2}{5}, \frac{1}{2}, \frac{4}{7}, \frac{5}{8}$ **11.** $1, \frac{1}{4}, \frac{1}{9}, \frac{1}{16}, \frac{1}{25}$

13. 2, 4, 8, 16, 32 **15.** $2, \frac{3}{2}, \frac{4}{3}, \frac{5}{4}, \frac{6}{5}$ **17.** $-2, 4, -8, 16, -32$ **19.** 3, 5, 3, 5, 3 **21.** $1, -\frac{2}{3}, \frac{3}{5}, -\frac{4}{7}, \frac{5}{9}$ **23.** $\frac{1}{2}, 1, \frac{9}{8}, 1, \frac{25}{32}$

25. $3, -9, 27, -81, 243$ **27.** 1, 5, 13, 29, 61 **29.** 2, 3, 5, 9, 17 **31.** 5, 11, 29, 83, 245 **33.** 4, 4, 4, 4, 4 **35.** $a_n = 4n$

37. $a_n = n^2$ **39.** $a_n = 2^{n+1}$ **41.** $a_n = \frac{1}{2^{n+1}}$ **43.** $a_n = 3n + 2$ **45.** $a_n = -4n + 2$ **47.** $a_n = (-2)^{n-1}$

49. $a_n = \log_{n+1}(n + 2)$ **51. a.** $28000, 29120, 30284.80, 31,496.19, 32756.04 **b.** $a_n = 28000(1.04)^{n-1}$

53. a. 16 ft, 48 ft, 80 ft, 112 ft, 144 ft **b.** 400 ft **c.** No **55. a.** $180°, 360°, 540°, 720°$ **b.** $3,240°$ **c.** Sum of interior angles would be $0°$ **57.** 27

59. 5 **61.** 1 **63.** 30 **65.** 40 **67.** 18 **69.** $-\frac{20}{81}$ **71.** $\frac{21}{10}$

73. $a_{100} \approx 2.7048, a_{1,000} \approx 2.7169, a_{10,000} \approx 2.7181, a_{100,000} \approx 2.7183$ **75.** 1, 1, 2, 3, 5, 8, 13, 21, 34, 55

Problem Set 13.2

1. 36 **3.** 11 **5.** 18 **7.** $\frac{163}{60}$ **9.** 60 **11.** 40 **13.** 44 **15.** $-\frac{11}{32}$ **17.** $\frac{21}{10}$ **19.** $5x + 15$

21. $(x - 2) + (x - 2)^2 + (x - 2)^3 + (x - 2)^4$ **23.** $\frac{x+1}{x-1} + \frac{x+2}{x-1} + \frac{x+3}{x-1} + \frac{x+4}{x-1} + \frac{x+5}{x-1}$

25. $(x + 3)^3 + (x + 4)^4 + (x + 5)^5 + (x + 6)^6 + (x + 7)^7 + (x + 8)^8$ **27.** $(x - 6)^6 + (x - 8)^7 + (x - 10)^8 + (x - 12)^9$

29. $\sum_{i=1}^{4} 2^i$ **31.** $\sum_{i=2}^{6} 2^i$ **33.** $\sum_{i=1}^{5}(4i + 1)$ **35.** $\sum_{i=2}^{5} -(-2)^i$ **37.** $\sum_{i=3}^{7} \frac{i}{i+1}$ **39.** $\sum_{i=1}^{4} \frac{i}{2i+1}$ **41.** $\sum_{i=6}^{9}(x - 2)^i$ **43.** $\sum_{i=1}^{4}\left(1 + \frac{i}{x}\right)^{i+1}$

45. $\sum_{i=3}^{5} \frac{x}{x+i}$ **47.** $\sum_{i=2}^{4} x^i(x + i)$

49. a. $0.3 + 0.03 + 0.003 + 0.0003 + \dots$ **b.** $0.2 + 0.02 + 0.002 + 0.0002 + \dots$ **c.** $0.27 + 0.0027 + 0.000027 + \dots$

51. seventh second: 208 feet; total: 784 feet **53. a.** $16 + 48 + 80 + 112 + 144$ **b.** $\sum_{i=1}^{5}(32i - 16)$ **55.** $3\log_2 x + \log_2 y$

57. $\frac{1}{3}\log_{10} x - 2\log_{10} y$ **59.** $\log_{10}\frac{x}{y^2}$ **61.** $\log_3 \frac{x^2}{y^3 z^4}$ **63.** 80 **65.** Possible solutions -1 and 8; only 8. **67.** 74 **69.** $\frac{55}{2}$

71. $2n + 1$ **73.** $\frac{13}{2}$

Problem Set 13.3

1. arithmetic; $d = 1$ **3.** not arithmetic **5.** arithmetic; $d = -5$ **7.** not arithmetic **9.** arithmetic; $d = \frac{2}{3}$

11. $a_n = 4n - 1$; $a_{24} = 95$ **13.** $a_{10} = -12$; $S_{10} = -30$ **15.** $a_1 = 7$; $d = 2$; $a_{30} = 65$ **17.** $a_1 = 12$; $d = 2$; $a_{20} = 50$; $S_{20} = 620$

19. $a_{20} = 79$; $S_{20} = 820$ **21.** $a_{40} = 122$; $S_{40} = 2{,}540$ **23.** $a_1 = 13$; $d = -6$ **25.** $a_{85} = -238$ **27.** $d = \frac{16}{19}$; $a_{39} = 28$

29. 20,300 **31.** $a_{35} = -158$ **33.** $a_{10} = 5$; $S_{10} = \frac{55}{2}$

35. a. $\$18{,}000, \$14{,}700, \$11{,}400, \$8{,}100, \$4{,}800$ **b.** $-\$3{,}300$ **c.**

d. $\$9{,}750$ **e.** $a_0 = 18000$; $a_n = a_{n-1} - 3300$ for $n \geq 1$ **37. a.** 1500 ft, 1460 ft, 1420 ft, 1380 ft, 1340 ft, 1300 ft
b. It is arithmetic because the same amount is subtracted from each succeeding term. **c.** $a_n = 1{,}540 - 40n$

39. a. 1, 3, 6, 10, 15, 21, 28, 36, 45, 55, 66, 78, 91, 105, 120 **b.** $a_1 = 1$; $a_n = n + a_{n-1}$ for $n \geq 2$
c. It is not arithmetic because the same amount is not added to each term.

41. 2.7604 **43.** -1.2396 **45.** 445 **47.** 4.45×10^{-8} **49.** $\frac{1}{16}$ **51.** $\sqrt{3}$ **53.** 2^n **55.** r^3 **57.** $\frac{2}{5}$

59. 255 **61.** $a_1 = 9$, $d = 4$ **63.** $d = -3$ **65.** $S_{50} = 50 + 1225\sqrt{2}$

Problem Set 13.4

1. geometric; $r = 5$ **3.** geometric; $r = \frac{1}{3}$ **5.** not geometric **7.** geometric; $r = -2$ **9.** not geometric

11. $a_n = 4 \cdot 3^{n-1}$ **13.** $a_6 = \frac{1}{16}$ **15.** $a_{20} = -3$ **17.** $S_{10} = 10{,}230$ **19.** $S_{20} = 0$ **21.** $a_8 = \frac{1}{640}$ **23.** $S_5 = -\frac{31}{32}$

25. $a_{10} = 32$; $S_{10} = 62 + 31\sqrt{2}$ **27.** $a_6 = \frac{1}{1000}$; $S_6 = 111.111$ **29.** ± 2 **31.** $a_8 = 384$; $S_8 = 255$ **33.** $r = 2$

35. 1 **37.** 8 **39.** 4 **41.** $\frac{8}{9}$ **43.** $\frac{2}{3}$ **45.** $\frac{9}{8}$ **51. a.** $\$450{,}000, \$315{,}000, \$220{,}500, \$154{,}350, \$108{,}045$ **b.** 0.7

d. Approximately $\$90{,}000$ **e.** $a_0 = 450000$; $a_n = 0.7a_{n-1}$ for $n \geq 1$

c.

53. a. $\frac{1}{2}$ **b.** $\frac{364}{729}$ **c.** $\frac{1}{1458}$ **55.** 300 feet **57. a.** $a_n = 15(\frac{4}{5})^{n-1}$ **b.** 75 feet **59.** 2.16 **61.** 6.36 **63.** $t = \frac{1}{5}\ln\frac{A}{10}$ **65.** 1

67. $x^2 + 2xy + y^2$ **69.** 15 **71.** $\frac{7}{11}$ **73.** $r = \frac{1}{3}$

75. a. stage 2: $\frac{3}{4}$; stage 3: $\frac{9}{16}$; stage 4: $\frac{27}{64}$ **b.** geometric sequence **c.** 0 **d.** increasing sequence

Problem Set 13.5

1. $x^4 + 8x^3 + 24x^2 + 32x + 16$ **3.** $x^6 + 6x^5y + 15x^4y^2 + 20x^3y^3 + 15x^2y^4 + 6xy^5 + y^6$ **5.** $32x^5 + 80x^4 + 80x^3 + 40x^2 + 10x + 1$

7. $x^5 - 10x^4y + 40x^3y^2 - 80x^2y^3 + 80xy^4 - 32y^5$ **9.** $81x^4 - 216x^3 + 216x^2 - 96x + 16$ **11.** $64x^3 - 144x^2y + 108xy^2 - 27y^3$

13. $x^8 + 8x^6 + 24x^4 + 32x^2 + 16$ **15.** $x^6 + 3x^4y^2 + 3x^2y^4 + y^6$ **17.** $16x^4 + 96x^3y + 216x^2y^2 + 216xy^3 + 81y^4$ **19.** $\frac{x^3}{8} + \frac{x^2y}{4} + \frac{xy^2}{6} + \frac{y^3}{27}$

21. $\frac{x^3}{8} - 3x^2 + 24x - 64$ **23.** $\frac{x^4}{81} + \frac{2x^3y}{27} + \frac{x^2y^2}{6} + \frac{xy^3}{6} + \frac{y^4}{16}$ **25.** $x^9 + 18x^8 + 144x^7 + 672x^6$

21. $\frac{x^3}{8} - 3x^2 + 24x - 64$ **23.** $\frac{x^4}{81} + \frac{2x^3y}{27} + \frac{x^2y^2}{6} + \frac{xy^3}{6} + \frac{y^4}{16}$ **25.** $x^9 + 18x^8 + 144x^7 + 672x^6$

27. $x^{10} - 10x^9y + 45x^8y^2 - 120x^7y^3$ **29.** $x^{25} + 75x^{24} + 2{,}700x^{23} + 62{,}100x^{22}$ **31.** $x^{60} - 120x^{59} + 7{,}080x^{58} - 273{,}760x^{57}$

33. $x^{18} - 18x^{17}y + 153x^{16}y^2 - 816x^{15}y^3$ **35.** $x^{15} + 15x^{14} + 105x^{13}$ **37.** $x^{12} - 12x^{11}y + 66x^{10}y^2$ **39.** $x^{20} + 40x^{19} + 760x^{18}$

41. $x^{100} + 200x^{99}$ **43.** $x^{50} + 50x^{49}y$ **45.** $51{,}963{,}120x^4y^8$ **47.** $3{,}360x^6$ **49.** $-25{,}344x^7$ **51.** $2{,}700x^{23}y^2$

53. $\binom{20}{11}(2x)^9(5y)^{11} = \frac{20!}{11!9!}(2x)^9(5y)^{11}$ **55.** $x^{20}y^{10} - 30x^{18}y^9 + 405x^{16}y^8$ **57.** $\frac{21}{128}$ **59.** $x \approx 1.21$ **61.** $\frac{1}{6}$ **63.** ≈ 7 years

65. 56 **67.** $125{,}970$

Chapter 13 Review

1. $7, 9, 11, 13$ **2.** $1, 4, 7, 10$ **3.** $0, 3, 8, 15$ **4.** $\frac{4}{3}, \frac{5}{4}, \frac{6}{5}, \frac{7}{6}$ **5.** $4, 16, 64, 256$ **6.** $\frac{1}{4}, \frac{1}{16}, \frac{1}{64}, \frac{1}{256}$ **7.** $a_n = 3n - 1$

8. $a_n = 2n - 5$ **9.** $a_n = n^4$ **10.** $a_n = n^2 + 1$ **11.** $a_n = (\frac{1}{2})^n = 2^{-n}$ **12.** $a_n = \frac{n+1}{n^2}$ **13.** 32 **14.** 25 **15.** $\frac{14}{5}$

16. -5 **17.** 98 **18.** $\frac{127}{30}$ **19.** $\sum\limits_{i=1}^{4} 3i$ **20.** $\sum\limits_{i=1}^{4}(4i - 1)$ **21.** $\sum\limits_{i=1}^{5}(2i + 3)$ **22.** $\sum\limits_{i=2}^{4} i^2$ **23.** $\sum\limits_{i=1}^{4}\frac{1}{i+2}$ **24.** $\sum\limits_{i=1}^{5}\frac{i}{3^i}$

25. $\sum\limits_{i=1}^{3}(x - 2i)$ **26.** $\sum\limits_{i=1}^{4}\frac{x}{x+i}$ **27.** geometric   **28.** arithmetic   **29.** arithmetic   **30.** neither   **31.** geometric

32. geometric   **33.** arithmetic   **34.** neither   **35.** $a_n = 3n - 1; a_{20} = 59$ **36.** $a_n = 8 - 3n; a_{16} = -40$

37. $a_{10} = 34; S_{10} = 160$ **38.** $a_{16} = 78; S_{16} = 648$ **39.** $a_1 = 5; d = 4; a_{10} = 41$ **40.** $a_1 = 8; d = 3; a_9 = 32; S_9 = 180$

41. $a_1 = -4; d = -2; a_{20} = -42; S_{20} = -460$ **42.** $20{,}100$ **43.** $a_{40} = -95$ **44.** $a_n = 3(2)^{n-1}; a_{20} = 1{,}572{,}864$

45. $a_n = 5(-2)^{n-1}; a_{16} = -163{,}840$ **46.** $a_n = 4(\frac{1}{2})^{n-1}; a_{10} = \frac{1}{128}$ **47.** -3 **48.** 8 **49.** $a_1 = 3; r = 2; a_6 = 96$ **50.** $243\sqrt{3}$

51. $x^4 - 8x^3 + 24x^2 - 32x + 16$ **52.** $16x^4 + 96x^3 + 216x^2 + 216x + 81$ **53.** $27x^3 + 54x^2y + 36xy^2 + 8y^3$

54. $x^{10} - 10x^8 + 40x^6 - 80x^4 + 80x^2 - 32$ **55.** $\frac{1}{16}x^4 + \frac{3}{2}x^3 + \frac{27}{2}x^2 + 54x + 81$ **56.** $\frac{1}{27}x^3 - \frac{1}{6}x^2y + \frac{1}{4}xy^2 - \frac{1}{8}y^3$

57. $x^{10} + 30x^9y + 405x^8y^2$ **58.** $x^9 - 27x^8y + 324x^7y^2$ **59.** $x^{11} + 11x^{10}y + 55x^9y^2$ **60.** $x^{12} - 24x^{11}y + 264x^{10}y^2$

61. $x^{16} - 32x^{15}y$ **62.** $x^{32} + 64x^{31}y$ **63.** $x^{50} - 50x^{49}$ **64.** $x^{150} + 150x^{149}y$ **65.** $-61{,}236x^5$ **66.** $5376x^6$

Chapter 13 Cumulative Review

1. 2 **2.** -5 **3.** 29 **4.** 12 **5.** $-2x - 15$ **6.** $2x + 20$ **7.** $24y$ **8.** $x^2 + x + 2$ **9.** $-\frac{1}{25}$ **10.** $\frac{y - 3x}{3y - x}$

11. $x^{13/15}$ **12.** $a^{13/5}$ **13.** $4x^2y\sqrt{3xy}$ **14.** $3xy\sqrt[3]{x}$ **15.** 1 **16.** 1 **17.** $-\frac{7}{3}, 3$ **18.** $\varnothing$ **19.** $-\frac{1}{2}, 3$

20. $-2, 2, 3$ **21.** $\frac{3}{2}$ **22.** 28 **23.** $\frac{-3 \pm i\sqrt{11}}{4}$ **24.** $\frac{3 \pm i\sqrt{23}}{4}$ **25.** 8 **26.** $\frac{2}{5}$ **27.** $-\frac{7}{3}$ **28.** 1.74

29. $(-1, 4)$ **30.** $\left(-\frac{31}{19}, -\frac{1}{19}\right)$ **31.** $\left(-\frac{19}{5}, -\frac{18}{5}, \frac{1}{5}\right)$ **32.** $(4, -2, 3)$ **33.** $x^{2/5} - 9$ **34.** $x^{6/5} - 4$

35.
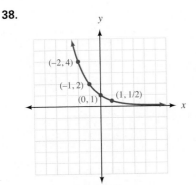

36.
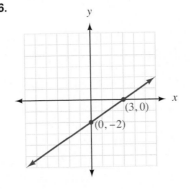

37.
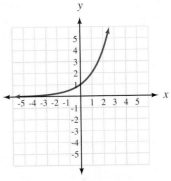

38.
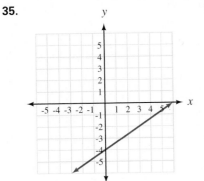

39.
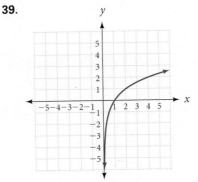

40.
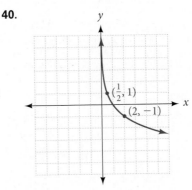

41.

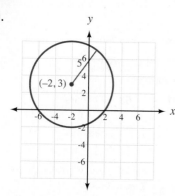

42.

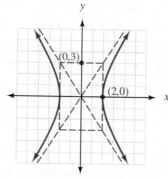

43. $a_n = 11n - 8$ **44.** $a_n = 16\left(\frac{1}{2}\right)^{n-1}$ **45.** $y = -\frac{4}{3}x - 9$ **46.** $y = -2x + 9$ **47.** $\sqrt{65}$ **48.** $2\sqrt{2}$

49. $\frac{1}{2}$ second, $\frac{5}{2}$ seconds   **50.** $\frac{1}{4}$ second, $\frac{11}{4}$ seconds   **51.** 25   **52.** $4x - 10$ **53.** 1.641×10^8 **54.** 4.99×10^7

Chapter 13 Test

1. $-2, 1, 4, 7, 10$ **2.** $2, 5, 10, 17, 26$ **3.** $2, 16, 54, 128, 250$ **4.** $2, \frac{3}{4}, \frac{4}{9}, \frac{5}{16}, \frac{6}{25}$ **5.** $3, 7, 11, 15, 19$

6. $4, -8, 16, -32, 64$ **7.** $a_n = 4n + 2$ **8.** $a_n = 2^{n-1}$ **9.** $a_n = \left(\frac{1}{2}\right)^n$ **10.** $a_n = (-3)^n$ **11.** 90   **12.** 28

13. 130   **14.** $3x + 12$ **15.** $\sum_{i=1}^{5}(3i + 1)$ **16.** $\sum_{i=1}^{4}\frac{i+1}{i+3}$ **17.** $\sum_{i=5}^{8}(x + 4)^i$ **18.** $\sum_{i=1}^{3}\frac{x^i}{x-i}$ **19.** geometric, $r = -5$

20. neither   **21.** arithmetic, $d = -\frac{1}{2}$ **22.** arithmetic, $d = 5$ **23.** $a_1 = 3$ **24.** $a_2 = -6$ or 6 **25.** $S_{10} = 320$

26. $S_{10} = 25$ **27.** $S_{50} = 3(2^{50} - 1)$ **28.** $S = \frac{3}{4}$ **29.** $x^4 - 12x^3 + 54x^2 - 108x + 81$

30. $32x^5 - 80x^4 + 80x^3 - 40x^2 + 10x - 1$ **31.** $x^{20} - 20x^{19} + 190x^{18}$ **32.** $-108,864x^3y^5$

33.

```
            1
          1   1
        1   2   1
      1   3   3   1
    1   4   6   4   1
  1   5  10  10   5   1
1   6  15  20  15   6   1
1  7  21  35  35  21  7  1
```

Index

W

X

Y

Z